SELECTED PHYSICAL QUANTITIES AND MEASUREMENT UNITS

Physical quantity	Quantity symbol	Measurement unit	Unit symbol	Unit dimensions
		Fundamental (base) units		
length	l	meter	m	m
mass	m	kilogram	kg	kg
time	t	second	s	s
electric charge*	Q	coulomb	c	c
temperature	T	degree Kelvin	°K	°K
amount of substance	n	mole	mol	mol
luminous intensity	I	candle	cd	cd
		Derived units		
acceleration	a	meter per second per second	m/s^2	m/s^2
area	A	square meter	m^2	m^2
capacitance	C	farad	f	$c^2\,s^2/kg\,m^2$
density	D	kilogram per cubic meter	kg/m^3	kg/m^3
electric current	I	ampere	a	c/s
electric field intensity	$\mathscr{E}$	newton per coulomb	n/c	$kg\,m/c\,s^2$
electric resistance	R	ohm	Ω	$kg\,m^2/c^2\,s$
emf	E	volt	v	$kg\,m^2/c\,s^2$
energy	E	joule	j	$kg\,m^2/s^2$
force	F	newton	n	$kg\,m/s^2$
frequency	f	hertz	hz	s^{-1}
heat	Q	joule	j	$kg\,m^2/s^2$
illumination	E	lumen per square meter	lm/m^2	cd/m^2
inductance	L	henry	h	$kg\,m^2/c^2$
luminous flux	Φ	lumen	lm	cd
magnetic flux	Φ	weber	wb	$kg\,m^2/c\,s$
magnetic flux density	B	weber per square meter	wb/m^2	kg/c s
potential difference	V	volt	v	$kg\,m^2/c\,s^2$
power	P	watt	w	$kg\,m^2/s^3$
pressure	p	newton per square meter	n/m^2	$kg/m\,s^2$
velocity	v	meter per second	m/s	m/s
volume	V	cubic meter	m^3	m^3
work	W	joule	j	$kg\,m^2/s^2$

*The SI fundamental (base) electric unit is the ampere. For pedagogical reasons the coulomb of electric charge is introduced in this book as the fundamental unit and the ampere as a derived unit having the dimensions, coulomb per second.

JOHN E. WILLIAMS
FREDERICK E. TRINKLEIN
H. CLARK METCALFE

MODERN PHYSICS

TEACHER'S EDITION

THE HOLT MODERN PHYSICS PROGRAM

Modern Physics (Student Text)

Modern Physics (Teacher's Edition)

Exercises and Experiments in Physics (Student's Edition)

Exercises and Experiments in Physics (Teacher's Edition)

Laboratory Experiments in Physics

Tests in Physics (duplicating masters)

HOLT, RINEHART AND WINSTON, PUBLISHERS

New York • Toronto • Mexico City • London • Sydney • Tokyo

ISBN 0-03-061937-8

345678 032 654321

CONTENTS

INTRODUCTION

The MODERN PHYSICS program is intended to help you, the teacher, in your effort to achieve the objectives of science teaching. This *Teacher's Edition of Modern Physics*, 1984, offers you aid in using the complete program of available text and supplementary materials.

Materials published by Holt, Rinehart and Winston
383 Madison Avenue, New York, NY 10017

Accompaniments to the Text: (By the authors of *Modern Physics*)
Laboratory Experiments in Physics
Exercises and Experiments in Physics
Teacher Edition of Exercises and Experiments in Physics
Tests in Physics (duplicating masters)

Materials by other authors:
Foundations of Physics, 1969, Lehrman, Robert L. and Clifford Swartz
Laboratory Experiments, 1969, Lehrman, Robert L. and Clifford Swartz
Foundations of Physics Program, Teacher's Guide, 1969, Lehrman and Swartz
Physics Problems, 1961, Castka, Joseph F. and Ralph W. Lefler
Modern Chemistry, 1974, Metcalfe, Williams, and Castka
Foundations of Chemistry, 1973, Toon, Ellis, and Brodkin
Radioactivity: Fundamentals and Experiments, 1963, Hermias and Joecile

The Project Physics Course provides the following components:
Project Physics Text, 1975
Project Physics Handbook, 1975
The Project Physics Course Readers, 1970:
1. *Concepts of Motion*
2. *Motion in the Heavens*
3. *The Triumph of Mechanics*
4. *Light and Electromagnetism*
5. *Models of the Atom*
6. *The Nucleus*
Project Physics Resource Book, 1975
Tests

Supplementary Materials: (not written specifically for high school use)
Fundamentals of Physics, Fourth Edition, 1966, Henry Semat
Great Experiments in Physics, edited by Morris H. Shamos
The Physics of Space, Richard M. Sutton
Introduction to Atomic and Nuclear Physics, 1962, Henry Semat

Materials available from other publishers:
  The books listed below will provide excellent sources of ideas for physics demonstrations and related activities.

Joseph, Alexander, Paul F. Brandwein, Evelyn Morholt, Harvey Pollack, and Joseph F. Castka, *A Sourcebook for the Physical Sciences* (Harcourt, Brace and Jovanovich, Inc.). A list of demonstration and laboratory procedures as well as numerous suggestions for projects and experiments. All the techniques have been tested in actual class situations.

Meiners, Harry F., ed., *Physics Demonstration Experiments* (The Ronald Press Company, 79 Madison Avenue, New York, NY 10016). A two-volume reference set that draws on the best demonstration techniques and equipment developed over the thirty years previous to 1970 and brings together the experiments that have proved most successful in helping students understand the principles of classical and modern physics. These books present a wide selection of ideas for demonstrations for each topic in the curriculum. The selections vary in sophistication for maximum teaching flexibility and are accompanied by lists of needed equipment and complete step-by-step instruction for effective presentation. Possibly one third of the ideas presented would be usable in any given high school.

Mlodzeevskii, A. B., ed., *Lecture Demonstrations in Physics* (M. V. Lomonosov State University, Moscow, USSR). Available in a condensation by H. A. Robinson (April, 1963) on Microfilm from the American Institute of Physics, 335 East 45th Street, New York, NY 10017.

Miller, Julius S., *Demonstrations in Physics* (Ure Smith Pty Limited, 155 Miller Street, North Sidney, N.S.W. 2060, Australia). This is a book on demonstrations in physics with a diagram or sketch revealing at a glance the physical arrangement of apparatus for each idea suggested. Each suggestion has associated with it a series of questions, exercises, and dilemmas that the student is to explore.

Nuffield Foundation, *Physics Teachers Guides*—5 volumes; *Physics Guide to Experiments*—5 volumes; *Physics Question Book*—5 volumes (Longmans, Green and Company, Ltd., London; Viking Penguin Inc., 425 Madison Avenue, New York, NY 10022).
  Science Masters Series, Part 1 of Series I, 1931; Part 1 of Series II, 1936; Part 1 of Series III, 1950; Part 1 of Series IV, 1961 (John Murray, Ltd., London).

Sutton, Richard M., *Demonstration Experiments in Physics* (McGraw-Hill Book Co., 1221 Avenue of the Americas, New York, NY 10020). This contains suggested experiments for both high school and elementary college physics, 238 in Mechanics, 153 in Sound, 182 in Heat, 269 in Electricity and Magnetism, 136 in Light, and 121 in Atomic and Electronic Physics.

References of a General Nature:
  Following is a list of recommended reference books. When ordering, ask for the latest edition.

American Radio Relay League, *The Radio Amateur's Handbook* (American Radio Relay League, Newington, CT 06111).

Arons, Arnold B., *Development of Concepts of Physics* (Addison-Wesley Publishing Co., Reading, MA 01867).

Beiser, Arthur, *Concepts of Modern Physics* (McGraw-Hill Book Co., 1221 Avenue of the Americas, New York, NY 10020).

Born, Max, *The Restless Universe* (Dover Publications, Inc., 180 Varick Street, New York, NY 10014).

Chedd, Graham, *Sound, From Communications to Noise Pollution* (Doubleday and Co., Inc., 501 Franklin Ave., Garden City, NY 11530).

Chemical Rubber Company, *Handbook of Chemistry and Physics* (Chemical Rubber Co., 18901 Cranwood, Cleveland, OH 44128).

Condon and Odishaw, *Handbook of Physics* (McGraw-Hill Book Co., 1221 Avenue of the Americas, New York, NY 10020).

Frisch, Otto R., *Atomic Physics Today* (Basic Books Inc., 10 East 53rd Street, New York, NY 10022).

Galilei, Galileo, *Dialogues Concerning Two New Sciences* (Dover Publications, Inc., 180 Varick Street, New York, NY 10014).

Gamow, George, *Mr. Tompkins in Paperback* (Cambridge University Press, 32 East 57th Street, New York, NY 10022). (*Mr. Tompkins in Wonderland* and *Mr. Tompkins Explores the Atom* are published in a single book.)

Gamow and Cleveland, *Physics, Foundations and Frontiers* (Prentice Hall, Inc., Englewood Cliffs, NJ 07632).

Ginzer and Youngner, *Physics* (Silver Burdett, Morristown, NJ 07960).

Haber-Schaim, Cross, Dodge, and Walter, *PSSC Physics Third Edition* (D.C. Heath and Company, Lexington, Mass.).

Halliday and Resnick, *Physics for Students of Science and Engineering* (John Wiley and Sons, 605 Third Avenue, New York, NY 10016).

Holton, Gerald J., *Introduction to Concepts and Theories in Physical Science* (Addison-Wesley Publishing Co., Reading, MA 01867).

Holton and Roller, *Foundations of Modern Physical Science* (Addison-Wesley Publishing Co., Reading, MA 01867).

Karplus, Robert, *Introductory Physics, A Modern Approach* (W. A. Benjamin, Inc., 2 Park Avenue, New York, NY 10016).

Lindsay, Robert Bruce, *Basic Concepts of Physics* (Van Nostrand Reinhold Co., 135 West 50th Street, New York, NY 10020).

Lenk, John D., *Handbook of Oscilloscopes, Theory and Application* (Prentice Hall, Inc., Englewood Cliffs, NJ 07632).

Lapedes, Daniel N., editor, *The McGraw-Hill Encyclopedia of Science and Technology* (McGraw-Hill Book Co., 1221 Avenue of the Americas, New York, NY 10020).

Millikan, R. A., *Electrons (+ and −)* (University of Chicago Press, 5750 Ellis Avenue, Chicago, IL 60637).

Radio Corporation of America, *RCA Receiving Tube Manual* (Radio Corporation of America, Harrison, NJ 07029).

Richard, Sears, Wehr, and Zemansky, *Modern College Physics* (Addison-Wesley Publishing Co., Reading, MA 01867).

Rogers, E. M., *Physics for the Inquiring Mind* (Princeton University Press, Princeton, NJ 08541).

Sears and Zemansky, *College Physics* (Addison-Wesley Publishing Co., Reading, MA 01867).

Spielman, Harold S., *Electronics Source Book for Teachers* (Hayden Book Company, 116 West 14th Street, New York, NY 10011). (Three volumes.)

Stollberg and Hill, *Physics, Fundamentals and Frontiers* (Houghton Mifflin Co., 2 Park Street, Boston, MA 02107). (Teacher's edition with useful notes available.)

Taylor, L. W., *Physics, The Pioneer Science* (Dover Publications, Inc., 180 Varick Street, New York, NY 10014).

Van Name, F. W., Jr., *Modern Physics* (Prentice Hall, Inc., Englewood Cliffs, NJ 07632).

Van Nostrand's Scientific Encyclopedia (Van Nostrand Reinhold Co., 135 West 50th Street, New York, NY 10020).

Weidner and Sells, *Elementary Modern Physics* (Allyn and Bacon Inc., 470 Atlantic Avenue, Boston, MA 02110).

Books dealing with the history of major developments in science and with some of the people who helped to shape science:

Bush, Vannevar, *Pieces of the Action* (Morrow, 105 Madison Avenue, New York, NY 10016).

Conant, James Bryant, *My Several Lives* (Harper and Row, Publishers, 10 East 53rd Street, New York, NY 10022).

Irwin, Keith Gordon, *The Romance of Physics* (Scribners, 597 Fifth Avenue, New York, NY 10017).

Magie, William F., *Source Book in Physics* (Harvard University Press, 79 Garden Street, Cambridge, MA 02138).

Seeger, Raymond J., *Men of Physics: Galileo Galilei, His Life and His Works* (Pergamon, Maxwell House, Fairview Park, Elmsford, NY 10523).

Segre, Emilio, *Enrico Fermi, Physicist* (University of Chicago Press, 5750 Ellis Avenue, Chicago, IL 60637).

Some ready sources of titles of physics or physics-related offprints and paperbacks:

AAAS Science Book List, Third Edition, AAAS Publications (Dept. BL-2), 1515 Massachusetts Avenue, Washington, DC 20005. Descriptive annotations for 2,441 titles.

Momentum Books, Published for the Commission on College Physics, Van Nostrand Reinhold Co., 135 West 50th Street, New York, NY 10020. Nineteen titles.

Science Study Series, Doubleday and Co., Inc., 501 Franklin Avenue, Garden City, NY 11530. More than 50 titles covering the most fundamental and exciting topics of physical science.

Scientific American Offprints, W. H. Freeman and Co., 660 Market Street, San Francisco, CA 94104.

Signet and Mentor Paperbacks in Science, The New American Library, Inc., 1301 Avenue of the Americas, New York, NY 10019.

Vistas of Science and other publications, The National Science Teachers Association, 1201 Sixteenth Street, N.W. Washington, DC 20036.

The following sections of the *Teacher's Edition of Modern Physics* provide a set of notes to accompany each of the twenty-five chapters of the text. The notes for each chapter are divided into:

1. **Synopsis**—a statement intended to provide suggestions for introducing, to the student, the material to be covered in the chapter. An historical approach is often provided.

2. **Comment**—suggestions for teaching the material; ideas for demonstrations, points to be emphasized, specific references to parts of books and articles in periodicals, general references, and a listing of appropriate films. A suggested instructional approach is often included.

3. **References by Text Section**—specific references are made to the corresponding presentations in other texts, overhead transparencies and 2″ × 2″ color slides, single concept film loops, demonstration ideas, and student laboratory experiments. Abbreviations are often used. These will be self-explanatory, especially when reference is made to the bibliographies given in this introduction.

4. **Questions**—brief answers are given for the questions included with each chapter. When questions are assigned as written work, the student should be expected to answer with complete statements in order to foster scholarly habits of writing. The *Teacher's Edition* provides short answers for economy of space. These answers are not intended as an example of the form that the teacher is to expect from the student.

5. **Problems**—solutions are shown and answers given for each problem included in the text. The last digit in many answers may vary by one digit, depending on whether or not intermediate numerical values were rounded off to the justified number of significant figures.

Student experiments are found in *Exercises and Experiments in Physics* and *Laboratory Experiments in Physics*. The experiments in the former are written in the "traditional" format. Teachers who prefer the "open-ended" approach will probably be happier with the latter or with Lehrman's *Laboratory Experiments*. Some teachers may wish to use the experiments prepared for the Project Physics Course and found in *Project Physics Handbook*. These too are referenced. Experiments in the Lehrman manual are of the PSSC type. It is suggested that teachers obtain copies of each of the aforementioned manuals or handbooks for examination to determine which is most suitable for their use. Some may wish to select the most appropriate experiment each time rather than be bound to one book. By doing this, the teacher will be able to present a course in physics best adapted to the abilities and needs of the students.

The unity of the chapter has not been broken by suggesting smaller divisions or lessons. The time required for instruction will vary from chapter to chapter and from teacher to teacher; however, time should be provided for significant student contact with the selected material through study of the text and references, through laboratory work, problem solving, and class and small group discussion. An appropriately selected demonstration, from those suggested, can form the basis for a valuable discussion.

No specific suggestion has been made regarding the selection of questions and problems to be assigned. This is a matter to be decided by the teacher and may well differ from one class to another and should take into consideration individual differences among students.

It is suggested that students be given appropriate assignments to be prepared in writing. In our work in physics we can help the student to develop powers of both written and oral expression.

References suggested vary considerably in reading difficulty and in mathematical sophistication. The teacher will have to decide whether a given reference will be appropriate for the class as a whole, for a single student, or a selected group of students. Some of the references should be retained for the teacher's personal use in preparing lessons.

The films suggested for use should probably be made a part of class instruction. However, they should also be made available for student use during their independent study time. Since these films are usually in the school on a rental basis, students should know in advance when they will be available for independent use so that they can properly schedule their time. Single concept films are listed. However, there are many available that are not listed. A study of the publication *Short Films for Physics Teaching*, Blum and Daugherty, The Commission on College Physics, Department of Physics and Astronomy, University of Maryland, College Park, Maryland 20742, and of catalogues provided by the suppliers will be useful.

Many schools are equipped with overhead projectors. These are useful tools in teaching some topics in physics. Teachers can prepare their own materials for projection. Do not use this as a time-saving device for the time you save may rob the student of necessary cogitation and incubation time. There are several companies that produce commercial transparencies for overhead projection. The Keuffel and Esser Company, Morristown, New Jersey 07960, has a book called *Diazo Transparency Master Book in Physics*. This book contains over 400 prepared tracing paper masters to produce (with overlays) over 80 diazo transparencies in color. The 3-M Company also produces masters from which transparencies may be made.

It will be noted that the bulk of audio-visual material that is referenced in this edition dates back to the period between the late 1950's and early 1970's. These have become increasingly difficult to obtain, but they have been retained here simply because their quality and effectiveness have not been improved upon. The PSSC films of the 1950's and the Project Physics film loops and overhead projector transparencies developed during the 1960's are especially useful in teaching the basic concepts of physics, and the effort to get them for your physics course is well worthwhile.

The authors recognize the increasing demand for programs in physics that have been prepared for the microcomputer. Many good programs have already been developed and new ones are appearing daily. These are largely the endeavor of individual physics teachers who have programmed materials for auto-tutorial use in their own classrooms. A compilation and evaluation of such software is now in progress and will eventually be included in future *Modern Physics* teacher editions, or in a supplement to this publication now in preparation.

The teacher should examine the exercises in *Exercises and Experiments in Physics* to determine how best to use them. It is not likely that all exercise items should be assigned to all students. Properly used, these exercises become effective instructional aids, but when applied in total to all students they become busy work for many. For this reason, the exercises are not annotated here; but the following table correlates the text sections with the exercise in *Exercises and Experiments* that is appropriate for those sections.

Text Sections	Exercise Number	Title
1.1–1.15	1	Physics: The Science of Energy
2.1–2.5	2A	Units of Measure
2.6–2.9	2B	Making and Recording Measurements
2.10–2.12	2C	Solving Problems
3.1–3.4	3A	Velocity
3.5–3.7	3B	Acceleration
3.8–3.10	3C	Newton's Laws of Motion
3.11–3.14	3C	Gravitation
4.1–4.3	4A	Composition of Forces
4.4–4.5	4B	Resolution of Forces
4.6–4.9	4C	Friction
4.10–4.13	4D	Parallel Forces
5.1–5.4	5A	Circular Motion
5.5–5.9	5B	Rotary Motion
5.10–5.12	5B	Harmonic Motion
6.1–6.6	6	Work, Machines, and Power
6.7–6.11	6	Energy
6.12–6.16	6	Momentum
7.1–7.5	7A	The Structure of Matter
7.6–7.11	7A	The Solid Phase
7.12–7.18	7B	The Liquid Phase
7.19–7.23	7B	The Gaseous Phase
8.1–8.3	8A	Units of Temperature and Heat
8.4–8.10	8A	Thermal Expansion
8.11–8.13	8B	Heat Exchange
8.14–8.21	8B	Change of Phase
9.1–9.5	9	Heat Engines—Heat and Work

Teachers should examine the testing program, which is an integral part of this *Modern Physics* package. The program consists of *Tests in Physics* duplicating masters.

A periodical that is essential to each physics teacher is *The Physics Teacher*. It is published by the American Association of Physics Teachers and dedicated to strengthening the teaching of introductory physics. It is published monthly by the Department of Physics, SUNY, Stony Brook, NY 11794. Each issue carries major articles, and includes sections entitled letters, notes, apparatus for physics teaching, book and film reviews, research frontier, and a calendar. Many publishers and apparatus manufacturers advertise in the journal. It should be a must in each physics teacher's library.

It is unlikely that all of the material in the complete *Modern Physics* program can be "covered" thoroughly in a school year of approximately 180 days. Therefore, it is wise for the teacher to plan the school year carefully in advance so that the available time is budgeted among those topics that will be of the greatest interest and value to the students. Thorough treatment of a limited number of topics is preferable to a hasty coverage of all the chapters in the text. The choice of topics should be based on the capabilities of the teacher and the needs of the students.

SYNOPSES, COMMENTARIES, AND REFERENCES

PHYSICS: THE SCIENCE OF ENERGY 1

Synopsis: The world today is dependent on technology and science. This basic idea is introduced by noting the nature of the world's increasing demand for energy and by indicating that solutions to our energy problems involve not only conservation but also intensive and extensive research and development.

As the world becomes more and more dependent on technology and science, the citizens should become more and more knowledgeable about both the nature of science and technology and the specific knowledge that science represents. New developments and elaboration of old techniques cannot be understood without knowing the present content of science. This is especially true in physics, which is the most basic science of all. Therefore, much of this chapter deals with a consideration of both the nature of science and the major concepts of physics.

First, the distinction between science and technology is made. Many people do not realize that there is a distinction. Recent attacks against technological innovation have been misdirected against science in general. Scientific research may result in a specific proposal, but its development depends on the support received by business, government, or both. The decision to use or reject a technological innovation does not rest with science.

A knowledge of the nature of scientific laws and theories is important to any educated individual. Examples of both are briefly presented, along with a comment on the usefulness of a theory.

The nature of scientific progress is indicated by consideration of the idea of scientific hypotheses. Although no step-by-step procedure will automatically lead to the solution of a problem in science, there are characteristics of scientific work that can be identified. Among these are the statement of an hypothesis, its confirmation, and its inclusion into a theory.

The uncertainty inherent in all science is then discussed. This uncertainty is attributed to observational methods and the individual scientist. The validity of science, therefore, is limited by both method and scientist. It is also dependent on an attitude that is both skeptical and questioning and, at the same time, characterized by curiosity and respect for logical thinking. It is pointed out that science cannot deal with all topics, especially those that concern human feelings or ultimate origins.

The next sections deal with the major ideas that form the content of physics: matter and energy. Matter and energy are fundamental physical entities that comprise the physical world that we perceive. In this course, students attempt to deepen and to refine their concept of matter and energy and to assimilate some understanding of the energy exchanges associated with changes that take place in the forms of matter. They will begin with operational definitions of concepts that can be used in describing matter and gradually proceed to an understanding that advances beyond the descriptive into the conceptual.

These two basic entities, matter and energy, along with their related concepts are presented. Stress is placed on the measurement of such concepts because the aim throughout the book is to use operational definitions of the ideas presented. That is, terms are defined by the way that they are measured. As an example, temperature is what is measured with a thermometer.

Two concepts that are related to the idea of mass are then presented from an operational viewpoint. These are inertia and mass density. This is followed by concepts related to energy and its conservation. Emphasis is given to the various forms of energy, the possibility of deriving useful work from energy, and the conversion of energy from one form to another. The two forms of mechanical energy especially important to a study of basic physics are discussed. These are potential and kinetic energy. The importance and challenge of the idea of conservation of energy is illustrated by use of an historical example.

The relationship between matter and energy is then discussed in terms of Einstein's mass-energy equivalence. This results in the matter and energy conservation statement, one of the most important laws of contemporary physics. Continually verified by laboratory work, this statement suggests that the sum total of mass and energy is not changed in any given chemical and physical system no matter how extensive the internal changes are.

The chapter concludes with a statement about the goal of physics and a final note about how this text is divided into four major topics for study.

Comment: Eric M. Rogers in "A Talk About Demonstrations," Chapter 4, pp. 42–60, *Physics Demonstration Experiments,* edited by Harry F. Meiners, 1970, states "Teaching physics is building new physics, new knowledge and understanding, in young minds. Our teaching duty goes far beyond supplying information: it is concerned with conveying to students the nature of physics—what it is, how it is done, how it grows in strength of knowledge, and even where it may lead." This chapter in the Meiner's book goes on to deal with the aims of demonstrations. Rogers states that although we should enliven our presentations with experiments, entertainment is not the goal. "Experiments that delight" go much deeper than entertainment. They provide for seeing a phenomenon, watching a test, or finishing an argument.

In this first chapter, some "experiments that delight" can be shown without explanation. Rogers suggests (pp. 45–46) a color-mixer, the monkey and hunter experiment, a loop-the-loop experiment, a Bernoulli paradox, and connecting a large soap bubble to a small one. These experiments can be used to start some thinking. Possibly one might create in the student a desire to understand why things in the physical world behave as they do.

This is a good time to give the student an idea of the general areas to be studied and the methods used to study them. Set the tone of the course by the manner of presentation.

Some additional suggestions for demonstrations are:

1. Cause the collision of dynamics carts along a straight line. The carts should be made to look alike but have greatly differing masses.
2. Roll two disks that look alike but with the mass of one concentrated near the center and the mass of the other concentrated near the periphery down an inclined plane.
3. Demonstrate a jumping ring by placing a closed aluminum ring over the extended iron pole of an electromagnet energized by an alternating current. The ring jumps as the current in the magnet is established.
4. A bologna bottle may be "exploded" by placing a small carborundum crystal in the bottle and shaking it to cause the crystal to scratch the interior.

Discuss the methods used by scientists, not as a hard and fast procedure, but as a logical way to go about the investigation of a scientific idea. Distinguish between science and technology. Comment on the problem of supplying the world with adequate energy, discussing Figure 1-1 and emphasizing the need for more research and development. It would also be pertinent to give reasons for knowing about basic physics in a world so dependent on technology and science. The nature of scientific work could be illustrated by the Project Physics film, "People and Particles."

Demonstrate the measurement of mass by use of an equal-arm and inertia balance. Perform a simple demonstration of inertia. Determine the mass and weight of a piece of brass. Demonstrate changes in physical properties.

Light an electric bulb to show the conversion of electric energy into light and heat. Discuss other transformations, that is, where electric energy comes from. Have a student push a book across the desk top and then push the desk in order to develop the concept of work. Bring out the interchangeability of matter and energy by showing that Einstein's prediction, proven by experiment, is now considered a law.

References: *Laboratory Experiments in Physics*, Holt, Rinehart and Winston, Publishers, New York, 1984.

Exercises and Experiments in Physics, Holt, Rinehart and Winston, Publishers, New York, 1984.

Brooks, Harvey and Raymond Bowers, "The Assessment of Technology," *Scientific American*, Feb. 1970.

Castka and Lefler, *Physics Problems*, Holt, Rinehart and Winston, Publishers, New York, 1961, Chapter 1, pp. 1–12.

Farre, G. L., "On the Problem of Scientific Discovery," *The Science Teacher*, Oct. 1966, pp. 26–29.

Gamow, George, *One, Two, Three . . . Infinity*, The Viking Press, New York, 1957, pp. 41–114.

Lehrman, Robert L., *Laboratory Experiments*, Holt, Rinehart and Winston, Publishers, New York, 1969.

Lehrman and Swartz, *Foundations of Physics*, Holt, Rinehart and Winston, Publishers, New York, 1969, pp. 1–5, 19–20, 32–36.

Lindsay, *Basic Concepts of Physics*, Van Nostrand Reinhold Co., New York, 1971, Chapter 1, The Nature of Science.

PSSC Physics, Third Edition, D. C. Heath and Co., Lexington, Mass., 1971, pp. 2–6.

Semat, Henry, *Fundamentals of Physics*, Holt, Rinehart and Winston, Publishers, New York, 1966, pp. 3–10.

Shamos, Morris H., *Great Experiments in Physics*, Holt, Rinehart and Winston, Publishers, New York, 1960, pp. 1–12.

The Project Physics Text, Holt, Rinehart and Winston, Publishers, New York, 1975.

Weisskopf, Victor F., *Knowledge and Wonder: The Natural World as Man Knows It*, Science Study Series. In this outstanding book—winner of the 1962 Edison Award—one of the world's most distinguished physicists brilliantly summarizes all that we have learned about ourselves and about our environment. (1966)

The Science Teacher, vol. 41, no. 8, Nov. 1974, pp. 12–28.

Films: *Introduction to Physics*, 11 min, s, bw/c, c. Shows the main divisions of physics and presents a basic vocabulary. Coronet Instructional Films, 65 East South Water Street, Chicago, IL 60601.

People and Particles, 28 min., black and white. A Project Physics film. A documentary in *cinéma verité* style, filmed over a $2\frac{1}{2}$ year period as student and faculty members worked together at the Cambridge Electron Accelerator on a problem in particle physics.

References for Chapter 1: Physics: The Science of Energy

1.2 Science and technology
Lindsay, *Basic Concepts of Physics*, Section 1.1, Physics and Human Curiosity.

1.4 Scientific hypotheses
Lindsay, *Basic Concepts of Physics*, Section 1.3, What is Science? Section 1.4, Science as Description. Section 1.5, Science as Creation. Section 1.6, Science as Understanding.

1.5 Certainty in science
Laboratory Exercise: Lehrman, *Laboratory Experiments*, Expt. 1, Determination of Volume, pp. 1–5.

1.7 Mass
Meiners, *Physics Demonstration Experiments*, 8–2, pp. 140–143, Weight and Mass.
Laboratory Exercise: Lehrman, Expt. 4, Mass and Weight, pp. 11–12.

Laboratory Exercise: *Project Physics Handbook*, Holt, Rinehart and Winston, 1975, Expt. 1–9, Mass and Weight, p. 33.

Lindsay, *Basic Concepts of Physics*, Chapter 4, p. 85ff. 4.1 The Concept of Mass. 4.2 Standards of Mass.

Sutton, M-2, p. 15, Error in Judgment of Weight.

1.8 Inertia

Suggested demonstrations:

Sutton, M-100, p. 46, Inertial Reaction. M-101, p. 47, Breaking Rope by Inertia. M-103, p. 47, Weight vs Inertial Reaction. M-104, pp. 47 & 48, Inertia Tricks. M-106, p. 49, Inertia Balance.

1.10 Energy

Lindsay, *Basic Concepts of Physics*, Chapter 5, p. 137ff.,

Section 5.5, p. 184ff., Generalization of the Concept of Energy.

1.14 Relationship between matter and energy

Schary, Stephen D., "Relativity in High School," *The Physics Teacher*, Jan. 1971, pp. 42–43.

Science Study Series: *Relativity and Common Sense* by Hermann Bondi, University of London (1964). *The Physical Foundations of General Relativity* by D. W. Sciama (1969).

Bronowski, J., "The Clock Paradox," *Scientific American*, Feb. 1963.

Gardner, Martin, "Can Time Go Backward," *Scientific American*, Jan. 1967.

2 MEASUREMENT AND PROBLEM SOLVING

Synopsis: The universe can be described by using certain physical quantities, such as length, mass, and time. How these quantities are described depends on the system of measurement being used. The metric system, based on the decimal system, is the official system of measurement for most of the world. It has proven to be a very suitable system for work in science and technology.

Units in any system of measurement are based on a few fundamental physical quantities. For example, in mechanics all physical quantities can be expressed in terms of mass, length, and time. How these basic quantities are actually defined, or measured, depends on the set of operations used to measure them. Other physical quantities, such as velocity, force, and momentum, are then defined in terms of these basic quantities.

The entire field of physics requires seven fundamental physical quantities. From these, all others are derivable. In this chapter, the basic concepts of length, mass, and time are examined in detail, with particular emphasis on distinguishing between the concepts of mass and weight.

The process of measurement is not simple. It involves both inherent uncertainties and instrument limitations, along with other variables that may influence the process. These ideas are considered in detail, along with the method of treating such inevitable errors in data recording. The ideas of accuracy, precision, and significant figures are used to describe these basic experimental errors, along with mathematical methods of determining them.

Scientific notation is introduced as a powerful device for comparing magnitudes of values obtained in calculations. How measured data are used is just as important as how they were obtained. Consequently, considerable attention is given to the recording and interpretation of graphed data, with two graphical examples presented and discussed thoroughly.

Physical quantities can be divided into scalars and vector quantities. Most students are familiar with scalars. The use of vectors is essential in many situations in physics and is introduced here in terms of velocities and graphical solutions.

The chapter concludes with a comment on what approach should be used in solving physics problems. Examples are presented to illustrate this approach, which consists of a sequence of separate mathematical steps.

Comment: An excellent reference for the nature of measurement appears in Chapter 2 of *Experimentation: An Introduction to Measurement Theory and Experiment Design*, D. C. Baird, Prentice-Hall, 1963. The SI system of units should be introduced with the use of measuring devices. Pass out meter sticks and have students measure lengths of objects and distances to nearby objects, as well as estimate distances or lengths in meters or centimeters. Consider also the measurement of areas and volumes using the metric system. Exhibit graduated cylinders, beakers, and flasks of marked capacity. Display a one-kilogram mass and submultiples thereof. Avoid allowing students to use the concepts of mass and weight interchangeably. Prepare a body with a weight of one newton and one with a weight of 10 newtons. Have students handle these to get a "feel" for the magnitude of these forces. Cover the face on a clock having a second hand. Have students estimate elapsed time and compare with clocked time. Exhibit the chart, "The Modernized Metric System," Superintendent of Documents, Government Printing Office, Washington, DC 20402.

Give specific attention to the section on accuracy and precision. Use these concepts in future problem work and in the discussion of experimental results until they become well understood by students. Develop the concept of significant figures by having students measure the length of suitable objects, compare different readings of identical objects, and arrive at the concept of uncertainty of measurement. Then lead to a discussion of significant figures. The rules for significant figures should be learned after illustration and discussion. The relationship between significant figures and scientific notation and the convenience of the latter should be explained. Refer to Fig. 2-10 to illustrate the utility of scientific notation and orders of magnitude. Develop and drill the arithmetic operations with significant figures. Discuss the recording and interpretation of data and the use of equations and graphs. Frequent reference to Table 2-3 will help reinforce the major ideas. Establish the need for vector quantities, distinguishing them from scalar concepts, and give considerable time to the graphical solution of vector problems. Reference to Figs. 2-13 through 2-17 will help clarify vector techniques.

Go through the procedure for solving problems, emphasizing the analytical approach as vital in dealing with situations in physics.

References: Astin, Allen V., "Standards of Measurement," *Scientific American*, June 1968.

Castka, Lefler, *Physics Problems*, Holt, Rinehart and Winston, Publishers, New York, 1961, Chap. 2.

Exercises and Experiments in Physics, Holt, Rinehart and Winston, Publishers, New York, 1984.

Laboratory Experiments in Physics, Holt, Rinehart and Winston, Publishers, New York, 1984.

Baird, *Experimentation: An Introduction to Measurement Theory and Experimental Design*, Prentice Hall, Inc., Englewood Cliffs, 1962, Chapter 2, The Nature of Measurement.

Furth, R., "The Limits of Measurement," *Scientific American*, July 1950.

Lehrman, *Laboratory Experiments*, Holt, Rinehart and Winston, Publishers, New York, 1969.

Lindsay, *Basic Concepts of Physics*, Van Nostrand Reinhold Co., New York, 1971, Chapter 2, The Concepts of Space and Time.

Richie-Calder, Lord, "Conversion to the Metric System," *Scientific American*, July 1970.

Films: *Precisely So.* 18 min, s, General Motors Corporation. To satisfy the high standards of measurement, it was necessary to develop delicate instruments of measurement: "split a hair's breadth measurement into hundreds of parts."

Understanding the Physical World Through Measurement. 35 min, free loan, National Bureau of Standards, Office of Technical Information, Washington, DC. Demonstrations include transformations between mechanical, chemical, heat, and other forms of energy.

PSSC Films: 0101. *Time and Clocks,* 28 min. Discusses concepts of time measurement and shows various devices used to measure and record time intervals from 1 second down to 10^{-9} seconds.

0119. *Short Time Intervals,* 21 min. Timing devices used include moving cameras, pen recorders, and the oscilloscope, with explanation of its use in measurements.

0102. *Long Time Intervals,* 25 min. Radioactive dating used in arriving at an estimate for the age of the earth.

0103. *Measuring Large Distances,* 29 min. Using models of earth, moon, and stars. Demonstrations point up the immensity of interstellar space and suggest complexities of measurement on this scale.

0104. *Measuring Short Distances,* 20 min. Starts with the centimeter scale, moves on to microscopic dimensions, and then to the dimensions of the atom by means of Erwin Mueller's field emission microscope. Shows calibration of instruments and how it can give us accurate knowledge of these small distances.

0105. *Measurement,* 21 min. The measurement of the speed of a rifle bullet is used as the basis for discussion of the art of measurement. Problems that are met and discussed include noise, bias, use of black boxes, and the element of decision in all measurements.

0106. *Changes of Scale,* 23 min. Demonstrates that change of size necessitates change in structure of objects: uses specially constructed props to emphasize scaling problems, then shows practical applications of scale models, i.e., movie-making.

0108. *Vectors,* 28 min. A lecture demonstration on vectors made to a high school class. The need for a quantity with direction and magnitude is pointed out. The meaning of addition and subtraction of vectors, and multiplication of vectors by a scalar is illustrated by demonstrations.

References for Chapter 2: Measurement and Problem Solving

2.1 The Metric System
Transparencies useful in presenting the work of this chapter: John Colburn Associates, 1122 Central Avenue, Wilmette, IL
Significant Figures.
Exponents.
Significant Figures, Order of Magnitude.
Linear Measurement, Order of Magnitude.
K & E Transparencies: Measurement. Coordinate Grid, 1 cm × 1 cm. Coordinate Grid, 1 cm × 1 cm, reversal. Coordinate Grid, 10 × 10 per inch. Coordinate Grid, 10 × 10 per inch, reversal. Scales, inch and centimeter; Protractor, 360 degrees. Slide Rule (Operable Transparency). Vernier Caliper (Operable Transparency). Micrometer Caliper (Operable Transparency). Units of Measurement.
Meiners: 6–1, pp. 85–87, Measurement. 6–2, pp. 87–90, Subjective Effects of Measurement.
Laboratory Exercise: *Project Physics Handbook,* Holt, Rinehart and Winston, 1975, Expt. 1-2, Regularity and Time, p. 16.
Exercises and Experiments in Physics, Expt. 3, Measuring Time.

2.2-2.5 Length, mass, time
Exercises and Experiments in Physics, Expt. 1, Measuring Length.
Lindsay, *Basic Concepts of Physics,* Section 2.3, p. 28ff., The Measurement of Space and Time Sutton, M-106, p. 49, Inertia Balance.
Exercises and Experiments in Physics, Expt. 2, Measuring Mass.
Lindsay, *Basic Concepts of Physics,* Section 4.2, p. 95ff., Standards of Measurement of Mass, Gravitational Mass.
Lehrman, Expt. 6, Period of a Pendulum, pp. 16–17.
Science Study Series; *Mathematical Aspects of Physics: An Introduction* by Francis Bitter, M.I.T. (1963).
Exercises and Experiments in Physics, Expt. 3, Measuring Time.

2.10 Data, equations, and graphs
Lindsay, *Basic Concepts of Physics,* Section 1.7, K & E Transparencies (cont'd.): Relationship between the Gram and the Cubic Centimeter. Use of Powers of 10. Systems of Units-Part I. Systems of Units-Part II.

VELOCITY AND ACCELERATION 3

Synopsis: Suppose a small object initially at point A is later found at point B. The path over which the particle moved might not be known but it could be assumed to have been along a straight line between the two points, the simplest mode of motion from A to B. Now we could ask: (1) How long did it take the particle to go from A to B? (2) How far is it from A to B? In answering these questions we are introducing a new concept, *speed*, and if we also ask what direction the particle traveled in going from A to B, we are introducing the concept *velocity*.

The nature of motion, the distinction between speed and velocity, the vector representation of a velocity, and the addition of velocity vectors are all considered. Space-time curves for moving bodies are plotted and interpreted.

It has been shown that a particle might have different instantaneous velocities at different points along its path. This is the type of motion usually encountered in real life situations. This type leads to the idea of *acceleration*. Just as velocity is the rate of change of position with time, accelerated motion approximating free fall is studied.

The text now leads from a purely descriptive study of motion called kinematics to a study of the relationship between forces and the motion of bodies called dynamics.

Forces are introduced as a push or pull, with more precise definitions to follow in the next chapter. (The term force is a commonly used word and has many meanings.)

Newton first clearly described the way in which forces control the motion of particles and objects in the realm of our experience. This description was in the form of three axioms or self-evident statements of fact now usually called "laws" although they are hypotheses about the fundamental behavior of moving bodies.

The first "law" is a statement of the existence of inertia. The second is equivalent to our equation of motion $F = ma$. The third law, which was Newton's principal contribution to the theory of motion, is generally called the law of action and reaction.

From the study of the laws of motion, the student proceeds to a study of Newton's *law of universal gravitation,* which eventually made it possible to calculate the mass of the earth.

The chapter closes with a consideration of gravity and an introduction of the concept of a force field, in this case a gravitational field.

Comment: The concepts of speed and velocity will need careful treatment. Sections 3.2 and 3.3 should serve as a base. Some teachers may wish to expand the graphical treatment.

See *PSSC Physics*, Third Edition, Chap. 9 for a series of graphical representations ranging from position-time graphs for motion along a straight line to velocity vs time graphs. Develop the relationship between motion and displacement. Bring out the difference between speed and motion using blackboard diagrams similar to those in Figure 3-2, page 41. Derive the equation for average velocity. Bring out the difference between uniform and variable motion. Devote time to velocity vectors, solving problems by application of vector principles previously derived and applied.

Develop the concepts of acceleration and average acceleration, basing the development on examples as in Section 3.5. Stress acceleration as a vector quantity and its units as containing two units of time. Demonstrate Galileo's experiment on uniform acceleration using a long inclined plane. A more sophisticated demonstration uses Fletcher's trolley and an inclined plane. This demonstration with details as to apparatus and its use is described in *Leybold Physics Leaflet*, DC 531:311; c. Tabulated data such as appears in Section 3.6 (or arrived at experimentally) is used as the basis for the development of significant relationships and representative equations and graphs. Special attention should be paid to concepts of average velocity, final velocity, and initial velocity. Refer to *Physics Problems*, pages 188–200, for a further discussion of accelerated motion, for the solution of sample problems, and for additional problems. For an excellent treatment of accelerated motion based on the work of Galileo Galilei see Chapter 2, pages 13–35 of *Great Experiments in Physics*. The PSSC film *Straight-Line Kinematics* (0107) develops the relationships between distance, speed, and acceleration. The kinematics of straight-line motion is developed by experiments using an automobile. Three graphs showing speed and acceleration curves are analyzed.

Lead up to the consideration of freely falling bodies along the lines of the first paragraph in Section 3.7. It may be worthwhile to demonstrate mathematically the effect of increasing the steepness of a plane on the magnitude of the vertical component of the force of gravity. Demonstrate the coin-and-feather-tube experiment. The equations employing g for freely falling bodies should be developed and related to those using a. Developing a table such as appears on page 192 of *Physics Problems* on the chalkboard is helpful in having students realize the various relationships represented by equations.

Introduce the idea of force by having the students pull or push carts, balls, pucks, etc. Load the carts and pucks with different masses. Use both "frictionless" and typical surfaces. Develop Newton's laws by intuitive discussions of these activities. Stress the fact that the word "force," like other scientific terms, is commonly and widely used, along with many definitions, but that in physics the term must be defined in a very precise way.

In developing the second law, bring out that a body is accelerated so long as an unbalanced force continues to act upon the body. If the unbalanced force is zero, the acceleration is zero and the body (if moving) is moving with uniform velocity. If the unbalanced force is constant, the body possesses uniform acceleration. A variable force produces a nonuniform acceleration. Emphasize the units related to $F = ma$. In developing the third law it is particularly important that students realize that the forces termed action and reaction never act on the same body and thus cannot establish equilibrium in a system.

Use Chapter 4, The Laws of Motion, Isaac Newton, *Great Experiments in Physics*, pages 42–58 as a reference or for a student report. Select appropriate PSSC films from Forces (0301), Inertia (0302), and Inertial Mass (0303).

Introduce the topic of gravitational attraction by considering a person's "weight" on the moon. Discuss and develop the relationship of variation of weight with distance from the center of the earth and variation with location on the earth's surface. Then after dealing with "facts," develop the law of universal gravitation and discuss the determination of G and the mass of the earth, M. Some teachers may desire to discuss inertial and gravitational mass (*PSSC Physics, Second Edition*, pages 325–328, Section 20-1, pages 335–336 and Section 21-8 through 21-11, pages 370–375), or demonstrate the inertial balance, *Laboratory Guide for Physics*, III-3. Gravitational forces associated with objects are now usually described in terms of fields. The intensity of a gravitational field is given by the magnitude and direction of the force on a one-kilogram mass placed at the point and is given in newtons. Stress the idea presented in Section 3.14. Note Fig. 3-20, which shows a diagram of the earth's gravitational field. Refer to *Physics Problems*, pages 20–21. A student report may be made on the Law of Gravitation, Henry Cavendish, *Great Experiments in Physics*, pages 75–92. Use PSSC film 0309, Universal Gravitation.

References: Cohen, I. Bernard, *The Birth of the New Physics*, PSSC Science Study Series, Doubleday and Co., Inc., Garden City, NY (Galileo & Newton).

Exercises and Experiments in Physics, Holt, Rinehart and Winston, Publishers, New York, 1984.

Lehrman, *Laboratory Experiments*, Holt, Rinehart and Winston, Publishers, New York, 1969.

Lindsay, *Basic Concepts of Physics*, Van Nostrand Reinhold Co., New York, 1971, selected readings from Chapter 3, Motion and Relativity, and Chapter 4, The Concepts of Mass and Force.

PSSC Physics, Third Edition, D. C. Heath and Co., Boston, 1971, selected readings from Chapter 10, Vectors; Chapter 11, Newton's Law of Motion; Chapter 12, Motion at the Earth's Surface; and Chapter 13, Universal Gravitation and the Solar System.

Semat, *Fundamentals of Physics, Fourth Edition*, Holt, Rinehart and Winston, Publishers, New York, 1966, selected readings from Chapter 3, Motion of a Particle, and Chapter 4, Force and Motion.

Project Physics Text, Holt, Rinehart and Winston, Publishers, 1975, selected topics from Chapter 3, The Birth of Dynamics-Newton Explains Motion, and Chapter 4, Understanding Motion.

Films: *Galileo's Laws of Falling Bodies*, 6 min, black and white, sound, Encyclopaedia Britannica Films, Inc. Experiments with the inclined plane and freely fallng bodies both in the atmosphere and in a vacuum.

Laws of Motion, sound and color, Encyclopaedia Britannica Films, Inc. Newton's three laws of motion are explained and dramatically demonstrated.

The ABC of Jet Propulsion, 18 min, sound and color, free loan, General Motors Corporation. Explains the basic principles of jet propulsion in animated form. Newton's third law of motion is illustrated.

PSSC Films: 0107. *Straight-Line Kinematics*, 34 min. Notions of distance, speed, and acceleration discussed; graphs of all three vs time are generated using special equipment in a real test car; relationships among them shown by measurements (and estimates) of slopes and areas from the graphs.

0109. *Vector Kinematics*, 16 min. Velocity and acceleration vectors are introduced and shown simultaneously for various 2-dimensional motions including circular and simple harmonic (you may wish to delay the use of this film until you are well into Chap. 5, Curvilinear Motion). The vectors are computed and displayed as arrows on a cathode ray tube screen by a digital computer, and their relationships are discussed.

0301. *Forces*, 23 min. (This film should probably be used with both Chap. 4, Resolution and Composition of Forces, and with this chapter, Section 3.12 dealing with gravitation). Although basically an introduction to mechanics, this film foreshadows later work with kinds of forces. A qualitative Cavendish experiment shows gravitational forces between small objects. By means of this experiment, gravitational force is compared with electrical force in a simple demonstration.

Galilean Relativity: Part I—Ball Dropped from Mast of Ship, Color, 2 min. 50 sec. This Film Loop is a partial realization of the experiment suggested by Galileo. A ball is dropped three times and shown in slow-motion. First the ball is dropped by a sailor from the mast, and the ball lands at the base of the mast of the moving ship. The ball is then tipped off a stationary support as the boat goes by. Finally, the sailor picks up the ball and holds it briefly before releasing it. Not only can the principle of relativity be discussed but students can also study superposition of motion and frames of reference.

Galilean Relativity: Part 2—Object Dropped from Aircraft, Color, 3 min. 50 sec. A flare is dropped from a moving airplane. Scenes are filmed showing a flare dropping toward the ground from the plane in both a side and head-on view. Because the side view is filmed at two different slow-motion factors with freeze-frames, the student can time the motion and compute both horizontal and vertical components of the flare's displacement, velocity, and acceleration.

References for Chapter 3: Velocity and Acceleration

3.1 The nature of motion
Sutton: M-72, p. 37, Velocity Magnitudes.

3.2 Speed
Meiners: 7-1.1, p. 99, Uniform Motion.

Lindsay, *Basic Concepts of Physics,* Chapter 3, p. 21ff., Motion and Relativity.

Meiners: Chapter 7, pp. 99–133, Kinematics covers Motion in One Dimension, Motion in Two Dimensions, Relative Velocity and Acceleration.

Laboratory Exercise: *Project Physics Handbook,* Expt. 1-4, Measuring Uniform Motion, pp. 18–21.

3.3 Velocity
Sutton: M-73 & 74, p. 38. Vector Addition of Displacements. M-75, p. 38, Vector Addition of Velocities.

3.4 Solving velocity vector problems
Meiners: 6-4.1, p. 92, Vector Addition of Velocities, Baby Bulldozers.

Meiners: 6-4.9, p. 97, Simultaneous Velocities, Vector Addition.

Meiners: 6-4.10, p. 97, Uniform Motion and Vector for Addition.

3.5 Acceleration
Meiners: Acceleration Carts and Track Experiments. 7-1.5 pp. 102–105, 1. Elementary Acceleration Meas. 2. Measuring Acceleration from the Relationship $S = \frac{1}{2}at^2$.

Laboratory Exercises: Lehrman, Expt. 9, Speed and Acceleration, pp. 22–25. Expt. 17, Acceleration Due to Uniform Force, pp. 37–40.

Exercises and Experiments in Physics, Expt. 4, Acceleration

3.6 Solving acceleration problems
Sutton: M-76, 78, 82 & 83, pp. 28–40, Uniform Acceleration.

Meiners: 7-1.6, p. 105, Acceleration on an Inclined Plane.

Laboratory Exercises: *Project Physics Handbook,* Expt. 1-5, A Seventeenth-Century Experiment, pp. 22–23; *Project Physics Handbook,* Expt. 1-6, Twentieth-Century Version of Galileo's Experiment, p. 24.

Freilich, Florence G., "A Modern Visit to Galileo," *The Physics Teacher,* Feb. 1970, pp. 63–66.

Project Physics Transparencies Unit 1, T2 Graphs of Various Motions, s vs t, v vs t. T_3 a vs t. T3 Instantaneous Speed. T4 Instantaneous Rate of Change. T6 Derivation of $s = v_i t + \frac{1}{2}a(t)^2$. T8 Tractor-log Problem—variation of a horse-cart problem. Unit 3, T19 One-Dimensional Collisions, T20 Equal Mass, Two-Dimensional Collisions. T21 Unequal Mass, Two-Dimensional Collisions. T22 Inelastic Two-Dimensional Collisions. T23 Slow Collisions.

K & E Transparencies: Uniform Acceleration. Newton's First Law of Motion. Newton's Second Law of Motion. Newton's Third Law of Motion.

3M Transparencies: Physics-Motion, Science Packet No.

26. Originals A-K, origin and meaning of motion equations. L-P, displacements and velocities with free fall. Difference between velocity and speed.

3.7 Freely falling bodies
Meiners: 7-1.12, 7-1.13, 7-1.14, pp. 113–115, Freely Falling Bodies.

Laboratory Exercise: Lehrman, Expt. 10, Acceleration Due to Gravity, pp. 25–26.

Pomeranz, Kalman B., "The Time of Ascent and Descent of a Vertically Thrown Object in the Atmosphere," *The Physics Teacher,* Dec. 1969, pp. 507–508.

Sutton: M-79, p. 39, Guinea-and-Feather Tube. M-8-&-81, p. 39, Rate of Fall Independent of Mass. M-84, p. 40, Freely Falling Bodies. M-85, p. 40, Freely Falling Slab, M-86, p. 41, Pendulum-times Free Fall.

Laboratory Exercise: *Project Physics Handbook,* Expt. 1-7, Measuring the Acceleration of Gravity, pp. 25–29.

3.8 Law of inertia
Sutton: M-100 through M-107, pp. 46–50.

See entry **3.9,** Meiners.

Laboratory Exercises: Lehrman, Expt. 19, Inertia, pp. 42–43.

Expt. 20, The Inertia Balance, pp. 43–44.

Science Study Series: *Sir Isaac Newton: His Life and Work* by E. N. da C. Andrade, University of London (1965).

3.9 Law of acceleration
Sutton: M-108 through M-117, pp. 50–53, Second Law Demonstrations.

Meiners: 8-1.2, p. 136, Demonstration of 1st and 2nd Law.

Meiners: 8-1.4, 8-1.5, and 8-1.6, p. 137, Second Law Demonstrations. Laboratory Exercise: *Project Physics Handbook,* Expt. 1-8, Newton's Second Law, pp. 30–32.

Dennis, J. and L. Choate, "Some Problems with Artificial Gravity," *The Physics Teacher,* Nov. 1970, pp. 441–444.

Gamow, George, "Gravity," *Scientific American,* March 1961.

Science Study Series; *Gravity* by George Gamow, University of Colorado (1962).

3.10 Law of interaction
Sutton: M-118 through M-123, pp. 53–55, Action and Reaction Demonstrations.

Meiners: 8-1.8, pp. 138–140, Several quick experiments to illustrate Newton's Third Law.

3.11 Newton's Law of Universal Gravitation
Sutton: M-128, p. 57, Gravitational Torsion Balance-Cavendish Experiment.

Meiners: Section 8-8, pp. 166–176 suggests demonstrations for study of gravitation.

3.12 The mass of the earth
Thorne, Kip S., "Gravitational Collapse," *Scientific American,* Nov. 1967.

CONCURRENT AND PARALLEL FORCES

4

Synopsis: The preceding chapter considered the effect of an unbalanced force on the motion of an object (Newton's laws of motion). This chapter is concerned with specific forces, such as friction and gravity, and with the analysis of forces.

The chapter begins by defining forces in terms of four very precise characteristics of force and how forces can be measured by use of a spring balance. The use of a realistic example is employed to illustrate how forces must be represented by vectors. The problem of determining the resultant force of concurrent forces is described by the process of vector addition. Both graphical and trigonometric solutions are presented.

The concept of equilibrium is introduced in response to the problem of constant velocity. The idea of the equilibrant is used to explain this situation, and the determination of the equilibrant is presented. Resolution of forces into their components is then considered, with the classic example of an object on an inclined plane being used to reinforce the idea of vector resolution.

Friction is introduced as a force opposing the motion of objects. Emphasis is given to the fact that the causes of friction are not well understood and several theories that may account for the observed principles of friction are discussed. These principles are presented as approximations and not as laws.

The chapter concludes with a study of parallel forces acting on an extended body. This is in contrast to forces acting at a point. The idea of center of gravity is used to introduce the concept of torque, and rotational equilibrium is presented in terms of torque. Examples involving rotational equilibrium and their analysis in terms of the concept of torque are presented.

Comment: Introduce the chapter by having students lift various masses, such as one kilogram, brick, bucket of water, or even a shotput. Use of various springs with different force constants will illustrate the relationship between the force of the earth (the weight) on the object and the actual extension of the spring. Newton scales should be used to indicate how gravitational forces can be measured. The various objects lifted by the students might be weighed with the scales. Try to have the students experience one (1) newton, ten (10) newtons, twenty (20) newtons of force by this method. Student experience with such scales is essential in order to acquire some conceptualization of a newton. Use vector notation to represent various forces, making sure proper and consistent scale constructions are used. Devise different physical situations, involving both horizontal and vertical force applications, so that the student must plot vectors to describe the force applied. Illustrate the process of constructing a vector diagram to scale and its use in solving vector problems graphically. The teacher may wish to state the law of sines and cosines and show their application. Given two angles and a side of a triangle, the third angle is easily computed from the relation: $A + B + C = 180°$; and the other two sides may be computed from the law of sines. Given two sides and the included angle of a triangle, the law of cosines may be used to compute the third side, and the law of sines may be applied to find the angles. Refer to the Mathematics Refresher: 12. Triangles in the back of the text. For further discussion see Unit 2, Part V, pages 84–97, *Physics Problems*, Castka and Lefler, Holt, Rinehart and Winston. See page 88, *Physics Problems* for the law of sines and cosines. Concentrate on the importance of establishing a suitable scale and writing down the scale unit. Solution by the parallelogram method is the basic skill required of all students. A protractor may be employed for angles other than 90°. Develop and drill trigonometric solution methods according to the dictates of student ability and/or syllabus requirement. Then develop the three-or-more-forces solution method. Use the force table to establish concepts of equilibrium and equilibrant. Establish generalities dealing with the magnitudes of varying angles for two fixed forces and the magnitudes of the resultants. Teach the resolution of a single force into components acting at right angles to each other and the necessity of knowing the direction of components to be determined for other than right angle situations. Develop and drill the necessary procedures for graphic and trigonometric solutions. Include the resolution of the force of gravity of an object on an inclined plane and the effects of increasing the steepness of the plane. See Chapter 5, Part A, *Physics Problems*, pages 166–174, for a treatment of concurrent forces.

Lay a block on the demonstration table as illustrated by Figure 4-15 (page 81). Show that an appreciable force must be applied to overcome friction. Try this experiment using other surfaces. Exhibit magnitude of friction with dry ice pucks on a flat glass surface (almost frictionless motion). Indicate how the coefficient of friction is computed. Discuss differences between starting and sliding friction. Discuss the effects of the nature of the surfaces, area of contact of the bodies, and the speed of sliding. Refer to *Physics Problems*, Castka and Lefler, pages 184–188 and 350. Use Sections 4-8 through 4-10, pages 69–74, *Fundamentals of Physics* by Henry Semat, fourth edition, Holt, Rinehart and Winston, 1966, as a reference.

Set up on the demonstration table a pivoted meter stick with necessary weight to show equilibrium condition. Point out by appropriate questions the meaning of torque and axis of rotation. Tilt the meter stick, and have the class discover the effect on the length of the torque arm(s). Emphasize the necessity of using the perpendicular distance from the axis of rotation to the line of action of the force as the length of the torque arm. Develop the equilibrium principles: the magnitude of the algebraic sum of the forces must be zero, and the sum of the clockwise torques must equal the sum of the counterclockwise torques. Illustrate a bridge problem and its solution. Develop the concept of a couple with illustrations as on page 89. Refer to *Physics Problems*, Castka and Lefler, pages 174–181, and *Fundamentals of Physics*, Semat, pages 25–31.

Have a student balance a meter stick and a baseball bat. Demonstrate the experimental method for finding the center of gravity by plumb line intersection. Attaching a lead weight to the back of an irregularly shaped board (unknown to the class) produces unexpected results.

References: *Exercises and Experiments in Physics*, Holt, Rinehart and Winston, Publishers, New York, 1984.

Lehrman, Robert L., *Laboratory Experiments*, Holt, Rinehart and Winston, Publishers, New York, 1969.

Lindsay, *Basic Concepts of Physics*, Van Nostrand Reinhold Co., New York, 1971, selected topics from Chapter 4, The Concepts of Mass and Force.

PSSC Physics, Third Edition, D. C. Heath and Co., Lexington, Mass., 1971, Chap. 11.

Semat, Henry, *Fundamentals of Physics*, Holt, Rinehart and Winston, Publishers, New York, 1966, Chap. 4, p. 58ff., Force and Motion.

Project Physics Text, Holt, Rinehart and Winston, Publishers, New York, 1975, selected topics from Chapter 3, The Birth of Dynamics—Newton Explains Motion.

PSSC Film: 0108, Vectors. This film is a record of a lecture demonstration on vectors made to a high school class. The need for a quantity with direction and magnitude is pointed out. The meaning of addition and subtraction of vectors, and multiplication of vectors by a scalar is illustrated by demonstrations.

References for Chapter 4: Concurrent and Parallel Forces

4.1 Describing forces
K&E Transparencies: Mechanics. Weight versus Mass. Composition of Vectors. Resolution of Vectors. Moment of a Force.

4.2 Combining force vectors
Sutton: M-4, p. 16, Resolution of Forces.
M-5, p. 16, Resolution of Forces.
Laboratory Exercise: Lehrman, Expt. 8, Addition of Forces, pp. 19–22.

4.3 The equilibrant force
Sutton: M-10, p. 18, Equilibrium of Two Forces. M-11 to 14, pp. 18–19, Equilibrium of Three Forces.
Exercises and Experiments in Physics, Expt. 6, Composition of Forces.

4.4 Components of force vectors
Sutton: M-3, p. 16, Components of a Vector.
Meiners: 6-4.6, p. 96, Resolution of Force Apparatus.
Exercises and Experiments in Physics, Expt. 5, Resolution of Forces.
Laboratory Exercise: Lehrman, Expt. 14. Rectangular Components of a Force, pp. 31–33.
Sutton: M-16, 17, p. 20, Components of Force.
M-18, p. 20, Car on Inclined Plane.

4.6 The nature of friction
For suggested demonstrations see Sutton, M-48 through 53, pp. 30–32.
Meiners: Section 8-4, pp. 151–157, Frictional Forces.
Laboratory Exercise: Lehrman, Expt. 12, Friction, pp. 28–29.

4.7 Measuring friction
Exercises and Experiments in Physics, Expt. 7, Coefficient of Friction.

4.10 Center of gravity
Sutton: M-31, p. 25, Center of Gravity.
Meiners: Section 14-2, p. 328, Center of Gravity.

4.11 Torques
Laboratory Exercise: Lehrman, Expt. 16, Center of Gravity, pp. 36–37.
Exercises and Experiments in Physics, Expt. 8, Center of Gravity and Equilibrium.

4.13 The second condition of equilibrium
Sutton: Demonstrations M-32 through M-38, pp. 25–27, Equilibrium.
Meiners: Section 14-3, pp. 329–332, Translational and Rotational Equilibrium.
Laboratory Exercise: Lehrman, Expt. 15, Rotational Equilibrium, pp. 33–35.

TWO-DIMENSIONAL AND PERIODIC MOTION

5

Synopsis: Motion in a curved path is more common than motion along a straight line. Curved motion also presents more complications than straight-line motion and is, therefore, more difficult to describe.

Two-dimensional motion is introduced by reference to an object that is projected horizontally. The study of such a path is a challenge since one does not intuitively perceive the independent horizontal and vertical components of the motion. This concept is also applied to objects projected at an angle to the horizontal.

Uniform circular motion is then introduced as a special case of two-dimensional motion. Both horizontal and vertical paths are considered. The explanation of such motions are then presented in terms of Newton's Laws of Motion.

Next, the idea of a reference frame is considered. A particular thought experiment seems to contradict Newton's laws until the point of view of the observer is considered. How one perceives a motion depends upon one's frame of reference. What may appear to be stationary to one observer is actually moving from the point of view of another. The detection of a curved path and, therefore, the inference of the existence of an unbalanced force, may appear to be a rectilinear motion to another observer. The importance of inertial and noninertial reference frames is established in such analyses.

Rotary motion is then introduced, using examples from everyday living. This type of motion, which is typified by the motion of a point on the periphery of a wheel rotating about a fixed axis through its center, can be described in two ways: (1) by the actual distance traveled along the circumference of the wheel and (2) by the angle generated by a line joining the axis and the point in question. This motion is described in terms of the familiar concepts of velocity, acceleration, time, displacement, and inertia. The relation between a force and the motion of a body in a circular path is further considered. In order to get a stationary wheel in rotation a force must be applied somewhere off center. The application of a torque will tend to produce an angular acceleration of the body to which the torque is applied. As in the case of linear motion where the acceleration was dependent on the mass of the body, the angular acceleration is dependent upon the rotational inertia. The parallelism between the concepts developed for motion in a straight line and rotational motion are everywhere evident. Precession then is analyzed in terms of the concepts of rotary motion.

Periodic motion, such as that of a mass hanging from an elastic spring, is noted to be a case where the motion of an object results from the action of forces that depend on the distance of the particle from a fixed point. The particle moves back and forth along a line in periodic motion about a fixed point as center, a motion called simple harmonic. This is a type of motion that is frequently found in nature, as for example, the swaying of a tree limb in the wind. The relationship between simple harmonic motion and circular motion is then developed mathematically. Finally, the pendulum is considered as an example of simple harmonic motion and is analyzed in terms of the variable period, length, and g.

Comment: Demonstrate the ballistics car (S-W 0742) and apparatus showing that, neglecting air friction, the gravitational force on a projectile acts in the same way as it does on a freely falling body (S-W 0877 or 0879) or the range prediction apparatus (Ealing 13-159) and/or the water drop parabola demonstrator (Cenco 76030). Refer to Figures 5-2 and 5-3 in developing the trajectories of projectiles having varying velocities.

Introduce uniform circular motion, then variable circular motion by swinging a ball at the end of a string, first in a horizontal plane and then in a vertical plane. Use centripetal force apparatus (S-W 0930 or 0932 or Macalaster 300).

In discussing centrifugal force, a pseudo-force, be certain that students understand that it can be observed only when the observer is in the same accelerated frame of reference as the body on which the centrifugal force acts. Centrifugal force is nonexistent for a person in an inertial frame of reference.

Distinguish between circular and rotary motion. The equations for uniformly accelerated rotary motion corresponding to those for linear motion should be developed. Note Table 5-1. See (Cenco 75256, 75260, 75305, and 77451) or (S-W 0915, 0925, or 0928) for examples of demonstration apparatus.

Discuss the principle of the gyroscope and the cause of precession. A bicycle wheel or a toy gyroscope will be useful.

Discuss and demonstrate periodic motion. Attach a mass to a spring as in Figure 5-15. Present in detail the characteristics of the motion of this mass when it is oscillating freely. Identify this type of motion as *simple harmonic*. Show the relationship between simple harmonic motion and the projected image of a bob in uniform circular motion as it is cast on a screen placed normal to the plane of the motion and parallel to a diameter of the circular orbit of the bob. Demonstrate a simple pendulum and show that the motion of the bob when the amplitude is small is simple harmonic. Terms such as amplitude, frequency, and period will be used many times in our study of physics.

References: Abramson, Morris B., and Solomon Schrier, "Simple Harmonic Motion and Sine Wave Generator," *The Science Teacher*, Vol. 30, 6, p. 22.

Beams, J. W., "High Centrifugal Fields," *The Physics Teacher*, Sept. 1963, pp. 103–108.

Bennett, D. M., "Apparatus for the Demonstration of Simple Harmonic Motion," *American Journal of Physics*, June 1962, p. 470.

Branley, Franklyn M., "Why Satellites Remain in Orbit," *The Science Teacher*, May 1959, p. 238.

Cise, John P., "Projectile Motion Using the Linear Air Track," *The Physics Teacher*, Feb. 1968, pp. 78–79.

Ellis, R. Hobart, Jr., "Some Different Views of Simple Harmonic Motion," *The Physics Teacher*, Oct. 1968, pp. 340–343.

Greenslade, Thomas B. Jr., "Damped Simple Harmonic Motion on a Linear Air Track," *The Physics Teacher*, Oct. 1969, pp. 395–396.

Haas, Douglas J., "Projectiles," *The Physics Teacher*, May 1971, pp. 261–262.

Project Physics Resource Book, Concepts of Motion, 1975.

Sawyer, Raymond B., "Projectile Motion in Different Frames of Reference," *The Physics Teacher*, April 1968, p. 178.

Sears, Francis W., "Weight and Weightlessness," *The Physics Teacher*, April 1963, pp. 20–26.

PSSC Films: 0304. *Free Fall and Projectile Motion,* 27 min. The behavior of freely falling bodies is explored from a dynamical point of view, leading to the proportionality of gravitational and inertial mass, the independence of perpendicular components of a motion, and the conclusion that Newton's Law is a vector relationship. Includes a slow-motion study of two balls simultaneously dropped and projected horizontally from the same height, and experiments with a large "monkey gun."

0305. *Deflecting Forces,* 29 min. Discusses nature of forces that produce curved paths, brings out concepts of centripetal vector acceleration, shows how knowledge of path and mass of an object give information on the force involved.

0306. *Periodic Motion,* 33 min. From a number of periodic motions, simple harmonic motion is selected for detailed examination: a pen moving in SHM plots its own displacement-time graph; graphs of velocity and acceleration versus time are derived from it. The formula for period of SHM is derived from the component of circular motion; the dynamics and period of SHM are checked experimentally by oscillation of a dry ice puck mounted between springs.

0307. *Frames of Reference,* 28 min. By means of a variety of experiments on frames of references moving at constant speed or at constant accelerations, this film demonstrates the distinction between an inertial and noninertial frame of reference and the appearance of fictitious forces in a noninertial frame.

McGraw-Hill Films: *Uniform Circular Motion,* b/w, 7 min.
Simple Harmonic Motion, b/w, 9 min.

8mm Film cartridges: *The Velocity Vector* — the first of a series of six films made from the 16mm film Vector Kinematics. Four of the other five are listed below. PSSC.
Velocity in Circular and Simple Harmonic Motion.

The Acceleration Vector.
Velocity and Acceleration in Simple Harmonic Motion.
Velocity and Acceleration in Circular Motion.

References for Chapter 5: Two-Dimensional and Periodic Motion

5.1 Motion in a curved path
Sutton: M-90, M-91, and M-92.
Lindsay, *Basic Concepts of Physics*, 1971, pp. 56–64, 110–113.
PSSC Physics, Third Edition, 1971, Section 10-5, Velocity Changes and Constant Vector Acceleration, pp. 205–207.
Meiners, pp. 122–131.
See *Project Physics Handbook*, pp. 46–47 Unit 1 for suggestions for demonstrations of projectile motion.

5.2 Motion in a circular path
Transparencies: Uniform circular motion and the origin of centripetal force, 3M Sci. Packet 26 Originals Q through U.
Lab Exercises: *Project Physics Handbook*, Expt. 1-12, Centripetal Force, p. 39, and Expt. 1-13, Centripetal Force on a Turntable, pp. 40–41.
Sutton, M-139, M-140, and M-141.
Western, 2 × 2 color transparencies, H-5 through 10, H-13 and 14.
Project Physics Handbook, Penny and Coat Hanger, p. 48 Unit 1.
PSSC Physics, Third Edition, 1971, Section 10-6 Changing Acceleration, and 10-7 Centripetal Acceleration, pp. 207–212.
Sutton, M-137 & M-177.
Meiners, 10-2.5, p. 229.
Lab Exercise: Lehrman, *Laboratory Experiments*, Expt. 21, Centripetal Force, pp. 45–47.
Laboratory Exercise: *Exercises and Experiments in Physics*, Expt. 9, Centripetal Force.

5.3 Motion in a vertical circle
Sutton, M-137.

5.4 Frames of reference
Halliday and Resnick, Wiley, 1966, pp. 120–121.
See *Project Physics Handbook*, pp. 47–48 for three ways to show how a moving object would appear in a rotating reference frame.
PSSC Physics, Third Edition, 1971, Section 10-8, The Description of Motion; Frames of Reference, pp. 212–215.
Section 12-9, Experimental Frames of Reference, pp. 262–263.
Section 12-10, Fictitious Forces in Accelerated Frames, pp. 263–264.

5.5 Motion around an axis
Sutton: Demonstrations associated with the study of rotation are found on pp. 61–68, M-137 through M-160.
Meiners, p. 271ff.

Semat, *Fundamentals of Physics*, Rotational Motion, Chapter 7, p. 123ff.

5.7 Angular acceleration
Refer to *Physics Problems*, Section IX, pp. 360–364.
Semat, *Fundamentals of Physics*, 1966, Chapter 7, pp. 123–133.

5.8 Rotational inertia
Sutton, M-170 & 171.
Meiners, Rotational Dynamics of a Rigid Body, Section 12.4, pp. 281–286.
Sutton, M-161 through M-165.
Lindsay, *Basic Concepts of Physics*, Section 4.6, Moment of Force and Angular Momentum, pp. 106–110.
Semat, *Fundamentals of Physics*, Section 7-6, pp. 127–129, Moments of Inertia.
Lindsay, *Basic Concepts of Physics*, pp. 106–110.

5.9 Precession
Sutton, M-178, M-187, and M-188.
Meiners, pp. 296–298.
Transparencies: 3M Sci. Packet 26, Originals V and W, Precession.
Semat, *Fundamentals of Physics*, Section 7-15, pp. 142–145, The Gyroscope.

5.10 Periodic motion
Semat, *Fundamentals of Physics*, 1966, Chapter 8, pp. 148–160.
Lindsay, *Basic Concepts of Physics*, pp. 113–120.
Sutton, S-1, S-2, S-3, and S-5.
Meiners, pp. 334ff.
Lab Exercise: Lehrman, *Laboratory Experiments*, Expt. 22, Simple Harmonic Motion, pp. 47–49.
Western, 2 × 2 color transparencies: H-17 through 20, and I-1 & 2.
PSSC Physics, Third Edition, 1971, Section 12-8, Simple Harmonic Motion, pp. 257–262.

5.11 Analyzing harmonic motion
Sutton, S-1 and S-2.
Project Physics Handbook, pp. 50–51 Unit 2, Forces on a Pendulum.
Project Physics Handbook, 1970, pp. 180–182 for the energy analysis of a pendulum swing.

5.12 The pendulum
Sutton, S-5.
Meiners, 15-5, Physical Pendulum, p. 353ff; 15-6, Center of Percussion, p. 355ff.
Laboratory Exercises: *Exercises and Experiments in Physics*, Expt. 10, The Pendulum.
Sutton, M-203, M-204, and M-205.

CONSERVATION OF ENERGY AND MOMENTUM

6

Synopsis: Work is defined in at least a dozen ways ranging from employment to the definition that we accept in physics as the product of the force and the displacement. The term power has nearly as many connotations ranging from political control or influence to the physical definition, the time rate of doing work.

The physical meaning of the term work, the units for the measurement of work, the calculation of the amount of work done under the action of steady and variable forces, and the work done in rotary motion are all considered. Power is defined and calculated for each instance.

Some of the meanings of *energy* are vitality, capacity for acting, forceful exertion, and our physical definition: the capacity for doing work. In this chapter, gravitational potential energy and the energy possessed by a moving body, whether in linear or rotational motion, are discussed, and the equations for the development of the magnitude of the energy are derived for each case. Further forms of energy such as that stored in a stretched spring or stored in a fossil fuel are considered.

People have long recognized the social significance of the fact that life is impossible without somebody's labor. The developments of science and technology have resulted in machines that diminish the "effort" involved in labor. Machines that make possible the exertion of a large force by the application of a smaller one have a compensating disadvantage. Even in a frictionless machine the effort force that is small must act through a great distance to cause the resistance force that is large to act through a short distance. Here we find a constancy (energy output never exceeds energy input) that is of enormous consequence in all science.

From the energy input and output relations, the efficiency of the machine is calculated.

We next consider the cumulative effect of a force applied to an object for a period of time. The result is a change in the quantity known as *momentum*, which is the product of the object's mass and velocity. Momentum is a vector quantity having the same direction as the velocity. The change in the momentum of the object is equal to the impulse, defined as the applied force multiplied by the time interval of application. Emphasis is then given to the law of conservation of momentum. Momentum conservation in collision experiments between both elastic and inelastic bodies is presented, along with a two-dimensional vector problem.

Finally, the concept of angular momentum is presented, with emphasis on its conservation also.

Comment: Using a spring balance, lift a block of wood from the floor. Using the spring balance, pull the block on the table surface with uniform velocity, applying the force horizontally. Repeat using the spring balance at an angle. Compare readings and point out that work was done in each case. Develop the work concept, $W = F \Delta d \cos \theta$. Introduce the joule (newton meter) as the SI work unit. Refer to Figure 6-2 in developing the concept of work done by a varying force. Develop the equations to compute the work done in rotary motion (see Section 6.3).

Discuss horsepower of automobile engines as an introduction to the concept of power. Emphasize that the watt (or kilowatt), frequently applied to electric power, is a fundamental SI unit of power. Devote some time to the consideration that if $P = W/t$ then $W = Pt$. Refer to *Physics*

Problems, Chapter 5, Section III, Work, Power, Energy, and Machines, pages 213–234 and to Chapter 11, Section VII, Work, Power, and Energy, pages 356–357, for added text material, solved problems, and problems to solve. Use PSSC film 0311, *Energy and Work,* which presents experimental evidence that the kinetic energy gained by an object is measured by the work done by the net force acting on it. Consider power in rotary motion. Review essential background material first presented in Chapter 5. Emphasize the similarity of relationships developed for the calculation of work and power in linear and rotary motion.

Referring to a pendulum bob, have the class explain what happens to the work done in raising the bob initially. Have the students trace the energy transformations that occur as the bob swings. By using other examples have the students arrive at definitions of potential and kinetic energy. Develop the equation for potential energy according to the treatment in Section 6.7, and the kinetic energy equation according to that in Section 6.8. Stress the importance of Newton's second law of motion and the relationship $w = mg$ in deriving and understanding the energy equations. Bring out that energy is expressed in work units. Show, for example, that $W = F\,\Delta d = ma \times \frac{1}{2}a\,\Delta t^2 = \frac{1}{2}ma^2\,\Delta t^2 = \frac{1}{2}mv^2$. Develop the relationships for the kinetic energy of rotation and not similarity to those for linear motion. Develop the relationship for the total energy of a wheel rolling down an incline, that is, potential energy plus rotational kinetic energy plus linear kinetic energy. Develop the relation for the energy of a stretched or compressed spring. Finally deal rather fully with Section 6.11 on conservation of energy.

Clarify the meaning in physics of the term *machine*. Discuss machines with regard to the six types of simple machines being variations of the two basic types — the lever and the inclined plane. The six types of simple machines listed must surely have been studied in some detail in earlier courses but an understanding of the meaning of efficiency of a machine should be considered with the solution of examples.

The concept of momentum can be introduced by involving the students in the motion of frictionless pucks, with either equal or different masses. An air track or air table could be used very effectively here. Stress the conservation of the concept in terms of the entire system. The Project Physics 8 mm film loops could also be used, both as discussion topics and/or as a student exercise. Demonstrations, including a compressed spring held closed between two pucks by a thread and then burning the thread will help emphasize the concepts involved.

For further discussion of momentum, see *PSSC Physics*, Third Edition, Chapter 14. Also, Fletcher's trolley experiments are described in Leybold Physics Leaflets: DC531 : 612;a (Impulse and Momentum), DC531 : 66;a (The Laws of Collisions Demonstrated with Fletcher's Trolley). Other demonstrations and experiments, which involve dynamics carts, timers, and timing tape, are described in *Laboratory Guide for Physics*, Second Edition, PSSC, Part III, pages 35–59.

An excellent test of the student's understanding would be the successful analysis of the Project Physics film loop, "Explosion of a Cluster of Objects."

The concept of angular momentum can be introduced from linear phenomena, stressing its conservation. Demonstrations with a rotating stool are very impressive, as are a bicycle wheel or toy gyroscope. Discuss the relationships between angular impulse and change in angular momentum.

References: Bauman, Robert F., "The Meaning of the Conservation Laws," *The Physics Teacher*, April 1971, pp. 186–189.

Harrison, George Russell, *The Conquest of Energy*, Morrow, 1968. Both theory and practical applications are considered in this unified picture of energy and its control.

Hartman, Claude C., "Momentum vs Kinetic Energy," *The Physics Teacher*, April 1968, p. 177.

Lindsay, Robert Bruce, *Basic Concepts of Physics,* Van Nostrand Reinhold Co., New York, 1971, Chapter 5, pp. 137–190.
Project Physics Resource Book. The Triumph of Mechanics, 1975.
  Project Physics Film Loops (Super 8 Cartridges):
  A Method of Measuring Energy, Nails Driven into Wood, 3 min. 20 sec.
  Gravitational Potential Energy, 3 min. 55 sec.
  Kinetic Energy, 3 min. 45 sec.
  Conservation of Energy, Pole Vault, 3 min. 55 sec.
  Conservation of Energy, Aircraft Takeoff, 3 min.
Semat, Henry, *Fundamentals of Physics,* Chap. 5, Work and Energy, pp. 88–104; Secs. 7-11, Work and Rotational Energy and 7-12, Power and Rotational Motion, pp. 135–137.
Strong, Lawrence, E., "Improving Concepts of Energy," *The Science Teacher,* Dec. 1968, pp. 8–10.

Films: PSSC, 0313. *Conservation of Energy,* 27 min. Energy traced from coal to electric output in a large power plant, quantitative data is taken in the plant; conservation law demonstrated for random and orderly motion.

PSSC, 0311. *Energy and Work,* 28 min. Shows that work, measured as the area under the force-distance curve, does measure the transfer of kinetic energy to a body, calculated from its mass and speed. A large-scale falling ball experiment, a nonlinear spring arrangement, and a "Rube Goldberg" graphically establish work as a useful measure of energy transfer in various interactions.

PSSC, 0318. *Elastic Collisions and Stored Energy,* 27 min. Various collisions between two dry ice pucks are demonstrated. Cylindrical magnets are mounted on the pucks producing a repelling force. Careful measurements of the kinetic energy of the pucks during an interaction lead to the concept of stored or potential energy.

PSSC, 0320. *Moving with the Center of Mass,* 26 min. Demonstrates the validity of the conservation of energy and momentum of several dry ice puck interactions as viewed in two different frames of reference.

PSSC, 0319. *Collisions of Hard Spheres,* 19 min. This is a laboratory instruction film dealing with conservation of momentum, primarily intended for teachers. The film is a demonstration of the adjustments and operation of the Collision in 2-D apparatus. The conservation of momentum is demonstrated for both equal and unequal mass spheres.

Filmstrip: *Conservation of momentum,* (fs), 41 frames, color, FOM, No. 626, McGraw-Hill Book Co., Text Film Dept., 1221 Avenue of the Americas, New York, NY 10020. Deals with impulse and momentum extending Newton's laws to conservation of momentum. Uses frictionless pucks for momentum and collision experiments.

Project Physics, 8 mm Film Loops: These loops could be used in the teaching of Momentum, Sections 6.12–6.14.

One-Dimensional Collisions: Part I, color, 3 min. 25 sec. The head-on collision of two steel balls of different masses hanging by long, thin wires is photographed in slow-motion. Different scenes are presented, each showing the ball at rest before collision. The student is able to time the motion and measure the distances of the moving balls just before and after collision. Since the masses are known, the student is able to make detailed analyses of the extent to which momentum and kinetic energy are conserved.

One-Dimensional Collisions: Part 2, color, 3 min, 35 sec. In slow-motion, two moving steel balls of different but known masses are shown colliding head-on. Since they are hanging by long-thin wires, their paths are very nearly straight lines. Two examples are shown. First the two balls collide approaching each other. Then they collide as one ball overtakes the other going in the same direction. The student can make detailed measurements of the total momentum and kinetic energy of the balls before and after collision.

Inelastic One-Dimensional Collisions, color, 3 min. 45 sec. Two sticky balls of known but different masses hanging by long, thin supports are photographed in slow-motion during head-on collision in one dimension. In one example, one ball is at rest before collision. In the second example, the balls move toward each other. In both cases, the balls stick together, providing a single moving system after collision. Detailed measurements and analyses can be made by the student before and after collision to explore the conservation of momentum and kinetic energy in an inelastic collision.

Two-Dimensional Collisions: Part 1, color, 2 min. 45 sec. Two hard, steel balls hanging from long, thin wires collide. The student sees the collision in slow-motion and can measure the velocities of the balls before and after the collision. The student can use the principles of conservation of momentum and conservation of energy to predict their behavior after collision. By projecting this Film Loop on a sheet of graph paper, a student can actually take the data and analyze it for this experiment.

Two-Dimensional Collisions: Part 2, color, 2 min. 40 sec. In two-dimensional collisions of two balls, both the direction and magnitude of the colliding balls have to be considered. By projecting the Film Loop onto a large piece of paper, the student can trace and time each ball's motion. The student can then determine the kinetic energy and momentum of each ball before and after collision. The analysis can be in arbitrary time units, or in clock time if the slow-motion rate is taken into consideration. Principles of conservation of momentum and energy can be studied.

Inelastic Two-Dimensional Collisions, color, 3 min. 35 sec. Two sticky balls of the same mass hanging from long, thin wires are photographed colliding inelastically in slow-motion. After collision, the balls stick together and move as one mass. The student can project this Film Loop on paper to mark the paths of motion, and by timing and measurement calculate the velocities. The kinetic energy and momentum can be found both before and after collision, and the student can check the usefulness of the principles of conservation of momentum and energy.

Scattering of a Cluster of Objects, color, 2 min. 10 sec. In another more complex example of a collision, a single ball of hardened steel collides with six clustered balls at rest. A student can easily determine if the total momentum of the system is conserved. This can be done by projecting the film on a piece of paper and noting the position of each ball at an initial time and at a later time. Connecting these two points and multiplying by each ball's mass will give a relative measure of the momentum of each ball. The student can then simply add vectors and compare the precollision and postcollision situations for momentum conservation.

Explosion of a Cluster of Objects, color, 2 min. 30 sec. Five balls of different masses, initially at rest, suspended independently from long, thin wires are photographed in slow-motion when a gunpowder charge in the center of the cluster is exploded. Each ball moves off in a different direction. Since the smoke of the explosion hides one of the balls at first, the student can try to predict where that "unknown" ball will be located when the smoke clears. To make this prediction, the student must make detailed measurements of the magnitude and direction of each ball's velocity by projecting the film on paper and measuring distances and time intervals. The student can then compute the total momentum of the system.

Recoil, color, 3 min. 55 sec. The recoil of a cannon upon firing at a fort in Montreal is presented in a preliminary scene. Then a laboratory set-up shows a bullet of known mass fired from a miniature brass "cannon" which is suspended by long, thin wires. After firing, the bullet and "cannon," projected in slow-motion, move away from each other in a straight line. The masses of the bullet and cannon are given. When the speed of the bullet has been measured, the speed of the cannon can be predicted from the conservation of momentum. The cannon's speed can now be checked by timing its motion.

Colliding Freight Cars, color, 2 min. 50 sec. Two colliding railroad freight cars are filmed. They briefly exert on each other a force of about 1,000,000 lb. The masses of the two freight cars are known, and the student can measure the velocities of the two cars before and after collision. Momenta and kinetic energies can be computed. Since the collision is partially elastic, and friction on both cars is appreciable after collision, the student is challenged to explain any differences in momentum and kinetic energy before and after collision.

Dynamics of a Billiard Ball, color, 3 min. 35 sec. In slow-motion, two billiard balls are photographed so that we can study the action resulting from a moving ball striking a stationary ball. Since the masses are equal and distances can readily be measured, the student can compute the velocities of the two balls after contact. But the results will indicate that applying conservation of momentum to the balls alone is not sufficient. The student is led to consider the elasticity of the balls, their contact with the supporting table, the spin of their motion, sliding friction, and rolling friction to explain discrepancies.

Finding the Speed of a Rifle Bullet: Method 1, color, 3 min. 10 sec. A rifle bullet of known mass is fired into a log of known mass and dimensions. The log, initially at rest, is suspended by four long, thin wires. After impact, the log and imbedded bullet move together and are photographed in extreme slow-motion. The student can measure the rate of this motion, and using the principle of conservation of momentum can compute the speed of the bullet before impact. The student can also compute the kinetic energy of the bullet before it strikes the log and the combined log-and-bullet mass afterwards.

Finding the Speed of a Rifle Bullet: Method 2, color, 2 min. 25 sec. A rifle bullet of known mass strikes a log of known mass and dimension that is suspended by long, thin wires so that it swings like a pendulum after impact. Slow-motion photography enables the student to observe the amount of rise of the log plus bullet. By using data from the Film Loop, the student can find the potential energy of the log at the top of its swing. This, by the principle of conservation of energy, will equal the kinetic energy of the moving mass at its lowest point, which in turn can be used to find the speed of the log. This speed can be checked by taking measurements from the Loop. The momentum of the log can now be calculated, and by using the principle of conservation of momentum, the student may find the speed of the bullet. The student may check this speed from his or her measurement in *Finding the Speed of a Rifle Bullet: Method 1*.

References for Chapter 6: Conservation of Energy and Momentum

See Sutton, pp. 59–60.

Semat, *Fundamentals of Physics*, Chapter 5, p. 88ff., Work and Energy.

Lindsay, *Basic Concepts of Physics*, Chapter 5, p. 137ff., The Concept of Energy.

PSSC *Physics, Third Edition*, Chapter 15, p. 326ff., Kinetic Energy, and Chapter 16, p. 344ff., Potential Energy.

Project Physics Text, Chapter 10, p. 29ff., Unit 3, Energy.

6.1 Definition of work
K & E Transparency: #22 Work.

6.4 Machines
Sutton: M-43 Lever. M-44 Windlass. M-45 Pulleys. M-46 Pulleys. M-47 Pulleys.
Lab Exercises: Lehrman, *Laboratory Experiments*, Expt. 24, Pulleys, pp. 52–54; Expt. 25, The Inclined Plane, pp. 55–56; Expt. 26, Efficiency of a C-Clamp, pp. 56–57.
K & E Transparencies: #24 Simple Machines, Part I. #25 Simple Machines, Part II.

Laboratory Exercise: *Exercises and Experiments in Physics*, Expt. 15, Efficiency of Machines.

6.5 Definition of power
Sutton, M-134 Brake Horsepower Test; M-135 Power Developed by a Person; M-136 Power Bicycle.
Laboratory Exercise: *Exercises and Experiments in Physics*, Expt. 12, Power.

6.6 Power in rotary motion
Semat, *Fundamentals of Physics*, Section 7-11, pp. 135–137,

Work and Rotational Energy. Section 7-12, p. 136, Power and Rotational Motion.

6.7 Gravitational potential energy
See Chapter 16, Potential Energy, *PSSC Physics, Third Edition,* 1971, pp. 344ff.

6.8 Kinetic energy in linear motion
Lab Exercise: Lehrman, *Laboratory Experiments,* Expt. 27, Kinetic Energy, pp. 57–58.
K & E Transparency: #23 Energy, Kinetic and Potential.

6.9 Kinetic energy in rotary motion
See Chapter 15, Kinetic Energy, *PSSC Physics, Third Edition,* 1971, pp. 326ff.

6.10 Elastic potential energy
Meiners, 9-1.4, The dependence of the kinetic energy on the square of the velocity.
Laboratory Exercise: *Exercises and Experiments in Physics,* Expt. 13, Elastic Potential Energy.

6.11 Conservation of mechanical energy
Sutton: M-132, Galileo's Pendulum Board. M-133, Pile Driver.

6.12 The nature of momentum
Sutton: M-124 through 127, pp. 55–57, Impact and Momentum.
Laboratory Exercise: Lehrman, Expt. 23, Momentum, pp. 50–52.
Meiners: 9-4, pp. 192–208, Impulse and Momentum.

Western, 2 × 2 Color Transparencies, I-3 through I-12, Momentum and Collisions, Conservation of Linear Momentum and Collision of a Sphere with a Wall.

6.13 The conservation of momentum
Meiners: 9-4.5, 9-4.7, p. 195, Conservation of Momentum.
Exercises and Experiments in Physics, Expt. 14, Conservation of Momentum.
Project Physics, 8mm Film Loop, Inelastic Two-Dimensional Collisions. Color, 3 min. 35 sec.

6.14 and 6.15 Inelastic and elastic collisions
Meiners: Section 9-5, pp. 208–219, Elastic and Inelastic Collisions.
Meiners: Chapter 10, pp. 221–249, Low Friction I: Air Supported Pucks and Air Tables and Chapter 11, pp. 250–269, Low Friction II: Air Troughs and Air Bearings.
Laboratory Exercise: *Project Physics Handbook,* Expt. 3-1, Collisions in One Dimension-I, pp. 4–5 Unit 3; Expt. 3-2, Collisions in One Dimension-II, pp. 5–12 Unit 3.
Project Physics Handbook, Expt. 3-3, Collisions in Two Dimensions-1, pp. 13–15 Unit 3; Expt. 3-4, Collisions in Two Dimensions-II, pp. 16–21 Unit 3.
Meiners, 9-1.2, 9-1.3, Conservation of Energy. 9-1.5 and 9-1.7, Interchange of Kinetic and Potential Energy.

6.16 Angular momentum
Semat, *Fundamentals of Physics,* Section 7-14, pp. 139–142, Angular Momentum.

7 PHASES OF MATTER

Synopsis: The kinetic theory of matter and its use in explaining specific phenomena are the major topics of this chapter. The text presentation illustrates how a scientific theory can be built on a body of indirect evidence. The chapter begins on an historical note by briefly commenting on the Greek attempt to conceptualize the structure of matter and indicating that it was another 2000 years before further serious inquiry resulted in Dalton's statement. Although the Greek concept was not quite the same as our modern concept, it was an intellectual attempt to solve a problem that was of great concern to them. This was the problem of continuity: is matter continuous throughout or is it discrete? Philosophers such as Aristotle, Anaxagoras, Leucippus, and Democritus were major contributors to the discussion of that question. Although Greek thought about the nature of matter never really evolved to a more acceptable theory, it did stimulate speculation.

Some of the supporting evidence for our belief in atoms and molecules is presented after the initial historical introduction. The two major tenets of the kinetic theory of matter are stated, and a brief comment follows regarding the motions of molecules, the forces acting between molecules, and the kinetic energy of molecules.

The kinetic theory of matter is then applied to various observed phenomena. The phenomena include the phases in which matter exists, cohesion, adhesion, malleability, and elasticity of solids. Additional observations pertaining to liquids are described. These include cohesion and adhesion, surface tension, capillarity, melting, and the effect of pressure and dissolved materials on the freezing point of a liquid. Observational evidence from the gaseous phase is also described, with emphasis on the processes of evaporation, sublimation, and boiling. The process of diffusion and the change of phase from a solid to a liquid to a gas are offered as additional evidence for the validity of the kinetic theory of matter. Plasma is briefly discussed as a special phase of matter at high temperatures.

Throughout the chapter, there is emphasis on the idea that a theory is an assumption made by scientists and that an acceptable theory will enable us to understand many seemingly unrelated phenomena.

Comment: This chapter presents many opportunities for consolidating, unifying, and applying previously learned concepts. Concepts such as forces, motion, and energy can easily be related to the structure of matter and, in particular, to the phenomenon of phase change.

One way of introducing the chapter would be through demonstrations such as a display of the fluid properties of mercury and its spherical droplets and the use of a molecular vibration tube. These phenomena pertain to many of the topics in the chapter, and they present an opportunity for the students to offer explanations for their observations. Along with this approach, you could easily lead into the question of the role of theory in science, especially if the demonstrations are presented as problem-solving situations.

Such pertinent demonstrations should permit you to develop the three basic assumptions of the kinetic theory of matter, stressing that the evidence is still of an indirect nature. Use of Experiment 15, The Size of a Molecule, *Exercises and Experiments in Physics,* will help to demonstrate that an indirect method can be used for obtaining an estimate of the size of a molecule.

In the study of solids, review crystal structure and types of forces involved. Demonstrate the magnitude of these forces by melting, or attempting to melt, samples of sodium chloride (M.P. is 801 °C), ice, and naphthalene (M.P. is 218 °C). Cohesion and adhesion can be demonstrated as suggested in Section 7.7 or by using experiments such as those in Sutton (M-259-261, pages 104–105). Demonstrate malleability by deforming a piece of lead. Sutton (M-63 through 69, pages 34–37) presents some good demonstrations on elasticity. Use any of the above demonstrations to develop the ideas of stress and strain. Demonstrate or assign Experiment 13, Elastic Potential Energy, *Exercises and Experiments in Physics.* Develop the concept of elastic modulus from it.

Introduce the study of liquids by comparing liquid structures with solid structures. Brownian motion, diffusion, cohesion, and adhesion should be demonstrated as suggested in Sections 7.12 and 7.13. Surface tension (Section 7.14) and capillarity (Section 7.15) should also be demonstrated, applying the kinetic interpretation to each. A highly readable, first-hand account of an investigation of surface tension and capillarity is included in *Soap Bubbles and the Forces that Mold Them Together,* by C. V. Boys (Science Study Series #S3).

References: Breck, D.W., and J.V. Smith, "Molecular Sieves," *Scientific American,* Jan. 1959, p. 85.
Derjaquin, Boris V., "The Force Between Molecules," *Scientific American,* July 1960, pp. 47–53.
Flowers, B. H., and E. Mendoza, *Properties of Matter,* Wiley, New York, 1970.
Holden, Alan, *Crystals and Crystal Growing,* Science Study Series, 1960.
———, *The Nature of Solids,* Columbia University Press, New York, 1965.

Kantor, Frederick W., "Floating Balloons and Soap Bubbles on an Air-Freon 12 Gas Interface," *The Physics Teacher*, May 1963, p. 82

Lehrman, Robert, and Clifford Swartz, *Foundations of Physics*, Holt, Rinehart and Winston, Publishers, New York, 1969, Chapter 10, Heat and Microstructure.

Lifshitz, Eugene M., "Superfluidity," *Scientific American*, June 1958, p. 30.

Lindsay, Robert Bruce, *Basic Concepts of Physics*, Van Nostrand Reinhold Co., New York, 1971, Chapter 6, The Atomic Concept.

PSSC Physics, Third Edition, 1971, Chapter 17, Heat Molecular and Conservation of Energy.

Renier, Marcus, "The Flow of Matter," *Scientific American*, December 1959, pp. 122–124.

Runnels, L. K., "Ice," *Scientific American*, Dec. 1966, pp. 118–124.

Schrödinger, Erwin, "What Is Matter," *Scientific American*, Sept. 1953.

Semat, *Fundamentals of Physics*, Holt, Rinehart and Winston, Publishers, New York, 1966, pp. 246–259.

Smith, Robert A., "Introduction to the Structure of Matter," *The Science Teacher*, Dec. 1969, pp. 74–75.

Trout, Verdine E., "Surface Tension," *The Science Teacher*, Oct. 1957, p. 285.

Zandy, Hassan F., "Plasma, The Fourth State of Matter," *The Physics Teacher*, Jan. 1970, pp. 27–31.

Films: PSSC, 0115. *Behavior of Gases*, 15 min. The Brownian motion of smoke particles is shown in photo-micrography and compared with a mechanical analogue. This evidence for molecules in chaotic motion is contrasted with the orderly behavior of gases as shown by Boyle's Law experiment. Animation and mechanical analogues are then used to develop a model for gas pressure based on chaotic molecular motion.

PSSC, 0113. *Crystals*, 25 min. Demonstrates the nature of crystals, how they are formed, and why they are shaped as they are. Shows actual growth of crystals under a microscope: discusses how they may be grown. Related these phenomena to the concept of atoms.

PSSC, 0312. *Mechanical Energy and Thermal Energy*.

References for Chapter 7: Phases of Matter

7.1 Early theories
Overhead Transparency K&E #79, Periodic Chart of the Elements.
Leherman and Swartz, *Fundamentals of Physics*, Chapter 21, pp. 634–665, Atomic and Molecular Systems.
Gamow and Cleveland, *Physics, Foundations and Frontiers*, Section 20-1, pp. 350–351, The Molecular Hypothesis. Section 20-2, pp. 351–355, Brownian Motion.

7.2 Molecules
Laboratory Exercise: *Exercises and Experiments in Physics*, Expt. 15, The Size of the Molecule.

7.3 Atoms
Lindsay, *Basic Concepts of Physics*, Section 9-1, pp. 343–345, The Continuum vs the Atomic Concept in Physics. Section 9.7, pp. 356–360, Atomic Structure.
Overhead Transparency K&E #78, Model of the Atom.
PSSC Physics, Third Edition, 1971, Chapter 24, pp. 554–570, The Rutherford Atom. Chapter 26, pp. 592–608, Atoms and Spectra.

Gamow and Cleveland, *Physics, Foundations and Frontiers*, Section 21-6, pp. 381–382, Thomson's Atomic Model. Section 21-7, pp. 382–384, Rutherford's Atomic Model. Chapter 23, pp. 408–417, The Bohr Atom. Chapter 24, pp. 419–433, The Structure of Atoms.
Beiser, *Concepts of Modern Physics*, Chapter 4, pp. 85–99, Atomic Structure, Chapter 5, pp. 101–122, The Bohr Model of the Atom.
Van Name, W. F. Jr., *Modern Physics*, Section 4.2, pp. 95–98, Atomic Models. Section 4.3, pp. 98–104, Bohr's Theory of the One-Electron Atom.

7.4 Kinetic theory of matter
See Sutton A-41 to A-56 for suggestions for demonstrations on kinetic theory and atomic structure, pp. 459–467.
Meiners, 27-2, p. 805ff., Kinetic Theory Models.
Lab Exercise: Lehrman, *Laboratory Experiments*, Expt. 36, pp. 74–75, The Size of the Molecule.
Semat, *Fundamentals of Physics*, Section 15-1, pp. 160–161, Phases of a Substance.

7.5 Forces between molecules
Semat, *Fundamentals of Physics*, Section 11-1, pp. 195–196, Internal Forces.

7.6 The nature of solids
See Sutton, M-267, Distinction Between Fluids and Solids.

7.7 Cohesion and adhesion
Sutton, M-259, Cohesion Plates. M-260, Cohesion of Water Column. M-261, Adhesion of Water to Glass.

7.8 Tensile strength
Semat, *Fundamentals of Physics*, Section 11-3, pp. 197–200, Tensile Stress and Strain.

7.10 Elasticity
See Sutton, M-63 through M-69 for demonstrations involving elasticity, deformation, and impact.
Sutton, H-100, A coiled lead wire becomes a spring when it is cooled in liquid air or nitrogen.
Semat, *Fundamentals of Physics*, Section 11-2, pp. 196–197, Elasticity.
Meiners, 18-1, p. 439ff., Elasticity.

7.11 Hooke's law
Sutton, M-63, Hooke's Law and Young's Modulus.
Meiners, 11-1.13, p. 259, Hooke's Law.
Western 2 × 2 transparencies, E-8 through E-10, Hooke's Law, 3 slides. E-11 Bending. E-12 & E-13 Strain and Photo-elasticity.
Laboratory Exercise: *Exercises and Experiments in Physics*, Expt. 13, Elastic Potential Energy.

7.12 The nature of liquids
See Sutton, M-268, Distinction Between Liquids and Gases.
Sutton, M-262 through M-266, Osmosis and Diffusion.
Meiners, 26-5.15, p. 776, Supercooling of Liquids.
Lindsay, *Basic Concepts of Physics*, Section 6.4, pp. 207–209, The Brownian Motion.
Meiners, 27-7.6, p. 813, Brownian Movement Analog.
Meiners, 17-6, p. 410, Diffusion of Liquids.

7.13 Cohesion and adhesion
Semat, *Fundamentals of Physics*, Section 14-7, p. 155, Diffusion of Liquids and Solids. Section 14-8, pp. 255–257, Brownian Motion.

7.14 Surface tension
See Sutton, pp. 92–105, for suggestions for demonstrations on surface tension.
See Sutton, H-51, H-52, and H-53, pp. 209 and 210, for suggestions for demonstrating sublimation.

Semat, *Fundamentals of Physics*, Section 11-8, pp. 207–209, Surface Tension.
Meiners, 16-5, p. 388ff., Surface Tension.
Physical Science Study Series: Soap Bubbles and the Forces which Mould Them by Sir Charles Vernon Boys.

7.15 Capillarity
Meiners, 16-5.2, p. 390, Capillarity.
Semat, *Fundamentals of Physics*, Section 11-9, pp. 209–211, Capillarity.

7.16 Melting
Semat, *Fundamentals of Physics*, Section 15-7, pp. 267–269, Melting, Fusion.
See Sutton H-43 through H-71, pp. 208–216, for demonstrations involving change of phase.
Meiners, 26-5.12, p. 775, Crystallization.
Semat, *Fundamentals of Physics*, Section 15-8, pp. 269–270, Dependence of Melting Point on Pressure.

7.17 Effect of pressure on the freezing point
Sutton, H-57, p. 211.

7.18 Effect of solutes on the freezing point
Sutton, H-66, p. 220, Comparison of Vapor and Gas.

7.19 The nature of gases
Sutton, M-268, p. 107, Distinction between Liquids and Gases.
Sutton, M-61, p. 33, showing viscosity in gases.
Meiners, 27-4, p. 796ff., Viscosity of a Gas.
Meiners, Gases and Kinetic Theory, Chapter 27, p. 783ff.
See *Project Physics Handbook*, 1970, Chapter 11, p. 195ff., The Kinetic Theory of Gases. Also Teachers Resource Book 3, Chapter 11, The Kinetic Theory of Gases.
Lindsay, *Basic Concepts of Physics*, Section 6.3, pp. 204–207, The Molecular Theory of Gases.
Sutton, M-117, p. 53, Gas-Pressure Analogy.
Semat, *Fundamentals of Physics*, Section 14-6, pp. 253–255, Diffusion of Gases.
Meiners, 27-9, p. 821ff., Diffusion.
Semat, *Fundamentals of Physics*, Section 15-6, pp. 265–267, Evaporation

7.20 Vaporization
Meiners, 26-5.7, p. 772, Sublimation.
See Sutton, pp. 216–224 for vapor pressure demonstrations: Pulse glass, vapor pressure of water at the boiling temperature, boiling water at reduced pressure, vapor pressure of water at room temperature, etc.

7.21 Equilibrium vapor pressure
Meiners, 27-3, p. 791ff., Vapor Pressure.

8 HEAT MEASUREMENTS

Synopsis: Heat is a form of energy that can be observed in many ways. The addition or removal of heat energy from a body may change its temperature, its linear dimensions, its volume, its color, its phase, the pressure exerted by the body on its surroundings, or other measurable physical properties, such as electrical resistance.

We are aware that some objects feel cold, cool, warm, or hot to the touch. Here we have the origin of the concept of temperature. We know, however, that we need to develop a more precise method for the measurement of temperature than to depend on "feel." The need for a device, a thermometer, by which a number can be assigned, for example, to the level of the mercury in a tube is introduced, and the result of this measurement is called temperature. We define the concept of temperature by the operation that is used to measure it.

Although human interest in heat phenomena dates well back in history, Galileo is credited with having invented the first thermometer, sometime between 1592 and 1598, a hundred years after the discovery of America by Columbus.

Temperature, as operationally defined, is not sufficient for the complete description of thermal phenomena. This need has brought about the development of a special discipline called calorimetry which involves experiments on the mixing of different masses of bodies at different temperatures.

In the study of thermal effects, we consider in some detail the specific effects of the addition or removal of heat on a body of gas, that is, the gas laws. Change of phase and the conditions required for a material to exist simultaneously with its three phases in equilibrium and in contact, that is, the triple point, is also considered.

The effects of heat on water, its peculiar properties, and the way in which water serves to stabilize our atmosphere are studied.

Comment: "Temperature is relative, and is like time in being difficult to define in terms of simple concepts. The word temperature means intensity of heat. The average temperature of a body is proportional to the average kinetic energy of the molecules of which the body is composed. Changes in the temperature of a body are accompanied by changes in the various properties, which can therefore be utilized in defining temperature as a number scale. . . . Early temperature measuring devices (thermometers) were based on . . . change of volume, and were exemplified by Galileo's air thermometer, and various liquid in glass thermometers." *Van Nostrand's Scientific Encyclopedia.* Carefully discuss the material covered in Sections 8.2 and 8.3. Focus attention on the fact that the triple point temperature of water (273.16 °K) is the single fixed point for defining temperature and also that the Celsius degree and the Kelvin degree were arbitrarily made equal in magnitude. Identify the absolute zero of temperature (0 °K) as a second fixed point on the Kelvin scale. Be certain that students understand that this is the temperature at which the molecular energy of a substance is at a minimum, but not zero. We are led by a study of the gas laws to regard 0 °K as the lowest temperature that can be reached, thus this is called the absolute zero of temperature. Section 8.3 will require focus of attention. Note that heat is defined as that which is transferred between a body and its surroundings as a result of temperature differences only and not as the energy that a body possesses at a given temperature. Refer to Section 22-1, pages 545–546 and Section 25-10, pages 640–641, *Physics*, Halliday and Resnick, Wiley, 1966; Part I, Chapter 6, pages 235–239, *Physics Problems*; and Sections 12-1 through 12-3, pages 219–222 and Section 13-1 and 2, pages 232–234, *Fundamentals of Physics*, Semat. Student reports may be based on the following sources, *Physics Problems:* The Scientific Method and the Caloric Theory, pages 2–6. *Great Experiments in Physics:* pages 166–168. *Harvard Case Histories in Experimental Science*, J. Conant, Ed., Harvard University Press, 1956, Case 3, The Early Development of the Concepts of Heat; The Rise and Decline of the Caloric Theory. An excellent reference is the article by Mark W. Zemansky, "The Use and Misuse of the Word 'Heat' in Physics Teaching," in the September 1970 issue of *The Physics Teacher*.

Note that there is no instrument that will directly measure the amount of thermal energy given out or absorbed by a body. Establish the definitions of kcal and cal. Perform the demonstration as in Figure 8-13 (paraffin may be used) to establish concepts of heat capacity and specific heat. Point out that water has one of the highest specific heats. Exhibit a calorimeter in preparation for a laboratory experiment on the measurement of heat capacity or specific heat. From the specific heat equation develop the equation for Q needed to raise the temperature of materials. Establish the law of heat exchange and go over the Examples on pages 185–186.

Approach the study of thermal expansion as an explanation of the behavior of thermometers. Demonstrate expansion and contraction of solids using the ball and ring apparatus, bimetal strip, and possibly its use in a thermostat demonstration. Also demonstrate a tube such as Figure 8-6, page 173, to show the volume expansion of liquids. A demonstration, showing the unequal volume expansion of liquids, consists of test tubes fitted with one-hole stoppers and inserted long glass tubes filled with water, mercury, and alcohol, and heated to the same temperature in a beaker of boiling water. Point out that both the glass and the liquids in thermometers expand, but that the coefficient of volume expansion is greater for the liquid. Emphasize the unusual expansion of water, using the graphs in Figures 8-7 and 8-8 on page 174.

The air thermometer operation may be demonstrated to indicate the magnitude of the volume expansion of gases as compared with liquids or solids. The effect of heat on the pressure of a confined gas may also be discussed in connection with this demonstration. Have students explain why all gases have approximately the same coefficient of volume expansion. You may wish to refer to the Avogadro hypothesis in this connection (equal volumes of all gases at the same temperature and pressure have equal numbers of molecules). Spend considerable time in developing the general gas law. Consider again the properties of an ideal gas.

Consider the effect of temperature and pressure on a pure substance, and show how these factors are related to the state of the substance. Study the changes of state that take place as a result of temperature and pressure changes. Refer to Figure 8-15, p. 188. Be certain that the triple point idea is understood so that students can interpret the graph along any vertical or horizontal line. Introduce the ideas of heat of fusion and heat of vaporization as energy either required or given off as the water undergoes a change of phase. Have students interpret Figure 8-16, p. 190. Discuss ways in which the properties of water contribute to our physical well-being.

Develop the difference between vaporization and boiling and refer to the graph, Figure 8-18, page 191. Demonstrate the influence of decreased pressure upon boiling point. Go over the Example on pages 193–194, emphasizing the two different aspects of the heat loss—by condensation and by temperature change. You may wish to demonstrate the liquefaction of an easily liquified gas (ammonia or sulfur dioxide) by passing the gas from a generator into a collecting vessel cooled by a dry ice-acetone mix. Demonstrate that evaporation is a cooling process by using cotton wads wound about thermometer bulbs, and compare evaporation effects of water, alcohol, and ether. Interpret this cooling effect in terms of the kinetic energy of the escaping and remaining molecules and the average kinetic energy of the nonevaporated molecules. Develop the concept of critical temperature.

An interesting method of controlling heat exchange involves the tendency of a fluid to cling to the surface over which the fluid is moving, even when the surface is curved. This is called the Coanda effect, in honor of the Romanian engineer Henri Coanda (1885–1972), who discovered it in 1928 when the hot exhaust from his experimental jet aircraft clung to the fuselage and burned off the tail section.

The United States Government used the Coanda effect to develop a foolproof method of switching hot rocket exhaust gases between two nozzles without the use of moving parts. The same principle is used in modern heating and air-conditioning systems to regulate the flow of hot and cold air in a building. A demonstration model of this application of the Coanda effect may be available from your local heating and air-conditioning contractor.

References: Brown, Sanford C., *Count Rumford, Physicist Extraordinary*, Science Study Series, 1962.

Buchhold, Theodore A., "Applications of Superconductivity," *Scientific American*, March 1960, pp. 74–82, (available as an offprint).

Castle, Jack, Jr., Werner Emmerich, Robert Heikes, Robert Miller, and John Rayne, *Science by Degrees: Temperature from Zero to Zero*. Walker, New York, 1965.

Conant, James B., Ed., *Harvard Case Histories in Experimental Science*, Case 3, Harvard University Press, Cambridge, 1956.

Davis, K. S., and J. R. Day, *Water, the Mirror of Science*, PSSC Science Study Series, Doubleday and Co., Inc., Garden City, N.Y., 1964.

Glover, Francisco, "Specific Heat Capacity, A Quantum Explanation," *The Physics Teacher*, March 1969, pp. 149–156.

Lehrman, Robert, and Clifford Swartz, *Foundations of Physics*, Holt, Rinehart and Winston, Publishers, New York, 1969.

Lindsay, Robert Bruce, *Basic Concepts of Physics*, Van Nostrand Reinhold Co., New York, 1971.

MacDonald, D. K. C., *Near Zero: The Physics of Low Temperature*, Science Study Series.

Mendelssohn, K., *The Quest for Absolute Zero: The Meaning of Low Temperature Physics*, McGraw-Hill Book Co., New York, 1966.

Ohring, George, *Weather on the Planets*, Science Study Series, 1966.

Osgood, Thomas H., "Is Your Idea of Temperature Measurement Up to Date?" *The Physics Teacher*, Oct. 1967, pp. 326–327.

PSSC Physics, Third Edition, 1971.

Semat, Henry, *Fundamentals of Physics*, Holt, Rinehart and Winston, Publishers, New York, 1966.

Wheatley, John C., and Voward J. Van Till, "Attaining Low Temperatures," *The Physics Teacher*, Feb. 1970, pp. 67–73.

Zemansky, Mark W., "The Use and Misuse of the Word 'Heat' in Physics," *The Physics Teacher*, Sept. 1970, pp. 295–300.

Zemansky, Mark W., "Classical and Modern Aspects of Low Temperature Physics," *The Science Teacher*, Oct. 1968, pp. 65–73.

Films: *Gas Laws and Their Application*, 14 min., black and white, EBF. Dramatizes early research that led to the discovery of the relationship between temperature, volume, and pressure of gases.

How We Measure Heat, fs, color, 45 frames, McGraw-Hill Book Co., Text Film Dept., 1221 Avenue of the Americas, New York, NY 10020. Explains how solids, liquids, and gases expand when heated. Shows how mercury thermometers are made.

Mechanical Energy and Thermal Energy, PSSC, 0312.

The Nature of Heat, 11 min., black and white, color, Coronet Films, 65 East South Water Street, Chicago, IL 60601. Describes heat as energy of molecular motion. Animated; transfer of heat is illustrated.

References for Chapter 8: Heat Measurements

See Meiners, Chapter 25, Temperature and Chapter 26, Heat and Thermodynamics, pp. 749–781.

Lehrman and Swartz, *Foundations of Physics*, Chapter 9, Heat Energy and Temperature.

PSSC Physics, Third Edition, 1971, Chapter 17, Heat, Molecular Motion, and Conservation of Energy, pp. 370–397.

8.1 Relationship Between Heat and Temperature
Sutton, H-1, Temperature Diagram.

Lindsay, *Basic Concepts of Physics*, 1971, Section 5.13, The Nature of Heat and Temperature, pp. 169–178.

Semat, *Fundamentals of Physics*, Chapter 12, Temperature, pp. 219–231.

Semat, *Fundamentals of Physics*, Chapter 12, pp. 219ff., Temperature. Section 12-1, Concept of Temperature. Section 12-2, Temperature Scales. Section 12-3, Celsius and Fahrenheit Scales.

Lindsay, *Basic Concepts of Physics*, Section 5.13, pp. 169–178, The Nature of Heat and Temperature, Section 5.14, pp. 178–184, The Mechanical Theory of Heat.

8.2 Temperature Scales
Sutton, H-2 Thermometers. H-3 Thermometers. H-4 Constant-Pressure Air Thermometer. H-5 Constant-Volume Air Thermometer. H-6 Lecture Table Thermometer. H-7 Lecture Table Thermometer. H-8 Optical Pyrometry.

Western, 2 × 2 Color Slides, F-15 Temperature Dependence on the Spectrum of Light Emitted from a Heated Body. F-16 & F-17 ditto above, three slides in the set. F-18 Color of Stars.

Sutton, H-1 Temperature Diagram.

See Meiners, Section 25-2 for demonstrations involving temperature change, pp. 750–854.

Lab Exercise: Lehrman, *Laboratory Experiments*, Expt. 29, Measurement of Temperature.

8.3 Heat Units
Semat, *Fundamentals of Physics*, Chapter 13, Heat and Work: Section 13-1, The Nature of Heat, Section 13-2, Heat and Internal Energy, Section 13-3, Heat Units, Section 13-4, Heat Capacity, Specific Heat, Section 13-5, Heat Measurements, pp. 232–243.

8.4 Thermal expansion of solids
For thermal expansion demonstrations see Sutton, H-9 through H-34, pp. 197–205.

Western, 2 × 2 Color Transparencies, F-9 Bimetal Expansion, F-10 Thermal Expansion.

Laboratory Exercise: Lehrman, *Laboratory Experiments*, Expt. 35, Expansion of a Solid.

Semat, *Fundamentals of Physics*, Section 12-4, Thermal Expansion of Solids.

Laboratory Exercise: *Exercises and Experiments in Physics*, Expt. 16, Coefficient of Linear Expansion.

8.5 Thermal expansion of liquids
Sutton, H-27 Expansion of Liquids, pp. 202–203. H-28 Expansion and Maximum Density of Water.

Sutton, H-32, Differential Expansion of Liquid and Container.

8.6 Abnormal expansion of water
Sutton, H-28, Expansion and Maximum Density of Water.

8.7 Charles' law
Sutton, H-33, Expansion of Gases, pp. 204–205.

Meiners, 27-2.7, Charles' Law, p. 790.

Laboratory Exercise: *Exercises and Experiments in Physics*, Expt. 17, Charles' Law.

Lab Exercise: Lehrman, *Laboratory Experiments*, Expt. 31, Temperature and Volume of a Gas.

Semat, *Fundamentals of Physics*, Section 12-7, pp. 228–229, Absolute Scale of Temperature.

8.8 Boyle's law
Sutton, M-319, Compressibility of Air Boyle's Law, p. 123

Western, 2 × 2 Color Transparencies F-5 through F-8, Boyle's Law.

Lab Exercises: Lehrman, *Laboratory Experiments*, Expt. 30, Pressure and Volume of a Gas.

Exercises and Experiments in Physics, Expt. 18, Boyle's Law.

8.10 The universal gas constant
Semat, *Fundamentals of Physics*, Section 14-1, pp. 246–247, General Gas Law.

8.12 Specific heat
See Sutton, H-35 through H-41, for demonstration ideas for the measurement of specific heat, pp. 205–208.

Laboratory Exercises: Lehrman, *Laboratory Experiments*, Expt. 32, Mechanical Determination of Specific Heat, Expt. 33, Calorimetric Determination of Specific Heat.

See Meiners, Section 26-2, for demonstrations involving specific heat, pp. 757–758.

Laboratory Exercise: *Exercises and Experiments in Physics,* Expt. 19, Specific Heat.

8.14 The triple point

Semat, *Fundamentals of Physics,* Chapter 15, pp. 260–276.

See Sutton, H-43 through H-71 for change of phase demonstrations, pp. 208–216.

Sutton, H-69, 70,, 71, p. 215 are demonstrations on the freezing of water by rapid evaporation—The Triple Point.

8.15 Heat of fusion

Sutton, H-54 Heat of Fusion of Ice, p. 210.

Meiners, Section 26-5.1, Heat of Fusion of Water, p. 770

Meiners, Section 26-5.2, Cooling Curve of a Melted Material (Lead), p. 770

Laboratory Exercise: *Exercises and Experiments in Physics,* Expt. 20, Heat of Fusion.

Lab Exercise: Lehrman, *Laboratory Experiments,* Expt. 34, The Latent Heats of Water.

8.17 The boiling process

For vapor pressure demonstrations see Sutton, H-72 through H-94, pp. 216–224.

Western, 2 × 2 Color Transparencies, F-11 & F-12, Vapor Pressure.

8.18 Heat of vaporization

Sutton, H-61, Heat of Condensation, p. 212.

Meiners, 26-5.5, Heat of Evaporation, p. 772.

Semat, *Fundamentals of Physics,* Section 15-4, p. 264, Heat of Vaporization.

Laboratory Exercise: *Exercises and Experiments in Physics,* Expt. 21, Heat of Vaporization.

8.20 The critical point

Semat, *Fundamentals of Physics,* Section 15-11, pp. 270–271, Triple Point.

9 HEAT ENGINES

Synopsis: Joule and Mayer, of different nationality and different professional backgrounds, were studying the relationship between heat and mechanical energy at about the same time. Mayer, a physician, took a biological approach although he understood the physics. Joule, an experimentalist, became intensely interested in the measurement of the relation connecting apparently different physical phenomena. Mayer interpreted the difference between the specific heat of a gas at constant pressure and that of a gas at constant volume to be due to the extra heat needed to produce the expansion required for maintaining constant pressure. Mayer provided one of the most fundamental derivations in thermodynamics.

Joule's experiments covered a wide range of physical phenomena, but he finally determined the numerical value of the equivalence of heat and energy by measuring the heat produced by the mechanical compression of air in a vessel and by using a paddle as a stirrer to heat water. The experimental work of Joule and the theoretical work of Mayer, after a time, formed the basis for the generalization of the concept of energy that includes heat as a form of energy.

Thermodynamics deals with the changes that take place in a system, as, for example, in a gas when it undergoes a change in temperature. Thermodynamics employs two fundamental concepts. The first states that in all transformations of energy there is no net gain or loss in a completely isolated system.

To say that heat, mechanical energy, and other forms of energy are equivalent—no matter what the nature of the transformation from one form to the other—overlooks the direction in which the energy transformations are observed to take place. In all naturally occurring processes, there is an increase in the unavailable energy (entropy) in the universe. This is to say, every time you get something by an energy transformation, you reduce by that amount the opportunity to get that something in the future. This is the second fundamental concept of thermodynamics.

Clausius envisioned the "heat death" of the universe when all parts have come to the same temperature level. The total energy of the universe will not have changed, but it will no longer be possible to produce a transformation of any kind.

Because of their interest in cars, most students find the discussion of the various types of engines in the last part of the chapter quite interesting. Stress the practical applications of each type of engine. In Section 9.16, emphasize particularly the cost efficiency and year-round applicability of the heat pump.

Comment: Have students test the temperature of a section of rubber band with their lips. Then have them rapidly extend the section by a quick pull and test the temperature again. This is one method of rapidly converting mechanical energy to heat and may serve as an introduction. Have one or more students report on the work of Rumford, Mayer, Joule, and Helmholtz. If time does not permit, the work of Joule should be emphasized and the details of the work of others filled in by the teacher.

Carefully consider the first law of thermodynamics and the conversion of heat into work. Place emphasis on Joule's experiment on the mechanical equivalent of heat in the light of its historical and scientific significance. The time lapse between the work of Rumford, the suggestion of Julius Mayer (1842) that heat and work were equivalent and interconvertible, and the actual measurement by Joule (reported in 1850) may be made the center of a discussion of the scientific method and the role of facts and ideas in the progress of science. (See References.) It might be pointed out that, prior to his work on the mechanical equivalent of heat, Joule worked out the relationship between heat and electric current and resistance: *Heat* $\propto I^2 R$.

Study carefully Sections 9.3 and 9.4, giving attention to the significance of isothermal and adiabatic expansion. Give attention to the graphical representations of work done by an expanding gas. Be certain that students understand why the specific heat at constant pressure is greater than the specific heat of a gas at constant volume. Point out that this difference exists for solids and liquids also; however, the amount of expansion is so small that the difference in the values of the two specific heats would only be noticed when working to a number of significant figures well beyond the three or four to which we work (see the next to the last paragraph in Section 9.5, page 199). Develop the theoretical basis for the maximum efficiency of a heat engine. Point out that a difference in temperature between that of the heat source and heat sink is necessary for the operation of a heat engine. Develop the second law of thermodynamics. Introduce the concept of entropy. Discuss the law of entropy as a restatement of the second law of thermodynamics.

An interesting method of controlling heat exchange involves the tendency of a fluid to cling to the surface over which the fluid is moving, even when the surface is curved. This is called the Coanda effect, in honor of the Romanian engineer Henri Coanda (1885–1972), who discovered it in 1928 when the hot exhaust from his experimental jet aircraft clung to the fuselage and burned off the tail section.

The United States Government used the Coanda effect to develop a foolproof method of switching hot rocket exhaust gases between two nozzles without the use of moving parts. The same principle is used in modern heating and air-conditioning systems to regulate the flow of hot and cold air in a building. A demonstration model of this application of the Coanda effect may be available from a local heating and air-conditioning contractor.

References: Brown, S. C., *Count Rumford: Physicist Extraordinary*, Science Study Series, Doubleday and Co., Inc., Garden City, N.Y.
Ehrenberg, W., "Maxwell's Demon," *Scientific American*, Nov. 1967.
Sandfort, John F., *Heat Engines: Thermodynamics in Theory and Practice*, Science Study Series, 1962.

Scientific American, Sept. 1954, symposium issue on heat.

Shamos, Morris H., *Great Experiments in Physics*, Holt, Rinehart and Winston, Publishers, New York, 1959, Chap. 12: The Mechanical Equivalent of Heat—James Joule, pp. 166–183.

Thompson, Paul D., *Gases and Plasmas*. (Introducing Modern Science Series.) Lippincott, Philadelphia, 1966.

Films: *An Introduction to Heat Engines*, 23 min., black and white, no charge except for return postage, Shell Oil Co., 50 West 50th Street, New York, NY 10020. Development of the heat engine from the earliest experiments to today's powerful and efficient turbine. Explains principles of diesel, gasoline, steam turbine, and gas turbine.

Mechanical Energy and Thermal Energy, 27 min., black and white, PSSC, 0312. Shows several models to help students visualize both bulk motion and random motion of molecules. It shows their interconnection as the energy of bulk motion turns into the thermal energy of random motion. Shows model of thermal conduction. Develops a temperature scale by immersing canisters of two gases in baths of various temperatures and reading the resultant pressures. Through this, the origin of the absolute temperature scale is explained.

Conservation of Energy, 27 min., black and white, PSSC, 0313. Energy traced from coal to electric output in a large power plant. Quantitative data are taken in the plant. Conservation law demonstrated for random and orderly motion.

References for Chapter 9: Heat Engines

Project Physics Text, Chapter 10, pp. 29–67, Energy. Sections of particular importance are 10.5, p. 39, Heat energy and the steam engine; 10.6, p. 46, James Watt and the Industrial Revolution; 10.7, p. 49, The experiments of Joule; 10.8, p. 51, Energy in biological systems; 10.9, p. 56, Arriving at a general law; 10.10, p. 60, A precise and general statement of energy conservation (the first law of thermodynamics); 10.11, p. 62, Faith in the conservation of energy.

See Meiners, Section 26-4, Heat and Work, pp. 766–770.

Semat, *Fundamentals of Physics*, Chapter 13, pp. 232–243, Heat and Work; Chapter 14, pp. 246–257, Kinetic Theory of Matter; Chapter 17, pp. 287–302, Laws of Thermodynamics.

9.1 Mechanical equivalent of heat
Meiners, 26-4.2, ''J''-Value Demonstration. 26-4.3, Mechanical and Thermal Energy. 26-4.4, Electrical Equivalent of Heat. 26-4.7, Mechanical Equivalent of Heat.

Lindsay, *Basic Concepts of Physics*, Section 5.15, pp. 184–186, Generalization of the Concept of Energy.

See Sutton H-169 through H-182 for demonstrations relating heat and work, pp. 243–248. H-169, Heat from Friction. H-170, Heat from Friction. H-171, Fire by Friction. H-172 through H-175, Heat from Work. H-176, Mechanical Equivalent of Heat—Shot Tube. H-177, Mechanical Equivalent of Heat. H-179, Heating by Compression—Fire Syringe. H-180, Heating by Compression. H-181, Temperature Changes Due to Expansion and Compression. H-182, Models, Engines.

Exercises and Experiments in Physics, Expt. 27, Mechanical Equivalent of Heat.

9.2 First Law of Thermodynamics
PSSC Physics, Third Edition, Chapter 17, p. 370ff., Heat, Molecular Motion and Conservation of Energy. See particularly 17-4, Mechanical Energy of Bulk Motion and Internal Energy, 17-5, The Quantitative Study of

the Conversion of Mechanical to Internal Energy, 17-7, Quantitative Relation of Energy Dissipation and Temperature Rise, 17-8, Conservation of Energy.

9.3 and 9.4 Isothermal and adiabatic expansion
Lindsay, *Basic Concepts of Physics*, Section 5.14, pp. 178–184, The Mechanical Theory of Heat.

See *Project Physics Text*, Chapter 11, The Kinetic Theory of Gases.

Meiners, 27-6, pp. 800–804, Adiabatic Processes. 27-6.1, Process of fuel ignition in a diesel engine.

9.5 Specific heats of gases
Semat, *Fundamentals of Physics*, Section 13.9, pp. 241–242, Specific Heats of Gases.

9.7 Second law of thermodynamics
Meiners, 3-1, p. 30, Quasi Static Expansion and Contraction of a Coil Spring. The second law of thermodynamics requires reversible changes of state, etc.

9.9 through 9.15 Conversion of heat into work
Sutton, H-182, Models, Engines.

Meiners, 26-4.5, Dust Explosion. 26-4.6, Work from Heat.

WAVES

10

Synopsis: Wave motion is a type of motion that is within the experience of every individual. It is one of the most important types of motion that we experience. It is fundamental to the understanding of the transfer of energy by radiation and constitutes one of the three means for the transfer of energy.

Students have observed surface waves on water. The waves that travel across a field of grain as the wind causes the individual stalks of the grain to oscillate back and forth and the waves sometimes formed in a wire such as a telephone line swinging in the wind are other examples of waves. Students have also noted the vibration of strings and have no doubt sent a wave down a long piece of rope by setting one end into vibratory motion. They may have observed waveforms on an oscilloscope tube in a radio repair shop.

Mechanical waves propagate through elastic media. For example, when one end of a steel bar is struck with a hammer, the effect can be felt at the other end of the bar. Earthquakes generate waves in the earth's crust. Matter media are not required for the propagation of electromagnetic waves.

In this chapter, students will study the nature and characteristics of wave motion as a basis for later study of sound and the various frequency bands of electromagnetic radiation.

Waves are described in terms of energy transfer, with both periodic longitudinal and periodic transverse waves being analyzed. The characteristics of waves in terms of speed, wavelength, frequency, and period are described. The relationship between the energy of a wave system and the amplitude is developed.

Associated with all types of wave motion are phenomena such as rectilinear propagation, reflection, refraction, impedance, and diffraction. These are presented separately, as is the superposition principle which is then used to describe interference. The phenomenon of standing waves is then explained in terms of the superposition principle, with emphasis on energy distribution.

Comment: Approach the study of waves by appealing to student experience. Just about every characteristic of wave motion probably has been experienced by most students. Using these experiences to point out wave properties is an excellent way to begin. For example, a pebble dropped into a basin of water could stimulate discussion. A short series of introductory experiments with the long coil spring and the ripple tank will provide a means for helping the students to visualize the type of wave motion and to observe wave phenomena. In addition, short demonstrations with such devices could be supplemented by the use of films and filmstrips.

Note that this chapter is entirely devoted to wave motion in general. Sound waves are treated separately in Chapter 11. After establishing that wave motion is a method of energy transfer, set up a spiral spring horizontally with paper riders placed according to numbered positions and representing particles corresponding to Figure 10-3. Develop the concept of an elastic medium on this model. Establish ideas of a transverse pulse and wave motion by using the model according to Figure 10-4. In a similar fashion treat longitudinal pulse and longitudinal wave motion, Figures 10-5 and 10-6. Set up and demonstrate periodic waves using apparatus as in Figures 10-7 and 10-8, stressing relationship to simple harmonic motion. Base treatment of wave characteristics on chalkboard diagrams according to Figure 10-9: speed, phase, frequency, period, wavelength, and amplitude. Establish the appropriate equations, particularly $V = f\lambda$. Discuss the relationship between energy transfer through wave motion and the characteristics of waves. Use diagrams for transverse waves to show varying frequencies and amplitudes. Wave machines, ripple tanks, films, and filmstrips may be used as desired.

Study the common properties of waves, rectilinear propagation, reflection, refraction, diffraction, and interference by use of chalkboard diagrams using colored chalk; with overlay transparencies for use with an overhead projector; and by use of the ripple tank (Figure 10-11) and accessories. The ripple tank is very useful for student laboratories as well as for demonstrations of wave phenomena. The wave phenomena illustrated in Figures 10-13 and 10-14, Figures 10-17 and 10-18, Figures 10-21 and 10-22, and Figure 10-24 can be demonstrated effectively by use of the ripple tank. Standing waves can be demonstrated by use of a mechanical vibrator or hand driven long coiled spring or piece of rope or rubber tubing. Illustrate production of standing waves by chalkboard diagrams as shown in Figure 10-29. Kundt's Apparatus may be used to demonstrate standing waves, nodes, and antinodes. A mechanical wave demonstrator is available from the apparatus companies and from the Bell Telephone Company. Make certain that students understand that the total energy of the two interfering waves in the case of standing waves remains unchanged, but that the energy distribution is different as a result of interference.

References: Bascom, Willard, *Waves and Beaches: The Dynamics of the Ocean Surface,* Science Study Series, Doubleday and Co., Inc., Garden City, N.Y., 1964.

Bond, B. Burtis, "A Ripple Tank Stroboscope," *The Science Teacher*, Vol. 30, No. 5, p. 48.

Strong, C. L., "The Amateur Scientist: How to Make a Ripple Tank to Examine Wave Phenomena," *Scientific American*, Oct. 1962, p. 171.

Waldron, R. A., *Waves and Oscillations,* Momentum Books, D. Van Nostrand Co., Inc., Princeton.

Films: *Similarities in Wave Behavior,* fs, 27 min., Local Bell Telephone Offices. This film, using a specially built torsion wave machine, discusses the reflection of waves from a free and clamped end, superposition, standing waves and resonance, energy loss by impedance mismatching, and other phenomena. The complete assembled kit with carrier may be purchased.

Simple Waves, 27 min., black and white, PSSC, 0204. Pulse propagation on ropes and Slinkies shows elementary characteristics of waves such as different speeds in different media. A torsion bar wave-machine is then used to repeat these experiments to demonstrate reflection and other phenomena.

Wave Motion—A Key to Modern Science, fs, color, 40 frames, FOM 600, McGraw-Hill Book Co., Text Film Dept., 1221 Avenue of the Americas, New York, N.Y. 10020. Shows wave phenomena using rubber tubes, "Slinkies," and ripple tanks and demonstrates wave properties and characteristics.

Project Physics Film Loops: *Vibrations of a Wire,* color, 3 min. 50 sec. Standing wave patterns are shown when a thin, stiff wire vibrates without the use of external tension. First, a long, straight wire, free at one end, but passing through a magnetic field, vibrates when the current direction in the wire is changed. Second, standing wave patterns appear in the same wire in a magnetic field when the adjustable frequency of the alternating current reaches certain values. Third, the wire, formed into a circular horizontal loop and fixed at one point in the magnetic field, develops nodes along the ring at certain current frequencies. This is an analogue to "stationary states" in an atomic model.

Vibrations of a Rubber Hose, color, 3 min. 50 sec. A long rubber hose hanging vertically from a fixed point is attached to a motor-driven vibrator. The student can observe the motion of the rubber hose as the motor is speeded up. As the frequency is increased, standing wave patterns appear only at certain frequencies. Up to 15 well-defined oscillating loops can be observed.

Vibrations of a Metal Plate, color, 3 min. 45 sec. A square metal plate, fixed at the center, is made to vibrate by a loudspeaker. The standing wave patterns created at certain frequencies are made visible by sprinkling sand on the plate. The sand assumes a series of beautiful patterns known as Chladni figures by collecting along nodal lines. Although the harmonic frequencies are too complex for simple mathematical analysis, the student can see that the distinct patterns are formed only at well-defined frequencies and that the patterns have symmetries. A student can compare the vibrations of this metal plate that is fixed at the center with the vibrations of the drum, as seen in the *Vibrations of a Drum* Film Loop, that was fixed along its periphery.

Vibrations of a Drum, color, 3 min. 25 sec. In this beautiful Film Loop, a rubber membrane stretched over a drum is made to vibrate by a loudspeaker. The vibrational modes of this membrane are varied by changing the frequency of the speaker. The vibrations of the membrane are shown in slow-motion by using stroboscopic illumination. This Loop is valuable for showing transverse wave motion on a two-dimensional surface as well as for showing the sequence of symmetrical patterns in phenomena controlled by eigen-values. This display will suggest to students other experiments to perform with a stroboscope and with objects undergoing harmonic motion.

Superposition, color, 3 min. 5 sec. On a cathode-ray tube, two sine waves are displayed together with their resultant superposition. The sine waves are changed in phase, frequency, and amplitude to produce changes in the resultant wave. The simultaneous display of the separate waves and of their resultant clarifies the addition of displacements as the fundamental process in superposition.

Standing Waves on a String, color, 2 min. 35 sec. Change of tension on a string vibrating continuously at 72 vibrations per second results in standing waves due to reflection from a fixed end. As the tension is changed, the student can see many wave patterns. The string vibrates successively in one, two, three, four, and more segments. Nodes and antinodes are clearly discernible. A stroboscopic illumination reveals in slow-motion the string going through its complete cycle of vibration.

Standing Waves in a Gas, color, 3 min. 45 sec. Standing waves in air are set up in a large glass tube using a loudspeaker and a movable piston. In the first half of the loop, sound waves are made visible by inserting cork dust into the tube. In the second half, a glowing wire is used to detect the wave motion. A student can "see" standing sound waves, their nodes and antinodes, and can estimate wavelength.

3M Science Packet Number 40 for Overhead Projection, *Waves.* The illustrations in this packet are designed to aid class discussion of wave phenomena seen on springs and in the ripple tank.

Visuals A and B illustrate the propagation of pulses as one-dimensional disturbances along springs. The top of Visual A emphasizes the distinction between wave motion and the motion of the conduction media. The bottom two figures on Visual A are before and after representations of the way pulses may pass through each other. Visual B concentrates on the interaction of the pulses as they come together and may be used to develop the technique of superposition. This technique of superposition will allow prediction of waveforms resulting from interference.

Visuals C and D review the boundary and junction effects as seen with waves on springs. Waves in the heavy spring illustrated in Visual D travel with half the speed of waves in the light medium. The difference of wave amplitude in the two media is intended. This is in accord with observed effects and the fact that the identical energy could not be associated with identical waveform in differing media.

Visuals E through I illustrate the action of pulses in a ripple tank. Visual E shows how the light projects the wave pattern below the tank through the converging and diverging action of wave crests and troughs. Visual F shows how the circular wave crest from a point source results from the equal wave speed in all directions. Visual G illustrates the mechanism of pulse reflection. The overlay H should be used to show the wave's reflection. Visual I shows several reflection phenomena that can be compared with image formation in plane and curved mirrors. The letter F represents the focus of the reflecting surface. The overlay J allows this to be used to quiz students about expected results. Visual K describes how the incident and reflection angles are related to the angles the wave fronts make with the barrier. You might prove geometrically that the angle i = angle i′ and that angle r = angle r′ by showing that in each case they are complements of the same angle. Simple measurement by cutting a colored overlay to fit or the use of a plastic protractor will bring home the same point.

Visual L can be helpful in establishing that the wave equation results as a direct consequence of the meaning of wavelength, period, and frequency. The fisherman follows the motion of a wave crest through a distance of one wavelength. (Fishing is poor.) The time for this motion is the period of the periodic waves. Since speed is distance traveled per time of travel and the reciprocal of the period of a periodic wave is its frequency (frequency = 1/period), then speed = (wavelength) (frequency).

Visuals M through P illustrate that wave refraction is a function of wavelength changes. In Visual M, attention can be drawn to the fact that the wavelengths in the first medium in each of the illustrations are the same, but that wavelengths in the second medium differ. This is the source of their different refraction angles. Visual N is intended to be cut so that the bottom part with the shorter wavelength can be moved around until the edges of the wave crests in the top half each intersect with a single wave crest of the bottom half. This can be done only at a particular angle. The meaning of this is that if waves in one medium each cause only one wave in the second medium (identical frequency), then the relative angles of the waves in the two media are predetermined by the relative wavelengths of the waves. Visual O is to be used with a proof that Snell's law (sin i/sin r = index of refraction) requires that the index of refraction is also the ratio of incident to refractive wavelength and the ratio of incident to refractive wave speed. A proof of this could follow a pattern such as this: angle i = angle i′ and angle r = angle r′. Measure with colored overlay that i and i′ are both complements of angle CBA while r and r′ are both complements of PBC′.

So,

$$\sin i = \sin i' = CB/BA \text{ and } \sin r = \sin r' = C'A/BA$$

then,

$$\frac{\sin i}{\sin r} = \frac{CB}{C'A} \text{ but } CB = \overset{\text{incident}}{\text{wavelength } (\gamma i)} \text{ and } C'A = \overset{\text{refractive}}{\text{wavelength } (\gamma r)}$$

Since the frequency of incident and refracted waves are identical, we have:

$$\frac{\sin i}{\sin r} = \frac{f\gamma i}{f\gamma r} \text{ or } \frac{\sin i}{\sin r} = \frac{\text{incident wave speed}}{\text{refracted wave speed}} = \text{refractive index}$$

Visual P illustrates dispersion, the changing of the refractive index with different frequencies. The diffraction of waves through apertures is shown in Visual O to be a function of the size of the aperture in comparison to the wavelength of the periodic wave. Visual R illustrates a similar relationship for barrier sizes and wavelength.

Visuals S through W are a set to be used to assist explanation of interference in waves from two point sources. On Visual S, the heavy lines may be thought to represent crests while the dotted lines represent wave troughs. The location of the point source is taken as the center of the circular wave pattern. Make two copies of Visual S. Place both copies of Visual S on the overhead projector. Slide one copy to the right so that the two patterns are not concentric. The overlapping of crests with crests and troughs with troughs traces out the locations of maxima, while the overlapping of crests and troughs traces out nodal lines. The interference pattern can be traced on a third clear overlay. Altering the position of the two S visuals allows study of the effect of wavelength and source separation on interference patterns. Visuals T, U, and V when used with one S visual illustrate the effect of phase changes on the interference patterns. Visual W is an interference pattern produced by waves in phase from the two point sources. The overlays from the bottom of this visual are sections of waves that formed this pattern. Punch out the dots labeled S_1 and S_2 on the overlays and Visual W. Attach the overlays with paper fasteners to Visual W so that these two strips can be swung around. Now, by intersecting the two movable overlays, you can illustrate how each maximum and minimum results from the differing distances waves travel from the two sources.

References for Chapter 10: Waves

Project Physics Text, Chapter 12, pp. 101–139 of Unit 3. Topics of particular interest: 12.2, Properties of waves; 12.3, Wave propagation; 12.4, Periodic waves; 12.5, When waves meet: the superposition principle; 12.6 A two-source interference pattern; 12.7, Standing waves; 12.8, Wave fronts and diffraction; 12.9, Reflection; 12.10, Refraction.

Sutton, S-22 through S-49, pp. 137–154 for suggested demonstrations of wave motion.

Meiners, Chapter 18, pp. 439–488, Waves in Elastic Media.

Lindsay, *Basic Concepts of Physics,* Sections 8.2–8.7, pp. 293–309.

Semat, *Fundamentals of Physics,* Chapter 18, pp. 307–323, Wave Motion.

10.1 Energy transfer
Meiners, 18-2, pp. 443–450, Traveling Waves, Transport of energy by waves.

Meiners, 18-8, Wave Models, pp. 476–488.

PSSC Physics, Third Edition, Chapter 5, pp. 90–105, Introduction to Waves.

Project Physics, Chapter 12, Waves.

Transparencies, 2 × 2, Color, Western J-1 through J-20, dealing with wave motion and the ripple tank. These deal with propagation of waves, reflection, superposition, stationary waves, refraction, and diffraction.

Transparencies: 3M Science Packet Number 40, Waves. The 23 overhead transparency masters in this packet are designed to aid class discussion of wave phenomena seen on springs and in the ripple tank.

10.2 Mechanical waves
Sutton, S-33, Wave Machine.

Project Physics Handbook, Experiment 3-15, Wave Properties, pp. 47–48; Experiment 3-16, Waves in a Ripple Tank, pp. 48–49; Experiment 3-17, Measuring Wavelength, pp. 49–51.

PSSC Physics, Third Edition, Chapter 5, pp. 90–105, Introduction to Waves. Chapter 6, pp. 108–125, Waves and Light. Chapter 7, pp. 128–144, Interference.

10.3 Transverse waves
Sutton, S-23, Transverse Waves on Strings.

Transparency: K & E, #37, Transverse Waves.

10.5 Periodic waves
Laboratory Exercise: Lehrman, *Laboratory Experiments,* Expt. 37, p. 75, Pulses and Waves in a Spring.

Exercises and Experiments in Physics or *Laboratory Experiments in Physics,* Expt. 23, Pulses on a Coil Spring.

Transparency: K & E, #38, Longitudinal Waves.

10.6 Characteristics of waves
Exercises and Experiments in Physics or *Laboratory Experiments in Physics,* Expt. 24, Wave Properties.

Meiners, 18-3, p. 450, Wave Speed.

10.7 Amplitude and energy
Sutton, S-22, Water Waves.

10.8 Properties of waves
Meiners, 18-6, pp. 466–473, Ripple Tank.

10.9 Rectilinear propagation
Laboratory Exercise: Lehrman, *Laboratory Experiments*,
    Expt. 38, p. 78, Two-Dimensional Waves.

10.10 Reflection
Laboratory Exercise: Lehrman, *Laboratory Experiments*,
    Expt. 39, p. 80, Reflection and Refraction of Waves.

10.11 Impedance
Sutton, S-24, Wave Reflection at Change of Media.
See Sutton, p. 142, Figs. 115a-d.

10.12 Refraction
Meiners, 19-8, p. 524, Refraction of Sound Waves by a
    Gas Lens.

10.13 Diffraction
Laboratory Exercise: Lehrman, *Laboratory Experiments*,
    Expt. 40, p. 82, Diffraction.
Meiners, 19-7, pp. 519–524, Diffraction of Sound Waves.

10.14 The superposition principle
Sutton, S-25 through S-28, Models of Wave Motion.

10.15 Interference
Meiners, 18-4, Interference of Waves, pp. 455–459.
Transparency: K & E, #41, Interference, Constructive
    and Destructive. #46, Overhead Projection Model of
    Interference.
Meiners, 19-5, pp. 508–510, Interference.
Laboratory Exercise: Lehrman, *Laboratory Experiments*,
    Expt. 63, pp. 129–132, Microwaves.
Project Physics Handbook, pp. 64–66 Unit 3, Moire Pat-
    terns.
Laboratory Exercise: Lehrman, *Laboratory Experiments*,
    Expt. 43, pp. 88–90, Interference from Two Point
    Sources.

10.16 Standing waves
Sutton, S-34 through S-37, Standing Waves in Strings.
Meiners, 18-5, Standing Waves, pp. 460–466.
Laboratory Exercise: Lehrman, *Laboratory Experiments*,
    Expt. 41, p. 84, Standing Waves.
Transparencies: K & E, #s 42-43, Beats, Parts I and II.

11 SOUND WAVES

Synopsis: Although we rely on all our senses for information about our environment, our eyes and our ears are the most important windows to our surroundings. The disturbances that are detectable by our ears (acoustical) are propagated by wave motion. Sound phenomena possess all the properties associated with longitudinal waves as set forth in the preceding chapter.

The sonic spectrum extends beyond the limits of audible sound. There is no clearly defined lower frequency limit of the spectrum—the length of an earthquake wave may be measured in kilometers, whereas the inter-particle spacing of the medium in which the wave is traveling establishes the lower wavelength limit of the sonic spectrum. The human ear is sensitive to a range of about ten octaves from 20 to 20,000 hertz.

There is currently considerable interest in sound: communication, speech and hearing, music, sonar, ultrasonics, noise and noise pollution. Ultrasonics is used in metallurgy to detect the presence of flaws; in medicine, to detect tumors and for therapy; in dentistry, for cleaning teeth; and in manufacturing, to remove contamination. The field of acoustics is important in oceanography, space exploration, aviation, architecture, noise control, and transportation.

Sound is produced by a vibrating source and transmitted as a vibrational disturbance at a definite velocity in each medium. Sound may be reflected (echoes), refracted (as by the atmosphere), and diffracted (as evidenced by the fact that we can hear a sound that is produced around the corner of a building). Interference of sound is evidenced in organ pipes and by production of beat tones.

Our study of sound touches on all the above and in addition considers the measurement of intensity, the effect on pitch produced by the relative motion of the source and the observer, the graphic representation of sound, and the characteristics of sound waves of concern in music.

Comment: Illustrate the production of sound by a variety of devices such as tuning forks, string instruments, reed instruments, and horns or whistles. Interpret Figure 11-3. Use a ringing bell inside an evacuated jar to demonstrate that sound transmission requires matter as a medium. Sound has a definite velocity in each medium in which it travels, this being determined by the density and elastic properties of the medium. Distinguish between intensity and loudness. Indicate how the intensity level of sound is measured. Interpret Figure 11-6, the range of audibility of the human ear.

Show the dependence of pitch on frequency. Introduce the Doppler effect by a discussion of familiar experiences and by use of the ripple tank. Develop the equations associated with the study of the Doppler effect.

Use an oscilloscope and associated equipment as well as chalkboard diagrams to produce a graphic representation of sound. If the oscilloscope and a loud speaker are connected in parallel to the audio system, one can hear the sound at the same time that the waveform is seen.

Stress the study of the characteristics of sound waves. Use a sonometer or stretched wire to illustrate the various modes of vibration shown in Figure 11-11. Support the study of harmonics with an oscillograph. Have students sound the same notes on various instruments to detect the meaning of "quality." Demonstrate and develop the equations for the laws of strings.

Discuss and demonstrate forced vibrations, resonance, and beats. Study resonant air columns, interpret Figures 11-20 and 11-21, and develop the equations associated with resonant air columns in open and closed tubes. Use two tuning forks to produce a beat tone and a chalkboard diagram to explain the formation of beats.

References: Amend, J. R., "The Velocity of Sound," *The Science Teacher*, Feb. 1964.

Benade, Arthur H., *Horns, Strings, and Harmony*, Science Study Series, Doubleday and Co., Inc., Garden City, N.Y.

Beranek, Leo L., "Noise," *Scientific American*, December 1966, pp. 66–74.

Bergeijk, Willem A. van, John R. Pierce, and Edward E. David, Jr., *Waves and the Ear*, Science Study Series, 1960.

Chedd, Graham, *Sound: From Communication to Noise Pollution*, Doubleday and Co., Garden City, N.Y., 1970.

Dwyer, Lester, "Demonstrations of Reflection and Diffraction of Sound," *The Physics Teacher*, May 1968, pp. 252–253.

Griffin, Donald R., *Echoes of Bats and Men*, Science Study Series, 1959.

Jordan, Byron E., "Hearing Is Believing or Is It?" *The Science Teacher*, Oct. 1968, pp. 80–81.

Kock, Winston E., *Sound Waves and Light Waves: The Fundamentals of Wave Motion*, Science Study Series, 1965.

Lindsay, R. B., "What Does Acoustics Have to Offer?" *The Physics Teacher*, Oct. 1963, pp. 159–163.

Pierce, John R., *Waves and Messages*, Science Study Series, 1967.

Stevens, S. S., Fred Warshofsky, and the editors of *Life*, *Sound and Hearing* (Life Science Library), Time, Inc., New York, 1965.

Films: *Approaching the Speed of Sound*, 27 min., color and sound, no charge except for return postage, Shell Oil Co., 50 West 50th St., New York. Describes and explains how sound waves travel through air and why the speed of sound affects high-speed aircraft. Striking views of shock waves building up are also shown in the film.

Sound Waves and Their Sources, 11 min., black and white, Encyclopaedia Brittanica Films, Inc., 1150 Wilmette Avenue, Wilmette, IL. Discusses vibrating columns of air, vibrating surfaces, and vibrating strings. Each is identified as a source of sound.

Sound Waves in the Air, 35 min., black and white, sound, PSSC, 0207. The wave characteristics of sound transmission are investigated with large-scale equipment using frequencies up to 5000 cycles. Experiments in reflection, diffraction, interference, and refraction are supplemented with ripple tank analogies. Standing waves.

The Sounds of Music, color, sound, Coronet Films, Inc., 65 East South Water Street, Chicago, IL 60601. Loudness, quality, and pitch are demonstrated through the use of string, wind, and percussion instruments.

References for Chapter 11:   Sound Waves

Semat, *Fundamentals of Physics*, Chapter 19, pp. 326–327, Vibrations and Sound.
Lindsay, *Basic Concepts of Physics*, Section 8.8, pp. 309–319, Sound Waves.

11.1 The sonic spectrum
For suggestions for demonstrations to accompany the work of this chapter see:
Sutton, S-50 through S-153, pp. 154 through 192, Sound.
Meiners, Chapter 19, pp. 590–629, Sound Waves.
Project Physics Text, Section 12.11, Sound Waves, Unit 3, pp. 128–129 and pp. 132–133. Noise and the Sonic Boom, pp. 130–131, Unit 3.
Karplus, *Introductory Physics, A Model Approach*, Chapter 7, pp. 173–196, Wave Models for Sound and Light, especially Section 7-1, pp. 173–179, Applications of Wave Theory to Sound.
Gamow and Cleveland, *Physics, Foundations and Frontiers*, Prentice-Hall, Englewood Cliffs, NJ, 1969, Section 9-2, pp. 135–136, Ultrasonics.

11.2 The production of sound
Sutton, S-50, S-51, pp. 154–155, A Sounding Body Is in Motion.
Transparency: 2 × 2 Color, Western, J-19 and J-20, Pressure vs Time Variation of Sound Waves, and Wave Form of a Tuning Fork.
Sutton, S-55 through S-68, pp. 157–162, Sources of Sound.
Sutton, S-69 through S-75, pp. 162–165, Sound Detectors.
Meiners, 19-4, pp. 496–508, Vibrating Systems and Sources of Sound.

11.3 Sound transmission
Sutton, S-80 through S-89, pp. 169–172, Transmission of Sound.
Lehrman and Swartz, *Foundations of Physics*, pp. 338–341, Speed of a Sound Wave.
Gamow and Cleveland, *Physics, Foundations and Frontiers*, Prentice-Hall, Englewood Cliffs, NJ, 1969, Chapter 9, pp. 132–141, Sound; Section 9-1, Sound Transmission.

Sutton, S-52, pp. 155–156, Material Medium Necessary for Transmission of Sound.

11.4 The speed of sound
Laboratory Exercise: Lehrman, *Laboratory Experiments*, Expt. 42, pp. 86–87, The Speed of Sound.
Sutton, S-54, pp. 156–157, Slow Mechanical Transmission of Sound Waves.
Meiners, 19-2, pp. 491–493, Velocity of Sound.

11.5 The properties of sound
Laboratory Exercise: *Project Physics Handbook*, Expt. 3-17, Measuring Wavelength, pp. 49–51 Unit 3; Expt. 3-18, Sound, pp. 51–53 Unit 3: Expt. 3-19, Ultrasound, pp. 53–55 Unit 3.
Sutton, S-90 through S-117, pp. 172–182, Wave Properties of Sound.
Lindsay, *Basic Concepts of Physics*, Intensity of Sound, p. 314 and Loudness, p. 317.

11.8 Frequency and pitch
Lehrman and Swartz, *Foundations of Physics*, pp. 343–345, Musical Sounds.

11.9 The Doppler effect
Meiners, 19-6, pp. 510–519, Doppler Effect.
Lehrman and Swartz, *Foundations of Physics*, pp. 345–347, Doppler Effect.
Gamow and Cleveland, *Physics, Foundations and Frontiers*, Prentice-Hall, Section 9-4, pp. 138–141, The Doppler Effect.

11.12 The quality of sound
Sutton, S-76 through S-79, pp. 165–169, Sound Analyzers.
Sutton, S-118 through S-153, pp. 182–192, Musical Sounds.

11.14 Forced vibrations
Sutton, S-110, p. 179, Forced Vibrations.

11.15 Resonance
Sutton, S-111 through S-117, pp. 179–182, Resonators.
Laboratory Exercise: *Exercises and Experiments in Physics*, Expt. 25, Resonance; The Speed of Sound.

Meiners, 19-3, pp. 493–496, Standing Longitudinal Waves.

11.16 Beats
Sutton, S-106 and 107, p. 178, Beats.
Meiners, 19-5, pp. 508–510, Interference; 19-5.4 and 5.5, p. 510, Beats.

THE NATURE OF LIGHT 12

Synopsis: The physicist's perception of "light" is likely to include electromagnetic radiation regions beyond that defined by the limits of vision. For example, the infrared region, the ultraviolet region, and even the X-ray region may be included. However, in its ordinary usage, *"light"* refers to the radiation frequencies (or wavelengths) that are associated with vision. When compared with the entire electromagnetic radiation spectrum, light occupies about one octave in a known spectral range of possibly ninety octaves. The strong absorption of radiation by most substances for radiations of longer or shorter wavelengths than light appears to place an inevitable limitation on the breadth of the optical spectrum.

It is interesting to observe that the ear and the eye each require about the same minimum amount of energy (about 10^{-18} j) for excitation. Both the ear and the eye are involved in normal life activities, but in physics, practically all instruments translate physical effects into visible, rather than audible, effects.

Certain optical phenomena such as reflection, refraction, interference, diffraction, and polarization are readily interpreted in terms of the wave theory and are difficult to explain satisfactorily in terms of the quantum theory. The photoelectric effect and emission or absorption spectra are readily interpreted in terms of the quantum theory. The measurement of luminous intensity is usually confined to visible light, although the physical ideas involved are not subject to this limitation.

Comment: The content of the first 14 sections of Chapter 12 involves the presentation and critical evaluation of theories dealing with the nature of light largely from the historical viewpoint. The use of student reports, of available films, and the previous overall treatment of wave motion may be combined to make this chapter a most stimulating and fruitful experience for students. Demonstrations should be employed wherever possible to supplement the discussions and present facts first hand. The inadequacies of the particle and wave models as well as the role of the quantum theory as a unifying theory should be clearly presented.

Review the five general properties of waves. Perform simple quick demonstration-experiments showing straight line (rectilinear) propagation (shadow formation); reflection (from a mirror); refraction (pencil in a cylinder filled with water); and possibly prism dispersion to review properties known to Newton. Have students attempt explanations according to the corpuscular theory, employing demonstrations suggested by Figures 12-2 and 12-3. Then have students attempt explanations of reflection and refraction according to the wave theory recalling (or redemonstrating) ripple tank experiments. Indicate the shortcoming of the wave theory in explanation of rectilinear propagation and its acceptance following the discovery of interference and

Foucault's determination of the relative speed of light in water and air. The film *Introduction to Optics* may be shown (see Films).

Present the electromagnetic theory as a solution to the problem of a medium that would transmit the energies of heat, light, and electric action (radiation phenomena). Discuss Maxwell's model as represented by Figure 12-6. Indicate the role of Hertz's experimental theory confirmation and later discovery of the absorption of light, which indicated the failures of the wave theory. You may wish to demonstrate how he detected electromagnetic waves using Hertz's ring and an induction coil. Acquaint students with the main parts of the electromagnetic spectrum, Figure 12-8. Demonstrate the photoelectric effect (see A. Ahlgren, "Inexpensive Apparatus for Studying the Photoelectric Effect and Measuring Planck's Constant," *The Physics Teacher*, Oct. 1963, pages 183–186; also see References). Discuss the laws of photoelectric emission, Figure 12-11, and Figure 12-12, and the consequent failures of the wave theory. For student reports on Maxwell and Hertz, see References.

Discuss the quantum theory and its resolution of the dual nature of light. Students can appreciate the summarizing statement in *italics* at the end of Section 12.10. Show how Einstein's photoelectric equation explained the photoelectric effect. Briefly consider the contributions of Bohr. Discuss de Broglie's matter waves and the representative equations as in Section 12.11. Deal with X rays as the converse of the photoelectric effect, and have students examine the table of photon energies (Table 12-1). Then consider the pressure of light. You may wish to show the film *Pressure of Light* (see Films). See References for student reports on the work of Einstein, Planck, Bohr, and Compton.

Discuss the two main sources of light. Demonstrate heated platinum or nichrome wire as an example of a luminous object and a mirror as an example of an illuminated object. Consider examples of reflection, absorption, and transmission as in the text. Use a light source as indicated in Figure 12-19 p. 304 to show pencils, beams, and the idea of rays. Show straight-line transmission by shadow formations as in Figures 12-20 and 12-21, page 305. Exhibit a pinhole camera. Discuss methods used to measure the speed of light. Display the authentic copy of the manuscript of an early experiment conducted by Michelson on the determination of the speed of light: "Velocity of Light," a reproduction of Michelson's handwritten report on his experiments in 1878, Honeywell Research, 2701 4th Avenue South, Minneapolis, Minnesota. You may wish to show the film *Speed of Light* (see Films).

Develop the definition of the *candle* as the luminous intensity of the source. Define luminous flux and the lumen. Bring out the fact that a luminous source having an intensity of one candle emits light at the rate of 4π lumens. Show how illumination on a surface varies inversely as the square of the distance from the source. Use a photocell and/or a wire frame model to illustrate inverse square law. Demonstrate operation of photometers.

References: Ainslie, Donald S., "Circuits for Demonstration Experiments with a Photoelectric Cell," *The Physics Teacher*, Nov. 1963, p. 229.

Brotherton, Manfred, *Masers and Lasers: How They Work, What They Do*, McGraw-Hill Book Co., New York, 1964.

Carroll, John M., *The Story of the Laser*, E. P. Dutton, New York, 1964.

Esterer, Arnulf K., *Discoverer of X Ray: Wilhelm Conrad Röntgen*, Messner, New York, 1968.

Froman, Robert, *Science, Art, and Visual Illusions*, Simon and Schuster, New York, 1970.

Gamow, George, *Thirty Years That Shook Physics, The Story of Quantum Theory*, Science Study Series, 1966.

Hendricks, Sterling B., "About Light and Its Action in Life," *The Science Teacher*, Dec. 1968, pp. 36–39.

Jaffee, Bernard, *Men of Science in America*, Simon and Schuster, New York, 1958, Chap. XV, Albert Michelson, America Participates in the Revolution in Modern Physics.

————, *Michelson and the Speed of Light*, Science Study Series.

Kock, Winston E., *Lasers and Holography: An Introduction to Coherent Optics*, Science Study Series, 1969.

Kock, Winston E., *Waves and Light Waves: The Fundamentals of Wave Motion*, Science Study Series, 1965.

Kreisler, Michael N., "Are There Faster Than Light Particles?," *The Physics Teacher*, Oct. 1969, pp. 391–394.

Leith, Emmett N., and Juris Upatnieks, "Photography by Laser," *Scientific American*, June 1965, pp. 24–35.

Michelson, Andrew A., *The Velocity of Light*, Honeywell, Inc. (Dept. 118, 2701 4th Avenue South, Minneapolis, MN 55408), 1965, 114 pp.

Miller, Stewart E., "Communication By Laser," *Scientific American*, Jan. 1966, pp. 19–27.

Newhall, Beaumont, *Latent Image*, Science Study Series, 1967.

Pierce, John R., *Quantum Electronics: The Fundamentals of Transistors and Lasers*, Science Study Series, 1966.

Pierce, John R., *Waves and Messages*, Science Study Series, 1967.

Pimentel, George C., "Chemical Lasers," *Scientific American*, April 1966, pp. 32–39.

Ready, John F., "Properties and Applications of Lasers," *The Physics Teacher*, Oct. 1968, pp. 344–351.

Schawlow, Arthur L., "Advances in Optical Masers," *Scientific American*, July 1963.

Scientific American, Readings from Lasers and Light. (Introduction by Arthur L. Schawlow.) Freeman, San Francisco, 1969.

Semat, Henry, *Fundamentals of Physics*, Holt, Rinehart and Winston, Publishers, New York, 1966, 4th Ed., Chap. 30, Light and Its Measurement, pp. 517–528.

Shamos, Morris H., *Great Experiments in Physics*, Holt, Rinehart and Winston, Publishers, New York, 1959; Chap. 13, Electromagnetic Waves—Heinrich Hertz; Chap. 14, X rays—Wilhelm C. Roentgen; Chap. 17, The Photoelectric Effect—Albert Einstein; Appendix 1, The Electromagnetic Field—James Clerk Maxwell; Appendix 2, the Quantum Hypothesis—Albert Einstein; Appendix 4, The Hydrogen Atom—Niels Bohr; Appendix 5, The Compton Effect—Arthur Compton.

Smith, G. R., and Victor, E., "Science Concepts of Light," *The Science Teacher*, Vol. 28, No. 1, p. 11.

Van Heel, A. C. S., and C. H. F. Velzel, *What is Light?* (Translated from the Dutch by J. L. J. Rosenfeld.) (World University Library Series.) McGraw-Hill Book Co., New York, 1968.

Films: *The Speed of Light*, 14 min., black and white, EBF. Shows how Galileo, Roemer, and Michelson developed the measurement of the speed of light and indicates some important applications.

The Nature of Light, 11 min., color, Coronet Films, 65 East South Water Street, Chicago, IL 60601. Shows light as a form of energy. Principles of reflection and refraction are discussed.

Principles of the Optical Maser, 30 min., color, sound, local Bell Telephone office. Shows how basic physical concepts are applied to make an optical maser oscillate. The optical maser is examined as a generator of electromagnetic energy in the optical range of frequencies, having many similar properties to standard radio and microwave oscillators. The principal types of gas and solid state optical masers are shown in the laboratory.

PSSC Films: 0201, *Introduction to Optics*, 23 min., color. Deals with the approximation that light travels in straight lines; shows the four ways in which light can be bent: diffraction, scattering, refraction, and reflection; refraction illustrated by underwater photography.

0202, *Pressure of Light*, 23 min., black and white. Light pressure on a tin foil suspended in a high vacuum sets the foil into oscillation. The film leads up to this by a discussion of the Crookes radiometer and the effect (not light pressure) that causes it to rotate. The role of light pressure in the universe is also briefly discussed.

0203, *Speed of Light*, 21 min., black and white. Outdoor at night, the speed of light over a 300 meter course using a spark-gap, parabolic mirrors, a photocell, and an oscilloscope. In the laboratory, the speed of light in air and in water is compared using a high-speed rotating mirror.

0418, *Photons*, 19 min. Photomultiplier and oscilloscope used to demonstrate that light shows particle behavior.

Project Physics Film Loop: Unit 4: Light and Electromagnetism. This Unit develops the field description for electricity and magnetism and discusses the phenomena of optics. It concludes with Maxwell's interpretation of light in terms of electromagnetic waves.

References for Chapter 12: The Nature of Light

Lindsay, *Basic Concepts of Physics*, Section 8.9, pp. 319–339, Light Waves.
Semat, *Fundamentals of Physics*, Chapter 30, pp. 517–528, Light and Its Measurement.

12.1 Properties of light
Gamow and Cleveland, *Physics, Foundations and Frontiers*, Prentice-Hall, Chapter 17, pp. 292–309, The Wave Nature of Light.
PSSC Physics, Third Edition, Chapter 1, pp. 8–22, How Light Behaves. Chapter 4, pp. 74–88, The Particle Model of Light.
Project Physics, Chapter 13, Light.
Meiners, Section 35-1, pp. 1064–1075, Nature and Propagation of Light.
Karplus, *Introductory Physics, A Model Approach*, Chapter 5, pp. 111–153, Models of Light and Sound; especially Sections 5.1, Properties of Light and Sound and 5.2, The Ray Model for Light, pp. 111–126. Chapter 12, pp. 139–167, The Wave Theory.

12.2 The corpuscular theory
PSSC Physics, Third Edition, Chapter 25, Photons, Section 25-1, p. 572ff., The Graininess of Light.
Karplus, *Introductory Physics*, p. 187, Newton's Corpuscular Model.

12.3 The wave theory
PSSC Physics, Third Edition, Chapter 27, pp. 610–626, Matter Waves.
Gamow and Cleveland, *Physics, Foundations and Frontiers*, Prentice-Hall, Section 17-2, pp. 295–296, Huygens' Principle. Section 17-6, pp. 305–308, The Nature of Light Waves.
Karplus, *Introductory Physics*, Section 6.3, pp. 153–156, Huygens' Principle.

12.4 The electromagnetic theory
Karplus, *Introductory Physics*, Section 7.3, pp. 192–196, The Electromagnetic Theory of Light.

12.5 The electromagnetic spectrum
PSSC Physics, Third Edition, Section 23-9, pp. 545–547, Evidence for Electromagnetic Radiation, The Electromagnetic Spectrum.

12.6 The photoelectric effect
Rogers, *Physics for the Inquiring Mind*, Princeton University Press, 1960, p. 185, The Electromagnetic Spectrum.
PSSC Physics, Third Edition, Sections 25-2 and 25-3, p. 575ff., The Photoelectric Effect and Einstein's Interpretation of the Photoelectric Effect.

12.7 Laws of photoelectric emission
Gamow and Cleveland, *Physics, Foundations and Frontiers*, Section 22-5, pp. 400–404, The Puzzle of the Photoelectric Effect.

12.9 The quantum theory
PSSC Physics, Third Edition, pp. 617–624, Sections 27-4, What is It That Waves, 27-5, Standing Waves, 27-6, A Particle in a "Box," 27-7, The Standing Wave Model of the Hydrogen Atom.
Gamow and Cleveland, *Physics, Foundations and Frontiers*, Prentice-Hall, Chapter 22, pp. 393–406, The Energy Quantum.

12.10 Einstein's photoelectric equation
PSSC Physics, Third Edition, pp. 575–581, Sections 25-2, The Photoelectric Effect, 25-3, Einstein's Interpretation of the Photoelectric Effect.

12.11 The quantized atom
Karplus, *Introductory Physics,* Section 8.3, pp. 212–215, Bohr's Model of the Atom.

12.14 The pressure of light
Use PSSC Film, 0202 Pressure of Light.
PSSC Physics, Third Edition, Sections 4-2, p. 77, Light Pressure, 4-6, p. 83, Some Difficulties with the Particle Theory, 14-9, p. 319, Light and the Conservation of Momentum, 25-4, p. 581, The Momentum of Photons.

12.15 Luminous and illuminated objects
PSSC Physics, Third Edition, Sections 1-1, p. 8, Sources of Light, 1-2, p. 9, Transparent, Colored and Opaque Materials.
Sutton, L-1 through L-4, pp. 367–369, Sources of Light. Sections L-5 through L-10, pp. 369–372, Light Paths Made Visible. L-15, p. 374, Rectilinear Propagation of Light. L-16, p. 375, Pinhole Camera.

Meiners, Section 35-4, pp. 1096–1097, Absorption of Light.
PSSC Physics, Third Edition, Section 1-8, p. 16, Light Beams, Pencils and Rays.

12.16 The speed of light
PSSC Physics, Third Edition, Section 1-6, p. 13, The Speed of Light.
Sutton, L-17, p. 376, Speed of Light.
Gamow and Cleveland, *Physics, Foundations and Frontiers,* Prentice-Hall, Section 16-1, pp. 276–279, The Velocity of Light.

12.17 Light measurements
Laboratory Exercise: Lehrman, *Laboratory Experiments,* Expt. 65, pp. 135–137, Photometry.
PSSC Physics, Third Edition, Section 4-1, p. 74, Source Strength and Intensity of Illumination.
Sutton, L-11 through L-14, pp. 372–374, Photometry.
Gamow and Cleveland, *Physics, Foundations and Frontiers,* Prentice-Hall, Section 16-2, pp. 279–281, Photometry.

12.18 The intensity of a source
Laboratory Exercise: *Exercises and Experiments in Physics,* Expt. 26, Photometry.

REFLECTION 13

Synopsis: As a beam of light passes through a medium, some of its energy is transformed into heat and some is scattered in directions other than the direction of the beam. The color of the sky is due to the scattering of sunlight by particles in the air from molecular size to the size of visible dust. Since light is electromagnetic in character, some energy may cause the electrons of the medium to vibrate. These molecules, in turn, may reradiate the energy at the same or a different frequency, as in the case of the phosphor of a fluroescent lamp. This reflected and reradiated energy is mostly in the blue region of the visible spectrum.

When a beam of light strikes a surface separating two media, air and glass for example, some of the light rays are reflected back into the first medium at the interface and the remaining rays are transmitted into the second medium. If the interface is smooth, the reflected rays exemplify *regular reflection*. If the interface is irregular, the reflected rays exemplify *diffusion*. If the transmitted rays pass obliquely through the interface into the second medium in which the propagation velocity is different, the rays are redirected at the interface and this change in direction exemplifies *refraction*.

Smooth, polished metal surfaces reflect about 90% of the incident light. Polished glass surfaces will reflect from 4% to 10% for angles of incidence from 0° to 60°.

In a metal, the light that is refracted through the surface is absorbed in a very thin layer of the material. Glass is transparent. Very little light is absorbed in its passage through reasonable thicknesses.

Reflection and refraction (Chapter 14) will be studied by assuming that light travels along straight lines called rays, in an optically homogeneous medium. Rays are at right angles to the wave front and indicate the direction of motion of the wave. The study of rays is called geometrical optics.

The laws of reflection, the formation of images by reflection, object-image relationships, and the principle of the reflecting telescope are developed in this chapter.

Comment: You may wish to begin the study of reflection and image formation by exhibiting and/or demonstrating the following: shaving mirror, auto rear view mirrors of various types, range finder, parallax viewer, optical micrometer, pinhole camera, or other device. As you proceed with the work, be certain that students understand that there are two types of images, the characteristics of each, how the laws of reflection are applied, how ray diagrams are constructed and interpreted, and the geometric relationships involved.

After using selected demonstrations to whet student curiosity and posing problems to be solved, the laws of reflection may be demonstrated in a dark room using an optical disk and a parallel-beam light source. Bouncing a ball at various angles on the floor may help students understand reflection. Diffuse reflection may be demonstrated by using white cardboard as a reflecting surface. Have diagrams similar to Figures 13-3 and 13-4 developed at the board. Exhibit types of mirrors and develop diagrams similar to Figure 13-5 for plane and curved surfaces. Discuss and demonstrate the formation of images. Project a slide to demonstrate the formation of a real image. Discuss virtual images and develop a diagram as in Figure 13-9. Demonstrate "sighting" technique to be used in laboratory experiments and develop ray diagrams that are appropriate to explain image formation.

Distribute concave mirrors to the students. Using one of these mirrors (or a cosmetic or shaving mirror) show how an image of a distant object such as a room window can be formed on a sheet of paper when held at the proper distance. The image of an object can be viewed in a convex mirror to discover the size and type of image formed. Using chalkboard diagrams, review curved-mirror terminology, as in Figures 13-11 and 13-20. Develop ray diagrams for image formation for the various cases for concave mirrors as in Figure 13-19. Call attention to spherical aberration. Develop and use the mirror equation for both concave and convex mirrors as in Section 13.12. Stress the significance and use of negative values in applying the equation. Explain the solution of the solved Example in Section 13.12. Assign or call for volunteers to demonstrate their solutions to the Practice Problems following the solved Example. Illustrate uses of curved mirrors and of the reflecting telescope.

References: Halliday and Resnick, *Physics*, John Wiley and Sons, New York, 1978, selected parts of Chap. 43, Reflection and Refraction—Plane Waves and Plane Surfaces.

Semat, Henry, *Fundamentals of Physics*, Holt, Rinehart and Winston, Publishers, New York, 1966, Chap. 31, Reflection and Refraction, Sec. 31-2 through 31-5, pp. 531–533.

References for Chapter 13: Reflection

Gamow and Cleveland, *Physics, Foundations and Frontiers*, Prentice-Hall, Chapter 15, pp. 247–271, Reflection and Refraction.

Project Physics Text, Unit 4, Chapter 13, Light, pp. 5–27. Meiners, Chapter 34, Section 34.1, pp. 1035–1059, Reflection and Refraction, 27 suggestions for demonstrations from which to choose.

13.1 Reflectance

PSSC Physics, Third Edition, Sec. 1-3, Reflection, pp. 10–12. Chapter 2, Reflections and Images, pp. 24–42.

Sutton: L-18, p. 376, Plane Mirror, L-19, p. 376, Reflection at Normal and Grazing Incidence. L-20, p. 377, Multiple Reflections in Plane Mirrors. L-21, p. 378, Plane Mirrors, Special Combinations. L-22, p. 378, Laws of Reflection.

2 × 2 color transparencies: Western, A-1, Reflection at the Surface of Water. A-2, Ditto at a Plane Mirror.

13.3 Mirrors as reflectors

Western: A-3, Diffuse Reflection. A-4, Reflection and Refraction at the Surface of Water. A-5, Ditto at a Glass Surface. A-13, Reflection of Converging Rays at a Plane Mirror.

PSSC Physics, Third Edition, Sec. 4-4, pp. 78–79, Reflection. Sec. 6-3, pp. 110–112, Reflection.

Laboratory Exercise: Lehrman, *Laboratory Experiments*, Expt. 66, pp. 138–140, Reflection of Light.

Semat, *Fundamentals of Physics*, Section 31-2, pp. 531–532, Laws of Reflection.

13.4 Images by reflection

Exercises and Experiments in Physics, Expt. 27, Plane Mirrors.

Semat, *Fundamentals of Physics*, Section 31-4, pp. 532–533, Images Formed by a Plane Mirror.

13.5 Images formed by plane mirrors

K&E Transparency: #55, Images from Mirrors.

Western, 2 × 2 color slides, A-14 through A-18, Images Formed by Plane Mirror or Mirrors.

13.7 Rays focused by spherical mirrors

Semat, *Fundamentals of Physics*, Section 32-8, pp. 556–557, Spherical Mirrors.

Western, 2 × 2 color slide, B-2, Focus of a Concave Mirror.

K&E Transparency: #56, Spherical vs Parabolic Mirrors.

13.8 Constructing the image of a point

Western, 2 × 2 color slide, B-3, Image Formed by a Concave Mirror. B-11, Aberration of a Concave Mirror.

Semat, *Fundamentals of Physics*, Section 32-9, pp. 557–558, Image Formation by Concave Mirrors.

13.9 Images formed by concave mirrors

Sutton: L-23 and L-24, pp. 378–379, Concave Mirror—Phantom Bouquet.

Laboratory Exercise: *Exercises and Experiments in Physics*, Expt. 28, Curved Mirrors.

13.10 Images formed by convex mirrors

Western, 2 × 2 color slides, A-19, A-20 and B-1, Convex Mirror.

Sutton: L-27, p. 380, Convex Mirror.

Semat, *Fundamentals of Physics*, Section 32-11, pp. 560–561, Image Formation by Convex Mirrors.

13.12 Object-image relationships

Semat, *Fundamentals of Physics*, Section 32-10, pp. 558–560, The Mirror Equation.

REFRACTION 14

Synopsis: The term refraction applies properly to the change in direction that radiation, especially light or sound, experiences in passing obliquely from one medium to another in which the velocity of propagation is different. The effect can be visualized in a ripple tank. Establish a train of straight parallel waves in such a way that they advance on an interface between water of one depth and water of a different depth. The velocity of the wave train changes as it crosses the interface. If it is made to progress perpendicularly across the interface, all points on each wave front change speed at the same time and there is no change in the direction of travel of the wave, that is, no refraction. If, however, the wave advances obliquely toward the interface, one end of each wave front will suffer a change in speed before the other end does, thus the direction of propagation is changed. The change in direction will be toward the normal to the interface if the wave is reduced in speed and away from the normal if it is increased in speed. A light beam will be affected similarly as it passes obliquely from one medium into another having different optical properties.

It can be shown that refraction is governed by a simple relation ($\sin i = n \sin r$) known as Snell's law. In this equation, n has the same value for all angles of incidence (i) and refraction (r)

for the passage of the radiation across the boundary between two specific substances. The constant n is known as the index of refraction. Physically, n represents the ratio of the velocity of the disturbance in the first medium to that in the second.

When light passes from one medium to another in which the velocity is greater ($n > 1$) and when the angle of incidence is sufficiently large, a phenomenon known as total reflection can be observed.

Refraction may occur in a single medium due to variations in its properties. The twinkling of stars is due to differences in the index of refraction of the atmosphere that result from temperature differences.

The index of refraction for all ordinary transparent substances is numerically greater than one. Since the index of refraction varies with the wavelength of the incident radiation, any exact statement of its value must specify the wavelength to which it refers. Tables of the index of refraction are usually given for yellow sodium light (5893 Å) since it is near the center of the visible spectrum. A prism, made of the material in question, is often used for the measurement of n. Dispersion is caused by the fact that the velocity of light in any given transparent medium is dependent on the wavelength of the light. In the visible part of the spectrum, red light is least changed in velocity and is therefore refracted the least amount while light of shorter wavelength is refracted the most.

In this chapter our knowledge of the refraction of light is applied to the formation of images by lenses, to the dispersion of light by prisms, and to the construction of optical instruments.

In addition this chapter deals with color of light, and the mixing of colored lights and pigments.

Comment: Demonstrate the floating-coin and bent-stick illusions. Use the optical disk and/or water in an aquarium to demonstrate refraction. Use the optical disk to demonstrate internal reflection, critical angle, properties of convex and concave lenses, and spherical aberration. Use a bent plastic rod to demonstrate internal reflection. Integrate laboratory experiments into the chapter development, particularly with the construction of ray diagrams. Emphasize the reversibility of object and image in lens diagram situations. Have students use the magnifier and microscope. Demonstrate projection of slides and use of film to form images. Redemonstrate refraction phenomena using the ripple tank.

After demonstrating illusions, use the optical disk and/or the water aquarium to provide data for understanding of refraction. You may wish to vary the angle of incidence or use a table of data (*PSSC Physics,* pages 46–47) to have students develop representative graphs of angle of incidence versus angle of refraction. Have students realize that some of the light is reflected as well as refracted (Figure 14-3, page 308). You may wish to develop Snell's law using the ratio of semi-chords (in effect what Descartes and Snell did) as in *PSSC Physics, Third Edition,* Section 3-3, pp. 48–51. Develop both equations for the index of refraction. Solicit questions about the solution to the solved Example in Section 14.3 from the class and elicit responses from other members of the class. Call for volunteers to demonstrate their solutions to the Practice Problems that follow the solved Example. Demonstrate and then have students construct ray diagrams as in Figures 14-7 and 14-8. Develop the concept of critical angle using Figure 14-9.

Develop the relationship between the index of refraction and the sine of the critical angle as in Section 14.5, page 312. Discuss applications of total reflection using the plastic rod transmission of light as a demonstration.

The work on lenses may be introduced by developing chalkboard ray diagrams of prisms placed base-to-base and then vertex-to-vertex to provide foundation for refraction by lenses and lens types. Show effect of lenses on plane waves. Develop definitions of important terms applicable to lenses, utilizing Figure 14-6 and Figure 14-7, pages 317–318. Demonstrate various cases by use of the optical disk. Compare image formation by lenses and mirrors. Demonstrate methods to be employed in construction of ray diagrams. Students should construct their own diagrams for the various cases and/or construct a table for the various cases, comparing d_o, d_i, and image characteristics as in *Physics Problems*, page 273. You may wish to show students how to find the focal length of a convex lens by finding the image of a window in the room on a piece of white paper mounted on an opposite wall or demonstrating Case 3, object at 2f (see Figure 14-20). The lens equation may be derived by dealing with the tabulated results of Experiment 30, Converging Lenses from *Exercises and Experiments in Physics* or *Laboratory Experiments in Physics*. The lens equation is also employed in Experiment 31, Focal Length of a Lens.

You may wish to apply the equation to Case 1 for convex lenses where d_o approaches infinity by relating to demonstration of simple method of finding focal length by projection of image on a paper screen or by use of a "burning" glass. For a discussion of the lens-maker's equation see pages 302–303, *PSSC Physics, First Edition.*

Point out the fact that 25 cm is the standard distance for most distinct *close* vision for the *average* eye. Define magnification and derive the equation for magnification. Set up an optical bench to demonstrate the compound microscope as illustrated in Figure 14-23 using a lighted candle as an object. Have students look through the eyepiece to see an enlarged image. The student may also observe a distorted image and chromatic aberration. You may wish to set up a refracting telescope (Figure 14-24) on the optical bench. Have ray diagrams drawn for microscope and telescope.

Demonstrate dispersion by the suggested methods or use a photographer's tungsten-filament flood lamp as suggested in Section 14.15. Use Table 14-1 to show variation of index of refraction with frequency (wavelength) of light (color). Demonstrate the color of opaque and transparent objects with colored lights. Use Van Nardroff or Singerman color apparatus or spinning adjustable color disks to show mixing. Marking colored crayons on white paper illustrates pigment mixing. The teacher should emphasize the difference between mixing the three colored lights (additive process) and mixing the three primary pigments (subtractive process). Discuss chromatic aberration. Have students view colored cloth and glass under varied colored light.

References: Ahlgren, Andrew, "Experiments with Simple Equipment—A Light Projector for Refraction Experiments," *The Physics Teacher*, May 1963, pp. 79–80.

Bond, Burtis B., "A Ripple Tank Stroboscope," *The Science Teacher*, Vol. 30, No. 5, p. 48.

Castka, J., and R. Lefler, *Physics Problems*, Holt, Rinehart and Winston, Publishers, New York, 1961, pp. 270–279.

Gluck, Irvin D., *Optics, The Nature and Applications of Light*, Holt, Rinehart and Winston, Publishers, New York, 1964, paperback, 144 pages.

Hilton, Wallace A., "Index of Refraction of Air," *The Physics Teacher*, April 1968, p. 176.

Julesz, Bela, "Texture and Visual Perception," *Scientific American*, Feb. 1965.

Lindsay, Robert Bruce, *Basic Concepts of Physics*, Van Nostrand Reinhold, New York, 1971, Section 8.9, Light Waves, pp. 319–339.

Pollack, Harvey, "Teaching the Law of Optics by the Ripple Tank Technique," *Harbrace Teacher's Notebook*, Spring, 1960, Harcourt, Brace and Jovanovich, New York.

Southall, James P. C., *Mirrors, Prisms and Lenses*, Dover, New York, 1964.

Sprecher, Roland O., "A Demonstration Lens," *The Science Teacher*, Dec. 1958, p. 470.

Free and Inexpensive Learning Materials: *Amateur Photography*, 47 pages, price 35¢, by Alexander J. Wedderburn, Haskin Service, 1200 I Street Northwest, Washington, DC. Contains information on cameras, film, and how to develop and print pictures.

Horizontal Section of Right Eyeball—Chart $8\frac{1}{2}''$ × 11″, free, undated, National Society for the Prevention of Blindness, Inc., 1790 Broadway, New York, NY.

Milestones in Optical History, 31 pages, free, 1954, Bausch and Lomb Optical Co., Rochester, NY. Brief biographical sketches accompanied by colored portraits of pioneers in eye hygiene and optics.

Photographic Optics, 64 pages, 1954, free, Bausch and Lomb Optical Co., Rochester, NY.

Films: *Eye to the Unknown*, 33 min., color, free loan, Modern Talking Picture Service, 3 East 44th Street, New York, NY. Presents problems in spectroscopy in science, medicine, and industry.

Fine Cameras and How They Are Made, 28 min., sound and color, free loan, Modern Talking Picture Service, 3 East 44th Street, New York, NY. Presents research and engineering behind cameras with diagrammatic explanations.

Introduction to Optics, 23 min., color, PSSC, 0201. Refraction illustrated by underwater photography to show how objects above water appear to a submerged skin diver.

NAA Optical Laboratory, 10 min., sound and color, free loan, North American Aviation, Inc., International Airport, Los Angeles, CA. Development of high-precision lenses in the optical laboratory.

The Nature of Color, 11 min., color, Coronet Films, Inc., 65 East South Water Street, Chicago, IL 60601. Principles of color reflection, absorption, mixing of colors by addition and subtraction. Application of color to printing and photography.

The Story of Lenses, fs, color, FOM 579, 42 frames, McGraw-Hill, Text Film Dept., 1221 Avenue of the Americas, New York, NY 10020. Principles of lenses.

References for Chapter 14: Refraction

Gamow and Cleveland, *Physics, Foundations and Frontiers*, Chapter 15, pp. 247–271, Reflection and Refraction of Light.

Semat, *Fundamentals of Physics*, Section 31-1, pp. 530–531, Passage of Light Through a Medium.

Project Physics Text, Unit 4, Chapter 13, pp. 5–27, Light.

14.1 The nature of optical refraction

Meiners: Chapter 34, Section 34.1, pp. 1035–1059, Reflection and Refraction, 27 suggestions for demonstrations from which to choose.

Laboratory Exercises: Lehrman, *Laboratory Experiments*, Expt. 67, pp. 140–142, Refraction of Light.

Laboratory Exercise: *Project Physics Handbook*, Expt. 4-1, pp. 4–6, Refraction of a Light Beam.

Sutton: L-28, p. 380, Refraction, L-29, p. 381, Refraction by Shadow with Cube of Glass. L-30, p. 381, Refraction Model. L-31, p. 381, Refraction by Gases.

14.2 Refraction and the speed of light

Semat, *Fundamentals of Physics*, Section 31-5, pp. 533–535, Refraction of Light.

Western, 2 × 2 color slides, A-4, Reflection and Refraction of Light at the Surface of Water. A-5, Ditto at the Surface of Glass.

14.3 The index of refraction

Sutton: L-33, p. 383, Refractive Index—Christiansen Filters.

Exercises and Experiments in Physics. Expt. 29, Index of Refraction of Glass.

PSSC Physics, Third Edition, Chapter 3, pp. 44–72, Refraction. Section 4-5, pp. 80–83, Refraction. Section 4-7, pp. 84–85, The Speed of Light and the Theory of Refraction. Section 6-6, pp. 118–120, Dispersion. Section 8-4, pp. 151–153, Color and Wavelength of Light. Section 1-2, pp. 9–10, Transparent, Colored and Opaque Materials.

Semat, *Fundamentals of Physics*, Section 31-7, pp. 537–539, Refraction Effects.

14.4 The laws of refraction

Western, 2 × 2 color slides, A-9, Path of a Ray Through a Plane-Parallel Glass Plate.

14.5 Total reflection

Sutton: L-34 through L-41, p. 385ff., Total Internal Reflection.

Western, 2 × 2 color slides, A-6, Reflection and Refraction of Light from a Glass-Air Surface. A-7, Ditto

Larger Angle of Incidence. A-8, Total Reflection. A-10, Totally-Reflecting Prism. A-11, Fiber Scope. A-12, Elevated Image of an Object in Water.

K & E Transparency: #57, Reflection and Refraction— Critical Angle.

Semat, *Fundamentals of Physics*, Sections 31-8, pp. 539–542, Critical Angle—Total Reflection; 31-9, Totally Reflecting Prisms; 31-10, Light Pipes, Fiber Optics.

14.6 Types of lenses

Semat, *Fundamentals of Physics*, pp. 545–547, Section 32-1, Lenses; 32-2, Thin Lenses.

Sutton: L-50, p. 390, Effect of Medium on Focal Length.

14.7 Ray diagrams

Western 2 × 2 color slides, B-4, Focus of a Convex Lens. B-5, Ditto. B-6, Focus of a Spherical Lens. B-9, Combination of two Convex Lenses.

Western, 2 × 2 color slides, B-7, Focus of a Concave Lens. B-8, Concave Lens and Converging Rays. B-10, Combination of Convex and Concave Lens.

14.8 Images by refraction

Laboratory Exercise: Lehrman, *Laboratory Experiments*, Expt. 69, pp. 146–149, Image Formation by Lenses.

Sutton: L-47, p. 388, Image Formation by Lenses. L-48, p. 388, Action of a Lens.

14.9 Images formed by converging lenses

Semat, *Fundamentals of Physics*, Section 32-3, pp. 547–548, Image Formation by Converging Lenses.

K & E Transparencies: #58, Images from Lenses. #59, Defects of Lenses.

Laboratory Exercises: *Exercises and Experiments in Physics*, Expt. 30, Converging Lenses. Expt. 31, Focal Length of a Diverging Lens.

Western, 2 × 2 color slides, B-12. Aberration of Lenses (1). B-13, Ditto (2).

14.10 Images formed by diverging lenses

Semat, *Fundamentals of Physics*, Section 35-2, pp. 550–551, Image Formation by Diverging Lenses.

14.11 Object-image relationships

Laboratory Exercise: Lehrman, *Laboratory Experiments*, Expt. 70, pp. 149–150, The Lens Equation.

Semat, *Fundamentals of Physics*, Section 32-4, pp. 548–550, The Lens Equation. Section 32-6, pp. 551–553, The Lensmaker's Equation.

14.12 The simple magnifier

Semat, *Fundamentals of Physics*, Section 33-4, pp. 570–572, The Magnifying Glass.

Laboratory Exercise: *Exercises and Experiments in Physics*, Expt. 32, Lens Magnification.

14.13 The microscope

K & E Transparency: #64, Microscope.

Semat, *Fundamentals of Physics*, Section 33-7, pp. 575–577, The Compound Microscope.

Sutton: L-54, p. 391, Model Microscope.

14.14 Refracting telescopes

Semat, *Fundamentals of Physics*, pp. 572–575, Sections 33-5, The Astronomical Telescope; 33-6, The Terrestrial Telescope.

K & E Transparency: #65, Telescope.

Sutton: L-55, p. 391, Telescope.

14.15 Dispersion by a prism

Western, 2 × 2 color slides, B-14, Dispersion of Light by Prisms (1). B-15, Ditto (2). B-18 through B-20, Rainbow.

K & E Transparency: #53, Light and the Prism.

Sutton: L-42, p. 386, Dispersion by a prism.

14.16 The color of light

Semat, *Fundamentals of Physics*, Section 31-6, pp. 535–537, Refraction and Dispersion.

Sutton: L-87 through L-100, pp. 404 through 411, Color.

14.17 The color of objects

Western, 2 × 2 color slide, C-15, Colors reflected at the surface of and transmitted through matter are compared.

Meiners: Demonstrations relating to color may be selected from Section 35-7, pp. 1113–1122, Color, eight suggestions.

14.18 Complementary colors

Western, 2 × 2 color slides, C-10, Mixing of Light Colors (1). C-11, Ditto (2). C-12 through C-14, Complementary Colors (1), (2), and (3).

14.19 The primary colors

Gamow and Cleveland, *Physics, Foundations and Frontiers*, Prentice-Hall, Section 16-3, pp. 282–284, Colored Light; Section 16-4, p. 285, Color in the Sky; Section 16-5, pp. 285–287, The Rainbow.

14.20 Mixing pigments

Western, 2 × 2 color slide, C-9, Mixing of Pigment Colors.

14.21 Chromatic aberration

Western, 2 × 2 color slides, B-16 and B-17, Chromatic Aberration.

Semat, *Fundamentals of Physics*, pp. 561ff., Sections 32-12, Spherical Aberration; 32-13, Chromatic Aberration.

Sutton: L-49, p. 389, Chromatic and Spherical Aberration.

15 DIFFRACTION AND POLARIZATION

Synopsis: Thus far in the study of light, reflection and refraction, while strictly wave phenomena, have been presented by using rays that are normals to the wave fronts. Diffraction, a property of light that can be best explained in terms of wave propagation will now be considered. It is possible to observe diffraction by looking at a distant light source through the space between two closely held fingers of the hand. When doing this, one sees a series of alternately dark and bright bands that can be explained in terms of wave theory as resulting from the bending of light wave fronts as they pass around the edges of the fingers.

Although diffraction may be easily observed, the explanation involves interference phenomena that can be understood after studying the double slit interference experiment performed by Thomas Young around 1800. Although Young's interpretation of the experiment was in terms of the wave theory, Newton's corpuscular theory was so firmly established that Young was bitterly attacked in Britain. It was not until after the work by the French physicists Fresnel, Malus, and Biot years later that Young was vindicated.

We shall study interference and diffraction phenomena and see how it is possible not only to observe and explain the phenomena but to use diffraction for measuring the wavelength of a given color of light.

An interesting property of light is known as polarization. One effect of polarized light can be observed by placing a sheet of Polaroid film in front of another sheet and rotating the sheets with respect to each other while viewing a source of light through them. The intensity of the transmitted light can be observed to vary from a maximum to near zero. The effect can be explained if the direction of vibration of the light waves is restricted to a single plane by its passage through the sheet of Polaroid. Light is transmitted through the second sheet when its polarizing plane is parallel to the polarizing plane of the first sheet and light is blocked when the polarizing plane of the second sheet is at 90° to the polarizing plane of the first sheet.

Comment: Students may give reports on the experiments of E. H. Land and on the contributions of Thomas Young and Augustin Fresnel. The multiplicity of interesting demonstrations and experiments related to interference, diffraction, and polarization phenomena should be used to maintain student interest. The studies of these phenomena should be explained on the basis of the wave theory of light.

Produce interference fringes (irregular) by illuminating two pieces of plate glass placed one on top of the other and viewed from the side. Pressing the plates together causes the fringes to spread out further. Microscope slides arranged in the same way give rings that are much closer together. Have students interpret formation of bands according to Figure 15-3. Have students report on ''The Interference of Light—Thomas Young'' from *Great Experiments in Physics*, pages 93–107. Discuss maximum constructive and destructive interference as in Section 15.2 and reproduce a chalkboard diagram for Figure 15-4 to aid in discussion.

Have students view spectra by looking at a light source through a grating and then also see the spectrum by placing the grating on a piece of white paper and viewing from appropriate angle. Discuss the determination of wavelength by diffraction according to treatment in Section 15.4 and explain the solution of the solved Example at the end of Section 15.4. You may wish to call upon students to demonstrate their solutions to the Practice Problems that accompany the solved Example.

Discuss single-slit diffraction as in Section 15.5 using Figure 15-8 as the basis. Explain scattering phenomena in terms of diffraction. You may wish to extend this lesson to include ripple tank demonstrations. Have students report on "The Diffraction of Light—Augustin Fresnel" from *Great Experiments in Physics*, pages 108–112. Place a slit in front of a flame and have students view it through a replica grating. Try some of the inert gas spectrum tubes as well for other bright-line spectra. Have them see the continuous spectrum emitted by a heated platinum wire.

Demonstrate polarization by double refraction with a crystal of Iceland spar, viewing letters on a printed page. Have students look at a light source with crossed and uncrossed Polaroid sheets. Shine a bright light on a shiny surface and have students observe through Polaroid sunglasses and describe glare reduction.

See Figure 15-9 and explain effects with mechanical analogy, Figure 15-10. Discuss examples of various processes producing polarization: selective absorption, reflection, and refraction. Students may wish to demonstrate plastic strain patterns. Discuss other uses of polarized light.

References: Barrow, Gordon, M., "Modern Spectroscopy," *The Science Teacher*, Feb. 1970, pp. 43–51.

Biser, Roy, H., "Modern Methods for the Study of Optical Diffraction," *The Physics Teacher*, March 1969, pp. 158–161.

Clark, Mary Lou, "Measuring the Wavelength of Light," *The Physics Teacher*, Feb. 1964, pp. 85–87.

Darby, William P., "Overhead Projection Demonstrations: A New Application of the Ripple Tank; Diameter of Molecule and Avogadro Number," *The Physics Teacher*, Feb. 1964, pp. 81–85.

Eisner, Leonard, "The Elastic String for Ray Indicators," ibid, p. 87.

Holden, Alan, and Phylis Singer, *Crystals and Crystal Growing*, Doubleday and Co., Inc., Garden City, 1960; Spectroscope, pp. 281–283; Polarimeter, pp. 244–248.

Lindsay, Robert Bruce, *Basic Concepts of Physics*, Van Nostrand Reinhold, New York, 1971, Section 8.9, Light Waves; pp. 319–339.

Long, Robert H., "An Apparatus Introducing the Principle of Polarized Light," *The Science Teacher*, Sept. 1956, p. 243.

Oster, Gerald, and Yasunori Nishijima, "Moire Patterns," *Scientific American*, May 1963, pp. 54–63.

Tomer, Darrell, "Strophotography," *The Science Teacher*, May 1960, pp. 47–52.

Williams, Dudley, "Infrared Radiation," *The Physics Teacher*, Nov. 1963.

Films: *Interference of Photons*, PSSC, 0419, 14 min., black and white. An experiment in which light exhibits both wave and particle characteristics. A very dim light source, a double slit, and a photo-multiplier are used in such a way that less than one photon (on the average) is in the apparatus at any given time. Characteristic interference pattern is pointed out by many individual photons hitting at places consistent with the interference pattern. Implications of this are discussed.

Polarized Light, fs, color, 41 frames, 617 FOM, McGraw-Hill Book Co., Text Film Dept., 1221 Avenue of the Americas, New York, NY 10020. Experiments with and nature of polarized light.

The Science of Color Photography, fs, color, 42 frames, 576 FOM, McGraw-Hill Book Co., Text Film Dept., 1221 Avenue of the Americas, New York, NY 10020. Explains color photographic principles and applications.

References for Chapter 15: Diffraction and Polarization

Project Physics Text, Unit 4, Chapter 13, pp. 5–27, Light.

Meiners: Section 35-2, pp. 1076–1087, Interference, 13 suggestions. Section 35-3, pp. 1087–1095, Diffraction, 10 suggestions.

15.1 Double-slit interference

PSSC Physics, Third Edition, Section 3-12, pp. 65–68, Light Pencils and Scaling. Chapter 8, pp. 146–167, Light Waves.

K & E Transparencies: #44, Young's Interference Expt. I. #45, Young's Interference Expt. II.

Exercises and Experiments in Physics, Expt. 33, Diffraction and Interference.

K & E Transparencies: #41, Interference, Constructive and Destructive. #46, Projection Model of Interference.

Laboratory Exercise: Lehrman, *Laboratory Experiments*, Expt. 43, pp. 88–90, Interferences from Two Point Sources.

Western, 2 × 2 color slides, K-8, Diffraction of Water Waves by a Double Slit. K-9 and K-10, Interference of Water Waves (1) and (2). K-11 through K-13, Young's Experiment (1), (2) and (3).

Semat, *Fundamentals of Physics*, Section 34-1, pp. 581–583, Interference of Light from Two Sources.

Laboratory Exercise: *Project Physics Handbook*, Expt. 4-2, pp. 6–8, Young's Experiment—The Wavelength of Light.

15.2 Interference in thin films

Semat, *Fundamentals of Physics*, Section 34-2, pp. 583–586, Interference from Thin Films.

Western, 2 × 2 color slides, L-1, Interference of Light in Soap Films (1). L-2, Ditto (2). L-3, Ditto (3). L-4, Apparatus for the Formation of Newton's Rings. L-5 through L-7, Newton's Rings (1) through (3). L-8, Interference of Light by a Wedge-like Air Film.

Sutton: L-67, p. 395, Interference-Color of Thin Films. L-68, p. 396, Ditto. L-70, p. 396, Color in Thin Air Films. L-71, p. 397, Newton's Rings.

15.3 Diffraction of light

Semat, *Fundamentals of Physics*, Section 34-3, pp. 586–587, Diffraction of Light.

Sutton: L-76, p. 399, Diffraction Box, L-73 through L-75 and L-77, pp. 398–400, Diffraction.

Karplus, *Introductory Physics*, Section 6.4, Diffraction of Waves.

15.4 Wavelength by diffraction

K & E Transparencies: #48, Diffraction Grating. #49, Gratings and Spectra.

Laboratory Exercise: Lehrman, *Laboratory Experiments*, Expt. 64, pp. 133–135, Wavelength of Light.

Laboratory Exercise: *Exercises and Experiments in Physics* or *Laboratory Experiments in Physics*, Expt. 34, Wavelength by Diffraction.

Sutton: L-82, p. 401, Diffraction by Single and Double Slits. L-83, p. 402, Projection of Single- and Double-Slit Patterns.

Western, 2 × 2 Color Slides, K-14 through K-17, Diffraction Grating (1) through (4). K-18, Diffraction by a Phonograph Record. K-19, Crossed Diffraction Gratings (1). K-20, Ditto (2).

15.5 Single-slit diffraction

Western, 2 × 2 color slides, K-5, Diffraction of Water Waves by a Single Slit. K-6, Diffraction of Light Waves by a Single Slit (1). K-7, Ditto (2).

Semat, *Fundamentals of Physics*, Section 34-4, pp. 587–589, Diffraction Through a Narrow Slit.

15.6 Polarization of transverse waves

Semat, *Fundamentals of Physics*, Section 35-1, p. 597, Polarization, Transverse Waves.

Sutton: L-116 through L-136, pp. 418–430, Polarization.

Meiners: Appropriate demonstrations may be selected from Section 35-6, pp. 1102–1113, Polarization, 18 suggestions.

Exercises and Experiments in Physics, Expt. 35, The Polarization of Light.

15.7 Selective absorption

Laboratory Exercise: Lehrman, *Laboratory Experiments*, Expt. 68, pp. 143–146, Polarization of Light.

15.9 Polarization by refraction

Semat, *Fundamentals of Physics*, Section 35-2, pp. 598–601, Polarization by Reflection, Section 35-4, pp. 602–604, Polarization by Double Refraction.

15.10 Interference patterns

Semat, *Fundamentals of Physics*, Section 35-7, pp. 605–606, Interference with Polarized Light.

15.11 Scattering

Western, 2 × 2 color slides, L-9 and L-10, Polarization of Light (1) and (2). L-11 and L-12, Polarization by Reflection (1) and (2). L-13 through L-16, Polarization by Double Refraction (1) through (4). L-17, Production of Colors by Polarized Light (1). L-18, Ditto (2). L-19, Photoelasticity (1). L-20, Ditto (2).

Semat, *Fundamentals of Physics*, Section 35-6, p. 605, Polarization by Scattering.

15.12 Optical rotation

Semat, *Fundamentals of Physics*, pp. 606–607, Sections 35-8, Photoelasticity; 35-9, Further Applications of Polarized Light.

ELECTROSTATICS 16

Synopsis: Common experiences with electrification are known to most of us. The crackling of hair when it is combed on a dry day or the shock that results when a metal door knob is touched after walking over a rug are two examples.

Some knowledge of these electrostatic effects dates back to about the same time in history that magnetic effects were first discovered. Miletus, an ancient Greek, is supposed to have called attention to the property of amber that, after it has been rubbed with fur, causes it to attract light objects such as small pieces of paper. Even the word *electricity* used to describe these effects, was derived from *elektron*, the Greek word for amber.

At the end of the 16th century, Gilbert in his *De Magnete* summarized knowledge to that time by comparing magnetic and electric action. Gilbert identified electrics as substances that could be electrified by rubbing and nonelectrics as substances that could not be. He theorized that some kind of effluvium emanated from a charged object that acted to draw light objects toward it. By this view Gilbert suggested the idea of a field although he thought that a real difference existed between electric attraction and magnetic attraction. He considered the attraction of iron particles to a magnet to be mutual, whereas he considered the electric attraction to be unilateral.

The repulsion of certain electrically charged objects was noted in the early 1600s by Cabeo, an Italian, and was studied in greater detail by Otto von Guericke who built one of the first electrostatic machines. Progress in the study of electrostatics continued. Gray in England found that conductors could be electrified, Du Fay postulated the existence of two kinds of electric charge, and Franklin established the identical character of frictional electricity and the electricity involved in lightning discharges. Franklin was the first to talk about electric charge as something the magnitude of which might be determined.

The modern theory that electric charges are discrete gradually evolved. Coulomb added a quantitative measure to the subject by his establishment of the inverse square law, which made possible the definition of a unit charge. There followed the discovery that free electric charges can exist and that such a charge is surrounded by an electric field that can be visualized by drawing lines of force. The graduated electric field about a point source gave rise to the idea of potential.

Although the historical development of knowledge in electrostatics is of great interest, our primary purpose in this chapter is to provide the student with a functional basis for understanding the modern concept of electric charge, the electric character of matter, the forces that exist between charges, the electric field and electric potential, and the distribution of charge on an isolated conductor.

The capacitor is introduced as a device designed for storing an electric charge. The quantity of stored charge is related to the materials of which the capacitor is constructed, its pertinent dimensions, and the potential difference that is developed across its terminals. The relation between the quantity of charge stored and the manner in which capacitors are combined into groups (in series or in parallel) is an important consideration.

Comment: Stress electrification as occurring through the transfer of electrons. Emphasize the similarity of Coulomb's law and Newton's law of universal gravitation, of gravitational field and electric field, of gravitational potential in mechanical energy transformations and potential difference in electric energy changes. Note the treatment of dielectric constant and of the nature and effect of dielectrics.

Use simple, quick demonstrations with which some students may be familiar. These include: having charged balloons stuck to the wall, rubbing a piece of mimeograph paper so that it sticks to the wall, picking up small bits of paper with a charged plastic pencil or pen. The large plastic bags used by cleaners to protect clothes may be slit along the vertical edges, supported by a meter stick inserted along the uncut edge, and charged by friction to show repulsion of like charges. An equivalent consists of strips of newspaper about one inch in width, cut from a double page so that an uncut portion joins the two strips. When the paper is pulled through the fingers, the two portions repel each other. A large test tube, dried by preheating and charged by stroking develops a positive charge much more readily than a glass or plastic rod. This may also be used to show that charge lies on the surface. Perform the demonstrations represented by the diagrams in the text to develop definitions of positive and negative charge, the basic law of electrostatics, electroscope construction and operation, and methods of charging an electroscope. Emphasize that, in general, electrification occurs through the transfer of electrons and use chalkboard diagrams extensively to represent the various demonstrations. Relate the differences between conductors and insulators to the structure of matter and presence of free electrons. Point out that electric charges are always conserved and that repulsion must be demonstrated to show that a body is charged. Where possible use the Braun type of electroscopes for demonstration purposes for visibility and charge retention.

In the treatment of Coulomb's law and electric fields of force, use supplementary demonstrations that emphasize concepts previously developed and lead to this topic. Such demonstrations include: deflection of a stream of water from the faucet by a charged rubber rod; insertion of a charged rubber rod into a box of sawdust, adherence of sawdust, and the "shooting" off of the charged sawdust particles. Students may report on Chapter 5, The Laws of Electric and Magnetic Force—Charles Coulomb, pages 59–74 from *Great Experiments in Physics*. Develop Coulomb's law as in Section 16.8 and use the solved Example and the Practice Problems at the end of this section. When discussing electric field intensity, point out that this quantity is a vector and that the test charge is positive by convention. Have students explain the solution to the solved Example at the end of Section 16.9 and demonstrate their solutions to the Practice Problems that follow.

Develop the concept of potential difference in terms of work and quantity of charge, and emphasize the analogy to gravitational potential in mechanical energy transformations. Demonstrate Faraday's ice-pail experiment and the principles of distribution of charge, Section 16-11. Demonstrations used should show shape and point effects.

Exhibit variable capacitors and rotate movable plates. Demonstrate parallel-plate condenser-electroscope to show effects of varying distance and insertion of dielectric: glass, paraffin, etc. Develop the concept of capacitance, and define the farad as unit of capacitance. Exhibit various types of capacitors. Students who have worked with radios and related electric devices may discuss use of capacitors, practical units, and capacitor arrangements. You may wish to perform Leyden jar demonstrations such as charging with electrophorus and discharging with discharge tongs. Discuss dielectric constant and effects of dielectrics as in Sections 16.15 and 16.16, and Figure 16-23. Develop equations for capacitors in series and parallel, and discuss the solution of the Example at the end of Section 16.17.

Students frequently build model Van de Graaff generators and may discuss and demonstrate their structure and operation. Operating models purchased from supply houses may also be used to perform interesting demonstrations.

References: Ainslie, D. S., "Construction and Uses of Equipment for Demonstrating the Fundamental Principles of Electrostatics," *The Physics Teacher*, Jan. 1964, pp. 32–33.

Barschall, H. H., "Electrostatic Accelerators," *The Physics Teacher*, Sept. 1970, pp. 316–322.

Jifmenko, Oleg, and David K. Walker, "Electrostatic Motors," *The Physics Teacher*, March 1971, pp. 121–129.

Moore, A. D., *Electrostatics*, Science Study Series, 1968.

Rosen, Louis, "Particle Accelerators: Instruments for Basic Research and Human Welfare," *The Physics Teacher*, Nov. 1970, pp. 432–440.

Sears, Francis W., "A Substitute for Pith Balls and Balloons in the Demonstration of Electrical Forces," *The Physics Teacher*, Nov. 1963, pp. 225–226.

Turner, R. P., *Basic Electricity*, Holt, Rinehart and Winston, Publishers, New York, 1963, Chapter 1.

Films: *Electric Fields*, fs, color, 41 frames, 60k FOM, McGraw-Hill Book Co., Text Film Dept., 1221 Avenue of the Americas, New York, NY 10020. Nature of and demonstrations on electric fields.

Electrostatics, 11 min., black and white, EBF. Shows the nature of attractive and repulsive forces between bodies. Demonstrates technique of charging by induction and the meaning of insulators and conductors.

PSSC Films: *Coulomb's Law*, 0403, 30 min., black and white. Demonstrates the inverse square variations of electric force with distance and also that electric force is directly proportional to charge. Introduces the demonstration with a thorough discussion of the inverse square idea. Also tests inverse square law by looking for electric effects inside a charged hollow sphere.

Coulomb Force Constant, 0405, 34 min., black and white. Shows a large-scale version of the Millikan Experiment (See also 0404 that title). The small charged plates of the original experiment are made very large, and the effects are shown of increasing plate area and separation and of adding more batteries to charge the plates. The same electric field strength used in the Millikan Experiment film enables the experimenter to count the number of elementary charges on an object and measure the constant in Coulomb's law of electric force.

Electric Fields, 0406, 25 min., black and white. An electric field discussed as a mathematical aid and as a physical entity. Experiments demonstrate vector addition of fields, shielding by closed metallic surfaces, the electric force that drives an electric current in a conductor for both straight and curved conductors. Physical reality of fields discussed briefly in terms of radiation.

Electric Lines of Force, 0407, 7 min., black and white. Shows how to produce electric field patterns using a neon sign transformer as a high voltage source. Indicates safety precautions. Fine grass seeds in the interface between freon film cleaner or carbon tetrachloride and mineral oil align themselves to form various electric field patterns.

References for Chapter 16: Electrostatics

Rogers, *Physics for the Inquiring Mind*, Princeton University Press, 1960, Chapter 33, pp. 533–565, Electric Charges and Fields: Electrostatics.

Semat, *Fundamentals of Physics*, Chapter 20, pp. 343–360, Electrostatics. (For Sections 16.1–16.13.)

Lindsay, *Basic Concepts of Physics*, Section 7.4, pp. 256–263, Electric Charges and the Electric Field.

Meiners, Section 19-1, pp. 835–856, 30 suggestions. *Very good.*

16.2 Two kinds of charge

Demonstration: *Project Physics, Teachers Resource Book IV*, 1970, D47, Some Electrostatic Demonstrations, pp. 59–61.

Meiners: Chapter 29, pp. 835–878, Electrostatics.

PSSC Physics, Third Edition, Chapter 18, pp. 398–414, Some Qualitative Facts about Electricity. Chapter 19, pp. 418–439. Coulomb's Law and the Elementary Electric Charge.

Project Physics Text, Section 14.2, The curious properties of lodestone and amber: Gilbert's *De Magnete*, pp. 30–33, Unit 4.

Sutton: E-1, p. 249, Electric Charges on Solids Separated after Contact.

16.3 Electricity and matter

Gamow and Cleveland, *Physics, Foundations and Frontiers*, Section 12-2, pp. 180–181, Elementary Atomic Structure.

Sutton: E-6, p. 253, Attraction and Repulsion. E-14 through E-27, pp. 258–263, A variety of suggestions for appropriate demonstrations in electrostatics. E-39, p. 266, Electric Chimes.

Project Physics Text, Section 14.5, The smallest charge, pp. 45–48, Unit 4. Section 14.6, The law of conservation of electric charge, pp. 48–50, Unit 4.

Gamow and Cleveland, *Physics, Foundations and Frontiers*, Chapter 12, pp. 178ff., Electrostatics; Section 12-1, pp. 178–179, Static Electricity.

PSSC Physics, Third Edition, pp. 400–406. Section 18-1, Electrified Objects. Section 18-2, Some Experiments with an Electroscope. Section 18-3, Electrostatic Induction. Section 18-4, A Model for Electric Charge.

Laboratory Exercise: *Exercises and Experiments in Physics*, Expt. 36. Electrostatics.

16.4 The electroscope

Western, 2 × 2 color transparency, M-3, Leaf-Electroscope.

Laboratory Exercise: Lehrman, *Laboratory Experiments*, Expt. 44, pp. 90–93, Electric Charge.

Sutton: E-2, p. 249, Large-Scale Electroscopes. E-3, p. 250, Projection Electroscopes. Figure 209, p. 252, Braun Electroscope.

Gamow and Cleveland, *Physics, Foundations and Frontiers*, Section 12-5, pp. 183–185, The Electroscope.

16.5 Conductors and insulators

Sutton: E-5, p. 252, Conductors and Insulators.

Gamow and Cleveland, *Physics, Foundations and Frontiers*, Section 12-3, p. 181, Conductors and Nonconductors.

16.7 Residual charge by induction

Western, 2 × 2 color transparency, M-2, Electrostatic Induction.

Sutton: E-8 and E-9, p. 254, Electrostatic Induction. E-10, p. 255, Electrophorus.

Gamow and Cleveland, *Physics, Foundations and Frontiers*, Section 12-4, Induced Charges.

16.8 The force between charges

K & E Transparency: #67, Coulomb's Torsion Balance.

Western, 2 × 2 color transparencies, M-1, Electrostatic Forces. M-4 through M-7, Coulomb's Law.

Laboratory Exercise: Lehrman, *Laboratory Experiments*, Expt. 45, pp. 93–95, Electrostatic Force.

Laboratory Exercise: *Project Physics Handbook*, Expt. 4-3, pp. 8–9, Electric Forces I. Expt. 4-4, pp. 10–12, Electric Forces II-Coulomb's Law.

Project Physics Text, Unit 4, Section 14.3, pp. 33–39, Electric Charges and Electric Forces.

Sutton: E-7, p. 253, Force between Charges, Coulomb's Law.

J. R. Zacharias, *Science*, March 8, 1957. ". . . Coulomb's law . . . in all of atomic and molecular physics, in *all* solids, liquids, and gases, and in *all* things that involve our relationship with our environment, the *only* force law, besides gravity, is some manifestation of this simple law. Frictional forces, wind forces, chemical bonds, viscosity, magnetism, . . . all these are nothing but Coulomb's law . . .''

PSSC Physics, Third Edition, Chapter 19, pp. 418–443, Coulomb's Law and the Elementary Electric Charge.

16.9 Electric fields

Laboratory Exercise: Lehrman, *Laboratory Experiments*, Expt. 46, pp. 96–98, Electric Fields.

Project Physics Text, Unit 4, Section 14.4, pp. 39–44, Forces and fields.

Sutton: E-42, p. 267, Movement of Charged Material along Lines of Force. E-43, p. 268, Movement of Charged Particles in Electric Field. E-44, p. 268, Mapping Field. E-52, p. 270, Extent of Electric Field.

Meiners: Section 29-2, pp. 856–865, Electric Fields, eight suggestions.

Western, 2 × 2 color transparencies, M-8 through M-18, Electrostatic Lines of Force.

Gamow and Cleveland, *Physics, Foundations and Frontiers*, Section 12-6, pp. 185–186, The Electric Field.

16.10 Electric potential

Sutton: E-56, p. 271, Electric Potential. E-57, p. 272, Mapping Potential Field. E-59, p. 273, Electroscope as a Potential Indicator. E-60, p. 273, Potential of Charged Sphere.

Meiners: Section 29-3, pp. 865–868, Electric Potential, 3 suggestions.

Gamow and Cleveland, *Physics, Foundations and Frontiers*, Section 12-7, pp. 189–191, Electric Potential.

Gamow and Cleveland, *Physics, Foundations and Frontiers*, Section 12-8, Practical Electrical Units.

16.11 Distribution of charges

Sutton: E-13, p. 257, Faraday's Ice-Pail Experiment. E-28, p. 263, Location of Charge, On Insulated Hollow Conductors.

16.12 Effect of the shape of conductors
Sutton: E-29, p. 264, Surface Distribution of Charge.
E-30, p. 264, Absence of Field within a Closed Conductor. E-31, p. 264, Electric Shielding.

16.13 Discharging effect of points
Sutton: E-32 through E-38, pp. 264–266. Discharge of Electricity from a Point.

16.14 Capacitors
For sections 16.14–16.17: Semat, *Fundamentals of Physics*, Chapter 21, Capacitors and Charges, pp. 363–378.
Sutton: E-40, p. 266, Effect of Intervening Medium on Force Between Charges. E-61, p. 274, Comparison of Charges on Spheres at Same Potential, Capacitance. E-62, p. 274, Elementary Condenser. E-63, p. 275, Leyden-Jar Condenser. E-64, p. 275, Dissectible Leyden Jar. E-65, p. 275, Bound Charge on Leyden Jar.

Meiners: Section 29-4, pp. 868–875, 13 suggestions from which to select.
Sutton: E-71, p. 277, Condensing Electroscope.
Gamow and Cleveland, *Physics, Foundations and Frontiers*, Section 12-9, pp. 191–196, Capacitance.

16.15 Dielectric materials
Sutton: E-40, p. 266, Effect of Intervening Medium on Force Between Charges.

16.16 The effect of dielectrics
Sutton: E-73 through E-75, p. 279, Dependence of Capacitance on Area.

16.17 Combinations of capacitors
Sutton: E-68, p. 276, Condensers in Series and Parallel. E-69, p. 276, Parallel-Plate Condenser. E-67, p. 267, Addition of Potentials.

DIRECT-CURRENT CIRCUITS 17

Synopsis: An electric charge of collection of charges in motion constitutes an electric current. The magnitude of the current is defined as the rate at which charge passes a given point in a circuit per unit of time. The existence of a direct current requires a closed circuit made up of a material capable of passing an electric charge and a device such as a dry cell or other source of continuous current capable of supplying the energy to cause the charge to move. Such a source must, of course, do work on the charge which, in turn, does an equal amount of work as it traverses the entire circuit or loop. Ohm's law applies to a complete circuit, including the source of electrical energy, and it may be written $E = V + Ir$. This equation states that the work done on a unit charge by the source of electrical energy (E) equals the work done by the charge in traversing the external circuit (V) plus the work done by the charge in traversing the internal circuit (Ir). The work done in the external circuit may all be converted into heat or to heat and other forms of energy such as mechanical energy if the external circuit should contain a motor. The work done in the internal circuit will all be converted to heat because of the internal resistance of the source.

Our study includes the effect of combining either cells or resistances in series, or parallel, or series-parallel; the effect that the physical and dimensional properties of conductors have on their resistance; and procedures for the measurement of resistance.

Comment: A variety of motivating demonstrations may be performed. A simple bell or buzzer circuit with an open switch that is subsequently closed may be used to develop concepts of an electric circuit. A simple voltaic cell, such as a penny and a dime separated by a piece of paper towel moistened with saliva, is connected to a galvanometer. The deflection is noted, and the leads reversed. Another simple cell consisting of strips of Cu and Zn in dilute acid is used to light a flashlight bulb. Use a photoelectric cell demonstration, a thermoelectric demonstration, and/or a piezoelectric demonstration.

A demonstration to show transient current, similar to that shown in Figure 17-2, may be demonstrated along with the circuit for a continuous current as in Figure 17-3. The ampere is defined, as are electric circuit and load. A circuit diagram is drawn and student attention directed to the use of the conventional symbols as in Figure 17-3. Various types of cells are exhibited and demonstrated as desired. Photoelectric cell kits are available from scientific supply houses. A solar cell kit along with text may be obtained from local offices of the Bell Telephone System. A student report may be given about the article, "The Revival of Thermoelectricity," Jaffee, Abram F., *Scientific American*, Nov. 1958, page 31. Exhibit a storage battery hydrometer and discuss 6- and 12-volt batteries. Have students examine storage cell series arrangement. A "safer" storage-cell experiment using sodium bicarbonate and magnesium sulfate as the electrolyte is described in the *American Journal of Physics*, June 1962, page 470, "Lead-Salt Storage Cell" by Siegfried S. Meyers. A demonstration fuel cell kit using alcohol (methanol) and hydrogen peroxide is commercially available (Allis-Chalmers, Milwaukee, WI). The cell is used to light a lamp and/or run a small motor.

Show a cutaway dry cell, Figure 17-9, and discuss the chemistry as outlined in Section 4 including an interpretation of Figure 17-10.

Cell combinations should be demonstrated and the characteristics of series and parallel arrangements discussed.

Exhibit materials having different resistances, rheostat, and carbon resistors. Develop Ohm's Law as in Section 17.6 page 395. Then treat internal resistance of a battery and its determination as in Section 17.7. Discuss the solution to the solved Example in Section 17.7 using the circuit diagram of Figure 17-17. Have students demonstrate their solutions to the accompanying Practice Problems.

Resistances in series and resistances in parallel may be demonstrated by using two different sets of Christmas-tree lights, one wired in series and the other in parallel. Investigate the effect of turning off one bulb. (Of course a-c is used but this need not interfere with the use of the bulbs at this time.) Another interesting parallel circuit demonstration uses absorbent paper and lines drawn in colored salt water as parallel branches in a circuit using a dry cell and milliammeter.

Develop the representative equations and rules, first for series circuits and then for parallel circuits. For the series circuit use Figure 17-18 and the related data representing application of the equations. Lamp-bank demonstrations with suitable measurements are often utilized during this lesson. Have students state and explain Kirchhoff's laws, Sections 17.8 and 17.9. Have students attempt a solution of the simple network, Figure 17-20. Teach them to make network and equivalent circuit diagrams, as in Figure 17-21. Practice in circuit-solving problems is essential to this study.

Point out the effect of length, cross-sectional area, material, and temperature on the resistance of metallic conductors. The resistance effect of increasing length may be related to the series-circuit concept while that of increasing cross-sectional area may be related to the parallel-circuit concept. Briefly mention super-conductivity and semiconductors. Assign and/or demonstrate the methods of determining resistance.

References: Joseph, Alexander, and Daniel J. Leahy, *Programmed Physics, Part II: Electricity and Magnetism*, Wiley, New York, 1965.

Long, Donald, "Electrical Conduction in Solids," *The Physics Teacher*, May 1969, pp. 264–270.

Semat, Henry, *Fundamentals of Physics, Fourth Edition*, 1966, Chapter 20, The Direct Current Circuit, pp. 381–393.

Suhr, Arthur G., "Ohm's Law," *The Science Teacher*, May 1957, p. 255.

Valasek, Joseph, "Piezoelectricity," *The Physics Teacher*, Nov. 1963, pp. 217–221.

Films: *Ohm's Law*, 6 min., color, Coronet Films, Inc., 65 East South Water Street, Chicago, IL 60601. Explains application of Ohm's law by analogy to water pressure, flow, and resistance.

Series and Parallel Circuits, 11 min., black and white, Encyclopaedia Britannica Films, Inc., 1150 Wilmette Avenue, Wilmette, IL. Shows the relationship between current, resistance, and electromotive force in series and parallel circuits.

Bell Solar Battery, 12 min., sound and color, free loan, local Bell Telephone System office. Conversion of sun's rays directly into usable energy.

PSSC Films: 0430, *EMF*, 20 min. Here is it shown that the energy transfers are independent of the geometry of the electrodes in the diode. It is further demonstrated that the energy per elementary charge delivered by a battery (its emf) depends only on the chemical constitution of the battery. The concept of emf is extended to describe any device that transforms energy by separating elementary charges. This discussion leads directly to the subject of the next film, *Electrical Potential Energy and Potential Difference.*

0431, 0432, *Electrical Potential Energy and Potential Difference* (Parts I and II) 54 min., black and white. In Part I, the mechanisms by which a battery establishes an electric field in a circuit is analyzed. The electric potential energy stored in such a system is measured experimentally and explained theoretically.

In Part II, it is shown how the energy transformations on a steady current-carrying circuit can be obtained from measurements of potential difference and electric current, including the energy dissipated internally in the batteries used.

0406, *Electric Fields*, 25 min. An electric field discussed as a mathematical aid and a physical entity; experiments demonstrate (1) vector addition of fields, (2) shielding effect by closed metallic surfaces, and (3) the electric force that drives an electric current in a conductor for both straight and curved conductors. Physical reality of fields is discussed briefly in terms of radiation.

References for Chapter 17: Direct-Current Circuits

Sutton, E-113 through E-130, Experiments to enable the instructor to show the relationship between static and current electricity.

Semat, *Fundamentals of Physics*, Chapter 22, p. 38lff., The Direct-Current Circuit.

17.1 Electric charges in motion
3M Brand Overhead Transparency Masters, Science Packet #43: 43K and 43L are intended to stress the meaning of the electric units, amperes, and volts.

Lindsay, *Basic Concepts of Physics*, Section 7.4, p. 256ff., Electric Charges and Electric Field.

Project Physics Test, Unit 4, Section 14.7, Electric currents.

Lehrman and Swartz, *Foundations of Physics*, Chapter 14, pp. 406–431, Electric Currents; covers the topics of: Production of Electric Currents, Currents Carried by Particles, Electric Current and Energy, Current in a Wire, Flow of Electrons, Drift Velocity, Resistance, and Series and Parallel Circuits.

Laboratory Exercise: Lehrman, *Laboratory Experiments*, Expt. 47, Electric Field in a Wire.

Sutton, E-113, Potential Drop along a Conductor with current from a Static Machine.

Karplus, *Introductory Physics*, Chapter 12, pp. 311–337, Electric Current and Energy Transfer; covers the following topics: Electric Current, Electric Circuits, Voltage, Analysis of Simple circuits. This will prove a valuable reference.

Rogers, *Physics for the Inquiring Mind*, Chapter 32, pp. 503–532, Electric Circuits—A Laboratory Chapter. Problems discussed are Short Circuit, Drawing Circuits, Resistance, Quantity of Electricity, Voltage, Current and Potential Difference, Measuring Resistance, and Ohm's Law. A well illustrated chapter.

17.2 Continuous current
PSSC Physics, Third Edition, Section 21-1, Conductors, Batteries and Potential Difference. Section 21-2, Measurement of Potential Difference. Section 21-3, Further Check on Potential Difference and Energy. Section

21-4, Current vs. Potential Difference. Section 21-5, An Overall View of a Circuit.

Project Physics Text, Unit 4, Section 14.8, Electric potential difference. Section 14.9, Electric potential difference and current.

3M Brand Overhead Transparency Masters, Science Packet #43: 43F and 43G, Common Electrical Symbols.

Meiners, Section 30-1, pp. 880–885, a series of nine suggestions under the title, Current and Resistance.

PSSC Physics, Third Edition, Section 18-8, Electrons in Metals. Section 19-4, Electric Potential. Section 19-6, The Elementary Charge. Section 20-2, Electric Current.

17.3 Sources of continuous current
Sutton, E-215, Direction of Induced Electromotive Force.

Sutton, A-89, Surface Photoelectric Effect.

Sutton, E-179, Thermoelectric Effect. E-182, Thermoelectric Magnet. E-183, Thermoelectric Magnet.

Meiners, Section 30-5, pp. 898–901, Thermoelectric Effects, two useful suggestions.

Sutton, E-186, Piezoelectric Effect in Rochelle Salt Crystals.

Meiners, Section 30-6, pp. 901–907, Piezoelectric Effects.

Laboratory Exercise: Lehrman, *Laboratory Experiments*, Expt. 48, The Elementary Charge, an Experiment in Electrochemical Reaction.

PSSC Physics, Third Edition, Section 20-5, EMF and Energy Supplied by a Battery. Section 20-6, Batteries, Volts and Amperes.

Sutton, E-119, Electric Currents from Voltaic Cells. E-120, Elementary Storage Cell. E-199, Cardboard Model to Illustrate Potential Differences and Electromotive Force in a Voltaic-Cell Circuit. E-200, Polarization of a Voltaic Cell. E-204, Secondary or Storage Cells. E-205, Large Current from Storage Cell.

Gamow and Cleveland, *Physics, Foundations and Frontiers*, Section 13-1, pp. 201–203, Electric Cells.

17.4 The dry cell
Sutton, E-116, Identification of Charge from Dry Cells.

17.5 Combinations of cells
Laboratory Exercise: *Exercises and Experiments in Physics*, Expt. 37, Combination of Cells; Internal Resistance.

17.6 Ohm's law for d-c circuits
Laboratory Exercise: Lehrman, *Laboratory Experiments*, Expt. 51, Factors Determining Current.

3M Brand Overhead Transparency Masters, Science Packet No. 43, 43O and 43P are intended for use in presenting Ohm's law. 43V and 43W present the common resistor code and an example of its use.

Sutton, E-153 through 166, series of demonstrations related to resistance and its measurement.

Gamow and Cleveland, *Physics, Foundations and Frontiers*, Section 13-3, pp. 205–206, Electrical Circuits.

17.7 Determining internal resistance
Laboratory Exercise: Lehrman, Expt. 53, Internal Resistance of a Battery.

See *Exercises and Experiments in Physics*, Expt. 37 suggested above.

17.8 Resistances in series
Laboratory Exercise: Lehrman, *Laboratory Experiments*, Expt. 52, Series and Parallel Circuits.

Meiners, 30-2.1, Apparatus to illustrate by analogy voltage drops across resistances in a parallel or series d-c circuit.

Gamow and Cleveland, *Physics, Foundations and Frontiers*, Section 13-4, pp. 206–207, Series Circuits.

17.9 Resistances in parallel
Meiners, Section 30-2.6, The Superposition Theorem.

Gamow and Cleveland, *Physics, Foundations and Frontiers*, Section 13-5, pp. 207–210, Parallel Circuits.

17.10 Resistances in simple networks
3M Brand Overhead Transparency Masters, Science Packet No. 43, 43S and 43T are intended to aid development of the concepts of the total effects of various circuits of resistors. 43U shows how a complex circuit can be reduced to a succession of equivalent resistances.

Laboratory Exercise: *Exercise and Experiments in Physics* or *Laboratory Experiments in Physics*, Expt. 41, Simple Networks.

17.11 The laws of resistance
Gamow and Cleveland, *Physics, Foundations and Frontiers*, Section 13-2, pp. 203–205, Electrical Resistance.

17.13 Measuring resistance
3M Brand Overhead Transparency Masters, Science Packet #43, 43M and 43N identify the common means of measuring electrical resistance. 43Q and 43R illustrate the Wheatstone Bridge method of measuring resistance.

Laboratory Exercise: *Exercises and Experiments in Physics*, Expt. 38, Measurement of Resistance: Voltmeter-Ammeter Method.

Laboratory Exercise: *Exercises and Experiments in Physics*, Expt. 39, Measurement of Resistance: Wheatstone Bridge Method. Expt. 40, Resistances in Series and Parallel, Expt. 41, Simple Networks.

K & E Overhead Transparency: #69 Wheatstone Bridge.

HEATING AND CHEMICAL EFFECTS 18

Synopsis: The concepts of work, power, and energy have important roles in any study of current electricity. The work W is done (energy E is expended) to move a charge q through a potential difference V between two points in an electric field in opposition to the force exerted on that charge by the field.

$$W = qV$$

One joule of work is required (one joule of energy is expended) as one coulomb of charge is moved through one volt of potential difference.

$$1\,\mathbf{j} = 1\,\mathbf{c} \times 1\,\mathbf{v}$$

The power P dissipated as work is done uniformly with time is the time rate of work done.

$$P = \frac{W}{t} = \frac{qV}{t}$$

One watt of power is dissipated as one joule of work is done (one joule of energy is expended) in one second, or as one coulomb of charge is moved through a potential difference of one volt in one second.

$$\mathbf{w} = \frac{\mathbf{j}}{\mathbf{s}} = \frac{\mathbf{cv}}{\mathbf{s}}$$

The electric current I in a conductor is the time rate of flow of electric charge through a cross section of the conductor.

$$I = \frac{q}{t}$$

and

$$P = \frac{q}{t} \times V = IV$$

From Ohm's law,

$$V = I \times R$$

Therefore,

$$P = IV = I^2R$$

Energy expended in an electric circuit is measured in terms of the units of work done and the units of heat produced. Useful work appears in a variety of forms depending on the nature of the load connected across the source of emf. Because of the inherent resistance of any circuit, some expended energy appears as heat. In a circuit that is pure Ohmic resistance, all of the electric energy is expended as heat.

Impedance matching as a condition for the maximum transfer of power from one element to another was first considered in the study of sound transmission. Impedance matching is again a concern in Section 18.5 relative to the electric circuit requirements for the transfer of maximum power from the source of emf to the external circuit (the load) connected across the source of emf.

Much of the early research in electrochemistry was carried out in England by Davy in the early 1800s. By passing d-c currents through fused salts of alkali metals such as sodium and potassium, he was able to recover these metals from their components in the pure metallic state. Around 1833 Faraday (Davy's pupil) formulated the quantitative relationships now known as Faraday's laws while continuing Davy's studies with water solutions of acids and salts. These laws provide a precise method for measuring the quantity of electric charge flowing in a circuit by determining the mass of silver plated out on a platinum rod in a silver coulombmeter.

Comments: Connect an electric heater or an electric iron into a regular service outlet. Disregard the fact that an a-c circuit, rather than a d-c circuit, is involved; the heating effects are the same. Have students compare the heat radiated by the appliance and by the electric cord. Exhibit lamps and other devices of various power ratings. Have students bring in electric bills and consider power versus energy and cost. Encourage students to construct their own electrolysis apparatus and demonstrate electroplating, electrolysis of water, and others that you might suggest. Refer your students to Chapter 10 of *Great Experiments in Physics*, pages 146–157, as the source of a timely report.

Discuss and/or demonstrate the energy conversions represented by Figure 18-2. Develop the concept of work in terms of current, potential difference, and time. The demonstration presented in Figure 18-3 may be performed. Paper riders may be used on the wires. Also, connect a piece of #28 copper wire (about 0.5 m long) to an equal length of #22 copper wire and place this momentarily across a dry cell. Using Ohm's law, develop equivalent expressions for work: $W = I^2Rt$ and $W = E^2t/R$. Develop Joule's law mathematically from the work equations and student laboratory experiences.

Demonstrate heating in series and parallel circuits with one-foot lengths of #28 copper wire and #22 chromel wire, using paper or cardboard riders. Relate to resistance and current calculations in series and parallel circuits. Demonstrate overloading by fusing a parallel circuit with heater element (550–600 watts) and lamp bank in parallel. Then by turning on the last lamp (small wattage such as 25 w), show that the fuse opens the circuit due to overload. Relate the blown fuse to safety considerations and to the use of present day circuit breakers. Relate electric power concept to mechanical power equation. Develop equations for total power and load power and relate to Joule's law equation. An exercise similar to that in Section 18.5 will serve to emphasize the condition necessary for maximum power transfer to a load. Emphasize that the watt-second, watt-hour, and kilowatt-hour are units of electric energy.

Demonstrate the electrolysis of water. Identify products by tests, and note relative volumes. Explain as an electron exchange phenomenon. You may try electrolysis of solution of sodium sulfate using pieces of litmus paper as indicators to show anode and cathode ion products. Demonstrate copper plating using a graphite cathode and 2 dry cells in series. Then reverse polarity of electrodes to remove the copper deposit. Develop the principles of electroplating based on the demonstration. The electrolytic refining of copper is explained on the basis of this demonstration. By coating a plastic object with graphite, it may also be electroplated. Discuss the conductivity of fused electrolytes (fused KNO_3 or $NaNO_3$) to establish understanding of metallurgical applications of electrolysis. Electrolysis of a concentrated solution of NaCl permits easy identification of chlorine as the anode product and of H_2 and NaOH as the cathode products.

Treat Faraday's laws from the qualitative point of view, particularly the first law. The second law may be based on student's laboratory experiences. Illustrate the meaning of chemical equivalent and electrochemical equivalent by examples.

References: Bennett, Wendell F., "A Classroom Demonstration of a Series-Parallel Circuit and the Heating Effects of an Electric Current," *The Science Teacher*, April 1958, p. 143.

Efron, Alexander, *Basic Physics*, John F. Rider Publisher, Inc., New York, 1957, pp. 277–285, 571–579.

Metcalfe, H. C., J. E. Williams, and J. F. Castka, *Modern Chemistry*, Holt, Rinehart and Winston, Publishers, New York, pp. 454–465.

Semat, Henry, *Fundamentals of Physics*, 1966, Section 22-4, Heating Effect of an Electric Current, pp. 383–384, Chapter 23, Electrochemical Effects, pp. 398–411.

Films: *Counting Electrical Charges in Motion*, 22 min., black and white, PSSC, 0408. The film shows how an electrolysis experiment enables us to count the number of elementary charges passing through an electric circuit in a given time, and thus to calibrate an ammeter. Demonstrates the random motion of elementary charges with a current of only a few charges per second.

Production of Sodium by Electrolysis, 3 min. 45 sec., color, 8 mm, Project Physics Film Loop. In 1807, Davy produced metallic sodium by electrolysis of sodium hydroxide. In a replication of Davy's experiment, sodium hydroxide is placed in an iron container and heated until molten. A rectifier connected to a power transformer supplies a steady current through the liquid solution with iron rods inserted in the melted sodium hydroxide. The sodium is seen to accumulate in a thin, shiny layer floating on the surface of the liquid sodium hydroxide. It is shown that metallic sodium has been produced when the shiny layer is removed and dropped into water with the resulting violent reaction.

References for Chapter 18: Heating and Chemical Effects

Sutton, E-171 through 178, pp. 319–322: E-171, Heating Effects of Electrical Current. E-172, Fuses and House-Lighting Circuits. E-174, Comparison of Heating in Various Conductors. E-175, Dependence of Resistance on Length and Area of Conductor. E-177, Resistances in Series and Parallel.

18.2 Energy conversion in resistance
Laboratory Exercise: Lehrman, *Laboratory Experiments*, Expt. 50, Energy Conversion in a Conductor.

18.3 Joule's law
Laboratory Exercise: *Exercises and Experiments in Physics*, Expt. 42, Electric Equivalent of Heat.

18.4 Power in electric circuits
Lindsay, *Basic Concepts of Physics*, Section 7.6, Production of Electric Currents, Ohm's Law, Electrical Energy and Power.
Project Physics Text, Section 14.10, Electric potential difference and power, p. 56, Unit 4.

18.7 Electrolytic cells
Semat, *Fundamentals of Physics*, Chapter 23, p. 398ff., Electrochemical Effects.

Sutton, E-191 through 214, pp. 329–338, Electrolytic Conduction.
Gamow and Cleveland, *Physics, Foundations and Frontiers*, Chapter 21, p. 370ff., The Electrical Nature of Matter, especially Section 21-1, Positive and Negative Ions and 21-2, The Laws of Faraday.
Rogers, *Physics for the Inquiring Mind*, Chapter 35, pp. 586–604, Chemistry and Electrolysis.

18.8 Electroplating metals
Sutton, E-195, Electroplating–Lead Tree. E-196, Electroplating–Tin Tree. E-197, Electroplating.

18.9 Faraday's laws of electrolysis
Project Physics Handbook, Expt. 5-1, pp. 4–6, Electrolysis.
Activities, pp. 20–22, Unit 5: Dalton's Puzzle, Electrolysis of Water, Periodic Table, and Single Electrode Plating.

19 MAGNETIC EFFECTS

Synopsis: The ancient Greeks were familiar with magnetic properties as early as 600 BC. Thales of Miletus, a Greek philosopher, was perhaps the first to call attention to the ability of magnetite, a natural oxide of iron, to attract bits of iron.

The first serious treatise on magnetism originating in Western Europe was written in 1269 by Petrus Peregrinus. He summarized the known properties of the lodestone (leading stone), the name given to a natural magnet composed of magnetite because of its self-orienting ability. Peregrinus observed that (1) a lodestone exhibits polarity; (2) when swinging freely about a vertical axis, its poles tend to become aligned in a north-south position; (3) like poles of two lodestones repel and unlike poles attract each other; (4) a piece of iron is magnetized by contact with a lodestone; and (5) there is a small angle between the north-south orientation of the lodestone when used as a compass and the true (geographic) north.

In his treatis, *De Magnete*, William Gilbert (1540–1603) summarized the then-current knowledge about magnets and added his own findings from his experiments. It was he who discovered that an artificial magnet could be made without direct contact between a natural magnet and a piece of iron (magnetic induction). This was the first indication of the ability of a magnet to exert an influence through space, the field concept in magnetism. Gilbert also realized that the earth acted as a huge magnet surrounded by a magnetic field.

The pattern of iron filings sprinkled on a glass plate placed over a magnet or a pair of magnets led Faraday to attach physical significance to the magnetic field, which he visualized in terms of "lines of force."

The concept of the magnetic field was originally related only to natural and permanent magnets. It was only after Oersted (1777–1851) discovered the link between electricity and magnetism that the mystery of the magnetic field of force began to unfold. Ampère (1775–1836) found that a current-carrying conductor exerts a force on another current-carrying conductor placed nearby. Ampère realized that a circular loop of wire carrying a current was the equivalent of a magnet with a north and a south pole. It became clear in time that an electric charge in motion is accompanied by a magnetic field of force. Ampère's discovery means that when a current-carrying conductor is placed in a magnetic field, the conductor is acted on by a force and if free to move, will move with accelerated motion. This action is the basis for many important applications from electric meters to electric motors.

Comment: Exhibit various magnets as available. Small ferrite magnets in a variety of shapes stir interest because of their exceptional magnetic strength. Demonstrate common applications such as magnetic latches, magnetic toys, etc. Emphasize that nature of the magnetic field.

Overhead projection of magnetic fields and their interaction is a recommended technique. Point out similarities (Coulomb's laws) and differences between charged bodies and magnetized bodies. Supplement the demonstrations suggested above by demonstrating magnetic properties of magnetite (lodestone); magnetization of a soft iron bar by holding parallel to flux lines of earth's magnetic field and tapping one end with a hammer, then dropping the magnetized bar to the floor, showing loss of magnetization; alternately use a permalloy bar placed in the earth's field; magnetize a glass tube full of iron filings, then demagnetize by shaking. Exhibit samples of magnetic, paramagnetic, and diamagnetic materials. Develop the present-day concept of the theory of magnetism using Figures 19-2 through 19-5. Stress electron motions and atomic structure. If time is available, a number of films, a filmstrip, and kits for demonstration may be obtained from the local Bell Telephone Co. One or more of these teaching aids should be used to establish the concept of ferromagnetic domains. A student report may be given on "The Laws of Electric and Magnetic Force—Charles Coulomb" from *Great Experiments in Physics*, pages 59–61, 66–74. Define unit pole, and demonstrate Coulomb's law qualitatively.

Demonstrate a magnetic field and permeability by having a steel paper clip attached by a thread from a bottom support apparently floating in air (a magnet is concealed in a platform above the clip). Placing a steel sheet between the suspended clip and the magnet causes the clip to fall. Use iron filings to demonstrate magnetic field and lines of flux as shown in Figures 19-10 through 19-12. If possible use an overhead projector. Another good technique is to make, or have students make, permanent displays using waxed paper that is heated slightly after the fields have been established. Demonstrate lines of flux by the experiment represented in Figure 19-8. Develop concepts of magnetic flux density and field intensity in terms of appropriate units and symbols as in Section 19.4. Insert a piece of soft iron between the N and S poles of adjacent magnets, and show resultant iron-filings pattern to establish idea of permeability and the need for a keeper. Demonstrate magnetization by induction using the demonstration represented by Figure 19-15. Mention retentivity, and show how breaking the contact at the magnet causes the nail string to drop. Some students may wish to report on terrestrial magnetism, or demonstrate and explain the behavior of the dipping needle.

Demonstrate Oersted's and Ampère's discoveries using Figure 19-19, Figure 19-21, and Figure 19-22 as guides. Specific details and suggestions are described in listed pages of the References at the end of this chapter. Base the definition of an ampere on the demonstration represented by Figure 19-21. Define the coulomb in terms of the ampere. Demonstrate Ampère's rules for a conductor and for a solenoid. By using a variety of turns and a variety of current values (use a number of dry cells), establish the dependence of electromagnetic strength on the nature of the core and the number of ampere-turns. Use the number of brads picked up as an index of relative strengths. Discuss and illustrate or exhibit applications and uses of the electromagnet. A student report may be based on Chapter 9, "Electromagnetism—Hans Christian Oersted," *Great Experiments in Physics*, pp. 121–127.

Exhibit various types of meters. Remove the cover cases from some damaged instruments and expose the permanent magnet, the moving coil between the magnetic poles, and the magnetic core. Commercial lecture-table galvanometers generally have such parts exposed so that they may be readily identified. The accessory resistance and shunts may be properly inserted to show their use in converting the galvanometer to a voltmeter and ammeter, respectively. Discuss current sensitivity and voltage sensitivity as in Sections 19.13 and 19.14. Have students study appropriate diagrams for voltmeter and ammeter conversion, Figures. 19-32 and 19-34. Some mention about the care and use of the instruments should be made. Refer students to Appendix A of *Exercises and Experiments in Physics* or *Laboratory Experiments in Physics*. Explain the basic circuit performance of the ohmmeter as illustrated by the simplified circuit shown in Figure 19-35.

References: Angrist, Stanley W., "Galvanomagnetic and Thermomagnetic Effects," *Scientific American*, Dec. 1961, p. 124.

Bleil, D. F., "Magnetics," *The Science Teacher*, Nov. 1967, pp. 17–23.

Ford, Kenneth W., "Magnetic Monopoles," *Scientific American*, Dec. 1963, p. 122.

Frank, N. F., "Faraday and the Field Concept of Physics," *The Physics Teacher*, May 1968, pp. 210–212.

Gross, Calvin F., "Construction of a Simple Meter," *The Science Teacher*, Sept. 1957, p. 232.

Haggerty, Brother Hugh, "Electric Forces—Laws of Ampère and Oersted," *The Science Teacher*, Vol. 28, No. 3, p. 46.

Hogan, Lester C., "Ferrites," *Scientific American*, June 1962, p. 92.

Jaffee, Bernard, *Men of Science In America*, Simon and Schuster, New York, 1958, Chapter VIII, Joseph Henry.

Kunzler, J. E., and Morris Tanenbaum, "Superconducting Magnets," *Scientific American*, June 1962.

Levine, Morton A., Harold P. Furth, and Ralph W. Waniek, "Strong Magnetic Fields," *Scientific American*, Feb. 1958, p. 28.

McKeehan, Louis W., *Magnets*, Van Nostrand, Princeton, 1967.

Nesbitt, E. A., *Ferromagnetic Domains*, Student's Edition and Teacher's Edition, Bell Telephone Laboratories, Inc., 1962.

Pohl, Herbert A., "Electrical Fields, Nonuniform," *Scientific American*, Dec. 1960, p. 106.

Semat, Henry, *Fundamentals of Physics, Fourth Edition*, Holt, Rinehart and Winston, Publishers, New York, 1966, Chapter 24, Magnetism and Chapter 25, Electromagnetic Effects, pp. 415 – 488.

Thumm, Walter, "Why Keepers on Magnets," *The Physics Teacher*, Jan. 1968, pp. 38–40.

Free and Inexpensive Learning Materials: *Magnetism and Electricity*, 34 pages, undated, free, Chrysler Corporation. P. O. Box 1919, Detroit, MI. How magnetism helps us to understand the electric conditions of our automobiles.

Bitter, Francis, *Magnets*, 65¢, Wesleyan University Press, Inc., Educational Center, Columbus, OH.

Bitter, E. A., *Ferromagnetic Domains*, free, Student's Edition, Teacher's Edition, Bell Telephone Laboratories, Inc., Murray Hill, NJ 07971. (To be used with kits and films below.)

Kits: *Seeing Magnetic Domains*, $25, Churchill Co., 468 West Broadway, New York, NY 10012. *Showing Hysteresis on the Oscilloscope*, Magnetico, Inc., 6 Richter Court, East Northport, NY. *Demonstrating the Barkhausen Effect*, Allegri-Tech Inc., 141 River Road, Nutley, NJ.

Films: *Magnetism*, 16 min., black and white, Encyclopaedia Britannica Films, Inc., 1150 Wilmette Avenue, Wilmette, IL. Illustrates the magnetic field, laws of polarity, and terrestrial magnetism.

DC Voltmeters and Ammeters, 41 frames, filmstrip, black and white, Long Film Service, 7505 Fairmont Avenue, El Cerrito 8, CA. Shows principles of voltmeter and ammeter designs of simple to complex stages.

A Magnet Laboratory, 20 min., black and white, PSSC, 0411. Professor Bitter's large magnet laboratory at MIT shows equipment used in producing strong magnetic fields; demonstrates magnetic effects of currents and the magnetism of iron.

Ferromagnetic Domains, Part I: Magnetism and Domains; Part II: How They Are Formed. 11 min. per part, color, Bell Telephone Business Office. Domain theory is treated comprehensively on a high school level. Early magnetization experiments are summarized, explanations of the magnetization curve, hysteresis loop, and Barkhausen effect are covered.

The Formation of Ferromagnetic Domains, 45 min., 35 mm color filmstrip with 33⅓ rpm record. Bell Telephone Business Office. Basic domain theory and techniques used in observing them are shown in this filmstrip. Emphasis on the energies involved in the formation of ferromagnetic domains and domain patterns that conform to theory are illustrated.

Domains and Hysteresis in Ferromagnetic Materials, 36 min., color, Bell Telephone Business Office. Introduction and development of fundamental theories necessary to an understanding of hysteresis in various magnetic materials. Designed to follow *The Formation of Ferromagnetic Domains.*

PSSC Films: 0407, *Electric Lines of Force,* 7 min. Shows how to produce electric field patterns using neon sign transformer as high voltage source. Indicates safety precautions. Fine grass seed in the interface between freon film cleaner (or carbon tetrachloride) and mineral oil align themselves to form various electric field patterns.

0412, *Electrons in a Uniform Magnetic Field,* 11 min. A spherical cathode-ray tube with a low gas atmosphere (Leybold) is used to measure the curvature of the path of electrons in a magnetic field and, with reference to the *Millikan Experiment,* the mass of the electron is determined. Arithmetic involved is worked out with the experiment.

References for Chapter 19: Magnetic Effects

Rogers, *Physics for the Inquiring Mind,* Chapter 34, pp. 568–585, Magnetism, Facts and Theory. A well illustrated, informative chapter.

Project Physics Text, Chapter 14, Electric and Magnetic Fields.

Gamow and Cleveland, *Physics, Foundations and Frontiers,* Chapter 14, pp. 216–242, Magnetism.

Semat, *Fundamentals of Physics,* Chapter 24, p. 415ff., Magnetism. Chapter 25, Electromagnetic Effects, p. 431ff. Chapter 27, Magnetic Properties of Matter, p. 470ff.

Lehrman and Swartz, *Foundations of Physics,* Chapter 15, pp. 435–457, Magnetism.

Project Physics Text, Section 14.13, Magnetic fields and moving charges.

19.1 Magnetic materials
Sutton, E-77, p. 280, Natural Magnets.

Meiners, Section 32-3, pp. 960–981, Ferromagnetism, 26 suggestions (this will provide interesting reading and suggestions for demonstrations that can be performed).

Meiners, Section 31-1, pp. 953–958, Poles and Dipoles, properties of a magnetic field demonstrated with an overhead projector; force between two small dipoles and force between a magnet and a piece of soft iron. Section 32-2, pp. 959–960, Diamagnetism and Paramagnetism.

Gamow and Cleveland, *Physics, Foundations and Frontiers,* Section 14-1, pp. 216–219, Magnets and Magnetic Fields.

Western, 2 × 2 Color Transparency, M-19, Magnetite (Lodestone).

PSSC Physics, Third Edition, Chapter 22, The Magnetic Field, 22-1, The Magnetic Needle.

Project Physics Text, Section 14.2, The curious properties of lodestone and amber: Gilbert's *De Magnete.*

19.2 The domain theory of magnetism
3M-34D, Alignment of Domains.

Sutton suggests a number of demonstrations for use in sections 1–10 of this chapter; see pages 280–296. See especially: E-78 and 79, Magnetization by Contact. E-80, Magnetization by Mechanical Disturbance in Earth's Field. E-81, Isolated Pole. E-82, Induced Magnetic Poles. E-83, Magnetization by Current. E-85, Which is the Magnet? E-86, Magnetic Balance. E-87, Determination of Pole Strength by Law of Force. E-88, Magnetic Induction. E-89, Magnetic Fields by Filings. E-90, Mapping Field of a Magnet. E-93, Breaking a Magnet. E-102, Paramagnetism and Diamagnetism. E-103, Effect of Temperature on the Magnetic Properties of Nickel. E-104, Iron Nonmagnetic at High Temperature. E-105 through 107, Magnetic Screening. E-110, Magnetic Heat Motor. E-111, Terrestrial Magnetism. E-112, Magnetic Induction in the Earth's Field.

19.3 Force between magnet poles
3M-34C, Identification of Poles. 3M-34B, Like and Unlike Poles.

Semat, Section 24-2, Coulomb's Law of Force.

19.4 Magnetic field of force
3M-34A, Magnetic Field, 3M-34E, Concentration of Lines of Flux.

Western, 2 × 2 Color Transparencies: M-20, Magnetic Field Produced by a Bar Magnet. N-1, Magnetic Lines of Force Around a Bar Magnet. N-2, Ditto Around a U-shaped Magnet. N-3, Magnetic Field Around a U-shaped Magnet. N-4, Magnetic Properties of Nickel Coins. N-5, Lines of Force Around a Nickel Coin.

Laboratory Exercises: Lehrman, *Laboratory Experiments,* Expt. 54, Measuring a Magnetic Field.

PSSC Physics, Third Edition, Section 22-2, Magnetic Fields of Magnets and Currents.

Project Physics Text, Section 14.4, Forces and Fields. Section 15.2, Faraday's First Electric Motor.

19.5 Magnetic permeability
Gamow and Cleveland, *Physics, Foundations and Frontiers*, Section 14-4, pp. 223–224, Magnetic Flux.
PSSC Physics, Third Edition, Section 22-11, Uniform Magnetic Fields.

19.6 Terrestrial magnetism
3M-34G, Earth's Magnetic Declination. 3M-34H, Magnetic Declination in the United States, 3M-34I, Magnetic Inclination.
K & E Overhead Transparency: #70, The Earth as a Magnet.
3M-34F, Earth's Magnetic Field.

19.8 The link between an electric current and magnetism
Sutton, E-121, Oersted's Experiment.
PSSC Physics, Third Edition, Section 22-3, The Vector Addition of Magnetic Fields.
Gamow and Cleveland, *Physics, Foundations and Frontiers*, Section 14-2, pp. 219–221, Currents and Magnetism.
Lindsay, *Basic Concepts of Physics*, Section 7.5, Electric Currents and Magnetic Fields, p. 264ff.
Project Physics Text, Section 14.11, Currents act on magnets.
Laboratory Exercise: *Exercises and experiments in Physics* or *Laboratory Experiments in Physics*, Expt. 43, Magnetic Field about a Conductor.

19.9 Magnetic field and a charge in motion
Project Physics Text, Section 14.12, Currents act on currents. Experiment 4-5, Forces on Currents.
Sutton, E-122, Magnetic Field Due to Current in a Long Straight Conductor.
Gamow and Cleveland, *Physics, Foundations and Frontiers*, Section 14-3, pp. 221–223, Force on a Moving Charge. Section 14-6, pp. 226–227, Currents in a Magnetic Field. Section 14-8, pp. 229–231, Interactions Between Currents.
Western, 2 × 2 Color Transparencies: N-6, Magnetic Field around a Straight Conductor. N-7, Magnetic Lines of Force around a Straight Conductor.
Laboratory Exercise: Lehrman, *Laboratory Experiments*, Expt. 55, Plotting Magnetic Fields.
PSSC Physics, Third Edition, Section 22-9, Magnetic Field Near a Long Straight Wire.
Lindsay, *Basic Concepts of Physics*, Section 7.10, the Electromagnetic Field, p. 286ff.
PSSC Physics, Third Edition, Section 22-4, Forces on Currents in Magnetic Fields-A Unit of Magnetic Field Strength. 3M-34J, Left Hand Rules.
Meiners, Section 31-1, pp. 908–931, Ampère's Law and the Magnetic Field, 30 suggestions.

19.10 Magnetic field and a current loop
K & E Overhead Transparency: #71, Magnetic Polarity of a Coil.
Western, 2 × 2 Color Transparencies: N-8, Magnetic Field Produced by a Circular Current. N-9, Magnetic Lines of Force Produced by a Circular Current.
Sutton, E-125, Field Produced by Current in a Solenoid; E-126, Electromagnet. 3M-34K, Electromagnet.
Western, 2 × 2 Color Transparency: N-10, Magnetic Field in a Solenoid. N-11, Magnetic Lines of Force in a Solenoid. N-12, Magnetic Lines of Force Due to a Toroidal Solenoid.

19.11 The electromagnet
Project Physics Handbook, Expt. 4-5, Forces on Currents, pp. 13-17; Expt. 4-6, Currents, Magnets, and Forces, pp. 18-20.
Gamow and Cleveland, *Physics, Foundations and Frontiers*, Section 14-5, Solenoids and Electromagnets.

19.12 The galvanometer
3M-34Q, Galvanometer.
PSSC Physics, Third Edition, Section 22-5, Meters and Motors. Section 22-6, Forces on Moving Charged Particles in a Magnetic Field.
Gamow and Cleveland, *Physics, Foundations and Frontiers*, Section 4-7, pp. 227–228, Galvanometer, Voltmeter, Ammeter.
K & E Overhead Transparency: #72, Electric Meters.
Project Physics Handbook, Expt. 4-6, Currents, Magnets, and Forces.
Sutton: E-131 through 152, Forces on Conductors in Magnetic Fields. 3M-43A through E: Visuals A through E show the construction of a galvanometer, d-c ammeter, and d-c voltmeter. Visual B shows an enlargement of the permanent field, movable coil, and cylindrical core of the galvanometer. When the coil carries an electric current, the coil produces a magnetic field. Visual C illustrates the individual effect of the coil and is intended to be used as an overlay on Visual B. This shows the origin of the force that gives a meter reading of current. It must be pointed out that the individual fields illustrated in Visual B and Visual C will result in a combined field that is not illustrated. 3M-43H through J: Visuals H, I, and J stress the cautions to be observed in using a galvanometer, d-c ammeter, and a d-c voltmeter. The visuals also contain the appropriate schematic diagram for a voltmeter and an ammeter circuit.

19.13 The d-c voltmeter
3M-34R, Voltmeter, 3M-34S, Ammeter.
3M Brand, Science Packet Number 34, Magnetism and Electromagnetism; transparencies for overhead projection.

ELECTROMAGNETIC INDUCTION 20

Synopsis: Oersted and Ampère established the foundations of electromagnetism with the discovery in about 1820 of the magnetic field associated with an electric current. Several experiments in the period from 1820 to 1830 suggested that since a current induces a magnetic field, there should be a mechanism by which a magnetic field induces a current. The solution was discovered at almost the same time by Faraday in England and Henry in the United States. Their simultaneous discovery of electromagnetic induction marks the beginning of the modern era in our civilization. Moreover, it strengthened the view that the field concept is one of the most important concepts in modern physics.

In our study of electromagnetic induction, we shall repeat Faraday's and Henry's experiments, study the cause, magnitude, and direction of the induced emf, and establish the energy relationships involved.

Having studied the physical basis for electromagnetic induction, we shall then consider machines such as generators, motors, and transformers that have been developed as a result of the accumulated knowledge of the basic principles of electromagnetic induction. In the remaining portion of the chapter, we shall investigate the effect of the magnetic field associated with a current in one circuit on another circuit poised within the magnetic field of the first.

Comment: Treat Faraday's discovery as the converse of Oersted's discovery. Demonstrate or have students demonstrate the variables involved. Reports on the original work of Faraday and Lenz may be given by students (see References). The basic understanding of alternating-current principles must be developed to provide background for subsequent chapters. Have students display related projects: homemade motors, Tesla coil, etc. Exhibit train transformers, automobile induction (spark) coil, electric clocks, and other related devices.

To establish Faraday's discovery, perform the demonstration shown in Figure 20-1. Establish the various factors that affect the induced emf. Use coils with varying numbers of turns, Figure 20-3. Show induction with an electromagnet. Have students explain the cause of induced emf in terms of Figure 20-4. Emphasize that the movement of the conductor must be opposed by a force if the free electrons in a conductor are to acquire potential energy. Develop Lenz's law on the basis of demonstrated change in current direction and coil polarity using chalkboard diagrams (Ampère's rule for a solenoid). Then analyze results as in Figures 20-6 and 20-7.

Use a magneto to operate a neon bulb, turning first rapidly and then slowly so that students can see that the plates glow alternately. Connect the neon lamp to the 120-volt a-c line and have students detect the same effect by holding the slot between the two semicircular segments horizontal and then swinging the lamp to and fro along a horizontal line. Slowly operate a demonstration a-c generator (Sargent-Welch 2408) so that students can detect deflections at the various angles of orientation of the armature coil. Using a chalkboard diagram, duplicate the sine curve of current or voltage, Figure 20-14. Develop the relationships for instantaneous voltage and current according to the treatment in Section 20.9, relating to Figure 20-11 and Figure 20-12. Have students point out the structural features of the demonstration generator: field coil and magnets, armature, slip rings, and brushes. Have them trace the armature circuit leading to the galvanometer. You may have them make a diagram for the simple generator, Figure 20-8. Develop the generator rule and relate to Lenz's law. Briefly discuss a-c generators and the pole-speed relationship for frequency.

Show how the demonstration a-c generator is converted to a d-c generator by replacing slip rings with a commutator. Use Figure 20-17 and student examination of the generator to show connections of ends of armature coil to commutator segments. Demonstrate generator operation with a lecture-table galvanometer; slowly turning to show the effect of a half-turn. Use diagrams of Figs. 20-18 and 20-19 to develop maximum-value and average-value effects of the number of coils and the number of commutator segments. Show how d-c generators are self-exciting. Point out, with the aid of Figures 20-20, 20-21, and 20-22, how the field magnets in a d-c generator may be connected. Discuss the advantages of using each circuit by analysis of the resistance networks in Figures 20-23, 20-24, and 20-25. Review the solution to the solved Example in Section 20.13 and call upon students to demonstrate their solutions to the Practice Problems that follow it.

The motor effect may be conveniently demonstrated by momentarily passing a strong current (a dry cell may be used) through a conductor placed in a magnetic field, as shown in Figure 20-26. Reverse the current, and note that the conductor moves in the opposite direction. Develop the relationship for torque according to Figures 20-27 and 20-28. Relate the results of the demonstration to the motor rule, Figure 20-29. Motor-generator relationships may be shown by using an electric swing composed of two connected coils suspended across the poles of two magnets. One coil set into motion acts as a generator causing the other to act as a motor (Sargent-Welch 2341). Demonstrate the back emf of a motor according to Figure 20-30. Emphasize the reason a starting resistance must be used in starting large d-c motors. Briefly mention types of d-c and a-c motors, exhibiting and demonstrating where possible.

Perform the demonstration represented by Figure 20-32, Section 20.18, even though it may have been shown in the introductory lesson. Show the effect of placing a soft iron core in the primary. Try moving it slowly through the opening with the switch key closed. Develop the concept of mutual inductance and discuss the solved Example at the end of Section 20.18. You may wish to assign students or call for volunteers to demonstrate their solutions to the Practice Problems that follow the Example. Perform the demonstration represented by Figure 20-34 and develop the concept of self-inductance with the support of Section 20.19. Emphasize the "flywheel effect" of inductance. Develop the relationships for the equivalent inductances of both series and parallel groupings of inductors. A dissectible model transformer can be used to show step-down operation with low voltage lamps and/or low voltage a-c voltmeter. Show relationships between turn ratio and current and voltage in primary and secondary. By using appropriate sets of lamp banks and meters, you can show the power output versus input. Discuss transformer losses and demonstrate effects of eddy currents. Transformer and inductance problems appear in *Physics Problems*, pages 328–329. Explain why electric power is transmitted at high voltage. To conclude the chapter, you may wish to get an electromagnetism demonstration kit, such as Sargent-Welch #2462, which may be used to show 20 interesting demonstrations such as the jumping ring, dancing coil, induction heating, induction motor, etc., as well as some of those already described.

References: Gould, Mauri, "C-Clamp into Transformer," *The Science Teacher*, Apr. 1968, pp. 45–47.

Phillips, Melba, "Electromotive Force and the Law of Induction," *The Physics Teacher*, Oct. 1963, pp. 155–159.

Ryan, Lawrence B., "Reactance Measurements in the Physics Lab," *The Physics Teacher*, Apr. 1968, pp. 176–177.

Semat, Henry, *Fundamentals of Physics, Fourth Edition*, Holt, Rinehart and Winston, Publishers, New York, 1966, Chapter 26, Electromagnetic Induction, pp. 452–466.

Sharlin, Harold, "Faraday to the Dynamo," *Scientific American*, May 1961, p. 107.

Trout, Verdine E., "The Motor Effect," *The Science Teacher*, Sept. 1968, pp. 74–76.

MacDonald, D. K. C., *Faraday, Maxwell and Kelvin*, Science Study Series, 1964.

Films: *Electrodynamics*, 11 min., black and white, Encyclopaedia Britannica Films, Inc., 1150 Wilmette Avenue, Wilmette, IL. Shows how moving electric charges create a magnetic field and how moving magnetic fields create an electric current.

AC and DC Generators, 29 frames, filmstrip, Long Film Slide Service, 7505 Fairmont Ave., El Cerrito 8, CA. Shows lines of magnetic flux, current in rotating coil, slip rings, brushes, operation of parts.

Transformers, 32 frames, filmstrip, Long Film Slide Service, 7505 Fairmont Avenue, El Cerrito 8, CA. Shows how transformer action takes place and how loss of energy is reduced to a minimum.

References for Chapter 20: Electromagnetic Induction

Project Physics Text, Chapter 15, Faraday and the Electric Age. See also the *Handbook*, pp. 40–44, Unit 4, for suggestions for activities.

PSSC Physics, Third Edition, Chapter 23, Electromagnetic Induction and Electromagnetic Waves.

Sutton, E-215 through 227, pp. 339–344, Electromagnetic Induction.

Lehrman and Swartz, *Foundations of Physics*, Chapter 16, pp. 472–501, Alternating Currents.

Semat, *Fundamentals of Physics*, Chapter 26, Electromagnetic Induction, p. 452ff. Chapter 28, p. 479ff., Alternating Currents.

20.1 Discovery of induced current
3M-34T, Induced Current.

20.2 Faraday's induction experiments
Laboratory Exercise: Lehrman, Expt. 56, Electromagnetic Induction.

Lindsay, *Basic Concepts of Physics*, Section 7.7, Electromagnetic Induction, Alternating Current, Electric Oscillations.

Meiners, Section 31-2, pp. 932–948, Faraday's Law of Electromagnetic Induction, 22 suggestions.

Project Physics Text, Section 15.3, The discovery of electromagnetic induction.

3M Brand, Science Packet #34, Magnetism and Electromagnetism; transparencies for overhead projection. Visuals T–W illustrate the production of an electric current from a magnetic field. The direction in which a current flows is dependent upon the direction in which the lines of force are cut (Visual T). Transformers (induction coils) are illustrated on Visual U. a-c and d-c generators differ only in the type of commutator used (Visual V). The production of current (sine curve) by the a-c generator is illustrated on Visual W.

PSSC Physics, Third Edition, Section 23-1, Induced Current. Section 23-2, Relative Motion. Section 23-3, Magnetic Flux Change.

Exercises and Experiments in Physics, Expt. 44, Electromagnetic Induction.

Western, 2 × 2 Color Transparencies: N-16 through N-20 and 0-1, a series of six slides on electromagnetic induction.

20.3 Factors affecting induced emf
PSSC Physics, Third Edition, Section 23-4, Induced emf.

Sutton, E-215, Direction of Induced Electromotive Force. E-216, Induced Current from Relative Motion of Current and Wire. E-217, Dependence of Induced Electromotive Force on Number of Turns. E-219, E-220, Current Induced When Change in Magnetic Flux is Caused by Changing Electric Current-Faraday's Experiment. E-221, Current Induced When Change in Magnetic Flux is Caused by Change in Magnetic Circuit. E-222, Earth Inductor.

Gamow and Cleveland, *Physics, Foundations and Frontiers*, Section 14-9, pp. 231–236, Generation of Electric Currents. Section 14-10, pp. 236–238, Changing Flux.

Lehrman and Swartz, *Foundations of Physics*, pp. 473–475, The Law of Induction.

20.4 The cause of an induced emf
PSSC Physics, Third Edition, See Sections 23-1 through 23-4 referred to earlier.

Rogers, *Physics for the Inquiring Mind*, pp. 655–660, Laboratory Work with Electrons; Electromagnetic Induction, Faraday's Discovery, Lenz's Law, and Transformers.

20.5 The direction of induced current
PSSC Physics, Third Edition, Section 23-5, Direction of Induced emf.

20.7 The generator principle
Lehrman and Swartz, *Foundations of Physics*, pp. 475–477, The Generator.

20.8 The basic a-c generator
K & E Overhead Transparency: #75. The AC Generator.
Project Physics Text, Unit 4, Section 15.4, Generating electricity by use of magnetic fields: the dynamo.

20.9 Instantaneous current and voltage
Lehrman and Swartz, *Foundations of Physics*, pp. 477–479, a-c in a Wire.

20.10 Practical a-c generators
3M-34V, a-c and d-c Generators.
Sutton, E-228, Waveform of Alternating Current.
3M-34W, Production of a-c Current.
Sutton, E-229, Alternating- and Direct-Current Generators.

20.11 The d-c generator
K & E Overhead Transparency: #76, The d-c Generator and d-c Motor.

20.14 The motor effect
Western, 2 × 2 Color Transparency: N-15, Magnetic Force on a Current.

20.15 Back emf
Sutton, E-230, Counter Electromotive Force in a Motor. See also E-231.
Meiners, 32-2.10, Demonstrates that back emf developed by electric motor running with no load almost equals driving voltage.

20.16 Practical d-c motors
K & E Overhead Transparency: #76, The d-c Generator and d-c Motor.
Exercises and Experiments in Physics, Expt. 45, The Electric Motor.

20.17 Practical a-c motors
3M-34O, Electric Motor.
Project Physics Text, Section 15.5, The electric motor.
Sutton, E-131 through 152, Forces on Conductors in Magnetic Fields. 3M-34P, Series and Shunt Wound Motors.
Laboratory Exercise: Lehrman, Expt. 57, Motors and Generators.

20.18 Mutual inductance
Meiners, Section 31-3, pp. 948–951, Inductance, six suggestions.
Lehrman and Swartz, *Foundations of Physics*, pp. 481–485, Inductance.
K & E Overhead Transparency: #74, The Induction Coil. 3M-34T, Induced Currents. 3M-34U, Induction Coil.
Western, 2 × 2 Color Transparency: 0-2, Current in an Induction Coil.
Lindsay, *Basic Concepts of Physics*, Section 7.8, The Concept of Inductance and Electric Oscillations.
Sutton, E-245, Induction Coil.

20.19 Self-inductance
K & E Overhead Transparency: #73, Self-Induction.
Western, 2 × 2 Color Transparencies: 0-13–0-15, the effect of self-inductance in a-c circuit.
Laboratory Exercise: Lehrman, *Laboratory Experiments,* Expt. 58, Self-Inductance.
Sutton, E-252, Self-Inductance. E-253, Self-Inductance. E-254, High Voltage from Low by Self-Induction.

20.21 The transformer
K & E Overhead Transparency: #77, The Transformer.
Sutton, E-235, Transformer. E-236, Jumping Ring. E-237, Large Currents in Step-Down Transformer. E-239, Dissectible Transformer. E-240, Toy Transformer. E-241, Reaction of Secondary Circuit on Primary. Also E-242. E-244, Use of Transformers for Transmission of Energy at High Voltage.
Gamow and Cleveland, *Physics, Foundations and Frontiers,* Section 14-11, pp. 238–242, Transformers and Alternating Current.
Lehrman and Swartz, *Foundations of Physics*, pp. 499–501, Transformers.

20.22 Transformer losses
Western, 2 × 2 Color Transparency: Damping by Eddy Current.
Sutton, E-255, Induced Current in a Metallic Disk, Eddy Currents. E-226, Arago's Disk. E-227, Magnetic Brake.

ALTERNATING-CURRENT CIRCUITS   21

Synopsis: We live in an age characterized by increasing energy consumption and decreasing energy reserves. With this imbalance of supply and demand, the rising cost of energy consumption with the present conversion technology is inevitable, as is our growing concern with energy conservation.

Because of its favorable energy conversion efficiency and convenient delivery system, we are deeply committed to the generation of electricity for most of our energy requirements. In the United States, electric energy is usually delivered to residential consumers as 60 hz, 120-vac service. Commercial and industrial consumers may require 240-vac and 440-vac service.

Incandescent lamps and various heating elements used in the home are typical resistive loads on the electric circuit. Appliances that are operated by electric motors present inductive loads. Electric power companies may add some capacitive loading to their delivery network as a means of maintaining the power factor near unity value, that is, to maintain a phase angle between the line current and voltage near zero. These power considerations in a-c circuits are considered in detail in Sections 21.3 through 21.8.

Comment: Use an oscilloscope to demonstrate the sine wave produced by a transformer. The material in the chapter is largely theoretical. The teacher should make continuous use of the phasor notation in showing magnitude and phase relationship between the resistive, inductive, and capacitive components of an a-c circuit. The rotating phasor is a convenient device for helping students understand the generation of a sine wave of a-c current (or voltage). In solving problems the student should draw both circuit diagrams and phase diagrams. It may be worthwhile to demonstrate again the resonance of two tuning forks of the same frequency as an analogy that, in a sense, establishes the problem to be presented and solved: Determination of conditions that result in circuits of identical frequency. The topic may be approached through a mechanical example using a bobbing weight at the end of a vertical coiled spring. This is described in Sutton, A22, page 444. Demonstrate tuning a radio or TV to a desired station as an application of resonance. Exhibit variable condensers used in such tuning to show how the process is usually performed. Most physics classes contain several students interested in radio and/or electronics as a hobby. Utilize their experiences to the fullest.

Review the generation of an alternating current (voltage) in a single conducting loop rotating at constant speed in a uniform magnetic field and the in-phase relationship according to Figure 21-2. Then compare d-c and a-c power according to Figure 21-3. Point out that the phase angle will affect a-c power. Use demonstrations to indicate that current and voltage measured by a-c instruments (or stated in problems) are effective or *rms* values. Derive the relationships for effective *versus* maximum values and illustrate as in Figure 21-4. Demonstrate home-made or student versions of the moving-vane and hot-wire types of meters. Discuss the electrodynamometer mechanism, and emphasize that the permanent magnet of the galvanometer movement of d-c meters is replaced by a pair of fixed coils. Discuss the other types of meters briefly.

You may wish to perform or repeat experiments dealing with inductance. Show with the aid of sine curves that in an a-c circuit consisting of pure inductance, the voltage leads the current by 90° (Figure 21-6). Then compare phase relation of an a-c current and voltage in pure resistive and pure inductive circuits (Figure 21-7). Go through the mathematics of the 10.0-v example that follows Figure 21-7. Interpret Figure 21-8 and show that the average power of a pure inductive a-c circuit is zero. You may wish to remind the students at this point that "pure inductance" is a theoretical consideration useful for analysis. The inherent resistance of the copper wire forming the loops of the inductor, however negligible, is present but is commonly neglected in the interest of simplicity. Show by vector addition how the resultant voltage of an a-c circuit consisting of resistance and inductance may be determined (Figures 21-9 and 21-10). Derive the expression for power factor. Define impedance, and show how an impedance diagram for an *L-R* series circuit is constructed. Emphasize those elements of trigonometry that will enable the student to analyze polar diagrams. Derive the relationships for phase angle and impedance, and analyze the solution to the Example at the end of Section 21.6. Have students demonstrate their solutions to the Practice Problems that follow this solved Example.

It may be useful to perform or repeat demonstrations on the capacitor and capacitance. Develop, with the aid of sine curves, how the voltage across a pure capacitor lags the current by 90°. Point out graphically (Figure 21-15) that in an a-c circuit consisting of a pure capacitance, the average power is zero. Define capacitive reactance and demonstrate it (Sutton: E-260, pages 360–361). Emphasize the representative circuit and diagrams of an *R-C* circuit. Derive the impedance diagram (Figure 21-7). Have students demonstrate the solution of the solved Example at the end of Section 21.8 and the solutions to the Practice Problems as applications of the principles learned.

You can now summarize by bringing together components of resistance, inductance, and capacitance. The circuit and phase diagrams for *R-L-C* in series, Figure 21-18, are useful to develop concepts of total reactance and impedance. Take the students through the solution using the values assigned in Section 21.9. In addition to the use of problems in the textbook, you may wish to refer to *Physics Problems*, page 329–340.

Review inductive reactance and capacitive reactance from results of demonstrations and individual student experiments. Point out through the use of Figures 21-19, 21-20, and 21-21, that the reactance of an a-c series circuit is inductive over the high-frequency range and is capacitive over the low-frequency range. Then demonstrate circuit resonance. This demonstration further emphasizes the energy transfer analogy to the bobbing weight-spring demonstration. By using Figure 21-23, show that at some intermediate frequency $X_L = X_c$ from which the formula for f_r is derived. Go through the calculations for the Figure 21-22 circuit. Demonstrate series resonance. Define selectivity and Q of a circuit. Use Figure 21-24 to show the influence of circuit resistance on selectivity. Discuss methods of tuning a resonant circuit.

References: Christenson, F. E., "The Oscilloscope and How it May be Used," *The Physics Teacher*, Oct. 1963, pp. 172–179.

Joseph, Alexander *et al.*, *A Sourcebook for the Physical Sciences*, Harcourt, Brace and Jovanovich, Inc., New York, 1961, Demonstrating Resonance with a Loop Oscillator and Receiver, p. 538.

Semat, Henry, *Fundamentals of Physics, Fourth Edition*, Holt, Rinehart and Winston, Publishers, New York, 1966, Chapter 28, Alternating Currents, pp. 479–493.

Stollberg and Hill, *Physics, Fundamentals and Frontiers*, Houghton Mifflin, Boston, 1965, Chapter 20, "Alternating Currents."

Turner, Rufus P., *Basic Electricity*, Holt, Rinehart and Winston, Publishers, New York, 1963, Chapters 6–9, and pp. 154–159.

Free and Inexpensive Learning Materials: *Alternating Current Simply Explained*, Timmerman, A. H., 16 pages, Wagoner Electric Corp., 6400 Plymouth Avenue, St. Louis, MO. A manual explaining alternating currents by means of graphs and equations. Free.

Films: *AC Voltmeters and Ammeters*, 30 frames, black and white, filmstrip, Long Film Slide

Service, 7505 Fairmont Avenue, El Cerrito 8, CA. Shows magnetic fields, a-c voltmeter, and a-c ammeter construction.

Alternating Current Circuits, 42 frames, color, filmstrip, 621 FOM, McGraw-Hill Book Co., Text Film Dept., 1221 Avenue of the Americas, New York, NY 10020. Describes the uses and principles of alternating current circuits.

The Versatile Oscilloscope, 41 frames, color, filmstrip, 624 FOM, McGraw-Hill Book Co., Text Film Dept., 1221 Avenue of the Americas, New York, NY 10020. Describes the uses and principles of the oscilloscope.

References for Chapter 21: Alternating-Current Circuits

Karplus, *Introductory Physics*, Section 12.4, pp. 331–336, Energy Transfer.
Project Physics Text, Section 15.7, pp. 86–90, a-c versus d-c, and the Niagara Falls power plant. Section 15.8, pp. 90–95, Electricity and society.

21.1 Power in an a-c circuit
Sutton, Section on Alternating Currents, E-228 through E-251, pp. 344–357. Covers waveform, generators, motors, transformers, and rotating magnetic fields.

Lehrman and Swartz, *Foundations of Physics*, Chapter 16, pp. 472–501, Alternating Currents.

Semat, *Fundamentals of Physics*, Chapter 28, pp. 479–493. Section 28-1, p. 479, Measurement of Alternating Current. Section 28-2, pp. 480–481, Instantaneous and Effective Values.

An extremely effective reference for this chapter is *The Radio Amateur's Handbook*, published by the American Radio Relay League. The present edition is available in radio shops. It may soon be replaced by a new edition; however, the chapter referred to here is not likely to be changed much. See particularly: Chapter 2, Electrical Laws and Circuits. This 44-page section deals, in language pitched to the needs of the amateur, with all topics covered in this chapter of the text.

Semat, *Fundamentals of Physics*, Section 28-5, pp. 484–485, Power in an a-c Circuit.

21.2 Effective values of current and voltage
Western, 2 × 2 color transparency, O-12, Alternating Current and Resistance.

21.3 Inductance in an a-c circuit
Western 2 × 2 color transparency: O-13, Alternating Current and a Coil, Alternating Current and a Capacitor.

Exercises and Experiments in Physics, Expt. 47, Inductance in a-c Circuits.

Lehrman and Swartz, *Foundations of Physics*, The Law of Induction, pp. 473–475. Inductance, pp. 481–485. Inductance in a-c Circuits, pp. 485–488.

Radio Amateur's Handbook, 1971, pp. 25–30, Inductance; Inductances in Series and Parallel.

Lindsay, *Basic Concepts of Physics*, Section 7.8, pp. 278–282, The Concept of Inductance and Electric Oscillations.

Semat, *Fundamentals of Physics*, Section 28-3, pp. 481–482, Inductance in an a-c Circuit.

21.4 Inductance and resistance
Lehrman and Swartz, *Foundations of Physics*, Inductance and Resistance, pp. 488–490.

Semat, *Fundamentals of Physics*, Section 28-4, pp. 482–484, Vector Diagram for an a-c Circuit.

21.6 Impedance
Radio Amateur's Handbook, 1971, pp. 32–37, Alternating Currents; Reactance; Impedance.

21.7 Capacitance in an a-c circuit
Meiners, Sections 33-1.2 and 1.3, pp. 991–992, Relaxation Oscillator.

Exercises and Experiments in Physics, Expt. 46, Capacitance in a-c Circuits.

Lehrman and Swartz, *Foundations of Physics*, Capacitance, pp. 490–492, Capacitance in an a-c Circuit, pp. 494–495.

Semat, *Fundamentals of Physics*, Section 28-6, pp. 485–487, Capacitance in an a-c Circuit.

Western, 2 × 2 color transparency: O-13, Alternating Current and a Coil. Alternating Current and a Capacitor.

21.8 Capacitance and resistance
Lehrman and Swartz, *Foundations of Physics*, pp. 492–494, Capacitance and Resistance.

Radio Amateur's Handbook, 1971, pp. 23–25, Capacitance and Capacitors in Series and Parallel.

21.9 L, R, and C in series
Meiners, Section 33-1.1, p. 990, Damped Oscillations.

Semat, *Fundamentals of Physics*, Section 28-7, pp. 487–489, Capacitance, Inductance and Resistance in Series. Section 28-8, p. 490, Resonance in an a-c Circuit.

Lehrman and Swartz, *Foundations of Physics*, pp. 495–498, Resonant Circuits (deals with both series and parallel circuits).

21.12 Series resonance
Laboratory Exercise: Lehrman, *Laboratory Experiments*, Expt. 60, Series Resonance.

Radio Amateur's Handbook, 1971, pp. 41–43, Resonance in Series Circuits.

22 ELECTRONIC DEVICES

Synopsis: This chapter opens with a brief history of the vacuum tube, and its role in the subsequent development of radio and television electronics as a viable technology. Thermionic emission, the boiling off of electrons from a heated (cathode) surface, is examined as is the function of an electrode (control grid) that regulates the magnitude of electron current passing from the cathode to the anode (plate) of the tube. In this sense, the vacuum tube functions as a versatile electron valve in an electronic circuit. The performance of the three-electrode (triode) vacuum tube circuit as a voltage amplifier is illustrated quantitatively. Special vacuum tubes that are essential components of television equipment are also described.

The main thrust of the chapter is concerned with the development of the transistor and its impact on electronic technology. The junction transistor is examined in some detail as a typical semiconductor product of research in solid state physics. The formation of P and N types of semiconductor materials is addressed. The diode characteristics of the P-N junction and the circuit characteristics of P-N-P and N-P-N junction transistors as current amplifiers are compared with the vacuum tube circuits which they have largely replaced.

Comment: The unit might be introduced by having students see "Microwaves in Action" a lecture-demonstration available through the local Bell Telephone Office. Class sets of the new instructional booklet "Satellite Communications Physics" are available from the same source. Hobby or project activities of students should be encouraged and reports or demonstrations by students should be made part of appropriate lessons. Various types of broken electronic tubes should be exhibited and their parts identified. A filmstrip, #628 FOM Vacuum Tubes, presents detailed structural and functional aspects of such tubes in color.

Discuss briefly the different functions of vacuum tubes, their inherent problems, and types of tubes. Emphasize the concept of thermionic emission and compare it with photoelectric emission. Demonstrate thermionic emission with a diode or adapted triode (see Sutton, A-78, page 481). Mention the various types of emitters, and on the chalkboard draw representative circuit symbols for cathode types, diodes, and triodes. The principle of the triode and function of the grid may be demonstrated.

Draw a simple diode circuit and point out the effect of plate voltage on plate current for different temperatures, as in Figure 22-8. Develop the concept of the space charge and the meaning of current saturation. Demonstrate and explain rectification by a diode, Sutton A-80, page 482. Discuss the addition of the third element to make a triode and explain how the grid acts as a gate in controlling the electron flow to the plate. Discuss its use as an amplifier and the meaning of cut-off and cut-off bias. The effects of control-grid bias on the plate current of a triode are illustrated in Figure 22-11. Amplification may be demonstrated. A mechanical amplifier model is suggested on page 486 (A-86) Sutton.

Go through the steps in the solution of a typical triode problem, Section 22.5 and Figure 22-13. Related problems appear on page 345 of *Physics Problems* and the preceding pages (342–344) show the solutions of Type Problems.

Emphasize that the plate voltage variations are opposite in phase to the voltage variations on the grid. For example, as the signal voltage on the grid reaches a maximum, the current in the tube reaches a maximum, and the plate voltage becomes a minimum, as shown in Figure 22-13. Point out that the elements of the tube form a capacitor. This becomes an important consideration when triodes are used as high-frequency amplifers.

Discuss the basic operation of a cathode-ray tube stressing the difference between electrostatic and magnetic deflection principles. Deflection of cathode rays may be shown using a Crookes tube. Call attention to the manner in which secondary emission of electrons is utilized in conjunction with the photoelectric effect to produce the photomultiplier tube.

Have a student demonstrate a crystal detector, possibly in a radio. A two-stage radio receiver consisting of a crystal detector and a triode amplifier may be constructed. The Bell Telephone Co. will furnish a kit, "From Sun to Sound," which provides materials and instructions for a student to construct a solar-powered transistor oscillator. An advanced science film, "Brattain on Semiconductor Physics," 30 min., is also available from the same source. Show the class samples of transistors. Explain how P- and N-type semiconductors have representative structures and how they function (Figures 22-19 and 22-20). Explain how P-N junction is a rectifying element in semi-conductor crystals. Explain how N-P-N and P-N-P junction transistors operate (Figures 22-22 through 22-25) and the representative circuit symbols. Have students display a transistor radio or a home-constructed version of the device.

References: Beck. A. H. W., *Thermionic Values*, Cambridge University Press, England, 1963, pp. 25–51.

Bennett, Alan, Robert Heikes, Paul Klemens, and Alexei Maradudin, *Electrons on the Move*, Walker, New York, 1964, 229 pp. Another excellent book by working scientists in the Westinghouse Research Laboratories who explore with the reader the properties of electrons, both free and in matter, and give particular emphasis to vacuum tubes and solid state devices. They also review some of the history of the electron and the scientific work that has led to its discovery, a determination of its properties and behavior. Some recent developments are discussed such as "picture tubes" and masers and lasers. The reader should know fundamentals of algebra.

Fink, Donald G., *The Physics of Television*, Science Study Series, 1960.

Freeman, Ira M., *Modern Introductory Physics*, McGraw-Hill Book Co., New York, 1957, pp. 346–354.

Pierce. John R., *Electrons and Waves, An Introduction to the Science of Electronics and Communications*, Science Study Series, 1964.

RCA Receiving Tube Manual, Electron Tube Division, Radio Corp. of America, Harrison, NJ.

Semat, Henry, *Fundamentals of Physics, Fourth Edition*, Holt, Rinehart and Winston, Publishers, New York, 1966, selected parts from Chapter 29, "Electromagnetic Waves and Electronics," pp. 496–512.

Spielman, Harold S., *Electronics Source Book for Teachers*, 3 vols., Hayden, New York, 1965, 1609 pp. in all, illus. This set presents, for instructor use primarily, a veritable encyclopedia of information in the broad subject area of electronics, representing a monumental effort on the part of the author. This compendium is arranged in progressive developmental order of subject matter, rather than alphabetically, and should be invaluable for those who must plan and teach various courses in electronics in secondary schools or technical institutes. The set is particularly useful for teachers who lack extensive background and practical experience in electronic technology—the books would probably not be suitable as textbooks in the hands of students, but can serve them as supplementary references. Many chapters, particularly in the second and third volumes, appear to be suitable frameworks of survey courses in specialized subjects, such as medical electronics. There are hundreds of diagrams, notably in Vol. 1, that are clearly slanted toward high school students, and all volumes are replete with materials around which lectures and classroom demonstrations or laboratory exercises can be planned by the teacher to meet the needs of almost any course of study.

Stollberg and Hill, *Physics: Fundamentals and Frontiers*, Houghton Mifflin Co., Boston, 1965, Unit Six, Electronics; Chapter 21, Electron Tubes and Electronic Circuits. Chapter 22, Communication Through Electronics. Chapter 23, Particles, Accelerators, and Holes, including semiconductors and some results of solid state research.

The Radio Amateur's Handbook, American Radio Relay League, West Hartford, CT.

Turner, R. P., *Semiconductor Devices*, Holt, Rinehart and Winston, Publishers, New York, 1961.

Wilcox, G., *Basic Electronics*, Holt, Rinehart and Winston, Publishers, New York, 1960.

Free and Inexpensive Learning Materials: *Adventures in Electronics*, N. 7, 16 pages, comic booklet, General Electric Co., Dept. 2-119, Schenectady, NY. Free.

Basic Theory and Application of Electron Tubes, #TM-11-662 and TO 16-1-255; *Basic Theory and Applications of Transistors*, #TM 11-690. Both from Department of the Army, Washington, DC.

Films and Filmstrips: *The Transistor*, 14 min., black and white, sound, Bell Telephone Offices, The story of the transistor. This device can do many of the jobs done by the vacuum tube and many that the vacuum tube cannot do. Describes transistor development and the part transistors play in communications.

Vacuum Tubes, 11 min., black and white, Encyclopaedia Britannica Films, Inc., 1150 Wilmette Ave., Wilmette, IL. Explains by animation the operation of a vacuum tube in terms of filament, plate, and grid circuits.

The Diode: Rectification, 60 frames, black and white, filmstrip, United States Office of Education, Washington, DC. Principles of operation of the diode as a rectifier.

The Triode: Amplification, 38 frames, black and white, filmstrip, United States Office of Education, Washington, DC. Explains the triode amplification and detection.

Vacuum Tubes, 37 frames, black and white, filmstrip, Photo Laboratory, Inc., 3825 Georgia Avenue, North West, Washington, DC. Shows the basic operation and theory of vacuum tubes.

References for Chapter 22: Electronic Devices

Meiners, Section 33-2, pp. 992–996, Electronics and a-c Circuits. 33-2.2, Current-Voltage Characteristics of Various Electrical Components. 33-2.3, Circuit to Demonstrate Phase Relationships. 33-2.4, Half- and Full-Wave Rectification. 33-2.6, Relationship between R, C and L. 33-2.10, 12-in. Scope for Demonstration Work.

22.2 Vacuum-tube applications

Section 33-3, p. 997, Forced Oscillations and Resonance. 33-3.1, Mechanical Analog of Parallel Resonance.

Radio Amateur's Handbook, 1971, Chapter 3, pp. 59–76, Vacuum-Tube Principles. (Several references to applications.)

See Spielman, *Electronics Sourcebook for Teachers*, Hayden Book Company, New York, 1965, three volumes, for applications for vacuum tubes. (This reference will hereafter be given as Spielman followed by chapter and page numbers.)

22.3 Vacuum tube development

Spielman, Chapter 3, pp. 108–122, Electron Emission.

22.4 Thermionic emission

Sutton, Section on Thermionic Emission, A-77 through A-88, pp. 480–488, covering thermionic effect in air and in a vacuum, rectifiers, triodes, amplifiers, and gas-filled tubes.

Exercises and Experiments in Physics, Expt. 48, Thermionic Emission.

22.5 Diode characteristics

Western, 2×2 color transparencies: 0-17, Experiment Diagram of $V_p - I_p$ Characteristics of a Diode. 0-18, $V_p - I_p$ Characteristics of a Diode. 0-19, Rectification and Smoothing by a Diode (1). 0-20. Rectification and Smoothing by a Diode (2).

Laboratory Exercise: Stollberg and Hill, Investigation 42, The Vacuum-Tube Rectifier.

Radio Amateur's Handbook, 1971, Figure 3-3, p. 60, Diode Characteristic.

Spielman, Chapter 4, pp. 123–145, High-Vacuum Diodes. Chapter 5, pp. 146–216, Gas-Filled Diodes.

Laboratory Exercise: Lehrman, *Laboratory Experiments*, Expt. 61, Full-Wave Rectification.

22.6 The triode amplifier

Radio Amateur's Handbook, 1971, p. 61, Triodes, Grid Control.

Laboratory Exercise: Stollberg and Hill, Investigation 43, The Three-Electrode Vacuum Tube.

Spielman, Chapter 7, pp. 249–260, Triodes. Chapter 8, pp. 261–312, Amplification.

22.7 The cathode-ray tube
PSSC Physics, Third Edition, Section 18-9, pp. 412–414, Electron Guns and Oscilloscopes.
Radio Amateur's Handbook, 1971, The Oscilloscope, pp. 544–545.
Spielman, Chapter 15, pp. 444–499, Cathode-Ray Tubes and Related Devices.
Laboratory Exercise: Stollberg and Hill, Investigation 44, The Oscilloscope.

22.8 Photomultiplier tube
Spielman, Section 13-4, pp. 364–366, Photoelectric Multiplier Tubes.

22.9 Crystal diodes
Radio Amateur's Handbook, 1971, pp. 81–92, Transistors.
Laboratory Exercise: Lehrman, *Laboratory Experiments*, Expt. 62, Solid-State Diode.
Spielman, Chapter 17, pp. 515–580, Transistors and Related Semiconductor Devices.

22.11 P- and N-type semiconductors
Ouseph, P. J. and Manuel Schwartz, "Semiconductor Detectors," *Scientific American*, Oct. 1970, pp. 374–379.

THE ATOMIC STRUCTURE 23

Synopsis: The idea that matter is made of four elements—earth, air, fire, and water—first proposed by Pythagoras and his disciples about 530 B.C., was believed commonly for over 2000 years. These four elements were supposed to have the properties of dryness, wetness, hotness, and coldness, which in various combinations could account for the observed properties of physical objects. It was eventually realized that very little that would account for observed properties of physical objects could be deduced from these four elements.

The Greeks of the time were bothered by the question of continuity. Aristotle championed the view that matter was continuous throughout; however, Anaxagoras suggested the notion that matter consists of combinations of small invisible particles, which are unchangeable and indestructible. Leucippus assumed the entire universe to be composed of a void and a large number of invisible and indivisible particles that differed from one another in shape and size. These particles were called *atoms* by the Greeks since the word *atom* in Greek means "that which cannot be cut."

Epicurus, in about 300 B.C., taught the atomic theory in Athens. It was clearly set forth that the universe consists of empty space and atoms, nothing more. It was thought that the collisions of atoms as they moved through the void brought about combinations that resulted in all physical objects and gave these objects their properties. Aristotle refused to accept the atomic theory, for he could not accept the idea of the existence of a totally empty space.

Because of the influence of Aristotle the atomic idea remained qualitative and quite dormant until the 1600s when it was revived again. At this time, density, the number of atoms per unit volume, became a numerical property. (A possible difference in the mass of different atoms was not then considered.)

Boyle (1626–1691) adopted a corpuscular hypothesis that did not include the idea of a void in which atoms could move in a purely random fashion. Boyle believed the corpuscles could form clusters that would provide the observed characteristics of objects. Predictions, though plausible, remained qualitative. Boyle's chief contribution to atomism was to make it a respectable theory.

Bernoulli (1700–1782) made the first analytical application of the atomic idea in his study of gases. His work led to the development of the molecular theory of gases, which was not considered seriously again until the 1800s.

Dalton (1766–1844) developed the idea that chemical elements taking part in a chemical reaction do so in the form of atoms. In Dalton's time two laws governing the formation and composition of chemical compounds, the law of definite proportions and the law of multiple proportions, followed directly from the assumption that the combining elements are groups of atoms that for a given element have the same mass. In 1811 Avogadro proposed that equal volumes of all gases at the same temperature and pressure contain the same number of molecules. He later developed the concept that a gram-molecule of any substance contains the same number of molecules. It follows that a gram-atom of any element contains the same number of atoms as a gram-atom of any other element. This number (6.03×10^{23}) is one of the most important constants in atomic theory.

In the early 1800s, Davy and later Faraday made extended measurements in electrolysis that gave reinforcement to the atomic hypothesis.

In the study of this chapter, we rather quickly assume the existence of atoms and molecules. After a consideration of the properties of atoms we consider the subatomic particles. Considerable attention is given to the existence and properties of the electron and to the atomic nucleus and the particles of which it is composed. We conclude with a study of isotopes and of nuclear binding energy.

Comment: Considerable time and emphasis may be given to the historical aspects of discoveries related to atomic structure. The Crookes tube should be demonstrated (Figure 23-2). Charts, transparencies, or chalkboard diagrams such as those of Figure 23-5 and Figure 23-9 should be used. Some chalkboard diagrams or transparencies (especially those with overlays) representative of atomic models may be used.

It is suggested that you prepare to discuss with students the material presented in Section 23.1 using this material as a "jumping off" point. You may wish to refer to Chapter 36, "Foundations of Atomics and Nucleonics," pages 611–624, *Fundamentals of Physics*, Semat, or Chapter 38, "Analyzing Atoms," pages 624–631 and Chapter 40, "Atoms: Experiment and Theory," pages 648–654, *Physics for the Inquiring Mind*, Rogers, 1960, Princeton University Press. Students may be advised to read selected topics from *PSSC Physics, Second Edition*, 1965. Chapter 8, Atoms and Molecules, pages 125–156. (Note, second edition.) This lesson should be used to establish the foundation for the study of atomic and nuclear structure. Insights that will encourage serious study of the remainder of this chapter should be given. The instructor will wish to use the references indicated in the previous lesson as well as Section 21-7 "Millikan's Determination of Electronic Charge," page 372, and Section 25-11, "Ratio of Charge to Mass of an Electron," page 433, from *Fundamentals of Physics*, Semat. If available, use a tube such as the Sargent-Welch 2145D or Cenco 71525 or 71535 for demonstrating the positive and negative charges produced when a high d-c voltage is impressed on a tube containing hydrogen at low pressure. Use the standard cathode-ray tubes to show effects such as those illustrated in Figures 23-2 and 23-3. Use a transparency with overlays to discuss the *e/m* experiment. Schools may have either the Sargent-Welch 0623 and 0623A or the Cenco 71267 or the Macalaster 1701 apparatus for measuring the *e/m* ratio. Student reports may be given from articles appearing in *Great Experiments in Physics*: "The Electron—J. J. Thomson" (pages 216–231), "The Elementary Electric Charge—Robert A. Millikan" (pages 238–249), and "The Neutron—James Chadwick" (pages 266–280). Schools should have the Macalaster 4900, Millikan apparatus for measuring the charge on the electron. This assignment can put the capstone on the study of the atom. The questions not already assigned can be used to focus attention on essential ideas associated with atomic structure. Problems may be assigned to give students experience with (a) determining mass numbers and atomic mass, (b) calculation of the charge on an ion, (c) determination of atomic number, (d) the determination of which particles and how many make up a given atom, (e) the calculation of the binding force per nucleon for an atom, (f) the determination of the velocity of particles resulting from the production of an electron pair produced by X rays of specific energy. The material presented in the text should be sufficiently considered to insure that students gain a "feel" for the ideas presented.

References: Andrade, E. N. da Costa, *Rutherford and the Nature of the Atom,* Science Study Series, 1964.

Barschall, H. H., "The Optical Model of the Nucleus," *The Physics Teacher*, Dec. 1969, pp. 481–485.

Calbick, C. J., "The Discovery of Electron Diffraction by Davisson and Germer," *The Physics Teacher*, May 1963, pp. 63–68.

Derjaguin, Boris V., "The Force Between Molecules," *Scientific American*, July 1960, 47–53.

Frisch, Otto R., *Atomic Physics Today*, Basic Books, New York, 1961.

_____, *Working with Atoms*, Basic Books, New York, 1966.

Glasstone, Samuel, *Sourcebook on Atomic Energy, Third Edition*, Van Nostrand, 1967.

Heckman, Harry H., and Paul W. Starring, *Nuclear Physics and Fundamental Particles*, Holt, Rinehart and Winston, Publishers, New York, 1963, Chapter 1–3.

Henry, Hugh F., "Radioactivity—Safer than You Think," *The Science Teacher*, Nov. 1970, pp. 29–32.

Hill, R. D., "Elementary Particles," *The Physics Teacher*, Jan. 1964.

Hofstadter, Robert, "The Atomic Nucleus," *Scientific American*, July 1956.

Hughes, Donald J., *The Neutron Story*, Science Study Series, 1959.

Lyons, Harold, "Atomic Clocks," *Scientific American*, Feb. 1957.

Marshak, Robert E., "The Nuclear Force," *Scientific American*, Mar. 1960.

Mayer, Maria G., "The Structure of the Nucleus," *Scientific American*, Mar. 1951.

Morrison, Phillip and Emily, "The Neutron," *Scientific American*, Oct. 1951.

Nier, Alfred O. C., "The Mass Spectrometer," *Scientific American*, Mar. 1953.

Peierls, R. E., "Models of the Nucleus," *Scientific American*, Jan. 1959.

Romer, Alfred, *The Restless Atom*, Science Study Series, 1960.

Schrödinger, Erwin, "What Is Matter?," *Scientific American*, Sept. 1953.

Shamos, Morris H., *Great Experiments in Physics*, Holt, Rinehart and Winston, Publishers, New York, 1959, "The Hydrogen Atom—Niels Bohr," pp. 329–347.

Stewart, Alec T., *Perpetual Motion: Electrons and Atoms in Crystals*, Science Study Series, 1965.

Swartz, Clifford E., *The Fundamental Particles*, Addison-Wesley, Reading, MA, 1965.

Thomson, George, *J. J. Thomson, Discoverer of the Electron*, Science Study Series, 1966.

Trower, W. P., and J. R. Ficenec, "Electron Scattering and Nuclear Structure," *The Physics Teacher*, Apr. 1971, pp. 175–184.

Watson, E. C., "Robert Millikan—A Physicist Who Changed the Course of History," *The Physics Teacher*, Jan. 1964, pp. 7–11.

Films: *Atomic Structure and Chemistry,* 41 frames, color, fs, FOM 616, McGraw-Hill Book Co., Text Film Dept., 1221 Avenue of the Americas, New York, NY 10020. Work of Rutherford and Bohr described. Orbits and quantum numbers treated.

The Structure of the Atom, 49 frames, black and white, fs, McGraw-Hill Book Co., Text Film Dept., 1221 Avenue of the Americas, New York, NY 10020.

What's In the Atom, 41 frames, color, fs, FOM 592, McGraw-Hill Book Co., Text Film Dept., 1221 Avenue of the Americas, New York, NY 10020. Standard treatment of particles with some frames devoted to mesons and anti-matter.

PSSC Films: 0111, *Elements, Compounds and Mixtures,* 28 min., color. Differences between elements, compounds, and mixtures demonstrated. Identification of components by means of physical properties.

0113, *Crystals,* 25 min., black and white. Demonstrates nature of crystals, their formation, growth, and shapes. Phenomena related to atoms.

0404, *Millikan Experiment,* 30 min., black and white. Simplified Millikan experiment photographed through the microscope. Emphasizes the evidence that charge comes in natural units that are all alike.

0416, *The Rutherford Atom,* 40 min., black and white. Uses a cloud chamber and gold foil to carry out the historic Rutherford experiment that led to the nuclear model of the atom.

0412, *Electrons in a Uniform Magnetic Field,* 10 min., black and white. A Leybold (spherical cathode-ray) tube with some gas to measure the curvature of path of electrons in a magnetic field and thus determine the mass of the electron. Arithmetic involved is worked out in the experiment.

0413, *Mass of the Electron,* 18 min. Using a cathode-ray tube encircled by a current carrying loop of wire, measurements are taken that enable the demonstrator to calculate the mass of the electron with reference to the Millikan experiment. The calculations are brought out in detail, step by step.

0408, *Counting Electrical Charges in Motion,* 22 min. This film shows how an electrolysis experiment enables us to count the number of elementary charges passing through an electric circuit in a given time and thus calibrate an ammeter. Demonstrates the random nature of motion of elementary charges, with a current of only a few charges per second.

References for Chapter 23: The Electron

23.1 Subatomic particles
Karplus, *Introductory Physics,* Chapter 8, Models for Atoms, pp. 203ff. Section 8.1, pp. 203–208, The Electrical Nature of Matter. Section 8.2, pp. 208–211, Early Models for Atoms. Section 8.3, pp. 212–215, Bohr's Model for the Atom.

23.2 Discovery of the electron
Laboratory Exercises: *Project Physics Handbook,* Expt. 5-2, p. 7, The Charge-to-Mass Ratio for an Electron. Expt. 5-3, p. 10, The Measurement of Elementary Charge.
Sutton, Properties of Electrons, A-71 and A-76, pp. 477–480, suggestion for demonstrating the *e/m* and the oil-drop experiments.
Karplus, *Introductory Physics,* pp. 207–208, Electrons.
Western, 2 × 2 Color Transparencies: Q-5 Fluorescence by Cathode Rays. Q-6 Pressure Exerted by Cathode Rays (1). Q-7 Ditto (2). Q-8 Rectilinear Motion of Cathode Rays. Q-9 Cathode Rays and Electric Field (1). Q-10 Ditto (2). Q-11 Cathode Rays and Magnetic Field (1). Q-12 Ditto (2). Q-13 Deflection of Cathode Rays (1). Q-14 Ditto (2). Q-15 Ditto (3). Q-16 Ditto (4).

23.3 Measuring the mass of the electron
Western, 2 × 2 Color Transparencies: Q-18 Measurement of *e/m*. Q-19 Magnetic Field Produced by *e/m* Apparatus. Q-20 Orbit of Electrons in *e/m* Apparatus (1). R-1 Ditto (2). R-2 Ditto (3).

Gamow and Cleveland, *Physics, Foundations and Frontiers,* Section 21-4, pp. 377–379, The Charge-to-Mass Ratio of an Electron. Section 21-5, pp. 380–381, The Charge and Mass of an Electron.
Weidner and Sells, *Elementary Modern Physics,* Allyn and Bacon, Boston, 1960. Section 3-5, pp. 111–116, Pair Production and Annihilation.
Semat, *Introduction to Atomic and Nuclear Physics, Fourth Ed.,* Holt, Rinehart and Winston, New York, 1963, Section 15-4, pp. 496–497, Production of Pairs of Charged Particles.

23.4 The electronic charge
Semat, *Introduction to Atomic and Nuclear Physics, Fourth Edition,* Holt, Rinehart and Winston, 1963, Section 3-6, pp. 61–65, Determination of the Charge of an Electron.

23.5 Size of the electron
Western, 2 × 2 Color Transparency: R-12 Electron Diffraction.

23.7 Discovery of the atomic nucleus
Lindsay, *Basic Concepts of Physics,* Section 9.11, pp. 379–383, Nuclear Physics.
Project Physics Supplemental Unit, The Nucleus. Project Physics Transparency, T41, Rutherford's Particle "Mousetrap." T38, Alpha Scattering.

Karplus, *Introductory Physics*, Section 8.6, pp. 220–228, The Atomic Nucleus. pp. 210–211, Rutherford's Nuclear Model.

Overhead Transparency K & E #80, Rutherford's Scattering Experiment.

Sutton, A-63, Alpha-Particle Scattering, Magnetic Model, A-64, Alpha-Particle Scattering-Electrostatic Model.

Gamow and Cleveland, *Physics, Foundations and Frontiers*, Chapter 28, pp. 479–491, The Structure of the Atomic Nucleus.

Project Physics 8mm Film: *Rutherford Scattering*. Color, 3 min. 50 sec.

Lehrman and Swartz, *Foundations of Physics*, Chapter 22, pp. 670–712, The Atomic Nucleus and Other Particles.

23.8 The proton

Karplus, *Introductory Physics*, pp. 221–222, Modern Nuclear Models.

Project Physics 8mm Film: *Collisions with an Unknown Object*. Color, 3 min. 35 sec.

23.9 The neutron

Project Physics Transparency, T47, Nuclear Equations.

23.10 Isotopes

Overhead Transparency K & E #82, Atomic Symbols. #81, Natural Radioactivity.

Overhead Transparency K & E #85, Mass Spectroscope.

23.11 The mass spectrograph

Project Physics Transparency, T45, Mass Spectrograph.

23.12 Nuclear binding

Project Physics Transparency, T48, Binding Energy Curves.

Gamow and Cleveland, *Physics, Foundations and Frontiers*, Section 28-3, pp. 482–485, Mass Defect and Nuclear Binding. Energy. Section 28-4, pp. 485–488, Mass Defect and Nuclear Reaction.

NUCLEAR REACTIONS 24

Synopsis: The nucleus provides the atom with a positive charge to balance the negative charge in the electron cloud surrounding it. It also provides the greater part of the mass of the atom and has a structure consisting of a variety of particles or entities that behave like particles when they are expelled from the nucleus.

We are concerned in this chapter with the particle emission from radioactive nuclei and with the devices that have been developed to detect and count the emitted particles. We are also concerned with the nature of the change that takes place in the nucleus as it emits a particle or particles in the process of radioactive disintegration.

The process of fission and of chain reactions and the means by which the energy released from the disintegration can be harnessed for useful purposes are considered. Fusion is studied as a reaction from which useful energy may someday be available. The chapter closes with a study of nuclear reactions and considers the production and use of natural and artificial radioactive isotopes. The concern in all of this is with the physics of the reactions, the conservation of mass and energy, and the evidence that we can gather in support of the theory that has been formulated.

Comment: Problems related to radioactive fallout, international attempts at controlling nuclear weapons, and research may be discussed to help students appreciate the international, political, social, and economic problems related to nuclear energy. Study the peacetime uses of nuclear energy. Films and/or visits may be used to acquaint students with some of the great centers for nuclear research. Students may be encouraged to construct and exhibit projects such as cloud chambers, Geiger counters, and Van de Graaff generators.

Use the Geiger counter as a black box (it will be discussed in Chapter 25) for the detection of radiation. Demonstrate that radiation can be detected from uranium ore, uranyl nitrate, and alpha and beta sources. The use of a radioactive source to bring about the rapid discharge of a charged electroscope is a stimulating demonstration. Discuss the properties of radioactive elements. Use a diagram like that of Figure 24-4, to show how the types of radioactivity are deflected by a magnetic or an electrostatic field. Introduce the half-life concept. Have a student report on "Natural Radioactivity—Henri Becquerel" (pages 210–215), *Great Experiments in Physics*. Differences in the penetrating ability of beta and gamma radiations may be demonstrated by using uranyl nitrate (primarily a beta emitter) and thorium nitrate (a gamma emitter) with various thicknesses of lead sheet and sheets of other materials: cardboard, paper, glass, aluminum. Use the Geiger counter as a detector.

Discuss the three types of changes in matter: physical, chemical, and nuclear. Practice writing equations using appropriate symbols for representing subatomic particles and radiation. Trace through selected portions of the uranium disintegration series, Figure 24-5, and write several equations for alpha and beta emission. Develop an understanding of and skill in writing nuclear equations, emphasizing here the fundamental ideas (conservation of mass number and conservation of nuclear charge). In alpha emission the mass number decreases by 4 and the nuclear charge decreases by 2. In beta emission the mass number is unchanged and the nuclear charge increases by 1. To help students understand the increase in the number of protons, suggest a mechanism, $_0^1n \times {}_1^1H + {}_{-1}^0e$, and draw diagrams of nuclei (parent and daughter). Teach the half-life concept, using illustrations such as ^{60}Co, ^{131}I, ^{90}Sr, which have social implications. Study man-made transmutations, fission, and fusion. Illustrate loss of mass and release of energy as in Section 24.7. Have students write representative equations for historically significant nuclear disintegrations and artificial radioactivity.

Make use of the many references given in this section for assigning supplementary reading and reports. The idea of a chain reaction and of a critical mass should be carefully considered. Film presentations should be useful in helping students understand how reactors are used to develop power. Give appropriate attention to the study of the uses of radioisotopes as tracers, in medicine, in industry, and in the dating of objects of historical significance.

References: Alexander, Peter, *Atomic Radiation and Life*, Penguin, Baltimore, 1965.

A collection of papers from the *Journal of Chemical Education* related to *Training and Experiments in Radioactivity*, Nuclear-Chicago Corporation, Des Plaines, 1959, free.

Anderson, David, *The Discovery of the Electron*, Momentum Book, D. Van Nostrand Co., Inc., Princeton.

Bauman, Robert F., "The Meaning of the Conservation Laws," *The Physics Teacher*, Apr. 1971, pp. 186–189.

Beeler, N. F., and F. M. Branley, *Experiments with Atomics*, Crowell, New York, 1954, Chapters 6, 7, 15, 17.

Burbidge, G., and F. Hoyle, "Anti-matter," *Scientific American*, Apr. 1958, p. 34.

Burwell, Calvin C., "Desalting and Nuclear Energy," *The Physics Teacher*, Feb. 1971, pp. 67–74.

Cloutier, R., "Radiation Safety," *The Science Teacher*, Dec. 1963, p. 35.

Cohen, Bernard L., *The Heart of the Atom, The Structure of the Atomic Nucleus*, Science Study Series, 1967.

Diehl, T. H., and C. D. Geilker, "Determination of Half-life," *The Science Teacher*, Dec. 1958, p. 442.

Frisch, David H., and Alan M. Thorndike, *Elementary Particles*, Momentum Book, D. Van Nostrand Co., Inc., Princeton.

Galley, E. L., "An Inexpensive Continuous Cloud Chamber," *The Physics Teacher*, May 1963, p. 80.

Ginzton, E. L., and W. Kirk, "The Two-mile Accelerator," *Scientific American*, Nov. 1961, p. 49.

Glasstone, Samuel, *Sourcebook of Atomic Energy, Third Edition*, D. Van Nostrand, Princeton, 1967.

Gottlieb, Herbert H., "Half-life Using Short-lived Radioisotopes," *The Physics Teacher*, Apr. 1968, p. 176.

Heckman, Harry H., and Paul W. Starring, *Nuclear Physics and the Fundamental Particles*, Holt, Rinehart and Winston, Publishers, New York, 1963, Chapters 4–13.

Hill, R. D., "Elementary Particles," *The Physics Teacher*, Jan. 1964, p. 22.

Hurley, P. M., "Radioactivity and Time," *Scientific American*, Aug. 1949.

Jaffee, Bernard, *Men of Science in America*, Simon and Schuster, New York, 1958, Chapter 19.

Jelley, J. V., "Cerenkov Radiation: Its Origin, Properties and Applications," *The Physics Teacher*, Nov. 1963, p. 203.

Post, Richard F., "Fusion Power," *Scientific American*, Dec. 1957.

Prowse, D. J., "Note on the Welch Scattering Apparatus," *American Journal of Physics*, Dec. 1961, p. 854.

Reynolds, John H., "The Age of the Elements in the Solar System," *Scientific American*, Nov. 1960.

Rogers, Eric M., *Physics for the Inquiring Mind*, Princeton University Press, Princeton, 1960, Part 5.

Schumar, J. F., "Reactor Fuel Elements," *Scientific American*, July 1959, p. 56.

Seaborg, Glenn T., and I. Perlman, "The Synthetic Elements I," *Scientific American*, Apr. 1950.

Seaborg, Glenn T., and Albert Ghiorso, "The Synthetic Elements II," *Scientific American*, Dec. 1956.

Seaborg, Glenn T., and Arnold R. Fritsch, "The Synthetic Elements III," *Scientific American*, Apr. 1963, p. 68.

Semat, Henry, *Fundamentals of Physics, Fourth Edition*, Holt, Rinehart and Winston, Publishers, New York, 1966, Chapters 39, 40.

Sisters Hermias and Joecile, *Radioactivity: Fundamentals and Experiments*, Holt, Rinehart and Winston, Publishers, New York 1963.

Skinner, S. Ballou, "A Simple Experiment to Illustrate Exponential Decay, Half-Life and Time Constant," *The Physics Teacher*, May 1971, pp. 269–270.

Spitzer, Lyman, "The Stellarator," *Scientific American*, Oct. 1958, p. 28.

Teller, Edward, "Atomic Explosives, Solved and Unsolved Problems," *The Physics Teacher*, May 1968, pp. 207–209.

Thomas, Miriam H., and Edward S. Josephson, "Radiation Preservation of Foods and Its Effect on Nutrients," *The Science Teacher*, March 1970, pp. 59–63.

Wahla, James, "A Simple Inexpensive Geiger Counter," *The Science Teacher*, May 1959, p. 255.

Wilson, Robert R., and Raphael Littauer, *Accelerators: The Machines of Nuclear Physics*, PSSC Science Study Series, Doubleday and Co., Inc., Garden City, NY.

Free and Inexpensive Learning Materials: *Atomic Energy in the Year 2000*, Standard Oil Co., 30 Rockefeller Center, New York, NY.

Experiments with Radioactivity, National Science Teachers Association, 1201 Sixteenth Street, Northwest, Washington 6, DC.

Laboratory Experiments with Radioisotopes (For High School Science Demonstrations), U.S. Atomic Energy Commission, Superintendent of Documents, U.S. Government Printing Office, Washington 25, DC.

Nuclear Research, Information Division, Radiation Laboratory, University of California, Berkeley, CA.

Films: *Atomic Energy*, 10 min., black and white, sound, U.S. Atomic Energy Commission, P.O. Box E, Oak Ridge, TN. Introduction to atomic principles including fission and chain reactions. Free loan.

Atomic Radiation, 12 min., black and white, sound, EBF. Story of atomic radiations explaining the roles of alpha, beta, and gamma radiations in radioisotope research.

Atoms and Industry, 10 min., black and white, sound, U.S. Atomic Energy Commission, P.O. Box E, Oak Ridge, TN. Atomic tracer studies in industrial processes. Free loan.

Basic Physics of an Atomic Bomb, f, color, sound, Dept. of the Army, Principles of the atom bomb. Free.

Carbon Fourteen, 12 min., black and white, EBF. Demonstrates the use of C-14 in determining the age of ancient objects and in tracing the processes of growth in living things.

Nuclear Reactors for Research, 20 min., f, sound, color, North American Aviation Inc., 12214 Lakewood Boulevard, Downey, CA. Design and development of atomic reactors and peacetime applications in medical, academic, and industrial research. Free loan.

Particle Accelerators, fs, color, 41 frames, FOM, McGraw-Hill Book Co., Text Film Dept., 1221 Avenue of the Americas, New York, NY 10020. Describes principles of operation of various types of accelerators.

The Strange Case of Cosmic Rays, 59 min., sound, color, Local Bell Telephone Business Office. Animated; behavior and characteristics of cosmic rays. Free loan.

References for Chapter 24: Nuclear Reactions

Beiser, *Concepts of Modern Physics*, Part 4, The Nucleus, pp. 291–391, consisting of Chapter 12, The Atomic Nucleus, Chapter 13, Nuclear Decay, Chapter 14, Nuclear Reactions, and Chapter 15, Elementary Particles.

Meiners, Section 41-2, pp. 1249–1266, Nuclear Reactions.

Lehrman and Swartz, *Foundations of Physics*, pp. 684–685, Radioactivity.

Semat, *Fundamentals of Physics*, Holt, Rinehart and Winston, Publishers, 1966, Chapter 39, pp. 676–689, Natural Radioactivity. Chapter 40, pp. 690–712, Nuclear Disintegration.

Project Physics Supplemental Unit, The Nucleus, Chapter 4, Nuclear Energy, Nuclear Forces.

24.1 Discovery of radioactivity

Project Physics Supplemental Unit, The Nucleus, Chapter 1, Radioactivity.

Karplus, *Introductory Physics*, pp. 224–225, Nuclear Reactions.

Rogers, *Physics for the Inquiring Mind*, Chapter 39, pp. 633–645, Radioactivity. Chapter 43, pp. 682–713, Nuclear Physics.

Gamow and Cleveland, *Physics, Foundations and Frontiers*, Chapter 26, pp. 447–463, Radioactivity and the Nucleus. Section 26-1, Discovery of Radioactivity.

24.2 Nature of radioactivity

Laboratory Exercises: *Project Physics Handbook, The Nucleus*, Expt. 1, Random Events, p. 110. Expt. 2, Range of α and β Particles, p. 114.

Exercises and Experiments in Physics, Expt. 50, Radioactivity.

Sutton, Radioactivity, A-111 to A-121, pp. 501–508. A-111, Sources of Radioactivity. A-112, Electroscope Experiments. A-113, Range of Alpha Particles. A-114, Scattering of Alpha Particles. A-115, Radioactive Decay. A-116, Wilson Cloud Chamber. A-118, Geiger-Muller Tube Counter.

24.3 Types of natural radioactivity

Project Physics Overhead Transparency, T40, Separation of α, β, and γ Rays.

Meiners, Section 21-1, pp. 1240–1249, Radioactivity.

Karplus, *Introductory Physics*, pp. 220–221, Radioactivity.

Gamow and Cleveland, *Physics, Foundations and Frontiers*, Section 26-4, p. 453, Alpha, Beta and Gamma Rays.

24.4 Nuclear symbols and equations

Rogers, *Physics for the Inquiring Mind*, p. 640, A "Family Tree" of Naturally Radioactive Elements.

Laboratory Exercise: Lehrman, *Laboratory Experiments*,

Expt. 72, p. 153, Intensity of Gamma Radiation from a Point Source.

Overhead Transparency K & E #82, Atomic Symbols.

Exercises and Experiments in Physics, Expt. 52, Simulated Nuclear Collisions.

24.5 Radioactive decay

Laboratory Exercise: Lehrman, *Laboratory Experiments*, Expt. 73, p. 154, Half-Life of a Radioactive Element.

Gamow and Cleveland, *Physics, Foundations and Frontiers*, Section 26-5, pp. 454–456, Families of Radioactive Elements. Section 26-6, p. 456, Decay Energies.

Overhead Transparency K & E #83, The Uranium Series of Radioactive Decay.

Project Physics Transparency, T42, Radioactive Disintegration Series.

Lehrman and Swartz, *Foundations of Physics*, pp. 685–689, Alpha Decay, Beta Decay.

Beiser, *Concepts of Modern Physics*, Chapter 13, pp. 318–343, Nuclear Decay.

General Reference for this Chapter: Weidner and Sells, *Elementary Modern Physics*, Allyn and Bacon, 1960, Chapter 9, pp. 313–363, Nuclear Structure and Chapter 10, pp. 369–418, Nuclear Reactions.

Semat, *Introduction to Atomic and Nuclear Physics*, Holt, Rinehart and Winston, Publishers, 1963, Part 3, Nuclear Physics consisting of Chapter 11, Natural Radioactivity, Chapter 12, Disintegration of Nuclei, Chapter 13, Fission and Fusion of Nuclei, Chapter 14, Nuclear Processes, Chapter 15, Fundamental Particles, and Chapter 16, Particle Accelerators.

Overhead Transparency K & E #84, Half-life and the Decay Curve.

Project Physics, T43, Radioactive Decay Curve. T44, Radioactive Displacement Rules.

Laboratory Exercise: *Project Physics Supplemental Unit, The Nucleus*, Expt. 3, Half-life I, p. 117. Expt. 4, Half-life II, p. 121.

Exercises and Experiments in Physics, Expt. 51, Half-Life.

Gamow and Cleveland, *Physics, Foundations and Frontiers*, Section 26-7, pp. 457–458, Half-Lifetimes. Section 26-8, pp. 458–459, Uranium-Lead Dating. Section 26-9, pp. 459–461, Carbon Dating. Section 26-10, pp. 461–463, Tritium Dating.

24.7 Fission

Lindsay, *Basic Concepts of Physics*, Section 9.12, pp. 383–386, Nuclear Energy.

Karplus, *Introductory Physics*, pp. 225–227, Nuclear Fission.

Gamow and Cleveland, *Physics, Foundations and Frontiers*, Chapter 29, pp. 493–507, Large-Scale Nuclear Reactions. Section 29-1, pp. 493–496, Discovery of Fission. Section 29-2, p. 496, Fission Neutrons. Section 29-3, p. 497, Fissionable Uranium-235.

Lehrman and Swartz, *Foundations of Physics*, pp. 689–692, Nuclear Fission, Controlled Nuclear Fission.

24.8 Fusion

Gamow and Cleveland, *Physics, Foundations and Frontiers*, Section 29-8, pp. 504–507, Fusion Reactors.

Lehrman and Swartz, *Foundations of Physics*, pp. 692–695, Nuclear Fusion.

24.10 Chain reactions

Overhead Transparency K & E #87, Chain Reaction.

Beiser, *Concepts of Modern Physics*, Section 14.6, pp. 358–361, The Chain Reaction.

24.11 Nuclear reactors

Overhead Transparency K & E #88, A Nuclear Reactor.

24.12 Nuclear power

Karplus, *Introductory Physics*, p. 227, Nuclear Reactors.

Gamow and Cleveland, *Physics, Foundations and Frontiers*, Section 29-4, pp. 497–500, The Fermi Pile and Plutonium. Section 29-5, pp. 500–501, Critical Size. Section 29-6, pp. 501–503, Nuclear Reactors.

24.13 Radioisotopes

Western, 2 × 2 Color Transparencies: R-16 Atomic Reactor (1). R-17 Ditto (2). R-18 Illustration of an Atomic Reactor. R-19 Core of Reactor. R-20 Critical Reactor.

Project Physics Supplemental Unit, The Nucleus, Chapter 2, Isotopes.

Laboratory Exercise: *Project Physics Supplemental Unit, The Nucleus*, Expt. 5, p. 123, Radioactive Tracers.

HIGH-ENERGY PHYSICS   25

Synopsis: This chapter is concerned with the methods for detecting and making appropriate measurements of nuclear particles and with a study of the procedures for accelerating subatomic particles to an energy level at which they will trigger a nuclear reaction in a target nucleus. The characteristics of the more important of the fundamental particles that have been identified as a result of the study of nuclear structure are introduced along with the nature of particle interactions and emphasis on the conservation principle.

The chapter closes with a consideration of the various atomic models, a brief but basic introduction to quantum mechanics, and a study of the mechanisms that bind atoms together to form the substances in our physical environment.

Comment: Consider the evolution of atomic models pointing out the reasons the earlier models were replaced by later ones. Indicate the effect that the announcement of the uncertainty principle had on the Bohr model. Discuss the model of the atom based on the principles of the quantum theory as presented in Section 25.2. Consider ionization energy. Discuss matter waves, Section 25.3.

Discuss the principles of the various types of particle accelerators that have been used or are now in use, as, for example, the Van de Graaff generator, betatron, cyclotron, synchrotron, and the linear accelerators. Demonstrate and explain the principle of a Van de Graaff type generator (Sargent-Welch 1910). Students may be asked to prepare reports on the various types of accelerators now in use or being planned. See Section 40-6, pages 697–699, *Fundamentals of Physics*, Semat.

Illustrate and discuss the principles of particle detection. Have students collect information about particle accelerators to be illustrated by pictures and diagrams. Selected parts from Part Five, Atomic and Nuclear Physics, pages 607–759, *Physics for the Inquiring Mind*, Rogers, Princeton University Press, 1960, will serve as an excellent reference source for both teacher and students. Have students develop some of the models of the atom that have been proposed. Then construct some molecular models such as those of water and sodium chloride.

Discuss the main characteristics of particles and rays emitted in nuclear reactions. Show how instruments to detect these rays and particles depend on these characteristics. Review the principle and purpose of the electroscope and show how it can be used to detect and measure the presence of ions in the surrounding air. See Sutton, A-112, page 502. Discuss the principle of the ionization chamber. Demonstrate a Geiger tube and counting circuit (Sargent-Welch 2192), and explain the principle of the Geiger tube. Build and operate a simple cloud chamber like the one shown in Figure 25-17, or use the Sargent-Welch 2158B or 2195 diffusion cloud chamber or equivalent. Emphasize the capabilities of the new electronic bubble chamber. Allow students to observe the scintillations caused by the impact of an alpha particle on a zinc sulfide screen using the spinthariscope (Sargent-Welch 5542). Discuss the use of photographic emulsions for the detection of subatomic particles. You may wish to use the Sargent-Welch 2164 Radioactive Film Loops. Refer to *A Sourcebook for the Physical Sciences*, Joseph *et al.*, Harcourt, Brace and World, Inc., 1961, pages 558–569 for suggestions for building and demonstrating devices for detecting radioactive substances.

Study the classification of the subatomic particles Section 25.14 and the accompanying chart. Study the forces involved in particle interactions, *e.g.*, strong nuclear interactions, electromagnetic interactions, and weak interactions. Study the nuclear conservation laws illustrating them by writing the interaction formulas. Introduce the idea of antiparticles, the principle of symmetry, and the parity law. Refer to appropriate sections of Chapter 40, Nuclear Disintegration, from Semat, *Fundamentals of Physics*, pages 690–711.

References: Alfvén, Hannes, "Antimatter and Cosmology," *Scientific American*, Apr. 1967. Asimov, Isaac, *The Neutrino: Ghost Particle of the Atom*, Doubleday and Co., Garden City, N.Y., 1966.

Beeler, N. F., and F. M. Branley, *Experiments with Atomics*, Crowell, New York, 1954, Chapters 6, 7, 15, 17.

Burbidge, G., and F. Hoyle, "Anti-matter," *Scientific American*, Apr. 1958, p. 34.

Casella, Russell C., "Time Reversal Symmetry and Its Breaking in Physics," *The Physics Teacher*, Mar. 1970. pp. 114–122.

Darrow, Karl K., "The Quantum Theory," *Scientific American*, Mar. 1952.

Fischer, Arthur, "Fundamental Particles—Children of the Atom," *Science World*, Apr. 17, 1963, pp. 12–15.

Frisch, David H., and Alan M. Thorndike, *Elementary Particles*, Momentum Book, D. Van Nostrand Co., Inc., Princeton.

Galley, E. L., "An Inexpensive Continuous Cloud Chamber," *The Physics Teacher*, May 1963, p. 80.

Gamow, George, "The Exclusion Principle," *Scientific American*, July 1959.

——— , "The Principle of Uncertainty, *Scientific American*, Jan. 1958.

Gell-Mann, Murray, and E. P. Rosenbaum, "Elementary Particles," *Scientific American*, July 1957.

Ginzton, E. L., and W. Kirk, "The Two-mile Accelerator," *Scientific American*, Nov. 1961 p. 49.

Glasstone, Samuel, *Sourcebook of Atomic Energy, Third Edition*, D. Van Nostrand Co., Inc., Princeton, 1967.

Heckman, Harry H., and Paul W. Starring, *Nuclear Physics and the Fundamental Particles*, Holt, Rinehart and Winston, Publishers, New York, 1963, Chapters 4–13.

Hill, R. D., "Elementary Particles," *The Physics Teacher*, Jan. 1964, p. 22.

Houston, Walter Scott, *Frontiers of Nuclear Physics*, A Science Unit Book, American Education Publications, Education Center, Columbus 16, OH.

Hughes, Donald, *"The Neutron Story,"* PSSC Science Study Series, Doubleday and Co., Garden City, N.Y.

Jaffee, Bernard, *Men of Science in America*, Simon and Schuster, New York, 1958, Chapter 19.

Marshak, Robert E., "Pions," *Scientific American*, Jan. 1957.

Morrison, Philip, "The Neutrino," *Scientific American*, Jan. 1956.

——— , "The Overthrow of Parity, *Scientific American*, Apr. 1957.

O'Neill, Gerard K., "Particle Storage Rings," *Scientific American*, Nov. 1966.

Panofsky, Wolfgang, "The Linear Accelerator," *Scientific American*, Oct. 1954.

Penman, Sheldon, "The Muon," *Scientific American*, July 1961.

Rosen, Louis, "Particle Accelerators: Instruments of Basic Research and Human Welfare," *The Physics Teacher*, Nov. 1970, pp. 432–440.

Rossi, Bruno, "Where Do Cosmic Rays Come From?," *Scientific American*, Sept. 1953.

Seaborg, Glenn T., and Justin L. Bloom, "Fast Breeder Reactors," *Scientific American*, Nov. 1970.

Sergre, Emilio, and Clyde C. Wiegand, "The Antiproton," *Scientific American*, June 1956.

Semat, Henry, *Fundamentals of Physics, Fourth Edition*, Holt, Rinehart and Winston, Publishers, 1966, Part Six: Atomics and Nucleonics, pp. 611–711.

Spitzer, Lyman, "The Stellarator," *Scientific American*, Oct. 1958, p. 28.

Trower, W. Peter, "The Accelerator in Particle Physics," *The Science Teacher*, Nov. 1969, pp. 25–30.

Wahla, James, "A Simple Inexpensive Geiger Counter," *The Science Teacher*, May 1959, p. 255.

Wiesner, Jerome B., and Herbert F. York, "National Security and the Nuclear Test Ban," *Scientific American*, Oct. 1964.

Wigner, Eugene P., "Violations of Symmetry in Physics," *Scientific American*, Dec. 1965.

Wilson, Robert R., and Raphael Littauer, *Accelerators: The Machines of Nuclear Physics*, PSSC Science Study Series, Doubleday and Co., Garden City, N.Y.

Wilson, Robert R., "Particle Accelerators," *Scientific American*, Mar. 1958.

Yagoda, Herman, "The Tracks of Nuclear Particles," *Scientific American*, May 1956.

Films: Films may be selected from the PSSC group distributed by Modern Learning Aids.

0416, *The Rutherford Atom:* qualitative experiments and appropriate models are used to indicate the ideas leading to Rutherford's nuclear concept of the atom.

0423, *Matter Waves:* the wave behavior of matter is illustrated by experiments which show that electrons display interference patterns.

0421, *Franck-Hertz Experiment:* shows the existence of discrete energy states in atoms.

Particle Accelerators, fs, color, 41 frames, FOM, McGraw-Hill Book Co., Text Film Dept. 1221 Avenue of the Americas, New York, NY 10020. Describes principles of operation of various types of accelerators.

The Strange Case of Cosmic Rays, 59 min., sound, color. Local Bell Telephone Business Office. Animated; behavior and characteristics of cosmic rays. Free loan.

References for Chapter 25: High-Energy Physics

General Reference: *Project Physics Supplemental Unit, The Nucleus*, Chapter 3, Probing the Nucleus.

Frisch, O. R., *Atomic Physics Today*, Basic Books Inc., 1961, New York.

Rogers, *Physics for the Inquiring Mind*, Chapter 43, pp. 682–713, Nuclear Physics.

25.2 Quantum numbers

Lindsay, *Basic Concepts of Physics*, Sections: 9.3, p. 348, Planck and the Invention of the Quantum Theory. 9.9, pp. 367–378, Quantum Mechanics.

Weidner and Sells, *Elementary Modern Physics*, Allyn and Bacon, Boston, 1960, p. 149, Quantum Mechanics.

Project Physics Transparency: T39, Energy Levels-Bohr Theory. General Reference: *PSSC Physics, Third Edition*, Chapter 27, pp. 610–626, Matter Waves.

25.5 Van de Graaff generators

Semat, *Introduction to Atomic and Nuclear Physics, Fourth Edition*, Holt, Rinehart and Winston, Publishers, 1963, Chapter 16, pp. 550–572, Particle Accelerators.

Lehrman and Swartz, *Foundations of Physics*, pp. 698–700, Accelerators.

Rogers, *Physics for the Inquiring Mind*, Chapter 42, pp. 672–680, Atom-Accelerators—The Big Machines.

Gamow and Cleveland, *Physics, Foundations and Frontiers*, Section 27-4, pp. 470–472, First Particle Accelerators. Section 27-5, pp. 472–474, The "Van de Graaff."

25.6 Circular accelerators

Overhead Transparency K & E #86, Cyclotron.

Gamow and Cleveland, *Physics, Foundations and Frontiers*, Section 27-6, pp. 474–475, The Cyclotron. Section 27-7, pp. 475–476, Other Particle Accelerators.

25.9 Low-energy detectors

Sutton, A-1, p. 431, Ionization in Air.

Meiners, Section 41-3, pp. 1266–1274, Particle Detectors.

Lehrman and Swartz, *Foundations of Physics*, pp. 673–676, Detection Devices for High Energy Particles.

Laboratory Exercise: Lehrman, *Laboratory Experiments*, Expt. 71, p. 150, The Geiger-Müller Counter.

Western, 2 × 2 Color Transparency: R-14, Wilson's Cloud Chamber. R-15, Diffusion Cloud Chamber.

Gamow and Cleveland, *Physics, Foundations and Frontiers*, Section 27-2, pp. 467–469, Photographing Nuclear Transformations, and Section 27-3, p. 470, Bubble Chambers.

Laboratory Exercise: Lehrman, *Laboratory Experiments*, Expt. 74, p. 155, Analysis of Particle Tracks.

25.14 Classification of subatomic particles

Lehrman and Swartz, *Foundations of Physics*, pp. 695–697, The Search for Fundamental Particles. Table 22-4, pp. 704–705, The Long-Lived Elementary Particles. pp. 708–710, The Elementary Particles. pp. 711–712, Structure of the Particles.

Gamow and Cleveland, *Physics, Foundations and Frontiers*, Chapter 30, pp. 509–521, Mystery Particles.

Beiser, Arthur, *Concepts of Modern Physics*, Revised Edition, McGraw-Hill, 1967, Chapter 14, pp. 345–364, Nuclear Reactions. Chapter 15, pp. 366–391, Elementary Particles.

Semat, *Introduction to Atomic and Nuclear Physics, Fourth Edition*, Holt, Rinehart and Winston, 1963, Chapter 15, pp. 489–546, Fundamental Particles.

Project Physics Transparency: T46, Chart of the Nuclides.

Weidner and Sells, *Elementary Modern Physics*, Allyn and Bacon, Boston, 1960, Section 10-10, pp. 414–417, The Elementary Particles.

25.15 Particle interactions

Exercises and Experiments in Physics, Expt. 52, Simulated Nuclear Interactions.

Lehrman and Swartz, *Foundations of Physics*, pp. 700–703, The Four Types of Physical Interaction.

25.16 Subatomic conservation laws

Lehrman and Swartz, *Foundations of Physics*, pp. 703–708, The Conservation Laws.

Beiser, *Concepts of Modern Physics*, Revised Edition, McGraw-Hill, 1967, Section 15-10, pp. 385–389, Symmetries and Conservation Principles. pp. 387–389, Parity. pp. 386–371, Antiparticle.

Lindsay, *Basic Concepts of Physics*, Sections: 9.7, p. 356, Atomic Structure, 9.8, p. 360, The Bohr Theory, and 9.10, p. 378, Review of Atomic Structure, Some Further Questions.

Project Physics 8mm Film: *Thomson Model of the Atom.* Color, 3 min. 50 sec.

ANSWERS
TO QUESTIONS
AND PROBLEMS

PHYSICS: THE SCIENCE OF ENERGY 1

Questions • Chapter 1/p. 6
GROUP A
1. Fossil fuels are limited. Students may have other answers, such as the rising cost of fossil fuels, the risk of spillage in transporting them in tankers, etc.
2. Lists will vary.
3. Students should ask store managers to explain UPC marks.
4. A theory is a reasonable explanation of a series of observed phenomena.
5. When a hypothesis is confirmed, it is incorporated into a theory or a law.
6. **(a)** A laboratory is a place where a scientist pursues an investigation. **(b)** The high school physics laboratory contains the facilities necessary to investigate the basic laws and theories of physics.
7. **(a)** Science is the study of the knowledge that explains and predicts the behavior of the observable universe. **(b)** Technology is the application of this knowledge to our needs and goals.
8. The school librarian can tell the students which sources are available in the school library.

GROUP B
9. Much more time and effort are required to test hypotheses than to formulate them.
10. Kepler formulated hypotheses that explained observed planetary motion.
11. There is no absolute certainty in science. The certainty of a scientific conclusion is always limited by the method of observation and by the person who made it.
12. Religion, morals, and human relations might be among the answers.

Questions • Chapter 1/p. 16
GROUP A
1. The mass is the same in both places.
2. **(a)** Mass is a measure of the quantity of matter that an object contains. Mass density is mass per unit volume. **(b)** Inertial mass is the property of matter that resists a change in its motion. Gravitational mass is the property of matter that causes a force of attraction between objects.
3. **(a)** Properties describe matter, while conditions describe its environment. **(b)** Mass, inertia, and mass density are properties. Temperature and pressure are conditions.
4. **(a)** The sun. **(b)** The answer should include the fact that gasoline, a fossil fuel formed with the help of solar energy, is mixed with air and exploded by an electric spark to furnish heat energy and mechanical energy. Also, as the car moves, some of its mechanical energy is transformed to heat energy through friction between the tires and the road. (There are many other energy transformations in a car, and the class should take some time discussing them.)
5. **(a)** potential **(b)** potential **(c)** kinetic **(d)** kinetic **(e)** potential
6. The amount of energy expended in the boring process.
7. The total amount of energy of all kinds in a given situation is constant.
8. Mass and energy are proportional to each other. Also, a given amount of mass is equivalent to a certain amount of energy.
9. Particles of matter exhibit wave characteristics.

10. Mechanics and heat: motion, forces, work, power, certain aspects of heat. Waves: sound and light. Electricity, including electronics. Nuclear and particle physics: composition of matter.

GROUP B

11. When there is no mass on the pan of the balance, it still has a vibration period of about 0.18 second. If the curve is extended backward until it crosses the horizontal axis (so that the vibration time is theoretically zero), the mass of the balance pan can be read on the leftward extension of the horizontal axis.
12. The shape of the resulting graph is a straight line.
13. No. Since the balance and the masses that are placed on it are both in free flight, the masses would not push down on the balance.
14. Energy is the capacity for doing work.
15. Answers will vary.
16. According to the theory of relativity, the mass of an object increases with speed. The effect is particularly noticeable as the speed approaches the speed of light. The greater the mass of an object, the more difficult it is to increase its speed.
17. (a) Expansion is a result of mechanical energy. Since heat produced the expansion, heat must be a form of energy also (in keeping with the law of conservation of energy).
   (b) Since some substances contract when heated, there must be different ways in which substances react to heat. In one of these ways, the space between the particles increases. This effect predominates in substances that expand when heated. In some substances, however, heat reduces the spaces between particles. This effect predominates in substances that contract when heated.
18. Count Rumford's observations showed that heat is proportional to mechanical work. Joule discovered the quantitative relationship between the two.

2   MEASUREMENT AND PROBLEM SOLVING

Questions • Chapter 2/p. 24
GROUP A

1. The metric system uses a decimal basis for multiples and subdivisions of the basic units of measure.
2. A physical quantity is a measurable aspect of the universe. A unit of measure is used to obtain the magnitude of a physical quantity.
3. Meter, kilogram, second, degree Kelvin, mole, candle, and coulomb.
4. Answers will vary.
5. SI stands for *Système International d'Unités,* the official name of the modern metric system.
6. The standard of length is the meter, which is defined as 1,650,763.73 wavelengths in a vacuum of the orange-red light of krypton-86.
7. A liter has a volume of 1000 cubic centimeters.
8. Masses can be compared with the standard kilogram with greater precision than is possible by other laboratory methods.
9. Mass is a measure of the quantity of matter that an object contains. Weight is the measure of the gravitational attraction between the earth and an object.

10. Cesium atoms are set in vibration by radio waves. The standard second is defined as 9,192,631,770 vibrations of cesium-133 atoms.

GROUP B
11. Answers will vary.
12. Answers will vary.
13. A cubic decimeter of water has a mass of approximately one kilogram.
14. The determination of the standard for each involves the use of the other.
15. Answers will vary.

Problems • Chapter 2/pp. 24–25
GROUP A
1. (a) 320 (b) 0.756 (c) 360,000 (d) 10,000 (e) 0.001 (f) 15,000 (g) 250 (h) 4.5 (i) 8
   (j) 67,000,000 (k) 24,000 (l) 30,000 (m) 10,000,000,000 (n) 7200 (o) 60,000
2. Area = (0.65 m)(0.92 m) = 0.60 m^2 (The reason for rounding off the answer will be explained in Section 2.8.)
3. Volume = (26 m)(14 cm)(5 cm) = 1820 cm^3 = 1820 mL
4. (a) Force = (100 g)(1 kg/1000 g)(9.8 n/kg) = 0.98 n
   (b) Force = (65 kg)(9.8 n/kg) = 640 n (Rounding off is explained in Section 2.8.)
   (c) Force = (2 tons)(1000 kg/ton)(9.8 n/kg) = 20,000 n (The reason for rounding off the answer is explained in Section 2.8.)

GROUP B
5. (a) Volume = total mass of water/mass of water per liter
   Volume = 1.00 kg/1.00 kg/L = 1.00 L
   (b) Volume in m^3 = volume in L × conversion factor
   Volume in m^3 = 1.00 L × 1.00 × 10^{-3} m^3/L = 1.00 × 10^{-3} m^3
   (c) Length of one side = cube root of volume = 0.100 m
6. Mass = mass of sea water per cm^3/volume
          = (1.04 g/cm^3)(1000 cm^3)
          = 1040 g = 1.04 kg
7. Mass = mass per unit length × length
   Mass = (4.52 g/cm)(1,500,000 cm) = 6,800,000 g (The reason for rounding off the answer will be explained in Section 2.8.)
8. Weight = (500 mL)(1 cm^3/mL)(1 g/cm^3) = 500 g

Questions • Chapter 2/p. 37
1. Accuracy is the closeness of a measurement to the accepted value of a physical quantity. Precision is the agreement among several measurements of a quantity that have been made in the same way.
2. Absolute error is the actual difference between a measured value and the accepted value. Relative error is the percentage error of a measurement.
3. Absolute deviation, relative deviation, tolerance, and significant figures.
4. The rightmost significant figure in a sum is determined by the leftmost place at which an uncertain figure occurs in the measurements being added.
5. Zero to the right of a nonzero figure but to the left of an understood decimal point, and where the zero is not indicated to be significant by a bar over it; a zero to the right of a decimal point but to the left of a nonzero figure; a single zero placed to the left of the decimal point to call attention to the decimal point.

6. One.
7. Scientific notation makes it easier to read very large and very small numbers, and it calls attention to the number of significant figures in a number.
8. **(a)** A straight line. **(b)** A hyperbola.
9. A basic equation relates the known and unknown quantities of a problem. A working equation expresses the unknown quantity of a problem in terms of the known quantities.
10. Dimensional analysis is the performance of the indicated mathematical operations in a problem with the units alone.
11. A check for reasonableness means to see whether the answer you obtained to a problem makes sense or not.
12. A scalar quantity can be expressed completely by a single number with appropriate unit. A vector quantity requires magnitude and direction for a complete description.
13. By the parallelogram method.
14. One way is to find the resultant of any two of the vectors and then use this resultant with a third vector to find a second resultant, and to continue in this way until all the component vectors have been used. The other way is to place the vectors head to tail. The resultant is the arrow that is made by starting at the tail of the first vector and drawing a straight line to the head of the last vector. (The sequence in which the vectors are placed in this method does not matter.)

Problems • Chapter 2/pp. 37–38
GROUP A
1. **(a)** 4 **(b)** 3 **(c)** 3 **(d)** 2 **(e)** 1 **(f)** 4
2. **(a)** 1.732×10^2 **(b)** 2.05×10^2 **(c)** 4.00×10^3 **(d)** 2.5×10^{-2} **(e)** 7×10^2
   (f) 9.050×10^{-2}
3. **(a)** 1.75×10^8 **(b)** 6.023×10^{23} **(c)** 4.7×10^{-5} **(d)** 1.67239×10^{-24}
4. **(a)** 10^8 **(b)** 10^{24} **(c)** 10^{-5} **(d)** 10^{-24}
5. Volume = mass/density
   Volume = $100 \text{ g}/0.20 \text{ g/cm}^3 = 500 \text{ cm}^3$
6. 67 cm^3
7. 65 km/hr
8. **(a)** 10^7 **(b)** 10^{-10} **(c)** 10^{-2} **(d)** 10^{-5}
9. **(a)** Student graph **(b)** 25 units, 12° south of west
10. **(a)** Student graph **(b)** Eastward 9.0 units, southward 21 units
11. **(a)** Absolute error = |observed value − accepted value|
       Absolute errors = 0.08 g/cm^3, 0.07 g/cm^3, 0.01 g/cm^3, 0.04 g/cm^3
    (b) Relative error = (absolute error/accepted value) × 100%
       Relative errors = 2%, 1%, 0.2%, 0.8%
    (c) Absolute deviation = |observed value − average value| Average value = 5.04 g/cm^3
       Absolute deviations = 0.09 g/cm^3, 0.06 g/cm^3, 0.02 g/cm^3, 0.03 g/cm^3
    (d) Relative deviation = (average of absolute deviations/average value) × 100%
       Relative deviation = $(0.05 \text{ g/cm}^3/5.04 \text{ g/cm}^3) \times 100\% = 1\%$
12. **(a)** A hyperbola **(b)** They are inversely proportional. **(c)** 112 s^{-1} **(d)** 0.19 m
13. Mass density = $\dfrac{\text{mass}}{\text{volume}} = \dfrac{(9.57 \text{ kg})(1000 \text{ g/kg})}{(6.24 \text{ L})(1000 \text{ cm}^3/\text{L})} = 1.53 \text{ g/cm}^3$

14. Mass density $= \dfrac{\text{mass}}{\text{volume}}$

$\text{Volume} = \dfrac{\text{mass}}{\text{mass density}} = \dfrac{(9.18 \text{ n})(1.00 \text{ mL/cm}^3)(1000 \text{ g/kg})}{(9.80 \text{ n/kg})(3.47 \text{ g/cm}^3)} = 27\bar{0} \text{ mL}$

15. Mass density $= \dfrac{\text{mass}}{\text{volume}} = \dfrac{(11.2 \text{ n})(1000 \text{ g/kg})}{(9.80 \text{ n/kg})(5.25 \text{ cm})^3} = 7.90 \text{ g/cm}^3$

16. Mass density $= \dfrac{\text{mass}}{\text{volume}} = \dfrac{795 \text{ g}}{(21.0 \text{ cm})(9.50 \text{ cm})(5.00 \text{ cm})} = 0.797 \text{ g/cm}^3$

17. $T = 2\pi \sqrt{\dfrac{l}{g}} \qquad T^2 = \dfrac{4\pi^2 l}{g} \qquad g = \dfrac{4\pi^2 l}{T^2}$

$g = \dfrac{4(3.14)^2(1.00 \text{ m})}{(2.01 \text{ s})^2} = 9.76 \text{ m/s}^2$

VELOCITY AND ACCELERATION **3**

Questions • Chapter 3/p. 47
GROUP A
1. (a) Displacement is a change of position in a particular direction. **(b)** Motion is the displacement of an object in relation to objects that are considered to be at rest.
2. (a) Speed is the time rate of motion. **(b)** Average speed is found by dividing the total path length by the elapsed time. Instantaneous speed is the speed at a specific instant. **(c)** Meters per second.
3. (a) Velocity is speed in a particular direction. **(b)** Speed is a scalar quantity whereas velocity is a vector quantity.
4. (a) The velocity of an object is constant when the determination over any randomly chosen interval gives the same value. **(b)** The velocity of an object is variable when the slope of the velocity graph changes.

GROUP B
5. Speed is completely described in terms of magnitude alone. The description of velocity includes direction as well as magnitude.
6. The resultant velocity is equal in magnitude and direction to the diagonal of the parallelogram of which the original velocities form the sides.
7. (a) Yes. **(b)** No. A change in speed is a change in the time rate of motion, which changes velocity; however, a change in direction is a change of velocity but not a change of speed.
8. The changing slope in a velocity graph signifies a change in instantaneous velocity.

Problems • Chapter 3/pp. 47–48
GROUP A
1. $v_{\text{av}} = \Delta d / \Delta t$
$v_{\text{av}} = 1712 \text{ km}/29.25 \text{ hr}$
$v_{\text{av}} = 58.53 \text{ km/hr}$
2. $25 \text{ km/hr} - 4 \text{ km/hr} = 21 \text{ km/hr}$ upstream

3. 825 km/hr − 35 km/hr = 79̄0 km/hr northward
4. $\tan\theta$ = 3.5 km/hr ÷ 5.0 km/hr = 0.700

 θ = 35°, the angle at which the boat heads downstream

 v_R = 3.5 km/hr ÷ $\sin\theta$ = 3.5 km/hr ÷ 0.574 = 6.1 km/hr, the resultant velocity of the boat

5. $\tan\theta$ = 40.0 km/hr ÷ 650 km/hr = 0.0615

 θ = 3.5° north of west, the direction of the resultant velocity

 v_R = 40.0 km/hr ÷ $\sin\theta$

 v_R = 40.0 km/hr ÷ 0.0610 = 656 km/hr, the resultant velocity

6. $\tan\theta$ = 50.0 km/hr ÷ 475 km/hr = 0.105

 θ = 6.0°

 45.0° − 6.0° = 39.0° east of north, direction of required velocity

 v_V = 475 km/hr ÷ $\cos\theta$

 v_V = 475 km/hr ÷ 0.995 = 477 km/hr, magnitude of required velocity

GROUP B

7. $c = \sqrt{a^2 + b^2 - 2ab\cos C}$

 $\cos C$ = $\cos 50.0°$ = 0.643

 $c = \sqrt{(75.0 \text{ m/s})^2 + (100.0 \text{ m/s})^2 - 2(75.0 \text{ m/s} \times 100.0 \text{ m/s} \times 0.643)}$

 c = 77.3 m/s, magnitude of resultant velocity

 b/sin B = c/sin C

 sin B = b sin C/c

 sin C = sin 50.0° = 0.766

 sin B = (100.0 m/s × 0.766)/77.3 m/s

 sin B = 0.991

 B = 82.3°. 82.3° + 25.0° = 107.3° − 90.0° = 17.3° south of east, direction of resultant velocity

8. 815 m/10.0 min = 4.89 km/hr

 $C = \sqrt{a^2 + b^2 - 2ab\cos C}$

 $\cos C$ = $\cos 110.0°$ = $-\cos 70.0°$ = −0.342

 $C = \sqrt{(4.89 \text{ km/hr})^2 + (3.0 \text{ km/hr})^2 - 2(4.89 \text{ km/hr} \times 3.0 \text{ km/hr} \times -0.342)}$

 C = 6.6 km/hr, required speed

 b/sin B = c/sin C

 sin B = b sin C/c

 sin C = sin 110.0° = sin 70.0° = 0.940

 sin B = (3.0 km/hr × 0.940)/6.6 km/hr

 sin B = 0.43   B = 25°

 25° + 20° = 45°, required heading upstream

9. $c = \sqrt{a^2 + b^2 - 2ab\cos C}$

 $\cos C$ = $\cos 130.0°$ = $-\cos 50.0°$ = −0.643

 $c = \sqrt{(25.0 \text{ km/hr})^2 + (425 \text{ km/hr})^2 - 2(25.0 \text{ km/hr} \times 425 \text{ km/hr} \times -0.643)}$

 c = 441 km/hr, required speed relative to the speed of the air

 b/sin B = c/sin C   sin B = b sin C/c

 sin C = sin 130.0° = sin 50.0° = 0.766

 sin B = (425 km/hr × 0.766)/434 km/hr

 sin B = 0.749 B = 48.5°

 180.0° − 48.5° = 131.5°

 131.5° + 40.0° = 171.5° = 3.5° east of south, required direction

Questions • Chapter 3/p. 55
GROUP A
1. **(a)** Acceleration is the rate of change of velocity with respect to time. **(b)** The units of acceleration are the units of velocity, m/s, divided by the unit of time, s, or m/s^2.
2. They act along the same line, and they have the same direction.
3. **(a)** Uniform acceleration is motion for which the velocity curve in a velocity-time graph is a straight line, and its slope is positive. **(b)** For variable acceleration, the velocity curve is not a straight line, but its slope is always positive. **(c)** The velocity curve is a straight line with a negative slope. **(d)** The velocity curve is not a straight line, but its slope is always negative.
4. The ball will be uniformly decelerated as it rises, momentarily come to rest, and then accelerate uniformly as it returns to the ground. The acceleration of gravity will always be in the direction of the weight force, toward the ground.
5. Instantaneous acceleration is the value of the slope of the tangent to the velocity-time curve at a given point.

GROUP B
6. $v_f = v_i + a \, \Delta t$
7. $v_{av} = (v_i + v_f)/2 \qquad \Delta d = v_{av} \, \Delta t \qquad \Delta d = (v_i + v_f/2) \, \Delta t$
8. A fast train on a straight track.
9. A racing car starting from rest.

Problems • Chapter 3/pp. 55–56
GROUP A
1. $a = v/\Delta t \qquad a = 29.31 \text{ m/s}/3.00 \text{ s} = 9.77 \text{ m/s}^2$
2. $a = v/\Delta t \qquad a = (142 - 95) \text{ km/hr}/8.1 \text{ s} = 5.8 \text{ km/hr/s, or } 1.6 \text{ m/s}^2$
3. **(a)** $v_i = 0$
   $\Delta t = 1.7 \text{ s}$
   Find v_f.
   $v_f = g \, \Delta t, \; v_f = 9.80 \text{ m/s}^2 \times 1.7 \text{ s} = 17 \text{ m/s}$
   (b) $v_i = 0, \; v_f = 17 \text{ m/s} \; \Delta t = 1.7 \text{ s}$

   $$\Delta d = v_{av} \, \Delta t, \; v_{av} = \frac{v_i + v_f}{2}$$

   $$\Delta d = \frac{17 \text{ m/s}}{2} \times 1.7 \text{ s} = 14 \text{ m}$$

4. **(a)** $v_i = 0 \qquad \Delta d = 145 \text{ m, acceleration} = g \text{ m/s}^2$
   $\Delta d = v_i \, \Delta t + \frac{1}{2}g(\Delta t)^2$

   $$\Delta t^2 = \frac{2 \, \Delta d}{g} = \frac{2 \times 145 \text{ m}}{9.80 \text{ m/s}^2} = 30 \text{ s}^2$$

   $t = \sqrt{30} \text{ s} = 5.5 \text{ s}$
   (b) $v_f = g \, \Delta t$
   $v_f = 9.8 \text{ m/s} \times 5.5 \text{ s}$
   $v_f = 54 \text{ m/s}$

5. **(a)** $v_i = 20.0 \text{ cm/s}, \; a = 8.0 \text{ cm/s}^2, \; t = 5.0 \text{ s}$
   $\Delta d = v_i \, \Delta t + \frac{1}{2}a(\Delta t)^2$
   $\Delta d = (20.0 \text{ cm/s} \times 5.0 \text{ s}) + \frac{1}{2} \times 8.0 \text{ cm/s}^2 \times (5.0 \text{ s})^2$
   $\Delta d = 100 \text{ cm} + 100 \text{ cm} = 200 \text{ cm}$

(b) At the end of any given second, v_f is

$v_f = v_i + a\,\Delta t$

v_f at end of 4th s $= 20.0$ cm/s $+ 8.0$ cm/s$^2 \times 4.0$ s

$v_{f_4} = 52$ cm/s

$v_{f_5} = 60$ cm/s

v_{av} for the fifth second $= \dfrac{52 \text{ cm/s} + 60 \text{ cm/s}}{2} = 56$ cm/s

$\Delta d_5 = v_{av_5} t = 56$ cm/s $\times 1.0$ s $= 56$ cm

6. $v_f = \sqrt{v_i^2 + 2a\,\Delta d}$ $\qquad \Delta d = 54$ m

$\qquad\qquad\qquad\qquad\quad v_i = 0.50$ m/s

$\qquad\qquad\qquad\qquad\quad a = 0.30$ m/s^2

$v_f = \sqrt{(0.50 \text{ m/s})^2 + 2 \times 0.30 \text{ m/s}^2 \times 54 \text{ m}}$

$v_f = \sqrt{33\,\dfrac{\text{m}^2}{\text{s}^2}} = 5.7$ m/s

7. (a) $v_i = \dfrac{157.0 \text{ km/hr} \times 1000 \text{ m/km}}{3600 \text{ s/hr}} = 43.6\,\dfrac{\text{m}}{\text{s}}$

$\qquad v_f = \dfrac{75.0 \text{ km/hr} \times 1000 \text{ m/km}}{3600 \text{ s/hr}} = 20.8\,\dfrac{\text{m}}{\text{s}}$

$\qquad \Delta t = \dfrac{v_i - v_f}{a} = \dfrac{(43.6 - 20.8) \text{ m/s}}{7.00 \text{ m/s}^2} = 3.26$ s

(b) $\Delta d = v_{av}\,\Delta t = \dfrac{v_i - v_f}{2}\,\Delta t$

$\qquad \Delta d = \left(\dfrac{43.6 + 20.8}{2}\right)$ m/s $\times 3.26$ s

$\qquad \Delta d = 105$ m

GROUP B

8. The height of the flagpole will be the distance that a ball will fall from rest in 2 s.

$\Delta d = \frac{1}{2}g(\Delta t)^2$

$\Delta d = \frac{1}{2}g(\Delta t)^2 = 9.80\,\dfrac{\text{m}}{\text{s}^2} \times (2.0 \text{ s})^2$

$\Delta d = 20$ m

9. The stone, which has an initial upward velocity of 5 m/s, will have an acceleration of g downward and will thus be brought to rest in 0.5 s.

During this time it will move upward a distance $\Delta d = \left(\dfrac{v_i + v_f}{2}\right)\Delta t = 2.5$ m/s $\times 0.5$ s

$\Delta d = 1.25$ m

Thus the stone will start to fall from rest from an elevation of 354 m

$\Delta d = \frac{1}{2}g(\Delta t)^2$

$\Delta t = \sqrt{\dfrac{2\,\Delta d}{g}} = \sqrt{\dfrac{2 \times 354 \text{ m}}{9.80 \text{ m/s}^2}} = 8.5$ s

10. (a) $v_f = \sqrt{v_i^2 + 2g\,\Delta d}$

$\qquad v_f = \sqrt{\left(25\,\dfrac{\text{m}}{\text{s}}\right)^2 + 2 \times 9.80\,\dfrac{\text{m}}{\text{s}^2} \times 155 \text{ m}}$ $\qquad v_f = 60.5$ m/s

(b) $v_f = v_i + g \, \Delta t$ $\Delta t = \dfrac{v_f - v_i}{g}$

$$\Delta t = \dfrac{(60 - 25) \text{ m/s}}{9.80 \text{ m/s}^2} = 3.6 \text{ s}$$

Questions • Chapter 3/pp. 62–63
GROUP A
1. (a) The force of gravity acting downward on the ball and the force exerted by the table acting upward on the ball. (b) No.
2. (a) It is accelerated in the direction of the push. (b) The force of gravity acting downward on the ball, the force exerted by the table acting upward on the ball, and the force of friction acting opposite to the direction of motion. (c) The force of friction is an unbalanced force. (d) The force of friction causes the ball to slow down and eventually stop. The other two forces balance each other.
3. (a) If there is no net force acting on a body, it will continue in its state of rest or will continue moving along a straight line with uniform velocity. (b) It is either at rest or moving with constant velocity.
4. (a) The acceleration is in direct proportion to the force. (b) The direction of the acceleration is the same as that of the force.
5. The acceleration is inversely proportional to the mass.
6. (a) The effect of an applied force is to cause the body to accelerate in the direction of the force. The acceleration is in direct proportion to the force and in inverse proportion to the mass of the body. (b) $F = ma$.
7. $F = F_w a / g$.
8. (a) The force of gravity. (b) The force of attraction of the brick for the earth.
9. (a) The force of the cord on the brick. (b) The force of the brick on the cord.

GROUP B
10. (a) An experiment that cannot actually be performed, as, for example, one requiring an object having zero mass. (b) See Galileo's inertia experiment.
11. It is constantly acted upon by an unbalanced force, the force of gravity.
12. (a) No. (b) When frictional forces are present, part of the applied force is used to overcome friction. The remaining force (if any) produces acceleration according to Newton's second law.
13. (a) No. (b) The tension exerted on the fishline is 400 n, which is less than the force required to break it. Each person exerts a 400-n force, which is balanced by the 400-n force of the other person.
14. Pull the 1-kilogram mass with the spring balance. Keep a constant force on the mass and measure the time required to pull the mass a given distance along the frictionless surface. Using the same force on the unknown mass, measure the time to pull it along the same distance. The masses will be directly proportional to the squares of the elapsed times, according to Newton's second law, when force and distance are kept constant.
15. (a) The principle of action and reaction applies. The forward pull of the tractor on the load is equal in magnitude to the backward pull of the load on the tractor. This is true whether the load is in motion or at rest. The actual acceleration of the tractor is due to the resultant, or unbalanced, force acting on it, that is, the forward pull of the tractor added vectorially to the backward pull of the ground. (b) Student diagram.
16. The reading on the center balance will be the same as that on the other two balances.

Problems • Chapter 3/p. 63
GROUP A
1. $F = ma$ $F = 3.0 \text{ kg} \times 5.0 \text{ m/s}^2$ $F = 15 \text{ n}$
2. $F = ma$ $m = F/a$ $m = 125 \text{ m}/3.00 \text{ m/s}^2$ $m = 41.7 \text{ kg}$
3. $F = ma$ $a = F/m$ $a = 25 \text{ m}/15 \text{ kg}$ $a = 1.7 \text{ m/s}^2$
4. $F_w = mg$ $F_w = 24 \text{ kg} \times 9.80 \text{ m/s}^2$ $F_w = 2.4 \times 10^2 \text{ n}$

GROUP B
5. $m = F_w/g$ $m = 485 \text{ n}/9.80 \text{ m/s}^2$ $m = 49.5 \text{ kg}$
6. $F = F_w a/g$ $F = (475 \text{ n} \times 3.00 \times 10^3 \text{ m/s}^2)/9.80 \text{ m/s}^2$ $F = 1.45 \times 10^5 \text{ n}$
7. $F = F_w a/g$ $F = (1.0 \times 10^5 \text{ n} \times 1.5 \text{ m/s}^2)/9.80 \text{ m/s}^2$ $F = 1.5 \times 10^4 \text{ n}$

Questions • Chapter 3/p. 68
GROUP A
1. **(a)** The force of gravitational attraction of the earth. **(b)** An equal force of attraction on the earth.
2. **(a)** The magnitude of the masses and the distances between their centers of mass. **(b)** It is directly proportional to the products of the masses and to G and inversely proportional to the square of the distance between their centers of mass.
3. The force required to overcome the force of gravity is directly proportional to the mass lifted.
4. Your weight becomes less.
5. Your weight becomes greater.
6. The answer should be a summary of the description in Section 3.11.

GROUP B
7. **(a)** The weight will decrease to zero during the first $\frac{5}{6}$ of the trip and then increase during the remaining $\frac{1}{6}$ to a value of about $\frac{1}{6}$ of the weight on the earth's surface. **(b)** Its weight decreases, becoming zero at the center of the earth.
8. Local variations in the composition of the earth's crust and the effect of the earth's rotation are not considered in the equation.
9. **(a)** A gravitational field is a region of space in which each point is associated with the value of g at that point. **(b)** The value of g, the acceleration due to gravity, at a particular point.

Problems • Chapter 3/pp. 68–69
GROUP A
1. $(1.62 \text{ m/s}^2/9.80 \text{ m/s}^2) \times 795 \text{ n} = 131 \text{ n}$
2. $\dfrac{F_1}{F_2} = \dfrac{d_2^2}{d_1^2}$ $F_2 = \dfrac{d_1^2}{d_2^2} \times F_1$

     $d_1 = 6.375 \times 10^3 \text{ km}$, radius of earth

     $d_2 = 6.375 \times 10^3 \text{ km} + 2.560 \times 10^4 \text{ km} = 3.198 \times 10^4 \text{ km}$

     $F_2 = \dfrac{(6.375 \times 10^3 \text{ km})^2}{(3.198 \times 10^4 \text{ km})^2} \times 1.058 \times 10^3 \text{ n}$

     $F_2 = 1.058 \times 10^3 \text{ n} \times 3.972 \times 10^{-2} = 4.202 \times 10^1 \text{ n}$
3. $F_w = mg$ $F_w = 0.250000 \text{ kg} \times 9.80336 \text{ m/s}^2$ $F_w = 2.45084 \text{ n}$

GROUP B
4. $F = Gm_1 m_2/d^2$

$F = (6.67 \times 10^{-11} \text{ n m}^2/\text{kg}^2 \times 10\bar{0} \text{ kg} \times 10\bar{0} \text{ kg}) \div (2.00 \text{ m})^2$
$F = 1.67 \times 10^{-7} \text{ n}$
5. $F = Gm_1m_2/d^2$
$F = (6.67 \times 10^{-11} \text{ n m}^2/\text{kg}^2 \times 5.00 \text{ kg} \times 10.0 \text{ kg}) \div (0.300 \text{ m})^2$
$F = 3.71 \times 10^{-8} \text{ n}$
6. $g = GM/d^2$
$g = (6.67 \times 10^{-11} \text{ n m}^2/\text{kg}^2 \times 5.96 \times 10^{24} \text{ kg}) \div (1.00 \times 10^7 \text{ m})^2$
$g = 3.98 \text{ m/s}^2$
7. $g = \dfrac{F_w}{m} = \dfrac{\text{n}}{\text{kg}} = \dfrac{\text{kg m/s}^2}{\text{kg}} = \text{m/s}^2$

CONCURRENT AND PARALLEL FORCES $\qquad$ 4

Questions • Chapter 4/p. 75
GROUP A
1. **(a)** Pushing a door shut with your hand or sitting in a chair. **(b)** Gravitational forces do not require physical contact.
2. A spring balance.
3. Magnitude and direction.
4. **(a)** A resultant force is a single force that has the same effect as two or more concurrent forces. **(b)** The resultant is found by algebraic addition.
5. The resultant is found by the parallelogram method and simple trigonometry.
6. The first condition of equilibrium is that there are no net forces acting on a body.
7. **(a)** The equilibrant force is the single force that will produce equilibrium. **(b)** At the same point as two or more concurrent forces. **(c)** It is equal and opposite to the resultant force.
8. If there are no net forces acting on a body, the body will remain in a state of rest or uniform motion, in accordance with Newton's first law of motion.
9. The resultant is found by the parallelogram method and the use of the law of sines and law of cosines.

Problems • Chapter 4/pp. 75–76
GROUP A
1. The two forces are acting along the same line and have the same direction, and therefore their sum 5.0 n + 15.0 n = 20.0 n is the magnitude of the resultant. The direction of the resultant is the same as that of each force, south.
2. **(a)** Force diagram includes two equilateral triangles. 20.0 n is magnitude of resultant. 30.0° north of east is direction. **(b)** The equilibrant is 20.0 n 60.0° west of south.
3. 1.0 m/1.5 m = 306 n/x $\quad$ x = 4.6 × 10² n, the force exerted by each hammock rope.

GROUP B
4. $\angle EOA = 140.0°$
$\angle OER = 40.0°$
By the cosine law,
$F_R = \sqrt{\overline{OE}^2 + \overline{ER}^2 - 2(OE)(ER)(\cos \angle OER)}$

$$\cos \angle OER = \cos 40.0° = 0.766$$
$$F_R = \sqrt{(10.0 \text{ n})^2 + (15.0 \text{ n})^2 - 2 \times 10.0 \text{ n} \times 15.0 \text{ n} \times 0.766}$$
$$F_R = 9.7 \text{ n}$$

By the sine law, $\dfrac{ER}{\sin \angle EOR} = \dfrac{OR}{\sin \angle OER}$

$$\sin \angle EOR = \dfrac{ER \times \sin \angle OER}{OR}$$

$$\sin \angle OER = \sin 40.0° = 0.643$$

$$\sin \angle EOR = \dfrac{15.0 \text{ n} \times 0.643}{9.7 \text{ n}} = 0.99$$

$\angle EOR = 98°$ ($\angle EOR$ is an obtuse angle)
The direction of F_R is 8° west of south. The data of the problem do not permit a more precise answer.

5. **(a)** $c = \sqrt{a^2 + b^2 - 2ab \cos C}$
$\cos C = \cos 30.0° = 0.866$
$c = \sqrt{(10.0 \text{ n})^2 + (30.0 \text{ n})^2 - 2(10.0 \text{ n} \times 30.0 \text{ n} \times 0.866)}$
$c = 21.9 \text{ n}$, magnitude of resultant force
$b/\sin B = c/\sin C$ $\sin B = b \sin C/c$ $\sin C = \sin 30.0° = 0.500$
$\sin B = (30.0 \text{ n} \times 0.500)/21.9 \text{ n}$
$\sin B = 0.685$ $B = 136.8°$ direction of resultant force $= 46.8°$ west of south
(b) The equilibrant is 21.9 n 46.8° east of north.

6. First determine the resultant of the two forces that are acting at right angles.
$c = \sqrt{a^2 + b^2}$ $c = \sqrt{(10.0 \text{ n})^2 + (15.0 \text{ n})^2}$
$c = 18.0 \text{ n}$, magnitude of resultant of 10.0-n and 15.0-n forces
$a/\sin A = c/\sin C$ $\sin A = a \sin C/c$
$\sin C = \sin 90.0° = 1.00$
$\sin A = (10.0 \text{ n} \times 1.00)/18.0 \text{ n}$
$\sin A = 0.556$
Direction of resultant of 10.0-n and 15.0-n forces $= 33.7°$ north of west
Now determine the resultant of the 18-n force and the 15-n force at 30.0° east of north
$c = \sqrt{a^2 + b^2 - 2ab \cos C}$ $\cos C = \cos 93.7° = -\cos 86.3° = -0.0645$
$c = \sqrt{(18.0 \text{ n})^2 + (15.0 \text{ n})^2 - 2(15.0 \text{ n} \times 18.0 \text{ n} \times -0.0645)}$ $c = 24.2 \text{ n}$, magnitude of resultant
$a/\sin A = c/\sin C$
$\sin A = a \sin C/c$
$\sin C = \sin 93.7° = \sin 86.3° = 0.998$
$\sin A = (18.0 \text{ n} \times 0.998)/24.2 \text{ n}$ $\sin A = 0.735$ $A = 47.9°$ direction of resultant $= 17.9°$ west of north

Questions • Chapter 4/p. 78
GROUP A
1. The resolution of forces is the opposite of the composition of forces.
2. 90°
3. A normal force is a force that acts perpendicularly to a surface.
4. **(a)** A component parallel to the hill and a component perpendicular to the hill. **(b)** The component parallel to the hill.

GROUP B
5. (a) The component that acts parallel to the plane and the component that acts perpendicular to the plane. (b) As the angle of the incline increases, the component parallel to the plane increases while the component perpendicular to the plane decreases in magnitude.
6. A component parallel to the ground and directed toward the tower and a component perpendicular to the ground and directed upward.

Problems • Chapter 4/pp. 78–79
GROUP A

1. Force diagram is a $45° - 45° - 90°$ triangle. $\dfrac{150.0\ n}{\sqrt{2}} = 106\ n$, magnitude of both the

    horizontal and vertical components

2. $F_w/F_p = l/h$ $F_p = F_w h/l$ $F_p = \dfrac{1.00 \times 10^5\ n \times 3.00\ m}{100\ m}$ $F_p = 3.00 \times 10^3\ n$

3. $\sin A = a/c$ $a = c \sin A$ $\sin A = \sin 35.0° = 0.574$ $a = 25.0\ n \times 0.574 = 14.4\ n,$
    the westward component $\cos A = b/c$ $b = c \cos A = \cos 35.0° = 0.819$
    $b = 25.0\ n \times 0.819 = 20.5\ n,$ the northward component

4. $\sin A = a/c$ $a = c \sin A$
    $\sin A = \sin 20.0° = 0.342$
    $a = 78.0\ n \times 0.342 = 26.7\ n,$ the vertical component
    $\cos A = b/c$ $b = c \cos A$
    $\cos A = \cos 20.0° = 0.940$
    $b = 78.0\ n \times 0.940 = 73.3\ n,$ the horizontal component

GROUP B
5. (a) $114\ kg \times 9.80\ n/kg = 1.12 \times 10^3\ n$
    $\cos A = b/c$ $b = c \cos A$
    $\cos A = \cos 12.0° = 0.978$
    $b = 1.12 \times 10^3\ n \times 0.978 = 1.10 \times 10^3\ n,$ component perpendicular to plane
    (b) $\sin A = a/c$ $a = c \sin A$
    $\sin A = \sin 12.0° = 0.208$
    $a = 1.12 \times 10^3\ n \times 0.208 = 2.33 \times 10^2\ n,$ component parallel to the plane
6. $\sin A = a/c$ $c = a/\sin A$ $\sin A = \sin 35.0° = 0.574$
    $c = 495\ n/0.574$
    $c = 813\text{-}n$ force exerted by the chain directed toward the building
    $\tan A = a/b$ $b = a/\tan A$ $\tan A = \tan 35.0° = 0.700$
    $b = 495\ n/0.700$
    $b = 707\text{-}n$ force exerted by the bracket away from the building
7. $a/\sin A = b/\sin B = c/\sin C$
    $2.00 \times 10^3\ kg \times 9.8\ n/kg/\sin 20.0° = b/\sin 30.0° = c/\sin 130.0°$
    $\sin 20.0° = 0.342$ $\sin 30.0° = 0.500$ $\sin 130.0° = \sin 50.0° = 0.766$
    $b = (2.00 \times 10^3\ kg \times 9.8\ n/kg \times 0.500)/0.342$
    $b = 2.87 \times 10^4\ kg,$ tie force
    $c = (2.00 \times 10^3\ kg \times 9.8\ n/kg \times 0.766)/0.342 = 4.39 \times 10^4\ n,$ thrust force
8. The force diagram formed by completing the force parallelogram contains two equilateral triangles. The force along each rafter is $6.00 \times 10^3\ n.$

Questions • Chapter 4/p. 84

GROUP A

1. **(a)** Friction is a force that resists the motion of objects that are in contact with each other and are sliding or rolling over each other. **(b)** As the surfaces of objects are rubbed together, they tend to interlock. Some cases of sliding friction may be caused by the forces that hold atoms and molecules together.
2. Walking, moving a car, braking, nailing, dishes on a table, etc.
3. Polished bearings, antifriction metals, ball bearings, roller bearings, lubricants, etc.
4. Sliding friction is usually less than starting friction.
5. **(a)** Materials and their surfaces, and the force pressing the two surfaces together. **(b)** The speed of sliding and the area of contact between the surfaces.
6. The ratio of the force needed to overcome friction to the normal force pressing the surfaces together.

GROUP B

7. It always acts parallel to the surface and in the direction opposing the motion.
8. Snow treads or other nonskid treads, chains, cinders, sand, salt, etc.
9. The lubricant separates the moving parts, thus substituting fluid friction, which is generally much less, for solid friction.
10. **(a)** If the force due to the weight of the object is normal to the surface; that is, the surface must be horizontal. **(b)** By determining the component of the weight normal to the surface.

Problems • Chapter 4/p. 85

GROUP A

1. $F = \mu F_N$ $\mu = F/F_N$
   $\mu = 1.00 \times 10^2 \text{ n}/2.00 \times 10^3 \text{ n}$ $\mu = 0.0500$
2. $F = \mu F_N$ $\mu = F/F_N$
   $\mu = 45 \text{ n}/125 \text{ n}$ $\mu = 0.36$
3. Inclined plane and force diagrams are $30° - 60° - 90°$ triangles.
   $1.25 \times 10^3 \text{ n} \div 2 = 6.25 \times 10^2 \text{ n}$, force parallel to plane
   $6.25 \times 10^2 \text{ n} \times \sqrt{3} = 1.08 \times 10^3 \text{ n}$, force normal to plane
   $\mu = F/F_N$
   $\mu = 6.25 \times 10^2 \text{ n}/1.08 \times 10^3 \text{ n}$
   $\mu = 0.579$

GROUP B

4. $80.0 \text{ kg} \times 9.80 \text{ n/kg} = 784 \text{ n} =$ weight of block
   $\sin A = a/c$ $a = c \sin A$
   $\sin A = \sin 20.0° = 0.342$
   $a = 784 \text{ n} \times 0.342$ $a = 268 \text{ n}$, force parallel to plane
   $\cos A = b/c$ $b = c \cos A$
   $\cos A = \cos 20.0° = 0.940$
   $b = 784 \text{ n} \times 0.940$ $b = 737 \text{ n}$, force normal to plane
   $F = \mu F_N$
   $F = 0.200 \times 737 \text{ n}$
   $F = 147 \text{ n}$, force to overcome friction
   $268 \text{ n} + 147 \text{ n} = 415 \text{ n}$, total force required

5. $F = (150 \text{ n})(\cos 45°) = 106 \text{ n}$, the horizontal component of the force exerted by the rope. Since the rope is at an angle of 45°, the normal component upward will also be 106 n.
$F_N = 400 \text{ n} - 106 \text{ n} = 294 \text{ n}$
$\mu = F/F_N$
$\mu = 106 \text{ n}/294 \text{ n} = 0.361$

6. $50.0 \text{ kg} \times 9.80 \text{ n/kg} = 490 \text{ n} =$ the weight of the box
Let $x =$ force required for 30.0° angle pull
$x \sin 30.0°$ is the vertical component of the force applied to the rope.
$F_N = 490 \text{ n} - x \sin 30.0° =$ the normal force between the box and the floor
$F = \mu F_N = 0.300 (490 \text{ n} - x \sin 30.0°)$
Also $F = x \cos 30.0° = 0.300 (490 \text{ n} - x \sin 30.0°)$
Solving $x = 144 \text{ n}$

7. **(a)** From similar triangle relationships between the force diagram and the inclined plane diagram, we find $\mu = \tan \theta$. Therefore
$\mu = \tan 10.0° = 0.176$
(b) $\mu = \tan 20.0° = 0.364$
(c) $\mu = \tan x°$

8. The force parallel to and down the plane equals the force parallel to and up the plane when the conditions of the problem are met. Therefore $F_w \sin \theta = \mu F_w \cos \theta + F_a$ where F_w is the weight of the block and F_a is the force applied to the block by the hanging weight. Substituting numerical values in the above equation,
$130 \text{ n} \sin \theta = 0.620 \times 130 \text{ n} \cos \theta + 45.0 \text{ n}$
and $\sin \theta = 0.620 \cos \theta + 0.346$
Squaring both sides of the equation,
$\sin^2 \theta = 0.384 \cos^2 \theta + 0.429 \cos \theta + 0.120$
and setting $\sin^2 \theta = 1 - \cos^2$ we have
$1 - \cos^2 \theta = 0.384 \cos^2 \theta + 0.429 \cos \theta + 0.120$ which set in the form of a quadratic becomes $1.384 \cos^2 \theta + 0.429 \cos \theta - 0.880 = 0$
$$\cos \theta = \frac{-0.429 \pm \sqrt{0.184 + 4.87}}{2.768}$$
$$\cos \theta = \frac{-0.429 \pm 2.25}{2.768} = 0.658 \text{ and } -0.968$$
$$\theta = 49°$$

Questions • Chapter 4/p. 90
GROUP A

1. The center of gravity is that point in an object at which all of its weight can be considered to be concentrated.

2. **(a)** Parallel forces are forces acting at different points but in the same or in opposite directions. **(b)** The resultant is equal in magnitude to the sum of the two forces and acts in the same direction. **(c)** The resultant has a magnitude equal to the difference in magnitude of the two forces and acts in the direction of the larger force.

3. **(1)** Move the pivot point of the seesaw to provide the lighter person with a longer torque arm. **(2)** Move the heavier person closer to the pivot point to shorten the torque arm.

4. **(a)** The torque equals the product of the force and the length of the torque arm, measured along the bar on which it acts. **(b)** The meter newton.

5. The magnitude of the algebraic sum of the forces must be zero, and the sum of the clockwise torques must equal the sum of the counterclockwise torques.

GROUP B
6. By choosing the point of application of one of the forces as the pivot point, the torque arm of that force becomes zero and the torque (which is also zero) drops out of the calculation.
7. The center of gravity is a specific point, whereas any point can be chosen as the pivot point in a torque problem.
8. A ring, a hollow sphere, etc.
9. The torque arm is first found by measuring the perpendicular distance from the pivot point to the line of action of the applied force. This is calculated by multiplying the distance between the pivot point and point of application of the force by the cosine of the angle between the lines. The length of the torque arm is then multiplied by the magnitude of the force, as before.
10. **(a)** A couple is a pair of forces of equal magnitude acting in opposite directions in the same plane, but not along the same line. **(b)** The torque of a couple is equal to the product of one of the forces and the perpendicular distance between them.

Problems • Chapter 4/pp. 90–91
GROUP A
1. Let x = force required to lift one end   clockwise torque = counterclockwise torque
   5.00 m $\times x$ = 2.5 m $\times$ 2.5 $\times$ 10^5 n
   x = 1.2 $\times$ 10^5 n
2. Equilibrant force = $40\bar{0}$ n + $30\bar{0}$ n + $50\bar{0}$ n = $120\bar{0}$ n directed upward
   Use the end at which the 300-n force is attached as a pivot point
   Let x = distance of equilibrant force from pivot point
   (1.5 m $\times$ $40\bar{0}$ n) + (4.0 m $\times$ $50\bar{0}$ n) = $120\bar{0}$ n $\times x$
   x = 2.2 m, distance from $30\bar{0}$-n force
3. Let x = force exerted by ladder A   1100 n $- x$ = force exerted by ladder B
   Clockwise torque = counterclockwise torque
   3.00 m $\times x$ = 875 n $\times$ 1.00 m + 223 n $\times$ 1.50 m
   x = 400 n, force exerted by ladder A
   $(110\bar{0} - 400)$ n = 700 n, force on ladder B
4. Let x = force exerted on end of scaffold near bricklayer
   8.00 $\times$ 10^2 n + 3.20 $\times$ 10^2 n + 7.50 $\times$ 10^2 n $- x$ = force on other end of scaffold
   Use end of scaffold farthest from bricklayer as a pivot point.
   Clockwise torque = counterclockwise torque
   3.00 m $\times x$ = (1.50 m $\times$ 3.20 $\times$ 10^2 n) + (2.00 m $\times$ 8.00 $\times$ 10^2 n) + (1.50 m $\times$ 7.50 $\times$ 10^2 n)
   x = 1.07 $\times$ 10^3 n, force exerted at end near bricklayer
   1.87 $\times$ 10^3 n $-$ 10.7 $\times$ 10^2 n = 8.00 $\times$ 10^2 n, force exerted at end farthest from bricklayer
5. Clockwise torque = counterclockwise torque.   Assuming that the center of gravity of the bench is at the geometric center, and using the position of the pair of legs near the 5.00 $\times$ 10^2-n person as the pivot point, (0.10 m)(5.00 $\times$ 10^2 n) + (0.90 m)(3.20 $\times$ 10^2 n + 7.50 $\times$ 10^2 n) + (1.70 m)(1.000 $\times$ 10^3 n) + (1.80 m)(15.0 n) = 1.80 nm $\times x$. x = 1520-n force exerted by the leg farthest from the 5.00 $\times$ 10^2 n person.   The total weight of

the system is 15.0 n $+ (5.00 \times 10^2$ n$) + (1.070 \times 10^3$ n$) + (1.00 \times 10^3$ n$) +$ 15.0 n $= 2.60 \times 10^3$ n.

$2\overline{6}00$ n $- 1520$ n $= 1080$-n force exerted by the legs nearest the 5.00×10^2-n person

6. Assume that the perpendicular distance between the forces comprising the couple is 2.0 m and that the pivot point is midway between these forces. Then the torque equals $F \times 1.0$ m $+ F \times 1.0$ m, or $F \times 2.0$ m. This is also the product of one force and the perpendicular distance between the lines of action of the forces.

7. Let $x =$ upward force exerted by pier A, and use pier B as the pivot point.
Clockwise torque $=$ counterclockwise torque
$x \times 36.5$ m $= 5.25 \times 10^4$ n $\times 26.3$ m
$x = 3.78 \times 10^4$ n, the upward force exerted by pier A
Let $y =$ upward force exerted by pier B, and use A as the pivot point.
Clockwise torque $=$ counterclockwise torque
5.25×10^4 n $\times 10.2$ m $= y \times 36.5$ m
$y = 1.47 \times 10^4$ n, the upward force exerted by pier B
Note that the sum $x + y =$ the weight of the truck.
The total force on each pier will be greater than the value shown above by an amount equal to $\frac{1}{2}$ the weight of the bridge.

GROUP B

8. Let $x =$ weight of bar. Using the edge of the block as the pivot point,
clockwise torque $=$ counterclockwise torque
2.0 m $\times x = 750$ n $\times 1.5$ m
$x = 5.6 \times 10^2$ n, weight of bar

9. $(2.50 \times 10^3$ n $+ 1.000 \times 10^4$ n $+ 3.50 \times 10^3$ n$) - (3.00 \times 10^3$ n $+ 4.00 \times 10^3$ n$) =$ 9.00×10^3 n, magnitude of equilibrant force, which acts upward
Let $x =$ distance of equilibrant force from B.   Use B as pivot point.
Clockwise torque $=$ counterclockwise torque
$(3.00 \times 10^3$ n $\times 4.0$ m$) + (4.00 \times 10^3$ n $\times 17.0$ m$) + (9.00 \times 10^3$ n $\times x) = (1.000 \times$ 10^4 n $\times 12.5$ m$) + (2.50 \times 10^3$ n $\times 25.0$ m$)$
$x = 12$ m, distance of equilibrant force from B

10. Since the door is in equilibrium, the net horizontal force must be zero. Hence, the horizontal components of the forces applied by the hinges must be equal and opposite, and these two forces constitute a couple. Let x represent the magnitude of the two equal horizontal components. Since the distance between the hinges is 1.90 m, the torque produced by the couple is $x \times 1.90$ m. Using either hinge (or any point on the line connecting them) as a pivot point,

clockwise torque $=$ counterclockwise torque
204 n $\times 0.50$ m $= x \times 1.90$ m
$x = 54$ n

The *vertical* components of the forces exerted by the hinges do not contribute any torque because their lines of action pass through the pivot point. The only torques that need to be considered are those caused by the weight of the door and by the couple.

11. (a) Torque of $10\overline{0}$-n force $= (10\overline{0}$ n$)(2.00$ m$)(\sin 45.0°) = 1.41 \times 10^2$ nm
Torque of $20\overline{0}$-n force $= (20\overline{0}$ n$)(3.00$ m$)(\sin 60.0°) = 5.20 \times 10^2$ nm
Torque of $10\overline{0}$-n force $= (10\overline{0}$ n$)(4.00$ m$) = 4.00 \times 10^2$ nm
Torque of $15\overline{0}$-n force $= 0$ nm

(b) Clockwise torque = 1.41×10^2 nm
    Counterclockwise torque = 5.20×10^2 nm + 4.00×10^2 nm = 9.20×10^2 nm
    Clockwise torque required to produce equilibrium =
    9.20×10^2 nm − 1.41×10^2 nm = 7.79×10^2 nm
    Downward force required at right end of bar =
    7.79×10^2 nm ÷ 4.00 m = 1.95×10^2 n

5

TWO-DIMENSIONAL
AND PERIODIC MOTION

Questions • Chapter 5/p. 100
GROUP A
1. Motion along a curved path results when an object moves with uniform velocity in one direction and uniformly accelerated motion in another.
2. (a) Because of its inertia, the projectile tends to maintain constant horizontal velocity and would do so except for the retarding force due to friction. It accelerates downward, as does any falling body, because of the force of gravity. (b) The constant horizontal motion accounts for the projectile's horizontal displacement only. The accelerated vertical motion due to gravitational force accounts for vertical displacement only and determines how long the projectile will be in flight.
3. (a) Motion of an object at constant speed along a curved path of constant radius. (b) Centripetal force. (c) Circular motion involves a change in direction. A change in direction is a form of acceleration. Acceleration requires a force, as stated in Newton's second law.
4. The mass of the object on which it is acting, the speed, and the radius of curvature of the path.
5. (a) Motion in a vertical circle involves a variable centripetal force. In uniform circular motion, the centripetal force is constant. (b) The force of gravity.
6. (a) The lowest velocity at which an object will describe a vertical circle. (b) It will fall out of its circular orbit.
7. (a) A system for specifying precisely the location of objects. (b) The earth, the satellite, the moon, the sun, etc.

GROUP B
8. (a) The sight is raised. (b) The sight would not have to be raised as much.
9. (a) The force required to produce the acceleration of a body is proportional to the magnitude of the acceleration; Newton's second law. (b) During reentry, air resistance causes the vehicle to decelerate at a rate equal to several times the magnitude of the acceleration due to gravity.
10. (a) In both cases, the force of gravity is all or part of the centripetal force. (b) In the motion of an earth satellite, the centripetal force is constant since the force of gravity is the only force acting. In the vertical circle, the centripetal force is variable since the force of gravity is only a part of the centripetal force and acts in one direction (downward).
11. (a) Strictly speaking, the statement is false. It would be better to say that the satellite has no "effective" weight since there is no force resisting it. (b) The statement is true.

12. From the frame of reference of the moving merry-go-round, a centrifugal force is pushing the merry-go-round and its occupant toward the outside of the curve. From the frame of reference of an observer standing on the ground next to the merry-go-round, a centripetal force is keeping the merry-go-round and its occupant in a curved path.

Problems • Chapter 5/p. 101
GROUP A

1. $a = v^2/r$ $a = \dfrac{(40.0 \text{ m/s})^2}{16.0 \text{ m}}$ $a = 1.00 \times 10^2 \text{ m/s}^2$

2. $F = ma$
   $F = 2.00 \text{ ton} \times 1000 \text{ kg/ton} \times 1.00 \times 10^2 \text{ m/s}^2$
   $F = 2.00 \times 10^5 \text{ n}$

3. $F = mv^2/r$ $F = \dfrac{2.5 \times 10^{-2} \text{ kg} \times (5.0 \text{ m/s})^2}{2.0 \text{ m}}$ $F = 0.31 \text{ n}$

4. (a) $F = mv^2/r$ $F = \dfrac{5.0 \text{ kg} \times (3.0 \text{ m/s})^2}{1.0 \text{ m}}$ $F = 45 \text{ n}$

   (b) $F = mv^2/r$ $F = \dfrac{5.0 \text{ kg} \times (3.0 \text{ m/s})^2}{3.0 \text{ m}}$ $F = 15 \text{ n}$

5. (a) $v = \sqrt{rg}$ $v = \sqrt{1.0 \text{ m} \times 9.8 \text{ m/s}^2}$ $v = 3.1 \text{ m/s}$
   $v = \sqrt{3.0 \text{ m} \times 9.8 \text{ m/s}^2}$ $v = 5.4 \text{ m/s}$

   (b) The value of g is inversely proportional to the square of the distance from the center of the earth, or $g_1 : g_2 = (r_2)^2 : (r_1)^2$ $g_2 = g_1 \times (r_1)^2/(r_2)^2$

   $g_2 = \dfrac{9.80 \text{ m/s}^2 \times (6370 \text{ km})^2}{(6370 \text{ km} + 6.0 \times 10^2 \text{ km})^2}$ $g_2 = 8.17 \text{ m/s}^2$
   $v = \sqrt{rg}$ $v = \sqrt{1.0 \text{ m} \times 8.17 \text{ m/s}^2}$ $v = 2.9 \text{ m/s}$
   $v = \sqrt{3.0 \text{ m} \times 8.17 \text{ m/s}^2}$ $v = 5.0 \text{ m/s}$

GROUP B

6. $F = mv^2/r$ $F = \dfrac{6.0 \times 10^4 \text{ n}/9.8 \text{ m/s}^2 \times (22 \text{ m/s})^2}{250 \text{ m}}$ $F = 1.2 \times 10^4 \text{ n}$

7. The value of g is inversely proportional to the square of the distance from the center of the earth, or $g_1 : g_2 = (r_2)^2 : (r_1)^2$
   $g_2 = g_1 \times (r_1)^2/(r_2)^2$
   $g_2 = \dfrac{9.80 \text{ m/s}^2 \times (6370 \text{ km})^2}{(6370 \text{ km} + 1.00 \times 10^3 \text{ km})^2}$ $g_2 = 7.32 \text{ m/s}^2$
   $v = \sqrt{rg}$
   $v = \sqrt{7370 \text{ km} \times 1000 \text{ m/km} \times 7.32 \text{ m/s}^2}$ $v = 7.34 \times 10^3 \text{ m/s}$

8. Circumference of the earth $= 2\pi r = 2 \times 3.14 \times 6370 \text{ km} \times 1000 \text{ m/km} = 4.00 \times 10^7 \text{ m}$

   Velocity of rotation $= \Delta d/\Delta t = \dfrac{4.00 \times 10^7 \text{ m}}{24 \text{ hr} \times 3600 \text{ s/hr}} = 4.63 \times 10^2 \text{ m/s}$

   $F = mv^2/r$
   $F = \dfrac{75.0 \text{ kg} \times (4.63 \times 10^2 \text{ m/s})^2}{6370 \text{ km} \times 1000 \text{ m/km}}$ $F = 2.52 \text{ n}$

9. The time required for the ball to rise to a height of 50.0 m and fall to the ground again will be 2 times the time required for the ball to fall 50.0 m from rest.

$\Delta d = \frac{1}{2}g \, \Delta t^2$; thus $\Delta t = \sqrt{\dfrac{2 \, \Delta d}{g}}$ and the time the ball will be in the air $= 2\sqrt{\dfrac{2 \, \Delta d}{g}}$.

$t = 2\sqrt{\dfrac{2 \times 50.0 \text{ m}}{9.80 \text{ m/s}^2}} = 2\sqrt{10.2 \text{ s}^2} = 6.39 \text{ s}$

10. (a) During the first second the ball will have $v_i = 0$ and $v_f = 9.8$ m/s

$\Delta d_i = v_{av} \times \Delta t = \dfrac{9.8}{2}$ m/s $\times$ 1.0 s = 4.9 m

During the second second
$v_i = 9.8$ m/s $v_f = 19.6$ m/s

$v_{av} = \left(\dfrac{9.8 + 19.6}{2}\right)$ m/s = 14.7 m/s

$\Delta d_2 = v_{av} \, \Delta t = 14.7$ m/s $\times$ 1.0 s = 14.7 m
During the third second
$v_i = 19.6$ m/s $v_f = 29.4$ m/s

$v_{av} = \left(\dfrac{19.6 + 29.4}{2}\right)$ m/s = 24.5 m/s

$\Delta d_3 = v_{av} \, \Delta t = 24.5$ m/s $\times$ 1.0 s = 24.5 m
(b) The horizontal component of the velocity is 25 m/s. Neglecting the effect of air resistance, this value remains constant. Thus the ball moves a horizontal distance of 25 m during each of the three seconds.

Questions • Chapter 5/p. 108
GROUP A
1. (a) The motion of a body turning about an axis. (b) The hands of an electric clock and the shaft of an electric motor are in uniform rotary motion while a car wheel and a spinning top are in variable rotary motion.
2. It is that angle which, when placed with its vertex at the center of a circle, subtends on the circumference an arc equal in length to the radius of the circle.
3. (a) The rate of change of angular displacement. (b) The rate of change of angular velocity.
4. (a) Because it includes the direction of the axis of rotation. (b) It is drawn parallel to the axis of rotation in the direction the thumb of the right hand points if the fingers of the right hand encircle the vector in the direction of rotation.
5. They are alike except for the substitution of θ for d, ω for v, and α for a.

GROUP B
6. Torques acting on it do not produce rapid changes in its angular momentum because its rotational inertia is large.
7. Motion about a third axis which is called precession.
8. The two vectors are perpendicular to each other.
9. The acceleration vector is at right angles to the velocity vector.
10. (a) See Fig. 5-12. (b) The earth precesses slightly, and thus the axis of rotation will not always be directed toward the present-day North Star (Polaris).

Problems • Chapter 5/p. 108

GROUP A

1. $4820/\text{min} \times 1 \text{ min}/60 \text{ s} \times 2\pi \text{ rad} = 504 \text{ rad/s}$
2. $625 \text{ rad/s} \times 1/2\pi \text{ rad} \times 60 \text{ s/min} = 5.97 \times 10^3 \text{ rev/min}$
3. $\alpha = Fr/I$

$$\alpha = \frac{10.0 \text{ n} \times 0.25 \text{ m}}{0.500 \text{ kg m}^2} \qquad \alpha = 5.0 \text{ rad/s}^2$$

4. $I = mr^2$
   $I = 5.7 \text{ kg} \times (0.15 \text{ m})^2 \qquad I = 0.13 \text{ kg m}^2$
5. $m = F_W/g$
   $m = 80.0 \text{ n}/9.80 \text{ m/s}^2 \qquad m = 8.16 \text{ kg}$
   $I = \frac{2}{5}mr^2$
   $I = \frac{2}{5} \times 8.16 \text{ kg} \times (0.050)^2$
   $I = 8.16 \times 10^{-3} \text{ kg m}^2$

GROUP B

6. $\alpha = \Delta\omega/\Delta t = \dfrac{545/\text{min} \times 2\pi \text{ rad} \times 1 \text{ min}/60 \text{ s}}{1 \text{ min} \times 60 \text{ s/min}} = 0.951 \text{ rad/s}^2$

7. (a) $\Delta\theta = \frac{1}{2}\alpha \Delta t^2$
   $\Delta\theta = \frac{1}{2} \times 0.951 \text{ rad/s}^2 \times (30.0 \text{ s})^2 \qquad \Delta\theta = 428 \text{ rad}$
   (b) $\omega_f = \omega_i + \alpha \Delta t$
   $\omega_f = 0 + 0.951 \text{ rad/s}^2 \times 30.0 \text{ s} \qquad \omega_f = 28.5 \text{ rad/s}$
   $\Delta\theta = \omega \Delta t + \frac{1}{2}\alpha \Delta t^2$
   $\Delta\theta = (28.5 \text{ rad/s} \times 30.0 \text{ s}) + [\frac{1}{2} \times 0.951 \text{ rad/s}^2 \times (30.0 \text{ s})^2]$
   $\Delta\theta = 1.28 \times 10^3 \text{ rad}$

8. (a) $I = \frac{1}{2}mr^2$
   $I = \frac{1}{2} \times 30.0 \text{ kg} = (0.200 \text{ m})^2 \qquad I = 0.600 \text{ kg m}^2$
   $\alpha = Fr/I$
   $\alpha = 25.0 \text{ n} \times 0.200 \text{ m}/0.600 \text{ kg m}^2 \qquad \alpha = 8.33 \text{ rad s}^2$
   $\omega_f = \alpha \Delta t$
   $\omega_f = 8.33 \text{ rad/s}^2 \times 12.0 \text{ s} \qquad \omega_f = 1.00 \times 10^2 \text{ rad/s}$
   (b) $\Delta\theta = \frac{1}{2}\alpha \Delta t^2 \qquad \theta = \frac{1}{2} \times 8.33 \text{ rad/s}^2 \times (12.0 \text{ s})^2$
   $\Delta\theta = 6.00 \times 10^2 \text{ rad}$

9. $I = \frac{1}{2}mr^2$
   $I = \frac{1}{2} \times 2.00 \times 10^3 \text{ n}/9.80 \text{ m/s}^2 \times (1.00 \text{ m})^2 \qquad I = 102 \text{ kg m}^2$
   $\alpha = Fr/I \qquad \alpha = 50.0 \text{ n} \times 1.00 \text{ m}/102 \text{ kg m}^2 \qquad \alpha = 0.490 \text{ rad/s}^2$
   $\omega_f = \alpha \Delta t$
   $\Delta t = \omega_f/\alpha$

   $$\Delta t = \frac{1200/\text{min} \times 2\pi \text{ rad} \times 1 \text{ min}/60 \text{ s}}{0.490 \text{ rad/s}^2} \qquad t = 256 \text{ s}$$

Questions • Chapter 5/p. 113

GROUP A

1. Simple harmonic motion is a type of periodic motion.
2. (a) A single back-and-forth movement. (b) The distance from the midpoint of the vibration. (c) The time required for a complete vibration. (d) The reciprocal of the period, or vibrations/time.

3. The square root of its length divided by the square root of the acceleration due to gravity.

4. **(a)** Every actual pendulum is a physical pendulum since the supporting string or rod has mass. A simple pendulum would have all its mass concentrated in a small bob. **(b)** A simple pendulum (thought experiment) can be constructed of such length that its period of vibration will be the same as that of the physical pendulum.

5. Determine the center of oscillation by making a simple pendulum with the same period of vibration; find the point on the bat where a ball can be hit without stinging the hands.

GROUP B

6. **(a)** The pendulum should be shortened. **(b)** Shortening the pendulum will make it swing faster.

7. Determine the period of vibration, measure the length of the pendulum, then substitute these values in the pendulum equation, and solve for g.

8. The displacement from the equilibrium position is directly proportional to the force exerted on the mass.

9. **(a)** Both undergo harmonic motion. **(b)** A quartz crystal and a simple pendulum both have definite periods of vibration.

Problems • Chapter 5/p. 113

GROUP A

1. $T_1 : T_2 = \sqrt{l_1} : \sqrt{l_2}$ $\quad T_2 = T_1\sqrt{l_2/l_1}$ $\quad T_2 = 1.1 \text{ s} \times \sqrt{0.10 \text{ m}/0.25 \text{ m}}$
   $T_2 = 0.70 \text{ s}$

2. $T = 2\pi\sqrt{\dfrac{l}{g}}$ $\quad T = 2 \times 3.14\sqrt{1.5 \text{ m}/9.8 \text{ m/s}^2}$ $\quad T = 2.5 \text{ s}$

GROUP B

3. $T = 2\pi\sqrt{\dfrac{l}{g}}$

   $g = \dfrac{4\pi^2 l}{T^2}$

   $g = \dfrac{4(3.14)^2(1.08 \text{ m})}{(2.05 \text{ s})^2}$

   $g = 10.1 \text{ m/s}^2$

4. $T = 2\pi\sqrt{\dfrac{l}{g}}$ $\quad l = \dfrac{T^2 g}{4\pi^2}$

   $l = \dfrac{(2.00000 \text{ s})^2 \times 9.79609 \text{ m/s}^2}{4 \times (3.14159)^2}$ $\quad l = 0.992553 \text{ m}$

5. $f_1^2 : f_2^2 = g_1 : g_2$

   $g_2 = \dfrac{g_1 f_2^2}{f_1^2}$

   $g_2 = \dfrac{(9.8 \text{ m/s}^2)(320 \text{ s}^{-1})^2}{(150 \text{ s}^{-1})^2}$

   $g_2 = 45 \text{ m/s}^2$

CONSERVATION OF ENERGY AND MOMENTUM

6

Questions • Chapter 6/p. 123
GROUP A
1. Work is done when a force affects the motion of an object moving in line with the direction of the applied force.
2. **(a)** If a graph is plotted in which the horizontal axis denotes the distance and the vertical axis denotes the force that acts in the direction of the motion, then the area under the curved line represents the work done by the force. **(b)** The graph is a straight line, and the area under the line is the area of a triangle.
3. No. The force acts at right angles to the displacement, and so there is no component of a centripetal force in the direction of the displacement.
4. **(a)** Lever, inclined plane, wheel and axle, wedge, pulley, screw. **(b)** Lever and inclined plane.
5. **(a)** The ratio of the useful work output to the total work input. **(b)** W_{output} (conservative)/W_{input} (total) × 100%.
6. The dissipative force of friction makes it impossible for the efficiency of a machine to be 100%.
7. Friction keeps the screw from falling out. Friction also makes it difficult to drive the screw.
8. **(a)** Power is the time rate of doing work. **(b)** The watt.

GROUP B
9. The area under the curve can be found by dividing the area into many adjacent rectangles, the areas of which are found by multiplying the ordinate by the small segments of abscissa.
10. Nine. One machine multiplies the effect of the other.
11. **(a)** Work. **(b)** Kilowatt-hours = watts × 10^3 × hours × 3600 s/hr = joule/s × 10^3 × 3600 s = 3.600 × 10^6 joules. Joules are work units.
12. Power = work/time = $W/\Delta t = T\theta/\Delta t$

Problems • Chapter 6/pp. 123–124
GROUP A
1. $W = F \Delta d$ $W = 750$ n × 5.0 m $W = 3.8 × 10^3$ j
2. $W = F \Delta d$ $W = 2.5$ kg × 9.8 m/s^2 × 2.2 m $W = 54$ j
3. $F_f = \mu F_N$ $F_f = 0.050 × 116$ kg × 9.8 m/s^2 $F_f = 57$ n
   $W = F \Delta d$ $W = 57$ n × 15.0 m $W = 8.5 × 10^2$ j
4. $W = F \Delta d \cos \theta$ $\cos 35.0° = 0.819$
   $W = 124$ n × 0.500 km × 1000 m/km × 0.819 $W = 5.08 × 10^4$ j
5. **(a)** $W = F \Delta d$ $W = \dfrac{18 \text{ n} × 0.25 \text{ m}}{2} = 2.2$ j   **(b)** $F = 18$ n × 2 = 36 n

   (c) $W = F \Delta d$ $W = \dfrac{36 \text{ n} × 0.5 \text{ m}}{2} = 9.0$ j

6. $\text{Efficiency} = \dfrac{F_w \, \Delta h}{F_a \, \Delta d}$ $\text{Efficiency} = \dfrac{2520 \text{ n} \times 1.50 \text{ m}}{545 \text{ n} \times 10.0 \text{ m}}$ $\text{Efficiency} = 0.694 = 69.4\%$

7. $\text{Efficiency} = \dfrac{F_w \, \Delta h}{F_a \, \Delta d}$ $F_a = \dfrac{F_w \, \Delta h}{\text{Eff.} \times \Delta d}$

$F_a = \dfrac{750 \text{ kg} \times 9.8 \text{ n/kg} \times 0.040 \text{ m}}{62\% \times 0.50 \text{ m}}$ $F_a = 950 \text{ n}$

Each boy exerts a force of 950 n/2 = 470 n.

8. $F_w \, \Delta h = F_a \, \Delta d$ $F_a = F_w \, \Delta h / \Delta d$
$F_a = 750 \text{ n} \times 0.90 \text{ m}/4.0 \text{ m}$ $F_a = 1.7 \times 10^2 \text{ n}$

9. $\text{Efficiency} = \dfrac{F_w \, \Delta h}{F_a \, \Delta d}$

$\text{Efficiency} = \dfrac{6.0 \times 10^3 \text{ kg} \times 9.8 \text{ m/s}^2 \times 0.010 \text{ m}/1.5}{320 \text{ n} \times 0.75 \text{ m} \times 2\pi}$

$\text{Efficiency} = 0.26 = 26\%$

10. $P = F \, \Delta d / \Delta t$ $P = \dfrac{47 \text{ kg} \times 9.8 \text{ m/s}^2 \times 12 \text{ m}}{15 \text{ s}} = 3.7 \times 10^2 \text{ w}$

11. $P = F \, \Delta d / \Delta t$

$P = \dfrac{2.50 \times 10^3 \text{ kg} \times 9.80 \text{ m/s}^2 \times 50.0 \text{ m}}{10.0 \text{ s}} \times \dfrac{1 \text{ kw}}{1000 \text{ w}} = 1.23 \times 10^2 \text{ kw}$

12. $P = F \, \Delta d / \Delta t$

$P = \dfrac{1500 \text{ kg} \times 9.8 \text{ m/s}^2 \times 33 \text{ m}}{30.0 \text{ s}} = 1.6 \times 10^4 \text{ w}$

GROUP B

13. $F_f = \mu F_N$ $F_f = 0.20 \times 35 \text{ kg} \times 9.8 \text{ m/s}^2$ $F_f = 69 \text{ n}$
$W = F \, \Delta d$ $W = 69 \text{ n} \times 8.0 \text{ m}$ $W = 5.5 \times 10^2 \text{ j}$
$W = F \, \Delta d$ $W = 35 \text{ kg} \times 9.8 \text{ m/s}^2 \times 1.3 \text{ m}$ $W = 4.5 \times 10^2 \text{ j}$
Total work = $5.5 \times 10^2 \text{ j} + 4.5 \times 10^2 \text{ j} = 9.9 \times 10^2 \text{ j}$

14. $F_w = mg$ $F_w = 75.0 \text{ kg} \times 9.80 \text{ m/s}^2$ $F_w = 735 \text{ n}$
$\sin A = a/c$ $a = c \sin A$
$\sin A = \sin 20.0° = 0.342$ $a = 735 \text{ n} \times 0.342$
$a = 251 \text{ n}$, force parallel to plane
$\cos A = b/c$ $b = c \cos A$ $\cos A = \cos 20.0° = 0.940$
$b = 735 \text{ n} \times 0.940$ $b = 691 \text{ n}$, force normal to plane
$\mu = F_f / F_N$ $F_f = \mu F_N$ $F_f = 0.150 \times 691 \text{ n}$
$F_f = 104 \text{ n}$, force to overcome friction
251 n + 104 n = 355 n, total force needed
$W = F \, \Delta d$ $W = 355 \text{ n} \times 3.00 \text{ m}$ $W = 1.07 \times 10^3 \text{ joules}$

15. $W = T \, \Delta \theta$ $W = 7.5 \text{ n} \times 0.25 \text{ m} \times 240 \text{ rev} \times 2\pi \text{ rad/rev}$ $W = 2.8 \times 10^3 \text{ joules}$

16. (a) $F = ma$ $a = F/m$ $a = 5.0 \text{ n} \div 2.0 \text{ kg}$ $a = 2.5 \text{ m/s}^2$
$\Delta d = \frac{1}{2} a \, \Delta t^2$ $\Delta d = \frac{1}{2} \times 2.5 \text{ m/s}^2 \times (2.0 \text{ s})^2$ $\Delta d = 5.0 \text{ m}$
$W = F \, \Delta d$ $W = 5.0 \text{ n} \times 5.0 \text{ m}$ $W = 25 \text{ joules}$

(b) $\Delta d = \frac{1}{2} a (2t - 1) \text{ s}$
$\Delta d = \frac{1}{2} \times 2.5 \text{ m/s}^2 (2 \times 10 \text{ s} - 1 \text{ s}) \text{ s}$ $\Delta d = 24 \text{ m}$
$W = F \, \Delta d$ $W = 5.0 \text{ n} \times 24 \text{ m}$ $W = 1.2 \times 10^2 \text{ joules}$

17. (a) $\omega_f = \alpha \, \Delta t \qquad \alpha = \omega_f / \Delta t$

$\alpha = (195 \text{ rev/min} \times 2\pi \text{ rad/rev} \times 1 \text{ min/60 s}) \div 60.0 \text{ s}$

$\alpha = 0.340 \text{ rad/s}^2$

$I = \frac{1}{2} m r^2 \qquad I = \frac{1}{2} \times 203 \text{ kg} \times (1.00 \text{ m})^2 = 101 \text{ kg m}^2$

$T = I\alpha = 101 \text{ kg m}^2 \times 0.340 \text{ rad/s}^2 = 34.3 \text{ m n}$

(b) $W = T \, \Delta\theta$ Since the wheel started from rest and had a speed of 195 rev/min at the end of 60 seconds, the average speed is 97.5 rev/min. In this time the wheel has turned 97.5 rev.

$W = 34.3 \text{ m n} \times 97.5 \text{ rev} \times 2\pi \text{ rad/rev} = 2.10 \times 10^4 \text{ joules}$

18. $60.0 \text{ kg} \times 9.80 \text{ n/kg} = 588 \text{ n}$ Let $x =$ force required for 45.0°-angle pull

$F_N = 588 \text{ n} - x \sin 45.0°$

$\mu = F_f / F_N \qquad F_f = \mu F_N$

$F_f = 0.020(588 \text{ n} - x \sin 45.0°)$

Also $F_f = x \cos 45.0° \qquad x \cos 45.0° = 0.020(588 \text{ n} - x \sin 45.0°)$

$x = 16 \text{ n}$

$W = F \, \Delta d \cos\theta \qquad W = 16 \text{ n} \times 1000 \text{ m} \times \cos 45.0°$

$W = 1.1 \times 10^4 \text{ joules}$

19. $P = F \, \Delta d / t$

$$P = \frac{20.0 \text{ liters} \times 0.700 \text{ kg/liter} \times 9.80 \text{ m/s}^2 \times 6.00 \text{ m}}{1.00 \text{ min} \times 60.0 \text{ s/min}} \times \frac{1 \text{ kw}}{1000 \text{ w}}$$

$P = 1.37 \times 10^{-2} \text{ kw}$

20. $P = F \, \Delta d / \Delta t \qquad \Delta d / \Delta t = P / F$

$\Delta d / \Delta t = (10.0 \text{ kw} \times 1000 \text{ w/kw}) \div (2.75 \times 10^4 \text{ kg} \times 9.80 \text{ m/s}^2)$

$\Delta d / \Delta t = 0.0371 \text{ m/s, or } 2.23 \text{ m/min}$

21. $F_w / F_a = l/h \qquad F_a = F_w h/l \qquad F_a = (1.00 \times 10^6 \text{ n} \times 1.50 \text{ m}) \div 10\overline{0} \text{ m}$

$F_a = 1.50 \times 10^4 \text{ n}$

$15.0 \times 10^3 \text{ n} + 2.00 \times 10^3 \text{ n} = 17.0 \times 10^3 \text{ n, total force required}$

$P = F \, \Delta d / \Delta t$

$P = 17.0 \times 10^3 \text{ n} \times 40.0 \text{ km/hr} \times 1000 \text{ m/km} \times 1 \text{ hr/3600 s}$

$P = 1.89 \times 10^5 \text{ w}$

Questions • Chapter 6/pp. 129–130

GROUP A

1. (a) Potential energy is stored energy. Kinetic energy is energy due to the motion of an object. **(b)** Gravitational potential energy is energy stored in an object because of work done on the object against the force of gravity. Elastic potential energy is stored energy in an elastic object because of work done in compressing or stretching the object.

2. $E_p = W = F \, \Delta d = mg \, \Delta h$

3. (a) Kinetic energy due to linear motion and kinetic energy due to rotary motion

(b) $E_k = \frac{1}{2} m v^2 \qquad E_k = \frac{1}{2} I \omega^2$

4. (a) The elasticity of the spring. **(b)** Newtons per meter.

5. (a) Mechanical energy is the sum of potential and kinetic energy. **(b)** Potential energy and kinetic energy interchange as the pendulum swings, but the sum remains constant.

GROUP B

6. (a) $v = \sqrt{2g \, \Delta d} \qquad \Delta d = v^2 / 2g$ Substituting this value for h in the E_p formula gives

$E_p = mg \, \Delta h = mg \times v^2 / 2g = \frac{1}{2} m v^2$

(b) $\frac{1}{2} m v^2 = \frac{1}{2} \times \text{kg} \times (\text{m/s})^2 = \text{kg m}^2/\text{s}^2 = \text{kg m/s}^2 \times \text{m} = \text{n m} = \text{j}$

7. **(a)** Conservative forces are forces involved in the conservation of mechanical energy. Dissipative forces produce deviations from the law of conservation of mechanical energy.
(b) Friction produces heat, which is not a form of mechanical energy.

8. As the man walks up the stairs, he has kinetic enegy because of his mass and velocity. His gravitational potential energy is increasing. As the elevator rises, the man and the elevator again have kinetic energy because of their mass and velocity. The gravitational potential energy of the elevator and the man are also increasing with respect to the bottom of the picture as a frame of reference.

Problems • Chapter 6/p. 130
GROUP A

1. $E_p = mg \, \Delta h$ $E_p = 2.00 \text{ kg} \times 9.80 \text{ m/s}^2 \times 0.80 \text{ m}$ $E_p = 16 \text{ j}$

2. $E_p = mg \, \Delta h$ $E_p = F_w \, \Delta h$
   $E_p = 1470 \text{ n} \times 40.0 \text{ m} = 5.88 \times 10^4 \text{ j}$

3. $E_k = \frac{1}{2}mv^2$
   $E_k = \frac{1}{2} \times 0.14 \text{ kg} \times (18 \text{ m/s})^2$ $E_k = 23 \text{ j}$

4. $E_k = \frac{1}{2}mv^2$ $E_k = F_w v^2/2g$
   $E_k = \dfrac{1860 \text{ n} \times (45.2 \text{ m/s})^2}{2 \times 9.80 \text{ m/s}^2} = 1.94 \times 10^5 \text{ j}$

5. **(a)** $E_p = \frac{1}{2}k \, \Delta d^2$
   $E_p = \frac{1}{2} \times 2.0 \times 10^2 \text{ n/m} \times (0.10 \text{ m} \div 4)^2$ $E_p = 6.2 \times 10^{-2} \text{ j}$
   (b) $E_p = \frac{1}{2}k \, \Delta d^2$
   $E_p = \frac{1}{2} \times 2.0 \times 10^2 \text{ n/m} \times (0.10 \text{ m})^2$ $E_p = 1.0 \text{ j}$

6. **(a)** $F = k \, \Delta d$ $F = 150 \text{ n/m} \times 0.25 \text{ m} = 38 \text{ n}$
   (b) $W = \frac{1}{2}k \, \Delta d^2 = \frac{1}{2} \times 150 \text{ n/m} \times (0.25 \text{ m})^2 = 4.7 \text{ j}$

GROUP B

7. **(a)** $E_p = mg \, \Delta h$ $E_p = 50.0 \text{ kg} \times 9.80 \text{ m/s}^2 \times 197 \text{ m}$ $E_p = 9.65 \times 10^4 \text{ j}$
   $E_k = 0 \text{ j}$
   (b) $\Delta d = \frac{1}{2}g \, \Delta t^2$ $\Delta d = \frac{1}{2} \times 9.80 \text{ m/s}^2 \times (1.00 \text{ s})^2$ $\Delta d = 4.90 \text{ m}$
   $E_p = mg \, \Delta h$ $E_p = 50.0 \text{ kg} \times 9.80 \text{ m/s}^2 \times (197 \text{ m} - 4.90 \text{ m})$
   $E_p = 9.41 \times 10^4 \text{ j}$
   $v_f = g \, \Delta t$ $v_f = 9.80 \text{ m/s}^2 \times 1.00 \text{ s}$ $v_f = 9.80 \text{ m/s}$
   $E_k = \frac{1}{2}mv^2$ $E_k = \frac{1}{2} \times 50.0 \text{ kg} \times (9.80 \text{ m/s})^2$ $E_k = 2.45 \times 10^3 \text{ j}$
   Note: $9.41 \times 10^4 \text{ j} + 2.45 \times 10^3 \text{ j} = 9.65 \times 10^4 \text{ j}$, since the total energy is constant
   (c) $\Delta d = \frac{1}{2}g \, \Delta t^2$ $\Delta d = \frac{1}{2} \times 9.80 \text{ m/s}^2 \times (5.00 \text{ s})^2$ $\Delta d = 123 \text{ m}$
   $E_p = mg \, \Delta h$ $E_p = 50.0 \text{ kg} \times 9.80 \text{ m/s}^2 \times (197 \text{ m} - 123 \text{ m})$
   $E_p = 3.63 \times 10^4 \text{ j}$
   $E_k = 9.65 \times 10^4 \text{ j} - 3.63 \times 10^4 \text{ j}$ $E_k = 6.02 \times 10^4 \text{ j}$
   (d) $v = \sqrt{2g \, \Delta d}$ $v = \sqrt{2 \times 9.80 \text{ m/s}^2 \times 197 \text{ m}}$ $v = 62.1 \text{ m/s}$
   $E_k = \frac{1}{2}mv^2$ $E_k = \frac{1}{2} \times 50.0 \text{ kg} \times (62.1 \text{ m/s})^2$ $E_k = 9.65 \times 10^4 \text{ j}$
   $E_p = 0 \text{ j}$

8. **(a)** $E_k = \dfrac{F_w(v_f^2 - v_i^2)}{2g} =$

   $\dfrac{2.00 \times 10^5 \text{ n} \times [(75.0 \text{ km/hr})^2 - (45.0 \text{ km/hr})^2] \times [(1000 \text{ m/km} \times 1 \text{ hr}/3600 \text{ s})^2]}{2 \times 9.80 \text{ m/s}^2}$

   $E_k = 2.83 \times 10^6 \text{ j}$

(b) $a = (v_f - v_i)/\Delta t \qquad a = \dfrac{(75.0 \text{ km/hr} - 45.0 \text{ km/hr}) \times 1000 \text{ m/km}}{11.0 \text{ s} \times 3600 \text{ s/hr}}$

$a = 0.758 \text{ m/s}^2 \qquad F = ma \qquad F = \dfrac{2.00 \times 10^5 \text{ n} \times 0.758 \text{ m/s}^2}{9.80 \text{ m/s}^2}$

$F = 1.55 \times 10^4 \text{ n}$

(c) $\Delta d = v_{av} \Delta t$

$\Delta d = (75.0 \text{ km/hr} + 45.0 \text{ km/hr})/2 \times 11.0 \text{ s} \times 1000 \text{ m/km} \times 1 \text{ hr}/3600 \text{ s}$

$\Delta d = 183 \text{ m} \qquad P = F \Delta d/\Delta t \qquad P = 1.55 \times 10^4 \text{ n} \times 183 \text{ m}/11.0 \text{ s} = 2.58 \times 10^5 \text{ w}$

Or: $P = \dfrac{E_k}{\Delta t} = \dfrac{2.83 \times 10^6 \text{ j}}{11.0 \text{ s}} = 2.57 \times 10^5 \text{ w}$

9. $E_k = \frac{1}{2}mv^2 \qquad E_k = \frac{1}{2} \times 9.1 \times 10^{-31} \text{ kg} \times (1.0 \times 10^7 \text{ m/s})^2$

$E_k = 4.6 \times 10^{-17} \text{ j}$

10. $E_k = \frac{1}{2}I\omega^2$

$E_k = \frac{1}{2} \times 45 \text{ kg m}^2 \times (1500 \text{ rev/min} \times 2\pi \text{ rad/rev} \times 1 \text{ min}/60 \text{ s})^2$

$E_k = 5.6 \times 10^5 \text{ j}$

11. (a) $E_k = \frac{1}{2}mv^2 \qquad E_k = \frac{1}{2} \times 8.00 \text{ kg} \times (7.00 \text{ m/s})^2 \qquad E_k = 196 \text{ joules}$

(b) For a sphere $E_k = \frac{1}{2}(\frac{2}{5}mv^2) \qquad E_k = \frac{1}{2}[\frac{2}{5} \times 8.00 \text{ kg} \times (7.00 \text{ m/s})^2]$

$E_k = 78.4 \text{ joules}$

(c) $E_k = 196 \text{ joules} + 78.4 \text{ joules} = 274 \text{ joules}$

12. $E_p = mg \Delta h \qquad E_p = 10.0 \text{ kg} \times 9.80 \text{ m/s}^2 \times 2.00 \text{ m} \qquad E_p = 196 \text{ joules}$

$E_k = \frac{1}{2}mv^2 + \frac{1}{2}I\omega^2 \qquad$ For a sphere, $I = \frac{2}{5}mr^2 \qquad \omega = \Delta\theta/\Delta t = (\Delta d/r)/\Delta t$

But $\Delta d/\Delta t = v$, and so $\omega = v/r$ and $\omega^2 = (v/r)^2 \qquad$ Substituting, $E_k = \frac{1}{2}mv^2 + \frac{1}{2}(\frac{2}{5}mv^2)$

$v = \sqrt{10 E_k/7 \text{ m}}$

$v = \sqrt{(10 \times 196 \text{ joules}/(7 \times 10.0 \text{ kg})} \qquad v = 5.29 \text{ m/s}$

13. $E_p = \frac{1}{2}k \Delta d^2 \qquad E_p = \frac{1}{2} \times 115 \text{ n/m} \times (0.200 \text{ m})^2$

$E_p = 0.230 \text{ j}$

$E_k = E_p = \frac{1}{2}mv^2 \qquad n = \sqrt{2E_p/m}$

$v = \sqrt{2 \times 0.230 \text{ j}/1.00 \text{ kg}}$

$v = 2.14 \text{ m/s}$

Questions • Chapter 6/p. 138
GROUP A

1. (a) The product of a force and the length of time it acts. **(b)** The product of the mass of an object and its velocity. **(c)** When an unbalanced force is applied to an object, the change in momentum that results is equal to the magnitude of the impulse.

2. In one case, the mass is very large while the velocity is small; in the other, the velocity is very large, while the mass is small. In both cases, the product of mass and velocity is large.

3. Vector quantities.

4. The total momentum of colliding bodies is not affected by the collision.

5. When the aerial rocket explodes, the sum of the momenta of the individual particles is equal to the momentum of the rocket just before the explosion.

6. In elastic collisions, objects rebound from each other. In inelastic collisions, the objects remain together after collision.

7. An isolated system is one that has no external forces, and therefore no torques, acting on it.

8. Answers will vary. A railroad car colliding with another railroad car on the same track is an example of a collision in one dimension, provided both cars stay on the track. Billiard balls colliding in such a way that one or both balls change direction after the collision is an example of a collision in two dimensions.

GROUP B

9. The great mass of the ocean liner and the great speed of the airplane give each of them a large momentum. Neither the water nor the air can exert a force sufficiently large to change this momentum suddenly.

10. The momentum of the car is decreased to zero. The force required to do this is that of the brakes against the wheel and, in turn, of the tires against the road. The total momentum of the earth and the car remains constant.

11. (a) The situation would be the same if the boy, instead of the man, did the pushing, provided that the magnitude of the push is the same in each case. (b) If the total magnitude of the push is the same in each case, then the situation is the same when both push as when only the man pushes.

12. The greater the mass of the spinning gyroscope and the greater its rotational velocity, the greater is the angular momentum of the gyroscope and the slower it will precess. The torque produced by gravity also increases with mass.

13. (a) Gravitation between the carts. (b) These forces are very small when compared to the force of friction.

14. No. Much or all of the kinetic energy in an inelastic collision is converted into heat or some other form of energy.

Problems • Chapter 6/pp. 138–139

GROUP A

1. $F \Delta t = m \Delta v$

$$F = \frac{m \Delta v}{\Delta t} = \frac{(50\bar{0} \text{ kg})(15\bar{0} \text{ m/s})}{25.0 \text{ s}}$$

$F = 30\bar{0}0 \text{ kg m/s}^2 = 30\bar{0}0 \text{ n}$

2. $mv_i + m'v_i' = mv_f + m'v_f'$ Both v_i and v_i' equal zero

$v_f' = -mv_f/m'$ 60.0 g = 0.0600 kg

$v_f' = -(0.0600 \text{ kg} \times 615 \text{ m/s})/5.00 \text{ kg}$

$v_f' = -7.38 \text{ m/s}$

3. $mv_i + m'v_i' = mv_f + m'v_f'$

$v_f = (mv_i + m'v_i' - m'v_f') \div m$

$v_f = [(1.50 \text{ kg} \times 8.00 \text{ m/s}) + (2.00 \text{ kg} \times 3.00 \text{ m/s}) - (2.00 \text{ kg} \times 7.29 \text{ m/s})]$
$\div 1.50 \text{ kg}$

$v_f = 2.28 \text{ m/s southward}$

4. $mv_i + m'v_i' = mv_f + m'v_f'$ Since $v_i' = 0$, this becomes $mv_i = mv_f + m'v_i'$

$$m' = \frac{mv_i - mv_f}{v_f'} = \frac{m(v_i - v_f)}{v_f'}$$

$$m' = \frac{1.67 \times 10^{-27} \text{ kg}[(6.00 \times 10^6 \text{ m/s}) - (-3.60 \times 10^6 \text{ m/s})]}{2.40 \times 10^6 \text{ m/s}}$$

$m' = 6.68 \times 10^{-27} \text{ kg}$

5. (a) The second object moves at an angle of 90° to the left of the path of the first object

after the collision. This follows from the discussion of the laws of conservation of energy and momentum in Section 6.15.

(b) m_1v_1 = momentum object 1 before collision = $30.0 \dfrac{\text{kg m}}{\text{s}}$

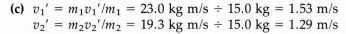

m_1v_1' = momentum object 1 after collision
m_2v_2' = momentum object 2 after collision

$\sin \overline{50}° = \dfrac{m_1v_1'}{m_1v_1}$

$m_1v_1' = (m_1v_1)(\sin \overline{50}°) = (30.0 \text{ kg m/s})(0.766)$
$\qquad = 23.0 \text{ kg m/s}$

$\sin \overline{40}° = \dfrac{m_2v_2'}{m_1v_1}$

$m_2v_2' = (m_1v_1)(\sin \overline{40}°) = (30.0 \text{ kg m/s})(0.643) = 19.3 \text{ kg m/s}$

(c) $v_1' = m_1v_1'/m_1 = 23.0 \text{ kg m/s} \div 15.0 \text{ kg} = 1.53 \text{ m/s}$
$v_2' = m_2v_2'/m_2 = 19.3 \text{ kg m/s} \div 15.0 \text{ kg} = 1.29 \text{ m/s}$

GROUP B

6. $m_1v_1 + m_2v_2 = m_1v_1' + m_2v_2'$
Since $v_2 = 0$ and $v_1' = v_2'$,

$m_1v_1 = v_1'(m_1 + m_2)$ and $v_1' = \dfrac{m_1v_1}{m_1 + m_2} = \dfrac{(3.0 \times 10^5 \text{ kg})(2.5 \text{ m/s})}{3.0 \times 10^5 \text{ kg} + 1.5 \times 10^5 \text{ kg}} = 1.7 \text{ m/s}$

7. $m_1v_1 = m_2v_2$
$m_1 = m_2v_2/v_1 = (10 \text{ kg})(100 \text{ m/s})/3 \times 10^8 \text{ m/s} = 3 \times 10^{-6} \text{ kg}$

8. (a) The average velocity during the impact, assuming that the person is decelerated at a constant rate, is $15 \text{ m/s} \div 2 = 7.5 \text{ m/s}$. The distance traveled into the drift is, therefore,
$\Delta d = v \, \Delta t = (7.5 \text{ m/s})(0.30 \text{ s}) = 2.2 \text{ m}$
(b) $F \, \Delta t = m \, \Delta v$
$F = m \, \Delta v/\Delta t = (65 \text{ kg})(15 \text{ m/s})/0.30 \text{ s} = 3.2 \times 10^3 \text{ n}$

9. $I\omega = \text{kg m}^2 \times \text{rad/s}$ Since the rad is a dimensionless quantity, $I\omega$ has the dimensions kg m^2/s

10. $m_1v_1 + m_2v_2 = m_1v_1' + m_2v_2'$ Assuming that m_1 is the more massive ball, $v_1 = -v_2$,
$m_1 = 3m_2$, $v_1' = 0$, and $v_2' = -2v_2$ Substituting, $-3m_2v_2 + m_2v_2 = 0 - 2m_2v_2$

11. Student diagram. The mass of the smaller ball is 63.3 g. Any answer between 60 g and 70 g should be accepted.

PHASES OF MATTER   **7**

Questions • Chapter 7/p. 150
GROUP A
1. Matter is composed of very tiny particles called molecules. Molecules are in constant motion. Momentum and kinetic energy are conserved.

2. **(a)** The phase of matter is the way in which particles are assembled in a substance.
   (b) In a solid, the particles are close together in a fixed pattern. In a liquid, the particles are not as close together and are not held in a fixed pattern. In a gas, the average separation of the particles is large and they are not held in a fixed pattern.

3. A molecule is the smallest particle of a substance that is capable of independent existence. An atom is the smallest particle of an element. Atoms can combine to make molecules.

4. **(a)** When molecules are farther apart than their normal spacing. **(b)** When molecules are closer together than their normal spacing.

5. **(a)** Crystalline and amorphous. **(b)** Crystalline: regular particle arrangement; amorphous: random particle arrangement.

6. **(a)** A vibratory motion. **(b)** The diffusion of solids.

7. **(a)** Cohesion. **(b)** Adhesion.

8. **(a)** The force required to break a rod or wire having a unit cross-sectional area. **(b)** Newtons per square meter.

9. The property of matter that requires a force to produce a distortion and, when the distorting force is removed, causes the matter to resume its original shape.

10. Stress is the ratio of distorting force to cross-sectional area of the distorted body. Strain is the ratio of change of length to the original length of the distorted body.

11. The smallest stress that produces permanent distortion.

12. Within the limits of perfect elasticity, strain is directly proportional to stress.

13. $Y = Fl/\Delta l A$, where Y is Young's modulus; F, the applied force; l, the original length; Δl, the change in length; A, the cross-sectional area.

GROUP B

14. Molecules have the closest spacing in solids, greater spacing in liquids, and the greatest spacing in gases.

15. **(a)** $1.6605655 \times 10^{-27}$ kg.   **(b)** It is 1/12th of the mass of a carbon-12 atom.

16. Large molecules have been photographed through an electron microscope.

17. The crystal pattern is unchanged. Layers of atoms are shifted past each other.

18. The force constant is the quantitative expression of Hooke's law for a specific substance.

Problems • Chapter 7/pp. 150–151
GROUP A

1. 0.050 m × 1.000 n/0.500 n = 0.10 m

2. **(a)** 5.00 cm × 125.0 n/40.0 n = 15.6 cm
   (b) 5.00 cm/40.0 n × 5.00 n = 0.625 cm separation of 5.00 n graduations

3. 489.0 n   The breaking force is dependent on the wire diameter and not on the length.

4. **(a)** Stress = F/A $F = 10.00$ kg × 9.80 n/kg = 98.0 n
   $A = \pi r^2 = 3.14 \times 0.100^2$ cm^2 = 0.0314 cm^2
   Stress = 98.0 n/0.0314 cm^2 = 3120 n/cm^2 = 3.12×10^7 n/m^2
   (b) Y = Stress/Stress   Strain = Stress/Y
   $Y_{copper} = 11.6 \times 10^{10}$ n/m^2
   Strain = 3.12×10^7 n/m^2/11.6×10^{10} n/m^2
   Strain = 2.69×10^{-4}

5. $F_{breaking}$ = (tensile strength)(area)   From Appendix B, Table 7, the tensile strength of

aluminum wire is 2.4×10^8 n/m^2. Hence, $F_b = (2.4 \times 10^8$ n/m$^2)(2.50 \times 10^{-3}$ cm$^2)/$ $(10^{-4}$ cm^2/m$^2) = 6\overline{0}$ n

6. $Y =$ Stress/Strain $Y = F/A \div \Delta l/l = Fl/A \Delta l$

$\Delta l = Fl/AY = mgl/\pi r^2 Y$

$\Delta l = 8.25$ kg $\times 9.80$ n/kg $\times 2.57$ m/$3.14 \times 1.00 \times 10^{-6}$ m$^2 \times 9.02 \times 10^{10}$ n/m^2

$\Delta l = 7.34 \times 10^{-4}$ m

7. From development in problem 6 solution $\Delta l = Fl/AY$

$\Delta l = 1.25$ n $\times 215$ cm/5.00×10^{-2} cm$^2 \times 9.02 \times 10^6$ n/cm^2

$\Delta l = 5.96 \times 10^{-4}$ cm

GROUP B

8. Tabulating the values given and those calculated, we have:

F	A	Stress F/A	Δl	Strain $\Delta l/l$
1 n	1.26×10^{-3} cm^2	794 n/cm^2	0.0224 cm	1.12×10^{-4}
2 n	1.26×10^{-3} cm^2	1590 n/cm^2	0.0464 cm	2.32×10^{-4}
3 n	1.26×10^{-3} cm^2	2380 n/cm^2	0.0681 cm	3.41×10^{-4}
4 n	1.26×10^{-3} cm^2	3170 n/cm^2	0.0897 cm	4.48×10^{-4}
5 n	1.26×10^{-3} cm^2	3970 n/cm^2	0.1128 cm	5.64×10^{-4}

Plot stress as the ordinate and strain as the abscissa. Draw the straight line that best fits the points. The slope of the curve will be stress/strain or Y, Young's modulus.

9. $Y = Fl/\Delta l A$ and $F = Y(\Delta l/l)A$

$F = 9.00 \times 10^{10}$ n/m$^2 \times 2.00 \times 10^{-4} \times \pi \times (3.00 \times 10^{-3}$ m$)^2$

$F = 5.10 \times 10^2$ n

Questions • Chapter 7/pp. 162–163

GROUP A

1. The constant, random vibration of molecules in a liquid.
2. A layer of water is spread out over a layer of copper (II) sulfate solution. Upon long standing, evidence of diffusion can be seen.
3. (a) Alcohol is more adhesive while mercury is more cohesive. (b) Cohesion causes viscosity.
4. It lowers the surface tension of the water and the needle will sink.
5. Surface tension causes a bubble to have the least surface area for its volume, which produces the spherical shape.
6. Whether or not a liquid wets (concave) or does not wet (convex) the container.
7. (a) No change. (b) Height increases. (c) Height decreases.
8. (a) Solid to liquid. (b) Melting point. (c) The change of phase from a liquid to a solid. (d) Freezing point.
9. (a) Water. (b) Ice has a crystal structure in which the molecules are farther apart than they are in water, which does not have a crystal structure.
10. The melting under pressure and the refreezing of a substance.
11. (a) Gas molecules are independent, rapidly moving particles. (b) The molecules constantly bombard the surface of their container. (c) Gas molecules are in constant and rapid motion.

GROUP B

12. Small groups of molecules may cling together, but the molecules can move over one

another easily. The force of attraction between liquid molecules is less than that for solids, although the spacing is about the same.

13. Force exerted by surface film counteracts weight of razor blade.
14. Surface tension will produce spheres.
15. Its cohesive force is greater than its adhesive force for glass.
16. The combined forces of adhesion and surface tension are greater than the weight of the elevated liquid.
17. In a crystalline solid, a definite amount of heat energy is necessary to overcome the bond forces, while in a noncrystalline solid the bond forces and the entangled chains of molecules are not overcome by a definite amount of energy.
18. **(a)** It raises the freezing point. **(b)** It lowers the freezing point.
19. The area of the skate runner in contact is least when the skate is sharp, thus the pressure (force/area) is greatest, and melting that will provide the lubricant is more likely to take place.
20. **(a)** Air would bubble from the end of the tube. **(b)** Liquid would rise in the tube.
21. **(a)** Plasma is a gas that is hot enough to become ionized and conduct electricity.
    (b) Most of the matter in the universe is in the form of stars, which consist almost entirely of plasma.

8   HEAT MEASUREMENTS

Questions • Chapter 8/pp. 170–171
GROUP A
1. Thermal energy is the total energy associated with the random motion and arrangement of particles. Heat is the thermal energy that is absorbed, given up, or transferred from one body to another.
2. See the definition for heat under question 1. The average temperature of a body is proportional to the average kinetic energy of the molecules of which the body is composed.
3. These two points may be easily determined with precision and are used as basic temperatures around which a temperature scale may be constructed.
4. The temperature of the triple point of water.
5. They have the same magnitude.
6. **(a)** $-273.16\,°C$. **(b)** At absolute zero a material has minimum thermal energy.
7. **(a)** The soldering iron has much more thermal energy. **(b)** The 10-section radiator has more thermal energy. **(c)** The kettle has more thermal energy. **(d)** The 20.0 kg of ice has more thermal energy. **(e)** The liter of liquid air has more thermal energy.
8. By the effects that they produce.
9. 10^3 calories, or 4.1868×10^3 joules

GROUP B
10. An increase in the average kinetic energy is proportional to an increase in temperature.
11. See Fig. 8-2.
12. The standard fixed point in thermometry is the triple point of water, $273.16\,°K$. The constant-volume gas thermometer is the standard thermometer. The extrapolated gas scale (identical with the Kelvin scale in the range in which a gas thermometer can be used) is used to define the ideal gas temperature. The use of the standard thermometer makes it

possible to experimentally determine reference points for temperature measurement, i.e., oxygen, normal boiling point: 90.18 °K; water, normal boiling point: 373.15 °K.
13. Slight differences in the diameter of the bore will alter the linear distance between the fixed points.

Problems • Chapter 8/p. 171
GROUP A
1. °K = °C + 273.2°   24.0 °C + 273.2° = 297.2 °K
2. °C = °K − 273.2°   77.0 °K − 273.2° = −196.2 °C
3. °K = °C + 273.2°   −183 °C + 273.2° = 9$\overline{0}$ °K
4. °C = °K − 273.2°   4.1 °K − 273.2° = −269.1 °C
5. $Q = mc\,\Delta T$ $Q = 50\overline{0} \text{ g} \times 1.00 \text{ cal/g C°} \times (10\overline{0} \text{ °C} - 2\overline{0} \text{ °C})$
   $Q = 4.00 \times 10^4$ cal
6. $8\overline{0} - 3\overline{0} = 5\overline{0}$ scale divisions between the ice point and the steam point. This means each scale division = 2 °C. $125 - 8\overline{0} = 45$ scale divisions above $10\overline{0}$ °C. $10\overline{0}$ °C + (2 C°/scale div. × 45 scale div.) = $19\overline{0}$ °C

Questions • Chapter 8/p. 181
GROUP A
1. The change in unit length of a solid when its temperature is changed one degree.
2. (a) Asphalt- or tar-filled joints. (b) A metal expansion joint. (c) The ring is cut.
3. The diameter of the hole increases.
4. The coefficients of expansion of platinum and glass are the same while the coefficients of expansion of copper and glass are different enough that expansion or contraction breaks the seal.
5. The coefficient of volume expansion of mercury is greater than the coefficient of volume expansion of glass.
6. Mercury and alcohol are liquids over a wide and useful temperature range and have a uniform rate of expansion. Alcohol remains a liquid at temperatures below −100 °C.
7. It cannot expand; therefore the increased kinetic energy increases its pressure.
8. Standard temperature and pressure.
9. (a) A mole is a specific amount of a substance. (b) A mole contains the Avogadro number of molecules of a substance.
10. The density varies directly with pressure.

GROUP B
11. No. The coefficient of linear expansion is the change in length per unit length per degree change in temperature. $\Delta l/l$ is a ratio that is independent of the unit used for the measure of l.
12. The increased energy increases the amplitude of the molecular vibrations.
13. (a) The closer movement of molecules due to the collapse of crystal structure predominates. (b) The increased amplitude of molecular vibration predominates.
14. The bulb or tube also expands or contracts.
15. It should be cooled.
16. Because the molecules are so widely separated that the attractive forces between them are negligible.
17. The distance separating the molecules makes their size and attractive force negligible.

Problems • Chapter 8/pp. 181–182

GROUP A

1. $\Delta l = \alpha l \, \Delta T$ $\Delta l = 16.8 \times 10^{-6}/\text{C}° \times 5.00 \text{ m} \times (70 \text{ °C} - 20 \text{ °C})$
                    $\Delta l = 4.20 \times 10^{-3} \text{ m}$

2. $\Delta l = \alpha l \, \Delta T$ $\Delta l = 18.8 \times 10^{-6}/\text{C}° \times 100.0 \text{ cm} \times (100.0 \text{ °C} - 0.0 \text{ °C})$
                    $\Delta l = 1.88 \times 10^{-1} \text{ cm}$

3. $\Delta l = \alpha l(T - T_0)$ $\Delta l = 10.5 \times 10^{-6}/\text{C}° \times 150\bar{0} \text{ m} \times (30.0 \text{ °C} - 10.0 \text{ °C})$
                    $\Delta l = 0.315 \text{ m}$ $150\bar{0} \text{ m} - 0.315 \text{ m} = 150\bar{0} \text{ m}$ to four significant
                    figures

4. $\Delta l = \alpha l(T - T_0)$ $\Delta l = 1.93 \times 10^{-5}/\text{C}° \times 1.500 \text{ cm} \times (15\bar{0} \text{ °C} - 2\bar{0} \text{ °C})$
$\Delta l = 3.76 \times 10^{-3} \text{ cm}$ $(1.500 + 0.0037) \text{ cm} = 1.504 \text{ cm}$

5. $\Delta V = \beta V(T - T_0)$
$\Delta V = 1.24 \times 10^{-3}/\text{C}° \times 500.00 \text{ mL} \times (45.0 \text{ °C} - 20.0 \text{ °C})$
$\Delta V = 15.4 \text{ mL}$ $500.00 \text{ mL} + 15.4 \text{ mL} = 515.4 \text{ mL}$

6. $\Delta V = \beta V(T - T_0)$
$\Delta V = (1.08 \times 10^{-3}/\text{C}°)(125.0 \text{ L})(25.0 \text{ °C} - 15.0 \text{ °C})$
$\Delta V = 1.35 \text{ L}$ $125.0 \text{ L} + 1.35 \text{ L} = 126.4 \text{ L}$

7. $V/T_K = V'/T_K'$ $V' = VT_K'/T_K$
$V' = [250.0 \text{ cm}^3 \times (67 \text{ °C} + 273°)]/(37 \text{ °C} + 273°)$
$V' = 274 \text{ cm}^3$

8. $pV = k$ $pV = p'V'$ $V' = Vp/p'$
$V' = 2.00 \text{ liters} \times 73.5 \text{ mm}/760 \text{ mm}$
$V' = 1.93 \times 10^{-1} \text{ liter}$

9. $V' = Vp/p'$ $V' = 465 \text{ cm}^3 \times 725 \text{ mm}/825 \text{ mm}$
$V' = 408 \text{ cm}^3$

10. $D/D' = p/p'$
$D' = Dp'/p$
$D' = 1.55 \text{ g/liter} \times 725 \text{ mm}/760 \text{ mm}$
$D' = 1.47 \text{ g/liter}$

GROUP B

11. $\Delta l = \alpha l(T - T_0)$ $\alpha = \Delta l(T - T_0)$
$\alpha = 0.0075 \text{ m}/12.0 \text{ m}[32.0 \text{ °C} - (-18.0 \text{ °C})]$ $\alpha = 1.2 \times 10^{-5}/\text{C}°$

12. $\Delta V = \beta V(T - T_0)$
$\Delta V = 1.82 \times 10^{-4}/\text{C}° \times 1.000 \text{ cm}^3 \times (100.0 \text{ °C} - 20.0 \text{ °C})$ $\Delta V = 0.0146 \text{ cm}^3$
$1.000 \text{ cm}^3 + 0.0146 \text{ cm}^3 = 1.015 \text{ cm}^3$ new volume
$D_m = m/V$ $D_m = 13.6 \text{ g}/1.015 \text{ cm}^3$ $D_m = 13.4 \text{ g/cm}^3$

13. $\Delta l = \alpha l \, \Delta T$ $\Delta l = 10.5 \times 10^{-6}/\text{C}° \times 100.0 \text{ m} \times 115 \text{ C}°$
$\Delta l = 0.121 \text{ m}$, the space that must be left for expansion

14. The distance between the marks was greater than 25.00 m by the amount of the expansion of a 25.00 m steel tape when heated through 25.00 C°, $\Delta l = \alpha l \, \Delta T$
$\Delta l = 10.5 \times 10^{-6}/\text{C}° \times 25.000 \text{ m} \times 25.00 \text{ °C}$ $\Delta l = 6.55 \times 10^{-3} \text{ m}$ **(b)** The distance between the lines is 25.007 m.   **(b)** The reading of the tape at 0.00 °C is 25.007 m

15. The diameter of the copper ring must be increased by 0.020 cm $\Delta l = \alpha l \, \Delta T$
$\Delta T = \Delta l/\alpha l$ $\Delta T = 0.020 \text{ cm}/16.8 \times 10^{-6}/°\text{C} \times 3.98 \text{ cm}$ $\Delta T = 30\bar{0} \text{ C}°$
Final temperature $= 20.\underline{0} \text{ °C} + 30\bar{0} \text{ C}° = 320 \text{ °C}$

16. $P' = PT'/T$ $P' = 60\bar{0} \text{ mm} \times 303 \text{ °K}/273 \text{ °K}$
$P' = 666 \text{ mm of mercury, pressure at } 30.0 \text{ °C}$

17. $V' = VT'/T = 60\overline{0}$ cm³ $\times$ 353 °K/273 °K
$V' = 776$ cm³, the volume at $8\overline{0}$ °C
18. $\Delta V_W = 2.07 \times 10^{-4}/$C° $\times$ 1000.0 cm³ $\times$ 60.0 C°
$\Delta V_W = 12.4$ cm³
$\Delta V_g = 3 \times 3.3 \times 10^{-6}/$C° $\times$ 1000.0 cm³ $\times$ 60.0 C°
$\Delta V_g = 0.59$ cm³
$\Delta V_W - \Delta V_g = 12.4$ cm³ $- 0.59$ cm³ $= 11.8$ cm³
The volume reading will be 1011.8 cm³
19. $T' = TP'V'/PV$
$T' = 300.0$ °K $\times$ 80.0 cm $\times$ 1.55 liters/76.0 cm $\times$ 1.25 liters
$T' = 392$ °K

20. $R = \dfrac{PV}{nT_K}$

$P = \dfrac{75.0 \text{ cm}}{76.0 \text{ cm}} = 0.987$ atm

$n = \dfrac{PV}{RT_K}$

$n = \dfrac{(0.987 \text{ atm})(1.75 \text{ L})}{(8.2057 \times 10^{-2} \text{ atm/mole °K})(298.2 \text{ °K})}$

$n = 7.06 \times 10^{-2}$ mole

Questions • Chapter 8/p. 187
GROUP A
1. The quantity of heat required to raise the temperature of a body 1 °C.
2. (a) The ratio of the heat capacity to the mass.   (b) $c = Q/m\ \Delta T$.
3. They are less.
4. $Q = mc\ \Delta T$
5. The heat given off by hot substances equals the heat received by cold substances.
6. Temperatures before and after, masses of all substances involved, and specific heats of all but one substance.
7. Law of heat exchange.

Problems • Chapter 8/p. 187
GROUP A
1. $Q = mc\ \Delta T$ $Q = 85$ g $\times$ 0.0305 cal/g C° $\times$ $(20\overline{0}$ °C $- 1\overline{0}$ °C) $Q = 4.9 \times 10^2$ cal
2. $Q_{lost} = Q_{gained}$ $m_W c_W (T_W - T_f) = m_{W'} c_{W'} (T_f - T_{W'})$
$T_f = (m_W c_W T_W + m_{W'} c_{W'} T_{W'})/(m_W c_W + m_{W'} c_{W'})$
$T_f = \dfrac{(20.0 \text{ g} \times 1.00 \text{ cal/g C°} \times 30.0 \text{ °C} + 10.0 \text{ g} \times 1.00 \text{ cal/g C°} \times 0.0 \text{ °C})}{(20.0 \text{ g} \times 1.00 \text{ cal/g C°} + 10.0 \text{ g} \times 1.00 \text{ cal/g C°})}$
$T_f = 20.0$ °C
3. $Q_{lost} = Q_{gained}$ $m_s c_s (T_s - T_f) = m_W c_W (T_f - T_W) + m_b c_b (T_f - T_b)$
$T_f = (m_s c_s T_s + m_W c_W T_W + m_b c_b T_b)/(m_s c_s + m_W c_W + m_b c_b)$
$T_f = (20\overline{0}$ g $\times$ .0562 cal/g C° $\times$ 100.0 °C $+ 135$ g $\times$ 1.00 cal/g C° $\times$ 21.0 °C $+$
45.0 g $\times$ 0.0917 cal/g C° $\times$ 21.0 °C)/200 g $\times$ .0562 cal/g C° $+ 135$ g $\times$
1.00 cal/g C° $+ 45.0$ g $\times$ 0.0917 cal/g C° $T_f = 26.9$ °C

4. $Q_{lost} = Q_{gained}$ $m_t c_t(T_t - T_f) = m_w c_w(T_f - T_w)$
$c_t = [m_w c_w(T_f - T_w)]/[m_t(T_t - T_f)]$
$c_t = [100\overline{0}\ g \times 1.00\ cal/g\ C° (20.0\ °C - 10.0\ °C)]/[225\ g\ (100.0\ °C - 20.0\ °C)]$
$c_t = 0.0556\ cal/g\ C°$

GROUP B

5. $Q_{lost} = Q_{gained}$ $m_m c_m(T_m - T_f) = m_w c_w(T_f - T_w) + m_c c_c(T_f - T_c)$
$c_m = [m_w c_w(T_f - T_w) + m_c c_c(T_f - T_c)]/[m_m(T_m - T_f)]$
$c_m = [100.0\ g \times 1.00\ cal/g\ C° \times (70.0\ °C - 0.0\ °C) + 50.0\ g \times$
$0.200\ cal/g\ C° \times (70.0\ °C - 0.0\ °C)]/[100\overline{0}\ g\ (300.0\ °C - 70.0\ °C)]$
$c_m = 3.35 \times 10^{-2}\ cal/g\ C°$

6. $Q_{lost} = Q_{gained}$ $m_m c_m(T_m - T_f) = m_w c_w(T_f - T_w) + m_c c_c(T_f - T_c)$
$c_m = [m_w c_w(T_f - T_w) + m_c c_c(T_f - T_c)]/[m_m(T_m - T_f)]$
$c_m = [300.0\ g \times 1.00\ cal/g\ C° \times (30.0\ °C - 20.0\ °C) + 75.0\ g \times$
$0.214\ cal/g\ C° \times (30.0\ °C - 20.0\ °C)]/[500.0\ g\ (100.0\ °C - 30.0\ °C)]$
$c_m = 9.03 \times 10^{-2}\ cal/g\ C°$

7. $Q_{lost} = Q_{gained}$ $m_m c_m(T_m - T_f) = m_t c_t(T_f - T_t)$
$c_t = [m_m c_m(T_m - T_f)]/[m_t(T_f - T_t)]$
$c_t = [95.3\ g \times 0.092\ cal/g\ C° \times (90.5° - 35.5\ °C)]/[75.2\ g\ (35.5\ °C - 20.5\ °C)]$
$c_t = 0.43\ cal/g\ C°$

8. $Q_{lost} = Q_{gained}$ $m_g c_g(T_f - T_g) + m_w c_w(T_f - T_w) = m_a c_a(T_a - T_f)$ $T_g = T_w$

$$T_f = \frac{T_g(m_g c_g + m_w c_w) + m_a c_a T_a}{m_g c_g + m_w c_w + m_a c_a}$$

$$T_f = \frac{20.0\ °C\ [(35\overline{0}\ g)(0.161\ cal/g\ C°) + (50\overline{0}\ g)(1.00\ cal/g\ C°)] + (40\overline{0}\ g)(0.581\ cal/g\ C°)(50.0\ °C)}{(35\overline{0}\ g)(0.161\ cal/g\ C°) + (50\overline{0}\ g)(1.00\ cal/g\ C°) + (40\overline{0}\ g)(0.581\ cal/g\ C°)}$$

$T_f = 29.1\ °C$

Questions • Chapter 8/pp. 196–197
GROUP A

1. See Fig. 8-15.
2. (a) The amount of heat needed to melt a unit mass of a substance at its normal melting point. (b) 80 cal/g at 0 °C
3. The cooling of a substance below its normal freezing point without a change of state.
4. (a) For the rapid cooking of food. (b) For the production of sugar crystals.
5. (a) The liquid to be distilled is boiled and the vapor condensed in a separate vessel.
(b) Distillation that makes use of the fact that different substances have different boiling points and may be separated because of these differences.
6. (a) The heat required to vaporize a unit mass of liquid at its normal boiling point.
(b) 539 cal/g at 100 °C
7. The steam has 539 cal/g more heat to give up due to the heat of vaporization.
8. The upper limit of the vaporization curve of a substance.
9. (a) −10 °C (b) 0 °C (c) 0 °C (d) 4 °C

GROUP B

10. The line TB slopes to the right of the vertical since an increase in temperature raises the freezing temperature.
11. (a) The solid and vapor changes to a liquid. (b) The solid and liquid vaporizes.
(c) The liquid and vapor solidifies. (d) The solid and liquid vaporizes.
12. The curve CT.

13. The solid vaporizes, i.e., sublimes.

14. **(a)** The surface pressure determines the energy needed for the molecules to escape during boiling. **(b)** The boiling temperature is higher with increased pressure.

15. The different materials condense at different temperatures and can therefore be collected separately.

16. The mass, temperature, and specific heat of steam and water.

17. A change in boiling temperature changes the energy required for molecules to escape. This increased or decreased energy requirement causes the latent heat to vary.

Problems • Chapter 8/pp. 197–198
GROUP A

1. $Q = mL_f$ $Q = 1.50 \text{ kg} \times 10^3 \text{ g/kg} \times 8\bar{0} \text{ cal/g}$ $Q = 1.2 \times 10^5 \text{ cal}$

2. $Q_{lost} = Q_{gained}$ $m_m c_m(T_m - T_f) = m_i L_f$ $T_m = (m_m c_m T_f + m_i L_f)/(m_m c_m)$
   $T_m = (2270 \text{ g} \times 0.1075 \text{ cal/g C°} \times 0.0 \text{ C°} + 1150 \text{ g} \times 80 \text{ cal/g})/2270 \text{ g} \times 0.1075 \text{ cal/g C°} = 380 \text{ °C}$

3. $Q_{lost} = Q_{gained}$ $m_m c_m(T_m - T_f) = m_i L_f$ $m_i = [m_m c_m(T_m - T_f)]/L_f$
   $m_i = [500 \text{ g} \times 0.217 \text{ cal/g C°} \times (35\bar{0} \text{ °C} - 0.00 \text{ °C})]/(8\bar{0} \text{ cal/g})$
   $m_i = 475 \text{ g}$

4. $Q_{lost} = Q_{gained}$ $m_c c_c(T_c - T_f) + m_w c_w(T_w - T_f) = m_i L_f + m_i c_w(T_f - 0.00 \text{ °C})$
   $m_i = [m_c c_c(T_c - T_f) + m_w c_w(T_w - T_f)]/[L_f + c_w(T_f - 0.00 \text{ °C})]$
   $m_i = [220.0 \text{ g} \times 0.924 \text{ cal/g C°} \times (21.0 - 5.00) \text{ °C} + 450.0 \text{ g} \times 1.00 \text{ cal/g C°} \times (21.0 - 5.00) \text{ °C}]/[8\bar{0} \text{ cal/g} + 1.00 \text{ cal/g C°} (5.00 - 0.00) \text{ °C}]$
   $m_i = 88.8 \text{ g}$

5. $Q_{lost} = Q_{gained}$
   $m_m c_m(T_m - T_f) = m_i c_i(0.0 \text{ °C} - T_i) + m_i L_f + m_i c_w(T_f - 0.0 \text{ °C})$
   $T_m = [m_m c_m T_f + m_i c_i(0.0 \text{ °C} - T_i) + m_i L_f + m_i c_w(T_f - 0.0 \text{ °C})]/m_m c_m$
   $T_m = \{500.0 \text{ g} \times 0.0917 \text{ cal/g C°} \times 20.0 \text{ °C} + 60.0 \text{ g} \times 0.530 \text{ cal/g C°} \times [0.0 \text{ °C} - (-20.0 \text{ °C})] + 60.0 \text{ g} \times 79.71 \text{ cal/g} + 60.0 \text{ g} \times 1.00 \text{ cal/g C°} \times (20.0 \text{ °C} - 0.0 \text{ °C})\}/(500.0 \text{ g} \times 0.0917 \text{ cal/g C°})$
   $T_m = 164 \text{ °C}$

6. $Q = mL_v$ $Q = 50.0 \text{ g} \times 538.7 \text{ cal/g}$
   $Q = 2.69 \times 10^4 \text{ cal}$

7. $Q = mL_v$ $Q = 2.00 \times 10^4 \text{ g} \times 538.7 \text{ cal/g}$ $Q = 1.08 \times 10^7 \text{ cal}$

8. $Q = mL_v + mc \, \Delta T + mL_f$
   $Q = 4.00 \times 10^3 \text{ g} \times 538.7 \text{ cal/g} + 4.00 \times 10^3 \text{ g} \times 1.00 \text{ cal/g C°} \times 10\bar{0} \text{ °C} + 4.00 \times 10^3 \text{ g} \times 79.71 \text{ cal/g}$
   $Q = 2.87 \times 10^6 \text{ cal}$

9. $Q = mL_v$ $m = Q/L_v$ $m = 1.00 \times 10^3 \text{ cal}/(70.6 \text{ cal/g})$ $m = 14.2 \text{ g}$

10. $Q_{lost} = Q_{gained}$ $m_s L_v + m_s c_w(100.0 \text{ °C} - T_f) = m_w c_w(T_f - T_w) + m_c c_c(T_f - T_c)$
    $T_f = (m_s L_v + 100.0 \text{ °C} \times m_s c_w + T_w m_w c_w + m_c c_c T_c)/m_s c_w + m_w c_w + m_c c_c$
    $T_f = (11.4 \text{ g} \times 538.7 \text{ cal/g} + 100.0 \text{ °C} \times 11.4 \text{ g} \times 1.00 \text{ cal/g C°} + 681.0 \text{ g} \times 1.00 \text{ cal/g C°} \times 25.0 \text{ °C} + 182.0 \text{ g} \times 0.220 \text{ cal/g C°} \times 25.0 \text{ °C})/11.4 \text{ g} \times 1.00 \text{ cal/g C°} + 681.0 \text{ g} \times 1.00 \text{ cal/g C°} + 182.0 \text{ g} \times 0.220 \text{ cal/g C°}$
    $T_f = 34.6 \text{ °C}$

GROUP B

11. $Q_{lost} = Q_{gained}$ $m_w c_w(T_w - T_f) = m_i L_f + m_i c_w(T_f - 0.00 \text{ °C})$
    $T_f = (m_w c_w T_w - m_i L_f)/(m_w c_w - m_i c_w)$
    $T_f = (50.0 \text{ g} \times 1.00 \text{ cal/g C°} \times 80.0 \text{ °C} - 50.0 \text{ g} \times 79.7 \text{ cal/g})/(50.0 \text{ g} \times$

1.00 cal/g C° − 50.0 g × 1.00 cal/g C°)

$T_f = 0.00$ °C

12. $Q_{lost} = Q_{gained}$ $m_c c_c(T_c − T_f) + m_w c_w(T_w − T_f) = m_i L_f + m_i c_w(T_f − 0.0$ °C$)$

$L_f = [m_c c_c(T_c − T_f) + m_w c_w(T_w − T_f) − m_i c_w T_f]/m_i$

$L_f = [200.0$ g $× 0.100$ cal/g C° $× (40.0$ °C $− 23.8$ °C$) + 300.0$ g $× 1.00$ cal/g C° $×$
$(40.0$ °C $− 23.8$ °C$) − 50.0$ g $× 1.00$ cal/g C° $× 23.8$ °C$]/50.0$ g

$L_f = 79.8$ cal/g

13. $Q_{lost} = Q_{gained}$ $m_s c_s(T_s − T_f) + m_c c_c(T_c − T_f) + m_w c_w(T_w − T_f) = m_i c_i(0$ °C $− T_i) +$
$m_i L_f + m_i c_w(T_f − 0$ °C$)$ $T_f = (m_s c_s T_s + m_c c_c T_c + m_w c_w T_w + m_i c_i T_i − m_i L_f)/$
$(m_s c_s + m_c c_c + m_w c_w + m_i c_w)$

$T_f = (500.0$ g $× 0.0562$ cal/g C° $× 100.0$ °C $+ 50.0$ g $× 0.100$ cal/g C° $× 30.0$ °C $+$
300.0 g $× 1.00$ cal/g C° $× 30.0$ °C $+ 50.0$ g $× 0.530$ cal/g C° $× (−10.0$ °C$) −$
50.0 g $+ 79.7$ cal/g$)/(500.0$ g $× 0.0562$ cal/g C° $+ 50.0$ g $× 0.100$ cal/g C° $+$
300.0 g $× 1.00$ cal/g C° $+ 50.0$ g $× 1.00$ cal/g C°$)$

$T_f = 20.1$ °C

14. $Q_{lost} = Q_{gained}$ $m_w c_w(T_i − T_f) + m_c c_c(T_i − T_f) = m_i L_f + m_i c_w T_f$

$T_f = m_w c_w T_i + m_c c_c T_i − m_i L_f/m_w c_w + m_c c_c + m_i c_w$

$T_f = 3400$ g $× 1.00$ cal/g C° $× 93.3$ °C $+ 1350$ g $× 0.090$ cal/g C° $× 93.3$ °C $− 900$ g $×$
79.7 cal/g$/3400$ g $× 1.00$ cal/g C° $+ 1350$ g $× 0.090$ cal/g C° $+ 900$ g $×$
1.00 cal/g C°

$T_f = 58.1$ °C

15. $Q_{lost} = Q_{gained}$

$m_s L_v + m_s c_w(100.0$ °C $− T_f) = m_w c_w(T_f − T_w) + m_c c_c(T_f − T_c)$

$L_v = [m_w c_w(T_f − T_w) + m_c c_c(T_f − T_c) − m_s c_w(100.0$ °C $− T_f)]/m_s$

$L_v = [150.0$ g $× 1.00$ cal/g C° $× (73.9$ °C $− 20.0$ °C$) + 75.0$ g $× 0.100$ cal/g C° $×$
$(73.9$ °C $− 20.0$ °C$) − 15.0$ g $× 1.00$ cal/g C° $× (100.0$ °C $− 73.9$ °C$)]/15.0$ g

$L_v = 540$ cal/g

16. $Q_{lost} = Q_{gained}$

$m_s L_v + m_s c_w(100.0$ °C $− T_f) = m_i L_f + m_w c_w(T_f − 0.0$ °C$) + m_c c_c(T_f − 0.0$ °C$)$

$m_i = [m_s L_v + m_s c_w(100.0$ °C $− T_f) − m_w c_w(T_f − 0.0$ °C$) − m_c c_c(T_f − 0.0$ °C$)]/L_f$

$m_i = [40.0$ g $× 538.7$ cal/g $+ 40.0$ g $× 1.00$ cal/g C° $× (100.0$ °C $− 60.0$ °C$) −$
200.0 g $× 1.00$ cal/g C° $× (60.0$ °C $− 0.0$ °C$) − 100.0$ g $× 0.200$ cal/g C° $×$
$(60.0$ °C $− 0.0$ °C$)]/(79.7$ cal/g$)$

$m_i = 125$ g

17. $Q_{lost} = Q_{gained}$

$m_s c_s(T_s − 100.0$ °C$) + m_s L_v + m_s c_w(100.0$ °C $− T_f) = m_a c_a(T_f − T_a) + m_b c_b(T_f − T_b) +$
$m_w c_w(T_f − T_w)$

$T_f = m_s c_s(T_s − 100.0$ °C$) + m_s L_v + 100.0$ °C $× m_s c_w + m_a c_a T_a + m_b c_b T_b +$
$m_w c_w T_w)/(m_s c_w + m_a c_a + m_b c_b + m_w c_w)$

$T_f = [25.0$ g $× 0.481$ cal/g C° $× (120.0$ °C $− 100.0$ °C$) + 25.0$ g $× 538.7$ cal/g $+$
100.0 °C $× 25.0$ g $× 1.00$ cal/g C° $+ 50.0$ g $× 0.220$ cal/g C° $× 20.0$ °C $+ 100.0$ g $×$
0.0917 cal/g C° $× 20.0$ °C $+ 250.0$ g $× 1.00$ cal/g C° $× 20.0$ °C$]/(25.0 ×$
1.00 cal/g C° $+ 50.0$ g $× 0.220$ cal/g C° $+ 100.0$ g $× 0.0917$ cal/g C° $+ 250.0$ g $×$
1.00 cal/g C°$)$

$T_f = 73.2$ °C

18. $Q_{lost} = Q_{gained}$ $m_c c_c(T_c − T_f) = m_w c_w(100.0$ °C $− T_w) + m_s L_v$

$T_c = (m_c c_c T_f + 100.0$ °C $× m_w c_w − m_w c_w T_w + m_s L_v)/m_c c_c$

T_c = 4540 g × 0.0924 cal/g C° × 100.0 °C + 100.0 °C × 1360 g × 1.00 cal/g C° −
1360 g × 1.00 cal/g C° × 22.0 °C + 450 g × 538.7 cal/g/4540 g ×
0.0924 cal/g C°

T_c = 929 °C

HEAT ENGINES   9

Questions • Chapter 9/pp. 205–206
GROUP A
 1. The study of the quantitative relationships between heat and other forms of energy.
 2. The quantity of energy supplied to any system in the form of heat is equal to the work done by the system plus the change in internal energy of the system.
 3. Q heat units are equivalent to JQ work units where J is 4.19 joules/calorie. J is the symbol for the mechanical equivalent of heat.
 4. 4.19 joules/calorie.
 5. (a) A process that occurs without a change in temperature. (b) A process that occurs without the addition or withdrawal of heat from the surroundings.
 6. (a) Expansion or contraction of an air mass when the atmospheric pressure changes. (b) A diesel engine.
 7. The specific heat of a gas varies, while the specific heat of a solid or liquid is relatively fixed.
 8. The specific heat at constant volume is the quantity of heat required to change the temperature of a unit mass of gas by 1 C° when the volume is held constant. The specific heat at constant pressure is the quantity of heat required to raise the temperature of a unit mass of gas by 1 C° when the pressure is held constant.

GROUP B
 9. It differs only in that the mechanical conservation of energy principle is a broader concept encompassing the first law of thermodynamics.
10. Joule measured the heat introduced into water by the friction of a paddle wheel driven by known weights moving through known distances.
11. Boyle's law.
12. Boyle's and Charles' laws.
13. (a) A heat source.   (b) A heat sink receives it.
14. (a) The thermal energy of the gas. (b) It increases the thermal energy of the gas.
15. (a) The average kinetic energy of the molecules remains constant since the temperature of a gas does not change during an isothermal expansion. (b) The gas must receive energy from an external heat source.
16. (a) Yes. The gas cools; therefore the average kinetic energy of its molecules is reduced. (b) Energy is taken from the gas.

Problems • Chapter 9/p. 206
Note: In all problems dealing with the mechanical equivalent of heat, dimensional analysis requires the use of the conversion factor 4.19 j/cal.
GROUP A
 1. W = 4.19 joules/cal × 1.00 × 10⁴ cal

$W = 4.19 \times 10^4$ joules
2. $W = 4.19$ joules/cal $\times$ 1.00 liter $\times$ 1000 cm³/liter $\times$ 0.700 g/cm³ $\times 1.15 \times 10^4$ cal/g
   $W = 3.37 \times 10^7$ joules
3. Assuming all potential energy of water is converted to heat at bottom of the fall,

   $E_p = Q$ $E_p = mgh$ $Q = mc\,\Delta T$ Thus $mgh = mc\,\Delta T$

   $$\Delta T = \frac{mgh}{mc} = \frac{gh}{c} = \frac{9.81 \text{ m/s}^2 \times 50.6 \text{ m}}{1.00 \text{ kcal/kg C}^\circ} \times \frac{\text{kcal}}{4.19 \times 10^3 \text{ j}}$$

   $\Delta T = 0.118$ C°, the change in temperature of the water

GROUP B

4. $P = \dfrac{W}{t} = 2.00$ g/s $\times$ 0.250 $\times$ 1.00 $\times 10^5$ cal/g $\times$ 4.19 joules/cal $\times$ 1 kw/1000 w

   $P = 2.10 \times 10^2$ kw

5. $Q = E_k = 1/2 mv^2$

   $$Q = \frac{1.25 \text{ kg} \times (26.6 \text{ m/s})^2}{2 \times 4.19 \text{ j/cal}} = 106 \text{ cal}$$

6. $E_k = \frac{1}{2}mv^2 = Q$

   $$v^2 = \frac{2Q}{m} \qquad v = \sqrt{\frac{2Q}{m}}$$

   $$v = \sqrt{\frac{2 \times 4.19 \text{ j/cal} \times 75\overline{0} \text{ cal} \times 0.300}{0.00500 \text{ kg}}} = 614 \text{ m/s}$$

7. $E_p = Q$ $E_p = mgh$ and $Q = mc\,\Delta T$ Thus $mgh = mc\,\Delta T$

   $$\Delta T = \frac{mgh}{mc} = \frac{gh}{c} = \frac{9.81 \text{ m/s}^2 \times 125 \text{ m}}{2 \times 0.0305 \text{ kcal/kg C}^\circ \times 4.19 \times 10^3 \text{ j/kcal}}$$

   $\Delta T = 4.79$ °C

8. $E_k = Q$ $E_k = \frac{1}{2}mv^2$ and $Q = mL_f$

   Thus $\frac{1}{2}mv^2 = mL_f$ and $v^2 = \dfrac{2mL_f}{m} = 2L_f$

   $v = \sqrt{2L_f J} = \sqrt{2 \times 79.7 \text{ kcal/kg} \times 4.19 \times 10^3 \text{ j/kcal}}$
   $v = 816$ m/s

9. $E_p = Q$ $E_p = mgh$ and $Q = (mc\,\Delta T)_w + (mc\,\Delta T)_c$
   Thus $mgh = [(mc\,\Delta T)_w + (mc\,\Delta T)_c]$

   $$= \frac{mgh}{(mc\,\Delta T)_w + (mc\,\Delta T)_c}$$

   $$= \frac{2.50 \text{ kg} \times 9.80 \text{ m/s}^2 \times 1.50 \text{ m} \times 25}{(70\overline{0} \text{ g} \times 1.00 \text{ cal/g C}^\circ \times 0.31 \text{ C}^\circ) + (15\overline{0} \text{ g} \times 0.0924 \text{ cal/g C}^\circ \times 0.31 \text{ C}^\circ)}$$

   $= 4.2$ j/cal

Questions • Chapter 9/p. 220
GROUP A
   1. **(a)** To furnish heat energy to a substance without undergoing any appreciable tempera-
      ture change itself.   **(b)** To receive heat energy from a substance without undergoing any
      appreciable temperature change itself.
   2. The Kelvin temperatures of the heat source and of the heat sink.
   3. It is not possible to construct an engine whose sole effect is the extraction of heat from a

heat source at a single temperature and the conversion of this heat completely into mechanical work.
4. Entropy is the amount of energy that cannot be converted into mechanical work. Entropy can also be defined as the measure of disordered energy.
5. In an external combustion engine, the fuel burns outside the engine. In an internal combustion engine, the fuel burns inside the engine.
6. Steam generated in a boiler passes into a sphere and out through nozzles. The reaction force rotates the sphere.
7. The pascal, or Pa, is equal to a newton per square meter.
8. Ship propulsion and electric power generation.
9. Compression ratio, piston displacement, and rate of fuel consumption.
10. Turbocharging is the use of a turbine to increase the pressure of the air before it enters an engine.
11. (a) With spark plugs. (b) By the heat of compression.
12. High temperatures, poor acceleration, and high emission levels.
13. A turbojet is a combination gas turbine and ramjet.
14. A jet engine carries fuel and uses the surrounding air as an oxidizer. A rocket carries both fuel and an oxidizer.
15. (a) Specific impulse is the product of the thrust-to-weight ratio and the burning time of a propellant. (b) thrust/weight × time = (n/n)s = seconds.
16. By using a heat pump.
17. (a) A heat pump can be used to heat a house in the winter or to cool it in the summer. (b) A heat pump moves heat from one place to another. This is more efficient than generating heat, as in an electric heater.

Problems • Chapter 9/p. 220
GROUP B

1. $e = \dfrac{T_1 - T_2}{T_1} = \dfrac{473\ °K - 373\ °K}{473\ °K} = 0.211$

$e = 21.1\%$

2. $e = \dfrac{T_1 - T_2}{T_1} \qquad T_1 = \dfrac{T_2}{1 - e} = \dfrac{373\ °K}{1 - 0.300}$

$T_1 = 533\ °K$

3. (a) $\Delta S_i = \dfrac{\Delta Q_i}{T_i} = \dfrac{m_i L_f}{T_i} = \dfrac{(5\bar{0}\ \text{g})(8\bar{0}\ \text{cal/g})}{273\ °K} = 15\ \text{cal/degree}$

(b) Since there is no change in the temperature of the water and no change of phase, there is no change in the entropy of the water. (c) The change in entropy of the universe is the sum of (a) and (b), or 15 cal/degree. That is, the melting of the ice increased the entropy of the universe.

WAVES   10

Questions • Chapter 10/p. 232
GROUP A
1. Energy may be transferred by the gross movement of masses of material, by the motion of particles, and by wave motion.

2. A mechanical wave is a disturbance that moves through a material medium. An electromagnetic wave does not require a material medium to transmit the wave energy, but moves through matter or empty space.

3. A *source* to produce a *disturbance* and a *medium* to transmit the disturbance.

4. **(a)** A displacement of some sort in the matter of the medium. **(b)** The displacement of molecules of water from their equilibrium positions. **(c)** The displacement of molecules of the air gases from their equilibrium positions.

5. An elastic medium behaves as if it were an array of particles connected by springs, with each particle occupying an equilibrium position.

6. **(a)** Transverse wave—A wave in which the particles vibrate at right angles to the path along which the wave travels. Longitudinal wave—A wave in which the particles vibrate to and fro along the path that the wave travels. **(b)** Transverse waves—vibrating string; longitudinal waves—sound waves.

7. **(a)** A pulse is a single nonrepeated disturbance. **(b)** A crest is an upward displacement in a transverse wave. **(c)** A trough is a downward displacement in a transverse wave. **(d)** A compression is a distortion in which the material of the medium is more compact than at equilibrium. **(e)** A rarefaction is a distortion in which the material of the medium is less compact than at equilibrium.

8. **(a)** Speed—The distance traveled per unit time. **(b)** Phase—The position and motion of a particle in wave motion. **(c)** Frequency—The number of waves passing a given point in unit time. **(d)** Period—The time required for a single wave to pass a given point. **(e)** Wavelength—The distance between any particle in a wave and the next particle that is in phase with it. **(f)** Amplitude—The maximum displacement from equilibrium of the vibrating particles.

9. It supplies energy.

10. Each advancing wave front is the circumference of an expanding circle and thus increases in length in proportion to r. Assuming no loss of energy, the energy in each unit length of crest is proportional to $1/r$. Since the wave energy is proportional to the square of the amplitude, the amplitude of the wave is proportional to $1/\sqrt{r}$.

11. No. The surface water wave advances as an expanding circle but the sound wave advances as an expanding sphere. Thus the amplitude of the sound wave diminishes more rapidly than that of the water wave.

12. **(a)** Heat transfer by radiation (heat waves). **(b)** Shock waves. **(c)** Shock wave (sound of explosion). **(d)** Heat transfer by convection (gross transport of matter). **(e)** Gross transport of matter. **(f)** Water-surface wave.

Problems • Chapter 10/pp. 232–233
GROUP A

1. $v = f\lambda$ $v = 2.5/s \times 0.60$ m $v = 1.5$ m/s

2. $v = f\lambda$ $\lambda = v/f$ $\lambda = 4.0$ m/s/0.50/s $\lambda = 8.0$ m

3. $v = f\lambda$ $\lambda = v/f$ $\lambda = 15$ m/s/5.0/s $\lambda = 3.0$ m

4. **(a)** 2×1.50 m $= 3.00$ m, wavelength
   (b) $v = f\lambda$ $v = 10.0/s \times 3.00$ m
   $v = 30.0$ m/s

5. $v = f\lambda$ $v = 20/s \times 0.40$ m $v = 8.0$ m/s

6. $v = f\lambda$ $\lambda = v/f$ $\lambda = 20$ m/s/8.0/s $\lambda = 2.5$ m

7. **(a)** $16/4.0$ s $= 1.0/0.25$ s $T = 0.25$ s

(b) $f = \dfrac{1}{T} = \dfrac{1}{0.25 \text{ s}} = 4.0/\text{s} = 4.0 \text{ hz}$

8. $v = f\lambda \qquad f = 1/T$
   $v = \lambda/T = 3.3 \text{ cm}/0.10 \text{ s} = 33 \text{ cm/s}$

Questions • Chapter 10/pp. 249–250
GROUP B

1. A reflected pulse has its displacement in the direction opposite to that of the incident pulse.
2. Rectilinear propagation—object location, aiming, pointing. Reflection—use of mirror, sound echoes. Refraction—camera lens, magnifying lens. Interference—dead spots in auditoriums. Diffraction—sound waves heard around corners.
3. Student diagram.
4. Student diagram.
5. Student diagram.
6. Student answer.
7. Standing waves are produced by two wave trains of the same wavelength, frequency, and amplitude that are traveling in opposite directions through the same medium.
8. The parts of the medium vibrating with a standing wave pattern that never move from their equilibrium positions are nodes. Halfway between the nodes, the amplitude of vibration is a maximum. These points are called loops.
9. None. Each wave is propagated through the medium as though the other were not present. The motion of the medium is the resultant of the motions imparted by the separate waves.
10. Energy is not lost. The total energy of the system remains the same but it is redistributed, the nodal regions having less energy and the antinodal regions having more.
11. A "free" termination of a medium allows unrestrained displacement of the medium at the boundary and the disturbance is reflected without a change in phase. A "fixed" termination restrains the medium allowing no displacement at the boundary. The disturbance is reflected with an inversion in phase.
12. The sound is diffracted around the edges of the door opening.
13. Experiments in diffraction and interference that would reveal the presence or absence of these wave properties.
14. Any device or technique that smooths over an impedance mismatch between two wave-transfer media.
15. The two impedances are badly mismatched.
16. **(1)** Match the impedance of the terminating medium to that of the transmitting medium. **(2)** Insert an appropriate impedance transformer between the two media.

Problems • Chapter 10/p. 250
GROUP B

1. (violet) $f = v/\lambda = \dfrac{3.0 \times 10^8 \text{ m/s}}{4.0 \times 10^{-7} \text{ m}} = 7.5 \times 10^{14} \text{ hz}$

   (red) $\quad f = v/\lambda = \dfrac{3.0 \times 10^8 \text{ m/s}}{7.6 \times 10^{-7} \text{ m}} = 3.9 \times 10^{14} \text{ hz}$

   (range) 3.9×10^{14} to 7.5×10^{14} hz
2. **(a)** 4.0 m

(b) $v = f\lambda = 7.0$ hz $\times$ 4.0 m $= 28$ m/s

3. $v = f\lambda = 15.0$ hz $\times$ 0.410 m $= 6.15$ m/s

4. $v = f\lambda = 1.1$ hz $\times$ 5.0 m $= 5.5$ m/s

5. **(a)** $T = 1/f = \dfrac{1}{11/19 \text{ s}} = 1.7$ s

(b) $\lambda = v/f = vT = \dfrac{24 \text{ m}}{6.5 \text{ s}} \times 1.7$ s $= 6.3$ m

11   SOUND WAVES

Questions • Chapter 11/pp. 264–265

GROUP A

1. By vibrating matter.
2. There is no medium through which sound waves may be transmitted.
3. **(a)** 331.5 m/s **(b)** 0.6 m/s/C°
4. Intensity: The rate at which the sound energy flows through a unit area. Loudness: The effect of the intensity of the sound waves on the ears.
5. Frequency: The number of waves passing a given point in unit time. Pitch: The identification of a certain sound with a definite tone. It depends on the number of vibrations per second that the ear receives.
6. The pitch is higher for the listener than for the engineer.
7. The vibrations are regular for musical tones; irregular for noise.
8. **(a)** 20 hz to 20,000 hz **(b)** Infrasonic. **(c)** Ultrasonic.
9. **(a)** A graph of the minimum intensity for an audible sound plotted as a function of the sound frequency. **(b)** A graph of the upper intensity for an audible sound plotted as a function of the sound frequency. **(c)** 10^{-16} w/cm² **(d)** 0 db **(e)** 10^{-4} w/cm² **(f)** 120 db

GROUP B

10. By successive compression and rarefaction of the molecules in the surrounding medium.
11. **(a)** It changes the amplitude of the sound waves produced. **(b)** It changes the frequency of the sound waves produced.
12. As the density of the gas decreases, less energy is transferred from the sources of the disturbance to the transmitting medium.
13. Because a sound must be 10 times as intense to be twice as loud, 100 times as intense to be three times as loud, etc.
14. The limiting condition is the interparticle spacing of the medium.
15. The sound intensity in the room is 10^4 times higher than the hearing threshold of 10^{-16} watt/cm², or is 10^{-12} watt/cm². This level of sound is about average for a residence.
16. Intensity I is the ratio of sound power P to the unit area A normal to the sound propagation. Relative intensity β is the ratio of the sound intensity I to the intensity of the threshold of hearing I_0.

$$\beta = 10 \log\frac{I}{I_0} = 10 \log\frac{\text{w/cm}^2}{\text{w/cm}^2} = \text{db}$$

The dimensional units w/cm² cancel, leaving the unit db dimensionless.

Problems • Chapter 11/pp. 265–266
GROUP A

1. (5.00 km × 1000 m/km)/[331.5 m/s + 0.6 m/s/C° × (10.0 °C − 0.0 °C)] = 14.8 s

2. $v = f\lambda$ $\lambda = v/f$
λ = [331.5 m/s + 0.6 m/s/C° × (15 °C − 0 °C)]/32$\overline{0}$/s = 1.06 m

3. $d = vt/2$
d = [331.5 m/s + 0.6 m/s/C° × (−10.0 °C − 0 °C)] × 5.0 s/2
d = 8.1 × 10² m

4. $d = vt/2$
d = [331.5 m/s + 0.6 m/s/C° × (20 °C − 0 °C)] × (4.0 s + 2.0 s)/2
d = 1.0 × 10³ m, or 1.0 km

5. $\beta = 10 \log I/I_0$ β = 10 log(10⁻¹¹ watt/cm²)/(10⁻¹⁶ watt/cm²)
β = 50 db

GROUP B

6. $d = \frac{1}{2}gt^2$ $t = \sqrt{2d/g}$
$t = \sqrt{2 \times 25\overline{0} \text{ m}/(9.80 \text{ m/s}^2)}$
t = 7.14 s, time for stone to fall
$v = d/t$
$t = d/v$
t = 25$\overline{0}$ m/[331.5 m/s + 0.6 m/s/C° × (5.0 °C − 0.0 °C)]
t = 0.747 s, time for sound to return
7.14 s + 0.747 s = 7.89 s, total time

7. Let x = time for stone to fall and 8.00 s − x = time for sound to return
$d = \frac{1}{2}gt^2 = vt$
(9.80 m/s² × x²/2 = [331.5 m/s + 0.6 m/s/C° × (25 °C − 0 °C)](8.00 s − x)
x = 7.25 s   8.00 s − 7.25 s = 0.75 s
0.75 s × 346.5 m/s = 260 m

8. $f_{LF} = f_S \dfrac{v}{v - v_S}$

v = 331.5 m/s + 0.6 m/s/C° × (15 °C − 0 °C) = 34$\overline{0}$ m/s
v_S = 95.0 km/hr × 10³ m/km × hr/3.6 × 10³ s = 26.4 m/s

$f_{LF} = 288 \text{ hz} \times \dfrac{34\overline{0} \text{ m/s}}{34\overline{0} \text{ m/s} - 26.4 \text{ m/s}} = 312 \text{ hz}$

9. $\Delta f = f_{LF} = f_{LB} = f_S \left[\dfrac{v}{v - v_S} - \dfrac{v}{v + v_S} \right]$

v = 331.5 m/s + (0.59 m/s × 25.0 °C) = 346 m/s
v_S = 145 km/hr × 10³ m/km × hr/3.6 × 10³ s = 40.3 m/s

$\Delta f = 32\overline{0} \text{ hz} \left[\dfrac{346 \text{ m/s}}{(346 - 40.3) \text{ m/s}} - \dfrac{346 \text{ m/s}}{(346 + 40.3) \text{ m/s}} \right]$

$\Delta f = 32\overline{0} \text{ hz} \left[\dfrac{346 \text{ m/s}}{306 \text{ m/s}} - \dfrac{346 \text{ m/s}}{386 \text{ m/s}} \right] = 32\overline{0}$ hz (1.13 − 0.896) = 76 hz

10. $\beta = 10 \log I/I_0$ 30 db = 10 log I/(10⁻¹⁶ watt/cm²) $I = 10^{-13}$ watt/cm²

11. $\beta = 10 \log \dfrac{I}{I_0} = 10 \log \dfrac{10^{-14} \text{ w/cm}^2}{10^{-16} \text{ w/cm}^2}$ β = 10 log 100 = 20 db

12. (a) $\beta = 10 \log \dfrac{I_1}{I_0} = 10 \log \dfrac{2.25 \times 10^{-6} \text{ w/cm}^2}{10^{-16} \text{ w/cm}^2} = 10 \log(2.25 \times 10^{10})$

$\beta = 10(\log 2.25 + \log 10^{10}) = 10(0.325 + 10) = 10\overline{0}$ db

(b) $\beta = 10 \log \dfrac{I_2}{I_0} = 10 \log \dfrac{1.17 \times 10^{-12} \text{ w/cm}^2}{10^{-16} \text{ w/cm}^2}$

$\beta = 10 \log(1.17 \times 10^4) = 10(\log 1.17 + \log 10^4)$
$\beta = 10(0.0682 + 4) = 40.7$ db

Questions • Chapter 11/pp. 277–278
GROUP A
1. **(a)** Its sound is reinforced by the forced vibrations of the table.
   (b) Vibrational energy is removed from the fork at a faster rate.
2. The natural vibration rates of two objects should be the same or the rate of one should be a multiple of the rate of the source.
3. One-fourth the wavelength of the sound wave.
4. One-half the wavelength of the sound wave.
5. The higher note has twice the frequency.
6. **(a)** The lowest pitch produced by a musical tone source. **(b)** As a single unit.
7. **(a)** Law of tensions. **(b)** Law of lengths.
8. **(a)** Constructive interference: Where the displacement of a particular particle of the medium caused by one wave at any instant is in the same direction as that caused by the other wave, the displacement of the particle at that instant is the sum of the separate displacements (superposition principle) and the resultant displacement is greater than that caused by either wave separately. **(b)** Destructive interference: Where the displacement effects of the two waves on the particle of the medium are in different directions, the resultant displacement is less than that caused by either wave separately. **(c)** The two wave trains would have slightly different frequencies and would be propagated through the medium in the same direction.

GROUP B
9. The sounds vary in the number and intensity of the overtones present.
10. The strings vary in length, mass per unit length, and tension.
11. The bridge transmits vibratory motion from the string to the body of the instrument.
12. The tuner should hear C, frequency 262 hz.
13. Move the forks either away from or toward one another.

Problems • Chapter 11/p. 278
GROUP A
1. $f = v/\lambda = \dfrac{331.5 \text{ m/s} + (0.6 \text{ m/s/C}° \times 2\overline{0} \text{ C}°)}{2(25.0 \text{ cm} + 0.8 \times 2.0 \text{ cm}) \times \text{m}/100 \text{ cm}}$

   $f = \dfrac{343.5 \text{ m/s}}{0.532 \text{ m}} = 646/\text{s} = 646$ hz

2. $v = f \times \lambda = f \times 4(l + 0.4d) = 384/\text{s} \times 4(20.0 \text{ cm} + 1.6 \text{ cm}) \times \text{m}/10^2 \text{ cm}$
   $v = 332$ m/s
3. $44\overline{0}/\text{s} - 5/\text{s} = 435/\text{s} = 435$ hz
4. 296 hz − 288 hz = 8 beats/s

5. $l' = fl/f' = 44\bar{0}$ hz $\times$ 25.4 cm/523.3 hz $= 21.4$ cm
   $l - l' = 25.4$ cm $- 21.4$ cm $= 4.0$ cm
6. $f/f' = l'/l \times d'/d = 10\bar{0}$ cm/25 cm $\times$ 0.25 mm/0.50 mm $= 2$
   The second string is one-half the frequency of the first.
7. $f_{av} = \frac{1}{2}(f_l + f_h) = \frac{1}{2}(32\bar{0}$ hz $+ 324$ hz$) = 322$ hz
   $f_h - f_l = 324$ hz $- 32\bar{0}$ hz $= 4$ beats/s
   The listener hears 4 beats per second of sound frequency 322 hz.
8. $l' = l \times f/f' = 26.7$ cm $\times$ 327.6 hz/293.7 hz $= 29.8$ cm
   $\Delta l = 29.8$ cm $- 26.7$ cm $= 3.1$ cm increase

GROUP B

9. $f/f' = l'/l \times \sqrt{F/F'}$ $f' = f \times l/l' \times \sqrt{F'/F}$
   $f' = 44\bar{0}/s \times 0.500$ m/0.400 m $\times \sqrt{5.00 \times 10^2 \text{ n}/2.50 \times 10^2 \text{ n}}$
   $f' = 778/s = 778$ hz
10. (a) $l' = l \times f/f' = 0.350$ m $\times$ 264 hz/297 hz $= 0.311$ m
       $\Delta l = 0.350$ m $- 0.311$ m $= 0.039$ m (decrease in length)
    (b) $F' = F(f'/f)^2 = 325$ n$(297$ hz/264 hz$)^2 = 411$ n
       $\Delta F = 411$ n $- 325$ n $= 86$ n (increase in tensional force)
11. $\beta = 10 \log I/I_0$
    110 db $= 10 \log I/10^{-14}$ w/cm^2 $I = 10^{-3}$ w/cm^2
12. (a) 261.6 hz $\times$ 16 $= 4186$ hz
    (b) $v = f\lambda$ $\lambda = v/f = (331.5$ m/s $+ 0.6$ m/s/C$^\circ \times 20.0$ C$^\circ)$/4186 hz
       $\lambda = 8.1 \times 10^{-2}$ m
13. (a) $f = v/\lambda$
       $\lambda = 2(l + 0.8d)$
       $$f = v/2(l + 0.8d) = \frac{331.5 \text{ m/s} + (0.6 \text{ m/s/C}^\circ \times 15 \text{ C}^\circ)}{2(1.23 \text{ m} + 0.8 \times 0.10 \text{ m})}$$
       $$f = \frac{331.5 \text{ m/s} + 9 \text{ m/s}}{2 \times 1.31 \text{ m}} = \frac{341 \text{ m/s}}{2.62 \text{ m}} = 13\bar{0} \text{ hz}$$
    (b) $2f = 26\bar{0}$ hz and $3f = 39\bar{0}$ hz
14. (a) $f = v/\lambda$
       $\lambda = 4(l + 0.4d)$
       $$f = v/4(l + 0.4d) = \frac{331.5 \text{ m/s} + (0.6 \text{ m/s/C}^\circ \times 12 \text{ C}^\circ)}{4(0.76 \text{ m} + 0.4 \times 0.050 \text{ m})}$$
       $$f = \frac{338 \text{ m/s}}{4 \times 0.78 \text{ m}} = \frac{338 \text{ m/s}}{3.1 \text{ m}} = 110 \text{ hz}$$
    (b) $3f = 330$ hz and $5f = 550$ hz

THE NATURE OF LIGHT   **12**

Questions • Chapter 12/pp. 301–302
GROUP A
1. Today scientists recognize the dual character of light; radiant energy is transmitted in photons, which are guided along their path by a wave field.

2. Gamma rays, X rays, ultraviolet radiations, visible light, infrared radiations, radio waves.

3. **(a)** *Sound*—vibrating matter. *Light*—electron transition between energy levels.
(b) *Sound*—solids, liquids, and gases. *Light*—vacuum, transparent materials, translucent materials.   **(c)** *Sound*—0.01 to 21 meters. *Light*—4000 to 7600 Å (4×10^{-7} to 7.6×10^{-7} m).   **(d)** *Sound*—longitudinal. *Light*—transverse.   **(e)** *Sound*—332 m/s at 0 °C. *Light*—3×10^8 m/s.

4. As an atom absorbs energy in a definite quantity, an electron moves out to a higher energy level, acquiring potential energy in the process. When the atom reverts to a lower energy level, it radiates energy in quantum units.

5. **(a)** No. The radiation from a flatiron at its normal operating temperature falls in the infrared region, well below the cutoff frequency of the photosensitive element of an ordinary photocell. The flatiron would probably have to be brought very near to a red heat for a cesium cell to sense the radiation.   **(b)** A special infrared-sensitive photographic film placed near a flatiron in a dark room would photograph the flatiron by its own radiation. Also, a thermometer with blackened bulb near the flatiron will show a rise in temperature.

GROUP B

6. Ordinary glass absorbs ultraviolet radiation, which is transmitted through quartz. The temperature changes suggest that the light source is a mercury-arc lamp, carbon arc, or some similar device having an emission spectrum that extends into the ultraviolet region.

7. **(a)** 1. The rate of emission of photoelectrons is directly proportional to the intensity of the incident light. 2. The kinetic energy of photoelectrons is independent of the intensity of the incident light. 3. Within the range of effective frequencies, the maximum kinetic energy of photoelectrons is directly proportional to the frequency of the incident light.   **(b)** An electromagnetic wave model of light is successful in dealing with the first law of photoelectric emission. However, the wave model requires that the kinetic energy of the photoelectrons be raised with an increase in light intensity and that photoemission occur at any frequency of incident light given enough time or high enough intensity. Thus the wave model fails with regard to the second and third laws.

8. All photoelectrons acquired energy hf from photons of the monochromatic source. Those ejected from the surface layer of the photosensitive material lose only an amount of energy equal to the work function of the metal and escape with the maximum velocity. Electrons ejected from subsurface layers lose varying amounts of energy through collisions before reaching the surface, plus energy equal to the work function, and escape with velocities ranging below this maximum.

9. Figure 12-11 shows that photoelectric currents produced by different intensities of illumination from a given source reach zero at the same collector potential, the cutoff potential for that cell and light source. Since cutoff potential measures the kinetic energy of the fastest photoelectrons (those that have lost only the energy equal to the work function of the photosensitive metal), these energies are shown to be the same for the different intensities of the light source.

10. When the photon energy hf imparted to electrons is just equal to the work function w of the emitter, the surface-layer electrons have just enough energy to penetrate the surface

barrier. No energy is left over to appear as velocity. Subsurface electrons may lose a portion of the energy hf in reaching the surface and will not overcome the surface barrier. The frequency, in this situation, is known as the cutoff frequency f_{CO}; and frequencies below this value cannot produce photoemission from this emitter material no matter how intense the light.

11. **(a)** Coherent light waves have the same frequency or phase, or both. **(b)** Laser beams consist of monochromatic (single frequency), parallel, and in-phase light waves.

12. In the photoelectric effect, photons of frequency f produce photoelectrons with maximum kinetic energy given by $\frac{1}{2}mv^2_{max} = hf - w$. Electrons accelerated to a velocity v should then cause the emission of photons of radiation having a maximum frequency given by $\frac{1}{2}mv^2 = hf_{max} - w$. In the production of X rays, high-speed electrons bombard a metal target, producing photon radiation (X rays) having a maximum frequency determined by the velocity of the electrons.

Questions • Chapter 12/p. 313
GROUP A

1. Luminous objects are visible because of the light they emit. Illuminated objects are visible because of the light they reflect.

2. **(a)** Transparent media are those which transmit light so that objects may be distinguished clearly through them. Examples: air, glass, water. **(b)** Translucent media are those which transmit light but diffuse it so that objects cannot be seen clearly through them. Examples: frosted glass, greased paper. **(c)** Opaque materials are those which do not transmit light. Examples: wood, cardboard, iron.

3. The quantities measured are luminous intensity, luminous flux, and illumination.

4. Luminous flux is that part of the total energy radiated per unit of time from a luminous source that is capable of producing the sensation of sight.

5. **(a)** The intensity of a light source is measured in candles, the candle being a fundamental unit. **(b)** Luminous flux and illumination are measured in derived units, the lumen and lumen/meter^2, respectively.

6. Light radiation from bodies that have been heated to incandescence. Fluorescent lighting is also commonly used and is the result of radiation of phosphors that are first irradiated with ultraviolet rays.

GROUP B

7. See Section 12-16.

8. The light meter is placed a measured distance from a light approximating a point source and an illumination reading is taken. Additional meter readings are then taken at half the original distance and at twice the original distance. Meter readings of 4 times and $\frac{1}{4}$ the original reading respectively will verify the inverse-square law.

9. The meter is first placed a measured distance from a known source and the reading noted. The known source is then replaced by the unknown source and the distance is adjusted to give the same scale reading as before on the meter. The intensity of the second source can be calculated since $\dfrac{I_2}{I_1} = \dfrac{r_2^2}{r_1^2}$.

10. The viewer in Chicago hears the performance first with a delay of approximately 10^{-3} s while the member of the studio audience hears the performer with a delay of approximately 10^{-1} s.

Problems • Chapter 12/pp. 313–314

GROUP A

1. $d = vt$ $t = \dfrac{d}{v} = \dfrac{1.5 \times 10^8 \text{ km}}{3.00 \times 10^5 \text{ km/s}} \times \dfrac{\text{min}}{60 \text{ s}} = 8.3 \text{ min}$

2. $E = I/r^2 = 125 \text{ cd}/(1.20 \text{ m})^2 = 86.8 \text{ lm/m}^2$

3. $E = I/r^2$ $r = \sqrt{I/E} = \sqrt{\dfrac{265 \text{ cd}}{540 \text{ lm/m}^2}} = \sqrt{0.49 \text{ m}^2} = 0.70 \text{ m}$

4. $E = I/r^2 = 150 \text{ cd}/(1.50 \text{ m})^2 = 67 \text{ lm/m}^2$

5. $\dfrac{I_1}{I_2} = \dfrac{r_1^2}{r_2^2}$ $I_2 = \dfrac{I_1 r_2^2}{r_1^2} = \dfrac{20 \text{ cd} \times (2.0 \text{ m})^2}{(3.0 \text{ m})^2} = 8.9 \text{ cd}$

6. $\dfrac{I_1}{I_2} = \dfrac{r_1^2}{r_2^2}$ $I_2 = \dfrac{I_1 r_2^2}{r_1^2} = \dfrac{16.0 \text{ cd} \times (100.0 \text{ cm} - 60.0 \text{ cm})^2}{(60.0 \text{ cm})^2} = 7.12 \text{ cd}$

7. $\dfrac{I_1}{I_2} = \dfrac{r_1^2}{r_2^2}$ $I_2 = \dfrac{I_1 r_2^2}{r_1^2} = \dfrac{40.0 \text{ cd} (30 \text{ cm})^2}{(100.0 \text{ cm} - 30 \text{ cm})^2} = 7.35 \text{ cd}$

8. $3 \text{ m} \times \sqrt{4} = 6 \text{ m}$

GROUP B

9. $E = I/r^2$ $r = \sqrt{I/E} = \sqrt{\dfrac{60.0 \text{ cd}}{10.0 \text{ lm/m}^2}} = 2.45 \text{ m}$

10. $4.3 \text{ yr} \times 365 \text{ da/yr} \times 24 \text{ hr/da} \times 3600 \text{ s/hr} \times 3.00 \times 10^5 \text{ km/s} = 4.1 \times 10^{13} \text{ km}$

11. Let r_1 = distance from the 30.0 cd lamp; then $r_2 = (10\overline{0} \text{ cm} - r_1)$

$\dfrac{I_1}{I_2} = \dfrac{r_1^2}{r_2^2}$ $r_1 = \sqrt{\dfrac{I_1 r_2^2}{I_2}} = \sqrt{\dfrac{30.0 \text{ cd} (10\overline{0} \text{ cm} - r_1)^2}{20.0 \text{ cd}}} = 55.1 \text{ cm}$

12. $\dfrac{I_1}{I_2} = \dfrac{r_1^2}{r_2^2}$ $I_1 = \dfrac{I_2 r_1^2}{r_2^2} = \dfrac{10 \text{ cd} (75 \text{ cm})^2}{(100 \text{ cm} - 75 \text{ cm})^2} = 90 \text{ cd}$ 90 cd/100 w = 0.9 cd/w

13. $\dfrac{299{,}729 \text{ km/s}}{70.8 \text{ km/}\frac{1}{8} \text{ rev}} = 529 \text{ rev/s}$

14. At the lower end: $\lambda = 7600 \text{ Å}$ $f = \dfrac{v}{\lambda} = \dfrac{3.0 \times 10^8 \text{ m/s}}{7.6 \times 10^3 \text{ Å} \times 10^{-10} \text{ m/Å}} = 4.0 \times 10^{14}\text{/s}$

At the upper end: $\lambda = 4000 \text{ Å}$ $f = \dfrac{v}{\lambda} = \dfrac{3.0 \times 10^8 \text{ m/s}}{4.0 \times 10^8 \text{ Å} \times 10^{-10} \text{ m/Å}} = 7.5 \times 10^{14}\text{/s}$

15. $E = \dfrac{I \cos \theta}{r^2}$ $\cos \theta = \dfrac{Er^2}{I} = \dfrac{0.025 \text{ lm/cm}^2 \times (75 \text{ cm})^2}{150 \text{ cd}} = 0.93$

$\theta = 21°$, the angle of the flux with the normal to the surface

13   REFLECTION

Questions • Chapter 13/pp. 329–330

GROUP A

1. The kind of material, the condition of the surface, and the angle of incidence.
2. (1) The incident ray, the reflected ray, and the normal to the reflecting surface lie in the

same plane.   (2) The angle of incidence is equal to the angle of reflection.
3. No, the image is reversed right and left.
4. Refer to Fig. 13-11 in the text.
5. A concave mirror produces real images of objects placed beyond the principal focus.
6. Use a mirror of small aperture or use a parabolic mirror.

GROUP B
7. The white paint is not as good a reflector of light as the mirror, and so the image of the word is darker.
8. (a) A mirror in which the reflecting surface is a portion of the outer surface of a right circular cylinder whose axis is horizontal.   (b) A mirror in which the reflecting surface is a portion of the outer surface of a right circular cylinder whose axis is vertical.
9. (a) Concave mirror, object between center of curvature and principal focus.   (b) Concave mirror, object a finite distance beyond center of curvature.   (c) Concave mirror, object at center of curvature.   (d) Concave mirror, object between principal focus and vertex.   (e) Convex mirror, object any distance from mirror.
10. (a) Student's diagram of Case 2, Fig. 13-19.   (b) The image is real, inverted, reduced, and located in front of the mirror between the principal (real) focus and the center of curvature.
11. (a) Student's diagram of Case 4, Fig. 13-19.   (b) The image is real, inverted, enlarged, and located in front of the mirror beyond the center of curvature.
12. (a) Student's diagram similar to Fig. 13-20.   (b) The image is virtual, erect, reduced, and located behind the mirror between the vertex and the principal (virtual) focus.

Problems • Chapter 13/pp. 330–331
GROUP A
1. Radius of curvature $= 2f = 2 \times 10.0$ cm $= 20.0$ cm
2. Focal length $= \frac{1}{2}$ radius of curvature $= \frac{8}{2}$ cm $= 4$ cm
3. $\dfrac{1.76 \text{ m}}{2} = 0.880$ m

4. Focal length $= \frac{1}{2}$ radius of curvature $= 150$ cm$/2 = 75$ cm

5. $\dfrac{1}{f} = \dfrac{1}{d_o} + \dfrac{1}{d_i}$ $d_i = \dfrac{d_o f}{d_o - f} = \dfrac{25.0 \text{ cm} \times 5.00 \text{ cm}}{25.0 \text{ cm} - 5.00 \text{ cm}} = 6.25$ cm

6. (a) $\dfrac{1}{f} = \dfrac{1}{d_o} + \dfrac{1}{d_i}$ $d_i = \dfrac{d_o f}{d_o - f} = \dfrac{91.5 \text{ cm} \times 15.0 \text{ cm}}{91.5 \text{ cm} - 15.0 \text{ cm}} = 17.9$ cm

   (b) $\dfrac{h_o}{h_i} = \dfrac{d_o}{d_i}$ $h_i = \dfrac{h_o d_i}{d_o} = \dfrac{25.4 \text{ cm} \times 17.9 \text{ cm}}{91.5 \text{ cm}} = 4.97$ cm

7. (a) $\dfrac{1}{f} = \dfrac{1}{d_o} + \dfrac{1}{d_i}$ $f = \dfrac{d_o d_i}{d_o + d_i} = \dfrac{50.0 \text{ cm} \times 33.3 \text{ cm}}{50.0 \text{ cm} + 33.3 \text{ cm}} = 20.0$ cm $r = 2f = 40.0$ cm

   (b) $\dfrac{h_o}{h_i} = \dfrac{d_o}{d_i}$ $h_o = \dfrac{h_i d_o}{d_i} = \dfrac{30.5 \text{ cm} \times 50.0 \text{ cm}}{33.3 \text{ cm}} = 45.8$ cm

8. Case 3.   36.4 cm $\div$ 2 $= 18.2$ cm, focal length

GROUP B

9. (a) $\dfrac{1}{d_o} + \dfrac{1}{d_i} = \dfrac{1}{f}$ $d_i = \dfrac{d_o f}{d_o - f} = \dfrac{5.0 \text{ cm} \times 15 \text{ cm}}{5.0 \text{ cm} - 15 \text{ cm}} = -7.5 \text{ cm}$

(b) $\dfrac{h_o}{h_i} = \dfrac{d_o}{d_i}$

$h_i = \dfrac{h_o d_i}{d_o} = \dfrac{2.0 \text{ cm} \times 7.5 \text{ cm}}{5.0 \text{ cm}} = 3.0 \text{ cm}$

10. (a) $\dfrac{1}{d_o} + \dfrac{1}{d_i} = \dfrac{1}{f}$

$d_i = \dfrac{d_o f}{d_o - f} = \dfrac{50.0 \text{ cm} \times (-25.0 \text{ cm})}{50.0 \text{ cm} - (-25.0 \text{ cm})}$

$d_i = -16.7 \text{ cm}$ (at 66.7-cm mark)

(b) $\dfrac{h_o}{h_i} = \dfrac{d_o}{d_i}$

$h_i = \dfrac{h_o d_i}{d_o} = \dfrac{5.00 \text{ cm} \times 16.7 \text{ cm}}{50.0 \text{ cm}}$

$= 1.67 \text{ cm}$

11. $\dfrac{h_o}{h_i} = \dfrac{d_o}{d_i}$ $h_i = \dfrac{h_o d_i}{d_o} = \dfrac{3480 \text{ km} \times 2.00 \text{ m}}{38\overline{0},000 \text{ km}} \times \dfrac{100 \text{ cm}}{\text{m}} = 1.83 \text{ cm}$

12. (a) $\dfrac{1}{f} = \dfrac{1}{d_o} + \dfrac{1}{d_i}$ $f = \dfrac{d_o d_i}{d_o + d_i}$

$d_o = \dfrac{h_o d_i}{h_i}$ Substituting for d_o and simplifying:

$f = \dfrac{h_o d_i}{h_o + h_i} = \dfrac{0.500 \text{ cm} \times 50\overline{0} \text{ cm}}{0.500 \text{ cm} + 50.0 \text{ cm}} = 4.95 \text{ cm}$

$2f = 9.90 \text{ cm}$, the radius of curvature

(b) $d_o = \dfrac{h_o d_i}{h_i} = \dfrac{0.500 \text{ cm} \times 50\overline{0} \text{ cm}}{50.0 \text{ cm}} = 5.00 \text{ cm}$

13. $\dfrac{1}{f} = \dfrac{1}{d_o} + \dfrac{1}{d_i}$ $d_i = \dfrac{d_o f}{d_o - f} = \dfrac{10.0 \text{ cm} \times 15.0 \text{ cm}}{10.0 \text{ cm} - 15.0 \text{ cm}} = -3\overline{0} \text{ cm}$

$h_i = \dfrac{d_i h_o}{d_o} = \dfrac{-3\overline{0} \text{ cm } h_o}{10.0 \text{ cm}} = -3.0 \, h_o \text{ cm}$

Magnification of virtual image = 3X

14. (a) $\frac{1}{2}(1\overline{0} \text{ cm} + 1.70 \text{ m}) = \frac{1}{2}(1.80 \text{ cm}) = 0.900 \text{ m}$

(b) No, the length of the mirror is independent of the separation distance

(c)

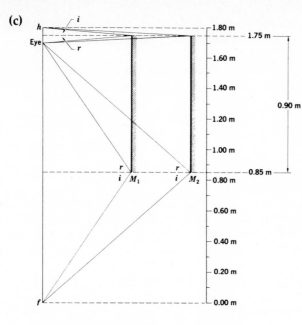

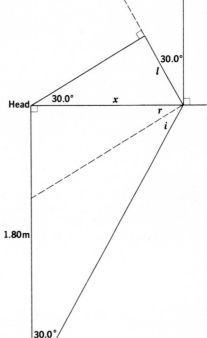

15. The feet are seen looking horizontally into the mirror through a minimum distance x. The head is seen looking perpendicularly into the mirror.

$l = x \sin 30.0°$

$x = 1.8 \text{ m}/\tan 30.0°$

$l = 1.80 \text{ m} \times \tan 30.0° \times \sin 30.0°$

$l = 1.80 \text{ cm} \times 0.577 \times 0.500 = 0.520 \text{ m}$

REFRACTION   14

Questions • Chapter 14/p. 340
GROUP A

1. **(a)** The bending of light rays as they pass obliquely from one medium into another of different optical density.   **(b)** The angle between the refracted ray and the normal drawn to the point of refraction.   **(c)** The ratio of the speed of light in a vacuum to the speed in another substance.
2. The change in the speed of light as the rays enter the substance.
3. First law: The incident ray, the refracted ray, and the normal to the surface at the point of incidence are all in the same plane.
   Second law: The index of refraction for any medium is a constant that is independent of the angle of incidence.
   Third law: When a ray of light passes obliquely from a medium of lesser to one of

greater optical density, it is bent toward the normal. Conversely, a ray of light passing obliquely from an optically denser to a rarer medium is bent from the normal to the surface.

4. **(a)** The particular angle of incidence at which the refracted ray makes an angle of 90° with the normal.   **(b)** The reflection that occurs when the angle of incidence equals or exceeds the critical angle.

5. For identification and determination of purity of substances.

GROUP B

6. Higher.

7. Refraction of light from the sun passing obliquely through the atmosphere causes the sun to appear above the horizon when it is actually below.

8. A right-angle prism is the better reflector since a beam may enter the glass perpendicular to the boundary with a minimum loss of energy and be totally reflected from the back of the prism. In order to produce the same change in beam direction with a plane mirror, some reflection would occur at the front surface, and the light entering the glass would be refracted, reflected, and refracted again as it leaves the glass.

9. The incidence angle is 0°, and therefore there is no refraction at the surface either where the light enters or leaves the prism.

10. Most of the light passing into the stone from above is totally reflected upward from the lower facets due to the high index of refraction of diamond, but not at the lower index of refraction of glass.

Problems • Chapter 14/p. 340
GROUP A

1. Index of refraction $= \dfrac{\text{speed in vacuum}}{\text{speed in chloroform}} = \dfrac{3.00 \times 10^8 \text{ m/s}}{1.99 \times 10^8 \text{ m/s}} = 1.51$

2. Index of refraction $= \dfrac{\text{speed in vacuum}}{\text{speed in diamond}}$

   speed in diamond $= \dfrac{\text{speed in vacuum}}{\text{index of refraction}} = \dfrac{3.00 \times 10^5 \text{ km/s}}{2.42} = 1.24 \times 10^5 \text{ km/s}$

3. Since the index of refraction of the water is $\frac{4}{3}$, the apparent depth of the coin would be $\frac{3}{4}$ as great as the actual depth of the water. 16.0 cm − 12 cm = 4 cm closer.

4. $n = \dfrac{\sin i}{\sin r}$ and $\sin r = \dfrac{\sin i}{n}$

   $\sin r = \dfrac{\sin(90.0° - 48.5°)}{n} = \dfrac{\sin 41.5°}{n} = \dfrac{0.663}{1.50} = 0.442$

   $r = 26.2°$

GROUP B

5. **(a)** Student's diagram using the procedure demonstrated in Fig. 14-7. The angle of refraction is found to be 39°.

   (b) $n_W = \dfrac{\sin i_a}{\sin r_W}$ thus $\sin i_a = n_W \sin r_W$

   Similarly,

   $n_g = \dfrac{\sin i_a}{\sin r_g}$ thus $\sin r_g = \dfrac{\sin i_a}{n_g}$

Substituting for sin i_a,

$$\sin r_g = \frac{n_W \sin r_W}{n_g} = \frac{1.33 \times \sin 45.0°}{1.50} = 0.627$$

$$r_g = 38.8°$$

6. $n = \dfrac{\sin i}{\sin r} = \dfrac{0.866}{0.656} = 1.32$

7. $\sin i_c = \dfrac{1}{n} = \dfrac{1}{1.46} = 0.685$ $i_c = \arcsin 0.685 = 43°$

8. (a) $n = \dfrac{\sin i}{\sin r} = \dfrac{\sin 35.0°}{\sin 20.9°} = \dfrac{0.574}{0.357} = 1.61$

   (b) flint.

Questions • Chapter 14/p. 353
GROUP A

1. (a) Light rays passing through the edge of a lens of large aperture do not focus at a point common to those traveling near the principal axis. (b) By use of a lens diaphragm to reduce the aperture.
2. The simple magnifier uses Case 6 (Fig. 14-20).
3. A real image of the object is formed by Case 4. This image is magnified by Case 6 (Fig. 14-20).
4. A real image of the object is formed by Case 2. This image is magnified by Case 6 (Fig. 14-20).
5. The terrestrial telescope contains an inverting lens so that the final image is erect. Otherwise they are similar.

GROUP B

6. (a) Converging lens, object between F' and 2F'. (b) Converging lens, object beyond 2F'. (c) Converging lens, object at 2F'. (d) Converging lens, object inside F'. (e) Diverging lens, object at any position.
7. The distance between the object and the lens must be decreased.
8. The focal length of the lens is increased. The amount of refraction for a given angle of incidence depends on the magnitude of the change in the speed of light at the refracting interface. The change in speed from water to glass is not as great as that from air to glass.
9. (a) If the medium in which the lens is immersed has a higher index of refraction than the lens, it will cause divergence. (b) Student's diagram that shows light ray passing from medium into lens bent away from the normal and passing from lens into medium bent toward the normal.
10. (a) Double concave shape. (b) Student's diagram that shows light ray bent away from the normal at the water-to-air interface and bent toward the normal at the air-to-water interface.

Problems • Chapter 14/p. 354
GROUP A

1. $\dfrac{1}{d_o} + \dfrac{1}{d_i} = \dfrac{1}{f}$ $\qquad d_i = \dfrac{d_o f}{d_o - f} = \dfrac{50.0 \text{ cm} \times 20.0 \text{ cm}}{50.0 \text{ cm} - 20.0 \text{ cm}} = 33.3 \text{ cm}$

2. $\dfrac{1}{d_o} + \dfrac{1}{d_i} = \dfrac{1}{f}$ $d_i = \dfrac{d_o f}{d_o - f} = \dfrac{\overline{10}\text{ cm} \times 5\text{ cm}}{\overline{10}\text{ cm} - 5\text{ cm}} = 10\text{ cm}$

Or, since $d_o = 2F$, $d_i = 2F = 10\text{ cm}$ (Case 3)

3. $\dfrac{1}{d_o} + \dfrac{1}{d_i} = \dfrac{1}{f}$ $d_o = \dfrac{d_i f}{d_i - f} = \dfrac{11.0\text{ cm} \times 10.0\text{ cm}}{11.0\text{ cm} - 10.0\text{ cm}} = 110\text{ cm}$

4. (a) $\dfrac{1}{d_o} + \dfrac{1}{d_i} = \dfrac{1}{f}$ $f = \dfrac{d_o d_i}{d_o + d_i} = \dfrac{\overline{20}\text{ cm} \times \overline{10}\text{ cm}}{\overline{20}\text{ cm} + \overline{10}\text{ cm}} = 6.7\text{ cm}$

   (b) $\dfrac{h_o}{h_i} = \dfrac{d_o}{d_i}$ $h_i = \dfrac{h_o d_i}{d_o} = \dfrac{3.0\text{ cm} \times \overline{10}\text{ cm}}{\overline{20}\text{ cm}} = 1.5\text{ cm}$

5. (a) $\dfrac{1}{d_o} + \dfrac{1}{d_i} = \dfrac{1}{f}$ $f = \dfrac{d_o d_i}{d_o + d_i} = \dfrac{\overline{30}\text{ cm} \times 60\text{ cm}}{\overline{30}\text{ cm} + 60\text{ cm}} = \overline{20}\text{ cm}$

   (b) $\dfrac{h_o}{h_i} = \dfrac{d_o}{d_i}$ $h_i = \dfrac{h_o d_i}{d_o} = \dfrac{5\text{ cm} \times 60\text{ cm}}{\overline{30}\text{ cm}} = 10\text{ cm}$

6. $\dfrac{1}{d_o} + \dfrac{1}{d_i} = \dfrac{1}{f}$ $d_i = \dfrac{d_o f}{d_o - f} = \dfrac{3.00\text{ m} \times 100\text{ cm/m} \times 5.00\text{ cm}}{3.00\text{ m} \times 100\text{ cm/m} - 5.00\text{ cm}} = 5.08\text{ cm}$

7. $\dfrac{1}{d_o} + \dfrac{1}{d_i} = \dfrac{1}{f}$ $f = \dfrac{d_o d_i}{d_o + d_i} = \dfrac{35.0\text{ cm} \times 10\text{ mm/cm} \times 19.0\text{ mm}}{35.0\text{ cm} \times 10\text{ mm/cm} + 19.0\text{ mm}} = 18.0\text{ mm}$

8. $\dfrac{1}{d_o} + \dfrac{1}{d_i} = \dfrac{1}{f}$ $f = \dfrac{d_o d_i}{d_o + d_i} = \dfrac{15.0\text{ m} \times 10^3\text{ mm/m} \times 19.0\text{ mm}}{15.0\text{ m} \times 10^3\text{ mm/m} + 19.0\text{ mm}} = 19.0\text{ mm}$

9. Magnification $= \dfrac{25\text{ cm}}{f} = \dfrac{25\text{ cm}}{\overline{10}\text{ cm}} = 2.5$

GROUP B

10. (a) $\dfrac{1}{d_o} + \dfrac{1}{d_i} = \dfrac{1}{f}$ $d_i = \dfrac{d_o f}{d_o - f} = \dfrac{16\text{ cm} \times 24\text{ cm}}{16\text{ cm} - 24\text{ cm}} = -48\text{ cm}$

    (b) $\dfrac{h_o}{h_i} = \dfrac{d_o}{d_i}$ $h_i = \dfrac{h_o d_i}{d_o} = \dfrac{3\text{ cm} \times 48\text{ cm}}{16\text{ cm}} = 9\text{ cm}$

11. (a) $\dfrac{1}{d_o} + \dfrac{1}{d_i} = \dfrac{1}{f}$ $f = \dfrac{d_o d_i}{d_o + d_i} = \dfrac{12\text{ cm} \times (-61\text{ cm})}{12\text{ cm} - 61\text{ cm}} = 15\text{ cm}$

    (b) $\dfrac{h_o}{h_i} = \dfrac{d_o}{d_i}$ $h_i = \dfrac{h_o d_i}{d_o} = \dfrac{5.0\text{ cm} \times 61\text{ cm}}{12\text{ cm}} = 25\text{ cm}$

12. (a) $\dfrac{1}{d_o} + \dfrac{1}{d_i} = \dfrac{1}{f}$ $f = \dfrac{d_o d_i}{d_o + d_i} = \dfrac{55\text{ cm} \times (-5.0\text{ cm})}{55\text{ cm} - 5.0\text{ cm}} = -5.5\text{ cm}$

    (b) $\dfrac{h_o}{h_i} = \dfrac{d_o}{d_i}$ $h_i = \dfrac{h_o d_i}{d_o} = \dfrac{1.0\text{ cm} \times 5.0\text{ cm}}{55\text{ cm}} = 0.091\text{ cm}$

13. Magnification $= \dfrac{25\text{ cm} \times l}{f_e \times f_o} = \dfrac{25\text{ cm} \times 15.0\text{ cm}}{2.00\text{ cm} \times 0.500\text{ cm}} = 375$

14. (a) $\dfrac{h_o}{h_i} = \dfrac{d_o}{d_i}$ $d_o = \dfrac{h_o d_i}{h_i} = \dfrac{6.4\text{ cm} \times 9.0\text{ m}}{1.5\text{ m}} = 38\text{ cm}$

    (b) $\dfrac{1}{d_o} + \dfrac{1}{d_i} = \dfrac{1}{f}$ $f = \dfrac{d_o d_i}{d_o + d_i} = \dfrac{38\text{ cm} \times 9.0\text{ m} \times 100\text{ cm/m}}{38\text{ cm} + 9.0\text{ m} \times 100\text{ cm/m}} = 36\text{ cm}$

15. Magnification $= \dfrac{25 \text{ cm} \times l}{f_e \times f_o} = \dfrac{250 \text{ mm} \times 16\overline{0} \text{ mm}}{30.0 \text{ mm} \times 5.00 \text{ mm}} = 267$

16. 15 cm + 3 cm + 6 cm + 3 cm + 5 cm = 32 cm (approximately)

Questions • Chapter 14/p. 360
GROUP A
1. The index of refraction of a substance is different for light of different colors.
2. The frequency (or wavelength) of light determines its color.
3. **(a)** Infrared rays.   **(b)** Ultraviolet rays.
4. The objects surrounding the black object reflect light showing its outline. Black surfaces are generally not completely nonreflective.
5. Black. The cloth absorbs all colors except red and reflects only red light. The green light is thus absorbed and no light is reflected.
6. **(a)** Red, green, and blue.   **(b)** A color that combines with another color to form white light.   **(c)** Cyan is the complement of red, magenta is the complement of green, and yellow is the complement of blue.

GROUP B
7. Mixing pigments is a subtraction process; the addition of pigments reduces the number of colors that they may reflect. White paint reflects all colors.
8. Disperse the light by means of a prism; one of the spectral colors will be red.
9. **(a)** The focal length for red light will be longer than that for green light.
(b) Student's diagram.
10. **(a)** The focal length for red light will be longer than that for green light.
(b) Student's diagram.

DIFFRACTION AND POLARIZATION   **15**

Questions • Chapter 15/p. 371
GROUP A
1. By interference of light reflected from the two surfaces of the film.
2. Different colors in white light have different wavelengths and therefore different diffraction angles.
3. Separation is greater in second-order spectrum.
4. More orders of images are given by the grating with fewer lines per centimeter.
5. Diffracted images are brighter and dispersion is greater.

GROUP B
6. **(a)** Emits red light.   **(b)** Reflects red light.   **(c)** Emits red light.   **(d)** Atmospheric scattering.   **(e)** Transmits only red light.
7. One of the two interfering waves experiences a phase inversion during reflection giving an extra phase difference of one-half wavelength beyond that normally expected.
8. The angular separation of orders will be greater when using the grating with the larger number of lines per centimeter. This can be shown by substituting arbitrary values in the expression $\sin \theta_n = n\lambda/d$, where $n = 1$ and 2.

9. Diffraction accounts for the light spreading out beyond the two slit openings. Interference accounts for the formation of light and dark bands of the double-slit image pattern.

10. Using the equation for finding wavelength by diffraction,

$$\lambda = \frac{d \sin \theta_n}{n}, \text{ solve for } d \qquad d = \frac{n\lambda}{\sin \theta_n}$$

Problems • Chapter 15/p. 372
GROUP B

1. $\lambda = \dfrac{d \sin \theta_n}{n}$

$$\sin \theta_n = n\lambda/d = 4.92 \times 10^3 \text{ Å} \times \frac{10^{-8} \text{ cm}}{\text{Å}} \times \frac{5.80 \times 10^3}{\text{cm}}$$

$\sin \theta = 0.285$

$\theta = 16.5°$

2. $\dfrac{\text{image distance}}{\text{screen distance}} = \tan \theta$

image distance = screen distance × tan θ

image distance = 34.5 cm × tan 16.5° = 34.5 cm × 0.296 = 10.2 cm

3. (a) $\lambda = \dfrac{d \sin \theta}{n} = \dfrac{\sin 38.0°}{2 \times 5.90 \times 10^3/\text{cm}} \times \dfrac{\text{Å}}{10^{-8} \text{ cm}}$

$$\lambda = \frac{0.616}{1.18 \times 10^4} \text{ cm} \times \frac{\text{Å}}{10^{-8} \text{ cm}} = 5.22 \times 10^3 \text{ Å}$$

(b) The color is green.

4. (a) $\lambda = d \sin \theta/n$

$\sin \theta_v = \lambda/d = 4.00 \times 10^3 \text{ Å} \times 10^{-8} \text{ cm/Å} \times 3.50 \times 10^3/\text{cm}$

$\sin \theta_v = 0.140$

$\theta_v = 8.0°$

image distance/screen distance = tan θ_v

image distance = screen distance × tan 8.0° = 50.0 cm × 0.141

image distance = 7.05 cm

(b) $\sin \theta_r = 7.60 \times 10^3 \text{ Å} \times 10^{-8} \text{ cm/Å} \times 3.50 \times 10^3/\text{cm}$

$\sin \theta_r = 0.266$

$\theta_r = 15.5°$

image distance = 50.0 cm × tan 15.5° = 50.0 cm × 0.277 = 13.8 cm

5. (a) $\lambda = d \sin \theta/n$

$\tan \theta = 9.25 \text{ cm}/50.0 \text{ cm} = 0.185$

$\theta = 10.5°$

$$\lambda = \frac{\sin 10.5°}{3500/\text{cm}} \times \frac{\text{Å}}{10^{-8} \text{ cm}} = 5.20 \times 10^3 \text{ Å}$$

(b) Green

Questions • Chapter 15/p. 378
GROUP A

1. Polarization.
2. Their transmission plane is vertical to eliminate the horizontally polarized glare from road and water surfaces.

3. A structural color results from the scattering of light by tiny particles suspended in a transmitting medium. A pigment color results from the absorption of certain wavelengths and the reflection or transmission of others.

4. A particular angle of incidence at which polarization of reflected light is complete.

5. Observe the light through a Polaroid film or a lens of Polaroid sun glasses and rotate the Polaroid through 90°. A variation in the amount of light observed through the Polaroid would indicate polarization.

6. Student responses may vary. A representative response: Observe light reflected from a flat, horizontal surface (such as a glass plate on a window ledge) through the Polaroid film. Rotate the film to find the position that reduces the reflected light to a minimum. The polarization axis is then vertical.

GROUP B

7. A Polaroid film absorbs light in planes of polarization other than that in which its polarization axis lies, the absorption being maximum for the polarization plane perpendicular to the polarization axis. When the axis of the analyzer disk is perpendicular to the plane of the polarized light, all incident light is absorbed.

8. No. Although emerging out of phase, they have polarization planes perpendicular to each other. For two light waves to interfere, their vibrations must be in the same plane.

9. The scattering of the shorter wavelengths of sunlight passing overhead causes the sky-light to appear light blue to the observer whose line of vision is perpendicular to the light path. The sun is near the horizon and the observer's line of vision is parallel to the light path. Scattering of the shorter wavelengths is maximum because the light must travel through a maximum amount of atmosphere. Thus, only the longer wavelengths remain in the light path to reach the observer and the sun appears red.

10. The magnitude of the angle of rotation depends on the length of the light path through the substance.

11. Scattered violet light is added to diminishing amounts of scattered blue, green, and yellow lights to give the additive resultant light blue light.

12. **(a)** Sucrose is optically active and rotates the plane of polarization of polarized light.
    (b) The analyzer disk is rotated to the right.
    (c) The analyzer disk is rotated to the left since the angle of rotation depends on the concentration of the optically active substance.

ELECTROSTATICS   16

Questions • Chapter 16/pp. 392–393
GROUP A

1. Student's answer.

2. **(a)** The rubber rod is charged negatively.   **(b)** The fur is charged positively.

3. **(a)** The glass rod is charged positively.   **(b)** The silk is charged negatively.

4. It does not indicate a charge on the pith ball. The pith ball may become charged temporarily by induction, or it may have a residual charge opposite that of the rod. In either case it is attracted to the rod.

5. Objects that are similarly charged repel each other; those with unlike charges attract each other.

6. The electroscope is charged temporarily by induction.

7. The electroscope is grounded temporarily to provide a sink for electrons to be repelled from the electroscope.
8. The number and freedom of motion of free electrons in a metallic substance determines a metal's property as a conductor.
9. Sulfur has few free electrons and strong nuclear charges that hamper the motions of those present.
10. The unit of charge is the coulomb.
11. The force between two point charges is directly proportional to the product of their magnitudes and inversely proportional to the square of the distance between them.
12. An electric field is a region in which an electric force is acting upon a charge brought into the region.

GROUP B

13. (a) Some electrons are transferred from the fur to the rubber rod, leaving the rod with an excess of electrons and thus negatively charged. (b) The fur is positively charged because of a deficiency of electrons.
14. A force of repulsion, since an uncharged ball may be attracted to an object possessing either kind of charge.
15. Charge a pith-ball electroscope by touching it to a charged rubber rod, and then bring it near the charged sphere. If the pith ball is repelled, the charge on the sphere is negative; if attracted, it may be positive. To verify, recharge the pith ball positively and bring it near the charged sphere. If repelled, the charge on the sphere is positive.
16. (a) The electroscope is charged positively. (b) As the negatively charged rod is brought near, electrons are repelled from the knob to the leaves, neutralizing the positive charge on the leaves. As the rod is brought nearer, enough electrons are repelled to the leaves to give them a temporary negative charge.
17. Dimensions of k are (a) $\dfrac{\text{n m}^2}{\text{c}^2}$ (b) $\dfrac{\text{n m}^2}{\text{c}^2} \times \dfrac{\text{c}^2}{\text{m}^2} = \text{n}$
18. Student's diagram. Lines are shown originating perpendicular to the circumference of a small circle at regular intervals and curving as necessary to terminate at the plate perpendicular to the plane.
19. The motion is similar to that of a projectile fired horizontally in the earth's gravitational field with a muzzle velocity v_m.

Problems • Chapter 16/p. 393
GROUP B

1. $F = k\dfrac{Q_1Q_2}{d^2} = 8.9 \times 10^9 \dfrac{\text{n m}^2}{\text{c}^2} \times \dfrac{2.0 \times 10^{-5}\,\text{c} \times -5.0 \times 10^{-6}\,\text{c}}{(0.10\ \text{m})^2}$

$F = 8.9 \times 10^9 \times \dfrac{-1.0 \times 10^{-10}}{1.0 \times 10^{-2}}\,\text{n} = -9\overline{0}\ \text{n}$, the attractive force

2. Resultant charge $= (+2\overline{0} - 5.0)\mu\text{c} = +15\ \mu\text{c}$
   When separated, $+15\ \mu\text{c} \div 2 = +7.5\ \mu\text{c}$ on each sphere

   $F = k\dfrac{Q_1Q_2}{d^2} = 8.9 \times 10^9 \dfrac{\text{n m}^2}{\text{c}^2} \times \dfrac{(7.5 \times 10^{-6}\,\text{c})^2}{(0.10\ \text{m})^2}$

   $F = 8.9 \times 10^9 \times 5.6 \times 10^{-9}\,\text{n} = 5\overline{0}\ \text{n}$, the repulsive force

3. $F = mg \tan 5° = 5.0 \times 10^{-5}\ \text{kg} \times 9.8\ \text{m/s}^2 \times 0.087 = 4.3 \times 10^{-5}\ \text{n}$, the repulsive force on each sphere

4. Each sphere is displaced 12 cm, giving a 3-4-5 right triangle for each sphere. Gravitational force, mg, acts as a vertical component of the force on the thread of each sphere and the repulsive force acts as a horizontal component. By similar triangles,

$$\frac{F}{mg} = \frac{12 \text{ cm}}{16 \text{ cm}}$$

$$F = 0.75mg$$
$$= 0.75 \times 1.0 \times 10^{-4} \text{ kg} \times 9.8 \text{ m/s}^2$$
$$F = 7.4 \times 10^{-4} \text{ n, the repulsive force}$$

$$F = k\frac{Q_1 Q_2}{d^2}$$

$$Q^2 = \frac{Fd^2}{k} = \frac{7.4 \times 10^{-4} \text{ n} \times (0.24 \text{ m})^2}{8.9 \times 10^9 \frac{\text{n m}^2}{\text{c}^2}}$$

$$Q^2 = 4.8 \times 10^{-15} \text{ c}^2$$
$$Q = 6.9 \times 10^{-8} \text{ c, the charge on each sphere}$$

5. $\mathscr{E} = \dfrac{F}{q}$

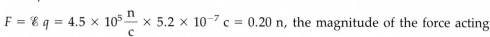

$$F = \mathscr{E} q = 4.5 \times 10^5 \frac{\text{n}}{\text{c}} \times 5.2 \times 10^{-7} \text{ c} = 0.20 \text{ n, the magnitude of the force acting}$$

6. $F = k\dfrac{Q_1 Q_2}{d^2} = k \times \dfrac{(25.5 \ \mu\text{c} \times \text{c}/10^6 \ \mu\text{c})^2}{(25.0 \text{ cm} \times \text{m}/10^2 \text{ cm})^2} = 8.93 \times 10^9 \dfrac{\text{n m}^2}{\text{c}^2} \times \dfrac{6.5 \times 10^{-10} \text{ c}^2}{6.25 \times 10^{-2} \text{ m}^2}$

$$F = 92.9 \text{ n, the attractive force}$$

7. (a) $F = k \times \dfrac{(25.5 \ \mu\text{c} \times \text{c}/10^6 \ \mu\text{c})(2.50 \ \mu\text{c} \times \text{c}/10^6 \ \mu\text{c})}{(12.5 \text{ cm} \times \text{m}/10^2 \text{ cm})^2}$

$$F = 8.93 \times 10^9 \times \frac{6.38 \times 10^{-11} \text{ c}^2}{1.56 \times 10^{-2} \text{ m}^2} = 36.5 \text{ n (each)}$$

(b) $F = 36.5$ n (repel left) + 36.5 n (attract left) = 73.0 n (left)
(c) $F = 92.9$ n (attract right) + 36.5 n (attract right) = 129.4 n (right)
(d) $F = 92.9$ n (attract left) − 36.5 n (repel right) = 56.4 n (left)

Questions • Chapter 16/pp. 407–408
GROUP A
1. The greatest charge is concentrated in the region of greatest curvature.
2. A difference of potential exists between two points if work is involved in moving a charge between these points.
3. (a) Potential difference is the work done per unit charge between two points. **(b)** The volt is the potential difference between two points such that 1 joule of work moves a charge of 1 coulomb between these points.
4. The earth is a limitless electron source, or sink, and is arbitrarily said to be at zero potential. A conducting body connected to earth is at zero potential, or "ground" potential.
5. No electric field exists within an isolated conductor, and thus no work is done in moving a charge from one point to another on its surface.
6. The air between two charged surfaces is subjected to an electric field. If the potential gradient becomes high enough, the air is ionized and a conducting path is formed for

ions and free electrons that move across to discharge the surfaces. Energy is dissipated as heat, light, and sound.

7. **(a)** Capacitance is the ratio of charge on either plate of a capacitor to the potential difference between the plates. **(b)** The farad is the unit of capacitance.

8. **(a)** The total capacitance of capacitors connected in parallel is the sum of the separate capacitances. **(b)** If capacitors are connected in series, the reciprocal of the total capacitance is equal to the sum of the reciprocals of the separate capacitances.

GROUP B

9. **(a)** $\text{volt} = \dfrac{j}{c} = \dfrac{n\,m}{c} = \dfrac{kg\,m^2}{cs^2}$

   (b) $\mathcal{E} = \dfrac{F}{q}$ so $q = \dfrac{F}{\mathcal{E}}$ and $V = \dfrac{W}{q}$ so $q = \dfrac{W}{V}$

   Thus, $\dfrac{F}{\mathcal{E}} = \dfrac{W}{V}$ whence $\mathcal{E} = \dfrac{FV}{W} = \dfrac{nv}{j} = \dfrac{nv}{n\,m} = \dfrac{v}{m}$

10. **(a)** Positively charged ions and free electrons in the air about the electroscope gradually remove the charge. **(b)** The knob should have a spherical shape.

11. The intensity of the electric field is increased by decreasing the radius of curvature at the tip. A very high potential gradient may result from a relatively small charge, ionizing the air and bleeding off the charge, thus preventing a dangerous accumulation of charge.

12. The building is well grounded through the steel framework, and the charge is conducted to ground.

13. No. It would be difficult to build up a high static charge because of the ease with which the charge leaks off through the film covering all contacting surfaces.

14. If a material of dielectric constant 2 were substituted for air (dielectric constant 1) separating two charges, the coulomb force would be reduced to 0.5 that for air. According to Coulomb's law, the proportionality constant k would then have to be 0.5 that for air.

15. The spheres will hold equal charge since the charge resides entirely on the surface and the spheres have equal surface areas.

Problems • Chapter 16/p. 408
GROUP A

1. $C_2 = KC_1 = 6.0 \times 150\ \text{pf} = 9\overline{0}0\ \text{pf}$

2. $Q_1 = C_1V = 5.0\ \mu f \times 28\ v = 140\ \mu c$
   $Q_2 = C_2V = 7.5\ \mu f \times 28\ v = 210\ \mu c$

3. **(a)** $C_T = C_1 + C_2 + C_3 = 2.00\ \mu f + 4.00\ \mu f + 5.00\ \mu f = 11.00\ \mu f$

   (b) $\dfrac{1}{C_T} = \dfrac{1}{C_1} + \dfrac{1}{C_2} + \dfrac{1}{C_3}$, solving for C_T,

   $C_T = \dfrac{C_1C_2C_3}{C_1C_2 + C_2C_3 + C_1C_3}$

   $C_T = \dfrac{2.00\ \mu f \times 4.00\ \mu f \times 5.00\ \mu f}{(2.00\ \mu f \times 4.00\ \mu f) + (4.00\ \mu f \times 5.00\ \mu f) + (2.00\ \mu f \times 5.00\ \mu f)}$

   $C_T = 1.05\ \mu f$

4. $Q = CV = 5.0\ \mu f \times 28\ v = 140\ \mu c$

5. $Q = CV = 2.70\ \mu f \times 15\overline{0}\ v = 40\overline{4}\ \mu c$

GROUP B

6. $V = \dfrac{W}{q}$ $V = \dfrac{3.2 \times 10^{-2}\, n \times 0.25\, m}{4.2 \times 10^{-5}\, c} = 1.9 \times 10^2\, v$

7. The charge on an electron is 1.60×10^{-19} coulomb.
   $W = qV = 1.60 \times 10^{-19}\, c \times 5.0 \times 10^6\, v = 8.0 \times 10^{-13}\, j$

8. (a) $\mathscr{E} = \dfrac{1.2 \times 10^3\, v}{5.0 \times 10^{-3}\, m} = 2.4 \times 10^5\, v/m$

   (b) $v = \dfrac{j}{c} = \dfrac{n\, m}{c}$ $\mathscr{E} = 2.4 \times 10^5 \dfrac{n\, m/c}{m} = 2.4 \times 10^5 \dfrac{n}{c}$

9. (a) $Q = CV = 1.50 \times 10^{-11}\, f \times 1.50 \times 10^2\, v = 2.25 \times 10^{-9}\, c$
   (b) The capacitance is increased 6 times.
       $C' = 6C = 6 \times 1.50 \times 10^{-11}\, f = 9.0 \times 10^{-11}\, f$
   (c) Since the total charge is 6 times the initial charge, the additional charge is 5 times
       the initial. Q added $= 5 \times 2.25 \times 10^{-9}\, c = 1.1 \times 10^{-8}\, c$

10. (a) $Q_1 = C_1 V = 0.15\ \mu f \times 240\ v = 36\ \mu c$
        $Q_2 = C_2 V = 0.22\ \mu f \times 240\ v = 53\ \mu c$
        $Q_3 = C_3 V = 0.47\ \mu f \times 240\ v = 110\ \mu c$
    (b) $C_T = C_1 + C_2 + C_3 = (0.15 + 0.22 + 0.47)\mu f = 0.84\ \mu f$
    (c) $Q_T = C_T V = 0.84\ \mu f \times 240\ v = 2.0 \times 10^2\ \mu c$

DIRECT-CURRENT CIRCUITS   17

Questions • Chapter 17/pp. 423–424
GROUP A
1. An electric current is the flow of charge past a given point in an electric circuit.
2. The unit of current is the ampere.
3. 1 ampere = 1 coulomb/1 second.
4. Electric resistance is the opposition to an electric current.
5. (1) Spontaneous oxidation-reduction reaction; (2) photoelectric effect; (3) thermoelectric effect; (4) piezoelectric effect; (5) electromagnetic induction.
6. (1) Primary (or storage) cell; (2) photoelectric cell; (3) thermocouple (or pyrometer); (4) crystal microphone (or phono pickup); (5) electric generator.
7. The parts of a primary cell are two electrodes and an electrolyte. One electrode must react chemically with the electrolyte.
8. An electrolyte is a substance whose water solution conducts an electric current.
9. (a) Cathode. (b) Anode.
10. A direct current is unidirectional.
11. An open circuit is an incomplete electric circuit that does not offer a continuous conducting path for electron flow. A closed circuit provides a complete conducting path through which an electron current may flow.
12. Open-circuit potential difference is the emf of the source.

GROUP B
13. (a) The potential energy of an electron is increased. (b) The potential energy of an electron is decreased.

14. The reacting materials in a primary cell must be replaced to reactivate the cell while those of a storage cell are renewable.
15. The conversion efficiency is a function of the temperature difference within the thermo-electric system.
16. Potential difference is the work done per unit of charge moved between any two points in a conducting circuit. The emf is the energy per unit charge supplied by a source and is the maximum potential difference developed across a source on open circuit.
17. **(a)** Four 1.5-volt dry cells connected in series. **(b)** Student's diagram showing 4 cells series connected.
18. No, a storage battery does not store electricity as such. Electricity is stored as an electric charge on the plates of a capacitor. In a storage battery, electric energy is converted to chemical energy during the charging cycle, and chemical energy is converted to electric energy during the discharging cycle.

Questions • Chapter 17/pp. 442–443
GROUP A

1. **(a)** The ratio of the emf applied to a closed circuit to the current in the circuit is a constant equal to the resistance of the circuit.

    (b) $\dfrac{E}{I} = R$ or $E = IR$

2. $V = IR$, where V is the potential difference across the part of circuit, I is the current in that part of the circuit, and R is the resistance of that part of the circuit.
3. There is but a single conducting path for current from the negative terminal of the source of emf through the external circuit to the positive terminal of the source.
4. (1) The current in all parts of a series circuit has the same magnitude. (2) The sum of all the separate drops in potential around a series circuit is equal to the applied emf. (3) The total resistance in a series circuit is equal to the sum of all the separate resistances.
5. A parallel circuit is one in which two or more components are connected across two common points in the circuit so as to provide separate conducting paths for current.
6. **(a)** The total resistance increases as resistances are added to the circuit in series.
    (b) The total resistance decreases as resistances are added to the circuit in parallel.
7. (1) The total current in a parallel circuit is equal to the sum of the currents in the separate branches. (2) The potential difference across all branches of a parallel circuit must have the same magnitude. (3) The reciprocal of the equivalent resistance is equal to the sum of the reciprocals of the separate resistances in parallel.
8. **(a)** The resistance of metallic conductors usually rises with rising temperature. **(b)** No, many conducting materials other than metals show a decrease in resistance with rising temperature.
9. The resistance of a conductor increases with its length. If the conductor is uniform in all other respects, the resistance is directly proportional to its length.
10. The resistance of a conductor varies inversely with its cross-sectional area.
11. Resistivity, ρ, is a proportionality constant equal to the resistance of a wire 1 cm long and having a cross-sectional area of 1 cm^2.
12. $R = \rho \dfrac{l}{A}$; so $\rho = \dfrac{RA}{l} = \dfrac{\Omega \text{ cm}^2}{\text{cm}} = \Omega \text{ cm}$

GROUP B

13. Energy is removed from the electrons flowing through the internal resistance, resulting in the fall of potential across the resistance that is subtracted from the emf (a rise in potential), and thus the potential drop is opposite in sign to the emf.

14. Assuming that the resistance of the voltmeter is very high compared to the internal resistance of the battery, the meter current is of negligible magnitude and produces a negligible potential drop across the internal resistance of the battery.

15. If the filament of one lamp opens, the entire series circuit is opened. All lamps in the circuit must operate in order for any one lamp to be used for lighting.

16. The 120-volt lamp would operate normally, but the 6-volt lamp would draw an excessive current and burn out.

17. Student's network diagram and analysis.

18. The current in the resistance is measured by an ammeter connected in series. The potential difference across the resistance is measured by a voltmeter connected in parallel. Resistance is then determined by the application of Ohm's law.

19. The voltmeter connected in parallel with the resistance to be measured has some loading effect on the circuit. The ammeter then measures the total current in the two parallel branches.

20. The fact that the resistance of each wire section is directly proportional to its length enables us to use the ratio of their lengths instead of the ratio of their resistances in solving the bridge problem.

21. **(a)** Student diagram. **(b)** B and C are of equal brightness but both are dimmer than A since the voltage drop across B and C in parallel is less than that across A (or the current in A is the sum of the currents in B and C). **(c)** With the filament of C open, A and B are in series and the potential difference across A decreases and across B increases, and lamp A dims and lamp B brightens until they achieve equal brightness.

Problems • Chapter 17/pp. 442–444
GROUP A

1. $I = \dfrac{E}{R} = \dfrac{4.5 \text{ v}}{18 \ \Omega} = 0.25$ a

2. **(a)** Student's diagram.
   (b) $R_T = R_1 + R_2 + R_3 = 15 \ \Omega + 21 \ \Omega + 24 \ \Omega = 60 \ \Omega$
   (c) $I = \dfrac{E}{R_T} = \dfrac{12 \text{ v}}{60 \ \Omega} = 0.20$ a

3. $V_1 = IR_1 = 0.20 \text{ a} \times 15 \ \Omega = 3.0$ v
   $V_2 = IR_2 = 0.20 \text{ a} \times 21 \ \Omega = 4.2$ v
   $V_3 = IR_3 = 0.20 \text{ a} \times 24 \ \Omega = 4.8$ v

4. $R = \dfrac{V}{I} = \dfrac{6.0 \text{ v}}{0.300 \text{ a}} = 20 \ \Omega$

5. **(a)** $V = I(R_1 + R_2) = 0.5 \text{ a} \times 6 \ \Omega = 3$ v
   (b) Four cells are used; two in series are connected in parallel with the other two in series. **(c)** Student's diagram showing a battery of two series-connected cells in parallel with two series-connected cells across the series combination of the ammeter, R_1, and R_2.

6. $I = \dfrac{V}{R} = \dfrac{115 \text{ v}}{230 \text{ }\Omega} = 0.50 \text{ a}$

7. $I = \dfrac{V}{R} = \dfrac{120 \text{ v}}{230 \text{ }\Omega} = 0.52 \text{ a}$

8. $R = \dfrac{V}{I} = \dfrac{230 \text{ v}}{0.50 \text{ a}} = 460 \text{ }\Omega$

9. $\dfrac{1}{R_{eq}} = \dfrac{1}{R_1} + \dfrac{1}{R_2} \qquad \dfrac{R_1 R_2 R_{eq}}{R_{eq}} = \dfrac{R_1 R_2 R_{eq}}{R_1} + \dfrac{R_1 R_2 R_{eq}}{R_2}$

$R_1 R_2 = R_2 R_{eq} + R_1 R_{eq} = R_{eq}(R_1 + R_2) \qquad R_{eq} = \dfrac{R_1 R_2}{R_1 + R_2}$

10. $R_{eq} = \dfrac{R_1 R_2}{R_1 + R_2} = \dfrac{12 \text{ }\Omega \times 18 \text{ }\Omega}{12 \text{ }\Omega + 18 \text{ }\Omega} = 7.2 \text{ }\Omega$

11. $\dfrac{1}{R_{eq}} = \dfrac{1}{R_1} + \dfrac{1}{R_2} + \dfrac{1}{R_3}$

$R_1 R_2 R_3 = R_2 R_3 R_{eq} + R_1 R_3 R_{eq} + R_1 R_2 R_{eq}$
$R_1 R_2 R_3 = R_{eq}(R_2 R_3 + R_1 R_3 + R_1 R_2)$

$R_{eq} = \dfrac{R_1 R_2 R_3}{R_1 R_2 + R_2 R_3 + R_1 R_3}$

12. By inspection: $R_{eq} = 12 \text{ }\Omega \div 3 = 4.0 \text{ }\Omega$

13. (a) $V = I_T R_{eq} = 1.5 \text{ a} \times 4.0 \text{ }\Omega = 6.0 \text{ v}$

(b) By inspection: $1.5 \text{ a} \div 3 = 0.5\overline{0} \text{ a}$ or, by each branch, $I = \dfrac{V}{R} = \dfrac{6.0 \text{ v}}{12 \text{ }\Omega} = 0.50 \text{ a}$

14. $R_{eq} = \dfrac{R_1 R_2 R_3}{R_1 R_2 + R_2 R_3 + R_1 R_3} = \dfrac{3 \text{ }\Omega \times 5 \text{ }\Omega \times 8 \text{ }\Omega}{(3 \text{ }\Omega \times 5 \text{ }\Omega) + (5 \text{ }\Omega \times 8 \text{ }\Omega) + (3 \text{ }\Omega \times 8 \text{ }\Omega)}$

$R_{eq} = \dfrac{120 \text{ }\Omega^3}{79 \text{ }\Omega^2} = 2 \text{ }\Omega$

15. (a) Student's diagram.

(b) $I_T = \dfrac{E}{R_{eq}} \qquad R_{eq} = \dfrac{R_1 R_2}{R_1 + R_2}$

$R_{eq} = \dfrac{4.0 \text{ }\Omega \times 6.0 \text{ }\Omega}{4.0 \text{ }\Omega + 6.0 \text{ }\Omega} = 2.4 \text{ }\Omega \qquad I_T = \dfrac{6.0 \text{ v}}{2.4 \text{ }\Omega} = 2.5 \text{ a}$

(c) $I_1 = \dfrac{E}{R_1} = \dfrac{6.0 \text{ v}}{4.0 \text{ }\Omega} = 1.5 \text{ a}$

(d) $I_2 = \dfrac{E}{R_2} = \dfrac{6.0 \text{ v}}{6.0 \text{ }\Omega} = 1.0 \text{ a}$ or, $I_2 = I_T - I_1 = 2.5 \text{ a} - 1.5 \text{ a} = 1.0 \text{ a}$

16. $I = \dfrac{V_{lamp}}{R_{lamp}} = \dfrac{24 \text{ v}}{8.0 \text{ }\Omega} = 3.0 \text{ a}$

$IR = V \text{ line} - V \text{ lamp} \qquad R = \dfrac{117 \text{ v} - 24 \text{ v}}{3.0 \text{ a}} = 31 \text{ }\Omega$

17. $\dfrac{l_1}{l_2} = \dfrac{R_1}{R_2}$

$R_2 = \dfrac{l_2 R_1}{l_1} = \dfrac{50\overline{0} \text{ m} \times 2.059 \text{ } \Omega}{10\overline{0} \text{ m}} = 10.3 \text{ } \Omega$

18. $\dfrac{R_1}{R_2} = \dfrac{d_2{}^2}{d_1{}^2}$ $R_2 = \dfrac{R_1 d_1{}^2}{d_2{}^2} = \dfrac{2.059 \text{ } \Omega \times (1.024 \text{ mm})^2}{(0.2546 \text{ mm})^2} = 33.31 \text{ } \Omega$

19. $R = \rho \dfrac{l}{A} = \rho \dfrac{l}{\pi d^2/4} = \rho \dfrac{4l}{\pi d^2}$

$R = \dfrac{2.83 \times 10^{-6} \text{ } \Omega \text{ cm} \times 4 \times 1.25 \times 10^4 \text{ cm}}{\pi (8.12 \times 10^{-2} \text{ cm})^2} = \dfrac{14.2 \times 10^{-2} \text{ } \Omega \text{ cm}^2}{2.07 \times 10^{-2} \text{ cm}^2}$

$R = 6.86 \text{ } \Omega$

20. $R = \rho \dfrac{l}{A}$ $l = \dfrac{RA}{\rho} = \dfrac{R\pi d^2}{4\rho}$

$l = \dfrac{10\overline{0} \text{ } \Omega \times 3.14 \times (5.11 \times 10^{-2} \text{ cm})^2}{4 \times 33 \times 10^{-6} \text{ } \Omega \text{ cm}} = \dfrac{8.25 \times 10^{-1} \text{ } \Omega \text{ cm}^2}{1.3 \times 10^{-4} \text{ } \Omega \text{ cm}}$

$l = 6.3 \times 10^3 \text{ cm}$

GROUP B

21. **(a)** Student's diagram.

(b) $R_T = r + R_L = 0.0300 \text{ } \Omega + 3.27 \text{ } \Omega = 3.30 \text{ } \Omega$

$I = \dfrac{E}{R_T} = \dfrac{6.60 \text{ v}}{3.30 \text{ } \Omega} = 2.00 \text{ a}$

(c) $V_L = IR_L = 2.00 \text{ a} \times 3.27 \text{ } \Omega = 6.54 \text{ v}$

22. $E = V_L + V_r = IR_L + Ir$ $Ir = E - IR_L$

$r = \dfrac{E - IR_L}{I} = \dfrac{3.1 \text{ v} - (0.50 \text{ a} \times 5.8 \text{ } \Omega)}{0.50 \text{ a}} = 0.40 \text{ } \Omega$

$r/\text{cell} = 0.20 \text{ } \Omega$

23. **(a)** Student's diagram.

(b) $I_T = \dfrac{E}{R_T}$ $R_T = R_1 + R_{eq} = R_1 + \dfrac{R_2 R_3}{R_2 + R_3}$

$R_T = 40.0 \text{ } \Omega + \dfrac{30.0 \text{ } \Omega \times 60.0 \text{ } \Omega}{30.0 \text{ } \Omega + 60.0 \text{ } \Omega} = 40.0 \text{ } \Omega + 20.0 \text{ } \Omega = 60.0 \text{ } \Omega$

$I_T = \dfrac{12.0 \text{ v}}{60.0 \text{ } \Omega} = 0.200 \text{ a}$

(c) $V_1 = I_T R_1 = 0.200 \text{ a} \times 40.0 \text{ } \Omega = 8.00 \text{ v}$

(d) $V_{eq} = I_T R_{eq} = 0.200 \text{ a} \times 20.0 \text{ } \Omega = 4.00 \text{ v}$

$I_2 = \dfrac{V_{eq}}{R_2} = \dfrac{4.00 \text{ v}}{30.0 \text{ } \Omega} = 0.133 \text{ a}$

24. **(a)** Student's diagram. The equivalent circuit shows the R_{eq} in series with r.

(b) $\dfrac{1}{R_{eq}} = \dfrac{1}{6.0\ \Omega} + \dfrac{1}{9.0\ \Omega} + \dfrac{1}{18.0\ \Omega}$ $R_{eq} = 3.0\ \Omega$

$R_T = r + R_{eq} = 0.10\ \Omega + 3.0\ \Omega = 3.1\ \Omega$

$I_T = \dfrac{E}{R_T} = \dfrac{6.2\ v}{3.1\ \Omega} = 2.0\ a$

(c) $V_L = I_T R_{eq} = 2.0\ a \times 3.0\ \Omega = 6.0\ v$

(d) In the 6.0-ohm resistance: $I = \dfrac{6.0\ v}{6.0\ \Omega} = 1.0\ a$

In the 9.0-ohm resistance: $I = \dfrac{6.0\ v}{9.0\ \Omega} = 0.67\ a$

In the 18.0-ohm resistance: $I = \dfrac{6.0\ v}{18.0\ \Omega} = 0.33\ a$

25. (a) Student's diagram.

(b) $R_{eq} = \dfrac{(R_1 + R_2) \times R_3}{R_1 + R_2 + R_3} = \dfrac{(15.0\ \Omega + 9.0\ \Omega) \times 8.0\ \Omega}{15.0\ \Omega + 9.0\ \Omega + 8.0\ \Omega} = 6.0\ \Omega$

$I_T = \dfrac{E}{R_{eq}} = \dfrac{6.0\ v}{6.0\ \Omega} = 1.0\ a$

(c) $I_3 = \dfrac{E}{R_3} = \dfrac{6.0\ v}{8.0\ \Omega} = 0.75\ a$

(d) $V_2 = I_2 R_2$ where $I_2 = I_T - I_3 = 0.25\ a$

$V_2 = 0.25\ a \times 9.0\ \Omega = 2.2\ v$

26. $I = \dfrac{V}{R} = \dfrac{9.0\ v}{3.0\ \Omega} = 3.0\ a$

$R_T = r + R_L = 1.5\ \Omega + 6.5\ \Omega = 8.0\ \Omega$

$E = I R_T = 3.0\ a \times 8.0\ \Omega = 24\ v$

27. (a) Student's diagram.

(b) $V_{eq} = I_2 R_2 = 0.800\ a \times 6.00\ \Omega = 4.80\ v$

$R_{eq} = \dfrac{R_1 R_2}{R_1 + R_2} = \dfrac{12.0\ \Omega \times 6.00\ \Omega}{12.0\ \Omega + 6.00\ \Omega} = 4.00\ \Omega$

$I_T = \dfrac{V_{eq}}{R_{eq}} = \dfrac{4.80\ v}{4.00\ \Omega} = 1.20\ a$

$E = I_T R_T = I_T \times (r + R_1 + R_{eq}) = 1.20\ a \times (0.250\ \Omega + 6.25\ \Omega + 4.00\ \Omega)$

$E = 1.20\ a \times 10.5\ \Omega = 12.6\ v$

28. $R = \rho \dfrac{l}{A}$

$\rho = \dfrac{RA}{l} = \dfrac{0.830\ \Omega}{5.00 \times 10^3\ cm} \times \dfrac{\pi(0.115\ cm)^2}{4}$ $\rho = 1.73 \times 10^{-6}\ \Omega\ cm$

29. $R = \dfrac{V}{I} = \dfrac{1.4\ v}{0.28\ a} = 5.0\ \Omega$

30. (a) (See two diagrams below.)

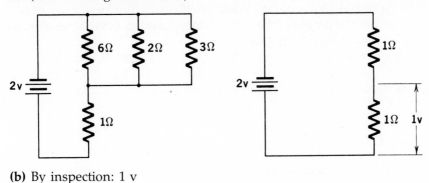

(b) By inspection: 1 v

HEATING AND CHEMICAL EFFECTS   **18**

Questions • Chapter 18/pp. 456–457
GROUP A

1. (1) Work done results in the production of light—electric lamp. (2) Work done results in the production of motion—electric motor. (3) Work done results in chemical action—electrolytic cell. (4) Work done results in the production of heat—electric resistance.

2. The iron wire becomes hotter because of its higher resistance.

3. The wire across the battery of two cells becomes hotter because of the larger current in it.

4. The heat developed in a conductor is directly proportional to the resistance of the conductor, the square of the current, and the time the current is maintained.

5. Electric power is the rate at which electric energy is delivered to an electric circuit.

6. **(a)** The heating effect is doubled. **(b)** Student's answer.

7. A short circuit results in a large current in the internal resistance of the battery. The I^2r power is dissipated as heat within the battery.

8. The maximum power is delivered to a load whose resistance is equal to the internal resistance of the source.

9. **(a)** Electric energy was purchased. **(b)** The energy used was measured in kilowatt-hours.

10. The load on a source of emf is increased by making the resistance of the external circuit smaller.

GROUP B

11. $Q = \dfrac{I^2 R t}{J} = \dfrac{a^2 \times \Omega \times s}{j/cal} = \dfrac{(a \times \Omega)(a \times s)}{j/cal}$

    $Q = \dfrac{v \times c}{j/cal} = \dfrac{j}{j/cal} = cal \qquad cal \times \dfrac{kcal}{1000\ cal} = kcal$

12. When two equal resistors are connected in parallel, their combined resistance is half that of either resistor. When connected in series, their combined resistance is twice that of either resistor. Thus the combined resistance changes by a factor of 4.

13. The starting motor on an automobile engine acts as a very low resistance across the battery during the starting operation. In order to deliver efficiently the large amount of power required to turn the engine over, the battery must have an internal resistance approximately equal to that of the starting motor.

14. High resistivity, high melting point, and be unreactive with oxygen at incandescent temperatures.

15. The rate of energy dissipation is an expression of power since $E/t = P$. **(a)** $P = VI$ **(b)** $P = I^2R$ **(c)** $P = V^2/R$

Problems • Chapter 18/pp. 457–458

GROUP A

1. $Q = It = 7.5 \text{ c/s} \times 15 \text{ min} \times 60 \text{ s/min} = 6800 \text{ c}$

2. $I = \dfrac{Q}{t} = \dfrac{4800 \text{ c}}{25 \text{ min} \times 60 \text{ s/min}} = 3.2 \text{ a}$

3. $W = QV = ItV = 0.52 \text{ a} \times 12 \text{ min} \times 60 \text{ s/min} \times 117 \text{ v} = 44{,}000 \text{ j}$

4. $w = \dfrac{j}{s} = \dfrac{44{,}000 \text{ j}}{12 \text{ min} \times 60 \text{ s/min}} = 61 \ w\text{(60-watt lamp)}$

5. $Q = \dfrac{I^2Rt}{J} = \dfrac{(V/R)^2Rt}{J} = \dfrac{(110 \text{ v}/55 \ \Omega)^2 \times 55 \ \Omega \times 1\overline{0} \text{ min} \times 60 \text{ s/min}}{4.19 \text{ j/cal} \times 10^3 \text{ cal/kcal}}$

   $Q = 32 \text{ kcal}$

6. $Q = \dfrac{I^2Rt}{J} = \dfrac{I^2(V/I)t}{J} = \dfrac{(9.00 \text{ a})^2 \times 117 \text{ v}/9.00 \text{ a} \times 30.0 \text{ min} \times 60 \text{ s/min}}{4.19 \text{ j/cal} \times 10^3 \text{ cal/kcal}}$

   $Q = 453 \text{ kcal}$

7. $Q = \dfrac{I^2Rt}{J} = \dfrac{(5.0 \text{ a})^2 \times 24 \ \Omega \times 45 \text{ min} \times 60 \text{ s/min}}{4.19 \text{ j/cal} \times 10^3 \text{ cal/kcal}} = 390 \text{ kcal}$

8. **(a)** $I = \dfrac{E}{R_L + r} = \dfrac{26.4 \text{ v}}{3.00 \ \Omega + 0.300 \ \Omega} = 8.00 \text{ a}$

   $P_L = I^2R_L = (8.00 \text{ a})^2 \times 3.00 \ \Omega = 192 \text{ w}$

   (b) $P_r = I^2r = (8.00 \text{ a})^2 \times 0.300 \ \Omega = 19.2 \text{ w}$

9. **(a)** For maximum power: $R_L = r = 0.300 \ \Omega$

   (b) $I = \dfrac{E}{R_T} = \dfrac{E}{R_L + r} = \dfrac{26.4 \text{ v}}{0.300 \ \Omega + 0.300 \ \Omega} = 44.0 \text{ a}$

   $P_L = I^2R_L = (44.0 \text{ a})^2 \times 0.300 \ \Omega = 582 \text{ w}$

   (c) $Pr = P_L = 582 \text{ w}$

10. $l = R \times \dfrac{m}{5.25 \ \Omega}$ where $R = V/I$

    $l = V/I \times m/5.25 \ \Omega = 12\overline{0} \text{ v}/8.70 \text{ a} \times m/5.25 \ \Omega = 2.63 \text{ m}$

GROUP B

11. $\text{cost} = \text{kw hr} \times \dfrac{\text{cost}}{\text{kw hr}} = mc \ \Delta T \times J \times \dfrac{\text{cost}}{\text{kw hr}}$

    $\text{cost} = 4.6 \text{ kg} \times \dfrac{1.00 \text{ kcal}}{\text{kg C}^\circ} \times 75 \text{ C}^\circ \times \dfrac{4.19 \text{ j}}{\text{cal}} \times \dfrac{10^3 \text{ cal}}{\text{kcal}} \times \dfrac{\text{kw hr}}{3.6 \times 10^6 \text{ w s}} \times \dfrac{9.2 \ \cent}{\text{kw hr}}$

    $\text{cost} = 3.7 \ \cent$

12. (a) $Q = \dfrac{I^2Rt}{J} \qquad \dfrac{Q}{t} = \dfrac{(V/R)^2 R}{J}$

$Q/t = \dfrac{(120 \text{ v}/20.0 \ \Omega)^2 \times 20.0 \ \Omega}{4.19 \text{ j/cal} \times 10^3 \text{ cal/kcal}} = 0.173 \text{ kcal/s, the heat liberated per second}$

(b) $E = Pt \qquad t = E/P$

$t = \dfrac{mc \ \Delta T}{Q/s \times 60 \text{ s/min}} = \dfrac{0.500 \text{ kg} \times 1.00 \text{ kcal/kg C}^\circ \times (10\bar{0} \ ^\circ\text{C} - 22.5 \ ^\circ\text{C})}{0.173 \text{ kcal/s} \times 60 \text{ s/min}}$

$t = 3.37 \text{ min}$

13. We are to assume that all the heat produced by the nichrome wire is taken up by the iron.

$\dfrac{I^2Rt}{J} = mc \ \Delta T \qquad \text{and} \qquad t = \dfrac{Jmc \ \Delta T}{I^2R}$

$I = V/R \qquad \text{and} \qquad R = \rho l/A$

Therefore, $t = \dfrac{Jmc \ \Delta T \ \rho l}{V^2 A}$

$t = \dfrac{(4.19 \times 10^3 \text{ j/kcal})(1.50 \text{ kg})(0.107 \text{ kcal/kg C}^\circ)(15\bar{0} \ ^\circ\text{C} - 20.0 \ ^\circ\text{C})(1.15 \times 10^{-4} \ \Omega \text{ cm})(2.00 \times 10^2 \text{ cm})}{(115 \text{ v})^2 \times 0.205 \times 10^{-2} \text{ cm}^2}$

$t = 74.2 \text{ j } \Omega/\text{v}^2 = 74.2 \text{ s}$

14. Efficiency $= \dfrac{W_{\text{out}}}{W_{\text{in}}} \times 100\%$

$W_{\text{out}} = mc \ \Delta T + mL_v$
$W_{\text{in}} = I^2Rt/J$ where $R = V/I$
$W_{\text{in}} = IVt/J$

Efficiency $= \dfrac{mc \ \Delta T + mL_v}{IVt/J} \times 100\%$

Efficiency $= \dfrac{(60\bar{0} \text{ g} \times 1 \text{ cal/g C}^\circ \times 80.0 \text{ C}^\circ) + (60.0 \text{ g} \times 539 \text{ cal/g})}{10.0 \text{ a} \times 12\bar{0} \text{ v} \times 7.00 \text{ min} \times 60 \text{ s/min} \times 1/4.19 \text{ j/cal}} \times 100\%$

Efficiency $= \dfrac{4.19 \text{ j/cal} \times 8.03 \times 10^4 \text{ cal}}{5.04 \times 10^5 \text{ j}} \times 100\% = 66.7\%$

15. (a) $P = V^2/R$
$V^2 = PR = 4.0 \text{ w} \times 2.0 \ \Omega = 8.0 \text{ v}^2$
$V = 2.8 \text{ v}$
(b) $V^2 = PR = 4.0 \text{ w} \times 20.0 \ \Omega = 8\bar{0} \text{ v}^2$
$V = 8.9 \text{ v}$

16. (a) $P = I^2R \qquad I^2 = P/R = 4.0 \text{ w}/10.0 \ \Omega = 0.40 \text{ a}^2$
$I = 0.63 \text{ a}$
(b) $V = P/I = 6.3 \text{ v}$

17. $R = \rho \dfrac{l}{A} \qquad \text{and} \qquad l = \dfrac{RA}{\rho}$

$P = \dfrac{V^2}{R} \qquad \text{and} \qquad R = \dfrac{V^2}{P}$

$l = \dfrac{V^2A}{P\rho} = \dfrac{(117 \text{ v})^2 \times 0.0509 \text{ mm}^2}{1.000 \times 10^3 \text{ w} \times 1.15 \times 10^{-4} \ \Omega \text{ cm}} \times \dfrac{\text{cm}^2}{10^2 \text{ mm}^2} \times \dfrac{\text{m}}{10^2 \text{ cm}}$

$l = 0.606 \text{ m}$

18. $P = VI$ $I = Q/t$ therefore, $P = VQ/t$ and $Q = Pt/V$

$$Q = \frac{100 \text{ w} \times 24.0 \text{ hr} \times 3.60 \times 10^3 \text{ s/hr}}{117 \text{ v} \times 1.60 \times 10^{-19} \text{ c/e}^-}$$

$Q = 4.61 \times 10^{23} \text{ e}^-$

Questions • Chapter 18/p. 463
GROUP A

1. Electron exchange reactions can be forced to occur by supplying electric energy from an external source.
2. An electrolytic cell is an arrangement in which a forced electron exchange reaction occurs.
3. The fused ionic compound may be decomposed.
4. An anode of silver, a cathode of the metal to be plated, and an electrolytic solution of a soluble silver salt.
5. Minute amounts of impurities in copper greatly increase its electric resistance. By refining copper electrolytically the purity is increased, electrolytic copper being over 99.99% pure.
6. (1) The mass of an element deposited during electrolysis is proportional to the quantity of charge that passes. (2) The mass of an element deposited during electrolysis is proportional to the chemical equivalent of the element.
7. The platinum dish serves as the cathode of the cell.

GROUP B

8. An electrochemical cell is one in which an electron exchange reaction occurs spontaneously, converting chemical energy to electric energy. An electrolytic cell is one in which an electron exchange reaction is forced to occur converting electric energy to chemical energy.
9. Electrode reactions in the two types of cells are reversed.
   Electrolytic cell: Cathode loses electrons to positive ions. Anode gains electrons from negative ions.
   Electrochemical cell: Cathode gains electrons during the chemical reaction between cathode and electrolyte. Anode loses electrons during this reaction.
10. Gold has the higher electrochemical equivalent since the chemical equivalent of Al is 9 g and of Au is 65.7 g and each is divided by the same constant, the faraday.
11. $m = I = \dfrac{\text{g}}{\text{c}} \times \dfrac{\text{c}}{\text{s}} \times \text{s} = \text{g}$
12. The ampere is the current that deposits 0.001118 g of silver in 1 second.

Problems • Chapter 18/pp. 463–464
GROUP A

1. $m = zIt$, where z for silver = 0.001118 g/c
   $m = 0.001118 \text{ g/c} \times 0.500 \text{ a} \times 8.00 \text{ min} \times 60 \text{ s/min} = 0.268$ g
2. $m = zIt$, where z for copper = 0.0003294 g/c

$$I = \frac{m}{zt} = \frac{0.750 \text{ g}}{0.0003294 \text{ g/c} \times 15.0 \text{ min} \times 60 \text{ s/min}}$$

$$I = \frac{0.750 \text{ g}}{0.296 \text{ g s/c}} = 2.53 \text{ c/s} = 2.53 \text{ a}$$

3. $m = zIt$, where z for oxygen $= 0.0000829$ g/c

$$t = \frac{m}{zI} = \frac{0.160 \text{ g}}{0.0000829 \text{ g/c} \times 0.300 \text{ a}} \times \frac{1 \text{ hr}}{3600 \text{ s}} = 1.79 \text{ hr}$$

4. $m = zIt = 1.04 \times 10^{-5}$ g/c $\times 25.0$ a $\times 24.0$ hr $\times 3.60 \times 10^3$ s/hr

$m = 22.5$ g

5. $m = zIt$. Thus, $t = m/zI = \dfrac{25.0 \text{ g}}{3.04 \times 10^{-4} \text{g/c} \times 3.40 \text{ a}} \times \dfrac{\text{hr}}{3.6 \times 10^3 \text{ s}}$

$t = 6.74$ hr

GROUP B

6. One electron is acquired per atom of hydrogen liberated.

No. H atoms $=$ No. $e^- = e^-/c \times$ c/faraday

6.25×10^{18} e^-/c $\times 9.65 \times 10^4$ c/faraday $= 6.03 \times 10^{23}$ e^-/faraday

6.03×10^{23} hydrogen atoms liberated

7. **(a)** $m = zIt$. But $I = Q/t$ Therefore, $m = zQ$ and $Q = m/z$

$Q = 1.01$ g/1.04×10^{-5} g/c $= 9.71 \times 10^4$ c (or by inspection: 1 faraday)

(b) $m = zQ = 8.29 \times 10^{-5}$ g/c $\times 9.71 \times 10^4$ c $= 8.05$ g (or by inspection: 8 g)

8. $m = zIt$ Therefore, $t = m/zI$

$D = m/V$ Therefore, $m = DV$ Substituting for m above

$$t = DV/zI = \frac{8.99 \times 10^{-2} \text{ g/L} \times 1.00 \text{ L}}{1.04 \times 10^{-5} \text{ g/c} \times 4.50 \text{ a}} = 1.92 \times 10^3 \text{ s}$$

9. $m = zIt = 1.12 \times 10^{-3}$ g/c $\times 2.00$ a $\times 2.00$ hr $\times 3.60 \times 10^3$ s/hr

$m = 16.1$ g

10. g-at wt $=$ chem eq $\times$ ionic charge

$$\frac{m_{Ag}}{\text{chem eq}_{Ag}} = \frac{m_x}{\text{chem eq}_x}   \text{Therefore, chem eq}_x = \frac{\text{chem eq}_{Ag} \times m_x}{m_{Ag}}$$

$$\text{g-at wt}_x = \frac{108 \text{ g} \times 2.73 \text{ g}}{16.1 \text{ g}} \times 3 = 54.9 \text{ g}$$

MAGNETIC EFFECTS   **19**

Questions • Chapter 19/p. 476

GROUP A

1. **(a)** A lodestone is a piece of magnetite that has magnetic properties. **(b)** Lodestones are found in nature possessing magnetic properties induced by the earth's magnetic field and thus are called natural magnets.

2. Ferromagnetic materials are strongly attracted by magnets.

3. Materials commonly referred to as nonmagnetic may be either feebly repelled or feebly attracted by strong magnets. Materials that are feebly repelled by strong magnets are said to be diamagnetic; those feebly attracted are said to be paramagnetic.

4. The two kinds of electron motion are (1) the revolving of an electron about the nucleus of the atom and (2) the spinning of an electron on its own axis.

5. Electron pairs consist of two electrons of opposite spin that, by pairing, neutralize their magnetic character.

6. (a) The important ferromagnetic metals are iron, cobalt, and nickel. (b) Alnico is a special alloy used to make strong magnets.

7. These elements have incomplete inner shells of electrons that contain unpaired electrons. The similarly oriented spins of these unpaired electrons in an appropriate crystal lattice account for the strong ferromagnetism of the metals of the iron family.

8. Domains are microscopic magnetic regions in ferromagnetic materials that are composed of atoms polarized parallel to the same crystal axis.

9. When all the domains are aligned in the same direction by a strong magnetizing force, the material is said to have reached magnetic saturation.

10. The force between two magnetic poles is directly proportional to the strengths of the poles and inversely proportional to the square of their distance apart.

11. (a) A repulsive force occurs between two similar poles. (b) An attractive force occurs between two dissimilar poles.

12. The magnets may be arranged under a glass plate and iron filings sprinkled over the glass. The alignment of the tiny iron fragments suggests the flux lines of the magnetic field.

13. By shaping the magnet like a horseshoe, the poles are brought close together, concentrating most of the magnetic flux in a small region and increasing the flux density.

14. Student's answer. A piece of soft iron near or in contact with a permanent magnet exhibits the properties of a temporary magnet.

15. When the magnetic pole region of the earth nearest the North Geographic Pole (actually an S pole) is designated the North Magnetic Pole, the pole of a small magnet or compass needle that is attracted by the North Magnetic Pole is called a north-seeking pole, or N pole.

16. The North Magnetic Pole does not coincide with the North Geographic Pole, and thus the compass needle does not point due north from most locations on the earth.

GROUP B

17. The atom structure has a magnetic property due to the movement of electrons about the nucleus. When the atoms of a substance are subjected to a strong magnetic force, this force opposes the revolving motion of the electrons and acts to repel the atoms.

18. Atoms have magnetic properties due to the orbital motions and the spinning motions of electrons. The permanent-magnet characteristic of some atoms may be due to an imbalance between orbits and spins. If this effect exceeds the diamagnetic effect common to all atoms, the material will be feebly attracted by strong magnets and is said to be paramagnetic.

19. The needle will not be attracted to the north edge of the pan of water since the S pole will be repelled by the North Magnetic Pole to the same extent that the N pole is attracted.

20. Use a small magnetic compass to detect a repulsion effect between one end of the bar and the compass needle.

21. The case should be made of ferromagnetic material so as to provide a highly permeable path for magnetic flux around the watch mechanism.

Questions • Chapter 19/pp. 487–488
GROUP A

1. Oersted discovered that a magnetic field is established about a conductor by a current in the conductor.

2. The N pole of the compass is deflected toward the east.

3. Student's answer: A current passing perpendicularly through a plane will enable the nature of the magnetic field around the current to be shown by means of iron filings or small compasses. The compasses will also indicate the direction of the field.
4. By Ampère's rule, the direction of the flux is clockwise around the current.
5. The strength of an electromagnet depends on the permeability of the core material, the magnitude of current, and the number of turns.
6. The torque produced by a current in the movable coil is opposed by the reaction of control springs against which the moving coil must do work. The coil moves until the torque acting on it is just neutralized by the reaction of the springs.
7. (a) An ammeter is connected in series with the load and therefore should add a negligible resistance to the circuit. (b) A voltmeter is connected across the load and therefore should have a negligible shunting effect on the circuit.

GROUP B
8. The coil will turn so that end A points in the direction of the North Magnetic Pole.
9. (a) The electron stream will be deflected upward. (b) The electron stream will be deflected downward.
10. The electron stream will be deflected to the right (looking from the source of electrons). The two fields aid each other on the left side of the beam, oppose each other on the right side, and are perpendicular above and below.
11. The meter is connected in parallel with the load resistance. If the meter resistance is not high with respect to the load resistance, the equivalent resistance will be appreciably smaller than the load. Since the internal resistance of the source remains the same, the indicated potential difference may be appreciably lower while the meter is connected.
12. Student's diagram: The ammeter is connected in series with the resistance and the voltmeter is connected across both the resistance and the ammeter. Since the resistance is high, the current in the voltmeter will be small. If the voltmeter were connected directly across the resistance, the small meter current would be an appreciable part of the total current measured by the ammeter. By connecting the voltmeter across both the resistance and the ammeter, the ammeter measures only the current in the resistance. Since the ammeter is a very low resistance instrument, the voltage drop across it is negligible and the voltmeter error is minimum.
13. Student's diagram: The ammeter is connected in series with the parallel combination of the resistance and voltmeter. Since the resistance is low, the current in it will be large and the small voltmeter current will be negligible. The voltage across the ammeter will be appreciable compared to that across the resistance. Thus the voltmeter should be connected across the resistance only to keep the voltmeter error to a minimum.

Problems • Chapter 19/p. 488
GROUP B
1. $F = \dfrac{2kI_1 I_2 l}{d}$

$I_1 I_2 = \dfrac{Fd}{2kl}$ but $I_1 = I_2$ so $I_1 I_2 = I_1^2$

$I_1^2 = \dfrac{Fd}{2kl} = \dfrac{1.6 \times 10^{-6}\, \text{n} \times 1.0\, \text{m}}{2 \times 10^{-7}\, \text{n/a}^2 \times 2.0\, \text{m}} = 4.0\, \text{a}^2$

$I_1 = I_2 = 2.0\, \text{a}$

2. $R_S = 0.1\ R_M$ $V_S = V_M$ $I_M = 0.1\ I_S$

I_T divides, 1 part in R_M and 10 parts in R_S

$$I_M = \frac{1\ a}{11}\,I_T = \frac{10\ a}{11} = 0.9\ a$$

3. (a) $k = \dfrac{I_M}{d} = \dfrac{375\ \mu a}{15.0\ \text{div}} = 25.0\ \mu a/\text{div}$

   (b) $I_M = kd = 25.0\ \mu a/\text{div} \times 20.0\ \text{div} = 50\bar{0}\ \mu a$

4. $V = I_M R_M + I_M R_S$

$$R_S = \frac{V}{I_M} - R_M = \frac{300.0\ v}{0.0750\ a} - 50.0\ \Omega$$

$$R_S = 40\bar{0}0 - 50.0\ \Omega = 3950\ \Omega$$

5. $V_M = I_S R_S$ where $I_S = I_T - I_M$

$I_M = V_M \div R_M = 0.050\ v \div 2.5\ \Omega$

$$R_S = \frac{V_M}{I_T - I_M} = \frac{0.050\ v}{7.5\ a - 0.02\ a} = 0.0067\ \Omega$$

6. $F = \dfrac{2kI_1 I_2 l}{d}$ and $d = \dfrac{2kI_1 I_2 l}{F}$

$$d = \frac{2 \times 10^{-7}\ n/a^2 \times 3.2\ a \times 3.2\ a \times 5.0\ m}{9.6 \times 10^{-4}\ n}$$

$$d = 1.1 \times 10^{-2}\ m, \text{ or } 1.1\ cm$$

20   ELECTROMAGNETIC INDUCTION

Questions • Chapter 20/p. 498–499

GROUP A

1. The essential condition for an induced emf is the changing flux linking the magnetic field and the conductor.
2. The magnitude of the emf is determined by the rate at which the flux linkage changes.
3. By increasing the rate of motion of the conductor. By increasing the density of the magnetic flux.
4. An induced emf is the energy per unit charge induced in a conductor on open circuit. An induced current is the rate of flow of charge past a point in a closed circuit across which an induced voltage is impressed.
5. The source of energy is the work done in moving the conductor in the magnetic field.
6. The direction of an induced current opposes, by its magnetic property, the change that induces it.
7. The negative sign indicates that the induced voltage opposes the change that induces it.

GROUP B

8. The induced current is in a clockwise direction around the loop (looking down on the loop).
9. The force acting on the loop would tend to push it downward and away from the approaching N pole of the magnet.
10. **(a)** magnetic flux, **(b)** flux density, **(c)** time rate of change of flux linkage, **(d)** potential difference.

11. The current induced in a conducting circuit provides a force in opposition to the motion producing it and work must be done against this force storing energy in the system.

12. If B is expressed in webers/square meter, l in meters, and v in meters/second:

$$Blv = \frac{wb}{m^2} \times m \times \frac{m}{s} = \frac{wb}{s} = \frac{n\ m/a}{s} = \frac{j}{c} = v$$

13. If Q is expressed in coulombs, v in meters/second, and B in wb/m^2:

$$QvB = c \times \frac{m}{s} \times \frac{wb}{m^2} = as \times \frac{m}{s} \times \frac{n\ m}{a\ m^2} = n$$

14. The magnetic field produced by the induced current in the solenoid must oppose the entry of the magnet. The end of the solenoid entered by the S pole of the magnet must be an S pole; the flux lines inside the solenoid must enter this end.

15. (a) Induced current is from a to b. (b) As S pole is produced at the end of the coil near the permanent magnet. (c) The direction of the induced current must be such that it opposes the motion of the magnet according to Lenz's law. Thus, an S pole must be produced at the end of the coil near the magnet to attract the retreating N pole of the magnet.

Problems • Chapter 20/p. 499
GROUP A

1. $E = -N\dfrac{\Delta\Phi}{\Delta t} = -325 \times \dfrac{1.15 \times 10^{-5}\ \text{weber}}{1.00 \times 10^{-3}\ s} = -3.74\ v$

2. $E = -N\dfrac{\Delta\Phi}{\Delta t}$ $\qquad$ and $\qquad$ $N = \dfrac{E}{\Delta\Phi/\Delta t}$

$N = \dfrac{0.25\ v}{5.0 \times 10^{-3}\ \text{weber/s}} = \bar{5}0\ \text{turns}$

3. $E = Blv = 0.025\ \text{weber/m}^2 \times 0.10\ m \times 0.75\ \text{m/s} = 0.0019\ v$, or 1.9 mv

4. $E = -N\dfrac{\Delta\Phi}{\Delta t}$ $\qquad$ and $\qquad$ $\Delta\Phi = BA$

$E = -N\dfrac{BA}{\Delta t}$

$E = -75 \times \dfrac{1.5\ \text{wb/m}^2 \times 4.0\ \text{cm}^2 \times m^2/10^4\ \text{cm}^2}{2.5 \times 10^{-2}\ s}$

$E = \dfrac{-4.5 \times 10^{-2}\ wb}{2.5 \times 10^{-2}\ s} = -1.8\ v$

5. $E = Blv$

$E = 4.5 \times 10^{-1}\dfrac{n}{a\ m} \times 15\ cm \times \bar{3}0\dfrac{cm}{s} \times \dfrac{m^2}{10^4\ \text{cm}^2}$

$E = 2.0 \times 10^{-2}\dfrac{n\ m}{a\ s} = 2.0 \times 10^{-2}\dfrac{j}{c} = 2.0 \times 10^{-2}\ v$

Questions • Chapter 20/p. 514
GROUP A

1. Armature, field magnet, and slip rings and brushes.

2. Extend the thumb, forefinger, and middle finger of the left hand at right angles to each other. Let the forefinger point in the direction of the flux and the thumb in the direction

the conductor is moving; then the middle finger points in the direction of the induced current.

3. A direct current is one in which electrons flow in only one direction through a conducting loop. An alternating current is one in which electrons flow alternately in opposite directions through a conducting loop.

4. The armature must rotate at a constant speed in a magnetic field of uniform flux density.

5. θ represents the displacement angle, the angle between the plane of the armature loop and the perpendicular to the flux of the magnetic field of a generator.

6. **(a)** A magneto is a small generator employing a permanent magnet to supply its field. **(b)** Used in the ignition systems of small gasoline engines.

7. An exciter is an auxiliary d-c generator used to supply direct current to the field magnet of an a-c generator.

8. **(a)** The frequency of an alternating current is the number of cycles the sine wave makes per second. **(b)** A two-pole generator has an output frequency the same as the armature rps.

9. By replacing the slip rings by a split-ring commutator.

10. The output of a d-c generator is pulsating, while that of a battery is a steady value for a given load.

11. **(a)** The field magnets are self-excited by connecting them in series, in parallel, or in series-parallel with the armature. **(b)** Student's diagram.

12. The armature resistance, the field windings, and the load.

13. α represents the angle between the plane of the armature loop and the magnetic flux of a motor.

14. Extend the thumb, forefinger, and middle finger of the right hand at right angles to each other. Let the forefinger point in the direction of the flux and the middle finger in the direction of the current; then the thumb points in the conductor's direction of motion.

15. **(a)** An emf that opposes the applied voltage. **(b)** It is induced by the generator effect of an operating motor.

16. The magnitude of armature current and the magnitude of the flux density.

17. Universal motor, induction motor, and synchronous motor.

GROUP B

18. The armature speed may be reduced to one-half or one-third the speed required for a 2-pole field magnet.

19. An increase in load means a reduction in the resistance of the external circuit and results in an increase in induced current. Since this larger current is in the field windings, the flux density is increased and the emf is increased.

20. An increase in load causes an increase in the current in the armature and a larger potential drop across the armature resistance. This results in a smaller potential difference across the field, and thus a smaller field current. The flux density decreases and the emf is decreased.

21. A current in a magnetic field produces a repulsion force according to Lenz's law that acts to expel the conductor from the field. The opposite conductors of a loop, having currents in opposite direction, have repulsion forces acting on them in opposite directions, producing a torque about the axis of the armature.

22. An operating motor has conducting loops moving in a magnetic field. Thus an emf is induced in the armature turns—the generator principle.

23. A load increase results in a reduction in speed of the armature, decreasing the back emf and increasing the armature current. The flux density of the series field increases, resulting in an increased torque.

24. The rotor must slip or lag behind the field if a torque is to be developed. This requires a rotating field. The field of a single-phase induction motor reverses periodically but does not rotate, and thus there can be no starting torque.

25. The synchronous motor, once started, runs in synchronism with the a-c generator supplying the current. Thus the clock accuracy is dependent on the accuracy with which the frequency of the generated current is controlled.

26. The silver loop requires the larger torque to turn it because the larger number of free electrons in the silver wire results in a larger induced current and, therefore, a larger opposition force to the motion producing it.

Problems • Chapter 20/pp. 514–515
GROUP A

1. $V = E - I_a R_a = 28\ \text{v} - (16\ \text{a} \times 0.25\ \Omega) = 28\ \text{v} - 4.0\ \text{v} = 24\ \text{v}$

2. $\text{Eff.} = \dfrac{\text{power Output}}{\text{power Input}} \times 100\% = \dfrac{15\bar{0}\ \text{a} \times 22\bar{0}\ \text{v}}{(15\bar{0}\ \text{a} \times 22\bar{0}\ \text{v}) + 32\bar{0}0\ \text{w}} \times 100\%$

$\text{Eff.} = \dfrac{33,\bar{0}00\ \text{w}}{36,200\ \text{w}} \times 100\% = 91.2\%$

3. **(a)** Student's diagram
   (b) $I_f = V/R_f = 24\bar{0}\ \text{v}/75.0\ \Omega = 3.20\ \text{a}$

   $I_a = I_L + I_f = \dfrac{1.80 \times 10^3\ \text{w}}{24\bar{0}\ \text{v}} + 3.20\ \text{a} = 75.0\ \text{a} + 3.20\ \text{a} = 78.2\ \text{a}$

   $E = V + I_a r_a = 24\bar{0}\ \text{v} + (78.2\ \text{a} \times 0.15\ \Omega) = 24\bar{0}\ \text{v} + 12\ \text{v} = 252\ \text{v}$
   (c) $P_T = EI_a = 252\ \text{v} \times 78.2\ \text{a} = 19,700\ \text{w}$, or 19.7 kw

4. **(a)** $T = F\ \omega \cos \alpha \qquad \alpha = 0° \qquad \text{and} \qquad \cos \alpha = 1$
   $T = 3.5\ \text{n} \times 0.15\ \text{m} \times 1 = 0.52\ \text{m n}$
   (b) $T = F\ \omega \cos \alpha \qquad \alpha = 30° \qquad \text{and} \qquad \cos \alpha = 0.866$
   $T = 3.5\ \text{n} \times 0.15\ \text{m} \times 0.866 = 0.46\ \text{m n}$

5. $T = F\ \omega \cos \alpha \qquad$ When torque is maximum, $\cos \alpha = 1 \qquad$ When torque is $\tfrac{1}{2}$ maximum,
   $\cos \alpha = 0.5\ \alpha = 60°,\ 120°,\ 24\bar{0}°,\ 300°$

6. $E = V - I_a r_a \qquad r_a = \dfrac{V - E}{I_a} = \dfrac{117\ \text{v} - 112\ \text{v}}{10\ \text{a}} = 0.5\ \Omega$

Questions • Chapter 20/p. 522
GROUP A

1. **(a)** Mutual inductance of two circuits is the ratio of the induced emf in one circuit to the rate of change of current in the other. **(b)** Self-inductance is the ratio of induced emf across a coil to the rate of change of current in the coil.

2. Copper losses and eddy-current losses.

3. The inductance of the coil is increased.

4. A closed core provides a continuous path for the magnetic flux of the primary to ensure maximum flux linkage with the secondary turns.

5. The secondary current is one-third the primary current.

GROUP B

6. Inductance opposes any change in circuit current. An increase is opposed and energy is stored in the magnetic field. A decrease in current is opposed also and energy is removed from the field to sustain the current.

7. Energy is transferred from the primary to the secondary circuit of a transformer only when the current in the primary varies in magnitude.

8. **(a)** Primary current is larger than secondary current. **(b)** Primary power and secondary power are equal. **(c)** Number of turns in the secondary is larger than in the primary.

9. A current is induced in the secondary only while the magnetic field of the primary is expanding or collapsing.

10. A voltage step-up is accompanied by a current step-down. Assuming no losses, the current induced in the secondary is one-tenth of that in the primary and no power gain can occur.

Problems • Chapter 20/p. 523
GROUP B

1. $E_S = -M \dfrac{\Delta I_P}{\Delta i} = 1.06 \text{ h} \times \dfrac{9.50 \text{ a}}{0.0336 \text{ s}} = 30\overline{0} \text{ v}$

2. $M = \dfrac{-E_S}{\Delta I_P/\Delta t} = \dfrac{-270 \text{ v}}{12 \text{ a}/0.048 \text{ s}} = 1.1 \text{ h}$

3. $\dfrac{V_S}{V_P} = \dfrac{N_S}{N_P}$ and $N_S = \dfrac{V_S N_P}{V_P}$

$N_S = \dfrac{2400 \text{ v} \times 75 \text{ turns}}{120 \text{ v}} = 1500 \text{ turns}$

4. (a) $L = \dfrac{e}{\Delta i/\Delta t} = \dfrac{16.5 \text{ v}}{7.5 \text{ a/s}} = 2.2 \text{ h}$ **(b)** $M = \dfrac{-e_S}{\Delta i_p/\Delta t} = \dfrac{5\overline{0} \text{ v}}{7.5 \text{ a/s}} \ 6.7 \text{ h}$

5. (a) $E = IR + L\dfrac{\Delta I}{\Delta t}$ Initially, $i = 0$ and $iR = 0$

$e = L\dfrac{\Delta i}{\Delta t}$ and $\dfrac{\Delta i}{\Delta t} = \dfrac{e}{L} = \dfrac{110 \text{ v}}{0.42 \text{ h}} = 260 \text{ a/s}$

(b) As $\dfrac{\Delta i}{\Delta t}$ approaches zero, $I = \dfrac{E}{R}$ (approximately)

$i = 0.85 \ I = 0.85 \dfrac{E}{R}$ $E = iR + L\dfrac{\Delta i}{\Delta t}$

$\dfrac{\Delta i}{\Delta t} = \dfrac{E - iR}{L} = \dfrac{E - \left(0.85\dfrac{E}{R}\right)R}{L} = \dfrac{110 \text{ v} - 94 \text{ v}}{0.42 \text{ h}} = 39 \text{ a/s}$

6. (a) $V_S = V_P N_S/N_P = 12\overline{0} \text{ v}/5 = 24.0 \text{ v}$
(b) $I_S = V_S/R_S = 24.0 \text{ v}/15.0 \ \Omega = 1.60 \text{ a}$
(c) $P_S = I_S^2 R_S = (1.60 \text{ a})^2 \times 15.0 \ \Omega = 38.4 \text{ w}$
(d) $I_p = P_p/V_p$ and $P_p = P_S$
$I_p = 38.4 \text{ w}/12\overline{0} \text{ v} = 0.320 \text{ a}$

7. $I_p = P_p/V_p$ and $P_p = P_S/\text{Eff.}$

$$I_P = \frac{38.4 \text{ w}}{0.925 \times 120 \text{ v}} = 0.346 \text{ a}$$

8. (a) $N_S = \dfrac{N_p V_S}{V_P} = \dfrac{400 \times 3000 \text{ v}}{120 \text{ v}} = 10000$

(b) $\text{Eff.} = \dfrac{P_S}{P_P} \times 100\% = \dfrac{V_S I_S}{V_P I_P} \times 100\% = \dfrac{3000 \text{ v} \times 0.0600 \text{ a}}{120 \text{ v} \times 1.85 \text{ a}} \times 100\% = 0.810,\ \text{or}\ 81.0\%$

9. (a) $V_R = IR = 0.0200 \text{ a} \times 270\ \Omega = 5.40 \text{ v}$

(b) $V_L = E - V_R = 12.0 \text{ v} - 5.40 \text{ v} = 6.6 \text{ v}$

(c) $\dfrac{\Delta I}{\Delta t} = \dfrac{V_L}{L} = \dfrac{6.6 \text{ v}}{2.50 \text{ h}} = 2.6 \text{ a/s}$

ALTERNATING-CURRENT CIRCUITS 21

Questions • Chapter 21/pp. 542–543

GROUP A

1. The current and voltage waveforms reach their positive and negative maxima and minima at the same instants of time.
2. A resistance load.
3. Instantaneous power is the product of the instantaneous values of current and voltage in an a-c circuit.
4. Upon its heating effect in a circuit.
5. The effective value of an alternating current is the number of amperes that, in a given resistance, product heat at the same average rate as that number of amperes of steady direct current.
6. (a) The current lags the applied voltage by 90°. (b) The current leads the induced voltage by 90°. (c) The voltages are 180° out of phase.
7. (a) The cosine of the phase angle. (b) The ratio of V_R to V. (c) pf is the cosine of the phase angle and indicates the relation between actual power and apparent power.
8. No power is consumed in a reactance.
9. (a) Impedance is the joint opposition of reactance and resistance to the current in an a-c circuit. (b) A diagram in which the magnitudes of resistance and reactance are plotted at right angles and are added vectorially to give impedance with a positive-direction angle. (c) A diagram in which the magnitudes of resistance and reactance are plotted at right angles and are added vectorially to give impedance with a negative direction angle.
10. In a d-c circuit, the capacitor, once charged to the applied voltage, acts as an open switch. In an a-c circuit, the capacitor reverses its charge as the current alternates and produces a lagging voltage.
11. (a) pf = 0 (b) pf = 1 (c) pf = 0
12. (a) $E = IZ$ (b) $V = IZ$

GROUP B

13. Since e and i are in phase, their product is positive when both are positive and when both are negative.

14. Power is dissipated in a resistance as heat.
$P = VI$
and $V = IR$
substituting, $P = I \times R \times I = I^2 R$
For a given resistance, $P \propto I^2$

15. **(a)** Student's graph similar to Figure 21-4. **(b)** $I = 0.707 I_{max}$, $I = 0.707 \times 5.7$ a $= 4.0$ a

16. The current in a pure inductance load lags the applied voltage by 90°. When the current is rising from zero to a maximum in either direction, energy is being taken from the source and stored in the expanding magnetic field. When the current is decaying from either maximum to zero, energy is being returned from the field to the source. The net energy transfer is zero.

17. **(a)** Student's graph of X_L vs f. **(b)** The curve is linear rising from zero at the origin in a positive sense. **(c)** Zero frequency is equivalent to a d-c value and, since reactance is an opposition to a-c currents, X_L must be zero for all values of L.

18. Zero frequency is equivalent to a d-c value, and any capacitor acts as an open switch in a d-c circuit.

19. **(a)** Whether X_L or X_C is larger. **(b)** If $X_L = X_C$, the load would be resistive.

20. As the current changes from zero to either maximum, energy in the form of capacitor charge is returned to the source.

Problems • Chapter 21/pp. 543–544
GROUP A

1. $I_{max} = 1.414 I = 1.414 \times 5.5$ a $= 7.8$ a

2. $E = 0.707 E_{max} = 0.707 \times 450$ v $= 318$ v

3. $E_{max} = 1.414 E = 1.414 \times 122.0$ v $= 172.5$ v
$e = E_{max} \sin \theta = 172.5$ v $\times \sin 50° = 172.5$ v $\times 0.7660 = 132.1$ v

4. **(a)** $P = I^2 R \qquad I = \sqrt{P/R} = \sqrt{250 \text{ w}/25.0 \ \Omega} = \sqrt{10.0 \text{ a}} = 3.16$ a
   (b) $V = IR = 3.16$ a $\times 25.0 \ \Omega = 79.0$ v

5. **(a)** $X_L = 2\pi f L = 2\pi \times 25.0/\text{s} \times 2.20$ h $= 345 \ \Omega$
   (b) Student's impedance diagram $\qquad Z = 405 \ \Omega \ \underline{|58°}$

6. **(a)** $X_C = \dfrac{1}{2\pi f C} = \dfrac{1}{2\pi \times 60.0/\text{s} \times 2.00 \times 10^{-6} \text{ f}} = 1320 \ \Omega$
   (b) $V = I X_C = 0.167$ a $\times 1320 \ \Omega = 220$ v

7. $X_C = V/I = 120$ v$/0.120$ a $= 1.00 \times 10^3 \ \Omega = \dfrac{1}{2\pi f C}$

$f = \dfrac{1}{2\pi C X_C} = \dfrac{1}{2\pi \times 2.65 \times 10^{-6} \text{ f} \times 1.00 \times 10^3 \ \Omega} = 60.2/\text{s, or } 60.2 \text{ hz}$

8. $V_R = IR = 20$ a $\times 4.0 \ \Omega = 80$ v
   $V_L = \sqrt{V^2 - V_R^2} = \sqrt{(110 \text{ v})^2 - (80 \text{ v})^2} = 75$ v
   $X_L = V_L/I = 75$ v$/20$ a $= 3.8 \ \Omega = 2\pi f L$

$L = \dfrac{X_L}{2\pi f} = \dfrac{3.8 \ \Omega}{2\pi \times 60/\text{s}} = 0.010$ h, or 10 mh

(Alternate solution)

$Z = \dfrac{V}{I} = \dfrac{110 \text{ v}}{20 \text{ a}} = 5.5 \ \Omega \qquad X_L = \sqrt{Z^2 - R^2} = \sqrt{(5.5 \ \Omega)^2 - (4.0 \ \Omega)^2}$

$X_L = 3.8 \ \Omega = 2\pi fL$

$L = \dfrac{X_L}{2\pi f} = \dfrac{3.8 \ \Omega}{2\pi \times 6\bar{0}/s} = 0.010 \ \text{h, or } 1\bar{0} \ \text{mh}$

GROUP B

9. **(a)** Student's diagram.

(b) $X_C = \dfrac{1}{2\pi fC} = \dfrac{1}{2\pi \times 8.0 \times 10^3/s \times 0.50 \times 10^{-6} \ \text{f}} = 4\bar{0} \ \Omega$

(c) Student's diagram.

$Z = \sqrt{R^2 + X_C{}^2} \ \underline{|\text{arctan} - X_C/R}$

$\quad = \sqrt{(3\bar{0} \ \Omega)^2 + (4\bar{0} \ \Omega)^2} \ \underline{|\text{arctan} - 4\bar{0} \ \Omega/3\bar{0} \ \Omega}$

$Z = \sqrt{2500 \ \Omega^2} \ \underline{|\text{arctan} - 1.3} = 5\bar{0} \ \Omega \ \underline{|-53^\circ}$

10. **(a)** $V = IZ = 0.050 \ \text{a} \times 5\bar{0} \ \Omega = 2.5 \ \text{v}$

(b) $\phi = -53^\circ$, V lags I by 53°

(c) $V_C = IX_C = 0.050 \ \text{a} \times 4\bar{0} \ \Omega = 2.0 \ \text{v}$

(d) $V_R = IR = 0.050 \ \text{a} \times 3\bar{0} \ \Omega = 1.5 \ \text{v}$

11. **(a)** $\text{pf} = \dfrac{\text{actual power}}{\text{apparent power}} = \dfrac{P}{V \times I} = \dfrac{40\bar{0} \ \text{w}}{117 \ \text{v} \times 4.75 \ \text{a}} = 0.719$

(b) $\phi = \arccos 0.719 = 44.0^\circ$

(c) $P = I^2 R$

$R = P/I^2 = 40\bar{0} \ \text{w}/(4.75 \ \text{a})^2$

$R = 17.7 \ \Omega$

(d) $X_L = 17.7 \ \Omega \times \tan 44.0^\circ = 17.7 \ \Omega \times 0.966 = 17.1 \ \Omega$

(e) $V_R = IR = 4.75 \ \text{a} \times 17.7 \ \Omega = 84.1 \ \text{v}$

(f) $V_L = IX_L = 4.75 \ \text{a} \times 17.1 \ \Omega = 81.2 \ \text{v}$

(g) Student's diagram $V = 117 \ \text{v} \ \underline{|44^\circ}$

12. **(a)** $X_L = 2\pi fL = 2\pi \times 1.0 \times 10^3/s \times 0.019 \ \text{h} = 120 \ \Omega$

$Z = \sqrt{R^2 + X_L{}^2} \ \underline{|\text{arctan} \ X_L/R}$

$Z = \sqrt{(9\bar{0} \ \Omega)^2 + (120 \ \Omega)^2} \ \underline{|\text{arctan} \ 120 \ \Omega/9\bar{0} \ \Omega} = 150 \ \Omega \ \underline{|53^\circ}$

(b) $I = V/Z = 6.0 \ \text{v}/150 \ \Omega = 0.040 \ \text{a}$

(c) $P = VI \cos 53^\circ = 6.0 \ \text{v} \times 0.040 \ \text{a} \times 0.602 = 0.14 \ \text{w}$

13. **(a)** $X_C = \dfrac{1}{2\pi fC} = \dfrac{1}{2\pi \times 6\bar{0}/s \times 5\bar{0} \times 10^{-6} \ \text{f}} = 53 \ \Omega$

$Z = \sqrt{R^2 + X_C{}^2} \ \underline{|\text{arctan} - X_C/R}$

$Z = \sqrt{(6\bar{0} \ \Omega)^2 + (53 \ \Omega)^2} \ \underline{|\text{arctan} - 53 \ \Omega/6\bar{0} \ \Omega} = 8\bar{0} \ \Omega \ \underline{|-41^\circ}$

$I = V/Z = 120 \ \text{v}/8\bar{0} \ \Omega = 1.5 \ \text{a}$

(b) $P = VI \cos (-41^\circ) = 120 \ \text{v} \times 1.5 \ \text{a} \times 0.75 = 140 \ \text{w}$

(c) $\text{pf} = \cos \phi = \cos (-41^\circ) = 0.75$

(d) $V_R = IR = 1.5 \ \text{a} \times 6\bar{0} \ \Omega = 9\bar{0} \ \text{v}$

(e) $V_C = IX_C = 1.5 \ \text{a} \times 53 \ \Omega = 8\bar{0} \ \text{v}$

(f) Student's diagram $\quad V = 120 \ \text{v} \ \underline{|-41^\circ}$

14. (a) $X_L = 2\pi f L = 2\pi \times 1.0 \times 10^3/s \times 4.8 \times 10^{-3}\, h = 30\, \Omega$

$$X_C = \frac{1}{2\pi f C} = \frac{1}{2\pi \times 1.0 \times 10^3/s \times 8.0 \times 10^{-6}\, f} = 20\, \Omega$$

$Z = \sqrt{R^2 + X^2} = \sqrt{R^2 + (X_L - X_C)^2} = \sqrt{(10\, \Omega)^2 + (30\, \Omega - 20\, \Omega)^2}$
$Z = 14\, \Omega$

(b) $\phi = 45°$ (by inspection)

(c) $I = V/Z = 6.0\, v/14\, \Omega = 0.43\, a$

(d) Student's diagram.

$V_L = IX_L = 0.43\, a \times 30\, \Omega = 13\, v$
$V_R = IR = 0.43\, a \times 10\, \Omega = 4.3\, v$
$V_C = IX_C = 0.43\, a \times 20\, \Omega = 8.6\, v$

15. (a) $V_X = V_L - V_C = 90\, v - 120\, v = -30\, v$

$V = \sqrt{V_R^2 + V_X^2} = \sqrt{(16\, v)^2 + (-30\, v)^2} = 34\, v$

(b) Student's diagram.

(c) $pf = \cos \phi = V_R/V = 16\, v/34\, v = 0.47$

Questions • Chapter 21/pp. 549–550
GROUP A

1. Students graph. **(a)** Each curve is linear. **(b)** The largest value of inductance yields the curve with the greatest slope. **(c)** If frequency equals zero, the value of inductive react-ance must equal zero for any value of inductance.
2. Series resonance is a condition in which the impedance of an L-R-C series circuit is equal to the resistance and the voltage across the circuit is in phase with the circuit cur-rent.
3. The voltage across the circuit lags the current since at frequencies below f_r, X_C is larger than X_L.
4. The voltage across the circuit leads the current since at frequencies above f_r, X_L is larger than X_C.
5. The quality factor, or Q, of the circuit determines the sharpness of the resonant peak, and Q is primarily dependent on the ratio of X_L to R.
6. A resonant circuit with a high Q is capable of discriminating between frequencies rela-tively close together; it is very selective.
7. **(a)** Decreased. **(b)** Increased. **(c)** Increased.

GROUP B

8. The resonant frequency can be varied by varying the inductance and having a fixed ca-pacitance or by varying the capacitance and having a fixed inductance. Inductance is usually varied by changing the position of a powdered-iron core in the coil. Capacitance is varied by changing the plate area or the spacing between the plates.
9. The lamp will dim. The current in the circuit will be reduced due to the inductive react-ance of the coil.
10. When the capacitor is adjusted, X_L and X_C are equal and cancel each other forming a series-resonant circuit. The current is then determined by the resistance of the lamp.
11. Student's graph showing circuit as a function of frequency (similar to Fig. 21-24).
12. **(a)** The circuit current lags behind the voltage and the inductive load draws a larger cur-rent than a load of the same power rating but with a power factor nearer unity. The large current results in excessive I^2R heat losses in the circuit. **(b)** Connect capacitance into the circuit to reduce the inductive effect of the load.

Problems • Chapter 21/p. 550
GROUP B

1. $f_r = \dfrac{1}{2\pi\sqrt{LC}} = \dfrac{1}{2\pi\sqrt{320 \times 10^{-6}\,h \times 8\bar{0} \times 10^{-12}\,f}}$

$f_r = \left(\dfrac{1}{2\pi \times 160 \times 10^{-9}}\right)\,hz = 1.0 \times 10^6\,cps$, or $1\bar{0}00\,khz$

2. Student's graphs similar to Fig. 21-23 showing f_r at 1000 khz.

3. $f_r = \dfrac{1}{2\pi\sqrt{LC}}$ $\qquad L = \dfrac{1}{4\pi^2 f_r^2 C}$

$L = \dfrac{1}{4\pi^2 \times (5.5 \times 10^5/s)^2 \times 350 \times 10^{-12}\,f} = 0.24\,mh$, or $240\,\mu h$

4. $f_r = \dfrac{1}{2\pi\sqrt{LC}} = \dfrac{1}{2\pi\sqrt{240 \times 10^{-6}\,h \times 15 \times 10^{-12}\,f}}$

$f_r = \left(\dfrac{1}{2\pi \times 6\bar{0} \times 10^{-9}}\right)\,hz = 2700\,khz$

ELECTRONIC DEVICES   **22**

Questions • Chapter 22/pp. 564–565
GROUP A

1. The vacuum tube.
2. The discovery of electrons.
3. The De Forest triode was able to amplify signals, an action not possible with diodes.
4. **(a)** Thermionic emission is the process whereby electrons are emitted from the surface of a hot body. **(b)** Oxide-coated emitters are used in receiving tubes.
5. The flow of electrons from cathode to plate may be controlled by a small potential difference between the grid and cathode.
6. The negative voltage on the control grid is high enough to prevent the flow of electrons from cathode to plate, thus reducing plate current in the circuit to zero.
7. If the grid becomes positive, there is some electron flow in the grid circuit. In the usual applications of vacuum tubes, this wastes power and is avoided by keeping the grid negative.

GROUP B

8. (1) He was preoccupied with the problem associated with the development of his electric lamp. (2) Electrons had not yet been discovered.
9. As the potential on the control grid becomes less negative, the plate current increases, causing a larger potential difference across the plate-load resistor and causing the plate potential to become less positive.
10. The electron gun consists of the cathode, control grid, focusing anode, and accelerating anode. Its function is to provide electrons and form them into a focused beam of the proper intensity.
11. In the electrostatic system the electron beam is passed through an electric field existing between two sets of deflecting plates set at right angles to each other inside the tube

envelope. In the electromagnetic system the electron beam is passed through a magnetic field existing between the coils of a deflection yoke located on the neck of the envelope.

12. (a) Photoelectrons emitted by the photoemission cathode must be accelerated toward the first secondary-emission electrode sufficiently to cause secondary emission on impact.
(b) Electrons from the preceding electrode must accelerate through a sufficient difference in potential to cause secondary emission on impact at each successive stage along the way to the collector plate.

Questions • Chapter 22/pp. 578–579
GROUP A

1. The crystal permits a large flow of electrons in one direction with a small applied voltage, and only a very small flow of electrons in the opposite direction even with a large applied voltage.
2. Minute traces of impurities having five valence electrons per atom are introduced into the germanium crystal structure. Each atom, on joining the structure, donates one electron to the crystal as a free electron. The impurity is called a donor and the crystal is called electron-rich germanium or N-type germanium.
3. Traces of impurities having three valence electrons per atom are introduced into the germanium crystal structure. Each of these atoms accepts an electron from a germanium atom leaving a hole in the crystal bond from which the electron is acquired. The impurity is called an acceptor and the crystal is called hole-rich germanium.
4. The P-N junction is an electric barrier formed when P- and N-type semiconductor crystals are joined. The plane at which the two structures meet is called the P-N junction.
5. As N-P-N and P-N-P transistors.
6. Student's diagram similar to that of Figure 22-23(A).
7. Student's diagram similar to that of Figure 22-23(B).
8. The arrowhead indicates the direction of the hole current, the electron current being in the opposite direction.
9. The emitter is equivalent to the cathode, the base to the control grid, and the collector to the plate.
10. Common-base circuit, common-emitter circuit, and common-collector circuit.

GROUP B

11. A germanium diode possesses the rectifying property of a diode because of the presence of a P-N junction across which electrons can easily pass from the N-type to the P-type crystal. A germanium transistor possesses the amplifying property of a triode because of the presence of two P-N junctions back to back.
12. Both are junction transistors, the P-N-P transistor consisting of a thin wafer of N-type germanium sandwiched between two sections of P-type germanium and the N-P-N transistor consisting of a thin wafer of P-type germanium sandwiched between two sections of N-type germanium.
13. The barrier potential at an N-P junction is opposite in polarity to that at a P-N junction. In both instances, the emitter bias opposes the barrier potential in order to move charges across the junction.
14. The alpha characteristic is the expression of the forward current gain of a transistor in the common-base circuit. It is the ratio of collector current to emitter current at a constant collector-to-base voltage.
15. Collector current is equal to the emitter current minus the small base current. See Figure 22-25.

16. The common-emitter circuit configuration. The emitter corresponds to the cathode as a source of charge carriers. The collector and the plate collect the charge for the output circuit. The base and the control grid control the output current. The input signal is applied to the base with respect to the emitter, to control the collector current. In the vacuum triode circuit, the input signal is applied to the grid with respect to the cathode to control the plate current.

17. The beta characteristic expresses the current gain across a transistor in the common-emitter circuit configuration. It is the ratio of collector current to base current at a constant collector-to-emitter voltage.

18. The collector current is slightly smaller than the emitter current; however, this current is in a high-impedance circuit and a signal gain is realized across the output impedance.

19. Photoelectric emission requires a vacuum environment and an external source of emf in the electron-emission circuit. Photovoltaic action does not require a vacuum environment and it generates its own emf.

20. Electron-hole pairs produced by photons in the junction region are separated by the electric field of this region before they have the opportunity to recombine. They are swept out of the region as additional charge carriers. Electron-hole pairs produced at a distance from the junction layer are likely to recombine before coming under the influence of the electric field.

ATOMIC STRUCTURE   23

Questions • Chapter 23/p. 590
GROUP A
1. **(a)** Cathode rays are so called because they emanate from the negative electrode, or cathode, of evacuated tubes. **(b)** Thomson found that cathode rays are streams of electrons.
2. **(a)** J. J. Thomson. **(b)** He showed that electrons have subatomic masses and that they are present in various kinds of matter.
3. **(a)** By the production of an electron and a positron from X rays and by using the electronic charge to find the charge-to-mass ratio of the electron. **(b)** The latter.
4. By friction in the inlet tube and by passing the oil drop through air ionized by X rays.
5. The electron has wave properties. The apparent size of the electron varies with its momentum. (See Section 23.5, pp. 588–589 of text.)
6. The electron has both wave and particle characteristics.
7. An orbital is a probable pattern of movement for an electron. Groups of orbitals are called shells.
8. **(a)** The rest mass of an electron is 1/1837th of the mass of a hydrogen atom.
   (b) Electrons do not account for a very large share of the mass of substances.

Questions • Chapter 23/p. 598
GROUP A

1.	*mass*	*charge*	*size*
molecule	variable	neutral	3 Å to 200 Å
atom	1.67343×10^{-27} kg to 3.95628×10^{-25} kg	neutral	1.0 Å to 5 Å
electron	9.109543×10^{-31} kg	negative	varies with momentum

1.

	mass	*charge*	*size*
proton	$1.6726485 \times 10^{-27}$ kg	positive	1.3×10^{-5} Å
neutron	1.674943×10^{-27} kg	neutral	1.3×10^{-5} Å
nucleus	variable	positive	1.3 to 8.0×10^{-5} Å
	(from 1 to over		
	238 nucleons)		

2. Atomic mass is the mass of an atom in kilograms or in atomic mass units (which is 1/12 the mass of a carbon-12 atom). Gram-atomic weight is the mass in grams of one mole of naturally occurring atoms of an element.

3. (a) All isotopes of an element have the same chemical properties. **(b)** Isotopes of an element have different numbers of neutrons in the nucleus, and hence isotopes have different atomic masses.

4. (a) Mass number; **(b)** angstrom unit; **(c)** atomic number; **(d)** number of neutrons; **(e)** atomic mass units; **(f)** mega electron volts.

5.

Isotope	Z	A	N	Binding energy/ nucleon (Mev/A)
deuterium	1	2	1	1.1
carbon-14	6	14	8	6.9
oxygen-18	8	18	10	8.1
sodium-23	11	23	12	8.2
sulfur-32	16	32	16	8.4
argon-40	18	40	22	8.6
uranium-235	92	235	143	7.5
uranium-238	92	238	146	7.4

GROUP B

6. (a) Rutherford used alpha particles emitted by radioactive polonium. He separated the alpha particles from the other emissions of polonium by using their ability to produce scintillations. **(b)** He detected the alpha particles by the scintillations they produced upon striking a fluorescent screen.

7. (a) Both experiments employed fluorescent screens. Both observed deflected particles. **(b)** Both bombarded elements with alpha particles from polonium.

8. Electrons do not have enough mass to deflect alpha particles.

9. In general, elements with odd atomic numbers exist naturally in only one or two isotopic forms. Those with even numbers exist in several forms.

10. (a) The lower mass of a nucleus as compared with the sum of the masses of its constituent particles. **(b)** The energy released in the formation of a nucleus. **(c)** Nuclear binding energy and nuclear mass defect are equivalent, in terms of the equation $E = mc^2$.

11. The element with a mass number of about 64.

Problems • Chapter 23/pp. 598–599
GROUP A
1. (a) $55.9349 \ u \times 1.6605655 \times 10^{-27}$ kg/u $= 9.28836 \times 10^{-26}$ kg $= 9.28836 \times 10^{-23}$ g
(b) 56

2. **(a)** 3.818×10^{-23} g $= 3.818 \times 10^{-26}$ kg $\times$ (1 u/1.6605655 $\times 10^{-27}$ kg) $= 23.00$ u
   (b) 23
3. (20 unit + charges) + (18 unit − charges) = (2 unit + charges)
   (2 unit + charges) $\times$ 1.6021892 $\times 10^{-19}$ c $= 3.2043784 \times 10^{-19}$ c
4. **(a)** 30 **(b)** 64
5. 47 protons   47 electrons $109 - 47 = 62$, number of neutrons

GROUP B
6. 6 protons $\times$ 1.007276470 u/proton $=$ 6.043658820 u
   6 neutrons $\times$ 1.008665012 u/neutron $=$ 6.051990072 u
   6 electrons $\times$ 0.00054858026 u/electron $=$ $\underline{0.00329148156\ u}$
                                          $\overline{12.098940373\ u}$
   Atomic mass of carbon-12 $=$ 12.00000 u
   Mass defect $=$ $\overline{0.09894\ u}$
   Mev/nucleon $= 0.09894$ u $\times$ 931 Mev/u $\times \frac{1}{12}$ nucleons $= 7.68$ Mev/nucleon
7. 16 protons $\times$ 1.007276470 u/proton $=$ 16.11642352 u
   16 neutrons $\times$ 1.008665012 u/neutron $=$ 16.13864019 u
   16 electrons $\times$ 0.00054858026 u/electron $=$ $\underline{0.0087772841\ u}$
                                          $\overline{32.26384099\ u}$
   Atomic mass of sulfur-32 $=$ $\underline{31.97207\ u}$
   Mass defect $=$ $\overline{0.29177\ u}$
   Binding energy/nucleon $= 0.29177$ u $\times$ 931 Mev/u $\times \frac{1}{32}$ nucleon $= 8.49$ Mev/nucleon
8. 0.754×34.96885 u $= 26.4$ u 0.246×36.96590 u $= 9.10$ u
   26.4 u + 9.10 u = 35.5 u, gram-atomic weight
9. 0.786×23.98504 u $= 18.9$ u 0.101×24.98584 u $= 2.52$ u
   0.113×25.98259 u $= 2.94$ u
   18.9 u + 2.52 u + 2.94 u = 24.3 u, gram-atomic weight
10. Mass of electron $= 0.00054858026$ u
    Mass of positron $= \underline{0.00054858026\ u}$
                       $\overline{0.0010971605\ u}$ $\times$ 931 Mev/u $= 1.02$ Mev
    Extra energy = 1.5 Mev − 1.02 Mev = 0.5 Mev = 8×10^{-14} j
    Hence, for each particle, $E_k = \frac{1}{2}mv^2 = 4 \times 10^{-14}$ j

    Solving for the velocity, $v = \sqrt{\dfrac{2E_k}{m}}$

    $$= \sqrt{\frac{2 \times 4 \times 10^{-14}\ \text{j}}{9.1 \times 10^{-31}\ \text{kg}}} = 3 \times 10^8 \text{ m/s}$$

    Note: This solution does not take relativity into account.

NUCLEAR REACTIONS   **24**

Questions • Chapter 24/pp. 611–612
GROUP A
1. **(a)** Radioactivity was discovered when uranium ore was stored in the same drawer with photographic plates, causing them to fog. **(b)** Henri Becquerel.

2. **(a)** The spontaneous, uncontrollable decay of the nucleus of an atom with the emission of particles and gamma rays. **(b)** All naturally occurring elements with an atomic number greater than 83.

3. (1) It affects light-sensitive emulsions. (2) It produces an electric charge in the surrounding air. (3) It produces fluorescence with certain compounds. (4) It has special physiological effects.

4. **(a)** Helium nuclei. **(b)** Electrons. **(c)** Electromagnetic waves.

5. **(a)** The chemical element. **(b)** Nuclear charge. **(c)** Number of nucleons.

6. **(a)** A change in the nuclear charge of an atom. **(b)** Emission of an alpha particle (alpha decay) or emission of a beta particle (beta decay).

7. **(a)** The length of time during which half a given number of atoms of a particular radioactive nuclide will decay. **(b)** The decay constant (the ratio between the number of nuclei decaying per second and the total number of original nuclei) is found by dividing 0.693 by the half-life. **(c)** The curie.

8. The resulting products of a nuclear bombardment have lower total mass numbers than the mass number of the target material. The difference shows up as a release of binding energy.

9. They have no electric charge, and hence they are not easily deflected by electrons and protons.

10. **(a)** To slow down fast neutrons. **(b)** Deuterium oxide and graphite.

11. **(a)** 105 **(b)** 92

12. **(a)** $4\,^1_1\text{H} \rightarrow\,^4_2\text{He} + 2\,^0_{+1}\text{e} + 25.7$ Mev **(b)** The mass of the star decreases.

13. The number of nucleons and the arithmetic sum of the electric charges.

14. **(a)** Cosmic rays are high-speed particles that strike the atmosphere. **(b)** Their high energy suggests that they come from beyond the solar system.

GROUP B

15. Alpha particles are deflected only slightly by a magnet: beta particles are deflected much more, and in the opposite direction from that of the alpha particles; gamma rays are not deflected at all.

16. (1) They have no electric charge. (2) They are more penetrating. (3) They are electromagnetic waves rather than particles.

17. **(a)** The mass defect. **(b)** The examples in Section 23.12 may be cited, or any of the mass defect problems in Chapters 23 or 24.

18. **(a)** The stability of the nucleus first increases with increasing mass. After reaching a maximum with nuclei having mass numbers around 70, the stability of the nucleus decreases again. **(b)** The stability of the nucleus is greatest when the ratio of protons to neutrons is about 1 for the lighter elements and about $1\frac{1}{2}$ for the heavier ones.

19. **(a)** The average life is the average time before a radioactive nuclide disintegrates. The half-life is the time during which half of a given number of atoms of a radioactive nuclide disintegrate. **(b)** The average life is 1.44 times the half-life.

20. **(a)** Remove an alpha particle and then a beta particle.
    (b) $^{200}_{80}\text{Hg} \rightarrow\,^{196}_{78}\text{Pt} +\,^4_2\text{He};\,^{196}_{78}\text{Pt} \rightarrow\,^{196}_{79}\text{Au} +\,^0_{-1}\text{e}$

21. **(a)** $^{234}_{90}\text{Th} \rightarrow\,^{230}_{88}\text{Ra} +\,^4_2\text{He}$ **(b)** $^{234}_{92}\text{U} \rightarrow\,^{230}_{90}\text{Th} +\,^4_2\text{He}$ **(c)** $^{214}_{83}\text{Bi} \rightarrow\,^{210}_{81}\text{Tl} +\,^4_2\text{He}$
    (d) $^{210}_{84}\text{Po} \rightarrow\,^{206}_{82}\text{Pb} +\,^4_2\text{He}$

22. $^{238}_{92}\text{U} +\,^1_0\text{n} \rightarrow\,^{239}_{92}\text{U}$ $^{239}_{92}\text{U} \rightarrow\,^{239}_{93}\text{Np} +\,^0_{-1}\text{e}$ $^{239}_{93}\text{Np} \rightarrow\,^{239}_{94}\text{Pu} +\,^0_{-1}\text{e}$

23. **(a)** $^2_1$H and $^3_2$He.   **(b)** Nuclides with mass numbers around 70.   **(c)** The greater the binding energy per nucleon, the more stable the nucleus.

24. The absorption of neutrons by ^{235}U produces fission and additional neutrons. The absorption of neutrons by ^{238}U results in the capture of the neutrons and the formation of ^{239}U.

25. **(a)** No; it tells when a given proportion of nuclei will disintegrate. Answers may vary, but the example of radium in Section 24.5 will probably be given.

26. The release of binding energy.

Problems • Chapter 24/pp. 612–613
GROUP A
1. $^{222}_{86}$Rn → $^{218}_{84}$Po + $^4_2$He + energy [222.0175 u − (218.0089 u + 4.00260 u)] × 931 Mev/u = 5.6 Mev

2. $^{214}_{82}$Pb → $^{214}_{83}$Bi + $^0_{-1}$e + energy
{[213.9982 u − 82 (0.000548 u)] − [213.9972 u − 83 (0.000548 u)] − 0.000548 u} × 931 Mev/u = 0.8 Mev

3. $\lambda = 0.693/T_{1/2}$
λ = 0.693/3.82 da × 24 hr/da × 3600 s/hr = 2.10 × 10^{-6}/s

4. $\lambda = 0.693/T_{1/2}$
λ = 0.693/4.49 × 10^9 yr × 365 da/yr × 24 hr/da × 3600 s/hr = 4.89 × 10^{-18}/s

GROUP B
5. **(a)** $\lambda = 0.693/T_{1/2}$ λ = 0.693/1.4 × 10^{10} yr × 365 da/yr × 24 hr/da × 3600 s/hr = 1.6 × 10^{-18}/s
100 g × (6.02252 × 10^{23} atoms/232.0382 g) × 1.6 × 10^{-18}/s × 1 curie/(3.70 × 10^{10} atoms/s) × 10^3 millicurie/curie = 1.1 × 10^{-2} mcurie

   (b) $^{232}_{90}$Th → $^{228}_{88}$Ra + $^4_2$He
[232.0382 u − (228.0303 u + 4.00260 u)] × 931 Mev/u × 100 g × (6.02252 × 10^{23} atoms/232.0373 g) × 1.6 × 10^{-18}/s = 2.0 × 10^6 Mev/s

6. **(a)** $\lambda = 0.693/T_{1/2}$ λ = 0.693/3.82 da × 24 hr/da × 3600 s/hr = 2.10 × 10^{-6}/s
5.00 × 10^{-3} g × (6.02252 × 10^{23} atoms/222.0175 g) × 2.10 × 10^{-6}/s × 1 curie/3.70 × 10^{10} atoms/s = 7.70 × 10^2 curie

   (b) From Problem 1, the energy evolution is 5.6 Mev/nucleus. 5.6 Mev/nucleus × 5.00 × 10^{-3} g × (6.02252 × 10^{23} atoms/222.0175 g) × 2.10 × 10^{-6}/s = 1.7 × 10^{14} Mev/s

7. {[239.0522 u + 1.0087 u)] − [136.91 u + 99.9076 u + (3 × 1.0087 u)]} × 931 Mev/u = 202 Mev, energy of fission from 240 u of reactants
[(2 × 2.0140 u) − 4.00260 u] × 931 Mev/u = 23.8 Mev, energy of fusion from 4.03 u of reactants

$$\frac{23.8\ \text{Mev}/4.03\ u}{202\ \text{Mev}/240\ u} = 7.02,$$ the ratio of energy of fusion to energy of fission

for the same mass of reactants

8. $$\frac{4000\ \text{kg/s} \times 3600\ \text{s/hr} \times 24\ \text{hr/da} \times 365\ \text{da/yr} \times 1000\ \text{yr}}{2 \times 10^{30}\ \text{kg}} = 6 \times 10^{-17}$$

9. $E = mc^2$

$$m = \frac{E}{c^2} = \frac{3 \times 10^{22} \text{ j/da} \times 365 \text{ da/yr}}{(3 \times 10^8 \text{ m/s})^2} = 1. \times 10^8 \text{ kg/yr}$$

1×10^8 kg/yr $\div 6 \times 10^{24}$ kg $= 2 \times 10^{-17}$/yr, the proportional increase in mass

10. 6 protons $\times$ 1.007276470 u/proton   =   6.043658820 u

7 neutrons $\times$ 1.008665012 u/neutron   =   7.060655084 u

6 electrons $\times$ 0.00054858026 u/electron   = $\dfrac{0.00329148156\ u}{13.107605386\ u}$

Atomic mass of $^{13}_{6}$C   = 13.00335 u

Mass defect = 13.107605386 − 13.00335 =   0.10426 u

Binding energy/nucleon = 0.10426 u × 931 Mev/u ÷ 13 nucleons =
7.46 Mev/nucleon

11. (a) 7 protons $\times$ 1.007276470 u/proton   =   7.050935290 u

5 neutrons $\times$ 1.008665012 u/neutron   =   5.043325060 u

7 electrons $\times$ 0.00054858026 u/electron = $\dfrac{0.0038400618\ u}{12.0981004118\ u}$

Atomic mass of $^{12}_{5}$N   = 12.0188 u

Mass defect   =   0.0793 u

Binding energy/nucleon = 0.0793 u × 931 Mev/u ÷ 12 nucleons =
6.15 Mev/nucleon

(b) $^{13}_{6}$C is more stable, since it has more binding energy per nucleon.

Questions • Chapter 24/p. 621
GROUP A

1. (a) The material surrounding the reaction must be pure enough as well as massive enough to make the reaction self-sustaining. **(b)** The amount of a fissionable material required to make the fission reaction self-sustaining.

2. Neutrons cause ^{235}U nuclei to undergo fission. Other neutrons strike ^{238}U nuclei to produce plutonium.

3. (a) The heat from a nuclear reaction is used to generate steam. The steam is used to drive conventional electric generators. **(b)** A critical reactor is one that contains a critical mass of fissionable material. **(c)** The ^{235}U in the natural uranium undergoes fission. The ^{238}U in the natural uranium changes to ^{239}Pu, which also undergoes fission when bombarded with slow neutrons.

4. In a breeder reactor fissionable material is produced at a greater rate than the fuel is consumed. This is not the case in a conventional reactor.

5. Nuclear and fossil-fuel plants both use steam to drive the turbines that turn electric generators. The difference is in the way in which the steam is produced. Conventional plants turn water into steam with the heat from fossil fuels. Nuclear plants use the heat from nuclear reactions to produce steam.

6. In a magnetic-confinement fusion reactor, plasma is concentrated by magnetic fields. In an inertial-confinement fusion reactor, lasers or heavy ions are directed against hydrogen isotopes. In both cases, the goal is to achieve the temperatures and densities necessary for controlled fusion reactions.

7. (a) Radioactive isotopes of elements. **(b)** They are produced naturally through radioactive decay, or they are produced through neutron bombardment in a nuclear reactor.

8. Radioisotopes could be mixed with the fertilizer. As the fertilizer is absorbed by the plant, the radioisotopes could be traced with a detecting device.
9. About 13,000 years.

HIGH-ENERGY PHYSICS   25

Questions • Chapter 25/pp. 629–630
GROUP A
1. (a) The method that is used to determine the position of the electron changes its momentum. Likewise, to determine the momentum, the electron's direction of motion is changed. (b) The uncertainty principle. (c) Werner Heisenberg.
2. (a) The branch of physics that deals with the behavior of particles whose specific properties are given by quantum numbers. (b) It describes the specific properties of subatomic particles in mathematical terms. This is not unique with quantum mechanics.
3. (a) The principal quantum number describes the radius of the orbit of the electron. The angular momentum quantum describes the magnitude of the angular momentum of the electron. The magnetic quantum number describes the direction of angular momentum. The spin magnetic quantum number describes the direction of spin. (b) The spin magnetic quantum number. (c) An electron can spin in one of two opposing directions.
4. (a) The splitting of the electron shells of an atom into two or more subshells by a magnetic field. (b) The Zeeman effect caused by electron spin does not require an external magnetic field.
5. (a) No two electrons in an atom can have the same set of quantum numbers.
    (b) Wolfgang Pauli. (c) The absence of certain spectral lines in atoms above hydrogen.
6. The greater the ionization energy of an electron, the greater is the stability of the atom of which it is a part.
7. The wavelength is inversely proportional to the mass of a moving object.
8. (a) Matter waves are mathematical concepts. There is no actual waving of matter as in a ripple tank. (b) There is none.

Problems • Chapter 25/p. 630
GROUP A
1. (a) $n = 1$, $\lambda = 0$, $m_\lambda = 0$, $m_s = +\frac{1}{2}$
   $n = 1$, $\lambda = 0$, $m_\lambda = 0$, $m_s = -\frac{1}{2}$

   (b)

n	λ	m_λ	m_s
1	0	0	$+\frac{1}{2}$
1	0	0	$-\frac{1}{2}$
2	0	0	$+\frac{1}{2}$
2	0	0	$-\frac{1}{2}$
2	1	+1	$+\frac{1}{2}$
2	1	+1	$-\frac{1}{2}$
2	1	0	$+\frac{1}{2}$
2	1	0	$-\frac{1}{2}$
2	1	−1	$+\frac{1}{2}$
2	1	−1	$-\frac{1}{2}$
3	0	0	$+\frac{1}{2}$

(c) Same as (b), plus the following:

n	λ	m_λ	m_s
3	0	0	$-\frac{1}{2}$
3	1	+1	$+\frac{1}{2}$
3	1	+1	$-\frac{1}{2}$
3	1	0	$+\frac{1}{2}$
3	1	0	$-\frac{1}{2}$
3	1	-1	$+\frac{1}{2}$
3	1	-1	$-\frac{1}{2}$

(d) Same as the first twelve electrons of (b) and (c)

2. (a) $E = -\dfrac{me^4}{8\varepsilon_0^2 h^2 n^2}$

$$= -\dfrac{9.1091 \times 10^{-31}\ kg \times (1.60210 \times 10^{-19}\ c)^4}{8 \times (8.854 \times 10^{-12}\ c^2/n\ m^2)^2 \times (6.6256 \times 10^{-34}\ j\ s)^2 \times 3^2}$$

$E = -2.41 \times 10^{-19}\ j$

(b) $\dfrac{-2.41 \times 10^{-19}\ j}{1.60 \times 10^{-19}\ j/ev} = -1.51\ ev$

The ionization energy of the outer electron of the neutral sodium atom is 5.1 ev.

GROUP B

3. (a) $\lambda = \dfrac{h}{mv} = \dfrac{6.6256 \times 10^{-34}\ j\ s}{0.15\ kg \times 26\ m/s} = 1.7 \times 10^{-34}\ m = 1.7 \times 10^{-24}\ \text{Å}$

(b) $\lambda = \dfrac{h}{mv} = \dfrac{6.6256 \times 10^{-34}\ j\ s}{9.1091 \times 10^{-31}\ kg \times 5.5 \times 10^6\ m/s} = 1.32 \times 10^{-10}\ m = 1.32\ \text{Å}$

4. The quantum numbers in Problem 1 show that 2 electrons have the principal quantum number 1, and 8 electrons have the principal quantum number 2. Continuation of the list shows that 18 electrons have the principal quantum number 3, 32 electrons have the principal number 4, etc. This agrees with the expression $2n^2$.

Questions • Chapter 25/p. 635
GROUP A

1. (a) It builds up a large difference of potential by means of a moving belt inside a spherical terminal. **(b)** Uncharged particles. **(c)** An uncharged particle is not attracted or repelled by an electric field.

2. (a) Electron volts. **(b)** An electron volt is the energy required to move an electron through a potential difference of one volt. **(c)** Van de Graaff, about 20 Mev; cyclotron, about 800 Mev; synchrotron, about 30 Bev.

3. In a cyclotron, particles are accelerated between two dees. In a synchrotron, the particles are confined to a doughnut-shaped enclosure.

4. (a) Particles are accelerated as they pass between oppositely charged drift tubes. **(b)** Because the speed of the particles is increasing.

5. The machine that accelerates the electrons. The rest mass of the proton is 1836 times as great as that of the electron. But at 99.9 percent of the speed of light, the relativistic mass of the electron is 22.4 times its rest mass (Section 1.14), while the relativistic mass of the proton at 10 percent of the speed of light is not significantly greater than its rest mass. Substituting these values in the kinetic energy equation, $E_k = \frac{1}{2}mv^2$, it is found that the electron accelerator requires about 22 percent more energy. A relativistic energy equation, $E_k = mc^2 - m_0c^2$, gives a more accurate answer of 20.6%.

6. **(a)** Protons are fed into the center of the cyclotron, between its two dees. As the dees are alternately and oppositely charged, the protons pass from one dee to the other with increasing velocity. **(b)** The magnetic field of the cyclotron keeps the protons inside the machine. **(c)** Neutrons are not attracted or repelled by electrically charged dees.

7. No energy is dissipated in pushing against the target material in an intersecting storage accelerator.

8. The particles move in a straight line.

Questions • *Chapter 25/pp. 640–641*
GROUP A

1. **(a)** The electroscope. **(b)** Madame Curie used the electroscope to study the radioactivity of uranium ore. This led to the discovery of radium.

2. **(a)** Spinthariscope comes from a Greek word meaning "spark." **(b)** Charged particles produce a spark when they hit a zinc sulfide screen in the instrument.

3. A scintillation counter provides information about the energy, and thus the identity, of the particles it detects.

4. They are both ionization devices. In the Geiger tube, a potential of about 1000 volts is applied to the electrodes, while the electroscope is not charged in this way.

5. See Section 25.9, part 6, for a description.

6. **(a)** Liquid hydrogen. **(b)** See Section 25.10.

7. **(a)** In an electronic bubble chamber, particles produce signals in a series of wire planes. The signals are analyzed in three dimensions by a computer. **(b)** It is faster and has better resolution.

8. Nuclear particles leave tracks in solid-state crystals that can be seen with an electron microscope. Each solid used as a detector has a threshold below which no tracks are produced. Thus, the dense background of light-particle tracks is eliminated.

Questions • *Chapter 25/pp. 649–650*
GROUP A

1. **(a)** M. Gell-Mann and K. Nishijima. **(b)** The periodic table of the chemical elements.

2. **(a)** A subatomic particle with a large mass. **(b)** A subatomic particle with medium mass. **(c)** A subatomic particle with a small mass. **(d)** Baryons and mesons. **(e)** A subatomic particle with zero rest mass.

3. **(a)** A substance composed of antiparticles. **(b)** Matter is annihilated.

4. **(a)** Gravitational, weak, electromagnetic, strong. **(b)** Graviton, W-particle, photon, meson. **(c)** Weak and strong.

5. Interaction is a broader concept. Interaction explains force by postulating the creation and annihilation of carrier particles.

6. **(a)** Diagrams that represent the production and exchange of particles in an interaction. **(b)** The carrier in a Feynman diagram transfers the energy that is responsible for the change in momentum of the colliding particles in an interaction.

7. **(a)** Baryons, leptons, hypercharge, and parity. **(b)** Hypercharge and parity.

8. It keeps the universe from deteriorating into low-mass mesons and leptons.

9. Parity is the mathematical expression of symmetry.

10. All forces in the universe are thought to be parts of a single concept.

11. Quarks were proposed because of the proliferation of subatomic particles. They were proposed in 1963 by Gell-Mann and Zweig.
12. Up, down, strange, charm, truth, and beauty.
13. By assigning color, flavor, and charm as properties of quarks.
14. A unified field theory combines two or more interactions. A grand unified field theory combines all four interactions.
15. The gluon is the carrier between quarks in the strong interaction.
16. To shield the experiment from cosmic rays.

JOHN E. WILLIAMS
FREDERICK E. TRINKLEIN
H. CLARK METCALFE

MODERN PHYSICS

THE HOLT MODERN PHYSICS PROGRAM

Modern Physics (Student Text)

Modern Physics (Teacher's Edition)

Exercises and Experiments in Physics (Student's Edition)

Exercises and Experiments in Physics (Teacher's Edition)

Laboratory Experiments in Physics

Tests in Physics (duplicating masters)

HOLT, RINEHART AND WINSTON, PUBLISHERS

New York • Toronto • Mexico City • London • Sydney • Tokyo

JOHN E. WILLIAMS, 2947 Java Road, Costa Mesa, California 92626; formerly physics and chemistry teacher at Newport Harbor High School, Newport Beach, California; and Head of the Science Department, Broad Ripple High School, Indianapolis, Indiana.

FREDERICK E. TRINKLEIN, 131 Brookville Road, Brookville, New York 11545; Dean of Faculty and physics teacher, Long Island Lutheran High School, Brookville, New York; and Adjunct Professor, Physical Science Department, Nassau Community College, Garden City, New York.

H. CLARK METCALFE, P.O. Box V2, Wickenburg, Arizona 85358; formerly chemistry teacher at Winchester-Thurston School, Pittsburgh, Pennsylvania; and Head of the Science Department, Wilkinsburg Senior High School, Wilkinsburg, Pennsylvania.

Cover: Spectacular rings. Dramatic differences in Saturn's rings are revealed by computer-enhanced colors. The picture was generated from two images transmitted by the interplanetary spacecraft Voyager 2 on August 17, 1981.

Photo credits appear on pages 719–720.
New art for this edition prepared by Vantage Art

ISBN 0-03-061936-X

3456 032 987654321

PREFACE

MODERN PHYSICS 1984 is the latest revision of a physics textbook first published in 1922. The many editions of MODERN PHYSICS have served more teachers and students than any other introductory physics text published in the United States. As with previous revisions, the aim of this edition is to increase its effectiveness with a changing population of teachers and students.

The basic concepts of physics are presented in a logical sequence believed to be the most productive for beginning students. Every effort has been made to present these concepts with such directness and simplicity that each student will achieve maximum comprehension. The organization of the text and the style of writing are designed to meet the requirements of today's physics students. A serious effort has been made throughout the text to improve readability without compromising precision.

MODERN PHYSICS 1984 strikes a realistic balance between theory and practical applications. The requirements for an outstanding college-preparatory program are fully met. The wide range of topics contains more material than is ordinarily included in a first year physics course. Thus, teacher's can exercise great flexibility in selecting appropriate subjects for students with varied interests and abilities and still offer a complete general course of study. The Teacher's Edition of MODERN PHYSICS offers many suggestions that can be helpful for implementing college-preparatory or general courses of study.

The authors believe that an effective presentation of physics must be based on a rigorous treatment of mechanics. Accordingly, the fundamental concepts of mechanics are developed early in the book. The introductory chapter is designed to speak directly to students about the importance of energy in their lives, the need for expanding knowledge of energy processes, and the recognition of physics as the science of energy. Measurement and problem-solving techniques are presented in the second chapter. Because mathematics is the language of physics, an adequate preparation in algebra and geometry is an essential prerequisite. The elementary trigonometry necessary for some problem solving is presented in the text where required.

The presentation of force and the results of balanced and unbalanced forces follow the description of motion. Thus the student has the advantage of having studied Newton's laws of motion before dealing with the topic of unbalanced forces. The chapters dealing with heat measurements and heat engines have been rewritten extensively and have been reorganized to give a more effective sequence of topics. The chapters on nuclear reactions and high-energy physics have been carefully updated in order to give the student an overview of recent developments in these important fields.

The International System of Units (SI) is used throughout MODERN PHYSICS with minor exceptions where modifications are deemed appropriate for beginning students. The SI rule that the symbol of a physical quantity is printed in italic type and the symbol of a measurement unit is printed in roman type is followed consistently. The principle that the unit symbol is, in general, lowercase is followed. However, the exception to this general principle, that the first letter of a unit symbol derived from a proper name is uppercase, is not generally followed in this book.

The techniques of dimensional analysis are emphasized in all quantitative work. Appropriate attention is given to the use of significant figures in the expression of measurement quantities and in computations involving measurement data. Furthermore, the derivation of each measurement concept is traced back to fundamental units. Such careful attention to the dimensional character of measurements throughout the book helps to unify the whole structure of physics.

A single-column format has been retained in this revision. Wide margins allow for flexible treatment of illustrative materials. A list of *Objectives* is given at the beginning of each chapter. *Marginal notes* are strategically placed throughout the text to reinforce important concepts, to provide supplementary information, and to assist students in developing productive study techniques. New terms, definitions, and principles are printed in *italic* and **boldface** type for emphasis and rapid review. *Example problems* with step-by-step solutions are used to illustrate important quantitative relationships and to ensure student comprehension of pertinent concepts. The solved *Examples* are followed by sets of related *Practice Problems*, with answers, to provide the student with opportunities for immediate reinforcement of these important quantitative concepts.

Chapters generally are divided into major subdivisions. *Question* and *Problem* sets are placed at the end of each subdivision to provide a liberal number and variety of exercises. The questions and problems are graded and differentiated as Group A and Group B. The A sets are constructed as essential drill exercises for *all* students. Vector analysis problems in these sets are limited to right triangles. The B sets are more difficult and challenging; they provide for individual and group differences. Vector analysis problems in these sets include non-right triangles requiring more sophisticated solutions. *Sections* within each chapter are identified by both chapter and section numbers for easy reference. For example, Section 4 of Chapter 10 is designated by the number 10.4 and is referred to in other parts of the text as Section 10.4. Each chapter concludes with a *Summary* and a *Vocabulary* which together provide a brief review of the major concepts and technical terms introduced in the chapter.

Red has been used as a second color and three-dimensional effects are used in diagrams. An eight-page *full-color insert* of photographs and drawings has been included between pages 336 and 337. Special *Photo Essays* provide supplementary information on relevant applications and frontiers of physics in a pictorial mode.

A *Mathematics Refresher* section precedes the Appendixes. This section briefly reviews the mathematical skills required for successful problem solving. *Appendix A* lists important equations presented in the sequence in which they appear in the text. *Appendix B* includes twenty-four tables of useful data, including a table of trigonometric functions and a table of logarithms. A *Glossary* of technical terms follows Appendix B.

The authors are especially indebted to Dr. David G. Haase, Associate Professor of Physics, North Carolina State University, Raleigh, North Carolina; Dr. John N. Shive, retired professional physicist with the Bell Telephone Laboratories; and Mr. Robert Roe, Jr., Science Instructor, Skyline High School, Dallas, Texas, all of whom critically reviewed the 1980 edition of MODERN PHYSICS. Their many valuable suggestions and criticisms were greatly appreciated. The authors are also indebted to Mr. William C. Metcalfe, Tampa, Florida, for his assistance in obtaining the postage stamps from various countries that are used throughout this book.

John E. Williams
Frederick E. Trinklein
H. Clark Metcalfe

CONTENTS

3 VELOCITY AND ACCELERATION

4 CONCURRENT AND PARALLEL FORCES

5 TWO-DIMENSIONAL AND PERIODIC MOTION

8   HEAT MEASUREMENTS

12   THE NATURE OF LIGHT

13   REFLECTION

14 REFRACTION

OPTICAL REFRACTION

LENS OPTICS

DISPERSION

15 DIFFRACTION AND POLARIZATION

INTERFERENCE AND DIFFRACTION

POLARIZATION

16   ELECTROSTATICS

17   DIRECT-CURRENT CIRCUITS

21   ALTERNATING-CURRENT CIRCUITS

22   ELECTRONIC DEVICES

25 HIGH-ENERGY PHYSICS

PHYSICS: THE SCIENCE OF ENERGY

Albert Einstein was a 25-year old clerk in a Swiss patent office when he published his theory on the equivalence of mass and energy. His concepts of space and time gave new dimensions to the study of the universe. In 1921, Einstein was awarded the Nobel Prize in Physics.

SCIENCE IN TODAY'S WORLD

1.1 The Energy Problem Modern civilization would be impossible without the use of large quantities of energy. In the last forty years, energy usage in the United States has increased more than 400% while the accompanying population increase has been only 72%. Thus the per capita use of energy in the United States has increased at a very fast rate.

Until recently, the world's energy demands were met almost entirely through the use of fossil fuels: wood, coal, petroleum, and natural gas. Since these are limited in quantity, however, other sources must be found. Nuclear and solar energy are two alternate sources, and neither of them is limited. It has been calculated, for instance, that the energy demands of the whole world could be met by utilizing the solar energy falling on a 125-mile square plot of ground near the equator. At the present time, however, nuclear and solar energy make up only a small fraction of the total energy used. See the graph of energy usage shown in Figure 1-1. As our knowledge and means of converting these ample energy supplies into usable forms increase, we shall be better able to keep pace with rising energy needs. However, it will take more than just the efforts of science to solve the energy problem. Fuel conservation by all segments of the population must accompany new scientific advances.

In this chapter you will gain an understanding of:

✔ the nature of science
✔ some basic properties of matter
✔ the relationship between matter and energy
✔ the subdivisions of physics

To prevent the interruption of solar energy at nighttime and during cloudy weather, huge solar satellites have been proposed.

The energy shortage and other world problems can be solved only if people in all specialties work together on them.

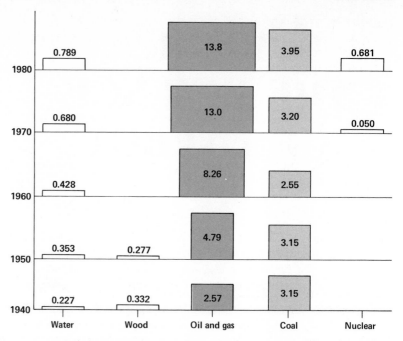

	Water	Wood	Oil and gas	Coal	Nuclear
1980	0.789		13.8	3.95	0.681
1970	0.680		13.0	3.20	0.050
1960	0.428		8.26	2.55	
1950	0.353	0.277	4.79	3.15	
1940	0.227	0.332	2.57	3.15	

Figure 1-1. Energy sources in the United States. The graph shows, in calories × 10^{18}, the changing amounts of energy that have been obtained from various fuels since 1940.

Figure 1-2. Magnified view of a modern computer chip containing hundreds of transistors. The chip is shown inside the eye of a needle.

1.2 Science and Technology

Briefly stated, *science* is the search for relationships that explain and predict the behavior of the universe. *Technology* is the application of these relationships to our needs and goals. Finding new energy sources falls into the realm of science. Developing and utilizing these discoveries are matters of technology.

The complementary work of scientists and technologists is also referred to, respectively, as *research* and *development*. Both government and private industry allocate large sums of money to support the research and development activities of scientists and engineers. Without such support, there would be little improvement in the processes, products, and services that are derived from these activities.

The development of the miniaturized computer is a good example of the result of cooperative effort between science and technology. At the heart of this amazing device is a tiny chip, such as the one shown in Figure 1-2, that contains thousands of transistors. This tiny chip has the calculating capacity of a vacuum tube computer that, twenty-five years ago, needed an entire room for space.

One application of the miniaturized computer is the supermarket electronic checkout. This system reads and records the names and prices of items labeled with the Universal Product Code (UPC). This code, a series of black lines underscored by numbers, is read by a laser beam scanner and the coded information is relayed to the store's computer. This coded information is processed by the computer, which then displays the product's name and price and provides an itemized receipt. While it is recording each purchase, the computer is also checking the store's inventory for reordering purposes.

Computer-assisted instruction demonstrates the application of computers in the field of education. Individualized instruction in any subject is available to the student. The student retrieves the stored information and then converses with the computer by means of a typewriter or television screen.

In banks computers already handle most of the billions of checks Americans write each year. In hospitals computers are used to analyze the medical histories and symptoms of patients as well as to detect and monitor their ailments. A computer program can order prescriptions and alert nurses to administer them at the right time and in the right dosage. A major advance is the use of the computer to assemble thousands of X rays of any part of the body into a single, two-dimensional view. See Figure 1-3.

In the home computer devices may someday be as commonplace as the kitchen sink. When mass production and new technology bring the cost within reach of most budgets, routine chores of all kinds can be handled quickly and accurately by the computer chip. In cars computers can automatically apply the brakes, determine gas mileage, and check the oil and tire pressure; a miniradar can warn the driver of impending collisions.

Many applications of computers still await discovery. Today's high school students, who have grown up in our highly technological society, can meet this challenge as they become the engineers and scientists of the future.

1.3 Scientific Laws and Theories In science *a **law** (or **principle**) is a statement that describes a natural event.* Unlike laws that *restrict* behavior (i.e., traffic laws), scientific laws *describe* behavior. For example, Jacques Charles (1746–1823) observed, under certain controlled conditions, a relationship between the temperature and the volume of most gases. This regular behavior of gases, which you will study in greater detail in Chapter 8, is now known as Charles' law. A scientific law may be stated in words. However, in physics a law is usually expressed by a mathematical equation relating the natural event to the various factors that cause it.

*A **theory** is a reasonable explanation of observed events that are related.* A theory often involves an imaginary *model* that helps scientists picture the way an observed event could be produced. A good example of this is our modern atomic

A system called PLATO (Programmed Logic for Automatic Teaching Operations) allows the student to "converse" with a computer by means of a typewriter and a TV screen.

In 1979, Allan M. Cormack of the United States and Godfrey N. Hounsfield of Great Britain received the Nobel Prize for developing the CAT scan.

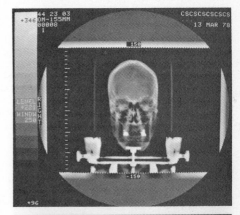
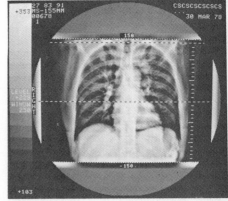

Figure 1-3. Computerized axial tomography (CAT) scan. Cross-sectional views of the head and chest cavity are obtained from the computer assembly of tens of thousands of X-ray readings.

theory, which you will study in Chapters 7, 23, 24, and 25. Another example is the kinetic molecular theory. In this theory gases are pictured as being made up of many small particles called molecules that are in constant motion. You will study the kinetic molecular theory in Chapter 7.

A scientific law describes a natural event. A theory is a reasonable explanation of a series of events.

A useful theory, in addition to explaining past observations, helps to predict events that have not as yet been observed. After a theory has been publicized, scientists design experiments to test the theory. For example, if matter is composed of small particles called atoms, it should behave in a certain way. If observations confirm the scientists' predictions, the theory is supported. If observations do not confirm the predictions, the scientists must search further. There may be a fault in the experiment, or the theory may have to be revised or rejected.

1.4 Scientific Hypotheses   Albert Einstein (1879–1955) was one of the greatest theoretical scientists of all time. Theoretical scientists work mostly with ideas. They use the results of other scientists' observations to develop their theories. Einstein's theories and predictions changed the course of modern science. Once he was asked to explain how a scientist works. "If you want to know the essence of scientific method," he said, "don't listen to what a scientist may tell you. Watch what he does."

Science is a way of doing things, a way that involves imagination and creative thinking as well as collecting information and performing experiments. Facts (observations, principles, laws, and theories) by themselves are not science, but science deals with facts. The mathematician Jules Henri Poincaré (1854–1912) said: "Science is built with facts just as a house is built with bricks, but a collection of facts cannot be called a science any more than a pile of bricks can be called a house."

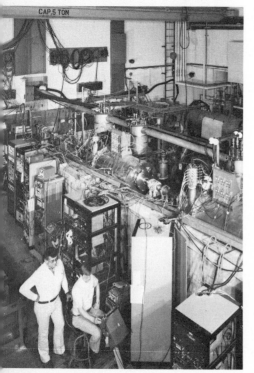

Figure 1-4. An industrial physics research laboratory. In this laboratory the world's first application of super conducting technology in the acceleration of heavy ions for nuclear research was performed.

CAP.5 TON

Most scientists start an investigation by finding out what other scientists have learned about the problem. After the known facts have been gathered, the scientist comes to the part of the investigation that requires considerable imagination. Possible solutions to the problem are formulated. These possible solutions are called *hypotheses.*

For example, at the time of Johannes Kepler (1571–1630), most scientists believed that the planets move in circular orbits. But the observed movements of the planets could not be satisfactorily explained by circular orbits. So Kepler formulated other hypotheses to explain planetary motion. The one that best fit the observations was: All the planets have elliptic orbits.

In a way, any hypothesis is a leap into the unknown. It

extends the scientist's thinking beyond the known facts. The scientist plans experiments, calculations, and observations to test hypotheses. For without hypotheses, further investigation lacks purpose and direction. In the case of Kepler, he continued his observations and calculations over a period of ten years after the announcement of his hypothesis about elliptic orbits. Eventually he was able to relate, in the form of an equation, the time required for one orbit of a planet with its distance from the sun.

When hypotheses are confirmed, they are incorporated into theories or laws. Thus Kepler's hypotheses have become parts of Kepler's laws of planetary motion.

A hypothesis is a possible solution to a problem.

The scientist's laboratory is any place where an investigation can be conducted. To the astronomer the starry skies are a laboratory. The biologist may do experimental work in a swamp or an ocean. The physicist and chemist are often surrounded by a maze of apparatus housed in buildings designed for specific purposes. See Figure 1-4.

1.5 Certainty in Science In a sense, there is no such thing as absolute certainty in science. The validity of a scientific conclusion is always limited by the method of observation and, to a certain extent, by the person who made it. If a nurse took a patient's temperature with a thermometer that read 1° too high, the physicians might reach an incorrect conclusion about the condition of the patient. The validity of every measurement is limited by the precision of the apparatus used. Furthermore, these limits are not always apparent to the experimenter. Galileo Galilei (1564–1642), the famous Italian scientist, thought that the motion of light was instantaneous, since he was unable to measure its speed. Later experiments showed that Galileo's conclusion was wrong.

Hence it is important to keep an open mind about the validity of scientific principles, theories, or hypotheses. No amount of experimentation can ever prove any one of them absolutely, whereas a single crucial experiment can disprove any of them. When a scientist is unwilling to question a statement in science, no matter how well established it is, that scientist has lost a very important attribute of a successful researcher. For this reason the famous physicist Niels Bohr (1885–1962) told his classes: "Every sentence I utter must be understood not as an affirmation but as a question."

Do you think it is frustrating to deal with scientific uncertainties?

Science has been so successful in answering questions about the physical universe and in making life more pleasant and productive that it has been proposed that the methods that scientists have found successful should be used in other areas as well. This means that the attitudes necessary for successful work in science should be universally productive. Among these attitudes are imagination, a thirst for knowledge, a regard for data and their meaning, a respect for logical thinking, patience in reaching conclusions, and the willingness to work with new ideas.

There are many topics, however, that do not fall into the realm of science. Among these are questions that concern human motives or the ultimate origin of matter and energy. In general, science deals with repeatable events and with answering "how" rather than "why" questions.

Questions

GROUP A

1. What are the disadvantages of using fossil fuels as energy sources?
2. List as many present and future applications of computers as you can. Compile a class list for the bulletin board and check it periodically to see whether any future applications become present ones during the school year.
3. Bring a UPC-marked item to class and compare it with those of your classmates for similarities and differences. Offer some explanations on the meaning of the lines and numbers.
4. What is meant by a theory?
5. How does a hypothesis become part of a law?
6. (a) Define the term "laboratory."
(b) How does your school's physics laboratory fit this definition?
7. Distinguish between science and technology.
8. Name the Nobel Prize winners in physics for the past five years and list the contribution for which each of them received the prize.

GROUP B

9. Edison once said: "Genius is 1% inspiration and 99% perspiration." Explain this remark in terms of the work of scientists and technologists.
10. How does the work of Kepler illustrate the scientific method?
11. Define the term "certainty" as it relates to science.
12. List questions and problems that, in your opinion, do not fall into the realm of science.

THE CONTENT OF PHYSICS

Words cannot fully describe an abstract idea.

1.6 Matter A good way to define matter is to describe its various forms. A description is not a definition in the real sense of the word, but it helps to bring an abstract idea down to familiar terms.

Matter can be described in terms of its measurable properties. In describing a person, you might refer to height, weight, eye color, hair color, etc. Similarly, all forms of matter possess properties. And just as a person can be identified by listing various characteristics, so a specimen of matter can be singled out by listing its properties.

The number of properties that can be measured for specimens of matter is very large. Entire handbooks of chemistry and physics are devoted to listings of various properties of matter.

In the study of physics it is important to recognize that *unless a property can be measured and compared with some kind of standard, it is of limited use to the scientist. Without measurement there can be no science.*

1.7 Mass An important property of matter is its *mass*. *Mass is a measure of the quantity of matter in an object.* This is another way of saying that matter takes up space. But the mass of an object cannot be determined merely by measuring its size. Two objects may have the same size and be composed of the same form of matter, yet one may contain more hollow spaces and thus have less mass. Or, one form of matter may have more mass in the same amount of space than does another form of matter. For example, a piece of brass and a piece of gold may have the same size but not the same mass. Or, a piece of lead and a piece of aluminum may have the same mass but not the same size, as shown in Figure 1-5.

The mass of an object is determined by comparing it with known masses. Mass measurements will be more fully described in Chapter 2.

1.8 Inertia The mass of an object can also be measured by using a property of matter called *inertia*. *Inertia is the property of matter that opposes any change in its state of motion.* Inertia shows itself when objects are standing still as well as when they are moving. A baseball resting on the ground will not start moving by itself. A baseball in flight will keep moving unless something stops it. This does not mean that there are two kinds of inertia—a stationary kind and a moving kind; the same property of matter is merely showing itself in different circumstances.

The inertia of matter can be used to measure mass with a device called an inertia balance, shown in Figure 1-6. One end of the balance is clamped to a table. The pan on the other end can be made to vibrate horizontally. The number of times this device vibrates in a second of time depends upon the length and stiffness of the two supporting metal blades and upon the total mass of the pan and any objects fastened to it. Since the blades vibrate horizontally, the action of the inertia balance is entirely independent of gravity, which acts vertically. The more massive the pan and its contents, the slower will be the changes in their motion and the longer will be the time they will take

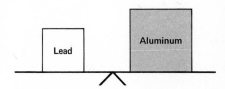

Figure 1-5. Distinction between mass and size. The piece of lead and the piece of aluminum have the same mass, but they do not have the same size.

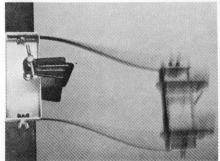

Figure 1-6. An inertia balance, as a person would view from above. When the instrument is set in horizontal motion (bottom), the period of vibration can be used to compare the magnitudes of masses attached to it.

Period and time are reciprocals.

to go through a complete vibration. The time required for a single vibration is called the *period*.

Unknown masses can be measured by comparing their periods with the periods of objects having known masses. First, several of the known objects are used and their periods are plotted as functions of their masses, as shown in Figure 1-7. Then the period of an unknown object is measured and its mass is read from the graph.

The graph in Figure 1-7 is not a straight line. But by careful analysis of the data the relationship can be expressed by the following equation:

$$\frac{m_1}{m_2} = \frac{T_1^2}{T_2^2}$$

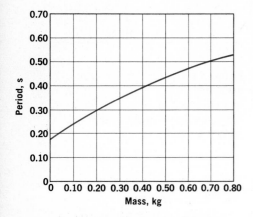

Figure 1-7. Graph of inertia balance data. If an object is attached to the pan of the balance and the period is measured, this graph can be used to find the mass of the object.

in which m_1 and m_2 are two masses (*including* the mass of the pan in each case) and T_1 and T_2 are their respective periods.

Masses can also be compared using either a spring balance or a platform balance, as shown in Figure 1-8. For comparing masses, these devices depend upon the pull of the earth's gravity. The stretch of a spring depends upon the amount of pull on the spring. The greater the pull, the greater is the amount of stretch. Since the pull of gravity is greater on a large mass than it is on a small mass, the spring in a spring balance stretches farther for a large mass than it does for a small mass. The amount of stretch for standard masses can be marked on a scale. The stretch for an unknown mass can then be compared with the stretch produced by the standard mass.

Figure 1-8. A spring balance (left) and a platform balance (right) in use.

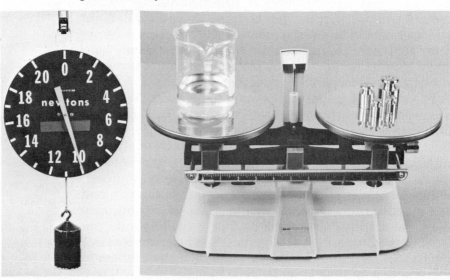

•

In the case of a platform balance, a uniform beam is balanced at its center on a sharp blade. The blade acts as a pivot. If a mass is added to either side of the beam, the beam tips. Platforms of equal masses are placed on the beam at equal distances from the pivot. If an unknown mass is placed on one platform (usually the left one), known masses can be added to the other platform until the beam again balances. At that point, the known and the unknown masses are equal.

When the mass of an object is measured by its resistance to a change in its motion, as in the inertia balance, the mass is sometimes referred to as *inertial mass*. Mass that is measured with a spring balance or a platform balance, on the other hand, is sometimes called *gravitational mass*. Many experiments have shown that the inertial mass and gravitational mass of a substance are always equal, even though they may seem to be unrelated. The single term "mass" can therefore be used to describe both forms.

Now that we have discussed two important properties of matter, we can use these properties to define matter. *Matter* is anything that has the properties of mass and inertia.

1.9 Mass Density   A property of matter that is closely related to mass is *mass density*. **Mass density** refers to the amount of matter in a given amount of space and is defined as the *mass per unit volume of a substance*. Thus if a substance has a mass of 45 g and occupies a space of 15 cm^3, its mass density is 3.0 g/cm^3. The mathematical equation is

$$\text{mass density} = \frac{\text{mass}}{\text{volume}}$$

In giving the mass density of a substance, it is important to include the units (kilograms per cubic meter, grams per cubic centimeter, or some other unit of mass per unit of volume) in order that it may be compared with other values of mass densities. For example, the mass densities of solids and liquids are usually compared to the mass density of water, while the mass densities of gases are compared to hydrogen. The ratio of such a comparison is called the *specific gravity* of the substance.

Mass density is a property that varies with the environment. The mass density of a gas increases when the gas is placed under pressure and decreases when the pressure is reduced. Such quantities as pressure and temperature, which describe the environment of an object, are called *conditions*. Very often a measurable property of an object will change if the conditions are changed.

Volume: quantity of space occupied by a substance, measured in cubic units of length.

Some stars have mass densities of thousands of kilograms per cubic centimeter!

1.10 *Energy* It is said that, as a youngster, James Watt (1736–1819) became interested in steam power by watching water boil in a teakettle over a fire. When he plugged the spout of the kettle, the lid rose. Perhaps this observation eventually led Watt to the improvements of steam engines for which he is famous. In any case, this observation is a good example of how energy is related to work.

We say that work is done when an object is lifted against the pull of the earth's gravity. Later we will define work more completely; for now this common use of the word will be adequate. We do work when we lift the lid of a teakettle. When we plug the spout, the steam lifts the lid. The work is done by the steam. So we reason that the steam in the kettle had the *capacity* for doing work before it lifted the lid. The steam had *energy*. **Energy** *is the capacity to do work,* if the conditions are right. Since energy can often be transformed into mechanical work, the units we use for measuring energy are the same as the units of work.

Energy is latent work.

Energy is sometimes quite noticeable because we have sense organs that are able to detect its presence in various forms. Our eyes respond to visible light energy. Our ears detect sound energy. Special nerves are sensitive to temperature, an indication of heat energy. Other nerves respond to electric energy.

Scientists have discovered forms of energy in addition to those that can be detected by our sense organs. Special detecting and measuring instruments had to be developed to record these forms of energy in a form that we can sense. For example, we cannot directly sense X rays, a form of energy similar to visible light. But X rays can affect special photographic film. Under visible light, we can look at a piece of exposed and developed X-ray film and tell where the X rays have affected the film. As another example, we cannot directly sense small amounts of infrared radiation, an invisible form of energy similar to visible light. (We can feel large amounts of infrared radiation as heat.) However, a special thermometer exposed to infrared radiation registers an increase in temperature. We can study the behavior of the thermometer and say that invisible radiation is present. By means of devices such as these, scientists have extended the range and sensitivity of the human senses. Among the forms of energy that fall into this "extrasensory" category that humans cannot directly detect are chemical energy, gravitational energy, nuclear energy, and forms of energy similar to visible light, such as X rays and infrared rays.

Energy is the concept that unifies physics. More specifically, physicists study and measure the transformation of energy, for the forms of energy are mutually interchangeable. For example, consider the energy you expend when you walk. Traced backward, it came from the food you ate. The food got its energy from nutrients in the ground and radiations from the sun. The sun got its energy from nuclear reactions in its interior, etc. Tracing your walking energy forward, the friction and motion of your feet on the ground heat the ground slightly. This heat radiates into space and helps to evaporate water from the earth, making rain possible, etc.

1.11 Potential Energy   Two important forms of energy that an object may have are its energy of position and its energy of motion. Energy of position is *potential energy.* Energy of motion is *kinetic energy.* A book resting on a desk top, as in Figure 1-9, acquired potential energy when someone did the work of lifting it to the desk top. The book has potential energy, or energy of position, while it is on the desk top. If the book is pushed off the desk top, it falls to the floor. While falling, it may strike some other object to which it gives some of its energy. This is kinetic energy, or energy of motion. When the book is stopped by the floor, the rest of its kinetic energy is transformed into other forms of energy, such as sound and heat.

In a way, the floor is an arbitrary zero level for our example. If a hole were cut in the floor under the book, the book would continue to fall and do more work when its motion is stopped. For each problem in *gravitational* potential energy, an arbitrary zero level must be specified.

An object may also have potential energy that is not connected with gravitation in any way. For example, a compressed spring gets its energy from the push that is exerted on it to get it into a compressed position. This energy is called *elastic* potential energy. Elastic potential energy is not as easy to compute as the gravitational variety, since one must be familiar with the transformations of various forms of energy. Some of these forms of energy, especially those concerning the interior of the atom, can take the problem to the very frontiers of modern physics.

1.12 Kinetic Energy   Every moving object has kinetic energy. This is the same as saying that everything has kinetic energy, because scientists believe that everything in the universe is moving in some way or other. Since the description of the motion of an object depends on another object in the universe designated as being "motionless," it is necessary to choose an arbitrary zero level or, in this

Figure 1-9. An example of gravitational potential energy. The amount of energy of the books depends upon the zero level chosen for the calculation.

case, an arbitrary stationary point. In ordinary, earthbound physics the surface of the earth is considered stationary. An object resting on the earth's surface is said to have zero kinetic energy even though the object rotates and revolves with the earth.

1.13 Conservation of Energy As Figure 1-10 shows, there is a constant interplay between potential energy and kinetic energy in physics. Consider the swinging of a mass at the end of a string. When the mass is at the top of its swing, it is momentarily stationary. At that point the energy of the mass is gravitational potential (except for internal energy). As the mass begins its downward swing, some of the gravitational potential energy changes into kinetic energy. At the bottom of the swing, which we will consider as the zero level for gravitational potential energy, the kinetic energy of the mass is at a maximum because it is moving at its maximum speed. As the mass swings up the other side of its arc, the energy interchange is reversed. The total amount of energy is always the same—it is merely changing form. This discussion dem-

Potential energy is energy due to the position of an object. Kinetic energy is energy due to the motion of an object.

Potential energy and kinetic energy are interchangeable. The total amount of energy in a situation remains constant.

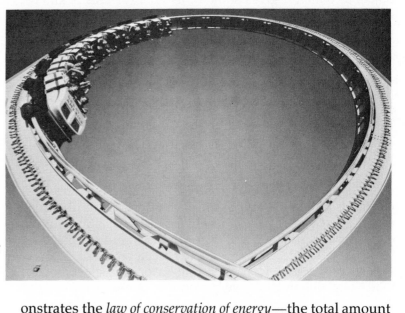

Figure 1-10. Energy transformation. As the roller coaster and its occupants descend, some of their gravitational potential energy changes into kinetic energy.

Count Rumford was a native American whose original name was Benjamin Thompson. He left for Europe at the start of the Revolutionary War because of his Tory sentiments. Yet in his will, which was signed by Lafayette, he left his estate to an American college.

onstrates the *law of conservation of energy*—the total amount of energy in a given situation is constant.

The conservation of energy was demonstrated in a striking way by the classical experiments with heat energy conducted by Count Rumford (1753–1814) and James Prescott Joule (1818–1889) in the early part of the nineteenth century. At that time most scientists thought of heat as a special kind of "fluid" called *caloric*, which was able to flow in

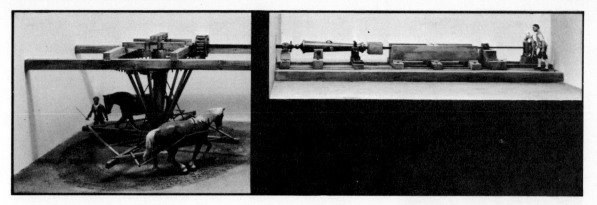

Figure 1-11. A model of Count Rumford's cannon-boring experiment. A team of horses furnished the energy to rotate the cannon barrel against a stationary boring tool. Water poured into the barrel during the boring process quickly came to a boil, especially when the boring tool became dull.

and out of objects without affecting their weights. Count Rumford was in charge of the boring of brass cylinders for the construction of cannon barrels in Bavaria. Figure 1-11 shows a model of Rumford's apparatus. During the boring process the barrels became so hot that water poured into them could be brought to a boil. According to the caloric theory, the brass contained a specific amount of caloric. This caloric was being released from the brass as the metal was broken down into chips by the boring tool. Count Rumford questioned this explanation since the amount of heat produced seemed limitless. As long as the horses provided the work to turn the boring tool, more heat developed. Rumford demonstrated that if the boring tools were dull and no metal were ground to chips that supposedly contained caloric, heat continued to pour out. In fact, the metal heated up even more. Rumford concluded that the heat produced was a result of the work done on the cannon barrel by the horses; it was not the result of a substance present in the brass. In spite of Rumford's experiments, the caloric theory existed for fifty more years.

Joule continued the investigation of heat by establishing the quantitative relationship between heat and mechanical energy. This relationship will be discussed more fully in Chapter 9. The equivalence between heat and mechanical energy provided strong evidence that heat is a form of energy. It also supported the theory that the total amount of energy remains constant when it is transformed from one kind to another.

1.14 Relationship between Matter and Energy   So far we have discussed matter and energy as though they were two entirely different entities. Actually, the two are inseparably related. All matter contains some form of energy. Ordinarily, the idea of energy is associated with a substance such as a hot gas or an electrified object. In some

Einstein's mass-energy equation led to the development of nuclear power.

Figure 1-12. Einstein did not spend all of his time thinking about the theory of relativity.

Table 1-1
RELATIVISTIC MASS

Speed of object (% of speed of light)	Relativistic mass (compared to rest mass)
0	1.00
25	1.03
50	1.15
75	1.51
80	1.67
85	1.88
90	2.30
95	3.16
99	7.07
99.9	22.4

cases it is necessary to associate energy with the empty space surrounding an electrified object.

In 1905 Einstein expressed the relationship between matter and energy with the following equation:

$$E = mc^2$$

in which E stands for units of energy (work units), m stands for mass, and c for the speed of light. Einstein (Figure 1-12) developed this equation entirely from theoretical considerations, and there was no way of verifying it in the laboratory at the time. Subsequent experiments, however, have shown that such a relationship does exist.

Einstein's equation states that mass and energy are directly proportional to each other. When one increases, the other increases; when one decreases, so does the other. The equation is also interpreted to mean that a given amount of mass is equivalent to a specific amount of energy.

A simple example will serve to illustrate the relationship between mass and energy. When you throw a ball, you impart energy to it. Energy is transferred from you to the ball. The mass-energy equation says that since c is constant, the increase in energy E will be accompanied by an increase in mass m. That is, while the ball is moving, its mass will be greater than it was while the ball was at rest. Both the energy and the mass of the ball have increased. The extra energy and mass of the ball came from you.

When an object is at rest with respect to the observer and the observer's measuring instruments, the mass of the object is said to be the *rest mass*. When the object is moving, its mass increases. This new mass, which increases rapidly as the object approaches the speed of light, is called the *relativistic mass* because it is in keeping with Einstein's *theory of relativity*. In the mass-energy equation, m is the relativistic mass. Table 1-1 shows the relationship between the speed of an object and its relativistic mass.

Actually, what Einstein's equation means is that matter is a form of energy and the two terms emphasize different aspects of the same quantity. This relationship is one of the most important concepts of contemporary physics.

It should not be assumed, however, that Einstein's mass-energy equation is a statement of the conservation of matter and energy. The equation has value even if the matter and energy of the universe were constantly changing. The conservation laws rest on repeated laboratory measurements that show mass and energy are not lost in chemical and physical changes. Some scientists think the law of conservation may not hold for large energies and masses in outer space, but this hypothesis has not been verified. The deviations predicted by scientists are too small for measurement with present instruments.

The relationship between matter and energy will become evident in the study of physics in still another way. We usually think of light energy in terms of waves and of matter in terms of particles. Actually, there are times when light acts as though it has granular properties. In other words, there are aspects of light that can be explained only by assuming that it is made up of discrete particles. Similarly, particles of matter exhibit wave properties. This wave-particle duality of nature is a key as well as a puzzle in physics, and scientists admit that they are far from a complete understanding of it. Perhaps you will play a role in unlocking this fundamental secret of nature.

1.15 Subdivisions of Physics The study of the physical universe is called *physical science. Chemistry* is a physical science. It deals with the composition of matter and reactions among various forms of matter. *Physics* is also a physical science. It is concerned with the relationship between matter and energy. The ultimate goal of physics is to explain the physical universe in terms of basic interactions and simple particles.

Since both chemistry and physics deal with matter, these two sciences overlap to some extent. But a distinguishing consideration in the study of physics is always the idea of energy: what it is, how it affects matter and how matter affects it, and how it can be changed from one form to another.

Classifying a concept is not as important as understanding it.

The study of physics can be subdivided in a number of ways. The sequence of topics followed in this book is one that has been used for many years because of its logical progression and because it leads from concepts developed many years ago to those at the very frontiers of physics.

The chapters in your text fall into four major categories: mechanics and heat, waves, electricity, and nuclear physics. The first category includes the study of motion, forces, work, power, and certain aspects of heat. These topics, and the ways in which they involve the transfer of energy, are considered in Chapters 3 through 9. (Chapter 2 describes measurement and problem solving, which are essential to the entire subject of physics.)

Waves are considered in Chapters 10 through 15. Sound energy and light energy are transmitted as waves. The nature of waves as a mechanism for energy transfer and the characteristics of sound and light comprise the subject matter of this section of your text.

Electricity involves a form of energy that is increasingly important to our way of life. In Chapters 16 through 22, the ways of generating and transmitting this vital commodity are explained. This section of the text also contains an introduction to the study of electronics, which is the basis for radio, television, and computers.

In the last three chapters of the book, the principles of nuclear and particle physics are introduced. It is here that the physicist tries to get to the very heart of the composition of matter and of the energy forms that bind the building blocks of the universe together. With topics ranging from "atom smashers" to the most distant star systems, this part of the course is usually called modern physics. The authors hope that as you reach this final section of the course, you will have caught the excitement of physics, a subject that features unlimited topics and possibilities.

Questions

GROUP A

1. How does the mass of an object on the earth compare with the mass of the same object on the moon?
2. (a) Distinguish between mass and mass density. (b) Distinguish between inertial and gravitational mass.
3. (a) Distinguish between properties and conditions. (b) Give examples of each.
4. (a) What is the source of energy in an automobile? (b) How is the energy transformed?
5. Classify the following as having mainly potential or mainly kinetic energy: (a) gasoline, (b) water behind a dam, (c) a moving car, (d) a spaceship in free flight, (e) gunpowder.
6. What limits the amount of heat energy that can be produced by the boring of a cannon barrel?
7. State the law of conservation of energy.
8. Give two meanings of the equation $E = mc^2$.
9. What is meant by the wave-particle duality of nature?

10. Briefly describe the four subdivisions of physics.

GROUP B

11. How can Figure 1-7 be used to find the mass of the balance pan used in the experiment?
12. Redraw Figure 1-7 using values of T^2 instead of T. What is the shape of the resulting graph?
13. Could a platform balance be used in a spaceship in free flight? Explain.
14. Why are work and energy measured in the same units?
15. State a fact to support the caloric theory.
16. Why does it become increasingly difficult to increase the speed of an object as it approaches the speed of light?
17. In general, objects expand when heated. (a) How does this phenomenon show that heat is a form of energy? (b) Modify your explanation to fit objects that contract upon being heated.
18. On the basis of Count Rumford's observations and conclusions about heat, can you explain how his experimentation led to the development by Joule of the mechanical equivalent of heat?

SUMMARY

The world's growing population re-quires ever-larger amounts of energy. By discovering and developing new energy sources, science and technology can help to meet this need. Science and technology are also called research and development. The miniaturized computer and its many applications are good examples of the impact of science and technology on our way of life. Even so, many applications of computers still await discovery.

Scientific laws, theories, and hypothe-ses are steps in the search for truth in the physical universe. A hypothesis is a pos-sible solution to a scientific problem. Con-firmed hypotheses become laws or theories, but even these are never abso-lutely certain.

Science deals with the "how" but not the "why" of the universe. Among the properties of matter are mass (the mea-sure of the quantity of matter), inertia (the resistance of matter to change in its state of motion), and mass density (mass per unit volume). The environment of a sample of matter is called a condition. Physics further deals with the various types of energy and their transformations. Energy is the capacity for doing work. Potential energy is energy of position. Gravitational energy and elastic energy are forms of potential energy. Kinetic energy is energy of motion. Potential and kinetic energy are readily interchanged. However, the law of conservation of en-ergy states that energy may be changed from one form to another without loss. The experiments of Count Rumford and James Prescott Joule verified this law.

Einstein's relativity equation, $E = mc^2$, states that energy and mass are directly proportional to each other and that they are interchangeable. Mass can thus be considered to be a form of energy.

Physics is the physical science that deals with the various forms of energy. Its four main subdivisions are mechanics and heat, wave motion, electricity, and nuclear and particle physics. The last of these is also referred to as modern physics.

VOCABULARY

caloric	law of conservation of	potential energy
conditions	energy	relativistic mass
energy	mass	rest mass
hypothesis	mass density	science
inertia	matter	specific gravity
kinetic energy	period	technology
law	physics	theory

Physics is more than a course of study. It is a foundation of information upon which careers are built. The people pictured here are applying the concepts and principles of physics daily in their work. Useful products and important solutions to interesting problems have resulted from their endeavors. Their careers supply the research and technological developments vital to the present needs and future goals of our society.

Careers

Scientist-astronaut Anna L. Fisher is pictured inside an orbiter one-g trainer. The trainer is part of the shuttle mock-up and integration laboratory at the Johnson Space Center. She is training for a future position as Shuttle Mission Specialist.

Scientist Janusz Wilczynski designed advanced optics for photolithography, work that has significantly increased the production volumes of very dense packaging used in many computers.

To try to understand more about the physical universe, the world in which we live, is a basic goal of the student and teacher in a physics classroom.

Studies in the fields of material composition and material phenomena, as well as in the physical sciences, are required to successfully employ a complex tool such as the electron-beam column shown here, which can be used in the manufacture of silicon semiconductor devices.

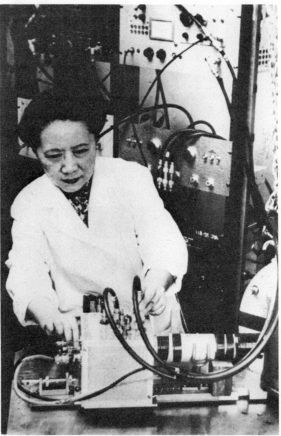

Chien Shiung Wu is professor of physics at Columbia University. In 1957 she performed an experiment which stunned the scientific community. Her research disproved what had seemed to be a basic law of physics—the law of conservation of parity. In 1975 she became the first woman president of the American Physical Society.

A scientist works with a Molecular Beam Epitaxy System in experiments to develop materials that can be used in advanced semiconductor components.

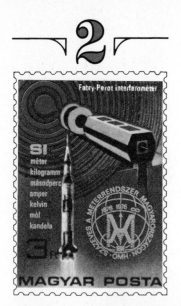

In 1976, Hungary issued this stamp to commemorate the 100th anniversary of the adoption of the metric system in that country. In addition to the seven fundamental SI units, the stamp depicts a space vehicle, a vacuum balance, and an interference pattern formed by a laser beam.

2 MEASUREMENT AND PROBLEM SOLVING

I n this chapter you will gain
an understanding of:

✓ the metric units of
measure
✓ accuracy and precision
in dealing with measurements
✓ mathematical operations
in solving physics problems

Figure 2-1. A highway sign that shows distances in both miles and kilometers.

Units of measure are used to describe physical quantities.

UNITS OF MEASURE

2.1 The Metric System The system of measurement that is best suited to scientific purposes is the *metric system*. Developed in France near the end of the eighteenth century, the metric system uses a decimal basis for multiples and subdivisions of the basic units of measure, which greatly simplifies calculations. In 1866 Congress legalized the metric system for use in the United States, and since 1893 the yard and the pound have been defined in terms of metric measures. In 1975, a law was passed to promote more extensive use of the metric system in the United States, but no date for complete conversion was specified. (See Figure 2-1.)

The modern metric system is officially called the International System of Units (abbreviated SI for *Système International d'Unités*). The International Bureau of Weights and Measures, which oversees the system, is located in Sèvres, France, a suburb of Paris. The property of the Bureau is internationally owned, much like the buildings and grounds of the United Nations in New York.

In a system of measurement, it is important to distinguish between *physical quantities* and *units of measure*. A physical quantity is a measurable aspect of the universe, such as *length*. The metric unit of measure for length is the *meter*.

There are seven *fundamental units* of measure. These seven units, the physical quantities they measure, and the

Table 2-1
THE FUNDAMENTAL UNITS OF MEASURE

Physical quantity	Quantity symbol	Unit of measure	Unit symbol
length	l	meter	m
mass	m	kilogram	kg
time	t	second	s
electric charge	Q	coulomb	c
temperature	T	degree Kelvin	°K
amount of substance	(none)	mole	mol
luminous intensity	I	candle	cd

symbol for each are shown in Table 2-1. (Because this in-formation is so important, it is also shown inside the front cover of the book and in Appendix B, Table 4.) The meter, kilogram, and second are basic to the study of mechanics. The degree Kelvin and the mole are used in the study of heat. The candle is used in measurements of light. The coulomb will be used in the section of the course on elec-tricity. Since the *m*eter, *k*ilogram, and *s*econd are three of the fundamental units, the modern metric system is some-times also called the MKS system.

Physical quantities other than those in Table 2-1 are measured by using combinations of the fundamental units of measure. For example, velocity is measured in meters per second and mass density in kilograms per cubic meter (or grams per cubic centimeter). Units that consist of com-binations of the fundamental units are called *derived units*.

The prefixes used in the metric system, with their corre-sponding symbols and values, are shown in Table 2-2. The most commonly used of these prefixes are kilo(k), centi(c), and milli(m).

The use of units and symbols in this book is based on the latest specifications of the International Bureau of Weights and Measures. But there are some exceptions. For example, Table 2-1 lists the coulomb as the fundamental unit of electricity, whereas SI specifies the ampere as the fundamental unit and the coulomb as a derived unit. SI specifies uppercase letters for several symbols that are not capitalized in this text, such as n for newton. These excep-tions to SI usage are made to avoid confusion between symbols or to simplify the understanding of new concepts.

2.2 *The Meter* The length of the meter was originally calculated to be one ten-millionth of the distance from the north pole to the equator along a line running through France and Spain. Until 1960, the standard for the meter

Table 2-2
METRIC SYSTEM PREFIXES

Prefix	Symbol	Factor
exa	E	10^{18}
peta (*pet*-ah)	P	10^{15}
tera (*tair*-ah)	T	10^{12}
giga (*jig*-ah)	G	10^{9}
mega	M	10^{6}
kilo	k	10^{3}
hecto	h	10^{2}
deka	da	10^{1}
deci	d	10^{-1}
centi	c	10^{-2}
milli	m	10^{-3}
micro	μ	10^{-6}
nano (*nan*-oh)	n	10^{-9}
pico (*pea*-koh)	p	10^{-12}
femto (*fem*-toh)	f	10^{-15}
atto (*at*-oh)	a	10^{-18}

In 1907, Michelson received the Nobel Prize in physics for mea-suring the standard meter bar with an interferometer.

Figure 2-2. A krypton-86 lamp. Electricity causes the krypton in the lamp to glow. The standard meter is based on the wavelength of a specific part of the light emitted in the tube.

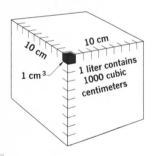

Figure 2-3. Relationship among the metric units of length, area, and volume.

Although SI does not include the liter as a unit, it has authorized its use as a special name for the cubic decimeter and has adopted both l and L as symbols to be used for the unit liter.

A standard is an object or a laboratory process that defines a fundamental unit of measure.

was a metal bar kept in a vault at the International Bureau of Weights and Measures.

Toward the end of the nineteenth century, the American physicist Albert A. Michelson (1852–1931) developed a very precise measuring device called an *interferometer*, which uses light waves to measure length. With an interferometer, light of only one color is used in the measuring process. The light is produced by sending electricity through a gas that is under low pressure and then filtering the resulting glow so that only a single color emerges.

As a result of the work of Michelson and others with interferometers, the standard meter was redefined in terms of light waves. The gas that is used in the modern definition of the meter is a form of the element krypton called krypton-86. Figure 2-2 shows a lamp in which the krypton is kept at low pressure while high-voltage electricity is passed through it. The specific wavelength of the glow of krypton-86 that is used to define the meter has an orange-red color. The definition adopted in 1960 is

1 meter (m) = 1,650,763.73 wavelengths in a vacuum of the orange-red light of krypton-86

The standard meter is accurate to about two parts in one billion, or about one meter of uncertainty in measuring the distance from the earth to the moon. Even greater accuracy can be achieved with the help of laser beams, and as a result, scientists expect that a new definition of the standard meter will be adopted.

The metric units for the measurement of area and volume are derived from the meter, as Figure 2-3 indicates. A rectangular floor 3 meters long and 2 meters wide has an area of 6 square meters. Similarly, a rectangular tank 4 meters long, 3 meters wide, and 2 meters deep has a volume of 24 cubic meters.

The liter (L) *is a special name for a cubic decimeter* (1000 cm^3). Since one milliliter (1 mL) is 1/1000th of a liter, 1 mL is equivalent to one cubic centimeter (1 cm^3).

2.3 The Kilogram The SI standard of mass is the *kilogram*. It is defined as the mass of the standard kilogram kept by the International Bureau of Weights and Measures, as shown in Figure 2-4. Duplicates of the standard kilogram are kept in various countries around the world. These duplicates are occasionally checked against the standard kilogram.

The standard for the kilogram is the only one of the seven standards that still consists of a natural object representing a unit of measure. One reason why the standard

kilogram is still a natural object is because masses can presently be compared with the standard with greater precision than is possible by means of other laboratory processes. All other units of measure are presently determined through reproducible processes in the laboratory, such as using an interferometer in determining the standard meter. See Section 2.2.

It is useful to know that one milliliter of water has a mass of approximately one gram and that a five-cent United States coin has a mass of about five grams. A metric ton contains 1000 kg.

2.4 Force and Weight The SI unit of force is the *newton* (n). The newton is a derived unit that will be more fully described in Chapter 4. But it is important at this point to know the relationship between a newton and a kilogram, namely that a force of about 9.8 newtons is required to lift a mass of one kilogram in most places on the earth. (The exact number of newtons required at a specific location depends on the pull of gravity at that location, but 9.8 is a useful approximation.)

In most countries that use the metric system, the weights of people and objects are commonly described in kilograms. Since weight is a force, it should be described in newtons, and the use of kilograms for weights is not consistent with the rules of the International System of Units. Perhaps a new definition of the kilogram will eventually be adopted to accommodate its widespread usage with weights. But in scientific literature at the present time, weights are correctly expressed in newtons.

2.5 The Second The unit of time is the *second* (s). Originally, the standard second was defined in terms of the earth's motion. This standard has been replaced by one that is based on atomic vibrations, which can be measured with an accuracy of one part in ten trillion.

Figure 2-4. The international standard of mass. Three evacuated bell jars cover the standard kilogram at the bottom center. Six comparison standards are under double bell jars on either side. The standards are cylinders made of platinum-iridium alloy, which does not oxidize and has a mass density almost twice that of lead. The cylinders are 3.9 cm high and 3.9 cm in diameter. (Equal height and diameter gives a cylinder the smallest possible area-to-volume ratio.) The standard meter bar used before 1960 is in a protective case on the shelf above the masses.

The lowercase symbol for the newton used in this text is an exception to the SI rule that symbols derived from proper names should be capitalized.

Figure 2-5. A decimal clock. This watch was made in anticipation of the adoption of decimal days, which never took place.

The standard second is based on the rate of vibration of the cesium atom. The vibration is produced by radio waves, and it is not affected by such external influences as temperature and gravity. The rate of vibration can be tuned precisely to the radio waves that are producing it. The definition adopted in 1967 is

1 second (s) = 9,192,631,770 vibrations of cesium-133 atoms

When the metric system was first proposed, an effort was made to redefine other time intervals as well. One proposal called for a month of three ten-day weeks. In this system, the day was divided into ten hours, the hour into 100 minutes, and each minute into 100 seconds. A decimal watch of this kind is shown in Figure 2-5. However, decimal days were never officially adopted in any country. Thus, the numbers on our clocks and calendars remain conspicuous exceptions to the metric system.

Questions

GROUP A

1. What makes the metric system convenient to use?
2. What is the relationship between a physical quantity and a unit of measure?
3. List the seven fundamental units of measure.
4. List three derived units of measure and their related physical quantities.
5. What is meant by SI?
6. Describe the standard of length.
7. How are the cubic centimeter and the liter related?
8. Why is the standard kilogram an object instead of a laboratory process?

9. Distinguish between the mass and weight of an object.
10. How is the standard for the second obtained?

GROUP B

11. List as many examples as you can of the use of the metric system in the United States.
12. List two advantages and two disadvantages of converting completely to the metric system in the United States.
13. How are the units of length and mass related?
14. What is the relationship between the second and the meter?
15. Why do you suppose decimal clocks and calendars were never adopted?

Problems

GROUP A

1. Convert each of the following quantities to the required units.
   (a) 32 cm = ___ mm
   (b) 756 mm = ___ m
   (c) 3.6 km = ___ cm
   (d) 10 cm^2 = ___ mm^2
   (e) 1000 cm^3 = ___ m^3
   (f) 15 L = ___ mL
   (g) 250 mL = ___ cm^3
   (h) 45 dL = ___ L
   (i) 8000 mg = ___ g
   (j) 67 kg = ___ mg
   (k) 0.024 g = ___ μg
   (l) 30 Mg = ___ kg

(m) 10 s = _____ ns
(n) 2 hr = _____ s
(o) 1 min = _____ ms

2. A table top measures 65 cm by 92 cm. What is the area of the table top in m²?

3. A container is 26 cm long, 14 cm wide, and 5 cm deep. Find the volume of the container in mL.

4. How much force is needed to lift each of the following masses at sea level? (a) 100 g, (b) 65 kg, (c) 2 metric tons.

GROUP B

5. A cubic container holds 1.00 kg of water. What is the length of one side of the container in m?

6. The mass of 1.00 cm³ of seawater is 1.04 g. What is the mass of 1.00 L of seawater?

7. A cable has a mass of 4.52 g/cm. What is the mass of 15 km of the cable?

8. A 500-mL flask is filled with water. How much does the water weigh?

MAKING AND RECORDING MEASUREMENTS

2.6 Accuracy The measurement of a physical quantity is always subject to some degree of uncertainty. There are several reasons for this: the limitations inherent in the construction of the measuring instrument or device, the conditions under which the measurement is made, and the different ways in which the person uses or reads the instrument, as you can see in Figure 2-6. Consequently, in reporting the measurements made during a scientific experiment, it is necessary to indicate the degree of uncertainty so far as it is known.

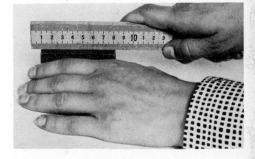

Figure 2-6. The accuracy of measurement depends upon the instrument used and the care with which the reading is made.

One way to express the uncertainty of a measurement is in terms of *accuracy*. **Accuracy** *refers to the closeness of a measurement to the accepted value for a specific physical quantity.* It is expressed as either an *absolute* or a *relative error*. *Absolute error* is the actual difference between the measured value and the accepted value. The equation for absolute error is

$$E_a = |O - A|$$

where E_a is the absolute error, O is the observed (measured) value, and A is the accepted value. (The vertical lines mean the absolute value is used, regardless of sign.)

Relative error is expressed as a percentage and is often called the percentage error. It is calculated as follows:

$$E_r = \frac{E_a}{A} \times 100\%$$

Accuracy is expressed as absolute error or relative error.

where E_r is the relative error, E_a the absolute error, and A the accepted value.

2.7 Precision In common usage accuracy and *precision* are often used synonymously. But in science it is important to make a distinction between them. You should learn

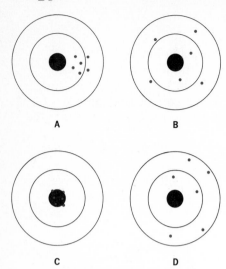

A   B

C   D

Figure 2-7. Distinction between accuracy and precision. Six shots were fired at each target. In A, the precision is good, but the accuracy is poor. In B, the average accuracy is good, but the precision is poor. In C, both the accuracy and precision are good. In D, both the average accuracy and the precision are poor.

to use the two terms correctly and consistently. ***Precision*** *is the agreement among several measurements that have been made in the same way.* It tells how reproducible the measurements are and is expressed in terms of *deviation*. Figure 2-7 should help you to distinguish between accuracy and precision as used in science. As in the case of accuracy, deviations in precision can be given as absolute or relative (percentage). *Absolute deviation* is the difference between a single measured value and the average of several measurements made in the same way, or

$$D_a = |O - M|$$

where D_a is the absolute deviation, O is the observed or measured value, and M is the mean, or average, of several readings.

Relative deviation is the percentage average deviation of a set of measurements and is calculated as follows:

$$D_r = \frac{D_a \text{ (average)}}{M} \times 100\%$$

where D_r is the relative deviation, D_a (average) is the average absolute deviation of a set of measurements, and M is the mean, or average, of the set of readings.

The precision of your laboratory measurements will be governed by the instruments at your disposal. In a measuring instrument the degree of precision obtainable is called the *tolerance* of the device. Any figure listed for the tolerance of an instrument indicates the limitations of the instrument. The instrument maker assumes that the instrument is used properly and that human errors are held to a minimum. The usual platform balance, for example, has a tolerance of 0.1 gram. Some modern research instruments, such as the precision balance shown in Figure 2-8, have tolerances of 0.00001 g or better!

2.8 Significant Figures Another way of indicating the precision of a measurement is by means of *significant figures*. ***Significant figures*** *are those digits in a number that are known with certainty plus the first digit that is uncertain.* For example, if the length of a laboratory table was measured with a meter stick graduated in millimeters and found to be 1623.5 millimeters, the measurement has five significant figures. The 5 is the uncertain digit. It was an estimation of the value between two millimeter marks on the meter stick.

In working with significant figures, you can usually assume to know the last figure of a number, the uncertain figure, to within 1. For instance, when the value for π is given as 3.14 you can assume that the true value lies between 3.13 and 3.15.

The following rules will help you to understand how significant figures are used in physics problems. Even these rules do not cover all cases. However, common sense usually shows how a specific case should be handled.

1. All nonzero figures are significant: 112.6 °C has four significant figures.

2. All zeros between nonzero figures are significant: 108.005 m has six significant figures.

3. Zeros to the right of a nonzero figure, but to the left of an understood decimal point, are not significant unless specifically indicated to be significant. In this book the rightmost such zero which is significant is indicated by a *bar* placed above it: 109,000 km contains *three* significant figures; 109,0$\overline{0}$0 km contains *five* significant figures.

4. All zeros to the right of a decimal point but to the left of a nonzero figure are not significant: 0.000647 kg has three significant figures. The single zero placed to the left of the decimal point in such an expression serves to call attention to the decimal point and is never significant.

5. All zeros to the right of a decimal point and following a nonzero figure are significant: both 0.07080 cm and 20.00 cm have four significant figures.

Further examples of these rules are given in Table 2-3.

6. Rule for addition and subtraction. Remember that the rightmost significant figure in a measurement is uncertain. The rightmost significant figure in a sum or difference will be determined by the leftmost place at which an uncertain figure occurs in any of the measurements being added or subtracted. The following example of addition illustrates this rule.

13.05 cm ⟶ leftmost place of uncertain
309.2   cm ⟶ figure in measurements being added
3.785 cm
326.035 cm ⟶ rightmost significant figure in the answer

Since the answer should have only one uncertain figure (the rightmost one), it should be recorded as 326.0 cm.

7. Rule for multiplication and division. Remember that when an uncertain figure is multiplied or divided by a number, the answer is likewise uncertain. Therefore the

Figure 2-8. A precision balance. With this instrument, differences of 10^{-8} kilogram can be measured.

Table 2-3
SIGNIFICANT FIGURES

Two significant figures	Three significant figures	Four significant figures
2300	5420	152$\underline{1}$
4$\overline{0}$00	60$\overline{0}$0	504$\overline{0}$
5.2	73.0	4.050
0.16	0.915	0.3780
0.0078	0.0467	0.01520

The rules for significant figures should be observed throughout the course.

product or quotient should not have more significant figures than the least precise factor. The following example illustrates the rule for multiplication.

3.54 cm × **4.8 cm** × **0.5421 cm** = **9.2113632 cm³**

least precise factor   **rightmost significant figure in the answer**

Consequently, the answer should be recorded as 9.2 cm³.

8. Rule for rounding. If the first figure to be dropped in rounding off is 4 or less, the preceding figure is not changed; if it is 6 or more, the preceding figure is raised by 1. If the figures to be dropped in rounding off are a 5 followed by figures other than zeros, the preceding figure is raised by 1. If the figure to be dropped in rounding off is a 5 followed by zeros (or if the figure is exactly 5), the preceding figure is not changed if it is even; but if it is odd, it is raised by 1.

EXAMPLE   Divide 5674 cm² by 85 cm.

SOLUTION   When this division is performed on an eight-place calculator, the numerical quotient is 66.752941. But since one of the factors in the problem, 85 cm, has only two significant figures, the correct answer is 67 cm.

Many students find it hard to believe that an answer with two figures can be more acceptable than an answer with eight figures. This difficulty may be the result of previous experience with pure-number problems. Pure numbers are not associated with units of measurement; hence the last figure of a pure number is not uncertain. In the example shown above, however, and in all problems dealing with measurements, the last figure of each number is assumed to be uncertain and the rules of significant figures apply to the solution.

PRACTICE   1. Add 132 g, 0.14 g, 6.2 g, and 24 g.   *Ans.* 162 g
PROBLEMS   2. Subtract 1.64 s from 19.56 s.   *Ans.* 17.92 s
3. Multiply 183 cm by 240 cm.   *Ans.* 44000 cm²
4. Divide 252 g by 42.19 cm³.   *Ans.* 5.97 g/cm³

2.9 Scientific Notation   Scientists frequently deal with very large and very small numbers. The velocity of light in air is about 300,000,000 meters per second. The mass of the earth is about 6,000,000,000,000,000,000,000,000 kilograms. The rest mass of an electron is 0.000,000,000,000,-000,000,000,000,000,000,910,953,4 kilogram. These numbers are inconvenient to write and difficult to read. Through the use of the mathematical rules of exponents, however, it is possible to express these numbers in a much simpler way. *Using powers of ten in writing a number is called* **scientific notation.** Such numbers have the form

$$M \times 10^n$$

where **M** is a number having a single nonzero figure to the left of the decimal point and **n** is a positive or negative exponent. To write a number in scientific notation:

1. Determine **M** by shifting the decimal point in the original number to the left or to the right until only one nonzero digit is to the left of it. **M** should contain only the significant figures of the original number.

2. Determine **n** by counting the number of places the decimal point has been shifted; if it has been moved to the left, **n** is positive; if to the right, **n** is negative.

For example, we see that the velocity of light can be written as 3×10^8 m/s, the mass of the earth as 6×10^{24} kg, and the mass of the electron as 9.109534×10^{-31} kg.

It is often helpful to know the magnitude of a number in terms of its powers of ten. In this form it can be compared quickly with other numbers. A numerical approximation to the nearest power of ten of a number is called its *order of magnitude* (Figure 2-9). The order of magnitude of the velocity of light in m/s is 10^8. The order of magnitude of the mass of the earth in kg is 10^{25}, since its mass is nearer to 10^{25} kg than it is to 10^{24} kg. Similarly, the order of magnitude of the mass of the electron in kg is 10^{-30}. Orders of magnitude of length of a number of objects in the universe are shown in Figure 2-10 on the next page.

Figure 2-9. The great spiral galaxy in the constellation Andromeda. This huge formation of stars, which is similar in size and shape to our own Milky Way galaxy, is estimated to be more than 10^{18} km in diameter and 2×10^{20} km from the earth. The mass of the Andromeda galaxy has been calculated to be 6×10^{35} kg, which is 3×10^{11} times the mass of the sun.

Scientific notation is sometimes called exponential notation. Can you explain why?

SOLVING PROBLEMS

2.10 Data, Equations, and Graphs   Scientists usually use the term "data" to refer to the numbers, units, and other information recorded when measurements are made. Much of your homework in physics will involve solving problems in which the necessary data are already provided for you. The problems at the ends of the chapters in this book are like that. Your objective is to use the data properly in order to find the quantity required.

The first step in solving a problem of this kind is to list carefully all the data provided. Be sure to include the correct symbol and unit for each physical quantity. If the symbol is not given in the problem, look it up inside the front cover of this book or in Appendix B, Table 4. Record these symbols correctly at the start. As you continue working the problem, be sure you use only these symbols. Also

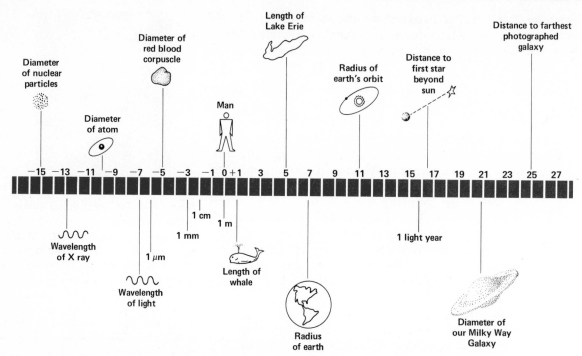

Figure 2-10. Orders of magnitude of length in meters of various objects in the universe. The numbers above the scale are powers of ten.

The Mathematics Refresher is a handy reference. Consult it frequently.

check to make sure that you have copied the right number of significant figures in each case.

In the laboratory most of the data needed must be obtained by direct measurement. In some experiments you will be told what to measure, what instruments to use, and even how many significant figures to use. But in other experiments you must make these decisions for yourself.

In recording and interpreting data in physics, you will use mathematical skills that you have acquired in previous courses. The Mathematics Refresher, following Chapter 25, contains a summary of information and skills from arithmetic, elementary algebra, plane geometry, and trigonometry that is useful in the study of physics. At this time you should review the information contained there and try some of the examples. Study carefully any procedure with which you are not familiar.

The purpose of a scientific experiment is to discover relationships that may exist among the quantities measured. The experimenter controls the conditions of the experiment in order to obtain the most accurate measurements. Uncontrolled conditions might make the measurements so confusing that relationships would not be discovered. As an example, suppose you wished to discover any relationships that might exist between the mass of a homogeneous substance (the same throughout) and its volume. You measure the volumes of several samples of different substances and determine the mass of each. Since temperature affects the volume of most substances, you control this condition by taking all of your measurements at the same temperature, such as 20 °C. Your data for three substances, with the volume measured in cm³ at 20 °C and the mass measured in grams, might look like Table 2-4.

Now you look for mathematical relationships among these numbers. As the volume of each substance increases, the mass also increases. But this statement is very general. You are looking for a specific equation. Sometimes, plotting the data on a graph reveals the equation that expresses the relationship among the measurements.

Table 2-4
VOLUME-MASS RELATIONSHIPS AT 20 °C

Volume (cm³)	Mass			Volume (cm³)	Mass		
	aluminum (g)	water (g)	maple wood (g)		aluminum (g)	water (g)	maple wood (g)
$1\overline{0}$	27	$1\overline{0}$	6	$6\overline{0}$	162	$6\overline{0}$	36
$2\overline{0}$	54	$2\overline{0}$	12	$7\overline{0}$	189	$7\overline{0}$	42
$3\overline{0}$	81	$3\overline{0}$	18	$8\overline{0}$	216	$8\overline{0}$	48
$4\overline{0}$	108	$4\overline{0}$	24	$9\overline{0}$	243	$9\overline{0}$	54
$5\overline{0}$	135	$5\overline{0}$	30	$10\overline{0}$	270	$10\overline{0}$	60

As you see, Figure 2-11 is a graph of the data in Table 2-4. The zero point is the origin of the graph. The horizontal line through the origin is called the X axis. On it are measured distances from the origin, or abscissas, equivalent to the various volumes. For convenience, use one small interval to represent $1\overline{0}$ cm³. The vertical line through the origin is called the Y axis. On it are measured distances from the origin, or ordinates, equivalent to the various masses. Use one small interval to represent $1\overline{0}$ g.

When the data for aluminum, water, and maple wood are plotted on the graph, the data for each substance can be connected by a straight line. The mathematical relationship between the mass and the volume is called a *direct proportion*. The symbol that means "is proportional to" is ∝. Hence the relationship between mass and volume is expressed mathematically as

mass *(m)* ∝ volume *(V)*

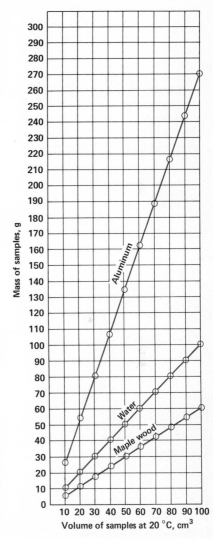

Figure 2-11. Graph of three direct proportions. Straight lines are typical of these relationships.

Table 2-5
SPEED-TIME RELATIONSHIPS

Speedometer reading (km/hr)	Time for 100-km trip (hr)
20.0	5.00
30.0	3.33
40.0	2.50
50.0	2.00
60.0	1.67
70.0	1.43
80.0	1.25
90.0	1.11
100.0	1.00

Plotting data that are **directly proportional** always results in the graph of a straight line.

To illustrate another kind of graph, let us plot the data given in Table 2-5, which shows the time required to make a trip of $10\overline{0}$ kilometers at various speeds. A line drawn through the plotted points, as shown in Figure 2-12, is a *hyperbola.* Furthermore, the product of the speed and time is constant. *Two quantities whose product is a constant are in* **inverse proportion.**

A proportion can be changed into an equation by introducing a constant. In the volume-mass relationship the constant for each substance can be calculated from the data in Table 2-4. (For example, if each measurement in the volume column is multiplied by 2.7, the measured mass of aluminum results.) The equation for this relationship is $m = DV$ where D is a constant. Solving the equation for D,

$$D = \frac{m}{V}$$

This is the equation for mass density that was given in Section 1.9. With this equation you can find any one of the three physical quantities (D, m, or V) provided the measurements for the other two are given. The mass densities of most substances range from 0.1 g/cm^3 to 10 g/cm^3. Mass densities are usually given in g/cm^3 instead of kg/m^3 to avoid cumbersome numbers. For example, 1 g/cm^3 becomes

$$\left(\frac{1 \text{ g}}{1 \text{ cm}^3}\right)\left(\frac{1 \text{ kg}}{10^3 \text{ g}}\right)\left(\frac{10^6 \text{ cm}^3}{1 \text{ m}^3}\right) = \frac{10^3 \text{ kg}}{\text{m}^3} = 1000 \text{ kg/m}^3$$

A carefully drawn graph provides you with another helpful tool. You can use it to estimate values between the plotted points (interpolation) and beyond the measurements made in the experiment (extrapolation).

Estimation of answers is not limited to interpolation and extrapolation. When you solve a problem in physics, you should make an intelligent guess about the reasonableness of your answer. Such a "check for reasonableness" will frequently disclose major errors in arithmetic or algebra.

2.11 Vectors Physical quantities such as length, area, volume, mass, density, and time can be expressed in terms of magnitude alone as single numbers with suitable units. For instance, the length of a table can be completely described as 1.5 m. The mass of a steel block is completely described as 54 kg. *Quantities, such as these, that can be*

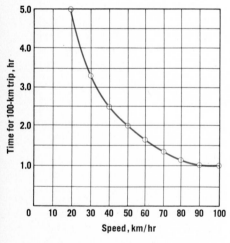

Figure 2-12. Graph of an inverse proportion. The curve is a hyperbola.

expressed completely by single numbers with appropriate units are called **scalar quantities,** *or simply,* **scalars.** ("Scalar" is derived from the Latin word *scala*, which means "ladder" or "steps" and implies magnitude.) Additional examples of scalars in physics are energy and temperature.

Other physical quantities, such as force, velocity, acceleration, electric field strength, and magnetic induction, cannot be completely described in terms of magnitude alone. In addition to magnitude these quantities always have a specific direction. *Quantities that require magnitude and direction for their complete description, and whose behavior can be described by certain mathematical rules, are called* **vector quantities.** *They can be represented by* **vectors.** ("Vector" is derived from a Latin word meaning "carrier," which implies displacement.) The ability to work with vectors is an important mathematical skill that is necessary in solving problems that deal with vector quantities.

A vector is usually shown as an arrow. The length of the arrow represents the magnitude of the vector, while the orientation of the arrow shows direction. North is usually toward the top of the page, east is toward the right, south is toward the bottom, and west is toward the left.

When two or more vectors act at the same point, it is possible to find a single vector that produces the same effect as the combination of separate vectors. In such a calculation, each separate vector is called a *component* and the single vector that produces the same result as the combined components is called the *resultant*. The following examples show basic rules in solving vector problems.

In Figure 2-13, a vector of 8.0 units and a vector of 3.0 units are shown, both directed northward. (These vectors might represent velocities, or forces, or any other vector quantities. But in this introductory treatment of vectors, we shall simply designate them as "units.") When two or more vectors act at the same point and in the same direction, the magnitude of the resultant is the algebraic sum of the magnitudes of the components. In Figure 2-13, the resultant has a magnitude of 11.0 units. (The vectors in the figure should be considered as acting at the same point, even though they are shown side by side for the sake of clarity.) The resultant has the same direction as the components, as shown in the figure.

A different situation is shown in Figure 2-14. This time the 8.0-unit vector acts northward, as before, but the 3.0-unit vector acts southward. When these two vectors act at the same point, the magnitude of the resultant is the algebraic difference of the component magnitudes, or 5.0 units. Another way of saying this is that the magnitude of

Figure 2-13. The magnitude of the resultant of two component vectors acting in the same direction is the sum of the magnitudes of the components. The resultant has the same direction as the components.

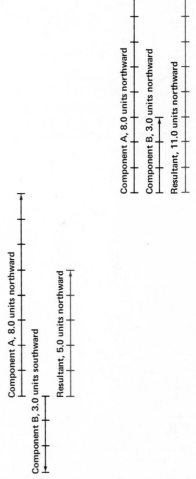

Figure 2-14. The magnitude of the resultant of two component vectors acting in opposite directions is the difference between the magnitudes of the components. The resultant has the same direction as the larger component.

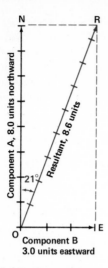

Figure 2-15. The resultant of two component vectors acting at right angles to each other is the diagonal of the parallelogram of which the components are the sides. The magnitude and direction of the resultant can be found either graphically or trigonometrically.

the resultant is the algebraic sum of the component magnitudes, with the 3.0-unit vector considered as negative, since it is acting in the opposite direction. The resultant has the direction of the larger component, or northward.

A third case is shown in Figure 2-15. The 8.0-unit vector is again acting northward, but the 3.0-unit vector is now directed eastward. In this case, the components do not act along a straight line, and the resultant cannot be found by simple addition or subtraction. Instead, the *parallelogram method* is used. (Experiments confirm that the parallelogram method accurately represents the behavior of vectors that do not act along a straight line.) To find the resultant, we construct the parallelogram **ONRE.** The diagonal of this parallelogram, **OR,** is the resultant.

The magnitude and direction of **OR** can be found in either of two ways. If the vector diagram is drawn accurately to scale, **OR** can be measured directly with a ruler and a protractor. This is called the *graphic solution*. Of course, in using a ruler it is important to use the same scale in measuring each vector. The ruler and protractor should be read to the limits of their precision and the rules of significant figures should be followed in rounding the answer.

A second way of finding **OR** is to use trigonometry. To understand this method, you should review the Mathematics Refresher, 12. Triangles (in the back of the book). The use of the Pythagorean theorem is not recommended for finding the resultant, since it gives only the magnitude and not the direction. The direction of a vector can be calculated only by the use of trigonometric ratios. This method of finding resultants is called the *trigonometric solution*. Be sure to follow the rules of significant figures.

By either the graphic or trigonometric method, **OR** in Figure 2-15 has a magnitude of 8.6 units and a direction of 21° east of north. The direction could also be designated as 69° north of east, but in the answers to the vector problems in this book, the smaller of the two complementary angles will always be given.

The parallelogram method can also be used when vectors act at angles other than right angles, as in Figure 2-16. The graphic solution is similar to the one in the previous example, but the trigonometric solution is lengthier, since it involves the law of cosines and law of sines. Examples and solutions of this kind are given in Chapter 3.

Vectors may also be combined by placing them head to tail, as shown in Figure 2-17. This method is especially helpful in finding the resultant of three or more vectors that act at the same point. If the parallelogram method

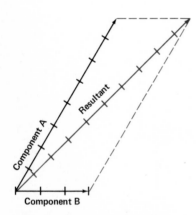

Figure 2-16. The parallelogram method can be used to find the resultant of two component vectors even though the components do not act perpendicularly to each other.

were used in such a case, it would first be necessary to find the resultant of any two of the vectors, then use this resultant with a third vector to find a second resultant, and continue in this way until all the component vectors have been applied. The final resultant would then be the required answer. The procedure shown in Figure 2-17 is really a shortcut for the parallelogram method. In using this method, it does not matter in what sequence the arrows are drawn, as long as the direction of each arrow is maintained when it is transposed. After all the components have been transposed, the resultant is the arrow that is made by starting at the tail of the first component and drawing a straight line to the head of the last component.

2.12 Rules of Problem Solving The problems in this text range from those that are quite simple and involve only a single computation to those that require several derivations and a number of separate mathematical steps. In general, problems of the first type will be found in Group A at the ends of the chapters. Examples of this type are given within the chapters themselves. Problems of greater difficulty involving several steps are included in Group B, and examples of these are also in the chapters.

The method used to solve physics problems consists of a number of logical steps. These steps are described below and they are illustrated in examples throughout the book.

1. Read the problem carefully and make sure that you know what is being asked and that you understand all the terms and symbols that are used in the problem. Write down all given data.

2. Write down the symbols for the physical quantity or quantities called for in the problem, together with the appropriate unit symbols.

3. Write down the equation relating the known and unknown quantities of the problem. This is called the *basic equation*. In this step you will have to draw upon your understanding of the physical principles involved in the problem. It is helpful to draw a sketch of the problem and to label it with the given data.

4. Solve the basic equation for the unknown quantity in the problem, expressing this quantity in terms of those given in the problem. This is called the *working equation*.

5. Substitute the given data into the working equation. Be sure to use the proper units and carefully check the significant figures in this step.

6. Perform the indicated mathematical operations with the units alone to make sure that the answer will be in the

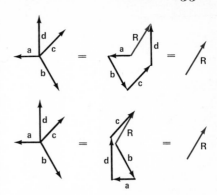

Figure 2-17. The resultant of component vectors acting at the same point can also be found by placing the components head to tail. The resultant is the vector drawn from the tail of the first component to the head of the last one. The sequence in which the components are transposed does not matter, but their directions must be maintained.

In this book the symbols for quantities are always italicized. The symbols for units are not.

units called for in the problem. This process is called *dimensional analysis.*

7. Perform the indicated numerical operations. Be sure to observe the rules of significant figures.

8. Check the answer for reasonableness. For example, if your answer shows a car moving at a speed of 1000 km/hr, you made a mistake in one of the previous steps.

9. Review the entire solution.

Some problems, especially those that deal with vectors, do not involve all of these steps and may require other skills instead. In every case, it is important to understand the basic principles of the problem. This is much more meaningful than merely "getting the right answer."

EXAMPLE   A rectangular metal block is 2.63 cm long, 2.15 cm wide, and 1.52 cm thick. The mass of the block is 56.0 g. What is the mass density of the metal?

SOLUTION   Given the length, width, thickness, and mass of a rectangular metal block, we are asked to calculate the mass density of the metal. Since this problem involves mass density, mass, and volume, the equation for mass density is the basic equation.

$$\text{Basic equation: } D = \frac{m}{V}$$

Since the volume is not given directly, however, we must use a working equation in which the given data for length, width, and thickness are shown.

$$\text{Working equation: } D = \frac{m}{lwh}$$

Now we are ready to substitute the given data into the working equation.

$$D = \frac{56.0 \text{ g}}{(2.63 \text{ cm})(2.15 \text{ cm})(1.52 \text{ cm})}$$

$$\text{Dimensional analysis: } D = \frac{\text{g}}{(\text{cm})(\text{cm})(\text{cm})} = \frac{\text{g}}{\text{cm}^3}$$

$$\text{Numerical operations: } \frac{56.0}{(2.63)(2.15)(1.52)} = 6.52 \text{ (rounded to three significant figures)}$$

Finally, we combine the numerical answer with the units obtained in the dimensional analysis, or $D = 6.52 \text{ g/cm}^3$. This answer is reasonable for the mass density of a metal, according to Section 2.10.

PRACTICE   **1.** A metal cube has a mass of 452 g. The mass density of
PROBLEMS   the metal is 11.3 g/cm³. Find the length of one side of the
metal cube.   *Ans.* 3.42 cm

2. A liquid has a volume of 2.47 L and a mass of 3.78 kg. Determine the
mass density of the liquid.   *Ans.* 1.53 g/cm³

Questions

GROUP A

1. Distinguish between the accuracy and precision of a measurement.
2. How does absolute error differ from relative error?
3. List four ways of expressing the precision of a measurement.
4. State the rule for determining the number of significant figures in the sum of added measures.
5. List three different cases in which a zero is not a significant figure.
6. How many uncertain figures should a correctly written measurement contain?
7. What is the advantage of using scientific notation?
8. (a) What is the shape of the graph of two quantities that are in direct proportion? (b) in inverse proportion?
9. Distinguish between a basic equation and a working equation.
10. Define "dimensional analysis."
11. What is meant by a "check for reasonableness" in solving a physics problem?
12. Distinguish between a scalar quantity and a vector quantity.
13. How is the resultant of two vectors determined when they are not acting along a straight line?
14. Describe two ways in which the resultant of three or more vectors can be found.

Problems

GROUP A

1. How many significant figures are there in each of the following numbers? (a) 173.2; (b) 205; (c) 4000; (d) 0.025; (e) 700; (f) 0.09050.
2. Write each of the numbers in Problem 1 in scientific notation form.
3. Write each of the following numbers in scientific notation form:
   (a) 175,000,000
   (b) 602,300,000,000,000,000,000,000
   (c) 0.000047
   (d) 0.000,000,000,000,000,000,000,001,672,39
4. What is the order of magnitude of each of the numbers in Problem 3?
5. A substance has a mass density of 0.20 g/cm³. What is the volume in cm³ of a 100-g sample of the substance?
6. A sample of maple wood has a mass of 40 g. Read the volume of the sample from Figure 2-11.
7. The time required for a 100-km trip is 1.5 hours. Read the average speed from Figure 2-12.
8. Give the order of magnitude of the answer for each of the following calculations:
   (a) $(6.1 \times 10^8)(3.2 \times 10^{-2})$
   (b) $\dfrac{5.92 \times 10^{-2}}{5.37 \times 10^8}$
   (c) $(65)(32 \times 10^{-5})$
   (d) $\dfrac{(2.35)(0.45)(546)}{(379)(4.3 \times 10^4)}$
9. A vector with a magnitude of 17 units is directed westward. A second vector

acts at the same point with a magnitude of 9.4 units and an angle of 35° south of west. Find the resultant (a) graphically and (b) trigonometrically.

10. A vector has a magnitude of 23 units and is directed 25° east of south. Find the eastward and southward components of this vector (a) graphically and (b) trigonometrically.

GROUP B

11. In a laboratory experiment a student obtained the following values for the mass density of a metal sample: 5.13 g/cm³, 4.98 g/cm³, 5.06 g/cm³, and 5.01 g/cm³. The accepted value for the mass density is 5.05 g/cm³. (a) List the absolute error of each measurement. (b) Determine the relative error of each measurement. (c) What is the absolute deviation of each measurement? (d) What is the relative deviation of the experiment?

12. If the tension of a vibrating string is kept constant, the length of the string is related to the frequency of the tone it emits. Plot a graph of the following data for a typical vibrating string, using lengths as abscissas and frequencies as ordinates.

Length (m)	Frequency (s⁻¹)	Length (m)	Frequency (s⁻¹)
0.075	1024	0.60	128
0.15	512	0.80	96
0.20	384	1.20	64
0.30	256	2.40	32
0.40	192		

(a) What is the shape of the graph? (b) What relationship exists between the length of the vibrating string and the frequency of the tone it emits? (c) From the graph interpolate the frequency the string would have if it were 0.75 m long. (d) How long must the string be to vibrate at the rate of $40\bar{0}$ s⁻¹?

13. A sample of liquid has a volume of 6.24 L and a mass of 9.57 kg. Find the mass density of the liquid.

14. A rock weighs 9.18 n. Its mass density is 3.47 g/cm³. How many mL does the rock displace when it is submerged in a container of water?

15. A cube of iron measures 5.25 cm on a side and weighs 11.2 n. Calculate the mass density of the iron cube.

16. A liquid with a mass of 795 g fills a container that is 21.0 cm long, 9.50 cm wide, and 5.00 cm high. Calculate the mass density of the liquid.

17. The equation for the period of vibration of a pendulum is $T = 2\pi\sqrt{\dfrac{l}{g}}$ where T is the period in s and l is the length of the pendulum in m. Find the value of g at a location where a pendulum 1.00 m long has a period of 2.01 s.

SUMMARY

The metric system of measurement is a decimal system well suited to scientific purposes. The United States is in the process of converting to the metric system.

Physical quantities are aspects of the universe. They are measured in units. All units of measure are derived from seven

fundamental units that are used to measure length, mass, time, electric charge, temperature, amount of substance, and luminous intensity. Many other units are derived from the fundamental units. The names and symbols for physical quantities and units have been standardized in the International System of Units (abbreviated SI).

The SI standard unit of length is the meter. It is based on a specific wavelength of the light emitted by krypton-86 atoms. The standard unit of mass is the kilogram, which is based on the mass of a cylinder kept in France. The unit of force is the newton, which is a derived unit. The unit of time is the second, which is based on the vibrations of cesium-133 atoms.

All measurements are uncertain to some degree. The closeness of a measurement to an accepted value is called accuracy. It is expressed as absolute error or relative (percentage) error. Precision is the agreement among measurements that have been made in the same way. Precision is expressed as absolute deviation, relative deviation, or tolerance. It may also be shown by significant figures, which are all the figures of a measurement that are known with certainty plus the first figure that is uncertain. Specific rules govern operations with significant figures.

Scientific notation is a convenient way of writing very large and very small numbers. Such numbers have the form $M \times 10^n$, in which M is a number with one nonzero figure to the left of the decimal point and n is a positive or negative exponent. The order of magnitude of a number is the approximation of the number to the nearest power of ten.

Data is the term used for the numbers, units, and other information obtained in experiments. Tables, graphs, and equations are used to express the relationships among data. A straight-line graph illustrates a direct proportion. A hyperbola is indicative of an inverse proportion, in which the product of two quantities is a constant.

A scalar is a quantity that needs only a number with appropriate units to describe it. A quantity that requires magnitude and direction for its description is a vector quantity. It is usually represented by an arrow in which the length represents magnitude and the orientation of the arrow specifies direction. Vectors in the same or opposite direction can be combined by simple addition or subtraction. The resultant of vectors at other angles is found by the parallelogram or the head-to-tail method.

The solving of a physics problem involves a number of logical steps. Among these are the use of a basic equation, which relates the known and unknown quantities of the problem, and a working equation, which expresses the unknown quantity in terms of the known ones. Another important step is dimensional analysis, in which the mathematical operations are performed with only the units.

VOCABULARY

absolute deviation	kilogram	relative deviation
absolute error	length	relative error
accuracy	liter	scalar quantity
basic equation	meter	scientific notation
component	metric system	second
derived unit	newton	significant figure
dimensional analysis	order of magnitude	standard
direct proportion	parallelogram method	vector quantity
fundamental unit	physical quantity	working equation
inverse proportion	precision	

3 VELOCITY AND ACCELERATION

Galileo's experiments with accelerating bodies provided the basis for a scientific understanding of motion. This stamp commemorates the 300th anniversary of the death of the famous Italian scientist. He is shown holding a telescope that he designed and used for astronomical observations.

In this chapter you will gain an understanding of:

✔ the nature of velocity
✔ the relationship between velocity and acceleration
✔ the effect of gravity on falling bodies
✔ the concept of inertia
✔ the relationship between force and acceleration
✔ the reaction between forces
✔ the force of gravitation between objects

VELOCITY

3.1 The Nature of Motion If you see a car in front of your house and later see it farther along the street, you are correct in assuming that the car has moved. To reach this conclusion, you observed two positions of the car and you also noted the passage of time. You might not know how the car got from one position to the other. It might have moved at a steady rate or it might have speeded up and then slowed down before it got to its second position. It might also have been moving when you first noticed it and might still have been moving at its second location. The car may have moved from one location to the other in a straight line, or it may not have. But none of these possibilities changes the truth of the statement that the car has moved.

Usually, when we see something move, we do not just make two observations. We observe the moving object continuously. However, even a case of continuous observation can be thought of as a series of observations in which the interval of time between two successive observations is small.

Figure 3-1 shows several important things about the nature of motion. First, it shows the changing positions of a moving object. Such *a change of position in a particular direction is called a **displacement.*** A displacement has magnitude, direction, and a point of origin. A displacement is

Figure 3-1. A study of motion. The disk was photographed every 0.1 s as it moved from left to right along the meterstick. Note that the displacement of the disk is equal in equal time periods.

Displacement is a change of position.

Motion is relative displacement.

Speed is the time rate of motion.

always a straight line segment from one point to another point, even though the actual path of the moving object between the two points is curved. Furthermore, displacements are vector quantities and can be combined like other vector quantities.

For example, the statement that an airplane flew 500 kilometers south from Chicago describes the airplane's displacement, since the statement specifies the point of origin, magnitude, and direction. Whether the airplane flew in a straight line or not, the airplane's displacement vector is a straight arrow directed south from Chicago and representing a magnitude of 500 kilometers.

Figure 3-1 also shows that relative motion can be detected by comparing the displacement of one object with respect to another. The reference object is usually considered to be stationary. Thus in the case of the moving car, the street and houses are the stationary references. So **motion** *may be defined as the displacement of an object in relation to objects that are considered to be stationary.*

This chapter deals with the simplest type of motion— the motion of an object along a straight line. Motion in a curved line will be discussed in Chapter 5.

3.2 Speed   *The time rate of motion is called* **speed.** For example, if you are walking at a speed of 3.0 m/s, in 1.0 second you travel 3.0 meters. In 2.0 seconds, you travel 6.0 meters, and in 3.0 seconds you travel 9.0 meters. A graph of this motion, and of several other motions with constant speeds, is shown in Figure 3-2. Notice that the lines that represent greater speeds are more steeply inclined. The *slope* of a line indicates how steep the line is. The slope is the ratio of the vertical component of a line to the corresponding horizontal component. For example, the red segment in line **A** has a vertical component of 3.0 units and a horizontal component of 1.0 unit. The slope of the line is $\dfrac{3.0}{1.0} = 3.0$.

This definition of the slope of a line on a graph may be used to represent a physical quantity such as speed. If the

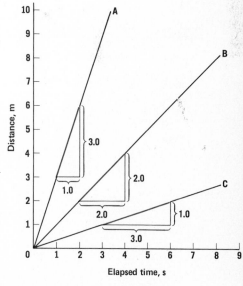

Figure 3-2. Three graphs of motion with different constant speeds. A is for 3 m/s, B is for 1 m/s, and C is for $\frac{1}{3}$ m/s. In each case, speed is represented by the slope of the line.

vertical axis represents distance and the horizontal axis represents elapsed time, then the slope of the line represents speed. For each of the speeds graphed in Figure 3-2 the data for computing the slope are given. A graph such as this and a single computation of the slope will determine the speed over the entire motion only if the speed is constant. *If the determination of speed over any randomly chosen interval gives the same value, the* **speed** *is* **constant.** The graph of constant speed is a straight line.

Average speed is found by dividing the total distance by the elapsed time. For example, if you run 112 m in 20.0 s, you might not run at constant speed. You might speed up and slow down. But as long as you move a total distance of 112 m in 20.0 s, your average speed is $\dfrac{112 \text{ m}}{20.0 \text{ s}} = 5.60$ m/s.

Suppose you ride a bicycle 120 m in $2\overline{0}$ s. Your average speed is $\dfrac{120 \text{ m}}{2\overline{0} \text{ s}} = 6.0$ m/s. A graph of your speed might look like Figure 3-3. You start at a speed of 0.0 m/s and increase your speed. Then from about 3 s to 15 s you travel at constant speed. Then at about 16 s you begin to tire and slow down until at $2\overline{0}$ s you stop. *Instantaneous speed is the slope of the line that is tangent to the curve at a given point.* Without a speedometer it is difficult to determine your instantaneous speed. It is the speed that you would have if your speed did not change from that point on.

Instantaneous speed can be measured approximately by timing the motion over a short distance. In Figure 3-3 instantaneous speed is found approximately by dividing a small amount of time, measured on the graph, into the corresponding distance for that time. The region of the curve marked **A** represents your change in speed as you move faster. An interval of time of 2.3 s is indicated on the horizontal axis. The corresponding distance for this time interval, approximated to the nearest 0.1 m, is 9.0 m. The slope of the red line changes over the interval **A,** but the average is $\dfrac{9.0 \text{ m}}{2.3 \text{ s}} = 3.9$ m/s. This method gives only an approximation of the instantaneous speed because we found the *average speed* for the time interval indicated. This value for the average speed is the same as the instantaneous speed (the slope of the tangent line) at some point in region **A,** but we cannot precisely determine that point.

Since the graph of your motion is a straight line in the region **B,** it represents constant speed. The elapsed time in region **B** is 9.0 s − 7.0 s = 2.0 s. The corresponding distance is 56 m − 41 m = 15 m. The instantaneous speed in

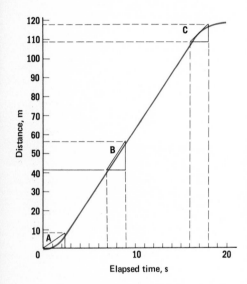

Figure 3-3. A graph of motion with variable speed.

region **B** can be determined precisely because it is the same as the average speed. It is $\dfrac{15 \text{ m}}{2.0 \text{ s}} = 7.5$ m/s.

Region **C** on the graph is curved. The elapsed time is 18.0 s − 16.0 s = 2.0 s. The corresponding distance is 118.0 m − 109.0 m = 9.0 m. The average speed over the interval **C** (as an approximation of the instantaneous speed) is $\dfrac{9.0 \text{ m}}{2.0 \text{ s}} = 4.5$ m/s.

Both instantaneous speed and average speed are completely described in terms of magnitude alone. Hence speed is a scalar quantity.

3.3 Velocity   When both speed and direction are specified for the motion, the term *velocity* is used. In other words, **velocity** is speed in a particular direction. A person walking eastward does not have the same velocity as a person walking northward, even though their speeds are the same. Two persons also have different velocities if they walk in the same direction at different speeds. As stated in Section 2.11, velocities are vector quantities. Magnitude and direction are necessary to describe velocities.

Velocity is speed in a given direction.

For the problems in this chapter, in which the motion is along a straight line, the word "velocity" and the symbol for velocity will be used without stating the direction angle. The values given for velocity will indicate only the magnitude and not the direction. In these cases the velocity may be either a positive or a negative number to indicate in which direction the object is moving. A negative sign as part of velocity, therefore, does not mean that the speed is negative; speed always represents the magnitude of the velocity. Hence, when the direction angle of velocity is not given, the equations for velocity and speed appear the same.

Average velocity is defined as the total displacement divided by the total elapsed time.

$$v_{av} = \frac{d_f - d_i}{t_f - t_i}$$

where v_{av} is the average velocity, d_i is the initial position at time t_i, and d_f is the final position at time t_f. This relationship may also be written

$$v_{av} = \frac{\Delta d}{\Delta t}$$

The Greek letter Δ in an equation means "change of."

where Δd represents the *change* of position (displacement)

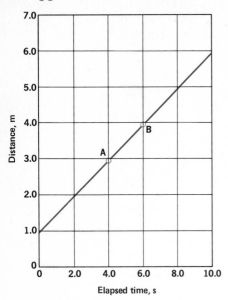

Figure 3-4. A graph of motion with constant velocity.

during the time interval Δt. The Greek letter Δ (delta) designates a change in the quantity of variables d and t.

Graphic representations of the velocity of motion along a straight line are shown in Figure 3-4 and Figure 3-5. These graphs indicate that the direction of motion does not change in either case because the magnitude of displacement is continuously increasing. Furthermore, since the graphs do not show the direction of motion, they have the same form as the speed graphs in Figure 3-2 and Figure 3-3. In Figure 3-4, Δt for the entire range of the graph is $1\overline{0}$ s and Δd is 5.0 m. Consequently, $v_{av} = \dfrac{5.0 \text{ m}}{1\overline{0} \text{ s}}$, or 0.50 m/s. The same value for v_{av} is obtained by choosing any other two points on the graph, such as **A** and **B**. In this case, Δt is 2.0 s and Δd is 1.0 m and v_{av} again equals 0.50 m/s. The velocity, represented by the slope of the straight line, is constant. The motion of the object is *uniform*.

In Figure 3-5, the motion is *variable*. The average velocity for the entire range of the graph $v_{av} = \dfrac{5.0 \text{ m}}{1\overline{0} \text{ s}}$, or 0.50 m/s, as before. However, the average velocity between points **C** and **D** is $\dfrac{2.0 \text{ m}}{2.0 \text{ s}}$, or 1.0 m/s.

Now suppose we wish to find the velocity at a specific point on the graph, such as point **C**. To do this, we draw a tangent to the line at **C** and measure the slope of the tangent. The slope is 0.5 m/s, the velocity at **C**. The velocity at a specific instant in variable motion is called instantaneous velocity to distinguish it from average velocity.

The reason that the slope method can be used to measure instantaneous velocity at a specific point on a curved line (just as it was used to measure instantaneous speed in the previous section) is that a very small segment of a curve is almost identical to a straight line segment. This becomes apparent when you look at the area around point **C** in Figure 3-5 and imagine it to be magnified 100 times. The red and black line segments near **C** would then virtually coincide.

Instantaneous velocity is an important concept in variable motion because it enables you to calculate the velocity of a moving object if its motion should stop changing at any point.

3.4 Solving Velocity Vector Problems

As vector quantities, velocities can be combined by the parallelogram method discussed in Section 2.11. The following examples show how to solve velocity problems of this kind.

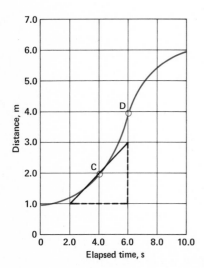

Figure 3-5. A graph of motion with variable velocity.

EXAMPLE   A person can row a boat at the rate of 8.0 km/hr in still water. The person heads the boat directly across a stream that flows at the rate of 6.0 km/hr. Find the resultant velocity (a) graphically and (b) trigonometrically.

SOLUTION   (a) To solve this problem graphically, use a protractor and a ruler with a millimeter scale. Draw a velocity vector, **OB**, to represent the velocity of the boat in still water, v_B, using an appropriate scale such as $1\bar{0}$ mm = 1.0 km/hr. (See Figure 3-6, but bear in mind that the scale in the figure is smaller than the one suggested for your paper.) Draw the velocity vector, **OC**, to the same scale and at right angles to **OB** to represent the velocity of the stream, v_c.

Complete the parallelogram **OBRC**, and draw in the diagonal **OR**. Place an arrowhead at **R**. The vector **OR** represents the resultant velocity, v_R. Using the same scale that you used in drawing **OB** and **OC**, read the magnitude of **OR** as precisely as possible. It should be about $1\bar{0}$ km/hr. Use the protractor to measure the direction of **OR**. It should be about $37°$ downstream from **OB**.

(b) The graphic solution of vector problems is limited by the precision with which the angles and lines are drawn and measured. This limitation is avoided by using the trigonometric solution. In this method tables or a calculator are used to determine the lengths of vectors. The precision of the tables or calculator is usually greater than that of the estimated answers in a graphic solution. Nevertheless, the graphic solution serves as an important check against errors in the trigonometric solution.

The trigonometric solution is found by using the relationship

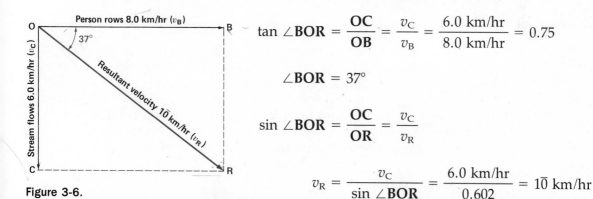

$$\tan \angle \mathbf{BOR} = \frac{OC}{OB} = \frac{v_C}{v_B} = \frac{6.0 \text{ km/hr}}{8.0 \text{ km/hr}} = 0.75$$

$$\angle \mathbf{BOR} = 37°$$

$$\sin \angle \mathbf{BOR} = \frac{OC}{OR} = \frac{v_C}{v_R}$$

$$v_R = \frac{v_C}{\sin \angle \mathbf{BOR}} = \frac{6.0 \text{ km/hr}}{0.602} = 1\bar{0} \text{ km/hr}$$

Figure 3-6.

Hence, $v_R = 1\bar{0}$ km/hr at an angle of $37°$ downstream.

EXAMPLE   In flying an airplane, a pilot wants to attain a velocity with respect to the ground of 485 km/hr eastward. A wind is blowing southward at 42.0 km/hr. What velocity must the pilot maintain with respect to the air to achieve the desired ground velocity?

SOLUTION   The situation in this example is shown in Figure 3-7. The desired velocity, v_R, is the diagonal of the parallelogram **OSRV**. We know one component of the resultant velocity, namely the velocity of the wind, v_S. We are required to find the other component, v_V, which is the velocity the pilot must maintain.

Figure 3-7. The vector representing the velocity that the airplane must maintain is a side of the parallelogram of which the desired velocity is the diagonal.

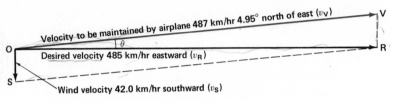

As with all problems involving the parallelogram method, v_V can be found either graphically or trigonometrically. The trigonometric solution is shown below.

$$\tan \theta = \frac{VR}{OR} = \frac{v_S}{v_R} = \frac{42.0 \text{ km/hr}}{485 \text{ km/hr}} = 0.0866$$

$$\theta = 4.95°$$

$$\sin \theta = \frac{VR}{OV} = \frac{v_S}{v_V}$$

$$v_V = \frac{v_S}{\sin \theta} = \frac{42.0 \text{ km/hr}}{0.0863} = 487 \text{ km/hr}$$

Hence, $v_V = 487$ km/hr 4.95° north of east.

EXAMPLE   The velocity of an airplane with respect to the air is 425 km/hr in a direction 30.0° north of east. A wind is blowing northward at 56.0 km/hr. What is the resultant velocity of the airplane?

SOLUTION   The graphic solution for this problem is shown in Figure 3-8. The trigonometric solution involves the law of cosines and law of sines, since the component vectors do not act at right angles.

$$\angle NOV = \angle NRV = 60.0°$$

$$\angle OVR = \angle ONR = \frac{360.0° - 2(60.0°)}{2} = 120.0°$$

$$ON = VR = v_N$$

$$NR = OV = v_V$$

$$OR = v_R = \sqrt{v_N^2 + v_V^2 - 2v_N v_V \cos \angle OVR}$$

$$v_R = \sqrt{(56.0 \text{ km/hr})^2 + (425 \text{ km/hr})^2 - 2(56.0 \text{ km/hr})(425 \text{ km/hr})(\cos 120.0°)}$$

$$v_R = 456 \text{ km/hr}$$

$$\frac{v_N}{\sin \theta} = \frac{v_R}{\sin \angle OVR}$$

$$\sin \theta = \frac{v_N \sin \angle OVR}{v_R}$$

$$\sin \theta = \frac{(56.0 \text{ km/hr})(\sin 120.0°)}{456 \text{ km/hr}}$$

$$\sin \theta = 0.106$$
$$\theta = 6.08°$$

Hence, the resultant velocity is 456 km/hr in a direction 30.0° + 6.08°, or 36.1°, north of east.

Figure 3-8.

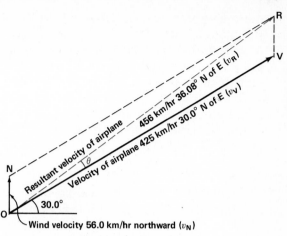

Wind velocity 56.0 km/hr northward (v_N)

Questions

GROUP A

1. (a) Define displacement. (b) Define motion.
2. (a) Define speed. (b) Distinguish between average and instantaneous speed. (c) What are the units of speed in the metric system?
3. (a) What is velocity? (b) Distinguish between speed and velocity.
4. (a) Describe the motion of an object that has uniform velocity. (b) Describe the motion of an object that has variable velocity.

GROUP B

5. Why is speed a scalar quantity but velocity a vector quantity?
6. How can the magnitude and direction of the resultant of two velocities be determined?
7. (a) Does a change in the speed of an object always produce a change in its velocity? (b) Does a change in the velocity of an object always indicate a change in its speed? Explain.
8. In a graph of variable velocity, what is the meaning of the changing slope of the line?

Problems

GROUP A

Note: Problems 2 through 6 should be solved (a) graphically and (b) trigonometrically.

1. The highway distance from Indianapolis to Denver is 1712 km. If the driving time is 29.25 hr, what is the average speed for the trip?
2. A motorboat travels 25 km/hr in still water. What will be the magnitude and direction of the velocity of the boat if it is directed upstream on a river that flows at the rate of 4 km/hr?

3. An airplane has a velocity, with respect to the air, of 825 km/hr northward. What is the velocity of the airplane with respect to the ground if the wind is blowing with a velocity of 35 km/hr southward?
4. A person can row a boat at a speed of 5.0 km/hr in still water. The person heads directly across a river that flows at a speed of 3.5 km/hr. Determine the magnitude and direction of the boat's resultant velocity.
5. An airplane flies westward at 650 km/hr. If the wind has a velocity of 40.0 km/hr northward, determine

the magnitude and direction of the resultant velocity.

6. An airplane is to maintain a velocity of 475 km/hr in a direction 45.0° north of east. If the wind velocity is 50.0 km/hr 45.0° south of east, what should be the magnitude and direction of the velocity of the airplane to offset the effect of the wind?

GROUP B

Note: Problems 7 through 9 should be solved (a) graphically and (b) trigonometrically.

7. Determine the magnitude and direction of the resultant velocity of

75.0 m/s, 25.0° east of north, and 100.0 m/s, 25.0° east of south.

8. A boat is to cross a river on a course directed 20.0° upstream to a dock 815 m distance in 10.0 min. If the river flows at a speed of 3.0 km/hr, at what angle upstream must the boat be headed and what must be its speed in km/hr relative to the moving water?

9. An airplane must fly at a ground speed of 425 km/hr in a direction of 10.0° east of south to be on course and on schedule. If the wind velocity is 25.0 km/hr 40.0° east of north, in what direction and at what speed relative to the air must the pilot fly?

ACCELERATION

Acceleration is a change in velocity. It can be positive or negative.

3.5 The Nature of Acceleration The gas pedal of a car is sometimes called the accelerator. When you drive a car along a level highway and increase the pressure on the accelerator, the speed of the car increases. The needle of the car's speedometer shows that for a certain amount of time the car's speed increases. We say that the car is *accelerating.* When you apply the brakes of the car that you are driving, the speed of the car decreases. The speedometer indicates that the car's speed is decreasing, and we say that the car is *decelerating.*

Acceleration is defined as the time rate of change of velocity. An automobile that goes from 0.0 km/hr to $6\overline{0}$ km/hr in $1\overline{0}$ seconds has greater acceleration than a car that goes from 0.0 km/hr to $6\overline{0}$ km/hr in 15 seconds. The first car has an average acceleration of $\dfrac{6\overline{0} \text{ km/hr}}{1\overline{0} \text{ s}} = 6.0$ (km/hr)/s. The second car has an average acceleration of 4.0 (km/hr)/s.

Acceleration, like velocity, is a vector quantity. In the case of a car moving along a straight road, acceleration can be represented by a vector in the direction the car is moving. Deceleration occurs in the opposite direction. For the cases in which the motion is known to be along a straight line, the direction of the acceleration is described as positive or negative. In Section 3.3 we saw that the equations for speed and velocity appear the same, since the equations do not specify the direction of displacement. For the same reason, the equations for rate of change of speed and acceleration appear the same. These equations involve only the magnitude and not the direction of the motion.

Average acceleration is found by the relationship

$$a_{\text{av}} = \frac{v_{\text{f}} - v_{\text{i}}}{t_{\text{f}} - t_{\text{i}}}$$

where a_{av} is the average acceleration, v_{i} is the initial velocity at time t_{i}, and v_{f} is the final velocity at time t_{f}. This relationship may also be stated

$$a_{\text{av}} = \frac{\Delta v}{\Delta t}$$

where Δv is the change in velocity for time interval Δt.

Figures 3-9 and 3-10 are velocity-time graphs for accelerated motion. In Figure 3-9 the velocity curve is a straight line. Hence the acceleration is a constant. This is an illustration of *uniformly accelerated motion.*

The motion illustrated in Figure 3-10 is an example of *variable acceleration.* As we mentioned in the case of finding instantaneous speed or instantaneous velocity, finding the average acceleration between two points on the curve only approximates the instantaneous acceleration. Instantaneous acceleration is equal to the slope of the line that is tangent to the curve at the given point.

Notice that the slope of the curve in Figure 3-10 is not always positive. That is, the velocity does not increase during all time intervals. For the interval between points **A** and **B**, Δt is 3.0 s and Δv is −2.0 m/s. Consequently,

$$a_{\text{av}} = \frac{-2.0 \text{ m/s}}{3.0 \text{ s}} = -0.67 \frac{\text{m/s}}{\text{s}}$$

This reduction in speed is called deceleration, or negative acceleration.

The dimensions of acceleration should be carefully examined. Acceleration is the rate of change of the rate of displacement, or $(\Delta d/\Delta t)/\Delta t$. If Δd is in meters and Δt is in seconds, acceleration is given in m/s/s, or m/s^2.

3.6 Solving Acceleration Problems Galileo Galilei (1564–1642) was the first scientist to understand clearly the concept of acceleration. He found that the acceleration of a ball rolling down an inclined plane and the acceleration of a falling object were both examples of the same natural phenomenon and were described by the same mathematical rules.

Let us consider one of Galileo's experiments. One end of a board is raised so that it is just steep enough to allow a ball to roll from rest to the 0.5-m position in the first second. The apparatus is shown in Figure 3-11. As the ball rolls down the board, its position is shown at one-second

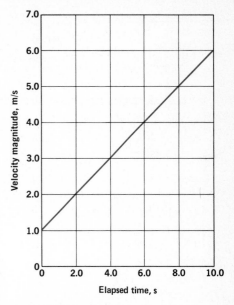

Figure 3-9. A graph of velocity versus time for uniformly accelerated motion.

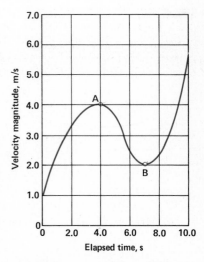

Figure 3-10. A graph of velocity versus time for motion with variable acceleration.

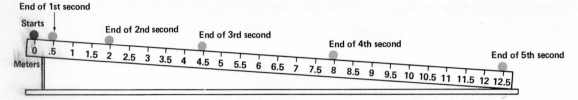

End of 1st second

Starts

End of 2nd second

End of 3rd second

End of 4th second

End of 5th second

Meters
0 .5 1 1.5 2 2.5 3 3.5 4 4.5 5 5.5 6 6.5 7 7.5 8 8.5 9 9.5 10 10.5 11 11.5 12 12.5

Figure 3-11. The ball rolls down the plane with uniformly accelerated motion.

intervals. The data obtained are shown in the first six columns of Table 3-1.

The data in the last four columns of Table 3-1 show that the velocity at the end of each second is 1.0 m/s greater than at the end of the previous second. Thus the acceleration is 1.0 (m/s)/s, or 1.0 m/s². The motion of a ball rolling down an inclined plane is an example of uniformly accelerated motion.

Table 3-1
UNIFORMLY ACCELERATED MOTION

Interval number	Time at start of interval	Time, t, at end of interval	Position at start of interval	Position at end of interval	Distance, Δd, traveled in interval	Average velocity, v_{av}, during interval	Instantaneous velocity, v_i, at start of interval	Instantaneous velocity, v_f, at end of interval	Average acceleration, a_{av}, during interval
	(s)	(s)	(m)	(m)	(m)	(m/s)	(m/s)	(m/s)	(m/s²)
1	0.0	1.0	0.0	0.5	0.5	0.5	0.0	1.0	1.0
2	1.0	2.0	0.5	2.0	1.5	1.5	1.0	2.0	1.0
3	2.0	3.0	2.0	4.5	2.5	2.5	2.0	3.0	1.0
4	3.0	4.0	4.5	8.0	3.5	3.5	3.0	4.0	1.0
5	4.0	5.0	8.0	12.5	4.5	4.5	4.0	5.0	1.0

Galileo's experiment verified the property of inertia.

The initial velocity of the ball is zero. Its acceleration, or rate of gain in velocity, is 1.0 m/s². Therefore its velocity at the end of the *first second* is 1.0 m/s. (If the ball at that instant were no longer accelerated, it would continue to move with a constant velocity of 1.0 m/s.) At the end of the next second, its velocity is 2.0 m/s. (2.0 s × 1.0 m/s² = 2.0 m/s.) Similarly the ball's velocity at the end of the fifth second is 5.0 m/s. These data show that *the final velocity of an object starting from rest and accelerating at a uniform rate equals the product of the acceleration and the elapsed time. If an object does not start from rest, its final velocity will equal the sum of its initial velocity and the increase in velocity produced by the acceleration.*

We may arrive at this same conclusion by mathematical reasoning. The equation defining acceleration is

Galileo was among the first scientists to use experiments in testing hypotheses.

$$a_{av} = \frac{v_f - v_i}{t_f - t_i}$$

For uniformly accelerated motion a_{av} becomes a. Then assuming that v_i is the velocity in the direction of the acceleration when $t_i = 0$ and t_f is elapsed time Δt,

$$a = \frac{v_f - v_i}{\Delta t}$$

Multiplying both sides of the equation by Δt,

$$a \, \Delta t = v_f - v_i$$

Isolating the term v_f yields

$$v_f = v_i + a \, \Delta t \qquad \text{(Equation 1)}$$

If the object starts from rest, $v_i = 0$

and $\qquad\qquad\qquad v_f = a \, \Delta t$

In Section 3.3 we saw that

$$v_{av} = \frac{\Delta d}{\Delta t}$$

or $\qquad\qquad\qquad \Delta d = v_{av} \, \Delta t \qquad \text{(Equation 2)}$

When an object is uniformly accelerated, the average velocity for any given interval of time can be found by taking one-half the sum of the initial and final velocities:

$$v_{av} = \frac{v_i + v_f}{2} \qquad \text{(Equation 3)}$$

By substituting the value of v_f (obtained in Equation 1) in Equation 3,

$$v_{av} = \frac{v_i + (v_i + a \, \Delta t)}{2} \qquad \text{(Equation 4)}$$

Now, substituting the value of v_{av} (obtained in Equation 4) in Equation 2 gives

$$\Delta d = v_i \, \Delta t + \tfrac{1}{2} a \, \Delta t^2 \qquad \text{(Equation 5)}$$

If the object starts from rest,

$$v_i t = 0$$

and $\qquad\qquad\qquad \Delta d = \tfrac{1}{2} a \, \Delta t^2$

From this equation we see that if an object starts from rest and travels with uniformly accelerated motion, the displacement of the object during the first second is numerically equal to one-half the acceleration.

Now let us solve Equation 1 for Δt, and substitute this value of Δt in Equation 5. From Equation 1,

$$\Delta t = \frac{v_f - v_i}{a}$$

The Mathematics Refresher may come in handy here.

Substituting in Equation 5,

$$\Delta d = v_i \left(\frac{v_f - v_i}{a} \right) + \tfrac{1}{2} a \left(\frac{v_f - v_i}{a} \right)^2$$

Multiplying by $2a$ and expanding terms,

$$2a\,\Delta d = 2v_i v_f - 2v_i^2 + v_f^2 - 2v_f v_i + v_i^2$$

Combining terms,

$$2a\,\Delta d = v_f^2 - v_i^2$$

Solving for v_f,

$$v_f = \sqrt{v_i^2 + 2a\,\Delta d} \qquad \text{(Equation 6)}$$

With uniform acceleration the final velocity equals the square root of the sum of the square of the initial velocity and twice the acceleration times the displacement. If the object starts from rest,

$$v_i^2 = 0$$

and

$$v_f = \sqrt{2a\,\Delta d}$$

Table 3-2
ACCELERATION EQUATIONS

With initial velocity	Starting from rest
$v_f = v_i + a\,\Delta t$	$v_f = a\,\Delta t$
$\Delta d = v_i\,\Delta t + \tfrac{1}{2}a\,\Delta t^2$	$\Delta d = \tfrac{1}{2}a\,\Delta t^2$
$v_f = \sqrt{v_i^2 + 2a\,\Delta d}$	$v_f = \sqrt{2a\,\Delta d}$

The equations just given are summarized in Table 3-2. They apply whether the uniform acceleration is positive or negative, hence it is important to note the sign of the displacement. When it is negative, the equations may be used to calculate the distance an object will travel when its motion is uniformly decelerated and the time that elapses in bringing the object to a stop. The following examples show how the equations are used in problems dealing with both positively and negatively accelerated motion.

EXAMPLE   A ball rolls down an inclined plane with a uniform acceleration of 1.00 m/s². At a certain instant its velocity is 0.50 m/s. (a) What is the velocity of the ball 10.0 s later? (b) How far has the ball traveled in that interval of 10.0 s? (c) How far did the ball travel during the eighth second?

SOLUTION   (a) Assuming that time is measured from the moment that the velocity of the ball is observed to be 0.50 m/s,

$$v_f = v_i + a \, \Delta t = 0.50 \text{ m/s} + (1.00 \text{ m/s}^2)(10.0 \text{ s})$$
$$v_f = 10.5 \text{ m/s}$$

(b) $\Delta d = v_i \, \Delta t + \frac{1}{2}a \, \Delta t^2$

$$\Delta d = (0.50 \text{ m/s})(10.0 \text{ s}) + \frac{1}{2}(1.00 \text{ m/s}^2)(10.0 \text{ s})^2$$
$$\Delta d = 55.0 \text{ m}$$

(c) The distance the ball travels during the eighth second equals the distance traveled in 8 s minus the distance traveled in 7 s.

$$\Delta d = (v_1 \, \Delta t_8 + \frac{1}{2}a \, \Delta t_8^2) - (v_i \, \Delta t_7 + \frac{1}{2}a \, \Delta t_7^2)$$
$$\Delta d = v_i(\Delta t_8 - \Delta t_7) + \frac{1}{2}a(\Delta t_8^2 - \Delta t_7^2)$$
$$\Delta d = (0.50 \text{ m/s})(8.00 \text{ s} - 7.00 \text{ s}) + \frac{1}{2}(1.00 \text{ m/s}^2)(64.0 \text{ s}^2 - 49.0 \text{ s}^2)$$
$$\Delta d = 8.00 \text{ m}$$

EXAMPLE   An automobile is traveling 65 km/hr. Its brakes decelerate it 6.0 m/s². (a) How long will it take to stop the car? (b) How far will the car travel after the brakes are applied?

SOLUTION

(a) $a = \dfrac{v_f - v_i}{\Delta t}$

$$\Delta t = \frac{v_f - v_i}{a} = \frac{(0 \text{ km/hr} - 65 \text{ km/hr})(1000 \text{ m/km})}{(-6.0 \text{ m/s}^2)(3600 \text{ s/hr})}$$

$$\Delta t = 3.0 \text{ s}$$

(b) $\Delta d = v_i \, \Delta t + \frac{1}{2}a \, \Delta t^2$

$$\Delta d = \frac{(65 \text{ km/hr})(3.0 \text{ s})(1000 \text{ m/km})}{3600 \text{ s/hr}} + \frac{1}{2}(-6.0 \text{ m/s}^2)(3.0 \text{ s})^2$$

$$\Delta d = 27 \text{ m}$$

(*Note:* Part B could also be solved by using the equation immediately preceding Equation 6.)

PRACTICE PROBLEMS   1. An object that is undergoing uniform acceleration from rest moves 7.4 m in 2.5 s. What is the acceleration of the object?   *Ans.* 2.4 m/s²

2. An automobile moves at an initial velocity of 12 m/s. Then it accelerates at the rate of 8.5 m/s² for a distance of 130 m. How fast is the automobile then moving?   *Ans.* 49 m/s

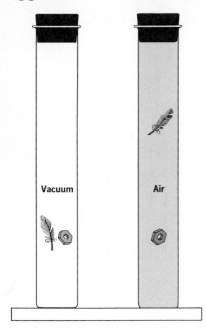

Figure 3-12. Objects fall with the same acceleration in a vacuum (left) but with different accelerations through air (right).

Vacuum

Air

Figure 3-13. Skydivers in free fall. Air resistance allows skilled divers to control the orientation and direction of their bodies even before the parachutes are opened.

3.7 Freely Falling Bodies In New York City if a body falls freely from rest, it reaches a velocity of 9.803 m/s in one second of time. Since $a = v_f/\Delta t$, the acceleration equals 9.803 m/s² at this location. We must consider the location because the value of the acceleration depends on the earth's gravity at that location. Since the gravity on the earth may vary from place to place, the value of the acceleration may also vary. (The nature of gravity will be discussed more fully in Sections 3.11–3.13.) A rounded value of 9.80 m/s² is approximately valid for locations in the United States.

The equations for accelerated motion apply to freely falling bodies. Since the acceleration due to gravity, g, is the same for all objects at a given location, we may substitute g for a in these equations. Thus

$$v_f = v_i + g\,\Delta t$$
$$\Delta d = v_i\,\Delta t + \tfrac{1}{2}g\,\Delta t^2$$
$$v_f = \sqrt{v_i{}^2 + 2g\,\Delta d}$$

When freely falling bodies start from rest, $v_i = 0$, and the motion is described by the following equations:

$$v_f = g\,\Delta t$$
$$\Delta d = \tfrac{1}{2}g\,\Delta t^2$$
$$v_f = \sqrt{2g\,\Delta d}$$

In the case of free-falling bodies, Δd is always a vertical distance. The vector quantities v and Δd in these equations are customarily assigned plus signs if they are directed downward and minus signs if they are directed upward. (The description of the motion is simplified by considering the downward direction of gravity as positive.) These equations apply only to objects that are falling freely in a vacuum because they do not take into consideration the resistance of air, as illustrated in Figure 3-12. These equations do not apply to objects whose air resistance is great compared to their mass, such as leaves falling in air.

An object thrown upward is uniformly decelerated by the force of gravity until it finally stops rising. Then as the object falls, it is uniformly accelerated by the force of gravity. If the effect of the atmosphere is neglected, then the time required for the object to fall is the same as the time required for it to rise. Study the following example below.

EXAMPLE An object is projected upward with a velocity of 125 m/s. (a) To what height will it rise? (b) How long will it take to reach that height? (c) What will be the total time elapsed until it strikes the earth?

SOLUTION

(a) $v_f = \sqrt{v_i^2 + 2g\,\Delta d}$

$$\Delta d = \frac{v_f^2 - v_i^2}{2g}$$

$$\Delta d = \frac{0 - (-125 \text{ m/s})^2}{2 \times 9.80 \text{ m/s}^2}$$ (The value for v_i is negative because the velocity is upward.)

$\Delta d = -797$ m   (The negative value obtained for Δd indicates an upward displacement.)

(b) $v_f = v_i + g\,\Delta t$

$$\Delta t = \frac{v_f - v_i}{g} = \frac{0 - (-125 \text{ m/s})}{9.80 \text{ m/s}^2} = 12.8 \text{ s}$$

(c) Neglecting the effect of the atmosphere, the same amount of time is required for the object to fall as for the object to rise. Thus the total time elapsed will be 2 × 12.8 s = 25.6 s.

PRACTICE PROBLEMS   **1.** A freely falling body starting from rest attains a velocity of 45 m/s. How long has it been falling?   *Ans.*   4.6 s

2. If a body falls freely from rest, how far does it move during the fifteenth second of its fall?   *Ans.*   140 m

---

Questions
GROUP A

1. (a) What is acceleration? (b) Why is the unit of time squared in measuring acceleration?
2. How do the directions of an acceleration vector and its corresponding velocity vector compare for the uniformly accelerated linear motion of an object starting from rest?
3. Define the following: (a) uniform acceleration; (b) variable acceleration; (c) uniform deceleration; (d) variable deceleration.

Problems
GROUP A

Note: Whenever necessary, use $g = 9.80$ m/s^2. Disregard the effect of the atmosphere.

4. What effect does the force of gravity have on a baseball thrown directly upward?
5. Explain instantaneous acceleration?

GROUP B
6. What equation relates final velocity to initial velocity, acceleration, and elapsed time?
7. How can displacement be calculated when the initial velocity, final velocity, and elapsed time are known?
8. Give an example of an object moving at high speed with zero acceleration.
9. Give an example of an object starting with zero speed and high acceleration.

1. If a ball is dropped and attains a velocity of 29.31 m/s in 3.00 s, what is the acceleration due to gravity?
2. An automobile can be accelerated from 95 km/hr to 142 km/hr in 8.1 s. What is the acceleration?

3. A large rock is dropped from a bridge into the river below. (a) If the time required for it to drop is 1.7 s, with what velocity, in m/s, does it hit the water? (b) What is the height, in meters, of the bridge above the water?

4. (a) How many seconds does it take for a metal ball to drop 145 m from rest? (b) What velocity does it attain?

5. An object with an initial velocity of 20.0 cm/s is accelerated at 8.0 cm/s^2 for 5.0 s. (a) What is the total displacement? (b) What is the displacement during the fifth second?

6. What velocity is attained by an object that is accelerated at 0.30 m/s^2 for a distance of 54 m if its initial velocity is 0.50 m/s?

7. (a) If the brakes of an automobile can decelerate it at 7.00 m/s^2, what time is required to reduce the velocity of the automobile from 157.0 km/hr to 75.0 km/hr? (b) How many meters does the car travel while decelerating?

GROUP B
Note: Whenever necessary, use g = 9.80 m/s^2. Disregard the effect of the atmosphere.

8. A ball is thrown from the ground to the top of the school flagpole. If it returns to the ground after 4.0 s, what is the height of the flagpole in meters?

9. A stone is released from a balloon while the balloon is ascending at the rate of 5.0 m/s and when the balloon is 353 m above the ground. What time is required for the stone to reach the ground?

10. A baseball is thrown vertically downward from the top of a 155-m tower with an initial velocity of 25 m/s. (a) With what velocity does it reach the ground? (b) What time is required for it to reach the ground?

NEWTON'S LAWS OF MOTION

3.8 Law of Inertia   Thus far in Chapter 3, we have discussed motion apart from its causes. Now we shall study the effects that forces have on motion. *A force is a physical quantity that can affect the motion of an object.* A push or a pull is a force. Since a force has magnitude, direction, and a point of application, it is a vector quantity.

The relationships among force, mass, and motion were described clearly for the first time by Sir Isaac Newton (1642–1727) in three laws of motion that bear his name. While some aspects of Newton's laws of motion can be tested only under carefully controlled conditions, repeated experiments and observations have led scientists to believe that the laws are universally true. That is, the laws apply not only to objects on or near the earth, but to objects throughout the universe.

Newton's first law of motion deals with the motion of a body on which no net force is acting. That is, either there is no force at all acting on the body or the *vector sum* of all forces acting on the body is zero. The word "net" refers to

Figure 3-14. Sir Isaac Newton formulated the law of gravitation and three laws of motion that describe how forces act on matter.

the second situation. Even though a body may have many forces acting on it, these forces may act against each other. They may balance each other in such a way that the body does not change its state of motion. If such a body is at rest, it will remain at rest. If it is in motion, it will continue in its motion *in a straight line with uniform speed.* **Newton's first law of motion** may therefore be stated as follows: *If there is no net force acting on a body, it will continue in its state of rest or will continue moving along a straight line with uniform speed.* As we saw in Section 1.8, this property of a body that opposes any change in its state of motion is called inertia. Hence Newton's first law of motion is known as the law of inertia.

The law of inertia states that unless a net force acts on an object, the motion of an object (or lack of it) does not change.

At first glance, this law seems to contradict our everyday experiences. If a car is to be kept moving with a constant velocity, the car's engine must apply a constant force to it. If the engine stops applying this force, the car comes to a stop. Only then does the car seem to obey the part of Newton's law that states that objects at rest will remain at rest unless acted upon by an unbalanced force.

Close study of a moving car shows, however, that it is the force of friction that brings the car to a stop and not the absence of the force provided by the engine. If it were possible to remove this friction, it would be reasonable to assume that the car would keep rolling without applying a constant force. This is, in fact, what happens in space travel. Once a spaceship is beyond the pull of the earth's gravity, it continues to move with constant velocity even without the thrust of its engines.

The study of the motion of a car in the absence of friction is an example of a *thought experiment* since the study cannot be performed under actual conditions. It was a thought experiment of this kind that led Galileo to an understanding of inertia even before Newton described it. Galileo noticed that if a ball rolls down one incline and up a second one, as shown in Figure 3-15, the ball will reach

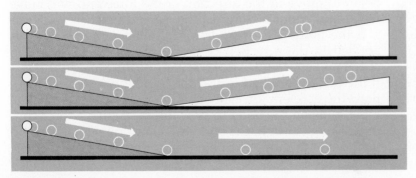

Figure 3-15. A diagram of Galileo's thought experiment. In the upper and middle drawings, the ball reaches almost the height at which it started. In the lower drawing, the ball continues to move indefinitely at constant velocity. In each drawing, the successive images show the positions of the ball after equal time intervals.

Inertia can be verified more easily in space than on the earth's surface, where friction and gravity intervene.

Figure 3-16. A demonstration of inertia. In the bottom photo, the table has been quickly pulled away, and the chicken dinner is momentarily suspended in midair.

almost the same height on the second incline as the height from which it started on the first incline. Galileo concluded that the difference in height is caused by friction and that if friction could be eliminated the heights would be exactly alike. Then he reasoned that the ball would reach the same height no matter how shallow the slope of the second incline. Finally, if the second slope were eliminated altogether, the ball would keep rolling indefinitely with constant velocity. This is the same idea as the one expressed in Newton's first law of motion. Inertia keeps a stationary object stationary and a moving object moving.

The greater the mass (and inertia) of an object, the greater is the force required to produce a given acceleration. A pencil lying on the floor has relatively little mass and therefore little inertia. You can produce acceleration easily by kicking it with your foot. A brick has much more mass and more inertia. You can easily tell its difference from the pencil if you try to kick it.

Newton's first law specifies which forms of motion have a *cause* and which forms do not. Uniform motion in a straight line is the only motion possible for an object far removed from other objects. Nonuniform motion is always *caused* by the presence of some other object.

3.9 Law of Acceleration Newton's first law of motion tells us how a body acts when there is no net applied force. Let us consider what happens when there *is* either a single applied force or two or more applied forces whose *vector sum* is not zero. (Forces are vector quantities since they have direction as well as magnitude.) In the following discussion, we shall use "applied force" or just "force" to mean the vector sum of all forces applied to the body.

Newton's second law of motion states: *The effect of an applied force is to cause the body to accelerate in the direction of the force. The acceleration is in direct proportion to the force and in inverse proportion to the mass of the body.*

If the body is at rest when the force is applied, it will begin to move in the direction of the force and will move faster and faster as long as the force continues.

If the body is moving in a straight line and a force is applied in the direction of its motion, it will increase in speed and continue to do so as long as the force continues. If the force is applied in the direction opposite to the motion, the acceleration will again be in the direction of the force, causing the body to slow down. If such a force continues long enough, the body will slow down to a stop and then begin to move with increasing speed in the opposite direction.

If the force applied to a moving object is not along the line of its motion, the acceleration will still be in the direction of the force. The effect in general will be to change both the direction and the speed of the motion. We shall postpone detailed consideration of this more complex case until Chapter 5. For the present we shall consider only the speed of an object initially at rest or one that is moving in a straight line that is changed only by a force acting along that line.

Because of friction, stepping on the accelerator of a car does not necessarily accelerate the car.

It should be emphasized again that a body to which no force is applied has zero acceleration. If such a body is in motion, its velocity does not change. If it is at rest, its velocity remains unchanged; its velocity stays at zero. And when there is no change in velocity, the acceleration is zero. When a force is applied to a body, the acceleration is not zero, the velocity changes, and the acceleration is in the direction of the applied force.

If different forces are applied to the same body, the magnitude of the acceleration is directly proportional to the amount of the force. That is, doubling the force causes the acceleration to double, and so on.

We can express such a direct proportion as an equation by saying that one of the quantities equals the other one times a constant, or

$$F = ka$$

in which F is the applied force, a is the resulting acceleration, and k is the constant of proportionality. Since F and a are both vectors, this equation also says that the applied force and the resulting acceleration are in the same direction (which is a restatement of the first part of Newton's second law of motion).

What happens when forces are applied to *different* objects? This question brings us to the third part of Newton's second law. It turns out that the equation $F = ka$ is always true but that the value of k is different for different objects. In fact, we find that if we properly define the units of force and mass, the value of k for any object is identical with the mass, m, of that object. Thus the equation is written

$$F = ma$$

In the metric system, the unit of mass is the kilogram and acceleration has the units of meters per second^2. (See Section 3.5.) The force required to accelerate 1 kilogram of mass at 1 meter per second^2 is 1 kg m/s^2. This relationship defines the newton in terms of fundamental units. That is,

$$1 \text{ n} = 1 \text{ kg m/s}^2$$

Since the acceleration acquired by a particular object is directly proportional to the amount of force applied, this law of acceleration can be expressed as the proportion

$$\frac{F}{F'} = \frac{a}{a'}$$

where F and F' are two different forces and a and a' are the corresponding accelerations.

In the case of a freely falling object, one of the forces is known. It equals the weight, F_w, of the object. The acceleration, g, is also known. It is the acceleration due to gravitation, 9.80 m/s². Making these substitutions, the proportion becomes

$$\frac{F}{F_w} = \frac{a}{g}$$

and

$$F = \frac{F_w a}{g}$$

Therefore, we can calculate the force needed to give any desired acceleration to an object of known weight.

Since $F = ma$, we can substitute this value for F in the above equation and get

Notice the similarity between $F_w = mg$ and $F = ma$. What is the reason for this similarity?

$$ma = \frac{F_w a}{g}$$

Dividing both sides of the equation by a, we have

$$m = \frac{F_w}{g} \quad \text{and} \quad F_w = mg$$

Provided the value of g is known, these two equations are used to convert from newtons to kilograms, and vice versa. The following examples show how Newton's second law is used in solving acceleration problems.

EXAMPLE   Neglecting friction and air resistance, what force is required to accelerate an automobile weighing 2.00×10^4 n from 30.0 km/hr to 70.0 km/hr in 10.0 s?

SOLUTION   *Basic equation:* $F = ma$

but

$$m = \frac{F_w}{g}$$

and

$$a = \frac{\Delta v}{\Delta t} = \frac{v_f - v_i}{\Delta t}$$

Substituting for m and a we get

Working equation:

$$F = \frac{F_w(v_f - v_i)}{g\,\Delta t}$$

$$F = \frac{2.00 \times 10^4 \text{ n } (70.0 \text{ km/hr} - 30.0 \text{ km/hr})(1000 \text{ m/km})}{(9.80 \text{ m/s}^2)(10.0 \text{ s})(3600 \text{ s/hr})}$$

$$F = 2.27 \times 10^3 \text{ n}$$

PRACTICE PROBLEMS **1.** A man has a mass of 75 kg. What is his weight at sea level?   *Ans.*   740 n

2. A force of 1250 n accelerates an automobile from rest to 60.0 km/hr. The automobile weighs 2.50×10^4 n. How much time is required for the acceleration?   *Ans.*   34.0 s

---

3.10 Law of Interaction   Newton's third law of motion may be stated as follows: *When one body exerts a force on another, the second body exerts on the first a force of equal magnitude in the opposite direction.* To illustrate this law, consider some of the forces that are exerted when a book is resting on the top of a level table. The book exerts a downward force against the table. The table top exerts an upward force on the book. These forces are equal in magnitude and opposite in direction. When you walk forward on a level floor your feet exert a horizontal force against the floor, and the floor pushes against your feet with a force of equal magnitude, but in the opposite direction.

For every action there is an equal and opposite reaction.

In each of these situations we have two objects. In the first instance, the objects are the book and the table. In the second instance, they are the foot and the floor. Two forces are involved in each situation. In the first, they are the force of the book against the table and the force of the table against the book. In the second, they are the force of the foot against the floor and the force of the floor against the foot. In cases such as these, one force may be called the *action*, while the second force may be called the *reaction.* Unaccompanied forces do not exist in nature.

The law of reaction holds true for all objects at all times, whether they are stationary or moving. Every force is resisted by an equal and opposite force, independent of the

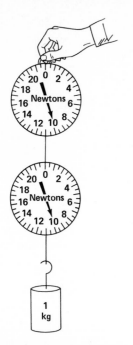

Figure 3-17. A demonstration of action-reaction. The kilogram mass exerts a downward force of 9.8 newtons, while the persons hand exerts an equal upward force of 9.8 newtons. (The weights of the balances are not considered in this illustration.)

motion of the objects involved. Let us consider another example. A boy rows a boat toward the shore of a lake and, when he is a meter or so from shore, he attempts to leap ashore. He exerts a force against the boat and the boat exerts an equal but opposite force against him. The force exerted by the boat against the boy accelerates him. The direction of his acceleration is opposite to the direction in which his force caused the boat to be accelerated. If we assume the resistance to the motion of the boat by the water to be negligible, the amount of acceleration of each object is inversely proportional to its mass. The boy might judge the force he must exert to reach shore on the basis of his experience in jumping the same distance from an object fixed on the earth. If so, when he jumps from the boat, he might not reach shore, but instead fall into the water.

Suppose a kilogram mass is suspended from two spring balances that are connected, as shown in Figure 3-17. As we learned in Section 2.4, a kilogram weighs about 9.8 newtons near the earth's surface. Then why do *both* balances have a reading of 9.8 newtons? The answer is Newton's third law. The downward force exerted by the mass is accompanied by an equal upward force exerted by the person's hand. Neither force can exist without the other. If additional balances are added in this experiment, each balance will still read 9.8 newtons (provided the weight of each balance is neglected). Can you explain why?

You may think that motion cannot occur if action and reaction are equal. If two people pull on a lightweight wagon with equal force in opposite directions, the wagon will not move. But this is not an example of action and reaction. There are two forces, it is true, and they have the same magnitude and opposite directions, *but they are both exerted on the same object.* Action and reaction apply when forces are exerted on different objects. For example, action and reaction are involved in the force that each one's feet exert against the ground and the equal but opposite force the ground exerts against the feet.

Questions
GROUP A

1. If a steel ball is placed on the top of a level table, it will remain there.
   (a) What forces are acting on the ball in this situation? (b) Are there any unbalanced forces acting on the ball?
2. (a) If the steel ball of Question 1 is given a slight push, what happens to the ball? (b) What forces act on the ball after it has been pushed? (c) Are any of these unbalanced forces? (d) What is the effect of the various forces on the motion of the ball?
3. (a) State Newton's first law of motion. (b) Describe the motion of a body in equilibrium.

4. (a) If other conditions are constant, how does the acceleration of an object vary with the amount of force applied? (b) How does the direction of the acceleration compare with the direction of the applied force?

5. How does the acceleration produced on different objects by equal forces vary with the mass of each object?

6. (a) State Newton's second law of motion. (b) What equation expresses the relationship of force to mass and acceleration?

7. How can the amount of force required to produce a certain acceleration of an object of known weight be calculated?

8. Suppose a brick is suspended from a rigid support by a suitable length of cord. (a) What downward force acts on the brick? (b) If this force is the action force, what force is the reaction force?

9. (a) What upward force acts on the suspended brick in Question 8? (b) If this force is the action force, what is the corresponding reaction force?

GROUP B

10. (a) What is meant by a thought experiment? (b) Give an example.

11. Why does a falling object in a vacuum undergo constant acceleration?

12. (a) Does Newton's second law hold only when frictional forces are absent? (b) Explain.

13. A fishline will break when a force of more than 600 n is exerted on it. Two people pull on the line in opposite directions with a force of 400 n each. (a) Will the fishline break? (b) Explain.

14. Explain the thought experiment you could devise to determine the mass of an object if you had only the following objects: a frictionless horizontal plane, a 1-kilogram standard, a meter stick, a spring balance in which the scale is marked in units unknown to you, and a stopwatch.

15. A tractor is pulling a heavy load. (a) If, according to Newton's third law, the load is pulling back as hard as the tractor is pulling forward, then why does the tractor move? (b) Make a drawing of the situation and show the appropriate force vectors.

16. If a third spring balance is placed between the other two in Figure 3-17 and in line with them, what will be the reading on the center balance? Test your answer in the laboratory.

Problems
GROUP A

Note: Use $g = 9.80$ m/s². All forces are net forces.

1. What force is required to accelerate a 3.0-kg object (which is free to move) at 5.0 m/s²?

2. What is the mass of an object that is accelerated 3.00 m/s² by a force of 125 n?

3. What acceleration does an object, mass 15 kg, undergo when a force of 25 n acts on it?

4. What is the weight of a 24-kg block of stone at sea level?

GROUP B

Note: Use $g = 9.80$ m/s². All forces are net forces.

5. What is the mass of a bag of cement that weighs 485 n at sea level?

6. What force is required to give a projectile weighing 475 n an acceleration of 3.00×10^3 m/s²?

7. A truck weighs 1.0×10^5 n. What force will give it an acceleration of 1.5 m/s²?

GRAVITATION

Newton developed his laws of motion and gravitation when he was only twenty-three years old and while he was home from college on an extended vacation during the Great Plague in 1665.

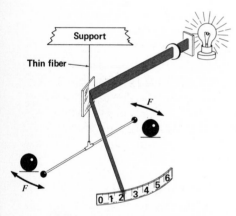

Figure 3-18. The Cavendish experiment. The force of gravitation, *F*, between the movable and stationary spheres can be measured by observing, with the help of a mirror and light beam, the amount of twist in the suspending fiber.

3.11 Newton's Law of Universal Gravitation   In addition to formulating his three laws of motion, Newton described the force that makes falling bodies accelerate toward the earth. In doing so, he made use of the laws of planetary motion that were developed by Johannes Kepler almost a century before Newton's time. See Section 1.4. In Newton's account of his study of falling bodies, he states that he wondered whether the force that makes an apple fall to the ground was related to the force that keeps the planets in their orbits. If so, a single law could be used to describe the attraction between objects in the entire universe.

From Kepler's laws, Newton deduced the fact that the *force of attraction between two objects is directly proportional to the product of the masses of the objects and inversely proportional to the square of the distance between their centers of mass.* (The center of mass of an object is that point at which all its mass can be considered to be concentrated. This concept is similar to that of "center of gravity," which will be discussed in Chapter 4.) Newton called this attractive force the *force of gravitation.* Hence the statement is known as Newton's **law of universal gravitation.** In equation form the law is

$$F = G\frac{m_1 m_2}{d^2}$$

where *F* is the force of gravitation, m_1 and m_2 are the masses of the attracting objects, *d* is the distance between their centers of mass, and *G* is a proportionality constant called the *universal gravitational constant.*

Newton was only able to confirm his theory with astronomical observations. He was unable to measure *G* with the laboratory instruments available at his time. The first determination of *G* was made in 1797 by the English scientist Henry Cavendish (1731–1810). A schematic diagram of this experiment is shown in Figure 3-18. Two small lead spheres are attached to the end of a lightweight rod that has a mirror attached to it. The rod is suspended by a thin quartz fiber. Two large lead spheres are placed in fixed positions near the small spheres. The gravitational force between the fixed and movable spheres draws the movable spheres toward the fixed ones. This motion causes the suspending fiber to twist. The fiber offers a slight resistance to twisting. The resistance increases as the twisting increases, and the angle of twist is proportional to the force of gravitation between the fixed and the movable spheres. The angle of twist could be measured directly from the rod's movement, but the sensitivity of the instrument can be increased by shining a light into the mirror. The light is reflected onto a distant scale, and small movements of the mirror result in large movements of the reflected light across the scale. Such an arrangement is called an optical lever. Since G has a very small value, extreme care must be used to isolate the Cavendish apparatus from outside forces such as those produced by air currents and electric charges. The value of G has been found to be 6.67×10^{-11} n m²/kg².

3.12 The Mass of the Earth   Once G has been measured, the equation for Newton's law of universal gravitation can be used to find the force of gravitation between any two objects of known mass with a known distance between their centers of mass. Or if the force of gravitation, the distance, and one of the masses are known, the equation can be used to find the value of the other mass. For example, the equation can be used to find the mass of the earth. In the equation, use m_e, the mass of the earth, for m_1. For m_2, use m_p, the mass of a particle on the earth's surface. Newton showed that, for a spherical mass, we can consider all of the mass to be located at the center. So in this case d will be the radius of the earth. The force of gravitation, F, will be the weight of the particle, so we shall replace F by F_w.

$$F_w = G\frac{m_e m_p}{d^2} \qquad \text{(Equation 1)}$$

Dividing both sides of the equation by m_p,

$$\frac{F_w}{m_p} = \frac{G m_e}{d^2} \qquad \text{(Equation 2)}$$

But
$$F_w = m_p g \qquad \text{(Equation 3)}$$

Or
$$\frac{F_w}{m_p} = g \qquad \text{(Equation 4)}$$

Substituting Equation 4 in Equation 2,

$$g = \frac{G m_e}{d^2} \qquad \text{(Equation 5)}$$

Since both G and m_e are constants, Equation 5 shows that the acceleration due to gravity in a given location depends only on the square of the distance from the center of the earth.

Compare this value with the one given in Section 2.9. Are they the same?

Figure 3-19. Saturn's rings. Each of the millions of particles circling the planet is held in a specific orbit by the force of gravitation between the planet and the particle.

Table 3-3
SURFACE GRAVITIES OF THE
MOON AND PLANETS

Body	Relative gravity (Earth = 1.00)
Jupiter	2.54
Neptune	1.20
Saturn	1.06
Uranus	0.92
Earth	1.00
Venus	0.92
Mercury	0.38
Mars	0.38
Pluto	uncertain
Moon	0.16

Equation 5 may be used to determine the mass of the earth. Solving this equation for m_e,

$$m_e = \frac{gd^2}{G}$$

Substituting 9.80 m/s² for g, 6.37×10^6 m for d, and 6.67×10^{-11} n m²/kg² for G, $m_e = 5.96 \times 10^{24}$ kg, the mass of the earth.

3.13 Relation Between Gravity and Weight *The term* **gravity** *is used to describe the force of gravitation on an object on or near the surface of a celestial body, such as the earth.* Thus we say that the moon has less gravity than the earth because the force of gravitation near the moon's surface is less than the force of gravitation near the earth's surface. A calculation of the gravity of a celestial body near the surface must take into consideration the size of the body since the force of gravitation varies inversely with the square of the distance to the center of mass of the body. See Table 3-3 for the surface gravities of the moon and planets.

The value for the mass of the earth can be used to determine the theoretical value of the gravity at any given location on or above the earth's surface. Such factors as local variations in the composition of the earth's crust and the effect of the earth's rotation must be considered to obtain the actual value of the gravity at a particular location.

Equation 1, Section 3.12 shows that the weight of an object is the measure of the force of attraction between the object and the earth. When we say that a person weighs 90$\overline{0}$ n, we mean that the force exerted on that person by the earth is 90$\overline{0}$ n. The person also exerts a force of 90$\overline{0}$ n on the earth. Equation 3 tells us that the weight of an object equals the product of its mass and the acceleration due to gravity. The mass of an object is constant. But Equation 5 indicates that the acceleration due to gravity is inversely proportional to the square of the distance from the center of the body to the center of the earth. All parts of the earth's surface are not the same distance from its center. The variations range from 393 m below sea level at the shore of the Dead Sea to 8848 m above sea level at the top of Mount Everest. An object at the top of a mountain, where gravity is less, will weigh less there than it does at sea level. Also, since the earth is slightly flattened at the poles and since the earth's rotation counteracts the force of gravitation, an object weighs a little more at the North Pole than it does at the equator.

3.14 Gravitational Fields The value of g, the acceleration due to gravity, at a particular point is sometimes called the *gravitational field strength* at that point. The effect of rotation, if any, is usually included in determining g. *A region of space in which each point is associated with the value of g at that point is called a* **gravitational field.** Since g is an acceleration, it is a vector. Hence a gravitational field is a region of space in which each point has associated with it a vector equal to the value of g at that point and which is called the gravitational field strength.

If an object of mass m is located at any point in a gravitational field, the force of gravitation on the object can be calculated by the equation in Section 3.9

$$F_w = mg$$

If the mass of the object is 1 kg, then F_w is numerically equal to g. Another way to think of the gravitational field strength at a point is that it equals the force of gravitation that would be exerted on a mass of one kilogram if the mass were located at that point, or

$$g = \frac{F_w}{m}$$

The value of g for the earth varies with the distance from the center of the earth and with certain other aspects of the earth's motion and composition. In Figure 3-20, vectors are used to represent the gravitational field strength at various altitudes. Table 3-4 lists the values of gravitational field strength for various locations on the earth's surface, as determined by careful measurements.

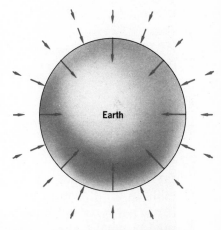

Figure 3-20. Simplified diagram of the earth's gravitational field. The inner circle of vectors represents the magnitude and direction of the field at various places on the earth's surface. The outer vectors represent the field at different altitudes.

Table 3-4
GRAVITATIONAL FIELD STRENGTH OF THE EARTH

Location	Latitude	Altitude (m)	Field strength (n/kg)
North Pole	90° N	0	9.832
Greenland	70° N	20	9.825
Stockholm	59° N	45	9.818
Brussels	51° N	102	9.811
Banff	51° N	1376	9.808
Chicago	42° N	182	9.803
New York	41° N	38	9.803
Denver	40° N	1638	9.796
San Francisco	38° N	114	9.800
Canal Zone	9° N	6	9.782
Java	6° S	7	9.782
New Zealand	37° S	3	9.800

The concept of force fields is useful in describing various natural phenomena.

The concept of a field is more complex than the idea of forces pushing on objects. Yet the field concept can explain more physical phenomena (such as electromagnetic waves) than can the simpler concept of forces. The field idea pictures objects as having associated regions such that a mass would be acted on by a force if it were placed in the region. Thus a ball falls to the earth because of interaction between the earth's gravitational field and the ball.

Another way to describe a gravitational field is to say that an object changes the properties of the space in its vicinity. However, the field concept does not explain the cause of gravitation any more than the idea of acceleration does. Even though scientists are able to express certain aspects of gravitation in terms of acceleration and force fields, the real nature and origin of the force of gravitation are still unknown.

Questions

GROUP A

1. (a) What force acts on an artificial satellite in orbit around the earth?
   (b) What force is the satellite exerting?
2. (a) On what factors does the magnitude of the force of gravitation depend? (b) How is the force of gravitation related to these factors?
3. Why must we exert five times as much force to lift a mass of 5 kg as we must to lift a mass of 1 kg?
4. Assuming your mass remains constant, how will your weight vary as you go from Colorado Springs, elevation 1800 m, to the top of Pikes Peak, elevation 4300 m?
5. Even though your mass remains constant, what happens to your weight as

you drive from the entrance to Death Valley National Monument, elevation 1600 m, down to the lowest spot in Death Valley, 85 m below sea level?
6. Describe the method used by Cavendish to measure the universal gravitational constant.

GROUP B

7. (a) How does the weight of an object vary as it is transported from the earth to the moon? (b) How would its weight vary if it were transported to the center of the earth?
8. Why does the equation $g = Gm_e/d^2$ give a theoretical and not an actual value for the acceleration due to gravity at a given location?
9. (a) What is meant by a gravitational field? (b) How is gravitational field strength defined?

Problems

GROUP A

1. The acceleration due to gravity on the moon is 1.62 m/s². If a person weighs 795 n on the earth, what will the person's weight be on the moon?
2. The instrument-carrying payload of a spaceship weighs 1058 n on the surface

of the earth. What does it weigh 2.560×10^4 km above the earth?
3. The acceleration due to gravity at Hartford, Connecticut, is 9.80336 m/s². What is the force of gravitation in newtons on a mass of 0.250000 kg at this location?

GROUP B

4. What is the force of gravitation between two spherical $10\overline{0}$-kg masses whose centers are 2.00 m apart?

5. A 5.00-kg and a 10.0-kg sphere are 0.300 m apart. Find the force of gravitation between them.

6. Calculate the theoretical value for the acceleration due to gravity at a point 1.00×10^7 m from the center of the earth.

7. Show that the dimensions of g in the equation $g = F_w/m$ are m/s^2.

SUMMARY

Displacement is a change of position of an object in a particular direction. Motion is the displacement of an object in relation to stationary objects. Speed is the time rate of change of position (displacement). Velocity is speed in a particular direction. Speed is a scalar quantity, but velocity is a vector quantity. Graphs can be used to determine average and instantaneous speed and velocity. Velocity problems can be worked either graphically or trigonometrically by use of the parallelogram method of vector addition.

Acceleration is the rate of change of velocity. The relationships among displacement, time, initial and final velocity, and acceleration can be expressed as equations. In the case of freely falling bodies, the acceleration caused by the earth's gravity is used.

A force is a physical quantity that can affect the motion of an object. Newton's laws of motion are (1) a body will con-

tinue in its state of rest or uniform motion in a straight line unless a net force acts on it; (2) the acceleration of a body is directly proportional to the net force exerted on it, inversely proportional to its mass, and in the same direction as the force; (3) when one body exerts a force on another body, the second body exerts an equal force on the first body but in the opposite direction.

The force of gravitation is the mutual force of attraction between bodies. Newton's law of universal gravitation states that the force of gravitation between two objects is directly proportional to the product of their masses and inversely proportional to the square of the distance between their centers of mass. The mass of the earth can be calculated by means of this law. The force of gravitation near the surface of a celestial body is called gravity. The values of gravity at various distances from a celestial body make up the gravitational field of the body.

VOCABULARY

acceleration	gravity	motion
average acceleration	instantaneous speed	slope
average speed	instantaneous velocity	speed
average velocity	law of acceleration	thought experiment
constant speed	law of inertia	variable acceleration
constant velocity	law of interaction	variable speed
displacement	law of universal	variable velocity
gravitational field	gravitation	velocity

Isaac Newton 1643-1727

3Ft MAGYAR POSTA

4

CONCURRENT AND PARALLEL FORCES

Isaac Newton's work on the nature of motion expanded on the ideas of Galileo, especially by defining force as the cause of acceleration. The lens and ray diagram on this stamp symbolizes the important discoveries that Newton also made about the nature and behavior of light.

In this chapter you will gain an understanding of:

✔ force vectors
✔ resultant and equilibrant forces
✔ the resolution of forces into components
✔ the nature and measurement of frictional forces
✔ torques
✔ the two conditions of equilibrium

COMPOSITION OF FORCES

4.1 Describing Forces   In Chapter 3, we studied the relationship between forces and motion. Now we shall take a look at some other characteristics of forces, and see how these characteristics are used in solving force problems.

When you push a door shut with your hand, your hand exerts a force on the door. The door also exerts a force on your hand. When you sit in a chair, you push on the chair and the chair pushes on you. In both of these cases, there is physical contact between the objects that are exerting forces on each other.

Forces can also be exerted without such physical contact. While an object is falling toward the earth, the earth exerts a gravitational force on the object and the object exerts a gravitational force on the earth. Yet there is no physical contact between the earth and the falling object.

These examples illustrate several important characteristics of forces:

1. A net force will change the state of motion of an object. The door moved because the force exerted by your hand was sufficient to overcome friction and other forces acting on the door. An object falls because a force is pulling it toward the earth. As we saw in Chapter 3, the application of a net force to an object always produces an acceleration.

2. Forces can be exerted through long distances. Gravitational and magnetic forces have this characteristic.

3. Forces always occur in pairs. When one object pushes

A force never exists by itself.

70

or pulls on another object, there is a force on *each* of the two objects. In the given examples, the two objects were your hand and the door, you and the chair, and the falling object and the earth.

4. *In each pair of forces, the two forces act in exactly opposite directions.* You pushed on the door, and the door pushed back. You pushed down, and the chair pushed up. The earth pulled the falling object toward the earth's center, and the object pulled the earth toward the object's center.

Now let us see how the magnitudes of forces are measured. (You will remember from Chapter 3 that forces are vector quantities and have both magnitude and direction.) When an object is suspended from a spring, it is pulled toward the earth by the force of gravitation. The spring stretches until the restoring force of the spring is equal to the force of gravitation on the object. Another object having the same weight stretches the spring by the same amount. Both objects together stretch the spring twice as far, and an object with three times the weight stretches it three times as far, etc. This characteristic of coiled springs, that the amount of stretch is proportional to the force pulling on the spring, means that we may use the amount of stretch to measure the size of a force. A device that measures forces in this way is called a *spring balance.* Forces may also be measured with devices other than spring balances. The results of such measurements can always be expressed in terms of newtons.

4.2 Combining Force Vectors

Since forces have both magnitude and direction, they are vector quantities. When a vector is used to represent a force, the magnitude of the force is represented by the length of the arrow. The direction of the force may be deduced from the physical situation. For example, suppose a barge is being towed through still water by a tugboat. The tugboat applies a force of 10,000 n to the barge through the towline. The length of the arrow representing the force is proportional to the magnitude of the force, 10,000 n. Figure 4-2 is a diagram of this example. The point of application of the force is the point at which the rope is attached to the barge. A long rope can transmit only a pull in a direction along its length. It cannot transmit a push or a sideways force. The rope is in the direction of the force. The arrow shows this direction.

Force vectors are treated like velocity vectors. For example, suppose two tugboats are attached to the same barge. Tugboat **A** is pulling with a force of 10,000 n and tugboat **B** is pulling with a force of 7,500 n in the same direction.

Figure 4-1. This telescoping boom crane lifts weights of more than 17,000 newtons to a height of sixteen stories.

Refer to Section 1.8 for a discussion of spring and platform balances.

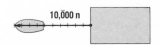

10,000 n

Figure 4-2. A vector diagram of a force of 10,000 n applied by a tugboat to a barge through a towline. In such a diagram, the direction and point of application of the vector represent the line of action and point of application of the force. These can be deduced from the direction and point of attachment of the towline.

Figure 4-3 represents this situation with vectors. The resultant force vector is 17,500 n in the direction of the towline. Figure 4-4 shows a vector diagram representing two tugboats pulling in opposite directions on a barge. Even though the forces act on extended objects such as opposite ends of the barge, the vectors may be considered as acting at the same point. The resultant force vector is 2,500 n in the direction of tugboat **A.**

Figure 4-3. A vector diagram of two tugboats applying forces to a barge in the same direction. Tugboat **A** exerts a force of 10,$\overline{0}$00 n, and tugboat **B** exerts a force of 7,500 n. The resultant force vector is 17,500 n in the direction both tugs are pulling.

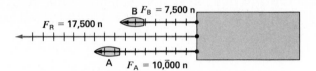

Figure 4-4. A vector diagram of two tugboats applying forces to a barge in opposite directions. Tugboat **A** exerts a force of 10,$\overline{0}$00 n. Tugboat **B** exerts a force of 7,500 n. The resultant force vector is 2,500 n in the direction of tugboat **A.**

The sum of two or more vectors is called the resultant.

Concurrent forces act through the same point at the same time.

Several concurrent forces can be combined into a single resultant that has the same effect.

*When two or more forces act on the same point at the same time, they are called **concurrent forces**. A **resultant force** is a single force that has the same effect as two or more concurrent forces.* When two forces act concurrently in the same or in opposite directions, the resultant has a magnitude equal to the algebraic sum of the forces and acts in the direction of the greater force.

When two forces act concurrently at an angle other than 0° or 180°, the resultant can be found by the parallelogram method, as in the velocity vector problems in Chapter 3. Suppose one force of 10.0 n, F_E, acts eastward upon an object at a point **O.** Another force of 15.0 n, F_S, acts southward upon the same point. Since these forces act concurrently upon point **O,** the vector diagram is constructed with the tails of both vectors at **O.** See Figure 4-5. F_E tends to move the object eastward. F_S tends to move the object southward. When the forces act simultaneously, the object tends to move along the diagonal of the parallelogram of which the two forces are sides. This is the vector F_R. The resultant force vector of two forces acting at an angle upon a given point is equal to the diagonal of a parallelogram of which the two force vectors are sides.

The graphic solution of the magnitude and direction of a resultant force consists of a diagram constructed to scale.

The trigonometric solution makes use of the facts that the opposite sides of a parallelogram are equal and that the diagonal of a parallelogram divides it into two congruent triangles. If the two forces act at right angles, the magnitude and direction of the resultant F_R are found as shown in the following example.

EXAMPLE   Calculate the magnitude and direction of the resultant of the two forces acting on point **O** as shown in Figure 4-5.

SOLUTION   Triangle **OSR** is a right triangle.

$$F_E = 10.0 \text{ n}$$
$$F_S = 15.0 \text{ n}$$

$$\tan \theta = \frac{F_E}{F_S} = \frac{10.0 \text{ n}}{15.0 \text{ n}} = 0.667$$

$$\theta = 33.7°$$

$$F_R = \frac{F_S}{\cos \theta} = \frac{15.0 \text{ n}}{0.833} = 18.0 \text{ n}$$

So $F_R = 18.0$ n 33.7° east of south.

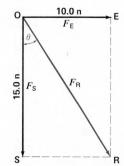

Figure 4-5. A diagram for the determination of the resultant force vector of two force vectors acting at right angles.

The angle between two forces acting on the same point is often not a right angle. Figure 4-6 represents the vectors for the two forces 10.0 n east and 15.0 n 40.0° east of south acting on point **O**. The parallelogram is completed as shown. Observe that the diagonal vector representing the resultant *is drawn from the point on which the two original forces are acting since the resultant will also act on this point.* The magnitude of F_R is found graphically to be 23 n. The direction is $\overline{30}°$ south of east.

The resultant is very different if the angle between the two forces is 140.0°. The parallelogram is constructed to scale in the same manner. The force vectors are the sides and the angle between them 140.0°. This parallelogram is shown in Figure 4-7. The diagonal must be drawn from **O**, the point at which the two forces act. The graphic solution for F_R yields $\overline{10}$ n 8° west of south.

The resultant of two forces acting at an acute or obtuse angle can also be found trigonometrically by means of the law of sines and the law of cosines. The use of these equations is shown in the following example.

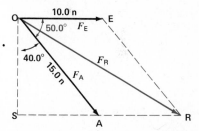

Figure 4-6. A diagram for the determination of the resultant of two force vectors acting at an acute angle.

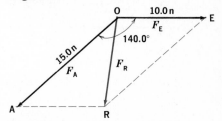

Figure 4-7. Diagram for the determination of the resultant of two force vectors acting at an obtuse angle.

EXAMPLE   Trigonometrically determine the magnitude and direction
of the resultant of the forces shown in Figure 4-6.

SOLUTION   The magnitude and direction of F_R can be found by the
application of the law of cosines and the law of sines.

$\angle EOA = 50.0°$
$\angle OER = 130.0°$
By the cosine law,

$$F_R = \sqrt{F_E^2 + F_A^2 - 2F_E F_A \cos \angle OER}$$
where
$$\cos \angle OER = \cos 130.0° = -\cos(180.0° - 130.0°) = -\cos 50.0° = -0.643$$

$$F_R = \sqrt{(10.0\text{ n})^2 + (15.0\text{ n})^2 - 2(10.0\text{ n})(15.0\text{ n})(-0.643)} = 22.8\text{ n}$$
The magnitude of F_R is 22.8 n.

By the sine law,
$$\frac{ER}{\sin \angle EOR} = \frac{OR}{\sin \angle OER}$$

$$\sin \angle EOR = \frac{(ER)(\sin \angle OER)}{OR}$$

$$\sin \angle OER = \sin 130.0° = \sin(180.0° - 130.0°) = \sin 50.0° = 0.766$$

$$\sin \angle EOR = \frac{(15.0\text{ n})(0.766)}{22.8\text{ n}} = 0.504$$

$\angle EOR = 30.3°$
The direction of F_R is 30.3° south of east.

PRACTICE PROBLEMS   1. A person weighing 730 n is sitting in a swing. The
swing is pulled sideways until the rope makes an angle of
35° with the vertical. What is the tension on the rope?   *Ans.* 890 n

2. A square picture frame weighing 4.0 n is supported by a wire attached
to its two top corners and passed over a nail in the wall. The top edge of
the picture frame is horizontal. What force is exerted by each half of the
wire if the angle between them is $6\bar{0}°$?   *Ans.* 2.3 n

4.3 The Equilibrant Force   *Equilibrium is the state of a
body in which there is no change in its motion.* A body in equilibrium is either at rest or moving at constant speed in a
straight line. In this section we shall discuss the conditions
for equilibrium of bodies at rest. The same conditions hold
for the equilibrium of bodies that are in motion. Examples
of these conditions will be discussed in Chapter 5.

*An equilibrant force is equal in
magnitude to the resultant of two
or more concurrent forces and
acts in the opposite direction.*

A body at rest must be in both translational and rotational equilibrium. *The first condition of equilibrium is that there are no unbalanced (net) forces acting on a body.* The second condition of equilibrium deals with rotation. It will be discussed in Section 4.12.

When there are no unbalanced forces acting on a body, the vector sum of all the forces acting on the body is zero. For example, if a person pulls on a rope with a force of $8\bar{0}$ n and another person pulls on the same rope in the opposite direction with a force of $8\bar{0}$ n, the vector sum of the two forces is zero and the system is in equilibrium. We can say also that each force is the *equilibrant* of the other.

To find the equilibrant of two concurrent forces, we first find their resultant. Then, since the equilibrant must balance the effect of this resultant, it must have the same magnitude but act in the opposite direction. See Figure 4-8.

The equilibrant of three or more concurrent forces can be found in a similar way. First find the resultant by vector addition. Then the equilibrant is graphically drawn from the origin of the forces so that it is equal in magnitude to the resultant but extends in the opposite direction. *When two or more forces act concurrently at a point, the **equilibrant force** is that single force that if applied at the same point would produce equilibrium.*

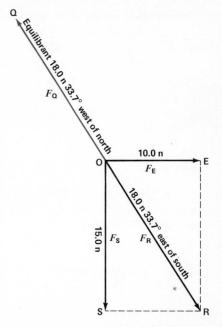

Figure 4-8. The equilibrant of two forces acting at a point is applied at the same point. It equals the magnitude of their resultant but acts in the opposite direction.

Questions
GROUP A

1. (a) Give examples of forces exerted by bodies in contact with each other.
   (b) What types of forces do not require such contact?
2. What device is frequently used to measure forces in physics?
3. What two properties of a force are represented by a force vector?
4. (a) What is a resultant force? (b) How is the resultant found when two concurrent forces act at an angle of 0° or 180°?

5. How is the resultant of two concurrent forces found that act at right angles to each other?
6. What is the first condition of equilibrium?
7. (a) What is an equilibrant force? (b) At what point must it be applied? (c) How does it compare with the resultant force?
8. What is the relationship between the first condition of equilibrium and Newton's first law of motion?
9. How is the resultant of two concurrent forces found that act at an angle to each other that is not a right angle?

Problems
GROUP A

Note: Solve the following problems (a) graphically and (b) trigonometrically.
1. Two forces act simultaneously on point A. One force is 5.0 n south; the other

is 15.0 n south. Determine the magnitude and direction of the resultant force vector.
2. Two forces act on point E. One is 20.0 n north; the other is 20.0 n 30.0° south of east. Find the magnitude and

direction of (a) the resultant, (b) the equilibrant.

3. A person weighs 612 n. If the person sits in the middle of a hammock that is 3.0 m long and sags 1.0 m below the points of support, what force would be exerted by each of the two hammock ropes?

GROUP B

Note: Solve the following problems (a) graphically and (b) trigonometrically.

4. Trigonometrically determine the magni-

tude and direction of the resultant of the two force vectors graphically shown in Figure 4-7.

5. Two forces of 10.0 n east and 30.0 n 30.0° south of west act on the same point. Determine the magnitude and direction of (a) the resultant, (b) the equilibrant.

6. Three forces act simultaneously on point J. One force is 10.0 n north; the second is 15.0 n west; the third is 15.0 n 30.0° east of north. Determine the magnitude and direction of the resultant.

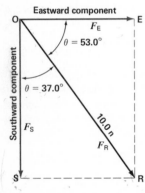

Figure 4-9. A diagram for the determination of the perpendicular components of a force.

A single force can be resolved into two or more components that have the same effect.

RESOLUTION OF FORCES

4.4 Components of Force Vectors Frequently a force acts on a body in a direction in which the body cannot move. For example, gravitational force pulls vertically downward on a wagon on an incline, but the wagon can move only along the incline. Finding the magnitude of the force that is pulling the wagon along the incline is an example of *resolution of forces.* Instead of a single force, two forces acting together can have the same effect as the original single force. One of the forces can be parallel to the surface of the incline and can pull the wagon along the incline. The other can be perpendicular to the surface of the incline. This force does not contribute to the force along the incline. These forces are at right angles to each other. As we saw in Section 2.11, two vectors that have the same effect as a single vector are called the *components* of the original vector. *This procedure of finding component forces is called* **resolution of forces.** Most of the examples we shall consider involve resolving a force into components that are at right angles to each other.

Before working problems with objects on an incline, study the following example, in which a force is resolved into two perpendicular components.

EXAMPLE A force of 10.0 n acts on point **O** at an angle of 37.0° east of south, as shown in Figure 4-9. Find the magnitudes of the southward and eastward components of its force vectors.

$$\cos \theta = \frac{F_S}{F_R}$$

SOLUTION $F_S = F_R \cos \theta$
$\cos \theta = \cos 37.0° = 0.799$
$F_S = (10.0 \text{ n})(0.799) = 7.99 \text{ n}$, the southward component

$$\sin \theta = \frac{F_E}{F_R}$$

$$F_E = F_R \sin \theta$$

$$\sin \theta = \sin 37.0° = 0.602$$

$$F_E = (10.0 \text{ n})(0.602) = 6.02 \text{ n, the eastward component}$$

4.5 Resolving Gravitational Forces

An object placed on an inclined plane is attracted by the earth. The force of attraction is the weight of the object. See Figure 4-10. The plane prevents the motion of the object in the direction of F_W, the direction in which the earth's attraction acts. The vector representing the force of attraction may, however, be resolved into two components. One component acts in a direction perpendicular to the surface of the plane. In physics, the term *normal* is often used to mean perpendicular. Hence we label the normal component F_N. The other component, F_P, acts parallel to the plane. We choose these two components because they have physical significance. The vector F_N represents the force exerted by the object perpendicular to the incline or the amount of the object's weight supported by the incline. The vector F_P represents the component that tends to move the object down the incline. (The plane is assumed to be frictionless.)

Using F_W as the diagonal, we can construct the parallelogram **ODEF** and find the relative values of the sides F_P and F_N by plotting to scale. We can also express these values trigonometrically. Since right triangles **ABC** and **OED** have mutually perpendicular, or parallel, sides, the triangles are similar and $\angle \textbf{EOD} = \theta$. Hence $\sin \theta$ can be expressed either as $\dfrac{\textbf{BC}}{\textbf{AB}}$ or $\dfrac{F_P}{F_W}$. This equation means that the force vector parallel to the plane, F_P, is smaller in magnitude than the weight vector, F_W, in the same ratio as the height of the plane, **BC**, is smaller than its length, **AB**.

By using $\cos \theta$, it may be similarly shown that the magnitude of the force vector perpendicular to the plane, F_N, is related to the weight vector of the object in the same way that the base of the plane is related to its length.

If θ and F_W are known,

$$F_P = F_W \sin \theta$$
and
$$F_N = F_W \cos \theta$$

Making the plane steeper increases the component F_P and decreases the component F_N. This steeper inclined plane is shown in Figure 4-11. The vector F_a, which is

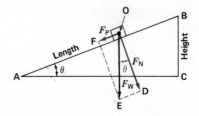

Figure 4-10. Resolution of gravitational force. One component acts parallel to the plane while the other component acts normal (perpendicular) to the plane.

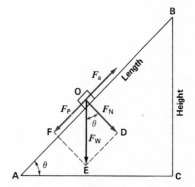

Figure 4-11. When the angle of the incline increases, the component of the weight acting parallel to the plane increases, while the component that acts normal (perpendicular) to the plane decreases.

equal and opposite to F_P, represents the applied force needed to keep the object from sliding down the plane. The steeper the plane, the greater this force becomes. It should be noted that in both Figure 4-10 and Figure 4-11, the normal force that the plane exerts on the object in reaction to force F_N is not shown, since it is not involved in the calculations for finding F_a.

Questions

GROUP A

1. How is the resolution of forces related to the composition of forces?
2. What is the angle between the components into which a force vector is usually resolved?
3. What is meant by a normal force?
4. A car is parked on a hill. (a) Into what two convenient components can the weight of the car be resolved? (b) Which component is counteracted by the brakes of the car?

GROUP B

5. (a) Describe two convenient components of the weight of an object lying on an inclined plane. (b) How are the magnitudes of these components related to the angle of the incline?
6. The transmitting tower of a TV station is held upright by guy wires that extend from the top of the tower to the ground. The wires make a 30.0° angle with the tower. Into what two components can the force that the tower exerts on each guy wire be resolved?

Problems

GROUP A

Note: Solve the following problems (a) graphically and (b) trigonometrically.

1. A person pushes with 150.0 n of force along the handle of a lawn roller. The angle between the handle and the ground is 45.0°. Determine the magnitudes of the horizontal and vertical components of this force.
2. A truck weighing 1.00×10^5 n is parked on a hill that rises 3.00 m in each $10\overline{0}$ m of road. What is the magnitude of the component of its weight that tends to make the truck roll down the hill?
3. A force of 25.0 n acts on point C at an angle of 35.0° west of north. What are the magnitudes of the northward and westward components of this force?
4. A person pulls a sled along level ground. The rope with which the person pulls the sled makes an angle of

20.0° with the ground as the person pulls with a force of 78.0 n. What are the horizontal and vertical components of the force?

GROUP B

Note: Solve the following problems (a) graphically and (b) trigonometrically.

5. A crate having a mass of 114 kg rests on an inclined plane that makes an angle of 12.0° with the horizontal. (a) What force does the crate exert perpendicular to the plane? (b) What force tends to make the crate slide down the plane?
6. A sign weighing 495 n is supported as shown in Figure 4-12. Determine the magnitudes and directions of the forces exerted by the chain and by the bracket. Assume that the horizontal rod cannot support vertical forces.
7. A block of stone having a mass of 2.00×10^3 kg is to be raised by a crane. The angle between the load

cable and the crane boom is 30.0°. The hinged boom is held in place by a tie cable that forms an angle of 20.0° with the boom. (See Figure 4-13.) Determine the magnitude of the thrust force (push) of the boom and of the tie force (pull) of the cable.

8. The rafters of a roof meet at an angle of 120.0°. What force is exerted along the rafters by an object weighing 6.00×10^3 n suspended from the peak?

FRICTION

4.6 The Nature of Friction   In Section 4.5 we discussed the resolution of the weight of an object resting on an incline. One of the components of the weight tends to pull the object down the incline. As the angle of the incline increases, this component also increases. The slightest angle of incline will produce the component that pulls the object down the incline *if* there is no restraining force on the object. However, in performing experiments of this type, we find that the object does not begin to slide until the component parallel to the incline reaches a certain value. This means that forces must exist between the object and the incline that prevent the object from sliding. These forces are called *forces of friction,* or simply *friction.* **Friction** *is a force that resists motion. It involves objects that are in contact with each other.*

The cause of friction is not simple. Some scientists believe that friction is caused mainly by the uneven surfaces of the touching objects. As the surfaces are rubbed together, they tend to interlock and thus offer resistance to being moved over each other. It has been shown that tiny particles are actually torn from one surface and become imbedded in the other.

From this theory of friction one would expect that if the two surfaces are carefully polished, sliding friction between them would be lessened. Experiments have shown, however, that there is a limit to the amount by which friction may be reduced by polishing the surfaces. If they are made very smooth, the friction between them actually increases. This observation has led to the theory that some cases of sliding friction may be caused by the forces of attraction between the molecules of substances.

In many instances, friction is very desirable. We would be unable to walk if there were no friction between the soles of our shoes and the ground. There must be friction between the tires of an automobile and the road before the automobile can move. When we apply the brakes on the automobile, the friction between the brake linings and

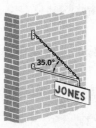

Figure 4-12.

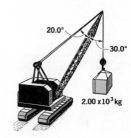

Figure 4-13.

Friction resists the motion of objects that are in contact with each other.

Some of the causes of friction are difficult to explain.

Without friction you couldn't write your homework.

the brake drums, or disks, slows down the wheels. Friction between the tires and the road brings the car to a stop. In a less obvious way, friction holds screws and nails in place and it keeps dishes from sliding off a table if the table is not perfectly level. On the other hand, friction can also be a disadvantage, as it is when we try to move a heavy piece of furniture by sliding it across the floor.

4.7 Measuring Friction Friction experiments are not difficult to perform but the results are not always easy to express as equations or laws. The following statements, therefore, should be understood as approximate descriptions only. Furthermore, they deal exclusively with solid objects. Frictional forces involving liquids and gases are beyond the scope of this book. Also, our discussion is restricted to starting and sliding friction. *Starting friction* is the maximum frictional force between stationary objects. *Sliding friction* is the frictional force between objects that are sliding with respect to one another. Static friction (which varies from 0 to the value of starting friction) and rolling friction are not considered at this time.

1. *Friction acts parallel to the surfaces that are in contact and in the direction opposite to the motion of the object or to the net force tending to produce such motion.* Figure 4-14 illustrates this principle. The weight of the block, F_W, is balanced by the upward force of the table, F_W'. The force F_a is sliding the block along the table top. In this case, F_a is parallel to the table top. The sliding frictional force, F_f, also parallel to the table top, resists the motion and is exerted in a direction opposite to that of F_a.

2. *Friction depends on the nature of the materials in contact and the smoothness of their surfaces.* The friction between two pieces of wood is different from the friction between wood and metal.

3. *Sliding friction is less than or equal to starting friction.* Starting friction prevents motion until the surfaces begin to slide. When the object begins to slide, less force is required to keep it sliding than was to start it sliding.

4. *Friction is practically independent of the area of contact.* The force needed to slide a block along a table is almost the same whether the block lies on its side or on its end. Figure 4-15 illustrates this principle. The surfaces are in contact in more places when the area of contact is large, but the pressure is greater when the area is small.

5. *Starting or sliding friction is directly proportional to the force pressing the two surfaces together.* It does not require as much force to slide an empty chair across the floor as it does

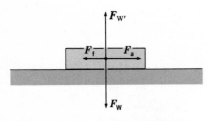

Figure 4-14. Forces on a block being pulled along a surface. F_f is the force of sliding friction. The block does not move until F_a exceeds the force of starting friction, which is usually greater than F_f.

to slide the same chair when a person is sitting on it. The reason for this is that the extra force actually deforms the surfaces to some extent and thus increases the friction.

A simple way to measure starting and sliding friction is with a spring balance, as shown in Figure 4-15. Blocks of

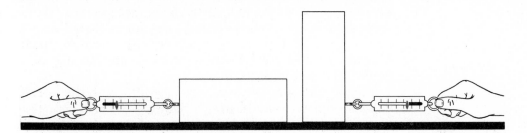

Figure 4-15. The force of friction does not vary significantly with the area of contact if the two blocks have the same weight and all surfaces are equally smooth.

the same substance, but with different sizes and shapes, are pulled along a smooth surface. If the surfaces in contact are consistently smooth, the ratio between the force of sliding friction and the weight of the block is the same in each trial. The ratio depends only on the substances used and not on the area of contact or the weight of the block. This ratio is called the *coefficient of sliding friction.* It may be defined as *the ratio of the force of sliding friction to the normal* (perpendicular) *force pressing the surfaces together.* As an equation, it can be written as

$$\mu = \frac{F_f}{F_N}$$

where F_f is the force of sliding friction, μ (the Greek letter mu) is the coefficient of sliding friction, and F_N is the normal (perpendicular) force between the surfaces. (Within certain limits, this equation is an approximate summary of friction measurements.) The coefficient of starting friction is determined in a similar fashion, except that F_f is the force of starting friction. The approximate values for the coefficients of friction of various surfaces in contact with each other are given in Table 4-1.

4.8 Changing Friction   In winter we sand icy sidewalks and streets in order to increase friction. Tire chains and snow tires are used for the same reason. In baseball, pitchers often use rosin to get more friction between their fingers and the ball. Many more examples could be given in which friction is purposely increased by changing the nature of the surfaces that are in contact.

The most common method of reducing sliding friction is by lubrication. The skier pictured in Figure 4-16 applied a layer of wax to the skis to reduce friction. A thin film of oil

Table 4-1
COEFFICIENTS OF FRICTION

Surfaces	Starting friction	Sliding friction
steel on steel	0.74	0.57
glass on glass	0.94	0.40
wood on wood	0.50	0.30
rubber tire on dry road		0.70
rubber tire on wet road		0.50
Teflon on Teflon	0.04	0.04

Figure 4-16. Friction between skis and snow is appreciably reduced with the application of a wax layer on the wood or metal surface of the skis.

between rubbing surfaces reduces friction. The lesser friction between a liquid and a solid has replaced the greater friction between two solids. Alloys have also been developed that are in effect self-lubricating. For example, when steel slides over an alloy of lead and antimony, the coefficient of friction is less than when steel slides over steel. Bearings lined with such an alloy reduce friction. From Table 4-1 it is also obvious that if a bearing is coated with a plastic such as Teflon, there is very little friction. Such bearings are used in electric motors where the use of a liquid lubricant is undesirable.

The squeaking wheel gets the grease.

Friction may also be greatly reduced through the use of ball bearings or roller bearings. Sliding friction is changed to rolling friction, which has a much lower coefficient. Using steel cylinders to roll a heavy box along the floor is another example of changing sliding to rolling friction.

4.9 Solving Friction Problems   The force required to slide an object along a level surface can be computed easily from the weight of the object and the coefficient of sliding friction between the two surfaces. However, when the force applied to the object is not applied in the direction of the motion, it is necessary to resolve forces in the calculation. This resolution of forces is illustrated in the following example.

In the case of an object resting on an incline, the forces of starting and sliding friction will determine whether the object remains at rest, slides down the incline with constant speed, or accelerates as it descends. How the angle of the incline and coefficient of sliding friction are used in such a problem is illustrated in another example.

EXAMPLE   A box weighing $45\overline{0}$ n is pulled along a level floor at constant speed by a rope that makes an angle of 30.0° with the floor, as shown in Figure 4-17. If the force on the rope is $26\overline{0}$ n, what is the coefficient of sliding friction?

SOLUTION   *Basic equation:* $\mu = \dfrac{F_f}{F_N}$

Before we can use this equation, we must calculate the component of the applied force that acts parallel to the floor. Also, we must calculate the net normal force acting on the floor while the box is being pulled. This is done by finding the appropriate vector components.

The force F_a has a horizontal component, F_h, which moves the object along the level floor. The force of friction, F_f, acts opposite to F_h. Since the speed is constant, $F_f = F_h$.

$$F_f = F_h = F_a \cos 30.0°$$

The force between the surfaces, F_N, is due to the downward action of the weight that is decreased by the vertical upward component, F_V, of the force F_a.

$$F_N = F_W - F_V$$
$$F_V = F_a \sin 30.0°$$
$$F_N = F_W - F_a \sin 30.0°$$

Since $\mu = \dfrac{F_f}{F_N}$ while $F_f = F_a \cos 30.0°$

and $F_N = F_W - F_a \sin 30.0°$,

substitution gives the

Working equation: $\mu = \dfrac{F_a \cos 30.0°}{F_W - F_a \sin 30.0°}$

$$\mu = \frac{(26\overline{0} \text{ n})(0.866)}{45\overline{0} \text{ n} - (26\overline{0} \text{ n})(0.500)} = \frac{225 \text{ n}}{32\overline{0} \text{ n}} = 0.703$$

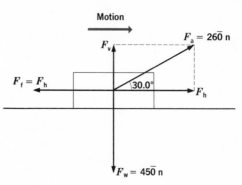

Figure 4-17.

EXAMPLE   A wood block weighing $13\overline{0}$ n rests on an inclined plane, as shown in Figure 4-18. The coefficient of sliding friction between the block and the plane is 0.620. Find the angle of the inclined plane at which the block will slide down the plane at constant speed once it has started moving.

SOLUTION   For the block to slide at constant speed, the net force in the direction of sliding must be zero. That is, the frictional force, F_f, must be just enough to equal and cancel the force F_P that tends to pull the block downward along the incline.

Basic equation: $F_P = F_f$

But we know that the frictional force equals the coefficient of friction times the normal force, or $F_f = \mu F_N$. Substituting in the basic equation, we have

$$F_P = \mu F_N \quad \text{or} \quad \frac{F_P}{F_N} = \mu$$

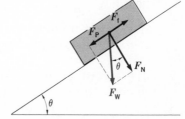

Figure 4-18.

From the diagram, we see that $\dfrac{F_P}{F_N} = \tan\theta$. Substituting in the previous equation, we now have $\tan\theta = \mu$. This is a useful result that holds for similar problems.

Since $\tan\theta = \mu$, then   Working equation: $\theta = \arctan\mu$

(This expression is read as θ is the angle whose tangent has a value of μ.)

Substituting the given value of μ in the equation,

$$\theta = \arctan 0.620$$

Looking up the angle whose tangent is 0.620 in Appendix B, Table 6, gives

$$\theta = 31.8°$$

In other words, the block will slide uniformly when the angle of the incline is 31.8°, no matter what the block weighs. (The weight of the block cancels out.) If the angle is less than 31.8°, the block will not slide at all. If the angle is greater than 31.8°, the block will accelerate as it slides down the plane. The angle at which the block slides depends only on the coefficient of friction.

PRACTICE PROBLEMS   1. The coefficient of sliding friction between a crate and a horizontal floor is 0.27. If the crate has a mass of 250 kg, what force is needed to slide it across the floor at constant velocity? *Ans.* 660 n

2. A crate weighing 1800 n is pulled along a floor by a force of 990 n. The floor is inclined 15° to the horizontal and the crate moves with constant velocity. What is the coefficient of sliding friction between the crate and the floor?   *Ans.*   0.31

Questions

GROUP A

1. (a) What is friction? (b) What ideas have scientists presented to explain friction?
2. Give several examples of ways in which friction is helpful.
3. What methods are used to reduce friction?
4. How does sliding friction compare with starting friction?
5. (a) On what does the amount of sliding friction usually depend? (b) Of what is it independent?
6. What is meant by the coefficient of friction?

GROUP B

7. What is the direction of the force of friction when two surfaces are moving over one another?
8. Name several devices that increase friction between the tires of a car and the pavement.
9. Why does a lubricant reduce friction in a bearing?
10. (a) Under what conditions is the weight of an object the same as the normal force pressing it to the surface over which it is moving? (b) How is the normal force component of the weight of an object determined under other conditions?

Problems
GROUP A

Note: For each problem, draw a force diagram using a suitable scale, and then perform the necessary calculations.

1. A block weighs 2.00×10^3 n. If a horizontal force of 1.00×10^2 n is required to keep it in motion with constant speed on a horizontal surface, what is the coefficient of sliding friction?
2. In a coefficient of friction experiment, a horizontal force of 45 n was needed to keep an object weighing 125 n sliding at constant speed over a horizontal surface. Calculate the coefficient of sliding friction.
3. A crate weighing 1.25×10^3 n slides down an inclined plane at constant speed. The plane is 6.0 m long. Its height is 3.0 m. What is the coefficient of sliding friction between the crate and the inclined plane?

GROUP B

4. The coefficient of sliding friction between a metal block and the inclined surface over which it will slide is 0.200. If the surface makes an angle of 20.0° with the horizontal and the block has a mass of 80.0 kg, what force is required to slide the block at constant speed up the plane?
5. A crate weighing $40\overline{0}$ n is pulled along a horizontal sidewalk at constant speed by a rope that makes an angle of 45.0° with the sidewalk. If a force of $15\overline{0}$ n is applied to the rope, what is the coefficient of sliding friction?
6. A box having a mass of 50.0 kg is dragged across a horizontal floor by means of a rope tied on the front of it. The coefficient of sliding friction between the box and the floor is 0.300. If the angle between the rope and the floor is 30.0°, what force must be exerted on the rope to move the box at constant speed?
7. (a) If a crate slides down a 10.0° incline at constant speed, what is the coefficient of sliding friction between the crate and the incline? (b) What is the coefficient of sliding friction when a crate slides down a 20.0° incline at constant speed? (c) What is the coefficient of sliding friction for a crate moving at constant speed down an incline of x°?
8. A block weighing $13\overline{0}$ n is on an incline. It is held back by a weight of 45.0 n hanging from a cord that passes over a frictionless pulley and is attached to the block as shown in Figure 4-19. Find the angle θ at which the block will slide down the plane at constant speed. The coefficient of friction is 0.620.

PARALLEL FORCES

4.10 Center of Gravity Thus far in our study of forces, we have been treating all the forces acting on a body as if they were acting at a single point. However, there may be many forces, each acting at a different point on the object. For example, Figure 4-20 represents a stone lying on the ground. Since every part of the stone has mass, every part is attracted to the center of the earth. Because of the large size of the earth, all the downward forces exerted on the stone are virtually parallel. The weight of the stone can be thought of as a force vector that is the vector sum, or resultant, of all these parallel force vectors. *Parallel forces*

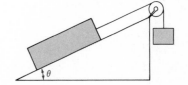

Figure 4-19.

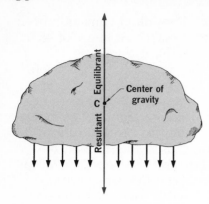

Figure 4-20. All the weight of the stone appears to be concentrated at the point called its center of gravity.

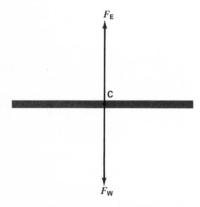

Figure 4-21. The weight of the bar, F_W, is apparently concentrated at the center of gravity, **C**, and can be balanced by an equal and opposite force, F_E, applied at **C**.

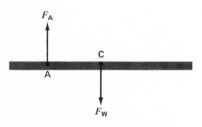

Figure 4-22. Two parallel forces, F_A and F_W, are applied at different points. A tendency to rotate results.

act in the same or in opposite directions at different points on an object. The resultant of parallel forces has a magnitude equal to the algebraic sum of all the forces. The resultant acts in the direction of this net force.

But where in the stone is this resultant force acting? Experiments show that if the proper point of application is chosen, the stone can be lifted without producing rotation. As shown in Figure 4-20, the equilibrant lifting force vector is then in line with the resultant (weight) vector of the stone. In other words, the stone acts as if all its weight were located at one point, which is called the *center of gravity. The **center of gravity** of any object is that point at which all of its weight can be considered to be concentrated.*

In Figure 4-21, the center of gravity of the bar is at **C**. In a bar of uniform construction, **C** is at the geometric center. But if the density or shape of the bar is not uniform, **C** is not at the geometric center.

Since the weight of the bar, F_W, can be considered to be acting at **C**, the bar can be suspended without changing its rotation by an equilibrant force, F_E, applied at **C**. Since F_W and F_E are equal but opposite vectors, they counterbalance each other. In this condition the bar is in both *translational and rotational equilibrium.* This expression means that the bar is not accelerating and its rotation (if any) is uniform.

4.11 Torques The two forces represented in Figure 4-22 by the vectors F_A and F_W are parallel. They do not act on the same point as did the concurrent forces we studied earlier in this chapter. To measure the rotating effect, or *torque,* of such parallel forces in a given plane, it is first necessary to choose a stationary reference point for the measurements. We shall refer to this stationary reference point as the *pivot point.*

Sometimes, as in the case of a seesaw, there is a "natural" point about which the rotating effects can be measured. However, such a pivot point is "natural" only when the seesaw is in motion. When it is motionless there is no "natural" pivot point. Any point on the seesaw, or even beyond it, may be chosen.

Once a suitable pivot point is chosen, a perpendicular line is drawn on the vector diagram from it to each of the lines along which force vectors act on the object. Each such line is called a *torque arm.* In some cases, the force vectors must be extended in order to meet the perpendicular. *Torque, T, is the product of a force and the length of its torque arm.* The unit of torque is the meter-newton.

To illustrate the concept of torque, consider the bar in Figure 4-22 with the application of an additional force, F_B, as shown in Figure 4-23. Choosing **A** as the pivot point, **CA** is the torque arm of F_W and **BA** is the torque arm of the additional upward force, F_B. The clockwise torque around **A** is the product of F_W and **CA**. The counterclockwise torque is the product of F_B and **BA**. Since F_A has a torque arm of zero, it produces no torque and does not enter into the calculations.

To identify a torque as clockwise or counterclockwise, imagine that the bar is free to rotate around a stationary pivot point. Further imagine that the force producing the torque is the only force acting on the bar. The direction in which the bar would rotate is the direction of the torque.

4.12 Rotational Equilibrium In Section 4.3 we discussed the conditions necessary for the translational equilibrium of an object. These conditions, however, do not prevent the *rotary motion* of an object that is subjected to torques. To prevent rotation in a given plane a second condition of equilibrium must be met. *The **second condition of equilibrium** in a given plane is that the sum of all the clockwise torques equals the sum of all the counterclockwise torques about any pivot point.*

Both conditions of equilibrium are illustrated in Figure 4-24. The sum of the force vectors is zero ($3\bar{0}$ n upward

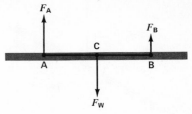

Figure 4-23. Rotational equilibrium results when the sum of the clockwise torques is equal to the sum of the counterclockwise torques. Any point may be chosen as the pivot point in making the computation provided the vector sum of the forces is zero.

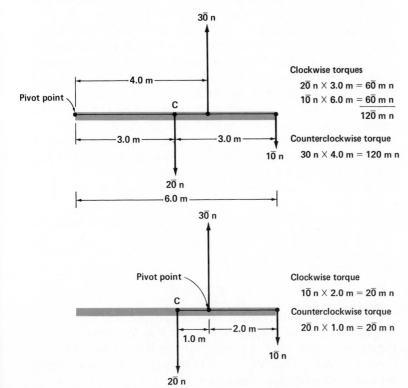

Clockwise torques

$2\bar{0}$ n $\times$ 3.0 m = $6\bar{0}$ m n
$1\bar{0}$ n $\times$ 6.0 m = $6\bar{0}$ m n
$12\bar{0}$ m n

Counterclockwise torque

30 n $\times$ 4.0 m = 120 m n

Clockwise torque

$1\bar{0}$ n $\times$ 2.0 m = $2\bar{0}$ m n

Counterclockwise torque

$2\bar{0}$ n $\times$ 1.0 m = $2\bar{0}$ m n

Figure 4-24. The calculation of torques can be simplified by setting one of the torque arms equal to zero. In the upper diagram, the left end of the bar is used as the pivot point. In the lower diagram, the point of application of the upward force is used, thereby reducing the number of torques.

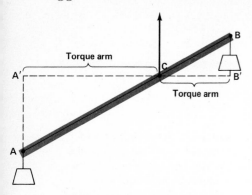

Figure 4-25. The torque arm is the perpendicular distance from the pivot point to the line indicating the direction of the applied force.

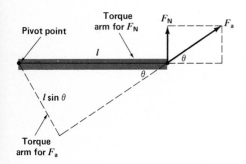

Figure 4-26. The resolution of forces is used to find the torque produced by a force acting on the bar at an angle other than perpendicular.

Figure 4-27. A pair of parallel forces of equal magnitude acting in opposite directions but not on the same point is called a couple. (The weight of the bar is not considered in this example.)

against $2\overline{0}$ n + $1\overline{0}$ n downward). The sum of the clockwise torques is equal to the sum of the counterclockwise torques. Two methods of computing the torques are shown. In the upper drawing of Figure 4-24, the left end of the bar is chosen as a pivot point. In the lower drawing, the point of application of the upward force is used as the pivot point. This simplifies the calculation.

When forces are applied to a bar at an angle other than perpendicular, the distances along the bar measured from the points of application of the forces cannot be used to measure the torque arms. *The torque arms must always be measured perpendicular to the directions of the forces.* Figure 4-25 shows such a situation. A meter stick is under the influence of three parallel forces. To find the torque arms it is necessary to draw a horizontal line through the pivot point. Since the weights hang vertically, a horizontal line is perpendicular to the force vectors of the weights. The problem is simplified by locating the pivot point at the center of gravity. If the bar is not homogeneous, this center of gravity may not be located at the geometric center of the bar. (Experimentally, the center of gravity can be approximately located by finding the point where the bar balances.) Locating the pivot point at the center of gravity eliminates two torques in the equation. It eliminates the torque produced by the weight of the bar located at some distance from the geometric center of the bar. And it also eliminates the torque produced by the force acting upward at **C**. The required torque arms, **CA′** and **CB′**, are then found by multiplying the distances **CA** and **CB** by the cosine of the angle **ACA′** or **BCB′**.

In Figure 4-26, F_a is applied to the right end of the bar at an angle other than perpendicular. In order to find the counterclockwise torque produced by F_a, we use the left end of the bar as the pivot point. Then we find the vertical component, F_N, of F_a. Then, by trigonometry,

$$F_N = F_a \sin \theta$$

Since F_N is perpendicular to the bar, the required torque is

$$T = F_N l$$

Substituting,

$$T = F_a l \sin \theta$$

Hence, $l \sin \theta$ is the torque arm of F_a. This result is further verified in Figure 4-26, where $l \sin \theta$ is the length of the perpendicular from the pivot point to the extended line of direction of F_a. This approach is in accord with the definition of torque arm in Section 4.11.

4.13 Coupled Forces The conditions of equilibrium hold true no matter how many forces are involved. An interesting example is one in which *two forces of equal magnitude act in opposite directions in the same plane, but not on the same point.* Such a pair of forces is called a **couple.** A diagram of a couple is shown in Figure 4-27. The torque is equal to the product of one of the forces and the perpendicular distance between them. (This can be proved by computing the sum of the torques produced by the action of the separate forces about any desired pivot point.) A good example of a couple is the pair of forces acting on the opposite poles of a compass needle when the needle is not pointing north and south.

A couple cannot be balanced by a single force since this single force would be unbalanced and would produce linear motion where it was applied. The only way to balance a couple is with another couple; the torques of the two couples must have equal magnitudes but opposite directions.

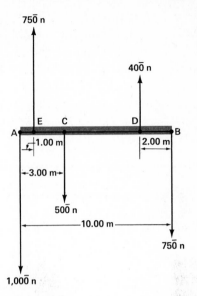

Figure 4-28.

EXAMPLE   A horizontal rod, **AB**, is 10.00 m long. It weighs $50\bar{0}$ n and its center of gravity, **C**, is 3.00 m from **A**. At **A** a force of $100\bar{0}$ n acts downward. At **B** a force of $75\bar{0}$ n acts downward. At **D**, 2.00 m from **B**, a force of $40\bar{0}$ n acts upward. At **E**, 1.00 m from **A**, a force of $75\bar{0}$ n acts upward. (a) What is the magnitude and direction of the force that must be used to produce equilibrium? (b) Where must it be applied?

SOLUTION   Construct a diagram as shown in Figure 4-28.

(a) Consider the known upward force vectors as being positive and the known downward force vectors as being negative. The algebraic sum of these force vectors will then be the resultant force vector. This resultant force vector must be counterbalanced by a force vector in the opposite direction in order to establish translational equilibrium.

$$75\bar{0} \text{ n} + 40\bar{0} \text{ n} - 100\bar{0} \text{ n} - 50\bar{0} \text{ n} - 75\bar{0} \text{ n} = -110\bar{0} \text{ n}$$

Therefore $110\bar{0}$ n must be applied upward to establish translational equilibrium.

(b) Use **A** as the pivot point, and let x be the distance from **A** to the point where the $110\bar{0}$-n force must be applied. To prevent rotary motion,

$$\text{clockwise torque} = \text{counterclockwise torque}$$
$$(50\bar{0} \text{ n})(3.00 \text{ m}) + (75\bar{0} \text{ n})(10.00 \text{ m}) = (75\bar{0} \text{ n})(1.00 \text{ m}) + (40\bar{0} \text{ n})(8.00 \text{ m}) + (110\bar{0} \text{ n})x$$
$$x = 4.59 \text{ m}$$

4.59 m is the distance from **A** to the point where the $110\bar{0}$-n upward force must be applied.

PRACTICE PROBLEMS   **1.** A nonuniform bar is 4.5 m long and weighs 640 n. It is balanced when a 270-n weight is hung 0.80 m from the light end of the bar and the bar is pivoted at its center. Where is the center of gravity of the bar?   *Ans.* 1.6 m from heavy end

2. Two workers carry a nonuniform beam weighing 730 n. The beam is 2.6 m long and its center of gravity is 1.1 m from one end. If the workers hold the beam at its ends, how much does each worker lift?   *Ans.* 310 n and 420 n

---

Questions

GROUP A

1. Define center of gravity.
2. (a) What are parallel forces? (b) How is the resultant of two parallel forces that act in the same direction calculated? (c) How is the resultant of parallel forces that act in opposite directions calculated?
3. Explain how two persons of unequal weight can be made to balance on a seesaw.
4. (a) A force acts at right angles to a pivoted bar. How is the torque produced by this force calculated? (b) What is the unit of torque?

5. If a bar on which parallel forces act is to be in equilibrium, what conditions must be met?

GROUP B

6. How can the choice of the pivot point simplify the calculations of the torques in a problem?
7. Distinguish between center of gravity and the pivot point.
8. Give an example of an object with a center of gravity that is not within the object.
9. A certain force does not act at right angles to a pivoted bar. How is the torque from this force calculated?
10. (a) What is a couple? (b) How is the torque of a couple calculated?

Problems

GROUP A

Note: For each problem, draw a force diagram using a suitable scale and perform the necessary calculations. Unless otherwise noted, the center of gravity is at the geometric center of the object.

1. A steel beam of uniform cross section weighs 2.5×10^5 n. If it is 5.00 m long, what force is needed to lift one end of it?
2. A bar 4.0 m long weighs $40\overline{0}$ n. Its center of gravity is 1.5 m from one end. A weight of $30\overline{0}$ n is attached at the heavy end and a weight of $50\overline{0}$ n is attached at the light end. What are the magnitude, direction, and point

of application of the force needed to achieve translational and rotational equilibrium of the bar?

3. A painter weighing 875 n stands on a plank 3.00 m long, which is supported at each end by a stepladder. The plank weighs 223 n. If the painter stands 1.00 m from one end of the plank, what force is exerted by each stepladder?
4. A bricklayer weighing 8.00×10^2 n stands 1.00 m from one end of a scaffold 3.00 m long. The scaffold weighs $75\overline{0}$ n. A pile of bricks weighing 3.20×10^2 n is 1.50 m from the other end of the scaffold. What force must be exerted on each end of the scaffold in order to support it?

5. A bench is 2.40 m long and weighs 3.20×10^2 n. The legs are attached 0.30 m from each end and weigh 15.0 n each. If three persons weighing, in order, 5.00×10^2 n, 7.50×10^2 n, and 1.000×10^3 n, sit 0.40 m, 1.20 m, and 2.00 m respectively from one end of the bench, what downward force must each set of legs exert on the floor?

6. Give an example showing the correctness of the rule for finding the torque of a couple by placing the pivot point somewhere between the two force vectors.

7. The bridge **AB** in Figure 4-29 is 36.5 m long. It weighs 2.56×10^5 n. A truck weighing 5.25×10^4 n is 10.2 m from one end of the bridge. Calculate the upward force that must be exerted by each pier to support this weight.

GROUP B

8. A bar 5.0 m long has its center of gravity 1.5 m from the heavy end. If it is placed on the edge of a block 1.5 m from the light end and a weight of 750 n is placed on the bar at the light end, it will be balanced. What is the weight of the bar?

9. A uniform bar 25.0 m long weighs 1.000×10^4 n. From end A a weight

of 2.50×10^3 n is hung. At B, the other end of the bar, there is a weight of 3.50×10^3 n. An upward force of 3.00×10^3 n is exerted 4.0 m from B, while an upward force of 4.00×10^3 n is exerted 8.0 m from A. Determine the magnitude, direction, and point of application of the force needed to establish translational and rotational equilibrium.

10. A door is 2.50 m high and 1.00 m wide. It weighs 204 n and its center of gravity is at its geometric center. The door is supported by hinges located 0.30 m from the top and bottom. If each hinge supports half the weight of the door, find the horizontal component of the force exerted by each hinge.

11. (a) Find the torques exerted on the rod in Figure 4-30. (b) Find the magnitude and direction of the additional force that must be exerted at the right end, perpendicular to the rod, to prevent rotational equilibrium.

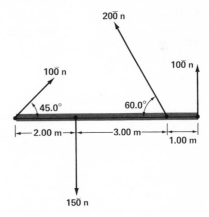

Figure 4-30.

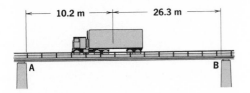

Figure 4-29.

SUMMARY

Forces are vector quantities. A net force will change the state of motion of an object. Forces can be exerted over long distances. Every force is opposed by an equal and opposite force.

A resultant force is a single force that produces the same effect as several forces acting along lines that pass through the same point. The equilibrant force is the single force that produces equilibrium when applied at a point at which two or more concurrent forces are acting. A single force may be resolved into two components, usually acting at right angles to each other. Resultant and component forces are found by the parallelogram method.

Friction is a force that resists the motion of objects that are in contact with each other. It acts parallel to the surfaces that are in contact, depends on the nature and smoothness of the surfaces, is independent of the area of contact, and is directly proportional to the force pressing the surfaces together. Starting friction is usually greater than sliding friction. The coefficient of friction is the ratio of the force of friction to the perpendicular force pressing the surfaces together. The resolution of forces is used in solving friction problems.

Parallel forces act in the same or in opposite directions. The center of gravity of an object is that point at which all of the object's weight can be considered to be concentrated. A stationary object is in translational and rotational equilibrium. The torque produced by a force is the product of the force and the length of the torque arm on which it acts. To produce equilibrium in parallel forces, the sums of the forces in opposite directions must be equal and the sum of all the clockwise torques must equal the sum of all the counterclockwise torques about a pivot point. Two forces of equal magnitude that act in opposite directions but not along the same line are called a couple.

VOCABULARY

center of gravity
coefficient of sliding
   friction
concurrent forces
couple

equilibrant force
equilibrium
friction
parallel forces
resolution of forces

resultant force
rotational equilibrium
torque
torque arm
translational equilibrium

Two-DIMENSIONAL AND PERIODIC MOTION

Christian Huygens developed the equation for centripetal acceleration. He also invented a pendulum-controlled clock and originated the wave theory of light, which was in opposition to the particle theory of light proposed by Newton. The two famous scientists were contemporaries.

CIRCULAR MOTION

5.1 Motion in a Curved Path In Chapters 3 and 4, we studied the motion of objects along a straight line. Now let us examine what happens to the motion of an object that moves horizontally with uniform velocity but is also accelerated vertically. That is, there is a right angle between the direction of the uniform velocity and the direction of the acceleration. What will be the path of such an object? The ball in Figure 5-1 illustrates such motion. It was released at the same time as the ball on the left, but had a uniform horizontal velocity. The horizontal distances between adjacent images of the right-hand ball are equal. This indicates that the ball is still traveling horizontally with uniform velocity. At the same time, the ball on the right covered the same vertical distance between each image as did the ball on the left, which had no horizontal velocity. Each position of the ball on the right is the resultant of motion with uniform velocity in one direction and with uniform acceleration in the other. This two-dimensional combination produces motion along a curved path.

Suppose a rifle bullet is fired horizontally with a velocity of 1250 m/s. Neglecting air resistance, the bullet travels 1250 m horizontally by the end of the first second. Immediately after it leaves the muzzle of the gun, the force of gravity begins to accelerate the bullet toward the earth's center of gravity. This force is vertical. During the first second, a freely falling body drops 4.90 m. The bullet

In this chapter you will gain an understanding of:

▷ motion in two dimensions
▷ the nature of centripetal acceleration
▷ the distinction between centripetal and centrifugal force
▷ radian measure
▷ vector representation of angular velocity and angular acceleration
▷ equations involving rotational inertia
▷ precession
▷ the pendulum as an example of harmonic motion

Two-dimensional motion has a curved path.

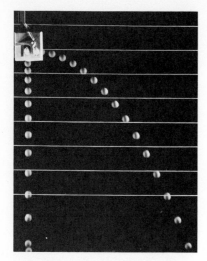

Figure 5-1. Two-dimensional motion. Both balls have the same vertical acceleration due to the force of gravity. But the ball at the right also has a uniform horizontal motion. The resultant of motion in two dimensions is a curved path.

drops 4.90 m while traveling the first 1250 m horizontally. See Figure 5-2. In two seconds the bullet travels 2500 m horizontally, at which time it also drops vertically through a distance of 19.6 m. Over short distances the path of a high-velocity projectile approximates a straight line, but over greater distances its path is noticeably curved.

If the line of sight to a target is horizontal, a projectile must be fired at a small upward angle in order to hit the target. This angle compensates for the downward acceleration due to the force of gravity on the projectile. Provided the same ammunition is used, the size of this angle depends upon the distance to the target. The front sight of most rifles is fixed at the end of the barrel while the rear sight is movable. Since the line of sight is a straight line to the target, the angle the barrel makes with this line is increased by raising the rear sight. When the rifle is aimed, its muzzle is directed upward at the predetermined angle. When the gun is fired, this angle gives the bullet the upward velocity component necessary to compensate for the bullet's drop on its way to the distant target.

The path a projectile takes if it is fired at an upward

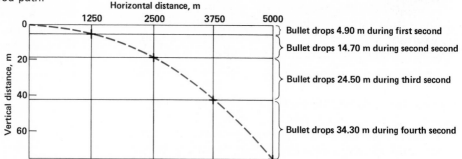

Figure 5-2. The path of a bullet is curved no matter how great the horizontal velocity is.

angle is shown in Figure 5-3. The muzzle velocity, v_m, is resolved into the horizontal component, v_h, and the vertical component, v_v. **AC** is called the range. The path of the projectile, **ABC**, is called the trajectory.

5.2 Motion in a Circular Path An important type of two-dimensional motion is motion in a circular path, as seen in Figure 5-4. Ball **A** is attached to the end of string **CA.** The string is fastened at **C.** If the speed of the ball in the circular path is constant, the ball is said to describe *uniform circular motion.* If the speed of the ball in the circular path varies, its motion is *variable circular motion.*

Imagine that you are at a location where there is no force of gravity. You attach a ball to a string as shown in Figure 5-4 and hold the string at **C.** Then you give the ball an

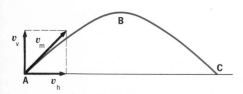

Figure 5-3. When air resistance is not considered, the path of a projectile is a parabola.

initial velocity in the direction of **B,** tangent to the circumference of the circle. The ball whirls in a circular path around **C.** You can feel yourself pulling continuously on the string to keep the ball in the circular path.

Let us analyze what you would observe in this situation. We know that the radius of a circle, **AC** for example, is perpendicular to the tangent drawn through the end of the radius. If the string is in the direction of the radius and the velocity is always directed along a tangent, your pull on the string is always directed perpendicularly to the velocity. Your pull accelerates the ball into a circular path, but the ball does not speed up or slow down. Your pull changes only the direction of the velocity but not the magnitude of the velocity. If your pull were not perpendicular to the velocity, a component of the acceleration in the direction of the ball's motion would exist and the speed of the ball would change.

In this example, the acceleration is directed toward the center of a circle. *Acceleration directed toward a central point is called* **centripetal acceleration.** (The word centripetal means "directed toward a center.")

This example can also be analyzed by means of velocity vectors. Even though the *speed* of the ball in its circular path is uniform, the *velocity* is constantly changing. In Figure 5-5(A), **A** and **B** represent two successive positions of an object moving with uniform circular motion about point **O.** The velocity vector v_i indicates the velocity of the object when at point **A,** and the velocity vector v_f indicates its velocity at point **B.** The velocity vectors, v_i and v_f, are tangent to the circle at **A** and **B** respectively, and are thus perpendicular to the respective radii **OA** and **OB.**

In order to study the change of velocity between v_i and v_f, a separate vector diagram, Figure 5-5(B), is drawn in which v_i and v_f originate at point **X.** If v_f is considered the resultant and v_i one of its components, then Δv is the other component. This vector Δv represents the change in velocity between v_i and v_f. Because the vectors v_i and v_f are equal in magnitude and are perpendicular to their respective radii, $\angle \theta$ in Figure 5-5(A) equals $\angle \theta$ in Figure 5-5(B) and triangle **ABO** is similar to triangle **XYZ.** Then

$$\frac{\Delta v}{v_i} = \frac{\text{chord } \mathbf{AB}}{r}$$

If angle θ is made smaller, arc Δd becomes more nearly equal to chord **AB** and can be substituted for it.

$$\frac{\Delta v}{v_i} = \frac{\Delta d}{r}$$

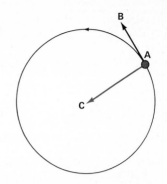

Figure 5-4. Uniform circular motion. When the force along **AC** stops acting, the ball moves in a direction that is indicated by the tangent **AB.**

Centripetal: toward a central point.

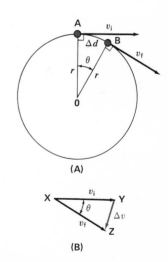

Figure 5-5. Centripetal acceleration. The change of velocity of an object in circular motion is directed toward the center of the circle.

But v_i and v_f are equal in magnitude and can be represented by v. If we let Δt represent the time between v_i and v_f and use the equation $\Delta d = v\,\Delta t$, we can write

$$\frac{\Delta v}{v} = \frac{v\,\Delta t}{r}$$

The equation $v = \dfrac{\Delta d}{\Delta t}$ is derived in Section 3.3. How is the form used here obtained?

Transposing terms

$$\frac{\Delta v}{\Delta t} = \frac{v^2}{r}$$

But $\Delta v/\Delta t$ is the centripetal acceleration directed along a radius of the circle and can be represented by a. Then

$$a = \frac{v^2}{r}$$

in which a is the centripetal acceleration of the object, v is its speed along the circular path, and r is the radius of its circular path.

Refer to Section 3.9.

By Newton's second law of motion, $F = ma$. Thus the force producing centripetal acceleration can be written

$$F_c = \frac{mv^2}{r}$$

and is called the *centripetal force*. The centripetal force is the net force directed toward the center of the circle. The equation for centripetal force indicates that it is directly proportional to the mass of the object, directly proportional to the square of the object's speed along the circular path, and inversely proportional to the radius of the object's circular path.

Because matter has inertia, a centripetal force is required to produce the centripetal acceleration that changes the direction of a moving object. If the centripetal force applied on the ball by the string ceases to act (for example, if the string breaks) the ball will move in a straight line tangent to the curved path at the point where the string breaks. This behavior is in accordance with Newton's first law of motion. When the string breaks, the centripetal force no longer acts on the ball; the ball has no unbalanced forces acting on it and thus moves in a straight line with constant speed.

EXAMPLE   An object weighing 14 n is swung at the end of a cord 0.80 m long. The object makes 2.0 revolutions per second. What is the centripetal force on the moving object?

SOLUTION

Basic equation: $F_c = \dfrac{mv^2}{r}$

$$m = \dfrac{F_w}{g}$$

$$v = 2.0 \text{ rev/s} = 2.0(2\pi r)/\text{s} = 4.0\pi r/\text{s}$$

$$F_c = \dfrac{(F_w)(4.0\pi r/\text{s})^2}{gr}$$

Working equation: $F_c = \dfrac{16\pi^2 F_w r}{g \text{ s}^2}$

$$F_c = \dfrac{16(3.14)^2(14 \text{ n})(0.80 \text{ m})}{(9.80 \text{ m/s}^2)(1.0 \text{ s})^2}$$

$$F_c = 180 \text{ n}$$

PRACTICE PROBLEMS   **1.** A mass moves in a circular path at a velocity of 2.0 m/s and a centripetal acceleration of 4.3 m/s². What is the radius of the circular motion of the mass?   *Ans.* 0.93 m

2. A cord 0.65 m long exerts a centripetal force of 23 n on a whirling 1.00-kg mass tied to the end of the cord. What is the velocity of the whirling mass?   *Ans.* 3.9 m/s

---

5.3 Motion in a Vertical Circle   Thus far, in describing the motion of an object in a circle, we have ignored the effect of the force of gravitation. This was done in order to simplify the analysis of the relationship between centripetal acceleration and the magnitude of the object's velocity. However, if the object moves in a vertical circle with gravity acting on it, the situation is not so simple.

   Suppose a ball on the end of a string moves along a circular path. The plane of the circle is vertical. Figure 5-6 is a diagram of the ball's motion. Because of the force of gravity, the speed of the ball in the circular path is not uniform. The ball accelerates on the downward part of its path and decelerates on the upward part. The speed of the ball is a minimum at the top of the circle and a maximum at the bottom. Consequently, the centripetal force is at a minimum at the top of the circle and at a maximum at the bottom. Let us see what forces comprise the centripetal force to make this so.

   At the top of the circle the ball has velocity v_{min}. The

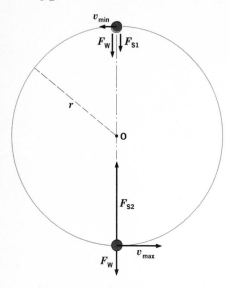

Figure 5-6. Motion in a vertical circle. When the ball is at its highest point, the centripetal force on the ball equals the tension of the string plus the weight of the ball. At its lowest point, the weight of the ball must be subtracted from the tension to find the centripetal force.

centripetal force, F_{c1}, is the sum of the force the string exerts on the ball, F_{s1}, and the weight of the ball, F_w, since these both act toward the center of the circle.

$$F_{c1} = \frac{mv_{min}^2}{r} = F_{s1} + F_w \qquad \text{(Equation 1)}$$

At the bottom of the circle, the ball has velocity v_{max}. The centripetal force, F_{c2}, is the difference between the magnitude of the force the string exerts on the ball, F_{s2}, and the weight of the ball, F_w, since these forces now act in opposite directions. The force the string exerts is inward, and the weight of the ball acts outward.

$$F_{c2} = \frac{mv_{max}^2}{r} = F_{s2} - F_w \qquad \text{(Equation 2)}$$

Observe that since v_{max} is greater than v_{min}, F_{c2} is greater than F_{c1}. Further, since the weight of the ball is added in Equation 1, but subtracted in Equation 2, F_{s2} is greater than F_{s1} by a ratio greater than the ratio F_{c2}/F_{c1}.

When a ball on a string moves in a vertical circle, there is a certain velocity below which the ball will not describe a circular path—the string slackens as the ball approaches its highest point. To find this minimum value of the velocity, we shall use Equation 1. When the string begins to slacken, $F_{s1} = 0$ and

$$\frac{mv_{min}^2}{r} = F_w$$

But
$$F_w = mg$$

So
$$\frac{mv_{min}^2}{r} = mg$$

Solving for v_{min}

$$v_{min} = \sqrt{rg}$$

A satellite in circular orbit around a planet moves with critical velocity.

This value of v_{min} is called the *critical velocity*. As the equation indicates, critical velocity depends only on the acceleration due to gravity and the radius of the vertical circle. The critical velocity does not depend on the mass of the object describing the motion. This is true because the force of gravity and the centripetal force are both proportional to the mass of the object.

5.4 Frames of Reference Figure 5-7 shows a chamber attached to the end of a long boom. The boom is designed to turn around a central point under the observation booth at the top of the picture. In operation, the chamber moves at constant speed in a large circle around the central point.

Suppose you are in the chamber while it is moving. You would still feel the pull of gravity. In addition, you would feel your body being pressed against the outside wall of the chamber. You would also observe that a ball placed on the floor would move across the floor toward the outside wall of the chamber. If you drop an object in the chamber, you would see it fall to the floor in a curved path.

From these observations it would seem that Newton's law of inertia does not hold true in the moving chamber. These observations can be explained, however, by attributing them to a force that tends to move all particles toward the outside wall of the moving chamber. This force moves the ball across the floor and accelerates the falling object so that its path is curved. *A force that tends to move the particles of a spinning object away from the spin axis is called* **centrifugal force.** Centrifugal force exists only for an observer in an accelerating system who is considered to be stationary.

An observer in the booth above the moving chamber in Figure 5-7 interprets the situation quite differently, however. To this observer, the chamber and its occupant tend to continue their motion in a straight line because of their inertia. A force along the boom, a centripetal force, causes the chamber to follow a circular path. And a force exerted by the wall of the chamber on its occupant causes the occupant also to follow a circular path. The observer in the booth can explain the behavior of objects in the chamber in terms of inertia and centripetal force. To this observer, no centrifugal force is involved.

To resolve this seeming contradiction, it is helpful to discuss this situation in terms of *frames of reference.* Each observer assumes that some surrounding objects are stationary because they do not move with respect to each other. To the observer in the moving chamber, the walls and floor of the chamber are stationary and are used as the basis of measurements. Other objects, such as the ball on the floor, move with respect to these stationary objects. The walls and floor of the chamber are the basis for this observer's *frame of reference.*

A **frame of reference** *is a system for describing the location of objects.* It is used to specify the positions and relative motions of objects. A frame of reference in which Newton's first law holds true is called an *inertial frame.* An accelerating frame of reference is *noninertial* because Newton's first law does not hold true.

Centrifugal: away from a central point.

Figure 5-7. This centrifuge is used to study the effects of angular acceleration on the human body. A volunteer is entering the test chamber of the centrifuge at the left.

The observer in the booth in Figure 5-7 is stationary with respect to the earth. Strictly speaking, however, a frame of reference based on the earth is not an inertial frame. The earth spins and orbits around the sun and describes continuous acceleration in doing so. However, the earth is so close to being an inertial frame that we can consider a frame of reference based on the booth as an inertial frame. (For practical purposes, we often neglect the accelerations of frames of reference that are stationary with respect to the earth.)

Centripetal and centrifugal are never used to describe motion in the same frame of reference.

The direction of motion of the moving chamber is continually changing. With respect to the observer in the booth, the moving chamber is accelerating. It is, therefore, a noninertial frame. Using such a frame, one can observe acceleration that is not attributed to such forces as friction or gravitation. The concept of centrifugal force is necessary. Centrifugal force is sometimes called fictitious because it is not involved when we decide to use an inertial frame to describe motion.

Questions
GROUP A

1. How is motion along a curved path produced?
2. (a) A projectile is fired horizontally. What motions can be used to explain its path? (b) What effect does each of these motions have on the path of the projectile?
3. (a) What is uniform circular motion? (b) What force causes circular motion? (c) Why must a force be used to produce circular motion?
4. What factors determine the magnitude of a centripetal force?
5. (a) How does motion in a vertical circle differ from uniform circular motion? (b) What causes this difference?
6. (a) What is meant by critical velocity? (b) What happens to an object that does not have critical velocity?
7. (a) What is meant by a frame of reference? (b) List the four frames of

reference that could be used to describe the motion of an earth satellite.

GROUP B
8. (a) What adjustment is made on the rear sight of a rifle when the distance to the target is increased? (b) How would the use of a higher velocity bullet affect the adjustment?
9. Why does an astronaut experience large forces (a) during the launching of a space vehicle? (b) during the reentry of the vehicle?
10. (a) In what way is the motion of a satellite in a circular orbit similar to motion in a vertical circle? (b) How are the two motions different?
11. Evaluate the statements: (a) "An earth satellite is weightless." (b) "An earth satellite is continuously falling around the earth."
12. Describe the motion of a person on a merry-go-round in terms of two different frames of reference.

Problems

GROUP A

1. What is the centripetal acceleration of an object moving along a horizontal circular path of 16.0-m radius with a speed of 40.0 m/s?

2. If the mass of the object in Problem 1 is 2.00 metric tons, what centripetal force is required to maintain it in a circular path?

3. A ball of mass 2.5×10^{-2} kg is swung at the end of a string in a horizontal circular path at a speed of 5.0 m/s. If the length of the string is 2.0 m, what centripetal force does the string exert on the ball?

4. Calculate the centripetal force exerted on a 5.0-kg mass that is moving at a speed of 3.0 m/s in a horizontal circular motion if the radius of the circle is (a) 1.0 m; (b) 3.0 m.

5. Find the critical velocity of the mass in Problem 4 for both radii if the mass is moving (a) in a vertical circle at the earth's surface; (b) in a vertical circle 6.0×10^2 km above the earth's surface. The radius of the earth is 6.37×10^6 m.

GROUP B

6. An automobile weighs 6.0×10^4 n. If it is driven around a horizontal curve that has a radius of 250 m at the rate of 22 m/s, what is the centripetal force of the road on the automobile?

7. What is the orbiting speed of a satellite moving uniformly in a circular path 1.00×10^3 km above the earth?

8. A person has a mass of 75.0 kg. If the person is standing on the equator, by how much is the person's weight changed because of the earth's rotation?

9. A baseball player hits a fly ball to a height of 50.0 m. After the bat strikes the ball, how much time does a fielder have to get into position to make the catch?

10. A person throws a ball from the roof of a building. The ball has a velocity of 25 m/s horizontally when it leaves the person's hand, and the ball hits the ground 3.0 s later. (a) How far does the ball fall during each second? (b) How far does the ball move horizontally during each second?

ROTARY MOTION

5.5 Motion Around an Axis
Rotary motion is the motion of a body about an internal axis. Rotary motion occurs in a spinning bicycle wheel, the spinning crankshaft of an automobile engine, and a wheel attached to the spinning shaft of an electric motor. Note the difference between circular motion and rotary motion. In circular motion, the axis of the motion is outside the object. In rotary motion, the axis of the motion is inside the moving object. A spinning wheel is in rotary motion; an object on the rim of the wheel describes circular motion.

For rotary motion to be *uniform*, the object must spin about a fixed axis at a constant rate. The movement of the hands of a clock is an example of uniform rotary motion. If either the direction of the axis or the rate of spin varies, the rotary motion is *variable*. The movements of automobile wheels as a car is driven at different speeds and the

Rotating and spinning are synonymous.

movements of a spinning top as it slows down are examples of variable rotary motion.

5.6 Angular Velocity For uniform linear motion, velocity is defined as the time rate of displacement. Similarly, for uniform rotary motion, **angular velocity** is defined as *the time rate of angular displacement*. Angular displacement is the angle about the axis of rotation through which the object turns. The symbol for angular velocity is the Greek letter ω (omega). The equation for angular velocity, ω, is

$$\omega = \frac{\Delta\theta}{\Delta t}$$

where $\Delta\theta$ is an angular displacement and Δt is the time interval in which the angular displacement occurs. The units of angular velocity are revolutions per second, degrees per second, or *radians per second*.

    An angle of one radian is the angle that, when placed with its vertex at the center of a circle, subtends on the circumference an arc equal in length to the radius of the circle. Thus in Figure 5-8, if radius r' is rotated from radius r until arc $\Delta d =$ radius r, the angle θ is 1 radian. Since the circumference of a circle is 2π times the radius

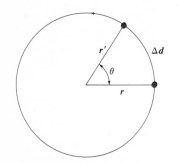

Figure 5-8. Radian measure. Since arc Δd equals radius r, angle θ is 1 radian. There are 2π radians in one revolution.

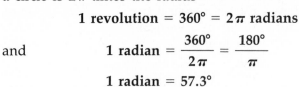

1 revolution = 360° = 2π radians

and   **1 radian $= \dfrac{360°}{2\pi} = \dfrac{180°}{\pi}$**

1 radian = 57.3°

Since an angle measured in radians equals the ratio of the length of the subtended arc to the length of the radius, the units cancel, and the angle in radians is a pure number.

    The concept of angular velocity includes both the rate of rotation and the direction of the axis of rotation. Angular velocity is a vector quantity represented by a vector along the axis of rotation. The length of the vector indicates the magnitude of the angular velocity. The direction of the vector is the direction in which the thumb of the right hand points when the fingers of the right hand encircle the vector in the direction in which the body is rotating. See Figure 5-9.

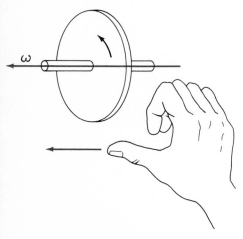

Figure 5-9. Right-hand rule of angular velocity. The vector representing angular velocity points in the direction of the thumb of the right hand when the fingers encircle the wheel in the direction of rotation.

5.7 Angular Acceleration From the study of linear motion we know that a change of velocity defines acceleration. The same is true for rotary motion. Changing either the rate of rotation or the direction of the axis involves a change of angular velocity and thus defines angular acceleration. The uniform rate of change of linear velocity is known as linear acceleration (Section 3.5). Similarly, the uniform rate of change of angular velocity is known as angular acceleration and designated as α (alpha).

$$\alpha = \frac{\Delta \omega}{\Delta t}$$

The equations for uniformly accelerated linear motion can be transformed into the corresponding equations for uniformly accelerated rotary motion by substituting $\Delta \theta$ for Δd, ω for v, and α for a. See Table 5-1.

Table 5-1
EQUATIONS FOR UNIFORMLY ACCELERATED MOTION

Linear	Rotary
$v_f = v_i + a\,\Delta t$	$\omega_f = \omega_i + \alpha\,\Delta t$
$\Delta d = v_i\,\Delta t + \frac{1}{2}a\,\Delta t^2$	$\Delta \theta = \omega_i\,\Delta t + \frac{1}{2}\alpha\,\Delta t^2$
$v_f = \sqrt{v_i^2 + 2a\,\Delta d}$	$\omega_f = \sqrt{\omega_i^2 + 2\alpha\,\Delta \theta}$

EXAMPLE Find the angular displacement in radians during the second 20.0-s interval of a wheel that accelerates from rest to 725 revolutions per minute in 1.50 min.

SOLUTION The angular displacement $\Delta \theta$ required for the second 20.0-s interval is equal to the difference between the angular displacement $\Delta \theta_2$ achieved during the first 40.0 s, Δt_2, and the angular displacement $\Delta \theta_1$ achieved during the first 20.0 s, Δt_1.

Basic equation: $\Delta \theta = \Delta \theta_2 - \Delta \theta_1$

Since $\omega_i = 0$ for each time interval,

$$\Delta \theta_2 = \tfrac{1}{2}\alpha\,\Delta t_2^2 \quad \text{and} \quad \Delta \theta_1 = \tfrac{1}{2}\alpha\,\Delta t_1^2$$

$$\Delta \theta = \tfrac{1}{2}\alpha\,\Delta t_2^2 - \tfrac{1}{2}\alpha\,\Delta t_1^2$$

Working equation: $\Delta \theta = \tfrac{1}{2}\alpha(\Delta t_2^2 - \Delta t_1^2)$

$$\Delta \theta = \frac{725/\text{min} \times 2\pi\ \text{rad}}{2 \times 1.50\ \text{min}} \times \left(\frac{\text{min}}{60\ \text{s}}\right)^2 \times [(40.0\ \text{s})^2 - (20.0\ \text{s})^2]$$

$$\Delta \theta = 506\ \text{rad}$$

EXAMPLE A flywheel accelerates uniformly from rest to an angular velocity of 15 revolutions per second in 6.0 s. What is the angular acceleration of the flywheel in rad/s^2?

SOLUTION *Basic equation:* $\alpha = \dfrac{\Delta\omega}{\Delta t}$

Since α is to be expressed in rad/s², ω must be converted from rev/s to rad/s,

or $1 \text{ rad/s} = \dfrac{1}{2\pi} \text{ rev/s}$ *Working equation:* $\alpha = \dfrac{\text{rev/s}}{2\pi\,\Delta t}$

$$\alpha = \frac{15/\text{s}}{2(3.14)(6.0 \text{ s})}$$

$$\alpha = 0.40 \text{ rad/s}^2$$

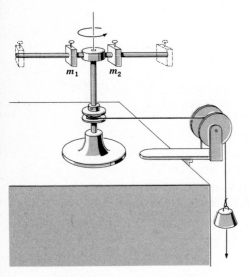

Figure 5-10. Rotational inertia. When the masses, m_1 and m_2, are moved outward on the rotating bar, the rotational inertia of the system increases.

Rotational inertia resists changes in rotation.

5.8 Rotational Inertia A wheel mounted on a shaft will not start to spin unless a torque is applied to the wheel. A wheel that is spinning will continue to spin at constant angular velocity unless a torque acts on it. In both cases the wheel is in equilibrium. Thus if we replace "force" with "torque," Newton's law of inertia also applies to rotary motion.

If we wish to change the rate of rotation of an object about an axis—that is, change its angular velocity—we must apply a torque about the axis. The angular acceleration that this torque produces depends on the mass of the rotating object and upon the distribution of its mass with respect to the axis of rotation.

In Figure 5-10 two masses, m_1 and m_2, are mounted on a bar that is fastened to an axle. The masses can be placed on the bar at varying distances from the axle. If a cord is wound around the axle and attached to a weight that falls under the influence of gravity, a constant torque is applied to the axle. If different weights are attached to the cord, different torques are applied. If the masses mounted on the bar remain in a fixed position, the greater the torque, the greater will be the angular acceleration. If the masses are placed near the axis of rotation, the acceleration produced by a given torque is greater than if they are placed at the ends of the bar. Moving the masses farther apart does not change the amount of mass that is rotated. It does, however, change the distribution of mass. We say that it increases the *rotational inertia*. **Rotational inertia** *is the resistance of a rotating object to changes in its angular velocity*. The angular acceleration, α, is directly proportional to the torque, T, but inversely proportional to the rotational inertia, I.

$$\alpha = \frac{T}{I}$$

Transposing $\qquad T = I\alpha$

which for rotary motion is analogous to $F = ma$ for linear motion. The torque, T, is computed by

$$T = Fr$$

where F is the force applied tangentially at distance r from the pivot point. Like a force, a torque has magnitude and direction. As in the case of angular velocity, the direction of the torque vector is the direction in which the thumb of the right hand points when the fingers of the right hand encircle the vector in the direction in which the mass is rotating. Torques behave like other quantities that can be represented by vectors.

Rotational inertia takes into account both the shape and the mass of the rotating object. Equations for the rotational inertia of certain regularly shaped bodies are given in Figure 5-11. Rotational inertia has the unit dimensions kg m². The following dimensional analysis verifies this.

$$I = \frac{T}{\alpha} = \frac{\mathbf{mn}}{1/s^2} = \frac{\mathbf{kg\ m^2/s^2}}{1/s^2} = \mathbf{kg\ m^2}$$

Figure 5-11. Rotational inertia equations for objects rotating about the indicated axes.

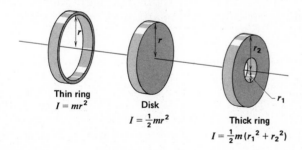

Thin ring
$I = mr^2$

Disk
$I = \frac{1}{2}mr^2$

Thick ring
$I = \frac{1}{2}m(r_1{}^2 + r_2{}^2)$

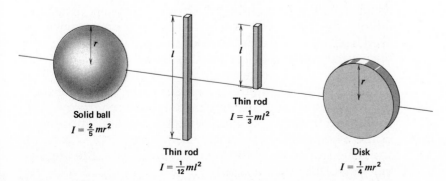

Solid ball
$I = \frac{2}{5}mr^2$

Thin rod
$I = \frac{1}{3}ml^2$

Thin rod
$I = \frac{1}{12}ml^2$

Disk
$I = \frac{1}{4}mr^2$

EXAMPLE   A solid ball is rotated by applying a force of 4.7 n tangentially to it. The ball has a radius of 14 cm and a mass of 4.0 kg. What is the angular acceleration of the ball?

SOLUTION   *Basic equation:* $T = I\alpha$

$$\alpha = \frac{T}{I}$$

$$T = Fr$$

From Figure 5-11, the value of I for a solid ball is $\frac{2}{5}mr^2$. Hence,

$$\alpha = \frac{Fr}{\frac{2}{5}mr^2} \quad \text{and}$$

Working equation: $\alpha = \dfrac{5F}{2mr}$

$$\alpha = \frac{5(4.7 \text{ n})}{2(4.0 \text{ kg})(0.14 \text{ m})}$$

$$\alpha = 21 \text{ rad/s}^2$$

PRACTICE PROBLEMS   1. What is the rotational inertia of a thick ring that is rotating about an axis perpendicular to the plane of the ring and passing through its center? The ring has a mass of 1.2 kg and a diameter of 45 cm. The hole in the ring is 15 cm wide.   *Ans.* 340 kg m^2

2. A flywheel in the shape of a thin ring has a mass of $3\bar{0}$ kg and a diameter of 0.96 m. A torque of 13 mn is applied tangentially to the wheel. How long will it take for the flywheel to attain an angular velocity of $1\bar{0}$ rad/s?   *Ans.* 5.3 s

---

A rotating wheel is sometimes called a gyroscope.

A gyroscope does not precess in the absence of gravity or in a freely falling spaceship.

5.9 Precession   Figure 5-12 shows a spinning bicycle wheel that seems to be defying gravity. The only means of support of the wheel is the cord that is holding up the left end of the shaft. Yet if the wheel is spinning fast enough, its shaft will remain almost horizontal in space. Instead of falling, the shaft will slowly rotate counterclockwise (as seen from above) around the axis marked by the cord. This horizontal rotation of the shaft is called *precession*.

To understand precession, we will need to study Figures 5-12 and 5-13 carefully. In Figure 5-12, **OA** is the shaft of a wheel spinning counterclockwise as seen from **A**. **OC** is the cord from which the shaft is suspended. The weight of the wheel produces a clockwise torque about **O** (as seen from the front). ω_S is the vector representing the angular velocity of the spinning wheel.

In Figure 5-13, α describes the angular acceleration due to the torque produced by F_w. (α is horizontal and directed into the paper.) Because of this angular acceleration there will be a change in the angular velocity during the short time Δt, or $\Delta\omega_\alpha = \alpha\,\Delta t$. This added vectorially to ω_S results in a new angular velocity ω_S' of the same magnitude but different direction. (Remember that the angular acceleration is small and can therefore be considered to be at right angles to both the initial and new angular velocities.) Because of the way the wheel is supported, if the new axis is now along ω_S', the shaft of the wheel has to turn toward this new position. The analysis has to be repeated for successive time intervals and the result is the precessional motion of the wheel.

The motion of the earth is a good example of precession since it spins about an axis that is tilted with respect to its plane of revolution around the sun. The rotation of the earth produces an equatorial bulge. The gravitational pull of the sun on the part of the bulge closest to the sun is stronger than the pull on the part of the bulge on the other side of the earth. Because of the resulting torque, the earth's axis goes through a precessional motion and does not point continually at the same place in space. See Figure 5-14. The earth's axis completes a single precessional cycle in 26,000 years.

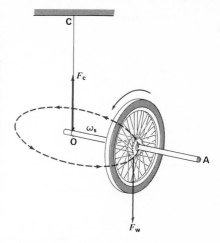

Figure 5-12. Precession. When the wheel rotates, the angular velocity combined with the torque produced by the weight of the wheel causes a precession.

Figure 5-13. Vector analysis of precession. All vectors are in the same horizontal plane. When vector $\Delta\omega_\alpha$ is small, vectors α, $\alpha\,\Delta t$, and $\Delta\omega_\alpha$ are all perpendicular to vectors ω_S and ω_S'.

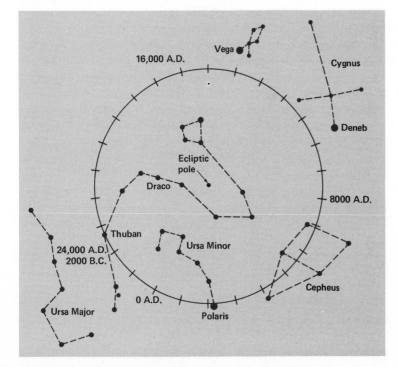

Figure 5-14. Precession in the northern sky. The present North Star is Polaris, because the northern end of the earth's axis points near that bright star at this time. As the earth precesses, however, other stars will become "north stars."

Questions
GROUP A

1. (a) What is rotary motion? (b) Give examples of uniform and variable rotary motion.
2. What is a radian?
3. (a) What is angular velocity? (b) What is angular acceleration?
4. (a) Explain why angular velocity is a vector quantity. (b) How is a vector representing an angular velocity drawn?
5. How do equations for uniformly accelerated rotary motions compare with those for uniformly accelerated linear motions?

GROUP B
6. Explain how a spinning flywheel maintains the shaft on which it is mounted at a constant angular velocity.
7. What is the result of the addition of two angular velocities about different axes?
8. For a rotating wheel, how do the directions of the linear velocity vector and the angular velocity vector compare at the same instant of time?
9. In the case of precession, how do the directions of an angular acceleration vector and the corresponding angular velocity vector compare? See Figure 5-13.
10. (a) What causes precession? (b) How does precession affect the position of the North Star?

Problems
GROUP A

1. What is the angular velocity in radians per second of a flywheel spinning at the rate of 4820 revolutions per minute?
2. If a wheel spins at a rate of 625 rad/s, what is its angular velocity in revolutions per minute?
3. A net force of 10.0 n is applied tangentially to the rim of a wheel having a 0.25-m radius. If the rotational inertia of the wheel is 0.500 kg m^2, what is its angular acceleration?
4. Calculate the rotational inertia of a thin ring whose mass is 5.7 kg and whose radius is 0.15 m and that rotates about an axis through the center perpendicular to the plane of the ring.
5. What is the rotational inertia of a solid ball 0.050 m in radius that weighs 80.0 n if it is rotated about a diameter?

GROUP B
6. Calculate the angular acceleration in radians per second^2 of a wheel that starts from rest and attains an angular velocity of 545 revolutions per minute in 1.00 minute.
7. (a) What is the angular displacement in radians of the wheel of Problem 6 during the first 0.500 min? (b) During the second 0.500 min?
8. (a) If a force of 25.0 n is applied tangentially to the rim of a disk whose radius is 0.200 m and mass is 30.0 kg, what angular velocity will the disk attain in 0.200 min? (b) What is the angular displacement during this acceleration?
9. A disk-shaped wheel 1.00 m in radius weighs 2.00×10^3 n. A force of 50.0 n is applied tangentially to the rim. After how many seconds does it attain an angular velocity of 1200 rev/min?

HARMONIC MOTION

Periodic: repeated in the same way in equal time intervals.

5.10 Periodic Motion When a body moves repeatedly over the same path in equal intervals of time, it is said to have

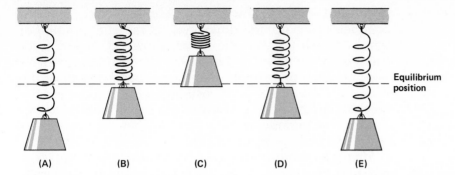

(A)   (B)   (C)   (D)   (E)

Equilibrium position

Figure 5-15. Simple harmonic motion. The frequency of vibration of the mass depends on the stiffness of the spring and the magnitude of the mass.

periodic motion. Figure 5-15 shows a mass attached to a spring that is supported from a horizontal beam. If we pull down on the mass and then let it go, it will vibrate up and down in periodic motion. Let us investigate quantitatively the periodic motion of this mass in Figure 5-15.

If we exert a force, F, acting downward on the mass, we displace it from its equilibrium position by a distance d. We also find by experiment that if we exert a force $2F$ we displace the mass a distance $2d$. The downward displacement from the equilibrium position is directly proportional to the downward force we exert. This is the principle of the spring balance described in Section 1.8.

In exerting this force downward we pull against the spring, and the spring also pulls against our hand with a force $-F$ (law of interaction). When we release the mass (A), the upward force of the spring exceeds the downward pull of gravity on the mass. An upward acceleration that is directly proportional to the excess force is produced (law of acceleration).

As the mass returns to its equilibrium position (B), less and less force is exerted on it by the spring. Consequently the acceleration becomes less. The net force is zero when the mass is at its equilibrium position. The acceleration is also zero. The mass has acquired its maximum velocity at this point. It moves past its equilibrium position because of its inertia. As this overshoot occurs, the upward pull of the spring becomes less than the downward force on the mass. There is a net downward force on the mass, and this downward force decelerates it. The mass stops (C). (During the upward movement of the mass, the spring's pull may change to a downward force due to compression.) The downward force then accelerates it again (D and E). As a consequence of the upward force exerted by the spring and the downward force due to gravity, the mass describes an up-and-down motion.

Assuming the ideal conditions of a frictionless spring, the mass moves above the equilibrium position the same distance it moves below the equilibrium position, and it completes each up-and-down cycle in the same amount of time. At each point in the up-and-down cycle, the force exerted on the mass, and therefore the resulting acceleration, are directly proportional to the displacement of the mass from the equilibrium position. Both the force and the acceleration are directed toward the equilibrium position. The type of periodic motion that has these characteristics is called *simple harmonic motion.*

Harmonic motion: acceleration directed toward an equilibrium position and proportional to the distance from that position.

Simple harmonic motion *is linear motion in which the acceleration is proportional to the displacement from an equilibrium position and is directed toward that position.*

5.11 Analyzing Harmonic Motion Simple harmonic motion can be analyzed in terms of circular motion. In Figure 5-16(A) a light is shining directly down on a rotating wheel. As the wheel spins, the handle on the wheel describes circular motion. The shadow of the wheel and handle falls perpendicularly on the horizontal surface below. The shadow of the handle moves back and forth along the shadow of the wheel. The shadow of the handle slows down when it is going away from the center of the wheel's shadow and speeds up when it approaches the center point. In other words, the handle's shadow describes simple harmonic motion.

The diagram in Figure 5-16(B) shows more clearly the relationship between a point on the handle, **P,** and its shadow, **P′**. **P** is moving with uniform speed in a circular path around point **O**. **P′** is the shadow of **P** on line **MN**. When **P** is at **a,** its shadow is at **a′;** when it is at **b,** its shadow is at **b′,** and so on. As **P** makes a single revolution, **P′** describes one *complete vibration.* **P′** moves with an acceleration proportional to **O′P′,** and the acceleration is always directed toward **O′.** Thus the motion of **P′** along **MN** is simple harmonic motion. When **P′** is at **b′,** the *displacement* of **P′** is the distance **b′O′**—its distance from the midpoint of its vibration at that particular instant. The *amplitude* of the vibration is the maximum displacement **O′M** or **O′N**—the radius of the reference circle. The *period* is the time of one complete vibration—the time required for the point to make one revolution on the reference circle. As we saw in Section 1.8, the *frequency* of a vibratory motion is the number of vibrations per second—the number of revolutions per second of a point on the reference circle. *The frequency is the reciprocal of the period.* The *equilibrium position* of an object that is describing simple harmonic motion is the midpoint of its path, **O′.**

To express the displacement of **P′** along **MN** in Figure

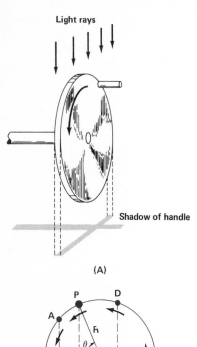

Light rays

(A)

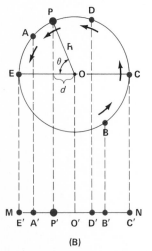

(B)

Figure 5-16. Simple harmonic motion can be analyzed in terms of circular motion.

5-16(B) as a function of time, we make use of the definition of angular velocity, $\omega = \theta/t$, or $\theta = \omega t$. If d is the displacement of **P′**, it can be seen in Figure 5-16(B) that

$$d = R \cos \theta$$

Substituting the value of θ above yields

$$d = R \cos \omega t$$

A graph of this relationship, which holds for all simple harmonic motion, is shown in Figure 5-17.

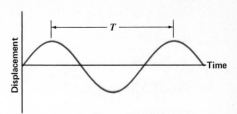

Figure 5-17. Graph of simple harmonic motion. *T* is the period of the motion.

5.12 The Pendulum *An object suspended so that it can swing back and forth about an axis is called a **pendulum**.* Figure 5-18 shows a simple pendulum in which a small dense mass, called the bob, is suspended by a cord. The mass of the cord is negligible in comparison to the mass of the bob. If the displacement of the bob is small in comparison to the length of the cord, the motion of a pendulum very closely approximates simple harmonic motion. As the pendulum bob moves from **A** to **B** and back again to **A,** it makes a complete vibration. **C** is the equilibrium position.

Galileo was probably the first scientist to make quantitative studies of the motion of a pendulum. It is said that he observed the gentle swaying of a chandelier in the cathedral at Pisa. Using his pulse as a timer, he found that successive vibrations of the chandelier were made in equal lengths of time, regardless of the amplitude of the vibrations. He later verified his observations experimentally and then suggested that a pendulum be used to time the pulse rates of medical patients.

In the simple pendulum shown in Figure 5-18, all the mass may be considered to be concentrated in the bob. For an ideal pendulum of this type, the following statements hold true.

1. The period of a pendulum is independent of the mass or material of the pendulum. This statement is strictly true only if the pendulum vibrates in a vacuum. Air resistance has more effect on a pendulum bob made of cotton than it does on a bob made of lead.

2. If the arc is small, the period of a pendulum is independent of the amplitude. The arc is considered small if 10° or less.

3. The period of a pendulum is directly proportional to the square root of its length. If we have a pendulum 25 cm long and one 100 cm long, the period of the longer pendulum will be twice that of the shorter one. Since the square roots of 25 and 100 are 5 and 10 respectively, the periods are in the ratio of 5 to 10, or 1 to 2.

4. The period of a pendulum is inversely proportional to the

The equation for the pendulum was developed by the Dutch physicist Christian Huygens (1629–1695).

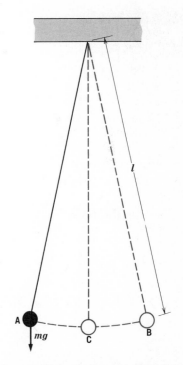

Figure 5-18. The simple pendulum. The frequency of vibration depends only on the values of *l* and *g*, provided that the displacement is small compared to the length of the cord and air resistance is negligible.

square root of the acceleration of free fall. Because the earth has an equatorial bulge, a pendulum vibrates slightly faster at the poles than at the equator. If we use the letter l to denote the length of a pendulum and g to denote the acceleration due to gravity, the period, T, is

$$T = 2\pi\sqrt{\frac{l}{g}}$$

Any real object that can vibrate like a pendulum is called a *physical pendulum*, as distinguished from a theoretical *simple pendulum*. For example, a baseball bat can be a physical pendulum as shown in Figure 5-19. When the bat is suspended from **O,** which is called the *center of suspension*, and set in vibration, its period will be the same as that of a simple pendulum with a length equal to the distance **OC.** Consequently, **C** is called the *center of oscillation*. **O** and **C** are interchangeable; that is, if the bat is suspended from **C,** the center of oscillation will be at **O.** (**G** is the center of gravity of the bat.)

An additional property of **C** is that it is the *center of percussion* of the bat. The batter's hands do not experience a "sting" if a ball strikes the bat at the center of percussion as the hands turn the bat about the center of suspension. If the ball hits the bat at a point other than the center of percussion, the bat rotates about some point other than the center of suspension. The part of the bat at the center of suspension then pushes against the hands, which accounts for the "sting."

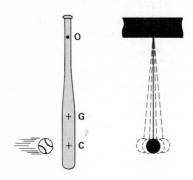

Figure 5-19. A physical pendulum and simple pendulum with the same period. **O** is the point of suspension, **G** is the center of gravity, and **C** is the center of oscillation and percussion.

EXAMPLE   A simple pendulum is 2.0 m long and has a period of 2.9 s. What is the acceleration of gravity at the location of the pendulum?

SOLUTION

$$\text{Basic equation: } T = 2\pi\sqrt{\frac{l}{g}}$$

$$\text{Solving for } g, \text{ or } \textit{Working equation: } g = \frac{4\pi^2 l}{T^2}$$

$$g = \frac{4(3.14)^2(2.0 \text{ m})}{(2.9 \text{ s})^2}$$

$$g = 9.4 \text{ m/s}^2$$

PRACTICE PROBLEM   What is the length of a pendulum with a period of 1.0 s at sea level?   *Ans.*   0.25 m

Questions

GROUP A

1. How is periodic motion related to simple harmonic motion?
2. Define the following terms as they apply to simple harmonic motion: (a) complete vibration; (b) displacement; (c) period; (d) frequency.
3. What factors determine the period of a pendulum?
4. (a) What is a physical pendulum? (b) How is it related to a simple pendulum?
5. Describe two ways of finding the center of percussion of a baseball bat.

GROUP B

6. (a) If a pendulum clock loses time, should the pendulum be made shorter or longer? (b) Explain.
7. Describe a method of using the simple pendulum to find the value of g for a given location.
8. Explain why less net force is exerted on the mass in Figure 5-15 as the mass returns to its equilibrium position.
9. (a) In what sense may the balance wheel of a wrist watch be compared with a simple pendulum? (b) Relate the vibration of the quartz crystal in some wrist watches to the simple pendulum.

Problems

GROUP A

1. A pendulum 0.25 m long has a period of 1.1 s. What is the period of a pendulum in the same location if it has a length of 0.10 m?
2. What is the period of a pendulum 1.5 m long at sea level?

GROUP B

3. What is the value of g for a location where a pendulum 1.08 m long has a period of 2.05 s?

4. A seconds pendulum is one with a period of exactly two seconds. Find the length of a seconds pendulum at Denver, Colorado ($g = 9.79609$ m/s^2).
5. A small pendulum is mounted in a space vehicle to measure acceleration. Before launch, the pendulum vibrates at the rate of 150 s^{-1}. At one point during the launch, the vibration rate is 320 s^{-1}. What is the acceleration of the vehicle at that point?

SUMMARY

Uniform circular motion is motion at constant speed along a curved path of constant radius. An object moving in a curved path undergoes centripetal acceleration. Centripetal acceleration is produced by a force known as centripetal force which is directed toward a central point. When gravity is acting on an object, a minimum velocity is required to keep the object moving in a vertical circle.

A frame of reference is a system for specifying precisely the location of objects. Centrifugal forces are required to explain the curved motions of objects in terms of a noninertial, or accelerating, frame of reference.

Rotary motion is the motion of a body about an internal axis. In such motion, the relationships among angular displacement, velocity, acceleration, and time are similar to the corresponding relationships for linear motion. The angular acceleration of a rotating body is directly proportional to the torque that produces motion and inversely proportional to the body's rotational inertia. The rotation of the shaft of a rotating body that is under the influence of gravity is called precession.

Periodic motion occurs when a body continually moves back and forth over a definite path in equal intervals of time. Simple harmonic motion is a special type of periodic motion. Simple harmonic motion can be analyzed in terms of circular motion. A swinging pendulum describes simple harmonic motion. The period of a pendulum is directly proportional to the square root of its length and inversely proportional to the square root of the acceleration of gravity.

VOCABULARY

angular acceleration
angular velocity
centrifugal force
centripetal acceleration
centripetal force

circular motion
critical velocity
frame of reference
pendulum
periodic motion

precession
radian
rotary motion
rotational inertia
simple harmonic motion

Conservation of Energy and Momentum

HERMANN VON HELMHOLTZ 1821-1894

25

DEUTSCHE BUNDESPOST BERLIN
1971

Hermann von Helmholtz published his statement on the conservation of energy in 1847. It was soon confirmed as a basic scientific law, equally as important as the law of the conservation of matter. Like Einstein, Helmholtz was only twenty-five years old when he proposed his theory.

WORK, MACHINES, AND POWER

6.1 Definition of Work   The word *work* has a specific meaning in physics. *Work is done when a force is exerted on an object causing the object to move in the direction of a component of the applied force.* Even if you hold a heavy load on your shoulder, as long as you do not move you are not doing any work on the load. You are merely exerting an upward force that counteracts the downward force of gravity on the load. You do work in a scientific sense when you raise the load to your shoulder, when you carry it up a flight of stairs, or when you pull it across the floor. In these cases, you exert a force that has a component in the direction in which the object moves.

Two factors must be considered in measuring work: the displacement of the object and the magnitude of the force in the direction of displacement. *The amount of work, W, equals the product of the force, F, in the direction of displacement and the displacement, Δd, of an object.*

$$W = F \, \Delta d$$

When the force is measured in newtons and the distance through which it acts is measured in meters, the work is expressed in *joules* (j). *A force of one newton acting through a distance of one meter does* **one joule** *of work.* This unit of work is named for the English physicist James Prescott Joule. Note that *a joule is a newton meter.*

For example, let us compute the work required to lift a

In this chapter you will gain an understanding of:

▷ the scientific definition of work
▷ simple machines
▷ the relationship between work output and work input in a machine
▷ the equations for calculating power in linear and rotary motion
▷ various forms of potential energy
▷ kinetic energy equations
▷ the law of conservation of mechanical energy
▷ the relationship between impulse and change of momentum
▷ the law of conservation of momentum
▷ energy and momentum in elastic and inelastic collisions

Work: the product of force and displacement.

115

1.0-kg mass to a height of 5.0 m. From the relationship between mass and weight discussed in previous chapters, we know that a force of about 9.8 n must be exerted to lift a mass of 1.0 kg at sea level. Thus, the amount of work involved is

The lower-case symbol for the joule used in this text is an exception to the SI rule that symbols derived from proper names should be capitalized.

$$W = F \Delta d$$

$$W = (9.8 \text{ n})(5.0 \text{ m})$$

$$W = 49 \text{ j}$$

If we merely slide the 1.0-kg mass at constant velocity along a horizontal surface having a coefficient of sliding friction, μ, of 0.30 for a distance of 5.0 m, the work required is

The relationship $F_f = \mu F_N$ is explained in Section 4.7.

$$W = F_f \Delta d$$

$$W = \mu F_N \Delta d$$

$$W = (0.30)(9.8 \text{ n})(5.0 \text{ m})$$

$$W = 15 \text{ j}$$

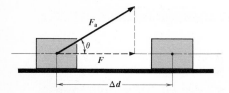

Figure 6-1. Definition of work. The work done by the applied force, F_a, is equal to the product of F, the component of F_a in the direction of the displacement, and Δd, the distance through which the mass moves.

In both instances, the force is applied in the direction in which the object moves.

Suppose, however, that the force is applied to the object in a direction other than that in which it moves. In that case, only the component of the applied force that acts in the direction the object moves is used to compute the work done on the object. Thus in Figure 6-1, the relationship between the applied force, F_a, and the component in the direction of motion, F, is

$$F = F_a \cos \theta$$

This relationship is used in the following example.

EXAMPLE   A box with a mass of 115 kg is pulled at constant velocity along an inclined plane 10.0 m long. The force is applied parallel to the plane, which makes an angle of 15.0° with the horizontal. How much work is done if the coefficient of sliding friction between the box and plane is 0.300?

SOLUTION   Since the box is moving at constant velocity, none of the applied force produces acceleration. Acceleration equations are, therefore, not involved in the solution.

Basic equation: $W = F \, \Delta d$

In this case, F is the sum of the force of friction, F_f, and the component of the weight of the box parallel to the plane, F_p, so that $F = F_f + F_p$.
The weight of the box, $F_w = mg$.
The force of friction, $F_f = \mu F_N$.
F_N is the component of the weight of the box normal to the plane, or $F_N = F_w \cos \theta = mg \cos \theta$. Thus $F_f = \mu mg \cos \theta$.
The component of the weight of the box parallel to the plane,
$F_p = F_w \sin \theta = mg \sin \theta$.
Substituting these values in the basic equation, we get the

Working equation: $W = (\mu mg \cos \theta + mg \sin \theta) \Delta d$, or
$$W = mg \, \Delta d (\mu \cos \theta + \sin \theta)$$
$$W = [(115 \text{ kg})(9.80 \text{ n/kg})(10.0 \text{ m})][(0.300)(0.966) + 0.259]$$
$$W = 6.20 \times 10^3 \text{ j}$$

PRACTICE PROBLEMS   1. A boy pulls a wagon with constant velocity along a level path for a distance of 85 m. He exerts a force of 110 n and the rope on which he pulls makes an angle of 25° with the horizontal. How much work does the boy do in pulling the wagon?   *Ans.* 8.5×10^3 j

2. How much work is done in pushing a crate weighing 970 n up an inclined plane 15 m long that makes an angle of 12° with the horizontal? The coefficient of sliding friction between the crate and the plane is 0.20.   *Ans.* 5.9×10^3 j

---

6.2 Work Done by Varying Forces   In the examples of work in Section 6.1, the forces involved did not vary. In many problems involving work, however, the forces may vary in direction, in magnitude, or in both during the time that they are acting on an object. For example, when a force is used to stretch a spring, the magnitude of the force increases as the spring gets longer.

An easy way to determine the amount of work done by a varying force is to use a graph. In Figure 6-2, the area under the curved line represents the work done by a force that varies in magnitude. The horizontal axis is the distance, Δd, in meters and the vertical axis is the force that acts in the direction of motion, F, in newtons. The amount of work required to provide the displacement indicated by the curve up to point **A**, for example, is equal to the area bounded at the top by the curve, at the right by a vertical line from **A** to the horizontal axis, and at the left and bottom by the two coordinate axes.

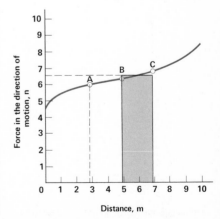

Figure 6-2. Work done by a variable force. The area under the curve represents the total work.

The mathematical term for finding the area under a curve is integration.

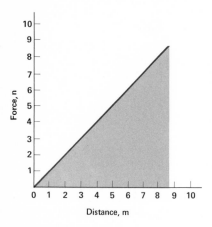

Figure 6-3. Work done by a constantly increasing force. The total work is represented by the red triangle.

To calculate the area of a geometric figure that is bounded by one or more curved lines requires a branch of mathematics called calculus. A good approximation is obtained, however, by the use of suitable rectangles. For example, the work indicated by the curve between points **B** and **C** is approximately equal to the area of the red rectangle in the figure. Thus, the work is approximately equal to the product of 2.0 m and 6.5 n, or 13 j.

The total amount of work that is represented by the area under the curve can be found by adding the areas of many rectangles formed in the same way as the one in the figure. If the rectangles are made very narrow, the answer will be more accurate. An interesting way to measure the area under a curve is to cut it out of a piece of paper, weigh it, and compare it with the weight of a known area.

The problem of finding the total work done by a varying force is somewhat simpler in the case of a stretching spring, as long as the force is constantly applied in the direction of the stretching. The force required to stretch a spring depends on the stiffness of the spring, but the force is directly proportional to the amount of stretching. (The limits within which this relationship is true will be discussed in Chapter 7.) Consequently, a graph of the work done in stretching a spring is shown in Figure 6-3. No approximations are required in computing the total work because the area of the triangle under the curve is exactly equal to one-half the product of its base and height.

6.3 Work in Rotary Motion To compute the work done in rotary motion, we make use of the principles of radian measure (Section 5.6). In Figure 6-4 the displacement of the rim of the wheel is designated by the arc Δd. If the angle $\Delta\theta$ is expressed in radians, then $\Delta d = r\,\Delta\theta$. Substituting this expression for Δd in the work equation, we get

$$W = Fr\Delta\theta$$

Furthermore, in Section 4.12 we saw that a torque, T, is equal to the product of a force and the length of its torque arm. In Figure 6-4, the torque arm is the radius of the circle, so $T = Fr$. The work equation can now be written

$$W = T\Delta\theta$$

which means that the work done in rotary motion can be computed by finding the product of the torque producing the motion and the angular displacement in radians.

For example, if the radius of the wheel in Figure 6-4 is

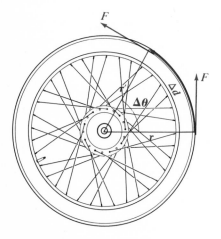

Figure 6-4. Work in rotary motion. The work done on the wheel is equal to the product of the applied force, *F*, the radius of the wheel, *r*, and the displacement of the rim, $\Delta\theta$.

2.0 m and a force of 12 n is applied tangentially to it, the work done in a single revolution is

$$W = T\Delta\theta$$

$$W = Fr\Delta\theta$$

$$W = 2\pi Fr$$

$$W = 2(3.14)(12 \text{ n})(2.0 \text{ m})$$

$$W = 150 \text{ j}$$

6.4 Machines Six types of simple machines are shown in Figure 6-5. Each one can be used to multiply force. Other machines are either modifications of these simple machines or combinations of two or more of them. The six machines shown are actually variations of two basic types: the pulley and the wheel and axle are forms of the lever, and the wedge and screw are modified inclined planes.

Figure 6-5. Simple machines. Each machine multiplies force at the expense of distance.

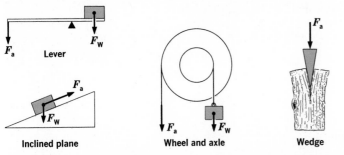

Lever

Inclined plane

Wheel and axle

Wedge

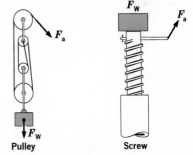

Pulley

Screw

Although a machine can be used to multiply force, it cannot multiply work. The work output of a machine cannot exceed the work input. In a frictionless machine, work output and work input would be exactly equal. In a machine that multiplies force, this equality means that the distance over which the input force moves is always greater than the distance over which the load moves.

The ratio of the useful work output of a machine to total work input is called the *efficiency*.

The ratio of the output force to the input force in a machine is called the mechanical advantage.

$$\text{Efficiency} = \frac{W_{output}}{W_{input}}$$

The efficiency of all machines is less than 100% because the work output is always less than the work input. This is due to the force of friction.

Thus in using a machine to lift an object, the efficiency equation becomes

$$\text{Efficiency} = \frac{F_w \, \Delta h}{F_a \, \Delta d}$$

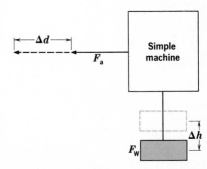

Figure 6-6. Principle of the simple machine. In the absence of friction, the work output, $F_w \, \Delta h$, is equal to the work input, $F_a \, \Delta d$.

where F_w is the weight of the object, Δh is the height through which it is lifted, F_a is the input force applied to the machine, and Δd is the distance through which F_a acts in the direction of the input motion.

The following example illustrates the use of the efficiency equation in solving problems dealing with simple machines.

EXAMPLE   An inclined plane is 6.00 m long and 1.00 m high. The coefficient of friction is 0.050. (a) What force is required to pull a weight of 1.00×10^4 n up the plane at constant velocity? (b) What is the efficiency of the plane?

SOLUTION   (a) The required force is the sum of the useful input force and the force required to overcome the force of friction.

Basic equation: $F = F_u + F_f$

$$W_{input} \text{ (useful)} = W_{output} \text{ (useful)}$$

$$F_u \, \Delta d = F_w \, \Delta h$$

$$F_u = \frac{F_w \, \Delta h}{\Delta d}$$

$$F_f = \mu F_N$$

Working equation: $F = \dfrac{F_w \, \Delta h}{\Delta d} + \mu F_N$

F_N is found by the method outlined in Section 4.5.
$F_N = 9860$ n

$$F = \frac{(1.00 \times 10^4 \text{ n})(1.00 \text{ m})}{(6.00 \text{ m})} + (0.050)(9860 \text{ n})$$

$$F = 2160 \text{ n}$$

(b) Working equation: Efficiency $= \dfrac{F_w \, \Delta h}{F_a \, \Delta d}$

$$\text{Efficiency} = \frac{(1.00 \times 10^4 \text{ n})(1.00 \text{ m})}{(2160 \text{ n})(6.00 \text{ m})}$$

$$\text{Efficiency} = 0.772, \text{ or } 77.2\%$$

PRACTICE PROBLEMS   1. How much work is required to raise a mass of 74 kg to a height of 4.2 m with a pulley system if the efficiency of the system is 82%?   *Ans.* 3.7×10^3 j

2. A force of 86 n is required to push a wagon up an inclined plane $2\overline{0}$ m long. The wagon has a mass of 81 kg and the end of the inclined plane rises 2.0 m off the ground. Calculate the efficiency of this application of a machine.   *Ans.* 0.92

6.5 Definition of Power   Like the term *work*, the term *power* has a scientific meaning that differs somewhat from its everyday meaning. When we say a person has great power, we usually mean that the person has great strength or wields great authority. In physics, the term ***power*** means *the time rate of doing work*.

You do the same amount of work whether you climb a flight of stairs in one minute or in five minutes, but your power output is not the same. See Figure 6-7. Power depends upon three factors: the displacement of the object, the force in the direction of the displacement, and the time required.

Since power is the time rate of doing work,

$$P = \frac{W}{\Delta t}$$

where P is power, W is work, and Δt is time. Or, because $W = F\Delta d$ and $v = \Delta d/\Delta t$,

$$P = \frac{F\Delta d}{\Delta t} = Fv$$

When work is measured in joules and time is measured in seconds, power is expressed in *watts* (w). *A **watt** is a joule per second.* This unit is named in honor of James Watt who designed the first practical steam engine. Since the watt is a very small unit, power is more commonly measured in units of 1000 watts, or kilowatts (kw).

Figure 6-7. A power comparison. In both cases, the girl does the same amount of work in climbing the stairs. However, in the lower drawing her output of power is greater because she gets to the top in less time. (Kinetic energy is disregarded in these examples.)

Figure 6-8. James Watt, the Scottish inventor for whom the metric unit of power is named, in his workshop.

The lower-case symbol for the watt used in this text is an exception to the SI rule that symbols derived from proper names should be capitalized.

The terms watt and kilowatt are used frequently in connection with electricity but they also apply to quantities of power other than electric power. The watt is the basic unit of power in the metric system and can be used to express quantities of mechanical as well as electric power. You may be familiar with the unit called the "horsepower." It is equal to 746 watts.

EXAMPLE A person's mass is 75.2 kg. If the person walks up a flight of stairs 12.6 m high in 28.0 s at constant velocity, what is the power output?

SOLUTION *Basic equation:* $P = \dfrac{F\Delta d}{\Delta t}$

In this case, F is the weight of the person, F_W, which is found from the person's mass by the relationship

$$F_w = mg$$

The distance Δd is the vertical height of the staircase.

Working equation: $P = \dfrac{mg\Delta d}{\Delta t}$

$$P = \frac{(75.2 \text{ kg})(9.80 \text{ m/s}^2)(12.6 \text{ m})}{28.0 \text{ s}}$$

$$P = 332 \text{ w}$$

6.6 Power in Rotary Motion By using radian measure, we can compute the power involved in rotary motion the same way we calculated the work done in rotary motion in Section 6.3. Substituting the expression for work in rotary motion, $T\Delta\theta$, the power equation becomes

$$P = \frac{T\Delta\theta}{\Delta t}$$

We saw in Section 5.6 that the time rate of angular displacement, $\Delta\theta/\Delta t$, is called the angular velocity, ω. If $\Delta\theta$ is measured in radians, then ω is in radians per second.

For rotary motion, the expression for power becomes

$$P = T\omega$$

which means that the power required to maintain rotary motion against an opposing torque is equal to the product of the torque maintaining the rotary motion and the constant angular velocity.

Questions

GROUP A

1. Give the scientific definition of "work."
2. (a) How is a graph used to find the work done by a varying force? (b) What method for finding the work done can easily be applied if the force varies constantly?
3. Is work done by the centripetal force that produces circular motion? Explain.
4. (a) List the six simple machines. (b) Into what two basic types can they be divided?
5. (a) Define efficiency for a machine. (b) How is it computed?
6. How does friction affect the efficiency of machines?
7. List the advantages and disadvantages of friction in the use of a screw.
8. (a) Give the scientific definition of "power." (b) What is the SI unit of power?

GROUP B

9. How are the coordinates of a graph used to find the area under a given curve?
10. Suppose that you have two simple machines, each of which multiplies a force by a factor of three. By what factor will the force be multiplied if the two machines are used in series? Explain.
11. (a) What physical quantity is expressed in kilowatt-hours? (b) Use dimensional analysis to verify your answer.
12. Show how the formula for power in rotary motion is derived.

Problems

GROUP A

1. A person weighing 750 n climbs a flight of stairs that is 5.0 m high. What work does the person do?
2. What work is done when a person lifts a 2.5-kg package and places it on a shelf 2.2 m high?
3. How much work must be done to roll a metal safe, mass 116 kg, a distance of 15.0 m across a level floor? The coefficient of friction is 0.050.
4. A sled is pulled over level snow a distance of 0.500 km by a force of 124 n applied to a rope that makes an angle of 35.0° with the snow. How much work is done?
5. A force of 18 n is required to stretch a spring 0.25 m from its equilibrium position. (a) Compute the amount of work done on the spring. (b) How much force is required to stretch the spring 0.50 m (assuming that the elastic limit of the spring is not exceeded)? (c) Compute the amount of work done.
6. A force of 545 n is exerted on the rope of a pulley system, and the rope is pulled in 10.0 m. This work causes an object weighing 2520 n to be raised 1.50 m. What is the efficiency of this machine?
7. Two people use a wheel and axle to raise a mass of 750 kg. The radius of the wheel is 0.50 m, and the radius of the axle is 0.040 m. If the efficiency of the machine is 62% and each person exerts an equal force, how much force must each apply?
8. The raised end of an inclined plane 4.0 m long is 0.90 m high. Neglecting friction, what force is required to push a steel box weighing 750 n up this plane?
9. A jackscrew has a lever arm 0.75 m long. The screw has 1.5 threads to the centimeter. If 320 n of force must be exerted in order to raise a load of 6.0×10^3 kg, calculate the efficiency.
10. What power is required to raise a mass of 47 kg to a height of 12 m in 15 s?

11. A loaded elevator has a mass of 2.50×10^3 kg. If it is raised in 10.0 s to a height of 50.0 m, how many kilowatts are required?

12. The mass of a large steel ball is 1500 kg. What power is used in raising it to a height of 33 m if the work is performed in 30.0 s?

GROUP B

13. A loaded trunk has a mass of 35 kg. The coefficient of friction between the trunk and the floor is 0.20. How much work is done in moving the trunk 8.0 m across a level floor and then lifting it into a truck 1.3 m above the floor?

14. A 75.0-kg crate is to be pushed up an inclined plane 3.00 m long that makes an angle of 20.0° with the horizontal. If the coefficient of friction between the crate and the inclined plane is 0.150, how much work must be done?

15. A tangential force of 7.5 n is applied to a wheel 0.25 m in radius to maintain it at a constant speed of 240 revolutions per minute. How much work is done per minute?

16. An unbalanced force of 5.0 n acts on a mass of 2.0 kg that is initially at rest. How much work is done (a) during the first two seconds;

(b) during the tenth second?

17. A flywheel, mass 203 kg, is a uniform disk 1.00 m in radius. (a) What constant torque must be applied to bring the flywheel from rest to an angular velocity of 195 revolutions per minute in 60.0 s? (b) How much work is done during this time?

18. A loaded sled is pulled by means of a rope that makes an angle of 45.0° with the horizontal. If the mass of the sled is 60.0 kg and the coefficient of friction is 0.020, how much work is done in pulling the sled along a level road for a distance of 1.000 km?

19. A pump can deliver 20.0 liters of gasoline per minute. What power, in kilowatts, is expended by the pump in raising gasoline a distance of 6.00 m? One liter of gasoline has a mass of 0.700 kg. (The kinetic energy acquired by the gasoline may be disregarded.)

20. A motor is rated to deliver 10.0 kw. At what speed in m/min can this motor raise a mass of 2.75×10^4 kg?

21. What power will be required to move a locomotive weighing 1.00×10^6 n up a grade that rises 1.50 m for each 100 m of track, so that it will move at a speed of 40.0 km/hr, if the frictional force opposing the motion is 2.00×10^3 n?

ENERGY

Potential and kinetic energy can each take a variety of forms.

6.7 Gravitational Potential Energy   In Chapter 1 we saw that there are two kinds of energy, potential and kinetic. In the following sections, we will deal quantitatively with these concepts and see how the various forms of energy are expressed in terms of work units.

The potential energy acquired by an object equals the work done against gravity or other forces to place it in position, as shown in Figure 6-9. The equation for calculating work when the force acts in the same direction as the displacement is

$$W = F\Delta d$$

Therefore the equation for potential energy is

$$E_p = F\Delta d$$

In lifting an object, F is its weight, which from Newton's second law of motion equals mg, and Δd is the vertical distance Δh through which it is lifted. Hence the potential energy equation can be written

$$E_p = mg\Delta h$$

The gravitational potential energy defined by this equation is expressed in relation to an arbitrary reference level where $h = 0$. It makes no difference what level is chosen. Sea level, street level, ground level, or floor level are all useful reference levels.

When the mass of an object is given in kilograms, the height in meters, and the acceleration due to gravity in m/s², the gravitational potential energy is expressed in joules. Thus if a $5\bar{0}$-kg mass of steel is raised 5.0 m, its gravitational potential energy is

$$E_p = mg\Delta h$$

$$E_p = 5\bar{0} \text{ kg} \times 9.8 \text{ m/s}^2 \times 5.0 \text{ m}$$

$$E_p = 2.5 \times 10^3 \text{ j}$$

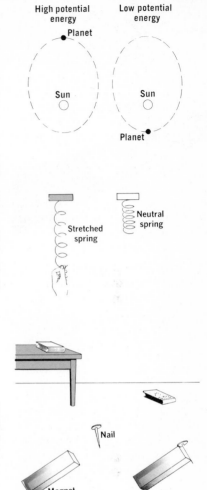

Figure 6-9. Examples of potential energy. Can you give the reason for the difference in energy for each pair of situations?

6.8 Kinetic Energy in Linear Motion   As we saw in Section 3.7, the velocity of a freely falling object that starts from rest, expressed in terms of the acceleration of gravity and the distance traveled, is given by the equation

$$v = \sqrt{2g\Delta d}$$

Solving for Δd, we obtain

$$\Delta d = \frac{v^2}{2g}$$

Since Δd in this equation corresponds to Δh in the equation $E_p = mg\Delta h$, let us substitute the expression we have just derived for Δd in place of Δh. The equation for the kinetic energy of a moving object then becomes

$$E_k = mg \times \frac{v^2}{2g}$$

$$E_k = \tfrac{1}{2}mv^2$$

Although this equation for kinetic energy was derived from the motion of a falling body, it applies to motion in any direction or from any cause.

As in the case of gravitational potential energy, kinetic

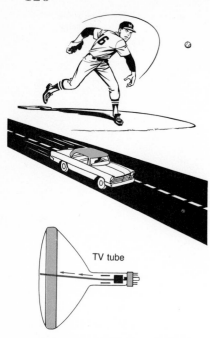

energy is expressed in joules if the mass is given in kilograms and the velocity in meters per second. Thus if a baseball has a mass of 0.14 kg and is thrown with a velocity of 7.5 m/s, its kinetic energy is

$$E_k = \tfrac{1}{2}mv^2$$

$$E_k = \frac{0.14 \text{ kg } (7.5 \text{ m/s})^2}{2}$$

$$E_k = 3.9 \text{ j}$$

6.9 Kinetic Energy in Rotary Motion As we saw in Section 5.6, the angular velocity, ω, of a rotating body that starts from rest is

$$\omega = \sqrt{2\alpha\Delta\theta}$$

where α is the angular acceleration and $\Delta\theta$ is the angular displacement. The equation given in Section 5.8 for the relationship among torque, rotational inertia, and angular acceleration is

$$T = I\alpha$$

Solving both of these expressions for angular acceleration,

$$\alpha = \frac{\omega^2}{2\Delta\theta} \text{ and } \alpha = \frac{T}{I}$$

Setting these two expressions equal,

$$\frac{T}{I} = \frac{\omega^2}{2\Delta\theta}$$

Solving for $T\Delta\theta$,

$$T\Delta\theta = \frac{I\omega^2}{2}$$

When only the net force and torque are considered in these equations, all of the work done to produce rotation appears as kinetic energy. Since for rotary motion $W = T\Delta\theta$, the equation for kinetic energy, E_k, is

$$E_k = \tfrac{1}{2}I\omega^2$$

Figure 6-10. Examples of kinetic energy. In each case, the energy is equal to the product of the mass of the moving particle and the square of its velocity.

The wheels of a moving car have both linear and rotational kinetic energy.

The wheel of a moving automobile has both linear motion and rotary motion. The wheel turns on its axle as the axle moves along parallel to the road. The kinetic energy of such an object is the sum of the kinetic energy due to linear motion and the kinetic energy due to rotary motion.

$$E_k = \tfrac{1}{2}mv^2 + \tfrac{1}{2}I\omega^2$$

EXAMPLE   A metal ring, mass 2.50 kg, rolls along a horizontal surface with a constant velocity of 5.25 m/s. What is the total kinetic energy of the moving ring?

SOLUTION   *Basic equation:* $E_k = \frac{1}{2}mv^2 + \frac{1}{2}I\omega^2$

The rotational inertia of a ring, $I = mr^2$ (See Figure 5-11.)

By definition, $\omega = \dfrac{\Delta\theta}{\Delta t}$ and $\theta = \dfrac{\Delta d}{r}$ and $v = \dfrac{\Delta d}{\Delta t}$

Consequently, $\omega = \dfrac{\Delta d/r}{\Delta t} = \dfrac{\Delta d}{r\Delta t} = \dfrac{v}{r}$

Substituting, $\frac{1}{2}I\omega^2 = \frac{1}{2}mr^2\,(v/r)^2 = \frac{1}{2}mv^2$

Working equation: $E_k = \frac{1}{2}mv^2 + \frac{1}{2}mv^2 = mv^2$

$$E_k = (2.50 \text{ kg})(5.25 \text{ m/s})^2$$

$$E_k = 68.9 \text{ j}$$

When energy is supplied to an object and simultaneously gives it both linear and rotary motion, the energy division depends on the rotational inertia of the object. If a ring and a solid disk of equal mass and diameter roll down the same incline, as in Figure 6-11, the disk will accelerate more. Its rotational inertia is less than that of the ring, thus its rotational kinetic energy is less. Its linear kinetic energy is therefore greater than that of the ring; the disk acquires a higher linear velocity and reaches the bottom of the incline first. However, at the bottom of the incline the total kinetic energy of the ring will be the same as that of the disk since both ring and disk have the same gravitational potential energy at the top.

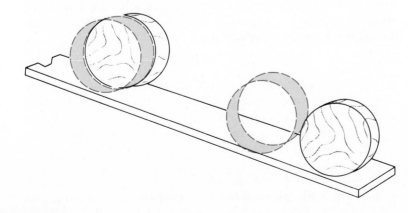

Figure 6-11. A kinetic energy race. If the ring and disk have equal masses and diameters, the rotational inertia of the disk is less than that of the ring. Hence the disk will accelerate more.

6.10 Elastic Potential Energy   When a spring is compressed, energy is stored in the spring. At any instant during the compression, the elastic potential energy in the spring is equal to the work done on the spring. The potential energy in a stretched or compressed elastic object is called *elastic potential energy*.

The work required to stretch or compress a spring does not depend on the weight of the spring. Consequently, gravity is not involved in the measurement of elastic potential energy. Instead, the work required to stretch or compress a spring is dependent upon a property of the spring known as the *force constant*. The force constant does not change for a specific spring so long as the spring is not permanently distorted.

Compare this equation with the one for kinetic energy in Section 6.8. Why do you suppose they are similar?

The force required to stretch a spring is written as

$$F = k\Delta d \qquad \text{(Equation 1)}$$

where k is the force constant of the spring and Δd is the distance over which F is applied. We noted in Section 6.2 that the force is directly proportional to the amount of stretching (within definite limits). Consequently, the work done on the spring also varies directly with the amount of stretching. The total amount of work, therefore, is given by the equation

$$W = \tfrac{1}{2}F\Delta d \qquad \text{(Equation 2)}$$

where F is the force exerted on the spring at the end of the stretch through distance Δd. Because the force varies from zero to F, the equation uses the average of these two values, or $\tfrac{1}{2}F$.

Substituting the expression for F from Equation 1 in Equation 2, we get

$$W = \tfrac{1}{2}k\,(\Delta d)^2 \qquad \text{(Equation 3)}$$

This equation applies to the compression of a spring as well as the stretching of a spring. Since the potential energy of an object is equal to the work done on the object, Equation 3 can be rewritten as a potential energy equation

$$E_\text{p} = \tfrac{1}{2}k\,(\Delta d)^2$$

where E_p is the elastic potential energy. This equation represents ideal conditions. In actual practice, a small fraction of the work of stretching or compression is converted into heat energy in the spring and thus does not show up as elastic potential energy.

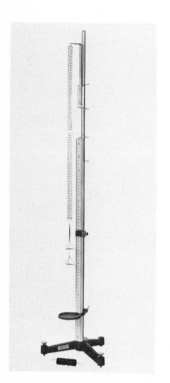

Figure 6-12. Apparatus for measuring force constants. If the weight is made to move up and down, the interchange of potential and kinetic energy can also be measured.

6.11 Conservation of Mechanical Energy   As we saw in Section 5.10, the vibration of a mass on a spring and the swinging of a pendulum are both examples of simple harmonic motion. Both can also be used to illustrate an important principle of physics called the ***law of conservation of mechanical energy.*** This law states that *the sum of the potential and kinetic energy of an energy system remains constant when no dissipative forces act on the system.*

Gravitational forces and elastic forces are called *conservative forces* because they conform to the law of conservation of mechanical energy. There are forces, however, that produce deviations from the law of conservation of mechanical energy. The force of friction is an example. Forces of this type are called nonconservative, or *dissipative forces.*

The reason that friction is a dissipative force is that it produces a form of energy (heat) that is not mechanical. Energy is lost to the system. When considered in light of the more comprehensive law of conservation of *total* energy, there is no "lost" energy, of course. The calculation of heat energy and its role in energy transformations will be discussed in Chapter 8.

Another way to distinguish between conservative and dissipative forces is to observe the relationship between the force and the path over which it acts. In the case of a conservative force, the work done and energy involved are completely independent of the length of the path, provided the paths have the same end point. For example, in the illustration of gravitational potential energy in Figure 6-13, the gravitational potential energy of both men is the same, provided they have equal masses, even though one of them traveled a greater distance than the other one. The mass of the men and their height above the reference level are the only factors necessary for the determination of gravitational potential energy.

The amount of heat energy lost through friction, a dissipative force, can be quite different for two objects moving through the same height, however. Where the path is longer, the amount of mechanical energy that is converted to heat energy will be greater. This is true even if the frictional forces were to remain constant. Thus the work done by a given dissipative force varies directly with the length of the object path.

This is a special case of the law of conservation of energy discussed in Section 1.13.

Figure 6-13. The work done by conservative forces is independent of the path. The potential energy of the man is, in each case, dependent only on his mass and his distance above the reference level.

Questions
GROUP A

1. Differentiate between (a) potential and kinetic energy; (b) gravitational potential energy and elastic potential energy.

2. Show how the formula for gravitational potential energy is derived from the work formula.

3. (a) Describe the kinetic energy of the wheel of a moving automobile.
   (b) Give the equations that can be used to calculate this energy.

4. (a) What determines the force required to stretch or compress a spring? (b) In what units is this factor expressed?

5. (a) Define mechanical energy. (b) How is it conserved in pendular motion?

GROUP B

6. (a) Show how the equation for kinetic energy is derived. (b) Use dimensional analysis to show that kinetic energy is expressed in work units.

7. (a) Distinguish between conservative and dissipative forces. (b) Why is the force of friction a dissipative force?

8. Describe the forms of energy involved in the two parts of Figure 6-13.

Problems

GROUP A

1. A mass of 2.00 kg is lifted from the floor to a table 0.80 m high. Using the floor as a reference level, what potential energy does the mass have because of this change of position?

2. A large rock with a weight of 1470 n rests at the top of a cliff 40.0 m high. What potential energy does it have using the bottom of the cliff as a reference level?

3. What is the kinetic energy of a baseball having a mass of 0.14 kg that is thrown with a speed of 18 m/s?

4. A meteorite weighing 1860 n strikes the earth with a velocity of 45.2 m/s. What is its kinetic energy?

5. A spring scale is calibrated from zero to 20 n. The calibrations extend over a length of 0.10 m. (a) What is the elastic potential energy of the spring in the scale when a weight of 5.0 n hangs from it? (b) What is the elastic potential energy when the spring is fully stretched?

6. The force constant of a spring is 150 n/m. (a) How much force is required to stretch the spring 0.25 m? (b) How much work is done on the spring in that case?

GROUP B

7. A stone having a mass 50.0 kg is dropped from a height of 197 m. What is the potential energy and the kinetic energy of the stone (a) at $t = 0$ s; (b) at $t = 1.00$ s; (c) at $t = 5.00$ s; (d) when it strikes the ground?

8. A car weighing 2.00×10^5 n is accelerated on a level road from 45.0 km/hr to 75.0 km/hr in 11.0 s. (a) What is its increase in kinetic energy? (b) What force produces the acceleration? (c) What power is required?

9. What is the kinetic energy, in joules, of an electron that has a mass of 9.1×10^{-31} kg and that moves at a speed of 1.0×10^7 m/s?

10. The rotor of an electric motor has a rotational inertia of 45 kg m². What is its kinetic energy if it turns at 1500 revolutions per minute?

11. A bowling ball has a mass of 8.00 kg. If it rolls down the alley, without slipping, at 7.00 m/s, calculate (a) the linear kinetic energy; (b) the rotational kinetic energy; (c) the total kinetic energy.

12. A solid ball, mass 10.0 kg, is at the top of an incline 2.00 m high and 10.00 m long. If it rolls down a frictionless incline, what is its linear speed as it reaches the bottom?

13. A 1.00-kg mass is placed at the free end of a compressed spring. The force constant of the spring is 115 n/m. The spring has been compressed 0.200 m from its neutral position. It is now released. Neglecting the mass of the spring and assuming that the mass is sliding on a frictionless surface, how fast will the mass move as it passes the neutral position of the spring?

MOMENTUM

6.12 The Nature of Momentum   More force is needed to stop a train than to stop a car, even though both are moving with the same velocity. A bullet fired from a gun has more penetrating power than a bullet thrown by hand, even though both bullets have the same mass. The physical quantity that describes this aspect of the motion of an object is called *momentum*. **Momentum is the product of the mass of a moving body and its velocity.** The equation for momentum is

$$p = mv$$

where p is the momentum, m is the mass, and v is the velocity of an object.

In the example of the car and train, the greater mass of the train gives it more momentum than the car. Consequently, a greater change of momentum is involved in stopping the train than in stopping the car. In the case of the bullets, the greater momentum of the fired bullet is due to its greater velocity; a large change of momentum takes place when the speeding bullet is stopped.

From Newton's second law of motion, we can derive an important relationship involving momentum. We know that over a short time interval $a_{av} = \Delta v/\Delta t$ (Section 3.5). If we substitute this value of acceleration for a in $F = ma$, we get

$$F = \frac{m\Delta v}{\Delta t} \quad \text{or} \quad F\Delta t = m\Delta v$$

The first of these equations states that force is the time rate of change of momentum.

The product of a force and the time interval during which it acts, $F\Delta t$, is called **impulse.** Hence from Newton's second law of motion we have established that impulse equals change in momentum. The equation $F = m\Delta v/\Delta t$ tells us that when a force is applied to a body, the body's rate of change of momentum is equal to the force. Since the equation is a vector equation, we also know that the body's rate of change of momentum is in the direction of the force.

A good example of the relationship between impulse and change in momentum is a bat hitting a baseball. See Figure 6-14. The impulse imparted to the ball depends on the force with which the ball is hit and the length of time during which the ball and bat are in contact. On leaving the bat, the ball has acquired a momentum equal to the product of its mass and its change of velocity. In a sense, the impulse produced the change of momentum; hence the two are equal.

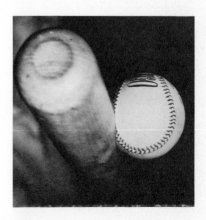

Figure 6-14. During the time of contact, much of the momentum of the bat is transferred to the baseball.

6.13 The Conservation of Momentum In Figure 6-15, a boy with a mass of $4\overline{0}$ kg and a man with a mass of $8\overline{0}$ kg are standing on a frictionless surface. When the man pushes on the boy from the back, the boy moves forward and the man moves backward.

Figure 6-15. Conservation of momentum. On a frictionless surface, the momentum of the boy toward the left is equal to the momentum of the man toward the right.

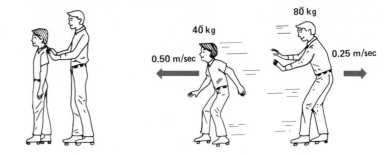

The velocities with which the boy and the man move are specified by one of the most important principles of physics, the *law of conservation of momentum.* This law states that *when no net external forces are acting on a system of objects, the total vector momentum of the system remains constant.*

Let us apply this law to the situation in Figure 6-15. Initially the man and the boy are at rest. The system, therefore, has zero momentum. When the man and the boy move apart, the law of conservation of momentum requires that the total vector momentum remain zero. Hence the momentum of the boy in one direction must equal that of the man in the other direction.

If the boy moves with a velocity of 0.50 m/s, the man will move with a velocity of 0.25 m/s since the mass of the man is twice as great as that of the boy. The momentum of the boy in one direction ($4\overline{0}$ kg $\times$ 0.5$\overline{0}$ m/s) must equal the momentum of the man in the opposite direction ($8\overline{0}$ kg $\times$ 0.25 m/s).

An important application of the law of conservation of momentum is the launching of a rocket. When a rocket fires, hot exhaust gases are expelled through the rocket nozzle. The gas particles have a momentum equal to the mass of the particles multiplied by their exhaust velocity. Momentum equal in magnitude is therefore imparted to the rocket in the opposite direction. See Figure 6-16. Newton's third law of motion (Section 3.10) is a special case of the law of conservation of momentum.

Figure 6-16. Conservation of momentum in a rocket. The Space Shuttle is launched when momentum equal in magnitude to that of the exhaust gases is imparted to the space vehicle.

6.14 Inelastic Collisions The law of conservation of momentum is very helpful in studying the motions of colliding objects. Collisions can take place in various ways, and we shall see how the momentum conservation principles apply in several such cases.

In Figure 6-17(A), two carts of equal mass approach each other with velocities of equal magnitude. A lump of putty is attached to the front of each cart so that the two carts will stick together after the impact. This situation is an example of *inelastic collision*. Since the carts are traveling along the same straight line, it is also an example of a *collision in one dimension*.

The momentum of cart **A** is $m_A v_A$. It is equal in magnitude to the momentum of cart **B**, $m_B v_B$. However, the direction of v_A is opposite to the direction of v_B, so

$$v_A = -v_B$$

Consequently,

$$m_A v_A = -m_B v_B$$

and

$$m_A v_A + m_B v_B = 0$$

This means that the total vector momentum of the system of two moving carts is zero. (We assume that the system is *isolated*, that is, there are no net external forces acting on it. In actual collision studies, the external force of friction is usually minimized by using rolling carts or air tracks, such as the one shown in Figure 6-18.)

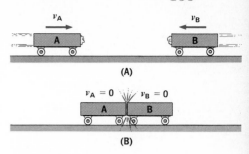

Figure 6-17. An inelastic collision. The carts have equal masses and approach each other with velocities of equal magnitude along the same straight line. The total momentum of the system is the same before and after the collision.

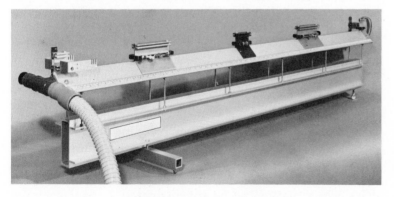

Figure 6-18. An air track designed for collision experiments. Air jets minimize the friction between the track and the masses placed on it.

After the carts in Figure 6-17(B) collide, they both come to rest. The cart velocities, v_A and v_B, are now both zero; the sum of the momenta of the two carts is zero, just as it was when the carts were in motion in opposite directions. Thus the total vector momentum of the system is unchanged by the collision.

If one of the carts has a greater mass than the other, although its velocity is still of equal magnitude but opposite sign, the outcome of the collision is different. After impact, the combined carts will move in the direction of the cart with the larger mass. The velocity of the combined carts will be such that the total momentum of the system remains unchanged.

For example, suppose

$$m_A = 2m_B$$

and $$v_A = -v_B, \text{ as before.}$$

Then, $$m_A v_A + m_B v_B \neq 0$$

Since $$m_B = \frac{m_A}{2}, \text{ by substitution,}$$

$$m_A v_A - 1/2\ m_A v_A = 1/2\ m_A v_A$$

This means that the total momentum of the system, before and after the collision, is $1/2\ m_A v_A$; the combined carts will move with this momentum in the direction of the original velocity of cart **A.** The velocity after the collision can be found by dividing the total momentum by the total mass

$$\frac{1/2\ m_A v_A}{3/2\ m_A} = 1/3\ v_A$$

When carts of equal and opposite momenta collide inelastically, they come to rest. Before the collision they have kinetic energy; after the collision, they do not. This is typical of all inelastic and partially elastic collisions. Much or all of the kinetic energy that the moving objects have before collision is converted into heat or some other form of energy. If all these forms of energy are taken into account, the law of conservation of energy holds true for inelastic collisions. But kinetic energy alone is not conserved.

Kinetic energy is not conserved in an inelastic collision.

6.15 Elastic Collisions   When colliding objects rebound from each other without a loss of kinetic energy, a perfectly *elastic collision* has just occurred. The only perfectly elastic collisions occur between atomic and subatomic particles. The situation can be approximated, however, by the use of hard steel balls or springs on an air track. The momentum conservation law holds for elastic collisions as well as inelastic ones and for collisions that are partly elastic and partly inelastic.

The law also holds for *collisions in two dimensions,* that is, when the colliding objects meet at an angle other than head-on. Figure 6-19(A) shows the elastic, two-dimensional collision of two balls of equal mass. Ball **B** collides at an angle with ball **R,** which is initially at rest. Figure

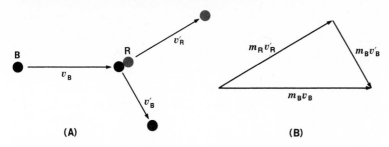

Figure 6-19. A vector diagram can be used to show that the total momentum and kinetic energy of a system both remain constant in a two-dimensional collision.

6-19(B) is the vector diagram representing the momenta of the balls before and after collision.

In Figure 6-19(B), $m_B v_B$ is the momentum of the black ball before the collision. Since the red ball is stationary before the collision, its momentum is zero and is not represented by a vector. The total momentum of the system before collision is

Review the parallelogram method of adding vectors explained in Section 2.11.

$$m_B v_B + 0 = m_B v_B$$

After the collision, the measured velocities of the black and red balls are v'_B and v'_R respectively, in the directions indicated. The total momentum now is the vector sum $m_B v'_B + m_R v'_R$. Since the momentum is conserved, the vector $m_B v_B$ must equal the vector sum of $m_B v'_B$ and $m_R v'_R$. The vector diagram appears as a closed triangle.

Since kinetic energy is also conserved in Figure 6-19,

$$1/2 m_B v_B{}^2 = 1/2 m_B v'_B{}^2 + 1/2 m_R v'_R{}^2$$

Since the two balls have equal masses, this equation can be simplified to

$$v_B{}^2 = v'_B{}^2 + v'_R{}^2$$

This is the equation that relates the sides and hypotenuse of a right triangle, as in Figure 6-19(B). So it follows from the laws of conservation of energy and momentum that, when a moving ball strikes a stationary ball of equal mass other than head-on in an elastic collision, the two balls move away from each other at right angles.

The conservation of momentum also holds for collisions involving more than two bodies and in three-dimensional situations. The examples that follow show how collision problems can be solved by the principles just discussed.

EXAMPLE   A ball having a mass of 0.25 kg and a velocity of 0.20 m/s eastward collides with another ball having a mass of 0.10 kg and a velocity of 0.10 m/s, also eastward and along the same straight line. After the collision the more massive ball has a velocity of 0.14 m/s eastward. What is the velocity of the less massive ball?

SOLUTION

Let m = the mass of the more massive ball
                m' = the mass of the less massive ball
        v_i and v_f = the initial and final velocities of the more massive ball
        v_i' and v_f' = the initial and final velocities of the less massive ball
The total vector momentum is the same after the collision as before the collision.

> *Basic equation:* $mv_i + m'v_i' = mv_f + m'v_f'$

> *Working equation:* $v_f' = \dfrac{mv_i + m'v_i' - mv_f}{m'}$

$$v_f' = \frac{(0.25 \text{ kg})(0.20 \text{ m/s}) + (0.10 \text{ kg})(0.10 \text{ m/s}) - (0.25 \text{ kg})(0.14 \text{ m/s})}{0.10 \text{ kg}}$$

$$v_f' = 0.25 \text{ m/s eastward}$$

The answer could also be obtained by using the conservation of kinetic energy as a basic equation.

EXAMPLE   Two balls with equal masses collide in the way shown in Figure 6-19. Before the collision, the black ball moves with a speed of 0.80 m/s. After the collision, the balls are observed to move along paths that are perpendicular to each other. If the angle between $m_R v_R'$ and $m_B v_B$ in the vector diagram is $40°$, calculate the speed of each ball after the collision.

SOLUTION   Since the balls have equal masses, we can ascribe any convenient mass to them for the purposes of the problem. If we choose 1.00 kg, then the numerical value of the momentum for each ball will equal the numerical value of the velocity. The vector diagram will then yield the velocities directly.

   v_B = 0.80 m/s (the speed of the black ball before the collision)
   v_R' = $v_B \times \cos 40° = (0.80 \text{ m/s})(0.77)$
       = 0.62 m/s (the speed of the red ball after the collision)
   v_B' = $v_B \times \sin 40° = (0.80 \text{ m/s})(0.64)$
       = 0.51 m/s (speed of the black ball after the collision)

6.16 Angular Momentum For rotary motion, the relationship between impulse and the change of angular momentum is similar to that for linear motion. Using the symbols for rotary motion, the equation becomes

$$T\Delta t = I\omega_f - I\omega_i$$

where $T\Delta t$ is the *angular impulse* and $I\omega_f - I\omega_i$ is the change in *angular momentum*. The dimensions of both angular impulse and angular momentum are kg m^2/s.

Just as the linear momentum of an object is unchanged unless a net external force acts on it, *the angular momentum of an object is unchanged unless a net external torque acts on it*. This is a statement of the *law of conservation of angular momentum*. A rotating flywheel, which helps maintain a constant angular velocity of the crankshaft of an automobile engine, is an illustration. The rotational inertia of a flywheel is large. Consequently torques acting on it do not produce rapid changes in its angular momentum. As the torque produced by the combustion in each cylinder tends to accelerate the crankshaft, the rotational inertia of the flywheel resists this action. Similarly, as the torques produced in the cylinders where compression is occurring tend to decelerate the crankshaft, the rotational inertia of the flywheel resists this action and the flywheel tends to maintain a uniform rate of crankshaft rotation.

If the distribution of mass of a rotating object is changed, its angular velocity changes so that the angular momentum remains constant. A skater spinning on the ice with arms folded, as in Figure 6-20, turns with relatively constant angular velocity. If she extends her arms, her rotational inertia increases. Since angular momentum is conserved, her angular velocity must decrease.

Figure 6-20. Conservation of angular momentum. When the skater spins (left), her rotational inertia is small and her rotational velocity is large. When she extends her arms (right), the situation is reversed.

Questions

GROUP A

1. (a) What is impulse? (b) What is momentum? (c) How are impulse and change of momentum related?
2. Why do both a slowly docking ferryboat and a speeding rifle bullet have a large amount of momentum?
3. Are impulse and momentum scalar or vector quantities?
4. What is the law of conservation of momentum?
5. How does an aerial display of fireworks illustrate conservation of momentum?
6. Distinguish between elastic and inelastic collisions.
7. What is meant by an isolated system? Consider torques as well as forces in your answer.
8. Give examples of one-dimensional and two-dimensional collisions.

GROUP B

9. Explain why it is impossible for a moving ocean liner or a speeding airplane to make an abrupt turn.
10. What happens to the momentum of a car when it stops?
11. Describe the situation in Figure 6-15 (a) if the boy, instead of the man, does the pushing and (b) if the boy and the man push simultaneously.
12. How can you explain the precession of a spinning gyroscope in terms of angular momentum and the torque produced by gravity?
13. (a) What external forces other than friction are acting on the moving carts in Figure 6-17? (b) Why can they be ignored in the experiment?
14. Does the law of conservation of energy hold for inelastic collisions? Explain.

Problems

GROUP A

1. A $50\overline{0}$-kg mass is at rest and is free to move. At a certain time, 25.0 s after a force acts on it, the mass has a velocity in the direction of the force of $15\overline{0}$ m/s. What is the magnitude of the force?
2. A bullet, mass 60.0 g, is fired from a gun that is suspended by wires and is free to move. The mass of the gun is 5.00 kg. The speed of the bullet is 615 m/s. What is the speed of recoil of the gun?
3. A 1.50-kg ball whose velocity is 8.00 m/s southward collides with a 2.00-kg ball traveling along the same path with a velocity of 3.00 m/s southward. If the velocity of the 2.00-kg ball is 7.29 m/s southward after impact, what is the velocity of the 1.50-kg ball?
4. A proton (mass = 1.67×10^{-27} kg) moves with a velocity of $6.00 \times$

10^6 m/s. Upon colliding with a stationary particle of unknown mass, the proton rebounds on its own path with a velocity of 3.60×10^6 m/s. The collision sends the unknown particle forward with a velocity of 2.40×10^6 m/s. What is the mass of the unknown particle?

5. An object with a mass of 15.0 kg and a velocity of 2.00 m/s collides with a stationary object having the same mass. After the collision the first object moves along a line that makes an angle of $4\overline{0}°$ to the right of its original path. (a) At what angle will the second object move after the collision? (b) Determine the momentum of each object after the collision. (c) Calculate the speed of each object after the collision.

GROUP B

6. A freight car with a mass of 3.0×10^5 kg travels at a velocity of

2.5 m/s. It collides with a stationary car having a mass of 1.5×10^5 kg on a horizontal track. The cars connect and roll together after impact. What is the velocity of the connected cars?

7. Benjamin Franklin once said that light could not consist of particles because if a particle would travel at such tremendous speeds (3×10^8 m/s), it would have the impact of a 10-kg cannon ball fired from a cannon with a velocity of 100 m/s. According to this comparison, what is the mass of Franklin's particle of light?

8. A person is skiing down a hill. The person's mass, including skis, is 65 kg. When the person reaches the bottom of the hill the speed is 15 m/s. The person hits a snowdrift and stops in 0.30 s with uniform deceleration. (a) How far does the person penetrate the snowdrift? (b) With what average force does the person hit the drift?

9. Show that angular momentum has the unit dimensions kg m^2/s.

10. Two balls collide head-on with equal velocity. One ball has three times the mass of the other. After the elastic collision, the more massive ball stops and the less massive one moves back along its original path with twice its original velocity. Show that momentum is conserved in the collision.

11. In the multiple-exposure photograph in Figure 6-21, a large ball approaches from the top and a smaller one from the bottom. The mass of the large ball is 15$\overline{0}$ g. The photo shows the balls at equal time intervals. By means of an appropriate vector diagram, find the mass of the smaller ball.

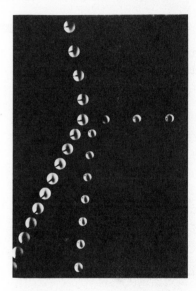

Figure 6-21.

SUMMARY

Work is the product of a displacement and the component of the force in the direction of the displacement. The unit of work is the joule. Work done by varying forces can be found by calculating the area under the curve of a graph in which the horizontal axis denotes the displacement and the vertical axis denotes the force. Radian measure is used to compute work done in rotary motion.

Machines can be used to multiply force at the expense of distance. The efficiency of a machine is the ratio of the useful work output to the total work input and is expressed as a percentage. Power is the time rate of doing work. It is measured in watts. Radian measure is used to compute power in rotary motion.

Gravitational potential energy is equal to the work done against gravity to place an object in position. Thus energy is measured in units of work. The kinetic energy of a body moving in a straight line is directly proportional to the mass of the body and the square of its velocity. In rotary motion, the kinetic energy is directly proportional to the rotational inertia and the square of the angular velocity.

Elastic potential energy is directly proportional to the force constant and the square of the amount of stretch. The law of conservation of mechanical energy states that the sum of the potential and kinetic energy of an ideal energy system remains constant.

Momentum is the product of the mass of a moving body and its velocity. The change of momentum of a moving body is equal to its impulse, which is the product of a force and the time interval during which the force acts. The momentum of a system of objects is conserved when no net external forces are acting on the system. This is the law of conservation of momentum. The reaction principle is an application of this law. The conservation of momentum also helps to describe the motion of objects colliding elastically or inelastically and in one, two, or three dimensions. Unless a net external torque acts on a rotating object, its angular momentum is also conserved.

VOCABULARY

angular impulse
angular momentum
efficiency
elastic collision
elastic potential energy
gravitational potential
  energy

impulse
inelastic collision
joule
kinetic energy
law of conservation of
  mechanical energy

law of conservation of
  momentum
machine
momentum
power
watt
work

PHASES OF MATTER

7

The idea that matter is composed of small, indivisible particles originated more than 2300 years ago with Greek philosophers like Democritus, who called these particles "atoms." Democritus reasoned that atoms differ in size and shape and combine with each other to form new substances.

THE STRUCTURE OF MATTER

7.1 Early Theories   When a lump of sugar is crushed into smaller pieces, each of these pieces is still a particle of sugar. Only the size of the piece of sugar is changed. If the subdividing process is continued by grinding the material into a fine powder, the results are the same. Even when the sugar is dissolved in water and the pieces are too small to be seen with a microscope, the taste of the sugar is retained. Evaporation of the water returns the solid sugar's other identifying properties, such as color, crystalline shape, and density. Observations such as these, together with the quantitative evidence supporting the laws of definite and multiple proportions, which you may have already studied in chemistry, lead to these conclusions about the nature of matter:

1. *Matter is composed of particles.*
2. *The ultimate particles of matter are extremely small.*

As early as 400 B.C., Greek philosophers formulated ideas about the ultimate composition of matter. Democritus (460–370 B.C.) believed that matter is indestructible and that the subdividing process mentioned above would reach a limit beyond which no further separation is possible. He called these ultimate particles *atoms*, after the Greek word *atomos* meaning "indivisible." This idea was largely the result of rational thinking. Democritus had no direct evidence to back up his concept of atoms.

In this chapter you will gain an understanding of:

▷ the relationship between atoms and molecules
▷ the kinetic theory of matter
▷ forces between molecules
▷ properties of matter in the solid phase
▷ Hooke's law relating stress and strain
▷ properties of liquids
▷ conditions affecting the boiling point of a liquid
▷ diffusion in liquids and gases
▷ vapor pressure and boiling

We still refer to wind and rain as the "elements of nature," especially during a storm.

The Greek philosopher Aristotle (384–322 B.C.) suggested that all matter can be reduced to four basic *elements:* air, earth, fire, and water. The reasoning that was used to support this idea was often quite complex, but it was a beginning in the attempt to list the ultimate particles of the universe.

It was a long time before the idea of atoms and ultimate particles was actively studied again. Near the beginning of the 19th century, the English scientist John Dalton (1766–1844) conducted a series of experiments with gases that added greatly to our knowledge of the nature of atoms. Dalton's *atomic theory* explained the mechanisms of known reactions between substances and made it possible to extend the list of known elements.

The atomic theory has been modified and expanded many times since Dalton's day. Almost two hundred years of investigation have added much information to our knowledge of the makeup and properties of atoms, the way in which they interact, and the nature of the compounds they form. Modern theory includes data concerning the mass and size of atoms as well as the energy relationships among atoms. Experiments have also revealed the fact that atoms are not ultimate particles of matter after all. A whole array of subatomic particles is presently being studied by scientists. (See Chapters 23–25.)

Figure 7-1. John Dalton postulated that each chemical element is composed of a different kind of atom. Dalton's ideas are the basis of modern atomic theory.

7.2 Molecules   *A **molecule** is the smallest chemical unit of a substance that is capable of stable, independent existence.* In the example in the previous section, the smallest particle of sugar that retains any identifying properties of the substance is a molecule of sugar. Not all substances, however, are composed of molecules. Some substances are composed of electrically charged particles known as ions.

A molecule of sugar is so small that it has never been seen, even with the help of the most powerful microscope. To get an idea of the extremely small size of molecules, imagine a drop of water magnified until it is as large as the earth. With this tremendous increase in size, a single molecule of water would be about one meter in diameter. A simple molecule, like that of water, is about 3×10^{-10} m, or 3 angstroms (Å) in diameter.

Molecules of more complex substances may have sizes of more than 200 Å. The electron microscope, which is capable of magnifications of several million times, can be used to photograph some of these "giant" molecules.

1 Å = 10^{-10} m or 10 nanometers. The angstrom is not an SI unit, but it is widely used with units of that system. It is named after the Swedish physicist Anders Ångström (1814–1874).

7.3 Atoms   If a molecule of sugar is analyzed, it is found to consist of particles of three simpler kinds of matter: carbon, hydrogen, and oxygen. These simpler forms of matter are called *elements*. *An* **atom** *is the smallest unit of an element that can exist either alone or in combination with other atoms of the same or different elements*. A molecule of sugar is made up of atoms of carbon, hydrogen, and oxygen.

Since atoms make up molecules, atoms are usually smaller than molecules. The smallest atom, an atom of hydrogen, has a diameter of about 0.6 Å. The largest atoms are a little more than 5 Å in size. The hydrogen atom is also the lightest atom. It has a mass of 1.673559×10^{-27} kg. The most common uranium atom, which is one of the heaviest atoms, has a mass of 3.952989×10^{-25} kg. There is about a tenfold range in the sizes of atoms and about a 250-fold range in their masses.

In 1970 the American scientist Albert Crewe (b. 1927) took the first photos of atoms. In 1976 Dr. Crewe obtained black and white movies of atoms. Two years later, Crewe used an electron microscope to take color movies of atoms.

Even with the use of scientific notation, it is difficult to express conveniently the masses of individual atoms. Consequently, scientists express the *atomic mass* of an atom in **atomic mass units** *(u)*. One *u* is equal to $1.6605655 \times 10^{-27}$ kg. This is $\frac{1}{12}$ the mass of a carbon-12 atom; carbon-12 is the most abundant form of carbon atoms. In other words, scientists use carbon-12 as the standard of mass for atoms.

In atomic mass units, the atomic mass of the most abundant form of hydrogen is 1.007825 *u*, while that of the most abundant form of uranium is 238.0508 *u*. You will notice that atomic masses are known very accurately. *The integer nearest to the atomic mass is called the* **mass number** *of an atom*. The mass number is represented by the symbol A. Thus for the common hydrogen atom, $A = 1$; for the common uranium atom, $A = 238$.

The concept of atoms is very useful in the study of chemical reactions. When substances interact chemically, atoms are regrouped, but the number of the various kinds of atoms does not change during the reaction.

7.4 Kinetic Theory of Matter So far in our study of physics, we have been concerned almost entirely with the behavior of solid particles of matter. A solid is one of the three *phases* (or states) of matter—the solid phase, the liquid phase, and the gaseous phase. In the description of matter *phase* indicates the way in which particles group together to form a substance. The structure of a substance can vary from compactly arranged particles to highly dispersed ones.

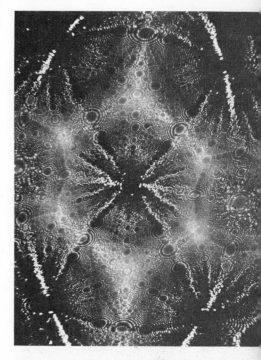

Figure 7-2. The point of a tungsten needle as seen through a field-ion microscope, which magnifies objects up to three million times. The photo shows the regular arrangement of tungsten atoms in the metal.

In a *solid*, the particles are close together in a fixed pattern. In a *liquid*, the particles are not usually as close together as in a solid and are not held in any fixed pattern. In a *gas*, the average separation of the particles is relatively large, and as in a liquid, the particles are not held in any fixed pattern. (Frequently the term *vapor* is used to describe a gas that under ordinary conditions of temperature and pressure is in the liquid phase.) Both liquids and gases are also called *fluids* (from Latin meaning "to flow").

To explain the motions of molecules and the energy molecules possess, particularly in the gaseous phase, scientists have developed the **kinetic theory of matter**. Two basic concepts of this theory are:

1. The molecules of a substance are in constant motion. The amount of motion depends upon the average kinetic energy of the molecules; this energy depends upon the temperature.

2. Collisions between molecules are perfectly elastic (except when chemical changes or molecular excitations occur).

The kinetic theory was developed independently of the atomic theory.

7.5 Forces between Molecules The forces required to pull a solid apart are generally much greater than the forces required to separate a similar amount of liquid. Liquids separate into drops. The forces of attraction between the molecules of liquids are not as great as the attractive forces of solids. Most liquids occupy a larger volume than the same mass of solid. Since the size of a molecule of a substance in one phase does not differ appreciably from its size in another phase, molecules of liquids must be farther apart than molecules of solids. In a gas, molecules separate from each other spontaneously; this accounts for the fact that a gas occupies a volume about 1000 times that of an equal mass of liquid. This spontaneous separation of gas molecules indicates that the kinetic energy of the molecules is great enough to keep them separated. We may conclude that forces between molecules decrease as the distance between them increases.

Solids and liquids are not easily compressed. Apparently, when molecules in solids and liquids are pushed closer together than their normal spacing, they repel each other. As the molecules are pushed still closer together, the repulsive forces become greater.

Intermolecular forces are mainly electric. They are about 10^{29} times as strong as the gravitational forces between molecules at the typical distances found in solids and liquids. Thus gravitational forces between molecules are negligible in comparison with intermolecular forces. Inter-

The forces that hold molecules together in crystals are known as van der Waals forces, after the Dutch scientist Johannes Diderik van der Waals (1837–1923).

molecular forces are small compared to the weight of objects we can see and handle; however, the masses these intermolecular forces act on—the masses of molecules—are small, too. These forces can impart instantaneous accelerations 10^{14} times the acceleration of gravity. Such accelerations last only a very short time since one molecule, so accelerated, moves quickly out of the range of another.

Figure 7-3 shows how the force of interaction between molecules varies with the distance between their centers. If we imagine one molecule to be fixed at the intersection of the axes, the other molecule will be repelled until the distance of separation is such that their outer charges do not overlap. This condition occurs at r_0 where no net force acts between the molecules. The distance to r_0 is often called the equilibrium distance. This distance (about 2.5×10^{-10} meter) is, therefore, the distance between the centers of two "touching" molecules and is the diameter of a single molecule. As the separation between the molecules increases, the force of attraction between opposite charges first increases and then approaches zero. Different molecules have different sizes and charge configurations, but they always show the qualitative behavior indicated by Figure 7-3.

In the solid phase, molecules vibrate about the equilibrium position r_0. They do not have enough energy to overcome the attractive force. The equilibrium positions are fixed. In a liquid the molecules have greater vibrational energy about centers that are free to move, although the average distance between the centers of the molecules remains nearly the same. The average distance of separation between gas molecules is considerably greater than the range of intermolecular forces, and the molecules move in straight lines between collisions.

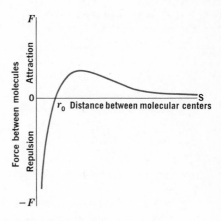

Figure 7-3. The force between molecules varies with the distance between molecular centers.

THE SOLID PHASE

7.6 The Nature of Solids   Solids have definite shapes and definite volumes. Scientists usually describe solids as either *crystalline* or *amorphous*. Crystalline solids have a regular arrangement of particles; amorphous solids have a random particle arrangement.

In addition to the forces that bind particles of a solid together, the motion of particles of a solid is an important consideration. Particles of a solid are held in relatively fixed positions by the binding forces. However, they do have a vibratory motion about their fixed positions. The amplitude of their vibration and their resulting vibratory

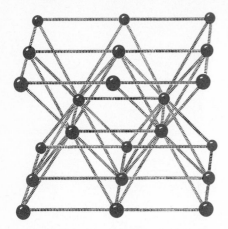

Figure 7-4. A crystalline solid. The particles of the solid are held in a specific pattern by intermolecular forces, which are shown here as springs. Each particle vibrates about its indicated position at a rate that depends on the temperature of the solid.

Why is the tape used to hold a bandage in place called adhesive tape?

Figure 7-5. Testing for tensile strength. The steel test rod elongates as large tensile forces act on it. The rod will break in the region where it has become noticeably thinner.

energy are related to the temperature of the solid. At low temperatures the kinetic energy is small, at higher temperatures it is larger.

Diffusion *is the penetration of one type of particle into a mass consisting of a second type of particle.* If a lead plate and a gold plate are in close contact for several months, particles of gold may be detected in the lead, and vice versa. This demonstrates that even solids may diffuse. Diffusion is slow in solids because of the limited motion of the particles and their close-packed, orderly arrangement.

7.7 Cohesion and Adhesion   The general term for *the force of attraction between molecules of the same kind is **cohesion.*** Cohesion, the force that holds the close-packed molecules of a solid together, is a short-range force. If a solid is broken, layers of gas molecules from the air cling to the broken surfaces. These gas molecules prevent the rejoining of the solid surfaces. The molecules of the broken surfaces are not close enough to have sufficient attraction to hold. However, if the surfaces of two like solids are polished and then slid together, cohesion causes the solids to stick together.

Molecules of different kinds sometimes attract each other strongly. Water wets clean glass and other materials. Glue sticks to wood. *The force of attraction between molecules of different kinds is called **adhesion.*** The forces of cohesion and adhesion have definite values for specific molecules.

7.8 Tensile Strength   Several properties of solids depend on cohesion; one of these is tensile strength. Suppose two wires of the same diameter, one copper and one steel, are put in a machine that pulls the wires until they break. When tested in this manner, steel wire proves stronger than copper wire of the same diameter. Therefore we say that steel has a higher *tensile strength* than copper. *The **tensile strength** of a material is the force per unit cross-sectional area applied perpendicularly to the cross section that is required to break a rod or wire of that material.* See Figure 7-5 and Appendix B, Table 7. Tensile strength is a measure of cohesion between adjacent molecules over the entire cross-sectional area.

7.9 Ductility and Malleability   If a metal rod can be drawn through a small opening, or die, to produce a wire, the metal is said to be *ductile* (*duk*-til) or to possess *ductility.* As the metal is pulled through the die, pressure decreases its diameter and increases its length and the rod becomes a wire. See Figure 7-6(A). In one industrial application of ductility, more than 1000 meters of wire are formed per

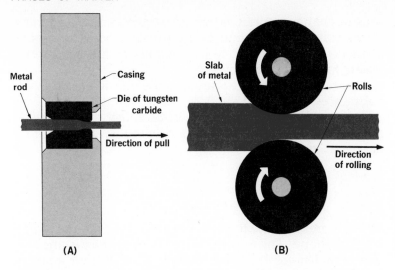

Metal rod
Casing
Die of tungsten carbide
Direction of pull

(A)

Slab of metal
Rolls
Direction of rolling

(B)

Figure 7-6. During the process of drawing (A) and rolling (B), the atoms of a metal are forced to move over each other from one position in the crystal pattern to another.

minute in a device in which the pressure is exerted by a fluid, and the wire never touches the solid die. Platinum is so ductile that a single gram of the metal can be drawn into a wire almost 600 kilometers long.

Metals that can be hammered or rolled into sheets are said to be *malleable* (*mal*-ee-uh-bul) or to have *malleability*. During the hammering or rolling, the shape of the metal is greatly changed. See Figure 7-6(B).

In the process of hammering, rolling, or drawing, layers of atoms of a metal are forced to slide over one another, thus changing their positions in the crystal pattern. Since the cohesive forces are strong and the atoms do not become widely separated from each other during their rearrangement, the metal holds together while its shape is being changed. Silver, gold, platinum, copper, aluminum, and iron are all highly malleable and ductile.

7.10 Elasticity When opposing forces are applied to an object, the size and shape of the object are changed. *The ability of an object to return to its original size or shape when external forces are removed* is described as its **elasticity.** However, there is a limit beyond which the change produced by the applied forces does not disappear when the forces are removed. Ductility and malleability are properties of substances that can undergo such permanent changes without fracturing. *When a substance is on the verge of becoming permanently changed*, we say it has reached its **elastic limit.**

At the elastic limit, molecular forces are overcome to such an extent that particles slide past each other, and the shape of the material is permanently changed. Such a

Elastic potential energy was discussed in Section 6.10.

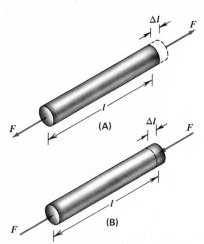

Figure 7-7. Two types of strain. Elongation strain (A) and compression strain (B) are both expressed by the ratio Δ*l*/*l*.

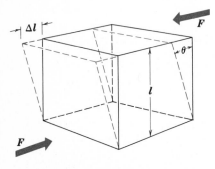

Figure 7-8. Shear strain. The amount of deformation produced by the forces, *F*, is equal to Δ*l*/*l*, or tan θ.

drastic change cannot be reversed by molecular forces. Every solid has a certain range through which it can be changed and yet return to its original condition before its elastic limit is reached.

Two terms are used to describe the elastic properties of substances: *stress* and *strain*.

*Stress is the ratio of the internal force, **F**, that occurs when the shape of a substance is changed to the area, **A**, over which the force acts, or*

$$\text{stress} = \frac{F}{A}$$

Stress represents the tendency of a substance to recover its original shape. The applied force changes the distance between molecules by either pulling them farther apart or pushing them closer together. When the force is removed, molecular forces restore the molecules to their normal spacing.

Strain is the relative amount of deformation produced in a body under stress. There are various kinds of strain, but its measure is always an absolute ratio—a number without units. When a rod is placed under tension, it stretches. *The ratio of the increase in length to the original length is called the **elongation strain** and is expressed as* Δ*l*/*l*. Linear compression strain is the reverse of elongation, as shown in Figure 7-7. Elongation and compression are accompanied by small changes in the cross-sectional area, but their effect is small in situations where elastic limits are not exceeded.

In an *elongation strain,* the particles have been moved in a direction perpendicular to the area over which the forces act. In a **shear strain,** the particles move in a direction parallel to the area over which the forces act. A cube will take the shape of a rhombic prism when shear forces are applied to it. *The measure of shear is the ratio of the amount the top of the cube is moved to the side to the length of one side of the cube.* This ratio, shear, is expressed as the tangent of the angle through which the oblique edges of the imaginary cube have been rotated from their original direction, as shown in Figure 7-8.

Volume strain is the ratio of the decrease in volume to the volume before the stress was applied.

Flexure (bending) and torsion (twisting) are combinations of elongation and compression strains. A straight beam bent into a plane curve undergoes compression on one side and elongation on the other side. The layer of material down the center of the beam undergoes neither compression nor elongation.

7.11 Hooke's Law Beams of buildings and bridges are often subjected to varying forces or stresses. It is important for engineers to know what deformation, or strain, these forces will produce. This involves the measurement of the elasticity of materials.

If a coiled spring is stretched by a weight, as shown in Figure 7-9(A), and then returns to its original form after the stretching force is removed, the spring is said to be *perfectly elastic*. If the spring is stretched too far, it remains permanently deformed; its elastic limit has been exceeded.

Suppose one end of a steel wire is fastened to a beam, as in Figure 7-9(B), and weights are gradually added to the hanger attached to the lower end. As the weights are added one by one, the wire stretches gradually. The wire stretches by an amount that is exactly proportional to the weight pulling on it, and it returns to its previous length when the weight is removed. If more weights are added, the elastic limit is eventually reached. Then when the weights are removed, the wire remains deformed; it does not return to its original length.

By making such measurements, the English philosopher and scientist Robert Hooke (1635–1703) found that the amount of elongation in elastic solids is directly proportional to the deforming force provided the elastic limit is not exceeded. The elongation also depends on the length and cross-sectional area of the wire or rod. All these facts were combined by Hooke into one simple law. *Hooke's law* states that *within certain limits strain is directly proportional to stress.*

The value of the ratio stress/strain is different for different solids. However, the ratio is constant for a given substance, even when the substance has different shapes. This ratio gives us a means of comparing the elasticity of various solids. The numerical value of Hooke's law is called *Young's modulus, Y,* and is defined by the equation

$$Y = \frac{\text{stress}}{\text{strain}}$$

From the definitions of stress and strain given in the previous section, we get

$$Y = \frac{F/A}{\Delta l/l}$$

and

$$Y = \frac{Fl}{\Delta l A}$$

As an example, suppose that the steel "A" string of a piano is 43.0 cm long and has a cross-sectional area of

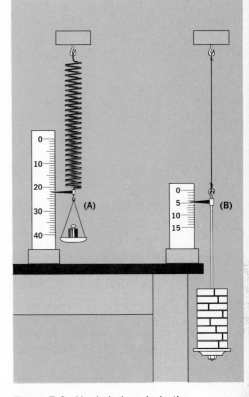

Figure 7-9. Hooke's law. In both cases, the pointers stood at zero before the weights were added, and the amount of stretch is proportional to the force exerted by the weights (provided that the elastic limit is not exceeded).

Young's modulus is named after the English physicist Thomas Young, who also did experiments with light. Values of Young's modulus for various substances are given in Appendix B, Table 8.

6.50×10^{-7} m². How much will the string stretch when a force of $71\bar{0}$ n is applied to it by a piano tuner to bring it up to pitch?

$$\Delta l = \frac{Fl}{YA} = \frac{(7.10 \times 10^2 \text{ n})(4.30 \times 10^{-1} \text{ m})}{(20.0 \times 10^{10} \text{ n/m}^2)(6.50 \times 10^{-7} \text{ m}^2)}$$

$$= 2.35 \times 10^{-3} \text{ m}$$

Questions

GROUP A

1. List the basic concepts of the kinetic theory.
2. (a) What is meant by the phase of matter? (b) Describe the three phases of matter.
3. Distinguish between a molecule and an atom.
4. Under what circumstances are the forces between molecules in a solid (a) attractive, (b) repulsive?
5. (a) What are the two classes of solids? (b) What is the particle arrangement in each?
6. (a) Describe the motion of the particles of a solid. (b) What experimental evidence supports this description?
7. What name is given to the force of attraction between (a) like molecules, (b) unlike molecules?
8. (a) What is tensile strength? (b) In what units is the value for tensile strength expressed?

9. Define elasticity.
10. Distinguish between stress and strain.
11. What is meant by the elastic limit?
12. State Hooke's law.
13. Give the equation for Young's modulus and identify each of its terms.

GROUP B

14. For a given mass of matter that can exist as a solid, a liquid, and a gas, what does the volume indicate about the spacing between molecules in these three phases?
15. (a) What is the magnitude of an atomic mass unit? (b) How is this value established?
16. What direct evidence is there that matter is composed of molecules?
17. What happens to the crystal pattern of an elastic material when its elastic limit is exceeded?
18. What is the relationship between Hooke's law and the force constant (Section 6.10)?

Problems

GROUP A

1. A coiled spring is stretched 0.050 m by a mass weighing 0.500 n that is hung from one end. How far will it be stretched by a mass weighing 1.000 n?
2. The hook of a spring balance is pulled down 5.00 cm by a mass weighing 40.0 n. (a) If a mass weighing 125.0 n is substituted for the previous mass, how far is the hook pulled down? (b) How far apart are the 5.00-n graduations on the scale?

3. Two identical wires are 125.0 cm and 375.0 cm long. The first wire is broken by a force of 489.0 n. What force is needed to break the other?
4. A copper wire 100.0 cm long and 0.200 cm in diameter is suspended from a solid support. A 10.00-kg mass is hung from the lower end of the wire. (a) Determine the stress in the wire. (b) Determine the strain in the wire.
5. The cross-sectional area of an aluminum wire is 2.50×10^{-3} cm². What force will break it?

6. A brass wire 2.57 m long and 2.00 mm in diameter is suspended from a fixed support. A mass of 8.25 kg is hung from the lower end. Determine the elongation of the brass wire.

7. A brass wire 215 cm long and 5.00×10^{-2} cm^2 in a cross-sectional area supports a mass weighing 1.25 n. How far is the wire stretched?

GROUP B

8. An aluminum wire 2.00 m long and 0.400 mm in diameter is attached to a firm support. Masses weighing 1.00 n each are added to it one at a time. The elongations of the wire as measured through a microscope and scale are 0.0224 cm, 0.0464 cm, 0.0681 cm, 0.0897 cm, and 0.1128 cm. Plot a graph with the stresses as ordinates and the strains as abscissas. From the information given above and the graph, determine Young's modulus.

9. A rod 6.00 mm in diameter is composed of a material having a Young's modulus of 9.00×10^{10} n/m^2. What force will stretch it by 0.02% of its length?

THE LIQUID PHASE

7.12 The Nature of Liquids In 1827 the English botanist Robert Brown (1773–1858) put some pollen grains in water and placed a bit of this suspension on a small glass slide. When he examined the suspension through a microscope, he found that the pollen grains moved in a very random way. The path of a single particle resembled that shown in Figure 7-10. This so-called *Brownian movement* is caused by the continual bombardment of the suspended particles by molecules of the surrounding liquid. Brownian movement shows that molecules of a liquid are in constant, rapid, and random motion. This is in keeping with the kinetic theory of matter described in Section 7.4.

Liquids diffuse. This can be shown by the experiment in Figure 7-11. Enough concentrated copper(II) sulfate solution is poured into a tall cylinder to form a layer several centimeters deep. A flat cork is floated on the surface of the solution. Water is carefully poured through a funnel tube onto the top of the cork. The water flows around the edge of the cork and spreads out over the surface of the copper(II) sulfate solution, producing two distinct layers. The water "floats" on the copper(II) sulfate solution because the water has a lower density. After the cylinder stands for a few days, the interface between the layers is less distinct. Some of the copper(II) sulfate solution diffuses into the water above, and some of the water molecules diffuse into the copper(II) sulfate solution below. Even though weeks may pass before the diffusion is complete, we can see that diffusion does occur in liquids despite the force of gravity. Because of the slightly more

Brownian movement is also exhibited by solid particles, such as smoke, suspended in the atmosphere.

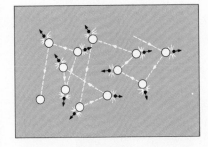

Figure 7-10. Brownian movement. The seemingly random motion of the white particle (starting at the upper right) is the result of collisions with molecules of the surrounding fluid. The directions of the fluid molecules after the collisions are pictured by the black arrows. (The distances between collisions are exaggerated for clarity.)

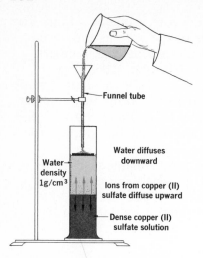

Figure 7-11. Diffusion in liquids. Ions of water diffuse downward into the heavier copper(II) sulfate solution, and vice versa.

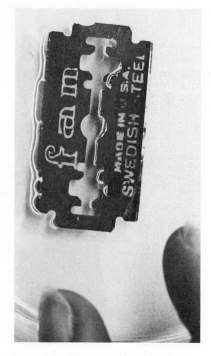

Figure 7-12. Surface tension keeps a razor blade afloat even though its mass density is much greater than that of water.

open molecular arrangement and greater molecular mobility of liquids, their rate of diffusion is considerably faster than that of solids.

7.13 Cohesion and Adhesion   If you push a spoon into a jar of honey and then pull it out, a certain amount of force is needed to pull apart the molecules of the honey. If you lick the honey from the spoon, you find that force is required to pull the molecules of the honey away from the molecules of the spoon. These are examples of cohesion within a liquid and adhesion between a liquid and a solid.

If a clean glass rod is dipped into water and then removed, some of the water clings to the glass rod. We say that the water wets the glass. The adhesion of water molecules to glass must therefore be greater than the cohesion between water molecules.

If the glass rod is dipped into mercury and then removed, the mercury does not cling to the glass because the cohesion between mercury molecules is greater than their adhesion to glass. Cohesive forces vary in different liquids. They are usually smaller in liquids than in solids.

If you examine the surface of water in a glass container, you will find that it is not completely level. It is very *slightly concave* when viewed from above. The edge of the surface where water comes in contact with the glass is lifted a little above the general level. *The crescent-shaped surface of a liquid column is called the **meniscus**.* The water rises at the edge because adhesion between water and glass is greater than cohesion between water molecules.

If the container contains mercury instead of water, the edges of the liquid are depressed and the surface is *slightly convex.* In this case, cohesion between mercury molecules is greater than adhesion between mercury and glass.

Cohesion is also responsible for a property of fluids called *viscosity.* **Viscosity** *is the ratio of shear stress to the rate of change of shear strain in a liquid or gas.* The viscosity of a fluid determines the rate of flow of the fluid. Viscosity is dependent on temperature. As the temperature increases, the viscosity of gases increases while the viscosity of liquids decreases. This explains the expression "as slow as molasses in January." Modern motor oils are specially formulated to minimize the effect of temperature on viscosity. Under normal conditions, gases are much less viscous than liquids.

7.14 Surface Tension   Have you ever seen a sewing needle or a razor blade floating on the surface of water? Even though they are about seven times as dense as water, if they are placed carefully on the surface, they remain there. A close look at the water surface shows that the needle or razor blade is supported in a hollow in the water surface, as seen in Figure 7-12. *The water acts as though it has a thin, flexible surface film.* The weight of the needle or razor blade is counterbalanced by the upward force that is exerted by the surface film. This property of liquids is due to *surface tension.*

All liquids show surface tension. Mercury has a very high surface tension. In many liquids the surface film is not as strong as that of water or mercury. Part of the cleaning action of detergents is due to their ability to lower the surface tension of water. This makes it possible for the water and detergent to penetrate more readily between the fibers of the articles being cleaned and the dirt particles.

Since particles in a liquid attract similar liquid particles that are nearby, they move as close together as possible. Hence the surface will tend to have a minimum area. The effect of this attraction is to make the liquid behave as if it were contained in a stretched elastic skin. The *tension* in this "skin" is the *surface tension.* When a force acts on a liquid surface film and distorts that film, the cohesion of the liquid molecules exerts an equal and opposite force that tends to restore the horizontal surface. Thus the weight of a supported needle produces a depression in the water surface film that increases the area of the film. In order to restore the surface of the liquid to its original horizontal condition, the cohesion of the water molecules exerts a counterbalancing upward force on the needle.

Surface tension produces contraction forces in liquid films. A liquid film has two free surfaces on which molecules are subject to an unbalanced force toward the inside of the film. Thus both free surfaces tend to assume a minimum area. The contraction of the film can be demonstrated by the device shown in Figure 7-13. A wire ring containing a loop of thread is dipped into a soap solution causing a film to form across the ring. If the film inside the loop of thread is broken with a hot wire the unbroken film outside the loop contracts and pulls the thread equally on all sides to form a circle.

Surface tension causes a free liquid to assume a spherical shape. A free liquid is one that is not acted upon by any external force. This condition can be approximated by small drops of mercury on a table top, as shown in Figure 7-14. A sphere has the smallest surface area for a given volume. The unbalanced force acting on liquid surface molecules tends to pull them toward the center of the liquid, reducing the surface area and causing the liquid to

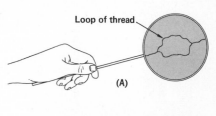

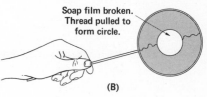

Figure 7-13. In (A), a soap film covers the entire area inside the ring. When the film inside the loop of thread is broken (B), the film outside the loop contracts and pulls the thread into a circle.

Figure 7-14. Small drops of mercury assume a spherical shape because of surface tension. Large drops are flattened because the downward force of gravitation is greater than the upward force produced by surface tension.

Water rises

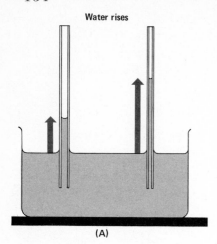

(A)

Mercury is depressed

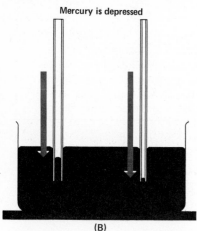

(B)

Figure 7-15. Capillarity. The elevation or depression is inversely proportional to the diameter of the tube. (The tube diameters and magnitudes of capillarity are exaggerated for clarity.)

Butter and glass do not have definite melting points.

assume a spherical shape. Since the cohesive force between mercury molecules is great and mercury has a very high surface tension, small drops of mercury are almost spherical. Larger drops of mercury, on which the effect of the force of gravity is greater, are noticeably flattened.

7.15 Capillarity When the lower ends of several tubes are immersed in water, the water will rise to the same height in each tube only if the tubes have large enough diameters so that the centers of the menisci are relatively flat. *Water does not rise to the same height in tubes of small diameters.* The height to which it rises increases as the diameter of the tube decreases. See Figure 7-15(A). When mercury is used, the depression of the surface is greater as the diameter of the tube is reduced. See Figure 7-15(B). *This elevation or depression of liquids in small-diameter tubes is called* **capillarity.**

Capillarity depends on both adhesion and surface tension. Adhesion between water and glass causes water to creep up the glass walls and produce a concave surface; surface tension tends to flatten this surface by contraction. The combined action of these two forces raises the water above its surrounding level. The water level rises until the upward force is counterbalanced by the weight of the elevated liquid.

Experiments have verified the following: *(1) Liquids rise in capillary tubes they wet and are depressed in tubes they do not wet. (2) Elevation or depression is inversely proportional to the diameter of the tube. (3) The elevation or depression decreases as the temperature increases. (4) The elevation or depression depends on the surface tension of the liquid.*

7.16 Melting *.The change of phase from a solid to a liquid is called* **melting.** Melting involves the breaking of bonds between the particles of a solid. The temperature at which this change occurs is called the **melting point.** Pure *crystalline* solids have definite melting points, and different solids have different melting points.

When a substance changes from a liquid to a solid, it is said to *freeze*. The temperature at which freezing occurs is known as the **freezing point.** For pure crystalline substances, the melting point and the freezing point are the same temperature at any given pressure.

Noncrystalline solids, like paraffin, have no definite melting point. When they are heated, they soften gradually. The temperature at which noncrystalline solids first soften and the temperature at which they flow freely are often greatly different.

In order for a solid to melt, energy must be supplied to it. This energy increases the energy of the particles of the solid and gives them the freedom of motion characteristic of the particles of a liquid. As the temperature of a solid increases, the vibrations of its particles increase in amplitude; thus more and more potential energy is stored in the average stretching of the bonds between the particles. Finally a point is reached at which the bonds between the particles cannot absorb any more energy without breaking. Thus crystalline solids have a definite melting point. When the liquid formed by melting a crystalline solid cools to a certain temperature, the energy of the liquid particles is reduced and the forces between the particles draw them into fixed positions in a crystal. Thus a liquid that forms a crystalline solid freezes at a definite temperature.

All the energy supplied to a substance during melting is used to increase the potential energy of the particles in changing from a crystal structure to a liquid. The kinetic energy of the particles does not change. Since average kinetic energy depends on temperature, the temperature is unchanged during the melting process.

Usually, the particles of noncrystalline solids are held together by forces of attraction and by the physical entanglement of long-chain molecules. The bonding combination is not of such definite strength that the bonds are broken when the particles acquire a fixed amount of energy. The energy required to overcome these bonds varies with the extent of the bonding and entanglement of each molecule. As they are heated, noncrystalline substances soften at a lower temperature first. At some higher temperature, they flow freely. Similarly, as they are cooled, the molecules become bonded at various kinetic energies, and the liquid does not solidify at a definite temperature.

The degree of separation of particles of a substance is different in the solid and liquid phases because of the difference in the potential energy of the particles. If melted paraffin is poured into a vessel and allowed to harden, the center becomes indented, or depressed, as shown in Figure 7-17(A). The paraffin cools and contracts as it solidifies. Both kinetic energy and potential energy are lost in the process. The loss of kinetic energy is indicated by the decrease in temperature. Loss of potential energy permits the particles to move closer together and take up less space. Almost all substances behave in this manner; the particles of most substances are closer in the solid phase than in the liquid phase.

Water is the most important exception to the rule that a

Figure 7-16. Noncrystalline solids, such as the glass in this building, do not have specific melting points.

The force of expansion that accompanies freezing can break water pipes in winter.

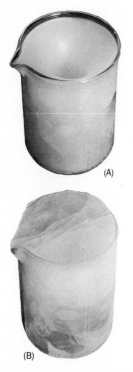

(A)

(B)

Figure 7-17. (A) Most substances, like paraffin, contract when they solidify. (B) Water is one of the few substances that expands on solidification.

substance contracts when it changes from a liquid to a solid. See Figure 7-17(B). When an ice cube tray is placed in the freezing compartment of a refrigerator, the level of the water in the sections of the tray is uniform. When the ice cubes are formed, however, each one has a slightly raised spot in the center. The volume occupied by ice is about 1.1 times that occupied by the water from which it was formed. The force of expansion when water freezes is enormous. Some calculations have placed the pressure as high as 83,000 newtons per square centimeter! Antifreeze is used in water-cooled cars that are driven in areas where the atmospheric temperature drops below the freezing point of water. As its name suggests, a mixture of antifreeze and water freezes at a lower temperature and does not expand significantly in freezing if the mixture is sufficiently concentrated.

Bismuth and antimony are two metals that expand rather than contract when they solidify. Substances such as ice, bismuth, and antimony have open crystal structures in which the particles are more widely separated than they are in the liquid phase. That is why the solid occupies a larger volume than the liquid.

7.17 Effect of Pressure on the Freezing Point

In substances that contract when they solidify, like the paraffin in Figure 7-17(A), the molecules of the solid are closer together than the molecules of the liquid. If additional pressure is exerted, such a liquid can be made to solidify at a higher temperature than its normal freezing point under atmospheric pressure. An increase in pressure raises the freezing (melting) point of most substances.

Some of the rock in the interior of the earth is hot enough to melt at normal pressure, but most of it remains solid because of the tremendous pressure. If the pressure is released, as when a volcano erupts, more of the rock melts and forms lava.

An increase in pressure has the opposite effect on the freezing point of a substance like water that expands as it freezes. In such a substance the molecules are farther apart in the solid than they are in the liquid. Since an increase in pressure makes formation of the solid more difficult, the freezing point is lowered if the pressure is raised.

We can illustrate this effect by suspending two weights over the surface of a block of ice by means of a strong wire. See Figure 7-18. The pressure of the wire on the ice lowers the melting point of the ice immediately below the wire. If the surrounding temperature is above the new melting point, this part of the ice melts, and the molecules of water are forced upward around the wire. When they reach a spot above the wire, the pressure returns to normal, the melting point rises, and the water freezes again. When the wire is embedded in the ice, the extra heat needed to melt the ice below the wire is supplied by the freezing of the water above the wire. In this way, the wire may cut its way through the block of ice, and yet leave the ice in one piece. This process of melting under pressure and freezing again after the pressure is released is called *regelation* (ree-jeh-*lay*-shun).

When the pressure of ice-skate blades on ice is sufficient to melt the ice at its existing temperature, the blades slide along with very little friction on a thin layer of water. However, if the ice is so cold that the pressure of the skate blades cannot melt the ice, the blades are retarded by the higher friction of steel on ice, and skating is difficult.

7.18 Effect of Solutes on the Freezing Point   The freezing point of a liquid is lowered whenever another substance is dissolved in it. The extent of the lowering depends on the nature of the liquid, the nature of the dissolved substance, and the relative amounts of each. The greater the amount of the solute in a fixed amount of liquid, the lower the freezing point of the liquid. We apply this principle when we use rock salt to prevent the formation of ice on roads and sidewalks, and to melt the ice that has formed. The dissolved substance interferes with crystal formation as the liquid cools. Before crystals form, the kinetic energy of the liquid particles must be reduced to a level below that at which they normally crystallize.

THE GASEOUS PHASE

7.19 The Nature of Gases   The following properties of gases show that gas molecules move independently of each other at high speed:

1. *Expansion.* A gas has neither a definite shape nor a definite volume; it expands and completely fills any container. This shows that gas molecules are independent particles.

2. *Pressure.* An inflated balloon may burst from the force that the air inside exerts on the balloon's inner surface. This force is a result of the continual bombardment of the inside surface by billions of moving molecules. If we increase the number of molecules within the balloon by blowing more air into it, the number of collisions against the inside surface increases, the pressure on the inside surface increases, and the balloon expands.

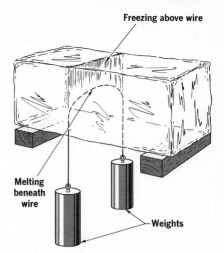

Figure 7-18. Regelation. The ice melts where pressure is applied by the wire and freezes again after the pressure is released.

For best results, the experiment in Figure 7-18 should be conducted at a temperature slightly below the freezing point.

Figure 7-19. Mobile regelation. A thin film of water forms as the skates press on the ice. As the skates pass, the water film freezes again.

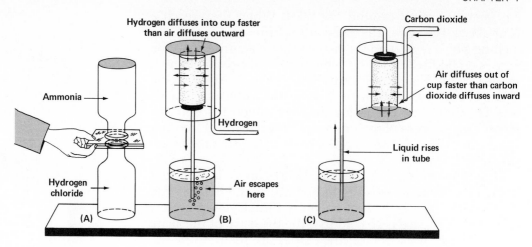

Figure 7-20. Diffusion in gases. (A) When the glass plates are removed, the gases mix by diffusion. (B) and (C) Differences in diffusion rates cause pressure differences between the upper and lower containers.

3. Diffusion. Hydrochloric acid is a water solution of the high-density gas hydrogen chloride; ammonia water is a water solution of the low-density gas ammonia. When these gases react chemically, they form a cloud of fine, white particles of solid ammonium chloride. Suppose we put a few drops of hydrochloric acid in a warm bottle and an equal amount of ammonia water in a second warm bottle. We then cover the mouth of each bottle with a glass plate and invert the bottle containing ammonia over the one containing hydrogen chloride, as shown in Figure 7-20(A). After the bottles have stood this way for a minute or two, we remove the glass plates so that the bottles are now mouth-to-mouth. The formation of white smoke indicates that the less dense ammonia descends and mixes with the more dense hydrogen chloride. The hydrogen chloride also rises and mixes with the ammonia in the upper bottle. The movement of each of these gases is opposite to that which would be caused by the density of the gases. The movement must be due to molecular motion. Thus diffusion of gases is evidence of the rapid movement of gas molecules.

Gases also diffuse through porous solids. In Figure 7-20(B), an inverted, unglazed earthenware cup is closed with a rubber stopper through which passes a glass tube that dips into a red liquid. When a low-density gas, such as hydrogen, is led into an inverted beaker placed over the cup, air immediately begins to bubble through the red liquid. Evidently the lighter molecules of hydrogen move in through the porous walls of the cup faster than the heavier molecules in the air move out. Thus there is an accumulation of hydrogen molecules inside the cup. This accumulation increases the pressure inside the cup, thus pushing the liquid down the tube and forcing some of the air and hydrogen mixture in the tube to escape.

In Figure 7-20(C), a porous cup is surrounded with a dense gas, such as carbon dioxide. Since carbon dioxide is denser than air, the apparatus must be modified; the beaker surrounding the porous cup must be positioned with the open side up. When the beaker is filled with carbon dioxide, molecules from the air flow out through the porous cup faster than carbon dioxide molecules enter. This difference reduces the pressure inside the cup, and the red liquid rises in the tube. Thus we see that there is an inverse relation between the rate of diffusion of a gas and its density.

Diffusion of gases through porous solids is an important process. Such diffusion occurs through the membranes of plants, animals, and humans, allowing oxygen to reach living cells and carbon dioxide to escape.

7.20 Vaporization When ether is placed in a shallow dish, within a short time the quantity of liquid decreases and the odor of ether becomes quite strong near the dish. Apparently molecules of ether liquid become molecules of ether vapor and mix with molecules of the gases in the surrounding air. In a similar manner, but at a much slower rate, moth balls left out in the air become smaller and smaller and eventually disappear while their characteristic odor is noticed in the air nearby. In this case, molecules of the solid moth balls turn into vapor and diffuse into the surrounding air. These are two examples of *vaporization, the production of a vapor or gas from matter in another phase.* See Figure 7-21.

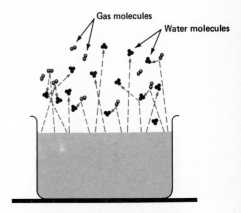

Figure 7-21. Vaporization. Water molecules escape into the air at the surface of the liquid. Some of them collide with molecules of gases in the air and rebound.

As in the case of melting (Section 7.16), vaporization is a constant-temperature process. All the energy supplied during vaporization goes to increase the potential energy of the particles. The average kinetic energy of the particles is unchanged.

The particles of all liquids and solids have an average kinetic energy that depends on the temperature of the liquid or solid. However, because of random collisions and vibratory motion, some particles have energies higher than average and others have energies lower than average. When the particle on the surface of a liquid or solid acquires enough energy to overcome the forces that hold it as part of the substance, the particle escapes and becomes a particle in the vapor phase. When vaporization occurs from liquids it is known as *evaporation;* when vaporization occurs from solids, it is known as *sublimation.* Sublimation is a direct change from solid to vapor without passing through the liquid phase.

The energy of the particles that evaporate from a liquid is obtained either from the surrounding atmosphere or from the remaining liquid. In either case, evaporation has a cooling effect on the environment. When perspiration evaporates from the skin, the process cools the skin. When rubbing alcohol is applied to the skin, the cooling effect is even greater because alcohol evaporates more rapidly than water.

Since the rate of evaporation or sublimation depends on the energy of the particles undergoing the change and in turn their energy depends upon their temperature, evaporation and sublimation occur more rapidly at higher temperatures and more slowly at lower temperatures.

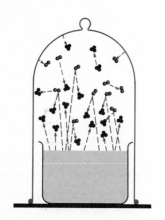

Figure 7-22. When the number of water molecules evaporating from the liquid equals the number of water vapor molecules returning to the liquid, the pressure exerted by the water vapor molecules against the walls of the bell jar is the equilibrium vapor pressure.

7.21 Equilibrium Vapor Pressure   A bell jar covering a container of water is shown in Figure 7-22. There are as many molecules of the gases of the air within the bell jar as there are in an equal volume of air outside it. The pressure (force per unit area) exerted by the gas molecules on the inside walls of the bell jar is the same as the pressure that such molecules exert on the outside walls.

When a molecule at the surface of the water inside the bell jar acquires sufficient kinetic energy, it escapes and becomes a water vapor molecule. As the water evaporates, water vapor molecules mix with the gas molecules in the bell jar. These water vapor molecules collide with gas molecules, with the walls of the bell jar, with the outside surface of the container of water, and with the surface on which the bell jar rests. They can also touch the water surface, be held by it, and become molecules of liquid again. The conversion of molecules of vapor to molecules of liquid is called **condensation.** Eventually the rate of evaporation equals the rate of condensation, and a condition of equilibrium prevails. At equilibrium, evaporation and condensation do not cease; they occur at the same rate. The number of water vapor molecules in the air in the bell jar remains constant. At equilibrium, the space above the water in the vessel is said to be *saturated* with water vapor.

The collision of water vapor molecules against the walls of the bell jar increases the pressure on the bell jar so that it exceeds the pressure exerted by the gases of the air. *Added pressure exerted by vapor molecules in equilibrium with liquid* is called **equilibrium vapor pressure.**

Since the kinetic energy of the particles of the liquid depends on both the temperature and mass of the particles, the equilibrium vapor pressure of a liquid depends on the composition as well as the temperature of the liquid. As the kinetic energy of the particles increases, so does the vapor pressure.

The ratio of the water vapor pressure in the atmosphere to the equilibrium vapor pressure at that temperature is called the *relative humidity*. Relative humidity is usually expressed as a percentage. For example, suppose that the air temperature on a given day is $3\overline{0}$ °C and that the water vapor pressure is 15.5 mm of mercury. The equilibrium vapor pressure at $3\overline{0}$ °C (as given in Appendix B, Table 12) is 31.8 mm of mercury. The relative humidity is

$$\frac{15.5 \text{ mm}}{31.8 \text{ mm}} \times 100\% = 48.7\%$$

In summer, the humidity is often more oppressive than the temperature. Why?

Relative humidity depends on both the temperature and the amount of water vapor in the atmosphere. The temperature at which a given amount of water vapor will exert equilibrium vapor pressure is called the *dew point*. In the above example, the dew point is 18 °C because at that temperature the equilibrium vapor pressure of water is 15.5 mm of mercury.

7.22 Boiling When water is heated sufficiently its vapor pressure will eventually equal the combined pressure of the atmosphere and the liquid pressure of the water. Vaporization then occurs at such a rapid rate throughout the water that the water becomes agitated. *Rapid vaporization that occurs when the vapor pressure of the liquid equals the pressure on its surface is called* **boiling.** If the pressure on the liquid surface is one atmosphere (760 mm of mercury), the temperature at which boiling occurs is called the *normal boiling point*. If the pressure on the liquid surface is greater than one atmosphere, boiling occurs at a higher temperature than the normal boiling point; if the pressure on the liquid surface is less than one atmosphere, boiling occurs at a lower temperature than the normal boiling point. Greater pressure is largely the result of the greater number of collisions of the particles against the surface of the liquid. Consequently, the kinetic energy of the liquid, and thus its temperature, must be raised to make boiling possible. The opposite is true when the pressure on a liquid is decreased.

Water or any other liquid that is boiling rapidly does not get hotter than when it is boiling slowly. While a liquid is boiling away, the boiling temperature remains constant until all the liquid has been vaporized.

Solids or gases dissolved in a liquid change the liquid's boiling temperature. For example, salt water boils at a

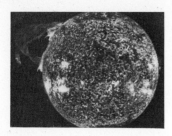

Figure 7-23. A solar eruption. A mass of plasma many times the size of the earth is ejected from the sun's surface at the upper left. The photo was taken in 1973 from the Skylab space station using an ultraviolet filter. The eruption is one of the largest ever photographed.

higher temperature than pure water. In general, solids dissolved in liquids raise the boiling temperature and gases dissolved in liquids lower the boiling temperature.

7.23 Plasma   A gas that is capable of conducting electricity is called a *plasma*. Gases do not ordinarily conduct electricity, but when they are heated to high temperatures gas molecules collide vigorously with each other to form electrically charged particles called ions. These ions give the plasma the ability to conduct an electric current.

Plasma is sometimes called the fourth phase of matter. Under normal conditions, matter exists only as a solid, liquid, or gas. But at the high temperatures that prevail in the sun and other stars, matter exists almost entirely in the plasma phase. Because there are so many stars, it is estimated that more than 99 percent of the matter in the universe is in the form of plasma.

Questions
GROUP A

1. What property of liquid molecules did Brown's experiment show?
2. Describe a demonstration that shows that molecules of liquids diffuse.
3. (a) In terms of adhesion and cohesion, explain why alcohol clings to a glass rod and mercury does not. (b) What is the relationship between cohesion and viscosity?
4. What is the effect of adding a detergent to the water on which a needle is supported?
5. Why is a soap bubble floating through the air spherical in shape?
6. What determines whether a meniscus is concave or convex when viewed from above?
7. Water rises to a certain height in a capillary tube of given diameter. What is the effect on the amount of rise if (a) the tube is lengthened, (b) the diameter of the tube is decreased, (c) the temperature of the water is raised?
8. (a) Which change of phase is called melting? (b) What name is given to the temperature at which melting occurs? (c) What is freezing? (d) What

name is given to the temperature at which freezing occurs?
9. (a) Which has a greater density, water or ice? (b) Explain in terms of molecular arrangement.
10. What is the meaning of the term "regelation"?
11. In terms of the kinetic theory, explain why gases (a) expand when heated, (b) exert pressure, (c) diffuse through porous substances.

GROUP B
12. Describe the properties of the particles of a substance that is liquid at room temperature.
13. Describe the forces that are acting when a razor blade is supported on water.
14. Why is lead shot made by allowing melted lead to fall through cool air?
15. In terms of adhesion and cohesion, explain why mercury is depressed in capillary tubes.
16. Describe the forces that cause water to rise in a capillary tube.
17. Why do crystalline solids have a definite melting point and noncrystalline solids do not?
18. What is the effect of increased pressure on the freezing point of (a) sub-

stances that contract as they freeze, (b) substances that expand as they freeze?

19. Why is it easier to skate with sharp ice skates than with dull ones?
20. What would happen if the gas introduced into the large beaker of the apparatus shown in Figure 7-20(B) were (a) helium, (b) argon?
21. (a) What is plasma? (b) Why is most of the matter in the universe in the form of plasma?

SUMMARY

The smallest particle of any substance that is capable of stable, independent existence is a molecule. Molecules are composed of atoms. Atoms are the smallest particles of elements that can exist either alone or in combination with other atoms of the same or different elements. The masses of individual atoms are usually expressed in atomic mass units, rather than kilograms. The atomic mass unit is based on the mass of the carbon-12 atom.

The kinetic theory of matter states that molecules of matter are in constant motion and undergo perfectly elastic collisions. The phase of matter (solid, liquid, or gas) is determined by the forces acting between molecules and the energy the molecules possess.

Properties of solids such as diffusion, cohesion, adhesion, tensile strength, ductility, malleability, and elasticity depend on molecular forces and/or molecular motion. Hooke's law states that, within limits, strain (the deformation of a solid) is directly proportional to the stress producing the strain. The ratio of stress to strain is called Young's modulus.

Molecular forces, molecular motion, and the weight of molecules are several factors on which liquid properties such as diffusion, cohesion, adhesion, viscosity, surface tension, and capillarity all depend. Melting is the change of phase from a solid to a liquid. Crystalline solids have characteristic melting points; noncrystalline solids do not have definite melting points. Most substances expand upon melting. Water is an important exception. Pressure and dissolved materials change the melting point and freezing point of substances.

Gases expand, exert pressure, and diffuse. Vaporization is the change of phase from a solid or liquid to a gas or vapor. The added pressure exerted by vapor molecules in equilibrium with liquid molecules is called equilibrium vapor pressure. Boiling is rapid vaporization that occurs when the vapor pressure of a liquid is equal to the pressure on its surface. In the stars, matter is in the form of plasma.

VOCABULARY

adhesion	elongation strain	plasma
atom	equilibrium vapor pressure	regelation
atomic mass unit	evaporation	shear strain
boiling point	freezing point	strain
Brownian movement	Hooke's law	stress
capillarity	kinetic theory	sublimation
cohesion	malleability	surface tension
condensation	mass number	tensile strength
diffusion	melting point	vaporization
ductility	meniscus	viscosity
elasticity	molecule	volume strain
elastic limit	phase	Young's modulus

Astrophysics is the study of the physical and chemical properties of astronomical objects and is a relatively recent development of twentieth century astronomy. It now flourishes as never before, identifying such celestial objects as quasars and pulsars and postulating the existence of singularities and black holes.

ASTROPHYSICS

Cosmologist Stephen W. Hawking sees science as an art in service of finding objective truths about the universe, for it appears that "the whole history of human thought has been to try to understand what the universe was like." Hawking is presently attempting to unify two great theoretical breakthroughs in twentieth century physics or "trying to find out whether there is some bigger law from which all other laws can be derived." Such a law would most likely involve "quantizing gravity." Dr. Hawking is shown here in his office at Cambridge University with an assistant.

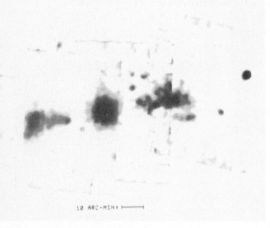

10 ARC-MIN

Einstein Observatory X-ray image of SS433 in the energy band 0.7 to 4 kev. The central object is the star SS433 and is greatly overexposed.

164

A

Saturn comes of age. Saturn, the most distant planet known to the ancients, has finally been reached by spacecraft. These three pictures show how our ability to see the details of the mysterious ringed planet has grown dramatically in the last few years as we have moved from earth-based telescopic observations (A) to computer-enhanced pictures returned from the spacecraft Voyager 1 (B) and Voyager 2 (C), which flew past Saturn. These computer-enhanced photos bring out faint details in Saturn's rings and show possible variations in the chemical composition of Saturn's ring system.

B

Beacons of distant space. An international team of astronomers discovered Quasar PKS 2000-330, shown here between the two black lines in the center of this astronomical photo. This is a negative of the star field—with the sky white and the stars dark. PKS 2000-330 has been determined to be the most distant object yet detected in the universe—12 billion light years away. The light from PKS 2000-330 left this distant object before the solar system, and probably the Milky Way galaxy, formed.

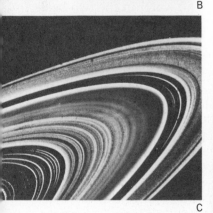

C

The galactic corona. Our Milky Way galaxy is surrounded by a galactic corona of thin, hot gas. The corona was discovered by the International Ultraviolet Explorer. This artistic illustration depicts the corona as a halo around our Milky Way galaxy.

HEAT MEASUREMENTS

Anders Celsius developed the first decimal scale of temperature. His was also the first scale based on the freezing and boiling points of water. At the left on this stamp is a drawing that Celsius used in describing his new instrument to the Swedish Academy of Sciences in 1742.

In this chapter you will gain an understanding of:

☞ the relationship between heat and temperature
☞ Celsius and Kelvin temperature scales
☞ the relationship between the calorie and the joule
☞ coefficients of thermal expansion
☞ Charles' law relating the temperature and volume of a gas
☞ Boyle's law relating the pressure and volume of a gas
☞ the universal gas constant
☞ the physical property called specific heat
☞ the law of heat exchange and the method of mixtures
☞ triple-point curves
☞ heat of fusion
☞ heat of vaporization

Temperature is the "hotness" of an object.

UNITS OF TEMPERATURE AND HEAT

8.1 Relationship Between Heat and Temperature Suppose we fill a large beaker and a small beaker with boiling water, as shown in Figure 8-1. The water in the two beakers is at the same temperature, but the water in the large beaker can give off more heat. It could, for example, melt more ice than could the water in the small beaker. It is possible, therefore, for a body to be at a high temperature and give off little heat; to be at a high temperature and give off a large quantity of heat; to be at a low temperature and give off little heat; or to be at a low temperature and give off a large quantity of heat.

When a material is hot, it has more *thermal energy* than when it is cold. ***Thermal energy** is the total potential and kinetic energy associated with the random motion and arrangements of the particles of a material.*

Temperature is the "hotness" or "coldness" of a material. The quantity of thermal energy in a body affects its temperature. The same quantity of thermal energy present in different bodies, however, does not give each the same temperature. The ratio between temperature and thermal energy is different for different materials.

We saw in Chapter 7 that the temperature of a substance will rise if the average kinetic energy of its particles is increased. If the average kinetic energy is decreased, the temperature goes down. On the other hand, when the potential energy of the particles is increased or decreased without a change in the average kinetic energy, a change of phase takes place without a change in temperature.

Heat *is thermal energy that is absorbed, given up, or transferred from one body to another.* The temperature of a body is a measure of its ability to give up heat to or absorb heat from another body. Thus, the temperature of a body determines whether or not heat will be transferred to or from any nearby body.

The experiments of Count Rumford and James Prescott Joule (Section 1.13) show that mechanical energy and heat are equivalent and that *heat must be a form of energy.*

Since thermal energy is also defined as a form of energy, you may wonder why two different terms—thermal energy and heat—are used. An example will illustrate the difference. The temperature of the air in a bicycle tire will rise when the tire is being pumped up. It will also rise when the tire is out in the sun. In both cases the thermal energy and the temperature of the air are increased. In the first case the work done in pumping was converted to thermal energy. In the second case the rise in temperature was due to energy transferred from the sun to the tire. The term *heat* is used when the transfer of thermal energy from one body to another body at a different temperature is involved.

Temperature is defined in terms of measurements made with thermometers that will be described later. But the following qualitative definition gives the relationship between temperature and energy. **Temperature** *is a physical quantity that is proportional to the average kinetic energy of translation of particles in matter.*

8.2 Temperature Scales The Celsius temperature scale is often heard during weather reports. In Section 7.21, we used the °C in the calculation of relative humidity. Now we will see how the Celsius scale was developed.

To measure temperature it is necessary to introduce a fourth fundamental unit. The unit of temperature difference, the degree, cannot be derived from length, mass, and time; a measurable physical property that changes with temperature must be used.

There are many physical properties that change with temperature. The length of a solid, the volume of a liquid, the pressure of a gas held at constant volume, the volume of a gas held at constant pressure, and the color of a solid heated to a high temperature are examples. Some of these properties of matter can be used in developing a temperature scale and constructing a thermometer.

To establish a temperature scale it is necessary to find a

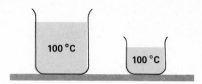

Figure 8-1. The water in the two beakers is at the same temperature, but the water in the large beaker can give off more heat.

Heat is thermal energy in motion.

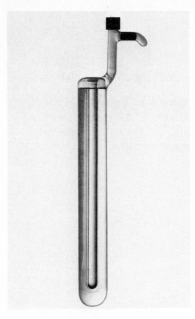

Figure 8-2. A triple-point cell. Pure water with air removed is sealed permanently in the cell, which is then immersed in a water-ice bath. The system is at the triple point when ice, water, and vapor are all present within the cell. A central well is provided for the insertion of a thermometer that is to be calibrated.

The SI symbol for the unit of temperature is K. But in this text the unit will be written as °K to avoid confusion. The Kelvin scale was invented by Sir William Thomson (1824–1907), who is better known by his title, Lord Kelvin.

0 °K (−273 °C) is the coldest possible temperature because at that temperature molecular energy is at a minimum.

process that occurs without a change in temperature. The temperature at which such a process takes place can then be used as a fixed point on a temperature scale. A change of phase of a substance, such as melting or boiling, can be used. The temperature at which the solid phase of a substance is in thermal equilibrium with its liquid phase is a fixed value at a given pressure. This is the melting point. Similarly, at the boiling point the liquid is in thermal equilibrium with its vapor, and this point has a fixed value at a given pressure. The boiling temperature of the substance is always higher than the melting temperature.

There is only one pressure at which the solid, liquid, and vapor phases of a substance can be in contact and in thermal equilibrium. This equilibrium occurs at only one temperature, which is known as the *triple-point temperature.* For example, there is only one pressure and temperature condition at which ice, liquid water, and water vapor can exist in a vessel in thermal equilibrium. The ice, liquid water, and vapor are all at the same temperature and can continue to exist indefinitely in the constant volume of the sealed vessel. A device for determining the triple point of water is shown in Figure 8-2.

The triple point of water is the SI standard for defining temperature. Its assigned value is 273.16 °K (Kelvin).

Originally, two fixed points were used to define the standard temperature interval. They were the *steam point* (the boiling point of water at standard atmospheric pressure) and the *ice point* (the melting point of ice when in equilibrium with water saturated with air at standard atmospheric pressure). The Celsius scale (formerly called the centigrade scale), devised by the Swedish astronomer Anders Celsius (1701–1744), assigned the value 0 °C to the ice point and 100 °C to the steam point, as shown in Figure 8-3. Thus the interval between the two fixed points was 100 C°. (It is interesting that Celsius originally assigned 0° to the steam point and 100° to the ice point, but the scale was changed to its present sequence within a year.) Note that specific temperatures on the Celsius scale are expressed in the unit °C. Temperature *differences* on the Celsius scale are expressed in the unit C°. Kelvin scale temperatures have the unit °K; Kelvin scale temperature *differences* are expressed in K°.

The magnitude of the Kelvin degree (K°) is the same as that of the Celsius degree (C°). These are arbitrarily established units for the measurement of temperature difference. 0 °K is called the *absolute zero of temperature.* Absolute zero should not be thought of as a condition of matter with zero energy and no molecular motion. Molecular action does not cease at absolute zero. The molecules of a substance at absolute zero have a minimum amount of kinetic energy, known as the zero-point energy. Molecular energy is a minimum, but not zero, at absolute zero.

Now let us consider the results obtained when a temperature is measured. To assign numbers on the temperature scale, it is assumed that a measurable physical property of a substance that changes with temperature, X, is proportional to the Kelvin temperature, T; thus $X \propto T$. If X_t is the measurement of the physical property at the standard fixed temperature, T_t, which is the triple point of water, and X is the measure of the same property of the same substance at the unknown temperature, T, then we can write the proportion

$$\frac{X}{X_t} = \frac{T}{T_t}$$

Since T_t is 273.16 °K, we can rewrite the proportion as

$$T = 273.16 \text{ °K} \frac{X}{X_t}$$

Because X and X_t are measurable quantities, the value of the temperature, T, can be computed.

The relationship between the Celsius and Kelvin temperature scales, rounded to three significant figures, is given by the equation

$$\text{°K} = \text{°C} + 273\text{°}$$

The degrees between the ice and steam points are numbered from zero degrees Celsius, 0 °C, to one hundred degrees Celsius, 100 °C. From one degree to the next is a temperature interval of one Celsius degree, 1 C°. Temperatures below 0 °C and above 100 °C are measured by extending the scale in 1 C° intervals. Temperatures below 0 °C are represented by negative values.

The most commonly used thermometers contain either mercury or alcohol. In both cases, the liquid volume increases rather uniformly with temperature over the useful range of the instruments.

8.3 Heat Units There is no instrument that directly measures the amount of thermal energy a body gives off or absorbs. Therefore, *quantities of heat must be measured by the effects they produce.* For example, the amount of heat given off when a fuel burns can be measured by measuring the temperature change in a known quantity of water that the burning produces. If one sample of coal warms 1.0 kg of water 1.0 C°, and another sample warms 1.0 kg

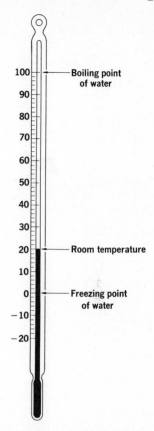

Figure 8-3. A Celsius thermometer of the type used in laboratory work. To convert to the Kelvin scale, add 273° to the Celsius reading.

Gabriel Fahrenheit (1686–1736) is credited with the first use of mercury in thermometers. He invented the temperature scale that bears his name.

of water 2.0 C°, then the second sample gives off twice as much heat.

In the past, water was the standard substance for defining heat units. In the metric system of units, the *kilocalorie* (kcal) was defined as the quantity of heat needed to raise the temperature of one kilogram of water one Celsius degree. The *calorie* (cal) was defined as the quantity of heat needed to raise the temperature of one gram of water one Celsius degree. Observe that a unit mass of water was used for defining each heat unit. Also note that the kilocalorie is one thousand times larger than the calorie. The kilocalorie is the "Calorie" used by biologists and dietitians to measure the fuel value of foods.

The Calorie used in dietary tables is equal to one thousand of the calories used by physicists.

The foregoing definition of the calorie is still used in many laboratory measurements. However, as thermal measurements increased in precision, these older definitions became inadequate. Another reason is that the quantity of heat required to raise the temperature of 1 gram of water through 1 Celsius degree varies slightly for different water temperatures.

The calorie is also defined as a specific number of joules.

$$1 \text{ calorie} = 4.1868 \text{ joules}$$

Defined in this way, the size of the calorie is nearly the same as the original calorie. When taken to three significant figures, 4.19 joules, this slight difference disappears. Thus the relationships stated in the original definition are still useful in measuring thermal properties.

The kilocalorie and the kilogram are consistent with the SI units of measurement we have stressed throughout this book. They are often inconveniently large for laboratory and discussion purposes, however. We shall use calorie and gram units in our consideration of the thermal properties of matter because of their more practical size.

Questions
GROUP A

1. Distinguish between thermal energy and heat.
2. How do heat and temperature differ?
3. Why were the temperature of the melting point of ice and the temperature of the boiling point of water originally called "fixed points" in defining a temperature scale?
4. What single fixed point is now used for the definition of temperature scales?
5. Compare the magnitude of a Celsius degree with that of a Kelvin degree.
6. (a) What is the Celsius temperature at absolute zero? (b) What would be the thermal energy of a material at absolute zero?
7. Compare the amount of thermal energy possessed by each of the following: (a) a soldering iron and a needle, both at $15\overline{0}$ °C; (b) a 4-section radiator and a 10-section radiator; (c) a kettle

of boiling water and cup of boiling water; (d) 20.0 kg of ice at -10.0 °C and 10.0 kg of ice at -10.0 °C; (e) a liter of liquid air and a milliliter of liquid air, both at -189 °C.
8. How are quantities of heat measured?
9. What is a kilocalorie?

GROUP B
10. How does the thermal energy of a material determine its temperature?

11. Describe a triple-point scale.
12. How can the boiling points of substances be determined from the triple point of water?
13. If you examine several similar laboratory thermometers, you may find that the distance between 0 °C and 100 °C is not exactly the same on all of them. Why is there a difference even when they are all the same make and model?

Problems
GROUP A

1. The temperature in a classroom is 24.0 °C. What is the Kelvin reading?
2. Liquid nitrogen boils at 77.0 °K. What is the reading on the Celsius scale?
3. The boiling point of liquid oxygen is -183 °C. What is this temperature in °K?
4. What is the boiling point of helium on the Celsius scale if its boiling point is 4.1 °K?

5. How many calories will be needed to change the temperature of $50\overline{0}$ g of water from $2\overline{0}$ °C to $10\overline{0}$ °C?
6. The stem of a thermometer is marked off in 150 equal scale divisions. When the bulb of this thermometer is placed in melting ice, the mercury stands at 30.0. When the bulb is suspended in the steam from water boiling at standard pressure, the mercury stands at 80.0. To what Celsius temperature does a reading of 125 on this thermometer correspond?

THERMAL EXPANSION

8.4 Thermal Expansion of Solids With few exceptions, *solids expand when heated and contract when cooled.* They not only increase or decrease in length but also in width and thickness. When a solid is heated, the increase in thermal energy increases the average distance between the atoms and molecules of the solid, and it expands.

The expansion of solids can be measured experimentally. A metal rod is heated in an apparatus that has a precise measuring device. If the temperature of an aluminum rod 1.0 m long is raised 1.0 C°, the increase in length is 2.3×10^{-5} m. An iron rod of the same length expands only 1.1×10^{-5} m when its temperature is raised 1.0 C°. For the same increase in temperature, different materials of the same length expand by different amounts. *The change in length per unit length of a solid when its temperature is changed one degree is called its* **coefficient of linear expansion.** See Figure 8-4. While the coefficient of linear expansion of most solids varies with temperature, the change is slight and we shall neglect it in our discussion.

As mentioned, 1.0 m of aluminum expands $2.3 \times$

Why should ovenware be made of a substance with a low coefficient of linear expansion?

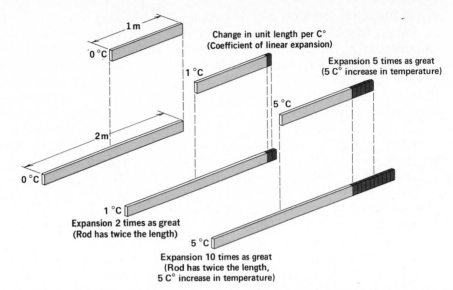

1 m

0 °C

Change in unit length per C°
(Coefficient of linear expansion)

1 °C

Expansion 5 times as great
(5 C° increase in temperature)

5 °C

2 m

0 °C

1 °C
Expansion 2 times as great
(Rod has twice the length)

5 °C
Expansion 10 times as great
(Rod has twice the length,
5 C° increase in temperature)

Figure 8-4. The change per unit length of a solid when its temperature is changed one degree is the coefficient of linear expansion. The change in length illustrated here is greatly exaggerated.

10^{-5} m when its temperature is raised 1.0 C°. The coefficient of linear expansion of aluminum is therefore 2.3×10^{-5}/C°. Likewise the coefficient of linear expansion of iron is 1.1×10^{-5}/C°. Since the coefficient of linear expansion is defined as *the change in length per unit length*, its value does not depend upon any particular length unit. Appendix B, Table 14 gives the value of the coefficient of linear expansion of several solids.

So far, we have been discussing 1.0-m lengths of aluminum and iron and a rise in temperature of 1.0 C°. If the temperature of 10.0 m of aluminum rod is raised 1.0 C°, the expansion is $1\overline{0}$ times as much as the expansion of the 1.0-m length: $1\overline{0} \times 2.3 \times 10^{-5}$ m = 2.3×10^{-4} m. If the temperature of this 10.0 m of aluminum is raised 10.0 C°, the increase is $1\overline{0}$ times as great as for 1.0 C°: $1\overline{0} \times 2.3 \times 10^{-4}$ m = 2.3×10^{-3} m. We can conclude from these observations that *the change in length of a solid equals the product of its original length, its change in temperature, and its coefficient of linear expansion.*

This can be given by the equation

$$\Delta l = \alpha l \Delta T$$

where Δl is the change in length, α is the coefficient of linear expansion, l is the original length, and ΔT is the difference between the final temperature, T_f, and the initial temperature, T_i.

In most practical situations, we are interested in the amount of *linear* expansion of solids. We must bear in mind, however, that when solids are heated they increase in all dimensions. The **coefficient of area expansion,** or *the*

How is the "clickety-clack" you usually hear when a train passes related to linear expansion?

change in area per unit area per degree change in temperature, is *twice* the coefficient of linear expansion. The *coefficient of cubic expansion,* or *the change in volume per unit volume per degree change in temperature,* is *three times* the coefficient of linear expansion.

Expansion of solids is considered in the design and construction of any structure that will undergo temperature changes. When a concrete highway is built, provision is made for expansion by pouring the concrete in sections that are separated by small spaces. Steel rails for railroads can be laid with small spaces between the ends of the rails for the same reason. Bridges are also built so that the sections can expand and contract without distorting the entire structure, as shown in Figure 8-5.

Suitable allowance must be made not only for changes in size due to expansion and contraction, but also for the different rates of expansion and contraction of different materials. For a tight seal, the wires that lead into the filament of an incandescent lamp must have the same coefficient of expansion as the glass of which the lamp is made. The principle of the expansion of solids is also applied in metallic thermometers, thermostats, and the balance wheels of some watches.

Figure 8-5. Expansion and contraction of the roadway is allowed for by this special joint.

8.5 Thermal Expansion of Liquids If the gasoline tank of an automobile is filled on a cool morning and the car is then parked in the sun, some of the gasoline may overflow the tank. Heat causes the gasoline to expand. Here again the increased thermal energy of the molecules and their resultant increase in amplitude of vibration cause them to move away from each other slightly. Thus the principle of expansion of liquids has many useful applications. Thermometers contain either mercury or colored alcohol because these liquids expand and contract quite uniformly as the temperature changes.

Since liquids do not have a definite shape, but take the shape of their container, we are concerned only with their volume expansion. An apparatus as shown in Figure 8-6 can be used to measure the volume expansion of a liquid.

Liquids have greater coefficients of volume expansion than solids have. If this were not so, the liquid in a thermometer would not rise. The coefficients of volume expansion for some common liquids are given in Appendix B, Table 15.

The change in volume of a liquid can be given by the formula

$$\Delta V = \beta V \Delta T$$

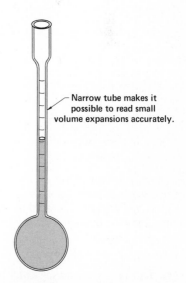

Narrow tube makes it possible to read small volume expansions accurately.

Figure 8-6. A tube for measuring the volume expansion of liquids. The scale takes into account the expansion of the container.

where ΔV is the change in the volume, β is the coefficient of volume expansion, V is the original volume, and ΔT is the difference between the final and initial temperatures.

PRACTICE PROBLEMS

1. What must be the length of a steel rod if its length increases by 0.00102 m as a result of a 20.0-C° change in temperature?   *Ans.*   4.86 m

2. What is the increase in volume of 15.0 L of ethyl alcohol when it is heated from 15.0 °C to 25.0 °C?   *Ans.* 0.168 L

8.6 Abnormal Expansion of Water   Suppose an expansion bulb, like that shown in Figure 8-6, is filled with pure water at 0 °C. As the bulb and water are warmed, the water gradually *contracts* until a temperature of 4 °C is reached. As the temperature of the water is raised above 4 °C, the water *expands*. Because the volume of water decreases as the temperature is raised from 0 °C to 4 °C, the mass density of the water increases. (The mass of the water is constant.) Above 4 °C, the volume of water increases as the temperature is raised, and the mass density decreases. Therefore, *water has its maximum mass density, 1.0000 g/cm³, at 4 °C.* The variation of the density of water with the temperature is shown in Figure 8-7. The temperature range of the graph was chosen to include the temperature at which the density of water is a maximum.

This unusual variation of the density of water with the temperature can be explained as follows. When ice melts to water at 0 °C, the water still contains groups of molecules bonded in the open crystal structure of ice. As the temperature of water is raised from 0 °C to 4 °C, these open crystal fragments begin to collapse and the molecules move closer together. The molecular speeds of the molecules also increase during the 0 °C to 4 °C interval, but the effect of the collapsing crystal structure predominates and the density increases. Above 4 °C the effect of increasing molecular speed exceeds the effect of collapsing crystal structure and the volume increases, as shown in Figure 8-8.

If water did not expand slightly as it is cooled below 4 °C and expand much more as it freezes, the ice that forms on the surface of a lake would sink to the bottom. During the cold winter months, ice would continue to form until the lake was frozen solid. In the summer months only a few feet of ice at the top of the lake would melt. However,

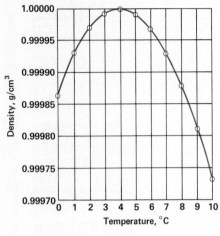

Figure 8-7. The density of water is greatest at 4 °C.

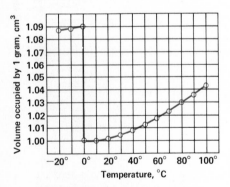

Figure 8-8. Graph of the volume of one gram of water between −20 °C and 100 °C. The abrupt change at 0 °C is due to the change in phase.

How does the expansion of water between 4 °C and 0 °C help to keep fish alive in the winter?

because of the unusual properties of ice and water, no ice forms at the surface of a pond until all the water in it is cooled to 4 °C. As the surface water cools below 4 °C, it expands slightly and floats on the 4 °C water. Upon freezing at 0 °C, further expansion takes place and the ice floats on the 0 °C water.

The graph in Figure 8-8 shows from right to left the gradual contraction of a given mass of water as it is cooled. Note the sharp expansion on freezing and the slight contraction as ice is cooled below 0 °C. The slight expansion that occurs during cooling from 4 °C to 0 °C is too small to show in this graph.

8.7 Charles' Law Gases expand when heated because the increase in the thermal energy of the gas molecules results in an increase in their kinetic energy.

Different solids and liquids have different coefficients of expansion, but *all gases have approximately the same coefficient of expansion.* Also, *the coefficient of expansion of gases is nearly constant at all temperatures,* except for those near the liquefying temperature of the gas. This was demonstrated experimentally in 1787 by French scientist Jacques Charles (1746–1823) in a manner similar to that described below.

In Figure 8-9, a column of air is trapped by a globule of mercury in a capillary tube that is sealed at one end. The length of the air column is measured when the tube is immersed in a mixture of ice and water. Then the length of the air column is measured when the tube is immersed in boiling water. Comparing these two readings, we find that the air column increases by $\frac{100}{273}$ of its original length when heated from 0 °C to 100 °C. For each degree of temperature change the expansion is $\frac{1}{273}$ of the volume at 0 °C. When gases other than air are used, similar results are obtained. The same fractional contraction occurs when gases are cooled below 0 °C.

The coefficient of volume expansion for gases is $\frac{1}{273}$ of the volume at 0 °C or $3.663 \times 10^{-3}/C°$. This is about 20 times the volume expansion of mercury and almost 60 times that of aluminum. All gases have approximately the same coefficient of expansion because they all consist of widely separated, exceedingly small molecules that are, in effect, independent particles. Gas molecules are separated by distances much greater than their molecular diameters. Consequently, the forces acting between them, and their volume in relation to the total gas volume, are negligible. For these reasons all gases have similar physical properties except at temperatures and pressures near those at which they liquefy.

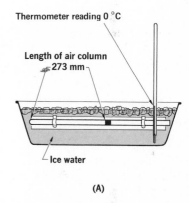

Thermometer reading 0 °C

Length of air column 273 mm

Ice water

(A)

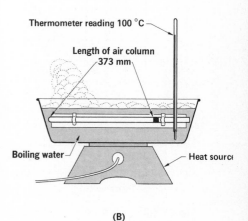

Thermometer reading 100 °C

Length of air column 373 mm

Boiling water   Heat source

(B)

Figure 8-9. Verifying Charles' law. For each degree rise in temperature, a gas expands 1/273rd of its volume at 0 °C.

Table 8-1
THE VOLUME-TEMPERATURE RELATIONSHIP OF A GAS

Volume (cm³)	Temperature (°C)	(°K)
373	100	373
323	50	323
273	0	273
223	−50	223
173	−100	173

Table 8-1 gives the volume occupied by a sample of gas at various Celsius and Kelvin temperatures. From the table we see that the volume of a gas varies directly with the Kelvin temperature. This relationship, which is known as *Charles' law,* can be stated thus: *The volume of a dry gas is directly proportional to its Kelvin temperature, provided the pressure remains constant.*

Unless a gas is under very high pressure or at very low temperature, Charles' law can be written as an equation in which the ratio of volume to Kelvin temperature is a constant,

$$\frac{V}{T} = k$$

or

$$\frac{V}{T_K} = \frac{V'}{T_K'}$$

and

$$V' = \frac{V T_K'}{T_K}$$

where V is the original volume, T_K is the original Kelvin temperature, V' is the new volume, and T_K' is the new Kelvin temperature.

Plotting the gas volumes given in Table 8-1 as a function of the Kelvin temperatures yields the curve shown in Figure 8-10. This linear relationship between the volume of a gas and its Kelvin temperature shows that these two quantities are directly proportional. Since 0 °C (273 °K) is frequently used as a reference point in calculations that involve temperature, this value is called *standard temperature.*

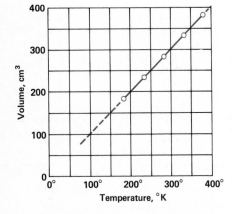

Figure 8-10. Graph of the data in Table 8-1. At constant pressure, the volume of a dry gas varies directly with the Kelvin temperature. (All gases liquefy before reaching 0 °K.)

8.8 Boyle's Law The English scientist Robert Boyle (1627–1691) was the first person to investigate what he called the "spring [elasticity] of the air." Other scientists at that time knew about compressed air, but none of them had performed experiments to learn how the volume of a gas is affected by the pressure exerted on it.

In his experiments, Boyle used a large J-shaped glass tube similar to that shown in Figure 8-11. He set up the apparatus in the stairwell of his laboratory at Oxford. The straight portion of the shorter arm was about one-third meter in length, while the longer arm had a length of about twenty-five meters. Enough mercury was poured into the tube to fill the bent portion. By tipping the tube to allow air to escape from the small arm, the mercury levels were then adjusted so that the mercury would stand at the same height in both arms. In this way Boyle trapped a column of air in the short arm of the tube. Next he measured the height of this air column. By assuming that the bore of the tube was uniform, Boyle used the height of the air column as a measure of the volume of air. The air in the short arm was at atmospheric pressure because the mercury levels were the same in both arms of the tube. See Figure 8-11(A).

In successive steps Boyle added more mercury to the long arm of the tube, as shown in Figure 8-11(B). By measuring the new height of the column of air in the short arm, he could determine each new volume. He found the corresponding pressure on this volume of air by measuring the height of the mercury column **ab** and adding that height to the atmospheric pressure as measured by a mercury barometer.

Boyle cooled the trapped, compressed air with a wet cloth, warmed it with a candle flame, and noted the resulting small changes in volume. These changes were so slight, however, that while it was obvious to Boyle that the temperature of the air during the experiment should be kept constant, small changes in temperature would not seriously affect the experimental results. Boyle's data, though not exceedingly accurate, convinced him and other scientists of his time that "the pressures and expansions . . . [are] in reciprocal proportion." Increasing the pressure on a column of confined air reduces its volume correspondingly. To reduce the volume to one-half, it is necessary to double the pressure; to reduce the volume to one-third, it is necessary to triple the pressure. Today we state **Boyle's law** as follows: *The volume of a dry gas varies inversely with the pressure exerted on it, provided the temperature remains constant.*

Boyle's data are plotted in Figure 8-12. The total pressures are the abscissas, and the heights of the air columns (volumes) are the ordinates. You will recall from Chapter 2 that a graph of this shape, a hyperbola, suggests an inverse proportion.

When two quantities are in inverse proportion, their product is a constant. At any given temperature, *the product of pressure and volume is always a constant.*

$$pV = k$$

Except during conditions of very high pressure, very low

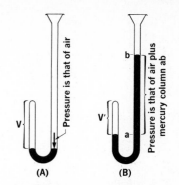

Figure 8-11. Verifying Boyle's law. The volume of the air enclosed in the short arm of the tube varies inversely with the pressure exerted on it.

The mercury barometer was invented by Evangelista Torricelli (1608–1647), a student of Galileo.

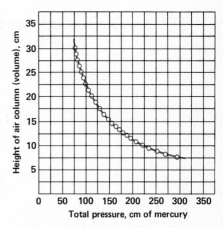

Figure 8-12. Graph of Boyle's original pressure-volume data. The shape of the curve is typical for an inverse proportion.

temperature, or both, it is true in all cases that

$$pV = p'V'$$

and

$$V' = V\frac{p}{p'}$$

Here p is the original pressure and V is the original volume; p' represents the new pressure and V' the new volume.

The pressure of a column of mercury exactly 760 mm high is called *standard pressure*. It is approximately equal to the average pressure of the atmosphere at sea level. Standard temperature, 0 °C, and standard pressure are frequently designated as STP.

There is no change in the mass of a gas when its volume is changed by a difference in the exerted pressure. Since an increase in pressure produces a decrease in the volume of a gas, it must also increase the density of the gas. *The density of a gas varies directly with the pressure exerted on it*, or

$$\frac{D}{D'} = \frac{p}{p'}$$

A cubic meter of air has a mass of 1.29 kg at a pressure of one atmosphere. Under a pressure of four atmospheres, a container having a volume of one cubic meter can hold four times 1.29 kg, or 5.16 kg, of air because the air is four times as dense.

PRACTICE PROBLEMS

1. A sample of oxygen occupies a volume of 5.00 L at 27 °C. If the pressure is unchanged, what volume does it occupy at 77 °C?   *Ans.*   5.83 L

2. A gas occupies 625 mL at 745 mm pressure. What is the volume at 815 mm pressure? The temperature remains constant.   *Ans.*   571 mL

8.9 The Combined Gas Equation   We can derive an equation that combines Boyle's and Charles' laws. At constant Kelvin temperature, T_{K1}, a certain mass of gas occupying volume V_1 is subject to a change in pressure from p_1 to p_2. The new volume, V_2, from Boyle's law is

$$V_2 = \frac{p_1 V_1}{p_2} \qquad \text{(Equation 1)}$$

Now if V_2 is subject to an increase in temperature from T_{K1} to T_{K2} at constant pressure, p_2, the new volume, V_3, from Charles' law is

$$V_3 = \frac{V_2 T_{K2}}{T_{K1}} \qquad \text{(Equation 2)}$$

Substituting the value of V_2 in Equation 1 into Equation 2, V_3 becomes

$$V_3 = \frac{p_1 V_1 T_{K2}}{p_2 T_{K1}}$$

and rearranging the terms, this expression becomes

$$\frac{p_1 V_1}{T_{K1}} = \frac{p_2 V_3}{T_{K2}}$$

But V_3 is the volume at pressure p_2 and temperature T_{K2}, and therefore we can write

$$\frac{pV}{T_K} = \frac{p'V'}{T_K'}$$

The laws of Charles and Boyle can be combined into a single equation.

where p, V, and T_K are original pressure, volume, and Kelvin temperature; and p', V', and T_K' are the new pressure, volume, and Kelvin temperature of a given mass of gas. You will note that when the pressure is constant in the above equation, it becomes Charles' law. When the temperature is constant, the equation becomes Boyle's law.

8.10 The Universal Gas Constant In 1811 the Italian physicist Amedeo Avogadro (1776–1856) recognized that at the same temperature and pressure, mass densities of different gases are proportional to their molecular weights. The molecular weight of a gas is the sum of the atomic weights of all the atoms comprising a molecule of that particular gas. Avogadro also postulated that the number of molecules in one gram-molecular weight (mole) of any substance was the same as that in a mole of any other substance. The number of molecules in one gram-molecular weight of a substance is called the *Avogadro number. A **mole** is the amount of a substance containing the Avogadro number of particles of that substance.*

The Avogadro number is 6.02 $\times$ 10^{23}.

The value of pV/T_K in the combined gas equation is designated by the symbol R and is the same for one mole of any gas. The value of R is independent of the chemical composition of the gas, except at very high pressure, and is known as the *universal gas constant.* If n moles are present in the sample we write

$$pV = nRT_K$$

This relationship defines the behavior of an ideal gas. An ideal gas is imagined to consist of infinitely small molecules that exert no forces on each other. The equation describes the behavior of real gases with reasonable accuracy except at low temperatures, extreme pressures, or both.

The numerical value of the universal gas constant R can be determined from the relationship $pV = nRT_K$. If the pressure is one atmosphere, one mole of an ideal gas will have a volume of 22.4 liters at the standard temperature of 273 °K. R is then found as follows.

$$R = \frac{pV}{nT_K} = \frac{(1\ \text{atm})(22.4\ \text{L})}{(1\ \text{mol})(273\ °\text{K})}$$

$$R = 8.21 \times 10^{-2}\ \text{L atm/mol }°\text{K}$$

We may calculate the pressure, volume, mass, molecular weight, or Kelvin temperature of a gas provided four of these five quantities are known. To do this we must recognize that $n = m/M$ where m is the mass and M is the gram-molecular weight. The following example illustrates a calculation involving the universal gas constant.

EXAMPLE   11.0 g of carbon dioxide (CO_2) occupy a volume of 10.0 L at a temperature of 350.0 °K. Find the pressure exerted by the CO_2.

SOLUTION   *Basic equation:* $pV = nRT_K$

The gram-molecular weight of CO_2 is $(12.0 + 16.0 + 16.0)$ g $= 44.0$ g; therefore, $n = 11.0$ g/44.0 g $= 0.250$ mol. $V = 10.0$ L and $T_K = 350.0$ °K. Solving for p,

Working equation: $p = \dfrac{nRT_K}{V}$

$$p = \frac{(0.250\ \text{mol})(8.21 \times 10^{-2}\ \text{L atm/mol }°\text{K})(350.0\ °\text{K})}{10.0\ \text{L}}$$

$$p = 0.718\ \text{atm}$$

PRACTICE PROBLEM   What volume will 5.00 moles of oxygen occupy at 77.0 °C and 80.0 cm pressure?   *Ans.* 137 L

Questions
GROUP A

1. What is meant by the coefficient of linear expansion?
2. What provision is made to allow for the expansion of (a) concrete highways, (b) bridges, (c) piston rings?
3. A hole 1.00 cm in diameter is drilled through a piece of steel at 20.0 °C. What happens to the diameter of the hole as the steel is heated to 100.0 °C?
4. A platinum wire can be easily sealed into a glass tube, but a copper wire does not form a tight seal with glass. Explain.
5. How does the coefficient of volume expansion of mercury compare with the coefficient of volume expansion of glass?
6. Why are mercury and alcohol used in making thermometers?
7. Why does the pressure of a confined gas increase as its temperature is raised?
8. What is the meaning of STP?
9. (a) What is a mole? (b) How is the mole related to Avogadro's number?

10. How does the density of a gas vary with the pressure exerted on it?

GROUP B
11. Does the coefficient of linear expansion depend on the unit of length used? Explain.
12. Why do solids expand when their temperature is raised?
13. (a) Why does water contract as its temperature is raised from 0 °C to 4 °C? (b) Why does it expand when heated above 4 °C?
14. Why does the measurement of the expansion of a liquid contained in a bulb or tube fail to give a true value for the expansion of the liquid?
15. A brass disk fits a hole in a steel plate snugly at 20.0 °C. In order to make the disk drop out of the plate, should the disk and plate combination be heated or cooled?
16. Why are the coefficients of volume expansion very nearly equal for all gases?
17. Why do real gases within the usual experimental ranges of temperature and pressure conform to the behavior of an ideal gas?

Problems
GROUP A

1. A piece of copper pipe is 5.00 m long at 20.0 °C. If it is heated to 70.0 °C, what is the increase in its length?
2. A rod of silver is 100.0 cm long at 0.0 °C. What is its increase in length when heated to 100.0 °C?
3. A steel pipeline is $150\overline{0}$ m long at 30.0 °C. What is its length when the temperature is 10.0 °C?
4. The diameter of a hole drilled through a piece of brass is 1.500 cm when the temperature is $2\overline{0}$ °C. What is its diameter when the brass is heated to $15\overline{0}$ °C?

5. A quantity of carbon tetrachloride occupies a volume of 500.00 mL at 20.0 °C. What is its volume at 45.0 °C?
6. The gasoline in the underground storage tanks of a service station is at a temperature of 15.0 °C. What is the volume of 125.0 L of the gasoline after it warms up to 25.0 °C in the fuel tank of a car?
7. A gas occupies a volume of 250.0 cm^3 at 37 °C. What is its volume at 67 °C if the pressure is not changed?
8. A gas occupies 2.00 L at 73.5 mm pressure. What is the volume at standard pressure if the temperature is unchanged?

9. A volume of 465 cm^3 of a gas is at a pressure of 725 mm of mercury. What volume will the gas occupy if the pressure is increased to 825 mm?

10. A certain gas has a density of 1.55 g/L at a pressure of $76\bar{0}$ mm. What is the density if the pressure is decreased to 725 mm?

GROUP B

11. The spaces between 12.0-m steel rails are 0.0075 m at −18.0 °C. If the rails close up at 32.0 °C, what is their coefficient of linear expansion?

12. Mercury has a mass density of 13.6 g/cm^3 at 20.0 °C. Find its mass density at 100.0 °C.

13. A steel steam pipe is 100.0 m long. How much room must be provided for expansion if its temperature change varies from $-1\bar{0}$ °C to 105 °C?

14. A steel tape is correctly calibrated in metric units at 0.00 °C. It is used at a temperature of 25.00 °C to measure the distance between two lines on a cement floor. The reading obtained is 25.000 m. (a) What is the actual distance between the lines? (b) What would the reading of the tape have been if its temperature had been 0.00 °C?

15. A copper ring has an inside diameter of 3.980 cm at 20.0 °C. To what temperature must it be heated to fit exactly on a shaft 4.000 cm in diameter?

16. The pressure exerted on a volume of gas at 0 °C is $60\bar{0}$ mm of mercury. The temperature of the gas is increased to $3\bar{0}$ °C, while the volume is held constant. What will the new pressure be?

17. A constant-pressure air thermometer contains a mass of air whose volume is $60\bar{0}$ cm^3 at 0 °C. What will be its volume at $8\bar{0}$ °C?

18. A calibrated Pyrex glass flask is filled to the 1000.0-cm^3 level with water. Both flask and water are at a temperature of 20.0 °C. The system is heated to 80.0 °C. Determine what the volume reading will be at this new temperature. (Bear in mind that both the water and the flask will expand.)

19. A certain mass of gas has a volume of 1.25 L at a pressure of 76.0 cm of mercury and a temperature of 27.0 °C. The gas expands to a volume of 1.55 L, at which time its pressure is 80.0 cm of mercury. What is the final temperature?

20. How many moles of argon occupy a volume of 1.75 L at 25.0 °C and 75.0 cm pressure? The gram-molecular weight of argon is 39.9 g.

HEAT EXCHANGE

8.11 Heat Capacity Blocks of five different metals—aluminum, iron, copper, zinc, and lead—are shown in Figure 8-13. They all have the same mass and the same cross-sectional area, but the pieces have different heights because the metals have different densities. First the blocks are put in a pan of boiling water to heat them all to the same temperature. Then they are transferred to a block of paraffin. The diagram shows the relative depths to which the metals melt the paraffin. The aluminum block melts the most paraffin, iron follows as a poor second, copper and zinc are tied for third, and lead melts the least

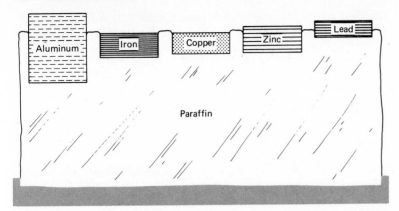

Figure 8-13. Because metals have different heat capacities, these blocks, all with equal masses and temperatures, melt the paraffin to different depths.

paraffin. This demonstration shows that different materials absorb or give off different amounts of heat, even though the materials have the same mass and undergo the same temperature change. Similarly, different amounts of heat are absorbed by blocks of the same material if their mass is different and their temperature change the same, if their mass is the same and their temperature change is different, or if they have different masses and undergo different temperature changes. Such objects are then said to differ in *heat capacity*. Those with a high heat capacity warm more slowly because they absorb a greater quantity of heat; they also cool more slowly because they give off more heat. *The **heat capacity** of a body is the quantity of heat needed to raise its temperature 1°.*

The calorie is not an SI unit, but it is widely used in heat measurements and in dietary literature.

$$\text{heat capacity} = \frac{Q}{\Delta T}$$

where Q is the quantity of heat needed to produce a change in the temperature of the body, ΔT. The units we shall use for heat capacity are cal/C°.

8.12 Specific Heat The heat capacity of an object does not describe the thermal properties of the material of which it is made. For example, the heat capacity of 1.0 kg of copper differs from that of 1.0 kg of aluminum, but the heat capacity of 1.0 kg of aluminum also differs from that of 2.0 kg of aluminum. In order to obtain a quantity that is characteristic of copper, aluminum, or any material, the heat capacities of *equal masses* of the materials must be compared. This comparison yields a more useful quantity known as *specific heat*.

Specific heat is the heat capacity of a material per unit mass. It is numerically equal to the quantity of heat that must be supplied to a unit mass of a material to raise its temperature one degree. If Q represents the quantity of heat needed to produce a temperature change, ΔT, in a quantity of material of mass m, the specific heat, c, is given by

$$c = \frac{Q/\Delta T}{m}$$

which when simplified yields

$$c = \frac{Q}{m\,\Delta T}$$

ΔT is always positive. Its value is found by subtracting the lower temperature from the higher temperature, regardless of which one is the initial or final temperature.

Since 1 calorie of heat raises the temperature of 1 g of water 1 C°, the specific heat of water is 1 cal/g C°. In SI units the specific heat of water is 1 kcal/kg C°. Appendix B, Table 13, shows that the specific heat of most substances is less than that of water.

Let us solve the specific heat equation for Q.

$$Q = mc\Delta T$$

Thus the quantity of heat needed to produce a certain temperature change in a body equals the product of the mass of the material, its specific heat, and its temperature change. If m is in g, c in cal/g C°, and ΔT in C°, Q will be expressed in calories.

8.13 Law of Heat Exchange Hot water can be cooled by the addition of cold water. As the two mix, the temperature of the hot water is lowered while the temperature of the added cold water is raised. The final temperature of the mixture lies between the original temperatures of the hot and cold water. Each time two substances of unequal temperature are mixed, the warmer one loses heat and the cooler one gains heat until both finally reach the same temperature. *A process that absorbs heat as it progresses is* **endothermic.** *An* **exothermic** *process gives off heat as it progresses.*

No thermal energy is lost when substances of unequal temperatures are mixed. *In any heat-transfer system, the heat lost by hot substances equals the heat gained by cold substances.* This is known as the *law of heat exchange.* The total number of heat units liberated by warmer substances equals the total number of heat units absorbed by cooler substances. This can be expressed as

$$Q_{lost} = Q_{gained}$$

This equality is the basis of a simple technique known as the *method of mixtures* for measuring a quantity of heat in transit from one substance to another. The method of mixtures and the law of heat exchange can be used to determine the specific heat of a solid, as shown in Figure 8-14.

Heat lost = heat gained.

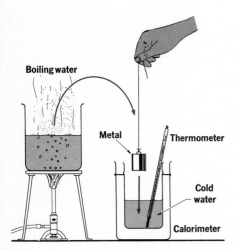

Figure 8-14. Apparatus for measuring the specific heat of a metal by the method of mixtures.

The hot solid of unknown specific heat, but of known mass and temperature, is "mixed" with water of known mass and temperature in a *calorimeter* (usually nested metal cups separated by an insulating air space) of known mass and temperature. The final temperature of the mixture is measured. All the data for the law of heat exchange equation are known except the specific heat of the solid; it can be calculated as in the example that follows.

In practical situations, some heat is usually transferred to the surroundings. This decrease in the amount of thermal energy that can be measured affects the accuracy of most heat experiments in the high school laboratory.

EXAMPLE A laboratory experiment to determine the specific heat of a sample of brass yielded the following data:

Mass of calorimeter	160.0 g
Specific heat of calorimeter	0.0924 cal/g C°
Mass of water	225.0 g
Specific heat of water	1.00 cal/g C°
Mass of brass	184.7 g
Initial temperature of water and calorimeter	18.0 °C
Initial temperature of brass	99.5 °C
Final temperature of water, calorimeter, and brass	23.5 °C

SOLUTION To calculate the specific heat of the brass from these data, first identify the substances losing and gaining heat. The brass cylinder lost heat while the water and the calorimeter gained heat. It is assumed that the inner cup of the calorimeter, which has a mass of 160.0 g and is in direct contact with the water, will always be at the temperature of the water. Since this cup is thermally insulated from the outer cup by a fiber ring and an air space, we can disregard at this level of accuracy any heat transfer to or from the inner cup system.

Now apply the basic idea that heat lost by the brass in cooling, Q_b, will equal heat gained by the water, Q_w, and the calorimeter, Q_c, in warming.

$$\text{Basic Equation: } Q_b = Q_w + Q_c$$

Since $Q = mc\Delta T$ in each case, we can substitute for Q_b, Q_w, and Q_c.

$$m_b c_b \, \Delta T_b = m_w c_w \, \Delta T_w + m_c c_c \, \Delta T_c$$

Solving for c_b

$$\text{Working Equation: } c_b = \frac{m_w c_w \, \Delta T_w + m_c c_c \, \Delta T_c}{m_b \, \Delta T_b}$$

Substituting the given data and solving

$$c_b = \frac{[(225.0 \text{ g})(1.00 \text{ cal/g C}°)(23.5 \text{ °C} - 18.0 \text{ °C})] + [(160.0 \text{ g})(0.0924 \text{ cal/g C}°)(23.5 \text{ °C} - 18.0 \text{ °C})]}{(184.7 \text{ g})(99.5 \text{ °C} - 23.5 \text{ °C})}$$

$$c_b = \frac{1240 \text{ cal} + 81 \text{ cal}}{14,\overline{0}00 \text{ g C}°} = \frac{1321 \text{ cal}}{14,\overline{0}00 \text{ g C}°}$$

$$c_b = 0.0944 \text{ cal/g C}°$$

EXAMPLE   In a laboratory experiment, 100.0 g of iron at 80.0 °C was added to 53.5 g of water at 20.0 °C. What is the final temperature of the mixture?

SOLUTION   Basic equation: $Q_l = Q_g$

From Appendix B, Table 13, we learn that the specific heat of iron is 0.108 cal/g C°, and the specific heat of water is 1.00 cal/g C°. Using T_f as the final temperature of the mixture, the heat lost by the iron is

$$Q_i = m_i c_i \Delta T_i = m_i c_i (T_i - T_f)$$

and the heat gained by the water is

$$Q_w = m_w c_w \Delta T_w = m_w c_w (T_f - T_w)$$

These expressions are equal. Therefore

$$m_i c_i (T_i - T_f) = m_w c_w (T_f - T_w)$$

Solving for T_f

$$\text{Working equation: } T_f = \frac{m_i c_i T_i + m_w c_w T_w}{m_i c_i + m_w c_w}$$

$$T_f = \frac{(100.0 \text{ g})(0.108 \text{ cal/g C}°)(80.0 \text{ °C}) + (53.5 \text{ g})(1.00 \text{ cal/g C}°)(20.0 \text{ °C})}{(100.0 \text{ g})(0.108 \text{ cal/g C}°) + (53.5 \text{ g})(1.00 \text{ cal/g C}°)}$$

$$T_f = 30.1 \text{ °C}$$

PRACTICE   **1.** An aluminum calorimeter has a mass of 60.0 g. Its tem-
PROBLEMS   perature is 25.0 °C. What is the final temperature attained
when 75.0 g of water at 95.0 °C is poured into it?   *Ans.* 84.5 °C

2. A metal cylinder, mass 450.0 g, temperature 100.0 °C, is dropped into
a 150.0-g iron calorimeter, specific heat 0.100 cal/g C°, that contains
300.0 g of water at 21.5 °C. If the resulting temperature of the mixture is
30.5 °C, what is the specific heat of the metal cylinder?   *Ans.* 9.1 ×
10^{-2} cal/g C°

Questions
GROUP A

1. What is the heat capacity of a body?
2. (a) Give a word definition for specific heat. (b) What is the formula that defines specific heat?
3. How do the specific heats of most common substances compare with the specific heat of water?

4. How is the amount of thermal energy required to produce a given temperature change in a substance calculated?
5. What is the law of heat exchange?
6. What data are required for the determination of specific heat by the method of mixtures?
7. On what basic law are experiments with calorimeters based?

Problems
GROUP A

1. How much heat is given out when 85 g of lead cools from $20\bar{0}$ °C to $1\bar{0}$ °C?
2. If 10.0 g of water at 0.0 °C is mixed with 20.0 g of water at 30.0 °C, what is the final temperature of the mixture?
3. What is the final temperature of a mixture of 135 g of water at 21.0 °C in a 45.0-g brass calorimeter and $20\bar{0}$ g of silver at 100.0 °C?
4. A piece of tin weighing 225 g and having a temperature of 100.0 °C is dropped into $10\bar{0}$ g of water at a temperature of 10.0 °C. If the final temperature of the mixture is 20.0 °C, what is the specific heat of the sample of tin?

GROUP B
5. A block of metal has a mass of $100\bar{0}$ g. It is heated to 300.0 °C and then put in 100.0 g of water at 0.0 °C in a calorimeter. The calorimeter's mass is 50.0 g and its specific heat is 0.200 cal/g C°. If

the final temperature is 70.0 °C, calculate the specific heat of the metal.
6. A block of brass, mass 500.0 g, temperature 100.0 °C, is put in 300.0 g of water, temperature 20.0 °C, in an aluminum calorimeter, mass 75.0 g. If the final temperature is 30.0 °C, what is the specific heat of the brass?
7. A cylinder of copper has a mass of 95.3 g and a specific heat of 0.092 cal/g C°. It is heated to 90.5 °C and then put in 75.2 g of turpentine, temperature 20.5 °C. The temperature of the mixture after stirring is 35.5 °C. State the specific heat of turpentine.
8. A glass beaker with a mass of $35\bar{0}$ g contains $50\bar{0}$ g of water. The beaker and the water are at a temperature of 20.0 °C. If $40\bar{0}$ g of ethyl alcohol at a temperature of 50.0 °C is poured into the beaker and thoroughly mixed with the water, what is the final temperature of the beaker and mixture? (The specific heats of alcohol and glass are given in Appendix B, Table 13.)

CHANGE OF PHASE

8.14 The Triple Point The phase and density of any pure substance are determined by its temperature and pressure. When temperature and pressure are controlled, it is possible to cause a pure substance to change from any one phase to either of the other two phases. The substance can make the phase change directly or by first passing through the third phase.

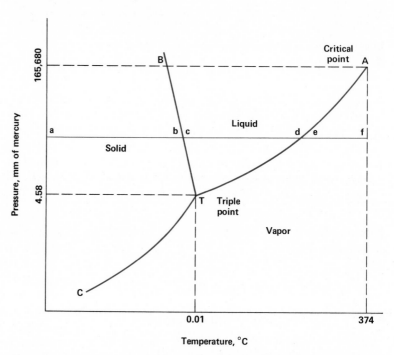

Figure 8-15. Temperature-pressure graph for water. The scales have been condensed in order to show the critical point.

Figure 8-15 is the temperature-pressure equilibrium graph for water. At the pressure and temperature of the triple point, **T,** the three phases of water can exist in equilibrium. It exists as a solid for temperature-pressure values lying in the area between the curves **TC** and **TB.** It is a liquid when these values lie in the area between **TB** and **TA.** In the area below the **ATC** curve it is a vapor. If the temperature-pressure values fall anywhere on the curve **TA,** water can exist with its vapor and liquid phases in equilibrium. Similarly, if these values fall on the curve **TB,** the solid and liquid phases can exist in equilibrium. For values on the curve **TC** the solid and vapor phases can exist in equilibrium. When there are two or more phases of a substance in equilibrium at any given temperature and pressure, there will always be interfaces separating the phases.

Figure 8-15 (with the solid-liquid curve **TB**, the liquid-vapor curve **TA**, and the solid-vapor curve **TC** plotted on a single graph) is typical of pure crystalline materials. However, each substance has a different set of curves. This is because the temperatures at which changes of phase occur (at a given pressure) are different for each substance.

Other features of the graph in Figure 8-15 will be explained in succeeding sections of this chapter. So it is important that you fully understand the meaning of the lines **TA**, **TB**, and **TC** in the figure.

8.15 Heat of Fusion   At the temperature and pressure indicated by point **a** in Figure 8-15, water exists as ice, a solid. Let us assume that we keep the pressure constant as we apply heat at a uniform rate. The ice will be warmed from its initial temperature to the temperature at point **b**. At this temperature the ice will begin to melt. Melting is an endothermic process. The application of more heat will melt more ice, but the temperature will not rise until all the ice is melted.

Following this change in phase, the temperature of the liquid water will rise. The horizontal line **abcdef** shows the temperature values as heat is applied first to the ice, then to the water, and finally to the vapor while the pressure on the system is held constant. This information can be shown more strikingly in another way. Assume that we start with a block of ice at $-2\overline{0}$ °C and add heat at a constant rate while holding the pressure constant at a value indicated by the line **af.** A plot of temperature readings against time during which heat is being applied at a uniform rate gives us the graph in Figure 8-16. The line **ab** in each figure represents the warming of the ice without change of state; **bc** represents the heat required to change the solid to a liquid without a change of temperature (note that **b** and **c** of Figure 8-16 are the same point); **cd** represents the warming of the water. The addition of heat to a solid at its melting point produces a change of phase instead of a rise in temperature. All the heat energy is used to increase the potential energy of the particles. The average kinetic energy is unchanged. Thus the addition of heat to ice at a pressure of one atmosphere and a temperature of 0 °C causes the ice to change into water at 0 °C.

The amount of heat needed to melt a unit mass of a substance at its melting point is called its **heat of fusion**, L_f. L_f for ice is approximately 80 cal/g at 0 °C, meaning that this quantity of heat must be added to each gram of ice at 0 °C to convert it to water at 0 °C. Heats of fusion of various substances are given in Appendix B, Table 13.

In melting, a substance absorbs heat without a rise in temperature.

The heat of fusion of ice is $8\overline{0}$ calories per gram at 0 °C.

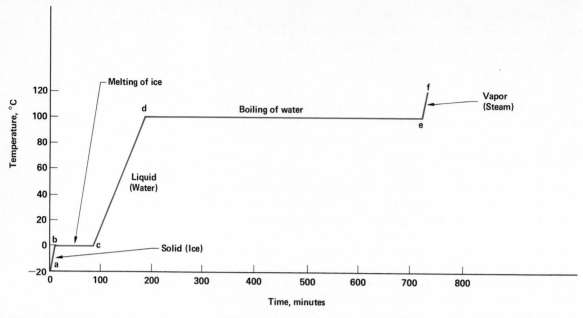

Figure 8-16. Time-temperature graph for the constant addition of heat to a mass of ice. The horizontal portions of the graph indicate phase changes.

The method of mixtures can be used to determine the heat of fusion of a solid. Suppose we wish to determine experimentally the heat of fusion of ice. Since hot water melts ice, we find the mass of ice at a known temperature that can be melted by a known mass of water at a known temperature. The following example illustrates this.

EXAMPLE   A calorimeter has a mass of 100.0 g and a specific heat of 0.900 cal/g C°. It contains 400.0 g of water at 40.0 °C. When 91.0 g of ice at 0.0 °C is added and completely melted, the temperature of the water is 18.2 °C. What is the heat of fusion of ice?

SOLUTION   *Basic Equation:* $Q_1 = Q_g$

Heat lost by water + heat lost by calorimeter = heat gained by melting ice + heat gained by ice water.

$$(mc\Delta T)_w + (mc\Delta T)_c = (mL_f)_i + (mc\Delta T)_{iw}$$

$$\textit{Working Equation: } L_f = \frac{(mc\Delta T)_w + (mc\Delta T)_c - (mc\Delta T)_{iw}}{m_i}$$

$(mc\Delta T)_w = [(400.0 \text{ g})(1.00 \text{ cal/g C°})(40.0 \text{ C°} - 18.2 \text{ C°})] = 8720 \text{ cal}$
$(mc\Delta T)_c = [(100.0 \text{ g})(0.0900 \text{ cal/g C°})(40.0 \text{ C°} - 18.2 \text{ C°})] = 196 \text{ cal}$
$(mc\Delta T)_{iw} = [(91.0 \text{ g})(1.00 \text{ cal/g C°})(18.2 \text{ C°} - 0.0 \text{ C°})] = 1660 \text{ cal}$
$m_i = 91.0 \text{ g}$

Then $L_f = \dfrac{8720 \text{ cal} + 196 \text{ cal} - 1660 \text{ cal}}{91.0 \text{ g}} = \dfrac{7260 \text{ cal}}{91.0 \text{ g}}$

$L_f = 79.8 \text{ cal/g}$

PRACTICE   What is the final temperature if 300.0 g of ice at 0.00 °C,
PROBLEM   500.0 g of ice water at 0.00 °C, and $120\overline{0}$ g of water at
100.0 °C are mixed?   *Ans.*   48.0 °C

8.16 The Freezing Process   The heat added to ice to
make it melt increases the thermal energy of its molecules
and changes it into water. This same amount of heat is
evolved when water freezes because this process is a reversible energy change. Freezing is an exothermic process.
Each gram of water at 0 °C that forms ice at 0 °C liberates
approximately 80 calories of heat. When the molecules of
water return to their fixed positions in ice, they give up in
the form of heat the energy that enabled them to slide over
one another. The reversible energy change between 1 g of
ice and 1 g of water at 0 °C is shown in Figure 8-17. To
change the ice to water, it is necessary to add $8\overline{0}$ cal; to
change the water to ice, it is necessary to take away $8\overline{0}$ cal.
A more precise value for the heat of fusion of ice is
79.71 cal/g.

If pure water is very carefully cooled without being
disturbed, it can reach temperatures as low as -20 °C
without freezing. Water that is cooled below the normal
freezing point is said to be *supercooled*. If a piece of ice or a
speck of dust is added to such water, freezing takes place
rapidly and the temperature rises to 0 °C, the normal
freezing point. The formation of ice takes place readily at
0 °C if there is some dust or other foreign matter on which
the first crystals of ice can form. Supercooling occurs when
no such foreign matter is present. Supercooling is of particular interest in meteorology because the process of supercooling takes place in the formation of some clouds and
when "freezing rain" turns to ice as it hits trees, telephone
wires, and other surfaces.

8.17 The Boiling Process   The equilibrium vapor pressure of a liquid is a characteristic of the liquid that depends
on the temperature only. Appendix B, Table 12, gives the
equilibrium vapor pressure of water at various temperatures. The vapor pressure curve for water is shown in Figure 8-18. This curve shows the relationship between the
pressure and temperature of water and its saturated
vapor. Any point on the curve or on **TA** (Figure 8-15) represents a definite temperature and pressure at which
water is in equilibrium with its saturated vapor. Figure
8-18 is simply a portion of the curve **TA** of Figure 8-15

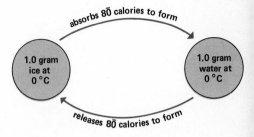

Figure 8-17. Heat of fusion. It
takes $8\overline{0}$ additional calories to
change 1.0 g of ice at 0 °C to
water. Water at 0 °C releases the
same number of calories upon
freezing.

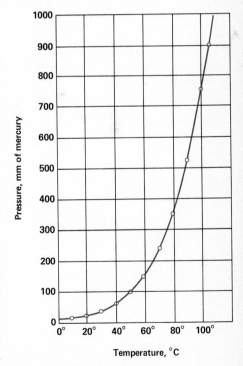

Figure 8-18. The equilibrium
vapor pressure curve for water. The
boiling point of water rises rapidly
as the temperature is raised.

drawn to specific temperature and pressure scales. Other liquids show vapor pressure curves that are similar.

Solids, like liquids, exert a vapor pressure. The equilibrium vapor pressure of ice at 0 °C is about 4.5 mm of mercury. The vapor pressure of solids is much less than that of liquids because solids sublime more slowly, if at all, at normal temperatures. Evaporation of both solids and liquids can occur at temperatures and pressures other than the equilibrium values that fall on the curves **TC** and **TA** of Figure 8–15. A solid at any temperature and pressure to the left of the curve **TC** can be evaporating to some degree depending upon the nature of the substance. Under these conditions the vapor pressure of the substance will be less than the pressure at saturation, however.

Equilibrium vapor pressure is defined in Section 7.21. Boiling point is defined in Section 7.22.

At 100.0 °C the vapor pressure of water is 760 mm of mercury. 100.0 °C is the *normal boiling point* of water. If the air pressure is reduced to 525.8 mm of mercury, water boils at 90.0 °C because at this temperature the vapor pressure of water is 525.8 mm of mercury. In order to make water boil at 50.0 °C, the pressure must be reduced to 92.5 mm of mercury. If the pressure is increased to 787.5 mm of mercury, water will not boil until 101 °C.

At very high altitudes, it is difficult to prepare a hard-boiled egg in an open pan. Can you explain why?

Strong-walled pressure cookers, in which water is boiled at pressures up to about 2 atmospheres and at temperatures up to about 120 °C, are useful for rapid cooking of foods. Special pans in which water is boiled at room temperature or slightly above are used in the production of sugar crystals.

If a mixture of ethyl alcohol and water is boiled, the boiling temperature is not the same as the boiling point of either liquid by itself. Alcohol boils at 78 °C and water at 100 °C; the boiling temperature of the mixture is between 78 °C and 100 °C, depending on the proportions of alcohol and water in the mixture. The boiling temperature of a mixture of two or more liquids each having different boiling points is different from that of any of the liquids used.

A liquid can be separated from a nonvaporizing dissolved solid by *distillation*. Distillation involves evaporation followed by condensation of the vapor in a separate vessel. Liquids that have different boiling points can be separated by *fractional distillation*. As the mixture boils, more of the component with the lower boiling point vaporizes and the boiling point of the resulting mixture rises. Samples collected at different temperatures are then redistilled.

8.18 Heat of Vaporization   If a liter of water at 0 °C is heated to 100 °C, each gram of water will absorb 10$\overline{0}$ cal. If heat is supplied at a constant rate, it takes more than five times as long to boil the water away as it did to heat it from 0 °C to 10$\overline{0}$ °C. (See Figure 8-16.) Each gram of water absorbs more than 500 calories of heat as it is changed into steam. The temperature of the water remains constant during boiling, and the steam produced has the same temperature as the boiling water. *The heat required per unit mass to vaporize a liquid at its boiling point is called its* **heat of vaporization,** L_v.

The heat required for vaporization gives the particles of liquid sufficient thermal energy to overcome the energy binding them to the liquid, and so enables them to separate from one another and move among the molecules of the gases above the liquid. As noted in Section 7.20, all the energy is used to increase the potential energy of the particles. Their average kinetic energy is not changed. Since the energy required for vaporization varies with temperature, the heat of vaporization varies with temperature. The heat of vaporization for water is about 539 cal/g at 100 °C. Water boiling under reduced pressure at a lower temperature has a heat of vaporization that is somewhat greater; at boiling temperatures above the normal boiling point, the heat of vaporization is smaller.

The method of mixtures is used to determine the heat of vaporization of water. As shown in Figure 8-19, a known mass of steam is passed into a known mass of cold water at a known temperature, and the increase in temperature is measured. To ensure that only steam enters the water in the calorimeter, a trap is used to catch any condensed water from the steam generator. The calculations for this method are given in the following example.

The heat of vaporization of water is 539 calories per gram at 10$\overline{0}$ °C.

Boiling is an endothermic process.

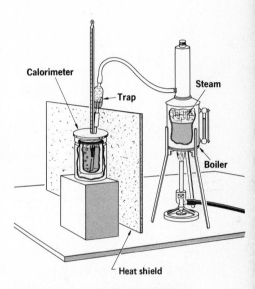

Figure 8-19. Laboratory apparatus that may be used for finding the heat of vaporization of water by the method of mixtures.

EXAMPLE Given the following data:

Mass of calorimeter	120.0 g
Specific heat of calorimeter	0.100 cal/g C°
Mass of water	402.0 g
Mass of steam	23.5 g
Initial temperature of cold water	6.0 °C
Temperature of steam	100.0 °C
Final temperature of water	40.0 °C

Calculate the heat of vaporization of water.

SOLUTION *Basic Equation:* $Q_1 = Q_g$

Heat lost by steam in condensing + heat lost by resulting water = heat gained by calorimeter + heat gained by water.

$$(mL_v)_s + (mc\Delta T)_{sw} = (mc\Delta T)_c + (mc\Delta T)_w$$

$$\text{Working Equation: } L_v = \frac{(mc\Delta T)_c + (mc\Delta T)_w - (mc\Delta T)_{sw}}{m_s}$$

$$(mc\Delta T)_c = [(120.0 \text{ g})(0.100 \text{ cal/g C}°)(40.0 \text{ C}° - 6.0 \text{ C}°)] = 408 \text{ cal}$$
$$(mc\Delta T)_w = [(402.0 \text{ g})(1.00 \text{ cal/g C}°)(40.0 \text{ C}° - 6.0 \text{ C}°)] = 13{,}700 \text{ cal}$$
$$(mc\Delta T)_{sw} = [(23.5 \text{ g})(1.00 \text{ cal/g C}°)(100.0 \text{ C}° - 40.0 \text{ C}°)] = 1410 \text{ cal}$$
$$m_s = 23.5 \text{ g}$$

$$\text{Then} \quad L_v = \frac{408 \text{ cal} + 13{,}700 \text{ cal} - 1410 \text{ cal}}{23.5 \text{ g}} = \frac{12{,}700 \text{ cal}}{23.5 \text{ g}}$$

$$L_v = 54\overline{0} \text{ cal/g}$$

PRACTICE PROBLEM A calorimeter contains 400.0 g of water at 20.0 °C. How many grams of steam at 100.0 °C are needed to raise the temperature of the water and calorimeter to 80.0 °C? The calorimeter has a mass of 100.0 g; its specific heat is 0.100 cal/g C°. *Ans.* 44.0 g

---

Condensation is an exothermic process.

8.19 The Condensing Process Heat is absorbed during vaporization. This increase in the thermal energy of the molecules of a liquid enables them to break away from the liquid and become molecules of vapor. When the vapor condenses to a liquid, this thermal energy is evolved as heat. This reversible energy change, shown in graphic form in Figure 8-20, is useful in a steam-heating system. The heat of vaporization changes water to steam in the boiler. The steam passes into radiators where it gives up its heat of vaporization and condenses to a liquid.

Even though steam and boiling water are at the same temperature, steam can produce a more severe burn. One reason is that steam at 100 °C has acquired about 540 more calories of heat per gram than water at 100 °C. When steam condenses, this heat of vaporization is given out. As it cools, the water that is formed gives out the same amount of heat that water at 100 °C does during cooling.

In Figure 8-16, the line **de** represents the heat required to change the liquid to a vapor (steam) without producing a change in temperature. Note that points **d** and **e** on Figures 8-15 and 8-16 represent the same values. After all the water is converted to steam, the addition of heat energy

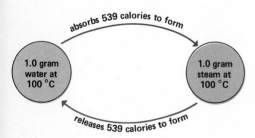

Figure 8-20. Heat of vaporization. It takes 539 additional calories to change 1.0 g of water at 10$\overline{0}$ °C to steam. Steam at 10$\overline{0}$ °C releases the same number of calories upon condensing.

causes the temperature of the steam to rise as indicated by **ef** in both figures.

In order to keep the pressure constant while heat is applied to change ice from $-2\overline{0}$ °C to steam at $12\overline{0}$ °C, the volume of the container must be greatly increased. In our previous discussions we noted that ice expands slightly as its temperature is raised to 0 °C, it contracts as it is melted to water at 0 °C, the water formed contracts to its minimum volume at 4 °C, and above 4 °C the water expands as its temperature is raised to $10\overline{0}$ °C. At constant pressure, as the water at $10\overline{0}$ °C is changed to steam at $10\overline{0}$ °C by the addition of heat the expansion is about 1700 times. As the vapor (steam) is heated further, it continues to expand.

8.20 The Critical Point The vaporization curve **TA** in Figure 8-15 is not unlimited in extent. The lower limit is the temperature and pressure of the triple point. The upper limit is the *critical point.* The temperature and pressure of the critical point are called the *critical temperature* and the *critical pressure.* A substance cannot exist as a liquid at a temperature above its critical temperature; no matter how great the pressure, it cannot be condensed to the liquid state. At the critical point the densities of the liquid and the vapor are equal and the heat of vaporization is zero. At temperatures above the critical temperature a substance is usually called a gas, while at temperatures below the critical temperature it is called a vapor. The critical temperature of water is 374 °C, and the critical pressure is 218 atmospheres. This means that the temperature of liquid water cannot be raised to 374 °C unless it is under a pressure of 218 atmospheres, and that at any higher temperature water can exist only in its gaseous phase, no matter how high the pressure.

Gaseous substances, such as oxygen and nitrogen, have very low critical temperatures. The critical temperature of oxygen is -119 °C and of nitrogen is -174 °C. These gases must first be cooled to their low critical temperatures before they can be liquefied. Helium has the lowest critical temperature, -268 °C.

8.21 Summary of Phase Changes We have described several effects of addition or loss of heat on water. In Figure 8-21 these are summarized in a somewhat different manner than has been shown previously. The graph shows the relationship among temperature, thermal energy, and phase as ice at $-2\overline{0}$ °C is heated to steam at $12\overline{0}$ °C. Since the specific heat of ice is 0.53 cal/g C°, 1.0 g of ice absorbs 11 cal in being warmed to 0 °C. As this ice

The lower limit of the vaporization curve of a substance is the triple point. The upper limit is the critical point.

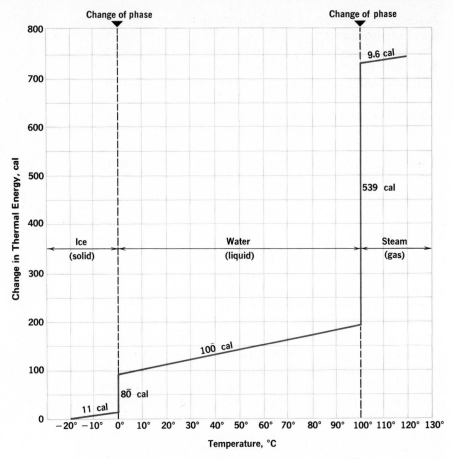

Figure 8-21. Temperature-energy graph for water.

melts there is *no temperature change* while $8\bar{0}$ cal of heat is being absorbed. There is a phase change only. As heating continues, the next $10\bar{0}$ cal increases the temperature to $10\bar{0}$ °C. As the water boils, 539 cal of heat converts the water into steam at 100 °C; during this change of phase there is *no temperature change*. If the steam is under one atmosphere pressure, its specific heat is 0.48 cal/g C°, and 9.6 cal is needed to heat it to 120 °C. Thus 740 cal is required to change 1.0 g of ice at $-2\bar{0}$ °C into 1.0 g of steam at $12\bar{0}$ °C. Conversely, 740 cal is evolved if 1.0 g of steam at $12\bar{0}$ °C is changed into 1.0 g of ice at $-2\bar{0}$ °C.

Questions
GROUP A

1. Draw the temperature-pressure equilibrium curves for water.
2. (a) What is heat of fusion? (b) What is the value for the heat of fusion of ice?

3. What is supercooling?
4. For what purposes are (a) pressure cookers, (b) vacuum pans used?
5. Explain the processes of (a) distillation, (b) fractional distillation.
6. (a) What is heat of vaporization? (b) What is the magnitude of heat of vaporization of water?

7. Why does steam at 100 °C produce a more severe burn to the skin than the same mass of water at 100 °C?
8. What is meant by the critical point?
9. The air above the ice of a pond is −10 °C. What is the probable temperature of (a) the upper surface of the ice; (b) the lower surface; (c) the water just beneath the surface; (d) the water at the bottom of the pond?

GROUP B

10. How do the curves drawn in Question 1 differ from those that would be drawn for a substance that contracts as it changes from a liquid to a solid?
11. Assume you have a sample of pure water at the triple point. What happens if you (a) increase the pressure while the temperature is held constant? (b) increase the temperature while the pressure is held constant? (c) reduce the temperature while the pressure is held constant? (d) reduce

the pressure while the temperature is held constant?
12. Refer to Figure 8-15. What is the location of the set of points that indicates the temperature-pressure values for situations where the solid and vapor of a substance are in equilibrium?
13. Isolate one point of the set referred to in Question 12. Assume that the pressure is held constant while the temperature of the substance is increased. What happens?
14. (a) Why does the boiling temperature of a liquid depend on the pressure exerted on its surface? (b) How does the boiling temperature of a liquid vary with the pressure exerted on its surface?
15. How can fractional distillation be used in separating the different liquids in petroleum?
16. What data are required in order to determine the heat of vaporization of water by the method of mixtures?
17. Why does the heat of vaporization vary with the boiling temperature?

Problems
GROUP A

1. How many calories will be absorbed by 1.50 kg of ice at 0.0 °C as it melts?
2. To what temperature must a 2270-g iron ball be heated so that it can completely melt 1150 g of ice at 0.0 °C?
3. A $50\overline{0}$-g aluminum block is heated to $35\overline{0}$ °C. How many grams of ice at 0.0 °C will the aluminum block melt on cooling?
4. A copper calorimeter has a mass of 220.0 g. It contains 450.0 g of water at 21.0 °C. How many grams of ice at 0.00 °C must be added to reduce the temperature of the mixture to 5.00 °C?
5. To what temperature must a 500.0-g brass weight be heated to convert

60.0 g of ice at −20.0 °C to water at 20.0 °C?
6. How many calories are given off by 50.0 g of steam at $10\overline{0}$ °C when it condenses?
7. How many calories are used to vaporize 20.0 kg of water at $10\overline{0}$ °C?
8. Calculate the number of calories evolved when 4.00 kg of steam at $10\overline{0}$ °C is condensed, cooled, and changed to ice at 0 °C.
9. How many grams of mercury can be vaporized at its boiling point, 356.58 °C, by the addition of 1.00×10^3 calories?
10. What is the final temperature attained by the addition of 11.4 g of steam at 100.0 °C to 681.0 g of water at 25.0 °C in an aluminum calorimeter having a mass of 182.0 g?

GROUP B

11. What is the final temperature of a mixture of 50.0 g of ice at 0.00 °C and 50.0 g of water at 80.0 °C?

12. A calorimeter, specific heat 0.100 cal/g C°, mass 200.0 g, contains 300.0 g of water at 40.0 °C. If 50.0 g of ice at 0.0 °C is dropped into the water and stirred, the temperature of the mixture when all the ice has melted is 23.8 °C. Calculate the heat of fusion of ice.

13. A block of silver, which has a mass of 500.0 g, temperature 100.0 °C, is put in a calorimeter with 300.0 g of water, temperature 30.0 °C. The mass of the calorimeter is 50.0 g, and its specific heat is 0.100 cal/g C°. A 50.0-g mass of ice at −10.0 °C is also put in the calorimeter. Calculate the final temperature.

14. What is the final temperature attained when $90\bar{0}$ g of ice at 0.0 °C is dropped into $340\bar{0}$ g of water at 93.3 °C in a calorimeter having a mass of $135\bar{0}$ g with specific heat 0.090 cal/g C°?

15. In an experiment to determine the heat of vaporization of water, 15.0 g of steam at 100.0 °C is added to 150.0 g of water at 20.0 °C in a calorimeter. The mass of the calorimeter is 75.0 g; its specific heat is 0.100 cal/g C°. The equilibrium temperature of the mixture is 73.9 °C. What is the heat of vaporization of water?

16. A mixture of ice and water, mass 200.0 g, is in a 100.0-g calorimeter, specific heat 0.200 cal/g C°. When 40.0 g of steam is added to the mixture, the temperature is raised to 60.0 °C. How many grams of ice were originally in the calorimeter?

17. An aluminum cylinder, mass 50.0 g, is placed in a 100.0-g brass calorimeter with 250.0 g of water at 20.0 °C. What equilibrium temperature is reached after the addition of 25.0 g of steam at 120.0 °C?

18. A copper ball with a mass of 4.54 kg is removed from a furnace and dropped into 1.36 kg of water, temperature 22.0 °C. After the water stops boiling, the combined mass of the ball and water is 5.45 kg. What is the furnace temperature?

SUMMARY

The thermal energy of a material is the potential and kinetic energy of its particles. Heat is the thermal energy that is absorbed, given up, or transferred from one material to another. Thus heat is a form of energy. Temperature is the physical quantity that is proportional to the average kinetic energy of the particles of a substance.

Temperature is measured in degrees. The Celsius and Kelvin temperature scales are based on the triple point of water, which is the temperature at which the solid, liquid, and vapor phases of water can coexist. Absolute zero (0 °K) is the temperature at which the kinetic energy of the molecules of a substance is at a minimum. Heat is measured in calories. A calorie is equivalent to 4.19 joules of energy.

The change in unit length of a solid when its temperature is changed one degree is its coefficient of linear expansion. The expansion of most liquids is proportional to their increase in temperature. The expansion of water is abnormal. Water reaches its maximum density at 4 °C. Gases expand uniformly except at very high pressures and very low temperatures. According to Charles' law, the volume of a dry gas is directly proportional to its Kelvin temperature provided the pressure is constant. Boyle's law states that if the temperature is constant, the volume of a dry gas is inversely proportional to the pressure. The universal gas constant relates the pressure, volume, Kelvin temperature, and number of moles of an ideal gas. Since the behavior of a real gas is similar to that of an ideal gas except at extreme pressures and low temperatures, the gas laws can be used with reasonable exactness with real gases.

The heat capacity of a body is the quantity of heat needed to raise its temperature one degree. Its specific heat is the ratio of its heat capacity to its mass. The heat given off by hot materials equals the heat received by cold materials.

The heat of fusion is the amount of heat needed to bring about fusion in a unit mass of a substance at its melting point. The heat required to vaporize a unit mass of liquid at its boiling point is the heat of vaporization. The temperature to which any gas must be cooled before it can be liquefied by pressure is called its critical temperature; the pressure needed to liquefy a gas at this temperature is called its critical pressure.

VOCABULARY

Boyle's law
calorie
calorimeter
Celsius scale
Charles' law
coefficient of area
  expansion
coefficient of cubic
  expansion
coefficient of linear
  expansion

critical point
critical pressure
critical temperature
endothermic
exothermic
heat
heat capacity
heat of fusion
heat of vaporization
Kelvin scale

law of heat exchange
mole
specific heat
standard pressure
standard temperature
temperature
thermal energy
triple point
universal gas constant

9 HEAT ENGINES

ROBERT H. GODDARD
U.S. AIR MAIL 8¢

Robert Goddard is known as the "father of modern rocketry." This stamp was issued on the 50th anniversary of Goddard's first rocket experiments in Massachusetts in 1914. In 1926 he successfully launched the world's first liquid-propellant rocket.

In this chapter you will gain an understanding of:

▶ the relationship between work and heat energy
▶ adiabatic and isothermal processes
▶ the two laws of thermodynamics
▶ specific heats of gases
▶ the effect of temperature difference on the efficiency of a heat engine
▶ the increase of entropy in the universe
▶ the principles and characteristics of various external and internal combustion engines
▶ heat pump operation and efficiency

Work and heat are interchangeable.

HEAT AND WORK

9.1 Mechanical Equivalent of Heat Since heat is a form of energy, it has the ability to do work. For example, you can warm your hands by briskly rubbing them together. You can heat up a nail by pounding it into a board with a hammer. And a tire pump gets hot when it is used vigorously.

As we noted in Section 1.13, Count Rumford and James Prescott Joule studied the relationship between heat and work in the 19th century. In experiments conducted between 1842 and 1870, Joule found that the heat produced by a given amount of mechanical energy is constant.

*The study of the quantitative relationships between heat and other forms of energy is called **thermodynamics.*** In this chapter we shall be concerned with the relationship of heat energy to mechanical energy. In discussing energy transformations, we shall also use the term *internal energy.* ***Internal energy** is the total potential and kinetic energy of the particles of a substance.* These particles include molecules, ions, atoms, and subatomic particles. Heat is one form of internal energy.

Joule used an apparatus similar to that shown in Figure 9-2 to determine the relationship between mechanical energy and heat energy. The system on which work was done was a mass of water in an insulated vessel designed to reduce the escape of heat from the vessel to a minimum. A set of movable paddles on a shaft was turned by a falling mass connected to the shaft by a cord. The paddles moved past fixed vanes, churned the water, and increased its temperature from T_i to some final value T_f. The work done on the water equals the loss of potential energy of mass m as it falls through a distance d, less the kinetic energy it possesses as it reaches the bottom. The loss in potential energy is mgd. The final kinetic energy of the mass is $\frac{1}{2}mv^2$. (In these experiments, v was small.) Hence, the amount of work, W, done on the system by the falling mass is

$$W = mgd - \tfrac{1}{2}mv^2$$

The amount of heat that would be needed to produce the observed temperature change in the water is

$$Q = m_w c_w (T_f - T_i)$$

Mechanical energy, or W, the work done on the system, is expressed in joules. Q is the equivalent heat energy expressed in calories. The results of many careful experiments have shown that

1 calorie = 4.19 joules

This equality is shown in Figure 9-3 and is known as the *mechanical equivalent of heat.*

Just as in the case of the hot brass chips in Count Rumford's cannon-boring experiments, it is possible to produce internal energy indefinitely in the Joule apparatus provided we continue to supply mechanical energy to it. Thus we must conclude that heat is a form of energy and not a substance.

9.2 First Law of Thermodynamics We can now broaden the statement of the conservation of energy in Section 1.13 to include internal energy as well as mechanical energy. This generalization is known as the ***first law of thermodynamics:*** *The quantity of energy supplied to any isolated system in the form of heat is equal to the work done by the system plus the change in internal energy of the system.* Thus the energy input to an isolated system equals the energy gained by the system plus the energy output in the form of work. We may also state the first law of thermodynamics as follows: *When heat is converted to another form of energy or when other forms of energy are converted to heat, there is no loss of energy.*

As an application of the first law of thermodynamics, let us assume that an amount of heat, Q, is added to a substance whose total internal energy is E_i. Generally we would find that the addition of this heat energy increases the internal energy of the substance to E_f and also causes

Figure 9-1. James Prescott Joule made the first determination of the mechanical equivalent of heat.

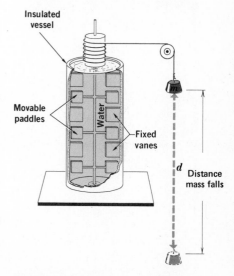

Figure 9-2. A simplified diagram of Joule's apparatus for observing the conversion of mechanical energy into internal energy.

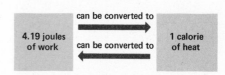

Figure 9-3. The relationship between heat and work.

the substance to do a quantity of work, W, on its surroundings. This can be stated algebraically as

$$Q = (E_f - E_i) + W$$

Q is positive when heat is added to the substance, and W is positive when the body does work on surrounding objects. All quantities must be expressed in the same units, either joules or calories.

In situations when no work is done by or on the substance, the change in internal energy equals the quantity of heat added to or removed from the substance. Adding heat to water that undergoes no change of phase will cause each gram of water to increase in temperature by one Celsius degree for each calorie of heat added. Here, we neglect the very small amount of work involved in changing the volume of the water. We know that the internal energy of the water is changed because the temperature of the water is changed.

Adiabatic: no heat is added or removed.

A process in which no heat is added to or removed from a substance is called an **adiabatic process.** In such a case, $Q = 0 = (E_f - E_i) + W$. Thus $W = E_f - E_i$. An example is the Joule experiment for determining the mechanical equivalent of heat described in the previous section. When the vessel containing the water permits no heat to enter or leave during the churning process, the work done on the water equals the change in its internal energy.

In another example let us assume that we have a quantity of air trapped in an *insulated* cylinder with a tight-fitting but freely moving piston. When the air is compressed by pushing the piston into the cylinder, the change in the internal energy of the air must equal the work done on it. This relationship is shown by the change in volume, pressure, and temperature of the air.

9.3 Isothermal Expansion In nearly all situations involving gases, the gases are confined by barriers on which they exert a force and which, in turn, exert an equal and opposite force on the gases. A gas may be under pressure in a storage tank or in a cylinder above a movable piston of an internal combustion engine. A quantity of air (an air mass) in the atmosphere is under pressure exerted by the air around it. In any of these cases, expansion of the gas requires it to do work against an external force; thus work is done on the external medium. On the other hand, the quantity of gas may be compressed by the action of an outside force; work may be done on the gas.

To calculate the work done by a gas in expanding, let us imagine the gas enclosed in a cylinder with a tight-fitting

piston, as illustrated in Figure 9-4. The piston rod is connected to a device on which it may exert a force. The force,
F, acting on the piston due to the pressure, p, exerted by
the gas on area A of the piston head is

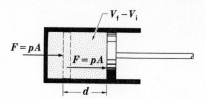

$$F = pA$$

Figure 9-4. Expansion of a gas at
constant pressure.

Suppose the piston is moved a distance d by the expanding gas in the cylinder. Suppose further that some heat is
applied to the gas while the pressure remains constant
during the expansion. The work done by the expanding
gas in moving the piston will be

Pressure = force per unit area

$$W = Fd = pAd$$

where the quantity Ad is the change in volume of the expanding gas. Thus the work done by the gas expanding at
constant pressure is

$$W = p(V_f - V_i)$$

where V_f and V_i are the final and initial volumes, respectively, of the confined gas. We may conveniently show the
work done by the gas expanding at constant pressure by
using a graph, as shown in Figure 9-5. On this graph volumes are plotted as abscissas and pressures as ordinates.
The expansion at constant pressure is represented by a
red horizontal line extending from V_i to V_f. The work, W,
done by the gas in expanding is given by the area between
this line and the X axis and between V_i and V_f (the red
portion shown in Figure 9-5). This graph is for the *expansion* of a gas at constant pressure. If the gas is *compressed* at
constant pressure, the work done is represented in the
same manner. It is considered negative, however, because
the volume is decreasing as a result of work being done on
the gas.

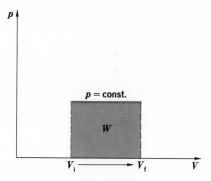

Figure 9-5. Graph of work done
by a gas expanding at constant
pressure.

   If during the expansion the pressure of the gas changes,
the calculation becomes somewhat more complicated. Let
us consider a situation in which sufficient heat is supplied
to keep the temperature constant during an expansion.
For a dry gas, the relationship between pressure and volume is given by Boyle's law

$$pV = \text{constant}$$

This relationship can be shown graphically, as in Figure
9-6. The area under the pV curve between V_f and V_i represents the work involved. This work can be computed by
dividing the total volume change into a number of very
small volume changes. Each of these volume changes

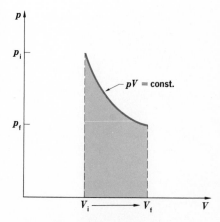

Figure 9-6. Graph of work done
during the expansion of a gas
kept at constant temperature.

must be so small that the pressure remains practically constant for the change. The work associated with each minute change in volume is determined by multiplying the change in volume by the pressure at the volume at which the change takes place. The sum of all these quantities of work equals the total work involved. *A process that takes place at constant temperature is known as an* **isothermal** **process.**

Isothermal: constant temperature

If the volume of an ideal gas increases at constant temperature, its pressure decreases. Similarly, if the volume of an ideal gas decreases at constant temperature, its pressure increases. Since the internal energy of an ideal gas determines the temperature, there can be no change in the internal energy of an ideal gas during isothermal processes. However, work is done by a gas during expansion. Since the volume occupied by the gas becomes greater, work is done on the gas molecules against external pressure. (In a real gas, some work is also done on the gas molecules, increasing their potential energy by moving them farther apart.) The heat equivalent of the work done isothermally *by* an ideal gas during expansion *must be absorbed from its surroundings*. In a similar manner, work must be done *on* an ideal gas during isothermal compression and the equivalent amount of heat *must be transferred to its surroundings*.

A sealed plastic bag containing air will increase in volume as the atmospheric pressure drops during an approaching storm, even if the room temperature remains constant. The air in the plastic bag pushes out against the atmosphere, yet there are no temperature changes.

9.4 Adiabatic Expansion During an adiabatic expansion of an ideal gas, work is done just as it is during an isothermal expansion. Now, however, the equivalent amount of heat is not withdrawn from the surroundings but is obtained at the expense of the thermal energy of the gas. Thus during adiabatic expansion the temperature as well as the pressure of the ideal gas is lowered. Similarly, when work is done on a gas as it is adiabatically compressed, the heat equivalent of the work is not lost to the surroundings but is used to increase the internal energy of the gas. Hence, during adiabatic compression both the pressure and the temperature of the gas increase.

Adiabatic processes do occur in fast compressions when there is not enough time for the resulting heat to escape. An example of this process is the diesel engine, which will be described later in this chapter.

9.5 *Specific Heats of Gases* While the specific heat of a solid or liquid is a fixed value to several significant figures for the substance at a given temperature, this is not true for gases. When heat is added to a mass of gas, its pressure, volume, or both may change, and its temperature may rise.

When the volume of a unit mass of gas is held constant as heat is added, the quantity of heat required to change the temperature of this unit mass of gas by 1 C° is called its *specific heat at constant volume*, c_v.

When the pressure of the same unit mass of gas is held constant as heat is applied to raise its temperature 1 C°, the quantity of heat required is called its *specific heat at constant pressure*, c_p.

The numerical value of c_v differs from that of c_p even though the same mass of the same gas is taken through the same temperature interval. The internal energy of the unit mass of gas must be changed by the same amount in each instance. Where the pressure is held constant, added energy must be provided to do the work required to produce the volume increase. That is, work must be done on the movable part of the gas container or the surrounding atmosphere. This added work equals the pressure times the change in volume. As a result c_p is greater than c_v. In addition to increasing the internal energy, as evidenced by the increased temperature of the gas, some of the added heat energy is transformed into mechanical energy as a result of the expansion.

The same situation exists for liquids and solids. However, the amount of expansion is so small that the two specific heats are numerically the same to the number of significant figures with which we work, even in most research situations.

The values of c_p and c_v for some common gases at room temperature and atmospheric pressure are given in Table 9-1.

Refer to Section 8.12 for a definition of specific heat.

Table 9-1
SPECIFIC HEATS OF GASES

Gas	c_p $\left(\dfrac{cal}{g\ C°}\right)$	c_v $\left(\dfrac{cal}{g\ C°}\right)$
air	0.242	0.173
ammonia	0.523	0.399
carbon dioxide	0.200	0.154
hydrogen	3.40	2.40
nitrogen	0.248	0.176
oxygen	0.218	0.156

Questions
GROUP A

1. What is thermodynamics?
2. State the first law of thermodynamics.
3. What is the meaning of the term *mechanical equivalent of heat*?
4. What is the metric value for the mechanical equivalent of heat?
5. What is (a) an isothermal process, (b) an adiabatic process?
6. Give an example of (a) an isothermal expansion, (b) an adiabatic expansion.
7. How does the specific heat of gases differ from that of a solid or liquid?
8. Define two kinds of specific heat for a gas.

GROUP B

9. How does the first law of thermodynamics differ from the mechanical conservation of energy principle?
10. Describe Joule's method for determining the mechanical equivalent of heat.
11. What physical law governs the isothermal expansion and compression of an ideal gas?
12. What physical laws govern the adiabatic expansion and compression of an ideal gas?
13. (a) What is the source of the heat equivalent of the work done by an ideal gas during isothermal expansion? (b) What happens to the heat equivalent of the work done on an ideal gas during isothermal compression?
14. (a) What is the source of the heat equivalent of the work done by an ideal gas during adiabatic expansion? (b) What happens to the heat equivalent of the work done on an ideal gas during adiabatic compression?
15. When an ideal gas expands isothermally, it does work on its surroundings. (a) Does the energy of the gas change during this process? (b) What is the source of the energy by which the gas can do this work?
16. (a) Does a gas do work on its surroundings when the gas expands adiabatically? (b) What is the source of energy by which the gas can do this work?

Problems

GROUP A

1. How many joules can ideally be obtained from 1.00×10^4 cal?
2. Gasoline, mass density 0.700 g/cm^3, liberates 1.15×10^4 cal/g when it is burned. How many joules of work can be obtained by burning 1.00 L of gasoline?
3. The water going over Niagara Falls drops 50.6 m. How much warmer is the water at the bottom of the falls than it is at the top? Disregard any possible effects of evaporation of water during the fall.

GROUP B

4. The natural gas burned in a gas turbine has a heating value of 1.00×10^5 cal/g. If 2.00 g of gas are burned in the turbine each second and the efficiency of the turbine is 25.0%, what is the output in kilowatts?
5. How much heat will be produced if a 1.25-kg mass moving at 26.6 m/s strikes a wall and all the energy is converted to heat?
6. The powder used in firing a 5.00-g bullet from a rifle produces $75\overline{0}$ calories of heat when burned. In the process of firing the rifle, 30.0% of the energy is converted into the kinetic energy of the bullet. Determine the muzzle speed of the bullet.
7. A lead block falls from a height of 125 m and strikes a cement block. Assume that half the energy of the lead block is converted into internal energy in the lead. Determine the rise in temperature of the lead block.
8. What must be the speed in meters per second of a snowball at 0 °C if it is completely melted by its impact against a wall? Assume that all the energy is absorbed by the snowball.
9. The following data were obtained by the use of an apparatus similar to that shown in Figure 9-2: descending mass, 2.50 kg; distance moved, 1.50 m; number of descents, 25; temperature rise of water and calorimeter, 0.31 C°; mass of water, 0.700 kg; mass of copper calorimeter, 0.150 kg. Calculate the mechanical equivalent of heat. (Assume that the mass has no kinetic energy when it stops descending.)

HEAT TRANSFER MECHANISMS

9.6 Efficiency of Ideal Heat Engines The processes occurring in the operation of an engine that converts heat energy into mechanical work are complex. However, we can simplify things by replacing the actual heat-engine cycle with an ideal cycle that can produce the same transformations of heat and work. We need not be concerned here with the fact that there are many different types of heat engines utilizing a variety of working substances such as steam, a mixture of fuel and air, or a mixture of fuel and oxygen.

In an engine, the working substance is taken through a series of operations known as a *cycle*. The result is that some of the heat supplied to the substance from a high temperature source is converted into work that is delivered to an external object. For example, high-temperature steam drives a turbine that in turn does work on an electric generator. Experimental and theoretical evidence indicate that not all the heat supplied to the engine can be converted into work. The heat that is not converted into work is delivered by the engine to some external reservoir at lower temperature. Such a reservoir is called a *heat sink*. It is a system that absorbs the exhausted heat, preferably without a significant increase in its own temperature.

In an ideal heat engine, the working substance is a gas that is returned at the end of the cycle to its original pressure, volume, and temperature conditions. There is no permanent loss or gain of internal energy; the internal energy of the working substance remains unchanged.

The operation of an ideal heat engine is schematically shown in Figure 9-7. A quantity of heat, Q_1, is delivered to the engine during the beginning of a cycle. This heat comes from a high-temperature heat source. The engine performs an amount of work, W, on some outside object and exhausts an amount of heat, Q_2, to a low-temperature heat sink. Low temperature in this case means any temperature below that of the heat source, and ideally much lower. Applying the first law of thermodynamics to this cycle,

$$W = Q_1 - Q_2$$

The thermal efficiency, e, of the heat engine is defined as

$$e = \frac{\textbf{work done during one cycle}}{\textbf{heat added during one cycle}}$$

or

$$e = \frac{W}{Q_1}$$

A heat sink is the opposite of a heat source.

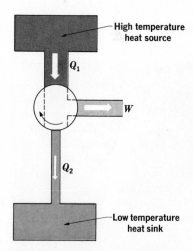

Figure 9-7. Diagram of the operation of a heat engine.

Since

$$W = Q_1 - Q_2$$

then

$$e = \frac{Q_1 - Q_2}{Q_1} = 1 - \frac{Q_2}{Q_1}$$

The efficiency of a heat engine is always less than 100%.

From this equation we see that the thermal efficiency of an operating heat engine must always be less than 100%.

It can be shown that the ratio Q_2/Q_1 is always equal to the ratio of the absolute temperatures T_2/T_1, provided that the engine is considered frictionless. Therefore we can write

$$e = 1 - \frac{T_2}{T_1}$$

From this equation we can see that the efficiency of a heat engine may be increased by making the temperature of the heat source as high as possible and the temperature of the heat sink as low as possible.

From experience we know that gasoline and steam engines expel some heat during the exhaust part of the cycle. The efficiency of any real heat engine will be less than the efficiency of an ideal heat engine because of heat loss from engine parts and friction.

It is interesting to note that the efficiency of steam engines has increased from 0.17% for the first steam engines of the seventeenth century to over 40% for the turbines used in modern power plants.

9.7 Second Law of Thermodynamics It was noted in the previous section that not all the heat supplied to a heat engine can be converted into mechanical work. This fact is true for all types of heat engines and is the basis for a far-reaching generalization known as the **second law of thermodynamics**, which was first formulated by the German physicist, Rudolf Clausius (1822–1888): *It is not possible to construct an engine whose sole effect is the extraction of heat from a heat source at a single temperature and the conversion of this heat completely into mechanical work.* Thus if a heat engine takes a quantity of heat from a heat source at a high temperature, it must transfer some of this heat to a heat sink at a lower temperature.

One interpretation of the second law of thermodynamics is that it is impossible to attain the absolute zero of temperature. There is simply no place to which heat can be transferred in order to cool a substance to exactly 0 °K.

Figure 9-8. Rudolf Clausius formulated the second law of thermodynamics that relates heat transfer and differences in temperature.

Values only a few thousandths of a degree above absolute zero have been attained. The experimental difficulties increase enormously as still lower temperatures are sought. The unattainability of absolute zero is sometimes called the third law of thermodynamics.

Some have considered extracting heat from the internal energy of the ocean and using this to operate the engines of a ship. If the water surrounding the ship is considered a heat source at a single temperature, the second law of thermodynamics indicates that it would be impossible to operate the engines of the ship from the internal energy of the sea water.

It also follows from the second law of thermodynamics that transfer of internal energy from a low-temperature heat source to a high-temperature heat sink requires work. This type of internal energy transfer takes place in refrigerators, air conditioners, and heat pumps.

In order to express the second law of thermodynamics in quantitative form, we must be able to measure the amount of mechanical work a system can do. For example, consider placing a hot and a cold body in thermal contact. After a time they will reach thermal equilibrium; the two bodies will come to the same temperature. There is no loss of energy in the process, but the system as a whole loses its capacity for doing work. (The temperature of the environment is disregarded in this discussion.) A heat engine connected between the two bodies before the bodies were placed in contact could have done useful work. After contact of the two bodies, when a temperature difference no longer exists between them, work cannot be done by an engine connected between the two.

9.8 Entropy   *The amount of energy that cannot be converted into mechanical work is related to* **entropy.** Entropy is a measurable property as important in the study of thermodynamics as is energy. As with potential energy or internal energy, it is the *difference* in entropy that is significant rather than the actual value of entropy. If an amount of heat, ΔQ, is added to a system that is at a Kelvin temperature, T, the *change in entropy*, ΔS, is

$$\Delta S = \frac{\Delta Q}{T}$$

If heat is removed from the system, the quantity ΔQ is negative and the change in entropy is also negative. Note that this equation defines the change in entropy and not entropy itself. Common units for denoting the change in entropy are calories per degree or joules per degree.

The second law of thermodynamics states that heat flows from objects with high temperatures to objects with lower temperatures.

EXAMPLE   The melting point of a given solid is 15 °C. In melting a sample of the solid, how much heat must be added in order to increase the entropy of the system by $100\overline{0}$ calories per degree?

SOLUTION

$$\text{Basic equation: } \Delta S = \frac{\Delta Q}{T}$$

In using the entropy equation, temperatures must be expressed in degrees Kelvin. Thus, 15 °C becomes $28\overline{8}$ °K, and

$$\text{Working equation: } \Delta Q = \Delta ST$$
$$\Delta Q = (100\overline{0} \text{ cal/°K})(28\overline{8} \text{ °K})$$
$$\Delta Q = 2.88 \times 10^5 \text{ cal}$$

Let us again refer to the heat transfer taking place between two bodies of different temperature that are brought into thermal contact. Assume that these bodies are insulated from their surroundings and the initial temperature, T_1, of the first body is greater than the initial temperature, T_2, of the second body. The moment they touch, a small quantity of heat, ΔQ, will be transferred from body 1 to body 2. The entropy change is $+\frac{\Delta Q}{T_2}$ for body 2 and $-\frac{\Delta Q}{T_1}$ for body 1. The net change in entropy for the system is $\frac{\Delta Q}{T_2} - \frac{\Delta Q}{T_1}$. Since $T_1 > T_2$, the change in entropy is positive; the entropy of the system increased. This is an example of an irreversible process in which a loss of capacity for work results in an increase in entropy for the system. It can be shown that for any transformation occurring in an isolated system, the entropy of the final state can never be less than that of the initial state. Only in cases where the transformation is reversible will the system undergo no change in entropy.

The *second law of thermodynamics* can now be restated as **the law of entropy:** *A natural process always takes place in such a direction as to cause an increase in the entropy of the universe. In the case of an isolated system it is the entropy of the system that tends to increase.*

The available energy of the universe is diminishing.

All natural processes are irreversible and involve increases in entropy. Thus the second law of thermodynamics is equivalent to the statement that the entropy of the universe is increasing, and the first law of thermodynamics is equivalent to the statement that the total energy of the universe is constant.

Heat can be considered disordered energy. An example of ordered energy is a flying rifle bullet that has kinetic energy. When the bullet is stopped suddenly by impact, the energy of its motion is transformed to random motion of atoms. This disordered energy is evidenced by the heating of both the bullet and the area of impact.

Work involves orderly motion. But when work is done against friction and is dissipated into internal energy, the disorderly motion of molecules is increased. There is an accompanying increase in disorder.

There are many examples in nature where energy processes are observed to go toward a state of greater disorder. The melting of an isolated crystalline solid is an example. When heat is added, the system goes from a well-ordered array of molecules in a crystal to a less well-ordered array of molecules in a liquid without a change in the temperature of the sample. Entropy may thus be thought of as a measure of the order-disorder in the system.

Most natural processes proceed toward a state of greater entropy as well as toward increased disorder. It is not surprising to note that physicists have found a connection between the thermodynamic concept of entropy and the statistical concept of disorder.

Two quantities can be used to describe heat. One quantity describes the energy and is measured in calories or joules. The other quantity describes the amount of disorder and is expressed as entropy, which is a mathematical concept.

Figure 9-9. As heat flows from the air and water into the ice, the ice melts and the entropy of the system increases.

9.9 Steam Engines The earliest known devices in which heat is converted to work were described by Hero of Alexandria in the first century A.D. In the so-called Hero's fountain, heated air was used to expel water vertically out of a container. In another device, known as the aeolipile, steam generated in a boiler passed through tubes into a sphere and out through nozzles that were mounted tangentially to the sphere, as shown in Figure 9-10. The reaction force of the steam leaving the nozzles made the sphere rotate. Neither of these devices had much practical value, however.

The first practical steam engine was built in England in 1712 by Thomas Newcomen (1663–1729). In the design of the Newcomen engine, steam was directed against a movable piston inside a cylinder. A rod that was connected to the moving piston activated a water pump or some

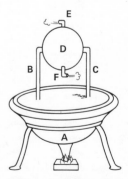

Figure 9-10. Hero's aeolipile. Steam produced in the boiler (A) passed through the hollow tubes (B and C), into the sphere (D), and out through the nozzles (E and F). The reaction force of the escaping steam made the sphere rotate.

212

It was noted in Section 6.5 that the metric unit of power is named in honor of James Watt.

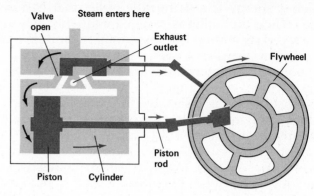

Figure 9-11. A double-acting steam engine. The sliding valve above the cylinder directs the steam first to one side and then the other side of the piston.

The pascal is named for the French mathematician Blaise Pascal (1623–1662).

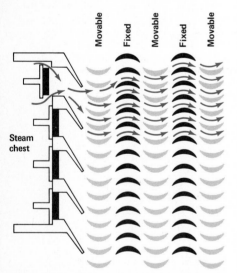

Figure 9-12. In a steam turbine, steam strikes a set of movable blades and exerts force against them. The steam is then deflected by a set of fixed blades to another set of movable blades, and so on.

other mechanical device. James Watt and others made important improvements on the steam engine, which became a major impetus for the Industrial Revolution.

Figure 9-11 is a cross-sectional view of a double-acting steam engine. As the name implies, steam enters the cylinder through a sliding valve and pushes the piston first in one direction and then in the opposite direction. In some modern steam engines, pressures of more than thirteen kilopascals are used. (The pascal, or Pa, is a derived SI unit that equals a newton per square meter.) The efficiency of the steam engine can be increased by arranging several cylinders in tandem, so that the steam leaving the first cylinder is used to push the piston in a second cylinder, and so on. Some compound steam engines of this type in large ocean liners have power outputs of more than 11,000 kilowatts.

Steam engines have been largely replaced by other power sources, such as the steam turbine and the gasoline engine. But a small steam engine is more efficient than a small steam turbine, and where large supplies of steam are required for other purposes, it can be economically feasible to use a steam engine to drive pumps and compressors.

9.10 Steam Turbines In one type of steam turbine, steam at a high temperature and high pressure is directed through nozzles against a set of cupped blades attached to a wheel. This arrangement is called an impulse turbine. The sideways component of the force of the steam against the blades causes the turbine wheel and the shaft on which it is mounted to rotate. Small impulse turbines may be made to move at rates of more than 10,000 revolutions per minute.

In a second type of steam turbine, high-pressure steam passes through alternating sets of movable and fixed vanes, as shown in Figure 9-12. This is called a reaction turbine. A single reaction turbine may contain forty or more pairs of movable and fixed sets of vanes. The fixed vanes are attached to the housing of the turbine, while the movable vanes are connected to a single shaft. The diameters of the turbine blades are made increasingly larger from the inlet to the outlet part of the engine, so that the energy of the expanding steam is transferred more efficiently. This arrangement can be seen in Figure 9-13.

Figure 9-13. A large steam turbine and electric generator during assembly. Notice that the turbine wheels have varying diameters.

Various combinations of steam turbines are used today for ship propulsion and for driving generators in electric power plants. In a large modern steam turbine, steam may enter at a pressure of more than 240 kilopascals and a temperature of over 560 °C. This obviously requires the use of special alloys that can withstand extreme conditions of temperature and pressure. As explained in Section 9.6, a large difference between the input and exhaust temperatures increases the efficiency of a heat engine. Such steam turbine-generator combinations with power outputs of more than 1,000,000 watts are in use today.

9.11 Gasoline Engines Steam engines and steam turbines are examples of *external combustion engines*. That is, the fuel burns outside the engine and the heat is transferred to a cylinder or turbine chamber by means of steam.

In an *internal combustion engine*, the fuel burns inside the cylinder or engine chamber. This requires the use of a fuel that does not leave appreciable amounts of solid combustion products in the engine. Combustible vapors are therefore used in most internal combustion engines.

The gasoline engine, which is the most widely used of all heat engines, is an internal combustion engine. Figure 9-14 shows the sequence of events in a four-stroke gasoline engine. As the piston moves down, (A), air and gasoline vapor are brought into the cylinder through a valve. As the piston moves up again, (B), the valve closes and the air-vapor mixture is compressed to a fraction of its original volume. This compression causes the temperature of the mixture to rise. A spark then starts a chemical reaction between the gasoline vapor and the oxygen in the air, (C).

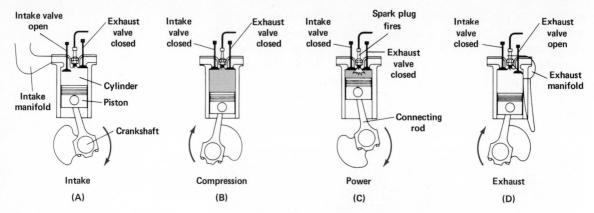

Figure 9-14. A four-stroke gasoline engine.

Heat produced by the chemical reaction increases the pressure of the gas in the cylinder. Changing some of the internal energy into mechanical energy causes the temperature of the gases to decrease, but not to a temperature as low as the temperature of the outside air. The heat that was added from the chemical reaction was partly converted to work and partly used to increase the internal energy of the gases that escape through the exhaust, (D).

The number of cylinders and their arrangement in a gasoline engine may vary, depending on the application. A lawn mower engine has only one cylinder, automobile engines have four or more cylinders, while some aircraft engines have as many as twenty-eight cylinders. In a V-8 gasoline engine, a row of four cylinders on one side of the engine is mounted opposite a second row of four cylinders, with the two rows of cylinders forming a V. This makes for compactness and freedom of vibration as the motion of the pistons is transferred to the crankshaft of the engine.

The firing sequence for the cylinders in an automobile engine is not the same for each model.

In two-stroke gasoline engines, ignition takes place during every revolution of the crankshaft and the intake and exhaust functions are carried out in a single operation. This increases the power-to-weight ratio of the engine.

The power of a gasoline engine depends on several factors. These include the compression ratio in the cylinder, piston size and displacement, and the amount of fuel the engine can burn in a given time period. In order to provide greater compression, especially in aircraft engines that are used at high altitudes, a turbine may be used to increase the pressure of the air before it enters the engine. This is called *turbocharging.* A turbocharged engine may have a power output as high as 2500 kilowatts.

The total piston displacement of an automobile engine is often given in liters.

Many types and sizes of gasoline engines have been developed for use in automobiles. One of these is called the Wankel engine, after its German inventor Felix Wankel (b. 1902). Instead of pistons, the Wankel engine has a triangular rotor that turns inside a specially shaped housing. See Figure 9-15. A Wankel engine has fewer moving parts and is quieter than a reciprocating engine. The power-to-weight ratio of a Wankel engine is also much higher than it is for a piston engine.

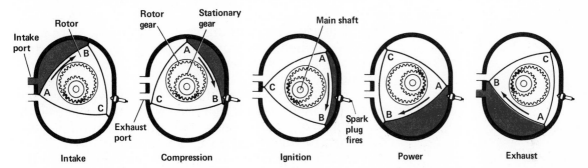

Intake — Compression — Ignition — Power — Exhaust

Figure 9-15. A Wankel engine.

9.12 Diesel Engines   In 1895, the German inventor Rudolf Diesel (1858–1913) developed an internal combustion engine that does not require spark plugs to ignite the fuel. The heat of compression of the air inside the cylinder is sufficient to produce ignition. The fuel is injected at the time of maximum compression. The heating of the air during compression in a diesel engine is an example of an adiabatic process.

See Section 9.4.

A diesel engine has a low power-to-weight ratio, but it is more efficient and can burn a lower grade of fuel than its gasoline counterpart. The diesel engine is also known for its reliability and durability. A small, single-cylinder diesel engine may have a power output of less than five kilowatts. Larger versions are used in buses, trucks, tractors, and locomotives. A diesel engine used for ship propulsion or in an electric-generating plant may have twelve cylinders, a piston diameter of more than seventy centimeters, a stroke of over one hundred centimeters, and develop 7500 kilowatts.

9.13 Gas Turbines   The gas turbine is another type of internal combustion engine. In a gas turbine, a large volume of air is continually drawn into a compressor, where its pressure is increased. This compressed air then flows into a combustion chamber, where the fuel is injected and burns with a continuous flame. Only part of the compressed air is needed for complete combustion of the fuel. From the combustion chamber the hot combustion gases

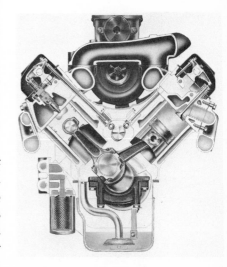

Figure 9-16. A diesel engine.

and heated air expand and move at great speed, first through a turbine and then through the exhaust nozzle.

Some of the energy imparted to the turbine is used to drive the compressor. The remaining energy may be taken from the shaft of the engine to run an electric generator or some other machine.

The efficiency of a gas turbine is rather low, partly because the temperature is not as high inside the engine as it is in a reciprocating engine. Very high temperatures would cause deterioration in the turbine blades. A gas turbine is also less efficient than a steam turbine, since more work is required to compress the air in a gas turbine than to pump the water needed for the steam in a steam turbine.

A gas turbine is most efficient at high velocities and under maximum load. Gas turbines are presently used to power fast ships and trains, to pump gas through long-distance pipelines, and to generate electricity. Used in conjunction with jet engines (which are described in the next section), gas turbines also power all but the lightest and slowest of today's aircraft.

Gas turbine automobiles are still in the developmental stage.

A practical gas turbine automobile engine still awaits development. Engineers estimate that good fuel efficiency will not be achieved in such an engine below an internal temperature of about 1350 °C. The metal components of an engine lose strength at about 1050 °C. Poor acceleration is another drawback of the gas turbine as a car engine.

While natural gas is the best fuel for a gas turbine, other fuels can also be used. In 1981, powdered coal was used as a fuel in gas turbines installed in test-model automobiles. Compressed air is used to pump the powdered coal into the turbine from a fuel tank over a front wheel. The fuel costs of this engine are comparatively low, but the emissions have not yet been reduced to acceptable levels.

9.14 Jet Engines   Hero's aeolipile (Figure 9-10) was the first jet engine. As the steam leaves the nozzles of the aeolipile, the reaction force turns the sphere on which the nozzles are mounted. The reaction force in a jet engine is called *thrust*. (Thrust is *not* produced, as is often imagined, by the push of the escaping gases against the surrounding atmosphere.)

The simplest type of modern jet engine is the ramjet. A ramjet has no moving parts. See Figure 9-17. Air is forced at increased pressure into the combustion chamber, where fuel is added and burns. The hot combustion gases and heated air rushing out through the jet nozzle produce the unbalanced reaction force that moves the engine forward.

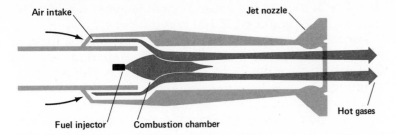

Figure 9-17. A ramjet. This engine has no moving parts and must be brought up to a certain speed before ignition.

Since a ramjet depends on its speed to compress the air necessary for combustion, it must have a high speed to operate efficiently. Therefore a ramjet must be boosted to the proper speed for ramjet operation. Rockets (which are described in the next section) are usually used for this purpose. Ramjets are used today mainly to propel guided missiles and are operated at speeds of up to five times the speed of sound.

The speed of sound is called Mach 1, in honor of the Austrian physicist Ernst Mach (1838–1916).

A turbojet engine is, as the name implies, a combination gas turbine and jet engine. As shown in Figure 9-18, the turbine turns a compressor at the front end of the engine. The compressor draws air into the engine and compresses the air before it reaches the combustion chamber.

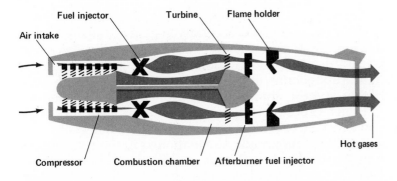

Figure 9-18. A turbojet engine. Part of the energy of the escaping gases is used to turn a turbine which, in turn, operates a compressor at the front of the engine.

Fuel, usually kerosene, is injected into the air stream just behind the compressor. Here it burns very rapidly at a high temperature, producing great pressure. The hot, high-pressure gases pass through the turbine that drives the compressor, and the unbalanced reaction force exerted by the gases as they escape through the nozzle pushes the engine forward at great speed.

The shaft in a turbojet engine rotates at about 12,000 revolutions per minute and internal temperatures can be as high as 1000 °C. The amount of thrust depends on the mass and velocity of the exhaust gases. A medium-sized turbojet engine may use as much as 4000 kilograms of air and 70 liters of fuel per minute! The exhaust velocity is about 6000 meters per second. Each engine on a large jet aircraft may produce 180,000 newtons of thrust.

Auxiliary power must be used to start turning a turbojet engine. The fuel is then ignited by a spark plug. The fuel mixture continues to burn as long as the engine is running.

A rocket engine carries an oxidizer as well as fuel.

Figure 9-19. Saturn V, the space vehicle that carried the first explorers to the moon. The rockets in the first stage of this vehicle were the most powerful ever built.

Figure 9-20. The Space Shuttle is launched with the help of two solid-propellant booster rockets. The Shuttle lands like an airplane and is then reused.

9.15 Rockets A rocket operates on the same principle as a jet engine. However, a rocket carries both fuel and the oxidizer needed to burn it, while a jet engine carries only fuel and uses oxygen from the air as the oxidizer. Thus a rocket may go beyond the earth's atmosphere. Rockets can move better through outer space than through air, because there is no friction.

Rocket engines can be classified as solid-propellant or liquid-propellant engines. The gunpowder rockets made by the Chinese in the thirteenth century were the first solid-propellant rockets. The first successful launch of a liquid-propellant rocket was made in 1926 by the American physicist Robert Goddard (1882–1945).

The performance of a rocket propellant is described in terms of *specific impulse*. **Specific impulse** *is the product of a propellant's thrust-to-weight ratio and its burning time.* Specific impulse is measured in seconds. A propellant consisting of liquid hydrogen and liquid oxygen has a specific impulse of 335 seconds. In general, as the temperature in the rocket chamber rises, so does the specific impulse. Also, as the definition of specific impulse suggests, light propellants like hydrogen-oxygen generally have higher specific impulses than do heavier propellants.

To get a rocket off the ground, the thrust of the propellant must exceed the total weight of the vehicle on which the rocket is mounted. As mentioned in the previous section, thrust depends in part on the velocity of the exhaust gases. In a typical rocket, exhaust velocities of 21,000 meters per second and internal pressures of 35 kilopascals are produced.

The most powerful rocket ever built was for Saturn V, which was used for the manned flights to the moon (Figure 9-19). The cluster of five engines in the first stage of Saturn V had a combined thrust of over 33 million newtons. The entire spacecraft was 132 meters tall and its fueled weight on the ground was slightly over 28 million newtons. The kerosene and liquid oxygen in the first stage burned at a rate of 14,100 liters per second for a total of 2.5 minutes. The maximum velocity attained by the spacecraft after the firing of its three stages was about eleven kilometers per second.

The Space Shuttle has three liquid-propellant engines that burn liquid hydrogen and liquid oxygen. At the time of launch, two solid-propellant booster rockets are also used. Each of these boosters weighs 5,750,000 newtons and has 12,900,000 newtons of thrust. The propellant is a mixture of powdered aluminum fuel and aluminum perchlorate oxidizer. The Shuttle boosters are the largest solid-propellant rockets ever built. Unlike previous space vehicles, the Space Shuttle and its booster chambers are reused.

The Space Shuttle is the first reusable spacecraft.

9.16 Heat Pumps A heat pump transfers heat from a low-temperature source to a high-temperature sink. This "uphill" flow of heat energy requires work, which is usually supplied in a modern heat pump by an electric motor. The principle of the heat pump was first proposed by Lord Kelvin in 1852, but it was not widely applied until electric refrigerators came into general use in the 1930's.

The fluid in a heat pump is a vapor that is easily condensed to a liquid when pressure is applied. The liquid, under pressure, gives up the heat developed during compression to the heat sink. The liquid is then released into a low-pressure zone where it is quickly evaporated, taking its heat of vaporization from the heat source. The vapor is then recompressed and the cycle is repeated.

In the winter, a house to be heated by a heat pump is made the high-temperature sink, as shown in Figure 9-21(A). The outside air serves as the low-temperature source. In the summer, Figure 9-21(B), these conditions

Figure 9-21. A heat pump. This heat transfer mechanism can be used to heat a home in the winter and cool it in the summer.

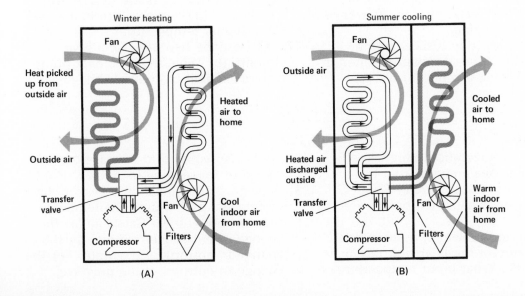

Winter heating
Summer cooling

(A) (B)

A heat pump can be used to heat a house in winter and to cool it in summer.

are reversed. The house is the low-temperature source and the outside air is the high-temperature sink.

Unlike other heating systems, a heat pump does not generate heat, but simply moves heat energy from one place to another. This makes the heat pump about twice as efficient as an electric heating system and about four times as efficient as a gas or oil furnace. However, a heat pump must be augmented by some other heating device when the outside temperature drops below −4 °C.

A refrigerator is a heat pump in which the freezing compartment is the low-temperature heat source and the air in the room is the high-temperature sink. Consequently, a refrigerator heats the room while at the same time cooling the food that is stored in the refrigerator.

Questions

GROUP A

1. What is the purpose of (a) a heat source, (b) a heat sink?
2. On what does the efficiency of an ideal heat engine depend?
3. State the second law of thermodynamics.
4. Give two definitions of entropy.
5. Distinguish between an external combustion engine and an internal combustion engine.
6. Describe Hero's aeolipile.
7. Define the metric unit of pressure.
8. List two uses for steam turbines.
9. What factors determine the power of a gasoline engine?
10. What is meant by turbocharging?
11. How is the air-fuel mixture ignited in (a) a gasoline engine, (b) a diesel engine?
12. Describe two difficulties in developing a practical gas turbine automobile engine.
13. What is the main difference between a ramjet and a turbojet engine?
14. What is the main difference between a jet engine and a rocket?
15. (a) Define specific impulse. (b) Use dimensional analysis to show that specific impulse is measured in seconds.
16. How can heat be made to flow from a body at a low temperature to a body at a higher temperature?
17. (a) Describe two uses for a heat pump. (b) What makes a heat pump more efficient than an electric heater?

Problems

GROUP B

1. What is the theoretically highest efficiency of a steam engine that has a steam input temperature of 200.0 °C and a steam exhaust temperature of 100.0 °C?
2. Referring to Problem 1, assume the temperature of the heat sink to remain at 100.0 °C. What input temperature would be required to increase the efficiency of this engine to 30.0%?
3. A $5\overline{0}$-g ice cube melts in a beaker containing $4\overline{0}0$ g of water. The temperature of both is 0 °C. Calculate (a) the change in entropy of the ice, (b) the change in entropy of the water, (c) the change in entropy of the universe.

SUMMARY

The study of the quantitative relationship between heat and other forms of energy is called thermodynamics. One calorie of heat is equal to 4.19 joules of mechanical energy. The first law of thermodynamics states that the quantity of heat supplied to a system is equal to the work done by the system plus its change in internal energy. The expansion of a gas may be an adiabatic or an isothermal process. Adiabatic processes are those in which no heat is added to or removed from a substance. Isothermal processes take place at constant temperatures. The specific heat of a gas varies with temperature.

The efficiency of an ideal heat engine increases as the difference between the temperatures of the heat source and heat sink increases. The second law of thermodynamics states that it is impossible to make an engine that will extract heat from a heat source at a single temperature and convert this heat completely to work.

The amount of energy that cannot be converted into mechanical work is related to entropy. The change in entropy of a system is equal to the amount of heat added to the system divided by its Kelvin temperature. The entropy of a system always tends to increase.

A heat engine converts heat energy into mechanical energy. Steam engines and steam turbines are external combustion engines. Gasoline engines, diesel engines, gas turbines, jet engines, and rockets are internal combustion engines. Unlike other engines, a rocket contains an oxidizer as well as fuel. The performance of rocket propellants is measured in terms of specific impulse. A heat pump transfers heat from a low-temperature source to a high-temperature sink. A heat pump is more efficient than most other heating systems.

VOCABULARY

adiabatic process
aeolipile
diesel engine
entropy
external combustion
   engine
first law of
   thermodynamics
gas turbine
gasoline engine
heat pump

heat sink
heat source
internal combustion
   engine
internal energy
isothermal process
jet engine
mechanical equivalent of
   heat
pascal
ramjet

rocket
second law of
   thermodynamics
specific impulse
steam engine
steam turbine
thermodynamics
turbocharging
turbojet
Wankel engine

WAVES

Max Planck originated the quantum theory, which states that electromagnetic waves consist of individual packets of energy. The theories of Planck and of his contemporary, Albert Einstein, are among the most revolutionary ideas in twentieth-century physics. Planck received the Nobel Prize in 1918.

In this chapter you will gain an understanding of:

▷ waves as an energy-transfer system
▷ the properties of an elastic medium through which mechanical waves are propagated
▷ the distinction between longitudinal and transverse wave systems
▷ periodic wave behavior
▷ characteristics and properties common to all waves
▷ the application of the superposition principle to wave trains traversing the same medium simultaneously
▷ the principles of wave interference
▷ the production of standing waves

THE NATURE OF WAVES

10.1 Energy Transfer   The locations of energy sources are often different from the locations of energy requirements. Clearly, some mechanism must be provided for the transport of energy from one place to another. One way of transporting energy is by the movement of materials or objects. Winds and projectiles in flight are well-known examples. When a baseball strikes a window, the glass is shattered by energy transferred from the ball during impact. Thermal convection is a familiar process for transferring heat energy from one location to another by the gross movement of quantities of heated gas or liquid between these locations.

A more interesting but more complicated way of transporting energy involves *waves*. Several natural phenomena that exemplify waves can be recognized. These phenomena are studied collectively because of an important simplifying fact: the ideas and language used to describe waves are the same, regardless of the kinds of waves involved. The basic concept in the use of the term *wave* is that the wave involves some quantity or disturbance that *changes in magnitude with respect to time at a given location* and *changes in magnitude from place to place at a given time*. We shall recognize that some wave disturbances occur only in material media and others are not restricted in this way. In general, *a **wave** is a disturbance that propagates through a medium or space.* All kinds of waves are characterized by their transport of energy without the bulk transport of matter.

A stone, or a single drop of water, falling into a quiet pond produces the familiar wave pattern on the surface of the water shown in Figure 10-1. A sound is heard because a wave travels from the source through the intervening atmosphere. The energy released by a great explosion can shatter windows far from its source because a wave of compression moves out from the source in all directions. The "shock wave" of a sonic boom can have similar destructive effects. *These waves are disturbances that move through a material medium.*

(A)

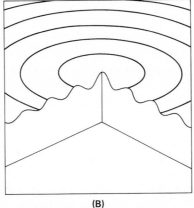

(B)

Figure 10-1. (A) A periodic circular wave train is generated by a drop of water falling on the surface of a quiet pool of water. (B) A cutaway diagram illustrates the succession of crests and troughs that characterize the surface wave train as it expands outward from the point of disturbance.

We are able to explain some properties of light by means of waves. Physicists have demonstrated that light waves, radio waves, infrared and ultraviolet waves, X rays, and gamma rays are fundamentally similar. They are *electromagnetic waves.* Although they can travel through matter, their transmission from one place to another does not require a material medium, that is, a medium composed of matter. The nature of electromagnetic waves will be discussed in detail in later chapters.

For the present (through Section 10.5) we shall be concerned with waves that require a *material medium* for their propagation. These waves are called *mechanical waves.* If the motion responsible for the wave disturbance is *periodic,* a periodic *continuous wave,* or *wave train,* is produced. Such wave motion is related to harmonic motion. Harmonic motion is described in Section 5.10 in terms of a single *object* vibrating about its equilibrium position. Here, we must consider many *particles* vibrating about their respective equilibrium positions as the wave train travels

Figure 10-2. Shock waves produced at supersonic speeds. This photograph of a Shuttle model in a wind-tunnel test shows flow lines indicating changes in air density.

through the medium. It is important to understand the behavior of waves because the language and ideas of wave motion are needed to correctly describe the motions of very small particles of matter.

As particles at some distance away from a source of vibrational energy are made to vibrate, their vibrations show that they have acquired energy. This energy has been transmitted to particles far from the energy source by the wave disturbance. Thus it is evident that waves provide a mechanism by which energy is transmitted from one location to another *without the physical transfer of matter between these locations.*

10.2 Mechanical Waves A **mechanical wave** *is a disturbance in the equilibrium positions of matter, the magnitude of which is dependent on location and on time.* To generate mechanical waves, a *source of energy* is required to cause a disturbance and an *elastic medium* is required to transmit the disturbance. An elastic medium behaves as if it were an array of particles connected by springs with each particle having an equilibrium position. A simple model of such a medium is shown in Figure 10-3.

Elastic forces obeying Hooke's law have the character that gives rise to simple harmonic motion.

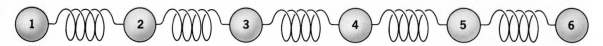

Figure 10-3. An elastic medium behaves as if it were an array of particles connected by springs, with each particle occupying an equilibrium position.

If particle **1** is displaced from its equilibrium position by being pulled away from particle **2**, it is immediately subjected to a force from particle **2** that attempts to restore particle **1** to its original position. At the same time, particle **1** exerts an equal but opposite force on particle **2** that attempts to displace it from its equilibrium position. Similar events occur but in opposite directions if particle **1** is displaced from its equilibrium position by being pushed toward particle **2**.

Suppose particle **1** is displaced by an energy source. Then particle **1** exerts a force that displaces particle **2**. Particle **2**, in turn being displaced from its equilibrium position, exerts a force on particle **3**, which is in turn displaced. In this way, the displacement travels along from particle to particle. Because the particles have inertia, the displacements do not all occur at the same time but successively as the disturbance affects particles farther and farther from the source. The energy initially imparted to particle **1** by the energy source is transmitted from particle to particle in the medium without motion of the medium as a whole.

10.3 Transverse Waves   A long spiral spring stretched between two rigid supports can represent an elastic medium, as shown in Figure 10-4(A). A portion of the spring is displaced at point **2** to form a *crest*, or upward displacement, as in Figure 10-4(B). To force the spring into this shape, point **2** must be pulled up while points **1** and **3** are held in place. When the spring is released at points **2** and **3** simultaneously, point **2** accelerates downward and point **3** accelerates upward. The crest moves toward the right as in Figure 10-4(C). Similarly, as the spring in the region of point **3** moves downward, the spring in the region of point **4** is displaced upward. In this manner the crest travels along the spring as shown in Figure 10-4(D) and (E). A *trough*, or downward displacement, formed at **2** travels along the spring in a similar fashion.

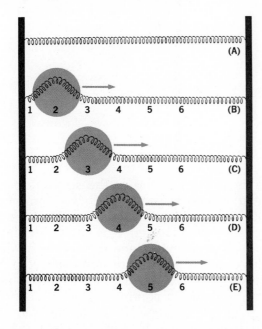

Figure 10-4. A transverse pulse traveling along an elastic medium.

A single nonrepeated disturbance such as a single crest or a single trough is called a single-wave pulse or, simply, a *pulse*. The displacement of the particles of the medium (the coiled spring) caused by the pulse is *perpendicular* to the direction in which the pulse travels. Such a pulse is said to be *transverse*. It is convenient to think of the crest as a *positive* pulse and the trough as a *negative* pulse.

When a periodic succession of positive and negative pulses is applied to the coiled spring, a series of crests and troughs, called a *continuous wave*, or *wave train*, travels through the medium. Again, the displacement of the particles of the medium is perpendicular to the direction the

If a pulse is produced in the middle of the spring, the disturbance will move in both directions.

wave train travels. The wave disturbance is *transverse*. A **transverse wave** *is one in which the displacement of particles of the medium is perpendicular to the direction of propagation of the wave.*

10.4 Longitudinal Waves Suppose a similar spiral spring is used, but instead of pulling the spring out of line, several coils are pinched closer together at one end, as in Figure 10-5(A). Such a distortion is called a *compression*. When these compressed coils are released, they attempt to spread out to their equilibrium positions. Thus, they compress the coils immediately to the right and the compression moves toward the right. See Figure 10-5(B)–(D).

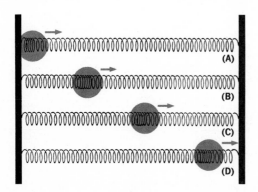

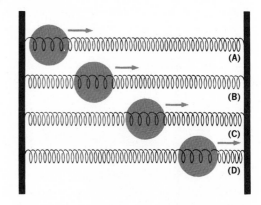

Figure 10-5. (Left) A compression pulse traveling along an elastic medium.

Figure 10-6. (Right) A rarefaction pulse traveling along an elastic medium.

Transverse displacements are across the path of propagation. Longitudinal displacements are along the path of propagation.

If the coils at the left end of the spring are stretched apart instead of compressed, a *rarefaction* is formed. When released, the rarefaction travels along the spring just as the compression did. This effect is shown in Figure 10-6.

These pulses propagate along the spring by displacing the particles of the spring in directions *parallel* to the direction the pulses are traveling. They are examples of *longitudinal pulses*. Similarly, a continuous wave disturbance of this type gives rise to *longitudinal wave motion*. Energy is transferred from particle to particle along the medium without motion of the medium as a whole. *A **longitudinal wave** is one in which the displacement of particles of the medium is parallel to the direction of propagation of the wave.*

10.5 Periodic Waves We have examined the effect of a single, nonrecurring wave disturbance on one end of a long, rigidly mounted spring. Now let us consider what happens if similar disturbances are repeated periodically.

Suppose we attach the left end of a long spring to a mass suspended by a second spring as in Figure 10-7. Assume that the mass can move up and down without friction between its vertical guides. If the mass is pulled down

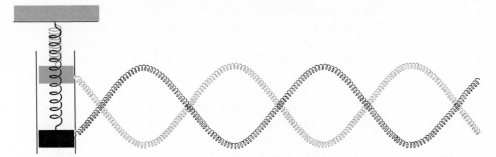

slightly and then released, it will vibrate within the guides with simple harmonic motion. Such a motion is *periodic*. That is, the mass repeats its motion once every certain time interval *T*, called the *period* of vibration. Since the vibrating mass is attached to the end of the long spring, it acts as a source of periodic disturbances. A *transverse wave train* is generated, which moves to the right along the spring. This wave is shown in Figure 10-7 at the instant the vibrating mass is at the upper limit of its excursion and again at the instant it is at the lower limit of its excursion.

This periodic transverse wave carries energy away from the vibrating mass. Unless energy is supplied to it, the mass loses amplitude and comes to rest. In order for a source to generate a continuous wave of uniform amplitude, energy must be supplied to the source at the same rate as the source transmits energy to the medium. Then

Figure 10-7. The oscillating mass generates a periodic transverse wave. The waves in color and the black waves represent a time difference of one-half period (*T*/2).

All simple harmonic motions are periodic; not all periodic motions are harmonic. For example, doing push-ups at a constant repetition rate is a periodic motion, but it is not simple harmonic motion.

Figure 10-8. (A) An apparatus for generating longitudinal waves. (B) A periodic longitudinal wave in a section of the spring.

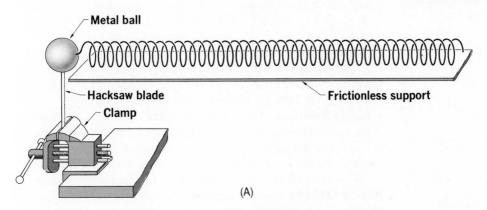

(A)

Compression   Rarefaction   Compression   Rarefaction   Compression   Rarefaction

(B)

the successive wave disturbances will be identical. When the periodic wave is a simple harmonic wave, it gives each particle of the medium a simple harmonic motion.

A periodic longitudinal wave can be generated by the apparatus shown in Figure 10-8(A) on the preceding page. The long spring is attached to a metal ball fastened to one end of a hacksaw blade. The other end of the blade is rigidly clamped. If the metal ball is displaced slightly to one side and released, it vibrates with simple harmonic motion. This motion produces a series of periodic compressions and rarefactions in the spring, and therefore a periodic longitudinal wave. This is illustrated in Figure 10-8(B). In order for the successive wave disturbances to be identical, energy must be provided to the source at the same rate it is transmitted by the wave.

10.6 Characteristics of Waves The *characteristics* and *properties* of waves discussed in the remaining sections of this chapter are descriptive of *all* waves. An elastic medium is required for the propagation of mechanical waves, but not for that of electromagnetic waves. The propagation of electromagnetic waves in free space is dramatically illustrated by the transmission of information back to earth by space probes from the outer regions of our solar system. For the transmission of electromagnetic waves, periodic oscillations of accelerating electrons in a conductor (such as an antenna) give rise to periodically reversing electric and magnetic fields that propagate in free space away from their source at the speed of light.

A graph of an electromagnetic wave is shown in Figure 12-6 of Chapter 12. The broad frequency range of the electromagnetic spectrum is evident in Figure 12-8 of Chapter 12. Observe the locations of familiar regions of the electromagnetic spectrum such as X rays, visible light, and radio waves. The *infrared* region includes ordinary *heat waves*, which account for the transfer of heat energy by *radiation*. This wave transfer of heat is evident when one stands directly in front of a log-burning fireplace and then moves aside to a position beyond the line of sight of the flame.

All waves have several common characteristics in addition to that of transferring energy without the transport of matter. As a periodic mechanical wave propagates through a medium, the particles of the medium vibrate about their equilibrium positions in identical fashion. Generally, however, the particles are in corresponding positions of their vibratory motion at different times. The *position* and *motion* of a particle indicate the *phase* of the wave. Particles that have the same displacement and are

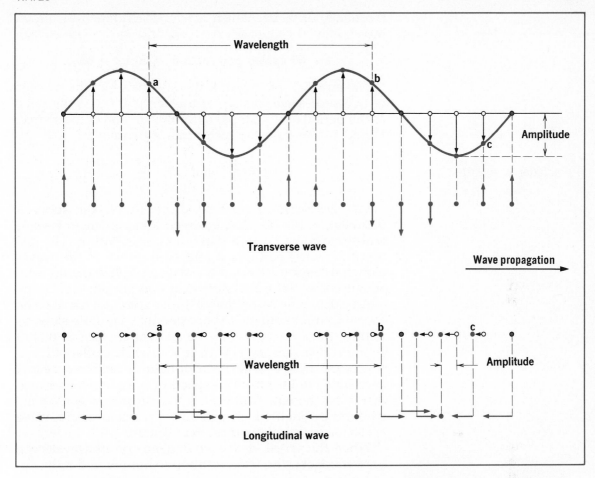

Transverse wave

Wave propagation ⟶

Longitudinal wave

moving in the same direction are said to be *in phase*. Particles **a** and **b** in both waves illustrated in Figure 10-9 are in phase. Those with opposite displacements and moving in opposite directions, such as particles **b** and **c** in both waves, are in *opposite phase*, or 180° out of phase.

The *frequency*, f, of a periodic wave is the number of crests (or troughs) passing a given point in unit time. It is the same as the frequency of the simple harmonic motion of the source and is conventionally expressed in terms of the vibrating frequency of the source. Thus the frequency of a motion is the number of vibrations, oscillations, or cycles per unit of time. The SI unit of frequency is the *hertz* (hz). One hz is equivalent to the expression "one cycle per second." Note that "cycle" is an event, not a unit of measure. Therefore it is not a part of the dimensional structure of the hertz. The dimension of the hertz is simply s^{-1}. For example, a wave generated at 60 cycles per second has a

Figure 10-9. Characteristics of transverse and longitudinal waves. The black open circles show the equilibrium positions of the particles of the media. The dots in color show their displaced positions. The black arrows indicate the displacements of the particles from their equilibrium positions. The arrows in color are the velocity vectors for the particles above them.

frequency of 60 hz, which is expressed dimensionally as 60/s.

$$f = \textbf{60 cycles per second} = \textbf{60 hz} = \textbf{60/s}$$

The *period*, T, of a wave is the time between the passage of two successive crests past a given point. It is the same as the period of the simple harmonic motion of the source. Period is therefore the reciprocal of frequency.

If $f = 100/s$, $T = 0.01$ s.

$$f = \frac{1}{T} \quad \text{and} \quad T = \frac{1}{f}$$

The *wavelength*, represented by the Greek letter λ (lambda), is the distance between any particle in a wave and the nearest particle that is in phase with it. The distance between particles **a** and **b** in either of the waves shown in Figure 10-9 is one wavelength. It is the distance advanced by the wave motion in one period, T.

An advancing wave has a finite *speed*, v, for a given transmitting medium. Wave speed may be quite slow, as that of water waves. It may be moderately fast, as that of sound waves that travel with speeds on the order of 10^2 to 10^3 m/s. It may be the speed of light or radio waves in a vacuum, 3×10^8 m/s. The speed of a wave depends primarily on the nature of the wave disturbance and on the medium through which it passes. In certain media, wave speed may also depend on wavelength.

When the speed of a wave depends on its wavelength (or frequency), the transmitting medium is said to be *dispersive*. A glass prism disperses (separates) a beam of white light into a color spectrum as the light of different wavelengths emerges along diverging paths. Glass is a dispersive medium for light waves. We can observe rainbows because water droplets in the atmosphere disperse sunlight in the same way. In fact, all electromagnetic waves propagating through matter, but not through a vacuum, show dispersion.

Water is highly dispersive for surface waves, and the wave speed can change significantly with wavelength. Ripples produced by a very gentle wind or vibrations of high frequency have very short wavelengths. The wave speed is governed by the surface tension of the liquid, which is a measure of the tendency of the surface to resist stretching. In this case, the *shorter* the wavelength, the higher is the wave speed.

A strong steady wind may produce surface waves of very long wavelength. Here, the wave speed is governed by gravity, which provides a restoring force. Such waves are called *gravity waves*. In this case, the *longer* the wavelength, the higher is the wave speed.

Since a wave travels the distance λ in the time T for a complete vibration of the source, its speed v is given by the expression

$$v = \frac{\lambda}{T}$$

and because

$$f = \frac{1}{T}$$

then

$$v = f\lambda$$

The wave equation $v = f\lambda$ is true for all wave systems.

This equation is true for all periodic waves, whether transverse or longitudinal, no matter what the medium is.

The maximum displacement of the vibrating particles of the medium from their equilibrium positions is called the *amplitude* of the wave. It is related to the energy flow in the system.

10.7 Amplitude and Energy   In all forms of traveling waves, energy is transmitted from one point to another. By expending energy, a vibrating source can cause a harmonic disturbance in a medium. The energy expended per unit of time depends on the amplitude, the frequency, and the mass of the particles of the medium at the source. Simple harmonic motion (vibrational motion) of a particle of the medium is characterized by an amount of energy, part kinetic and part potential, that the particle has at any instant. The disturbance is transferred through the medium because of the influence that a particle has on other particles adjacent to it. Thus the wave carries energy away from the source.

Assuming no losses in the system, the energy transported by the advancing wave during a given time is the same as the energy expended by the source during that time. The energy transported, or the energy expended, per unit of time is the *power* transmitted by the wave.

If the wave amplitude is doubled, the vibrational energy is increased fourfold. Doubling the vibrating frequency has the same effect. *The rate of transfer of energy, or the power transmitted by a wave system, is proportional to the square of the wave amplitude and also to the square of the wave frequency.*

Surface waves on water that emanate from a point source undergo a decrease in amplitude as they move away from their source. Each advancing wave crest is an expanding circle. The energy contained in each crest is a fixed amount, and the crest is expanding. The energy per unit length of crest must decrease. Thus the amplitude of

Figure 10-10. Energy is transported by an ocean wave.

the wave must diminish. This effect is greater for sound waves as the advancing crests are expanding spheres.

We know from experience that the amplitude of a vibrating pendulum or spring gradually decreases with time and the vibrations eventually stop. Frictional and resistive effects that oppose the motion slowly remove energy from the vibrating system. Similarly, in a wave system energy is dissipated and the wave amplitude gradually diminishes. *The reduction in amplitude of a wave due to the dissipation of wave energy as it travels away from the source is called damping.* Damping effects may be quite small over relatively short distances.

Questions

GROUP A

1. In what ways may energy be transferred?
2. Distinguish between mechanical waves and electromagnetic waves.
3. What are three requirements for the production of mechanical waves?
4. (a) What is a disturbance? (b) What constitutes the disturbance when a wave moves across the surface of a pond? (c) What constitutes the disturbance when a longitudinal wave travels through the air?
5. How does an elastic material behave as a wave medium?
6. (a) Distinguish between transverse and longitudinal waves. (b) Give an example of each.
7. Define (a) pulse; (b) crest; (c) trough; (d) compression; (e) rarefaction.
8. For wave motion, define (a) speed; (b) phase; (c) frequency; (d) period; (e) wavelength; (f) amplitude.
9. What does a wave source supply to the medium through which the wave passes?

10. A pebble is dropped into a quiet pond of water. Neglecting damping losses, how does the amplitude of the resulting wave disturbance vary with the distance from the source?
11. Neglecting damping losses, would you expect the amplitude of an advancing surface wave on a pond to diminish with distance from its point source at the same rate as an advancing sound wave diminishes with distance from its point source in air? Justify your conclusion qualitatively.
12. A charge of dynamite is exploded at a construction site overlooking a lake. (a) A grass fire is ignited nearby. (b) A window in a cottage nearby is shattered. (c) A flock of birds fly frantically out of the trees in the surrounding area. (d) A smoke cloud rises from the site of the explosion. (e) A huge boulder falls into the lake causing (f) a small sailboat that is passing nearby to capsize. Identify the energy-transport mechanism responsible for each event.

Problems

GROUP A

1. What is the speed of a periodic wave disturbance, frequency 2.5 hz and wavelength 0.60 m?

2. Calculate the wavelength of water waves that have frequency 0.50 hz and speed 4.0 m/s.
3. The speed of transverse waves in a string is 15 m/s. If a source produces a disturbance, frequency 5.0 hz, what is the wavelength of the wave produced?

4. A periodic transverse wave, frequency 10.0 hz, travels along a string. The distance between a crest and either adjacent trough is 1.50 m. What is (a) the wavelength, (b) the speed of the wave?

5. A periodic longitudinal wave, frequency 2̄0 hz, travels along a coil spring. If the distance between successive compressions is 0.40 m, calculate the speed of the wave.

6. What is the wavelength of a periodic longitudinal wave in a coil spring, frequency 8.0 hz and speed 2̄0 m/s?

7. A wave generator produces 16 pulses in 4.0 seconds. (a) What is its period? (b) What is its frequency?

8. One pulse is generated every 0.10 s in a tank of water, and the wavelength of the surface wave measures 3.3 cm. What is the propagation speed?

WAVE INTERACTIONS

10.8 Properties of Waves In any study of wave behavior, the various properties common to all kinds of waves should be recognized. These common properties are: *rectilinear* (straight line) *propagation, reflection, refraction, diffraction,* and *interference.* Being familiar with the properties of waves helps to determine whether an observed phenomenon involves a wave or not.

Surface waves on water are particularly useful in studying wave properties. A surface wave is easily generated, its speed is quite slow, and the transparency of the water to light allows the wave pattern to be projected on a screen. A continuous wave can be established on the water surface by allowing a probe to dip into the surface periodically. An apparatus for observing wave motion on the surface of water is shown in Figure 10-11. It is called a ripple tank.

Figure 10-11. (Left) The ripple tank provides a means of observing the properties of waves in the laboratory.

Figure 10-12. (Right) A functional diagram of a ripple tank.

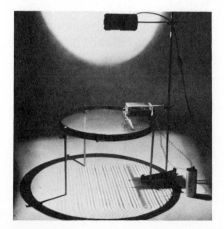

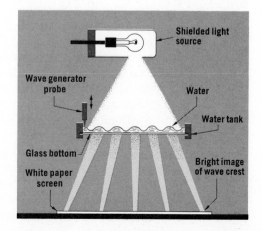

Shielded light source

Wave generator probe

Water

Water tank

Glass bottom

White paper screen

Bright image of wave crest

The convex surface of a wave crest serves to converge light.

The concave surface of a wave trough serves to diverge light.

The diagram in Figure 10-12 shows how images of surface wave crests can be projected on a screen. The ripple tank has a glass bottom that allows light to be projected down through the water and the glass onto a screen below the tank. The crests appear on the screen as bright regions, and troughs appear as dark regions.

10.9 Rectilinear Propagation   We can generate a train of straight waves in a ripple tank by placing a long, straight edge, like that of a ruler, along the surface of the water and causing it to vibrate up and down. A photograph of the image of such a continuous wave is shown in Figure 10-13. The wave crests are the parallel white lines. The wave troughs are the dark spaces separating the crests. The portions of water surface whose particles are all in the same phase of motion are called *wave fronts*. The paths of the two points **a** and **b** show that *the direction of propagation of the advancing straight wave is perpendicular to the wave front.*

Figure 10-13. (Left) Periodic straight wave train moving across a ripple tank. The wavelength is the distance between adjacent crests or adjacent troughs.

Figure 10-14. (Right) Periodic circular wave train generated by a point source in a ripple tank. The wave train moves radially away from the disturbance point in ever-widening circles.

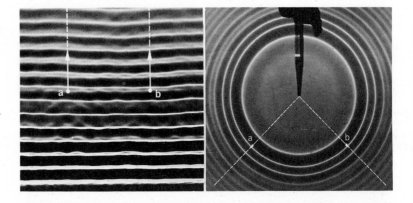

By changing from a straight to a pointed probe, such as the tip of a pencil, we can generate a train of circular waves. See Figure 10-14. Here the advancing wave crests are expanding circles; the wave fronts are moving out in all directions from the center of disturbance. The letters **a** and **b** locate two points on an expanding crest. Their paths show that *the directions of propagation of these two segments of the advancing circular wave lie along radial lines away from the center of disturbance.* This is true of all other points on any circular wave crest. The radial lines are perpendicular to the segments of the wave front through which they pass.

Light regions of a ripple-tank image relate to wave crests; dark regions relate to wave troughs.

From these experiments with the ripple tank we observe that *continuous waves traveling in a uniform medium propagate in straight lines perpendicular to the advancing wave fronts.* The uniform spacing between the wave crests suggests that the wave speed in the medium is constant. Considering the speed and direction of propagation of traveling waves, we can state that *the wave velocity at every point along the advancing wave front is perpendicular to the wave front.*

10.10 Reflection Familiar examples of reflection are the sound echoes that return from a distant canyon wall, the reflection of light from a mirror, and the reflection of waves from the edge of a pool. Our image as seen in a mirror appears to be reversed, right for left. The crest of a water wave is reflected from the pool side as a crest traveling in a different direction. The echo of a sound retains the character of the original sound.

The illusion of reversal occurs because of the left-right symmetry of the body.

About the only inference we can draw from these casual observations is that *a wave is turned back, or **reflected,** when it encounters a barrier that is the boundary of the medium in which the wave is traveling.* A close examination of the boundary conditions can reveal much more about the nature of reflection.

Suppose we generate a single straight pulse in a ripple tank that has a straight barrier placed across the tray parallel to the advancing wave front. When the pulse reaches the barrier, it is reflected back in the direction from which it came. No disturbance appears in the water behind the barrier. The paths of the incident pulse (approaching the barrier) and the reflected pulse lie perpendicular to the surface of the barrier.

The angle formed by the path of incidence and the perpendicular (normal) to the reflecting surface at the point of incidence is called the *angle of incidence, i.* The angle formed by the path of reflection and the normal is the *angle of reflection, r.* In this instance both the incident and reflected angles are 0°, and i is equal to r.

Certainly i and r equal 0° when the incident wave approaches the barrier along a line perpendicular to it. We may then ask whether the equality between i and r is coincidental in this situation, or whether it is characteristic of reflection in general. To investigate the relationship of i and r further, we need to change the position of the barrier so that it is no longer parallel to the wave fronts and then send more pulses against it. Figure 10-15 shows the reflection of a single pulse from a diagonal barrier.

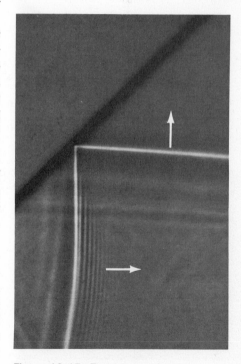

Figure 10-15. The image of a single straight pulse being reflected from a diagonal barrier in a ripple tank. The incident segment of the pulse is moving toward the top of the picture. The reflected segment is moving toward the right.

Reasonably good measurements can be made of the angles that the incident and reflected pulses make with the barrier by placing suitable markers on the screen parallel to the projected images of the pulse. We shall call these angles i' and r' respectively. Measurements of i' and r' for several positions consistently indicate that these angles

Figure 10-16. The geometry of reflection. A straight pulse is reflected from a straight barrier in a ripple tank. The angle of incidence, *i*, is shown to be equal to the angle of reflection, *r*.

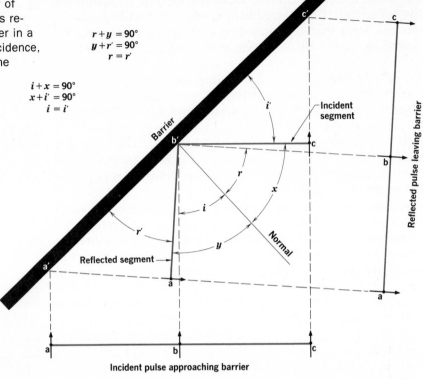

$$r + y = 90°$$
$$y + r' = 90°$$
$$r = r'$$

$$i + x = 90°$$
$$x + i' = 90°$$
$$i = i'$$

are equal. From Figure 10-16 it is evident that angles $i + x = 90°$ because the incident path **bb'** is perpendicular to the incident segment **b'c** of the pulse. Also, angles $x + i' = 90°$ because they form the angle between the barrier and its normal. Thus i and i' are equal angles. Similar reasoning shows that r and r' are also equal angles. From measurement, i' and r' are equal. Thus the angle of incidence, i, and the angle of reflection, r, are equal.

$$i = r$$

We can state this relationship as a ***law of reflection.*** *When a wave disturbance is reflected at the boundary of a transmitting medium, the angle of incidence, i, is equal to the angle of reflection, r.* A photograph of a periodic straight wave train reflected from a diagonally placed barrier is shown in Figure 10-17.

The reflection of a circular wave from a straight barrier is shown in Figure 10-18. Each segment of the expanding wave front reflected from the barrier surface rebounds according to the principle just stated. Observe that the reflected portions of the wave crests are arcs of circles. The apparent center of these reflected wave crests is as far behind the reflecting surface as the real center of disturbance is in front of it.

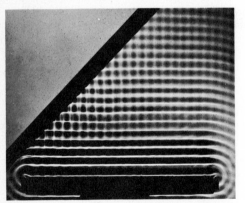

Figure 10-17. Reflection of a periodic straight wave train from a diagonal barrier. The incoming waves are generated by the black bar at the bottom and are moving toward the top. The left segment of each wave has already been reflected toward the right.

Reflection may be partial or complete depending on the nature of the reflecting boundary. If there is no boundary mechanism for extracting energy from the wave, all the energy incident with the wave is reflected back with it.

When a surface wave in the ripple tank encounters the straight barrier, which is a rigid vertical bulkhead, the vertical transverse component of the wave is unrestrained. The wave crest is reflected *as a crest,* and the trough is reflected *as a trough.* The surface wave is reflected without a change in phase. *A boundary that allows unrestrained displacement of the particles of a medium reflects waves with no change in the direction of the displacement, which is equivalent to no change in phase.* The bulkhead acts as a *free-end* or *open-end* termination to the wave medium.

Suppose we send a transverse pulse traveling along a stretched string of which the far end is free. Having such a string presents no problem in a "thought" experiment. In practice, the free-ended termination might be approximated as shown in Figure 10-19. To the extent that the termination is frictionless, the end of the string is free to move in the plane of the pulse.

After arriving at the free end of the string, the pulse is reflected as shown in Figure 10-19. Observe that it has not been inverted. Displacement of the particles of the string is in the same direction as that for the incident pulse. During the time interval that the pulse is arriving at the free end of the string and the reflected pulse is leaving it, the combined effect of the two pulses causes the free end to experience twice the displacement as from either pulse alone. *Reflection at the free-end termination of a medium occurs without a change in phase.*

Now suppose we clamp the far end of the string in a fixed position and send a transverse pulse toward this termination. See Figure 10-20. When the pulse arrives at the fixed end, it applies an upward force to the clamp. An equal but opposite reaction force is applied to the string generating a reflected pulse that is inverted with respect to the incident pulse. *Reflection at the fixed termination of a medium occurs with a reversal of the direction of the displacement, or a phase shift of 180°.* In general, we may conclude that *a boundary that restrains the displacement of the particles of a medium reflects waves inverted in phase.*

Some examples of reflecting terminations for different kinds of wave disturbances are given in Table 10-1. The first three systems listed summarize the phase relations discussed in this section.

Generally, something less than total reflection occurs at the termination of a transmitting medium. The "free" end

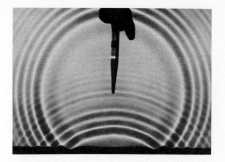

Figure 10-18. A periodic circular wave train generated at the center of the photograph is reflected from a straight barrier located at the bottom of the photograph.

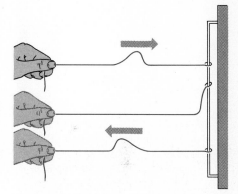

Figure 10-19. Reflection of a pulse at the "free" end of a taut string. The string is shown terminated in a "frictionless" ring that allows the end segment unrestrained displacement in the plane of the pulse.

Table 10-1
REFLECTION BEHAVIOR

Type of wave	Example	Termination	
		For in-phase reflection	For out-of-phase reflection
transverse (mechanical)	stretched string	free-ended	fixed-ended
longitudinal (sound)	air motion in organ pipe	open pipe	closed pipe
liquid surface (vertical component)	water	solid bulkhead	(no simple analog)
optical (electromagnetic)	light	(no simple analog)	mirror
electric	transmission line	short circuit	open circuit

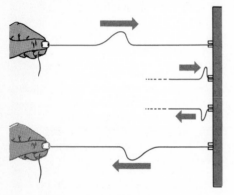

Figure 10-20. Reflection of a pulse at the clamped end of a taut string. The string is terminated in a fixed position and displacement of the end segment is restrained.

of a vibrating string, as well as the fixed end, must be supported in some manner, and some of the wave energy is transferred to the medium providing this support.

The speed of the wave disturbance depends on the properties of the transmitting medium. Waves often traverse the boundary between two media, the second of which can be thought of as the *terminating medium.* Thus waves in the string and in the terminating medium travel at different speeds. If this difference in speed is very great, the portion of wave energy reflected at the boundary will be very large. If the two wave speeds are very similar, little reflection will occur at the boundary. An intermediate effect is seen in Figure 10-22, where partial reflection is evident at the boundary of the deep and shallow water. Light is partially transmitted and partially reflected when it passes from air into glass because of the difference in the transmission speeds in the two media.

10.11 Impedance If we were to apply the same wave-producing force to a light string and to a heavy rope, two entirely different *displacement velocities* would result. The string would experience a very large displacement velocity compared with that of the rope. We could say that the ratio of the *wave-producing force* to the *displacement velocity* is quite small for the string and quite large for the rope. In the study of waves, this *ratio of the applied wave-producing force to the resulting displacement velocity is called the **impedance** of the medium.*

$$\text{Impedance} = \frac{\textbf{wave-producing force}}{\textbf{resulting velocity within the medium}}$$

By this definition, the string is said to have a relatively low impedance and the rope a relatively high impedance. The impedance of a wave medium denotes the ease or difficulty with which a wave disturbance of given amplitude can be launched in it.

The concept of impedance is quite general and can be applied to wave media of all kinds. For a stretched string, the wave-producing force is the transverse oscillatory force applied to the end of the string. The displacement velocity is the oscillating transverse velocity imparted to the particles of the first portion of the string. In an organ pipe, the ratio of the alternating compressional force applied to the molecules of the air to the longitudinal oscillating velocity acquired by these molecules defines the acoustic impedance of the air column.

"Impedance" is a common concept in alternating-current electricity and electronics.

In cases of *total* in-phase reflection discussed in Section 10.10, the impedance of the terminating medium may be considered to be zero. For *total* out-of-phase reflection, the impedance of the terminating medium is infinite. In systems having infinite or zero impedance terminations, reflection is complete and no wave energy is transferred to the terminating medium.

If the terminating impedance exactly matches that of the transmitting medium, the wave energy is completely transferred to the terminating medium and there is no reflection. What about systems that fall in between the extreme impedance mismatch, zero or infinite impedance terminations, and the perfect impedance match? In these cases some energy is transferred to the terminating medium and some is reflected back through the medium of the incident wave.

In practical energy-transfer systems, an abrupt impedance discontinuity between two wave propagation media causes unwanted wave reflection that wastes energy. Numerous devices and techniques are used to smooth over an impedance mismatch and reduce or prevent reflections. They are called *impedance transformers.*

An impedance transformer is an energy-transfer device.

A cheerleader's megaphone is a simple approximation of an acoustic impedance transformer. It effectively matches the impedance of the air column of the throat and mouth to the free air. The coating on an optical lens acts as an impedance transformer. It reduces reflection at the air-glass boundary and causes a greater portion of the total incident light energy to be transmitted through the lens. There are many applications of impedance transformers in electric and electronic systems. In general, whenever a device intended to receive wave energy efficiently cannot be designed to have the same impedance as the transmitting medium, it must be coupled to the medium by means of an impedance transformer.

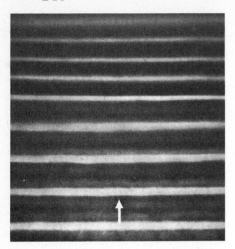

Figure 10-21. The passage of a surface wave from deep to shallow water. The shallow water is at the top of the picture.

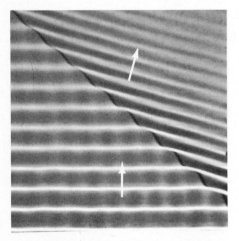

Figure 10-22. Refraction of a surface wave at the boundary of deep and shallow water. The shallow water is above the boundary in the picture. Observe the slight reflection of the incident wave to the left at the boundary.

10.12 Refraction The properties of the medium through which a certain wave disturbance moves determine the propagation speed of that disturbance. It is not surprising therefore to find that traveling waves passing from one medium into another experience a change in speed *at the interface (boundary) of the two media.*

From the wave equation, $v = f\lambda$, we see that the wavelength, λ, for a wave disturbance of a given frequency, f, is a function of its speed, v, in the medium through which it is propagating. If the speed decreases when the wave enters a second medium, the wavelength is shortened proportionately. If the speed increases, the wavelength is lengthened proportionately.

A useful aspect of water waves is the relation of the speed of surface waves to the depth of the water. Such wave disturbances travel faster in deep water than in shallow water. For a given wave frequency, the wavelength in deep water is longer than it is in shallow water. Therefore water of two different depths acts just like two different transmitting media for wave propagation.

This difference is easily observed in the ripple tank by arranging the tray so that it has two different depths of water. Then surface waves are generated that travel across the boundary of the two regions. In Figure 10-21 the deep water representing the medium of higher propagating speed is in the lower portion of the photograph. The wave moves toward the shallow water in the upper portion. The boundary between the deep and shallow water is parallel to the advancing wave front. Thus each incident wave crest approaches the shallow region along the normal to the boundary.

Observe that the change in wavelength is abrupt. It occurs simultaneously over the entire wave front at the boundary of the deep and shallow water. All segments of the advancing wave front change speed at the same time, and the wave continues to propagate in its original direction.

Suppose we adjust the deep and shallow regions of the tray so that the boundary is no longer parallel to the advancing straight wave but cuts diagonally across its path. This arrangement is shown in Figure 10-22. Each advancing wave crest now approaches the boundary obliquely. Adjacent segments of the wave front pass from deep to shallow water successively rather than simultaneously. At the boundary, the direction of the wave front changes; the advancing wave has undergone *refraction.* **Refraction** is the *bending of the path of a wave disturbance as it passes obliquely from one medium into another of different propagation speed.*

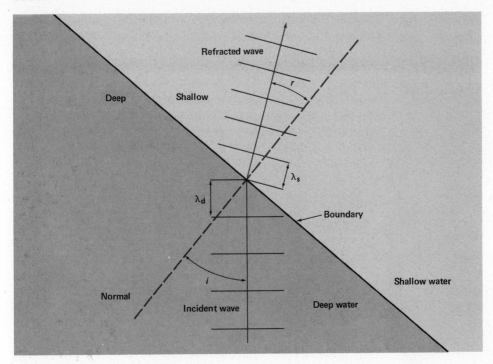

The refraction of surface waves on water is pictured in Figure 10-23. As the wave passes into the shallow water where its speed is less and its wavelength is shorter, it is refracted *toward* the normal drawn to the deep-shallow boundary. The angle of refraction, r, is smaller than the angle of incidence, i. (The angle r is the angle between the path of the refracted wave and the normal.)

Had we generated the wave in the shallow region and directed it obliquely toward the boundary, it would have been refracted *away from* the normal on entering the deep water. Angle r would have been larger than i. The change in direction and the change in speed occur simultaneously. How are these two changes related? We shall investigate this question quantitatively later when we study optical refraction.

Figure 10-23. The geometry of refraction. Water of two depths in the ripple tank represents two media of different transmission speeds. Compare this diagram with Figure 10-22.

10.13 Diffraction Sounds can be heard even though they may originate around the corner of a building. However, the source of the sound disturbance cannot be seen around the corner. Rectilinear propagation appears to hold in this situation for light but not for sound. The building is an obstruction that prevents the straight-line transmission of light from the source of the sound disturbance to the observer.

Perhaps, where sound waves are concerned, the corner of the building creates a discontinuity in the transmitting

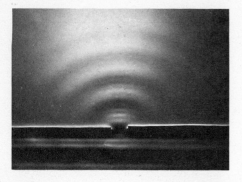

Figure 10-24. The diffraction of a periodic straight wave as it passes through a small aperture. Observe the decrease in the diffraction effect as the wavelength of the disturbance sent against the barrier is shortened.

An open doorway diffracts sound, but it does not diffract light.

medium that causes the waves to *spread*. Even if we find support for this idea, the question of the different behavior of light remains to be answered.

We will again use the ripple tank to observe wave behavior, this time with obstructions inserted into the transmitting medium. This may be done by placing two straight barriers across the tray on a line parallel with the straight-wave generator. An *aperture*, or opening, is left between them approximately equal to the wavelength of the wave to be used. When a periodic straight wave is sent out, the wave pattern beyond the barrier opening appears as shown in Figure 10-24. As a segment of each wave crest passes through the aperture, it clearly spreads into the region beyond the barriers. *This spreading of a wave disturbance beyond the edge of a barrier is called* **diffraction.**

Sound waves spread around the edges of doorways in the same way that water waves spread around the edges of obstructions. Audible sounds have wavelengths in air ranging from a few centimeters to several meters. The doorway of a room or building has dimensions within this range. This suggests, but does not prove, that a discontinuity such as an aperture in the path of an advancing wave will diffract the wave if its dimensions are comparable with the wavelength of the oncoming wave.

We produced the diffraction pattern in the ripple tank with a barrier aperture approximately the same width as the wavelength of the surface wave used. Suppose we shorten the wavelength by increasing stepwise the frequency of the wave generator. We would observe a series of patterns with waves of decreasing wavelengths.

The spreading of the wave at the edges of the aperture diminishes as the wavelength of the wave sent against them is shortened. When the wavelength is a small fraction of the width of the opening between the barriers, the wave segments passing through show little tendency to spread into the shadow regions beyond the barriers. These observations suggest that if the wavelength is much smaller than the width of the aperture, no diffraction occurs. The part of the straight wave that passes through the opening will continue in straight-line propagation.

How are these observations related to the fact that light is not diffracted at the edges of a doorway as is sound? The speed of light in air is high, being on the order of 3×10^8 m/s. We have not yet dealt with light frequencies, but if they were extremely high, light wavelengths (v/f) would be very short. Then it would not be surprising to find that light is diffracted only by apertures having very small dimensions.

Indeed, the wavelength of visible light ranges from approximately 7.5×10^{-7} m down to about 4×10^{-7} m. This range of wavelengths suggests that the diffraction of light occurs in the realm of extremely small dimensions. Diffraction phenomena associated with light waves will be presented in Chapter 15.

10.14 The Superposition Principle   We have described the passage of a single pulse and a continuous wave through a medium. It is possible for two or more wave disturbances to move through a medium at the same time. Sound waves from all the musical instruments of an orchestra move simultaneously through the air to our ears. Electromagnetic waves from many different radio and television stations travel simultaneously through regions of space to our receiving antennas. Yet, we still can listen to the sound of a particular musical instrument or select a particular radio or television station. These facts suggest that each wave system proceeds independently along its pathway as if the other wave systems were not present.

Let us consider the effects of two periodic transverse waves traveling along a taut string in the same direction. The displacements they produce *at a particular instant* are shown in Figure 10-25. These two waves have different amplitudes and frequencies. The displacements y_1 produced by one wave are represented by the black solid curve. The displacements y_2 produced by the other wave are represented by the black dashed curve. The resultant displacements **Y** all along the string at this instant are represented by the colored curve. These resultant displacements are determined by algebraically adding the displacements y_1 and y_2 for every point along the string.

In effect, the displacement of any particle of the medium by one wave at any instant is superimposed on the displacement of that particle by the other wave at that instant. The action of each wave on a particle is independent

Figure 10-25. An instantaneous view of the superposition of two periodic transverse waves of different amplitudes and frequencies. The component waves y_1 and y_2 are shown in black; the resultant wave Y is in color.

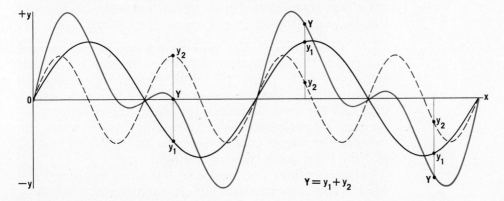

of the action of the other, and the particle displacement is the resultant of both wave actions. This phenomenon is known as *superposition*. The two waves are said to *superpose*.

The superposition principle: *When two or more waves travel simultaneously through the same medium, (1) each wave proceeds independently as though no other waves were present and (2) the resultant displacement of any particle at a given time is the vector sum of the displacements that the individual waves acting alone would give it.*

Providing the displacements are small, this principle holds for light and all other electromagnetic waves as well as for sound waves and waves on a string or on a liquid surface. For example, this principle does not hold for "shock waves" produced by violent explosions.

The component waves y_1 and y_2 in Figure 10-25 are simple sine waves. Observe that the frequency f_2 of wave y_2 is twice the frequency f_1 of wave y_1. That is,

$$f_2 = 2f_1$$

Either periodic wave alone would cause each particle of the string to undergo simple harmonic motion about its equilibrium point as the wave traveled along the string. The resultant wave is also periodic. However, it has a complex, or complicated, wave form. The superposition principle makes it possible to analyze a complex wave in terms of a combination of simple component waves that are frequency multiples of the lowest frequency wave.

Actual wave disturbances generally have complex wave forms. For example, the sound waves produced by musical instruments may be very complicated compared with simple sine wave variations. Wave forms of musical tones from a French horn and a trumpet are shown in Figure 11-15 in Chapter 11. Each of these complex waves can be represented by a certain combination of simple waves that are whole number-multiples of the lowest frequency and that by superposition yield the complex wave form.

10.15 Interference The general term *interference* is used to describe the *effects* produced by two or more waves that superpose while passing through a given region. Special consideration is given to *waves of the same frequency*, particularly in the case of sound and light. Interference phenomena are exclusively associated with waves. In fact, the existence of interference in experiments with light first established the wave character of light.

Let us consider two waves of the same frequency traversing the same medium simultaneously. *Each particle* of the medium is affected by *both* waves. Suppose the displacement of a particular particle caused by one wave at any instant is in the same direction as that caused by the other wave. Then the total displacement of *that particle at that instant* is the *sum* of the separate displacements (superposition principle). The resultant displacement is *greater* than either wave would have caused separately. This effect is called *constructive interference.*

On the other hand, if the displacement effects of the two waves on the particle are in opposite directions, they tend to cancel one another. The resultant displacement of *that particle at that instant* is the *difference* of the two separate displacements and is in the direction of the larger (superposition principle). The resultant displacement is *less* than one of the waves would have caused separately. This effect is called *destructive interference.*

If two such opposite displacement effects are equal in magnitude, the resultant displacement is zero. The destructive interference is complete. The particle is not displaced at all but is in its equilibrium position *at that instant.*

Notice that we have been considering the effect of two waves on the position of just *one particle* of the medium at *one particular instant.* We shall continue to fix our attention on one instant in time as we consider the displacements of many *different* particles of the medium in which the two

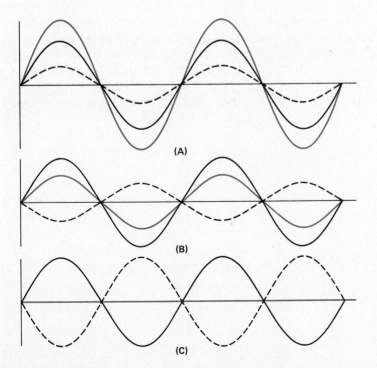

(A)

(B)

(C)

Figure 10-26. Interference of two periodic waves of the same frequency and traveling in the same direction. The component waves are in black; the resultant wave is in color. (A) Constructive interference. (B) Destructive interference. (C) Complete destructive interference.

waves are propagating. We find that the interference effects on different particles are different. There will appear constructive interference at some locations and destructive interference at others.

In Figure 10-26(A) two periodic waves with the same frequency and in phase are traveling in the same direction. They interfere constructively. The resultant periodic wave, shown in color, has the same frequency as the component waves but has an amplitude equal to the sum of the component amplitudes.

In Figure 10-26(B) the two periodic waves have the same frequency but are opposite in phase, that is, 180° out of phase. The displacements of the two waves are opposite in sign, and they interfere destructively. In Figure 10-26(C) the amplitudes of the two waves are equal and the destructive interference is complete.

The arrangement of interference effects (called an *interference pattern*) depends on the relative characteristics of the interfering waves. Figure 10-27 is a photograph of an interference pattern produced in a ripple tank by two identical circular waves emanating from two points. The two waves are in phase at their sources. That is, the two probes that generate the waves by their vibrations are moving up and down together.

This interference pattern is diagrammed in Figure 10-28. Points **A** and **B** are the sources of two periodic circular waves of the same frequency and amplitude. The sources are acting in phase. Solid circular lines represent wave crests and dashed circular lines represent wave troughs.

At each point similar to those marked **C** the crest of a

Figure 10-27. A photograph of the interference pattern of water waves from two point sources. Locate the nodal and antinodal regions.

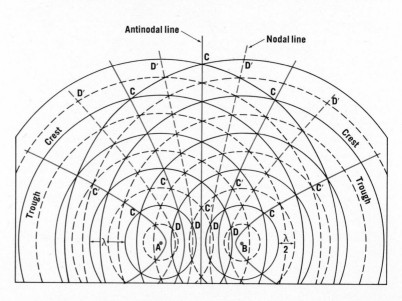

Figure 10-28. A diagram of the interference pattern of two periodic circular waves of the same frequency emanating in phase from points A and B.

wave from one source is superposed with the crest of a wave from the other source. At points similar to **C′** the troughs are superposed. A half period later the crests are at **C′** and the troughs are at **C**. Constructive interference occurs along the lines **CC′** giving the resultant wave an amplitude twice the amplitudes of the individual waves.

At points similar to those marked **D** the crest of a wave from one source is superposed with the trough of a wave from the other source. Destructive interference occurs along the lines **DD′**, displacements being reduced to zero.

Points of zero displacement in the interference pattern are called *nodes*. The lines **DD′** along which they occur are *nodal lines*. Similarly, points of maximum displacement are called *antinodes* (or loops) and the lines **CC′** along which they occur are *antinodal lines*. Observe these nodal and antinodal regions in the photograph of the interference pattern shown in Figure 10-27.

At any nodal point where the destructive interference is complete, there is no motion of the medium and no energy. This does not mean that the energy of the two interfering waves has been destroyed or otherwise lost. Instead, it appears at points of constructive interference, such as points **C** in Figure 10-28. Recall from Section 10.7 that where the amplitude of the resultant wave crest is double that of a single interfering wave, the energy is four times greater. The total energy of the two wave systems remains unchanged, but the *energy distribution* resulting from the interference is different.

10.16 Standing Waves As seen in Section 10.10, a transverse pulse traveling on a taut string is inverted when reflected at the clamped end. This inversion upon reflection at a clamped end occurs also with a train of transverse waves. If no energy is lost in reflection at the fixed end, a continuous wave will be reflected back upon itself giving two wave trains of the same wavelength, frequency, and amplitude but traveling in opposite directions.

Portions of two such wave trains are shown at a certain instant in time in Figure 10-29(A) and (B). If their displacements are added at this instant, as in (C), the resultant displacement is zero. When the wave patterns have moved $\frac{1}{4}$ wavelength, each in its own direction of travel, superposition gives the resultant shown in (D). Another $\frac{1}{4}$ wavelength of movement gives the resultant in (E), and an additional movement of $\frac{1}{4}$ wavelength gives the resultant displacement in (F). Such a wave pattern is a *standing wave*. The particles in a standing wave vibrate in simple harmonic motion with the same frequency as each of the

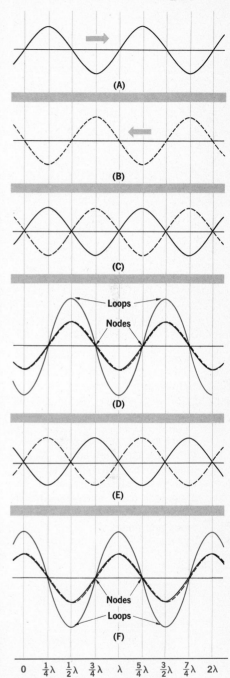

Figure 10-29. A standing wave (shown in color) produced by the interference of two periodic waves of the same frequencies and amplitudes traveling in opposite directions.

component waves. The amplitude of their motion is not the same for all points along the string. It varies from a minimum of zero amplitude at scale positions $\frac{1}{4}\lambda$, $\frac{3}{4}\lambda$, $\frac{5}{4}\lambda$, $\frac{7}{4}\lambda$, etc., to a maximum of twice the amplitude of one component wave at 0λ, $\frac{1}{2}\lambda$, 1λ, $\frac{3}{2}\lambda$, etc.

In the displacement shown in Figure 10-29(F) the motions of all the particles between 0λ and $\frac{1}{4}\lambda$ are exactly in phase with one another and with the motion of particles between $\frac{3}{4}\lambda$ and $\frac{5}{4}\lambda$. They are 180° out of phase with the motion of particles between $\frac{1}{4}\lambda$ and $\frac{3}{4}\lambda$ and with the motion of particles between $\frac{5}{4}\lambda$ and $\frac{7}{4}\lambda$. When particles in a length of string equal to one-half wavelength are going up, the particles in the immediately adjacent parts of the string are going down.

Certain parts of a string vibrating with a standing wave pattern never move from their equilibrium positions. These parts are the *nodes*. Halfway between the nodes, where the amplitude of vibration is maximum, are the *loops*. The wavelength of the component periodic transverse waves that produce the standing wave is *twice* the distance between the adjacent nodes or the loops in a standing wave.

A standing wave is produced by the interference of two periodic waves of the same amplitude and wavelength traveling in opposite directions. No standing wave can be produced if the two waves have different wavelengths.

When a stretched string is clamped at both ends, a standing wave pattern can be formed only for certain definite wavelengths. In such a vibrating string both ends must be nodes. Four possible standing wave patterns in a stretched string are shown in Figure 10-30. Loops are indicated by the letter **L** and nodes by the letter **N.** The wavelength, λ, for each standing wave is expressed in terms of

Figure 10-30. Standing wave patterns in a stretched string. (A) A time-exposure view. (B) A strobe-flash view. Wavelength λ is expressed in terms of the length l of the string. Labels L and N identify loops and nodes.

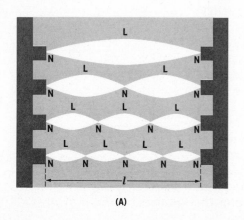

(A)

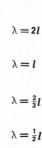

$$\lambda = 2l$$

$$\lambda = l$$

$$\lambda = \tfrac{2}{3}l$$

$$\lambda = \tfrac{1}{2}l$$

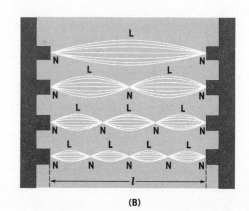

(B)

the length of the string, *l*. Figure 10-30(A) illustrates a time-exposure view showing the envelopes of the standing waves. Part (B) is a strobe-flash view showing positions of the strings at different instants of time.

In a standing wave, particles of the string at the nodes are continuously at rest. Energy is not transported along the string, but remains "standing" in the string. The energy of each particle remains constant as that particle executes its simple harmonic motion. When the string is straight, or undistorted, particles have their maximum velocities and the energy is all kinetic. The energy is all potential when the string has its maximum displacement and all particles are momentarily at rest.

Standing waves are established in water, in air, or in other elastic bodies, just as they are in taut strings. Standing waves can also be produced with electromagnetic waves. A standing wave pattern can occur in a circular ripple tank in which a periodic circular wave is generated at the center of the tank. Nodal rings appear where the surface of the water is at rest. The surface oscillates up and down between the nodal rings. Many vibrating objects vibrate normally in a way that establishes standing waves in the object. The strings or the air columns of musical instruments establish various modes of standing waves.

Questions
GROUP B

1. How does a pulse on a taut string differ from the incident pulse after being reflected at a fixed termination?
2. List the common properties of waves and give an illustration of each.
3. Make a drawing showing two transverse waves of the same wavelength but of different amplitudes that are in phase. Graphically find the wave pattern that results if these two waves are superposed.
4. Repeat Question 3 with waves of the same amplitude but with one having twice the wavelength of the other. At $t = 0$, the displacement of both waves is zero. At $t = T/4$ later for each, the displacement is maximum positive.
5. Repeat Question 4, but take the displacement as zero for the wave of

longer wavelength and as positive maximum for the wave of shorter wavelength at $t = 0$ while the reverse applies at $t = T/4$.

6. In your drawings of Questions 4 and 5 mark the points of maximum displacement and zero displacement. Do they occur at the same points at $t = 0$ as at $t = T/4$? Would you consider the resulting pattern as a standing wave or a progressing wave?
7. What are the requirements for production of a standing wave pattern?
8. What are *loops* and *nodes* in a standing wave pattern?
9. When two waves interfere, what influence does each exert on the progress of the other? Explain.
10. When two waves interfere, is there a loss of energy in the system? Explain.
11. Distinguish between the effects of a fixed termination and a free termination of a wave-transmission medium.

12. Two rooms are separated by a special soundproof partition except for a connecting doorway. How can you explain the fact that a sound produced anywhere in one room can be heard anywhere in the other?

13. A new phenomenon has been observed in which there is a transfer of energy. What type of experiments would you propose to determine whether it is a particle or wave phenomenon?

14. What is an impedance transformer?

15. A wave disturbance travels through a uniform medium to its termination, where a small part of the wave energy is transferred to the terminating medium and the remainder is reflected. How would you describe the impedance of the terminating medium relative to the impedance of the transmitting medium?

16. Suggest two possible ways of modifying the wave-transfer system described in Question 15 to eliminate the reflection of wave energy.

Problems
GROUP B

1. Electromagnetic waves travel through space at a speed of 3.0×10^8 m/s. The visible region of the spectrum has wavelengths ranging from about 4.0×10^{-7} m in the violet region to about 7.6×10^{-7} m in the red region. What is the frequency range of visible light?

2. A string stretched between two clamps is 2.0 m long. When plucked at the center, a standing wave is produced that has nodes at the clamped ends and a single loop at the center. By means of a stroboscope, the string is observed to complete 7.0 vibrations per second. (a) What is the wavelength of the component traveling waves? (b) What is the speed of the transverse waves in the string?

3. Periodic longitudinal waves are generated in a coil spring from a vibrating source at one end of the spring as shown in Figure 10-8. The frequency of the vibrating source is 15.0 hz. The distance between successive compressions is 41.0 cm. What is the speed of each compression moving along the spring?

4. An observer determined that 2.5 m separated a trough and an adjacent crest of surface waves on a lake and counted 33 crests passing in $3\overline{0}$ s. What was the speed of these surface waves?

5. Two physics students were fishing from a boat anchored in a lake 24 m from shore. One student observed that the boat rocked through 11 complete oscillations in 19 s and that one wave crest passed the boat with each oscillation. The other student noted that each crest required 6.5 s to reach the shore. (a) What was the period of the surface wave? (b) What was the wavelength?

SUMMARY

Energy can be transferred by the movement of materials or objects, by the motion of particles, by the flow of gases or liquids, and by waves. Energy is transferred by waves without the transport of matter. Waves that require a material medium for their transmission are called mechanical waves. Electromagnetic waves can travel through matter but a material medium is not required for their transmission. A pulse is a single nonrecurring wave disturbance. A continuous wave results from a wave disturbance that recurs periodically.

Mechanical waves in which the particles of the transmitting medium are displaced perpendicular to the direction in which the wave travels are called transverse waves. Those in which the particles of the medium are displaced parallel with the wave direction are longitudinal waves.

Waves have common characteristics such as phase, frequency, period, wavelength, speed, and amplitude. A wave's velocity is the product of its frequency and its wavelength in the medium. Transmission media in which the speed of a wave is dependent on its frequency are said to be dispersive.

Some of the common properties of waves are rectilinear propagation, reflection, refraction, diffraction, and interference. Superposition occurs when two or more wave disturbances move through a medium at the same time. Standing waves are produced by the interference of two periodic waves of the same amplitude and wavelength and traveling in opposite directions.

VOCABULARY

amplitude	interference	simple harmonic motion
angle of incidence	longitudinal wave	standing wave
angle of reflection	mechanical wave	superposition
damping	period	transverse wave
diffraction	periodic motion	wave
dispersive medium	phase	wave crest
elastic medium	rectilinear propagation	wave front
frequency	reflection	wavelength
impedance	refraction	wave trough

11 SOUND WAVES

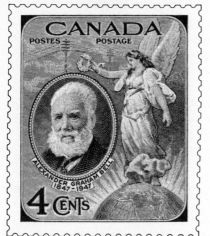

Alexander Graham Bell was a pioneer in the science of sound and in the improvement of education for the deaf. Bell invented the telephone in 1875. The world's first long-distance phone call (thirteen kilometers) was made the following year in Canada, where Bell had his summer home.

In this chapter you will gain an understanding of:

▷ the sonic spectrum
▷ the production, properties, and transmission of sound waves
▷ the measurement of sound levels
▷ the Doppler effect
▷ harmonic content of musical sounds
▷ the laws of vibrating strings and their application to stringed instruments
▷ forced vibrations and beats

THE NATURE OF SOUND

11.1 The Sonic Spectrum   The energy-transfer mechanism of compression waves is a major interest of physicists. In a medium having elastic and inertial properties, compression waves propagate as longitudinal disturbances. The disturbances consist of *compressions* and *rarefactions* that give rise to elastic forces in the propagating (transmitting) medium. Through these elastic forces the energy of the wave is transferred to the particles of the medium that are next in line. The energy exchange is continuous as particles receive wave energy and pass it along.

The frequency range over which such longitudinal waves occur is very large. It is called the *sonic spectrum.* There is no clearly defined lower limit of frequency of the sonic spectrum. However, the frequency of an earthquake wave may be a fraction of a cycle per minute and its wavelength may be measured in kilometers.

The upper limit of the sonic spectrum is well defined. For a constant-velocity condition, periodic waves of increasing frequency have proportionally decreasing wavelength. A wave is not transported by a medium in which the wavelength is small compared to the inter-particle spacing of the medium. For a gas the mean free path of the molecules is the limiting dimension. Thus at ordinary temperatures and pressures the upper range of sonic frequencies is of the order of 10^9 hertz in a gaseous medium. In liquids and solids the upper frequency limit is higher because of the smaller inter-particle spacing.

Within the sonic spectrum lies the region of **sound,** *a range of compression-wave frequencies to which the human ear is sensitive.* This audible range of frequencies, called the *audio range,* or the *audio spectrum,* extends from approximately 20 to 20,000 hertz. Compression waves at frequencies above the audio range are referred to as *ultrasonic;* those below the audio range are referred to as *infrasonic.*

11.2 The Production of Sound A sound is produced by the initiation of a succession of compressive and rarefactive disturbances in a medium capable of transmitting these vibrational disturbances. Particles of the medium acquire energy from the vibrating source and enter the vibrational mode themselves. The wave energy is passed along to adjacent particles as the periodic waves travel through the medium.

Such vibrating elements as reeds (clarinet, saxophone), strings (guitar, vocal cords), membranes (drum, loudspeaker), and air columns (pipe organ, flute) initiate sound waves. Sound waves are transmitted outward from their source by the surrounding air. When they enter the ear, they produce the sensation of sound.

Suppose we clamp one end of a thin strip of steel in a vise to serve as a vibrating reed. When struck sharply, the free end vibrates to and fro. See Figure 11-2. If the reed vibrates rapidly enough, it produces a humming sound.

When the reed vibrates, its motion approximates simple harmonic motion. See Figure 11-3(A). A graph of its displacement with time produces the sine wave shown in Figure 11-3(B). As the vibrating reed moves from **a** to **b,** it compresses the gas molecules of the air immediately to the right. The reed transfers energy to the molecules in the direction in which the compression occurs. At the same time the air molecules to the left expand into the space behind the moving reed and become rarefied. Thus the movement of the reed from **a** to **b** produces a compression and rarefaction simultaneously with a net transfer of energy to the right, the direction of the compression.

As the reed moves in the reverse direction from **b** to **a,** it compresses the gas molecules to the left, while those to the right become rarefied. The combined effect of this simultaneous compression and rarefaction transfers energy to the left, again in the direction of the compression.

If we consider just the series of compressions and rarefactions produced to the right, it is apparent that the maximum compression occurs as the reed moves through

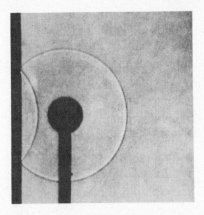

Figure 11-1. Photograph of a sound, circa 1930. Dr. A. L. Foley devised a technique for photographing the sound from a loud electric spark. Light from a second spark discharge casts a shadow on photographic film of the shock wave traveling outward in the air as a spherical shell. This photograph shows the wave being reflected from a plane surface.

Recall that sounds are produced by vibrating matter.

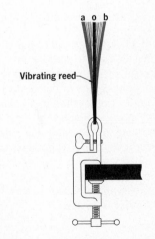

Figure 11-2. Sounds are produced by vibrating matter.

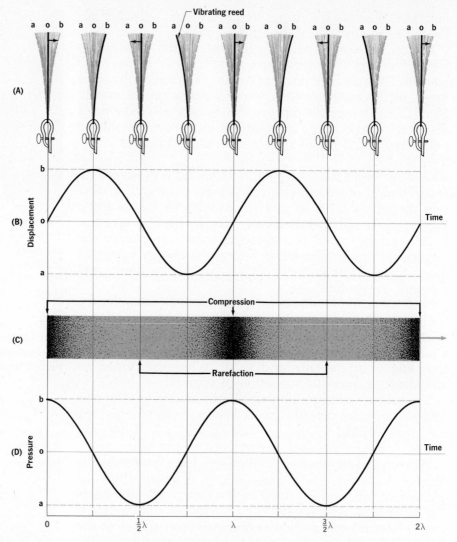

Figure 11-3. Variations of the vibrating reed in (A) produce the displacement variations with time in (B). The regions of compression and rarefaction of the longitudinal sound wave produced to the right of the reed are shown in (C). The pressure, or density variations, of the sound wave with time are plotted in (D). Observe that the displacement and pressure curves are 90° out of phase.

its equilibrium position **o** traveling from **a** to **b.** See Figure 11-3(C). The maximum rarefaction occurs as the reed moves through its equilibrium position **o** traveling from **b** to **a.** This is shown in (D) as a plot of the variation of pressure with time.

Simultaneously, of course, a corresponding series of rarefactions and compressions is produced to the left. The vibration of the reed thus generates longitudinal trains of waves. In these longitudinal waves, vibrating gas molecules move back and forth along the path of the traveling waves, receiving energy from adjacent molecules nearer the source and passing it on to adjacent molecules farther from the source. *Sound waves are longitudinal waves.*

Variations in the work done in setting the reed in vibration alter the amplitude of the vibration but not the frequency. The greater the energy of the moving reed, the greater is the amplitude of its vibrations and the amplitude of the resulting longitudinal waves.

Figure 11-4. In this specially designed room, only 0.1% of the sound is reflected. The acoustical conditions here are similar to those in the atmosphere 1.6 km above the earth's surface.

11.3 Sound Transmission To produce sound waves, we must have a source that initiates a mechanical disturbance and an elastic medium through which the disturbance can be *transmitted*. Most sounds come to us through the air, and it is the air that acts as the transmitting medium. At low altitudes, we usually have little difficulty hearing sounds. At higher altitudes, where the density of the air is lower, less energy may be transferred from the source to the air. Dense air is a more efficient transmitter of sound than rarefied air.

The following experiment provides evidence that a material is required for the transmission of sound. As shown in Figure 11-5, an electric bell under a bell jar is connected to a source of energy so that it rings. While it is ringing, air is removed from the bell jar. As the remaining air becomes less and less dense, the sound becomes fainter and fainter. When air is allowed to reenter, the sound becomes louder. These effects show that it is increasingly difficult to effect the transfer of sound energy to the air as the density of the air decreases. This demonstration suggests that if all the air could be removed from the bell jar, no sound would be heard. *Sound does not travel through a vacuum; it is transmitted only through a material medium.*

The sound of two rocks being struck together under water can be heard quite plainly by a swimmer submerged in the water. If the swimmer is near the source of the disturbance, the effect can be painful. Liquids are efficient transmitters of sound.

A loose faucet washer may vibrate when water is drawn from a water pipe. The sound of this vibration is carried to

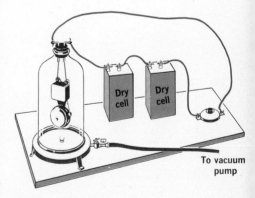

To vacuum pump

Figure 11-5. As air is pumped from the bell jar, the sound of the bell becomes fainter. Such a demonstration is evidence that a material medium is needed for the transmission of sound.

all parts of the house by water pipes and wood framework. Many solids are efficient sound transmitters. In long rods or pipes, the sound propagation may be restricted to one dimension instead of spreading in space.

11.4 The Speed of Sound   During a thunderstorm, a distant lightning flash can be seen several seconds before the accompanying thunder is heard. The timer at the finish line during a track meet may see the smoke from the starter's gun before he hears the report. Over short distances, light travels practically instantaneously. Therefore, the time that elapses between a lightning flash being seen and the thunder being heard or between a gun being fired and the report being heard must be the time required for the sound to travel from its source to the listener. The speed of sound in air is 331.5 m/s at 0 °C. This speed increases with temperature about (0.6 m/s)/C°.

The speed of sound in water is about four times the speed in air. In water at 25 °C sound travels about 1500 m/s. In some solids, the speed of sound is even greater. In a steel rod, for example, sound travels approximately 5000 m/s—about 15 times the speed in air. In general the speed of sound varies with the temperature of the transmitting medium. For gases the change in speed is rather large. For liquids and solids, however, this change in speed is small and is usually neglected. Representative values for the speed of sound waves in various media are given in Table 11-1. (See also Appendix B, Table 17.)

Table 11-1
SPEED OF SOUND
(25 °C)

Medium	Speed (m/s)
air	346
hydrogen	1339
alcohol	1207
water	1497
glass	4540
aluminum	5000
iron (steel)	5200

11.5 The Properties of Sound   A nearby clap of thunder is loud; a whisper is soft. A cricket has a shrill, high chirp; a bulldog has a deep growl. A violin string bowed by a virtuoso produces a note of fine quality and is far more pleasing to hear than that produced by the novice. Each of these sounds has characteristics clearly associated with it. Sounds differ from each other in several fundamental ways. We shall consider three physical properties of sound waves: *intensity, frequency,* and *harmonic content.* The effects of these properties on the ear are called, respectively, *loudness, pitch,* and *quality.*

11.6 Intensity and Loudness   The **intensity** of a sound is *the time rate at which the sound energy flows through a unit area normal to the direction of propagation.* Intensity thus has the dimension of power/area.

Recall that $P = \dfrac{E}{t}$.

$$I = \frac{P}{A}$$

Where P is sound power in watts and A is area in square

centimeters, the sound intensity I is expressed in watts per square centimeter. The intensity of a sound wave of given frequency is dependent on its amplitude; it is proportional to the square of the amplitude.

Review Section 10.7.

Sound waves that emanate from a vibrator approximating a point source and that travel in a uniform medium spread out in a spherical pattern. Thus the area of the expanding wave front is directly proportional to the square of its distance from the source. Since the total power of the wave is constant, the intensity of the wave diminishes as it moves away from the source. The sound produced by a whistle is only one-fourth as intense at a distance of one kilometer as it is at a distance of half a kilometer from the source. The intensity of sound in a uniform medium is inversely proportional to the square of its distance from the point source.

We interpret the intensity of a sound in terms of its loudness.

The **loudness** *of a sound depends on an auditory sensation in the consciousness of a human listener.* In general, sound waves of higher intensity are louder, but the ear is not equally sensitive to sounds of all frequencies. Thus a high- or a low-frequency sound may not seem as loud as one of mid-range frequency having the same intensity.

The ear is a nonlinear device.

An increase in the intensity of a sound of fixed frequency causes it to seem louder to the listener. Suppose a 1000-hz tone is generated of such low intensity that it is barely audible. The intensity can be increased approximately 10^{12} times before the tone becomes so loud that it is painful to the listener. While the *intensity* of a sound can be directly measured with instruments, the *loudness* can not as it depends on the ear and subjective judgment of the listener.

11.7 Relative Intensity Measurements Intensity is measured with acoustical instruments and does not depend on the hearing of a listener. At 1000 hz the intensity of the average faintest audible sound, called the *threshold of hearing,* is 10^{-16} w/cm^2. When expressing the intensity of a sound, it is often compared with the intensity of the threshold of hearing and this ratio of intensities yields an expression of the *relative intensity* of the sound. Because of the great range of intensities over which the ear is sensitive, a *logarithmic* rather than an arithmetic intensity scale is used. On this kind of scale the relative intensity of a sound is given by the equation

$$\beta = 10 \log \frac{I}{I_0}$$

where β (the Greek letter beta) is the relative intensity in

Table 11-2
RELATIVE INTENSITIES OF
SOUNDS (at 1000 hz)

Type of sound	Intensity (db)
threshold of hearing	0
whisper	10–20
very soft music	30
average residence	40–50
conversation	60–70
heavy street traffic	70–80
thunder	110
threshold of pain	120
jet engine	170

decibels (db) of a sound of intensity I. I_0 is the intensity of the threshold of hearing. I and I_0 have the same dimensions, usually w/cm². Therefore the decibel is a dimensionless unit. Relative intensities of several familiar sounds are listed in Table 11-2. The following example illustrates the use of the relative intensity equation.

EXAMPLE   Sound energy is radiated uniformly in all directions from a small source at a rate of 1.2 w. (a) What is the intensity of the sound at a point 25 m from the source? (b) What is the relative intensity (the decibel measure) at this reception point?

SOLUTION   Assuming no absorption of sound energy by the transmitting medium, the 1.2 w of sound power is associated with a spherical area having a radius of 25 m.

(a) $I = \dfrac{P}{A} = \dfrac{P}{4\pi r^2} = \dfrac{1.2 \text{ w}}{4\pi(2500 \text{ cm})^2} = 1.5 \times 10^{-8} \text{ w/cm}^2$

(b) $\beta = 10 \log \dfrac{I}{I_0} = 10 \log \dfrac{1.5 \times 10^{-8} \text{ w/cm}^2}{10^{-16} \text{ w/cm}^2} = 10 \log (1.5 \times 10^8)$

$\beta = 10(\log 1.5 + \log 10^8) = 10(0.1761 + 8) = 82 \text{ db}$

PRACTICE PROBLEMS   **1.** A source emits sound energy at a rate of $1\overline{0}$ w. (a) What is the intensity of the sound heard at a distance of $2\overline{0}$ m? (b) What is the relative intensity of the sound at the listener's position?   *Ans.*   (a) 2.0×10^{-7} w/cm² (b) 93 db

2. A jet plane in a landing approach is found to have a noise intensity at ground level of 1.1×10^{-5} w/cm². Determine the relative intensity at this location.   *Ans.*   110 db

---

Table 11-3
RELATIVE INTENSITIES

I/I_0	bel	db
10	1	10
4		6
2		3
1	0	0
0.5		−3
0.25		−6
0.1	−1	−10

The *decibel* is the practical unit for the relative intensity of sound. It is equal to 0.1 *bel*, the original unit named in honor of Alexander Graham Bell (1847–1922), who invented the telephone.

$$\textbf{Relative intensity in bels} = \log \dfrac{I}{I_0}$$

In the human sensory response, a change in the sound power level of 1 db is just barely perceptible. This change of 1 db represents a change in sound intensity of 26%. Table 11-3 shows several power ratios and their decibel equivalents.

We stated in Section 11.1 that the audio spectrum extends from approximately 20 to 20,000 hz. The lower frequency is called the *lower limit of audibility* and the higher frequency the *upper limit of audibility*. However, there is considerable variation in the ability of individuals to hear sounds of high or low frequencies.

The graph in Figure 11-6 shows the characteristics of audible sound waves. Since the graph is a composite of results obtained by testing many persons, it refers to the performance of the "average" ear.

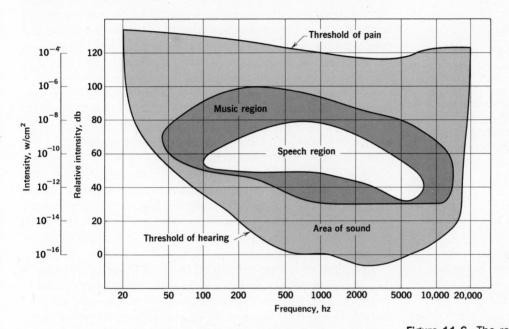

Figure 11-6. The range of audibility of the human ear.

The lower curve, the *threshold of hearing,* shows the minimum intensity level at which sound waves of various frequencies can be heard. Observe that the lowest intensity that will produce audible sound is for frequencies between 2000 and 4000 hz. The ear is most sensitive in this frequency range. At the lower and upper limits of audibility the intensities of sound waves must be greater to be heard.

The upper curve, the *threshold of pain,* indicates the upper intensity level for audible sounds. Sounds of greater intensity produce pain rather than hearing.

The graph shows the general frequency limits for the audio range to be between 20 and 20,000 hz. The relative intensity limits in the region of maximum sensitivity are between 0 and 120 decibels, corresponding to intensities of from 10^{-16} w/cm² to 10^{-4} w/cm².

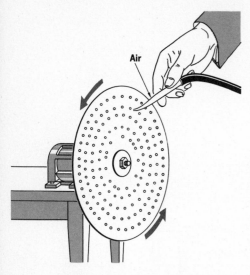

Air

Figure 11-7. A siren disk perforated to demonstrate the relationship between pitch and frequency and the difference between musical tones and noise.

We interpret the frequency of a sound in terms of its pitch.

Noise: sounds with nonperiodic waveforms.

Music: sounds with approximate periodic waveforms, unless their intensities are too high.

11.8 Frequency and Pitch   A perforated disk, as shown in Figure 11-7, can be used to show that the frequency of a sound determines its *pitch*. The disk has five concentric rings of holes. The holes of the innermost ring are irregularly spaced. The holes in the other four rings are regularly spaced, and, respectively, there are 24, 30, 36, and 48 holes in them.

A stream of air directed against a ring of holes is interrupted by the metal between the holes as the disk rotates rapidly at constant speed. The interruptions give rise to a series of air pulses that produce a sound characteristic of the frequency of the pulses. A sound of constant pitch is heard if the holes are evenly spaced. The pitch rises as the disk rotation is accelerated; it falls as the disk rotation is decelerated. If the stream of air is directed at the 24-, 30-, 36-, and 48-hole rings successively as the disk rotates at a constant rate, tones of successively higher pitch are produced. *Pitch is the characteristic of a sound that depends on the frequency the ear receives.* It is the *pitch* that allows us to assign the sound its place in the musical scale.

We usually distinguish between a musical tone and a noise by the fact that the former is a pleasing sound, and the latter is a disagreeable one. When the stream of air is directed at the innermost ring of holes, an unpleasant noise is heard. The noise is characterized by a random mixture of frequencies and is not easily identified in terms of pitch.

The pitch of a tone produced by directing the air against one of the regularly spaced rings of holes is appropriately described in terms of the frequency of the sound produced. If the disk is rotating at the rate of 20 revolutions per second, the frequency of the tone produced by the ring with 24 holes is 480 hertz. The ring with 30 holes produces a tone having a frequency of 600 hertz; the 36-hole ring, a frequency of 720 hertz; and the 48-hole ring, 960 hertz.

The two tones produced by the outermost and innermost of these four regularly spaced rings are an *octave* apart, their frequencies being in the ratio of 2:1. The four tones have frequencies in the ratios 4:5:6:8. When these tones are produced simultaneously, they comprise a musical sound known as a *major chord*.

Tone is associated with the pleasing quality of the sound that reaches the ear. Tone quality is enhanced by the complexity of the sound wave, that is, by the number and distribution of harmonics of the fundamental frequency superposed in the complex sound wave. *Pitch* is associated principally with the fundamental frequency of the complex sound wave. This distinction is important when considering the quality of musical sounds. For example, notes of the same pitch sounded by the pianist and the violinist are very different in tonal quality.

11.9 The Doppler Effect   In our discussion of frequency and pitch in Section 11.8, both the source of the emission and the listener were assumed to be stationary. In that case the pitch of the sound heard is characteristic of the frequency of the emission. If the source emits 1000 vibrations per second, the listener hears a 1000-hz tone.

When there is relative motion between the source of sound and the listener, the pitch of the sound heard is not the same as that when both listener and source are stationary. Two situations are common: The source may be moving and the listener is stationary, and the listener may be moving and the source is at rest.

The first situation prevails when the listener stands at a railroad crossing while a train passes. The pitch of the warning horn of the locomotive drops abruptly as the train passes the listener's position. The second situation prevails when the listener rides in the train as it passes through a highway intersection. In this instance the pitch of the warning bell at the crossing drops abruptly as the listener passes through the intersection.

In both situations the steady pitch of the sound heard by the observer is *higher* than the actual frequency of the source would indicate as the distance between the listener and the source *decreases* at a constant rate. Also, the steady pitch of the sound heard is *lower* than the actual frequency of the source would indicate as the distance between the listener and the source *increases* at a constant rate. *The change in pitch produced by the relative motion of the source and the observer is known as the* **Doppler effect.** Of course, the frequency of the sound emitted by the source remains unchanged, as does the velocity of the sound in the transmitting medium. Figure 11-8 illustrates this principle.

The Doppler effect is named for the Austrian physicist Christian Johann Doppler (1803–1853).

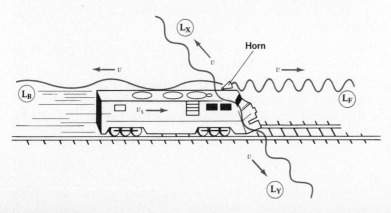

Horn

Figure 11-8. Doppler effect. The pitch of the horn on the speeding diesel locomotive sounds normal to the listeners L_X and L_Y, off to the sides, higher to the listener L_F, in front, and lower to the listener L_B, in back.

Suppose that a source of sound is stationed at **S** in Figure 11-9 and that one listener is at **L$_F$** and another is at **L$_B$**. If the source emits a sound of frequency f_S and the velocity of sound in the medium is v, the wavelength λ is

$$\lambda = \frac{v}{f_S} \qquad \text{(Equation 1)}$$

The time t in this diagram is unit time of 1 second. Thus the distance **SL$_F$** = **SL$_B$** = vt, that is, the distance the sound travels in 1 second. Then there are f_S waves in both distances **SL$_F$** and **SL$_B$,** each of wavelength λ.

Figure 11-9. Sound waves emitted in unit time by the source **S** during its movement from **S** to **S$_1$** with speed v_s are contained within the distance **S$_1$L$_F$** in the direction of listener **L$_F$** in front of the source, and within the distance **S$_1$L$_B$** in the direction of the listener **L$_B$** behind the source.

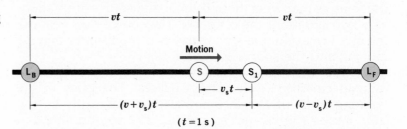

Now suppose that the source is moving toward **L$_F$** and away from **L$_B$** with speed v_S and in 1 second travels from **S** to **S$_1$**, a distance $v_S t$. During this time, f_S vibrations are emitted by the source. The first vibration is emitted at **S.** At the end of the 1-second interval this vibration is at **L$_F$** in front of the source, and also at **L$_B$** behind the source. Remember that the advancing wave fronts are spherical. The last vibration just emitted at the end of the 1-second interval is at **S$_1$**. Therefore, the same number of vibrations, f_S, are in the front region **S$_1$L$_F$** and the back region **S$_1$L$_B$** in the 1-second interval. These two distances are respectively

$$\mathbf{S_1L_F} = (v - v_S)t \qquad \text{(Equation 2a)}$$
and
$$\mathbf{S_1L_B} = (v + v_S)t \qquad \text{(Equation 2b)}$$

The wavelength λ' of the sound reaching the stationary listener in front of the moving source is shorter than λ of Equation 1 and is given by

$$\lambda' = \frac{v - v_S}{f_S} \qquad \text{(Equation 3)}$$

The velocity of the sound in the medium, v, is independent of the motion of the source. Thus the frequency of the sound reaching the stationary observer in front of the moving source, f_{LF}, is

$$f_{LF} = \frac{v}{\lambda'} \qquad \text{(Equation 4)}$$

Substituting the expression for λ' given in Equation 3 in Equation 4 yields

$$f_{LF} = \frac{v}{\dfrac{v - v_S}{f_S}}$$

or

$$f_{LF} = f_S \frac{v}{v - v_S} \qquad \text{(Equation 5)}$$

Equation 5 shows us that *the frequency of the sound reaching the stationary listener in front of the moving source is higher than the frequency of the source*. Therefore, the sound that is heard has a higher pitch than that emitted by the source.

We shall now turn to the other listener, $\mathbf{L_B}$, in Figure 11-9 and show how the pitch of the sound heard is related to the frequency of the source. *The wavelength λ' of the sound reaching the stationary listener behind the moving source is longer than λ of Equation 1* and can be expressed as

$$\lambda' = \frac{v + v_S}{f_S} \qquad \text{(Equation 6)}$$

As in Equation 4 the frequency of the sound reaching the observer behind the moving source, f_{LB}, is

$$f_{LB} = \frac{v}{\lambda'} \qquad \text{(Equation 7)}$$

Substituting the expression for λ' given in Equation 6 in Equation 7, we get

$$f_{LB} = f_S \frac{v}{v + v_S} \qquad \text{(Equation 8)}$$

Equation 8 indicates that *the frequency of the sound reaching the stationary listener behind the moving source is lower than the frequency of the source*. Thus the sound that is heard has a lower pitch than that emitted by the source.

If the listener is moving toward a stationary source, the pitch of the sound heard is higher than f_S, but it is not exactly that indicated by Equation 5. If the listener has a closing velocity v_{LC}, then v_{LC}/λ waves per second are received in addition to the f_S emitted by the source. From Equation 1,

$$f_S = \frac{v}{\lambda}$$

so,

$$f_{LC} = \frac{v}{\lambda} + \frac{v_{LC}}{\lambda}$$

or

$$f_{LC} = \frac{v + v_{LC}}{\lambda} \qquad \text{(Equation 9)}$$

Substituting the expression for λ given in Equation 1 in Equation 9, we have

$$f_{LC} = f_S \frac{v + v_{LC}}{v} \qquad \text{(Equation 10)}$$

Equation 10 shows the frequency of the sound reaching the listener moving toward the stationary source. Thus a sound is heard that has a higher pitch than that emitted by the source but not the same pitch as given in Equation 5.

The listener moving away from the stationary source with an opening velocity v_{LO} will receive $v_{LO}/λ$ waves per second fewer than f_S emitted by the source.

$$f_{LO} = \frac{v - v_{LO}}{λ} \qquad \text{(Equation 11)}$$

Substituting for λ as before,

$$f_{LO} = f_S \frac{v - v_{LO}}{v} \qquad \text{(Equation 12)}$$

From Equation 12 we can determine the frequency of the sound reaching a listener moving away from the stationary source. Observe that the pitch is lower than that emitted by this source, but is not the same pitch as given in Equation 8.

We have considered the special cases in which the velocities of the source and the observer lie along the line common to both. If the medium has a velocity along this line joining the source and the listener, it must be considered also.

The Doppler effect is most easily observed in connection with sound waves. Astronomers have observed slight changes in the wavelength of light received from distant stars and interpret this as being a Doppler effect caused by the relative motions of the stars toward or away from the earth. Light from many distant stars is shifted to longer wavelengths or toward the red region of the light spectrum. This effect, called the *red shift*, suggests that the universe is expanding. Doppler radar techniques can be used to detect moving objects, and some satellite tracking systems also use the Doppler principle.

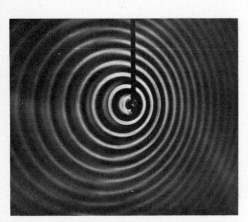

Figure 11-10. Doppler effect due to motion of the source. This effect is observed in a ripple tank in which the source of the disturbance is moving from left to right. Observe the crowding effect on the waves in front of the source and the spreading effect on those behind the source.

Questions

GROUP A

1. How are sounds produced?
2. Why is sound not transmitted through a vacuum?
3. (a) State the speed of sound in air at 0 °C in metric units of measure. (b) What is the rate of increase of speed with rise in temperature?
4. Distinguish between intensity and loudness.
5. Distinguish between frequency and pitch.

6. The engineer of a diesel locomotive sounds the horn as the train approaches you. How does the pitch you hear compare with the pitch the engineer hears?

7. How do the vibrations of the source of a musical tone differ from the vibrations of the source of a noise?

8. (a) What is the range of audio frequencies? (b) What name is applied to sound vibrations below the audio range? (c) What name is applied to sound vibrations above the audio range?

9. Referring to Figure 11-6, interpret (a) the *threshold of hearing* curve and (b) the *threshold of pain* curve. At a sound frequency of 1000 hz, what is (c) the intensity of the threshold of hearing, (d) the relative intensity of the threshold of hearing, (e) the intensity of the threshold of pain, and (f) the relative intensity of the threshold of pain?

GROUP B

10. Describe how a vibrating reed sets up longitudinal waves in the air surrounding it and thereby transfers its energy to the surrounding air.

11. (a) How does variation in the work done in setting a reed in vibration affect the sound produced? (b) How does variation in the length of the vibrating portion of the reed affect the sound?

12. Explain why the audibility of a sound disturbance in an enclosed gaseous medium diminishes as the density of the gas is lowered.

13. Why is the relative intensity of sound measured on a logarithmic scale?

14. What condition determines the upper limit of the sonic spectrum in a given transmitting medium?

15. Explain the meaning of the statement, "the relative intensity of sound in the room is 40 db."

16. Sound intensity, I, is a measured quantity having dimensions of power per unit area, P/A, expressed in watts/centimeter^2. The relative intensity of sound, β, is also a measured quantity expressed in decibels, the decibel, db, being a dimensionless unit. Explain the dimensionless character of the decibel.

Problems
GROUP A

1. What time is required for sound to travel 5.00 km if the temperature of the air is 10.0 °C?

2. What is the wavelength, in meters, of the sound produced by a tuning fork that has a frequency of $32\overline{0}$ hz? The temperature of the air is 15 °C.

3. The echo of a ship's foghorn, reflected from an iceberg, is heard 5.0 s after the horn is sounded. The temperature is −10.0 °C. How many meters away is the iceberg?

4. A rifle is fired in a valley with parallel vertical walls. The echo from one wall is heard in 2.0 s; the echo from the other wall 2.0 s later. The temperature is 20 °C. What is the width of the valley?

5. What is the relative intensity in decibels of a sound that has an intensity of 10^{-11} w/cm^2?

GROUP B

6. A stone is dropped into a mine shaft $25\overline{0}$ m deep. The temperature is 5.0 °C. How many seconds pass before the stone is heard to strike the bottom? (Neglect the effect of air resistance on the stone.)

7. A stone is thrown out horizontally over a cliff and is heard to strike the ground after 8.00 s. The temperature is 25 °C. What is the height of the cliff in meters? (See note in Problem 6.)

8. A locomotive approaches a crossing at 95.0 km/hr. Its horn has a frequency of 288 hz and the temperature is 15 °C. What is the frequency of the sound heard by the guard at the crossing?

9. What is the drop in frequency of the sound a listener hears as a train with its horn sounding passes at 145 km/hr? The frequency of the horn is $32\overline{0}$ hz; the temperature is 25.0 °C.

10. What is the intensity in w/cm² of a sound that has a relative intensity of

$3\overline{0}$ db?

11. The intensity of a sound is determined to be 10^{-14} w/cm². What is its relative intensity?

12. An explosion occurs and the intensity I_1 of the sound arriving at a nearby observation point is 2.25×10^{-6} w/cm². The intensity of the sound arriving at a more distant observation point I_2 is 1.17×10^{-12} w/cm². (a) Determine the relative intensity at the nearby observation point, (b) at the more distant point.

CHARACTERISTICS OF SOUND WAVES

11.10 Fundamental Tones Suppose a piano wire is stretched about a meter in length between two clamps and drawn tight enough to vibrate when plucked. If plucked in the middle, the wire vibrates *as a whole*, as shown in Figure 11-11(A). *A taut wire or string that vibrates as a single unit*

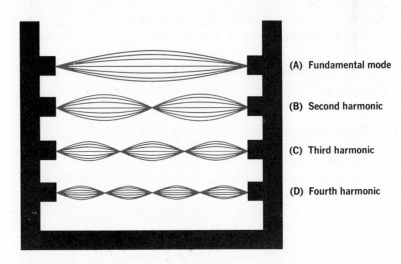

(A) Fundamental mode

(B) Second harmonic

(C) Third harmonic

(D) Fourth harmonic

Figure 11-11. A strobe-flash view of some vibration modes of a single string.

produces its lowest frequency, called its **fundamental.**

A vibrating string, having a small surface area, cuts through the air without disturbing it effectively. The string itself transfers very little of its vibrational energy to the air and therefore produces only a feeble sound. If, however, its vibrations are transferred to a larger surface, such as a sounding board, the air will be disturbed more effectively and a louder sound will be produced. The sounding board enables a substantial quantity of the vibrational energy to be transferred to the air. In this sense, the sounding board can be thought of as an impedance transformer (Section 10.11), which reduces the impedance discontinuity between the vibrational source and the air. These considerations are enlarged upon in Section 11.14.

The properties of vibrating strings and the sounds they produce can be studied using an instrument called a *sonometer*, which is shown in Figure 11-12. The sonometer may have two or more wires or strings stretched over a sounding board, which reinforces the sound produced by the wires. Because wires may vary in diameter, tension, length, and composition, a sonometer is useful for studying the effects of these variables on the vibrational characteristics of the wires.

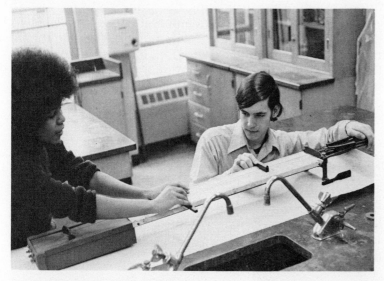

Figure 11-12. The sonometer can be used to study the vibrational characteristics of strings and wires.

By plucking a string, energy is transferred to the string and causes it to vibrate in transverse wave motion. Just as a vibrating reed transfers energy to molecules of air gases and causes them to exhibit longitudinal wave motion, so a vibrating string transfers energy to these gas molecules and causes them to vibrate and transmit a longitudinal wave. The frequency of the longitudinal vibration of the molecules, which is the frequency of the sound wave, is the same as the frequency of the transverse vibration of the string. Since the production of sound by a vibrating string dissipates energy, energy must be continuously supplied to the string by plucking or bowing to maintain the sound.

11.11 Harmonics A string may vibrate as an entire unit and may also vibrate in two, three, four, or more segments depending on the standing wave pattern that is established on the string. See Figure 11-11(B–D). When a string is plucked or bowed, not only the fundamental mode but other, higher modes of vibration may be present.

Suppose a string vibrates in two segments with a node in the middle, as in Figure 11-11(B). Its vibrating frequency is twice that of the string vibrating as a single segment. Doubling the frequency of a sound wave raises the pitch one *octave*. When the string vibrates in four segments, its vibrating frequency (and that of the sound waves produced) is four times its fundamental frequency. The pitch of the sound produced is then two octaves above the fundamental. *The fundamental and the vibrational modes having frequencies that are whole-number multiples of the fundamental are called* **harmonics.** Only these frequencies can be produced by plucking or bowing the string.

By this definition the fundamental is the *first harmonic* because it is *one* times the fundamental. The vibrational mode having a frequency twice that of the fundamental is the *second harmonic.* The vibrational mode of frequency *three* times that of the fundamental is the *third harmonic,* and so on. Oscillograms of a fundamental mode and several harmonics are shown in Figure 11-13.

Figure 11-13. Oscillograms of (A) a fundamental, (B) the second harmonic, (C) the third harmonic, and (D) the fourth harmonic. These oscillograms are traces of light on the screen of a cathode-ray tube (similar to a television picture tube) that presents a graph of the instantaneous values of the amplitude of the sound wave as a function of time. The instrument that contains the cathode-ray tube is called an oscilloscope.

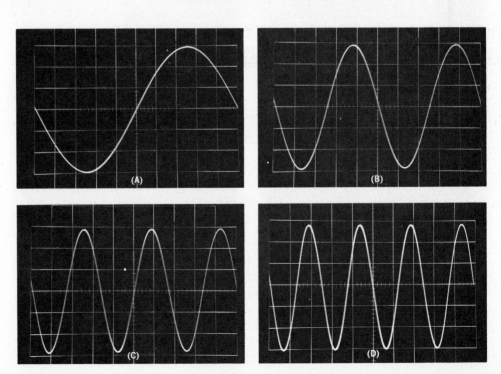

11.12 The Quality of Sound It is not difficult to pick out the sounds produced by different instruments in an orchestra, even though they may be producing the same tone with equal intensity. The difference is a property of sound called *quality*, or *timbre*.

If the string of a sonometer is touched very lightly in the middle while being bowed near one end, it will vibrate in two segments and as a whole at the same time, as Figure 11-14 illustrates. The fundamental is audible, but added to it is the sound of the second harmonic. The combination is richer and fuller; the quality of the sound is improved by the addition of the second harmonic to the fundamental. See Figure 11-15(A). *The quality of a sound depends on the number of harmonics produced and their relative intensities.* When stringed instruments are played, they are bowed, plucked, or struck near one end to enhance the production of harmonics that blend with the fundamental and give a richer sound. See Figure 11-15(B).

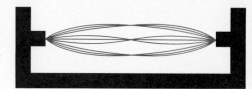

Figure 11-14. A strobe-flash view of a single string vibrating in its fundamental and second harmonic modes simultaneously.

Figure 11-15. Oscillograms of (A) a fundamental and second harmonic, (B) a fundamental and fourth harmonic, (C) the sound of a French horn, and (D) the sound of a trumpet.

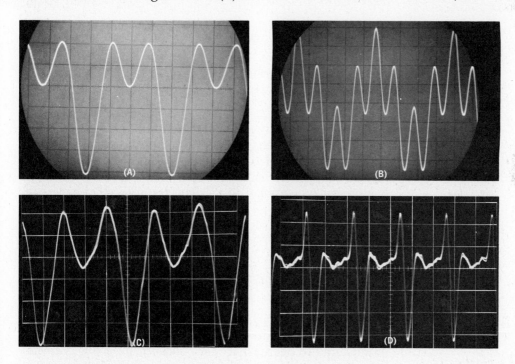

The quality of the tones produced by orchestral instruments varies greatly. For example, the tone produced by a French horn consists almost entirely of the fundamental and the second harmonic. This harmonic content can be recognized by comparing Figures 11-15(A) and 11-15(C). The tonal quality of the trumpet is due to the intensity of its high-frequency harmonics. See Figure 11-15(D).

We interpret the harmonic content of a sound in terms of its quality.

Figure 11-16. The laws of strings are used by the guitarist to produce a wide range of musical tones.

11.13 The Laws of Strings The frequency of a vibrating string is determined by its *length, diameter, tension,* and *density*. The strings of a piano produce tones with a wide range of frequencies. If we examine them we find that the strings that produce low-frequency tones are long, thick, and loose, while the strings that produce high-frequency tones are short, thin, and taut. One string produces a tone loud enough for the low-pitched notes, but three strings are needed to produce high-pitched notes of comparable loudness.

The conditions affecting the frequency of vibrating strings are summarized in four *laws of strings*.

1. Law of lengths. A musician may raise the pitch produced by a stringed instrument by shortening the vibrating length of the string. A violinist shortens the length of the A string about 2.5 cm to produce the note B. *The frequency of a string is inversely proportional to its length if all other factors are constant.*

$$\frac{f}{f'} = \frac{l'}{l}$$

Here f and f' are the frequencies corresponding to the lengths l and l'.

2. Law of diameters. In a piano and other stringed instruments like the cello and the guitar, the strings that produce higher frequencies have smaller diameters. A string with a diameter of 0.1 cm has twice the frequency of a similar string 0.2 cm in diameter. *The frequency of a string is inversely proportional to its diameter if all other factors are constant.*

$$\frac{f}{f'} = \frac{d'}{d}$$

Here f and f' are the frequencies corresponding to the diameters d and d'.

3. Law of tensions. When stringed instruments are tuned, the strings are tightened to increase their frequency or loosened to decrease it. *The frequency of a string is directly proportional to the square root of the tension on the string if all other factors are constant.*

$$\frac{f}{f'} = \frac{\sqrt{F}}{\sqrt{F'}}$$

Here f and f' are the frequencies corresponding to the tensions F and F'. See the following example.

4. *Law of densities.* The more dense a string is, the lower its frequency. Usually three of the strings on a violin are of plain gut; the fourth is wound with fine wire to increase its density so that it can produce the low-frequency tones. *The frequency of a string is inversely proportional to the square root of its density if all other factors are constant.*

$$\frac{f}{f'} = \frac{\sqrt{D'}}{\sqrt{D}}$$

Here f and f' are the frequencies corresponding to the densities D and D'.

EXAMPLE A string stretched with a force of 50.0 n produces the note C, 261.6 hz. What force must be applied to this string to produce the note C', one octave higher in frequency?

SOLUTION The frequency of C is 261.6 hz; that of C' is 523.2 hz.

$$\text{Basic equation: } \frac{f}{f'} = \frac{\sqrt{F}}{\sqrt{F'}}$$

$$\text{Working equation: } F' = \frac{f'^2 F}{f^2} = \frac{(523.2 \text{ hz})^2 \times 50.0 \text{ n}}{(261.6 \text{ hz})^2} = 20\bar{0} \text{ n}$$

PRACTICE PROBLEMS 1. A string under a tensional force of 71.2 n has a fundamental frequency of 261.6 hz. What tensional force is required to yield a fundamental frequency of 329.6 hz? *Ans.* 113 n

2. A string has a fundamental frequency of $44\bar{0}$ hz under a tensional force of 204 n. What is the fundamental frequency after the tension is reduced to 115 n? *Ans.* $33\bar{0}$ hz

11.14 Forced Vibrations When a tuning fork is struck with a rubber hammer, it vibrates at its fundamental frequency together with some low-order harmonics. The fundamental has a natural frequency that depends upon the fork length, thickness, and composition. When a key on a piano is struck, the piano string vibrates at its fundamental frequency and at harmonics of this frequency. The only external forces that affect these natural rates of vibration are friction and gravitation.

Suppose we strike a tuning fork and then press its stem against a table top. The tone becomes louder when the fork is in contact with the table because the fork *forces* the table top to vibrate with the same frequency. Since the table top has a much larger vibrating area than the tuning fork, these *forced vibrations* produce a more intense sound.

A vibrating violin string stretched tightly between two clamps does not produce a very intense sound. When the string is stretched across the bridge of a violin, however, the wood of the violin is forced to vibrate in response to the vibrations of the string; the intensity of the sound is increased by these forced vibrations. The sounding board of a piano acts in the same way to intensify the sounds produced by the vibrations of its strings.

11.15 Resonance The two tuning forks in Figure 11-17 have the same frequency. They are mounted on sounding boxes that increase the intensity of the sound through forced vibrations. One end of each sounding box is open.

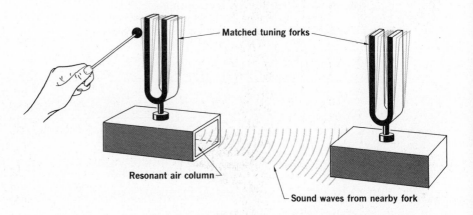

Matched tuning forks

Resonant air column

Sound waves from nearby fork

Figure 11-17. Resonance between two matched tuning forks.

Suppose we place these forks a short distance apart, with the open ends of the boxes toward each other. Now let us strike one fork and, after it has vibrated several seconds, touch its prongs to stop them. We find that the other fork is vibrating weakly. The compressions and rarefactions produced in the air by the first tuning fork act on the second fork in a regular fashion, causing it to vibrate. Such action is called *resonance,* or *sympathetic vibration.* A person who sings near a piano may cause the piano strings that produce similar frequencies to vibrate. Resonance occurs when the natural vibration rates of two objects are the same or when the vibration rate of one of them is equal to one of the harmonics of the other. If we changed the frequency of one of the tuning forks by adding mass to one of its prongs, we would alter the conditions of the experiment, and no resonance would be evident. Both forks should have the same natural frequency to produce this resonant effect.

Figure 11-18 illustrates a method of producing resonance between a tuning fork and a column of air. The vibrating fork is held above the hollow cylinder immersed in water. The length of the air column is gradually increased by raising the cylinder. As seen in Figure 11-18(A), there is a marked increase in the loudness of the sound. Here the column of air in the tube vibrates vigorously at the frequency of the tuning fork, and the two are in resonance.

This behavior is similar to the production of a standing wave on a stretched string. During resonance, a compression of the reflected wave unites with a compression of the direct wave, and a rarefaction of the reflected wave unites with a rarefaction of the direct wave. The constructive superposition of the two waves in phase amplifies the sound.

The resonant air column is simply a standing longitudinal wave system. The water surface in Figure 11-18(A) closes the lower end of the tube and prevents longitudinal displacement of the molecules of the air immediately adjacent to it. This termination effectively clamps the air column at the closed end and gives rise to a displacement node. Because the upper end of the tube is open, it provides a free-ended termination for the air column and gives rise to a displacement antinode, or loop. These are, respectively, shown in Figure 11-18(A) as **N** and **L**.

Because the air particles at the closed end of the tube are unable to undergo longitudinal displacement, they experience maximum changes in pressure. This position of a displacement node is also a pressure antinode, or a point of maximum pressure change. (Refer to Figure 11-3 for the phase relationship between displacement and pressure waves.) Compressions are reflected as compressions, and rarefactions are reflected as rarefactions. This behavior amounts to a change in phase upon reflection from a rigid termination in the sense that the displacement vector of the air particles is inverted. See Figure 11-19(A).

The open end of the tube is a displacement loop and, consequently, a pressure node. Here the pressure remains at the constant value of the outside atmosphere, and air rushes into and out of the tube as the column vibrates. It is this large movement of air at the open end of the tube that

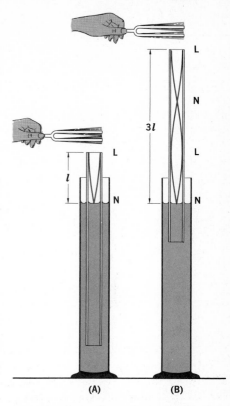

(A)   (B)

Figure 11-18. Resonance between a tuning fork and a vibrating air column. (A) and (B) give greater sound reinforcement than would any water-level positions in between.

Figure 11-19. A compression traveling from left to right is reflected at the closed end in (A) as a compression and at the open end in (B) as a rarefaction. The particle displacement vector in the closed tube is inverted by reflection, but it remains unchanged in the open tube because the displacement of air layers is in the same direction for a rarefaction traveling to the left as for a condensation traveling to the right.

Particle displacement:	Particle displacement:
before reflection ⟶	before reflection ⟶
after reflection ⟵	after reflection ⟶
(A) Closed tube	**(B) Open tube**

transfers energy to the atmosphere and provides the reinforcement of the sound produced by the tuning fork.

The fundamental frequency of the resonant column corresponds approximately to a displacement node at the closed end and an adjacent displacement loop at the open end, as shown in Figure 11-18(A). Since the distance separating a node and an adjacent loop of a standing wave is one-fourth wavelength, *the length of the closed tube is approximately one fourth the wavelength of its fundamental resonant frequency.*

$$\lambda \simeq 4l$$

By applying a small empirical correction proportional to the diameter of the tube (because the motion of molecules of the air at the open end of the tube is not strictly in one dimension), we can state the relationship more precisely as follows:

$$\lambda = 4(l + 0.4d)$$

λ is the wavelength of the fundamental resonant frequency, l is the length of the closed tube, and d is its diameter.

If the cylinder of Figure 11-18(A) is lifted higher out of the water, resonance will occur again when the length is three quarters of a wavelength. This length of air column also corresponds to a displacement node at the closed end and a loop at the open end and allows a standing wave to develop. See Figure 11-18(B). For a tube that is long enough, successively weaker resonance points could be found at $5/4\lambda$, $7/4\lambda$, etc. Thus *a closed tube is resonant at odd quarter-wavelength intervals.*

The quarter-wave resonant column for the fundamental frequency of the tuning fork of Figure 11-18(A) is three quarters of a wavelength long for the third harmonic of this frequency. It is $5/4\lambda$ for the fifth harmonic and is $7/4\lambda$ for the seventh harmonic. Therefore *the resonant frequencies of a closed tube are harmonics, but only odd harmonics of the fundamental mode are present.* Figure 11-20 illustrates this.

Many musical instruments employ vibrating air columns open at one end and closed at the other *(closed tube)* or open at both ends *(open tube).* From the above discussion we have shown that the normal modes of oscillation of air columns are characterized by

1. *a displacement node (or pressure loop) at a closed end,* and
2. *a displacement loop (or pressure node) at an open end.*

Compressions traveling in an open tube are reflected at the open ends as rarefactions, and rarefactions are reflected as compressions. In such reflections, the longitudi-

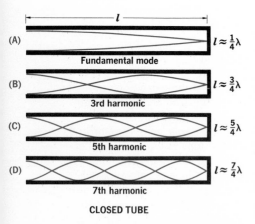

Figure 11-20. The normal modes of oscillation of a closed tube. The fundamental frequency is half that of an open tube of the same length, and only odd harmonics are produced as shown by the standing wave patterns.

nal displacement vector of the air particles is *not* inverted, and in this sense there is no change in phase at the open termination of the air column. This condition is illustrated in Figure 11-19(B).

The fundamental frequency of a resonant air column in an open tube corresponds approximately to displacement loops at opposite ends and a displacement node in the middle, as shown in Figure 11-21(A). Since adjacent loops of a standing wave are one-half wavelength apart, *the length of an open tube is approximately one half the wavelength of its fundamental resonant frequency.*

$$\lambda \simeq 2l$$

Again, by applying a small empirical correction proportional to the diameter of the tube, we can state the relationship more precisely as follows:

$$\lambda = 2(l + 0.8\,d)$$

λ is the wavelength of the fundamental resonant frequency, l is the length of the open tube, and d is its diameter.

An open tube, which is a half-wave resonant column at its fundamental frequency, is a full wavelength long at the second harmonic of this frequency. It is $3/2\lambda$ at the third harmonic. In each case displacement loops occur at both open ends, and the column resonates in these modes. Therefore *the resonant frequencies of an open tube are harmonics, and all harmonics of the fundamental mode are present.* See Figure 11-21.

The quality of a musical tone is generally enhanced by its harmonic content. The open tube resonator has both odd and even harmonics of the fundamental mode present, whereas the closed tube resonator has only odd harmonics. Thus the quality of the sounds from open tubes and closed tubes is not the same.

11.16 Beats We have already discussed interference of transverse waves in strings and of water waves in the ripple tank in Chapter 10. We were concerned with standing waves in resonant air columns in the preceding section. There is abundant experimental evidence that two or more wave disturbances can travel through the same medium independently of one another. The superposition principle shows us that the displacement of a particle of a medium at any time is the vector sum of the displacements it would experience from the individual waves acting alone.

A standing wave is formed by two wave trains of the same frequency and amplitude traveling through a medium in opposite directions. We are generally concerned

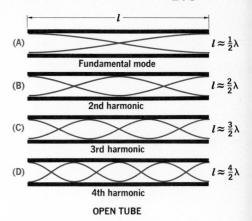

Figure 11-21. The normal modes of oscillation of an open tube. The fundamental frequency is twice that of a closed tube of the same length, and all harmonics are produced.

with the conditions or behavior of the space in which the standing wave exists. This generalization is true for resonant air columns used in musical instruments. The standing wave is characterized by an amplitude *that varies with distance or position in space.*

Two wave trains of slightly different frequencies traveling in the same direction through a medium will interfere in a different way. At any fixed point in the medium through which the waves pass, their superposition gives a wave characterized by an amplitude *that varies with time.*

In Figure 11-22 the resultant displacement of two wave trains of slightly different frequencies is plotted as a function of time. In curve (A) the frequency is 8 hz, and in (B) it is 10 hz. Curve (C) shows the combined effect of these two waves at a fixed point in their pathway. This resultant wave varies periodically in amplitude with time. Such amplitude pulsations are called *beats.*

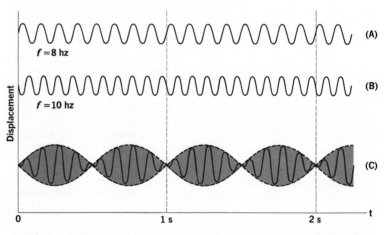

Figure 11-22. Beats. Waves (A) and (B) of slightly different frequencies combine to give a wave (C) that varies in amplitude with time.

Observe that waves (A) and (B) come into phase two times each second and out of phase the same number of times each second. The *beat frequency* can be described as *two beats per second. The number of beats per second equals the difference between the frequencies of the component waves.*

When the interfering frequencies are audible sounds, the amplitude variations, or beats, are recognized as variations in loudness. The average human ear can distinguish beats up to a frequency of approximately ten per second.

The beat phenomenon is frequently used to tune vibrating systems with great precision. Two vibrating strings can be tuned to the same frequency by adjusting the tension of one until the beats disappear. This procedure is called "zero beating." The piano tuner adjusts frequencies precisely by systematically beating harmonics of notes against each other.

To illustrate, assume that two vibrating strings have fundamental frequencies of 165 hz and 325 hz respectively. A person (who possesses a "trained" ear) may distinguish 5 beats per second as a consequence of the 325-hz tone beating against the second harmonic of the 165-hz tone.

A precision tuning fork vibrates only in its fundamental mode and produces a tone that is free of harmonics. Suppose a tuning fork of frequency 260 hz is sounded together with one of 264 hz. A listener will hear 4 beats each second of sound having an average frequency of 262 hz.

If the lower frequency is represented by f_1 and the higher frequency by f_h, the average frequency f_{av} can be expressed as follows:

$$f_{av} = 1/2(f_1 + f_h)$$

The sound is a sine wave of average frequency f_{av} and varying amplitude that reaches a maximum $(f_h - f_1)$ times each second (the beat frequency). When the frequency separation of the two tones is too large for the beats to be distinguished by the listener, a difference *tone* of frequency $(f_h - f_1)$ may be heard if the difference frequency falls within the audio range.

Questions
GROUP A

1. (a) Why does a tuning fork sound louder when its stem is pressed against a table top? (b) Why doesn't its sound last as long?
2. What conditions are necessary to produce resonance?
3. To produce the best resonance, how must the length of a closed tube compare with the wavelength of the sound?
4. To produce the best resonance, how must the length of an open tube compare with the wavelength of the sound?
5. How do the frequencies of notes an octave apart compare?
6. (a) What is a fundamental? (b) How does a string vibrate to produce the fundamental?
7. Which law of strings is used (a) in tuning a guitar; (b) in playing the instrument?
8. Given two sound wave trains of the same frequency traversing the same medium simultaneously, describe the conditions for (a) constructive interference; (b) destructive interference. (c) What changes and/or additions in the above wave specifications are required for the interfering wave trains to produce a beat note?

GROUP B
9. Why is it easy to distinguish between the sound produced by a piano and a trombone even if both play the same note?
10. How do the strings on a piano illustrate the laws of strings?
11. Suggest a function of the bridge between the strings and the sounding board of a musical instrument such as a cello.
12. Suppose a piano tuner has tuned the

notes middle C and G to frequencies of 262 hz and 392 hz respectively. The note G is then beat against low C, which is one octave below middle C. What should be heard to indicate that

low C has been adjusted to its proper frequency?

13. Given two tuning forks that vibrate at the same frequency, suggest a way to hear beats between them.

Problems
GROUP A

1. What is the frequency of a tuning fork that resonates with an open tube 25.0 cm long and 2.0 cm in diameter when the temperature is $2\overline{0}$ °C?

2. A tuning fork, frequency 384 hz, produces resonance with a closed tube 20.0 cm long and 4.0 cm in diameter. What is the speed of sound?

3. A tuning fork has a frequency of 440 hz. If another fork of slightly lower pitch is sounded at the same time, the beats produced are 5/s. What is the frequency of the second tuning fork?

4. How many beats will be heard each second when a string, frequency 288 hz, is plucked simultaneously with another string, frequency 296 hz?

5. A violin string is 25.4 cm long and produces a fundamental frequency of $44\overline{0}$ hz. What change in length is required to produce a frequency of 523.3 hz?

6. Compare the frequency of one string 25 cm long and 0.50 mm in diameter with that of another $10\overline{0}$ cm long and 0.25 mm in diameter, assuming all other factors are constant.

7. Two tuning forks of $32\overline{0}$ hz and 324 hz are sounded simultaneously. What sound will the listener hear?

8. A violin string is 26.7 cm long and has a resonant frequency of 327.6 hz. What change in length is required to lower the resonant frequency to 293.7 hz?

GROUP B
9. When a string 0.500 m long is stretched with a force of 2.50×10^2 n,

its frequency is $44\overline{0}$ hz. If the string is shortened to 0.400 m and the stretching force is increased to 5.00×10^2 n, what is the new frequency?

10. A string 0.350 m long and under a tension of 325 n has a resonant frequency of 264 hz. (a) What change in the length of the string is required to raise the resonant frequency to 297 hz? (b) If the length is not changed, what change in the tensional force is required to effect this change in frequency?

11. Sound intensity at the hearing threshold is approximately 10^{-14} w/cm^2 for a sound frequency of 200 hz. The pain threshold is reached for sound at this frequency when the intensity is increased by 110 db. Calculate the intensity of this sound at the pain threshold.

12. Middle C on the piano keyboard is tuned to 261.6 hz when the temperature is 20.0 °C. Calculate (a) the frequency, and (b) the wavelength in air of the highest note on the keyboard that is exactly 4 octaves above middle C.

13. An organ pipe open at both ends is 1.23 m long and has a diameter of 10 cm. (a) What is its fundamental frequency when the air temperature is 15 °C? (b) What are the frequencies of the two lowest harmonics produced along with the fundamental tone?

14. An organ pipe closed at one end is 0.76 m long and has a diameter of 5.0 cm. The air temperature is 12 °C. (a) Determine its fundamental frequency. (b) What are the frequencies of the two lowest harmonics produced along with this fundamental tone?

SUMMARY

Sound is a longitudinal disturbance consisting of a succession of compressions and rarefactions to which the ear is sensitive. The term also applies to similar disturbances above and below the normal range of hearing. A medium having elastic and inertial properties is required for its propagation. The intensity of a sound is the rate at which the wave energy flows through a unit area. The sensory response to sound intensity is loudness, the sensory response to frequency is pitch, and the sensory response to harmonic content is tonal quality.

The sensory response to sounds is nonlinear. Minimum intensity levels for hearing various sound frequencies define the threshold of hearing. Upper intensities that produce pain rather than increased loudness define a threshold of pain. The change in pitch heard when there is relative motion between an observer and a constant-frequency source is called the Doppler effect.

When set in motion, sound generators such as taut strings and air columns vibrate at their fundamental and various harmonic frequencies. The greater the harmonic content of a sound produced by a musical instrument, the higher is its quality and the more pleasing is the auditory response. The frequency of a vibrating string is determined by its length, diameter, tension, and density. The laws of strings enunciate the relationships between frequency and these physical factors.

The intensity of a sound generated by a vibrating source can be increased by a sounding board forced to vibrate at the sound frequency. When forced vibrations are at the natural frequency of the reinforcing body (resonance), maximum reinforcement occurs.

Resonant air columns in the form of open and closed tubes are standing longitudinal wave systems. The resonant length of a closed tube is approximately one-fourth the wavelength of its fundamental frequency, and it resonates at odd harmonics of the fundamental mode. The resonant length of an open tube is approximately one-half the wavelength of its fundamental frequency, and it resonates at all harmonics of the fundamental mode.

Two wave trains of slightly different frequencies traveling in the same direction in a medium interfere in a way characterized by an amplitude that varies with time at any fixed point in the medium. The amplitude pulsations are called beats. The number of beats per second is the difference between the frequencies of the interfering waves.

VOCABULARY

audio spectrum
beat frequency
beats (sound)
bel
compression
decibel
Doppler effect
forced vibration

frequency
fundamental (sound)
harmonic
infrasonic
intensity (sound)
loudness
pitch
propagation (wave)

quality (sound)
rarefaction
resonance
resonant frequency
sonic spectrum
threshold of hearing
threshold of pain
ultrasonic

THE NATURE OF LIGHT

In 1865, James Clerk Maxwell proposed the electromagnetic theory of light on the basis of extensive mathematical calculations. Twenty years later Heinrich Hertz accidentally discovered radio waves, which verified Maxwell's theory. The metric unit of frequency is named in honor of Hertz.

In this chapter you will gain an understanding of:

▷ the properties of light
▷ the historical development of the wave theory of light
▷ the range of the electromagnetic spectrum
▷ the photoelectric effect and its importance in the development of the quantum theory
▷ the basic principle of the laser
▷ the parameters that characterize photometry, the quantitative study of light

WAVES AND PARTICLES

12.1 Properties of Light The general properties of waves are described in Chapter 10. They are restated and briefly summarized as follows:

1. *Propagation* within a uniform medium is along straight lines.

2. *Reflection* occurs at the surface, or boundary, of a medium.

3. *Refraction,* or bending, may occur where a change of speed is experienced.

4. *Interference* is found where two waves are superposed.

5. *Diffraction,* or bending around corners, takes place when waves pass the edges of obstructions.

These properties are easily recognized in the behavior of sound and water waves. Such disturbances occur in matter, the particles of the medium being set in motion about their equilibrium positions by the passing waves.

Matter is not required for the propagation of light. While light does pass readily through certain kinds of matter, its transmission is unhindered in interstellar space or through an evacuated vessel.

The regular reflection of light from smooth surfaces was known in the time of Plato, almost twenty-four hundred years ago. (Reflection is the subject of Chapter 13.)

The refraction of light at the interface (boundary) of two transparent media of different optical densities was observed by the Greeks as early as the second century A.D.

The Arabian mathematician Alhazen (965–1039) studied the refraction of light and disputed the ancient theory that visual rays emanated from the eye. He demonstrated the refractive behavior of light as it passed from one medium into another of greater optical density. See Figure 12-1. He believed that the angles of incidence and refraction are related, but was unable to determine how they are related. This relationship, now known as *Snell's law,* was established six hundred years later. (Refraction and Snell's law are discussed in Chapter 14.)

The shapes of shadows and the practical use of sight lines for placing objects in a straight line give evidence of the straight-line, or rectilinear, propagation of light. Greek philosophers were familiar with this property of light. Sir Isaac Newton conducted experiments on the separation of light into colors by means of a prism and was aware of the colors produced by thin films.

Rectilinear propagation, reflection, and refraction are the principal light phenomena that were familiar to observers in the seventeenth century. It must have been evident to Newton and his contemporaries that since the transmission of energy from one place to another was involved, only two general theories could explain these properties of light.

In general, energy can be propagated either by particles of matter or by wave disturbances traveling from one place to another. A window pane may be shattered by a moving object, such as a baseball thrown from a distance, or by the concussion ("shock wave") from a distant explosion.

Arguments favoring both a particle (corpuscular) theory and a wave theory were plausible when applied to the properties of light observed in the seventeenth century. The principal advocate of the corpuscular theory was Newton, whose arguments were supported by the French mathematical physicist and astronomer Laplace (1749–1827). The wave theory was upheld principally by Christian Huygens (*hi*-ganz) (1629–1695), a Dutch mathematician, physicist, and astronomer. He was supported by Robert Hooke of England. Because of the plausibility of both theories, a scientific debate concerning the nature of light developed between the followers of Newton and the followers of Huygens and continued unresolved for more than a century. What are some of the arguments that support each of these classical theories of light?

12.2 The Corpuscular Theory
Sir Isaac Newton believed that light consists of streams of tiny particles, which he called "corpuscles," emanating from a luminous

Figure 12-1. The Arabian mathematician Alhazen studied the refraction of light nearly a thousand years ago.

The light phenomena known in Newton's time were rectilinear propagation, reflection, and refraction.

Sir Isaac Newton lived from 1642 to 1727.

source. Let us examine the arguments used by those who believed the particle theory best explained the various light phenomena known to them.

1. Rectilinear propagation. A ball thrown into space follows a curved path because of the influence of gravity. Yet if the ball is thrown with greater and greater speed, we know that its path curves less and less. We can easily imagine minute particles traveling at such enormous speed that their paths essentially form straight lines.

Newton experienced no difficulty in explaining the rectilinear propagation of light by means of his particle model. In fact, this property of light provided the supporters of the corpuscular theory with one of their strongest arguments against a wave theory. How, they asked, could waves travel in straight lines? A sound can easily be heard around the corner of an obstruction, but a light certainly cannot be seen from behind an obstruction. The former is unquestionably a wave phenomenon. How can the latter also be one?

The simple and direct explanation of rectilinear propagation provided by the particle model of light, together with the great prestige of Sir Isaac Newton, were largely responsible for the preference shown for the corpuscular theory during the seventeenth and eighteenth centuries.

2. Reflection. Where light is incident on a smooth surface, such as a mirror, we know that it is regularly reflected. How do particles behave under similar circumstances? Steel ball bearings thrown against a smooth steel plate rebound in much the same way light is reflected. Perfectly elastic particles rebounding from a resilient surface, then, could provide a suitable model for the reflection of light. See Figure 12-2.

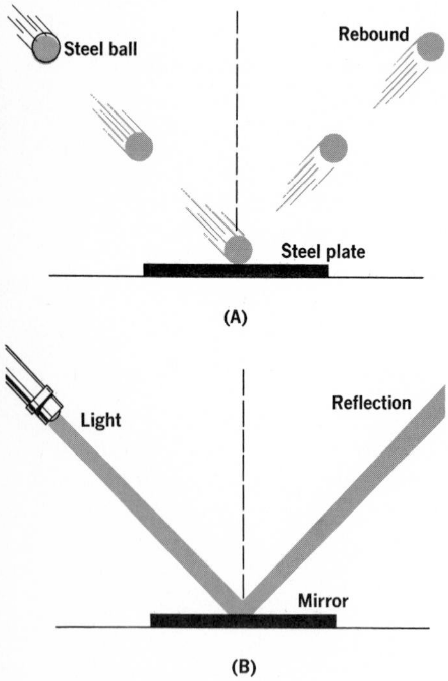

3. Refraction. Newton was able to demonstrate the nature of refraction by means of his particle model. We can duplicate this experimentally by arranging two level surfaces, one higher than the other, with their adjacent edges joined by an incline, as shown in Figure 12-3. A ball may be rolled across the upper surface, down the incline, and across the lower surface. Of course, it will experience an acceleration due to the force of gravity while rolling down the incline and will move across the lower surface at a higher speed than it had initially.

Suppose the ball is set rolling on the higher surface toward the incline at a given angle with the normal to the edge. At the incline, the accelerating force exerts a pull on the ball causing it to roll across the lower surface at a smaller angle with the normal to its edge. Now if we think of the upper surface as representing air, the lower surface

Figure 12-2. The rebound of a steel ball from a resilient surface resembles the reflection of light from a mirror surface. This example was used as an early argument to "prove" that light rays were streams of tiny particles.

as an optically more dense medium like water, and the incline as the interface of the two transmitting media, the rolling ball behaves as particles of light being redirected, or refracted, as they pass from air into water.

By varying the grade of the incline while maintaining both constant rolling speed and constant angle with the normal on the upper surface, the refractive characteristics of different transmitting media can be illustrated by the rolling-ball model.

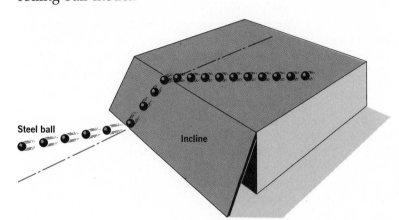

Steel ball

Incline

Figure 12-3. A ball bearing rolling from a higher to a lower surface illustrates one aspect of the refraction of light. In what way does this rolling-ball model of refraction fail?

Newton believed that water attracted the approaching particles of light in much the same way gravity attracts the rolling ball on the incline. The rolling-ball experiments imply, as they did to Newton, that light particles accelerate as they pass from air into a medium of greater optical density like water or glass. The corpuscular theory required that the speed of light in water be higher than the speed of light in air. Newton recognized that if it should ever be determined that the speed of light in water is lower than the speed of light in air, his corpuscular theory would have to be abandoned. This uncertainty about the speed of light in water remained unresolved for one hundred twenty-three years after Newton's death. In 1850 the French physicist Jean Foucault (foo-*koh*) (1819–1868) demonstrated experimentally that the speed of light in water is indeed lower than the speed of light in air. This condition was predicted by the wave theory.

Physicist Foucault is recognized primarily for his pendulum experiment, which demonstrates that the earth rotates on its axis in space. The Foucault pendulum experiment is now repeated daily in science museums throughout the world.

12.3 The Wave Theory Christian Huygens is generally considered to be the founder of the wave theory of light. Although somewhat different in its modern form, Huygens' basic concept is still very useful to us in predicting and interpreting the behavior of light. Let us recall a familiar characteristic of water waves as an introduction to this important principle.

If a stone is dropped into a pool of quiet water, it creates a disturbance in the water and a series of concentric waves travels out from the disturbance point. The stone quickly comes to rest on the bottom of the pool, so its action on the water is of short duration. However, wave disturbances persist for a considerable time thereafter and cannot reasonably be attributed to any activity on the part of the stone. It must be that the disturbances existing at all points along the wave fronts at one instant of time generate those in existence at the next instant.

Huygens recognized this logical deduction as a basic aspect of wave behavior and devised a geometric method of finding new wave fronts. His concept, published in 1690 and now recognized as *Huygens' principle,* may be stated as follows: *Each point on a wave front may be regarded as a new source of disturbance.*

According to this principle, a wave front originating at a source **S** in Figure 12-4(A) arrives at the position **AB.** Each point in this wave front may be considered as a secondary source sending out wavelets. Thus from points **1, 2, 3,** etc., a series of wavelets develops simultaneously. After a time *t* these wavelets have a radius equal to *vt,* where *v* is the velocity of the wave.

The principle further states that the surface **A′B′,** tangent to all the wavelets, constitutes the new wave front. It is apparent from Figure 12-4 that spherical wave fronts are

Figure 12-4. By Huygens' principle, every point on an advancing wave front is regarded as a source of disturbance.

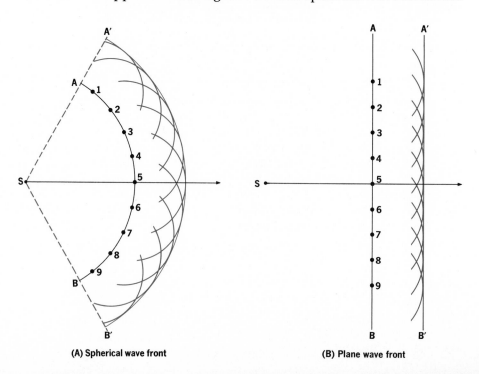

(A) Spherical wave front

(B) Plane wave front

propagated from spherical wavelets and planar wave fronts from planar wavelets.

The wave theory treats light as a train of waves having wave fronts perpendicular to the paths of the light rays. In contrast to the particle model discussed earlier, the light energy is considered to be distributed uniformly over the advancing wave front. Huygens thought of a ray merely as a line of direction of waves propagated from a light source.

The supporters of the wave theory were able to satisfactorily explain reflection and refraction of light. The explanation of refraction required that the speed of light in optically dense media, such as water and glass, be *lower* than the speed of light in air. They had trouble, however, explaining rectilinear propagation. This was the primary reason Newton rejected the wave theory.

Before the nineteenth century, interference of light was unknown and the speed of light in such media as water and glass had not been measured. Diffraction fringes or shadows had been observed as early as the seventeenth century. In the absence of knowledge of interference, however, neither Newton nor Huygens attached much significance to this diffraction phenomenon.

In 1801 the interference of light was discovered. This was followed in 1816 by the explanation of diffraction based on interference principles, as discussed in Chapter 15.

These two phenomena imply a wave character and cannot be satisfactorily explained by the behavior of particles. Thus despite the great prestige of Sir Isaac Newton, the corpuscular theory was largely abandoned in favor of the wave theory. The final blow to the corpuscular theory came when Foucault found that the speed of light in water was lower than the speed of light in air. Through the remainder of the nineteenth century the wave concept supplied the basic laws from which came remarkable advances in optical theory and technology.

The corpuscular theory was largely abandoned in favor of the wave theory following the discovery of interference and diffraction of light. Note that the speed of light in water had not yet been measured.

12.4 The Electromagnetic Theory Hot objects transfer heat energy by radiation. If the temperature is high enough, these objects radiate light as well as heat. If a light source is blocked off from an observer, its heating effect is cut off as well. For this reason, a cloud that obscures the sun's light cuts off some of the sun's heat at the same time.

The English physicist Michael Faraday (1791–1867) became concerned with the transfer of another kind of energy while investigating the attraction and repulsion of electrically charged bodies. In 1831 these experiments led him to the principle of the electric generator.

Figure 12-5. Before he was fifteen years old, James Clerk Maxwell (1831–1879), the Scottish theoretical physicist, wrote papers that were recognized for their scientific value. His laws of electromagnetic waves (Maxwell's equations) have the same role in electromagnetism that Newton's laws of gravitation and motion have in mechanics.

Faraday's practical mind required a model to interpret and explain physical phenomena. It was difficult for him to visualize electrically charged objects attracting or repelling each other at some distance with nothing taking place in the intervening space. Thus he conceived a space under stress and visualized *tubes of force* between charged bodies.

Faraday was not an astute mathematician and so did not put his model for this "transmission of electric force" into abstract mathematical form.

The Scottish mathematical physicist James Clerk Maxwell (1831–1879) set out to determine the properties of a medium that would transmit the energies of heat, light, and electricity. By the year 1865 he had developed a series of mathematical equations from which he predicted that all three are propagated in free space at the speed of light as *electromagnetic disturbances.* This unification, the *electromagnetic theory,* brought into common focus the various phenomena of radiation. Maxwell determined that the energy of an electromagnetic wave is equally divided between an electric field and a magnetic field, each perpendicular to the other, and both perpendicular to the direction of propagation of the wave. In Section 10.1 we defined a wave as a periodic disturbance that propagates itself through a medium or space. In this sense, *an electromagnetic wave is a periodic disturbance involving electric and magnetic forces.* A model of an electromagnetic wave at a given instant is shown in Figure 12-6.

Figure 12-6. Electromagnetic wave. The electric and magnetic fields are each at right angles to the direction of propagation. Shown is a graph of the two fields at a given instant.

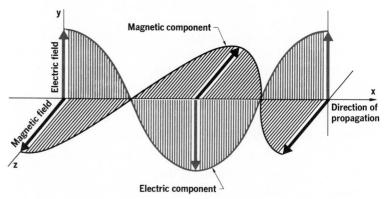

By 1885 experimental confirmation of the electromagnetic theory was achieved by the German physicist Heinrich Rudolf Hertz (1857–1894). Hertz showed that light transmissions and electrically generated waves are of the same nature. Of course, many of their properties are quite different because of great differences in frequency.

Maxwell's theory of electromagnetic waves seemed to provide the final architecture for optical theory; all known optical effects could now be fully explained. Many physicists felt at this time that all the significant laws of physics had been discovered and that there was little left to do other than develop new and more sophisticated techniques for measuring everything more accurately. In this connection Hertz stated, "The wave theory of light is, from the point of view of human beings, a certainty." Ironically, it was Hertz who was soon to discover a most important phenomenon having to do with the absorption of light energy, one that would create a dilemma involving the wave theory and at the same time set the stage for the *new physics* that was to emerge in the early years of the twentieth century.

Figure 12-7. Heinrich Hertz, the German experimental physicist whose research provided the experimental confirmation of the electromagnetic waves predicted by Maxwell.

12.5 The Electromagnetic Spectrum   Electromagnetic energy can be detected and measured by physical means only when it is intercepted by matter and changed into another form of energy such as thermal, electric, kinetic, potential, or chemical energy. Today, the electromagnetic spectrum is known to consist of a tremendous range of radiation frequencies extending from about 10 hz to more than 10^{25} hz. All electromagnetic radiations travel in free space with the constant velocity of 3×10^8 meters per second. From the wave equation ($v = f\lambda$) of Section 10.6, it is evident that the wavelengths (λ) of these radiations are inverse functions of their respective frequencies (f). Therefore, the range of the electromagnetic spectrum in terms of radiation wavelengths is from about 3×10^7 meters in the low-frequency region to less than 3×10^{-17} meter in the high-frequency region. See Figure 12-8. In terms of the angstrom (Å), a unit frequently used to express the wavelengths of electromagnetic radiations, the range is from about 3×10^{17} Å to 3×10^{-7} Å, an angstrom being equal to 10^{-10} meter.

Eight major regions of the electromagnetic spectrum are commonly recognized. These regions are based on the general character of the radiations and are identified in Figure 12-8. All kinds of electronic transmissions are accommodated in the *radio-wave* region. Commercial electricity falls within the *power* region. Observe the very small region occupied by the *visible spectrum*.

The optical spectrum includes those radiations, commonly referred to as light, that can be detected visually. Their wavelengths range from approximately 7600 Å to 4000 Å. Accordingly, *light may be defined as radiant energy*

Figure 12-8. The electromagnetic spectrum.

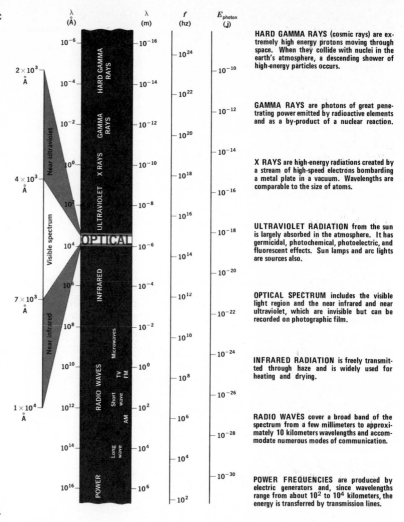

λ (Å)		λ (m)	f (hz)	E_{photon} (J)	

HARD GAMMA RAYS (cosmic rays) are extremely high energy protons moving through space. When they collide with nuclei in the earth's atmosphere, a descending shower of high-energy particles occurs.

GAMMA RAYS are photons of great penetrating power emitted by radioactive elements and as a by-product of a nuclear reaction.

X RAYS are high-energy radiations created by a stream of high-speed electrons bombarding a metal plate in a vacuum. Wavelengths are comparable to the size of atoms.

ULTRAVIOLET RADIATION from the sun is largely absorbed in the atmosphere. It has germicidal, photochemical, photoelectric, and fluorescent effects. Sun lamps and arc lights are sources also.

OPTICAL SPECTRUM includes the visible light region and the near infrared and near ultraviolet, which are invisible but can be recorded on photographic film.

INFRARED RADIATION is freely transmitted through haze and is widely used for heating and drying.

RADIO WAVES cover a broad band of the spectrum from a few millimeters to approximately 10 kilometers wavelengths and accommodate numerous modes of communication.

POWER FREQUENCIES are produced by electric generators and, since wavelengths range from about 10^2 to 10^4 kilometers, the energy is transferred by transmission lines.

Figure 12-9. X-ray photograph of a sea horse. Locate the X-ray region of the electromagnetic spectrum in Figure 12-8.

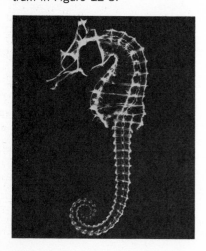

that a human observer can see. The optical spectrum also extends into the near infrared and into the near ultraviolet. Although our eyes cannot see these radiations, they can be detected by means of photographic film.

12.6 The Photoelectric Effect As mentioned in Section 12.4, Hertz conducted experiments that led to confirmation of the electromagnetic theory. While studying the radiation characteristics of oscillatory discharges, he observed that a spark discharge occurred more readily between two charged spheres when they were illuminated by another spark discharge. At about the same time, other investigators found that negatively charged zinc plates lost their charge when illuminated by the ultraviolet radiations from an arc lamp. Positively charged plates were not discharged when similarly illuminated.

Observations of the peculiar effects of ultraviolet radiation on metal surfaces led to the discovery of the photoelectric effect, a phenomenon that defied explanation based on the electromagnetic wave theory of light.

Figure 12-10 shows two freshly polished zinc plates **A** and **B** that are sealed in an evacuated tube having a quartz window and are connected externally to a battery and galvanometer (a sensitive current-indicating meter). The quartz window transmits ultraviolet radiation, which does not pass through glass. The galvanometer indicates a small current in the circuit when ultraviolet light falls on the negative plate **A**. If a sensitive electrometer circuit is substituted for the battery, it may be shown that the plate exposed to the ultraviolet light acquires a positive charge.

The results of these experiments imply that the action of the light on the zinc plate causes it to lose electrons. The German physicist Philipp Lenard (1862–1947) published the results of the first quantitative studies of the photoelectric phenomenon in 1902. By measuring the charge-to-mass ratio of the negative electricity derived from an aluminum plate illuminated by ultraviolet light, he was able to prove that electrons were ejected from the metal surface. Such electrons are called *photoelectrons.* Subsequent investigations have shown that all substances exhibit photoemission of electrons. *The emission of electrons by a substance when illuminated by electromagnetic radiation is known as the* **photoelectric effect.**

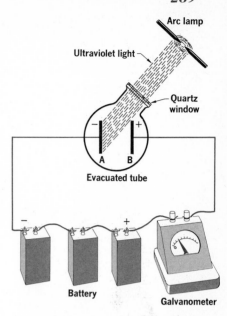

Figure 12-10. Apparatus for demonstrating the photoelectric effect.

12.7 Laws of Photoelectric Emission Plate **B** of Figure 12-10, having a positive potential V with respect to the emitter plate **A,** acts as a collector of photoelectrons ejected from the emitter. If the positive potential is increased enough, all photoelectrons are collected by plate **B** and the photoelectric current I reaches a certain limiting, or *saturation,* magnitude.

Curve **a** of Figure 12-11 is a graph of photoelectric current as a function of collector plate potential for a given source of light. Curve **b** shows the result of doubling the intensity of the light. Observe that the magnitude of the saturation current is doubled. This means, of course, that the rate of emission of photoelectrons is doubled. Here we have evidence of the *first law of photoelectric emission: The rate of emission of photoelectrons is directly proportional to the intensity of the incident light.*

For an electron to escape through the surface of a metal, work must be done against the forces that bind it within the surface. This work is known as the *work function.* The

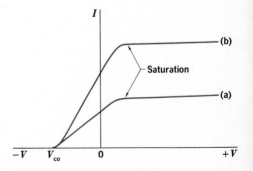

Figure 12-11. Photoelectric current as a function of collector plate potential. Curve (b) shows the effect of doubling the intensity of the same incident light.

photoelectrons must acquire from the incident light radiation the energy needed to overcome this surface barrier. If electrons acquire less energy than the work function of the metal, they cannot be ejected. On the other hand, if they acquire more energy than is required to pass through the surface, the excess appears as kinetic energy and consequently as velocity of the photoelectrons.

We may assume that light penetrates a few atom layers into the metal and that the photoelectric effect will occur at varying depths beneath the surface. Photoelectrons ejected from atom layers below the surface will lose energy through collisions in reaching the surface and must then give up energy equal to the work function of the metal in escaping through the surface. Photoelectrons ejected from the surface layer of atoms lose only the energy necessary to overcome the surface attractions. Thus in any photoelectric phenomenon we should expect photoelectrons to be emitted at various velocities ranging up to a maximum value possessed by electrons having their origin in the surface layer of atoms.

We can test the logic of these deductions by experimenting further with the photoelectric cell of Figure 12-10. A positive potential of a few volts on the collector plate **B** produces a saturation current as shown in Figure 12-11. If the collector plate potential is lowered towards zero, the photoelectric current decreases slightly, but at zero potential may still be close to the saturation magnitude.

As the collector plate potential is made slightly negative with respect to the emitter, the photoelectric current decreases also. By increasing this negative potential a value is reached where the photoelectric current drops to zero. This is called the *stopping*, or *cutoff*, potential V_{co} and is shown in Figure 12-11.

The cutoff potential "cuts off" the photoelectric current at the collector plate.

A negative potential on the collector plate repels the photoelectrons, tending to turn them back to the emitter plate. Only those electrons having enough kinetic energy and velocity to overcome this repulsion reach the collector. As the cutoff potential is approached, only those photoelectrons with the highest velocity reach the collector. These are the electrons with the maximum kinetic energy that are emitted from the surface layer of the metal. At the cutoff potential even these electrons are repelled and turned back to the emitter.

Thus the negative collector potentials, which repel rather than attract photoelectrons, reveal something about the kinetic energy distribution of the ejected electrons. As this potential is made more negative, a nearly linear decrease in photoelectron current shows us that the photo-electrons do have a variety of velocities. The cutoff potential measures the kinetic energy of the *fastest* photoelectrons.

The cutoff potential measures the maximum kinetic energy of photoelectrons.

From curves **a** and **b** shown in Figure 12-11 observe that the photoelectric currents produced by different intensities of incident light from a certain source reach zero at the same collector potential. Thus the cutoff potential for a given photoelectric system and *the velocity of the electrons expelled* are independent of the intensity of the light source. From these observations we can formulate a **second law of photoelectric emission:** *The kinetic energy of photoelectrons is independent of the intensity of the incident light.*

The second law of photoelectric emission makes trouble for the wave theory.

The American physicist Robert A. Millikan (1868–1953) performed many experiments with various emitters and light sources. By illuminating emitters made of sodium metal with light radiations of different frequencies, Millikan found that the cutoff potential had different values for the various frequencies of incident light. By plotting the cutoff potentials V_{co} as a function of light frequencies f, the straight line shown in Figure 12-12 is obtained. Millikan demonstrated that *the cutoff potential depends only on the frequency of the incident light.*

Recalling that the cutoff potential measures the kinetic energy of the fastest photoelectrons ejected in a particular photoelectric system, we must conclude that the *maximum kinetic energy of photoelectrons increases with the frequency of the light illuminating the emitter.*

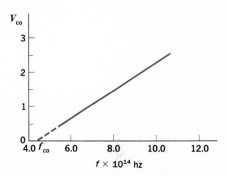

Figure 12-12. A plot of cut-off potentials for sodium as measured by Millikan at various frequencies of illumination. The cut-off frequency of sodium is 4.4×10^{14} hz.

Other physicists, notably A. L. Hughes at the Cavendish Laboratory in England and K. T. Compton at Princeton University, obtained similar results in numerous experiments. All these experiments further showed that *for each kind of surface there is a characteristic threshold or cutoff frequency f_{co} below which the photoelectric emission of electrons ceases* no matter how intense the illumination.

Only a few elements, in particular the alkali metals, exhibit photoelectric emission with ordinary visible light. The cutoff frequency of cesium is in the infrared region of the electromagnetic spectrum. Millikan determined the f_{co} of sodium to be 4.4×10^{14} hz, which is red light with a wavelength of about 6800 Å. The cutoff frequency of potassium is in the visible green region. Those of copper and platinum are in the ultraviolet and deep ultraviolet.

The number of materials available as photoelectric emitters with visible light is limited.

For photoelectric emission from any surface, the incident light radiation must contain frequencies higher than the cutoff frequency characteristic of that surface. The maximum kinetic energy of photoelectrons emitted from any surface can be increased only by raising the frequency of the light illuminating the surface. These facts provide us with a ***third law of photoelectric emission:*** *Within the region of effective frequencies, the maximum kinetic energy of photoelectrons varies directly with the difference between the frequency of the incident light and the cutoff frequency.*

12.8 Failures of the Wave Theory The experiments of Hertz, together with those which followed, confirmed that Maxwell's electromagnetic theory correctly describes the transmission of light and other kinds of radiation. This wave theory requires the radiated energy to be distributed uniformly and continuously over a wave front. The higher the intensity of the radiating source, the greater should be the energy distributed over the wave front.

The first law of photoelectric emission does not imply any deviation from the electromagnetic theory since the magnitude of the photoelectric current is proportional to the incident light intensity. The surprising thing is that the velocity of the photoelectrons is not raised with an increase in the intensity of the illumination on the surface of an emitter, contrary to what the wave theory suggests. Measurements of emission cutoff potentials over very large ranges of intensities show that the maximum kinetic energy of photoelectrons is independent of the light intensity.

According to the wave theory, light of any frequency should cause photoelectric emission provided it is intense enough. Experiments show, however, that all substances have characteristic cutoff frequencies below which emission does not occur, no matter how intense the illumination. On the other hand, a very feeble light containing frequencies above the threshold value causes the ejection of photoelectrons.

Again, from the wave theory we could argue that given enough time an electron in an area illuminated by a very feeble light would "soak up" enough energy to escape from the surface. No such time lag between the illumination of a surface and the ejection of a photoelectron has ever been measured. If emission of photoelectrons occurs at all, it begins simultaneously with the illumination of a photosensitive surface.

The wave theory does not provide an explanation for the photoelectric effect.

The wave theory, which serves so well in the explanation of radiation transmission phenomena, is incapable of describing the processes of radiation absorption observed in the photoelectric effect. In this instance a particle model of light appears to be useful. Had Newton perhaps been on the right track after all? The classic particles of Newton's theory could explain the relation between photoelectron emission rate and intensity of illumination but not that between photoelectron velocities and frequency of illumination.

The discovery of the photoelectric effect, coming as it did from the experiments that established the correctness of the electromagnetic wave theory, presented a paradox to twentieth-century physicists. Was there no theory to explain these widely divergent phenomena?

12.9 The Quantum Theory   Aside from the disturbing implications that the photoelectric effect introduced into twentieth-century physics, physicists were much concerned about discrepancies between their experimental studies of absorption and emission of radiant energy and the requirements derived from the electromagnetic radiation theory.

The first great step toward the resolution of these discrepancies was taken by the German theoretical physicist Max Planck (1858–1947). See Figure 12-13. In 1900 Planck was investigating the spectral distribution of electromagnetic radiation from a hot body. Classic electromagnetic theory predicted that the emission intensity would increase continuously as the radiation frequency increased. The spectral distribution of radiated energy measured experimentally was very different from that predicted by theory at the higher frequencies. It approached zero rather than infinity as the frequency increased without limit. Clearly something was wrong in this prediction that the radiated energy would keep on increasing with frequency without limit.

Planck found that he could bring radiation theory and experiment into agreement by assuming the energy emitted by the radiating sources to be an integral multiple of a fundamental quantity hf, where h is a universal constant now known as *Planck's constant*, and f is the frequency of these radiating sources. To explain why this bold assumption worked, he postulated that light is radiated and absorbed in indivisible packets, or quanta. We now call these packets, or quanta, *photons.*

The amount of energy comprising a photon is determined by the frequency of the radiation. It is directly proportional to this frequency since

$$E = hf$$

When f is given in hertz and h in joule seconds ($h = 6.63 \times 10^{-34}$ j s), the energy, E, of a photon is expressed in joules.

Figure 12-13. Max Planck, the German physicist, developed the initial assumptions on which the quantum theory is based while studying radiation phenomena.

Planck published his quantum hypothesis in 1901 and, although not immediately accepted, it was destined to profoundly influence the *new physics* emerging with the twentieth century. Certainly there is nothing in the classical physics of Newton to suggest that certain values of energy should be allowed and others should not.

Albert Einstein recognized the value of Planck's quantum hypothesis in connection with the photoelectric effect. He reasoned that since emission and absorption of light radiation occur discontinuously, certainly the transmission field should be discontinuous. Proceeding with this hypothesis, in 1905 Einstein published a very simple and straightforward explanation of the photoelectric effect. A few years later, the accurate experimental work of Millikan, Hughes, and Compton established the correctness of Einstein's explanation. The bold extension of Planck's quantum ideas by Einstein firmly proved *the **quantum theory**, which assumes that the transfer of energy between light radiations and matter occurs in discrete units called quanta, the magnitude of which depends on the frequency of the radiation.*

12.10 Einstein's Photoelectric Equation   According to Einstein, light illuminating an emitter surface consists of a stream of photons. When a photon is absorbed by the emitter, its quantum of energy *hf* is transferred to a single electron within the surface. If the acquired energy is sufficient to overcome the surface barrier and the electron is moving in the right direction, the electron will escape from the surface. In penetrating the surface, the electron must give up a certain energy *w*, the work function of the substance. If the energy *hf* imparted by the photon is greater than *w*, the electron will have kinetic energy as indicated by its velocity after leaving the surface.

The maximum kinetic energy possessed by photoelectrons ejected from an emitter illuminated by light of frequency *f* was given by Einstein as

$$\tfrac{1}{2}mv^2_{\text{max}} = hf - w$$

where *m* and *v* are the mass and velocity, respectively, of the photoelectrons, *h* is Planck's constant, *f* is the frequency of the impinging light radiation (the product *hf* is the energy of the photon), and *w* is the work function of the emitting material.

Now, recalling the failures of the wave theory to explain photoelectric emission, we find no such difficulties in the application of Einstein's photon hypothesis. A lower light intensity means fewer photons impinging on the emitter surface in a given period of time and fewer photoelectrons ejected. However, as long as the frequency of the incident light remains unchanged, each photon transfers the same energy hf to the electron when a collision occurs.

Einstein's photoelectric equation shows clearly why light of too low a frequency will not cause photoelectric emission from a given material no matter how intense the illumination. Photon energy is a linear function of the frequency of the light since it equals the product hf. Now if $\frac{1}{2}mv^2_{max} = 0$, it follows that

$$hf_{co} = w$$

Recall that f_{co} for an emitting surface is the illumination frequency below which the photoemission of electrons ceases.

Here the photon can impart to the electron just enough energy for it to penetrate the surface barrier. No energy is left over to appear as electron velocity. The frequency in this situation is the cutoff frequency f_{co}. Illumination of any lower frequency on this emitter will consist of photons with energy $hf < w$, and photoelectric emission cannot occur no matter how many photons there are.

If photons have enough energy to eject electrons, no "soaking up" time is required before a feeble light can start the photoelectric emission process. This is true because the ejection energy for each photoelectron is delivered in a single concentrated bundle.

The quantum theory, which provides a particle model of light, meets every objection raised when the electromagnetic wave theory is employed to interpret photoelectric emission phenomena. Yet many experiments have proved the correctness of the wave theory. When we consider the full range of radiation phenomena, we are persuaded that light must have a dual character. In some circumstances it behaves like waves, and in other circumstances it behaves like particles. *The modern view of the nature of light recognizes this dual character: Radiant energy is transported in photons that are guided along their path by a wave field.*

Light exhibits a dual character: that of waves and of particles.

12.11 The Quantized Atom

Striking evidence favoring the quantum theory was given by Niels Bohr (1885–1962), who in 1913 devised a model of the atom based on quantum ideas. Ernest Rutherford (1871–1937), the English physicist, had conducted experiments (described in Section 23.7) that in 1911 led him to postulate the nuclear atom with its planetary arrangement of electrons. This model of Rutherford's immediately encountered difficulties when subjected to the classical laws of physics, that is, the known laws of mechanics and electromagnetism as formulated by Newton and Maxwell.

The electromagnetic theory predicts that charged particles undergoing acceleration must radiate energy. Indeed,

Figure 12-14. Niels Bohr, the Danish physicist, demonstrating a point on his blackboard. Bohr received the Nobel Prize in 1922 for his work in atomic physics.

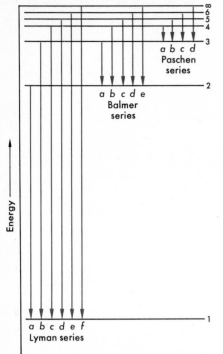

Figure 12-15. This electronic energy-level diagram for hydrogen shows some of the transitions that are possible in this atom.

Figure 12-16. Representative lines in the hydrogen spectrum. The small letter below each line indicates which of the energy-level transitions shown in Figure 12-15 produces it.

this accounts for the radiation of energy from a radio transmitting antenna. A planetary electron moving about the nucleus of an atom experiences acceleration and accordingly should radiate energy. As energy is drained from the electron, it should spiral in toward the nucleus causing the atom to collapse. Of course atoms do not collapse, and so this classical model cannot be correct.

Bohr assumed that an electron in an atom can move about the nucleus in certain discrete orbits without radiating energy. Such orbits represent "allowable" energy levels in the Planck concept. The level closest to the nucleus corresponds to the orbit of lowest energy. This postulate could account for the stability of an atom. However, atoms do radiate energy. Bohr assumed further that an electron may "jump" from one discrete orbit to another of lower energy. In the process a photon is emitted. Its energy represents the difference between the energies of the levels involved in the transition. The frequency of the emission is proportional to this energy difference, the proportionality factor being Planck's constant h. Thus the frequency of the photon emission depends on the magnitude of this energy change ΔE and can be expressed as $\Delta E/h$.

The return of excited atoms to their stable state results in the emission of *bright line spectra*. (See Color Plate VII, Chapter 14.) Excited hydrogen atoms produce the simplest such atomic spectra. Hydrogen lines appear in several series in and beyond the visible region of the radiation spectrum. These include the *Balmer series* in the visible region, the *Lyman series* in the ultraviolet region, and the *Paschen series* in the near infrared region. See Figures 12-15 and 12-16.

Numerous attempts had been made to account for the discrete frequencies of the bright lines observed in these spectra. It was Bohr who first associated spectral lines

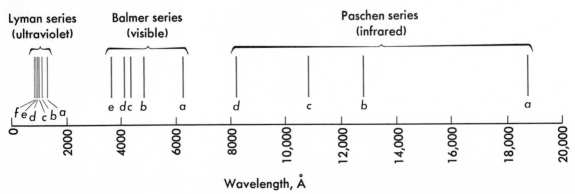

Wavelength, Å

with pairs of energy levels, thus marking the initial step in the development of the *quantum mechanics* of atomic structure.

Bohr's atom model provided no information about the mechanics of photon emission when an electron passes from a higher to a lower energy level; but his concept of energy levels provided a great stimulus for theoretical studies in this field. In about 1924 the French physicist Louis de Broglie (b. 1892) suggested that the dual particle-and-wave nature of light provided evidence of the wave nature of particles as well. Accepting Einstein's idea of the equivalence between mass and energy, he postulated that *in every mechanical system, waves are associated with matter particles.* The application of this concept of matter waves to the study of the structure of matter is known as *wave mechanics.* Just as the ordinary laws of mechanics are essential to any explanation of the behavior of objects of large dimensions, *wave mechanics* is necessary in dealing with objects that have atomic and subatomic dimensions.

A photon has energy that is the product of Planck's constant and its radiation frequency.

$$E = hf$$

In Einstein's equation $E = mc^2$ for mass-energy equivalency, m is the mass of the particle while in motion. When at rest, its mass m_0, known as the rest mass, is smaller than m.

If we consider the energy of the photon to be equivalent to that of a moving particle, then

$$hf = mc^2$$

and we could solve for m. However, this solution would also imply a certain rest mass for the photon. But a photon is never at rest; it always moves with the speed c. Photons do have momentum and can transfer it to any surface on which they impinge. Calculating this momentum mc, we get

$$mc = \frac{hf}{c}$$

Since
$$c = f\lambda$$

by substitution,
$$mc = \frac{h}{\lambda}$$

or
$$\lambda = \frac{h}{mc}$$

The wavelength of the photon may be expressed in terms of its momentum mc. Similarly, the wavelength of any matter particle having a velocity v may be expressed in terms of its momentum mv.

$$\lambda = \frac{h}{mv}$$

There is abundant evidence today of the wave nature of subatomic particles. Accelerated electrons have been found to behave like X rays. The electron microscope is an application of electron waves. The development of the quantum theory and the system of wave mechanics have provided physicists with their most powerful means of studying the structure and properties of matter. Matter waves will be discussed more extensively in Chapter 25.

12.12 Coherent Light: the Laser Common light sources such as fluorescent and incandescent lamps radiate light in all directions over a wide range of frequencies with random phase relationships. Even groups of waves of the same frequency have random phase relationships. These disordered light waves are said to be *incoherent*.

A stationary interference pattern is described in Section 10.15 and represented in Figures 10-27 and 10-28. To produce such a pattern, it is necessary to employ two wave trains with identical wavelengths and a constant phase relationship. Such wave trains have *coherence;* they are said to be *coherent* waves.

It is not difficult to obtain coherence from two sources of sound waves. However, a stationary interference pattern from light waves is ordinarily produced in a high school laboratory by means of a subtle arrangement using a single source of incoherent light.

Coherent light is produced by a remarkable instrument called a *laser*. It is a light amplifier that functions because of stimulated emissions of radiation. The term "laser" is an acronym derived from the description of its function: **L**ight **A**mplification by **S**timulated **E**mission of **R**adiation. The laser output is a single-frequency (monochromatic) radiation of parallel waves with a constant phase relationship. An example of a laser application is shown in Figure 12-17. See also Color Plate VI B, Chapter 14.

The first successful laser produced a pulsed coherent red beam from a ruby crystal. The ruby laser was followed by gaseous lasers which produced continuous coherent beams. Today there are many different types of lasers using gases, liquids, crystals, semiconductors, glass, and plastics. These lasers produce coherent beams of different

Incoherent waves differ in frequency or phase, or both.

Coherent waves have the same frequencies and a constant phase relationship.

Figure 12-17. A laser used in conjunction with an aircraft instrument landing.

spectral colors. Laser research is continuing on a broad front. Industry, engineering, medicine, communications, photography, and the military are areas in which lasers are being used. Scientists and engineers are constantly finding new applications.

The underlying principle of laser technology can be illustrated with a model of the basic ruby laser system shown in Figure 12-18. The model consists of a pencil-shaped ruby crystal and an external flash lamp with its power source. The ends of the ruby crystal are optically flat, silvered reflecting surfaces, one end being partly silvered to allow some light to emerge. The flash lamp surrounds the ruby crystal as illustrated. When this lamp flashes, photons enter the ruby crystal and initiate the laser action.

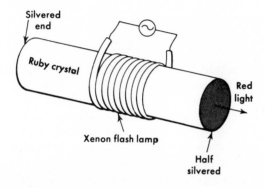

Figure 12-18. Simplified model of a ruby laser.

The effect of the flash lamp is to "pump" energy into the crystal. Some atoms of the crystal become excited by absorbing photons and are elevated to a higher energy state. They may then emit some of the energy and make a transition to an intermediate state. From this intermediate state an atom may return to its stable state either by spontaneous or stimulated emission of a photon of light energy equal to its energy level difference.

Suppose one atom emits a photon spontaneously. If traveling in the proper direction, the photon is reflected at the end of the ruby crystal. As it travels back through the crystal, the photon stimulates another atom in the intermediate state to emit a photon. This process continues, and the intensity of the photon beam in the crystal is *amplified*.

With the reflective ends of the crystal the proper distance apart, a standing wave of single-frequency red light is set up in the ruby rod. This particular standing light wave is analogous to the standing sound wave in an organ pipe. As the photon beam is amplified, photons pass through the partially reflective end of the rod in the form of intense pulses of coherent red light. Gaseous lasers produce a continuous but less intense beam of coherent light. The laser provides practical evidence of the quantized energy levels of atoms as proposed by Bohr in 1913. See Section 12.11.

12.13 Production of X Rays In the photoelectric effect, photons of frequency f cause the ejection of electrons from atoms of a substance with a maximum velocity given by the equation

$$\tfrac{1}{2}mv^2_{\text{max}} = hf - w$$

It is not surprising that the converse of this phenomenon can be produced also. High-speed electrons with velocity v projected against the surface of a target cause the emission of photons of radiation having a maximum frequency given by the photoelectric equation

$$\tfrac{1}{2}mv^2 = hf_{\text{max}} - w$$

The work function w of target material is negligible compared to the photon energy at X-ray frequencies and is usually omitted from the above equation.

This phenomenon is sometimes referred to as the *inverse photoelectric effect*. An example is the production of accelerated electrons. Each electron that strikes the target loses its energy, and this energy reappears as photons of radiation. Neglecting the work-function energy, as previously indicated, no more than the full kinetic energy of the electron can appear in the X-ray photons as the energy hf. Since h is constant, the maximum frequency f_{max} of the X-ray radiation is determined by the velocity v of the electron.

The penetrating properties of X rays and their many practical uses stemming from these properties are well known. X rays lie deep in the high frequency end of the electromagnetic spectrum beyond the ultraviolet region. The energy hf of X-ray photons is very high, being of the order of 10^4 times the energy of photons in the visible region. As a consequence of this high photon energy and the very short wavelengths, X rays are used extensively in numerous areas of research. Table 12-1 gives photon energies for important regions of the spectrum.

12.14 The Pressure of Light The electromagnetic theory of Maxwell predicted that incident light should exert a pressure applied in the direction of the radiation. According to Maxwell, the pressure on a totally reflecting surface should be twice that on a totally absorbing surface illuminated by the same radiation.

Table 12-1
PHOTON ENERGIES

Radiation	Typical value of f (hz)	hf (j)
gamma rays	3.0×10^{19}	2.0×10^{-14}
X rays	3.0×10^{18}	2.0×10^{-15}
visible light	6.0×10^{14}	4.0×10^{-19}
heat waves	3.0×10^{13}	2.0×10^{-20}
radio waves	3.0×10^{6}	2.0×10^{-27}

In 1901 Nichols and Hull in the United States and Lebedev in Russia succeeded experimentally in measuring the pressure of light. Their results were in such close agreement with theory as to provide proof of the existence of light pressure.

What does the particle model tell us about the pressure of light? Our common experiences with pressure are those involving collisions of particles in motion. Since particle collisions are momentum problems, conservation of momentum applies.

A perfectly elastic ball rebounding from an object transfers twice the momentum that a perfectly inelastic ball having the same mass and velocity would transfer. In Section 12.11 we expressed the momentum of a photon in terms of its frequency as

$$mc = \frac{hf}{c}$$

When a beam of photons is incident on a totally absorbing surface, the pressure exerted depends on the rate of change of photon momentum per unit area of illuminated surface.

$$p = \frac{F}{A} = \frac{\Delta(hf/c)}{\Delta t \times A} = \frac{\Delta hf}{c \Delta t A}$$

The pressure, p, has the dimensions

$$p = \frac{\text{j s/s}}{\text{m/s} \times \text{s} \times \text{m}^2} = \frac{\text{j}}{\text{m}^3} = \frac{\text{n}}{\text{m}^2}$$

If the photon energy hf is totally reflected from the surface of a body, the momentum change is $2hf/c$ and the pressure of light is twice that for the totally absorbing surface illuminated by the same beam.

The pressure of light is exceedingly small in comparison to pressures we commonly experience. The pressure of sunlight on the earth is approximately 4×10^{-11} standard atmosphere. Comet tails always point away from the sun, apparently due to the pressure of sunlight and solar particles on the extremely diffuse matter in them.

Pressure, p, is force per unit area. Force, F, is time rate of change of momentum.

Questions

GROUP A

1. What is the modern view of the nature of light?
2. Arrange the following in order of their increasing wavelength: visible light, infrared radiations, ultraviolet radiations, X rays, radio waves, gamma rays.
3. Compare visible light and sound as to: (a) origin, (b) transmitting media, (c) wavelength, (d) type of wave, (e) speed.

4. How does an atom of a substance radiate energy according to the quantum theory?

5. The flatiron is a source of radiations that cannot be detected by the eye. (a) Would you expect an ordinary photoelectric cell to detect these radiations? Explain. (b) Suggest a way to prove that such radiations are emitted.

GROUP B

6. A beam of light passes through a small aperture and illuminates the blackened bulb of a thermometer. A piece of ordinary glass is placed over the aperture and the thermometer reading drops two degrees. The glass is replaced by a quartz window and the thermometer reading returns approximately to the original value. What do the results suggest concerning the nature of the light source? Explain.

7. (a) State the three laws of photoelectric emission.

(b) To what extent is a wave model of light successful in explaining these laws?

8. Given a monochromatic (single frequency) light source to illuminate a photocell, how can you explain the fact that photoelectrons are ejected at various velocities ranging up to a maximum value?

9. How does the pair of curves shown in Figure 12-11 show that the velocity of photoelectrons is independent of the intensity of the light illuminating the photocell?

10. What special significance can you attach to the photoelectric situation in which the photon energy hf transferred to an electron is just equal to the work function w of the emitter?

11. (a) Distinguish between coherent and incoherent light waves. (b) Describe the light emitted by a laser.

12. How would you defend the assertion that the production of X rays illustrates the inverse photoelectric effect?

ILLUMINATION

12.15 Luminous and Illuminated Objects Nearly all the natural light we receive comes from the sun. Distant stars account for an extremely small amount of natural light. Of course, moonlight is sunlight reflected from the surface of the moon.

We produce light from artificial sources in several ways. Materials may be heated until they glow, or become incandescent, as in an electric lamp. Molecules of a gas at reduced pressure may be bombarded with electrons to produce light, as in a neon tube. Most of the visible light from fluorescent tubes results from the action of ultraviolet radiations on phosphors that coat the inside surface of the glass tubes. The firefly produces light by means of complex chemical reactions.

Most artificial sources of light are hot bodies that radiate energy in the infrared region of the spectrum as well as visible light. The energy of the *thermal radiation* emitted by hot bodies depends upon the temperature of the body and the nature of its surface.

At a temperature of 300 °C the most intense radiation emitted by a hot body has a wavelength of about 5×10^{-4} cm, well down in the infrared region of the electromagnetic spectrum. As the temperature is raised to 800 °C, enough radiation is emitted in the visible region to cause the body to appear "red hot" although the bulk of the energy radiated is still in the infrared region. If the temperature of the body is raised to 3000 °C, the most intense emission remains in the near infrared region; however, there is enough blue visible radiation to cause the body to appear "white hot."

Visible spectrum: 7.6×10^{-5} cm to 4.0×10^{-5} cm.

This last temperature is near that of the filament of an incandescent lamp. Hence such lamps have low efficiencies as producers of visible radiation. Generally the efficiency improves as the filament temperature is raised.

The white-hot filament of an incandescent lamp is said to be *luminous*. It is visible primarily because of the light it emits. *An object that gives off light because of the energy of its accelerated particles is said to be **luminous**.* The sun and the other stars are luminous objects.

Just as radiant heat may be reflected, light may be reflected from the surfaces of objects. Mirrors reflect a beam of light in a definite direction. Other surfaces scatter the light that is incident on them in all directions. *An object that is seen because of the light scattered from it is said to be **illuminated**.* The moon is illuminated, for it reflects radiant energy from the sun. Some of the light energy arriving at the surface of an object is reflected, some of it is transmitted, and some of it is absorbed by the substance.

Any dark-colored object absorbs light, but a black object absorbs nearly all the light it receives. When the rays of the sun strike a body of water vertically, most of them are either absorbed or transmitted. Most of the rays are reflected, however, when they strike the water at an oblique angle. For this reason the image of the sun can be seen in the water without discomfort when the sun is directly overhead, but not when the sun is near the horizon.

Air, glass, and water transmit light readily and are said to be *transparent*. Other substances transmit light but scatter or diffuse it so that objects seen through them cannot be identified. Such substances are *translucent*. Typical examples are frosted electric lamps and parchment lampshades. *Opaque* substances do not transmit light at all.

From a luminous point source, light waves travel outward in all directions. If the medium through which they pass is of the same nature throughout, the light progresses along the normal to the wave front. A single line of light from a luminous point is called a *ray;* a group of closely spaced rays forms a *beam* of light. Small beams are referred to as *pencils.* See Figure 12-19. Light coming from the sun is in rays so nearly parallel that it may be considered as a parallel beam. When several rays of light come from a point, they are called a *diverging* pencil, while rays proceeding toward a point form a *converging* pencil. When the sun's rays pass through a magnifying glass, they converge at a point called a *focus.*

Figure 12-19. The formation of beams and pencils of light.

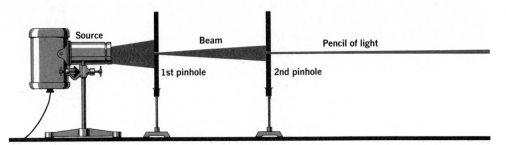

Because an opaque object absorbs light, it casts a shadow in the space behind it. When the source of light is a point, as in Figure 12-20(A), an opaque ball, **B,** cuts off all the rays that strike it and produces a shadow of uniform darkness on the screen, **S.** If the light comes from an extended source, the shadow varies in intensity, as shown in Figure 12-20(B). The part from which all the rays of light are excluded is called the *umbra;* the lighter part of the shadow is the *penumbra.* Within the region of the penumbra the luminous source is not entirely hidden from an observer. The umbral and penumbral regions of the moon's shadow cast during a solar eclipse are shown in Figure 12-21.

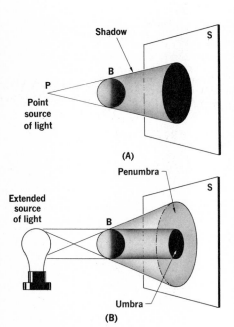

Figure 12-20. An obstruction in the path of light from a point source (A) casts a shadow of uniform density. If the light is from an extended source (B), the shadow is of varying density.

12.16 The Speed of Light The speed of light is one of the most important constants used in physics, and the determination of the speed of light represents one of the most precise measurements ever achieved. Before 1675 light propagation was generally considered to be instantaneous, although Galileo had suggested that a finite time was required for it to travel through space. In 1675 the Danish astronomer Olaus Roemer (1644–1710) determined the first value for the speed of light to be 140,000 miles per second. This is approximately equivalent to 225,000 kilometers per second. He had been puzzled by a variation in his calculations of the time of eclipse of one of Jupiter's satellites as seen from different positions of the earth's orbit about the sun. Roemer concluded that the variation was due to differences in the distance that the light traveled to reach the earth.

Very precise modern measurements of the speed of light are made using laboratory methods. The most notable experiments were performed by Albert A. Michelson (1852–1931), a professor of physics at the University of Chicago, who measured the speed of light in air and in a vacuum with extraordinary precision.

1. The speed of light in air. Michelson measured the speed of light over the accurately determined distance between Mt. Wilson and Mt. San Antonio, California. His method is illustrated in principle by Figure 12-22. The light source, octagonal mirror, and telescope were located on Mt. Wilson and the concave mirror and plane mirror were located on Mt. San Antonio, 35.4 km (approximately 22 mi) away. The octagonal mirror, **M**, could be rotated rapidly under controlled conditions and was timed very accurately.

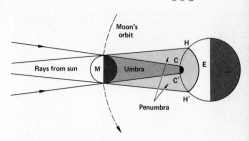

Figure 12-21. A diagram of a total solar eclipse showing the umbral and penumbral regions of the moon's shadow. CC' is the region of totality on the earth; CH and C'H' are the regions from which a partial eclipse is seen. (Sizes and distances are not to scale.)

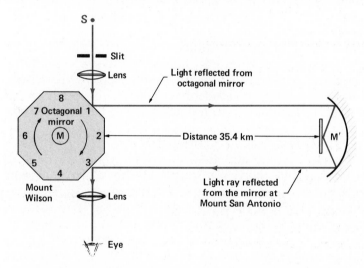

Foucault first used a rotating mirror in 1850 to compare the speed of light in water and in air.

Figure 12-22. A diagram of Michelson's octagonal-mirror method for measuring the speed of light.

With mirror **M** stationary, a pencil of light from the slit opening was reflected by **M₁** to the distant mirror **M'**, from which it was returned to **M₃**. The image of the slit in the mirror at the **M₃** position could be observed accurately through the telescope. The octagonal mirror was then set in motion and the speed of rotation brought up to the value that moved **M₂** into the position formerly occupied by **M₃** during the time required for the light to travel from **M₁** to Mt. San Antonio and return. The slit image was again seen in the telescope precisely as it was when the octagon was stationary. The light traveled twice the optical path of 35.4 km in one eighth of the time of one revolution of the octagonal mirror. Thus

$$c = \frac{2MM'}{t}$$

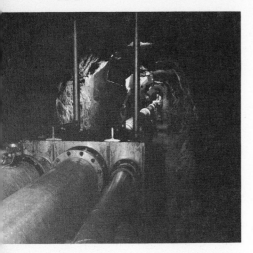

Figure 12-23. A laser beam is directed down a 30-meter vacuum tube in an experiment to measure the speed of light. The apparatus is installed in an abandoned gold mine and is so sensitive that it records "tides" in the earth's crust caused by the sun and moon.

A proposal is now under consideration within the International Bureau of Weights and Measures which could fix the speed of light in a vacuum as a constant at 299,792,458 m/s. The standard meter could then be defined as the distance traveled by light in a vacuum during the time interval of 1/299,792,458 s.

where MM' is the optical path, t is the time of one-eighth revolution of the octagon, and c is the speed of light in air.

Michelson's investigation of the speed of light in air required several years to complete and extremely high precision was attained in making the necessary observations. The optical path was measured by the U.S. Coast and Geodetic Survey and found to be 35,385.5 meters, accurate to about one part in seven million. The rate of the revolving mirror was measured by stroboscopic comparison with an electric signal of standard frequency. The average of a large number of determinations yielded a speed of light in air of 299,729 km/s.

2. The speed of light in a vacuum. Michelson conducted similar experiments using an evacuated tube one mile long to eliminate the problems of haze and variations in air density. In these investigations he determined the speed of light to be 299,796 km/s, which he believed to be accurate to within 1 km/s.

Modern laboratory methods of measuring the speed of light require very complex apparatus and are considered to be more accurate than the methods used by Michelson. In some experiments, electromagnetic waves much longer than light waves have been used and good agreement has been found between their speed and that of visible light. These findings provide experimental confirmation of Maxwell's electromagnetic theory, which requires that all electromagnetic waves throughout the electromagnetic spectrum have the same speed c in free space. Thus we must consider the speed of light within the larger framework of the speed of electromagnetic radiations in general.

As Michelson's figures show, the speed of light is slightly higher in a vacuum than in air. The speed in free space generally accepted as most accurate is

$$c = 2.99792458 \times 10^8 \text{ m/s}$$

with an uncertainty of 1.2 m/s. Physicists now conduct laser experiments (Figure 12-23) in their quest for improved accuracy in the determination of this important constant.

12.17 Light Measurements   *The quantitative study of light is called* **photometry.** *Three quantities are generally measured in practical photometry: the* luminous intensity *of the source; the* luminous flux, *or light flow, from a source; and the* illumination *on a surface.*

1. *Luminous intensity.* In order to derive a set of photometric units it is necessary to introduce an additional fundamental unit into our measurement system. The unit for
luminous intensity, *I*, is the arbitrary choice. It is the *candle*, cd. The candle is known internationally as *candela*.

Originally the light from a certain type of candle was
used as a standard, but this has been replaced by a more
readily reproducible source of luminous intensity. *The
candle is one sixtieth of the luminous intensity of a square centimeter of a black-body radiator maintained at the temperature of
freezing platinum (2046 °K).* In practice it is convenient to
use incandescent lamps that have been rated by comparison with the standard.

Incandescent lamps used for interior lighting generally
have an intensity ranging from a few candles to several
hundred candles. The common 40-watt lamp has an intensity of about 35 candles; a 100-watt lamp gives about 130
candles. Note that a 40-watt fluorescent lamp has an intensity of about 200 candles. The intensity of any lamp depends on the direction from which it is measured; the average candle power in all directions in space (spherical
candle power) is often given.

2. *Luminous flux.* Not all of the energy radiated from a
luminous source is capable of producing a visual sensation. Most of the radiation is in the infrared region, and a
small amount is in the ultraviolet region. The rate of flow
of visible radiation is called *luminous flux*, the symbol for
which is the Greek letter Φ (phi). ***Luminous flux** is that part
of the total energy radiated per unit of time from a luminous
source that is capable of producing the sensation of sight.* The
unit of luminous flux is the *lumen*, lm.

Suppose a standard light source of 1 candle is placed at
the center of a hollow sphere having a radius of 1 meter, as
illustrated in Figure 12-24. The luminous source is pre-
sumed to radiate light equally in all directions and to have
such small dimensions that it may be termed a "point
source." The area of the surface of a sphere of radius *r* is
equal to $4\pi r^2$. Since the radius of the unit sphere is 1
meter, its surface area is $4\pi \times (1 \text{ m})^2$. One lumen of flux is
radiated by the 1-candle source to each square meter of
inside surface of the sphere. *The **lumen** is the luminous flux
on a unit surface all points of which are at unit distance from a
point source of one candle.* The lumen is not a measure of a
total quantity of luminous energy but a *rate* at which luminous energy is being emitted, transmitted, or received.

The unit surface area of the unit sphere is intercepted by
a solid angle, *ω*, of 1 *steradian*, sr. The unit surface area of a

*The original standard candle was
one made of spermaceti (wax obtained from the oil of sperm
whales) that burned 120 grains
per hour.*

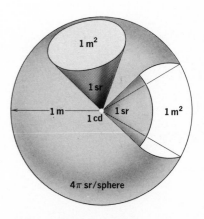

Figure 12-24. A 1-candle point
source radiates luminous flux at
the rate of 4π lumens.

sphere of radius r is intercepted by a solid angle of $1/r^2$ steradians. *The ratio of the intercepted surface area of a sphere to the square of the radius is the measure of the solid angle in steradians.*

$$\omega = \frac{A}{r^2} \text{ (in sr)}$$

Observe that the steradian is dimensionless.

The luminous flux of the standard light source of 1 candle is 1 lumen per steradian.

Since the unit sphere has 4π unit areas of surface, there are 4π steradians per sphere. The total luminous flux emitted by a point source is therefore 4π lumens per candle of luminous intensity.

$$\Phi = \frac{4\pi \text{ lm}}{\text{cd}} \times I \text{ cd}$$

$$\Phi = 4\pi I \qquad \text{(in lumens)}$$

Thus a luminous source having an intensity of 1 candle emits light at the rate of 4π lm, or 12.57 lumens. In fact, light sources are usually rated in terms of the total flux emitted, with 12.57 lumens being radiated by 1 candle. A 40-watt incandescent lamp is rated at about 450 lumens and a 40-watt fluorescent lamp at about 2600 lumens.

3. *Illumination.* In Figure 12-24 it is evident that as the intensity of the source is increased, the luminous flux transmitted to each unit area of surface and the flux on each unit area are similarly increased. The *illumination* on the surface is said to be increased. ***Illumination*** *is the density of the luminous flux on a surface.* When the surface is uniformly illuminated, the illumination E is the quotient of the flux on the surface divided by the area of the surface and is expressed as the luminous flux per unit area.

$$E = \frac{\Phi}{A}$$

Φ is the luminous flux in lumens and A is the area in square meters. The illumination E is in *lumens per square meter.*

Suppose the radius of the unit sphere shown in Figure 12-24 is increased to 2 meters. The surface area is $4\pi(2 \text{ m})^2$, or approximately 50 m^2, and is *four* times the area of the unit sphere. Similarly, a radius of 3 meters gives an area of $4\pi(3 \text{ m})^2$, which is *nine* times the area of the unit sphere. Thus as the radial distance from the luminous source is increased, the area illuminated is increased in proportion to the *square* of the distance.

If a point source of constant intensity is located at the center of the sphere, *the illumination decreases as the square of the distance from the source.* See Figure 12-25. If the intensity of the source is doubled, of course the luminous flux transmitted to the surface is doubled and the illumination is doubled. Thus the illumination E on a surface perpendicular to the luminous flux falling on it is dependent on the intensity I of the source and its distance r from the source.

The distance r from the point source is the radius of a spherical surface of area $4\pi r^2$ illuminated by the source I at its center. Thus the proper dimensions of E become apparent from the following:

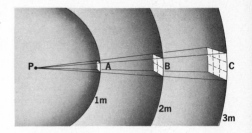

Figure 12-25. The illumination on a surface varies inversely as the square of the distance from the luminous point source.

$$E = \frac{\Phi}{A} = \frac{\cancel{4\pi}I}{\cancel{4\pi}r^2}$$

$$E = \frac{I}{r^2}$$

If r is expressed in meters and I in candles, then

$$E = \frac{\cancel{4\pi}\ \text{lm/}\cancel{\text{cd}} \times I\ \cancel{\text{cd}}}{\cancel{4\pi}r^2\text{m}^2}$$

$$E = \frac{I}{r^2}\ (\text{in lm/m}^2)$$

The *inverse square law* is used to calculate the illumination from an individual point source on planes perpendicular to the beam. In practice, if the dimensions of the source are negligible compared with its distance from the illuminated surface, it is considered a point source. For long fluorescent tubes, the illumination varies inversely as the distance, but only for distances somewhat smaller than the length of the tube.

When a surface is not perpendicular to the beam of light illuminating it, the luminous flux spreads over a greater area and the level of illumination is reduced. In Figure 12-26 the perpendicular surface **ABC** illuminated by the beam is square, having an area equal to **AB** × **BC**. As the surface is tilted away from the source, the illuminated area becomes rectangular, the width remaining the same and the length increasing to **BD**. In the right triangle **ABD**

$$BD = \frac{AB}{\cos\theta}$$

where θ equals the tilt angle of the surface from its perpendicular position with respect to the beam. (It is evident that θ also represents the angle that the light beam makes with the perpendicular to the illuminated surface.)

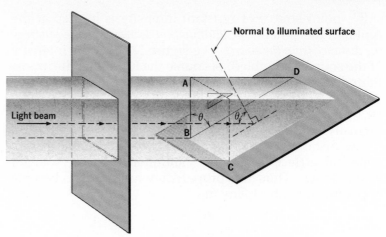

Figure 12-26. The illumination on a surface varies directly with the cosine of the angle between the luminous flux and the normal to the surface.

As θ reaches 60°, cos θ equals 0.5 and the length **BD** becomes twice the width **AB.** Thus the area is doubled and, since the same flux is spread over twice the area, the illumination is reduced to half the original level. In the general case,

$$E = \frac{I \cos \theta}{r^2}$$

The illumination on a surface varies inversely with the square of the distance from the luminous source and directly with the cosine of the angle between the luminous flux and the normal to the surface. A surface that is perpendicular to the luminous flux is simply a special case of the general expression for illumination in which the angle θ becomes zero and cos θ = 1. See the following example.

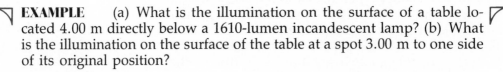

EXAMPLE   (a) What is the illumination on the surface of a table located 4.00 m directly below a 1610-lumen incandescent lamp? (b) What is the illumination on the surface of the table at a spot 3.00 m to one side of its original position?

SOLUTION

(a) $E = \dfrac{I}{r^2}$ information given in lm

$\Phi = 4\pi I$ $I = \dfrac{\Phi}{4\pi}$ express I in terms of Φ

$E = \dfrac{\Phi}{4\pi r^2} = \dfrac{1610 \text{ lm}}{4\pi \times (4.00 \text{ m})^2}$ substitute $\Phi/4\pi$ for I

$E = 8.00 \text{ lm/m}^2$

(b) $E = \dfrac{I \cos \theta}{r^2}$

$I = \dfrac{\Phi}{2\pi}$

$E = \dfrac{\Phi \cos \theta}{4\pi r^2}$

$E = \dfrac{1610 \text{ lm} \times 0.800}{4\pi \times (5.00 \text{ m})^2}$ $\cos \theta = 4.00 \text{ m}/5.00 \text{ m } (3 - 4 - 5 \text{ rt } \Delta)$

$E = 4.10 \text{ lm/m}^2$

PRACTICE PROBLEMS   **1.** (a) What is the illumination on a horizontal surface at a spot located 2.50 m directly below a 945-lumen incandescent lamp? (b) What is the illumination on this horizontal surface at a second spot 2.50 m to one side of the original spot?   *Ans.* (a) 12.0 lm/m² (b) 4.25 lm/m²

2. Referring to Problem 1, what is the illumination on the horizontal surface at a third spot 3.00 m to one side of the original spot? (Determine angle θ to the nearest 0.5°.)   *Ans.*   3.20 lm/m²

12.18 The Intensity of a Source   The candle power of a light source can be measured by comparing its intensity with that of a standard light source having the same color quality. This is done by using an instrument called a *photometer*. See Figure 12-27. Photometric measurements are made in a darkened room.

Figure 12-27. A photoelectric photometer.

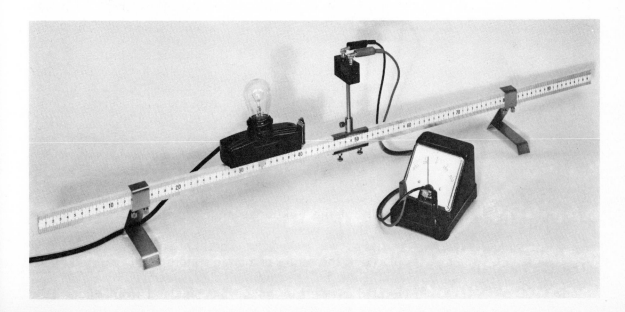

1. *Bunsen photometer.* The Bunsen photometer is sometimes called the grease-spot photometer. If a disk of white paper with a grease spot in the center is held toward a light, the grease spot will appear lighter than the rest of the paper because it transmits more light. On the other hand, because it is a poorer reflector of light than the paper, the grease spot will appear darker than the paper when held away from a light. In that case it is seen by reflected light. A laboratory model of the Bunsen photometer consists of such a paper screen supported on a meter stick between a standard lamp and a lamp of unknown candle power. The paper screen is moved back and forth along the meter stick until it is equally illuminated on both sides. At this point the grease spot seems to disappear. When the screen is correctly positioned, illuminations E_1 and E_2 of the two sides of the screen are equal.

Since $\qquad E_1 = E_2 \qquad$ then $\qquad \dfrac{I_1}{r_1^2} = \dfrac{I_2}{r_2^2}$

I_1 and I_2 are the intensities of the two sources producing illuminations E_1 and E_2, and r_1 and r_2 are their respective distances from the screen. Each distance r is a radial distance from a source.

2. *Joly photometer.* The Joly photometer gives more satisfactory results than the grease-spot photometer. It consists of two blocks of paraffin separated by a thin sheet of metal. The light from either side is transmitted by the paraffin, but is stopped by the metal. By looking at the edges of the blocks of paraffin it is easy to adjust the photometer so that both sides are equally illuminated. Then distances from the sheet of metal to the lamps are measured and calculations made as with the Bunsen photometer.

3. *Photoelectric photometer.* The Bunsen or Joly photometer head can be replaced by a photoelectric cell connected to a meter suitable for measuring the photoelectric current of the cell. The photoelectric cell is then placed a given distance from a standard lamp of known luminous intensity, and the photocurrent is measured. An unknown lamp is then substituted for the standard lamp, and the position of the photocell is adjusted to give the same photocurrent as for the standard lamp. The luminous intensity of the unknown lamp can then be calculated as with the Bunsen photometer.

4. *Spherical photometer.* The light source being tested is placed in the center of a large sphere that is painted white on the inside. The luminous flux is received by a photocell located inside the sphere but shielded from direct light

from the source. The light on this cell is equal to that received by any other similar portion of the sphere interior, due to cross-reflections, and is therefore proportional to the total light emitted by the test source. A meter outside the sphere is connected to the cell and is calibrated to read the mean spherical candle power or lumens directly. This accurate photometer is used commercially.

Questions

GROUP A

1. Distinguish between luminous and illuminated objects.
2. Define and illustrate the following terms: (a) transparent, (b) translucent, (c) opaque.
3. What are the physical quantities generally measured in practical photometry?
4. Define luminous flux.
5. (a) Which of the quantities from Question 3 is measured in fundamental units? (b) Which are measured in derived units?
6. What is our chief method of providing artificial light?

Problems

GROUP A

Note: Assume filament lamps to be point sources.

1. The distance from the earth to the sun is approximately 1.5×10^8 km. What time, in minutes, is required for light from the sun to reach the earth?
2. What is the illumination on the page of a book 1.20 m directly below a source whose intensity is 125 cd?
3. It is recommended that the illumination be 540 lm/m^2 for newspaper reading. How far from the paper should a 265-cd source be placed to provide this illumination? (Assume the paper to be perpendicular to the luminous flux reaching it.)
4. What is the maximum illumination 1.50 m from a lamp whose intensity is 150 cd?

GROUP B

7. By using Figure 12-22, explain how Michelson determined the speed of light in air.
8. Devise an experiment that would enable you to verify the inverse-square law with a light meter. Describe it.
9. Suggest a procedure by which a light meter with an arbitrary scale could be used in the absence of a photometer to determine the intensity of an unknown light source.
10. A member of a television studio audience in New York sits 30 m from the performer. A viewer observes the performance at home in Chicago, 1200 km away. Who hears the performer first?

5. The amount of illumination thrown on a screen by two sources of light is the same when the distances from the lamps to the screen are 3.0 m and 2.0 m, respectively. If the intensity of the first lamp is $2\overline{0}$ cd, what is the intensity of the second lamp?
6. A lamp, intensity 16.0 cd, is placed at the 0.0-cm mark on a meter stick. A lamp of unknown intensity is placed at the 100.0-cm mark. If a Bunsen photometer is equally illuminated at the 60.0-cm mark, what is the intensity of the unknown lamp?
7. A photometer is in balance on a meter stick when a 40.0-cd lamp and a second lamp are 100.0 cm apart, and the grease spot is 30.0 cm from the second lamp. Find the intensity of this lamp.
8. The illumination on a screen located 3 m from a source of light is 4 times

as much as that on a second screen that is illuminated by the same source. The intensity of the source is 25 cd. How far is the second screen from the source?

GROUP B

9. The intensity of a lamp is 60.0 cd. At what distance does it provide an illumination of 10.0 lm/m^2 on a table located directly beneath it?
10. How far away is the nearest star if it takes 4.3 years for the light from the star to reach the earth?
11. An incandescent lamp of 30.0 candles is placed at the 0-cm mark on a meter stick and a lamp of 20.0 candles is placed at the opposite end. Where must the screen of a photometer be placed so that both sides are equally illuminated?
12. A 100-watt lamp placed at one end of a meter stick and a 10-candle source placed at the other end equally illuminate a photometer that is 75 cm from the 100-watt lamp. How many candles per watt does the lamp supply?
13. How many revolutions per second did Michelson's octagonal mirror make if light traveled 70.8 km while the mirror made one eighth of a revolution?
14. What is the range of frequencies of visible light?
15. A surface is 75 cm from a luminous source of 150 cd. At what angle can it be tilted and still have an illumination of 0.025 lm/cm^2?

SUMMARY

A particle model and a wave model were developed in the seventeenth century to explain the then known properties of light. The particle model failed after additional properties of light were discovered. The wave model prevailed, and with the experimental confirmation of the electromagnetic theory the wave model of light was thought to be complete. The discovery of the photoelectric effect, the failure of the electromagnetic theory, and the success of the quantum theory in explaining photoemission established the dual character of light.

An object is luminous if it is visible because of the light it emits. An object is illuminated if it is visible because of the light it reflects. An object may reflect, absorb, or transmit the light it receives. Opaque objects cast shadows when illuminated. Shadows may consist of umbras and penumbras.

Photometry is the quantitative study of light. Three quantities measured in practical photometry are luminous intensity, luminous flux, and illumination. The intensity of a light source can be measured by means of a photometer.

VOCABULARY

coherent light	incoherent light	photometer
converging pencil	inverse square law	photometry
corpuscular theory	laser	Planck's constant
diverging pencil	lumen	radiant energy
electromagnetic spectrum	luminous flux	rectilinear propagation
electromagnetic wave	luminous object	Snell's law
Huygens' principle	penumbra	umbra
illuminated object	photoelectric effect	visible spectrum
illumination	photoelectron	

REFLECTION

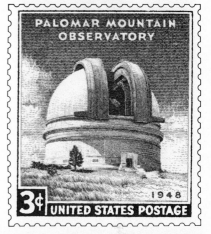

PALOMAR MOUNTAIN OBSERVATORY
1948
3¢ UNITED STATES POSTAGE

This dome in California houses one of the largest optical telescopes in the world. The main mirror of the telescope is 5.08 meters in diameter and has a focal length of 16.764 meters. Its light-gathering power is 640,000 times greater than that of the unaided human eye.

13.1 *Reflectance* A light beam passing through any material medium will become progressively weaker due to two effects. Part of the light beam's energy will be *absorbed* by molecules of the medium and part will be *scattered* in all directions by the molecules. A light beam striking the boundary between two media can be partly *transmitted* and partly returned to the first medium. See Figure 13-1. Light returned to the first medium is said to be *reflected* at the boundary. Reflection is described as a wave property in Section 10.10.

Part of the reflected light proceeds in a common direction but part can be reflected in all possible directions as scattered light. See Figure 13-2 for these effects. The fact that a portion returns in a common direction provides us

In this chapter you will gain an understanding of:

▷ regular and diffused reflection of light
▷ the two laws of reflection
▷ the formation of images by plane mirrors
▷ the terminology of curved mirrors
▷ the location of image points formed by concave and convex mirrors
▷ problem solving involving the mirror equation, the relationship between focal length of a mirror and object and image locations

Reflection
Regular
Scattered
Scattered
Light rays
Absorbed
Boundary
Transmitted

Figure 13-1. A diagram showing the effects on light rays of a material medium and an interface, or boundary, between two media.

315

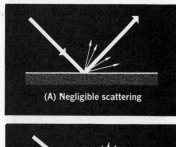

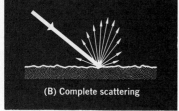

Figure 13-2. Reflecting surfaces vary in the extent to which reflected rays are scattered.

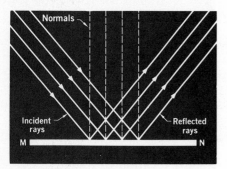

Figure 13-3. Regular reflection of light from a specular surface.

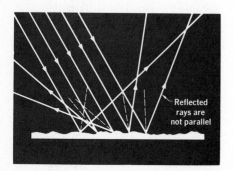

Figure 13-4. Irregular reflection promotes the diffusion of light.

with a method for controlling the reflection of a light beam.

The amount of light reflected at the surface of an object, whether by scattering or by unidirectional reflection, depends on the kind of material underlying the surface, the smoothness of the surface, and the angle at which the light strikes it. *The ratio of the light reflected from a surface to the light falling on the surface is called **reflectance**.* This ratio is commonly expressed as a percentage. Materials differ widely in their reflectance. The material of highest reflectance is magnesium oxide, a white chalky substance that reflects about 98% of incident light with practically complete scattering. The reflectance of a smooth surface of silver is about 95% with negligible scattering. Some black surfaces have reflectances of 5% or less.

13.2 Regular and Diffused Reflection The reflection of sunlight by a mirror produces a blinding glare. The rays of the sun reaching the earth are practically parallel and thus have the same angle of incidence. They remain practically parallel after being reflected from a plane mirror. Such reflection, in which scattering of the reflected rays is negligible, is called *regular* reflection. Polished, or *specular*, surfaces cause regular reflection and the image of the luminous source is sharply defined. The regular reflection of a narrow beam of light is in one direction with no appreciable loss of definition or intensity. The nature of regular reflection is shown in Figure 13-3.

Regular reflection from highly polished surfaces provides for relatively accurate control of light rays. Searchlights, beacons, and automobile spotlights use concentrated light sources of high intensity and highly-polished regular reflectors to redirect light rays in the desired direction.

In Figure 13-4 we observe what happens to a beam of light incident on an irregular surface. The laws of reflection hold true for each particular ray of light, but the normals to the surface are not parallel and the light is reflected in many directions. Such scattering, or *diffusion*, of light is extremely important.

If the sun's rays were not diffused by rough and irregular surfaces and by dust particles in the air, the corners of a room and the spaces under shade trees would be in almost total darkness and the glare would be dazzling in sunlit areas. Astronauts have found that it is possible for them to see in the shadows on the lunar surface where there is no air, but they cannot see as easily as we can on the surface of the earth.

13.3 Mirrors as Reflectors The reflection of light is similar to the reflection of sound or to the rebound of an elastic ball. The line **MN** in Figure 13-5(A) represents a reflecting surface; **AD** is a ray of light incident upon the reflector at **D; DB** is the path of the reflected ray. The line **CD,** perpendicular to the reflecting surface at the point of incidence, is called a *normal*. In Figure 13-5(B) the normal is perpendicular to the tangent to the curved reflecting surface at the point of incidence. Recall from Section 10.10 that the *angle of incidence (i)* is the angle between the incident ray and the normal at the point of incidence [angle **ADC** in Figure 13-5(A)]. In Chapter 10 we examined the nature of reflection by sending water waves against a barrier in the ripple tank. At that time we recognized *the first law of reflection: The angle of incidence, i, is equal to the angle of reflection, r*. The *angle of reflection, r*, is the angle between the reflected ray and the normal at the point of incidence [angle **BDC** in Figure 13-5(A)].

The relationships between an incident ray of light and the reflected ray and between the angles they form with the normal are easily determined in the laboratory. Critical observation of the specular reflection of light reveals a *second law of reflection: The incident ray, the reflected ray, and the normal to the reflecting surface lie in the same plane.* As mentioned in Section 10.10, these laws are true for all forms of wave propagation.

Any highly polished surface that forms images by the regular reflection of light can act as a mirror. A *plane mirror* of plate glass is silvered on one surface to reflect the light efficiently. Its reflectance may be nearly 100%. The plane parallel surfaces of the plate glass allow an essentially distortion-free image to be observed.

A *spherical mirror* is a small section of the surface of a

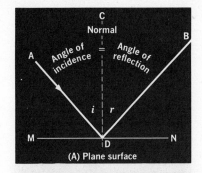

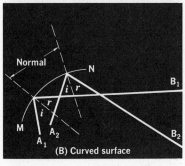

Figure 13-5. Reflections from plane and curved surfaces.

The reflective coating is commonly placed on the back surface of a plane mirror to protect the coating from damage.

Figure 13-6. Spherical mirrors are circular sections of spheres.

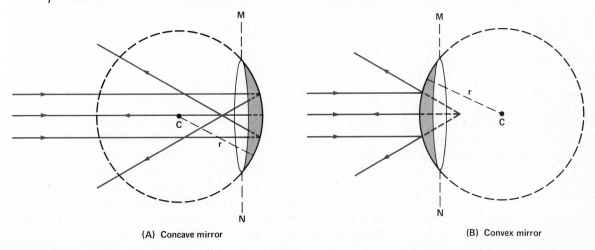

(A) Concave mirror　　　　　　　　　　　　　　(B) Convex mirror

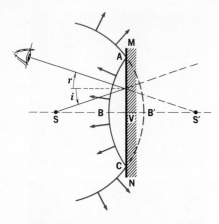

Figure 13-7. A wave-front diagram of reflection of light from a point source by a plane mirror.

Figure 13-8. The image formed by a plane mirror appears to be reversed right and left. Observe that the image of the right hand has the symmetry of the left hand.

sphere. One side of the section is polished or coated with a reflective material. When viewed from the *inside*, the mirror is called a *concave* mirror, as indicated in Figure 13-6(A). When viewed from the *outside*, it is called a *convex* mirror, as indicated in Figure 13-6(B).

The laws of reflection hold for both spherical mirrors and plane mirrors. However, the size and position of the images formed by spherical mirrors are quite different from the size and position of images formed by plane mirrors.

13.4 Images by Reflection Light rays reflected from a concave mirror may meet in front of the mirror and form an image of the object from which the light comes. Such an image can be projected on a screen placed at the image location. *An image formed by converging rays of light actually passing through the image point is called a **real image**.* Real images are inverted relative to the object and can be larger or smaller than the object.

Your image observed in a plane mirror seems to be behind the mirror, although the rays forming it actually are reflected forward from the mirror surface. *This image is called a **virtual image**. Rays of light that appear to have diverged from the image point do not actually pass through that point.* Virtual images cannot be projected on a screen. They are erect with respect to the object and can be enlarged or reduced in size.

In Figure 13-7 light traveling from a point source in spherical wave fronts is reflected from a plane mirror. At the instant that points **A** and **C** on the wave front reach the mirror, point **B** has been reflected back towards the source at **S,** having traveled the distance **VB**. The reflected wave front **ABC** has an apparent source **S'** at an apparent distance **VS'** behind the mirror. Note that **VS'** is equal to the distance **VS** of the source in front of the mirror. The reflected wave front approaches an observer, the eye, as though its source were **S'**, **S'** being the virtual image of the source **S**. The real image of an object is composed of many image points, each being the point image of the corresponding point on the object.

13.5 Images Formed by Plane Mirrors The image formed by a plane mirror is neither enlarged nor reduced but is always virtual, erect, and appears to be as far behind the mirror as the object is in front of it. The image differs from the object in that right and left appear to have been interchanged. In Figure 13-8 the image of the right hand has the appearance of a left hand. If an object is spinning clockwise about its axis, its image appears to be spinning counterclockwise about its image axis.

The ray diagram in Figure 13-9 shows a simple graphical method of locating the image formed by a plane mirror. The triangle **ABC,** drawn on paper, is used as the object in front of a mirror, **MN.** A pin is placed at each vertex of the triangle. Rays AO_1 and AO_2 from pin **A** strike the mirror so as to be reflected to the two eyes, E_1 and E_2, of the observer. When the sight lines E_1O_1 and E_2O_2 are produced (extended) behind the mirror, they intersect at **A'**, which is the location of the image of pin **A.** A comparison of triangles AO_2O_1 and $A'O_2O_1$ makes it apparent that **A'** is as far behind the mirror as **A** is in front of it.

The images of **B** and **C** are located in similar fashion at **B'** and **C'**. In each instance *the image appears to be as far behind the mirror as the object is in front of it.* When the image points **A'B'C'** are joined, it becomes evident that *the image is the same size as the object.*

This method of constructing images is based on the laws of reflection. The light ray AO_1 is reflected at point O_1 and is sighted at E_1. By drawing a normal to the mirror at O_1, the angle AO_1E_1 is bisected and the angle of reflection E_1O_1R is equal to the angle of incidence AO_1R. The incident ray AO_1, the reflected ray O_1E_1, and the normal O_1R all lie in the same plane.

Sound waves bound and rebound between parallel cliffs or walls and produce multiple echoes. In a similar fashion, light reflects back and forth between parallel mirrors or mirrors set at an acute angle and forms multiple images. The image formed in one mirror appears to act as the object forming the image in the next mirror. Because some light energy is absorbed and some is scattered at each mirror, the succeeding images become fainter. A thick plate-glass mirror produces multiple reflection, as shown in Figure 13-10, because some light is reflected each time a boundary is encountered.

Multiple reflections can be avoided by aluminizing or silvering the front surface of a glass mirror. Mirrors for astronomical telescopes and other precision instruments are usually coated with aluminum in this manner. Such reflecting surfaces are exposed and can be easily damaged.

13.6 Curved-Mirror Terminology

In order to discuss image formation by spherical mirrors, several terms associated with this process must be identified. In this discussion, reference is made to the diagrams of mirrors shown in Figures 13-11 and 13-12.

The arc **MVN** is the portion of the spherical surface separated from the sphere by a cutting plane represented by the line **MN.**

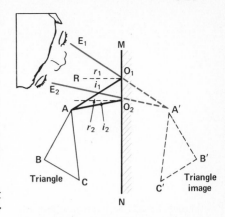

Figure 13-9. A ray diagram used to construct an image formed by a plane mirror.

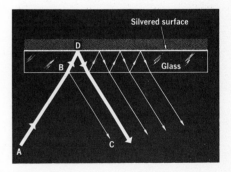

Figure 13-10. Multiple reflections that result from the use of a back-silvered mirror.

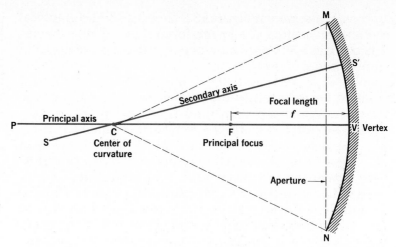

Figure 13-11. A plane diagram for defining terms used with curved mirrors.

1. The *center of curvature,* **C,** is the center of the original sphere. The radius of the sphere is also the *radius* **CV** of the spherical mirror.

2. The *vertex,* **V,** is the center of the mirror.

3. The *principal axis* of the mirror is the line **PV** drawn through the center of curvature and the vertex.

4. A *secondary axis* is any other line drawn through the center of curvature to the mirror: **SS'**, for example.

5. A *normal* to the surface of a spherical mirror is a radius drawn from the point of incidence of a light ray. Therefore the normal is perpendicular to the mirror surface at the incidence point. (Normals are illustrated in Figure 13-12.)

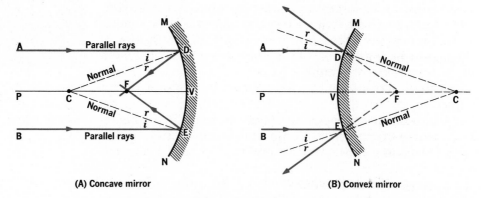

(A) Concave mirror   (B) Convex mirror

Figure 13-12. Locating the principal focus of spherical mirrors.

The aperture size relative to focal length in mirror diagrams is exaggerated for the sake of clarity.

6. The *principal focus,* **F,** is the point on the principal axis where light rays close to and parallel to the principal axis converge, or from which they appear to diverge.

7. The *focal length, f,* is the distance from the vertex of the mirror to its principal focus.

8. The *aperture* determines the amount of light intercepted by the mirror. The length **MN** is called the *linear*

aperture of the mirror, and the angle **MCN** is called the *angular aperture.* For a spherical mirror to be capable of forming sharp images, its aperture must be small compared with its focal length.

13.7 Rays Focused by Spherical Mirrors In Figure 13-12(A) the rays **AD** and **BE,** parallel to the principal axis, are shown incident on a concave spherical mirror at points **D** and **E** respectively. The normals **CD** and **CE** are drawn to the points of incidence. By the second law of reflection, the reflected rays converge on the principal axis at **F,** the principal focus. Because the incident rays converge upon being reflected, concave mirrors are also known as *converging* mirrors. The distance **FV** is the *focal length* of the mirror. For spherical mirrors of small aperture, the focal length is one-half the radius of curvature.

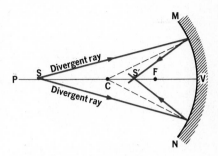

Figure 13-13. The reflection of diverging rays by a concave mirror.

Parallel rays incident on the surface of a convex spherical mirror are shown in Figure 13-12(B). The normals **CD** and **CE** produced beyond the mirror surface show the reflected rays to be divergent along the lines **FD** and **FE** produced, **F** being the principal focus of the mirror. Rays parallel to the principal axis appear to diverge from the principal focus of a convex mirror. Because incident rays diverge on being reflected, convex mirrors are known as *diverging* mirrors.

Diverging rays from a point **S** incident on a concave mirror, Figure 13-13, converge to a focus at some point **S'** beyond the principal focus of the mirror. If **S** is located on the principal axis, the image **S'** is also on the principal axis. Suppose the source of diverging rays is now moved to position **S'**. The image formed by the mirror will be at **S.**

Converging rays are also focused by concave mirrors. As shown by Figure 13-14, the point of focus is between the vertex and the principal focus. In every instance the direction of a reflected ray is determined by the laws of reflection.

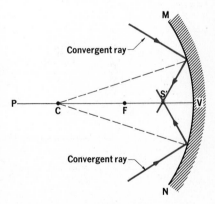

Figure 13-14. The reflection of converging rays by a concave mirror.

As observed in Figure 13-12(A), the incident parallel rays that are also parallel to the principal axis of the concave mirror are reflected and converge at the principal focus. The image is ideally a point of light at the principal focus. If the parallel rays are not parallel to the principal axis but are incident at a small angle with the principal axis, they do not converge at the principal focus. Instead they are focused at a point in a *plane* that contains the principal focus and is perpendicular to the principal axis. This plane is called the *focal plane* of the mirror. Figure 13-15 illustrates the reflection of a bundle of parallel light rays incident on a concave mirror. In Figure 13-15(A) the rays are parallel to the principal axis of the mirror, and in

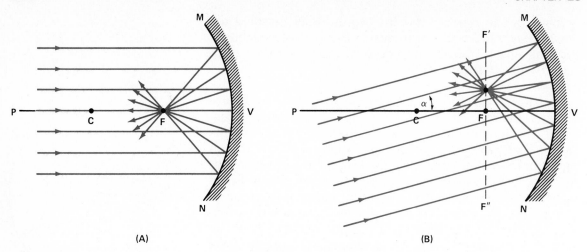

(A)   (B)

Figure 13-15. In (A) parallel rays are shown incident on a concave mirror parallel to its principal axis. The reflected rays converge at the principal focus. In (B) the parallel rays are incident at a small angle α (alpha) with the principal axis. Here the reflected rays are focused at a point in the focal plane F'F″ of the mirror.

Figure 13-15(B) the rays are at a small angle α (the Greek letter alpha) with the principal axis.

If the aperture of a spherical mirror is large, the parallel rays of light striking the mirror near its edge are not reflected through the principal focus but are focused at points nearer the mirror, as shown in Figure 13-16(A). Only those parallel rays that are incident on the mirror near its vertex are reflected to the principal focus. This characteristic of spherical mirrors of large aperture is known as *spherical aberration* and results in the formation of fuzzy images.

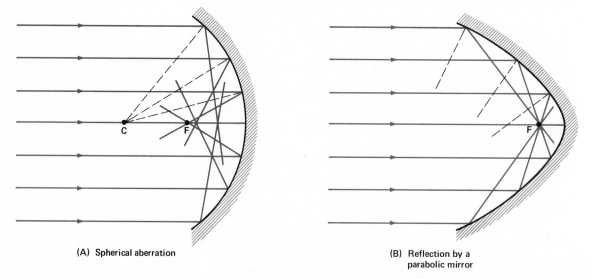

(A) Spherical aberration   (B) Reflection by a
parabolic mirror

Figure 13-16. Spherical aberration is a characteristic of spherical mirrors with large apertures (A). This defect is avoided by the use of parabolic mirrors (B).

In order to reduce spherical aberration, the aperture of the spherical mirror must be small. For angular apertures no larger than about 10°, the distortion of the image is negligible. Spherical aberration can be avoided for parallel rays if the mirror surface is made parabolic instead of spherical. See Figure 13-16(B). The parabolic mirror shown in Figure 13-17 reflects light rays originating at the focus as parallel rays. Parabolic reflectors are used in automobile headlights, microwave communication relays, radar, and radio telescopes. Parabolic mirrors are also used in large reflecting telescopes to collect and focus light rays from distant objects.

Parabola: the curve of intersection of a plane parallel to a straight line in the surface of a right circular cone.

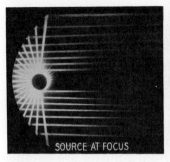
SOURCE AT FOCUS

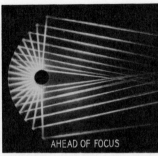
AHEAD OF FOCUS

BEHIND FOCUS

13.8 Constructing the Image of a Point In Figure 13-18 the image of point **S** formed by a concave mirror is to be located. First the principal axis **PV** is drawn through **C**. Because light is given off in all directions from point **S**, *any two lines* can be drawn from **S** to represent rays of light incident on the mirror. The point of intersection of the *reflected* rays locates the image of **S** at **S'**.

The construction is greatly simplified if incident rays are selected for which the directions of the reflected rays are known from the geometry of the system. Three such special rays are shown in Figure 13-18. Ray **SA**, *parallel to the principal axis,* is reflected through the principal focus. Ray **SB** *passes through the principal focus* and is reflected parallel to the principal axis. Ray **SC** *lies along the secondary axis.* It is reflected back along its path of incidence because the

Figure 13-17. Reflections from a parabolic mirror.

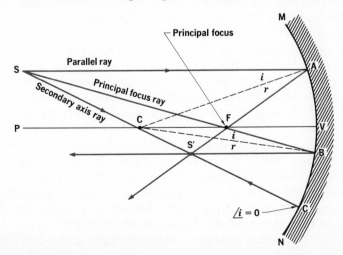

Figure 13-18. Locating an image formed by a concave mirror.

angle of incidence is zero. *Note that the convergence of any two of these three reflected rays at a common point* **S'** *forms the image of* **S.**

If the object were located at **S'** the image would be formed at **S.** The object and image are thus interchangeable, or *conjugate.* The positions of the object and its image, **S** and **S',** are called conjugate points.

Conjugate points: any two points so situated that light from one is concentrated at the other.

13.9 Images Formed by Concave Mirrors The image formed by a concave mirror can be constructed by locating the images of an adequate number of object points. Suppose the object is an arrow. The entire image can be located by constructing the image points of the head and the tail of the arrow. Concave mirror images can be grouped into six cases based on the position of the object in front of the mirror.

Case 1. Object at an infinite distance. Rays emanating from an object at an infinite distance from a mirror are parallel at the mirror. If the rays are parallel to the principal axis, they are reflected through the principal focus. See Figure 13-19(A). We therefore conclude that *the image formed by a concave mirror of an object an infinite distance away is a point at the principal focus.* The rays of the sun reaching the earth are practically parallel and provide a simple method for finding the approximate focal length of a concave mirror.

Case 2. Object at a finite distance beyond the center of curvature. In Figure 13-19(B) the image of the object arrow can be located using the method illustrated in Section 13.8. Two known rays, **1** and **2,** emanating from the head of the arrow, intersect after reflection to locate the image of the arrowhead. The image of the tail lies on the principal axis as shown earlier in Figure 13-13. *The image in this case is real, inverted, reduced in size, and located between the center of curvature and the principal focus.* The nearer the object approaches the center of curvature, the larger the image becomes and the nearer the image approaches **C.**

Case 3. Object at the center of curvature. When the object arrow is at the center of curvature, as shown in Figure 13-19(C), the image of the arrowhead is found inverted at **C.** *When the object is at the center of curvature, the image is real, inverted, the same size as the object, and located at the center of curvature.*

Case 4. Object between the center of curvature and principal focus. This case is the converse of Case 2 and is shown in Figure 13-19(D). *The image is real, inverted, enlarged, and located beyond the center of curvature.* The nearer the object approaches the principal focus of the mirror, the larger the image becomes and the farther it is beyond **C.**

Case 5. Object at principal focus. This case is the converse

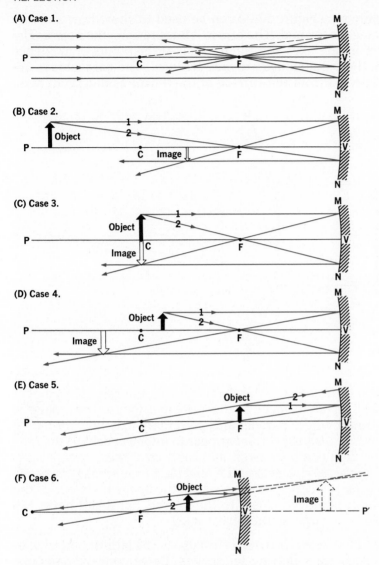

(A) Case 1.

(B) Case 2.

(C) Case 3.

(D) Case 4.

(E) Case 5.

(F) Case 6.

Figure 13-19. Ray diagrams of image formation by concave mirrors.

of Case 1; all rays originating from the same point on the object are reflected from the mirror as parallel rays. See Figure 13-19(E). *When the object is at the principal focus, no image is formed.* If the object is a point source, all reflected rays are parallel to the principal axis.

Case 6. *Object between principal focus and mirror.* The reflected rays from any point on the object are divergent; they can never meet to form a real image. They appear to meet behind the mirror, however, to form a virtual image as shown in Figure 13-19(F). In this case, *the image is virtual, erect, enlarged, and located behind the mirror.*

13.10 Images Formed by Convex Mirrors A convex mirror forms an erect image of reduced size. The diagram

Figure 13-20. The only image formed by a convex mirror is virtual, erect, reduced, and located behind the mirror.

shown in Figure 13-20 can be used to show how such an image is formed. The arrow **AB** represents the object. The secondary axes and the normals at the points of incidence of the parallel rays are *radii produced*. The parallel rays are reflected from the surface of the mirror as divergent rays.

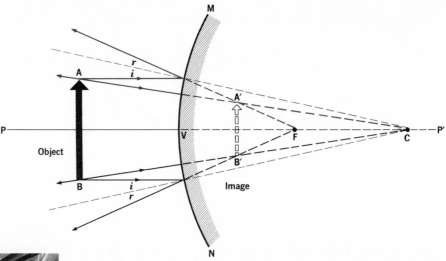

When produced behind the mirror, the reflected parallel rays *appear* to meet at the principal focus. The image is formed where parallel and secondary-axis rays *emanating from the same object point* appear to intersect behind the mirror. In a convex mirror, *all images are virtual, erect, smaller than the object, and located behind the mirror between the vertex and the principal focus.* The size of the image increases as the object is moved closer to the mirror, but it can never become as large as the object itself.

13.11 *The Reflecting Telescope* The largest optical telescopes are reflecting telescopes. Large concave parabolic mirrors are used to collect and focus light rays. See Figure 13-21. Reflecting telescopes can be made much larger than refracting (lens) telescopes. It is difficult to obtain a large piece of glass of sufficiently high optical quality for a lens, to grind and polish two surfaces with precision, and to support it appropriately.

When objects being viewed with a reflecting telescope are at a finite distance beyond the center of curvature of the mirror, Case 2 (Section 13.9) applies. The image is real, inverted, and located near the principal focus of the mirror. This real image can be reflected out of the path of the rays of light incident on the mirror so that it can be viewed or photographed from a position outside the telescope.

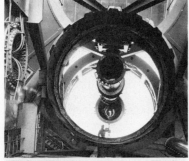

Figure 13-21. (A) The 200-inch mirror of the Mt. Palomar, California, reflecting telescope. (B) An 8-inch reflecting telescope ideally suited for amateur astronomers.

The inversion of the image is not objectionable when reflecting telescopes are used for astronomical viewing.

13.12 Object-Image Relationships   A simple relationship exists among the distance of an object d_o from a curved mirror, the distance of its image d_i from the mirror, and the focal length f of the mirror. This relationship, known as the *mirror equation,* can be derived by considering the diagram shown in Figure 13-22, in which a real

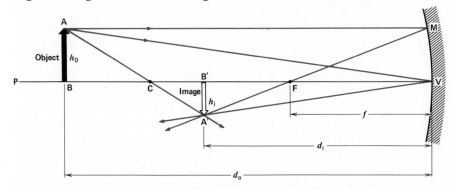

image of the object **AB** is formed by a concave mirror. (In this derivation we will consider as insignificant the very slight error introduced by the curvature **MV.**)
In similar triangles **A′B′F** and **MVF,**

Figure 13-22. A diagram for the derivation of the mirror equation for curved mirrors.

$$\frac{\mathbf{A'B'}}{\mathbf{MV}} = \frac{\mathbf{B'F}}{\mathbf{VF}} \qquad \text{(Equation 1)}$$

Taking **MV** = **AB**, **VF** = f, and **B′F** = $d_i - f$, we have

$$\frac{\mathbf{A'B'}}{\mathbf{AB}} = \frac{d_i - f}{f} = \frac{d_i}{f} - 1 \quad \text{(Equation 2)}$$

In similar triangles **ABV** and **A′B′V,**

$$\frac{\mathbf{A'B'}}{\mathbf{AB}} = \frac{\mathbf{B'V}}{\mathbf{BV}} \qquad \text{(Equation 3)}$$

Taking **BV** = d_o and **B′V** = d_i,

$$\frac{\mathbf{A'B'}}{\mathbf{AB}} = \frac{d_i}{d_o} \qquad \text{(Equation 4)}$$

Substituting in Equation 2

$$\frac{d_i}{d_o} = \frac{d_i}{f} - 1$$

Dividing by d_i,

$$\frac{1}{d_o} = \frac{1}{f} - \frac{1}{d_i} \qquad \text{(Equation 5)}$$

Or,

$$\frac{1}{f} = \frac{1}{d_o} + \frac{1}{d_i} \qquad \text{(Equation 6)}$$

where d_o is the distance of the object from the mirror, d_i the image distance, and f the focal length of the mirror.

Equation 6, the mirror equation, is applicable to all concave and convex mirrors. When an object is located less than one focal length in front of a concave mirror, the image is virtual. It appears to be located behind the mirror, and the image distance d_i is a *negative* quantity. In the case of convex mirrors, the radius of curvature is negative and the image is virtual. Therefore, both the focal length f and the image distance d_i are *negative* quantities. For concave mirrors, the radius of curvature and the focal length are *positive*. The image distance can be either positive or negative—being *positive* for real images and *negative* for virtual images. The significance of these negative quantities is illustrated in the following example.

EXAMPLE   An object located $2\overline{0}$ cm in front of a convex mirror forms an image $1\overline{0}$ cm behind the mirror. What is the focal length of the mirror?

SOLUTION

$$\frac{1}{d_o} + \frac{1}{d_i} = \frac{1}{f}$$

Solving for f:

$$f = \frac{d_o d_i}{d_o + d_i} = \frac{2\overline{0}\text{ cm} \times (-1\overline{0}\text{ cm})}{2\overline{0}\text{ cm} + (-1\overline{0}\text{ cm})} = -2\overline{0}\text{ cm}$$

PRACTICE PROBLEMS   **1.** An object is placed 25.0 cm in front of a convex mirror and the image appears to be 5.0 cm behind the mirror. What is the focal length of the mirror?   *Ans:* −6.25 cm

2. The object in Problem 1 is moved to a position 15.0 cm in front of the convex mirror. Where does the image appear to be located?   *Ans:* −4.42 cm (behind)

3. An object is located 14.0 cm in front of a concave mirror having a focal length of 7.0 cm. Where is the image located?   *Ans:* 14 cm (in front)

The relative heights of an object and its image formed by a curved mirror depend on their respective distances from the center of curvature of the mirror. This can be seen from similar triangles **ABC** and **A'B'C** of Figure 13-22.

We can demonstrate that for mirrors of small aperture the relative heights of the object and image also depend on their respective distances from the vertex of the mirror. If the height of object **AB** in Figure 13-22 is represented by h_o and the height of the image **A'B'** is represented by h_i, Equation 4 of Section 13.12 becomes

$$\frac{h_i}{h_o} = \frac{d_i}{d_o}$$

The sign conventions for spherical mirror computations using the mirror equation are summarized as follows:

1. The numerical value of the focal length f of a concave mirror is positive.

2. The numerical value of the focal length f of a convex mirror is negative.

3. The numerical value of the real object distance d_o is positive.

4. The numerical value of the real image distance d_i is positive.

5. The numerical value of the virtual image distance d_i is negative.

Questions
GROUP A

1. What three factors determine the amount of light an object will reflect?
2. What are the two laws of reflection?
3. When you look in a plane mirror, do you see yourself as others see you? Explain.
4. Illustrate the following by diagram as they are related to curved mirrors: center of curvature, vertex, principal axis.

5. What type of mirror produces real images?
6. In what ways can spherical aberration in mirrors be reduced?

GROUP B

7. Suppose we write the word "light" on a mirror with white paint. When the mirror is placed in a beam of sunlight, the reflection of the sunlight on a smooth wall consists of a bright area in which the letters of the word "light" appear dark. Explain.

8. What kind of trick mirror produces (a) a short, fattened image; (b) a tall, thin image?

9. What kinds of mirrors could be used and where should the object be placed to produce (a) an enlarged real image, (b) a reduced real image, (c) a real image the same size as the object, (d) an enlarged virtual image, (e) a reduced virtual image?

10. (a) Construct a ray diagram to show the formation of an image by a concave mirror when the object is at a finite distance beyond the center of curvature. (b) Describe fully the image that is formed.

11. (a) Construct a ray diagram to show the formation of an image by a concave mirror when the object is between the principal focus and the center of curvature. (b) Describe the image formed.

12. (a) Construct a ray diagram to show the formation of an image by a convex mirror. (b) Describe fully the image that is formed.

Problems
GROUP A

1. A concave mirror has a focal length of 10.0 cm. What is its radius of curvature?

2. If the radius of curvature of a curved mirror is 8 cm, what is its focal length?

3. While you are looking at the image of your feet in a plane vertical mirror, you see a scratch in the glass. Assuming your height to be 1.76 m, what is the approximate height of the scratch from the floor?

4. The light from a distant star is collected by a concave mirror. If the radius of curvature of the mirror is 150 cm, how far from the mirror is the image?

5. An object is placed 25.0 cm from a concave mirror whose focal length is 5.00 cm. Where is the image located?

6. An object 25.4 cm high is located 91.5 cm from a concave mirror, focal length 15.0 cm. (a) Where is the image located? (b) How high is it?

7. An object placed 50.0 cm from a spherical concave mirror gives a real image 33.3 cm from the mirror. (a) What is the radius of curvature of the mirror? (b) If the image is 30.5 cm high, what is the height of the object?

8. An object and its image in a concave mirror are the same height when the object is 36.4 cm from the mirror. What is the focal length of the mirror?

GROUP B

9. An object is placed 5.0 cm from a concave mirror whose focal length is 15 cm. (a) Where is the image located? (b) If the object is 2.0 cm high, what is the height of the image?

10. An object 5.00 cm tall stands at the 0.0-cm mark on a meter stick. (a) If a convex mirror having a focal length of 25.0 cm is placed at the 50.0-cm mark, where is the image? (b) How tall is it?

11. The image of the moon is formed by a concave mirror whose radius of curvature is 4.00 meters at a time when the distance to the moon is 380,000 km. What is the diameter of the image in centimeters if the diameter of the moon is 3480 km?

12. The image of an incandescent lamp filament is to be formed on a screen 5.00 m from a concave mirror. The lamp filament is 5.00 mm long and the image is to be 50.0 cm long. (a) What should be the radius of curvature of the mirror? (b) How far in front of the vertex should the filament be?

13. A spherical concave mirror has a radius of curvature of 30.0 cm. When your face is 10.0 cm from the vertex of the mirror, what is the image magnification?

14. A physics student 1.80 m tall stands in front of a mirror admiring the reflected image. Assume the student's eyes are 10 cm below the top of the head. (a) What is the smallest vertical mirror that will enable the entire image to be seen? (b) Is the length of the mirror dependent upon the distance between the student and the mirror? (c) Construct a ray diagram that supports your answers.

15. The mirror for Problem 14 is hung at an angle of 30.0° from the vertical toward the student. What will be the minimum-length mirror that will allow the student to see the full image if the line of vision cannot be depressed below the horizontal? Assume that the eyes are at the top of the head.

SUMMARY

The amount of incident light reflected from the surface of an object depends on the composition of the object, the character of its surface, and the angle at which the light rays strike the surface. The ratio of the reflected light to the incident light is known as reflectance. The angle of reflection and the angle of incidence are equal, and both lie in the same plane.

Images may be either real or virtual. Real images are inverted with respect to the object and can be projected onto a screen. Virtual images are erect and cannot be projected onto a screen. Images formed by plane mirrors are virtual, erect, and the same size as the object. Images formed by plane mirrors appear to be the same distance behind the mirror as the object is in front of the mirror.

Concave mirrors cause reflected rays of light to converge. These mirrors can form real images. The size and position of an image are related to the size of the object and its position relative to the mirror. A virtual image is formed by a concave mirror if the object is less than a focal length away from the mirror. Convex mirrors cause reflected rays of light to diverge. They form virtual images of objects irrespective of the object distance from the mirror.

Spherical mirrors of large aperture form indistinct images. The spherical-mirror defect responsible for fuzzy images is known as spherical aberration. Parabolic mirrors do not show this aberration defect. When mirrors of large aperture are required, parabolic surfaces rather than spherical surfaces are used to reflect incident light.

VOCABULARY

angle of incidence
angle of reflection
angular aperture
aperture
concave
conjugate points
converge
convex
diffusion (light)
diverge

focal length
focal plane
linear aperture
mirror equation
normal (surface)
plane mirror
principal axis
principal focus
real image

reflectance
regular reflection
scattering (light)
secondary axis
specular (surface)
spherical aberration
spherical mirror
vertex (mirror)
virtual image

14 — REFRACTION

Alhazen Ibn-al-Haitham, an Arabian mathematician and astronomer, studied the reflection and refraction of light more than 900 years ago. The drawings on this stamp symbolize Alhazen's work with the magnifying properties of glass hemispheres, which led to the use of lenses for the correction of faulty vision.

In this chapter you will gain an understanding of:

➤ the relation between optical refraction and the wave character of light
➤ the relation between optical refraction and the speed of light
➤ light control with lenses
➤ image formation by ray diagrams
➤ problem solving with object-image relationships
➤ image magnification with optical instruments
➤ light dispersion by prisms
➤ color as a property of light; light spectra
➤ primary and complementary colors
➤ the additive process with colored lights and the subtractive process with colored pigments

OPTICAL REFRACTION

14.1 The Nature of Optical Refraction   In Section 10.12 we discussed refraction as a property of waves. Now we shall examine the refractive behavior of light and relate this behavior to its wavelike nature.

When aiming a rifle at a target, one relies on the common observation that light travels in straight lines. It does so, however, only if the transmitting medium is of the same *optical density* throughout. **Optical density** *is a property of a transparent material that is an inverse measure of the speed of light through the material.*

Consider a beam of light transmitted through air and directed onto the surface of a body of water. Some of the light is reflected at the interface (boundary) between the air and water; the remainder enters the water and is transmitted through it. Because water has a higher optical density than air, the speed of light is reduced as the light enters the water. This change in the speed of light at the air-water interface is diagrammed in Figure 14-2.

A ray of light that strikes the surface of the water at an *oblique* angle (less than 90° to the surface) changes direction abruptly as it enters the water because of the change in speed. The reason for this change in direction with a change in speed can be illustrated if we redraw the wavefront diagram of Figure 14-2(A) to make the angle of the incident ray oblique, as in Figure 14-2(B). When interpreting this diagram, you should remember that a light ray indicates the direction the light travels and is perpendicular to the wave front. We have already defined refraction as a bending of a wave disturbance. (See Section 10.12.) This bending of a light ray is called *optical refraction*. **Optical refraction** *is the bending of light rays as they pass obliquely from one medium into another of different optical density.*

Because of refraction, a fish observed from the bank appears nearer to the surface of the water than it actually is. A teaspoon in a tumbler of water appears to be bent at the surface of the water. A coin in the bottom of an empty teacup that is out of the line of vision of an observer may become visible when the cup is filled with water.

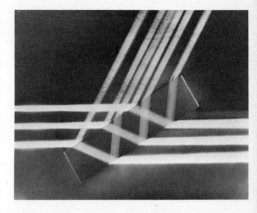

Figure 14-1. Parallel beams of light incident upon a rectangular glass plate. The beams are mainly refracted at both surfaces. Some reflection occurs at each surface, however.

14.2 Refraction and the Speed of Light Line **MN** of Figure 14-3 represents the surface of a body of water. The line **AO** represents a ray of light through the air striking the water at **O**. Some of the light is reflected along **OE**. Instead of continuing in a straight line along **OF**, the light ray entering the water is bent as it passes from air into water, taking the path **OB**.

The incident ray **AO** makes the angle **AOC** with the normal. Angle **AOC** is the *angle of incidence, i*. Recall that the angle of incidence is defined as the angle between the incident ray and the normal at the point of incidence. The refracted ray **OB** makes the angle **DOB** with the normal produced. *The angle between the refracted ray and the normal*

Figure 14-2. Wave-front diagrams illustrating (A) the difference in the speed of light in air and in water and (B) the refraction of light at the air-water boundary.

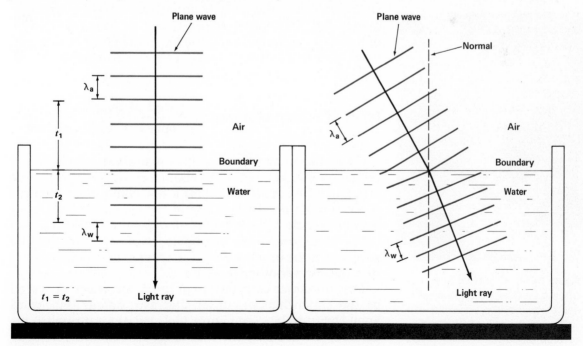

(A)   (B)

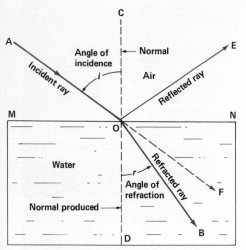

Figure 14-3. A ray diagram of refraction showing the angle of incidence and the angle of refraction.

at the point of refraction is called the **angle of refraction,** *r.*

In the examples given so far, the light rays have passed from one medium into another of *higher* optical density, with a resulting reduction in speed. When a ray enters the denser medium normal to the interface, no refraction occurs. When a ray enters the denser medium at an oblique angle, refraction does occur and the ray is bent *toward* the normal.

What is the nature of the refraction if the light passes obliquely from one medium into another of *lower* optical density? Suppose the light source were at **B** in Figure 14-3. Light ray **BO** then meets the surface at point **O,** and angle **BOD** is the angle of incidence. Of course some light is reflected at this interface. However, on entering the air the refracted portion of the light takes the path **OA.** The angle of refraction in this case is angle **COA.** It shows that the light is bent away from the normal. When a light ray enters a medium of lower optical density at an oblique angle, the ray is bent *away* from the normal. Had the light ray entered this less dense medium normal to the interface, no refraction would have occurred.

14.3 The Index of Refraction   The speed of light in a vacuum is approximately 300,000 kilometers per second. The speed of light in water is approximately 225,000 kilometers per second, or just about three fourths of that in a vacuum. The speed of light in ordinary glass is approximately 200,000 kilometers per second or about two thirds of that in a vacuum. *The ratio of the speed of light in a vacuum to its speed in a substance is called the* **index of refraction** *for that substance.* For example:

$$\text{Index of refraction (glass)} = \frac{\text{speed of light in vacuum}}{\text{speed of light in glass}}$$

Using the approximate values just given, the index of refraction for glass would be about 1.5 and for water about 1.3. The index of refraction for a few common substances is given in Appendix B, Table 18. The speed of light in air is only slightly different from the speed of light in a vacuum. Therefore, with negligible error, we can use the speed of light in a vacuum for cases where light travels from air into another medium.

The fundamental principle of refraction was discovered by the Dutch mathematician and astronomer Willebrord Snell (1580–1626). See Figure 14-4. He did not publish his discovery, but his work was taught at the University of Leyden, where he was a professor of mathematics and

Figure 14-4. Willebrord Snell became a professor of mathematics and physics at the University of Leyden at the age of 21. He discovered the law of refraction now known as Snell's law.

physics. The French mathematician and philosopher René Descartes (1596–1650) published Snell's work in 1637.

Snell's discoveries about refraction were not stated in terms of the speed of light. The speed of light in empty space was not determined until 1676, and the speed in water was not measured until 1850. From his observations, however, Snell defined the index of refraction as the ratio of the sine of the angle of incidence to the sine of the angle of refraction. This relationship is known as Snell's law. If n represents the index of refraction, i the angle of incidence, and r the angle of refraction,

$$n = \frac{\sin i}{\sin r}$$

The sines of angles from 0° to 90° are given in Appendix B, Table 6.

The relationship between Snell's law and the ratio of the speeds of light in air and a refracting medium can be recognized from Figure 14-5. Rays of light travel through the first medium (air) with a speed v_1 and enter the refracting medium in which their speed is v_2.

A wave front **MP** approaches the refracting interface **MN** in the first medium at an incident angle i. The wave front at **P** travels to **N** in the time t with a speed v_1. Simultaneously the wave front at **M** travels in the second medium to **Q** in the same time t but with a speed v_2. The new wave front is **NQ,** which travels forward in the second medium at a refractive angle r. The distances **PN** and **MQ** are respectively $v_1 t$ and $v_2 t$, so that

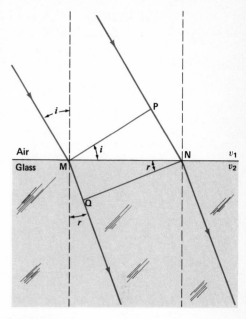

Figure 14-5. A wave-front diagram of refraction.

$$\frac{\mathbf{PN}}{\mathbf{MQ}} = \frac{v_1 t}{v_1 t} = \frac{v_1}{v_2}$$

In triangle **MNP**

$$\sin i = \frac{\mathbf{PN}}{\mathbf{MN}}$$

In triangle **MNQ**

$$\sin r = \frac{\mathbf{MQ}}{\mathbf{MN}}$$

The ratio

$$\frac{\sin i}{\sin r} = \frac{\mathbf{PN/MN}}{\mathbf{MQ/MN}} = \frac{\mathbf{PN}}{\mathbf{MQ}} = \frac{v_1}{v_2}$$

From Snell's law

$$n = \frac{\sin i}{\sin r}$$

Therefore

$$n = \frac{v_1}{v_2}$$

These relationships are illustrated in the following example.

EXAMPLE   A ray of light travels from air into water at an angle with the surface of 60.0°. The index of refraction of the water is 1.33. Find (a) the angle of refraction and (b) the speed of light (in m/s) in the water.

SOLUTION

(a) From Snell's law,

$$n = \frac{\sin i}{\sin r} \text{ and } \sin r = \frac{\sin i}{n}$$

($i = 90.0° - 60.0° = 30.0°$ and $n = 1.33$)

$$\sin r = \frac{\sin 30.0°}{1.33} = \frac{0.500}{1.33} = 0.376$$

$$r = 22.1°$$

(b) From the definition of the index of refraction,

$$n = \frac{v_1}{v_2} \text{ and } v_2 = \frac{v_1}{n}$$

($v_1 = 3.00 \times 10^8$ m/s and $n = 1.33$)

$$v_2 = \frac{3.00 \times 10^8 \text{ m/s}}{1.33} = 2.26 \times 10^8 \text{ m/s}$$

PRACTICE PROBLEMS   1. A ray of light passes from air into water at an angle to the surface of the water of 50.0°. What is (a) the angle of refraction and (b) the speed of light in water?   *Ans:* (a) 28.9° (b) 2.26×10^8 m/s

2. A ray of light passes from air into crown glass at an angle to the normal to the glass surface of 35.0°. (a) Determine the angle of refraction in the glass. (b) Determine the speed of light in the glass. (Index of refraction values are listed in Appendix B, Table 18.)   *Ans:* (a) 22.2° (b) 1.97×10^8 m/s

3. A ray of light in air is incident on the surface of a diamond gemstone at an angle of 42.0° with the gemstone surface. The angle of refraction is found to be 17.9°. What is (a) the index of refraction of the diamond and (b) the speed of light in the diamond?   *Ans:* (a) 2.42 (b) 1.24×10^8 m/s

---

The index of refraction of a homogeneous substance is a constant quantity that is a definite physical property of the substance. Consequently, the identity of such a substance can be determined by measuring its index of refraction with an instrument known as a *refractometer*. For example,

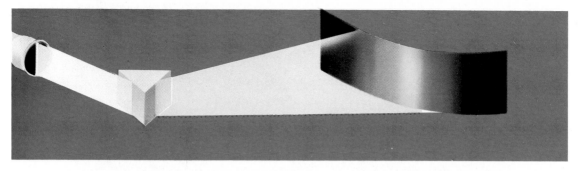

A. The prism spreads a narrow beam of white light out into the visible spectrum, the shorter wavelengths being bent more than the longer wavelengths.

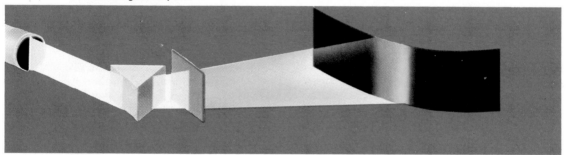

B. A red filter between the prism and the screen allows only light of the longer wavelengths to pass.

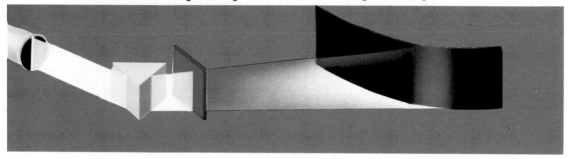

C. A green filter allows only light from the middle region of the spectrum to pass.

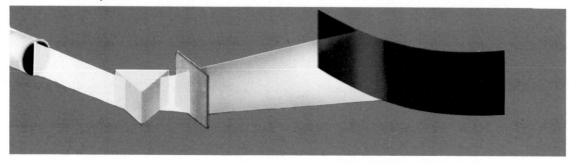

D. A blue filter passes only light of the shorter wavelengths.

(Adapted from COLOR AS SEEN AND PHOTOGRAPHED, Eastman Kodak)

Plate I

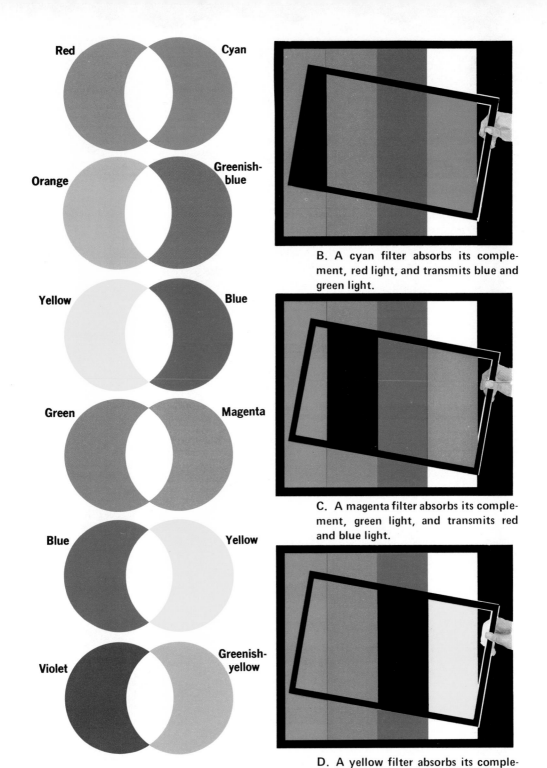

Red Cyan

Orange Greenish-blue

Yellow Blue

Green Magenta

Blue Yellow

Violet Greenish-yellow

A. Complementary colors.

B. A cyan filter absorbs its complement, red light, and transmits blue and green light.

C. A magenta filter absorbs its complement, green light, and transmits red and blue light.

D. A yellow filter absorbs its complement, blue light, and transmits red and green light.

(Adapted from COLOR AS SEEN AND PHOTOGRAPHED, Eastman Kodak)

Plate II

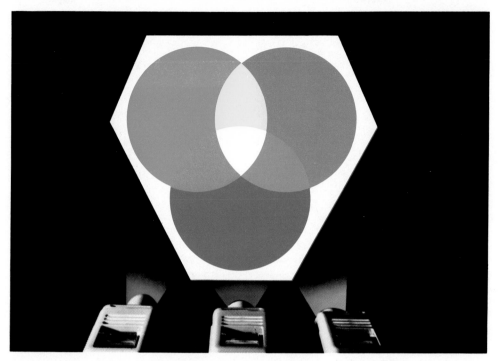

A. Primary colors. The additive mixture of red, green, and blue lights produces white light. Combined in pairs, two primaries give the complement of the third: cyan, magenta, and yellow. (Adapted from COLOR AS SEEN AND PHOTOGRAPHED, Eastman Kodak)

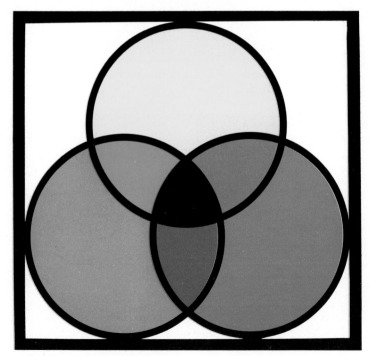

B. Primary pigments. The subtractive combination of cyan, magenta, and yellow filters transmits no light. Combined in pairs, two primary pigments give the complement of the third by subtraction: red, green, and blue.

Plate III

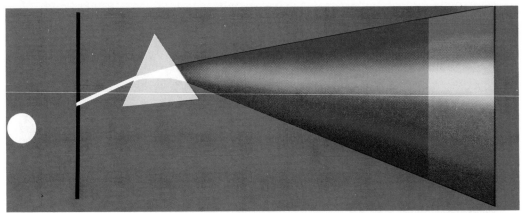

A. A continuous spectrum is produced by incandescent solids, liquids, and gases under high pressure. The very dense incandescent gases in the main body of the sun and stars produce continuous spectra.

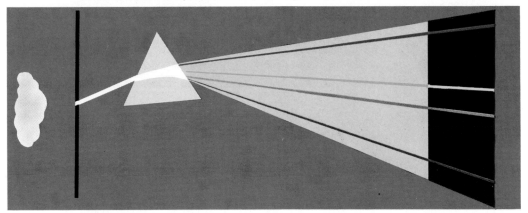

B. A bright-line spectrum is produced by incandescent gases of low density. Each chemical substance yields a characteristic pattern of lines that differ from all others.

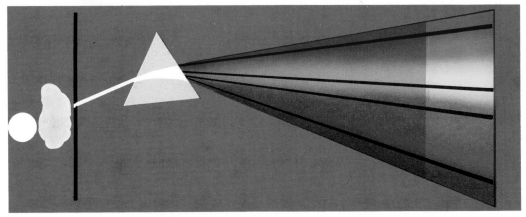

C. A dark-line spectrum is produced by a nonluminous (cooler) gas in front of an incandescent source of a continuous spectrum. The cooler gas absorbs light energy from the parts of the spectrum where it would emit bright lines if heated to incandescence. (Adapted from THE UNIVERSE, Time, Inc.)

Plate IV

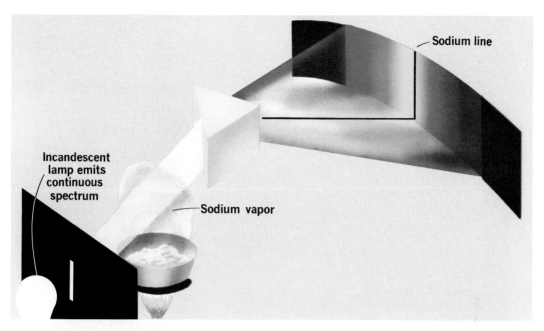

A. Nonluminous sodium vapor absorbs yellow light of the same wavelength as that emitted by luminous sodium vapor.

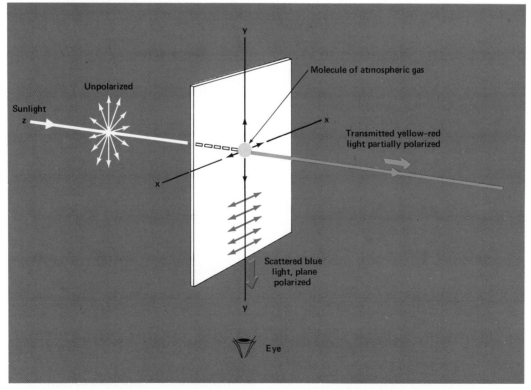

B. The appearance of the blue sky is due to the scattering of the shorter wavelengths of sunlight passing overhead through the atmosphere.

Plate V

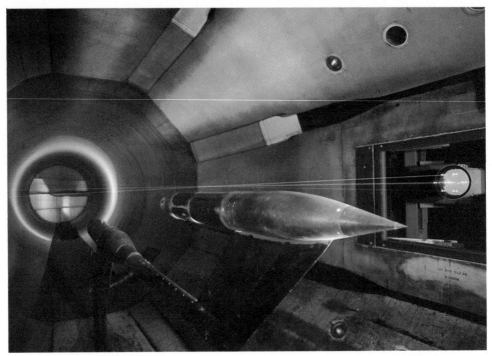

A. Laser beam in a transonic wind tunnel measures the relative air velocity of the model under test.

B. Interference produced by reflecting white light from a soap film. The picture on the right shows the fringes produced by red light.

Plate VI

EMISSION SPECTRA

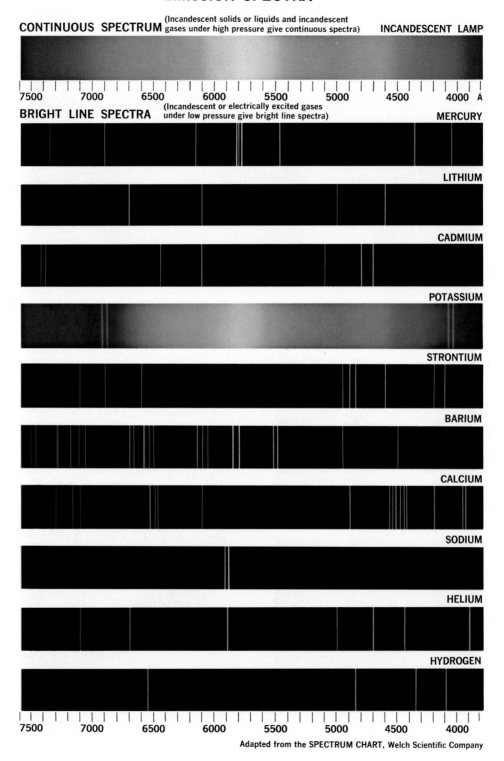

CONTINUOUS SPECTRUM (Incandescent solids or liquids and incandescent gases under high pressure give continuous spectra) INCANDESCENT LAMP

7500 7000 6500 6000 5500 5000 4500 4000 Å

BRIGHT LINE SPECTRA (Incandescent or electrically excited gases under low pressure give bright line spectra) MERCURY

LITHIUM

CADMIUM

POTASSIUM

STRONTIUM

BARIUM

CALCIUM

SODIUM

HELIUM

HYDROGEN

7500 7000 6500 6000 5500 5000 4500 4000

Adapted from the SPECTRUM CHART, Welch Scientific Company

Plate VII

A. Interference pattern produced by light of a single spectral color passing through two narrow slits.

B. Passing white light through a single slit produces this diffraction pattern.

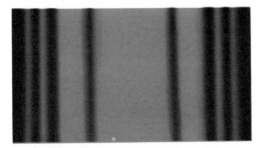

Red light passed through the same slit produces this pattern.

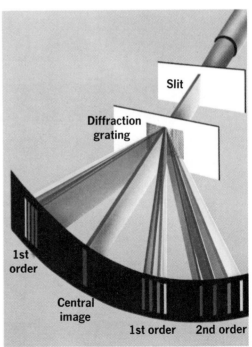

C. Mercury light through a diffraction grating. First and second order spectra appear to the right of the central image.

Plate VIII

butterfat and margarine have different indexes of refraction. One of the first tests made in a food-testing laboratory to determine whether butter has been mixed with margarine is the measurement of the index of refraction. The high index of refraction of a diamond furnishes one of the most conclusive tests for its identification.

Because light travels very slightly faster in outer space than it does through air, light from the sun or the stars is refracted when it enters the earth's atmosphere obliquely. Since the atmosphere is denser near the earth's surface, a ray of light from the sun or a star striking the atmosphere obliquely follows a path suggested by the curve shown in Figure 14-6. There is no abrupt refraction such as that which occurs at the interface between two media of different optical densities.

Atmospheric refraction prevents the sun and stars from being seen in their true positions except when they are directly overhead. In Figure 14-6, the sun appears at **S'** instead of **S,** its true position. Since the index of refraction from outer space to air is only 1.00029, the diagram is greatly exaggerated to show the bending. Refraction of sunlight by the earth's atmosphere causes the sun, when geometrically on the horizon, to appear about one diameter ($\frac{1}{2}°$) higher than it really is. Around 2 minutes are required for the earth to rotate through $\frac{1}{2}°$ of arc. Thus, we gain about 4 minutes of additional daylight each day because of atmospheric refraction at sunrise and sunset.

14.4 The Laws of Refraction   If the index of refraction of a transparent substance is known, it is possible to trace the path that a ray of light will take in passing through the substance. In Figure 14-7 rectangle **ABCD** represents a piece of plate glass with parallel surfaces. The line **EO** represents a ray of light incident upon the glass at point **O**. From point **O** as a center, two arcs are drawn. The radii of these arcs are in the ratio 3/2 based on the index of refraction of the glass being 1.5 and that of the air being 1. The normal **OF** is drawn. Then a line is drawn parallel to **OF** through the point **H** where the incident ray intersects the *smaller* arc. This line intersects the *larger* arc at point **G**. The line **OP**, determined by points **G** and **O**, marks the path of the refracted ray through the glass.

If the ray were not refracted as it leaves the glass, it would proceed along the line **PK**. To indicate the refraction at point **P**, this point is used as a center and arcs are drawn having the same ratio, 3/2, as before. The normal **PL** and a line **MN** parallel to the normal are drawn. The parallel line is drawn this time through point **N** where the

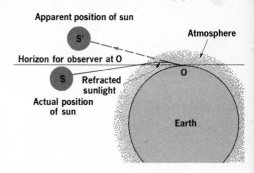

Figure 14-6. The sun is visible before actual sunrise and after actual sunset because of atmospheric refraction.

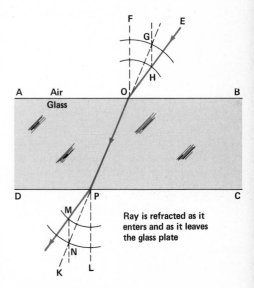

Figure 14-7. A method of tracing a light ray through a glass plate.

Test the validity of this construction using Figure 14-5 and Snell's law.

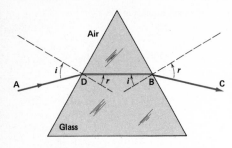

Figure 14-8. The path of a light ray through a prism. As the ray AD passes from the optically less dense to the optically more dense medium at D, its path is refracted toward the normal. As the ray passes from the optically more dense to the optically less dense medium at B, its path is refracted away from the normal.

Figure 14-9. (Left) The critical angle i_c is the limiting angle of incidence in the optically denser medium that results in an angle of refraction of 90°.

Figure 14-10. (Right) Total reflection at the water-air interface occurs when the angle of incidence exceeds the critical angle.

larger arc is intersected by the extension of refracted ray **OP.** This line intersects the smaller arc at point **M.** Points **P** and **M** determine the path of the refracted ray as it enters the air.

In Figure 14-8 line **AD** represents a light ray incident upon a triangular glass prism at point **D.** Passing obliquely from air into glass, the ray is refracted toward the normal along line **DB.** Observe that at point **D** angle i in air is larger than angle r in glass. As light ray **DB** passes obliquely from glass into air at point **B,** it is refracted away from the normal along line **BC.** At point **B** angle i in glass is smaller than angle r in air.

Refraction of light can be summarized in *three laws of refraction:*

1. The incident ray, the refracted ray, and the normal to the surface at the point of incidence are all in the same plane.

2. The index of refraction for any homogeneous medium is a constant that is independent of the angle of incidence.

3. When a ray of light passes obliquely from a medium of lower optical density to one of higher optical density, it is bent toward the normal to the surface. Conversely, a ray of light passing obliquely from an optically denser medium to an optically rarer medium is bent away from the normal to the surface.

14.5 Total Reflection   Suppose an incident ray of light, **AO,** passes from water into air and is refracted along the line **OB,** as shown in Figure 14-9. As the angle of incidence, i, is increased, the angle of refraction, r, also increases. When this angle approaches the limiting value, $r_l = 90°$, the refracted ray emerges from the water along a path that gets closer to the water surface. As the angle of incidence continues to increase, the angle of refraction finally equals 90° and the refracted ray takes the path **ON**

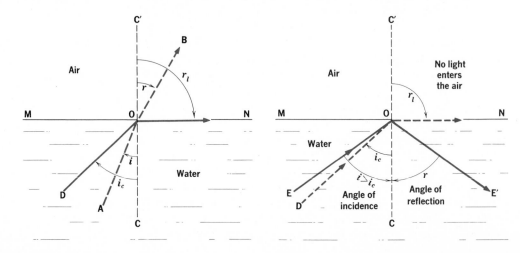

along the water surface. *The limiting angle of incidence in the optically denser medium that results in an angle of refraction of 90° is known as the **critical angle,** i_c.* The critical angle for water is reached when the incident ray **DO** makes an angle of 48.5° with the normal; the critical angle for crown glass is 42°, while that for diamond is only 24°.

In Section 14.3 we defined the index of refraction of a material as the ratio of the speed of light in a vacuum (air) to the speed of light in the material or as the ratio sin *i*/sin *r*. In Figure 14-9 the light passes from the optically denser water to the air. Thus in the form of Snell's law, the roles of the angles of incidence and refraction are reversed. Here the angle of refraction, *r*, is related to the speed of light in air. The angle of incidence, *i*, is associated with the speed of light in water. The index of refraction of the water in this instance is

$$n = \frac{\sin r \ (\text{air})}{\sin i \ (\text{water})}$$

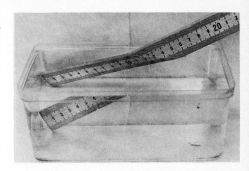

Figure 14-11. The meter stick appears to be bent at the surface of the water as a result of the refraction of light. Can you account for the second image of the end of the meter stick at the lower left?

At the critical angle, i_c, *r* is the limiting value r_l, which equals 90°. Thus

$$n = \frac{\sin r_l}{\sin i_c} = \frac{\sin 90°}{\sin i_c} = \frac{1}{\sin i_c}$$

Therefore, in general

$$\sin i_c = \frac{1}{n}$$

where *n* is the index of refraction of the optically denser medium relative to air and i_c is the critical angle of this medium.

If the angle of incidence of a ray of light passing from water into air is increased beyond the critical angle, no part of the incident ray enters the air. The incident ray is *totally reflected* from the water interface. In Figure 14-10 **EO** represents a ray of light whose angle of incidence exceeds the critical angle, the angle of incidence **EOC** being greater than the critical angle **DOC**. The ray of light is reflected back into the water along the line **OE′**, a case of simple reflection in which the angle of incidence **EOC** equals the angle of reflection **E′OC**. Total reflection always occurs when the angle of incidence exceeds the critical angle.

Some reflection (but not total reflection) occurs for angles of incidence equal to or less than i_c.

A diamond is a brilliant gem because its index of refraction is exceedingly high and its critical angle is therefore correspondingly small. Very little of the light that enters the upper surface of a cut diamond passes through the diamond; most of the light is reflected internally (total reflection), finally emerging from the top of the diamond. The faces of the upper surface are cut at such angles as to ensure that the maximum light entering the upper surface is reflected back to these faces.

Questions
GROUP A

1. Define (a) refraction, (b) angle of refraction, and (c) index of refraction.
2. What property of a transparent substance causes the refraction of light rays that strike it at an oblique angle?
3. What are the three laws of refraction?
4. What is meant by the terms (a) critical angle and (b) total reflection?
5. What practical use is made of the index of refraction of a substance?

GROUP B
6. You are looking diagonally down at a fish in a pond. To the fish, assuming that it can see you, does your head appear higher or lower than it actually is?
7. Explain why we see the sun before it actually rises above the horizon in the morning, and why we see it after it has dropped below the horizon in the evening.
8. Which is the better reflector of light, a right-angle prism or a plane, silvered mirror? Explain.
9. Why are totally reflecting prisms usually designed so that light enters and leaves the prism at an angle 90° with respect to the surface?
10. Why is a diamond more brilliant than a glass imitation cut the same way?

Problems
GROUP A

1. The speed of light in chloroform is 1.99×10^8 m/s. What is its refractive index?
2. What is the speed of light in kilometers per second in a diamond that has a refractive index of 2.42?
3. A penny at the bottom of a glass cylinder is 30.0 cm below the eye. If water is poured into the cylinder to a depth of 16.0 cm, how much closer does the coin appear?
4. A light ray passes from air into the surface of a Lucite block at an angle with the Lucite surface of 48.5°. What is the angle of refraction? (Index of refraction values are listed in Appendix B, Table 18.)

GROUP B
5. A ray of light passes from a medium of refractive index 1.33 into a medium of refractive index 1.50. The angle of incidence is 45.0°. (a) By the construction method, determine the angle of refraction. (b) Compute the angle of refraction.
6. The angles of incidence and refraction of a ray of light passing from air into water are 60.0° and 41.0°. Find the index of refraction of the water.
7. Find the critical angle for a carbon-tetrachloride-to-air surface.
8. A physics student wished to know whether a glass cube to be used in an experiment was composed of crown or flint glass. A light ray passing from air into the cube was traced and the angles of incidence and refraction were determined to be 35.0° and 20.9° respectively. (a) What index of refraction did these data yield? (b) What kind of glass did the student have?

LENS OPTICS

14.6 Types of Lenses   A lens is any transparent object having two nonparallel curved surfaces or one plane surface and one curved surface. The curved surfaces can be spherical, parabolic, or cylindrical. Lenses are usually made of glass but can be made of other transparent materials. There are two general classes of lenses based upon their effects on incident light.

   1. *Converging lenses.* Cross sections of converging lenses are shown in Figure 14-12(A). All are thicker in the middle than at the edge. The concavo-convex lens is known as a meniscus lens.

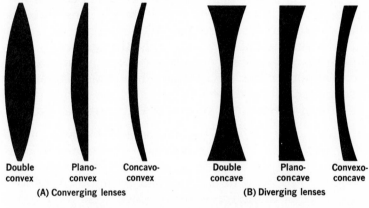

| Double convex | Plano-convex | Concavo-convex | Double concave | Plano-concave | Convexo-concave |

(A) Converging lenses   **(B) Diverging lenses**

Figure 14-12. Common lens configurations.

   Recall that light travels more slowly in glass and other lens materials than it does in air. Light passing through the thick middle region of a converging lens is retarded more than light passing through the thin edge region. Consequently, a wave front of light transmitted through a converging lens is bent as shown in Figure 14-13(A). The plane wave in this diagram is incident on the surface of the converging lens parallel to the lens plane. The wave is refracted and converges at the point **F** beyond the lens.

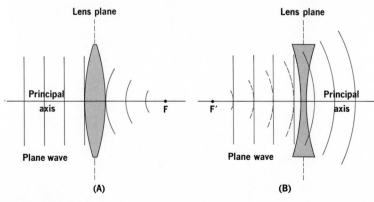

Figure 14-13. Refraction of a plane wave (A) by a converging lens and (B) by a diverging lens. The incident wave is perpendicular to the lens axis.

2. Diverging lenses. Lenses that are thicker at the edges than in the middle are diverging lenses. Their cross sections are shown in Figure 14-12(B). The convexo-concave lens is a meniscus lens.

Light passing through the thin middle region of a diverging lens is retarded less than light passing through the thick edge region. A wave front of light transmitted through a diverging lens is bent as shown in Figure 14-13(B). Observe that the plane wave in this diagram is also incident on the surface of the lens parallel to the lens plane. In this case, however, the refracted wave diverges in a way that makes it appear to come from the point **F′** in front of the lens.

A spherical lens usually has two centers of curvature— the centers of the spheres that form the lens surfaces. These centers of curvature determine the principal axis of the lens. The radii of curvature of the two surfaces of double convex and double concave lenses are not necessarily equal, although they are so drawn in the ray diagrams in this chapter. The geometry of lens surfaces having different radii of curvature is shown in Figure 14-14.

Figure 14-14. The geometry of lens surfaces that have different radii of curvature. Observe that the surfaces of the double convex lens are sections of intersecting spheres.

14.7 Ray Diagrams The nature and location of the image formed by a lens is more easily determined by a ray diagram than by a wave-front diagram. The plane waves of Figure 14-13 can be represented by light rays drawn perpendicular to the wave fronts. Since these wave fronts

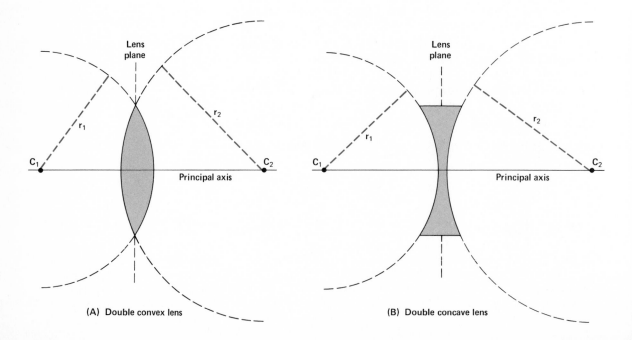

(A) Double convex lens   (B) Double concave lens

approaching the lens are shown parallel to the lens plane, their ray lines are parallel to the *principal axis* of the lens. Observe that the principal axis passes through the center of the lens and is perpendicular to the lens plane.

In Figure 14-15(A) these parallel rays (which are also parallel to the principal axis) are shown incident on a converging lens. They are refracted as they pass through the lens, and they converge at a point on the principal axis that locates the *principal focus* **F** of the lens. At point **B** the incident ray is refracted toward the normal drawn to the front surface of the lens. At point **D** the incident ray is refracted away from the normal drawn to the back surface of the lens. Because the rays of light actually pass through the principal focus **F**, it is called the *real focus*. *Real* images are formed on the same side of the lens as the real focus. Had these rays approached the lens from the right parallel

Real images can be projected on a screen.

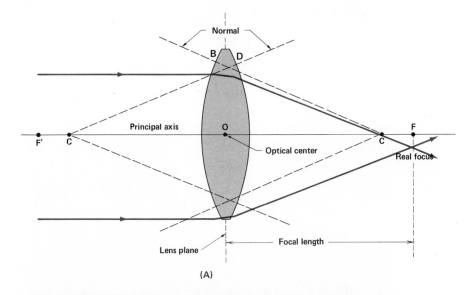

(A)

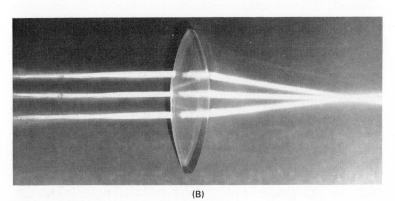

(B)

Figure 14-15. The converging lens. (A) A ray diagram that illustrates the refractive convergence of incident light rays parallel to the principal axis of the lens. (B) A photograph using light pencils shows a similar convergence.

to the principal axis, they would have converged at a point on the principal axis that located the principal focus **F′**.

Rays parallel to the principal axis are shown incident on a diverging lens in Figure 14-16(A). The rays of light are refracted as they pass through the lens and diverge as if they had originated at the principal focus **F′** located in front of the lens. Because the rays do not actually pass through this principal focus, it is called a *virtual focus*. *Virtual* images are formed on the same side of the lens as the virtual focus.

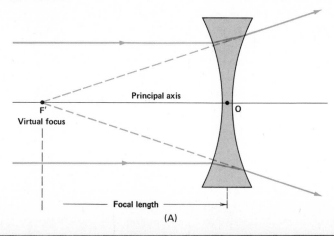

(A)

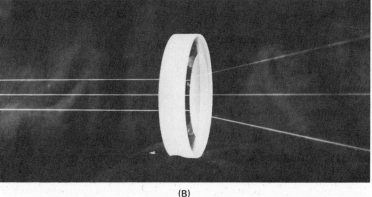

(B)

Figure 14-16. The diverging lens. (A) A ray diagram illustrates the refractive divergence of incident light rays parallel to the principal axis of the lens. (B) A photograph using light pencils shows a similar divergence.

In general, lenses refract rays that are parallel to the principal axis and the refracted rays either converge at the real focus behind the (converging) lens or diverge as if they originated at the virtual focus in front of the (diverging) lens. However, these foci are not midway between the lens and the center of curvature as they are in spherical mirrors. *The positions of the foci on the principal axis depend on the index of refraction of the lens.* A common double convex lens of crown glass has principal foci and centers of curvature that practically coincide and a focal length that is approximately equal to its radius of curvature. The *focal length, f,* of a lens is the distance between the optical center of the lens and the principal focus. The focal length of any lens depends on its index of refraction and the curvature of its surfaces. The higher its index of refraction, the shorter is its focal length. The longer its radius of curvature, the longer is its focal length.

The focal length of a lens is the same for light traveling in either direction even if the two lens surfaces have different radii of curvature.

A "bundle" of parallel light rays incident on a converging lens parallel to its principal axis is refracted and converges at the principal (real) focus behind the lens. The image is ideally a point of light at the real focus. This refraction is shown in Figure 14-17(A).

Bundles of parallel rays are not always incident on a lens parallel to its principal axis. If such rays are incident on a converging lens at a small angle with the principal axis, they converge at a point in the plane that contains the principal focus of the lens and is perpendicular to the principal axis. This plane is called the *focal plane* of the lens. A bundle of parallel rays incident on a converging lens at a small angle α (alpha) with its principal axis is shown in Figure 14-17(B).

Bundle of rays: the essentially parallel rays in a beam or a pencil of light.

The focal plane of a concave mirror is described in Section 13.7.

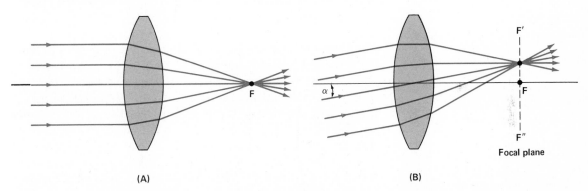

(A)   (B)

14.8 Images by Refraction   The image of a point on an object is formed by intersecting refracted rays emanating from the object point. In Section 13.8 we recognized that the task of locating graphically an image formed by reflection is simplified if we select incident rays that have known paths after reflection. These rays, called *principal rays,* are shown with their *reflected* paths in Figure 13-17. Their paths after refraction are also known. They are shown with their *refracted* paths in Figure 14-18.

The image **S′** of point **S** (the object in Figure 14-18) is formed by the converging lens when the refracted rays from point **S** intersect behind the lens. Ray **1,** *parallel* to the principal axis, is refracted through the real focus, **F.** See Figure 14-15. Ray **2,** along the *secondary axis,* passes

Figure 14-17. In (A) parallel rays are shown incident on a converging lens parallel to its principal axis. The refracted rays converge at its principal focus. In (B) the rays are incident at a small angle α with the principal axis. Here the refracted rays are focused at a point in the focal plane F′F″ of the lens.

Review Figure 13-17.

through the optical center of the lens, **O**, without being appreciably refracted. Ray **3** passes through the principal focus, **F'**, and is refracted parallel to the principal axis. The three refracted rays intersect at the image point **S'**. Observe that any two of the principal rays *emanating from the same object point* are able to locate the image of that point.

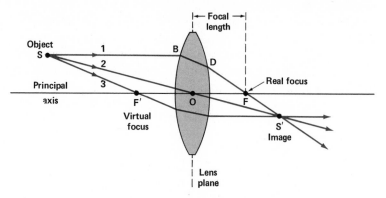

Figure 14-18. The principal rays used in ray diagrams. Any two of these three rays from the same point on the object locate the image of that point.

Lenses and mirrors differ in several ways.

1. Secondary axes pass through the optical center of a lens and not through either of its centers of curvature.

2. The principal focus is usually near the center of curvature, depending on the refractive index of the glass from which the lens is made. Thus the focal length of a double convex lens is about equal to its radius of curvature.

3. Since the image produced by a lens is formed by rays of light that actually pass through the lens, a *real* image is formed on the side of the lens opposite the object. *Virtual* images formed by lenses appear to be on the same side of the lens as the object.

4. Convex (converging) lenses form images in almost the same manner as concave mirrors, while concave (diverging) lenses are like convex mirrors in the manner in which they form images.

Spherical lenses, like spherical mirrors, have aberration defects. See Figure 14-19. When such lenses are used with large apertures, images formed by rays passing through the central zones of the lens are generally sharp and well-defined, while images formed by rays passing through the edge zones are fuzzy. This defect of lens images is called *spherical aberration.*

Similarly, rays of light coming from an object point not on the principal axis are not brought to a sharp focus in the image plane. This defect of lenses is known as *lens astigmatism.* By using a combination of lenses of suitable refractive indexes and focal lengths, lens makers produce *anastigmatic* lenses. Such lenses give good definition over the entire image even when used with large apertures.

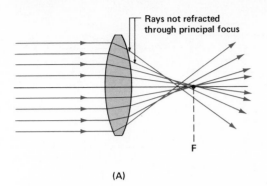

Rays not refracted
through principal focus

F

(A)

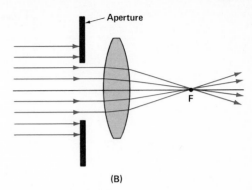

Aperture

F

(B)

A lens of short focal length that can be used with a large aperture has a large light-gathering capability. It is said to be a "fast" lens. The light-gathering power of a camera lens is given in terms of its **f-number.** This number is determined by the focal length of the lens and its effective diameter. The effective diameter is the diameter of the camera aperture (diaphragm) that determines the useful lens area. The light-gathering power, or "speed," of a lens is expressed as the ratio of its focal length to its effective diameter. If the speed of a lens is given as f/4, it means that its focal length is 4 times its effective diameter.

Because the useful area of a lens is proportional to the square of its effective diameter, the light-gathering power of a lens increases four times when its effective diameter is doubled. An f/4 lens is 4 times as fast as an f/8 lens, and is 16 times as fast as an f/16 lens. It follows that the required time of exposure increases as the square of the f-number.

It should be recognized that all lenses having the same f-number give the same illumination in the image plane, regardless of their individual diameters. Therefore, a lens having twice the diameter of another lens of the same f-number will have four times the light-gathering power, but this light will be spread over an image having four times the area and will give the same image brightness.

14.9 Images Formed by Converging Lenses We shall consider six different cases of image formation. These cases are illustrated in Figure 14-20.

Case 1. Object at an infinite distance. The use of a small magnifying glass to focus the sun's rays upon a point approximates this first case. While the sun is not at an infinite distance, it is so far away that its rays reaching the earth are nearly parallel. When an object is at an infinite distance and its rays are parallel to the principal axis of the lens, *the image formed is a point at the real focus.* See Figure 14-20(A). This principle can be used to find the focal length of a lens by focusing the sun's rays on a white

Figure 14-19. (A) Rays parallel to the principal axis but near the edge of a converging lens are not refracted through the principal focus. (B) An aperture can be used to block these rays from the lens.

The area of a circle:
$$A = \pi r^2 = \frac{\pi d^2}{4}. \; \textit{Thus, } A \propto d^2.$$

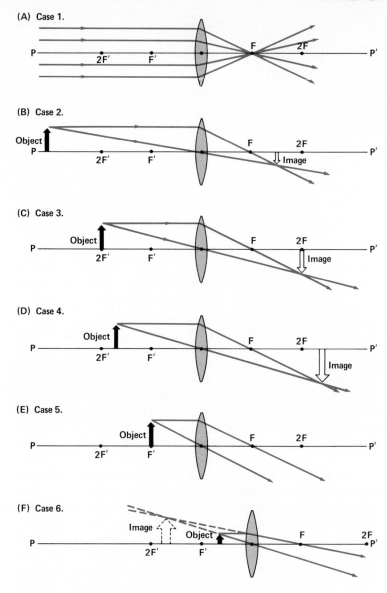

Figure 14-20. Ray diagrams of image formation by converging lenses.

screen. The distance from the screen to the optical center of the lens is the focal length of the lens.

Case 2. Object at a finite distance beyond twice the focal length. Case 2 is illustrated in (B). Rays parallel to the principal axis, along the secondary axis, and emanating from a point on the object are used to locate the corresponding image point. *The image is real, inverted, reduced, and located between F and 2F on the opposite side of the lens.* The lenses of the eye and the camera, and the objective lens of the refracting telescope are all applications of this case.

Case 3. Object at a distance equal to twice the focal length. The construction of the image is shown in (C). *The image is real, inverted, the same size as the object, and located at 2F on the opposite side of the lens.* An inverting lens of a field telescope, which inverts an image without changing its size, is an application of Case 3.

Case 4. Object at a distance between one and two focal lengths away. This is the converse of Case 2 and is shown in (D). *The image is real, inverted, enlarged, and located beyond 2F on the opposite side of the lens.* The compound microscope, slide projector, and motion picture projector are all applications of a lens used in this manner.

Case 5. Object at the principal focus. This case is the converse of Case 1. *No image is formed,* since the rays of light are parallel as they leave the lens (E). The lenses used in lighthouses and searchlights are applications of Case 5.

Case 6. Object at a distance less than one focal length away. The construction in (F) shows that the rays are divergent after passing through the lens and cannot form a real image on the opposite side of the lens. These rays appear to converge behind the object to produce *an image that is virtual, erect, enlarged, and located on the same side of the lens as the object.* The simple magnifier and the eyepiece lenses of microscopes, binoculars, and telescopes form images as shown in Case 6.

14.10 Images Formed by Diverging Lenses   The only kind of image of a real object that can be formed by a diverging lens is one that is *virtual, erect, and reduced in size.* Diverging lenses are used to neutralize the effect of a converging lens, or to reduce its converging effect to some extent. The image formation is shown in Figure 14-21.

Figure 14-21. The image of an object formed by a diverging lens.

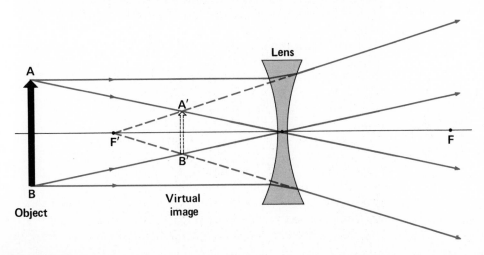

14.11 Object-Image Relationships For thin lenses, the ratio of object size to image size equals the ratio of the object distance to image distance. This rule is the same as the rule for curved mirrors. Thus

$$\frac{h_i}{h_o} = \frac{d_i}{d_o}$$

Observe that the lens equations are the same as the mirror equations of Section 13.12.

where h_o and h_i represent the heights of the object and the image respectively, and d_o and d_i represent the respective distances of the object and image from the optical center of the lens.

The equation used to determine the distances of the object and image in relation to focal length for curved mirrors applies also to lenses. It can be restated here as

$$\frac{1}{f} = \frac{1}{d_o} + \frac{1}{d_i}$$

The lens equation is valid providing these sign conventions are followed:

d_o *is* $\begin{bmatrix} \textit{positive for real objects} \\ \textit{negative for virtual objects} \end{bmatrix}$

d_i *is* $\begin{bmatrix} \textit{positive for real images} \\ \textit{negative for virtual images} \end{bmatrix}$

f *is* $\begin{bmatrix} \textit{positive for converging lenses} \\ \textit{negative for diverging lenses} \end{bmatrix}$

where d_o represents the distance of the object from the lens, d_i the distance of the image from the lens, and f the focal length.

The numerical value of the focal length f is *positive* for a converging lens and *negative* for a diverging lens. For real objects and images, the object and image distances d_o and d_i have *positive* values. For virtual objects and images, d_o and d_i have *negative* values.

14.12 The Simple Magnifier A converging lens of short focal length can be used as a simple magnifier. The lens is placed slightly nearer the object than one focal length and the eye is positioned close to the lens on the opposite side. This is a practical example of Case 6; the image is virtual, erect, and enlarged as shown in Figure 14-22. A reading glass, a simple microscope, and an eyepiece lens of a compound microscope or telescope are applications of simple magnifiers.

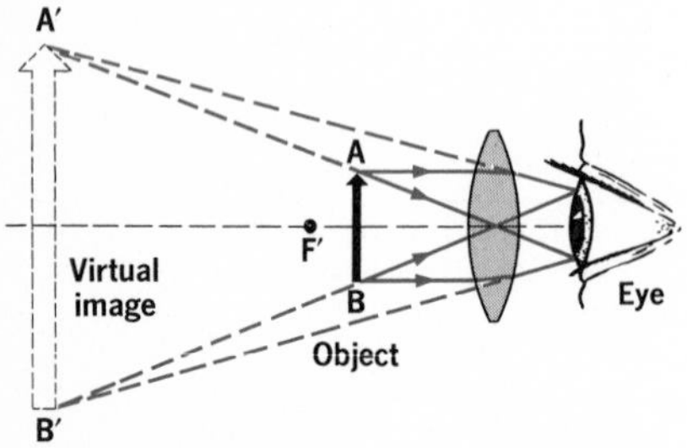

Magnification M is simply the ratio of the image height to the object height.

$$M = \frac{h_i}{h_o}$$

But,

$$\frac{h_i}{h_o} = \frac{d_i}{d_o}$$

So,

$$M = \frac{d_i}{d_o}$$

Figure 14-22. The simple magnifier.

Suppose an object is viewed by the unaided eye. As it is moved closer and closer to the eye, the image formed on the retina becomes larger and larger. Eventually, a *nearest* point is reached for the object at which the eye can still form a clear image. This minimum distance for distinct vision is approximately 25 cm from the eye. Although this nearest point varies among individuals, 25 cm is taken as the standard distance for most distinct vision; it is called the *near point*. As a person grows older the muscles of the eye, which thicken the lens and thus increase its convergence (shorten its focal length), gradually weaken. Consequently the near point moves out with aging.

If a converging lens is placed in front of the eye as a simple magnifier, the object can be brought much closer and the eye focuses on the virtual image. When the lens is used in this way, the object is placed just inside the principal focus $(d_o \simeq f)$. The image is then formed approximately at the near point. Magnification, shown above to be equal to the ratio d_i/d_o, can now be expressed for a simple magnifier as the ratio of the distance for most distinct vision to the focal length of the lens. When f is given in centimeters, the magnification becomes approximately

$$M = \frac{25 \text{ cm}}{f}.$$

Magnifiers are labeled to show their magnifying power. Thus a magnifier with a focal length of 5 cm would be marked 5X. One with a focal length of 2.5 cm would be marked 10X, etc. Observe that the shorter the focal length of a converging lens, the higher is its magnification.

14.13 The Microscope The compound microscope, thought to be invented in Holland by Zacharias Janssen about 1590, uses a lens, the *objective*, to form an enlarged image as in Case 4. This image is then magnified, as in Case 6, by a second lens, called the *eyepiece*.

In Figure 14-23 a converging lens is used as the objective, with the object **AB** just beyond its focal length. At **A'B'**, a distance greater than twice the focal length of the objective lens, an enlarged, real, and inverted image is formed. The eyepiece lens acts as a simple magnifier to enlarge this image.

The magnifying power of the objective is approximately equal to the length of the tube, l, divided by the focal length, f_o, of the objective, or l/f_o. The magnifying power of the eyepiece, acting as a simple magnifier, is approximately 25 cm/f_e. The total magnification is the product of

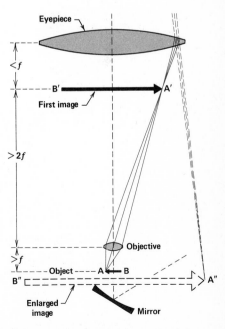

Figure 14-23. Image formation by a compound microscope.

the two lens magnifications. (The equation is again an approximation.)

$$M = \frac{25 \text{ cm} \times l}{f_e \times f_o}$$

14.14 Refracting Telescopes A refracting astronomical telescope has two lens systems. The *objective lens* is of large diameter so that it will admit a large amount of light. The objects to be viewed in telescopes are always more distant than twice the focal length of the objective lens. As a consequence the image formed is smaller than the object. The *eyepiece lens* magnifies the real image produced by the objective lens. The magnifying power is approximately equal to the focal length of the objective, f_o, divided by the focal length of the eyepiece, f_e, or f_o/f_e.

The lenses of a *terrestrial*, or field, telescope form images just as their counterparts do in the refracting astronomical telescope. Since it would be confusing to see objects inverted in a field telescope, another lens system is used to reinvert the real image formed by the objective. This additional inverting lens system makes the final image erect, as shown in Figure 14-24. The inverting lens system does not magnify the image because the lens system is placed exactly its own focal length from the image formed by the objective.

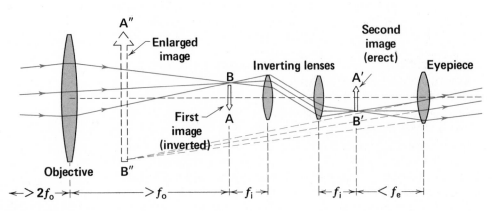

Figure 14-24. Image formation by a terrestrial telescope.

The *prism binocular* is actually a double field telescope that uses two sets of totally reflecting prisms instead of a third lens system to reinvert the real images formed by the objective lenses. This method of forming final images that are upright and correctly oriented also has the effect of folding the optical path, thus making binoculars more compact and easier to use than telescopes.

The prism binocular customarily has descriptive markings stamped on its case, such as 7×35, 8×50, etc. The first number gives its magnification, and the second number gives the diameters (in mm) of its objective lenses.

Figure 14-25. A pair of totally reflecting prisms reinvert the image in binoculars.

Questions
GROUP A

1. (a) What causes spherical aberration in lenses? (b) How can it be remedied?
2. Of which case of converging lenses is the simple magnifier an application?
3. Explain the lens system and image formation of a compound microscope in terms of the lens cases used.
4. Explain the lens system and image formation of a refracting astronomical telescope in terms of the lens cases used.
5. How does a terrestrial telescope compare with a refracting astronomical telescope?

GROUP B

6. What kinds of lenses could be used and where should the object be placed to produce (a) an enlarged real image, (b) a reduced real image, (c) a real image the same size as the object, (d) an enlarged virtual image, (e) a small virtual image?
7. When the distance between a projector and the screen is increased, what adjustment must be made in the distance between the film and the objective lens to bring the image back into focus?
8. Is the focal length of a crown-glass diverging lens changed when the lens is immersed in water? Justify your answer.
9. A double convex lens in air acts as a converging lens. (a) Can it act as a diverging lens in another medium? (b) Construct a ray diagram to illustrate your response to (a).
10. Suppose you were going to use a clear plastic bag inflated with air as an underwater converging lens. (a) What shape should the inflated bag have? (b) Illustrate your response to (a) with an appropriate ray diagram.

Problems

GROUP A

1. A converging lens has a focal length of 20.0 cm. If it is placed 50.0 cm from an object, at what distance from the lens will the image be?

2. If an object is $1\overline{0}$ cm from a converging lens of 5 cm focal length, how far from the lens will the image be formed?

3. The focal length of the lens in a box camera is 10.0 cm. The fixed distance between the lens and the film is 11.0 cm. If an object is to be clearly focused on the film, how far must it be from the lens?

4. An object 3.0 cm tall is placed $2\overline{0}$ cm from a converging lens. A real image is formed $1\overline{0}$ cm from the lens. What is (a) the focal length of the lens, (b) the size of the image?

5. An object $3\overline{0}$ cm from a converging lens forms a real image $6\overline{0}$ cm from the lens. (a) Find the focal length of the lens. (b) What is the size of the image if the object is 5 cm high?

6. The focal length of a camera lens is 5.00 cm. How far must the lens be from the film to produce a clear image of an object 3.00 m away?

7. What is the focal length of the image-forming optics (lens and cornea) in your eye when you read a book 35.0 cm from your eye? The effective length of the ray path in the eyeball is 19.0 mm.

8. What is the focal length of the image-forming optics in your eye when you are looking at a person standing 15.0 m away? The effective length of the ray path in the eyeball is 19.0 mm.

9. What is the magnifying power of a simple magnifier whose focal length is $1\overline{0}$ cm?

GROUP B

10. An object 3 cm tall is placed 16 cm from a converging lens with a focal length of 24 cm. (a) Find the location of the image. (b) What is its size?

11. When an object 5.0 cm tall is placed 12 cm from a converging lens, an image is produced on the same side of the lens as the object but 61 cm away from the lens. What is (a) the focal length of the lens, (b) the size of the image?

12. An optical bench pointer 1.0 cm tall is placed at the 55.0-cm mark on the meter stick. When a diverging lens is placed at the 0.0-cm mark, an image is formed at the 5.0-cm mark. What is (a) the focal length of the lens, (b) the size of the image?

13. The objective lens of a compound microscope has a focal length of 0.500 cm. The eyepiece has a focal length of 2.00 cm. If the lenses are 15.0 cm apart, what is the magnifying power of the microscope?

14. The dimensions of the picture on a slide are 6.4 cm by 7.6 cm. This slide is to be projected to form an image 1.5 m by 1.8 m at a distance of 9.0 m from the objective lens of the projector. (a) What is the distance from the slide to the objective lens? (b) What focal length objective lens must be used?

15. The tube of a microscope is $16\overline{0}$ mm long. If the focal length of the eyepiece is 30.0 mm and the focal length of the objective is 5.00 mm, find the magnifying power.

16. What is the distance between the objective lens and the eyepiece lens of a terrestrial telescope when an object 1.5 kilometers away is being observed? The focal length of the objective lens is 15 cm and the focal length of the eyepiece lens is 5 cm. Each lens of the two-lens inverting system has a focal length of 3 cm, and these two lenses are 6 cm apart.

DISPERSION

14.15 Dispersion by a Prism Suppose a narrow beam of sunlight is directed onto a glass prism in a darkened room. If the light that leaves the prism falls on a white screen, a band of colors is observed, one shade blending gradually into another. *This band of colors produced when sunlight is dispersed by a prism is called a* **solar spectrum.** The dispersion of sunlight was described by Newton, who observed that the spectrum was "violet at one end, red at the other, and showed a continuous gradation of colors in between."

We can recognize six distinct colors in the visible spectrum. These are *red, orange, yellow, green, blue,* and *violet.* Each color gradually blends into the adjacent colors giving a *continuous* spectrum over the range of visible light. A continuous spectrum is shown in **Plate VII** of the color insert at the end of this chapter. *Light consisting of several colors is called* **polychromatic light;** *light consisting of only one color is called* **monochromatic light.** All colors of the spectrum are present in the incident beam of sunlight. White light is a mixture of these colors.

The dispersion of light by a prism is shown in **Plate I** of the color insert. It is evident that the refraction of red light by the prism is not as great as that of violet light; the refractions of other colors lie between these two. Thus the index of refraction of glass is not the same for light of different colors. If we wish to be very precise in measuring the index of refraction of a substance, monochromatic light must be used and the monochrome color must be stated. Some variations in the index of refraction of glass are given in Table 14-1.

Table 14-1
VARIATION OF THE INDEX
OF REFRACTION

Color	Crown glass	Flint glass
red	1.515	1.622
yellow	1.517	1.627
blue	1.523	1.639
violet	1.533	1.663

14.16 The Color of Light A hot solid radiates an appreciable amount of energy that increases as the temperature is raised. At relatively low temperatures, the energy is radiated only in the infrared region. As the temperature of the solid is raised, some of the energy is radiated at higher frequencies. These frequencies range into the red portion of the visible spectrum as the body becomes "red hot." At still higher temperatures the solid may be "white hot" as the major portion of the radiated energy shifts toward the higher frequencies.

Suppose we have a clear-glass tungsten-filament lamp connected in an electric circuit so that the current in the filament, and thus the temperature of the filament, can be controlled. A small electric current in the filament does not

change the filament's appearance. As we gradually increase the current, however, the filament begins to glow
with a dark red color. To produce this color electrons in
the atoms of tungsten must have been excited to sufficiently high energy levels so that upon de-excitation they
emit energy with wavelengths of about 7600 Å. Even before the lamp filament glows visibly, experiments show
that it radiates infrared rays that can be detected as heat.
As the current is increased further, the lamp filament
gives off orange light in addition to red; then the filament
adds yellow, and finally at higher temperatures it adds
enough other colors to produce white light.

A photographer's tungsten-filament flood lamp operates at a very high temperature. If the white light of such a
lamp is passed through a prism, a band of colors similar to
the solar spectrum is obtained. Figure 14-26 shows the distribution of radiant energy from an "ideal" radiator for
several different temperatures.

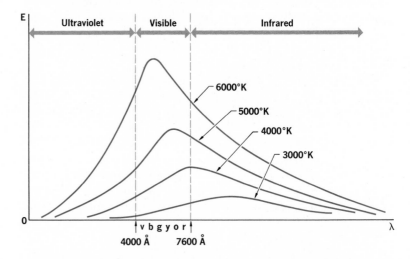

Figure 14-26. Distribution of
radiant energy from an ideal
radiator.

Considering the wavelengths of various colors shown in
Plate I, it is evident that our eyes are sensitive to a range of
frequencies equivalent to about one octave. The wavelength of the light at the upper limit of visibility (7600 Å) is
about twice the wavelength of the light at the lower limit
of visibility (4000 Å). We use the word *color* to describe a
psychological sensation through the visual sense related
to the physical stimulus of light. The color perceived for
monochromatic light depends on the wavelength of the
light. For example, when light of about 6500 Å enters the
eye, the color perceived is red.

14.17 The Color of Objects   Color is a property of the
light that reaches our eyes. Objects may absorb certain

wavelengths from the light falling upon them and reflect other wavelengths. For example, a cloth that appears blue in sunlight appears black when held in the red portion of a solar spectrum in a darkened room. A red cloth held in the blue portion of the solar spectrum also appears black. *The color of an opaque object depends upon the frequencies of light it reflects.* If all colors are reflected, we say it is *white*. It is *black* if it absorbs all the light that falls upon it. It is called *red* if it absorbs all other colors and reflects only red light. The energy associated with the colors absorbed is taken up as heat.

Actually, in order to be reflected the red light must interact with the object. The interaction consists of a resonant absorption by electrons of the object's atoms. The energy is immediately re-emitted or "reflected."

A piece of blue cloth appears black in the red portion of the spectrum because there is no blue light there for it to reflect and it absorbs all other colors. For the same reason, a red cloth appears black in the blue portion of the spectrum. *The color of an opaque object depends on the color of the light incident upon it.*

Ordinary window glass, which transmits all colors, is said to be colorless. Red glass absorbs all colors but red, which it transmits. The stars of the United States flag would appear red on a black field if viewed through red glass. *The color of a transparent object depends upon the color of the light that it transmits.*

14.18 Complementary Colors Because polychromatic light can be dispersed into its elementary colors, it is reasonable to suppose that elementary colors can be combined to form polychromatic light. There are three ways in which this can be done.

The term "color" refers to the color of light.

1. A prism placed in the path of the solar spectrum formed by another prism will recombine the different colors to produce white light. Other colors can be compounded in the same manner.

2. A disk that has the spectral colors painted on it can be rotated rapidly to produce the effect of combining the colors. The light from one color forms an image that persists on the retina of the eye until each of the other colors in turn has been reflected to the eye. If pure spectral colors are used in the proper proportion, they will blend to produce the same color sensation as white light.

Combining colored lights is an additive process.

3. Wavelengths from the middle region of the visible spectrum combined with wavelengths from the two end regions produce white light. This method is described in Section 14.19.

Two prisms are used as described above, but with the red light from the first prism blocked off from the second prism. The remaining spectral colors are recombined by the second prism to produce a blue-green color called

*Become familiar with the six elementary colors and their complements illustrated in **Plate II.***

cyan. Red light and cyan should therefore combine to produce white light, and a rotating color wheel shows this to be true. *Any two colors that combine to form white light are said to be* **complementary.**

In similar fashion it can be shown that blue and yellow are complementary colors. White fabrics acquire a yellowish color after continued laundering. A blue dye added to laundry detergents neutralizes the yellow color and the fabrics appear white. Iron compounds in the sand used for making glass impart a green color to the glass. Manganese gives glass a *magenta,* or purplish-red, color. However, if both these elements are present in the right proportion, the resulting glass will be colorless. Green and magenta are complementary colors. The complements of the six elementary spectral colors are shown in **Plate II.**

14.19 The Primary Colors The six regions of color in the solar spectrum are easily observed by the dispersion of sunlight. Further dispersion within a color region fails to reveal any other colors of light. We generally identify the range of wavelengths comprising a color region by the color of light associated with that region. These are the six *elementary* colors of the visible spectrum; they combine to produce white light. However, the complement of an elementary color is not monochromatic but is a mixture of all of the elementary colors remaining after the one elementary color has been removed.

Experiments with beams of different colored lights have shown that most colors and hues can be described in terms of three different colors. Light from one end of the visible spectrum combined with light from the middle region in various proportions will yield all of the color hues in the half of the spectrum that lies in between them. Light from the opposite end, when combined with light from the middle region, will also yield all hues in the half of the spectrum that lies in between them. Colored light from the two end regions and the middle region can be combined to match most of the hues when mixed in the proper proportions. The three colors that can be used most successfully in color matching experiments of this sort are *red, green,* and *blue.* Consequently these have been called the *primary colors.*

Suppose we project the three primary colors onto a white screen as shown in **Plate III(A).** The three beams can be adjusted to overlap, producing additive mixtures of these primary colors. Observe that green and blue lights combine to produce cyan, the complement of red; green and red lights combine to produce yellow, the complement of blue; and red and blue lights combine to produce magenta, the complement of green. Thus two primary colors combine to produce the complement of the third primary color. Where the three primary colors overlap, white light is produced.

Combining primary colors is an additive process.

14.20 Mixing Pigments When the complements blue light and yellow light are mixed, white light results by an additive process. If we mix a blue pigment with a yellow pigment, a green mixture results. This process is subtractive since each pigment subtracts or absorbs certain colors. For example, the yellow pigment subtracts blue and violet lights and reflects red, yellow, and green. The blue pigment subtracts red and yellow lights and reflects green, blue, and violet. Green light is the only color reflected by both pigments; thus the mixture of pigments appears green under white light.

Combining primary pigments is a subtractive process.

The subtractive process can be demonstrated by the use of various color filters that absorb certain wavelengths and transmit others from a single white-light source.

When pigments are mixed, each one subtracts certain colors from white light, and the resulting color depends on the light waves that are not absorbed. *The **primary pigments** are the complements of the three primary colors.* They are cyan (the complement of red), magenta (the complement of green), and yellow (the complement of blue). When the three primary pigments are mixed in the proper proportions, all the colors are subtracted from white light and the mixture is black. See **Plate III(B).**

*Compare the additive combinations of primary colors in **Plate III(A)** with the subtractive combinations of primary pigments in **Plate III(B).***

14.21 Chromatic Aberration Because a lens configuration has some similarity to that of a prism, some dispersion occurs when light passes through a lens. Violet light is refracted more than the other colors and is brought to a focus by a converging lens at a point nearer the lens than the other colors. Because red is refracted the least, the focus for the red rays is farthest from the lens. See Figure 14-27(A). Thus images formed by ordinary spherical lenses are always fringed with spectral colors. *The nonfocusing of light of different colors is called **chromatic aberration**.* Sir Isaac Newton developed the reflecting telescope to avoid the objectionable effects of chromatic aberration that occur when observations are made through a refracting telescope.

The English optician John Dollond (1706–1761) discovered that the fringe of colors could be eliminated by means of a combination of lenses. A double convex lens of crown glass used with a suitable plano-concave lens of flint glass

Figure 14-27. (A) Chromatic aberration is caused by unequal refraction of the different colors. (B) A two-lens combination of crown and flint glass corrects chromatic aberration.

corrects for chromatic aberration without preventing refraction and image formation. A lens combination of this type is shown in Figure 14-27(B). It is called an *achromatic* (without color) lens.

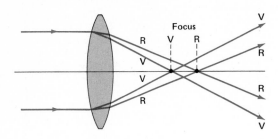

(A) Chromatic aberration

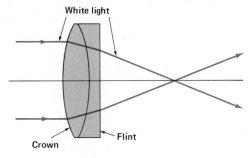

(B) Achromatic lens

Questions

GROUP A

1. Why does a prism disperse sunlight into a band of colors?
2. What property of light determines its color?
3. What name is given to electromagnetic radiations having wavelengths slightly (a) longer and (b) shorter than visible light?
4. If a black object absorbs all light rays incident upon it, how can it be seen?
5. What is the appearance of a red dress in a closed room illuminated only by green light? Explain.
6. (a) What are the primary colors? (b) Define a complementary color. (c) Name the complement of each primary color.

GROUP B

7. Why can't white paint be made from orange paint by adding a pigment of another color?
8. How could you demonstrate that a piece of white-hot iron gives off red light?
9. The focal length of a converging lens is determined experimentally using red light. (a) How will this focal length compare with that for green light? (b) Construct a ray diagram to support your conclusion in (a).
10. Suppose a diverging lens is substituted for the converging lens in Question 9. (a) How will the focal length of the lens for red light compare with that for green light? (b) Construct an appropriate diagram to support your conclusion.

SUMMARY

The laws of refraction describe the behavior of light rays that pass obliquely from one medium into another of different optical density. The index of refraction of any transparent material is defined in terms of the speed of light in a vacuum and the speed of light in the material. The index of refraction is also expressed in terms of Snell's law. Total reflection is explained on the basis of the critical angle of a material and the limiting value of the angle of refraction.

Converging lenses have convex surfaces and form images in a manner similar to that of concave mirrors. Diverging lenses have concave surfaces and form images in a manner similar to that of convex mirrors. The general lens equations correspond to those for curved mirrors. These equations relate object and image sizes with their respective distances from the lens and object and image distances with the focal length of the lens.

A converging lens forms either real or virtual images of real objects depending on the position of the object relative to the principal focus of the lens. A diverging lens forms only virtual images of real objects. Lens functions in common types of refractive instruments such as microscopes and telescopes can be analyzed by considering each lens separately.

Sunlight is composed of polychromatic light that undergoes dispersion when refracted by a prism. Six elementary colors are recognized in dispersed white light. The visual perception of color is related to the wavelength of visible light.

The removal of an elementary color from white light leaves a polychromatic color that is the complement of the color removed. Addition of complementary colors produces white light. White light is produced by adding the primary colors. Any two of the three primary colors will combine to produce the complement of the third. The primary pigments are complements of the primary colors. Their combination is considered to be a subtractive process.

VOCABULARY

achromatic lens
angle of incidence
angle of refraction
chromatic aberration
complementary color
converging lens
critical angle
diverging lens
elementary color
eyepiece
focal length
focal plane

index of refraction
lens equation
monochromatic
near point
objective
optical density
polychromatic
primary color
primary pigment
principal axis
principal focus
principal ray

real focus
real image
refraction
refractometer
secondary axis
Snell's law
solar spectrum
spherical aberration
total refraction
virtual focus
virtual image

DIFFRACTION AND POLARIZATION

The equation for the diffraction of light was derived by William Henry Bragg and his son, William Lawrence Bragg, by studying crystal structures with X rays. In 1915, the Braggs jointly received the Nobel Prize in Physics for this work, which provided a valuable new tool for the study of atomic structure.

In this chapter you will gain an understanding of:

☛ solving problems based on the relationship between wavelength and diffraction angle
☛ how polarization confirms the transverse-wave character of light
☛ different ways light becomes plane-polarized
☛ double refraction
☛ polarized light and strain patterns

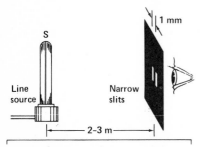

Figure 15-1. A method of observing double-slit interference.

INTERFERENCE AND DIFFRACTION

15.1 Double-Slit Interference The superposition of two identical wave trains traveling in the same or opposite direction illustrates the phenomenon of *interference*. Sound waves of the same amplitude and wavelength projected in the same direction by two loudspeakers provide an interference pattern in which alternate regions of reinforcement and cancellation can be found. Two identical waves traveling in opposite directions on a stretched string interfere and produce a standing wave pattern.

The interference of two water waves of identical wavelength is easily observed in the laboratory using the ripple tank. More precise studies of such wave behavior are made by photographing light reflections from the surface of a mercury ripple tank. Interference and superposition topics can be reviewed by referring to Sections 10.14 through 10.16.

The superposition principle holds for these wave disturbances in material media and for light waves and other electromagnetic waves in free space. Thus light shows *interference: the mutual effect of two beams of light that results in a loss of intensity in certain regions (destructive interference) and a reinforcement of intensity in other regions (constructive interference).*

The English physician and physicist Thomas Young (1773–1829) first demonstrated interference of light in 1801 and showed how this phenomenon supports the wave theory of Huygens. Two narrow slits, as shown in Figure 15-1, about 1 mm apart in a piece of black paper are used to observe a narrow source of light, **S,** placed 2 or 3 meters away. A series of narrow bands alternately dark and light is seen in the center of a fairly wide band of light. See Figure 15-2. If a red filter is placed between the source and the slits so that a single spectral color is the light source, a series of red and black bands is seen. A double-slit interference pattern from a monochromatic light is shown in **Plate VIII(A).**

A wave-front diagram of double-slit interference with a monochromatic light source is shown in Figure 15-3. Slits S_1 and S_2 are equidistant from the source **S.** As light from **S** reaches S_1 and S_2, each slit serves as a new light source producing new wave fronts in phase with each other. These waves travel out from S_1 and S_2 producing bright bands of light due to reinforcement where constructive interference occurs and dark bands due to cancellation where destructive interference occurs.

Figure 15-2. A double-slit interference pattern.

Compare the diagram in Figure 15-3 with the photograph in Figure 10-27.

Figure 15-3. A wave-front diagram of double-slit interference.

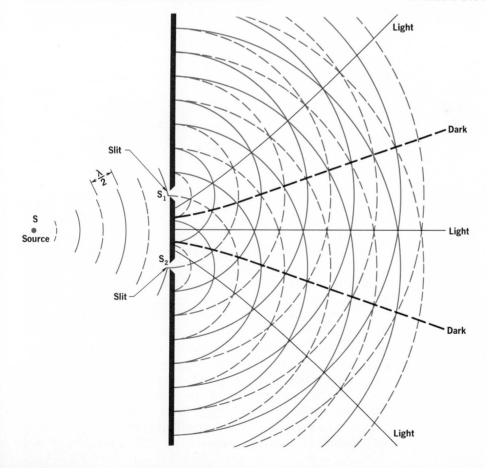

15.2 Interference in Thin Films   Sir Isaac Newton knew about the color fringes produced by thin films. He devised experiments to determine the film thickness that corresponds to a specific color. Interference of light was unknown in his time, however. His corpuscular explanation of the partial reflection and partial refraction of light at the interface of two different transmitting media is inadequate in view of modern theories of light.

When viewed by reflected white light, thin transparent soap films, oil slicks, and wedge-shaped films of air show varying patterns of colors. When illuminated by monochromatic light, alternate bright and dark regions are observed, and the positions of the bands shift as the color of the monochromatic light is changed. Thomas Young explained this thin-film phenomenon in terms of the interference of light waves.

A ray of monochromatic light is incident on a thin transparent film at a small angle i, as shown in Figure 15-4.

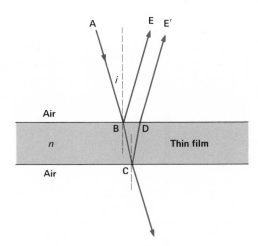

Figure 15-4. Partial reflection and refraction of light in a thin film results in interference.

The two reflected waves, BE and DE', are coherent because they both originate from the same point A on the monochromatic source.

Some light is reflected at **B** and some is refracted toward **C**. The index of refraction, n, of the transparent film is higher than that of air. The refracted light is then partially reflected at **C,** emerging from the film along **DE'** parallel to **BE**. If the wave fronts traveling along **BE** and **DE'** reach the eye, interference is observed. This is because the paths **ABE** and **ACE'** differ in length and the partial path **BCD** of **ACE'** is in a medium of different optical density. The nature of the interference depends on the phase relation between the waves arriving at the eye.

For the monochromatic light that is used, assume that the incidence angle i is very small and the film at **B** has an optical thickness of a quarter wavelength. Thus the optical path *in the thin film* from **B** to **C** is one-quarter wavelength. From **C** to **D** is another one-quarter wavelength. Light emerging from the film at **D** is therefore one-half wavelength behind that reflected at **B** and will be expected to interfere destructively causing the film to appear dark.

Now if we assume the film has an optical thickness of a half wavelength, light reflected from the lower surface of the film will emerge at **D** a whole wavelength behind that reflected from the upper surface at **B**. The waves will be expected to interfere constructively causing the film to appear bright at this point. *Observation of reflected light from thin films shows that a reverse effect actually occurs.* Thus, the quarter-wave point appears *bright* and the half-wave point appears *dark*.

This anomalous situation prevails whenever air is on both sides of the film. It is easily observed in soap films. If a soap film is supported vertically, as it drains its upper region will become quite thin and a very small fraction of a wavelength in thickness just before it breaks. If observed by reflected monochromatic light at this moment, this upper region appears dark. See **Plate VI(B).**

Applying the same line of argument to this film of almost negligible thickness, we would expect the light reflected from the front surface and that reflected from the back surface to reach the eye so nearly in phase that this upper region of the film would appear bright by reflected light. Here is a clear-cut discrepancy between observation and theory. How can it be reconciled?

Thomas Young resolved the discrepancy by suggesting that *one of the interfering waves undergoes a phase change of 180° during reflection.* This phase change is in addition to the phase change that results from the unequal lengths of their optical paths. The phase inversion occurs during one reflection and not the other because the two reflections are opposite in kind. One reflection takes place at an interface at which the medium beyond has the higher index of refraction. This reflection corresponds to that at point **B** in Figure 15-4. *The reflected wave is inverted in phase.* The other reflection occurs at an interface at which the medium beyond has the lower index of refraction. This reflection corresponds to that at point **C** in Figure 15-4. Here, *the reflected wave experiences no change in phase.*

Because of the phase-inverting effect of one of the two reflecting interfaces, the rules for constructive and destructive interference for thin films *bounded on both sides with a medium of lower index of refraction* are just the opposite from those we might expect. *Maximum constructive interference occurs with such thin films if the optical path difference is an*

The wavelength referred to is the wavelength of the monochromatic light in the thin film. Since the index of refraction of the film is higher than that of air, the wavelength of the light is shorter in the film than in the air.

Camera lenses are often coated with transparent thin films to reduce stray reflected light and increase image contrast.

See Section 10.10 for a review of incident and reflected waves at a reflecting interface.

odd number of half wavelengths. This happens where the film thickness is an *odd* number of quarter wavelengths. *Maximum destructive interference occurs if the optical path difference is a whole number of wavelengths*. Here the film thickness is an *even* number of quarter wavelengths.

The vertically supported soap films of **Plate VI(B)** gradually increase in thickness toward the bottom as they drain. A vertical cross section of such a film would have a wedge-like appearance. As the thickness increases, the odd $\lambda/4$ requirement for the reinforcement of different colors and the even $\lambda/4$ requirement for their cancellation are met at successive intervals down the film. The soap film shown on the left of **Plate VI(B)** therefore reflects white light as a succession of colors at intervals down the film.

An air film between two optically flat glass plates produces a regular pattern of interference fringes. Irregular surfaces produce irregular patterns of interference fringes. Using interference patterns, inspection techniques have been developed that make extremely high precision measurements possible. Test plates can be polished optically flat with a tolerance of approximately 5×10^{-7} cm. Interference techniques are used to establish standards of measurement in many mechanical processes. The *interferometer* is an instrument that uses interference in the measurement of distance in terms of known wavelengths of light or in the measurement of wavelengths in terms of a standard of length.

Figure 15-5. Interference patterns. (Left) Straight interference fringes from an optical flat. (Right) An interference pattern obtained from a spherical mirror being tested in an interferometer. The white spots are reflections from the interferometer light source.

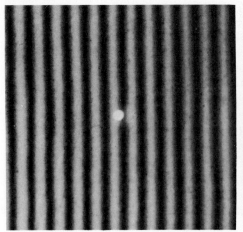 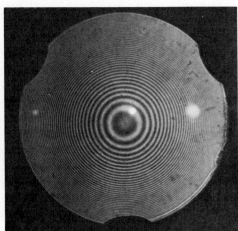

15.3 Diffraction of Light According to the wave theory, light waves should bend around corners although our common experience with light shows that it travels in straight lines. Under some conditions light waves do bend out of their straight paths. When light waves encounter an obstruction with dimensions comparable with their wavelengths, the light spreads out and produces spectral colors due to interference. *The spreading of light into a region behind an obstruction is called* **diffraction.** A slit opening, a fine wire, a sharp-edged object, or a pinhole can serve as a suitable obstruction in the path of a beam of light from a point source. It is possible to see diffraction fringes if one peers between two fingers at a distant light source. With the fingers held close to the eye and brought together to form a slit opening, dark fringes will be seen just before the light is shut out.

Diffraction effects, such as indistinct edges of shadows and shadow fringes, are known to have been observed as early as the seventeenth century. However, before the discovery of interference in 1801, neither the wave theory nor the corpuscular theory could offer a suitable explanation for these diffraction effects. In 1816 the French physicist A. J. Fresnel (fray-*nel*) (1788–1827) demonstrated that the various diffraction phenomena are fully explained by the interference of light waves.

Very useful diffraction patterns can be produced by illuminating an optical surface, either plane or spherical concave, that has many thousands of straight, equally-spaced parallel grooves ruled on it. These ruled surfaces are known as *diffraction gratings.* Light is diffracted when it is transmitted through or reflected from the narrow spaces between the ruled lines.

Standard gratings can have as many as 12,000 ruled lines per centimeter of grating surface. X rays are investigated with ordinary diffraction gratings set with their surface at a low angle to the rays. The gratings thus have the equivalent of many lines per centimeter. Gratings are generally superior to optical prisms for displaying the length or spread of spectra. However, grating spectra tend to be less intense than those formed by prisms.

15.4 Wavelength by Diffraction

A transmission grating placed in the path of plane waves disturbs the wave front because the ruled lines are opaque to light and the narrow spacings between the lines are transparent. These spaces provide a large number of fine, closely-spaced transmission slits. New wavelets generated at these slits interfere in such a way that several new wave fronts are established. One wave front travels in the original direction and the other wave fronts travel at various angles from this direction depending on their wavelength.

Suppose a single narrow slit is illuminated by white

Review diffraction as observed in the ripple tank in Section 10.13.

Figure 15-6. A master diffraction grating being ruled by a diamond-pointed scribe.

Recall Huygens' principle explained in Section 12.3: Each point on a wave front may be regarded as a new source of disturbance.

light and viewed through a transmission grating. A white image of the illuminated slit is seen directly in line with the slit opening. In addition, pairs of continuous spectra are observed, each pair equally spaced on opposite sides of the principal image.

If we illuminate the slit with monochromatic light, successive pairs of slit images of decreasing intensity will appear on opposite sides of the principal image. The two images forming the first pair are known as *first-order* images, those forming the second pair as *second-order* images, etc.

In Figure 15-7, **A** and **B** are parallel spaces between the ruled lines on a diffraction grating. They act as adjacent transmission slits, being uniformly separated by the distance *d*, called the *grating constant*. Monochromatic light from a distant illuminated slit traveling normal to the grating surface produces secondary wavelets simultaneously at **A** and **B**. A new wave front of these wavelets proceeds along **MN** and produces the principal image of the distant slit at **N**.

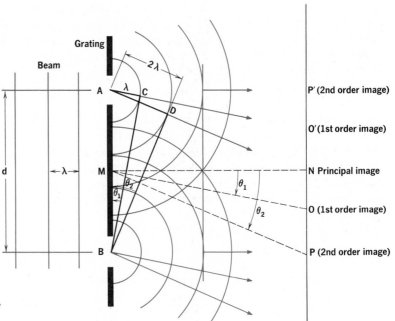

Figure 15-7. The optical geometry of a transmission grating.

A given wavelet from **B** and the first preceding wavelet from **A** produce a wave front **CB** that travels along **MO** and gives a *first-order* image of the slit at **O**. Similarly, a given wavelet from **B** and the second preceding wavelet from **A** give a wave front **DB** that yields a *second-order* image at **P**, etc. Of course, corresponding images appear at **O′**, **P′**, etc., on the other side of **N**.

In the right triangle **ABC**, side **AC** equals the wavelength λ of the incident light and angle θ_1 is the angle of the first-order diffracted wave front from the grating plane, the *diffraction angle*.

It is evident that

$$\lambda = d \sin \theta_1$$

Side **AD** of triangle **ABD** is equal to 2λ, and θ_2 is the second-order diffraction angle. Thus for second-order images,

$$\lambda = \frac{d \sin \theta_2}{2}$$

In the general case, for any order n, the grating equation becomes

$$\lambda = \frac{d \sin \theta_n}{n}$$

The diffraction angle θ can be determined experimentally. Knowing the order of image n observed, the diffraction angle θ, and the grating constant d, the wavelength of the light can be calculated from this equation. Of course, if the grating constant of a particular diffraction grating is not known, it can be calculated from an experimental determination of θ_n in which a monochromatic light of known wavelength is used. In the example, the grating equation is used to determine the wavelength of a monochromatic light from experimental data.

Reliable measurements of the wavelengths of light from various sources can be made in the laboratory using this diffraction method.

EXAMPLE   An optical grating that has 6.00×10^3 lines/cm gives a second-order image at a diffraction angle of $44.8°$. Calculate the wavelength of the light used.

SOLUTION

$$\lambda = \frac{d \sin \theta_n}{n}$$

where $n = 2$ and $d = \dfrac{1}{6.00 \times 10^3/\text{cm}} = 1.67 \times 10^{-4} \text{ cm}$

$$\lambda = \frac{1.67 \times 10^{-4} \text{ cm} \times \sin 44.8°}{2}$$

$$\lambda = \frac{1.67 \times 10^{-4} \text{ cm} \times 0.705}{2} \times \frac{\text{Å}}{10^{-8} \text{ cm}}$$

$$\lambda = 5890 \text{ Å}$$

PRACTICE   **1.** An illuminated grating with 4.85×10^3 lines/cm gives a
PROBLEMS   second-order image at a diffraction angle of 37.4°. Calculate the wavelength, in angstroms, of the monochromatic light.
Ans: 6260 Å

2. A monochromatic light of 4550 Å illuminates a transmission grating having 5600 lines/cm and produces first- and second-order images. Determine the diffraction angle of (a) the first-order image and also (b) the second-order image.   *Ans:*   (a) 14.7°; (b) 30.5°

15.5 Single-Slit Diffraction   According to Huygens' principle, every point on an advancing wave front can be regarded as a new source of disturbance from which secondary waves spread out as spherical wavelets. If a barrier with a narrow slit opening is placed in the path of advancing plane waves, the disturbance will be transmitted by the slit to the region beyond the barrier. See Figure 15-8(A). When the width of the slit opening is reduced to only a few wavelengths of the light, a broad region **MN** is illuminated by the slit. Experiments show that the central portion of this region is always brighter than the remote portions that reveal diffraction fringes of diminishing intensity. Examples of single-slit diffraction patterns are shown in **Plate VIII(B).**

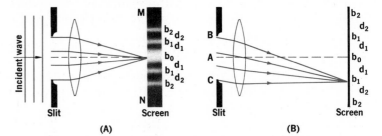

Figure 15-8. Single-slit diffraction. Slit size is exaggerated relative to the lines and screen.

In Figure 15-8(B) the point b_0 lies on the perpendicular bisector of the slit and is equidistant from points **B** and **C**. Because the distance from the slit to the screen is very large compared with the width of the slit, point b_0 is essentially equidistant from all points along the line **BAC**. Thus the wavelets originating simultaneously from all points along **BA** and from all points along **AC** will reach b_0 in phase and the screen in this region will be bright.

At points d_1, above and below b_0, where the distance from **B** and **C** is different by a whole wavelength, the distance from **A** and **C** is different by a half wavelength. For every point along **CA** there is a corresponding point along **BA** that is a half wavelength different in distance from d_1.

Therefore wavelets arriving at d_1 from one half of the slit will be annulled by the wavelets arriving from the other half of the slit. The region of d_1 will be dark.

Beyond points d_1 there are points b_1 where the distance from **B** and **C** is different by 1.5 wavelengths. Now we can think of the slit opening as being divided into three equal parts. Wavelets from one part arrive at b_1 a half wavelength behind wavelets from the second part. Wavelets from these parts of the slit opening interfere destructively, while wavelets arriving from the third part produce brightness at b_1. However, the illumination at b_1 will be lower than that at b_0 since only one-third of the slit opening contributes to the brightness of these regions.

By the same reasoning the regions of d_2, where the difference in distance from **B** and **C** is 2 wavelengths, will be dark. Similarly, the regions of b_2 will be bright. Thus a series of alternate bright and dark regions appears on either side of b_0. The intensity of these regions decreases as the distance from b_0 increases.

Questions
GROUP A

1. How do the colors of a soap bubble originate?
2. Why are the various colors of white light separated by a diffraction grating?
3. Will the angular separation between red and blue rays be greater in the first-order or second-order spectrum of white light produced by a diffraction grating?
4. You are given two diffraction gratings, one with 400 lines per centimeter and the other with 4000 lines per centimeter. Which grating yields more orders of images of an illuminated slit?
5. What advantage is realized by increasing the number of ruled lines per centimeter of grating surface?

GROUP B
6. Explain why the following appear red: (a) glowing charcoal, (b) a ripe cherry, (c) a neon sign, (d) the sunrise, (e) objects viewed through red sunglasses.
7. How can you explain that in observations of thin films by reflected light the interference effects are the reverse of those we would normally expect?
8. You are given two optical gratings, one with fewer lines per centimeter than the other, and a single monochromatic light source. For which grating will the angular separation of orders be greater? Justify your conclusion.
9. What part of Young's double-slit experiment depended on diffraction and what part depended on interference?
10. Monochromatic light of known wavelength was transmitted by a grating and the diffraction angle of a first-order image was determined. Derive the equation for the grating constant of the transmission grating.

Problems
GROUP B

1. A transmission grating with $58\overline{0}0$ lines/cm is illuminated by monochromatic light with a wavelength of 4920 Å. What is the diffraction angle for the first-order image?

2. In Problem 1, the perpendicular distance from the grating to the image screen was 34.5 cm. How far from the principal image was the first-order image found?

3. Monochromatic light illuminates a grating having $59\overline{0}0$ lines/cm. The diffraction angle for a second-order image is 38.0°. (a) Determine the wavelength of the light in angstroms. (b) What is its color?

4. White light falls on a grating that has $35\overline{0}0$ lines/cm. The perpendicular distance from the grating to the image screen is 50.0 cm. (a) Find the distance of the near edge of a first-order image from the principal image on the screen. (b) Find the distance of the far edge of the same image.

5. In the experiment of Problem 4, a marker was placed in the first-order image on the screen 9.25 cm from the principal image. (a) Determine the wavelength (in angstroms) at the marker position. (b) What color corresponds to this wavelength?

POLARIZATION

15.6 Polarization of Transverse Waves The general wave properties of rectilinear propagation in a homogeneous medium—reflection, refraction, interference, and diffraction—are clearly recognized in the behavior of light as well as in sound and water disturbances.

In Chapter 12 we considered a particle model of light that at one time was as successful as the wave model in explaining rectilinear propagation, reflection, and refraction. Only interference and diffraction failed to accommodate the particle model. They required a wave model for explanation. Thus interference and diffraction give the best evidence that light has wavelike characteristics.

Before the introduction of the electromagnetic theory, light was assumed to be a longitudinal wave disturbance. Electromagnetic theory predicts that light is a transverse wave. Sound waves are longitudinal disturbances and water waves have both transverse and longitudinal characteristics. From interference or diffraction experiments, we can infer nothing concerning the transverse nature of light waves since both sound and water waves show these properties. What experimental basis do we have that supports the theoretical prediction that light waves are transverse?

Fresnel observed that a beam of light falling on a calcite crystal was separated into two beams that were incapable of producing interference fringes. Young suggested that this could be explained by assuming that the light consisted of transverse waves that were separated into component waves having oscillating planes at right angles to each other. He called this *a plane-polarization effect.*

Many experiments with calcite and other similar materials have demonstrated the correctness of Young's polarization hypothesis. **Plane-polarized light** *is light in which the oscillations are confined to a single plane that includes the line of propagation.* Transverse waves but not longitudinal waves can be polarized. This is because the oscillations of transverse waves are independent and randomly oriented. Longitudinal waves vibrate along only one line of direction, and that direction is parallel to its propagation.

Light radiated by ordinary sources is unpolarized since the primary radiators, the atoms and molecules of the light source, oscillate independently. Beams of unpolarized light are made up of independent wave trains with oscillation planes oriented in a random manner about the line of propagation. Diagrammatically, we can represent unpolarized light being propagated in space by a system of *light vectors* as shown in Figure 15-9(A). It is customary to resolve these vectors into vertical and horizontal components as at (B), which is a convenient but entirely arbitrary orientation. Figure 15-9(C) represents a side view of (B) in which the horizontal component vectors are perpendicular to the page. It is customary for the light vectors of Figure 15-9 to represent the oscillating *electric components* of the electromagnetic (light) waves.

Ordinary light can become plane-polarized through interactions with matter. The scattering effect of small particles (see Section 15.11) is accompanied by a polarization effect. Polarization can also result from reflection of light from various surfaces, from refraction of light through some crystals, and from selective absorption of light in some crystals.

A simple mechanical model can be used to illustrate the polarization concept. Suppose we set up transverse waves in a rope passed through a slot as shown in Figure 15-10. We can readily see that vibrations are transmitted beyond the slot only when the vibrating plane of the rope and the plane of the slot are aligned. We shall call this frame of slots the *polarizer.* A second slot, parallel to the first, will transmit the waves. See Figure 15-10(A). If the second slot is perpendicular to the first slot, it obstructs them. See

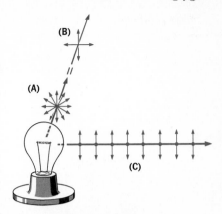

Figure 15-9. The vectors representing random oscillation planes of unpolarized light (A) may be arbitrarily resolved into vertical and horizontal component vectors as in (B) for an end-on view and as in (C) for a side view.

Review Section 12.4.

Figure 15-10. A mechanical analogy of polarization.

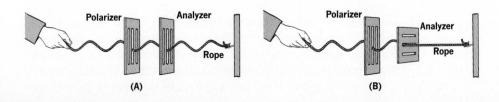

(A)   (B)

Figure 15-10(B). We shall call this second frame of slots the *analyzer*.

If the rope is replaced by a long coiled spring, longitudinal waves set up in the spring will pass through both slots regardless of their orientation. *Polarization is a property of transverse waves.*

Polaroid sunglasses are perhaps the most familiar optical polarizing device.

15.7 Selective Absorption It is known that certain crystalline substances transmit light in one plane of polarization and absorb light in other polarization planes. Tourmaline is such a material. Unpolarized light incident on a tourmaline crystal emerges as green, plane-polarized light of low intensity. This property of crystals in which one polarized component of incident light is absorbed and the other is transmitted is called *dichroism*. See Figure 15-11.

Dichroic crystals of quinine iodosulfate transmit plane-polarized light very efficiently but the crystals are too small for practical use. In 1935 Edwin H. Land developed a method of imbedding these crystals in cellulose film so that the dichroic properties of the crystals were retained. This highly efficient polarizing film is known commercially as Polaroid. Improved Polaroid sheets have now been developed in which polarizing molecules, rather than crystals, are imbedded in appropriate films.

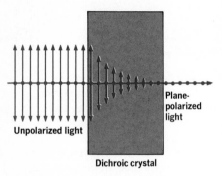

Figure 15-11. Polarization by selective absorption.

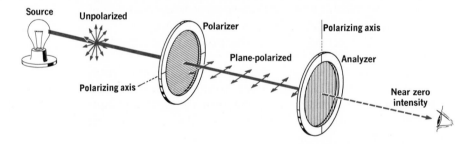

Figure 15-12. Polaroid disks in use. One acts as a polarizer and the other as an analyzer. Notice that the polarizing axis of each disk is marked on the rim.

15.8 Polarization by Reflection Sunlight reflected from the surface of calm water or from a level highway can be quite objectionable to the observer. Sunglasses made of polarizing films reduce the intensity of these reflections. By rotating the lenses slightly from side to side the reflections are seen to pass through a minimum, indicating that these rays are partially polarized.

Ordinary light incident obliquely on the surface of a glass plate is partly reflected and partly refracted. Both the transmitted and the reflected beams are partly polarized. This can be verified by observing the light through a polarizing disk used as an analyzer. By rotating the disk in a plane perpendicular to the light axis, one can determine the direction of the dominant light vectors in each beam.

Polarization by reflection does not occur for metals and other good conductors. Polarizing glasses can reduce glare from painted surfaces of automobiles, but not from chrome-plated bumpers.

The component of the incident light lying in the plane parallel to the surface of the glass is largely reflected. The component lying in the plane perpendicular to the surface is largely refracted. A particular angle of incidence at which polarization of the reflected light is complete, known as the *polarization angle,* can be found experimentally. Polarization by reflection is shown in Figure 15-13.

At the polarizing angle the reflected beam is of low intensity and the reflectance is about 15%. The refracted beam, which is not completely polarized, is bright. The intensity of the plane-polarized reflected beam can be increased by combining reflections as a result of stacking several plates. The combined refracted beam becomes less intense but more completely plane-polarized.

15.9 Polarization by Refraction If a thick glass plate is placed on a printed page, the print viewed through the glass may appear displaced because of refraction. A natural crystal of calcite placed on the page shows *two* refracted images of the print, as shown in Figure 15-14. Calcite and many other crystalline materials exhibit this property of *double refraction.*

Upon entering a doubly refracting crystal such as calcite, a beam of unpolarized light can divide into two beams at the crystal surface. This separation is shown diagrammatically in Figure 15-15. Analysis of these separate beams with polarizing disks reveals that they are plane-polarized with their planes of polarization perpendicular to each other.

Experimentally, one of the polarized beams can be shown to follow Snell's law; the other beam does not. Using monochromatic sodium light of 5893 Å and measuring the angles of incidence and refraction, we find that one beam yields a constant index of refraction of 1.66. However, the other beam shows an index of refraction that varies from 1.49 to 1.66 depending on the angle of incidence. This difference suggests that the light energy is propagated through the crystal at different speeds that are determined by the orientation of the light planes with the crystal lattice.

Calcite crystals are sometimes polished, cut through, and cemented back together in such a way that one of the polarized beams is totally reflected at the cemented face. Such a crystal, known as a *Nicol prism,* can be used to produce a beam of completely polarized light.

15.10 Interference Patterns The two plane-polarized beams of light that emerge from a doubly refracting crystal

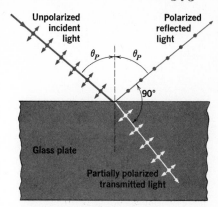

Figure 15-13. At the polarizing angle the reflected light is completely polarized but of low intensity.

Figure 15-14. Double refraction in calcite.

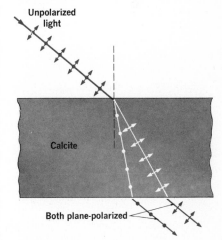

Figure 15-15. Double refraction. Light in one polarization plane conforms to Snell's law; light in the other plane does not.

cannot be made to interfere with each other even though they are transmitted through the crystal at different speeds. This is because their planes of polarization are perpendicular to each other. It follows that *if two light waves interfere, their oscillations (or components of these oscillations) must lie in the same plane.*

If *polarized light* is incident on a doubly refracting crystal at the proper angle, the emerging beams *are* found to interfere as they pass through an analyzer disk. The analyzer passes only the components of these perpendicularly polarized rays that lie in its transmission plane. Those waves from the two beams having a phase difference of an odd number of half wavelengths interfere destructively, the corresponding color is removed, and its complement is observed.

Materials, such as glass and Lucite, that become doubly refracting when subjected to mechanical stress are said to be *photoelastic*. When a photoelastic material is placed between polarizing and analyzing disks, the strain patterns (and thus the stress distributions) are revealed by interference fringes, as shown in Figure 15-16.

Figure 15-16. Strain patterns in a cylindrical disk subjected to diametrical compression.

15.11 Scattering A beam of light traveling in dust-free air cannot be seen even in a darkened room. If, however, there is dust in the path of the beam, it becomes visible by reflection of light from the surface of the particles.

Suppose a beam of white light travels in a medium containing suspended particles that have diameters smaller than the mean wavelength of visible light. When observed at right angles to its path, the beam has a bluish cast.

When observed in its path, the remaining transmitted beam has a red-orange cast. With certain concentrations of suspended particles, the bluish light sent off radially becomes quite intense and the transmitted light acquires a more intense red-orange color. An excess of shorter wavelengths is emitted at right angles to the path of the beam and an excess of longer wavelengths is transmitted along the path of the beam. This phenomenon, known as *scattering*, occurs when a beam of light encounters suspended particles with dimensions that are small compared to the wavelengths of the light.

If we look at blue sky light through a polarizing disk, we observe that it is plane polarized. By selecting a point in the blue sky near the horizon with the sun overhead, we find the scattered light to be horizontally polarized.

In the late afternoon near the time of sunset, sunlight must travel a maximum distance through the atmosphere to reach an observer. The setting sun has a yellow to red hue. The sunlight reflected to an observer from clouds near the sight line of the observer has the yellow or red hue characteristic of the sunset. Much of the energy in the blue region of the sunlight has been removed by the scattering effect of the atmospheric gas molecules. Under these conditions the transmitted light is largely yellow-red. This accounts for the blue appearance of the sky and the reddish appearance of the setting sun. See **Plate V(B).**

Atmospheric scattering accounts for our colorful sunsets and our blue sky.

If the earth had no envelope of atmospheric gases, the sky would appear black to an observer on the surface of the earth since there would be no scattering of sunlight passing overhead. Astronauts have observed the blackness of outer space while in flight beyond the earth's atmosphere.

Figure 15-17. Scattering of light by particles in the air brightens the daytime sky on earth (left). An observer on the lunar surface, however, sees a dark daytime sky (right) because the moon has almost no atmospheric particles.

Figure 15-18. A modern polarimeter in use.

Scattering is responsible for certain other natural color phenomena known as *structural colors* (in contrast to the pigment colors that we have already considered). The colors of many minerals are structural. The blue color of a bluejay's feathers results from scattering of light by tiny bubbles of air dispersed through the feather structure. The blue color in the eyes of a person is also a structural color; there is no blue pigment in the irises of blue eyes.

15.12 Optical Rotation A number of substances, such as quartz, sugar, tartaric acid, and turpentine, are said to be *optically active* because they rotate the plane of polarized light. A water solution of cane sugar (sucrose) rotates the plane of polarization to the right. For a given path length through the solution, the angle of rotation is proportional to the concentration of the solution.

In analytical procedures chemists find numerous applications of this property of optically active substances. Instruments for measuring the angle of rotation under standard conditions are known as *polarimeters;* those used specifically for sugar solutions are called *saccharimeters.*

Questions

GROUP A

1. What phenomenon provides evidence of the transverse wave character of light?
2. How do Polaroid sunglasses reduce the glare of bright sunlight?
3. Distinguish between a structural color and a pigment color.
4. Define the polarizing angle.
5. How could you determine whether a beam of light is plane polarized or unpolarized?
6. Suggest a way to identify the polarizing axis of a sheet of Polaroid film.

GROUP B

7. Explain why a Polaroid disk used as an analyzer blocks the beam of light transmitted by the polarizer disk when it is properly oriented.
8. The two beams of light that emerge from a doubly refracting crystal travel through the crystal at different speeds. Can they be made to produce an interference pattern? Explain.

9. A physics student observes that the setting sun near the horizon is red while the skylight overhead is blue. Explain.
10. For a given optically active substance, upon what does the magnitude of the rotation angle for the plane of polarized light passing through the substance depend?
11. If scattering of sunlight is more pronounced in the shorter wavelength regions, violet light must be scattered more than blue. Why does the sky appear to be light blue rather than violet?
12. A tube filled with a sucrose solution is placed between two crossed polarizing disks. (a) Explain why some light is transmitted through the analyzer. (b) In what direction must the analyzer be rotated to reduce the light intensity to a minimum? (c) The tube is now filled with a more dilute solution of the same substance. How must the analyzer be adjusted to reestablish the light minimum?

SUMMARY

Interference of light can be compared with interference in other wave disturbances. It is interpreted in terms of the superposition principle. Thomas Young first demonstrated interference of light and related double-slit interference and thin-film interference to Huygens' wave theory.

The analysis of thin-film interference requires an understanding of the nature of wave reflections at interfaces where the medium beyond has the higher or the lower index of refraction.

The various diffraction phenomena are explained by the interference of light waves. Diffraction gratings produce diffraction patterns by the reflection or transmission of incident light. The wavelength of light can be determined experimentally from measurements of grating spectra.

Polarization of light depends on the transverse nature of light waves. Polarization is a property of transverse waves. Light may be plane polarized by the selective absorption of certain crystals, by reflection from certain smooth surfaces, and by doubly refracting crystals. Photoelastic substances become doubly refracting when subjected to mechanical stress.

The scattering of sunlight accounts for the blue appearance of sky light and the reddish appearance of sunset and sunrise. Substances that rotate the plane of polarized light are said to be optically active.

VOCABULARY

constructive interference
destructive interference
dichroism
diffraction
diffraction angle
diffraction grating

first-order image
grating constant
interference (light)
interferometer
Nicol prism
optically active substance

plane-polarized light
polarimeter
saccharimeter
scattering (light)
second-order image
structural color

U.S. POSTAGE 3¢

BENJAMIN FRANKLIN 250TH ANNIVERSARY

16

ELECTROSTATICS

Benjamin Franklin performed many experiments on the nature of electricity. In one of these, he flew a kite during a thunderstorm to show that clouds are electrically charged and that he could draw sparks from a key tied to the kite string—at the risk of his life.

In this chapter you will gain an understanding of:

✔ the two kinds of electrostatic charge and methods of identifying and transferring them

✔ Coulomb's law of electric fields of force

✔ the concept of potential difference

✔ mapping and describing electric fields quantitatively

✔ the properties of electric conductors as well as nonconductors

✔ how capacitors store electric charge

✔ methods for solving problems involving electric fields of force and potential difference

✔ the relationship between potential difference, capacitance, and quantity of charge

ELECTRIC CHARGE

16.1 Charges at Rest We can observe about us a number of physical effects that are sometimes produced by rubbing pieces of dry matter. Sometimes an annoying shock is felt when the door handle of an automobile is touched after one slides over the plastic-covered seat. We may feel a shock after we walk on a woolen carpet and then touch a doorknob or other metal object. The slight crackling sound heard when dry hair is brushed and the tendency of thin sheets of paper to resist separation are other common observations of these physical effects.

When an object shows effects of the type we have described, we say that it has an *electric charge. The process that produces electric charges on an object is called* **electrification.** Electrification is most apparent when the air is dry. An object that is electrically charged can attract small bits of cork, paper, or other lightweight particles. Because the electric charge is confined to the object and is not moving, it is called an *electrostatic charge.* Thus **static electricity** *is stationary electricity in the form of an electric charge at rest.* Static electricity is commonly produced by friction between two surfaces in close contact.

16.2 Two Kinds of Charge We can detect the presence of an electrostatic charge by means of an instrument called an *electroscope.* The simplest kind of electroscope is a small ball of wood pith or Styrofoam suspended by a silk thread. This electroscope is more sensitive if the pith ball is coated

380

with aluminum or graphite. Such an instrument is shown in Figure 16-1.

Suppose a hard rubber or Bakelite rod is charged by stroking it with flannel or fur. If the end of the charged rod is then held near a simple electroscope, the pith ball is attracted to the rod.

If the pith ball is allowed to come in contact with the charged rod, it immediately rebounds and is then repelled by the rod. We can reasonably assume that some of the *charge* has been transferred to the pith ball so that both the rod and the ball are now similarly charged.

Now suppose a glass rod is charged by stroking it with silk. If the glass rod is held near the charged electroscope, the pith ball is attracted rather than repelled as it was with the charged rubber rod. These effects of attraction and repulsion can be explained if we assume that there are two kinds of electric charge.

The electric charge produced on the rubber rod when the rod is stroked with flannel or fur is called a *negative* charge. The rod is said to be charged negatively. The electric charge produced on the glass rod when the rod is stroked with silk is called a *positive* charge. The rod is said to be charged positively.

From a study of atomic structure it is known that all matter contains both positive and negative charges. For simplification, however, the diagrams that follow show only the excess charge. If an object has an excess negative charge, the *net* charge will be indicated by negative ($-$) signs. If it has an excess positive charge, the *net* charge will be indicated by positive ($+$) signs. A neutral object will have no sign because it has the same amount of each kind of charge and therefore *a net charge of zero.*

Two pith balls that are negatively charged by contact with a charged rubber rod repel each other. Similarly, two pith balls that are positively charged by contact with a charged glass rod also repel each other. However, if a negatively charged pith ball is brought near a positively charged pith ball, they attract each other. See Figure 16-2. These observations are summarized in *a **basic law of electrostatics:** Objects that are similarly charged repel each other; objects that are oppositely charged attract each other.*

16.3 Electricity and Matter
As an understanding of the nature of static electricity requires a knowledge of the basic concepts regarding the structure of matter, we will briefly discuss these concepts now. (These concepts will be discussed more fully in Chapter 23.) All matter is composed of atoms, of which there are many different kinds.

A pith ball coated with aluminum or graphite is not appreciably more massive, but it can acquire a higher electric charge.

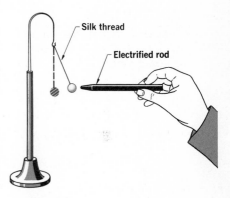

Figure 16-1. The pith-ball electroscope can be used to detect an electrostatic charge.

Objects that are electrically neutral have equal numbers of protons and electrons.

Figure 16-2. Like charges repel each other and unlike charges attract each other.

Each atom consists of a positively charged nucleus surrounded by negatively charged electrons.

Protons and neutrons are tightly packed into the very dense nucleus. Because each proton possesses a single unit of positive electric charge and neutrons, as their name suggests, are neutral particles, the nucleus is positively charged. This charge is determined by the number of protons the nucleus contains.

All electrons surrounding the nucleus are alike. They all carry the same amount of negative electric charge—one unit of negative charge per electron. Because an atom is electrically neutral, we know that the nucleus has a positive charge equal in magnitude to the total negative electronic charge. Thus the number of electrons of a neutral atom equals the number of protons.

Objects that are electrically neutral have zero net charge.

The rest mass of an electron is 9.1095×10^{-31} kg. The mass of a proton is 1.6726×10^{-27} kg and that of a neutron is 1.6750×10^{-27} kg. Both the proton and the neutron have masses that are nearly 2000 times greater than the mass of an electron. The mass of an atom is almost entirely concentrated in the nucleus.

Protons and neutrons are bound together within the nucleus by strong forces acting through very short distances. By comparison, the repulsions between protons that are due to their similarity of charge are weak forces.

The electrons are retained in the atom structure by the electric attraction exerted by the positive nuclear charge. In general the outermost electrons of higher energies are held less firmly in the atom structure than inner electrons of lower energies. The outer electrons of the atoms of metallic elements in particular are loosely held and are easily influenced by outside forces.

A net negative charge results from electrons being deposited on an isolated neutral body.

When two appropriate materials are in close contact, some of the loosely held electrons can be transferred from one material to the other. If a hard rubber rod is stroked with fur, some electrons can be transferred from the fur to the rod. The rubber rod becomes negatively charged because it has a *net excess of electrons*, and the fur becomes positively charged because it has a *net deficiency of electrons*.

Similarly, when a glass rod is stroked with silk, some electrons are transferred from the glass to the silk. The glass is positively charged because it has a *net deficiency of electrons*, and the silk is negatively charged because it has a *net excess of electrons*. All these charged states result from *the transfer of electrons*.

Electric charge is a scalar quantity. The net charge on an object is the sum of its positive charges minus the sum of its negative charges.

A net positive charge results from electrons being removed from an isolated neutral body.

16.4 The Electroscope Two common types of electroscopes that are more sensitive than the simple pith-ball device are shown in Figure 16-3.

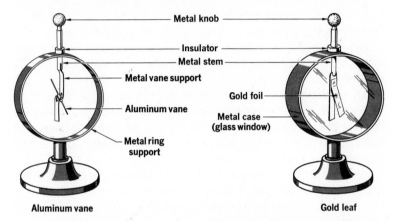

Metal knob

Insulator
Metal stem
Metal vane support

Gold foil

Aluminum vane

Metal case (glass window)

Metal ring support

Aluminum vane

Gold leaf

Figure 16-3. Two types of sensitive electroscopes.

The *vane electroscope* consists of a light aluminum rod mounted by means of a central bearing on a metal support that is insulated from its metal stand. When charged, the vane is deflected at an angle by electrostatic repulsion. The angle of deflection depends on the magnitude of the charge.

The *leaf electroscope* consists of very fragile strips of gold leaf suspended from a metal stem that is capped with a metal knob. The leaves are enclosed in a metal case with glass windows for their protection, and the metal stem is insulated from the case. When electrified, the leaves diverge because of the force of repulsion due to their similar charge. Good sensitivity is realized because of the very low mass of the gold leaf.

A *proof plane* is frequently used with an electroscope to test or transfer charges. The proof plane is a small metal disk with an insulating handle. One can easily be made by cementing a small coin to a glass rod. When the proof plane is used to transfer a charge, the metal disk is brought in contact with the charged object and then with an electroscope.

When used to test a charge, the charged disk of the proof plane is brought in contact with an electroscope previously electrified with a known charge. If the charge on the electroscope increases, the test charge is of the same sign as that on the electroscope. If the charge on the electroscope decreases, the test charge is of the opposite sign as that on the electroscope.

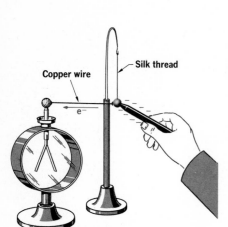

Copper wire

Silk thread

e⁻

Figure 16-4. A charge is conducted to the electroscope by the copper wire.

16.5 Conductors and Insulators Suppose that an aluminum-coated pith ball is suspended by a silk thread and the ball is connected to the knob of a leaf electroscope by means of a copper wire, as shown in Figure 16-4. When a charge is placed on the pith ball, the leaves of the electroscope diverge. Apparently the charge on the ball is transferred to the electroscope by the copper wire.

Now suppose a silk thread is substituted for the copper wire. When a charge is placed on the pith ball, the leaves of the electroscope do not diverge. The charge has not been conducted to the electroscope by the silk thread.

*A **conductor** is a material through which an electric charge is readily transferred.* Most metals are good conductors. At normal temperatures, silver is the best solid conductor; copper and aluminum follow in that order.

*An **insulator** is a material through which an electric charge is not readily transferred.* Good insulators are such poor conductors that for practical purposes they are considered to be nonconductors. Glass, mica, paraffin, hard rubber, sulfur, silk, dry air, and many plastics are good insulators.

Liquid solutions and confined gases conduct electricity in a different way than solids. At this time we are interested in the conductivity of solids. We can use our knowledge of the structure of matter in describing why some materials are conductors and why some are insulators.

A few grams of matter contain a very large number of atoms and, except for hydrogen, an even larger number of electrons. For example, 27 g of aluminum consists of 6.02×10^{23} aluminum atoms containing 7.83×10^{24} electrons. On an atomic scale, there is one electron in approximately 2.2 atomic mass units of matter. (One atomic mass unit = 1.66×10^{-24} g of matter.)

Metals have close-packed crystal structures. The crystal lattice consists of positively charged particles permeated by a cloud of *free electrons*. This cloud of free electrons is commonly referred to as the *electron gas*. The binding force in such structures is the attraction between the positively charged metal ions and the electron gas. The loosely held outermost electrons of the metal atoms have been "donated" to the electron gas and belong to the crystal as a whole. These electrons are free to migrate throughout the crystal lattice. Their migration gives rise to the high electric conductivity commonly associated with metals.

A good conductor contains a large number of free electrons whose motions are relatively unimpeded within the material. Since like charges repel, the free electrons spread throughout the material in order to relieve any local concentration of charge. If such a material is in contact with a charged body, the free electrons surge in a common direction. If the charged object is deficient in electrons (positively charged), this surge is in the direction of the object. If the charged object has an excess of electrons (negatively charged), the surge is away from the object. See Figure 16-5. In either case a transfer of electric charge continues until the repulsive forces between the free electrons are in equilibrium throughout the entire system.

An insulator is characterized by a lack of free electrons because even the outermost electrons are rather firmly held within the atom structure. Thus the transfer of charge through an insulator is usually negligible. If an excess of electrons is transferred to one particular region of such a material, the extra electrons remain in that region for some time before they gradually leak away.

16.6 Transferring Electrostatic Charges Suppose a rubber rod is charged negatively by stroking it with fur. If the rod is brought near the knob of an electroscope, the leaves diverge. If the rod is removed, the leaves collapse; no charge remains on the electroscope. The charge that makes the leaves diverge is called an *induced* charge. The electroscope is said to be charged temporarily by *induction*.

We can reason that the negative charge on the rod when brought near the metal knob of the electroscope repels free electrons in the knob and metal stem and forces them down to the leaves. See Figure 16-6(A). The force of repulsion of the extra electrons on the leaves causes them to diverge. The knob is then deficient in electrons and is positively charged. As soon as the force of repulsion exerted by the charged rod is withdrawn, the excess free electrons on the leaves scatter throughout the stem and knob, restoring the normal uncharged state throughout the electroscope.

Similarly, a glass rod that has been stroked with silk temporarily induces a positive charge on the leaves of an electroscope by attracting electrons up through the stem to the knob. See Figure 16-6(B).

Electrostatic experiments are best performed in dry air.

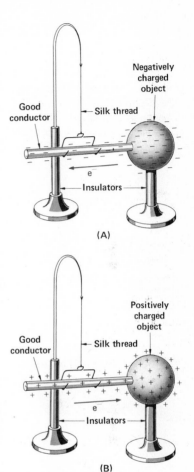

Figure 16-5. Free electrons of a conductor surge in the direction that reduces the net charge.

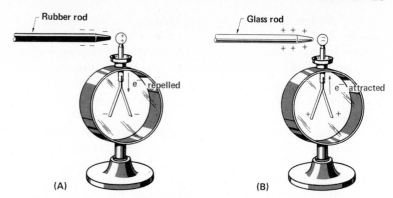

Figure 16-6. An electroscope may be charged temporarily by induction because of a redistribution of the free electrons of the metallic conductor.

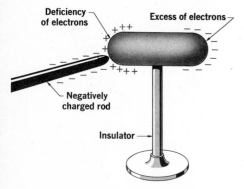

Figure 16-7. A charged rod brought near an isolated conductor induces electric charge of the same sign on the far end of the conductor.

In moist air an invisible film of water condenses on the surfaces of objects, including those of charged insulators. Dissolved impurities in the film make these surfaces conductive, and an isolated charge cannot be maintained for any length of time.

Any conducting object when properly isolated in space can be temporarily charged by induction. The region of the object nearest the charged body will acquire a charge of the *opposite* sign; the region farthest from the charged body will acquire a charge of the same sign. See Figure 16-7. This statement is consistent with the basic law of electrostatics stated in Section 16.2.

A charge can be transferred by *conduction*. If we touch the knob of an electroscope with a negatively charged rubber rod, the leaves diverge. When the rod is removed, the leaves remain apart, indicating that the electroscope retains the charge. How can we determine the nature of this residual charge on the electroscope?

We can reason that some of the excess electrons on the rod have been repelled onto the knob of the electroscope. This would be true only for the region of the rod immediately in contact with the electroscope since rubber is a very poor conductor and excess electrons do not migrate freely through it. Any free electrons thus transferred to the electroscope, together with other free electrons of the electroscope itself, would be repelled to the leaves by the excess electrons remaining on the parts of the rod not in contact with the electroscope.

When the rod is removed, and with it the force of repulsion, the electroscope is left with a residual negative charge of a somewhat lower density. This deduction can be verified by bringing a positively charged glass rod near the electroscope to induce a positive charge on the leaves. The leaves collapse and diverge again when the glass rod is removed. See Figure 16-8. *Any conducting object, properly isolated in space and charged by conduction, acquires a residual charge of the same sign as that of the body touching it.*

An electroscope with a known residual charge can be used to identify the nature of the charge on another object. This second object merely needs to be brought near the knob of this charged electroscope.

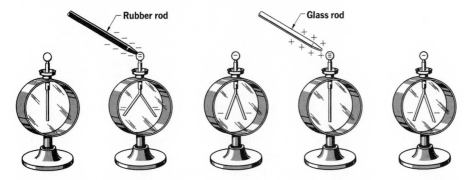

16.7 Residual Charge by Induction When a charged rubber rod is held near the knob of an electroscope, there is no transfer of electrons between the rod and the electroscope. If a path is provided for electrons to be repelled from the electroscope while the repelling force is present, free electrons escape. Then if the escape path is removed before removing the repelling force, the electroscope is left with a deficiency of electrons, giving it a residual positive charge. We can verify this conclusion by bringing a positively charged glass rod near the knob of the charged electroscope. The leaves of the electroscope diverge even more. When the charged rod is withdrawn, the leaves fall back to their original divergence and remain apart. The steps in placing a residual charge on an electroscope by induction are shown in Figure 16-9.

Figure 16-8. The residual charge on an electroscope, when charged by conduction, is of the same sign as the charge on the object that touches it.

Figure 16-9. Steps in placing a residual charge on an electroscope by induction.

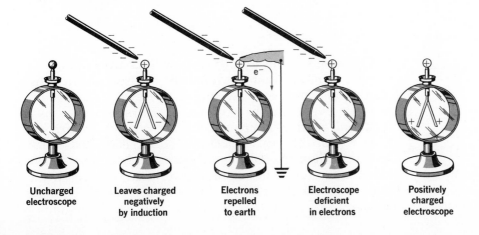

| Uncharged electroscope | Leaves charged negatively by induction | Electrons repelled to earth | Electroscope deficient in electrons | Positively charged electroscope |

Similarly, we can induce a residual negative charge on an electroscope using a positively charged glass rod. *When an isolated conductor is given a residual charge by induction, the charge is opposite in sign to that of the object inducing it.*

16.8 The Force Between Charges   From the basic law of electrostatics stated in Section 16.2, we recognize that like charges repel and unlike charges attract. If a charge is uniformly dispersed over the surface of an isolated sphere, its influence on another charged object some distance away is the same as if the charge were concentrated at the center of the sphere. Thus the charge on such an object is considered to be located at a particular point and is called a *point charge.*

The SI symbol for the coulomb is C. Because of its similarity to the symbol for capacitance, C, the older lowercase form, c, is used for the coulomb in this book.

The quantity of charge on a body, represented by the letter Q, is determined by the number of electrons in excess of (or less than) the number of protons. The quantity of charge is measured in coulombs (c), named for the French physicist Charles Augustin de Coulomb (1736–1806).

$$\textbf{1 coulomb} = \textbf{6.25} \times \textbf{10}^{\textbf{18}} \textbf{ electrons}$$

Thus the charge on one electron, expressed in coulombs, is the reciprocal of this number and the sign of Q is $-$.

$$\textbf{e}^- = \textbf{1.60} \times \textbf{10}^{-\textbf{19}} \textbf{ c}$$

Similarly, the charge on one proton is 1.60×10^{-19} coulomb and the sign of Q is $+$.

The coulomb is a very large unit of charge for the study of electrostatics. Frequently it is convenient to work with a fraction of this unit called the microcoulomb (μc).

$$\textbf{1 } \mu\textbf{c} = \textbf{10}^{-\textbf{6}} \textbf{ c}$$

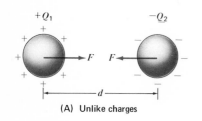

(A) Unlike charges

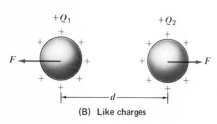

(B) Like charges

Figure 16-10. When Q_1 and Q_2 are of opposite sign, F is negative and is interpreted as a force of attraction. When Q_1 and Q_2 are of the same sign, F is positive and is interpreted as a force of repulsion.

Coulomb's many experiments with charged bodies led him to conclude that the forces of electrostatic attraction and repulsion obey a law similar to Newton's law of universal gravitation. We now recognize his conclusions as *Coulomb's law of electrostatics: The force between two point charges is directly proportional to the product of their magnitudes and inversely proportional to the square of the distance between them.* Charged bodies approximate point charges if they are small compared to the distances separating them. See Figure 16-10.

Coulomb's law can be expressed as

$$F \propto \frac{Q_1 Q_2}{d^2}$$

If we use a proportionality constant that takes into account the properties of the medium separating the charged bodies and that has the proper dimensions, Coulomb's law becomes

$$F = k \frac{Q_1 Q_2}{d^2}$$

The proportionality constant k has the numerical value 8.987×10^9 for vacuum and 8.93×10^9 for air. The dimensions of k are n m^2/c^2. The point charges Q_1 and Q_2 are in coulombs and are of proper sign to indicate the nature of each charge. As d is distance in meters, F is expressed in newtons of force.

In Figure 16-10(A) the charges Q_1 and Q_2 have opposite signs and the force F acts on each charge to move it toward the other. In Figure 16-10(B), charges Q_1 and Q_2 have the same sign and the force F acts on each charge to move it away from the other. The force between the two charges is a vector quantity that acts on each charge.

Suppose the objects in Figure 16-10(B) are charged to 0.01 coulomb each and placed 10 meters apart. Since the charges are of like sign, the force between them is one of repulsion. Using k for air (to one significant figure it is 9×10^9 n m^2/c^2), the Coulomb's law expression for this force becomes

$$F = 9 \times 10^9 \frac{\text{n m}^2}{\text{c}^2} \times \frac{(10^{-2} \text{ c})(10^{-2} \text{ c})}{(10 \text{ m})^2}$$

$$F = 9 \times 10^9 \frac{\text{n m}^2}{\text{c}^2} \times \frac{10^{-4} \text{ c}^2}{10^2 \text{ m}^2}$$

$$F = 9 \times 10^3 \text{ n of repulsive force}$$

EXAMPLE   Find the force between charges of $+100.0$ μc and -50.0 μc located 50.0 cm apart in air.

SOLUTION   Two unlike charges are separated by a given distance in air. We are asked to determine the force acting on them through this distance. Coulomb's law applies.

$$F = k \frac{Q_1 Q_2}{d^2}, \text{ where } k \text{ is the constant for air.}$$

Note that Coulomb's law of electrostatics has the same form as Newton's law of universal gravitation, Section 3.11.

Since k has the dimensions n m²/c², the quantities of charge must be converted to coulombs and the distance to meters. Force will then be calculated in newtons.

$$F = 8.93 \times 10^9 \frac{\text{n m}^2}{\text{c}^2} \times \frac{(100.0 \ \mu c \times c/10^6 \ \mu c)(-50.0 \ \mu c \times c/10^6 \ \mu c)}{(50.0 \ \text{cm} \times \text{m}/10^2 \ \text{cm})^2}$$

$$F = 8.93 \times 10^9 \frac{\text{n m}^2}{\text{c}^2} \times \frac{(1.000 \times 10^{-4} \ c)(-5.00 \times 10^{-5} \ c)}{(5.00 \times 10^{-1} \ \text{m})^2}$$

$$F = 8.93 \times 10^9 \frac{\text{n m}^2}{\text{c}^2} \times \frac{-5.00 \times 10^{-9} \ c^2}{2.50 \times 10^{-1} \ \text{m}^2}$$

$F = -1.79 \times 10^2$ n (the negative sign indicating *attraction* between the two charges)

PRACTICE PROBLEMS   **1.** What force acts on two electrostatic charges of 60.8 μc and 76.5 μc isolated in air 42.5 cm apart?   *Ans:* $23\overline{0}$ n
2. Two small isolated spheres are 20.0 cm apart. The sphere on the left receives a charge of +10.8 μc and the sphere on the right receives a charge of +12.2 μc. What force acts on each charge and in what (left/right) direction?   *Ans:*   left charge, 29.4 n right; right charge, 29.4 n left

---

The idea of a field of force was introduced by Faraday.

16.9 Electric Fields   The concept of a field of force will be helpful as we consider the region surrounding an electrically charged body. A second charge brought into this region experiences a force according to Coulomb's law. Such a region is an *electric field. An **electric field** is said to exist in a region of space if an electric charge placed in that region is subject to an electric force.*

Let us consider a positively charged sphere +Q of Figure 16-11(A) isolated in space. A small positive charge +q, which we shall call a *test charge*, is brought near the surface of the sphere. Since the test charge is in the electric field of

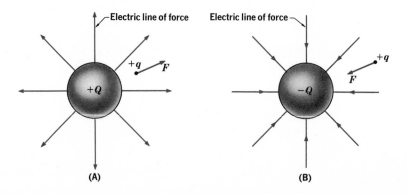

Figure 16-11. The electric field surrounding a charged sphere isolated in space.

(A)   (B)

the charged sphere and the charges are similar, it experiences a repulsive force directed radially away from $+Q$. Were the charge on the sphere negative, as in Figure 16-11(B), the force acting on the test charge would be directed radially toward $-Q$.

An *electric line of force is a line so drawn that a tangent to it at any point indicates the orientation of the electric field at that point.* We can imagine a line of force as the path of a test charge moving slowly in a very viscous medium in response to the force of the field. By convention, electric lines of force *originate* at the surface of a positively charged body and *terminate* at the surface of a negatively charged body, each line of force showing the direction in which a positive test charge would be accelerated in that part of the field. A line of force is *normal* to the surface of the charged body where it joins that surface.

The *intensity*, or strength, of an electrostatic field, as well as its direction, can be represented graphically by lines of force. *The electric field intensity is proportional to the number of lines of force per unit area normal to the field.* Where the intensity is high, the lines of force will be close together. Where the intensity is low, the lines of force will be more widely separated in the graphical representation of the field.

In Figure 16-12(A), electric lines of force are used to show the electric field near two equally but oppositely charged objects. At any point in this field the resultant force acting on a test charge $+q$ can be represented by a vector drawn tangent to the line of force at that point.

The electric field near two objects of equal charge of the same sign is shown by the lines of force in Figure 16-12(B). The resultant force acting on a test charge $+q$ placed at the midpoint between these two similar charges would be zero.

Figure 16-12. Lines of force show the nature of the electric field near two equal charges of opposite sign (A), and near two equal charges of the same sign (B).

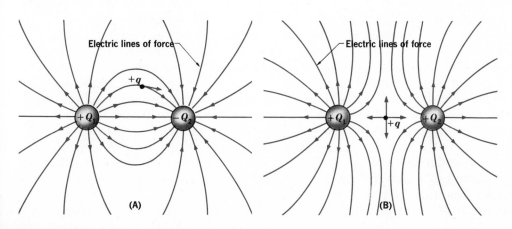

Electric lines of force

$+q$

$+Q_1$ $-Q_2$

(A)

Electric lines of force

$+Q_1$ $+q$ $+Q_2$

(B)

The *electric field intensity,* $\mathcal{E}$, *at any point in an electric field is the force per unit positive charge at that point.* The electric field intensity has the dimensions *newton per coulomb.* Thus,

$$\mathcal{E} = \frac{F}{q}$$

where $\mathcal{E}$ is the electric field intensity and F is the force in newtons acting on the test charge q in coulombs.

The following example will illustrate the use of this equation.

EXAMPLE   A charge of 2 μc placed in an electric field experiences a force of 0.08 n. What is the magnitude of the electric field intensity?

SOLUTION   The charge of 2 μc expressed in coulombs becomes 2 μc × $c/10^6$ μc or 2×10^{-6} c. The force of 0.08 n is most conveniently expressed as 8×10^{-2} n.

$$\mathcal{E} = \frac{F}{q} = \frac{8 \times 10^{-2} \text{ n}}{2 \times 10^{-6} \text{ c}}$$

$\mathcal{E} = 4 \times 10^4$ n/c, the electric field intensity.

PRACTICE PROBLEMS   **1.** A force of 2.5×10^{-3} n acts on a charge of 3.8×10^{-7} c in an electric field. Determine the magnitude of the electric field.   *Ans:* 6.6×10^3 n/c

2. An electric charge of 1.9 μc is placed at a point in an electric field where the field intensity has a magnitude of 1.7×10^5 n/c. To what force is the charge subjected?   *Ans:* 3.2×10^{-1} n

Questions

GROUP A

1. List five examples (other than those given in this chapter) in which electrification occurs.
2. A hard rubber rod is rubbed with fur. What kind of charge is acquired (a) by the rubber rod, (b) by the fur?
3. A glass rod is rubbed with silk. What kind of charge is acquired (a) by the glass rod, (b) by the silk?
4. A pith ball suspended on a silk thread is attracted to a charged rubber rod. Does this indicate that the pith ball is oppositely charged? Explain.

5. State the law of electrostatics that reveals the nature of the attraction and repulsion of charged objects.
6. Why do the leaves of an uncharged electroscope diverge when a charged object is brought near?
7. Why is it necessary to ground an electroscope temporarily while inducing a residual charge on it with a negatively charged object?
8. What determines the property of a metal as a conductor of electricity?
9. Explain why sulfur is a very poor conductor.
10. What is the unit of electric charge in the metric system?

11. State Coulomb's law of electrostatics.
12. Define an electric field.

GROUP B

13. How can you explain (a) the presence of a charge on a rubber rod after it has been rubbed with fur, (b) the charge remaining on the fur?
14. Which do you think offers the more conclusive proof of the presence of a charge on a pith-ball electroscope: an observed force of repulsion on it or one of attraction? Explain.
15. Given a charged sphere, describe a simple experiment that would enable you to determine conclusively the nature of the charge on the sphere.
16. A negatively charged rod is brought near the knob of a charged electroscope. The leaves first collapse and then as the rod is brought nearer they

again diverge. (a) What is the residual charge on the electroscope? (b) Explain the action of the leaves.

17. (a) In order for a coulomb force to be expressed in newtons, what must be the dimensions of the proportionality constant in the Coulomb's law equation, $F = k\,Q_1Q_2/d^2$? (b) Verify your answer.
18. A very small sphere is given a positive charge and is then brought near a large negatively charged plate. Draw a diagram of the system showing the appearance of the electric lines of force.
19. An electron of mass m and charge e^- is projected into a uniform electric field with an initial velocity v_i at right angles to the field. Describe its motion in the electric field.

Problems

GROUP B

1. A small sphere is given a charge of $+2\bar{0}$ μc and a second sphere of equal diameter located $1\bar{0}$ cm away is given a charge of -5.0 μc. What is the force of attraction between the charges?
2. The two spheres of Problem 1 are allowed to touch and are again spaced $1\bar{0}$ cm apart. What force exists between them?
3. Two small spheres each having a mass of 0.050 g are suspended by silk threads from the same point. When given equal charges, they separate, the threads making an angle of $1\bar{0}°$ with each other. What is the force of repulsion acting on each sphere? (Suggestion: Construct a vector diagram and consider the horizontal component.)
4. Two small spheres each having a mass of 0.10 g are suspended from the same point on silk threads $2\bar{0}$ cm long. When given equal charges, they repel each other, coming to rest 24 cm apart. Find the charge on each sphere. (Suggestion: Construct a vector diagram and consider similar triangles.)

5. A charge of 0.52 μc is placed in an electric field where the field intensity is 4.5×10^5 n/c. What is the magnitude of the force acting on the charge?
6. Two identical spheres are separated by 25.0 cm in air. The sphere on the left receives a charge of -25.5 μc. The sphere on the right receives a charge of $+25.5$ μc. What force acts? (Suggestion: Sketch a working diagram after the style of Figure 16-10. Provide enough sphere separation to accommodate Problem 7.)
7. A third sphere with a charge of $+2.5$ μc is now placed halfway between the two spheres of Problem 6. (a) What force does each of the original charges exert on this third charge? (b) What total force acts on the third charge and in which (left/right) direction? (c) What net force acts on the -25.5-μc charge and in which (left/right) direction? (d) What net force acts on the $+25.5$-μc charge and in which (left/right) direction? [Suggestion: Add all force vectors (left/right) to your diagram.]

POTENTIAL DIFFERENCE

16.10 Electric Potential   Let us consider the work done by gravity on a wagon coasting down a hill. The wagon is within the gravitational field of the earth and experiences a gravitational force causing it to travel downhill. Work is done by the gravitational field, so the energy expended comes from within the gravitational system. The wagon has less potential energy at the bottom of the hill than it had at the top; in order to return the wagon to the top, work must be done on it. However, in this instance, the energy must be supplied from an outside source to pull against the gravitational force. The energy expended is stored in the system, imparting to the wagon more potential energy at the top of the hill than it had at the bottom.

Similarly, a charge in an electric field experiences an electric force according to Coulomb's law. If the charge moves in response to this force, work is done by the electric field. Energy is removed from the system. If the charge is moved against the coulomb force of the electric field, work is done on it using energy from some outside source, this energy being stored in the system.

If work is done as a charge moves from one point to another in an electric field or if work is required to move a charge from one point to another, these two points are said to *differ in electric potential. The magnitude of the work is a measure of this difference of potential.* The concept of potential difference is very important in the understanding of electric phenomena. *The **potential difference**, V, between two points in an electric field is the work done per unit charge as a charge is moved between these points.*

$$\text{potential difference } (V) = \frac{\text{work } (W)}{\text{charge } (q)}$$

The unit of potential difference is the *volt (v)*. One **volt** is *the potential difference between two points in an electric field such that 1 joule of work is done in moving a charge of 1 coulomb between these points.*

$$1 \text{ volt} = \frac{1 \text{ joule}}{1 \text{ coulomb}}$$

Small differences of potential are commonly expressed in *millivolts* (mv) or *microvolts* (μv). Large differences of potential are measured in *kilovolts* (kv) and *megavolts* (Mv).

$$1 \ \mu v = 10^{-6} \ v$$
$$1 \ mv = 10^{-3} \ v$$
$$1 \ kv = 10^{3} \ v$$
$$1 \ Mv = 10^{6} \ v$$

Figure 16-13. A strand of copper wire is completely vaporized as a potential difference of 5,000,000 volts is placed across it.

The SI symbol for the volt is V. Because of its similarity to the symbol for potential difference, V, the older lowercase form, v, is used for the volt in this book.

Suppose the potential difference between two points in an electric field is 6.0 v. The work required to move a charge of 3.00×10^2 μc between these points can be determined as follows.

$$V = \frac{W}{q}$$

$W = Vq = 6.0 \text{ v} \times 3.00 \times 10^2 \ \mu c \times c/10^6 \ \mu c$
$W = 6.0 \text{ v} \times 3.00 \times 10^{-4} \text{ c}$
$W = 1.8 \times 10^{-3} \text{ j}$

Since a joule of work is a force of 1 newton applied through a distance of 1 meter, it follows that

$$v = \frac{j}{c} = \frac{n \ m}{c}$$

and

$$\frac{v}{m} = \frac{n}{c}$$

In Section 16.9 it was shown that $\mathscr{E}$, the electric field intensity, is expressed in newtons per coulomb. Thus,

$$\mathscr{E} = \frac{n}{c} = \frac{v}{m}$$

The electric field intensity is commonly expressed in terms of *volts per meter* and can be referred to as the *potential gradient*. The **potential gradient** *of an electric field is the change in potential per unit of distance.*

We can consider the earth to be an inexhaustible *source* of electrons, or limitless *sink* into which electrons can be "poured" without changing its potential. For practical purposes, the potential of the earth is arbitrarily taken as *zero*. Any conducting object connected to the earth must be at the same potential as the earth; that is, the potential difference between them is zero. Such an object is said to be *grounded*.

The potential at any point in an electric field is the potential difference between the point and earth taken as zero. This potential can be either positive or negative depending on the nature of the charge producing the electric field.

16.11 Distribution of Charges

Michael Faraday performed several experiments to demonstrate the distribution of charge on an isolated object. He charged a conical silk bag like that shown in Figure 16-14 and found that the charge was on the outside of the bag. By pulling on the

Figure 16-14. The type of conical silk bag used by Faraday to demonstrate that electric charges reside on the outside.

Silk thread

Electric charge on outside

Insulator

silk thread, he turned the bag inside out and found that the charge was again on the outside. The inside of the bag showed no electric charge in either position.

Faraday connected the outer surface of an insulated metal pail to an electroscope by means of a conducting wire, as shown in Figure 16-15(A). He then lowered a positively charged metal ball supported by a silk thread into the pail. The leaves of the electroscope diverged (B), indicating a charge by induction. The positive charge on the ball attracted free electrons in the pail to the inner surface, leaving the outside of the pail and the electroscope positively charged. When the ball was allowed to come in contact with the inside of the pail (C) and was then removed, the leaves remained apart without change. It was found that the ball no longer had a charge after its removal. The fact that the electroscope remained unchanged after the ball was removed (D) indicated (1) that there was no redistribution of positive charge on the outside surface of the pail and (2) that the outside of the pail (and the electroscope) had acquired a net charge equal to the charge originally placed on the ball.

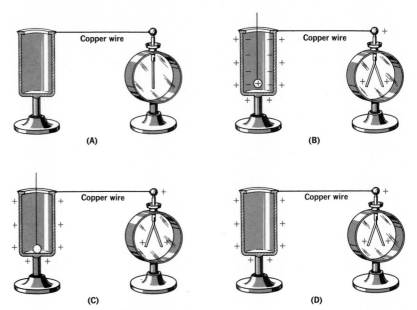

Figure 16-15. Faraday's ice-pail experiment.

From these and other experiments with isolated conductors we can conclude that:

1. All the static charge on a conductor lies on its surface. Electrostatic charges are at rest. If the charge were beneath the surface so that an electric field existed within a conductor, free electrons would be acted upon by the coulomb force of this field. Work would be done, the electrons would move because of a difference of potential, and energy would be given up by the field. Since movement is not consistent with a static charge, the charge must be on the surface and the electric field must exist only externally to the surface of a conductor. See Figure 16-16.

2. There can be no potential difference between two points on the surface of a charged conductor. A difference of potential is a measure of the work done in moving a charge from one point to another. As no electric field exists within the conductor, no work is done in moving a charge between two points on the same conductor. No difference of potential can exist between such points.

3. The surface of a conductor is an equipotential surface. All points on a conductor are at the same potential and no work is done by the electric field in moving a charge residing on a conductor. If points of equal potential in an electric field near a charged object are joined, an *equipotential line* or *surface* within the field is indicated. No work is done when a test charge is moved in an electric field along an equipotential surface.

4. Electric lines of force are normal to equipotential surfaces. A line of force shows the direction of the force acting on a test charge in an electric field. It can be shown that there is no force acting normal to this direction. Thus no work is done when a test charge is moved in an electric field normal to the lines of force. See Figure 16-17.

5. Lines of force originate or terminate normal to the conductive surface of a charged object. Since the surface of a conductor is an equipotential surface, lines of force must start out perpendicularly from the surface. For the same reason, a line of force cannot originate and terminate on the same conductor.

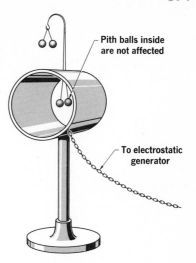

Figure 16-16. When the metal cylinder is charged by an electrostatic generator, the pith balls outside diverge while those inside are not affected.

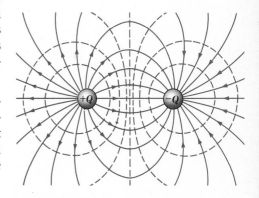

Figure 16-17. Lines of force (solid lines) and equipotential lines (dashed lines) define the electric field near two equal but opposite charges.

16.12 Effect of the Shape of a Conductor   A charged spherical conductor perfectly isolated in space has a uniform charge density, or charge per unit area, over the outer surface. Lines of force extend radially from the surface in all directions and the equipotential surfaces of the electric field are spherical and concentric. Such symmetry is not found in all cases of charged conductors.

A charge acquired by a nonconductor such as glass is confined to its original region until it gradually leaks away. The charge placed on an isolated metal sphere quickly spreads uniformly over the entire surface. If the

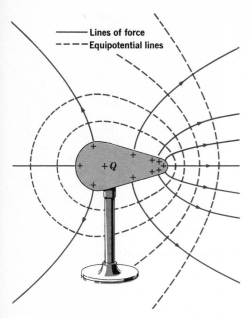

——— Lines of force
- - - - Equipotential lines

+Q

Figure 16-18. The charge density is greatest at the point of greatest curvature.

A lightning bolt transfers an average of about 1 coulomb of charge between a cloud and the earth.

conductor is not spherical, the charge distributes itself according to the surface curvature, concentrating mainly around points. The pear-shaped conductor illustrated in Figure 16-18 shows the charge more concentrated on the curved regions and less concentrated on the straight regions. If the small end is made more pointed, the charge density will increase at that end.

16.13 *Discharging Effect of Points* In Figure 16-18 the lines of force and equipotential lines are shown more concentrated at the small end of the charged conductor. This geometry indicates that the intensity of the electric field, or potential gradient, in this region is greater than elsewhere around the conductor. If this surface is reshaped to a sharply pointed end, the field intensity can become great enough to cause the gas molecules in the air surrounding the pointed end to *ionize*. Ionized air consists of gas molecules from which an outer electron has been removed, thus allowing both the positively charged *ion* and the freed electron to respond to the electric force. When the air is ionized, the point of the conductor is rapidly discharged.

There are always a few positive ions and free electrons present in the air. The intense electric field near a sharp point of a charged conductor will set these charged particles in motion such that the electrons are driven in one direction and the positive ions in the opposite direction. Violent collisions with other gas molecules will knock out some electrons and produce more charged particles. In this way air can be ionized quickly when it is subjected to a sufficiently large electric stress.

In dry air at atmospheric pressure, a potential gradient of 30 kv/cm between two charged surfaces is required to ionize the intervening column of air. When such an air gap is ionized, a *spark discharge* occurs. There is a rush of free electrons and ionized molecules across the ionized gap, discharging the surfaces and producing heat, light, and sound. Usually the quantity of static electricity involved is quite small and the time duration of the spark discharge is very short. Atmospheric lightning, however, is a spark discharge in which the quantity of charge is great.

The intensity of an electric field near a charged object can be sufficient to produce ionization at sharp projections or sharp corners of the object. A slow leakage of charge can occur at these locations producing a *brush* or *corona discharge*. A faint violet glow is sometimes emitted by the ionized gases of the air. A glow discharge known as St. Elmo's fire can sometimes be observed at night at the tips of ship masts and at the trailing edges of wing and tail surfaces of aircraft. The escape of charges from sharply pointed conductors is important in the operation of electrostatic generators and in the design of lightning arresters.

Lightning is a gigantic electric discharge in which electric charges rush to meet their opposites. The interchange can occur between clouds or between a cloud and the earth. See Figure 16-19. According to the Lightning Protection Institute, one hundred bolts of lightning strike the earth every second, each bolt initiated by a potential difference of millions of volts.

There are no known ways of preventing lightning. However, there are effective means of protection from its destructiveness. Lightning rods, invented by Benjamin Franklin, are often used to protect buildings made of nonconducting materials from lightning damage. Sharply pointed rods are strategically located above the highest projections of the building, and by their discharging effect, they normally prevent the accumulation of a dangerous electrostatic charge. Lightning rods are thoroughly grounded. If lightning strikes, the rods provide a good conducting path into the ground. Being well grounded, the steel frames of large buildings offer excellent protection from lightning damage.

Television receiving antennas, even though equipped with lightning arresters, do not protect a building from lightning. Without lightning arresters they are a distinct hazard because they are not grounded.

Figure 16-19. Streaks of lightning brighten the nighttime sky. Each flash requires a potential difference of millions of volts.

16.14 Capacitors Any isolated conductor is able to retain an electrostatic charge to some extent. If we place a positive charge on such a conductor by removing electrons, the potential is raised to some positive value with respect to ground. Conversely, a negative charge placed on the conductor results in a negative potential with respect to ground. By increasing the charge, we increase the potential of the conductor since the potential of an isolated conductor is a measure of the work done in placing a charge on the conductor. It is evident that we can continue to increase the charge until the potential with respect to ground or other conducting surface becomes so high that corona or spark discharges occur. However, if the conductor is in an evacuated space, it could be raised to a much higher potential by continuing the addition of charge.

Suppose a charged conductor is connected to an electroscope, as shown in Figure 16-20(A). The leaves of the electroscope will diverge indicating the potential of the charged conductor. The electroscope is now a part of the

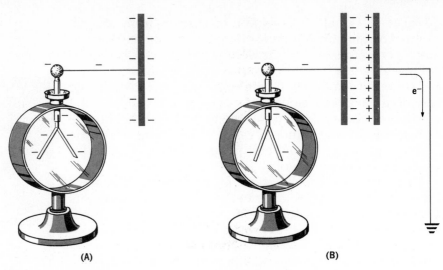

(A)   (B)

Figure 16-20. The principle of operation of a capacitor.

conducting surface and there can be no difference of potential between different regions of a single charged conducting surface.

Now suppose a grounded conductor is brought near the charged conductor, as shown in Figure 16-20(B). A positive charge is induced on this second conductor as free electrons are repelled to ground by the negative field of the first conductor. The leaves of the electroscope partially collapse. The closer we move the grounded plate to the charged conductor, the more pronounced is this effect. Because of the attractive force of the induced charge on the grounded plate, less work is required to place the same negative charge on the first conductor. Its potential is consequently reduced accordingly. A greater charge can now be placed on this conductor to raise the potential back to the initial value.

*A combination of conducting plates separated by an insulator that is used to store an electric charge is known as a **capacitor**.* The area of the plates, their distance of separation, and the character of the insulating material separating them determine the charge that can be placed on a capacitor. The larger the charge, the greater is the potential difference between the plates of a capacitor. The ratio of charge Q to the potential difference V is a *constant* for a given capacitor and is known as its *capacitance, C.* **Capacitance** *is the ratio of the charge on either plate of a capacitor to the potential difference between the plates.* Thus

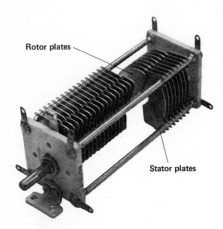

Figure 16-21. A two-section, air-dielectric variable capacitor. The capacitance of each section is varied by rotating the shaft to which the rotor (movable) plates are attached, thereby altering the area of rotor plates that engage the stator (fixed) plates.

$$C = \frac{Q}{V}$$

where C is the capacitance of a capacitor, Q is the quantity of charge on either plate, and V is the potential difference between the conducting plates.

The unit of capacitance is the *farad* (f), named in honor of Michael Faraday. *The capacitance is 1 **farad** when a charge of 1 coulomb on a capacitor results in a potential difference of 1 volt between the plates.* The farad is an extremely large unit of capacitance. Practical capacitors have capacitances of the order of *microfarads* (μf) or *picofarads* (pf).

The SI symbol for the farad is F. Because of its similarity to the symbol for force, F, the older lowercase form, f, is used for the farad in this book.

$$1 \; \mu f = 10^{-6} \; f$$
$$1 \; pf = 10^{-12} \; f$$

16.15 Dielectric Materials Faraday investigated the effects of different insulating materials between the plates of capacitors. He constructed two capacitors with equal plate areas and equal plate spacing. Using air at normal pressure in the space between the plates of one and an insulating material between the plates of the other, he charged both to the same potential difference.

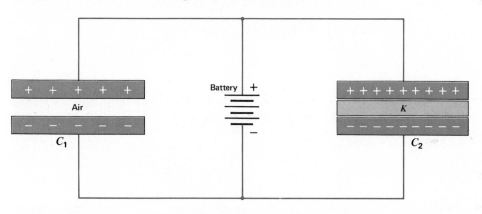

These capacitors are shown as C_1 and C_2 respectively in Figure 16-22. Faraday measured the quantity of charge on each capacitor and found that C_2 had a greater charge than C_1 by a factor K.

Figure 16-22. Capacitors C_1 and C_2 have identical dimensions and are charged to the same potential difference by the battery. C_2 accumulates the larger quantity of charge.

$$Q_2 = KQ_1$$

Except for the material separating the plates, the two capacitors are identical and have the same potential difference across their plates. The factor K by which the charge on C_2 exceeds the charge on C_1 must be due to a property of the insulating material of C_2. The ratio Q_2/V is larger than Q_1/V by this factor K.

$$\frac{Q_2}{V} = K \frac{Q_1}{V}$$

Thus the capacitance of C_2 is larger than the capacitance of C_1 by the factor K.

$$C_2 = KC_1$$

Many materials such as mica, paraffin, oil, waxed paper, glass, plastics, and ceramics can be used instead of air in the space between the plates of a capacitor. For each material, the resulting capacitance Q/V will have a different value.

Materials used to separate the plates of capacitors are known as *dielectrics*. *The ratio of the capacitance with a particular material separating the plates of a capacitor to the capacitance with a vacuum between the plates is called the* **dielectric constant,** *K, of the material.* Dry air at atmospheric pressure has a dielectric constant of 1.0006. In practice this is taken as *unity* (the same as vacuum) for dielectric constant determinations. Dielectric constants are dimensionless numbers ranging from 1 to 10 for materials commonly used in capacitors. The dielectric constant, K, of the dielectric material used in Figure 16-22 is

$$K = \frac{C_2}{C_1}$$

Typical dielectric constants for some common dielectric materials are given in Table 16-1.

Table 16-1
DIELECTRIC CONSTANTS

Dielectric material	Dielectric constant K	Proportionality constant k (n m²/c²)
air	1.0	8.9×10^9
paper (oiled)	2.0	4.5×10^9
paraffin	2.2	4.1×10^9
polyethylene	2.3	3.9×10^9
polystyrene	2.5	3.6×10^9
hard rubber	2.8	3.2×10^9
mica	6.0	1.5×10^9
glass	8.0	1.1×10^9

Instead of charging the identical capacitors of Figure 16-22 to the same potential difference, suppose we place the same charge Q on them. The experiment now shows that the potential difference across C_2 is smaller than that across C_1 by the factor $1/K$.

$$V_2 = \frac{V_1}{K}$$

This leads us to our previous conclusion about the influence of the dielectric material separating the plates of a capacitor. Since $C = Q/V$ and Q_1 and Q_2 are equal,

$$\frac{Q_2}{V_2} = \frac{Q_1}{V_1/K} \quad \text{and} \quad C_2 = KC_1$$

16.16 The Effect of Dielectrics   The molecules of some dielectrics have a permanent separation of their positive and negative centers of charge. This property is described as a permanent *electric dipole moment*. These molecules are called *polar* molecules, or *dipoles*. When placed in an electric field, polar molecules tend to become aligned with the external field to a degree characteristic of the molecules.

Other dielectrics are composed of molecules that are essentially *nonpolar* and that show no permanent electric dipole moments. When placed in an electric field, these molecules can acquire a temporary polar character by *induction*. While in the electric field, they have induced electric dipole moments.

The space between the plates of a charged capacitor is permeated by a uniform electric field. If a dielectric slab is inserted into such an electric field denoted by the potential gradient $\mathscr{E}_0$, the slab as a whole becomes polarized by induction. The surface near the positive plate of the capacitor acquires a negative charge and the surface near the negative plate acquires a positive charge. See Figure 16-23. These surface charges result from the dipole moments of the dielectric molecules and not from the transfer of electrons as in the case of metallic conductors.

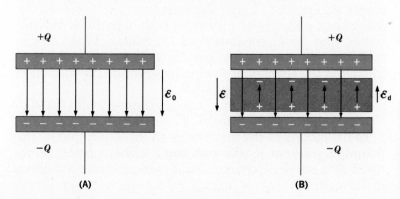

(A)   **(B)**

Figure 16-23. A dielectric slab placed in an electric field tends to weaken the field within the dielectric. What should be the effect on the potential difference across the plates?

The electric field established in the dielectric slab by the surface charge *opposes* the external field $\mathscr{E}_0$ of the capacitor. This opposing field is shown in Figure 16-20(B) as $\mathscr{E}_d$. Its effect is to weaken the original field within the dielectric. The net electric field $\mathscr{E}$ is the vector sum of the two fields $\mathscr{E}_0$ and $\mathscr{E}_d$ and is always a weaker field in the direction of $\mathscr{E}_0$.

$$\mathscr{E} = \mathscr{E}_0 + \mathscr{E}_d$$

The net effect of a dielectric between the plates of a charged capacitor is to lower the potential gradient of the electric field. If a parallel-plate capacitor with an air dielectric is charged and then isolated, a potential difference of Q/C volts remains across the plates. The introduction of a dielectric slab between the plates causes a decrease in this potential difference and reveals its weakening effect on the electric field of the capacitor.

If the charged spheres of Figure 16-10 are immersed in some dielectric medium, the medium would also become polarized by induction. Clustered around each of the charged spheres would be opposite charges from the medium. The result would be an effective reduction in the values of Q_1 and Q_2. A rigorous analysis shows that the force between the two spheres is reduced in value by the factor K, which is the dielectric constant of the medium. Thus in the equation for Coulomb's law

In Table 16-1 observe that the value of k for air divided by K for another dielectric material yields k for that material.

$$F = k\frac{Q_1Q_2}{d^2}$$

the value of k is 8.987×10^9 n m^2/c^2 for vacuum but is reduced by the factor K for material media. Its value for various materials is given in the last column of Table 16-1.

Capacitors find wide application in all kinds of electronic circuits, ignition systems, and telephone and telegraph equipment. Large capacitances are achieved by using large plate areas, insulators with high dielectric constants, and small separation of the plates. Physical size limits the plate area while cost limits the choice of dielectric. *Dielectric strength* of the insulator limits the reduction in spacing between the plates.

Dielectric strength should not be confused with the dielectric constant of a material. The dielectric strength defines the quality of the material as an insulator; that is, the potential gradient it will withstand without being punctured by a spark discharge. Some typical values are given in Table 16-2.

Table 16-2
DIELECTRIC STRENGTHS

Dielectric material	Dielectric strength (kv/cm to puncture)
air	30
oil	75
paraffin	350
paper (oiled)	400
mica	500
glass	1000

16.17 Combinations of Capacitors Suppose three capacitors of capacitances C_1, C_2, and C_3 are connected in *parallel*, that is, with one plate of each capacitor connected to one conductor while the other plate is connected to a second conductor. See Figure 16-24(A).

The plates connected to the + conductor are parts of one conducting surface. Those connected to the − conductor form the other conducting surface. If the three capacitors are charged, it is apparent that they must have the same difference of potential, V, across them. The quantity of charge on each must be respectively

$$Q_1 = C_1 V, \; Q_2 = C_2 V, \; \text{and} \; Q_3 = C_3 V$$

The total charge, Q_T, must be the sum of the separate charges on the three capacitors,

$$Q_T = Q_1 + Q_2 + Q_3$$

then

$$Q_T = C_1 V + C_2 V + C_3 V$$

Because the total charge is equal to the product of the total capacitance, C_T, and the potential difference, V, it is evident that

$$Q_T = C_T V$$

Substituting,

$$C_T V = C_1 V + C_2 V + C_3 V$$

and

$$C_T = C_1 + C_2 + C_3$$

For capacitors connected in parallel, the total capacitance is the sum of all the separate capacitances.

Now suppose we connect the three capacitors in *series*, as shown in Figure 16-24(B). A positive charge placed on C_1 from the + source induces a negative charge on the second plate as the electrons are attracted away from the plate C_2, which is connected to C_1. A positive charge of the same magnitude is left on the + plate of C_2. Similarly the plates of C_3 acquire the same magnitude of charge. Thus

$$Q = Q_1 = Q_2 = Q_3$$

The negative plate of C_1 must be at the same potential with respect to ground as the positive plate of C_2 since they are connected and are parts of the same conducting surface. Similarly the negative plate of C_2 must be at the same potential as the positive plate of C_3. The total difference of potential, V_T, across the three series capacitors

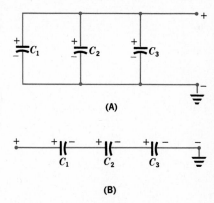

(A)

(B)

Figure 16-24. Capacitors connected in parallel (A) and in series (B).

must be equal to the sum of the separate potential differences across each capacitor: V_1, V_2, and V_3.

$$V_T = V_1 + V_2 + V_3$$

Since the charge received from the source is Q, then by definition

$$V_T = \frac{Q}{C_T}$$

and

$$V_1 = \frac{Q}{C_1}, \ V_2 = \frac{Q}{C_2}, \ V_3 = \frac{Q}{C_3}$$

Substituting,

$$\frac{Q}{C_T} = \frac{Q}{C_1} + \frac{Q}{C_2} + \frac{Q}{C_3}$$

or

$$\frac{1}{C_T} = \frac{1}{C_1} + \frac{1}{C_2} + \frac{1}{C_3}$$

For capacitors connected in series, the reciprocal of the total capacitance is equal to the sum of the reciprocals of all the separate capacitances. See the following example.

EXAMPLE   Three capacitors have capacitances of 0.200 μf, 0.300 μf, and 0.500 μf. (a) If they are connected in parallel and charged to a potential difference of $10\overline{0}$ v, what is the charge on each capacitor? (b) What is the total charge acquired? (c) If these three capacitors are discharged, then connected in series, and a charge of +30.0 μc is transferred to the ungrounded terminal, what potential difference appears across each? (d) What total potential difference exists across the three capacitors?

SOLUTION   (a) The diagram shown in Figure 16-24(A) applies to this part of the problem. In parallel,

$$V = V_1 = V_2 = V_3 = 10\overline{0} \text{ v}$$

Then

$$Q_1 = C_1 V = 0.200 \ \mu\text{f} \times 10\overline{0} \text{ v} = 20.0 \ \mu\text{c}$$
$$Q_2 = C_2 V = 0.300 \ \mu\text{f} \times 10\overline{0} \text{ v} = 30.0 \ \mu\text{c}$$
$$Q_3 = C_3 V = 0.500 \ \mu\text{f} \times 10\overline{0} \text{ v} = 50.0 \ \mu\text{c}$$

(b) In parallel,

$$Q_T = Q_1 + Q_2 + Q_3 = (20.0 + 30.0 + 50.0) \ \mu\text{c}$$
$$Q_T = 100.0 \ \mu\text{c}$$

(c) The diagram shown in Figure 16-24(B) applies to this part of the problem. In series,

$$Q = Q_1 = Q_2 = Q_3 = 30.0 \ \mu c$$

Thus each capacitor is charged to 30.0 μc, the ungrounded terminal being positive. Then,

$$V_1 = \frac{Q}{C_1} = \frac{30.0 \ \mu c}{0.200 \ \mu f} = 15\bar{0} \ v$$

$$V_2 = \frac{Q}{C_2} = \frac{30.0 \ \mu c}{0.300 \ \mu f} = 10\bar{0} \ v$$

$$V_3 = \frac{Q}{C_3} = \frac{30.0 \ \mu c}{0.500 \ \mu f} = 60.0 \ v$$

(d) In series,

$$V_T = V_1 + V_2 + V_3 = 15\bar{0} \ v + 10\bar{0} \ v + 60.0 \ v$$
$$V_T = 31\bar{0} \ v \ (\text{positive with respect to ground})$$

PRACTICE PROBLEMS   **1.** Two capacitors, C_1 and C_2, with capacitances of 0.25 μf and 0.50 μf respectively are connected in parallel as in Figure 16-24(A) across an $8\bar{0}$-volt circuit. (a) When fully charged, what is the potential difference across each capacitor? (b) What is the charge on each capacitor? (c) What is the total capacitance in the circuit? (d) What is the total charge stored in the capacitors?   *Ans:* (a) $8\bar{0}$ v; (b) $Q_1 = 2\bar{0} \ \mu c$, $Q_2 = 4\bar{0} \ \mu c$; (c) 0.75 μf; (d) $6\bar{0} \ \mu c$.

2. Capacitors C_1 and C_2 of Problem 1 are fully discharged and then connected in series as in Figure 16-24(B). A charge of $+25 \ \mu c$ is transferred to the positive (ungrounded) terminal as shown in Figure 16-24(B). (a) What is the charge on each capacitor? (b) What is the total capacitance of the series circuit? (c) What is the potential difference across each capacitor? (d) What is the potential difference across the two capacitors in series?   *Ans:* (a) 25 μc; (b) 0.17 μf; (c) $V = 10\bar{0}$ v, $V = 5\bar{0}$ v; (d) 150 v

---

Questions

GROUP A

1. In what region of an insulated ellipsoidal conductor is the greatest charge density concentrated?

2. What determines whether a difference of potential exists between two points?

3. Define (a) potential difference, (b) the unit of potential difference.

4. A conducting object connected to the earth is at zero potential. Explain.

5. Is work required to move a charge along the surface of a charged conductor isolated in space? Explain.

6. How does a spark discharge occur between two charged surfaces?

7. (a) Define capacitance. (b) What is the unit of capacitance?

8. What is the effect of connecting capacitors (a) in parallel, (b) in series?

GROUP B

9. (a) Express the dimensions of the volt

in fundamental units. (b) Demonstrate that the electric field intensity can have the dimensions of volt per meter.

10. (a) Why is it not possible to maintain a charge on an electroscope indefinitely? (b) What shape should the knob have for a minimum rate of loss of charge?

11. Why are the tips of lightning rods shaped into sharp points?

12. Why are the occupants of a steel-frame building not harmed when the building is struck by lightning?

13. Would you expect to be very successful in conducting experiments with static electricity on a humid day? Explain.

14. The electrostatic force between two charges immersed in oil is found to be one half of the electrostatic force when the same two charges are in air. Explain.

15. A solid metal sphere and a hollow metal sphere have the same dimensions, both perfectly insulated. Which will hold the larger maximum charge in air without breaking down?

Problems
GROUP A

1. A capacitor with air as its dielectric has a capacitance of 150 pf. What is its capacitance when the air is replaced by mica? (See Table 16-1 for values of dielectric constants.)

2. A 5.0-μf capacitor, C_1, and a 7.5-μf capacitor, C_2, are connected in parallel across a 28-v battery. What charge is stored in each capacitor?

3. (a) Capacitors C_1, C_2, and C_3, of capacitances 2.00 μf, 4.00 μf, and 5.00 μf, respectively, are connected in parallel. What is their total capacitance? (b) When these three capacitors are connected in series, what is their total capacitance?

4. A 5.0-μf capacitor is charged to a potential difference of 28 v. What is the charge on the capacitor?

5. The potential difference across the plates of a 2.70-μf capacitor is $15\overline{0}$ v. What charge is stored in the capacitor?

GROUP B

6. A force of 0.032 n is required to move a charge of 42 μc in an electric field between two points 25 cm apart. What potential difference exists between the two points?

7. An electron is accelerated in a machine in which it is subjected to a potential difference of 5.0 megavolts. What energy has the electron acquired?

8. (a) What is the potential gradient between two parallel plates 0.50 cm apart when charged to a potential difference of 1.2 kv? (b) Convert your result to an expression of electric field intensity in terms of n/c.

9. A capacitor consisting of two parallel plates separated by a layer of air 0.3 cm thick and having a capacitance of 15.0 pf is connected across a $15\overline{0}$-volt source. (a) What is the charge on the capacitor? The air dielectric is replaced by a sheet of mica 0.3 cm thick. (b) What is the capacitance with the mica dielectric? (c) What additional charge does the capacitor take up?

10. Three paper capacitors having capacitances of 0.15 μf, 0.22 μf, and 0.47 μf are connected in parallel and charged to a potential difference of 240 volts. (a) Determine the charge on each capacitor. (b) What is the total capacitance of the combination? (c) What is the total charge acquired?

SUMMARY

Electric charges are of two kinds, negative and positive. Like charges repel and unlike charges attract. Electrification is the process that produces electric charges on an object. The process involves the transfer of free electrons. An electroscope may be used to indicate the presence of an electric charge and to provide a rough measurement of its magnitude.

Substances are classified as conductors or insulators based on their ability to conduct an electric charge. Most metals are good conductors. Most nonmetals are good insulators because they are poor conductors of electric charges. The conduction of electric charge by solids is related to the availability of free electrons in their structures. Isolated bodies with conducting surfaces can be charged by either induction or conduction.

The force of attraction or repulsion between point charges is enunciated in terms of Coulomb's law. Charged bodies approximate point charges if they are small in comparison with the distance separating them. The region about a charged body in which the coulomb force exists is an electric field. The path along which a positive test charge is driven by an electric field is called a line of force. Lines of force are used to represent both the direction and intensity of the field.

The difference in potential between two points in an electric field is defined in terms of work expended in moving a charge between two points and quantity of charge moved between these points.

Experiments with isolated conductors have shown that (1) the residual charge on a conductor resides on its surface; (2) there can be no difference of potential between two points on the surface of a charged conductor; (3) the surface of a conductor is an equipotential surface; (4) electric lines of force are normal to equipotential surfaces; and (5) lines of force originate and terminate normal to the conductive surface of a charged object.

Charge density on the conductive surface of a charged body varies with the surface curvature. The intensity of the electric field near sharp points may be great enough to ionize the surrounding air, thus discharging the body.

An arrangement of conducting surfaces separated by insulators and used to store electric charge is called a capacitor. The capacity for storing electric charge is its capacitance. The dielectric constant of an insulating material is determined by comparing the material's effect on capacitance with the effect of air. Capacitors may be connected in series or in parallel. The total capacitance of the capacitor network is dependent upon the kind of connection used.

VOCABULARY

capacitance
capacitor
conductor (electric)
corona discharge
coulomb
Coulomb's law of
    electrostatics
dielectric
dielectric constant
electric field

electric field intensity
electrification
electron gas
electroscope
farad
free electron
induced charge
insulator (electric)
line of force (electric)
microcoulomb

microfarad
picofarad
point charge
potential difference
potential gradient
proof plane
residual charge
static charge
volt

DIRECT-CURRENT CIRCUITS

The SI unit of electric current is named in honor of André Marie Ampère, the French physicist, mathematician, philosopher, and teacher. Ampère, whom James Clerk Maxwell called the "Newton of electricity," was self-taught, since there was no school in the village in which he grew up.

In this chapter you will gain an understanding of:

▶ the nature of an electric current in terms of electric charge

▶ the common dry cell as a source of electric current

▶ current and voltage characteristics of cells connected in series and in parallel

▶ applications of Ohm's law to series and parallel circuits

▶ the effect of internal resistance of a source of emf in electric circuits

▶ measurements of voltage, current, and resistance in series and parallel circuits

▶ the analysis of simple networks consisting of combinations of resistances in series and parallel

▶ the laws of resistance

▶ the Wheatstone bridge circuit and its use

SOURCES OF DIRECT CURRENT

17.1 Electric Charges in Motion The quantity of charge on a capacitor is indicated by the potential difference between the plates.

$$Q = CV$$

If the capacitance, C, for a given capacitor is a constant, then the charge, Q, is proportional to the potential difference, V, across it.

$$Q \propto V \quad (C \text{ constant})$$

A potential difference across a charged capacitor, and therefore a charge on the capacitor, can be indicated by an electroscope connected across the capacitor plates. See Figure 17-1. Suppose the two plates of the capacitor are now connected by a heavy copper wire as in Figure 17-2(A). The leaves of the electroscope immediately collapse, indicating that the capacitor has been discharged rapidly. Free electrons flow from the negative plate of the capacitor and from the electroscope through the copper wire to the positive plate. This action quickly establishes the normal distribution of electrons characteristic of an uncharged capacitor.

Considering the charged capacitor of Figure 17-1 again, suppose the plates are connected by means of a long, fine wire made of nichrome, which has few free electrons compared with copper. The leaves of the electroscope collapse more gradually than before. See Figure 17-2(B). This delay means that a longer time is required to discharge the capacitor completely. When any conductor connects the plates of the capacitor, electrons move from the negative plate through the conductor to the positive plate. This transfer decreases the charge on each plate and the difference of potential between the plates.

The charge, Q, on either plate of a capacitor is proportional to the potential difference, V, between the plates. The *rate* at which the charge decreases, in c/s, is proportional to the rate at which the potential difference decreases, in v/s. Now, the rate of decrease of charge on the capacitor must represent the rate of flow of the charge, in c/s, through the conductor. Thus the rate at which the leaves of the electroscope collapse indicates the rate of flow of the charge through the conductor. During the time the capacitor is discharging, a *current* is said to exist in the conducting wire. *An **electric current**, I, in a conductor is the rate of flow of charge through a cross section of the conductor.*

$$\text{current } (I) = \frac{\text{charge } (Q)}{\text{time } (t)}$$

The unit of current is the *ampere* (a), named for the French physicist André Marie Ampère (1775–1836). *One ampere is a current of 1 coulomb per second.*

$$1 \text{ a} = \frac{1 \text{ c}}{\text{s}}$$

Recall that 1 coulomb is the charge of 6.25×10^{18} electrons or a like number of protons. Small currents can be expressed in *milliamperes* (ma) or *microamperes* (μa).

$$1 \text{ ma} = 10^{-3} \text{ a}$$
$$1 \text{ }\mu\text{a} = 10^{-6} \text{ a}$$

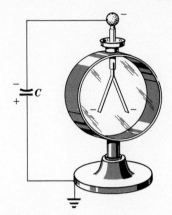

Figure 17-1. The charge on a capacitor can be indicated by an electroscope.

The SI symbol for the ampere is A. Because of its similarity to the symbol for (conductor) cross-sectional area, A, the older lower-case form, a, is used in this book.

Figure 17-2. The rate of discharge of a capacitor depends on the conducting path provided.

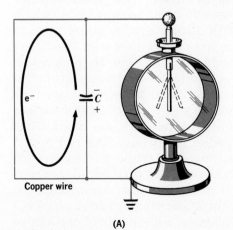

Copper wire

(A)

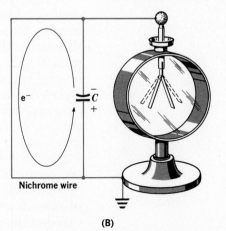

Nichrome wire

(B)

A moving charge is basic for an electric current. The character of a conducting substance determines the nature of the charge set in motion in an electric field. In solids the charge carriers are electrons; in gases or electrolytic solutions the charge carriers can be positive ions, negative ions, or both. *An **electrolyte** is a substance whose aqueous solution conducts an electric current.*

A negative charge moving in one direction in an electric circuit is equivalent to a positive charge moving in the opposite direction insofar as external effects are concerned. Negative charges departing from a negatively charged surface and arriving at a positively charged surface leave the former surface less negative and the latter surface less positive. Positive charges departing from the positively charged surface and arriving at the negatively charged surface would have a similar effect, leaving the positively charged surface less positive and the negatively charged surface less negative.

If both positive and negative charges are set in motion in a conducting medium, the current in the medium is the sum of the currents due to the motion of the positive charges in one direction and the negative charges in the opposite direction.

In an electric circuit, the directional sense of electron flow through an electric field in a metallic conductor is from negative to positive. The directional sense of a positive ion flow in an ionized gas is from positive to negative. In our study of electric circuits we will be concerned generally with current in metallic conductors and will deal mainly with electron flow in which the directional sense is from negative to positive. In any case, confusion about current direction can be avoided by indicating the directional sense of the charge carriers that are in motion.

The above reference to an electric field in a conductor does not contradict the assertions of Section 16.11, which require the electric field in a conductor to equal zero. In this earlier discussion we were concerned with static electric charges on an isolated conductor. The charges reside on the surface of such a conductor and there is no net charge motion on this equipotential surface. We are now dealing with moving charges and with conductors across the ends of which potential differences are deliberately maintained. Thus this earlier restriction does not apply.

Free electrons in empty space are accelerated by an electric field. The effect of an electric field on the free electrons of a metallic conductor is quite different. An acceleration of extended duration of these electrons is not realized because of their frequent collisions with the fixed particles of

The electric current in metallic conductors is an electron flow from negative to positive.

the conductor, the metallic ions. With each collision the electron loses whatever velocity it had acquired in the direction of the accelerating force, transfers energy to the fixed particle, and makes a fresh start.

The kinetic energy transferred in these collisions increases the vibrational energy of the conductor particles and heat is given up to the surroundings. The free electrons move in the direction of the accelerating force with an average velocity called the *drift velocity*. This drift velocity has a constant average value for a given conductor carrying a certain magnitude of current. It is much smaller than the speed with which *changes in the electric field* propagate through the conductor, which is approximately 3×10^8 m/s.

It has been estimated that the free electrons of a copper conductor of 1-mm^2 cross section have an average drift velocity of about 1 mm/s when a current of 20 a is in the conductor. The situation is somewhat analogous to that of a ball rolling down a flight of stairs. With the proper conditions, the successive collisions of the ball with the steps effectively cancel the acceleration it acquires between steps and it rolls down with a constant average speed.

As we saw, collision between the free electrons and the fixed particles of a conductor is an energy-dissipating process that opposes the flow of charge through the conductor. *This opposition to the electric current is called* **resistance, R.** The practical unit of resistance is the *ohm*, symbolized by the Greek letter Ω (omega). The laws of electric resistance are discussed in Section 17.11.

17.2 Continuous Current   The current from a discharging capacitor persists for a very short interval of time; it is known as a *transient* current. The effects of such a current are also transient, but those effects are very important in certain types of electronic circuits. In the general applications of electricity, more continuous currents are required.

The effects of current electricity are quite different from those of static electricity. The electric current is one of our most convenient means of transferring energy. A *closed-loop* conducting path is needed if energy is to be utilized outside the source; this conducting loop is known as an *electric circuit*. The device or circuit component utilizing the electric energy is called the *load*. The basic components of an electric circuit are illustrated in three different ways in Figure 17-3. Conventional symbols used in schematic (circuit) diagrams are shown in Figure 17-4.

A capacitor would be useful as a source of continuous current over a prolonged period of time only if some

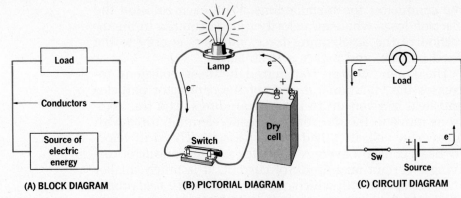

(A) BLOCK DIAGRAM   (B) PICTORIAL DIAGRAM   (C) CIRCUIT DIAGRAM

Figure 17-3. An electric circuit is a conducting loop in which a current can transfer electric energy from a suitable source to a useful load.

means were available for keeping it continuously charged. We would need to supply electrons to the negative plate of the capacitor as rapidly as they were removed by the current in the conducting loop. Similarly, we would need to remove electrons from the positive plate of the capacitor as rapidly as they were deposited by the current. We must

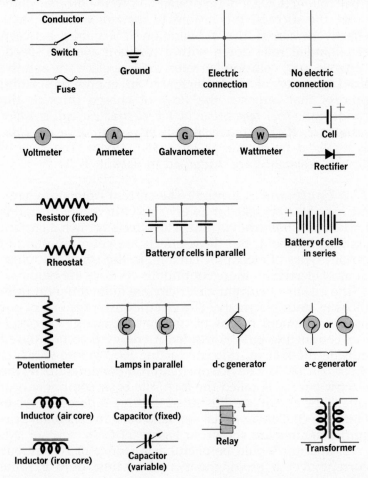

Figure 17-4. Conventional symbols used in schematic diagrams of electric circuits.

maintain the potential difference across the source during the time a continuous current is in the circuit.

Maintaining a potential difference across a source of current in an operating circuit is an energy-consuming process; work is done on the electrons in opposition to the force of an electric field. In the process the potential energy of electrons is raised, and this energy is available to do useful work in the external circuit. What is needed is some kind of "electron pump." Several practical energy sources are available to maintain a difference of potential across an operating electric circuit.

17.3 Sources of Continuous Current   The transformation of available energy into electric energy is accomplished by one of two basic methods: *dynamic conversion* or *direct conversion*. Dynamic converters are rotating machines, and it is these machines that supply most of our requirements for electric energy today. Direct conversion methods produce energy transformations without moving parts. We shall consider the following sources of continuous current:

1. *Electromagnetic.* The major source of electric current in practical circuits today depends on the principle of *electromagnetic induction*. If a conducting loop is rotated in a magnetic field, the free electrons of the conductor are forced to move around the loop and thus constitute a current. This is the basic operating principle of electric generators. It is an example of the transformation of mechanical energy into electric energy, a dynamic conversion process. Electromagnetic induction and the electric generator are discussed in Chapter 20.

2. *Photoelectric.* The photoelectric effect was first observed by Heinrich Hertz. He noticed that a spark discharge occurred more readily when charged metal spheres were illuminated by another spark discharge.

Electrons are emitted from the surface of a metal illuminated by light of sufficiently short wavelength. The alkali metals are photosensitive to light in the near ultraviolet region of the radiation spectrum. Two of these metals, potassium and cesium, emit photoelectrons when exposed to ordinary visible light. In the case of most metals, the incident light for photoemission corresponds to radiations deep in the ultraviolet region. When a photon of the incident light is absorbed by the metal, its energy is then transferred to an electron. If the motion of this electron is toward the surface and its energy at the surface is sufficient to overcome the forces that bind it within the surface of the metal, it is emitted as a photoelectron.

The energy supply for the dynamic conversion process comes mainly from burning fossil fuel, from falling water, and from nuclear fission reactions.

Review the laws of photoelectric emission in Section 12.7.

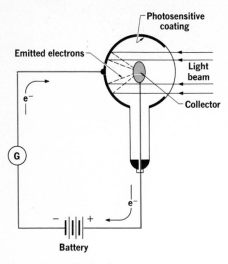

Figure 17-5. The circuit of a photoelectric cell.

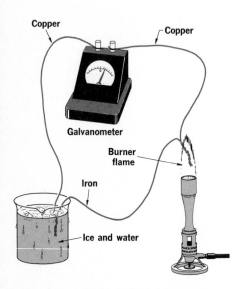

Figure 17-6. A thermocouple.

If properly isolated, the metal surface acquires a positive charge as photoelectrons are emitted. As the charge increases, other electrons are prevented from escaping from the metal and the action stops. If, however, the metal is placed in an evacuated tube and made part of a circuit, as in Figure 17-5, it can act as a source of current whenever light is incident upon its surface. The metal thus serves as the electron pump mentioned in Section 17.2. The evacuated tube containing the metallic emitter is called a *photoelectric cell*.

The photoelectric cell shown in Figure 17-5 has a coating of potassium metal deposited on the inner surface of a glass tube, leaving an aperture through which light can enter. A metallic ring near the focus of the emitter surface acts as a collector of photoelectrons. The emitter is connected externally to the negative terminal of a battery, and the collector is connected to the positive terminal. The collector is thus maintained at a positive potential with respect to the emitter. The force of the resulting electric field acts on the emitted electrons, driving them to the collector. The small photoelectron current can thus be maintained in the external circuit as long as light photons are incident on the emitter surface. A *galvanometer,* an electric meter sensitive to very feeble currents, can be placed in the circuit to indicate the relative magnitude and direction of the current.

A bright light of effective frequency causes the ejection of many electrons from the surface of the photosensitive metal, while light of the same frequency but of a lower intensity produces fewer photoelectrons. Thus the current produced by the photoelectric cell is determined by the intensity of the incident light; this is a case in which radiant energy is transformed into electric energy. Some form of photosensitive cell is used in devices that are controlled or operated by light.

3. Thermoelectric. Suppose we form a conducting loop circuit that consists of a length of iron wire, a length of copper wire, and a sensitive galvanometer, as shown in Figure 17-6. One copper-iron junction is put in a beaker of ice and water to maintain a low temperature; the other junction is heated by a gas flame. As the second junction is heated, the galvanometer indicates a current in the loop. An electric circuit incorporating such junctions is known as a *thermocouple.*

The magnitude of the current is related to the nature of the two metals forming the junctions and the temperature difference between the junctions. A thermocouple can be used as a sensitive thermometer to measure radiant energy. The iron-copper junction is useful for temperatures up to about 275 °C. Junctions of copper and constantan, a copper-nickel alloy, are widely used in lower temperature ranges. Junctions of platinum and rhodium are used to measure temperatures ranging up to 1600 °C.

Since heat energy is transformed directly into electric energy, the thermocouple is a thermoelectric source of current. If a sensitive galvanometer is properly calibrated, temperatures can be read directly. The hot junction of the thermocouple can be placed in a remote location where it would be impossible to read an ordinary thermometer.

Recall that the first law of thermodynamics tells us that heat can be converted into other forms of energy, but energy can neither be created nor destroyed. Thus in a closed thermodynamic system no energy will be gained or lost. Thermoelectric converters are fundamentally heat engines and, as for all other heat engines, are limited to the efficiency of the conversion process. The actual efficiency with which heat is converted to electricity is a function of the temperature difference within the operating system.

The heating effect of an electric current in the resistance of a conductor is an irreversible phenomenon. On the other hand, the basic thermoelectric effects discussed in this section are reversible. The best known of these effects is described in this section: when the junctions of two dissimilar metals are subjected to a difference in temperature, a current is set up in the loop. Known as the *Seebeck effect*, it is named after its discoverer who first observed the phenomenon in 1821. See Figure 17-7(A).

The *Peltier effect* is produced when a direct current is passed through a junction formed by two dissimilar metals. The junction becomes warmer or cooler depending on the direction of the current, as shown in Figure 17-7(B).

Another thermoelectric effect, called the *Thomson effect*, is observed in certain homogeneous materials. If a single solid is subjected to a temperature gradient between opposite faces, a difference of potential is established across it. This potential difference can be used to supply an electric current in an external circuit connected to the solid.

Certain semiconductor materials show thermoelectric properties and perform the heat-to-electricity conversion more efficiently than metals. In addition to the application of semiconductor converters as electric generators, the Peltier effect makes possible their use as cooling devices.

Thermocouples in which heat from radioactive isotopes

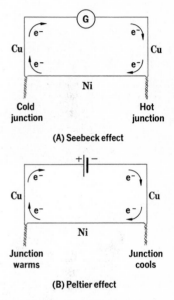

Figure 17-7. The Seebeck and Peltier effects for copper and nickel. One is the inverse of the other.

is converted to electric energy are used in some unmanned satellites. Nuclear heat sources that produce an electric current by thermoelectric conversion were used by astronauts on the surface of the moon to provide power for their lunar experiments. These energy converters (Figure 17-8) used plutonium-238 as the heat source.

4. *Piezoelectric.* When certain crystals, such as Rochelle salt and quartz, are subjected to a mechanical stress, the opposite surfaces will become electrically charged. The difference of potential between the stressed surfaces is proportional to the amount of stress applied to the crystal. Reversing the stress reverses the charges also.

If the crystal is suitably supported and a conducting circuit is connected between surfaces having a difference of potential, electricity will flow through the circuit. Thus mechanical energy is transformed into electric energy. This transformation is known as the *piezoelectric effect,* and the crystal with its supporting mechanism is called a *piezoelectric cell.*

Crystal microphones use the piezoelectric effect to transform the alternating pressure of sound waves into an electric current that varies as the pressure of the sound waves varies. A crystal phonograph pickup responds in a similar manner to the varying stress caused by the needle riding in the groove of the record.

The piezoelectric effect is reversible; that is, the crystal experiences a mechanical strain when subjected to an electric stress. An alternating difference of potential properly placed across the crystal will cause mechanical vibrations proportional to the variations of the voltage. Crystal headphones use this reciprocal effect.

5. *Chemical.* There are certain chemical reactions that involve a transfer of electrons from one reactant to the other. The reactions that occur *spontaneously* (that is, of and by themselves) can be used as sources of continuous current. During the chemical reaction, electrons are removed from one reactant. A like number of electrons is added to the other reactant. If the reactants are separated in a conducting environment, the transfer of electrons will take place through an external circuit connecting the reactants. This electron transfer will continue for as long as reactants are available. Such an arrangement is known as an *electrochemical cell.* Chemical energy is transformed into electric energy during the chemical reaction.

Electrochemical cells include primary, storage, and fuel cells. A primary cell, a flashlight cell for example, is ordinarily replaced when its reactants are used up. Storage cells, as in an automobile battery, are reversible. Electric

Figure 17-8. A thermoelectric generator on the surface of the moon. Heat is supplied by nuclear energy. The shadow of an astronaut is seen at the right.

energy is initially stored as chemical energy, and the cell can be repeatedly "recharged" with energy. If the reactants in a primary cell are supplied continuously, the arrangement is called a fuel cell. Fuel cells have been used as the source of electric current in spacecraft.

17.4 The Dry Cell The essential components of a primary cell are an *electrolyte* and two *electrodes* of unlike materials, one of which reacts chemically with the electrolyte. Let us examine the carbon-zinc dry cell as a common example of the primary cell. A cut-away diagram of this cell is shown in Figure 17-9. The electrolyte is a moist paste of ammonium chloride. One electrode is a carbon rod. The other electrode, a zinc cup, is consumed as the cell is used.

When an external circuit is connected across the cell, zinc atoms react with the electrolyte and lose electrons, which remain with the zinc cup. This reaction leaves the zinc electrode *negatively charged*. We shall refer to the negative electrode as the **cathode.** *It is the electron-rich electrode.*

Simultaneously, the electrolyte removes a like number of electrons from the carbon electrode. This reaction leaves the carbon electrode positively charged. We shall call the positive electrode the **anode.** *It is the electron-poor electrode.*

The electrons move through the external circuit from cathode to anode because of the difference in potential between the electrodes. Thus the cell acts as a kind of electron pump. See Figure 17-10. It removes electrons with low potential energy from the anode and supplies electrons with high potential energy to the cathode while current is in the external circuit. Electrons flow through the external circuit from cathode to anode. This unidirectional current is an example of *direct current*, dc.

If we open the circuit and stop the current, the *maximum* potential difference is quickly established across the cell. The magnitude of this *open-circuit* potential difference is dependent solely on the materials that make up the electrodes. For the carbon-zinc cell it is about 1.5 volts.

The open circuit potential difference can be referred to as the *emf*, or electromotive force, of the cell. The symbol for emf is E and the unit is the *volt* (v). The cell itself must supply a definite amount of energy for each unit of charge that is stored on either electrode; *the **emf** of a source is the energy per unit charge supplied by the source.* Thus the potential difference across the cell on an *open* circuit is equal to the emf of the cell.

$$V_{oc} = E$$

As shown in Section 17.6, the potential difference, V, across a source of emf on a closed circuit is less than the emf, E.

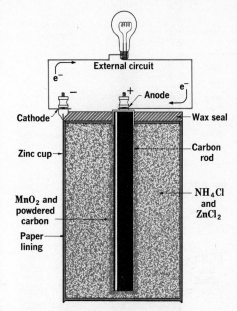

Figure 17-9. The dry cell is a practical source of direct current in the physics laboratory.

The definitions for electrodes are sometimes based on oxidation-reduction reactions rather than on their electric charge. In that case their names in primary cells become the reverse of those used here.

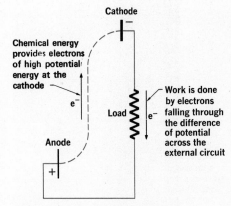

Figure 17-10. A voltaic cell provides electrons of high potential energy, the result of chemical reactions within the cell.

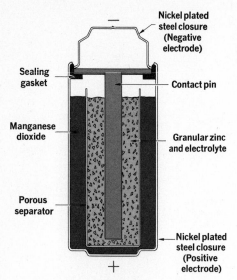

Figure 17-11. Cross section of an alkaline-manganese cell. An outer jacket (not shown) and appropriate terminal connections provide the conventional appearance of the carbon-zinc cell.

The carbon-zinc cell has an emf of 1.5 volts, regardless of the size of the cell. However, the larger the electrode surfaces, the larger the current the cell can deliver. The standard No. 6 dry cell is capable of supplying a *momentary* current up to 35 amperes. It should not be required to furnish more than 0.25 ampere of continuous current, however. The internal resistance can vary from less than 0.1 ohm as in the No. 6 cell to over 1 ohm as in small special-purpose cells.

Alkaline-manganese cells can be substituted for carbon-zinc cells in many applications. While more expensive to produce, alkaline cells have lower internal resistance, higher efficiency, and the ability to deliver from 50% to 100% more current than their carbon-zinc counterparts.

A sectional view of an alkaline-manganese cell is shown in Figure 17-11. The positive electrode consists of manganese dioxide in conjunction with a steel can that serves as the electrode terminal. The negative electrode is granulated zinc mixed with a highly conductive potassium hydroxide electrolyte. The nominal potential difference across the cell is 1.5 volts.

17.5 Combinations of Cells Energy-consuming devices that form the loads of electric circuits are designed to draw the proper current for their operation when a specific potential difference exists across their terminals. For example, a lamp designed for use in a 6-volt electric system of an automobile would draw an excessive current and burn out if placed in a 12-volt system. A lamp that is intended for use in a 12-volt system, on the other hand, would not draw sufficient current at 6 volts to function as intended.

Electrochemical cells are sources of emf. Each cell furnishes a certain amount of energy to the circuit for each coulomb of charge that is moved. It is often necessary to combine cells in order to provide either the proper emf or an adequate source of current. Groups of cells can be connected in *series*, in *parallel*, or in *series-parallel* combinations. *Two or more electrochemical cells connected together form a battery.* The emf of a battery depends on the emf of the individual cells and the way in which they are connected.

A *series* combination of cells is shown in Figure 17-12. Observe that the positive terminal of one cell is connected to the external circuit, and the negative terminal is connected to the positive terminal of a second cell. The negative terminal of this cell is connected to the positive terminal of a third cell, and so on. The negative terminal of the

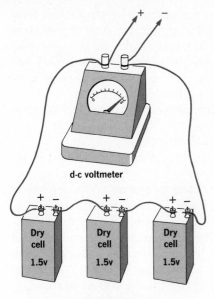

(A) Pictorial diagram

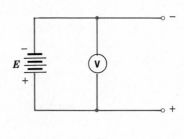

(B) Circuit diagram

last cell in the series is connected to the load to complete the circuit. Thus the cells are joined end to end and *the same quantity of electricity must flow through each cell.*

Suppose the battery is thought of as a group of electron pumps connected in *series* (adding). Each electron pump (cell) removes electrons of low potential energy from its anode and supplies electrons of high potential energy to its cathode. The first pump lifts electrons to a certain level of potential energy, the second pump takes them from this level to the next higher level, the third lifts them to a still higher potential energy level, and so on. Similarly, the emfs of the cells are added to give the emf of the battery.

A battery made up of cells connected in *series* has the following characteristics:

1. *The emf of the battery is equal to the sum of the emfs of the individual cells.*

2. *The current in each cell and in the external circuit has the same magnitude throughout.*

3. *The internal resistance of the battery is equal to the sum of the internal resistances of the individual cells.*

In a *parallel* combination of cells, all negative terminals are connected together. One side of the external circuit is then connected to any one of these common terminals. The positive terminals are connected together and the other side of the external circuit can be connected to any one of these common terminals to complete the circuit. See Figure 17-13. This arrangement has the effect of increasing the total cathode and anode surface areas. The

Figure 17-12. Dry cells connected in series to form a battery. The same quantity of electric charge flows through each cell.

Resistances in series are treated analytically in Section 17.8.

battery is then roughly the equivalent of a single cell having greatly enlarged electrodes in contact with the electrolyte and a lower internal resistance. Each cell of a group in parallel merely furnishes its proportionate share of the total circuit current. Of course, cells should not be connected in parallel unless they have equal emfs.

A battery of *identical* cells connected in *parallel* has the following characteristics.

1. The emf is equal to the emf of each separate cell.

2. The total current in the circuit is divided equally among the cells.

Resistances in parallel are treated analytically in Section 17.9.

3. The reciprocal of the internal resistance of the battery is equal to the sum of the reciprocals of the internal resistances of the cells.

The lead-acid storage battery for automobiles with 6-volt electric systems consists of three cells connected in series. The emf is approximately 6.6 volts. Six cells are connected in series to make up the battery for 12-volt systems.

As mentioned previously, a No. 6 dry cell should not be required to deliver more than 0.25 ampere of continuous current. Suppose we have an electric device designed to perform with a 0.75-ampere continuous current in a 3-volt circuit. How should the 1.5-volt dry cells be grouped to make up the battery for operating the device?

Certainly, two dry cells in series would provide an emf of 3 volts for the circuit. Three cells in parallel would require no more than 0.25 ampere from each cell. Thus a series-parallel arrangement of dry cells, as shown in Figure 17-14, is appropriate.

Figure 17-13. Dry cells connected in parallel to form a battery.

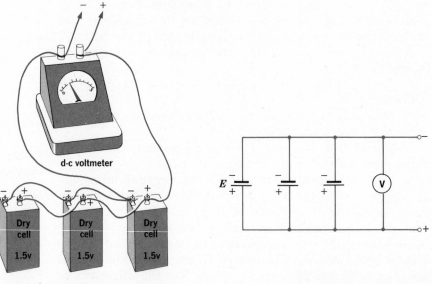

(A) Pictorial diagram   **(B) Circuit diagram**

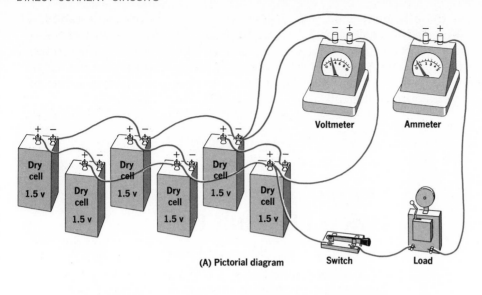

(A) Pictorial diagram

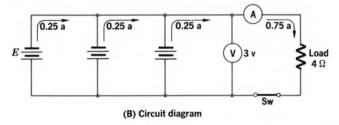

(B) Circuit diagram

Figure 17-14. A circuit using a battery that consists of a series-parallel arrangement of cells.

Questions

GROUP A

1. What is an electric current?
2. What is the unit of current?
3. Define the unit of current in terms of fundamental units.
4. What is electric resistance?
5. Name five basic sources of continuous current.
6. Name a current-producing device for each of these sources.
7. What are the essential parts of a primary cell?
8. Define an electrolyte.
9. What term is used to identify (a) the negative electrode of an electric cell, (b) the positive electrode?
10. What is the distinguishing characteristic of direct-current electricity?
11. Distinguish between the terms "open circuit" and "closed circuit."
12. What is the open-circuit potential difference of a source of current called?

GROUP B

13. What change in the energy of an electron is produced (a) by a source of emf, (b) by a load connected across a source of emf?
14. Both primary and storage cells are sources of emf. What distinguishes one cell from the other?

15. What determines the efficiency with which heat is converted to electricity in thermoelectric converters?

16. Distinguish between potential difference and emf.

17. (a) How would you connect 1.5-volt dry cells to provide a 6-volt battery to supply a continuous current of 0.25 ampere to a load? (b) Draw a circuit diagram of the battery you describe.

18. Does a storage battery store electricity? Justify your answer.

SERIES AND PARALLEL CIRCUITS

17.6 Ohm's Law for d-c Circuits Resistance, *R*, is defined as the opposition to the flow of electric charge. Metallic substances in general are classed as good conductors of electricity. Silver and copper, because of their abundance of free electrons, are excellent conductors; yet even these metals offer some opposition to the current in them. Through its normal range of operating temperatures, every conductor has some inherent resistance. Thus every device or component in an electric circuit offers some resistance to the current in the circuit.

Copper wire is used to connect various circuit components in electric circuits because ordinarily its resistance is low enough to be neglected. Certain metallic alloys offer unusually high resistance to the flow of electricity; *nichrome* and *chromel* are notable examples. Spools wound with wire made from these alloys can be used in an electric circuit to provide either *fixed* amounts of resistance or *variable* resistance in definite amounts.

Carbon granules can be mixed with varying amounts of clay and molded into cylinders having a definite resistance. These devices, known as *carbon resistors*, are commonly used in electronic circuits.

The German physicist Georg Simon Ohm (1789–1854) discovered that *in a closed circuit the ratio of the emf of the source to the current in the circuit is a constant.* This constant is the *resistance* of the circuit. The above statement, known as **Ohm's law of resistance,** describes a basic relationship in the study of current electricity. It can be stated mathematically:

$$\frac{E}{I} = R \quad \text{or} \quad E = IR$$

Figure 17-15. Georg Simon Ohm, the German physicist for whom the unit of resistance is named.

The units are the volt, ampere, and ohm respectively. The *ohm* (Ω) has the following dimensions:

$$\Omega = \frac{v}{a} = \frac{j/c}{c/s}$$

$$\Omega = \frac{j\ s}{c^2} = \frac{n\ m\ s}{c^2}$$

$$\Omega = \frac{kg\ m^2\ s}{s^2\ c^2} = \frac{kg\ m^2}{c^2\ s}$$

Suppose there is an emf of 12 v across an electric circuit that has 33 Ω resistance. The current in the circuit is

$$E = IR \quad \text{so} \quad I = E/R$$

$$I = \frac{12\ v}{33\ \Omega} = 0.36\ a$$

Now suppose the circuit resistance is doubled. In this situation the opposition to the current is doubled but the emf remains the same, and so we should expect the current in the circuit to be reduced to one-half the former value.

$$I = \frac{12\ v}{66\ \Omega} = 0.18\ a$$

The emf, E, applied to a circuit equals the drop in potential across the *total* resistance of the circuit, but Ohm's law applies equally well to *any part* of a circuit that does not include a source of emf. Suppose we consider the circuit shown in Figure 17-16. Points **A** and **B** are the terminals of the battery that supply an emf of 12.0 volts to the circuit. The total resistance of the circuit is 12.0 ohms. Of this total, 0.20 ohm is *internal resistance*, r, of the battery and is conventionally represented as a resistor in series with the battery inside the battery terminals. The remaining 11.8 ohms of resistance consists of a *load resistor*, R_L, in the external circuit.

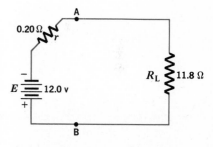

Figure **17-16.** Ohm's law applies to an entire circuit or to any part of the circuit.

For the entire circuit, the total resistance, R_T, is equal to the sum of internal and external resistances.

$$R_T = R_L + r$$

The current in the entire circuit is

$$I = \frac{E}{R_T} = \frac{E}{R_L + r}$$

$$I = \frac{12.0\ v}{11.8\ \Omega + 0.20\ \Omega} = 1.00\ a$$

When we apply Ohm's law to a part of a circuit that does not include a source of emf, we are concerned with the potential difference, V, the current, I, and the resistance, R, that apply only to that part of the circuit. Thus Ohm's law for a part of the circuit becomes

$$V = IR$$

The drop in potential across the external circuit of Figure 17-16 must be

$$V_L = IR_L$$
$$V_L = 1.00 \text{ a} \times 11.8 \text{ } \Omega = 11.8 \text{ v}$$

The drop in potential across the internal resistance of the battery due to current in the battery can be found by a similar application of Ohm's law.

$$V_r = Ir$$
$$V_r = 1.00 \text{ a} \times 0.20 \text{ } \Omega = 0.20 \text{ v}$$

This is a drop in potential across the battery resistance that removes energy from the electrons. It is, therefore, opposite in sign to the emf of the battery. The potential difference across the battery terminals of a closed circuit with a current I is determined by subtracting this drop in potential from the emf.

$$V = E - Ir$$

The closed-circuit potential difference across the terminals of a source of emf that is applied to the external circuit will always be less than the emf by the amount Ir volts. This is not an important difference for a source of emf of low internal resistance unless an excessive current is in the circuit. (In the discussion of Sections 17.8–17.10 the internal resistance of a source of emf will be neglected and the emf will be assumed to be applied to the external circuit.)

17.7 Determining Internal Resistance   A good quality voltmeter is an instrument of very high resistance. When such a voltmeter is connected across the terminals of a source of emf, even a source with low internal resistance, negligible current is drawn from the source. The meter registers an "open circuit" voltage that, for all practical purposes, is the emf of the source. Such a measurement of the emf of a cell or battery provides a convenient method of determining its internal resistance. The following example, based on the circuit shown in Figure 17-17, illustrates this method of determining the internal resistance, r, of a dry cell.

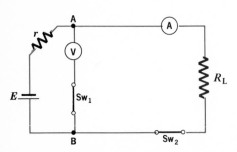

Figure 17-17. A circuit for measuring the internal resistance of a cell.

EXAMPLE   A dry cell gives an open-circuit voltmeter reading of 1.5 v with **Sw₁** closed and **Sw₂** open. The voltmeter, **V**, is then removed from the circuit by opening switch **Sw₁**, and an external load of 2.8 ohms, **R_L**, is connected across the source of emf by closing switch **Sw₂**. The ammeter, **A**, in the external circuit reads 0.50 a. (a) Find the internal resistance, *r*, of the cell. (b) What would the voltmeter have read if it had been left connected while the current was in the load?

SOLUTION   (a) We are given the following information: the emf of the cell, $E = 1.5$ v; $I = 0.50$ a; and $R_L = 2.8 \ \Omega$. We are asked to find r. We know that E is the work done per coulomb of charge moved through the total resistance of the circuit. Thus

$E = IR_T.$ But $R_T = R_L + r$

Substituting for R_T: $E = I(R_L + r)$

Solving for r: $r = \dfrac{E - IR_L}{I}$

$r = \dfrac{1.5 \ \text{v} - (0.50 \ \text{a} \times 2.8 \ \Omega)}{0.50 \ \text{a}} = 0.20 \ \Omega$, the internal resistance

(b) $V = E - Ir = 1.5 \ \text{v} - (0.50 \ \text{a} \times 0.20 \ \Omega)$
$V = 1.4$ v, the potential difference across the circuit

PRACTICE PROBLEMS   1. A resistance load, R_L, of 7.0 Ω is connected across a battery with an emf of 4.5 v. An ammeter in the external circuit shows a current of 0.60 a. (a) What is the potential difference, V_L, across the load resistance, R_L? (b) What is the internal resistance, r, of the battery?   *Ans.* (a) 4.2 v; (b) 0.50 Ω

2. A resistance load, R_L, of 3.7 Ω is connected across a source of emf with an internal resistance, r, of 0.30 Ω. The current in the circuit is determined to be 4.0 a. (a) What is the potential difference across the load, R_L? (b) What is the potential difference across the internal resistance, r? (c) What is the emf of the source?   *Ans.* (a) 14.8 v; (b) 1.2 v; (c) 16.0 v

17.8 Resistances in Series   If several electric devices are connected end to end in a circuit, electricity must flow through each in succession. There is a single path for the moving charge. The same current must be in each device or else there would be an accumulation of charge at different points around the conducting circuit. We know that a charge can be accumulated on a conductor only if it is iso-lated. *An electric circuit with components arranged to provide a single conducting path for current is known as a **series circuit**.*

All elements connected in series in a circuit have the same magnitude of current.

Series-circuit operation has several inherent characteristics. If one component of a series circuit fails to provide a conducting path, the circuit is opened. Each component of the circuit offers resistance to the current, and this resistance limits the current in the circuit according to Ohm's law. When components are connected in series, their resistances are cumulative. The greater the number of resistive components, the higher is the total resistance and the smaller is the current in the circuit for a given applied emf. Obviously, since the current must be the same at all points in the circuit, *all devices connected in series must be designed to function at the same current magnitude.*

From Ohm's law, the drop in potential across each component of a series circuit is the product of the current in the circuit and the resistance of the component. Electrons lose energy as they fall through a difference of potential; in a series circuit these losses occur in succession and are therefore cumulative. *The drops in potential across successive components in a series circuit are additive, and their sum is equal to the potential difference across the whole circuit.*

These observations are summarized by stating three cardinal rules for resistances in series:

1. *The current in all parts of a series circuit has the same magnitude.*

$$I_T = I_1 = I_2 = I_3 = \text{etc.}$$

2. *The sum of all the separate drops in potential around a series circuit is equal to the applied emf.*

$$E = V_1 + V_2 + V_3 + \text{etc.}$$

3. *The total resistance in a series circuit is equal to the sum of all the separate resistances.*

$$R_T = R_1 + R_2 + R_3 + \text{etc.}$$

These rules have been applied to a circuit consisting of several resistors connected in series as shown in Figure 17-18. For simplicity, the internal resistance of the source is assumed to be negligible and is omitted from the circuit diagram. The computations are shown at the right of the

Figure 17-18. A series circuit with Ohm's law relationships calculated.

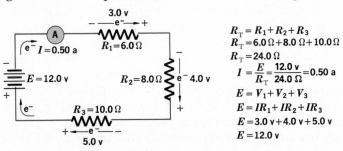

$$R_T = R_1 + R_2 + R_3$$
$$R_T = 6.0\,\Omega + 8.0\,\Omega + 10.0\,\Omega$$
$$R_T = 24.0\,\Omega$$
$$I = \frac{E}{R_T} = \frac{12.0\,\text{v}}{24.0\,\Omega} = 0.50\,\text{a}$$
$$E = V_1 + V_2 + V_3$$
$$E = IR_1 + IR_2 + IR_3$$
$$E = 3.0\,\text{v} + 4.0\,\text{v} + 5.0\,\text{v}$$
$$E = 12.0\,\text{v}$$

circuit. Observe that the "positive" side of R_1 is 3.0 volts *positive with respect to the cathode of the battery* and is 9.0 volts *negative with respect to the anode of the battery.* The three voltage drops around the external circuit are in series, and their sum equals the applied emf in magnitude, but their signs are opposite.

Stated in another way, *the algebraic sum of all the changes in potential occurring around the complete circuit is equal to zero.* This is a statement of **Kirchhoff's second law,** credited to Gustav Robert Kirchhoff (1824–1887), an eminent German physicist. Since potential difference is an expression of work (or energy) per unit charge involved in transporting a charge around a closed circuit, Kirchhoff's second law can be recognized as an application of the law of *conservation of energy* to electric circuits.

Kirchhoff's first law is given in Section 17.9.

17.9 Resistances in Parallel Some of the characteristics of series operation of electric devices were mentioned in the preceding section. It would be quite disconcerting to have all the lights go out in your home each time one lamp were turned off. For this reason each lamp is on a separate circuit so that each is operated independently. Such lamps are connected in parallel. *A **parallel circuit** is one in which two or more components are connected across two common points in the circuit to provide separate conducting paths for current.*

Since there can be only one difference of potential between any two points in an electric circuit, *electric devices or appliances connected in parallel should usually be designed to operate at the same voltage.* The currents in the separate branches must vary inversely with their separate resistances since the same potential difference exists across each branch of the parallel circuit.

By Ohm's law, which can be applied to any part of a circuit, the current in each branch is

$$I = \frac{V}{R}$$

But V is constant for all branches in parallel. Then

$$I_{\text{(for each branch)}} \propto \frac{1}{R_{\text{(of that branch)}}}$$

In Figure 17-19 it is apparent that more electric charge can flow between points **A** and **B** with the two parallel paths R_1 and R_2 in the circuit than with either path alone. The current I entering junction **A** must be the same as the current $(I_1 + I_2)$ leaving junction **A**. Similarly, the current $(I_1 + I_2)$ entering junction **B** must be the same as the current I leaving junction **B**. Then I must be equal to $I_1 + I_2$.

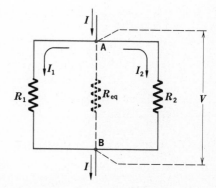

Figure 17-19. Two resistances in parallel and the single equivalent resistance, R_{eq}, that can replace them with no change in the circuit parameters.

All elements connected in parallel in a circuit have the same voltage across them.

Let us consider in another way the currents entering and leaving junctions **A** and **B**. Electric current is the rate of transfer of charge, $I = Q/t$. In the steady-state operation of the circuit, the charge of **A** can neither increase nor decrease. Current I carries charge toward this junction and I_1 and I_2 carry charge away from it. Thus

$$I - I_1 - I_2 = 0$$

Similarly, currents I_1 and I_2 carry charge toward junction **B** and I carries charge away from it. Using the same sign convention for current direction as before,

$$I_1 + I_2 - I = 0$$

These expressions tell us that *the algebraic sum of the currents at any circuit junction is equal to zero.* This is known as **Kirchhoff's first law.** It states the law of *conservation of charge* in electric circuits. Kirchhoff's two laws define the basic principles we must employ in our interpretation of electric circuits: *that energy and charge are conserved.*

In Figure 17-19, assume that $R_1 = 1\bar{0}\ \Omega$, $R_2 = 15\ \Omega$, and $V = 3\bar{0}$ v. From Ohm's law, the current in R_1 is

$$I_1 = \frac{V}{R_1} = \frac{3\bar{0}\text{ v}}{1\bar{0}\ \Omega} = 3.0\text{ a}$$

The current in R_2 is

$$I_2 = \frac{V}{R_2} = \frac{3\bar{0}\text{ v}}{15\ \Omega} = 2.0\text{ a}$$

and

$$I = I_1 + I_2 = 5.0\text{ a}$$

According to Ohm's law, the current from **A** to **B**, which is I, is the quotient of the potential difference across **A-B** divided by the effective resistance between **A** and **B**. It is convenient to think of the effective resistance due to resistances in parallel as a single *equivalent resistance, R_{eq},* that, if substituted for the parallel resistances, would provide the same current in the circuit.

An equivalent resistance is shown in Figure 17-19.

$$I = \frac{V}{R_{eq}}$$

Then

$$R_{eq} = \frac{V}{I} = \frac{3\bar{0}\text{ v}}{5.0\text{ a}} = 6.0\ \Omega$$

The equivalent resistance for the parallel combination of

R_1 and R_2 is *less than* either resistance present. As the current approaching junction **A** *must be larger* than either branch current, clearly the equivalent resistance of the parallel circuit *must be smaller* than the resistance of any branch.

Since $$I = \frac{V}{R_{eq}}, \; I_1 = \frac{V}{R_1}, \; I_2 = \frac{V}{R_2}$$

and $$I = I_1 + I_2$$

then, $$\frac{V}{R_{eq}} = \frac{V}{R_1} + \frac{V}{R_2}$$

Dividing by V, we get

$$\frac{1}{R_{eq}} = \frac{1}{R_1} + \frac{1}{R_2} \qquad \text{(Equation 1)}$$

Equation 1 shows that the sum of the reciprocals of the resistances in parallel is equal to the reciprocal of their equivalent resistance. Then solving Equation 1 for R_{eq},

$$R_1 R_2 = R_{eq}(R_1 + R_2)$$

and $$R_{eq} = \frac{R_1 R_2}{R_1 + R_2}$$

$$R_{eq} = \frac{1\overline{0} \; \Omega \times 15 \; \Omega}{1\overline{0} \; \Omega + 15 \; \Omega} = \frac{150 \; \Omega^2}{25 \; \Omega}$$

$$R_{eq} = 6.0 \; \Omega$$

These observations are summarized by stating three cardinal rules for resistances in parallel.

1. The total current in a parallel circuit is equal to the sum of the currents in the separate branches.

$$I_T = I_1 + I_2 + I_3 + \text{etc.}$$

2. The potential difference across all branches of a parallel circuit must have the same magnitude.

$$V = V_1 = V_2 = V_3 = \text{etc.}$$

3. The reciprocal of the equivalent resistance is equal to the sum of the reciprocals of the separate resistances in parallel.

$$\frac{1}{R_{eq}} = \frac{1}{R_1} + \frac{1}{R_2} + \frac{1}{R_3} + \text{etc.}$$

17.10 Resistances in Simple Networks Practical circuits can be quite complex compared to the series and parallel circuits we have considered. There can be resistances

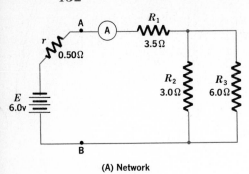

(A) Network

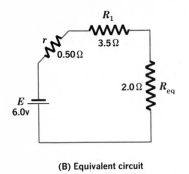

(B) Equivalent circuit

Figure 17-20. A simple network and its equivalent circuit.

in series, other resistances in parallel, and different sources of emf. Such complex circuits are commonly called *networks*.

A simple network is shown in Figure 17-20(A). There is a single source of emf and one resistor in series with the combination of two resistors in parallel. We will use this circuit to demonstrate the steps taken to simplify a network and reduce circuit problems to simple relationships.

We are given the emf and internal resistance of the battery and the resistance of each component resistor in the circuit. The first step is to find the total current in the circuit. Since the emf, E, is known, the total resistance, R_T, must also be known in order to calculate the total current, I_T. To find the total resistance, we must first reduce the parallel resistances to an *equivalent* value.

$$\frac{1}{R_{eq}} = \frac{1}{R_2} + \frac{1}{R_3} \quad \text{and} \quad R_{eq} = \frac{R_2 R_3}{R_2 + R_3}$$

$$R_{eq} = \frac{3.0 \ \Omega \times 6.0 \ \Omega}{3.0 \ \Omega + 6.0 \ \Omega}$$

$$R_{eq} = 2.0 \ \Omega, \text{ the equivalent resistance}$$

A simpler *equivalent* circuit can now be drawn showing this equivalent resistance in series with R_1 and r. See Figure 17-20(B). The total resistance is now readily seen to be 6.0 ohms.

$$R_T = R_1 + R_{eq} + r = 6.0 \ \Omega$$

And

$$I_T = \frac{E}{R_T} = \frac{6.0 \text{ v}}{6.0 \ \Omega} = 1.0 \text{ a}$$

The drop in potential across R_1 is

$$V_1 = I_T R_1 = 1.0 \text{ a} \times 3.5 \ \Omega$$

$$V_1 = 3.5 \text{ v}$$

Since R_2 and R_3 are in parallel, the same potential drop must appear across each. This is found by considering the total current in the equivalent resistance of the parallel segment.

$$V_{eq} = I_T R_{eq} = 1.0 \text{ a} \times 2.0 \ \Omega$$

$$V_{eq} = 2.0 \text{ v}$$

The potential difference across the terminals of the battery is the difference between the emf and the $I_T r$ drop.

$$V = E - I_T r = 6.0 \text{ v} - (1.0 \text{ a} \times 0.50 \text{ } \Omega)$$

$$V = 5.5 \text{ v}$$

This result is verified by the fact that V must equal $V_1 + V_{eq}$.

The current in R_2 may be found since the potential difference, V_{eq}, is known.

$$I_2 = \frac{V_{eq}}{R_2} = \frac{2.0 \text{ v}}{3.0 \text{ } \Omega}$$

$$I_2 = 0.67 \text{ a}$$

The current in R_3 is

$$I_3 = \frac{V_{eq}}{R_3} = \frac{2.0 \text{ v}}{6.0 \text{ } \Omega}$$

$$I_3 = 0.33 \text{ a}$$

This is verified by the fact that I_3 must equal $I_T - I_2$.

A more complex network is shown in Figure 17-21(A).

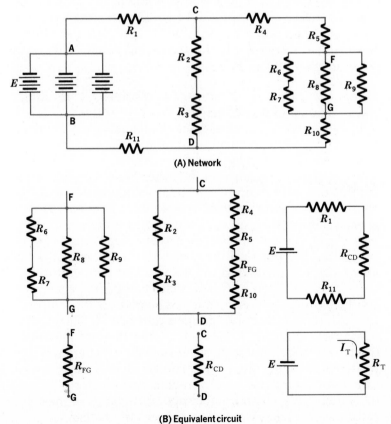

(A) Network

(B) Equivalent circuit

Figure 17-21. A complex network can be reduced stepwise to its simple equivalent circuit by applying the equivalent-circuit analysis techniques.

The battery is a series-parallel arrangement of similar cells, and so the magnitude of the emf will depend on the number of cells in *series*. The total current must be known. The approach is the same as that in our earlier example. First, reduce the segment **F-G** to its equivalent value. We will label this R_{FG}. Now R_{FG} is in series with R_4, R_5, and R_{10} to form one branch of the segment **C-D**. When the two branches of **C-D** are reduced to their equivalent, R_{CD}, we can consider R_1, R_{CD}, and R_{11} to be in series. This reduction provides R_T, and I_T may be calculated. See the diagrams in Figure 17-21(B).

Suppose we need to know the magnitude of current in R_7 and the drop in potential across this resistor. Now the current in R_{CD} must be I_T and the drop in potential across **C-D** must be $I_T R_{CD}$, or V_{CD}. The current in the branch of **C-D** containing R_{FG} is the quotient of V_{CD} divided by the sum $(R_4 + R_5 + R_{FG} + R_{10})$. This current (we will label it I_{FG}) is in R_{FG}, and the potential drop, V_{FG}, across R_{FG} is the product $I_{FG} R_{FG}$.

Since R_6 and R_7 are in series, the resistance of this branch of **F-G** is equal to their sum. The current in R_7 is V_{FG} divided by $(R_6 + R_7)$. With I_7 known, the drop in potential across R_7 is the product $I_7 R_7$. Observe that we have depended on Ohm's law and the six extensions of Ohm's law listed as cardinal rules for series and parallel circuits in Sections 17.8 and 17.9.

17.11 The Laws of Resistance Various factors affect the resistance of a conductor.

1. Temperature. The resistance of all substances changes to some degree with changes in temperature. *In the case of pure metals and most metallic alloys, the resistance increases with a rise in temperature.* On the other hand, carbon, semiconductors, and many electrolytic solutions, show a decrease in resistance as their temperature is raised. A few special alloys, such as constantan and manganin, show very slight changes in resistance over a large range of temperatures. For practical purposes, the resistance of these alloys can be considered independent of temperature.

Thermal agitation of the particles composing a metallic conductor increases as the metal is heated. As the temperature is lowered, thermal agitation diminishes. The electric resistance of metals is related to this thermal agitation; their resistance approaches zero as the temperature approaches absolute zero (0 °K). The graph of resistance as a function of temperature of a metallic conductor is essentially linear over the normal temperature range of the solid state of a metal. Thus when the electric resistance of a conductor is stated, the temperature of the material at which this resistance applies should be indicated. See the graph in Figure 17-22.

2. Length. Resistances in series are added to give the total resistance of the series circuit. From this fact we can conclude that 10 meters of a certain conductor should have a resistance 10 times that of 1 meter of the same conductor. Experiments prove this relationship to be true: *The resistance of a uniform conductor is directly proportional to the length of the conductor, or*

$$R \propto l$$

where l is the length of the conductor.

3. Cross-sectional area. If three *equal* resistances are connected in parallel in an electric circuit, the equivalent resistance is one-third the resistance of any one branch. The equal currents in the three branches add to give the total current in the circuit.

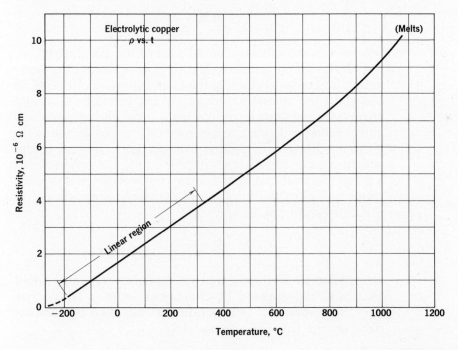

Figure 17-22. The resistivity of copper as a function of temperature.

Assuming the three resistances to be three similar wires 1 meter in length and each having a cross-sectional area of 0.1 cm², we can think of the equivalent resistance as a similar wire of the same length but having a cross-sectional area of 0.3 cm². This deduction is reasonable since a wire carrying a direct current of constant magnitude has a constant *current density* throughout its cross-sectional area. If the parallel resistances are designated R_1, R_2, and R_3, then

$$\frac{1}{R_{eq}} = \frac{1}{R_1} + \frac{1}{R_2} + \frac{1}{R_3}$$

But

$$R_1 = R_2 = R_3$$

So

$$\frac{1}{R_{eq}} = \frac{3}{R_1}$$

And

$$R_1 = 3R_{eq}$$

We can conclude that the resistance of the large wire, R_{eq}, is one-third the resistance of the small wire, R_1, and experiments prove this conclusion to be true. *The resistance of a uniform conductor is inversely proportional to its cross-sectional area,* or

$$R \propto \frac{1}{A}$$

where A is the cross-sectional area of the conductor.

4. *Nature of the materials.* A copper wire having the same length and cross-sectional area as an iron wire offers about one-sixth the resistance to an electric current. A similar silver wire presents even less resistance. *The resistance of a given conductor depends on the material of which it is made.*

Taking these four factors into account, several useful conclusions become apparent. For a uniform conductor of a given material and temperature, the resistance is directly proportional to the length and inversely proportional to the cross-sectional area.

$$R \propto \frac{l}{A}$$

Suppose we introduce a proportionality constant, called *resistivity,* represented by the Greek letter ρ (rho). The laws of *resistance* for wire conductors can then be summarized by the following expression:

$$R = \rho \frac{l}{A}$$

Resistivity depends on the material composing the wire and its temperature, as shown in Figure 17-22. If R is expressed in ohms, l in centimeters, and A in centimeters, then ρ has the dimensions *ohm centimeter* (Ω cm).

$$\rho = \frac{RA}{l}$$

$$\rho = \frac{\Omega \ cm^2}{cm} = \Omega \ cm$$

Resistivity, ρ, is equal to the resistance of a wire 1 cm long and having a uniform cross-sectional area of 1 cm². (The resistivities of various materials are listed in Appendix B, Table 20. The wire gauge scale and the dimensions of wire by gauge number are given in Table 21.) If the resistivity of a material is known, the resistance of any other wire composed of the same material can be calculated. See the following example.

EXAMPLE What is the resistance of a copper wire $2\overline{0}$ m long and 0.81 mm in diameter at 20 °C? The resistivity of annealed copper at 20 °C is 1.72×10^{-6} Ω cm.

SOLUTION We are given the resistivity of copper at 20 °C, $\rho = 1.72 \times 10^{-6}$ Ω cm. The length of wire can be converted to centimeters and expressed as $l = 2.0 \times 10^3$ cm. The diameter of the wire in millimeters can be converted to 0.081 cm. The area, A, of a circle with diameter

$$d = \frac{\pi d^2}{4}.$$

$$R = \rho \frac{l}{A}$$

$$R = 1.72 \times 10^{-6} \ \Omega \ cm \times \frac{2.0 \times 10^3 \ cm}{\dfrac{\pi(0.081 \ cm)^2}{4}}$$

$$R = \frac{1.72 \times 10^{-6} \ \Omega \ cm \times 4 \times 2.0 \times 10^3 \ cm}{\pi \times 6.56 \times 10^{-3} \ cm^2}$$

$R = 0.67$ Ω, the resistance of the wire

PRACTICE PROBLEMS 1. The resistivity of German silver at 20 °C is given in Appendix B, Table 20 as 33×10^{-6} Ω cm. Determine the resistance of a German silver wire 0.40 mm in diameter and $1\overline{0}$ m long at $2\overline{0}$ °C. *Ans.* 26 Ω

2. A resistance wire 15.0 m in length and 0.361 mm in diameter is found experimentally to have a resistance of 72.1 Ω at $2\overline{0}$ °C. (a) Determine the resistivity of the metal. (b) By consulting Appendix B, Table 20, identify the material under test. *Ans.* (a) 49.2×10^{-6} Ω cm; (b) nickel silver

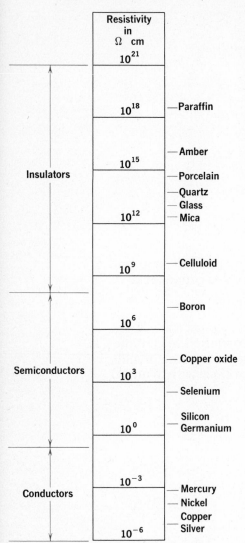

Figure 17-23. A resistivity spectrum. Observe the enormous range of resistivities between the best conductor and the best insulator.

17.12 Range of Resistivities The enormous range of resistivities of common materials is shown in Figure 17-23. The resistance of the best insulator can be greater than that of the best conductor by a factor of approximately 10^{25}. Insulators and semiconductors can decrease in resistivity with increasing temperature and with increasing potential gradient. We can interpret this latter behavior to mean that these materials do not follow Ohm's law.

A small group of *semiconductors* has assumed considerable importance because of unusual variations in resistance with changes in temperature and potential gradient. Copper oxide *rectifiers* will pass a current when a potential difference is applied in one direction but not in the opposite direction. Silicon and germanium *transistors* are substituted for vacuum tubes in many electronic circuits.

At extremely low temperatures, some metals exhibit the unusual property of zero resistance. The production of very low temperatures and the study of the properties of materials at low temperatures is called *cryogenics*. Cryogenic temperatures are generally considered to be those below the boiling point of liquid oxygen, which is 80 °K (−193 °C). Ultra-low temperature research began with the liquefaction of helium in 1908 by the Dutch physicist Heike Kamerlingh Onnes (1853–1926). The boiling point of liquid helium is about 4 °K (−269 °C).

We have learned that the resistivity of a conductor is a function of its temperature. We would expect that the resistivity would become less and less as the temperature of a conductor drops, until the conductor offers no resistance at all to the passage of electrons. In 1911 Kamerlingh Onnes found that resistivity did indeed continue to drop with falling temperature. However, instead of gradually approaching zero resistivity, some materials had a specific temperature at which the resistivity dropped suddenly to zero. This temperature is called the *transition temperature*. *The condition of zero resistivity below the transition temperature of a material is called* **superconductivity.** Temperatures at which some materials become superconductive are shown in Figure 17-24.

The region of very low temperatures at which superconductivity prevails poses a severe limitation for the utilization of superconductors. Obviously superconductors offer great promise for such applications as power transmission, electric energy storage, and superelectromagnets that operate without power loss. However, the problem of keeping a transmission line below the transition temperature is a formidable one.

17.13 Measuring Resistance There are several methods of measuring the resistance of a circuit component. Two methods utilizing apparatus normally available in the laboratory will be discussed here. These are the *voltmeter-ammeter* method and the *Wheatstone bridge* method.

1. *The voltmeter-ammeter method.* Ohm's law applies equally well to any part of an electric circuit that does not contain a source of emf and to an entire circuit to which an emf is applied. For a part of a circuit

$$V = IR \quad \text{and} \quad R = \frac{V}{I}$$

If the potential difference, V, across a resistance component is measured with a voltmeter and the current, I, in the resistance is measured with an ammeter, the resistance, R, can be computed.

Suppose we wish to measure the resistance of R_1 in the circuit shown in Figure 17-25. The ammeter is connected in series with the resistance R_1 to determine the current in R_1. The voltmeter is connected in *parallel* with R_1 to determine the potential difference across R_1.

If the resistance of R_1 is very low, it can have an excessive loading effect on the source of emf, causing an excessive current in the circuit. This difficulty can be avoided by placing a variable resistance, or *rheostat*, in series with R_1. The rheostat provides additional series resistance and limits the current in the circuit. The rheostat can be adjusted to provide an appropriate magnitude of current in the circuit. See Figure 17-26. Observe that the voltmeter is connected across that part of the circuit for which resistance is to be determined.

The voltmeter-ammeter method of measuring resistance is convenient but not particularly precise. A voltmeter is a high-resistance instrument. However, as it is connected in parallel with the resistance to be measured, there is a small current in it. The ammeter reading is actually the sum of the currents in these two parallel branches. Precision is limited by the extent to which the resistance of R_1 exceeds the equivalent resistance of R_1 and the voltmeter in parallel.

Of course the ammeter can be placed in the R_1 branch in series with R_1. In this location the ammeter reading is the current in R_1. However, the voltmeter reading is now the sum of the voltage drop across the ammeter and R_1 in series.

2. *The Wheatstone bridge method.* A precise means of determining resistance utilizes a simple *bridge* circuit known

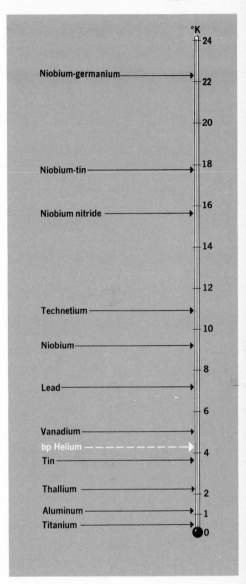

Figure 17-24. Kelvin temperatures at which various materials become superconductive.

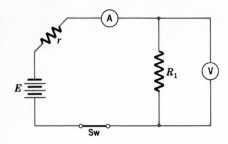

Figure 17-25. Determining resistance by the voltmeter-ammeter method.

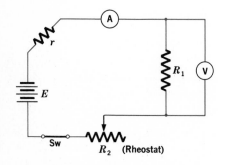

Figure 17-26. A rheostat provides a means of limiting the current in a circuit.

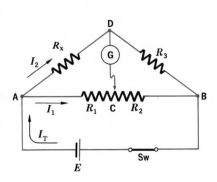

Figure 17-27. The circuit diagram of the Wheatstone bridge set up to measure the resistance R_x.

as the *Wheatstone bridge*. It consists of a source of emf, a galvanometer, and a network of four resistors. If three are of known resistance, the resistance of the fourth can be calculated.

The circuit diagram of a Wheatstone bridge is shown in Figure 17-27, in which an unknown resistance, R_x, may be balanced against known resistances R_1, R_2, and R_3. A galvanometer is *bridged* across the parallel branches **ADB** and **ACB.** By adjusting R_1, R_2, and R_3, the bridge circuit can be balanced and the galvanometer will show zero current.

When the bridge is balanced in this manner, there can be no potential difference between **C** and **D**, since the galvanometer indicates zero current. Thus

$$V_{AD} = V_{AC}$$

and

$$V_{DB} = V_{CB}$$

The current in the branch **ACB** is I_1, and the current in **ADB** is I_2. Then

$$I_2 R_x = I_1 R_1$$

and

$$I_2 R_3 = I_1 R_2$$

Dividing one expression by the other, we have

$$\frac{R_x}{R_3} = \frac{R_1}{R_2}$$

or

$$R_x = R_3 \frac{R_1}{R_2}$$

In the laboratory form of the Wheatstone bridge, resistances R_1 and R_2 usually consist of a uniform resistance wire, such as constantan, mounted on a meter stick. See Figure 17-28. The length of wire comprising R_1 and the length comprising R_2 are determined by the position of the contact **C.** If l_1 is the length of wire forming R_1 and l_2 is the length of wire forming R_2, it is apparent that

$$\frac{l_1}{l_2} = \frac{R_1}{R_2}$$

We do not need to know the resistances of R_1 and R_2, since the lengths of wire l_1 and l_2 can be taken directly from the meter stick. Our Wheatstone bridge expression

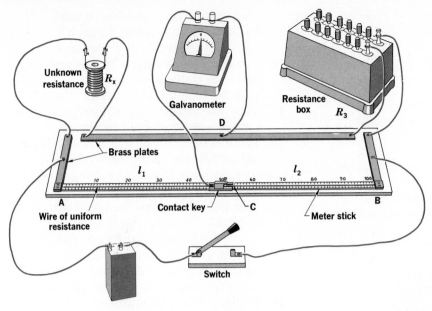

Figure 17-28. Pictorial diagram of the Wheatstone bridge circuit for measuring resistance R_x shown in Figure 17-27.

then becomes
$$R_x = R_3 \frac{l_1}{l_2}$$

A precision resistance box of the type shown in Figure 17-29 is often used to provide a suitable known resistance R_3 in the bridge circuit. The internal arrangement is shown in the diagram. Resistance coils are wound with manganin or constantan wire to minimize changes in resistance with temperature. The free ends of each resistance coil are soldered to heavy brass blocks that can be electrically connected by inserting brass plugs between them. Each plug in place provides a low-resistance path across a resistance coil and thus effectively removes the coil from the circuit. Each plug removed places a coil in the circuit. The resistance between the terminals of the box is the sum of the resistances of all coils whose plugs are removed.

Figure 17-29. A plug-type resistance box that provides resistance between 0.1 ohm and 111.0 ohms in 0.1-ohm steps. The total resistance across the terminals will depend on which plugs are removed.

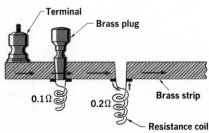

Questions
GROUP A

1. State Ohm's law (a) verbally, (b) mathematically.
2. What is the mathematical expression for Ohm's law as it applies to a part of a circuit that does not contain a source of emf?
3. What is the identifying characteristic of a series circuit?
4. State the three cardinal rules for circuits in which resistances are in series.
5. Define a parallel circuit.
6. What change occurs in the total resistance of a circuit as additional resistances are added (a) in series, (b) in parallel?
7. State the three cardinal rules for circuits in which resistances are connected in parallel.
8. (a) What is the usual relationship between resistance and temperature of metallic conductors? (b) Is this true of all conducting materials in general?
9. How does the resistance of a conductor vary with its length?
10. What is the relationship between the resistance of a conductor and its cross-sectional area?
11. Define the resistivity constant, rho, that appears in the mathematical expression for the laws of resistance.
12. Demonstrate that resistivity, rho, has the dimensions "ohm-centimeter."

GROUP B
13. Explain why the fall of potential that appears across the internal resistance of a source of emf when a current is in the circuit is always opposite in sign to the emf of the source.
14. What condition must be assumed in order to determine the emf of a battery by placing a voltmeter across it?
15. Why are lamps in a house-lighting circuit not connected in series?
16. What would happen if a 6-volt lamp and a 120-volt lamp were connected in parallel across a 120-volt circuit?
17. Draw a diagram of an electric circuit that would be classified as a simple network. Include the internal resistance of the source of emf. Assign values to all resistances and to the source of emf, and be prepared to analyze your network for the class.
18. Explain how the voltmeter-ammeter method is used for measuring resistance.
19. Why is the voltmeter-ammeter method not a particularly precise way to measure resistance?
20. In the laboratory form of the Wheatstone bridge, in Figure 17-28, resistors R_1 and R_2 are lengths of wire. Why can we use the lengths of wire rather than their actual resistances in determining the unknown resistance once the bridge is balanced?
21. Lamps A, B, and C have equal resistances and have been connected in series-parallel across an appropriate source of emf, A being in series with the parallel combination of B and C. (a) Draw the circuit diagram. (b) Discuss the relative brightness of the three lamps. (c) If C burns out, what change in relative brightness occurs in the remaining lamps? Explain.

Problems
GROUP A
(Refer to Appendix B, Tables 20 and 21)
1. A resistance of 18 ohms is connected across a 4.5-volt battery. What is the current in the circuit?
2. The load across a 12-volt battery consists of a series combination of three resistances, R_1, R_2, and R_3, that are 15 ohms, 21 ohms, and 24 ohms respectively. (a) Draw the circuit diagram. (b) What is the total resistance

of the load? (c) What is the magnitude of the circuit current?

3. What is the potential difference across each of the three resistances of Problem 2?

4. A small lamp is designed to draw $30\overline{0}$ ma in a 6.0-v circuit. What is the resistance of the lamp filament?

5. Resistances R_1 of 2 ohms and R_2 of 4 ohms are connected in series across a combination of 1.5-volt dry cells. An ammeter in the circuit reads 0.5 a. (a) What potential difference is applied across the circuit by the battery? (b) If the continuous current of each cell is limited to 0.25 a, how many cells are used and how are they arranged? (c) Draw the circuit diagram.

6. The resistance of an electric lamp filament is $23\overline{0}$ ohms. The lamp is switched on when the line voltage is 115 volts. What current is in the lamp circuit?

7. If the line voltage of Problem 6 rises to $12\overline{0}$ volts, what current is in the lamp circuit?

8. What would be the filament resistance of a lamp designed to draw the same current as the lamp in Problem 6 but in a 230-volt circuit?

9. Prove algebraically that the equivalent resistance of two resistances, R_1 and R_2, connected in parallel is

$$R_{eq} = \frac{R_1 R_2}{R_1 + R_2}$$

10. Two resistances, one 12 ohms and the other 18 ohms, are connected in parallel. What is the equivalent resistance of the parallel combination?

11. Solve the following expression algebraically for R_{eq}:

$$\frac{1}{R_{eq}} = \frac{1}{R_1} + \frac{1}{R_2} + \frac{1}{R_3}$$

12. Three resistances of 12 ohms each are connected in parallel. What is the equivalent resistance?

13. (a) What potential difference must be applied across the parallel combination of resistances in Problem 12 to produce a total current of 1.5 amperes? (b) What is the current in each branch?

14. A 3.0-ohm, a 5.0-ohm, and an 8.0-ohm resistance are connected in parallel. What is the equivalent resistance?

15. A 4.0-ohm resistance R_1, and a 6.0-ohm resistance R_2, are connected in parallel across a 6.0-volt battery. (a) Draw the circuit diagram. (b) What is the total current in the circuit? (c) What is the current in R_1? (d) What is the current in R_2?

16. A 24-volt lamp has a resistance of 8.0 ohms. What resistance must be placed in series with it if it is to be used on a 117-volt line?

17. At 20 °C, 100 m of No. 18 gauge copper wire has a resistance of 2.059 ohms. What is the resistance of $50\overline{0}$ m of this wire?

18. The diameter of the copper wire in Problem 17 is 1.024 mm. What is the resistance of $10\overline{0}$ m of No. 30 gauge copper wire that has a diameter of 0.2546 mm?

19. What is the resistance of 125 m of No. 20 gauge aluminum wire at 20 °C?

20. No. 24 gauge German silver wire has a diameter of 0.511 mm at 20 °C. How many centimeters are needed to make a resistance spool of $10\overline{0}$ ohms?

GROUP B

21. Three cells of a storage battery, each with a resistance of 0.0100 ohm and an emf of 2.20 volts, are connected in series with a load resistance of 3.27 ohms. (a) Draw the circuit diagram. (b) Determine the current in the circuit. (c) What is the potential difference across the load?

22. Two dry cells connected in series provide an emf of 3.1 volts. When a load

of 5.8 ohms is connected across the battery, an ammeter in the external circuit reads 0.50 ampere. Find the internal resistance of each cell.

23. A load connected across a 12.0-volt battery consists of the resistance R_1, 40.0 ohms, in series with the parallel combination of R_2 and R_3, 30.0 ohms and 60.0 ohms respectively. (a) Draw the circuit diagram. (b) What is the total current in the circuit? (c) What is the potential difference across R_1? (d) What is the current in the R_2 branch?

24. A 6.0-ohm, a 9.0-ohm, and an 18.0-ohm resistance are connected in parallel and the combination is connected across a battery having an internal resistance of 0.10 ohm and an emf of 6.2 volts. (a) Draw the circuit diagram and the equivalent circuit. (b) Determine the total current in the circuit. (c) What is the potential difference across the load? (d) What is the current in each of the three load resistances?

25. Resistors R_1, R_2, and R_3 have resistances of 15.0 ohms, 9.0 ohms, and 8.0 ohms respectively. R_1 and R_2 are in series and the combination is in parallel with R_3 to form the load across two 6-volt batteries connected in parallel. (a) Draw the circuit diagram. (b) Determine the total current in the circuit. (c) What is the current in the R_3 branch? (d) What is the potential drop across R_2?

26. A battery with an internal resistance of 1.5 ohms is connected across a load consisting of two resistances, 3.0 ohms and 3.5 ohms, in series. The

potential difference across the 3.0-ohm resistance is 9.0 volts. What is the emf of the battery?

27. Two resistances, R_1 and R_2, of 12.0 ohms and 6.00 ohms are connected in parallel, and this combination is connected in series with a 6.25-ohm resistance, R_3, and a battery that has an internal resistance of 0.250 ohm. The current in R_2, the 6.00-ohm resistance, is 800 milliamperes. (a) Draw the circuit diagram. (b) Determine the emf of the battery.

28. The resistance of a uniform copper wire 50.0 m long and 1.15 mm in diameter is 0.830 Ω at 20 °C. What is the resistivity of the copper at this temperature?

29. In determining the resistance of a component of an electric circuit by the voltmeter-ammeter method, the voltmeter reads 1.4 volts and the ammeter reads 0.28 ampere. What is the resistance of the component?

30. (a) Reduce the accompanying circuit to its series equivalent circuit. (b) Determine the potential difference across the 1-Ω resistance.

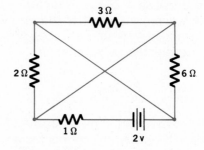

Figure 17-30.

SUMMARY

Electric current is the rate of flow of charge in a circuit. The charge carriers in metallic conductors are electrons. The current in these conductors is an electron current and its directional sense is from negative to positive. A continuous current requires a source of emf. There are several sources of emf and continuous current, each of which transforms energy of some form to electric energy.

The dry cell is a common source of emf that supplies direct current to a circuit. It is a primary cell that is a form of voltaic cell in which chemical energy is converted to electric energy. Electric cells can be combined in series, in parallel, or in series-parallel to form batteries of suitable characteristics for a given circuit.

All circuit components offer some resistance to the current in a closed circuit. The relationship between resistance, current, and potential difference is expressed by Ohm's law. Ohm's law applies to an entire circuit or to any part of a circuit.

Resistances can be connected in series, in parallel, or in series-parallel combinations called networks. Networks can be reduced to simple equivalent circuits in which those components in parallel are reduced to their series equivalents. Equivalent circuit problems are solved by the application of Ohm's law for series circuits.

The resistance of a metal conductor depends on the temperature, length, cross-sectional area, and identity of the metal. From a knowledge of these factors, the laws of resistance are formulated and a property of materials called resistivity is defined. Resistivity depends on only the identity of the material and its temperature.

The resistance of a circuit element can be measured directly by the voltmeter-ammeter method or by the Wheatstone bridge method. The voltmeter-ammeter method is rapid, and the Wheatstone bridge method is more precise.

VOCABULARY

anode
battery
cathode
closed circuit
cryogenics
drift velocity
electric current
electrode
electrolyte
emf

equivalent resistance
external resistance
galvanometer
internal resistance
load resistor
open circuit
parallel circuit
Peltier effect
photoelectric effect
piezoelectric effect

primary cell
resistance
resistance network
resistivity
Seebeck effect
semiconductor
series circuit
thermocouple
Wheatstone bridge

Highly technical calibration stations are designed to fix, check, or correct the graduations of a measuring instrument.

Ultra High Vacuum Calibration System

X-Band Microwave Calibration Station

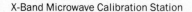

Automated Calibration System for
Precision AC and DC Instruments

Automated Pressure Calibration Station

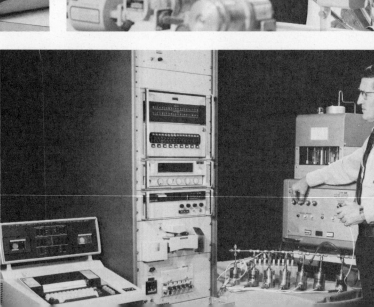

446

Metrology is the study of measurement. In the field of physics it is important to recognize that unless a property can be measured and compared with some kind of standard, it is of limited use to the scientist. Scientists usually use the term "data" to refer to the numbers, units, and other information recorded when measurements are made. The validity of every measurement is limited by the method of observation of the person who made it and by the precision of the apparatus used. In this sense, there is no such thing as absolute certainty in measurement, and yet without skillful measurement there can be no basis for science. Technological advances in making more precise measuring apparatus provide us with the necessary tools to advance in research and development vital to the future progress of our society.

Metrology

Metrology in outer space. The Viking spacecraft consists of an Orbiter and Lander. The Lander is an automated scientific laboratory designed to obtain data about the structure, surface, and atmosphere of Mars. Its data is radioed to the Orbiter from the surface of Mars through an antenna located on one of the Orbiter's four solar panels.

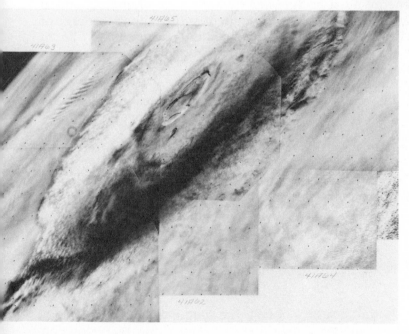

The great Martian volcano, Olympus Mons, was photographed by the Viking 1 Orbiter from a distance of 8000 kilometers. The mountain is seen in midmorning wreathed in clouds. The clouds are thought to be composed principally of water ice condensed from the atmosphere as it cools while moving up the slopes of the volcano.

447

18

HEATING AND CHEMICAL EFFECTS

Thomas Edison was a prolific inventor, with 1033 patents to his credit over a period of fifty years. His inventions include the phonograph (1877), the incandescent lamp (1879), and the motion picture camera (1891). A favorite expression of Edison's was: "There is no substitute for hard work."

In this chapter you will gain an understanding of:

▶ the transfer of energy in an electric circuit

▶ the relationship between the heat energy developed in an electric circuit and the resistance of the circuit

▶ the condition under which maximum power is dissipated in the external load connected across a source of emf

▶ how the cost of electric energy used in your home is determined

▶ the distinction between electrolytic and electrochemical cells

▶ electroplating processes

▶ two important relationships in electrochemistry, Faraday's laws of electrolysis

HEATING EFFECTS

18.1 Energy of an Electric Current   The relationships among objects, time, and space are basic concerns of physics. Work, power, and energy are important physical concepts involved in these relationships. We have described energy as the ability to do work and have measured it in units of work and heat. To maintain an electric current, energy must be supplied by the source of emf. Chemical energy, which can be thought of as the potential energy of chemical bonds, is transformed into electric energy in an electrochemical cell. Thus a charge acquires energy within the cell or other source of emf and expends this energy in the electric circuit.

In Section 16.10 we recognized that one joule of work is done when one coulomb of charge is moved through a potential difference of one volt.

$$W = qV$$

One coulomb of charge transferred in one second constitutes one ampere of current.

$$I = \frac{q}{t} \quad \text{and} \quad q = It$$

Therefore

$$W = VIt \qquad \text{(Equation 1)}$$

Within a source of emf, the charge is moved in opposition to the potential difference, and work, W, is done on the electrons *by the source of emf.* As the charge moves through the external circuit in response to the potential difference across the circuit, work, *W*, is done *by the electrons* on the components of the circuit. It is this electric energy, expended in the external circuit, that is available to perform useful work.

We can observe the work done by an electric current in a variety of ways. If a lamp is in the circuit, part of the work appears as light; if an electric motor is in the circuit, part of the work appears as rotary motion; if an electroplating cell is in the circuit, part of the work promotes chemical action in the cell. In all cases, *part* of the work, *W*, appears as *heat* because of the inherent resistance of the circuit components. If an ordinary resistor comprises the circuit, *all* the work goes to the production of heat since no mechanical work or chemical work is done. See Figure 18-2.

18.2 Energy Conversion in Resistance Suppose short lengths of No. 30 gauge copper wire and German silver wire are connected in series and placed in the circuit shown in Figure 18-3. When the switch is closed, the German silver wire may become red hot and melt. The copper wire will become warm but is less likely to melt. The melting point of German silver is about 30 C° higher than that of copper, but its resistivity is nearly 20 times higher than that of copper. Since the two wires were in series, the same magnitude of current was in each. We may infer that *a greater quantity of electric energy was converted to heat in the wire having the higher resistance.*

Figure 18-1. The electric energy expended in the lamp filament by the electron-charge carriers heats the filament to incandescence, and hence the name "incandescent lamp."

Figure 18-2. Work is done when a charge is moved through a potential difference.

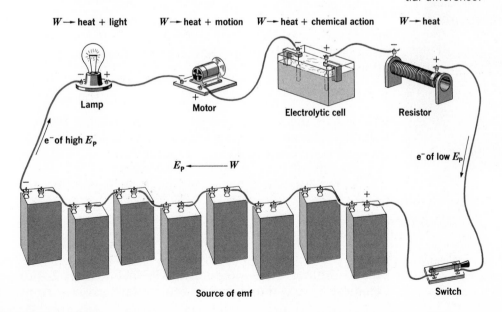

$W \rightarrow$ heat + light $W \rightarrow$ heat + motion $W \rightarrow$ heat + chemical action $W \rightarrow$ heat

Lamp   Motor   Electrolytic cell   Resistor

e^- of high E_P

$E_P \longleftarrow W$

e^- of low E_P

Source of emf   Switch

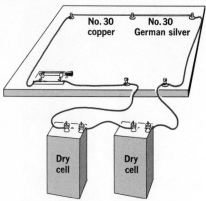

Figure 18-3. A circuit for studying the heating effect of an electric current.

Now suppose a single piece of No. 30 copper wire is connected in the circuit in place of the copper–German silver link. When the switch is closed, the copper wire may warm to a red heat and even melt. Because the resistance of this copper wire is less than that of the copper–German silver link, the current in the circuit is higher. We may infer that *the quantity of electric energy converted to heat in the wire is greater when the current is increased.* We can reason also that *the longer the switch is closed* (assuming the wire does not melt), *the greater will be the quantity of electric energy converted to heat.*

From Ohm's law, the potential difference across the resistance of an external circuit is the product of the current in the circuit and the resistance of the circuit.

$$V = IR$$

Substituting this product for V in Equation 1, we have an expression for the energy transferred to the resistance of a circuit.

$$W = I^2Rt \qquad \text{(Equation 2)}$$

18.3 Joule's Law James Prescott Joule, the English physicist who first studied the relationship between heat and work, also pioneered the quantitative study of the heating effect of an electric current. From the results of his experiments he formulated the generalization known as *Joule's law: The heat developed in a conductor is directly proportional to the resistance of the conductor, the square of the current, and the time the current is maintained.*

Since all of the work done by a current in a resistance appears as heat, the electric energy expended in the resistance is directly proportional to the heat energy that appears.

$$W = JQ \qquad \text{(Equation 3)}$$

where W is the electric energy in joules, J is the proportionality constant we have known previously as the mechanical equivalent of heat (4.19 joules/calorie), and Q is the quantity of heat given up in calories.

Solving Equation 3 for Q we have an expression for Joule's law.

$$Q = \frac{W}{J} \qquad \text{(Equation 4)}$$

By substituting the expression for W in Equation 2 into Equation 4, a useful form of Joule's law evolves from

Figure 18-4. Electric energy is converted to heat in an electric furnace for the production of special grades of steel.

which the quantity of heat developed in the resistance of an electric circuit can be calculated directly.

Electric power is dissipated in a resistance as heat.

$$Q = \frac{I^2Rt}{J} \qquad \text{(Equation 5)}$$

Observe that the quantity of heat Q is expressed in kilocalories by introducing the conversion factor $kcal/10^3$ cal.

$$\frac{I^2Rt}{J} = \frac{(a \times \Omega)(a \times s)}{j/cal} \times \frac{kcal}{cal}$$

$$= \frac{v \times c \times kcal}{j} = \frac{j \times kcal}{j}$$

$$= kcal$$

The following example illustrates an application of Joule's law.

EXAMPLE   The heating element in a percolator has a resistance of 22.0 Ω and is designed for use across a 117-v circuit. How long will it take to heat 1.00 kg of water at 20.0 °C to the boiling point, 100.0 °C, assuming no loss of heat to surroundings?

SOLUTION   The problem is concerned basically with the heating effect of an electric current—Joule's law. However, the specific heat of water is a factor in determining the time of the circuit operation. The basic equation is

$$Q = \frac{I^2Rt}{J}$$

The quantity of heat, Q, required to warm a given mass of water through a change of temperature, ΔT, is

$$Q = mc\,\Delta T$$

From Ohm's law: $$I = \frac{V}{R}$$

Substituting for Q and I in the basic equation,

$$mc\,\Delta T = \frac{V^2Rt}{R^2J} = \frac{V^2t}{RJ}$$

We now have an expression that makes use of the quantities given in the problem. Since we are solving the problem for time, t, our working equation is

$$t = \frac{mc\,\Delta TRJ}{V^2}$$

Substituting and solving:

$$t = \frac{1.00 \text{ kg} \times 1 \text{ kcal/kg C}° \times 80.0 \text{ C}° \times 22.0 \text{ } \Omega \times 4.19 \text{ j/cal} \times 10^3 \text{ cal/kcal}}{(117 \text{ v})^2}$$

$$t = 539 \frac{\Omega j}{v^2} = 539 \text{ s}$$

Dimensional analysis:

$$\frac{\Omega j}{v^2} = \frac{v}{a} \times \frac{v \, c}{v^2} = \frac{c}{a} = \frac{a \, s}{a} = s$$

PRACTICE PROBLEMS   **1.** The heating element of an electric tea kettle with a resistance of 24.0 Ω is connected across a 12$\bar{0}$-v circuit. How long will it take to heat 50$\bar{0}$ g of water at 10.0 °C to its boiling point, 100.0 °C, assuming no loss of heat to surroundings?   *Ans.* 314 s

2. An electric hot plate with a resistance of 28.8 Ω is connected across a 12$\bar{0}$-v circuit. How many minutes are required to heat 75$\bar{0}$ g of water at 4.5 °C to its boiling point, assuming no heat loss to surroundings? *Ans.* 10.0 min

---

The energy delivered to a circuit element is the product of power and time.

18.4 Power in Electric Circuits   Power is defined as the rate of doing work. The symbol for power is P and the unit of power is the watt (w). The watt is a *rate* of one joule of work per second. In an electric circuit, it is convenient to think of power as *the rate at which electric energy is delivered to the circuit.*

$$P = \frac{W}{t} \qquad \text{(Equation 6)}$$

Recalling Equation 1 of Section 18.1, $W = VIt$, let us solve this expression for W/t.

$$\frac{W}{t} = VI \qquad \text{(Equation 7)}$$

Combining Equations 6 and 7 yields a basic power equation for electric circuits.

$$P = VI \qquad \text{(Equation 8)}$$

If I is in amperes and V is in volts, then P is in watts.

$$P = VI = \frac{\text{joule}}{\text{coulomb}} \times \frac{\text{coulomb}}{\text{second}}$$

$$P = \frac{\text{joule}}{\text{second}} = \text{watt}$$

By Ohm's law, the potential difference across a load resistance in a circuit is equal to the product IR_L. Substituting for V in Equation 8,

$$P_L = I \times IR_L$$

or

$$P_L = I^2 R_L \qquad \text{(Equation 9)}$$

This shows us that *the power expended by a current in a resistance is proportional to the square of the current in the resistance.*

Similarly, the power dissipated in the internal resistance of a source of emf is

$$P_r = I^2 r \qquad \text{(Equation 10)}$$

The total power consumed in a circuit must be the sum of the power dissipated within the source of emf and the power expended in the load of the external circuit.

$$P_T = I_T^2 R_T = I_T^2 (R_L + r)$$

Recall that

$$E = I_T(R_L + r)$$

Therefore

$$P_T = I_T E \qquad \text{(Equation 11)}$$

There are instances in which the current in a circuit is unknown or of no interest. Power can then be expressed in terms of voltage and resistance since, by Ohm's law,

$$I_T = \frac{E}{R_T}$$

Substituting for I_T in Equation 11, we get

$$P_T = \frac{E^2}{R_T} \qquad \text{(Equation 12)}$$

and for the external circuit,

$$P_L = \frac{V^2}{R_L} \qquad \text{(Equation 13)}$$

18.5 Maximum Transfer of Power There are some electric-circuit applications in which the transfer of maximum power to the load is important. One example is the circuit of the starter motor of an automobile engine. To operate the starter, a large amount of electric energy must be supplied by a battery during a short interval of time. Other, less familiar examples would be the transfer of

Can you derive Equation 9 from Equations 2 and 6 in just three steps?

Power dissipated in the internal resistance of a source of emf is wasted as heat.

audio-frequency power from the output stage of a radio to the speaker and the transfer of radio-frequency power from the transmission line to the antenna.

Consider the circuits of Figure 18-5 in order to determine the circumstances under which maximum power from a given source of emf is expended in the load. Observe that the same source of emf is used in all three circuits and loads greater than, less than, and equal to the internal resistance of the source are supplied. In each circuit I_T, P_L, P_r, and P_T can be determined by using equations derived in Section 18.4. For comparison these values are listed below.

Circuit **A**	Circuit **B**	Circuit **C**
$I_T = 2.0$ a	5.0 a	3.0 a
$P_L = 4.0$ w	2.5 w	4.5 w
$P_r = 2.0$ w	12.5 w	4.5 w
$P_T = 6.0$ w	15.0 w	9.0 w

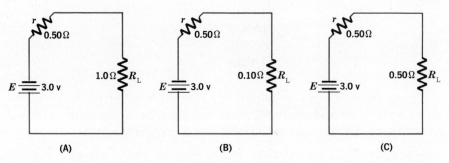

(A)   (B)   (C)

Figure 18-5. In which circuit is the most power expended in the load?

From such calculations it can be shown that the power expended in the load of circuit **C**, 4.5 w, is the maximum that can be transferred from this source of emf. *Maximum power is transferred to the load when the load resistance is equal to the internal resistance of the source of emf.* The source and the load are said to be *matched*. In circuit **B**, the load resistance is less than the internal resistance of the source and most of the power generated is dissipated as heat in the source itself. This power is wasted. In circuit **A**, the load resistance is greater than the internal resistance of the source and little power is lost in the internal resistance of the source. However, the power expended in the load is not as high as in circuit **C**.

18.6 Purchasing Electric Energy An operating circuit that requires 100 watts of power converts 100 joules of

energy per second. Obviously, the longer this circuit continues to operate, the larger will be the total quantity of
energy transformed. Since

$$\text{power} = \frac{\text{energy}}{\text{time}}$$

then

$$\text{energy} = \text{power} \times \text{time}$$

In an electric circuit the current, potential difference,
and resistance are measured in amperes, volts, and ohms
respectively. If time is measured in seconds, then power is
expressed in watts. Energy can thus be expressed in *watt-
seconds* (w s). The watt-second is inconveniently small for
electric energy sold on a commercial scale. For example, a
100-watt lamp operated for 1 hour transforms 3.6×10^5
watt-seconds of energy. Dividing by 3.6×10^3 seconds/
hour, this quantity is converted to 1×10^2 watt-hour
(w hr). Dividing again by 10^3 watts/kilowatt, the energy
expression is 1×10^{-1} kilowatt-hour (kw hr), or simply
0.1 kilowatt-hour. The kilowatt-hour is the practical electric energy unit in common use. Observe the magnitude of
this combined conversion constant.

Your electric utility statement is based on the number of kilowatt-hours of energy used and the cost per kilowatt-hour.

$$\text{kw hr} \times \frac{10^3 \text{ w}}{\text{kw}} \times \frac{3.6 \times 10^3 \text{ s}}{\text{hr}} = 3.6 \times 10^6 \text{ w s}$$

The domestic cost of electric energy may be in the range
of 6¢ to 12¢ per kilowatt-hour, depending on location and
primary energy source. The cost of electric energy is calculated in the example that follows.

EXAMPLE A small electric furnace connected across a 117-v line dissipates 2.00 kw of power as heat. (a) What current is in the circuit?
(b) What is the resistance of the furnace? (c) What is the cost of operation
for 24.0 hr at 10.0¢ per kw hr? (d) What quantity of heat, in kcal, is developed in 1.00 hr?

SOLUTION (a) The potential difference across the furnace and the
power expended in the furnace are given. The current can be calculated
knowing that $P = IV$. Therefore,

$$I = \frac{P}{V} = \frac{2.00 \text{ kw}}{117 \text{ v}} \times \frac{10^3 \text{ w}}{\text{kw}} = 17.1 \text{ a}$$

(b) Both the current in the circuit and the potential difference across the furnace are known; therefore, Ohm's law can be applied to find the resistance.

$$V = IR \quad \text{and} \quad R = \frac{V}{I} = \frac{117 \text{ v}}{17.1 \text{ a}} = 6.84 \ \Omega$$

(c) The power dissipation in the furnace, the cost per kilowatt-hour, and the time of operation are given. The conversion factor, $\$1/10^2 ¢$, is useful here to express cost in appropriate terms.

$$\text{cost} = \text{kw hr} \times \frac{\text{cost}}{\text{kw hr}} = \frac{2.00 \text{ kw} \times 24.0 \text{ hr} \times 10.0¢}{\text{kw hr}} \times \frac{\$1}{10^2 ¢} = \$4.80$$

(d) Both the time of operation and the power dissipated as heat are given. Recognizing that $P = I^2R$, the expression for Q can be simplified.

$$Q = \frac{I^2 Rt}{J} = \frac{Pt}{J}$$

$$= \frac{2.00 \text{ kw} \times 1.00 \text{ hr}}{4.19 \text{ j/cal}} \times \frac{10^3 \text{ w}}{\text{kw}} \times \frac{3.6 \times 10^3 \text{ s}}{\text{hr}} \times \frac{\text{kcal}}{10^3 \text{ cal}}$$

$$Q = 1.72 \times 10^3 \text{ kcal}$$

(Proof of answer unit is left to the student.)

PRACTICE PROBLEMS   **1.** The heating element in an electric toaster dissipates 1.37 kw of power as heat when connected across a $12\bar{0}$-v line. (a) What current is in the heating element? (b) What is the resistance of the heating element? (c) What is the cost for 1.25 hours of operation at 8.75¢ per kw hr? (d) How many kilocalories of heat are produced during the 1.25-hour period of operation?   *Ans.* (a) 11.4 a; (b) 10.5 Ω; (c) 15.0¢; (d) 1.47×10^3 kcal

2. An electric heater has an element resistance of 7.80 Ω and when connected across a source of emf draws a current of 15.0 a. (a) What is the potential difference across the heater element? (b) How many kilocalories of heat are produced per hour? (c) What is the cost for 8.00 hours of operation at 9.25¢ per kw hr?   *Ans.* (a) 117 v; (b) 1.51×10^3 kcal; (c) $1.30

Questions

GROUP A

1. State four ways in which the work done by an electric current in a load may be observed, and name the kind of load responsible in each instance.
2. Equal lengths of silver wire and iron wire having the same diameters are connected in series to form the external circuit across a dry cell. Which wire becomes hotter? Explain.
3. One length of platinum wire is connected across the terminals of a single dry cell. A second platinum wire, identical with the first, is connected

across a battery of two dry cells in series. Which wire becomes hotter? Explain.
4. State Joule's law.
5. How may electric power be defined?
6. (a) What change occurs in the heating effect of an electric circuit if the source of emf remains the same but the total resistance is cut in half? (b) Devise a simple circuit problem to prove your answer.
7. If a battery is short-circuited by means of a heavy copper wire, its temperature rises. Explain.
8. Under what circumstance is the power expended in a load the maximum the source is able to deliver?
9. Suppose you have just paid the electric utility bill for your home. (a) What did you purchase? (b) In what units was it measured?
10. What change occurs in the resistance

in the external circuit connected to a source of emf when the load across the source is increased?

GROUP B
11. Demonstrate that Q in Joule's law can be expressed in kilocalories.
12. Two resistors of equal resistance were originally connected in parallel. By what factor does their combined resistance change when they are connected in series instead?
13. Why is it important that an automobile battery have a very low internal resistance?
14. What special characteristics should a toaster wire have?
15. Express the rate of energy dissipation in a resistance in terms of (a) current and potential difference, (b) current and resistance, (c) potential difference and resistance.

Problems
GROUP A

1. The current in an electric heater is 7.5 a. What quantity of electric charge flows through the heater in 15 min?
2. Determine the current in a lamp circuit if 4800 c of electric charge flows through the lamp in 25 min.
3. An electric lamp connected across a 117-v line has a current of 0.52 a in it. How much work is done in 12 min?
4. What is the approximate wattage rating of the lamp in Problem 3?
5. How many kilocalories of heat are produced by a resistance of 55 Ω connected across a 110-v line for 10 min?
6. A heating coil across a 117-v line draws 9.00 a. How many kilocalories are liberated if the heater operates for 30.0 min?
7. The heating element of an electric iron has a resistance of 24 Ω and draws a current of 5.0 a. How many kilocalories are developed if the iron is used for 45 min?

8. A battery has an emf of 26.4 v and an internal resistance of 0.300 Ω. A load of 3.00 Ω is connected across the battery. How much power is (a) delivered to the load, (b) dissipated in the battery?
9. (a) To what value would the load resistance of Problem 8 have to be changed in order that maximum power is delivered to the load using the same source of emf? (b) What is the magnitude of the maximum power delivered to this load? (c) What power is now dissipated in the battery?
10. A wire for use in an electric heater has a resistance of 5.25 Ω/m. What length of this wire is needed to make a heating element for a toaster that will draw 8.70 a across a 12$\bar{0}$-v line?

GROUP B
11. If electric energy costs 9.2¢ per kw hr, what is the cost of heating 4.6 kg of water from 25 °C to the boiling point, assuming no energy is wasted?

12. A percolator with a heating coil of 20.0-Ω resistance is connected across a 120-v line. (a) What quantity of heat is liberated per second? (b) If it contains 500 g of water at 22.5 °C, how much time is required to heat the water to boiling, assuming no loss of heat?

13. An electric iron has a mass of 1.50 kg. The heating element is a 2.00-m length of No. 24 gauge nichrome wire. What time is required for the iron to be heated from 20.0 °C to 150 °C when connected across a 115-v line, assuming no loss of heat?

14. An electric hot plate draws 10.0 a on a 120-v circuit. In 7.00 min the hot plate can heat 600 g of water at 20.0 °C to boiling and boil away 60.0 g of water. What is the efficiency of the hot plate?

15. Each coil of a plug-type resistance box is capable of dissipating heat at the rate of 4.0 w. What is the maximum voltage that should be applied across (a) the 2.0-Ω coil? (b) the 20.0-Ω coil?

16. A dial-type rheostat of 10.0-Ω resistance is capable of dissipating heat at the rate of 4.0 w. (a) What is the maximum current the rheostat can carry? (b) What is the maximum voltage that can be applied across it?

17. How many meters of No. 30 gauge nichrome wire are needed to form the heating element of a 1000-w electric iron used on a 117-v line?

18. A lamp operates continuously across a 117-v line dissipating 100 w. How many electrons flow through the lamp per day?

ELECTROLYSIS

18.7 Electrolytic Cells   The close relationship between chemical energy and electric energy has already been discussed. Spontaneous reactions that involve the transfer of electrons from one reactant to the other were described in Section 17.3. Such reactions can be sources of emf. In the electrochemical cell chemical energy is converted to electric energy as the spontaneous electron-transfer reaction proceeds. The products of the reaction have less energy than did the reactants.

Similar reactions *that are not spontaneous* can be forced to occur if electric energy is supplied from an external source, as in charging a storage battery. In such forced reactions, the products have more chemical energy than did the reactants. This means electric energy from the external source of emf is transformed into chemical energy as the reaction proceeds. An arrangement in which a forced electron-transfer reaction occurs is known as an *electrolytic cell*.

The basic requirements for an electrolytic cell are shown in Figure 18-6. The conducting solution contains an electrolyte that furnishes positively and negatively charged ions. Two electrodes are immersed in the electrolytic solution. The negative terminal of a battery or other source of direct current is connected to one electrode to form the *cathode* of the cell; the positive terminal of the source is

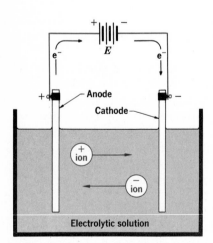

Figure 18-6. The essential components of an electrolytic cell.

The cathode is the electron-rich electrode.

connected to the other electrode to form the *anode* of the cell.

When the circuit is closed, the cathode becomes negatively charged and the anode positively charged. Positive ions migrate to the cathode where they acquire electrons of high potential energy and are then discharged. Negatively charged ions migrate to the anode and are discharged by giving up electrons of low potential energy to the anode. Note that the electrode reactions occurring in electrochemical cells, presented in Sections 17.3 and 17.4, are just the reverse of those in an electrolytic cell.

The loss of electrons by the cathode and the acquisition of a like number of electrons by the anode constitute the conduction of an electric charge through the cell. *The conduction of an electric charge through a solution of an electrolyte or through a fused ionic compound, together with the resulting chemical changes, is called* **electrolysis.**

The end result of electrolysis depends on the nature of the electrolyte, the kinds of electrodes, and to some extent the emf of the source. Electrolytic cells are useful in decomposing compounds, plating metals, and refining metals.

18.8 Electroplating Metals An electrolytic cell is set up with a metallic copper anode and positively charged copper(II) ions, Cu^{++}, in solution. When a small potential difference is placed across the cell, the Cu^{++} ions acquire electrons from the cathode and form copper atoms. The following chemical equation represents this reaction.

$$Cu^{++} + 2e^- \rightarrow Cu^0 \text{ (Cathode reaction)}$$

These copper atoms plate out on the cathode surface.

Simultaneously, atoms of the copper anode give up electrons to this positively charged anode and form Cu^{++} ions that go into the solution to replace those being plated out as copper atoms. The chemical reaction is

$$Cu^0 \rightarrow Cu^{++} + 2e^- \text{ (Anode reaction)}$$

These Cu^{++} ions pass into the solution from the anode at the same rate as similar Cu^{++} ions leave the solution and form the plate on the cathode. To maintain the concentration of Cu^{++} ions in the electrolytic solution, Cu^{++} ions must continuously pass into the solution from the anode. The result is that the anode is used up in the plating process. A simplified diagram of a copper-plating cell is shown in Figure 18-7.

The reaction at the cathode makes an electrolytic cell useful for the electrolytic deposition, or *electroplating,* of

The anode is the electron-poor electrode.

Chemical reactions in electro-chemical cells are spontaneous; those in electrolytic cells are driven.

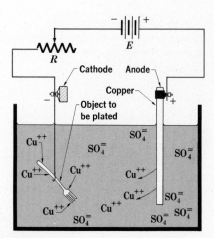

Figure 18-7. An electrolytic cell used for copper plating.

one metal on the surface of another. The object to be plated is used as the cathode of the cell. The plating metal is used as the anode. The conducting solution contains a salt of the metal to be plated out onto the cathode.

Metallic copper of the highest purity is required for electric conductors. Even very small amounts of impurities in copper cause a marked increase in its electric resistance. These impurities can be removed almost completely from the copper by electrolysis. For this purpose an electroplating cell is used in which the anode is composed of impure copper. During electrolysis, refined copper is plated onto the cathode made of pure copper. This electrolytically refined copper metal deposited on the cathode is more than 99.99% pure.

18.9 Faraday's Laws of Electrolysis Michael Faraday investigated the distribution of charge on isolated bodies as well as the nature of electric fields. He also performed experiments on the conduction of electric charge by solutions. Faraday discovered that a current that deposited a half-penny-weight of silver in ten minutes would deposit one penny-weight of silver in twenty minutes. This discovery suggests an important relationship between the quantity of electric charge that passes through an electrolytic cell and the quantity of a substance liberated by the chemical action.

Suppose we consider the quantitative significance of Faraday's discovery by imagining several cells connected in series and supplied by a source of emf. Each cell has a pair of inert platinum electrodes immersed in a water solution of an electrolyte. See Figure 18-9. Each cell has a different electrolyte: cell No. **1** has sulfuric acid (hydrogen sulfate), cell No. **2** has silver nitrate, cell No. **3** has copper(II) sulfate, and cell No. **4** has aluminum sulfate. Since this is a "thought" experiment we need not be concerned with the possibility of running out of ions in the cells.

Figure 18-8. Michael Faraday, the English scientist who formulated the laws of electrolysis and for whom the unit of electric capacitance is named. His lectures were popular events.

Figure 18-9. Electrolytic cells in series must have the same magnitude of current for the same length of time.

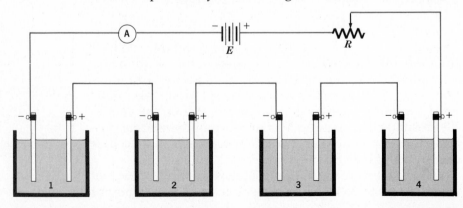

Since the cells are in series, a certain constant current in the circuit for a given time interval signifies that *the same quantity of charge has passed through each cell.*

Hydrogen, silver, copper, and aluminum are liberated at the respective cathodes. The quantity of each liberated substance must be proportional to the quantity of charge that passed through each cell: that is, to the product of the current and the time.

Faraday's first law: The mass of an element deposited or liberated during electrolysis is proportional to the quantity of charge that passes.

$$m \propto Q$$

where

$$Q = It$$

By varying either current or time interval, we merely vary the quantity of charge passing through each cell. By Faraday's first law, the mass of the element deposited or liberated varies accordingly. It follows that whatever the charge through the cells, the relationship among the quantities of the elements deposited remains *constant*. Such quantities of the elements are said to be *electrochemically equivalent*. The **electrochemical equivalent,** z, of an element is the mass in grams of the element deposited or liberated by 1 coulomb of electric charge. See Table 18-1. A more extensive table of electrochemical equivalents of common elements is given in Appendix B, Table 19.

Recall that there can be only one magnitude of current in a series circuit because there is only one conducting path.

Table 18-1
ELECTROCHEMICAL EQUIVALENTS

Element	g-at wt	Ionic charge (oxidation state)	z
	(g)		(g/c)
aluminum	27.0	3	0.0000932
copper	63.5	2	0.0003294
hydrogen	1.01	1	0.0000104
silver	108	1	0.0011179

Suppose we allow the charge to flow through our cells until 1 g of hydrogen has been liberated in cell No. **1.** We would find 108 g of silver deposited in No. **2,** 31.8 g of copper in No. **3,** and 9 g of aluminum in No. **4.**

The hydrogen ion, H^+, and the silver ion, Ag^+, are each deficient by 1 electron and carry a net positive charge of 1 unit. The elements hydrogen and silver are said to have an ionic charge of 1, which means they lose 1 electron per

The number of electrons of an atom of an element participating in a chemical process is sometimes referred to as the valence of the element.

atom in a chemical reaction. Similarly the charge of the copper(II) ion, Cu^{++}, is 2, and that of the aluminum ion, Al^{+++}, is 3.

Using this information and the information previously obtained in the thought experiment, we can make two very interesting and important observations. *First, for a given quantity of charge, the quantity in grams of each element deposited or liberated is proportional to the ratio of its gram-atomic weight (g-at wt) to its ionic charge.* This ratio is known as the ***chemical equivalent*** of an element.

$$\text{chemical equivalent} = \frac{\text{g-at wt}}{\text{ionic charge}}$$

Thus the chemical equivalents of hydrogen, silver, copper, and aluminum are respectively 1, 108, 31.8, and 9 grams.

Second, the product of the current in amperes and the time in seconds during which the cells operated to deposit the chemical equivalents of these elements is found to be approximately 96,500 coulombs. This product is a constant and is called a *faraday*.

1 faraday = 96,500 coulombs

*One **faraday**, 96,500 coulombs, is the quantity of electric charge required to deposit one chemical equivalent of an element.* This unit has many significant applications in electrolysis and electrochemistry.

Faraday's second law: *The mass of an element deposited or liberated during electrolysis is proportional to the chemical equivalent of the element.*

$m \propto$ chemical equivalent

From our definition of the electrochemical equivalent of an element, we can see that

$$z = \frac{\text{chemical equivalent}}{\text{faraday}}$$

and has the dimensions g/c. The electrochemical equivalent z is a constant for a given element but is different for different elements.

We can combine Faraday's laws of electrolysis into the following equation:

$$m = zIt$$

where m is the mass in grams of an element deposited or liberated, z is the electrochemical equivalent in g/c of the element, I is the current in amperes, and t is the time in seconds.

Faraday's laws give us a precise method of measuring the quantity of electric charge flowing through a circuit. If we know the length of time the charge flows, we can determine the average current magnitude as shown in Figure 18-10. A platinum dish is connected with the negative terminal of a source of about 2 volts emf, and then the dish is partly filled with a solution of silver nitrate. A platinum spiral rod is connected to the positive terminal of the source of current and dipped into the silver solution.

One coulomb of charge (one ampere-second) flowing through such a silver solution will deposit on the walls of the dish 0.001118 g of silver. If the average current in the circuit is one ampere for one hour, 3600 coulombs will deposit 4.025 g of silver. From the mass of the silver deposited, the number of coulombs of electric charge can be determined with good precision.

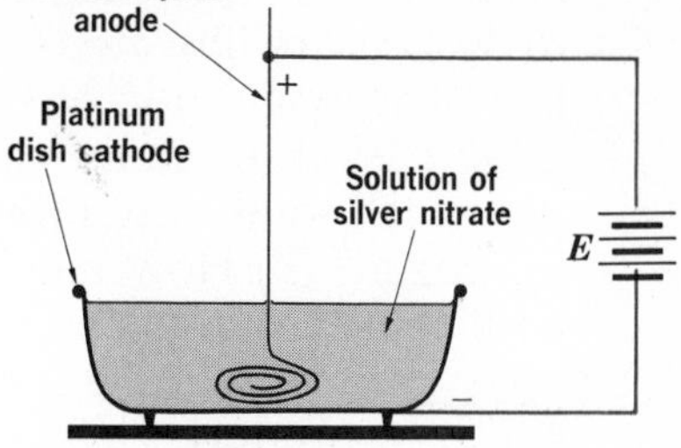

Figure 18-10. A silver coulombmeter.

Questions

GROUP A

1. How can electron-transfer reactions that are not spontaneous be forced to occur?
2. What is an electrolytic cell?
3. What may be the effect of the conduction of electric charge through a fused ionic compound?
4. What are the basic requirements for a silver-plating cell?
5. Why is electrolytically refined copper used to make electric conductors?
6. State Faraday's laws of electrolysis.
7. What part of a silver coulombmeter is used as the cathode?

GROUP B

8. Distinguish between an electrochemical cell and an electrolytic cell.
9. Compare the cathode and anode reactions in an electrolytic cell with the corresponding electrode reactions in an electrochemical cell.
10. The g-at wt of aluminum is 27, and the g-at wt of gold is 197; both ions have a charge of 3. Which has the higher electrochemical equivalent?
11. Show that the mass of an element deposited according to Faraday's law can be expressed in grams.
12. How could you define the ampere in relation to Faraday's laws and the silver coulombmeter?

Problems

GROUP A

1. How much silver can be deposited in 8 min by making use of a constant current of 0.500 a?
2. An electroplating cell connected across a source of direct current for 15.0 min deposits 0.750 g of copper. What is the average current in the cell?
3. How many hours are required for an electrolytic cell to liberate 0.160 g of oxygen using a constant current of 0.300 a?
4. A water-decomposition cell is maintained in continuous operation with a constant current of 25.0 a. How much hydrogen is recovered per day?
5. How many hours would it take to plate 25.0 g of nickel onto an automobile bumper if the current in the plating bath is 3.40 a?

GROUP B

6. An electrolysis-of-water cell is operated until 1 faraday of electricity has

been passed. How many atoms of hydrogen are liberated?

7. (a) What magnitude of electric charge must pass through an electrolysis-of-water cell to liberate one g-at wt of hydrogen? (b) How much oxygen will be liberated during the same time?

8. How long will it take for a current of 4.50 a to produce 1.00 L of hydrogen by the electrolysis of water? Assume STP conditions.

9. Two coulombmeters, one of silver and the other of an unknown metal that forms ions having a +3 charge, are connected in series. A constant current of 2.00 a is maintained in the circuit for 2.00 hr and 2.73 g of the unknown metal are deposited. How much silver is deposited?

10. What is the g-at wt of the unknown metal that was deposited in the above problem?

SUMMARY

The electric energy expended in a resistance is converted to heat. The relationship among the heat developed in a circuit element, the resistance, and the time the current is maintained is expressed in terms of Joule's law.

Electric power dissipated in the resistance of a circuit is proportional to the square of the circuit current. Power dissipated in the internal resistance of the source of emf is wasted. When it is important to deliver maximum power to the load, the load resistance and that of the source of emf must be matched.

Electrolytic cells are driven by an external source of emf. Electric energy is then transformed to chemical energy through electron-transfer reactions at the cell electrodes. The products have higher chemical energy than the reactants. Electrolytic cells are useful in decomposing compounds, plating metals, and refining metals.

Faraday's laws of electrolysis provide an insight into the quantitative significance of energy transformations in chemical reactions. These laws relate the mass of an element deposited during electrolysis to the quantity of charge that passes through the cell and to the chemical equivalent of the element. The silver coulombmeter provides a precise method of measuring the quantity of charge in an electric circuit.

VOCABULARY

anode	electrochemical equivalent	faraday
cathode	electrolysis	Faraday's first law
chemical equivalent	electrolyte	Faraday's second law
coulombmeter	electrolytic cell	Joule's law
electrochemical cell	electroplating	spontaneous reaction

MAGNETIC EFFECTS

In 1820, Hans Christian Oersted discovered that an electric current is surrounded by a magnetic field. This discovery stimulated the investigation of electricity by scientists in many other countries. The SI unit of magnetic intensity is named in honor of Oersted.

MAGNETISM

19.1 Magnetic Materials Physicists believe that all magnetic phenomena result from forces between electric charges in motion. Vast quantities of electric energy are now generated as a consequence of relative motion between electric conductors and magnetic fields. Electric energy is transformed into mechanical energy by relative motion between electric currents and magnetic fields. The function of many electric measuring instruments depends on the relationship between electricity and magnetism.

The basic theory of electric generators and motors is presented in Chapter 20. Electric measuring instruments are discussed later in this chapter. Before undertaking the study of magnetic effects of electric currents, we shall examine the magnetic properties of substances and learn of the nature of magnetism and magnetic fields.

Deposits of a magnetic iron ore were discovered many centuries ago by the Greeks in a section of Turkey. The region was then known as Magnesia and the ore was called *magnetite*. Deposits of magnetite are found in the Adirondack Mountains of New York and in other regions of the world. Pieces of magnetite are known as *natural magnets*. A suspended piece of magnetite aligns itself with the magnetic field of the earth. These natural magnets, known as lodestones (leading stones), were first used as magnetic compasses during the twelfth century.

A few materials, notably iron and steel, are strongly attracted by magnets; cobalt and nickel are attracted to a

In this chapter you will gain an understanding of:

▷ the domain theory of magnetism
▷ magnetic force and the nature of this force between magnetic poles
▷ the concept of the magnetic monopole
▷ the characteristics of a magnetic force
▷ techniques for mapping magnetic fields
▷ the nature of magnetic induction
▷ the earth's magnetic field and its magnetosphere
▷ the link between moving charges and magnetic fields of force
▷ the magnetic field of a current in a straight conductor and in a solenoid
▷ electric meters and their use in d-c circuits

The Latin word for iron is "ferrum"; thus the name "ferromagnetic."

Alnico (Al Ni Co) consists mainly of aluminum, nickel, and cobalt plus iron.

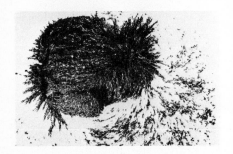

Figure 19-1. Iron filings attracted to a lodestone give evidence of the field of force surrounding the natural magnet.

The orbiting electron produces a magnetic field because, in this motion, it constitutes an electric current.

lesser degree. These substances are said to have *ferromagnetic* properties. Special alloys such as *permalloy* and *alnico* have extraordinary ferromagnetic properties. Physicists have shown much interest in the structure of materials possessing the property of *ferromagnetism.*

Today very strong magnets are made from ferromagnetic substances. Alnico magnets may support a weight of over 1000 times that of the magnets themselves. Ferromagnetic substances are commonly referred to simply as "magnetic substances."

Materials are commonly classified as magnetic or nonmagnetic. Those that do not demonstrate the strong ferromagnetism of the Iron Family of metals are said to be "nonmagnetic." However, if these materials are placed in the field of a very strong magnet, some are observed to be slightly repelled by the magnet while others are very slightly attracted.

Zinc, bismuth, sodium chloride, gold, and mercury are a few of the substances that are feebly repelled; they are *diamagnetic.* The property of *diamagnetism* is an important concept in the modern theory of magnetism, as we shall see in Section 19.2.

Wood, aluminum, platinum, oxygen, and copper(II) sulfate are examples of substances that are very slightly attracted by a strong magnet. Such materials are *paramagnetic,* and this magnetic behavior is called *paramagnetism.*

19.2 The Domain Theory of Magnetism   William Gilbert's report on his experiments with natural magnets, published in 1600, probably represents the first scientific study of magnetism. In the years that followed, discoveries by Coulomb, Oersted, and Ampère added to our knowledge of the behavior of magnets and the nature of magnetic forces. Physicists believe, however, that it is only within this century that they have begun to understand the true nature of magnetism. The present view is that the magnetic properties of matter are electric in origin and result from the movements of electrons within the atoms of substances. Since the electron is an electrically charged particle, this theory suggests that *magnetism is a property of a charge in motion.* If so, we can account for the energy associated with magnetic forces by using known laws of physics.

Two kinds of electron motion are important in this modern concept of magnetism. *First, an electron revolving about the nucleus of an atom imparts a magnetic property to the atom structure.* See Figure 19-2. When the atoms of a substance are subjected to the magnetic force of a strong magnet, the force affects this magnetic property, opposing the motion of the electrons. The atoms are thus repelled by the magnet. This is diamagnetism. If the electron's only motion were its movement about the nucleus, all substances would be diamagnetic. Diamagnetic repulsion is quite feeble in its action on the total mass of a substance.

The second kind of motion is that of the electron spinning on its own axis. Each spinning electron acts as a tiny permanent magnet. Opposite spins are designated as + and − spins; electrons spinning in opposite directions tend to form *pairs* and so neutralize their magnetic character. See Figure 19-3. The magnetic character of an atom as a whole may be weak because of the mutual interaction between the electron spins.

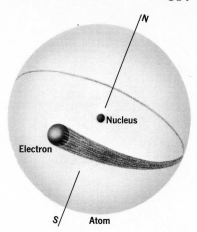

Figure 19-2. Revolving electrons impart a magnetic property to the atom.

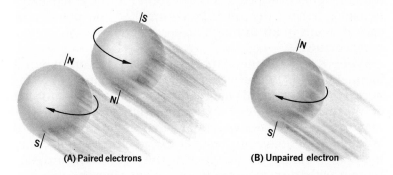

(A) Paired electrons   (B) Unpaired electron

Figure 19-3. Ferromagnetism in matter from the spin of electrons.

Magnetic properties are associated with both kinds of electron motion. The atoms of some substances may possess permanent magnet characteristics because of an imbalance between orbits and spins. These atoms act like tiny magnets, called *dipoles,* and are attracted by strong magnets. Substances in which this attractive effect exceeds the diamagnetism common to all atoms show the property of paramagnetism.

In the atoms of ferromagnetic substances there are unpaired electrons whose spins are oriented in the same way. The common metals iron, cobalt, and nickel and the rare earth elements gadolinium and dysprosium show strong ferromagnetic properties. Some alloys of these and other elements, as well as certain metallic oxides called ferrites, also exhibit strong ferromagnetic properties.

The inner quantum levels, or shells, of the atom structures of most elements contain only paired electrons. The highest quantum level, or outer shell, of each of the noble gases (except helium) consists of a stable octet of electrons made up of four electron pairs. The atoms of other elements achieve this stable configuration by forming chemical bonds. Only in certain transition elements that have

Each iron atom has four unpaired inner-shell electrons.

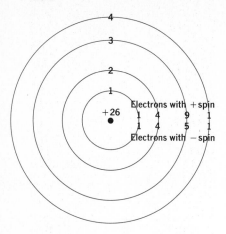

Figure 19-4. The iron atom has strong ferromagnetic properties.

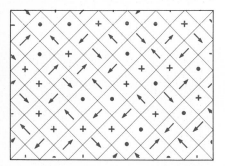

Figure 19-5. The domains of an unmagnetized ferromagnetic substance are polarized along the crystal axes. Dots and plus signs represent arrows going out of and into the page, respectively.

Table 19-1
CURIE POINTS OF
FERROMAGNETIC ELEMENTS

Element	Curie point
iron	770 °C
cobalt	1131 °C
nickel	358 °C
gadolinium	16 °C

incomplete inner shells do unpaired electrons result in ferromagnetic properties. The electron configuration of the iron atom, Figure 19-4, shows four unpaired electrons in the third principal quantum level. The similarly oriented spins of these electrons, enhanced by the influence of nearby atoms in the metallic crystal, account for iron's strong ferromagnetism.

From the preceding discussion, it would seem that every piece of iron should behave as a magnet. However, such is not the case. Atoms are grouped in microscopic magnetic regions called *domains.* The atoms in each domain are magnetically polarized parallel to a crystal axis. In a polycrystalline specimen, ordinarily these axes (and the domains) are oriented in all possible directions. The domains effectively cancel one another and the net magnetism is essentially zero. In Figure 19-5 the polarity of each domain in an unmagnetized material is represented by an arrow.

When a ferromagnetic material is placed in an external magnetic field, two effects occur. The domains more favorably oriented in this magnetic field may increase in size at the expense of less favorably oriented adjacent domains. Other domains may rotate in order to become more favorably oriented with respect to the external field. The material becomes magnetized. If the domain boundaries remain extended to some degree even after the external magnetizing force is removed, the material is said to be "permanently" magnetized. When the direction of magnetization of a magnetic domain is rotated by an external magnetic field, it must be understood that the *material* of the domain does not change its position in the specimen. It is only its *direction* of magnetization that changes.

When the temperature of a ferromagnetic material is raised above a certain critical value, the domain regions disappear and the material becomes paramagnetic. This temperature is known as the *Curie point.* It is usually lower than the melting point of the substance. The Curie points for some ferromagnetic substances are given in Table 19-1.

When a single crystal of iron is sprinkled with colloidal particles of iron oxide, the microscopic domains become visible. Using this technique, physicists are able to photograph magnetic domains and observe the effects of external magnetic fields on them. Typical photomicrographs of magnetic domains are shown in Figure 19-6.

A recent magnet technology that makes use of a group of ferromagnetic substances known as *ferrites* yields strong hard magnets with unique properties. Ferrites are iron oxides combined with oxides of other metals such as manga-

IRON   COBALT

Figure 19-6. Photomicrographs of magnetic domains.

nese, cobalt, nickel, copper, or magnesium. The combined oxides are powdered, formed into the desired shape under pressure, and fired. The ferrites have very high electric resistance, a property that is extremely important in some applications of ferromagnetic materials. The original lodestone, commonly called magnetic iron oxide, is a material of this type. Chemically it is a combination of iron(II) oxide, FeO, and iron(III) oxide, Fe_2O_3. Its formula is considered to be $Fe(FeO_2)_2$.

19.3 Force Between Magnet Poles The fact that iron filings cling mainly to the ends of a bar magnet indicates that the magnetic force acts on the filings primarily in these regions, or *poles;* it does not mean that the middle region of the magnet is unmagnetized. The pole that points toward the north when the magnet is free to swing about a vertical axis is commonly called the *north-seeking pole,* or **N** pole. The opposite pole, which points toward the south, is called the *south-seeking pole,* or **S** pole.

Suppose a bar magnet is suspended as shown in Figure 19-7. When the **N** pole of a second magnet is brought near the **N** pole of the suspended magnet, the two repel each other. A similar action is observed with the two **S** poles. When the **S** pole of one magnet is placed near the **N** pole of the other magnet, they attract each other. Such experiments show that *like poles repel and unlike poles attract.*

Magnets usually have two well-defined poles—one **N** and one **S**. Sometimes long bar magnets acquire more than two poles, and an iron ring may have no poles at all when magnetized. Physicists have long speculated about the existence of single-pole magnetic particles called *monopoles*. Known magnetic poles, however, always come in pairs called *dipoles*. The most elementary magnet has an **S** pole and an **N** pole. If cut in half, each half is found to be dipolar. A magnet has an **S** pole for every **N** pole. An isolated **N** pole of unit strength is sometimes *assumed* in "thought" experiments. *A unit pole may be thought of as one that repels an exactly similar pole, placed one centimeter away, with a force of one dyne.* (1 dyne = 10^{-5} newton.)

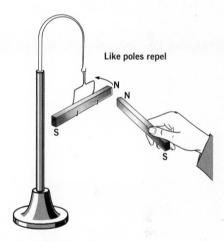

Like poles repel

Figure 19-7. Like poles repel. Unlike poles attract.

Experimental evidence of the possible existence of magnetic monopoles has been reported but not verified. Physicists believe that proof of the existence of monopoles could help verify some of the basic concepts of physics.

The quantitative expression for Coulomb's law of magnetism is

$$F = k \frac{M_1 M_2}{d^2}.$$ *Compare this*

equation with those in Sections 3.11 and 16.8.

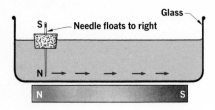

Figure 19-8. The path followed by the floating magnet in this experiment is approximately that of an independent N pole.

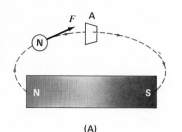

(A)

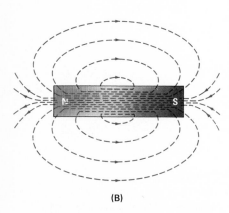

(B)

Figure 19-9. (A) The path taken by an independent N pole in a magnetic field suggests a line of flux. (B) Magnetic flux about a bar magnet.

The first quantitative study of the force between two magnetic poles is generally credited to Coulomb. He found this magnetic force governed by the same inverse-square relationship that applies to gravitational force and electrostatic force. **Coulomb's law of magnetism is:** *The force between two magnetic poles is directly proportional to the product of the strengths of the poles and inversely proportional to the square of the distance between them.* The force is one of repulsion or attraction, depending on whether the magnetic poles are alike or different.

19.4 Magnetic Fields of Force In Section 16.9 we described the electric field of force near an electrically charged object. Electric forces are not the only forces that act on charged particles. Sometimes we observe the effect of a force that is both *perpendicular* and proportional to the velocity of a *moving* charge. This force identifies a *magnetic field.* A dipole magnet in such a region of space experiences a torque. We speak of a magnetic field in the space around a bar magnet in the same way we speak of an electric field around a charged rod. Furthermore, *we can represent a magnetic field by lines of flux,* just as we represented an electric field by lines of force.

The behavior of our imaginary independent **N** pole in a magnetic field can be approximated by using a magnetized darning needle as illustrated in Figure 19-8. The needle is supported by cork so that it floats with the **N** pole extended below the surface of the water. The **S** pole is far enough removed to have negligible influence on the movement of the needle. A bar magnet placed under the glass dish with its **N** pole near the needle causes the floating magnet to move along a path that approximates the path an isolated **N** pole would follow.

The path of an independent **N** pole in a magnetic field suggests a *line of flux. A **line of flux** is a line so drawn that a tangent to it at any point indicates the direction of the magnetic field.* Flux lines are assumed to emerge from a magnet at the **N** pole and to enter the magnet at the **S** pole. Every flux line is a closed path running from **S** pole to **N** pole within the magnet. See Figure 19-9.

The lines of flux perpendicular to a specified area in the magnetic field are collectively called the *magnetic flux,* for which the Greek letter Φ (phi) is used. The unit of magnetic flux is the *weber* (wb).

*The **magnetic flux density, B,** is the number of flux lines per unit area that permeates the magnetic field.* The flux density B is a vector quantity; the direction of B at any point in the magnetic field is the direction of the field at that point.

$$B = \Phi/A$$

Flux density is expressed in *webers per square meter* (wb/m^2). The flux density determines the *magnetizing force* at any point in the magnetic field. The weber per meter^2 is also called the *tesla*.

$$1 \text{ weber/meter}^2 = 1 \text{ tesla}$$

The measurement of these quantities is in Section 19.9.

Flux lines drawn to indicate how tiny magnets would behave when placed at various points in a magnetic field provide a means of *mapping* the field. A line drawn tangent to a flux line at any point indicates the direction a very small magnet would assume if placed there. An arrowhead can be added to the tangent line to indicate the direction in which the **N** pole of the tiny magnet would point, thus giving the direction of the magnetic field, and the *B* vector, at that point.

Imaginary lines of magnetic flux are useful for mapping magnetic fields.

Using a suitable scale of flux lines per unit area perpendicular to the field, the flux density, *B*, at any point can be illustrated. Selection of a number of lines to represent a unit of magnetic flux is arbitrary. Usually, one flux line per square meter represents a flux density of 1 wb/m^2. In this sense, one line of flux is a weber.

The magnetic field near a single bar magnet is suggested by the pattern formed by iron filings sprinkled on a glass plate laid over the magnet. A photograph of this field pattern is shown in Figure 19-10. Using a similar technique, the magnetic fields near the *unlike* poles and near the *like* poles of two bar magnets are illustrated in Figure 19-11. Observe that the magnetic force acting on the two unlike poles is one of attraction and that acting on the two like poles is one of repulsion. Figure 19-12 similarly illustrates an end-on view of the magnetic field between the poles of a horseshoe magnet.

19.5 Magnetic Permeability   In Section 19.4 we described the effect of a magnetic field of force on iron filings and on a magnetized needle as experienced through glass

Figure 19-10. Iron filings near a single bar magnet.

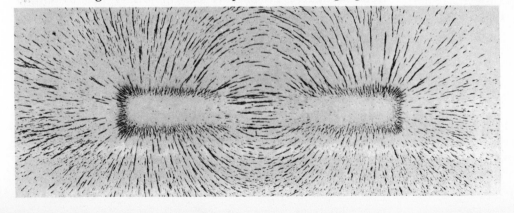

Figure 19-11. Iron filings are shown near unlike poles of two bar magnets in (A) and near their like poles in (B).

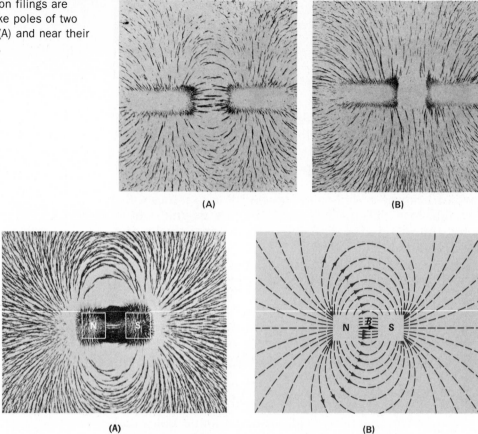

(A)   (B)

(A)   (B)

Figure 19-12. (A) Iron filings near the poles of a horseshoe magnet, end view. (B) An idealized drawing of (A) showing lines of flux.

In a practical sense, all materials except those that are ferromagnetic can be considered magnetically inert. In a magnetic field, they behave just like air.

and water. Nonmagnetic materials in general are transparent to magnetic flux; that is, their effect on the lines of flux is not appreciably different than that of air. *The property of a material by which it changes the flux density in a magnetic field from the value in air is called its **permeability**, μ.* Permeability is a ratio of flux densities and is without dimension. The permeability of empty space is taken as unity and that of air as very nearly the same. The permeabilities of diamagnetic substances are slightly less than unity; permeabilities of paramagnetic substances are slightly greater than unity. Permeabilities of ferromagnetic materials are many times that of air.

If a sheet of iron covers a magnet, there is little magnetic field above the sheet. The flux enters the iron and follows a path within the iron itself. Similarly, an iron ring placed between the poles of a magnet provides a better path than air for the magnetic flux. This effect is illustrated in Figure 19-13. The flux density in iron is greater than it is in air; therefore, iron is said to have a high permeability. The permeabilities of other ferromagnetic substances are also very high.

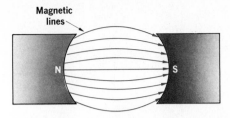

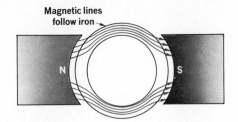

Suppose a bar of soft iron lies in a magnetic field, as in Figure 19-14. Because of the high permeability of the iron, the field is distorted and the magnetic flux passes through the iron in preference to the air. Under these circumstances the soft iron bar becomes a magnet with end **A** as the **S** pole and end **B** as the **N** pole. Such a bar is said to be magnetized by *induction. Magnetism produced in a ferromagnetic substance by the influence of a magnetic field is called* **induced magnetism.**

Figure 19-13. At left, magnetic flux crosses the air gap between the poles of a magnet. At right, magnetic flux follows the soft iron ring, which is more permeable than air.

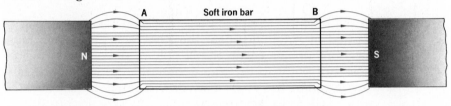

A   Soft iron bar   B

If the magnetic field is removed by withdrawing the two bar magnets, most of the induced magnetism will be lost. Magnets produced by induction are known as temporary magnets. A piece of hardened steel is not so strongly magnetized by induction but retains a greater *residual* magnetism when removed from the induction field.

There is no significant difference in the process if the iron bar in Figure 19-14 is brought into contact with one of the magnet poles. The magnetization process is somewhat more efficient due to the reduction of the air gap. See Figure 19-15.

Figure 19-14. An iron bar magnetized by induction.

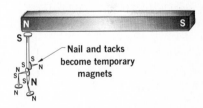

Nail and tacks become temporary magnets

Figure 19-15. The nail becomes a magnet by induction. Are the tacks also magnets?

19.6 Terrestrial Magnetism Suppose the earth contained a great bar magnet. See Figure 19-16. It would produce a magnetic field similar to its actual field. Over most of the earth's populated surface the north-seeking pole of a compass points northward. Although it is the south pole of our fictitious magnet that attracts the **N** pole of the compass, the pole region is conventionally called the *north magnetic pole* because it is located in the northern hemisphere. Similarly, the pole region in the southern hemisphere is called the *south magnetic pole.*

The earth's magnetic axis does not coincide with its polar (geographic) axis, but is inclined to the polar axis at a small angle. The north magnetic pole, at latitude 73°N and longitude 100°W, is about 2000 km (1200 miles) south

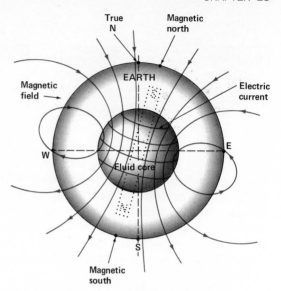

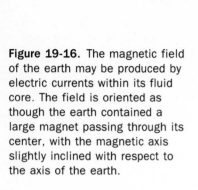

Figure 19-16. The magnetic field of the earth may be produced by electric currents within its fluid core. The field is oriented as though the earth contained a large magnet passing through its center, with the magnetic axis slightly inclined with respect to the axis of the earth.

of the north geographic pole. The south magnetic pole is located in Antarctica near the Ross Sea. Thus from most locations on the earth, the **N** pole of the compass needle does not point to the true geographic north. At any surface location the angle between magnetic north and the true north is called the *declination*, or *variation*. In the region of Los Angeles the compass variation is about 15°E. That is to say a compass needle points about 15° to the east of true north. In the region of Boston the variation is about 15°W. Cincinnati, Ohio, is located very near the line of zero declination. Here the compass needle points to the true north and the variation is 0°.

A compass needle mounted on a horizontal axis and provided with a means of measuring the angle the needle makes with the horizontal plane is called a *dipping* needle. At certain places on the earth's surface, about midway between the magnetic poles, the angle of dip is zero and the needle is horizontal. A line drawn through a succession of such points identifies the *magnetic equator*. The angle of dip is 90° at the magnetic pole. The dip, or deviation between the equilibrium position of a dipping needle and the horizontal, is known as the *magnetic inclination*.

In 1600 the English physicist William Gilbert (1540?–1603) published his scientific treatise *De magnete*, which deals with the magnetism of the earth. This is one of the earliest publications on the experimental treatment of a scientific topic. Gilbert inferred that the earth behaved as a large magnet because the interior consisted of permanently magnetic material. Today scientists believe the core of the earth is too hot to be a permanent magnet and is fluid rather than solid.

The German physicist Karl Friedrich Gauss (1777–1855) showed that the magnetic field of the earth must originate inside the earth. In 1939 the American theoretical physicist Walter M. Elsasser suggested that the earth's magnetic field results from electric currents generated by the flow of matter in the earth's fluid core. See Figure 19-16. Today physicists believe that the magnetic field is due primarily to electric currents within the earth, but they have not yet established the origin of these currents.

Electric current loops inside the earth are responsible for its magnetic field.

19.7 The Magnetosphere Because space vehicles now travel to the outer limits of the earth's atmosphere and beyond, there is a growing interest in a region of the outer atmosphere known as the *magnetosphere*. Located beyond 200 km, the magnetosphere is the region in which the motion of charged particles is governed primarily by the magnetic field of the earth. At lower altitudes, where the density of the atmosphere is much greater, the motion of charged particles is controlled largely by collisions.

The magnetosphere on the side facing the sun extends beyond the earth's surface approximately 57,000 km, or about 10 earth radii. On the side away from the sun, the magnetosphere probably extends outward for hundreds of earth radii. See Figure 19-17. The elongated shape results from the influence of the onrushing *solar wind*, or *solar plasma*. The solar wind, consisting mainly of protons and electrons emitted by the sun, compresses the magnetosphere on the side nearest the sun.

In 1958 regions of intense radiation were discovered within the magnetosphere by a team of physicists headed

Figure 19-17. The magnetosphere of the earth. The overall radiation regions are shown in color. The inner and outer Van Allen belts of intense radiation are the dark regions ranging outward to approximately 4 earth radii.

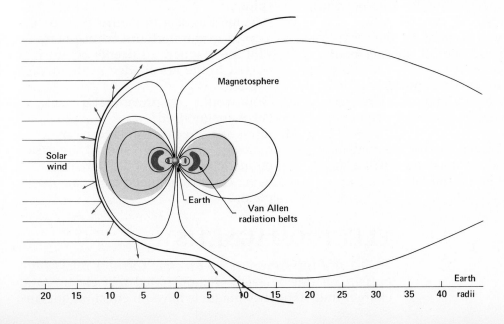

by Dr. J. A. Van Allen (b. 1914). These regions, now known as the Van Allen radiation belts, contain energetic protons and electrons trapped by the earth's magnetic field. Those trapped in the inner belts probably originate in the earth's atmosphere; those trapped in the outer belts probably have their origin in the sun. When these intense radiation belts were first discovered, scientists were concerned about the serious threat they appeared to present to space travel. Today, astronauts journeying into outer space are able to pass quickly through these regions with adequate protection from the Van Allen radiation.

Auroral displays over the polar regions are related to the escape of energetic particles from the radiation belts.

Questions
GROUP A

1. (a) What is a lodestone? (b) Why are lodestones sometimes called natural magnets?
2. What is the distinguishing property of ferromagnetic materials?
3. Distinguish between diamagnetic and paramagnetic materials.
4. What two kinds of electron motion are important in determining the magnetic property of a material?
5. What are electron pairs and what is their significance in regard to magnetic character?
6. (a) What three metals are the most important ferromagnetic materials? (b) What special alloy is often used to make very strong magnets?
7. How can you account for the ferromagnetic properties of the metals of the Iron Family?
8. What are magnetic domains?
9. Describe the condition of a ferromagnetic material in a state of magnetic saturation.
10. State Coulomb's law for magnetism.
11. Under what circumstances will a magnetic force be (a) one of repulsion, (b) one of attraction?

12. Describe a simple way to examine the fields of small magnets.
13. What is the advantage in making a magnet in the shape of a horseshoe?
14. Describe an experiment that illustrates induced magnetism.
15. Distinguish between a north-seeking pole and the north magnetic pole.
16. Why is a declination angle involved in the use of a compass over most of the earth's surface?

GROUP B
17. Explain, on the basis of atomic structure, the property of diamagnetism.
18. Explain paramagnetism on the basis of imbalance between orbits and spins.
19. Assume the dish in Figure 19-8 to be located in the northern hemisphere. Will the magnetized needle be attracted to the north edge of the dish? Explain.
20. How would you prove that a steel bar is magnetized?
21. If a watch mechanism is to be magnetically "insulated," should the case be made of diamagnetic, paramagnetic, or ferromagnetic material? Explain.

ELECTROMAGNETISM

19.8 The Link Between an Electric Current and Magnetism It can be easily demonstrated that electrostatic charges and stationary magnets have no effect on one another. However, in 1820 Hans Christian Oersted (*er*-stet) (1777–1851), a Danish physicist and professor of physics at the University of Copenhagen, observed that a small compass needle is deflected when brought near a conductor carrying an electric current. This was the first evidence of a long-suspected link between electricity and magnetism. Oersted discovered that forces exist between a magnet and electric charges in motion. His famous experiment is so significant that a brief description of it is in order.

A dry cell, compass, switch, and conducting wire are arranged as shown in Figure 19-19(A). With the switch open, a straight section of the conductor is supported *above* the compass in the vertical plane of the compass needle. In Figure 19-19(B) the dry-cell connection is such that the electron flow will be from north to south. When the switch is closed, the **N** pole of the compass is deflected toward the west. When the dry-cell connections are reversed so electron flow is from south to north, the **N** pole of the compass is deflected to the east, as in Figure 19-19(C). It is evident that *a magnetic field exists in the region near the conductor when the circuit is closed*. Furthermore, *the direction of the field is dependent on the direction of the current in the conductor*.

Figure 19-18. Hans Christian Oersted studied medicine before becoming professor of physics at the University of Copenhagen in 1806. Several years before he performed his famous experiment, he predicted that a link between electricity and magnetism would be found.

Figure 19-19. The Oersted experiment as viewed from above. In each diagram the compass needle is located below the conductor.

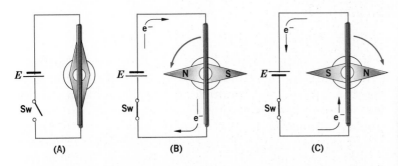

(A)   (B)   (C)

If the experiment is repeated with the conductor placed *below* the compass needle, the compass deflection is opposite to that in the first experiment. This suggests, but does not prove, that the magnetic field encircles the conductor.

19.9 Magnetic Field and a Charge in Motion Shortly after Oersted's discovery, the French physicist Ampère determined the shape of the magnetic field about a conductor carrying a current. He had discovered that forces exist between two parallel conductors in an electric circuit. If the two currents are in the same direction, the force is one of attraction; the force is one of repulsion if the currents are in opposite directions. See Figure 19-21.

In a quantitative sense, two long, straight, parallel conductors of length *l* separated by a distance *d* and carrying

Figure 19-20. André Ampère, the French physicist for whom the unit of electric current is named, did fundamental work in electromagnetism.

currents I_1 and I_2 will each experience a force F of magnitude

$$F = \frac{2k\,l\,I_1 I_2}{d}$$

The constant k is exactly 10^{-7} n/a². If I_1 and I_2 are expressed in amperes and l and d in meters, the force F is given in newtons.

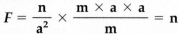

$$F = \frac{\mathbf{n}}{\mathbf{a}^2} \times \frac{\mathbf{m} \times \mathbf{a} \times \mathbf{a}}{\mathbf{m}} = \mathbf{n}$$

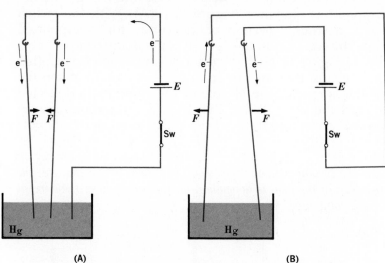

Figure 19-21. Forces between parallel currents (A) in the same direction and (B) in opposite directions.

(A)   (B)

Because these attractive and repulsive forces between current-carrying conductors are directly proportional to the currents in the conductors, they provide a precise method of defining the unit of current, the ampere. In this sense, the **ampere** may be defined as *the current in each of two long parallel conductors spaced one meter apart that causes a magnetic force of 2×10^{-7} newton per meter length of conductor.*

Following this scheme, the **coulomb** as a quantity of charge (an ampere-second) may be defined as *the quantity of electric charge that passes a given point on a conductor in one second when the conductor carries a constant current of one ampere.*

Ampère investigated the magnetic fields about conductors to find an explanation of the magnetic forces. Suppose a heavy copper wire passes vertically through the center of a horizontal sheet of stiff cardboard. When the ends of the vertical conductor are connected to a dry cell, iron filings sprinkled over the surface of the cardboard form a pattern of concentric circles around the conductor. See Figure 19-22. If a small compass is placed at successive points around a circle of filings, the needle always comes to rest

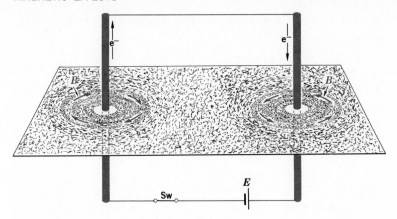

Figure 19-22. The magnetic field encircling a current in a straight conductor.

tangent to the circle and with the same tangential orientation of its **N** pole.

If the direction of current in the vertical conductor is reversed, the compass needle again becomes aligned tangent to the circle of filings, but with its **N**-pole orientation reversed. From these observations we conclude that *a magnetic field encircles an electric charge in motion.* The lines of flux are closed concentric circles in a plane perpendicular to the conductor with the axis of the conductor as their center. The direction of the magnetic field is everywhere tangent to the flux and is dependent on the direction of the current.

Ampère devised a rule, known today as *Ampère's rule,* for determining the direction of the magnetic field around a current in a straight conductor when the direction of the electron flow is known.

Ampère's rule for a straight conductor: Grasp the conductor in the left hand with the thumb extended in the direction of the electron current. The fingers then will circle the conductor in the direction of the magnetic flux. See Figure 19-23.

The flux density, *B,* also called the *magnetic induction,* at any point in the magnetic field of a long straight conductor carrying a current, *I, is directly proportional to the current in the conductor and inversely proportional to the radial distance, r, of the point from the conductor.*

$$B = 2k\frac{I}{r}$$

The constant *k* again is 10^{-7} n/a². When *I* is given in amperes and *r* is in meters, *B* is expressed in newtons per ampere meter, which is equivalent to webers per square meter.

In Section 19.4 flux density is defined in terms of the lines of flux per unit area that permeate the magnetic field.

Magnetic phenomena are interpreted in terms of the forces associated with electric charges in motion.

The left-hand rule for a straight conductor indicates the direction of the magnetic flux surrounding the conductor.

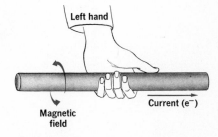

Figure 19-23. Ampère's rule for a straight conductor.

In this sense the expression for flux density is $B = \Phi/A$. Thus

$$\Phi = BA$$

When B is expressed in newtons per ampere meter, the unit for Φ, the weber, can be shown to be 1 newton meter per ampere. Dimensionally

$$wb = \frac{n}{a\ m} \times m^2 = \frac{n\ m}{a}$$

Whether Φ is expressed in webers or newton meters per ampere and B is expressed in webers per square meter or newtons per ampere meter is a matter of convenience in each situation.

Observe that the definition of B given in Section 19.4 is based on the force exerted on an isolated unit pole. An isolated pole exists only in the fiction of a thought experiment; consequently, measurements based on this definition lack precision. The more practical definition given above involves quantities that can be measured precisely and is therefore generally preferred.

19.10 Magnetic Field and a Current Loop Keeping Ampère's rule in mind, let us consider a loop in a conductor carrying a current. The magnetic flux from all segments of the loop must pass through the inside of the loop in the same direction; that is, *the face of the loop must show polarity.* See Figure 19-24.

Figure 19-24. The magnetic field through a current loop.

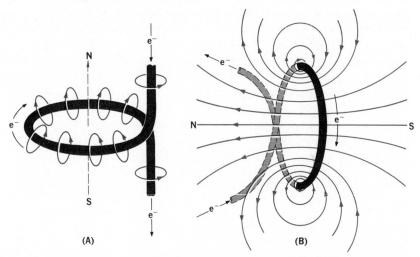

(A)   (B)

A magnetic tube is an imaginary tube bounded by magnetic lines. It always links the current that produces the field.

This loop magnet can be made stronger if the flux density can be increased. Because the magnetic field around a conductor varies with the current, the flux density can be increased by increasing the magnitude of the current in the conducting loop, by forming additional loops in the conductor, or by both.

A linear coil of such conducting loops takes the form of a helix and is called a *solenoid*. The cylindrical column of air inside the loops, extending the length of the coil, is called the *core*. When a current is in a solenoid, the core of each *turn* (loop) becomes a magnet; the core of the solenoid is a magnetic tube through which practically all the magnetic flux passes. See Figure 19-25.

Because a solenoid conducting an electric current has the magnetic properties of a bar magnet, its polarity can be determined by means of a compass. However, the magnetic flux in the core of the solenoid is derived from the magnetic field of each turn of the conductor. Thus Ampère's rule is modified to adapt it to this special case of the solenoid.

Ampère's rule for a solenoid: Grasp the coil in the left hand with the fingers circling the coil in the direction of the electron current. The extended thumb will point in the direction of the **N** pole of the core. See Figure 19-26.

19.11 The Electromagnet A solenoid with a core of air, wood, or some other nonmagnetic material does not produce a very strong electromagnet because the permeability of all nonmagnetic substances is essentially equal to that of air—unity. Substitution of such materials for air does not appreciably change the flux density.

Soft iron, on the other hand, has a high permeability. If an iron rod is substituted for air as the core material, the flux density is greatly increased. Strong electromagnets therefore have ferromagnetic cores with high permeability. For a given core material, the strength of the electromagnet depends on the magnitude of the current and the number of turns. In other words, its strength is determined by *the number of ampere-turns*.

19.12 The Galvanometer Suppose we form a wire loop in a vertical plane, place a compass needle (free to rotate in a horizontal plane) in the center of it, and then introduce a current into the loop. The needle will be deflected. If we increase the number of turns sufficiently, even a feeble current will produce a deflection of the needle. Such a device, called a *galvanoscope*, may be used to detect the presence of an electric current or to determine its direction. A simple galvanoscope is shown in Figure 19-28.

A more versatile instrument for detecting feeble currents is the *galvanometer*, the essential parts of which are shown in Figure 19-29. A coil of wire wound on a soft iron core is pivoted on jeweled bearings between the poles of a

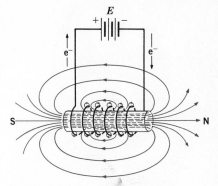

Figure 19-25. Magnetic field about a solenoid.

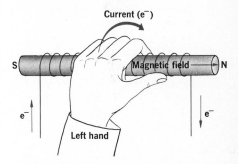

Figure 19-26. Ampère's rule for a solenoid.

Figure 19-27. A superconducting electromagnet. In operation, the eight-foot cylinder is immersed in liquid helium (−232 °C). At this temperature the niobium-titanium strips imbedded in the copper coils lose all electric resistance and the magnet produces a force field up to 5000 times greater than that of the earth.

Figure 19-28. A simple galvano-scope.

Because a magnetic field exerts forces on moving charges, it exerts torques on current-carrying coils.

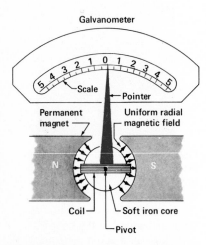

Galvanometer

Figure 19-29. The basic components of a moving-coil galvanometer.

permanent horseshoe magnet. The coil becomes a magnet when current is in it. The instrument then has two magnets: a permanent horseshoe magnet in a fixed position and an electromagnet free to turn on its axis. Electric connections to the coil are made through two control springs (not shown), one above and one below the coil. These coiled springs also restrain the rotational motion of the coil so that the attached pointer returns to the zero scale position when no current is present in the coil. This zero position is often located at the midpoint of the scale, as in Figure 19-29.

When there is a current in the movable coil, its core is magnetized. The poles of the core are then attracted and repelled by the poles of the permanent magnet. A torque acts upon the coil and the coil rotates in an attempt to align its plane perpendicular to the line joining the poles of the permanent magnet. As the coil rotates, however, it does work against the two control springs. Its final position is reached when the torque acting on it is just neutralized by the reaction of the springs. Since the permanent field flux is constant, the torque on the coil is proportional to the current in it. We may assume, for small movements of the coil, that the reaction of the springs is proportional to the deflection angle. When the coil reaches its equilibrium position, these two opposing torques are equal, and the deflection angle of the coil is therefore proportional to the current in it.

The scale of the galvanometer is marked at intervals on either side of the zero center. Readings are made on this scale by means of a small, lightweight pointer attached to the coil. For a coil current in one direction, the needle deflection is to the left. If the current direction is reversed, the needle is deflected to the right.

The galvanometer is a sensitive instrument for detecting feeble currents of the order of microamperes. For translation of a reading into absolute current values, the *current sensitivity* of the specific instrument must be known. Current sensitivity is usually expressed in *microamperes per scale division*.

The pointer deflection, d, of a galvanometer is proportional to the current, I_M, in the coil

$$I_M \propto d$$

or

$$I_M = kd$$

and

$$k = \frac{I_M}{d}$$

where k is the current sensitivity in microamperes per scale division.

The moving coil of the galvanometer has resistance. By Ohm's law, a potential difference appears across the resistance of the meter when a current is in the coil. We can express the *voltage sensitivity* of the instrument since it must be equal to the product of the meter resistance and the current per scale division.

$$\text{Voltage sensitivity} = kR_{M} = \frac{I_{M}}{\text{div}} \times R_{M}$$

where R_M is the resistance of the meter movement. Voltage sensitivity is given in microvolts per scale division.

If provision is made to prevent excessive currents from entering the coil, the galvanometer can be adapted for service as either a d-c ammeter or d-c voltmeter.

The following examples illustrate calculations involving the galvanometer.

Figure 19-30. Sensitive galvanometers use a mirror to indicate the position of a suspended coil by producing an image of a scale viewed through a telescope or by reflecting a beam of light onto a scale.

EXAMPLE   What current is required for full-scale deflection of a galvanometer having a current sensitivity of 50.0 μa per scale division? The meter has exactly 50 divisions on either side of the mid-scale index.

SOLUTION   We are given the current sensitivity, $k = 50.0\ \mu\text{a/div}$, and the number of divisions on either side of mid-scale (zero) position, $d = 50$ div.

$$I_{M} = kd = \frac{50.0\ \mu\text{a}}{\text{div}} \times 50\ \text{div}$$
$$I_{M} = 25\overline{0}0\ \mu\text{a, or } 2.50 \times 10^{-3}\ \text{a, the required current}$$

EXAMPLE   What potential difference appears across the galvanometer described in the previous example when the pointer is fully deflected? The meter resistance is 10.0 Ω.

SOLUTION   We are given the current sensitivity, $k = 50.0\ \mu\text{a/div}$, the resistance of the meter, $R_M = 10.0\ \Omega$, and the number of divisions of deflection, $d = 50$ div.

$$V_{M} = I_{M}R_{M}$$

But
$$I_{M} = kd$$

Then $$V_{M} = kdR_{M} = \frac{50.0\ \mu\text{a}}{\text{div}} \times 50\ \text{div} \times 10.0\ \Omega$$

$V_{M} = 2.50 \times 10^{4}\ \mu\text{v, or } 2.50 \times 10^{-2}\ \text{v, the potential difference}$ across the meter.

PRACTICE **1.** What magnitude of current will produce a full-scale
PROBLEMS deflection of a galvanometer with exactly 40 scale divisions
and a sensitivity of 30.0 μa per scale division?   *Ans.* $1.20 \times 10^3 \mu$a

2. What potential difference is across the galvanometer in Problem 1
when the pointer is fully deflected? The meter resistance is 8.50 Ω.
Ans. $1.02 \times 10^4 \mu$v

3. The galvanometer in Problem 1 is connected in a circuit with a current
of 675 μa. Assuming a linear scale calibration, what scale pointer deflection is observed?   *Ans.* 22.5 div

19.13 The d-c Voltmeter The potential difference
across a galvanometer is quite small even when the needle
is fully deflected. If a galvanometer is to be used to measure voltages of ordinary magnitudes, we must convert it to
a high-resistance instrument. The essential parts of a d-c
voltmeter are shown in Figure 19-31.

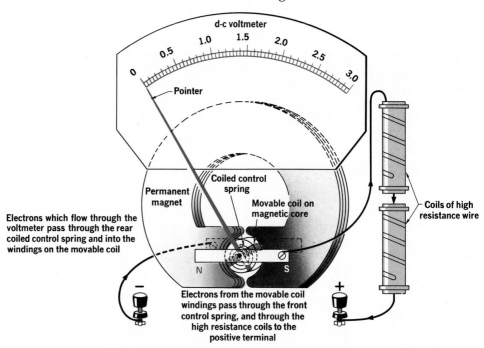

The high resistance coils permit only a few electrons
to flow through the movable coil of a voltmeter

Figure 19-31. The construction of
a d-c voltmeter showing the high
resistance in series with the windings of the movable coil.

If a high resistance is added in series with the moving
coil, most of the potential drop appears across this series
resistor, or *multiplier*. Since *a voltmeter is connected in parallel*
with the part of a circuit across which the potential difference is to be measured, a high resistance prevents an appreciable loading effect. By the proper choice of resistance,

the meter can be calibrated to read any desired voltage.

Suppose we convert the galvanometer used in the examples in Section 19.12 to a voltmeter reading 15.0 volts on full-scale deflection. The current required for full deflection has been found to be 2.50×10^{-3} ampere, and the resistance of the meter coil is 10.0 ohms. We must determine the value of the resistor R_S to be placed in series with the moving coil. Figure 19-32 illustrates this problem.

Since R_M and R_S (of Figure 19-32) are in series,

$$V = I_M R_M + I_M R_S$$

Then $\quad R_S = \dfrac{V}{I_M} - R_M$

$$R_S = \dfrac{15.0 \text{ v}}{2.50 \times 10^{-3} \text{ a}} - 10.0 \text{ } \Omega$$

$R_S = 5990 \text{ } \Omega$, value of the series resistor

Observe that the total resistance between the terminals of the meter is $60\overline{0}0$ ohms.

The voltmeter sensitivity is frequently expressed in terms of *ohms per volt*. When the ohms-per-volt sensitivity of a voltmeter is known, we can quickly estimate the loading effect it will have when placed across a known resistance component of a circuit. For example, at 400 ohms per volt, our meter, which reads from 0 to 15 volts, has $60\overline{0}0$ ohms between the terminals. If it were placed across resistances greater than about 600 ohms, the loading effect would result in serious meter errors.

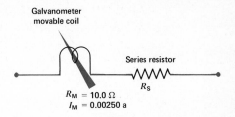

Galvanometer movable coil

Series resistor

R_S

$R_M = 10.0 \text{ } \Omega$
$I_M = 0.00250 \text{ a}$

Figure 19-32. Converting a galvanometer to a voltmeter.

19.14 The d-c Ammeter We could use the basic galvanometer in Section 19.12 as a microammeter by calibrating the graduated scale to read directly in microamperes. However, the meter would not be useful in circuits in which the current exceeded 2500 microamperes. Current in the resistance of a galvanometer coil produces I^2R heating, and an excessive current would burn out the meter.

To convert the galvanometer to read larger currents, an alternate (parallel) low-resistance path for current, called a *shunt*, must be provided across the terminals. By the proper choice of shunt resistance, we can calibrate the meter to read over the required range of current magnitudes. See Figure 19-33.

Suppose we wish to convert the same galvanometer to an ammeter reading 10.0 amperes full scale. As before, the current required for the full deflection of the moving coil is 2.50×10^{-3} ampere and the resistance is 10.0 Ω. We

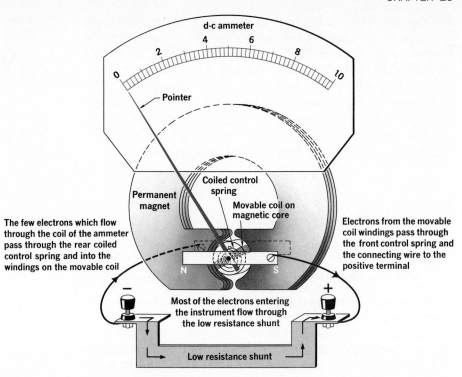

Figure 19-33. The construction of a d-c ammeter showing the low resistance in parallel with the windings of the movable coil.

must determine the resistance of the shunt to be used across the coil. Figure 19-34 applies.

Since R_M and R_S are in parallel,

$$I_M R_M = I_S R_S$$

But

$$I_S = I_T - I_M$$

Then

$$R_S = \frac{I_M R_M}{I_T - I_M}$$

$$R_S = \frac{2.50 \times 10^{-3} \text{ a} \times 10.0 \text{ }\Omega}{10.0 \text{ a} - 0.00250 \text{ a}}$$

$$R_S = 0.00250 \text{ }\Omega, \text{ the value for the shunt resistor}$$

The total resistance of the ammeter is the equivalent value for 10.0 ohms and 0.00250 ohm in parallel; it will be less than 0.00250 ohm. This exercise demonstrates clearly why *an ammeter must be connected in series* in a circuit, and why it does not materially alter the magnitude of current in the circuit.

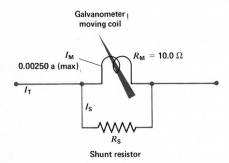

Figure 19-34. Converting a galvanometer to an ammeter.

19.15 The Ohmmeter An ohmmeter provides a convenient means of measuring the resistance of a circuit component. A basic ohmmeter circuit is shown in Figure 19-35. Its accuracy limitation is approximately the same as that of the voltmeter-ammeter method of measuring resistance; actually, it is a modified version of this method. The ohmmeter must be used only on a completely de-energized circuit.

The ohmmeter circuit of Figure 19-35 shows a milliammeter requiring 1 ma for full-scale deflection. With an emf of 4.5 volts, by Ohm's law, 4500 ohms of resistance will provide 1 ma of current when terminals **A-B** are short-circuited. A fixed resistor, R_2, of 4000 ohms and a rheostat, R_1, of 0-1000 ohms are provided.

To use this ohmmeter, **A** and **B** are short-circuited and R_1 is adjusted to give full deflection. If the emf is 4.5 v, R_1 will be set at 500 ohms. The pointer position at full-scale deflection is now marked as zero ohm (0 Ω). The rest position of the pointer is the open-circuit position with infinite resistance between **A** and **B**. This position is marked as ∞ Ω. Other resistance calibrations may be made from Ohm's law applications. For example, 4500 ohms between **A** and **B** will mean a total of 9000 ohms in the circuit and 0.5 ma of current. This mid-scale position of the pointer can be marked 4500 Ω. The meter, when recalibrated, will read the resistance between terminals **A-B**.

Each time the ohmmeter is used, it is first shorted across **A-B** and R_1 is adjusted to "zero" the meter. This operation calibrates the meter and accommodates any decrease in the terminal voltage of the battery with age. The resistance R_1 allows the ohmmeter to be used until E drops below 4.0 volts.

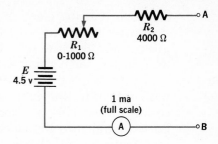

Figure 19-35. A basic ohmmeter circuit.

Questions

GROUP A

1. What important discovery was made by Oersted?
2. A conductor carrying a current is arranged so that electrons flow in one segment from north to south. If a compass is held over this segment of the wire, in what direction is the needle deflected?
3. Describe a simple experiment to show the nature of the magnetic field about a straight conductor carrying a current.
4. Suppose an electron flow in a conductor passing perpendicularly through this page is represented by a dot inside a small circle when the direction of flow is up out of the page. What is the direction of the magnetic flux about this current?
5. Upon what factors does the strength of an electromagnet depend?
6. What prevents the movable coil of a galvanometer from aligning its magnetic field parallel to that of the permanent magnet each time a current is in the coil?
7. (a) Why is it necessary that an ammeter be a low-resistance instrument? (b) Why must a voltmeter be a high-resistance instrument?

GROUP B

8. A solenoid with ends marked A and B is suspended by a thread so that the core can rotate in the horizontal plane. A current is maintained in the

coil such that the electron flow is clockwise when viewed from end A toward end B. How will the coil align itself in the earth's magnetic field?

9. A stream of electrons is projected horizontally to the right. A straight conductor carrying a current is supported parallel to the electron stream and above it. (a) What is the effect on the electron stream if the direction of the current in the conductor is from left to right, (b) if the current is reversed?

10. If the conductor in Question 9 is replaced by a magnet that produces a magnetic field directed downward. What is the effect on the electron stream?

11. Why might the potential difference indicated by a voltmeter placed across a circuit load be different from the potential difference with the meter removed?

12. Suppose the resistance of a high-resistance load is to be determined using the voltmeter-ammeter method. Considering the design characteristics of ammeters and voltmeters, how would you arrange the meters in the circuit to reduce the error to a minimum? Draw your circuit diagram and justify your arrangement.

13. Assume that the resistance of a low-resistance load is to be determined using the voltmeter-ammeter method. How would you arrange the meters in this circuit to reduce the error to a minimum? Draw your circuit diagram and justify your arrangement.

Problems
GROUP B

1. Two parallel conductors 2.0 m long and 1.0 m apart and carrying equal currents experience a total force of 1.6×10^{-6} n. What magnitude of current is in each conductor?

2. An ammeter that has a resistance of 0.01 ohm is connected in a circuit and indicates a current of 10 amperes. A shunt having a resistance of 0.001 ohm is then connected across the meter terminals. What is the new reading on the meter? Assume the introduction of the shunt does not affect the total circuit current.

3. A galvanometer has a zero-center scale with 20.0 divisions on each side of zero. The pointer deflects 15.0 scale divisions when a current of 375 μa is in the movable coil. (a) What is the current sensitivity of the meter? (b) What current will produce a full-scale deflection?

4. A galvanometer has a resistance of 50.0 Ω and requires 75.0 ma to produce a full-scale deflection. What resistance must be connected in series with the galvanometer in order to use it as a voltmeter for measuring a maximum of 300.0 v?

5. A galvanometer movement has a resistance of 2.5 ohms and when fully deflected has a potential difference of $\overline{5}0$ millivolts across it. What shunting resistance is required to enable the instrument to be used as an ammeter reading 7.5 amperes full scale?

6. A repulsive force of 9.6×10^{-4} n is experienced by each of two parallel conductors 5.0 m long when a current of 3.2 a is in each conductor. By what distance are they separated?

SUMMARY

Magnetite, a magnetic iron ore, is a natural magnet. Metals of the Iron Family and special metallic alloys and oxides are strongly attracted by magnets; they have ferromagnetic properties. Very strong magnets are made from ferromagnetic substances. Materials that are not ferromagnetic are commonly said to be nonmagnetic. Ferromagnetic materials have high permeabilities. Nonmagnetic materials in general are transparent to magnetic flux. These materials may be very feebly diamagnetic or paramagnetic. Magnetism is explained by the domain theory.

Coulomb's law for magnetism is a quantitative expression for the force acting between two magnetic poles. A magnetic field, and its influence on a fictitious N pole, shows similarities to an electric field and its influence on a positive test charge.

A charge in motion is surrounded by a magnetic field. The core of a coil carrying an electric current becomes a magnet. Strong electromagnets are produced by winding a conducting coil around a ferromagnetic core. The strength of an electromagnet depends on the number of turns of coil and the magnitude of the current in the coil.

The galvanometer is the basic meter for d-c measurements. The galvanometer can be calibrated as a voltmeter by placing a high resistance in series with the galvanometer coil. It can be calibrated as an ammeter by placing a very low resistance shunt across the galvanometer coil. An ohmmeter requires a source of emf, an adjustable resistance, and a sensitive ammeter. It is essentially a voltmeter-ammeter method of measuring the resistance of a circuit component.

VOCABULARY

ammeter	electron pair	magnetosphere
Ampère's rule	ferromagnetism	ohmmeter
Coulomb's law of	flux density	paramagnetism
magnetism	galvanometer	permeability
Curie point	induced magnetism	solar wind
declination	line of flux	solenoid
diamagnetism	magnetic domain	voltmeter
dipping needle	magnetic induction	

ELECTRO-MAGNETIC INDUCTION

Werner von Siemens and his brother, William, developed the first electric generator that did not use permanent magnets. This marked the commercial beginning of the electric industry. Firms headed by Siemens also laid the first transatlantic cable in 1874 and built the world's first public street railway in 1881.

In this chapter you will gain an understanding of:

✔ the distinction between an applied emf and an induced emf

✔ the importance of Faraday's experiments in the discovery of electromagnetic induction

✔ the factors that affect an induced emf

✔ the effect of a magnetic force on a moving electric charge

✔ the basic principle of operation of the electric generator

✔ the concept of instantaneous current and voltage

✔ the "motor effect" and the principle of the electric motor

✔ the concepts of mutual inductance as well as self-inductance

✔ the transformer and its uses

INDUCED CURRENTS

20.1 Discovery of Induced Current In Section 19.8 we discussed Oersted's discovery of the link between magnetism and electricity. Soon after Oersted's work, other scientists attempted to find out whether an electric current could be produced by the action of a magnetic field. In 1831 Michael Faraday discovered that *an emf is set up in a closed electric circuit located in a magnetic field whenever the total magnetic flux linking the circuit is changing.* The American physicist Joseph Henry (1797–1878) made a similar discovery at about the same time. This phenomenon is called *electromagnetic induction.* The emf is called an *induced emf,* and the resulting current in the closed conducting loop is called an *induced current.*

The production and distribution of ample electric energy are essential functions of a modern technological economy. The invention of electric generators and transformers has made it possible to provide these essential services. The discoveries of Faraday and Henry represent the initial step in the development of the broad knowledge base in electromagnetic induction that makes such inventions possible.

20.2 Faraday's Induction Experiments We shall examine some of Faraday's experiments in order to understand their significance. Suppose a sensitive galvanometer is connected in a closed conducting loop as shown in Figure 20-1. A segment of the conductor is poised in the field flux of a strong magnet. As the conductor in Figure 20-1(A) is moved down between the poles of the magnet, there is a momentary deflection of the galvanometer needle, indicating an induced current. The needle shows no deflection when the conductor is stationary in the magnetic flux. This observation suggests that *the induced current is related to the motion of the conductor in the magnetic flux.*

As the conductor is raised between the poles of the magnet, as in Figure 20-1(B), there is another momentary deflection of the galvanometer, but in the opposite direction. This suggests that *the direction of the induced current in the conductor is related to the direction of motion of the conductor in the magnetic field.* The emf induced in the conductor is of opposite polarity to that in the first experiment.

Faraday found that he could induce an emf in a conductor either by moving the conductor through a stationary field or by moving the magnetic field near a stationary conductor. He observed that the direction of the induced current in the conducting loop is reversed with a change in either the direction of motion or the direction of the magnetic field.

Supporting the conducting loop of Figure 20-1 in a fixed position and lifting the magnet result in a deflection similar to that of Figure 20-1(A). When the magnet is lowered, the galvanometer needle is momentarily deflected as in

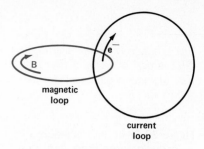

Magnetic lines of flux always link the current loop that sets up the magnetic field. Review Section 19.9.

Figure 20-1. A current is induced in a closed conducting loop when the magnetic flux linked through the circuit is changing.

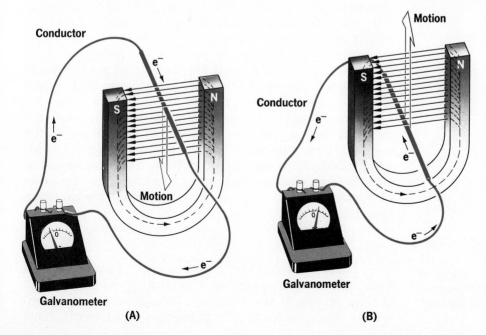

(A) (B)

The conducting segment of Figure 20-2 is part of a conducting loop that includes a galvanometer as shown in Figure 20-1.

Figure 20-2. An emf is induced in a conductor when there is a change of flux linked by the conductor.

Figure 20-1(B). The relative motion between the conductor and the magnetic flux is the same whether the conductor is raised through the stationary field or the field is lowered past the stationary conductor.

Observe that the relative motion of the conductor in each of these experiments is *perpendicular* to the magnetic flux. If the conductor in the magnetic field is now moved *parallel* to the lines of flux, the galvanometer shows no deflection. A conductor that moves perpendicular to the magnetic flux can be construed to "cut through" lines of flux and experience *a change in flux linkage.* Conversely, a

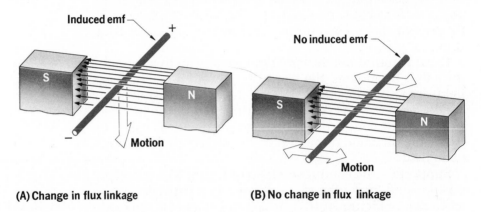

(A) Change in flux linkage   **(B) No change in flux linkage**

conductor that moves parallel to the magnetic field does not "cut through" lines of flux and does not experience a change in flux linkage. See Figure 20-2. These observations suggest that *electromagnetic induction results from those relative motions between conductors and magnetic fields which are accompanied by changes in magnetic flux linkage.*

Suppose the conductor is looped so that several turns are poised in the magnetic field, as in Figure 20-3. When the coil is moved down between the poles of the magnet as before, there is a greater deflection on the galvanometer. By increasing the rate of motion of the coil across the magnetic flux or by substituting a stronger magnetic field, greater deflections are produced. In each of these cases the effect is to increase the number of flux lines "cut" by turns of the conductor in a given length of time. We can then state that *the magnitude of the induced emf, or of the induced current in a closed loop, is related to the rate at which the flux linked by the conductor changes.*

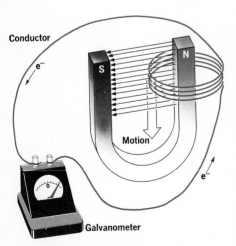

Figure 20-3. A greater change in flux linkage occurs when several turns of a conductor cut through the magnetic flux.

20.3 Factors Affecting Induced emf Faraday found that an emf is induced in a conductor whenever any change occurs in the magnetic flux (lines of induction) linking the conductor. Very precise experiments have shown that *the emf induced in each turn of a coiled conductor is proportional to the time rate of change of magnetic flux linking each turn of the coil.*

The total magnetic flux linking the coil is designated by the Greek letter Φ (phi). If $\Delta\Phi$ represents the *change* in magnetic flux linking the coil during the time interval Δt, the emf E induced in a single turn of the coil can be expressed as

$$E \propto -\frac{\Delta\Phi}{\Delta t}$$

By introducing a proportionality constant k, the value of which depends on the system of units used, the expression can be written as

$$E = -k\frac{\Delta\Phi}{\Delta t}$$

Where E is measured in volts and Φ in webers, the numerical value of k is unity and the equation becomes

$$E = -\frac{\Delta\Phi}{\Delta t}$$

Thus, a change in the magnetic flux linking a coil occurring at the rate of 1 weber per second induces an emf of 1 volt in a single turn of the coil.

A coil consists of turns of wire that are, in effect, connected in series. Therefore, the emf induced across the coil is simply the sum of the emfs induced in the individual turns. A coil of **N** turns has **N** times the emf of the separate turns. This relationship can be represented by the following equation

$$E = -\text{N}\frac{\Delta\Phi}{\Delta t}$$

The negative sign merely indicates the relative polarity of the induced voltage. It expresses the fact that *the induced emf is of such polarity as to oppose the change that induced it,* a basic energy conservation principle discussed in detail in Section 20.6.

To illustrate this concept, suppose a coil of 150 turns linking the flux of a magnetic field uniformly is moved perpendicular to the flux and a change in flux linkage of

3.0×10^{-5} weber occurs in 0.010 second. The induced emf is then

$$E = -N\frac{\Delta\Phi}{\Delta t}$$

$$E = -150 \times \frac{3.0 \times 10^{-5} \text{ wb}}{1.0 \times 10^{-2} \text{ s}}$$

$$E = -0.45 \text{ v}$$

$$\left(\frac{\text{wb}}{\text{s}} = \frac{\text{n m/a}}{\text{s}} = \frac{\text{j}}{\text{c}} = \text{v}\right)$$

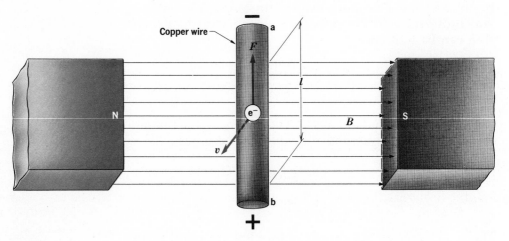

Figure 20-4. A force acts on a moving charge in a magnetic field.

20.4 The Cause of an Induced emf   A length of conductor moving in a magnetic field has an emf induced across it that is proportional to the rate of change of flux linkage. However, an induced current persists *only* if the conductor is a part of a closed circuit. In order to understand the cause of an induced emf, we shall make use of several facts that have already been established.

A length of copper wire poised in a magnetic field, as shown in Figure 20-4, contains many free electrons, and these moving charges constitute an electric current. In Section 19.12 we recognized that a force acts on the movable coil of a galvanometer in a magnetic field when a current is in the coil. This force, in effect, acts on the moving charges themselves. We shall call this force a *magnetic force* to distinguish it from the force exerted on free electrons by the electrostatic field of a stationary charge.

Suppose the copper wire of Figure 20-4, at right angles to the uniform magnetic field, is pushed downward (into the page) with a velocity v through the magnetic field of flux density B. As a consequence of the motion of the wire, the free electrons of the copper conductor may be considered to move perpendicular to the flux with the speed v. A magnetic force, F, acts on them in a direction perpendicular to both B and v. In response to the magnetic force, these electrons move toward end **a** and away from end **b.** Because the two ends of the conductor are not connected in a circuit that would provide a closed path for the induced electron flow, end **a** acquires a growing negative charge while a residual positive charge builds up on end **b.** Thus a difference of potential is established across the conductor with **a** the negative end and **b** the positive end. If either the motion of the wire or the magnetic field is reversed, the direction of F in Figure 20-4 is reversed and an emf is induced so that end **a** is positive and end **b** is negative.

The accumulations of charges at the ends of the conductor establish an electric field that increasingly opposes the movement of electrons through the conductor. The force of this electric field acting on the electrons from **a** toward **b** soon balances the magnetic force arising from the motion of the conductor, and the flow of electrons ceases.

The equilibrium potential difference across the open conductor is numerically equal to the induced emf and depends on the length, l, of wire linking the magnetic flux; the flux density, B, of the field; and the speed, v, with which the conductor is moved through the field.

$$E = Blv$$

When B is in newtons/ampere meter (or webers/meter^2), l in meters, and v in meters/second, E is given in volts.

Assume that the length of the conductor linking the magnetic field in Figure 20-4 is 0.075 m and the flux density is 0.040 n/a m. If the conductor is moved down through the flux with a velocity of 1.5 m/s, the induced emf is

$E = Blv$
$E = $ **0.040 n/a m** $\times$ **0.075 m** $\times$ **1.5 m/s**
$E = $ **0.0045 v or 4.5 mv**

In Figure 20-4 a magnetic force equal but opposite to F will act on the positively charged protons in the copper nuclei. Since these are in the bound parts of the copper atoms, they will not move in response to this force. But, positively charged ions in a liquid or gas would move.

The vectors B, v, and F are all mutually perpendicular as shown in Figure 20-5. If a charge Q moves with a velocity v

A changing magnetic field induces an electric field (Faraday's law).

The movement of electrons through the open conductor of Figure 20-4 comprises a "transient" induced current of extremely short duration.

Flux density, B, is also called magnetic induction.

To compare the response of a moving electric charge to an electric force and to a magnetic force, see Figures 16-11 and 20-5.

Verify that if B is expressed in n/a m, Φ can be expressed in n m/a.

through a magnetic field of flux density B, the force, F, acting on the charge becomes

$$F = QvB$$

The force F is in newtons, Q is in coulombs, v is in meters/second, and B is in newtons/ampere meter (or webers/meter^2).

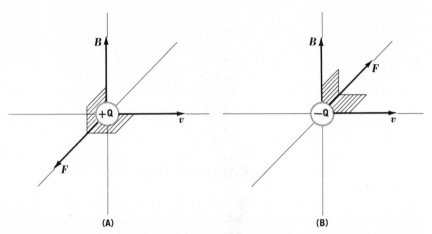

(A)   (B)

Figure 20-5. The magnetic force acting on a charge moving in a magnetic field. (A) The charge is positive and the force is directed out of the page. (B) The charge is negative and the force is directed into the page.

20.5 The Direction of Induced Current   Now consider that a length of conductor is used outside the magnetic flux to connect ends **a** and **b** of the copper wire in Figure 20-4, thus providing a closed-loop path for electron current. We shall call the length of wire linked with magnetic flux the *internal circuit* and the rest of the conducting path the *external circuit*.

When the straight conductor is pushed downward, as in Figure 20-4, electrons move from the negative end **a** through the external circuit to the positive end **b** and through the internal circuit from end **b** to end **a**. Over the external path from **a** to **b**, electrons transform the potential energy acquired in the internal circuit into kinetic energy that they expend in the external circuit. In the internal path, the electric force acting on the electrons from **a** to **b** is reduced below its equilibrium value because of the partial depletion of accumulated charges at **a** and **b**. A net magnetic force pumps electrons from **b** to **a**, maintaining a potential difference across the internal circuit. The value of this potential difference is less than the open-circuit potential difference which, you will recall, was numerically equal to the induced emf.

20.6 Lenz's Law   The downward motion of the copper wire in Figure 20-4 is maintained by exerting a force on it in the direction of v. This force may be thought of as the action that generates the induced electron current in the closed circuit.

The relationship between the direction of an induced current and the action inducing it was recognized in 1834 by the German physicist Heinrich Lenz (1804–1865). His discovery, now referred to as *Lenz's law,* is true of all induced currents. Because Lenz's law refers to induced currents, it applies only to closed circuits. ***Lenz's law*** states that *an induced current is in such a direction that it opposes the change that induced it.* Thus, the magnetic effect of the induced current in Figure 20-4 must be such that it opposes the downward force (the pushing action) applied to the conductor segment in the external field of the magnet.

Faraday discovered how to determine the direction of an induced current also, but he did not express it as clearly as Lenz.

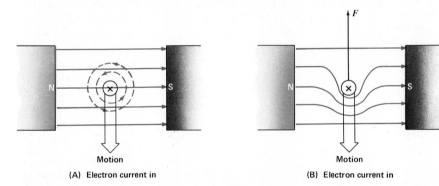

    (A) Electron current in   (B) Electron current in

We can readily visualize what this effect must be. A cross-sectional view of a conductor poised in a magnetic field is shown in Figure 20-6(A). The conductor corresponds to end **b** of the wire in Figure 20-4. It is part of a closed circuit that cannot be seen in this cross-sectional diagram. As the wire is pushed downward through the magnetic field, the induced electron current is directed into the page as indicated by the × symbol (tail of the arrow). By Ampère's rule for a straight conductor, the magnetic field of this induced current encircles the current in the counterclockwise sense. Similar diagrams in Figure 20-7 represent the effects of the wire being pulled up through the magnetic field. Here the induced electron current is directed out of the page as indicated by the · symbol (head of the arrow). The composite, or resultant, fields are shown in Figures 20-6(B) and 20-7(B). They suggest that the agent moving the conductor always experiences an opposing force.

The fact that an induced current always opposes the motion that induces it illustrates the conservation-of-energy principle. Work must be done to induce a current

Figure 20-6. Two interacting magnetic fields. The separate magnetic lines of the fields of the magnet and of the induced current (into the page) are shown in (A). The lines of the resultant field are shown in (B).

See Section 19.9 for Ampère's rule.

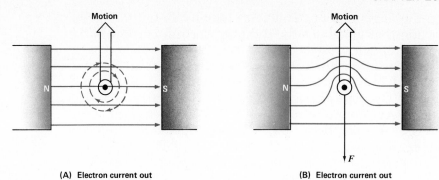

(A) Electron current out   (B) Electron current out

Figure 20-7. Two interacting magnetic fields. The separate magnetic lines of the fields of the magnet and of the induced current (out of the page) are shown in (A). The lines of the resultant field are shown in (B).

in a closed circuit. The energy expended comes from outside the system and is a result of work done by the external force required to keep the conductor segment moving. The induced current can produce heat or do mechanical or chemical work in the external circuit as electrons of high potential energy fall through a difference of potential.

Questions
GROUP A

1. What is the essential condition under which an emf is induced in a conductor?
2. What determines the magnitude of the emf induced in a length of conductor moving in a magnetic field?
3. In what two ways may the rate of change of magnetic flux linking a conductor be increased?
4. Distinguish between an induced emf and an induced current.
5. What is the source of the energy expended as work is done in a load by an induced current resulting from the movement of the conductor in a magnetic field?
6. State Lenz's law.
7. What is the meaning of the negative sign in the expression $E = -N\,\Delta\Phi/\Delta t$?

GROUP B
8. If a bar magnet is held in a vertical position with the **N** pole down and is dropped through a closed-loop coil whose plane is horizontal, what is the direction of the induced current as the **N** pole approaches the loop?
9. What is the direction of the force acting on the loop in Question 8 as the **N** pole approaches?
10. What physical quantities are measured in (a) webers, (b) webers per meter^2, (c) webers per second, (d) joules per coulomb?
11. Explain how Lenz's law illustrates conservation of energy.
12. Demonstrate that the product Blv is properly expressed in volts.
13. Demonstrate that the product QvB is properly expressed in newtons.
14. The **S** pole of a bar magnet is entering a solenoid. What is the direction of the flux lines of the induced current inside the solenoid?
15. A permanent magnet is moved away from a stationary coil as shown in the accompanying diagram. (a) What is

the direction of the induced current in the coil? (Indicate direction of electron flow in the straight section of conductor below the coil.) (b) What magnetic polarity is produced across the coil by the induced current? (c) Justify your answers to (a) and (b).

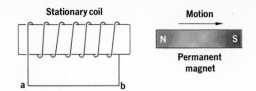

Stationary coil

Motion

N   S

Permanent magnet

a   b

Problems
GROUP A

1. A coil of 325 turns moving perpendicular to the flux in a uniform magnetic field experiences a change in flux linkage of 1.15×10^{-5} weber in 0.00100 s. What is the induced emf?
2. How many turns are required to produce an induced emf of 0.25 volt for a coil that experiences a change in flux linkage at the rate of 5.0×10^{-3} weber per s?
3. A straight conductor $1\overline{0}$ cm long is moved through a magnetic field perpendicular to the flux at a velocity of 75 cm/s. If the flux density is 0.025 weber/m², what emf is induced in the conductor?
4. A coil of 75 turns and an area of 4.0 cm² is removed from the gap between the poles of a magnet having a uniform flux density of 1.5 wb/m² in 0.025 s. What voltage is induced across the coil?
5. A rod 15 cm long is perpendicular to a magnetic field of 4.5×10^{-1} n/a m and is moved at right angles to the flux at the rate of $3\overline{0}$ cm/s. Find the emf induced in the rod.

GENERATORS AND MOTORS

20.7 The Generator Principle An emf is induced in a conductor whenever the conductor experiences a change in flux linkage. When the conductor is part of a closed circuit, an induced current can be detected in the circuit. By Lenz's law, work must be done to induce a current in a conducting circuit. Accordingly, this is a practical source of electric energy.

Moving a conductor up and down in a magnetic field is not a convenient method of inducing a current. A more practical way is to shape the conductor into a loop, the ends of which are connected to the external circuit by means of *slip rings*, and rotate it in the magnetic field. See Figure 20-8.

Such an arrangement is a basic *generator*. The loop across which an emf is induced is called the *armature*. The ends of the loop are connected to slip rings that rotate as the armature is turned. A graphite *brush* rides on each slip ring, connecting the armature to the external circuit. *An **electric generator** converts mechanical energy into electric energy.* The essential components of a generator are a *field magnet, an armature,* and *slip rings and brushes.*

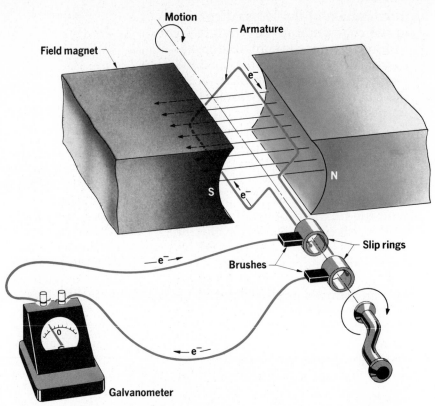

Figure 20-8. The essential components of an electric generator.

Figure 20-9. The left-hand generator rule.

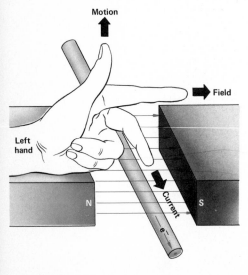

The induced emf across the armature and the induced current in the closed circuit result from relative motion between the armature and the magnetic flux that effects a change in the flux linkage. Thus either the armature or the magnetic field may be rotated. In some commercial generators the field magnet is rotated and the armature is the stationary element.

According to Lenz's law, an induced current will appear in such a direction that the magnetic force on the electrons comprising the induced current opposes the motion producing it. The direction of induced current in the armature loop of a generator can be easily determined by use of a *left-hand rule* known as the *generator rule.* This rule takes into account Lenz's law and the fact that an electric current in a metal conductor consists of a flow of electrons. See Figure 20-9.

The generator rule: Extend the thumb, forefinger, and the middle finger of the left hand at right angles to each other. Let the forefinger point in the direction of the magnetic flux and the thumb in the direction the conductor is moving; the middle finger points in the direction of the induced electron current.

20.8 The Basic a-c Generator   The two sides of the conducting loop in Figure 20-8 move through the magnetic flux in opposite directions when the armature is rotated. By applying the generator rule to each side of the loop, the direction of the induced current is shown to be toward one slip ring and away from the other. Thus a single-direction current loop is established in the closed circuit. As the direction of each side of the loop changes with respect to the flux, the direction of the induced current is reversed. As the armature rotates through a complete cycle, there are two such reversals in direction of the induced current.

In Figure 20-10 one side of a conducting loop rotating in a magnetic field is shown cross-sectionally in color while the other side is shown in black. In (A) the colored side of the loop is shown moving down and cutting through the flux. By the generator rule, the induced current is directed into the page. (The white × represents the tail of the arrow.) The motion of the black side of the loop induces a current directed out of the page. (The white dot represents the head of the arrow.)

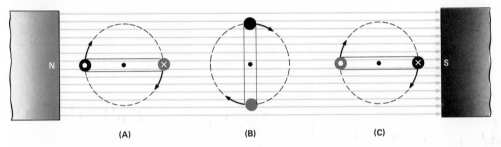

(A)   (B)   (C)

An emf is induced in a conductor as a result of a change in the flux linking the conductor. In (B) both sides of the loop are moving parallel to the flux and there is no change in linkage. Therefore, no emf is induced across the loop and no current is in the closed circuit.

In (C) the black side of the loop is moving down and cutting through the flux so that the induced current is directed into the page. The colored side is moving up through the flux, and thus the induced current is directed out of the page. The direction of the current in the circuit is the reverse of that in (A). A quarter cycle later, the emf again drops to zero and there is no current in the circuit. The emf induced across the conducting loop reaches a maximum value when the sides of the loop are moving perpendicular to the magnetic flux. See Section 20.2.

We can see from Figure 20-10 that the magnitude of the induced emf across the conducting loop must vary from zero through a maximum and back to zero during a half

Figure 20-10. An emf is induced only when there is a change in the flux linking a conductor.

cycle of rotation. The emf must then vary in magnitude in a similar manner, from zero through the same magnitude maximum and back to zero, during the second half cycle. However, polarity across the loop is opposite. The emf across the loop thus *alternates* in polarity.

Similarly, the current in a circuit connected to the rotating armature by way of the slip rings alternates in direction: electrons flowing in one direction during one half cycle and in the opposite direction during the other half cycle. *A current that has one direction during part of a generating cycle and the opposite direction during the remainder of the cycle is called an* **alternating current,** *ac.* A generator that produces an alternating current must, of course, produce an alternating emf. See Figure 20-11.

Figure 20-11. One cycle of operation of an a-c generator.

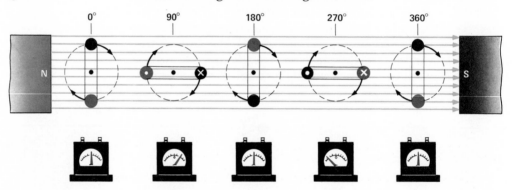

Commercial electric power is usually supplied by the generation of alternating currents and voltages. Such power is referred to as *a-c power.* The expressions *a-c* and *d-c* are often used to distinguish between alternating-current and direct-current properties: a-c voltage and d-c current are two examples.

20.9 Instantaneous Current and Voltage The open-circuit voltage across a battery has a constant magnitude characteristic of the chemical makeup of the battery. The voltage across an armature rotating in a magnetic field, however, has no constant magnitude. It varies from zero through a maximum in one direction and back to zero during one half cycle. It then rises to a maximum in the opposite direction and falls back to zero during the other half cycle of armature rotation. At successive instants of time, different magnitudes of induced voltage exist across the rotating armature. *The magnitude of a varying voltage at any instant of time is called the* **instantaneous voltage,** *e.*

The maximum voltage, E_{max}, is obtained when the conductor is moving perpendicular to the magnetic flux because the rate of change of flux linking the conductor is

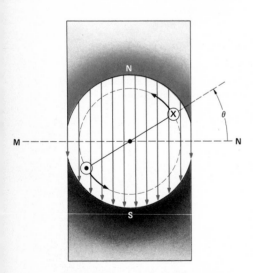

Figure 20-12. The instantaneous value of an induced voltage varies with the sine of the displacement angle of the loop in the magnetic field.

maximum during this time. *If the armature is rotating at a constant rate in a magnetic field of uniform flux density, the magnitude of the induced voltage varies sinusoidally (as a sine wave) with respect to time.*

In Figure 20-12 a single loop is rotating in a uniform magnetic field. When the plane of the loop is perpendicular to the flux (**MN** of Figure 20-12), the conductors are moving parallel to the flux lines; the displacement angle of the loop is said to be zero. We shall refer to this angle between the plane of the loop and the perpendicular to the magnetic flux as θ (theta). When $\theta = 0°$ and $180°$, $e = 0$ v. When $\theta = 90°$, $e = E_{max}$; and when $\theta = 270°$, $e = -E_{max}$. These relationships are apparent from Figure 20-11. In general, the instantaneous voltage, e, varies with the sine of the displacement angle of the loop.

$$e = E_{max} \sin \theta$$

The current in the external circuit of a simple generator consisting of pure resistance will vary in a similar way; the *maximum current, I_{max},* occurs when the induced voltage is maximum. From Ohm's law

$$I_{max} = \frac{E_{max}}{R}$$

The *instantaneous current, i,* is accordingly

$$i = \frac{e}{R}$$

but

$$e = E_{max} \sin \theta$$

so

$$i = \frac{E_{max}}{R} \sin \theta$$

and

$$i = I_{max} \sin \theta$$

20.10 Practical a-c Generators   The simple generator consists of a coil rotating in the magnetic field of a permanent magnet. Any small generator employing a permanent magnet is commonly called a *magneto.* Magnetos are often used in the ignition systems of gasoline engines for lawn mowers, motorbikes, and boats.

The generator output is increased in a practical generator by increasing the number of turns on the armature or increasing the field strength. The field magnets of large generators are strong electromagnets; in a-c generators they are supplied with direct current from an auxiliary d-c generator called an *exciter.* See Figure 20-13.

The performance of large a-c generators is generally more satisfactory if the armature is stationary and the field rotates inside the armature. Such stationary armatures are

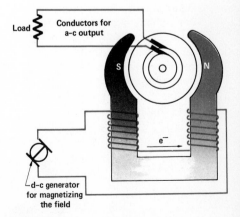

Figure 20-13. The field of a large alternating-current generator is produced by an electromagnet.

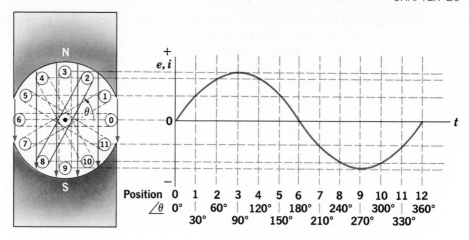

Figure 20-14. A sine curve representing current or voltage generated by a single-loop armature rotating at a constant rate in a uniform magnetic field. Positions 0 through 12 on the graph correspond to positions of the rotating armature in the magnetic field, as shown to the left.

The "frequency" of a d-c voltage or current is considered to be zero.

Figure 20-15. The principle of the three-phase alternator. Armature coils schematically diagrammed in (A) are spaced 120° apart and generate peak voltage output in three phases as shown in (B). In (C) the output is shown as a voltage diagram.

referred to as *stators* and the rotating field magnets as *rotors.* Circuit current is taken from the stator at the high generated voltage without the use of slip rings and brushes. The exciter voltage, which is much lower than the armature voltage, is applied to the rotor through slip rings and brushes.

In a simple two-pole generator one cycle of operation produces one cycle or two alternations of induced emf, as shown in Figure 20-14. If the armature (or the field) rotates at the rate of 60 cycles per second, the frequency, f, of the generated voltage sine wave is 60 hertz and the period, T, is $\frac{1}{60}$ second.

*The **frequency** of an alternating current or voltage is the number of cycles of current or voltage per second.* If the generator has a 4-pole field magnet, 2 cycles of emf are generated during 1 revolution of the armature or field. Such a generator turning at 30 rps would generate a 60-hz voltage. In general

$$f = \textbf{No. of pairs of poles} \times \textbf{revolution rate}$$

Practically all commercial power is generated by *three-phase* generators having three armature coils spaced symmetrically and producing emfs spaced 120° apart. The coils are usually connected so that the currents are carried by three conductors.

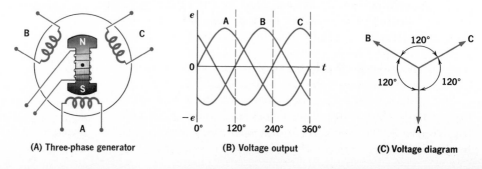

(A) Three-phase generator　　　**(B) Voltage output**　　　**(C) Voltage diagram**

It is evident from Figure 20-15(B) that three-phase power is smoother than the single-phase power of Figure 20-14. Electric power is transmitted by a three-phase circuit but it is commonly supplied to the consumer by a single-phase circuit. A modern center for the control of production and distribution of electric energy is shown in Figure 20-16.

20.11 The d-c Generator The output of an a-c generator is not suitable for circuits that require a direct current. However, the a-c generator can be made to supply a *unidirectional* current to the external circuit by connecting the armature loops to a *commutator* instead of slip rings. *A **commutator** is a split ring, each segment of which is connected to an end of a corresponding armature loop.*

The current and voltage generated in the armature are alternating, as we would expect. By means of the commutator, the connections to the external circuit are reversed at the same instant that the direction of the induced emf reverses in the loop. See Figure 20-17. The alternating current in the armature appears as a *pulsating* direct current in the external circuit, and a pulsating d-c voltage appears across the load. A graph of the instantaneous values of the pulsating current from a generator with a two-segment commutator plotted as a function of time is shown in Figure 20-18. A graph of the voltage across a resistance load would have a similar form. Compare this pulsating d-c output to the a-c output of the generator in Figure 20-14.

To secure from a d-c generator a more constant voltage and one having an instantaneous value that approaches the average value of the emf induced in the entire armature, many coils are wound on the armature. Each coil is connected to a different pair of commutator segments. The two brushes of each commutator segment are positioned so that they are in contact with successive pairs of commutator segments at the time when the induced emf in their respective coils is in the E_{max} region. See Figure 20-19.

20.12 Field Excitation Most d-c generators use part of the induced power to energize their field magnets and are said to be *self-exciting*. The field magnets may be connected in series with the armature loops so all of the generator current passes through the coil windings. In the *series-wound* generator an increase in the load increases the magnetic field and hence the induced emf. See Figure 20-20.

The field magnets may also be connected in parallel with the armature so only a portion of the generated current is used to excite the field. In this *shunt-wound* generator, Figure 20-21, an increase in load results in a decrease in the field and hence a decrease in the induced emf.

Figure 20-16. An electric system operating center utilizing computer technology for monitoring and controlling the production and transmission of electric power from generating stations to customers.

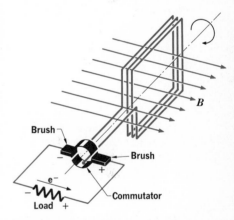

Figure 20-17. A split-ring commutator of two segments.

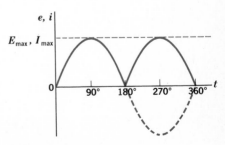

Figure 20-18. The variation of current or voltage with time in the external circuit of a simple generator with a two-segment commutator.

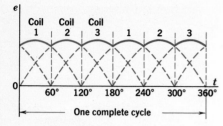

Figure 20-19. The output of a d-c generator having three armature coils and a six-segment commutator is fairly constant.

By using a combination of both series and shunt windings to excite the field magnets, the potential difference across the external circuit of a d-c generator may be maintained fairly constant; an increase in load causes an increase in current in the series windings and a decrease in current in the parallel windings. With the proper number of turns of each type of winding, a constant flux density can be maintained under varying loads. A compound-wound generator is shown in Figure 20-22.

20.13 Ohm's Law and Generator Circuits We can think of the armature turns as the source of emf in the d-c generator circuit. If the field magnet were separately excited, this emf would appear across the armature terminals on open-circuit operation since there would be no induced armature current.

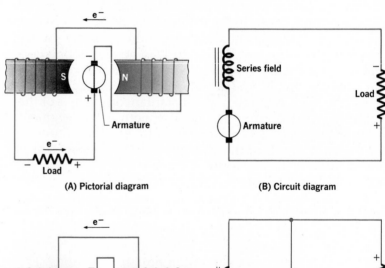

Figure 20-20. A series-wound d-c generator.

(A) Pictorial diagram   **(B) Circuit diagram**

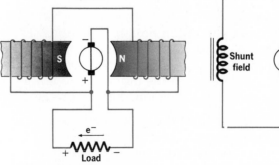

Figure 20-21. A shunt-wound d-c generator.

(A) Pictorial diagram   **(B) Circuit diagram**

However, in a self-excited generator the armature circuit is completed through the field windings. The resistance of the armature turns, r_a, is in this current loop, and a situation analogous to that of a battery with internal resistance furnishing current to an external circuit results. A potential drop, $I_a r_a$, of opposite polarity to the induced

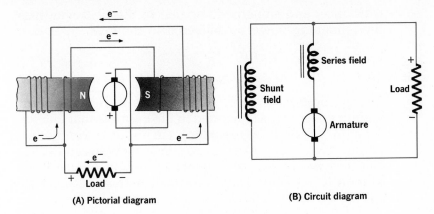

(A) Pictorial diagram   (B) Circuit diagram

emf, must appear across the armature. The armature potential difference, V, is

$$V = E - I_a r_a$$

Figure 20-22. A compound-wound d-c generator.

Figure 20-23 is a resistance circuit of a series-wound generator. The resistance of the field windings, R_f, is in series with the internal resistance of the armature and the load resistance so that the armature current is present in all three resistances.

In Figure 20-21, the resistance of the shunt windings is in parallel with the load. The resistance circuit of a shunt-wound generator is diagrammed in Figure 20-24, with the resistance of the shunt windings shown as R_f. The total current of the circuit, I_a, must be in r_a, producing an $I_a r_a$ drop across the armature. The potential difference, V, which has been shown to be $E - I_a r_a$, then appears across the network consisting of R_f and R_L in parallel. The following Ohm's law relationships hold:

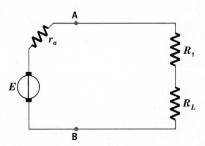

Figure 20-23. The resistance circuit of a series-wound d-c generator.

$$I_f = \frac{V}{R_f}$$

$$I_L = \frac{V}{R_L}$$

and $$I_a = I_f + I_L$$

The total electric power in the generator circuit, P_T, derived from the mechanical energy source that turns the armature, is the product of the armature current, I_a, and the induced emf, E.

$$P_T = EI_a$$

Some of this power is dissipated as heat in the armature

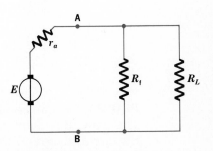

Figure 20-24. The resistance circuit of a shunt-wound d-c generator.

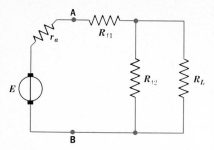

Figure 20-25. The resistance circuit of a compound-wound d-c generator.

resistance, and some is dissipated in the field windings; the remaining power is delivered to the load.

$$P_T = P_a + P_f + P_L$$

or

$$EI_a = I_a^2 r_a + I_f^2 R_f + I_L^2 R_L$$

The application of these relationships to the d-c shunt-wound generator is illustrated in the example that follows.

Figure 20-25 is a resistance circuit of a compound-wound generator. Observe the similarity between this circuit and the resistance network shown in Figure 17-21, Section 17.10. How would you analyze this resistance circuit of the compound-wound d-c generator?

EXAMPLE   The armature resistance of a shunt-wound d-c generator is 0.50 ohm, the resistance of the shunt winding is 60.0 ohms, and 0.960 kilowatt is delivered to a 15.0-ohm load. (a) What is the potential difference across the load? (b) What is the magnitude of the armature current? (c) What is the emf of the generator?

SOLUTION   The circuit of Figure 20-24 applies.

(a) $P_L = I_L^2 R_L = \dfrac{V_L^2}{R_L}$

$V_L = \sqrt{P_L R_L} = \sqrt{960 \text{ w} \times 15.0 \ \Omega}$
$V_L = 12\overline{0} \text{ v}$

(b) $I_a = I_f + I_L$

$I_f = \dfrac{V_f}{R_f}$ and $I_L = \dfrac{V_L}{R_L}$

But $V_L = V_f = V$

$I_a = \dfrac{V}{R_f} + \dfrac{V}{R_L} = \dfrac{12\overline{0} \text{ v}}{60.0 \ \Omega} + \dfrac{12\overline{0} \text{ v}}{15.0 \ \Omega}$

$I_a = 2.00 \text{ a} + 8.00 \text{ a} = 10.00 \text{ a}$

(c) $V = E - I_a r_a$
$E = V + I_a r_a = 12\overline{0} \text{ v} + (10.00 \text{ a} \times 0.50 \ \Omega)$
$E = 125 \text{ v}$

PRACTICE PROBLEMS   1. A shunt-wound d-c generator has an armature resistance of 0.35 Ω and a field-windings resistance of 75.0 Ω. The generator delivers 1.44 kw to a load resistance of 22.5 Ω. (a) Draw the circuit diagram. (b) What is the potential difference across the load? (c) What is the magnitude of the armature current? (d) What is the emf of the generator?   Ans.   (a) As Figure 20-24; (b) 18$\overline{0}$ v; (c) 10.40 a; (d) 184 v

2. A series-wound d-c generator has an armature resistance of 0.25 Ω and a field-windings resistance of 2.50 Ω. When the generator is connected to a load resistance of 7.50 Ω, the circuit current is 4.80 a. (a) Draw the circuit diagram. (b) What is the potential difference across the external circuit? (c) What power is dissipated in the load resistance? (d) What is the emf of the generator?   *Ans.*   (a) As Figure 20-23; (b) 48.0 v; (c) 173 w; (d) 49.2 v

20.14 The Motor Effect   We have learned that a current is induced in a conducting loop when the conductor is moving in a magnetic field so that the magnetic flux linking the loop is changing. This is the generator principle. By Lenz's law, we have seen that work must be done against a magnetic force as an induced current is generated in the conducting loop. Recall that an induced current always produces a magnetic force that opposes the force causing the motion by which the current is induced.

Figure 20-26. A visual representation of the motor effect.

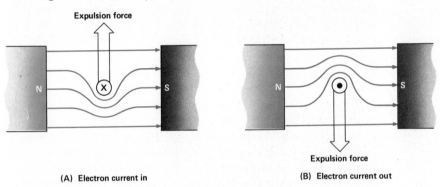

(A) Electron current in   (B) Electron current out

Instead of employing a mechanical effort to move a conductor poised in the magnetic field, suppose a current is *supplied* to the conductor from an external source. The magnetic field is distorted, as discussed in Section 20.6, and the resulting magnetic force tends to expel the conductor from the magnetic field. This action is known as the *motor effect*, and it is illustrated in Figure 20-26.

If a current is supplied to an armature loop poised in a uniform magnetic field, as in Figure 20-27, the field around each conductor is distorted. A force acts on each side of the loop proportional to the flux density and the current in the armature loop. These two forces, equal in magnitude but opposite in direction, constitute a *couple* and produce a torque, *T*, that causes the armature loop to rotate about its axis. The magnitude of this torque is equal to the product of the force and the *perpendicular* distance between the two forces.

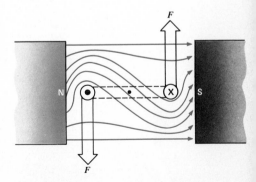

Figure 20-27. The resultant magnetic field of a current loop in the external field of a magnet gives a visual suggestion of the forces acting on the loop.

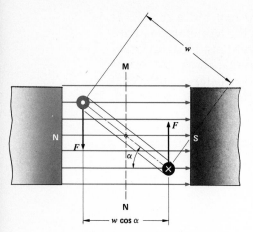

Figure 20-28. The torque on a current loop in a magnetic field is proportional to the perpendicular distance between the forces acting on the conductors.

It is evident from Figure 20-27 that the perpendicular distance between the torque-producing forces is maximum when the conductors are moving perpendicular to the magnetic flux. This distance is equal to the width of the armature loop. When the loop is in any other position with respect to the flux lines, the perpendicular distance between the forces is less than the width of the loop, and consequently the resulting torque must be less than the maximum value. See Figure 20-28.

We shall refer to the angle between the plane of the loop and the magnetic flux as angle α (alpha). When the angle is zero, the plane of the loop is parallel to the flux lines and the torque is maximum.

$$T_{max} = Fw$$

where F is the magnetic force acting on either conductor and w is the width of the conducting loop.

As the armature turns, the angle α approaches 90° and the torque diminishes since the perpendicular distance between the couple approaches zero. In general, the perpendicular distance between the forces acting on the two conductors is equal to $w \cos \alpha$, as shown in Figure 20-28. Hence

$$T = Fw \cos \alpha$$

When the plane of the loop is perpendicular to the magnetic flux, angle α is 90°, the cosine of 90° = 0, and the torque is zero. As the inertia of the conductor carries it beyond this point, a torque develops that reverses the motion of the conductor and returns it to the zero-torque position. In order to prevent this action, the direction of the current in the armature loop must be reversed at the proper instant. To reverse the current when the neutral position is reached, the conducting loop terminates in a commutator.

An electric motor performs the reverse function of a generator. *Electric energy is converted to mechanical energy using the same electromagnetic principles employed in the generator.* We can determine the direction of the motion of the conductor on a motor armature by use of a *right-hand rule* known as the *motor rule*. It is illustrated in Figure 20-29.

The motor rule: Extend the thumb, forefinger, and middle finger of the right hand at right angles to each other. Let the forefinger point in the direction of the magnetic flux and the middle finger in the direction of the electron current; the thumb points in the direction of the motion of the conductor.

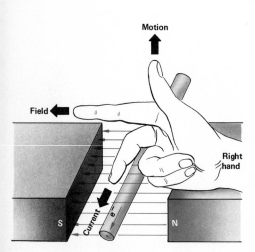

Figure 20-29. The right-hand motor rule.

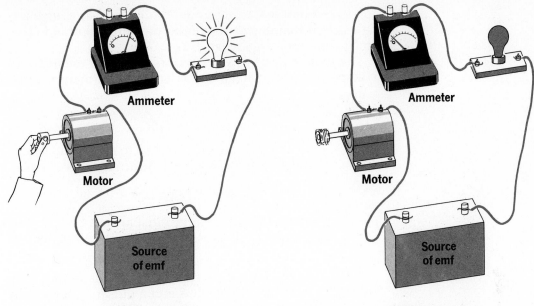

(A) Motor stalled   **(B) Motor running**

Figure 20-30. Demonstrating the back emf of a motor.

20.15 Back emf The simple d-c motor does not differ essentially from a generator; it has a field magnet, an armature, and a commutator ring. In fact, *the operating motor also acts as a generator.* As the conducting loop of the armature rotates in a magnetic field, an emf is induced across the armature turn. The magnitude of this emf depends on the speed of rotation of the armature.

Suppose that an incandescent lamp and an ammeter are connected in series with the armature of a small battery-driven motor, as illustrated in Figure 20-30. If the motor armature is held so that it cannot rotate as the circuit is closed, the lamp glows and the circuit current is indicated on the meter. Releasing the armature allows the motor to gain speed, and the lamp dims; the ammeter indicates a smaller current.

According to Lenz's law, the induced emf must oppose the motion inducing it. The emf induced by the generator action of a motor consequently opposes the voltage applied to the armature. *Such an induced emf is called the **back emf** of the motor.* The difference between the applied voltage and the back emf determines the current in the motor circuit.

A motor running at full speed under no load generates a back emf nearly equal to the applied voltage; thus a small current is required in the circuit. The more slowly the armature turns, the smaller is the back emf and consequently the larger is the voltage difference and the circuit current. A motor starting under a full load has a large initial current that decreases due to the generation of a back emf as the motor gains speed.

The induced emf in a generator is equal to the terminal voltage *plus* the voltage drop across the armature resistance.

$$E = V + I_a r_a$$

In the motor, the induced emf is equal to the terminal voltage *minus* the voltage drop across the armature resistance.

$$E = V - I_a r_a$$

Hence, the back emf in a motor must always be less than the voltage impressed across the armature terminals.

20.16 Practical d-c Motors A simple motor with a single armature coil would be impractical for many purposes because it has neutral positions and a pulsating torque. In practical motors a large number of coils is used in the armature; in fact there is little difference in the construction of motor and generator armatures. Multiple-pole field coils can be used to aid in the production of a uniform torque. The amount of torque produced in any given motor is proportional to the armature current and to the flux density.

Depending on the method used to excite the field magnets, practical d-c motors are of three general types. These are *series, shunt,* and *compound* motors. The excitation methods are similar to those of the d-c generators discussed in Section 20.12.

20.17 Practical a-c Motors Nearly all commercial distributors of electric power supply alternating-current power. Except for certain specialized applications, a-c motors are far more common than d-c motors. Much of the theory and technology of a-c motors is complex. We will briefly summarize the general characteristics of three common types of a-c motors: *the universal motor, the induction motor,* and *the synchronous motor.*

1. The universal motor. Any small d-c series motor can be operated from an a-c source. When this is done, the currents in the field windings and armature reverse direction simultaneously, maintaining torque in the same direction throughout the operating cycle. Heat losses in the field windings are extensive, however, unless certain design changes are incorporated into the motor. With laminated pole pieces and special field windings, small series motors operate satisfactorily on either a-c or d-c power. They are known as *universal* motors.

2. *The induction motor.* Induction motors are the most widely used a-c motors because they are rugged, simple to build, and well adapted to constant speed requirements. They have two essential parts—a stator of field coils and a rotor. The rotor is usually built of copper bars laid in slotted, laminated iron cores. The ends of the copper bars are shorted by a copper ring to form a cylindrical cage; this common type is known as a *squirrel-cage* rotor.

By using three pairs of poles and a three-phase current, the magnetic field of the stator is made to rotate electrically and currents are induced in the rotor. In accordance with Lenz's law, the rotor will then turn so as to follow the rotating field. Since induction requires relative motion between the conductor and the field, the rotor must *slip* or lag behind the field in order for a torque to be developed. An increase in the load causes a greater slip, a greater induced current, and consequently a greater torque.

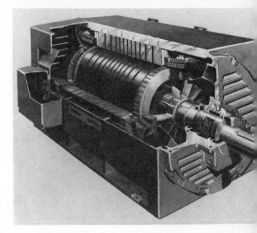

Figure 20-31. A cut-away view of an induction motor.

For its operation, an induction motor depends on a rotating magnetic field. Thus a single-phase induction motor is not self-starting because the magnetic field of its stator merely reverses periodically and does not rotate electrically. Single-phase induction motors can be made self-starting by a *split-phase winding*, a *capacitor*, a *shading coil*, or a *repulsion winding*. The name of the motor usually indicates the auxiliary method used for starting it.

3. *The synchronous motor.* We can illustrate the principle of synchronism by placing a thoroughly magnetized compass needle in a rotating magnetic field. The magnetized needle aligns itself with the magnetic field and rotates in synchronization with the rotating field.

The synchronous motor is a constant speed motor, running in synchronism with the a-c generator that supplies the stator current. However, it is not self-starting; the rotor must be brought up to synchronous speed by auxiliary means.

Electric clocks are operated by small single-phase synchronous motors that start automatically as a form of induction motor. Once synchronism is attained, they run as synchronous motors. Electric power companies maintain very accurate control of the 60-hz frequency of commercial power now that electric clocks are in general use.

Large industrial synchronous motors use an electromagnetic rotor supplied with d-c power, and a stator supplied with three-phase a-c power. The synchronous motors are used where the speed requirements are very exacting.

Questions
GROUP A

1. What are the essential components of an electric generator?
2. State the rule that helps us determine the direction of the induced current in the armature loops of a generator.
3. Distinguish between a direct current and an alternating current.
4. Under what circumstances does a simple generator produce a sine-wave variation of induced voltage?
5. What does the term θ (theta) represent in the expression $e = E_{max} \sin \theta$?
6. (a) What is a magneto? (b) For what is it used?
7. What is the function of an *exciter* in the generation of a-c power?
8. (a) What is meant by the frequency of an alternating current? (b) Under what circumstances will the frequency of a generated current be the same as the rps of the armature?
9. How can a generator be made to supply a direct current to its external circuit?
10. In what way is the output of a d-c generator different from the d-c output of a battery?
11. (a) What methods are used to energize the field magnets of d-c generators? (b) Draw a circuit diagram of each method.
12. What are the three power-consuming parts of a d-c generator circuit?
13. What is meant by the term α (alpha) in the torque expression $T = Fw \cos \alpha$?

14. State the rule that helps us determine the direction of motion of the armature loops of a motor.
15. (a) What is meant by back emf? (b) How is it induced in an electric motor?
16. What two quantities influence the amount of torque produced in a motor?
17. What are the three common types of a-c motors?

GROUP B

18. What is the advantage of having a 4- or 6-pole field magnet in a generator producing a 60-hz output?
19. How does an increase in the load on a series-wound d-c generator affect the induced emf? Explain.
20. How does an increase in the load on a shunt-wound d-c generator affect the induced emf? Explain.
21. How is torque produced on the armature loops of a motor?
22. Why is it true that an operating motor is also a generator?
23. The torque in a series motor increases as the load increases. Explain.
24. Explain why a single-phase induction motor is not self-starting.
25. Why are synchronous motors used in electric clocks?
26. Two conducting loops, identical except that one is silver and the other aluminum, are rotated in a magnetic field. In which case is the larger torque required to turn the loop?

Problems
GROUP A

1. A series-wound d-c generator turning at its rated speed develops an emf of 28 v. The current in the external circuit is 16 a and the armature resistance is 0.25 Ω. What is the potential drop

across the external circuit?
2. A d-c generator delivers $15\overline{0}$ a at $22\overline{0}$ v when operating at normal speed connected to a resistance load. The total losses are $320\overline{0}$ w. Determine the efficiency of the generator.
3. A shunt-wound generator has an armature resistance of 0.15 Ω and a

shunt winding of 75.0-Ω resistance. It delivers 18.0 kw at 24$\overline{0}$ v to a load. (a) Draw the circuit diagram. (b) What is the generator emf? (c) What is the total power delivered by the armature?

4. A magnetic force of 3.5 n acts on one conductor of a conducting loop in a magnetic field. A second force of equal magnitude but opposite direction acts on the opposite conductor of the loop. The conducting loop is 15 cm wide. Find the torque acting on the conducting loop (a) when the plane of the loop is parallel to the magnetic flux, (b) after the loop has rotated through 30°.

5. The maximum torque that acts on the armature loop of a motor is 10 m n. At what positions of the loop with respect to the magnetic flux will the torque be 5 m n?

6. A shunt-wound motor connected across a 117-v line generates a back emf of 112 v when the armature current is 1$\overline{0}$ a. What is the armature resistance?

INDUCTANCE

20.18 Mutual Inductance An emf is induced across a conductor in a magnetic field when there is a change in the flux linking the conductor. In our study of the generator we observed that an induced emf appears across the armature loop whether the conductors move across a stationary field or the magnetic flux moves across stationary conductors. *In either action there is relative motion between conductors and magnetic flux.*

This relative motion can be produced in another way. By connecting a battery to a solenoid through a contact key, an electromagnet is produced that has a magnetic field similar to that of a bar magnet. When the key is open, there is no magnetic field. As the key is closed, the magnetic field builds up from zero to some steady value determined by the number of ampere-turns. The magnetic flux spreads out and permeates the region about the coil. *An expanding magnetic field is a field in motion.* When the key in the solenoid circuit is opened, the magnetic flux collapses to zero. *A collapsing magnetic field is also a field in motion.* Its motion, however, is in the opposite sense to that of the expanding field.

Suppose the solenoid is inserted into a second coil whose terminals are connected to a galvanometer as in Figure 20-32. The coil connected to a current source is the *primary coil;* its circuit is the *primary circuit.* The coil connected to the galvanometer (the load) is the *secondary coil;* its circuit is the *secondary circuit.* At the instant the contact key is closed, a deflection is observed on the galvanometer. There is no deflection, however, while the key remains closed. When the key is opened a galvanometer deflection again occurs, but in the opposite direction. An emf is induced across the secondary turns whenever the

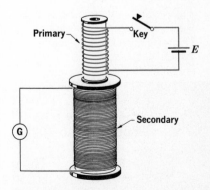

Figure 20-32. Varying the current in the primary induces an emf in the secondary.

flux linking the secondary is increasing or decreasing.

The relative motion between conductors and flux is in one direction when the field expands and in the opposite direction when the field collapses. Thus the emf induced across the secondary as the key is closed is of opposite polarity to that induced as the key is opened. The more rapid this relative motion, the greater the magnitude of the induced emf; the greater the number of turns in the secondary, the greater the magnitude of the induced emf. A soft-iron core placed in the primary greatly increases the flux density and the induced emf.

The SI symbol for the henry is H. Because of its similarity to the symbol, H, for magnetic force (also called magnetic field strength and magnetic intensity), the older form, h, is used in this book.

Two circuits arranged so that a change in magnitude of current in one causes an emf to be induced in the other show *mutual inductance*. The **mutual inductance,** M, of two circuits is the ratio of the induced emf in one circuit to the rate of change of current in the other circuit.

The unit of mutual inductance is called the *henry,* h, after the American physicist Joseph Henry. *The mutual inductance of two circuits is one henry when one volt of emf is induced in the secondary as the current in the primary changes at the rate of one ampere per second.*

$$M = \frac{-E_S}{\Delta I_P / \Delta t}$$

Here, M is the mutual inductance of the two circuits in henrys, E_S is the average induced emf across the secondary in volts, and $\Delta I_P / \Delta t$ is the time rate of change of current in the primary in amperes per second. The negative sign indicates that the induced voltage opposes the change in current according to Lenz's law. From this equation the emf induced in the secondary becomes

$$E_S = -M \frac{\Delta I_P}{\Delta t}$$

Figure 20-33. Joseph Henry, the American physicist for whom the unit of inductance is named.

See the following example.

EXAMPLE   Two coils have a mutual inductance of 1.25 henrys. Find the average emf induced in the secondary if the current in the primary builds up to 10.0 amperes in 0.0250 second after the switch is closed.

SOLUTION   The primary current must build up from zero to 10.0 amperes during 0.0250 s, so

$$\frac{\Delta I_P}{\Delta t} = \frac{10.0 \text{ a}}{0.0250 \text{ s}}$$

The mutual inductance is given as 1.25 h

$$E_S = -M\frac{\Delta I_P}{\Delta t} = -1.25 \text{ h} \times \frac{10.0 \text{ a}}{0.0250 \text{ s}}$$

$E_S = -50\overline{0} \text{ v}$, the emf induced in the secondary

Observe that since $\text{h} = \dfrac{\text{v}}{\text{a/s}}$, the units $\text{h} \times \dfrac{\text{a}}{\text{s}} = \dfrac{\text{v} \times \text{s} \times \text{a}}{\text{a} \times \text{s}} = \text{v}$

PRACTICE PROBLEMS　**1.** Two adjacent coils have a mutual inductance of 0.880 h. What is the average induced emf in the secondary if the current in the primary changes from 0.00 a to 18.0 a in 0.0550 s?　*Ans.* −288 v

2. Two coils have a mutual inductance of 1.70 h. The current in the primary reaches 11.5 a in 0.0326 s. What is the average induced emf in the secondary?　*Ans.* −60$\overline{0}$ v

3. The current in the primary of two adjacent coils changes from 0.0 a to 14 a in 0.025 s and the average induced emf in the secondary is 320 v. What is the mutual inductance of the two circuits?　*Ans.* 0.57 h

---

20.19 Self-Inductance　Suppose a coil is formed by winding many turns of insulated copper wire on an iron core and connected in a circuit to a 6-volt battery, a neon lamp, and a switch. See Figure 20-34. Since the neon lamp requires about 85 v-dc to conduct, it acts initially as an open switch in a 6-volt circuit.

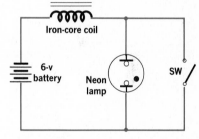

Figure 20-34. A circuit to demonstrate self-inductance.

When the switch is closed, a conducting path is completed through the coil; the fact that the lamp does not light is evidence that it is not in the conducting circuit. If now the switch is quickly opened, the neon lamp conducts for an instant, producing a flash of light. This means that the lamp is subjected to a potential difference considerably higher than that of the battery. What is the source of this higher voltage?

Any change in the magnitude of current in a conductor causes a change in the magnetic flux about the conductor. If the conductor is formed into a coil, a changing magnetic flux about one turn cuts across adjacent turns and induces a voltage across them. According to Lenz's law, the polarity of this induced voltage acts to oppose the motion of the flux inducing it. The sum of the induced voltages of all the turns constitutes a *counter* emf across the coil.

If a rise in current with its expanding magnetic flux is responsible for the counter emf, this rise in current will be opposed. Therefore, the counter emf is opposite in polarity to the applied voltage. When the switch in Figure 20-34 was closed, the rise in current from zero to the steady-state magnitude was opposed by the counter emf induced across the coil. Once the current reached a steady value, the magnetic field ceased to expand and the opposing voltage fell to zero.

If a fall in current with its collapsing flux is responsible for the induced voltage across the coil, the fall in current is opposed. In this case the induced emf has the same polarity as the applied voltage and tends to sustain the current in the circuit. A very rapid collapse of the magnetic field may induce a very high voltage. *It is the change in current, not the current itself, that is opposed by the induced emf.* It follows, then, that the greater the rate of change of current in a circuit containing a coil, the greater is the magnitude of the induced emf across the coil that opposes this change of current.

It is not the magnitude of current but the rate at which current is changing that relates to the magnitude of an induced emf.

The property of a coil that causes a counter emf to be induced across it by the change in current in it is known as *self-inductance*, or simply *inductance*. *The **self-inductance**, L, of a coil is the ratio of the induced emf across the coil to the rate of change of current in the coil.*

The unit of self-inductance is the *henry*, the same unit used for mutual inductance. Self-inductance is *one henry* if *one volt of emf* is induced across the coil when the current in the circuit changes at the rate of *one ampere per second.*

$$L = \frac{-E}{\Delta I / \Delta t}$$

Here L is the inductance in henrys, E is the emf induced in volts, and $\Delta I / \Delta t$ is the rate of change of current in amperes per second. The negative sign merely shows that the induced voltage opposes the change of current. From this expression, the induced emf is expressed by

$$E = -L \frac{\Delta I}{\Delta t}$$

The equation for inductance, L, is analogous to the defining equation for capacitance, C, as expressed in Section 16.14.

$$C = \frac{Q}{V}$$

The presence of a magnetic field in a coil conducting an electric current corresponds to the presence of an electric field in a charged capacitor.

Once a coil has been wound, its inductance is a constant property that depends on *the number of turns, the diameter of the coil, the length of coil,* and *the nature of the core.* Because of its property of inductance, a coil is commonly called an *inductor.*

Inductance in electricity is analogous to inertia in mechanics: the property of matter that opposes a *change* of velocity. If a mass is at rest, its inertia opposes a change that imparts a velocity; if the mass has a velocity, its inertia opposes a change that brings it to rest. The flywheel in mechanics illustrates the property of inertia. We know that energy is stored in a flywheel as its angular velocity is increased and is removed as its angular velocity is decreased.

In an electric circuit inductance has no effect as long as the current is steady. An inductance does, however, oppose any change in the circuit current. An increase in current is opposed by the inductance; energy is stored in its magnetic field since work is done by the source against the counter emf induced. A decrease in current is opposed by the inductance; energy is removed from its field, tending to sustain the current. *We can think of inductance as imparting a flywheel effect in a circuit having a varying current.* This concept is very important in alternating-current circuit considerations.

20.20 Inductors in Series and Parallel The total inductance of a circuit consisting of inductors in series or parallel can be calculated in the same manner as total resistance. When inductors are connected in series, the total inductance, L_T, is equal to the sum of the individual inductances providing there is no mutual inductance between them.

$$L_T = L_1 + L_2 + L_3 + \text{etc.}$$

However, when two inductors are in series and arranged so that the magnetic flux of each links the turns of the other, the total inductance is

$$L_T = L_1 + L_2 \pm 2M$$

The $\pm$ sign is necessary in the general expression for the counter emf induced in one coil by the flux of the other may either aid or oppose the counter emf of self-induction. The two coils can be connected either in series "aiding" or series "opposing," depending on the manner in which their turns are wound.

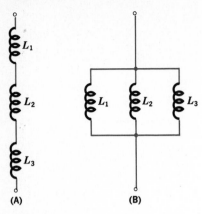

Figure 20-35. Inductors in series (A) and in parallel (B).

Inductors connected in parallel so that each is unaffected by the magnetic field of another provide a total inductance according to the following general expression:

$$\frac{1}{L_T} = \frac{1}{L_1} + \frac{1}{L_2} + \frac{1}{L_3} + \text{etc.}$$

See Figure 20-35 for schematic representation of these series and parallel connections for inductors.

20.21 The Transformer In principle the transformer consists of two coils, a primary and a secondary, electrically insulated from each other and wound on the same ferromagnetic core. Electric energy is transferred from the primary to the secondary by means of the magnetic flux in the core. A transformer is shown in its simplest forms in Figure 20-36.

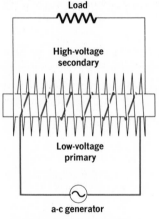 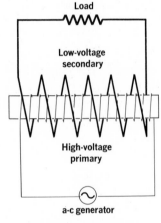

Figure 20-36. The transformer in its simplest forms.

Refer to Figure 20-14, in which the current sine wave produced by the a-c generator is given. The current is changing at its maximum rate as it passes through zero. Consequently, the emf induced across the secondary winding of a transformer is maximum as the primary current passes through zero. The polarity of the secondary emf reverses each time the primary current passes through a positive or negative maximum, since at these instants the current is changing at its minimum rate.

In a power transformer, a *closed* core is used to provide a continuous path for the magnetic flux, ensuring that practically all the primary flux links secondary turns. Since the same flux links both primary and secondary turns, the same emf per turn is induced in each and the ratio of secondary to primary emf is equal to the ratio of secondary to primary turns. Neglecting losses, we may assume the terminal voltages to be equal to the corresponding emfs. Thus

$$\frac{V_S}{V_P} = \frac{N_S}{N_P}$$

where V_S and V_P are the secondary and primary terminal voltages and N_S and N_P are the number of turns in the secondary and primary windings respectively.

If there are 20 turns in the secondary winding for every turn of the primary, the transformer is said to have a *turns ratio* of 20.

$$\textbf{Turns ratio} = \frac{N_S}{N_P} = \frac{20}{1} = 20$$

If the primary voltage is 110 volts, the secondary terminal voltage is 2200 volts and the transformer is called a *step-up transformer*. If the connections are reversed and the coil with the larger number of turns is made the primary, the transformer becomes a *step-down transformer*. The same primary voltage now produces a voltage across the secondary of 5.5 volts.

It can be shown that when power is delivered to a load in the secondary circuit, the product of the secondary current and secondary turns is essentially equal to the product of the primary current and primary turns.

$$I_S N_S = I_P N_P$$

or

$$\frac{I_S}{I_P} = \frac{N_P}{N_S}$$

Thus when the voltage is stepped up, the current is stepped down; there is no power gain as a result of transformer action. Ideally primary and secondary power are equal, but actually there are power losses as in any machine.

The efficiencies of practical transformers are high and constant over a wide power range. Transformer efficiencies above 95% are common. Efficiency can be expressed as the ratio of the power dissipated in the secondary circuit to the power used in the primary.

$$\textbf{Efficiency} = \frac{P_S}{P_P} \times \textbf{100\%}$$

20.22 Transformer Losses While transformer efficiencies are high, the transfer of energy from the primary circuit to the secondary circuit does not occur without some

Figure 20-37. An external view, low-voltage side of a power transformer.

loss. We shall consider two types: *copper losses* and *eddy-current losses*. They represent wasted energy and appear as heat.

1. *Copper losses.* These losses result from the resistance of the copper wires in the primary and secondary turns. Copper losses are I^2R heat losses. They cannot be avoided.

2. *Eddy-current losses.* When a mass of conducting metal is moved in a magnetic field or is subjected to a changing magnetic flux, induced currents circulate in the mass. These closed loops of induced current circulating in planes perpendicular to the magnetic flux are known as *eddy currents*.

Eddy currents in motor and generator armatures and transformer cores produce heat due to the I^2R losses in the resistance of the iron. They are induced currents that do no useful work, and they waste energy by opposing the change that induces them according to Lenz's law. Eddy-current losses are reduced by *laminating* the armature frames and cores. Thin sheets of metal with insulated surfaces are used to build up the armatures and cores. The laminations are set in planes parallel to the magnetic flux so that the eddy-current loops are confined to the width of the individual laminations. The high resistance associated with the narrow width of the individual laminations effectively reduces the induced currents and thus the I^2R heating losses.

Recall that resistance varies inversely with cross-sectional area.

Questions
GROUP A

1. Define (a) mutual inductance, (b) self-inductance.
2. What types of losses occur in a transformer?
3. What is the effect on the inductance of a coil when an iron core is inserted?
4. Why is a closed core used in a power transformer?
5. If an ideal transformer triples the voltage, how does the secondary current compare with the primary current?

GROUP B
6. Why is an inductor in an electric circuit said to impart a flywheel effect?

7. Why is a transformer considered to be an a-c circuit device?
8. The primary of a step-up transformer with negligible losses is connected to a source of emf and the secondary is connected to a resistance load. What is the relationship between primary and secondary (a) current, (b) power, and (c) number of turns?
9. Explain why the galvanometer in Figure 20-32 indicates a current in the secondary at the time the key in the primary circuit is closed or opened but not while it remains closed.
10. The voltage induced across the secondary of an ideal transformer is ten times the voltage applied across the primary. Explain why there is no power gain in the secondary circuit.

Problems

GROUP B

1. A pair of adjacent coils has a mutual inductance of 1.06 h. Determine the average emf induced in the secondary circuit when the current in the primary circuit changes from 0.00 a to 9.50 a in 0.0336 s.

2. When the primary circuit of a pair of adjacent coils is activated, the current surges to 12 a in 0.048 s and the emf induced in the secondary circuit is 270 v. Determine the mutual inductance of the pair of inductors.

3. A step-up transformer is used on a 120-v line to provide a potential difference of 2400 v. If the primary has 75 turns, how many turns must the secondary have (neglecting losses)?

4. An initial rise in current in a coil occurs at the rate of 7.5 a/s at the instant a potential difference of 16.5 v is applied across it. (a) What is the self-inductance of the coil? (b) At the same instant a potential difference of 5$\overline{0}$ v is induced across an adjacent coil. Find the mutual inductance of the two coils.

5. A coil with an inductance of 0.42 h and a resistance of 25 Ω is connected across a 110-v d-c line. What is the rate of current rise (a) at the instant the switch is closed; (b) at the instant the current reaches 85% of its steady-state value?

6. A 5:1 step-down transformer with negligible losses is connected to a 12$\overline{0}$-v a-c source. The secondary circuit has a resistance of 15.0 Ω. (a) What is the potential difference across the secondary? (b) What is the secondary current? (c) How much power is dissipated in the secondary resistance? (d) What is the primary current?

7. Assume that the transformer of Problem 7 is replaced by one having an efficiency of 92.5%. What is the primary current?

8. A transformer with a primary of 40$\overline{0}$ turns is connected across a 12$\overline{0}$-v a-c line. The secondary circuit has a potential difference of 300$\overline{0}$ v. The secondary current is 60.0 ma and the primary current is 1.85 a. (a) How many turns are in the secondary winding? (b) What is the transformer efficiency?

9. A 27$\overline{0}$-Ω resistor, a 2.50-h coil, and a switch are connected in series across a 12.0-v battery. At a certain instant after the switch is closed the current is 20.0 ma. What is the potential difference (a) across the resistor; (b) across the inductor? (c) What is the rate of change of current at this instant?

SUMMARY

An emf is induced in a conductor when relative motion between the conductor and a magnetic field produces a change in the flux linkage. The greater the rate of relative motion, the greater is the magnitude of the induced emf. If the conductor is part of a closed circuit, an electron current is induced in the circuit. The direction of an induced current is always in accord with Lenz's law.

An electric generator converts mechanical energy into electric energy. The direction of induced electron current in the armature turns is determined by use of the left-hand generator rule. The generator induces a sinusoidal emf across the armature turns. The current in the armature circuit alternates. The frequency of the generated current is expressed in hertz; one hertz is equivalent to one cycle per second. An a-c generator may be modified for a pulsating d-c output. The d-c generator is self-excited.

Electric motors convert electric energy

into mechanical energy. The motor effect is the result of an electric current in a magnetic field; it is the reverse of the generator effect. The direction of motion of the armature turns is determined by the use of the right-hand motor rule. An electric motor produces a back emf that subtracts from the applied voltage. Practical d-c motors are of three types: series, shunt, and compound wound. Three common types of a-c motors are the universal motor, the induction motor, and the synchronous motor. Of these, the induction motor is most widely used.

If a change in current in one circuit induces an emf in a second circuit, the two have a property of mutual inductance. An emf is induced across a coil by a change of current in the coil; this property is known as self-inductance. The unit of inductance is the henry. The induced emf across a given inductance depends on the time rate of change of current in the inductance. The property of inductance is described as a kind of electric inertia.

Transformers are alternating-current devices. Primary and secondary windings have a common core. For a given primary voltage, the turns ratio determines the secondary voltage. The transformer may either step up or step down the a-c voltage of the primary circuit. Efficient and economical distribution of a-c power is possible through the use of the transformer principle.

VOCABULARY

alternating current
back emf
commutator
compound-wound
   generator
eddy currents
exciter
force couple
generator rule
induced current

induced emf
instantaneous current
instantaneous voltage
Lenz's law
magnetic force
magnetic induction
magneto
motor effect
motor rule

mutual inductance
rotor
self-inductance
series-wound generator
shunt-wound generator
slip ring
stator
synchronous motor
three-phase generator

ALTERNATING-CURRENT CIRCUITS

POSTE ITALIANE

CENT. 50

Guglielmo Marconi developed the first practical wireless telegraph system in 1895 and successfully transmitted the first transatlantic radio signals in 1901. Later he also did important pioneering work with short-wave radio transmission. Marconi won the Nobel Prize in Physics in 1909.

a-c MEASUREMENTS

21.1 Power in a-c Circuits A single conducting loop rotating at a constant speed in a uniform magnetic field generates an alternating emf. The magnitude of the emf varies with the sine of the angle that the plane of the loop makes with the perpendicular to the magnetic flux. The instantaneous value of the emf is

$$e = E_{max} \sin \theta$$

The instantaneous current in a pure resistance load comprising the external circuit of the generator is similarly expressed as

$$i = I_{max} \sin \theta$$

This current is a consequence of the alternating voltages impressed across the load, and its maxima and minima occur at the same instants as the voltage maxima and minima. The alternating current and voltage are said to be *in phase*. This in-phase relationship is characteristic of a resistance load across which an alternating voltage has been applied. Sine curves of in-phase alternating current and voltage are shown in Figure 21-2.

In a d-c circuit, electric power, P, is the product $E \times I$. In an a-c circuit, the instantaneous power, p, is the product $e \times i$. These relationships are shown in Figure 21-3.

The voltage and current in Figure 21-3(B) are shown in phase as they are in a resistive load. The instantaneous

In this chapter you will gain an understanding of:

➤ the distinction between a-c power and d-c power
➤ the significance of effective values of alternating current and voltage
➤ the effects of inductance and capacitance in a-c circuits
➤ impedance in a-c circuits containing inductive and capacitive reactance
➤ the effect of circuit impedance on the phase relation between current and voltage
➤ applications of Ohm's law in a-c circuits
➤ the special condition of resonance in a-c circuits
➤ the importance of resonant circuits

power curve varies between some positive maximum and zero and has an average value P. Observe that the instantaneous power curve, p, has a frequency twice that of e and i. The ordinates of this power curve are always positive since e and i are in phase. During the first half-cycle, both e and i are positive, so their product is positive. During the second half-cycle, both e and i are negative, but their product is positive.

Figure 21-1. Thomas A. Edison and Charles P. Steinmetz in Steinmetz's laboratory in Schenectady, New York, in 1922. Steinmetz was a brilliant mathematician and electrical engineer. He developed the mathematical analysis for alternating-current circuits and provided the firm mathematical base of electrical engineering.

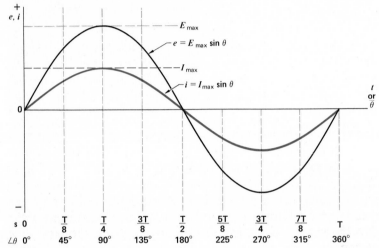

Figure 21-2. One cycle of alternating current and voltage in phase.

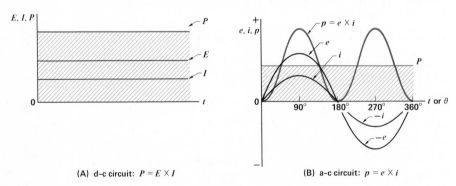

(A) d-c circuit: $P = E \times I$ (B) a-c circuit: $p = e \times i$

Figure 21-3. A comparison of d-c and a-c power.

21.2 Effective Values of Current and Voltage If a zero-centered d-c ammeter were placed in an a-c circuit, the pointer would tend to swing alternately in the positive and negative directions. At all but the very lowest frequencies, the inertia of the meter movement would cause the pointer to remain at zero. This, in effect, averages out the sine-wave variations. The same would be true of an a-c voltage measurement. *The average value of a sine curve of alternating current or voltage is zero.* This is true regardless of the magnitude of the maximum values attained.

The most useful value of an alternating current is based on its heating effect in an electric circuit and is commonly referred to as the *effective value. The **effective value** of an alternating current is the number of amperes that, in a given resistance, produces heat at the same average rate as that number of amperes of steady direct current.*

Let us assume that a resistance element is immersed in water in a calorimeter and that a steady direct current of 1 ampere in the circuit raises the temperature of the water 25 C° in 10 minutes. An alternating current in the same resistance that would raise the temperature the same amount in the same time is said to have an effective value of 1 ampere. The symbol for the effective value of an alternating current is I, the same as that for steady direct current. When the magnitudes of alternating currents are given in amperes, they are understood to be effective values unless other values are clearly stated.

From Joule's law, $Q = I^2Rt/J$, we know that the heating effect of an electric current is proportional to I^2. The average rate of heat production by an alternating current is proportional to the mean (average) of the instantaneous values of current squared. From Figure 21-4 it is evident that the mean value of I^2 is $\frac{1}{2}I_{max}^2$ because the square of the instantaneous values of the alternating current of frequency f gives a curve of frequency $2f$ that varies between I_{max}^2 and zero. Thus

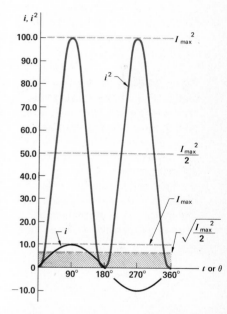

Figure 21-4. Current and current squared curves showing the effective value of an alternating current.

$$I^2 = \frac{I_{max}^2}{2}$$

and

$$I = \sqrt{\frac{I_{max}^2}{2}} = \frac{I_{max}}{\sqrt{2}}$$

$$I = 0.707\ I_{max}$$

Then

$$I_{max} = 1.414\ I$$

A sine curve of alternating current having a maximum value of 10.0 amperes is plotted in Figure 21-4. *The effective value is the square root of the mean of the instantaneous values squared* and is frequently called the *root-mean-square, rms,* value. In the example shown, the rms, or effective, value of current is

$$I = \sqrt{\frac{I_{max}^2}{2}} = \sqrt{\frac{100\ a^2}{2}} = \sqrt{50.0\ a^2} = 7.07\ a$$

The rms, or effective, value of an a-c emf is expressed similarly.

$$E = 0.707\ E_{\text{max}}$$

Then
$$E_{\text{max}} = 1.414\ E$$

For a potential difference that does not include a source of emf, these expressions become

$$V = 0.707\ V_{\text{max}}$$

and
$$V_{\text{max}} = 1.414\ V$$

In special applications a-c meters can be calibrated to indicate maximum values.

Thus a house-lighting circuit rated at 120 volts (effective) varies between +170 volts and −170 volts during each cycle. An appliance that draws 10.00 amperes (effective) when connected across this circuit has a current that varies between +14.14 amperes and −14.14 amperes. The meters used for measurements in a-c circuits are usually designed to indicate effective values rather than maximum values. Representative a-c meter movements are shown in Figure 21-5, (A, B, C, D).

If the load in an a-c circuit is a pure resistance, the current and voltage are in phase and the average power consumed is

$$P = I^2R = VI$$
$$P = 120\ \text{v} \times 10.00\ \text{a}$$
$$P = 1200\ \text{w}$$
$$(\text{v} \times \text{a} = \text{j/c} \times \text{c/s} = \text{j/s} = \text{w})$$

If, however, the current and voltage are not in phase, *the average power of the circuit is not a simple product of the*

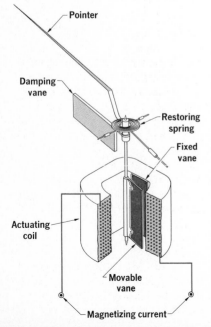

Figure 21-5(A). Moving iron-vane meter. When a magnetizing current is in the actuating coil, a repelling force develops between the fixed and movable iron vanes. The moving vane exerts a force against the restoring spring, and the final pointer position is a measure of the current in the coil. The moving-vane mechanism is used in inexpensive a-c ammeters and voltmeters.

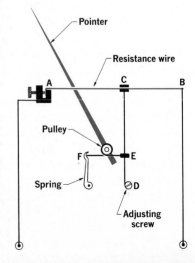

Figure 21-5(B). Hot-wire meter. When a current is carried in the meter circuit, the temperature of the platinum-alloy resistance wire **AB** rises because of I^2R heating. The resulting expansion reduces the tension on wire **CD** and allows the spring to pull thread **EF** to the left. The pulley rotates and moves the pointer slowly to the final deflection. The hot-wire mechanism is sensitive to temperature variations. A low-resistance shunt is added for ammeter functions. A high resistance is added in series with wire **AB** for voltmeter functions.

Figure 21-5(C). Electrodynamometer. The electrodynamometer is similar to the galvanometer movement of d-c meters except that it does not include a permanent magnet. The moving coil rotates in the magnetic field of a pair of fixed coils carrying magnetizing current. As a current in the fixed coils produces magnetic flux **B**, a current in the moving coil produces a flux along its axis and the coil tends to align the two magnetic fields. The torque produces a pointer deflection related to the product of the two coil currents. The dynamometer mechanism is used in quality ammeters, voltmeters, and wattmeters.

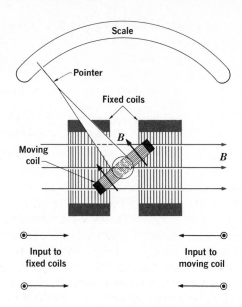

effective values of voltage and current. The products I^2R and VI are no longer the same; I^2R gives the *actual* power dissipation and VI gives the *apparent* power. This is a basic difference between d-c and a-c circuits, and it stems from the fact that phase differences between current and voltage may be produced by capacitance and inductance in an a-c circuit. These effects have added a new dimension to alternating-current circuits known as *impedance*, which is considered in Section 21.6.

21.3 Inductance in an a-c Circuit In Section 20.19 inductance was described as an inertia-like property that opposes any *change* in current in a coil and causes a counter emf proportional to the rate of change of current to be induced across the coil. Figure 21-6 illustrates the waveforms characteristic of a circuit consisting of pure inductance to which an alternating voltage is applied. The instant the current is passing through zero in a positive direction, it is changing at its maximum rate. Thus the counter emf induced across the coil at this instant must be at its negative maximum value. When the current reaches the positive maximum, the rate of change of current is zero and the induced emf is zero. Again 90° later, the current is changing in a negative direction at its maximum rate and the counter emf is at its positive maximum.

The counter emf follows behind the current inducing it by 90° and is opposite in polarity to the applied voltage, as shown in Figure 21-6. The current in the circuit must then

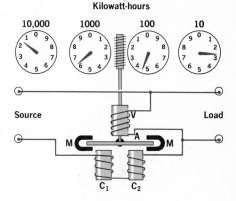

Figure 21-5(D). The induction watt-hour meter is used to record electric energy consumption. It is a small single-phase induction motor that turns at a rate proportional to the power being used. A coil **V** is connected in parallel with the circuit load and acts as a voltage winding. The coils C_1 and C_2 are connected in series with the load and act as current windings. An aluminum disk **A** turns as an induction motor armature to follow the sweeping flux set up by the combination of coils. Permanent magnets **M** induce eddy currents in the rotating disk that oppose the motion of the disk and produce the slippage required for operation.

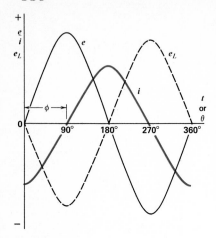

Figure 21-6. Current and voltages in an a-c circuit containing pure inductance.

lag behind the applied voltage by 90°. It is just as appropriate to consider that the voltage leads the current by 90°; the *difference in phase* is 90°.

Alternating currents and voltages do not have *direction* in the sense of vector quantities. They are not representable by directional components as are vector quantities such as force, velocity, momentum, and electric field intensity. These currents and voltages can be regarded, however, as rotors (rotating vectors), or *phasors*, rotating counterclockwise as spokes of a wheel and denoting quantities related in time or phase. *A **phasor** is a representation of the concepts of magnitude and direction in a reference plane.*

Phasors are used in Figure 21-7 to show the phase relation between current and voltage in a pure resistance (**A**) and in a pure inductance (**B**). Phasors usually represent effective values of current and voltage in *phase* diagrams such as those shown at the right in Figure 21-7. These diagrams are analyzed and solved by geometric or vector methods.

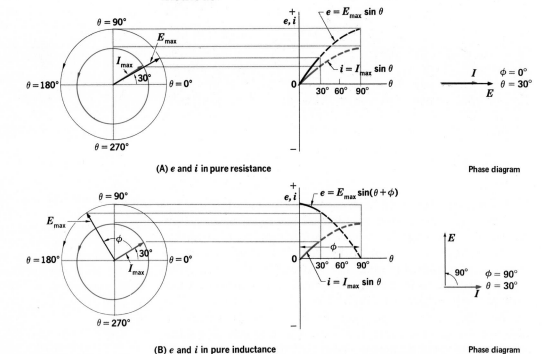

(A) *e* and *i* in pure resistance

(B) *e* and *i* in pure inductance

Figure 21-7. Phase relation of an a-c current and voltage in a pure resistance circuit (**A**) and in a pure inductance circuit (**B**).

The phase angle between voltage and current is commonly referred to as the angle ϕ (the lowercase Greek letter phi). This relation in an a-c circuit can be expressed as follows:

$$i = I_{max} \sin \theta$$
$$e = E_{max} \sin (\theta + \phi)$$

The phase angle, ϕ, of a leading voltage is *positive*, and it indicates that the voltage is ahead of the current in phase. Suppose the effective value of voltage represented in Figure 21-7(B) is 10.0 volts. When the displacement angle, θ, is 30°, the instantaneous value e will be

$$e = E_{max} \sin (\theta + \phi)$$
$$e = E_{max} \sin (30° + 90°)$$
$$e = E_{max} \sin 120°$$

But
$$E_{max} = 1.414\ E = 1.414 \times 10.0\text{ v}$$
$$E_{max} = 14.1\text{ v}$$

Then
$$e = 14.1\text{ v} \times \sin 120°$$
$$\sin 120° = \sin 60° = 0.866$$
$$e = 14.1\text{ v} \times 0.866$$
$$e = 12.2\text{ v}$$

Plotted as a function of time, the products of the instantaneous values of voltage and current in a pure inductance yield an instantaneous power curve as shown in Figure 21-8. Observe that the average power is zero. While the current is changing from zero to a positive maximum, energy is taken from the source and stored in the magnetic field of the inductor. These portions of the power curve are positive. When the current is changing from a positive maximum to zero, all the energy stored in the magnetic field is returned to the source. These portions of the power curve are negative. Therefore, the net energy removed from the source during a cycle by a pure inductance is zero; the product of the rms voltage and current can indicate only an *apparent power*.

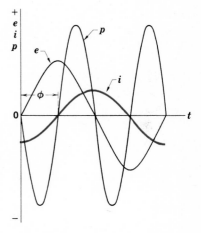

Figure 21-8. Power curve in a pure inductance.

21.4 Inductance and Resistance
An inductor cannot be entirely without resistance since it is not possible to have pure inductance in a circuit. The ordinary resistance of the conductor is an inherent property of any coil. We can treat the resistance of an inductive circuit as a lumped value in series with a pure inductance. This arrangement is shown in Figure 21-9(A).

The same current must be present in all parts of a series circuit. The circuit current is thus used as a reference to show the phase relation between current and voltages in the resistance and inductance. The voltage across the resistance is in phase with the circuit current; this in-phase relation is shown in Figure 21-9(B). The voltage across the inductance leads the circuit current by 90°, as shown in Figure 21-9(C). The voltage across the series combination of R and L then leads the circuit current by some angle between 0° and 90°.

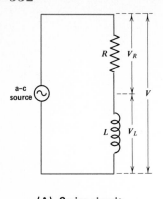

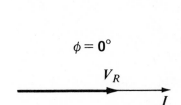

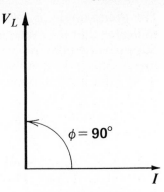

(A) Series circuit   (B) Voltage and current in phase   (C) Voltage and current 90° out of phase

Figure 21-9. An *L-R* circuit showing the phase relation between voltage and current in the resistance and the inductance.

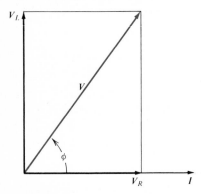

Figure 21-10. A phase diagram showing the addition of voltage phasors in a series circuit containing inductance and resistance.

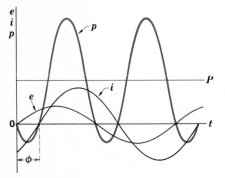

Figure 21-11. Power in an a-c circuit containing resistance and inductance.

With the circuit current of Figure 21-9 as a reference, it is apparent that V_R and V_L have a phase difference of 90°. These two voltages in series cannot be added algebraically but must be added vectorially to give the circuit voltage, V. This voltage V across the external circuit is the *resultant*, or vector sum, of voltages V_R and V_L shown in the phase diagram, Figure 21-10. The phase angle, ϕ, has a positive value less than 90°.

No power is dissipated in the inductance. The average power dissipated in the resistance is equal to the product of the effective values of the circuit current and the voltage across the resistance.

$$P_R = V_R I$$

From the voltage diagram shown in Figure 21-10 it is apparent that

$$V_R = V \cos \phi$$

Thus the actual power consumed in the external circuit is

$$P = VI \cos \phi$$

The total power consumed in the internal and external circuits is, of course,

$$P_T = EI \cos \phi$$

When the phase angle, ϕ, is *zero* (a pure resistive load), then $\cos \phi = 1$ and the product $VI \cos \phi = VI$. When the phase angle, ϕ, is +90° (a pure inductive load), $\cos \phi = 0$ and the product $VI \cos \phi = 0$ (Figure 21-8). The cosine of ϕ is known as the *power factor*, pf, of an a-c circuit. See Figure 21-11.

$$pf = \cos \phi = \frac{V_R}{V}$$

Electric power companies strive to maintain a power factor near unity in their distribution lines. Why?

21.5 Inductive Reactance The potential drop across the resistance in the circuit of Figure 21-9 is represented by the phasor V_R in Figure 21-10. This potential difference is equal to the product IR, where R is the common resistance to current in the external circuit. As a consequence of its nonresistive opposition to the current in the circuit, the inductance has a potential difference across it. This opposition is *nonresistive* because no power is dissipated in the inductance and is called *reactance*, X. Reactance is expressed in *ohms*. When reactance is due to inductance, it is referred to as *inductive reactance*, X_L. **Inductive reactance, X_L, is the ratio of the effective value of inductive potential, V_L, to the effective value of current, I.**

$$X_L = \frac{V_L}{I}$$

Thus the voltage V_L of Figure 21-10 is equal to the product IX_L.

The inductive reactance of a coil is directly proportional to the inductance and to the frequency of the current in the circuit. This is true because the rate of change of a given current increases with frequency. The larger the inductance, the greater is the opposition to the change in current at a given frequency.

$$X_L = 2\pi f L$$

When f is in hertz and L is in henrys, X_L is expressed in ohms of inductive reactance; 2π is a proportionality constant.

21.6 Impedance The vector sum of IX_L and IR is equal to the voltage, V, applied to the external circuit, as in Figure 21-10. This potential difference is the product of the circuit current, I, and the combined effect of the inductive reactance, X_L, and the resistance, R, in the load. The joint effect of reactance and resistance in an a-c circuit is called *impedance*, Z. It is expressed in ohms. Thus V is equal to the product IZ. See Figure 21-12(A). If each voltage is divided by the current I, IX_L yields inductive reactance X_L, IR yields resistance R, and IZ yields impedance Z. This is illustrated in the *impedance diagram* that is shown in Figure 21-12(B).

The tangent of the phase angle ϕ is defined by the ratio of the side opposite ϕ to the side adjacent to ϕ. In the impedance diagram these are the sides X and R, respectively, of the right triangle. Therefore

$$\frac{X}{R} = \tan \phi$$

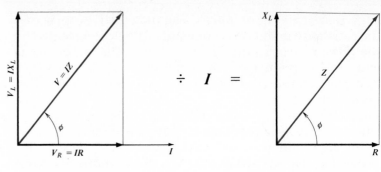

Figure 21-12. The derivation of an impedance diagram of an *L-R* series circuit.

(A) Voltage diagram   (B) Impedance diagram

If reactance is inductive, the phase angle is positive.

This equation gives X/R *as a function of* ϕ. It also enables us to determine ϕ *as a function of* X/R; that is, ϕ *is an angle whose tangent is* X/R. This expression, "ϕ is an angle whose tangent is X/R," gives the *inverse function* to the tangent in the equation. The inverse function to the tangent is called the **arc tangent.** The symbol for arc tangent is **arctan.** The inverse function to the tangent of the phase angle ϕ in the above equation is commonly written as

The term "arctan" is sometimes written as "$\tan^{-1}$." They are equivalent notations for the arc tangent.

$$\phi = \arctan \frac{X}{R}$$

The impedance, Z, has a magnitude equal to $\sqrt{R^2 + X^2}$ and a phase angle ϕ (direction angle with respect to R) whose tangent is X/R. It can be written in the polar form $Z \underline{|\phi}$ as

Relate $Z = \sqrt{R^2 + X^2}$ to the Pythagorean theorem of geometry.

$$Z = \sqrt{R^2 + X^2} \ \underline{|\arctan X/R}$$

When the reactance is inductive, $X = X_L$ and ϕ is a positive angle whose tangent is X_L/R. In this polar expression, $\sqrt{R^2 + X^2}$ yields the *magnitude* of the impedance and the symbol $\underline{|\quad}$ is interpreted to mean "*at an angle of* — °."

We can now rewrite the Ohm's-law expression for d-c circuits in a form that applies to a-c circuits. For the entire circuit:

$$E = IZ$$

For the external circuit:

$$V = IZ$$

EXAMPLE   A coil with a resistance of $3\overline{0}$ ohms and an inductance of 0.12 henry is connected across a 120-volt, $6\overline{0}$-hertz source. (a) What is the magnitude of the circuit current? (b) What is the phase angle? (c) What power is expended in the external circuit?

SOLUTION   The circuit shown in Figure 21-9 and the phasor diagram shown in Figure 21-10 apply to this problem.

(a) The magnitude of Z will be determined since only the magnitude of the circuit current is required in part (a).

By Ohm's law, $I = \dfrac{V}{Z}$

where the magnitude of $Z = \sqrt{R^2 + X_L^2}$ and $X_L = 2\pi fL$

So, $Z = \sqrt{R^2 + (2\pi fL)^2}$

Thus, $I = \dfrac{V}{\sqrt{R^2 + (2\pi fL)^2}} = \dfrac{120 \text{ v}}{\sqrt{(3\overline{0} \text{ } \Omega)^2 + (2\pi \times 6\overline{0}/\text{s} \times 0.12 \text{ h})^2}}$

$I = \dfrac{120 \text{ v}}{54 \text{ } \Omega} = 2.2 \text{ a},$ the circuit current

(b) $\phi = \arctan \dfrac{X_L}{R} = \arctan \dfrac{45 \text{ } \Omega}{3\overline{0} \text{ } \Omega} = \arctan 1.5$

From Appendix B, Table 6, $\phi = 56°$, the phase angle of a leading voltage.

(c) $P = VI \times \text{pf} = VI \cos \phi = 120 \text{ v} \times 2.2 \text{ a} \times \cos 56° = 120 \text{ v} \times 2.2 \text{ a} \times 0.56$
$P = 150 \text{ w}$

PRACTICE PROBLEMS   1. A coil has an inductance of 0.0664 h and a resistance of 20.0 Ω. It is connected across a $12\overline{0}$-v, 60.0-hz source. (a) Determine the magnitude of the impedance in the circuit. (b) Determine the circuit current. (c) Determine the phase angle. (d) What power is expended in the external circuit?   *Ans.* (a) 32.0 Ω; (b) 3.75 a; (c) 51.3°; (d) 281 w

2. A coil with resistance of 34.7 Ω and inductance of 0.0850 h is connected across a 60.0-hz source of emf and draws a current of 0.637 a. (a) What is the potential difference across the circuit? (b) What is the phase angle? (c) What power is expended in the external circuit? *Ans.* (a) 30.1 v; (b) 42.7°; (c) 14.0 w

---

21.7 *Capacitance in an a-c Circuit*   A capacitor in a d-c circuit charges to the applied voltage and effectively opens the circuit. The potential difference across a capacitor can change only as the charge on the capacitor changes. Since $Q = CV$, the charge Q, in coulombs, is directly proportional to the potential difference V, in volts, for a given capacitor. Thus we can refer to the charge in terms of the voltage across the capacitor.

Because $Q \propto V$, we describe the charge on a capacitor in terms of its voltage.

An uncharged capacitor offers no opposition to a charging current from a source of emf. However, as a charge builds up on the capacitor plates, a voltage develops across the capacitor that opposes the charging current. When the charge is such that the voltage across the capacitor is equal to the applied voltage, the charging current must be zero. Why? These two conditions are illustrated in Figure 21-13. We can conclude that *the current in a capacitor circuit is maximum when the voltage across the capacitor is zero and is zero when the capacitor voltage is maximum.*

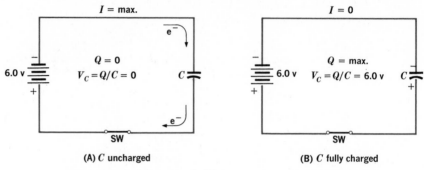

(A) *C* uncharged   (B) *C* fully charged

Figure 21-13. As a capacitor is charged, a voltage develops across it that opposes the charging current.

Let us examine the effect of pure capacitance in an a-c circuit. When the initial charging current is a positive maximum, the voltage across the capacitor plates is zero. As the capacitor charges, the potential difference builds up and the current decays; the charging current reaches zero when the capacitor voltage is maximum.

As the charging current reverses direction and increases toward a negative maximum, electrons flow from the negative plate of the capacitor through the circuit to the positive plate, thereby removing the charge. The capacitor then charges in the opposite sense, and the potential difference reaches a maximum with opposite polarity as the current returns to zero. Thus *the voltage across the capacitor lags the circuit current by 90°.* See Figure 21-14.

Figure 21-14. Pure capacitance in an a-c circuit.

The phase angle, ϕ, in an a-c circuit with a pure capacitance load is $-90°$, and the circuit has a *lagging* voltage.

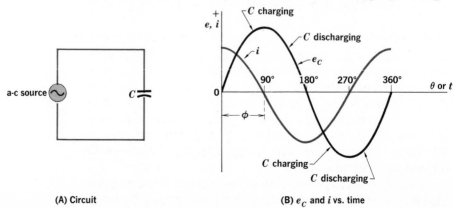

(A) Circuit   (B) e_C and i vs. time

The phase relationship is expressed as

$$i = I_{max} \sin \theta \quad \text{and} \quad e = E_{max} \sin (\theta + \phi)$$

These expressions are similar to those given for i and e in Section 21.3. However, the phase angle of a lagging voltage is *negative*. Hence the sign of the phase angle, ϕ, must be introduced properly.

Suppose the effective value of voltage across a capacitor is 10.0 volts. When the displacement angle, θ, is 30°, the instantaneous value of e is

$$e = E_{max} \sin (\theta + \phi)$$
$$e = E_{max} \sin [30° + (-90°)]$$
$$e = E_{max} \sin (-60°)$$

Now
$$E_{max} = 1.414 \ E = 1.414 \times 10.0 \ \text{v}$$
$$e = 14.1 \ \text{v} \times \sin (-60°)$$

and
$$\sin (-60°) = -0.866$$

Therefore
$$e = 14.1 \ \text{v} \times (-0.866)$$
$$e = -12.2 \ \text{v}$$

Compare this result with the similar computation in Section 21.3.

Plotted as a function of time, the product of the instantaneous values of current and voltage in a circuit of pure capacitance results in the instantaneous power curve shown in Figure 21-15. Similar to a circuit with pure inductance, the average power is zero. As the current changes from each maximum to zero, energy is taken from the source of emf and stored in the electric field between the capacitor plates. All this energy is returned to the source as the current rises from zero to either maximum. Thus the product of the effective values of current and voltage indicates an *apparent power*. The phase angle of −90° yields a power factor equal to zero [$\cos (-90°) = 0$], and so the *actual* power dissipated must be zero.

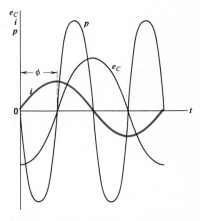

Figure 21-15. Power curve in a pure capacitance.

21.8 Capacitance and Resistance

In an a-c circuit the opposition to the current due to capacitance is *nonresistive* since no power is dissipated in the capacitor. This nonresistive opposition is called *capacitive reactance*. The symbol for capacitive reactance is X_C and the unit is the *ohm*.

If reactance is capacitive, the phase angle is negative.

The inherent resistance in any practical circuit containing capacitance can be represented in series with X_C. An R-C series circuit is shown in Figure 21-16. The potential difference, V, across the external circuit is the vector sum of V_C and V_R; the series current is used as the reference. There is, of course, no electron flow through the capacitor;

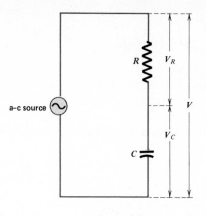

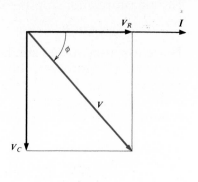

(A) Circuit diagram

(B) Voltage diagram

Figure 21-16. An *R-C* series circuit showing the phase relations of voltages V_R and V_C with the circuit current.

it is alternately charged, first in one sense and then in the other, as the charging current alternates. The phase angle, ϕ, is negative, which is characteristic of a lagging voltage.

The corresponding impedance diagram can be produced by dividing each voltage phasor by the circuit current. See Figure 21-17. Just as V_C is negative with respect to V_L in Figure 21-12(B), so X_C is plotted in the opposite direction to X_L.

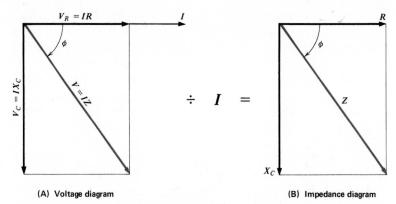

(A) Voltage diagram

(B) Impedance diagram

Figure 21-17. The derivation of an impedance diagram of an *R-C* series circuit.

Because $V_C = Q/C$, the higher the capacitance C, the lower is the potential difference V_C that develops across the capacitor for a given charge Q. *Thus the capacitive reactance to a circuit current varies inversely with the capacitance in the circuit.* As the frequency of a charging current is increased, a shorter time is available during each current cycle for the capacitor to charge and discharge. Smaller changes in voltage occur across the capacitor. *Hence the capacitive reactance to the circuit current varies inversely with the current frequency.*

$$X_C = \frac{1}{2\pi f C}$$

When f is in hertz and C is in farads, the capacitive reactance, X_C, is expressed in ohms.

The impedance of a circuit containing capacitance and resistance in series has a magnitude equal to $\sqrt{R^2 + X^2}$ and a phase angle, ϕ, whose tangent is $-X_C/R$. The impedance, in its general polar form $Z\lfloor\phi$, is expressed as $\sqrt{R^2 + X^2}\lfloor\text{arctan } X/R$, where $X = X_C$. See the following example.

EXAMPLE   A capacitance of 20.0 μf and a resistance of 100 Ω are connected in series across a $12\bar{0}$-v, 60.0-hz source. (a) Find the impedance of the circuit. (b) What is the magnitude of the circuit current? (c) What is the potential difference across the capacitor?

SOLUTION   (a) The circuit and diagrams of Figures 21-16 and 21-17 apply.

$$Z = \sqrt{R^2 + X^2}\lfloor\text{arctan } X/R$$

But

$$X = X_C = \frac{1}{2\pi fC} = \left(\frac{1}{2\pi \times 60.0 \times 20.0 \times 10^{-6}}\right)\Omega$$

$$X_C = 133\ \Omega$$

Magnitude of Z

$$Z = \sqrt{(10\bar{0}\ \Omega)^2 + (133\ \Omega)^2}$$
$$Z = 166\ \Omega$$

Phase angle

$$\phi = \text{arctan } \frac{-X_C}{R} = \text{arctan } \frac{-133\ \Omega}{10\bar{0}\ \Omega}$$

$$\phi = \text{arctan } -1.33$$
$$\phi = -53°, \text{ the phase angle of a lagging voltage}$$

Therefore the impedance can be expressed as

$$Z = 166\ \Omega\ \lfloor-53°$$

(b) $I = \dfrac{V}{Z} = \dfrac{12\bar{0}\text{ v}}{166\ \Omega}$

$$I = 0.723\text{ a, the circuit current}$$

(c) $V_C = IX_C = 0.723$ a $\times$ 133 Ω
$$V_C = 96.2\text{ v, the potential difference across } C$$

PRACTICE PROBLEMS   1. A 15.0-μf capacitor and a $20\bar{0}$-Ω resistor are connected in series across a 115-v, 60.0-hz source. (a) Determine the circuit impedance. (b) Determine the magnitude of the circuit current. (c) What is the potential difference across the capacitor? *Ans.* (a) 267 $\Omega\lfloor-41.5°$; (b) 0.431 a; (c) 76.3 v

2. A 30.0-μf capacitor and a 15$\bar{0}$-Ω resistor are connected in series across a 90.0-v, 60.0-hz source. (a) Find the impedance of the circuit. (b) What is the magnitude of the circuit current? (c) What is the potential difference across the capacitor? (d) What is the potential difference across the resistor?   *Ans.*   (a) 174 $\Omega\lfloor-30.5°$; (b) 0.517 a; (c) 45.7 v; (d) 77.6 v

3. A 0.500-μf capacitor and a 30$\bar{0}$-Ω resistor are connected in series across a 30$\bar{0}$-v, 1.00-khz source. (a) What is the load impedance? (b) Determine the magnitude of the circuit current. (c) What is the potential difference across the capacitor? (d) What is the potential difference across the resistor? (e) What power is expended in the resistor? (f) Demonstrate that the correct answer unit in (e) is the watt.   *Ans.*   (a) 437 $\Omega\lfloor-46.7°$; (b) 0.686 a; (c) 218 v; (d) 206 v; (e) 141 w; (f) student's proof

---

If reactance is zero, the phase angle is zero.

21.9 L, R, and C in Series   Inductance produces a leading voltage and a positive phase angle in an a-c circuit. Capacitance produces a lagging voltage and a negative phase angle. Of course, the voltage across a resistance must be in phase with the current in the resistance.

Suppose values of L, R, and C are connected in series across an a-c source of emf as in Figure 21-18(A). The *total reactance* of the circuit is

$$X = X_L - X_C$$

since X_L and X_C are 180° apart.

The impedance has the general form $Z\lfloor\phi$.

$$Z = \sqrt{R^2 + X^2}\ \lfloor\underline{\text{arctan } X/R}$$

However, X can be either positive or negative (X^2 is always positive) since

$$X = X_L - X_C = 2\pi fL - \frac{1}{2\pi fC}$$

Therefore the magnitude of the impedance can be expressed as

$$Z = \sqrt{R^2 + (X_L - X_C)^2}$$

$$Z = \sqrt{R^2 + \left(2\pi fL - \frac{1}{2\pi fC}\right)^2}$$

The calculation of X_L is shown in Section 21.5.

Suppose the resistance R in Figure 21-18(A) is 16.0 Ω and the a-c source is a 60.0-hz generator that applies a potential difference of 10$\bar{0}$ v across the external circuit. At this frequency, X_L is determined to be 22.0 Ω and X_C to be 10.0 Ω. The impedance, Z, of the circuit is calculated from these known values of R, X_L, and X_C as illustrated in the following example.

The calculation of X_C is shown in Section 21.8.

$$Z = \sqrt{R^2 + X^2}\ \lfloor\underline{\text{arctan } X/R}$$

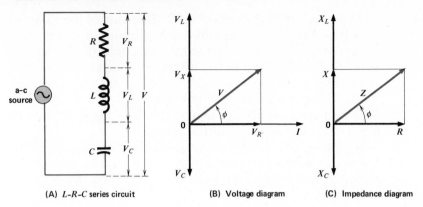

(A) *L–R–C* series circuit   (B) Voltage diagram   (C) Impedance diagram

Figure 21-18. An *L-R-C* series circuit with voltage and impedance diagrams.

But

$$X = X_L - X_C = 22.0\ \Omega - 10.0\ \Omega = 12.0\ \Omega$$
$$Z = \sqrt{(16.0\ \Omega)^2 + (12.0\ \Omega)^2}\ \underline{|\text{arctan } 12.0\ \Omega/16.0\ \Omega}$$
$$Z = 20.0\ \Omega\ \underline{|37.0°}$$

By calculating $Z\underline{|\phi}$ we found that the magnitude of the impedance is 20.0 Ω and its direction angle, ϕ, is 37.0° positive with respect to R. The impedance diagram is shown in Figure 21-18(C). This angle ϕ is positive because X_L is larger than X_C. Thus the voltage across the external circuit must *lead* the circuit current, I, by an angle of 37.0°. This voltage was given as $10\overline{0}$ v. Therefore the magnitude of the current is

Positive phase angle: voltage leads current.
Negative phase angle: voltage lags current.

$$I = \frac{V}{Z} = \frac{10\overline{0}\text{ v}}{20.0\ \Omega} = 5.00\text{ a}$$

Knowing the circuit current, we can determine the magnitudes of the voltages across L, R, and C.

$$V_L = IX_L = 11\overline{0}\text{ v}$$
$$V_R = IR = 80.0\text{ v}$$
$$V_C = IX_C = 50.0\text{ v}$$

Observe that the vector sum of V_L, V_R, and V_C must be equal to V.

From the voltage diagram of Figure 21-18(B) it is evident that the magnitude of the voltage V across the external circuit is

$$V = \sqrt{V_R^2 + V_X^2}$$

where

$$V_X = V_L - V_C$$

Then

$$V = \sqrt{(80.0\text{ v})^2 + (60.0\text{ v})^2}$$
$$V = 10\overline{0}\text{ v}$$

This result is in agreement with the value of the external circuit voltage V assumed at the beginning of this circuit-computation exercise. The direction angle of V must be $37.0°$ since this is the phase angle, ϕ, using I as the reference. This example of a series L-R-C circuit illustrates the fact that the magnitude of the voltage across a reactance in an a-c circuit can exceed the magnitude of the applied voltage across the entire external circuit.

Resistive and reactive components may also be connected in parallel in practical a-c circuits. The analysis of their parallel impedance networks is beyond the scope of this book.

Questions

GROUP A

1. What is the significant relationship between an alternating current and alternating voltage that are in phase?
2. In a-c circuits what type of load has a current in phase with the voltage across it?
3. What is meant by instantaneous power?
4. Upon what property of an alternating current is its effective value based?
5. Define the effective value of an alternating current.
6. What is the phase relation between (a) the current in a pure inductance and the applied voltage, (b) the current and the induced voltage, (c) the applied voltage and the induced voltage?
7. Express power factor (a) in terms of the phase angle, (b) in terms of V and V_R. (c) What is the significance of the power factor pertaining to an a-c circuit?
8. Why is reactance described as being nonresistive?
9. (a) Define impedance. (b) Draw an impedance diagram for an inductor with reactance X_L and resistance R. (c) Draw an impedance diagram for a capacitor with reactance X_C and resistance R.

10. Describe the difference in the performance of a capacitor in a d-c circuit and in an a-c circuit.
11. What is the power factor of an a-c circuit containing (a) pure inductance, (b) pure resistance, (c) pure capacitance?
12. Express Ohm's law for an a-c circuit (a) that includes a source of emf, (b) that does not include a source of emf.

GROUP B

13. Explain why the instantaneous power curve for current in a resistive load varies between some positive maximum value and zero.
14. Demonstrate algebraically that the heating effect of an electric current is proportional to the effective value of current squared.
15. (a) Plot the current and current squared curves for an alternating current having a maximum value of 5.7 amperes. Use cross-section paper and a suitable scale for the coordinate axes to produce a graph similar to Figure 21-4. (b) Show that the rms value of the current is 4.0 amperes.
16. If an a-c circuit has a pure inductance load, the average power delivered to the load is zero. Does this mean that there is no energy transfer between the source and the load? Explain.

17. (a) Considering the expression for X_L in terms of frequency and inductance, draw a rough graph of X_L as a function of frequency for a given inductance. (b) What is the nature of the curve? (c) What is the significance of the fact that X_L is zero when f is zero, regardless of the value of L?

18. Considering the expression for X_C in terms of frequency and capacitance, what is the significance of the fact that when the frequency is zero, X_C is infinitely high for any capacitance?

19. (a) In an L-R-C series circuit operating at a certain frequency, what determines whether the load is inductive or capacitive? (b) Suggest a circumstance in which the load would be resistive.

20. What is the significance of the negative portions of the power curve shown in Figure 21-15?

Problems
GROUP A

Note: Compute angles to the nearest 0.5 degree.

1. A current in an a-c circuit measures 5.5 a. What is the maximum instantaneous magnitude of this current?

2. A capacitor has a voltage rating of 45$\bar{0}$ v maximum. What is the highest rms voltage that can be impressed across it without danger of dielectric puncture?

3. The emf of an a-c source has an effective value of 122.0 v. Determine the instantaneous value when the displacement angle, θ, is 50.0°?

4. An alternating current in a 25.0-Ω resistance produces heat at the rate of 25$\bar{0}$ w. What is the effective value (a) of current in the resistance, (b) of voltage across the resistance?

5. An inductor has an inductance of 2.20 h and a resistance of 22$\bar{0}$ Ω. (a) What is the reactance if the a-c frequency is 25.0 hz? (b) Draw the impedance diagram and determine graphically the magnitude and phase angle of the impedance.

6. A 2.00-μf capacitor is connected across a 60.0-hz line and a current of 167 ma is indicated. (a) What is the reactance of the circuit? (b) What is the voltage across the line?

7. A capacitance of 2.65 μf is connected across a 12$\bar{0}$-v line and is found to draw 12$\bar{0}$ ma. What is the frequency of the source?

8. When a resistance of 4.0 Ω and an inductor of negligible resistance are connected in series across a 110-v, 6$\bar{0}$-hz line, the current is 2$\bar{0}$ a. What is the inductance of the coil?

GROUP B

9. A 0.50-μf capacitor and a 3$\bar{0}$-Ω resistor are connected in series across a source of emf whose frequency is 8.0×10^3 hz. (a) Draw the circuit diagram. (b) Calculate the capacitive reactance. (c) Draw the impedance diagram and determine the magnitude and phase angle of the impedance.

10. The current in the circuit of Problem 9 is found to be 5$\bar{0}$ ma. (a) What is the magnitude of the voltage across the series circuit? (b) What is the phase relation of this voltage to the circuit current? (c) What is the voltage across the capacitor? (d) What is the voltage across the resistor?

11. A 60.0-hz circuit has a load consisting of resistance and inductance in series. A voltmeter, ammeter, and wattmeter, properly connected in the circuit, read respectively 117 v, 4.75 a, and 40$\bar{0}$ w. (a) Determine the power factor. (b) What is the phase angle? (c) What is the resistance of the load?

(d) What is the inductive reactance of the load? (e) Determine the voltage across the resistance. (f) Determine the voltage across the inductance. (g) Draw the voltage diagram with the circuit current as the reference phasor. What is the applied voltage as found graphically and in polar form?

12. A coil has a resistance of $9\bar{0}$ Ω and an inductance of 0.019 h. (a) What is the impedance (complex) at a frequency of 1.0×10^3 hz? (b) Determine the magnitude of the current when a potential difference of 6.0 v at this frequency is applied across it? (c) How much power is delivered to the coil?

13. A capacitance of $5\bar{0}$ μf and a resistance of $6\bar{0}$ Ω are connected in series across a 120-v, $6\bar{0}$-hz line. (a) What is the magnitude of current in the circuit? (b) What power is dissipated? (c) What is the power factor? (d) Determine the voltage across the resistance. (e) Determine the voltage across the capacitance. (f) Draw the voltage diagram with the circuit current as

the reference phasor. What is the applied voltage found graphically? Express in polar form.

14. An inductance of 4.8 mh, a capacitance of 8.0 μf, and a resistance of $1\bar{0}$ Ω are connected in series, and a 6.0-v, 1.0×10^3-hz signal is applied across the combination. (a) What is the magnitude of the impedance of the series circuit? (b) What is the phase angle? (c) Determine the magnitude of the current in the circuit. (d) Find the potential drop across each component of the load, and draw the voltage diagram.

15. An inductor, a resistance, and a capacitor are connected in series across an a-c circuit. A voltmeter reads $9\bar{0}$ v when connected across the inductor, 16 v across the resistor, and 120 v across the capacitor. (a) What will the voltmeter read when placed across the series circuit? (b) Draw the voltage diagram, and graphically verify your answer to (a). (c) Compute the power factor.

RESONANCE

21.10 Inductive Reactance vs. Frequency

An inductor in a d-c circuit has no effect on the steady, direct current except for that due to the ordinary resistance of the wire forming the coil. A steady, direct current can be thought of as an alternating current having a frequency of zero hz. Since $X_L = 2\pi fL$, inductive reactance in the d-c circuit must be zero.

An alternating current changes from its positive maximum to its negative maximum in a time equal to one-half the period, T; if the frequency of this current is increased, the time rate of change of current is increased accordingly. See Figure 21-19. The inductive reactance presented to the current in an a-c circuit by an inductor increases as the frequency of this current is increased.

Consider the reactance of a certain inductance over a range of frequencies. A plot of the values of inductive reactance as ordinates against frequencies as abscissas produces the linear curve shown in Figure 21-20. Any value of

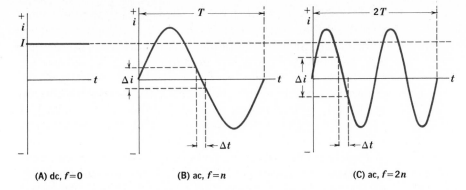

(A) dc, $f=0$ (B) ac, $f=n$ (C) ac, $f=2n$

inductance yields a reactance curve that starts at the origin of the coordinate axes; the larger the value of inductance, the greater is the slope of the curve.

21.11 Capacitive Reactance vs. Frequency When a capacitor in a d-c circuit is charged to the applied voltage, it acts as an open switch and the current in the circuit is zero. While the capacitor remains in this fully charged state, the capacitive reactance in the circuit is infinitely high regardless of the value of C.

A capacitor in an a-c circuit has a very different effect on the alternating current and voltage. Electrons flow in one direction during one half-cycle, charging the capacitor in one sense. The electrons flow in the opposite direction during the other half-cycle, reversing the charge on the capacitor.

If the current has a very low frequency, the capacitive reactance is very large since $X_C = 1/(2\pi fC)$. This large reactance effectively limits the circuit current. At higher frequencies, less change occurs in the charge on the capacitor during each half-cycle and the reactive opposition to the circuit current is lower. At extremely high frequencies, the change in capacitor charge becomes negligible and the capacitive reactance in the circuit becomes a negligible factor in limiting the circuit current. The larger the capacitor in the circuit, the more rapidly the capacitive reactance decreases with increasing frequency. Observe that in the above equation for X_C, the magnitude of X_C varies inversely with frequency f, with capacitance C, and with their product fC.

In Figure 21-21 values of reactance for a given capacitor are plotted as a function of frequency. The slope of the curve is hyperbolic, which is typical of the inverse relation between capacitive reactance, X_C, and frequency, f. A larger capacitor will yield a reactance-frequency curve with a greater slope than that shown.

Figure 21-19. The time rate of change of current increases with frequency.

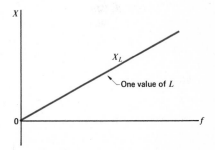

Figure 21-20. The reactance of a given inductor plotted as a function of frequency.

X_C for a given capacitor varies inversely with frequency.

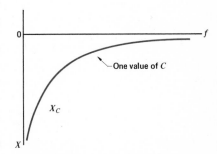

Figure 21-21. The reactance of a given capacitor plotted as a function of frequency. X_C is plotted as $-[1/(2\pi fC)]$ for comparison with X_L in Figure 21-20.

21.12 Series Resonance

In Section 21.9 it was stated that the voltages across an inductance and a capacitance in a series circuit are of opposite polarity. If X_L is larger than X_C, the circuit is inductive and the voltage across the circuit leads the current. If X_C is the larger reactance, the circuit is capacitive and the circuit voltage lags behind the current. In either instance the impedance, Z, has a magnitude $\sqrt{R^2 + (X_L - X_C)^2}$ and a phase angle, ϕ, whose tangent is $(X_L - X_C)/R$.

The frequency of an a-c voltage applied to a series circuit with fixed values of L, R, and C determines the reactance of the circuit. If Figures 21-20 and 21-21 are combined, they will show that the reactance of the circuit is inductive over a range of high frequencies and capacitive over a range of low frequencies. At some intermediate frequency, the inductive and capacitive reactances are equal; the reactance of the circuit, $X_L - X_C$, equals zero. At this frequency the impedance, Z, is at its minimum value for the series circuit. It is equal to the inherent circuit resistance, R. As is true of all circuits with pure ohmic resistance, the voltage across the circuit and the current in the circuit are in phase. This special circuit condition is known as *series resonance*. The frequency at which resonance occurs is called the *resonant frequency*, f_r. *Series resonance is a condition in which the impedance of an L-R-C series circuit is equal to the circuit resistance, and the voltage across the circuit is in phase with the circuit current.*

Suppose an inductor of $1\overline{0}$-millihenrys inductance and 25-ohms resistance is placed in series with an ammeter and a capacitor of 1.0-microfarad capacitance. A signal of $1\overline{0}$ volts rms from a variable frequency source, such as an audio signal generator, is applied. The circuit is shown in Figure 21-22. When the frequency of the applied signal is

Figure 21-22. An *L-R-C* series circuit that has a resonant frequency of 1600 hz.

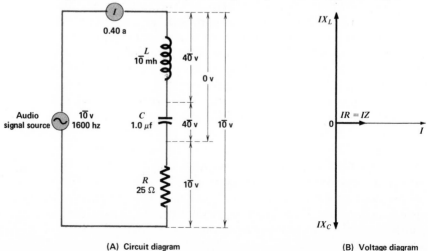

(A) Circuit diagram

(B) Voltage diagram

varied through the audio range, a distinct rise in the current occurs as the frequency approaches 1600 hz. The current magnitude quickly falls away as the frequency of the signal increases beyond this value. We can conclude that the series circuit is resonant at 1600 hz.

At resonance, $X_L = X_C$ and the value of each is found as follows:

$$X_L = 2\pi f_r L$$
$$X_L = (2\pi \times 1.6 \times 10^3 \times 1.0 \times 10^{-2})\ \Omega$$
$$X_L = 1.0 \times 10^2\ \Omega$$

$$X_C = \frac{1}{2\pi f_r C}$$

$$X_C = \left(\frac{1}{2\pi \times 1.6 \times 10^3 \times 1.0 \times 10^{-6}}\right)\ \Omega$$

$$X_C = 1.0 \times 10^2\ \Omega$$

The impedance at resonance is

$$Z = \sqrt{R^2 + X^2}\ \underline{|\arctan X/R}$$

But

$$X = X_L - X_C = 0\ \Omega$$

and the phase angle $\phi = \arctan \dfrac{X}{R} = 0°$

Thus

$$Z = R = 25\ \Omega$$

The impedance, Z, of the series-resonant circuit is the minimum value and is equal to R. The current, I, is the maximum value and is equal to V/R.

$$I_r = \frac{V}{R} = \frac{1\bar{0}\ \mathrm{v}}{25\ \Omega} = 0.40\ \mathrm{a}$$

$$V_L = I_r X_L = 0.40\ \mathrm{a} \times 1.0 \times 10^2\ \Omega$$
$$V_L = 4\bar{0}\ \mathrm{v}$$
$$V_C = I_r X_C = 0.40\ \mathrm{a} \times 1.0 \times 10^2\ \Omega$$
$$V_C = 4\bar{0}\ \mathrm{v}$$
$$I_r X_L - I_r X_C = 4\bar{0}\ \mathrm{v} - 4\bar{0}\ \mathrm{v} = 0\ \mathrm{v}$$
$$V_R = I_r \times R = 0.40\ \mathrm{a} \times 25\ \Omega = 1\bar{0}\ \mathrm{v}$$

The phase angle, ϕ, at resonance is $0°$ and the power factor of the circuit, which is the cosine of ϕ, is 1. The series-resonant circuit has unity power factor. Thus maximum power, P, is dissipated in the circuit at resonance.

$$pf = \cos \phi = \cos 0° = 1$$

$$P = VI_r \cos \phi = VI_r$$

The frequency at which a series circuit resonates is determined by the combination of L and C used. For a particular combination, there is one resonant frequency, f_r. At series resonance

$$X_L = X_C$$

$$2\pi f_r L = \frac{1}{2\pi f_r C}$$

Solving for f_r, $$f_r^2 = \frac{1}{4\pi^2 LC}$$

$$f_r = \frac{1}{2\pi\sqrt{LC}}$$

When L is in henrys and C is in farads, f_r is expressed in hertz.

At frequencies below the resonance point, the series circuit is capacitive, performing as one with an X_C and R in series. At the resonant frequency, f_r, the circuit performs as one containing pure resistance. Above the resonant frequency, it is inductive. These characteristics of the L-R-C series circuit are shown by the reactance curve X in Figure 21-23.

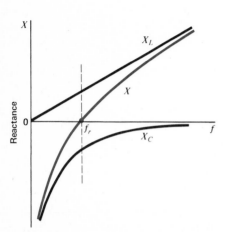

Figure 21-23. The reactance characteristic of an **L-R-C** series circuit on either side of the resonant frequency.

21.13 Selectivity in Series Resonance

A resonant circuit responds to impressed voltages of different frequencies in a very selective manner. The circuit presents a high impedance to a signal voltage of a frequency above or below the resonant frequency of the circuit. The resulting current is correspondingly small. This same circuit presents a low impedance, consisting of the circuit resistance, to a signal voltage at the resonant frequency. The current is correspondingly large.

A tuned circuit thus *discriminates* among signal voltages of different frequencies. This property of signal discrimination is known as the circuit's *selectivity*. The lower the resistance of the series circuit, the higher is the resonant current and the more sharply selective is the circuit. The family of current-resonance curves shown in Figure 21-24 illustrates the influence of circuit resistance on selectivity. Curves are shown for three different values of circuit resistance.

The usual series-resonant circuit consists of an inductor and a capacitor. The only resistance is that inherent in the circuit, which is almost entirely the resistance of the coil. The characteristics of the series-resonant circuit depend primarily on the ratio of the inductive reactance to the circuit resistance, X_L/R. This ratio is called the Q, quality factor, of the circuit. It is essentially a design characteristic of

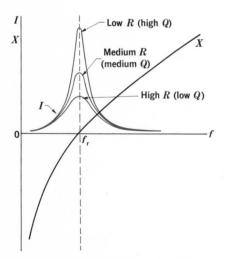

Figure 21-24. The magnitude of current in a series-resonant circuit plotted as a function of frequency.

the inductor. The higher the Q, the more sharply selective is the series-resonant circuit.

Resonant circuits can be tuned over a range of frequencies by making either the inductance or the capacitance variable. The position of a powdered-iron core in a coil can be varied to change the coil's inductance. A capacitor can be made variable by providing a means of changing its plate area or its plate separation. A variable capacitor of the type used in the tuning circuits of small radios is shown in Figure 21-25. Resonance is particularly important in communication circuits.

Figure 21-25. A two-gang variable capacitor of the type used in the tuning circuits of radio receivers. Each capacitor section has a set of stator plates and a set of rotor plates. The rotor plates are attached to the shaft. The capacitances are varied simultaneously by rotating the shaft and thus varying the amounts of effective plate areas of the two capacitor sections.

Questions
GROUP A

1. Select three different values of inductance, and determine the reactance of each at four different frequencies. Plot a curve of inductive reactance as a function of frequency for each using the same coordinate axes. (a) What is the shape of each curve? (b) Which curve has the greatest slope? (c) Why should each curve start at the origin of the coordinate axes?
2. Define series resonance.
3. What is the relation between the voltage across a series L-R-C circuit and the current in the circuit at a frequency below the resonant frequency? Explain.
4. What is the relation between the voltage across a series L-R-C circuit and the circuit current at a frequency above the resonant frequency? Explain.

5. What factor largely determines the sharpness of the rise of current near the resonance frequency in a series-resonant circuit?
6. What is the advantage of a resonant circuit having a high Q?
7. How is the Q of a resonant circuit affected by an increase in (a) resistance, (b) inductance, (c) frequency?

GROUP B

8. How can the resonant frequency of an L-R-C circuit be varied?
9. What would be the effect of having an inductor connected in series with a lamp in an a-c circuit? Explain.
10. A variable capacitor is connected in series in the circuit of Question 9 and when adjusted, the lamp glows normally. Explain.
11. A series L-R-C circuit is connected across an a-c signal source of variable frequency. Draw a curve to show the way in which the circuit current varies with the frequency of the source.

12. An a-c circuit has a low power factor because of an inductive load.
(a) What is the effect of this low power factor on the circuit current?
(b) Suggest a way to raise the power factor closer to unity.

Problems
GROUP B

1. A resonant circuit has an inductance of 320 μh and a capacitance of $8\bar{0}$ pf. What is the resonant frequency?

2. Assume that the frequency of the signal voltage applied to the resonant circuit of Problem 1 is varied from a value of 25 khz below the resonant frequency to 25 khz above. (a) Obtain values of inductive reactance for every 5 khz over this range of frequencies. Plot a curve showing inductive reactance as a function of frequency, using the frequency as the reference abscissa. (b) Obtain values of capacitive reactance for every 5 khz, and plot a curve that shows the capacitive reactance as a function of frequency, using the same reference abscissa. (c) Determine the resonant frequency graphically from your curves by plotting the values for $(X_L - X_C)$.

3. The variable capacitor used to tune a broadcast receiver has a maximum capacitance of 350 pf. The lowest frequency to which we wish to tune the receiver is 550 khz. What value of inductance should be used in the resonant circuit?

4. The minimum capacitance of the variable capacitor of Problem 3 is 15 pf. What is the highest resonant frequency that may be obtained from the circuit?

SUMMARY

The power expended in the load of an a-c circuit is determined by the average of the instantaneous values over the period of one cycle. Instantaneous power is the product of the instantaneous current and the instantaneous voltage. In a resistive load, the average power is the product of the effective values of current and voltage. In either a pure inductive load or a pure capacitive load, the average power is zero. Power in an a-c circuit is equal to the product of the effective value of the current, the effective value of the voltage, and the cosine of the phase angle.

An a-c circuit containing both inductance and resistance produces a current that lags behind the voltage. Such a load presents an impedance to the source. This impedance is the vector sum of the resistance and the inductive reactance in the circuit. The phase angle is the angle at which the current lags the voltage.

Capacitance and resistance in an a-c circuit produce an impedance equal to the vector sum of the capacitive reactance and resistance in the circuit. The phase angle in this case is due to a lagging voltage. When inductance and capacitance are both present, the phase angle is determined by the relative magnitudes of the two reactances. Inductive reactance is directly proportional to both frequency and inductance. Capacitive reactance is inversely proportional to both frequency and capacitance.

Series circuits containing inductance, capacitance, and resistance can be analyzed by the application of Ohm's law for a-c circuits.

At zero frequency the reactance of an inductor is zero, with a linear increase as the frequency increases. The reactance of a capacitor is infinitely large at zero frequency and decreases nonlinearly with increasing frequency. The resonant frequency of the circuit is the frequency at which the inductive and capacitive reactances of a circuit are equal.

At the resonant frequency the impedance of a series-resonant circuit is at a minimum and is equal to the circuit resistance. The resonant current is the max-imum current possible in the series circuit and is in phase with the applied voltage. If the Q of the circuit is high, the resonant current is very large compared with the average circuit current over a range of frequencies. The current curve is sharply peaked. The selectivity of such a resonant circuit is high.

VOCABULARY

actual power
apparent power
arc tangent
capacitance
capacitive reactance
capacitor
effective value (of current
　or voltage)

impedance
impedance diagram
inductance
inductive reactance
inductor
phase angle
phase diagram
phasor

power factor
reactance
resonance
resonant frequency
root-mean-square
selectivity
series resonance

In the early 1800's, the English mathematician Charles Babbage realized that if a machine capable of storing numbers and performing simple arithmetic tasks could be built, then eventually a machine could be programmed to do even more complex calculations automatically. His "difference engine" and "analytic engine" were the first steps in the development of today's high-speed computers. Computer technology has come a long way since then; vacuum tubes and semiconductors have been followed by integrated circuits. Computer applications have grown to include fields as diverse as astronomy, medicine, and athletics! The world is experiencing a computer revolution.

COMPUTERS

Forecasters of the National Weather Service provide nearly two million predictions per year to the public and to commercial interests. This meteorologist is operating a new computerized communications system designed to speed weather data handling.

Computer-assisted instruction (CAI) enables students to receive individualized instruction. Here students converse with a computer by means of a video display screen and keyboard during a classroom activity.

Nan Davis, a paralyzed Wright State University student, is shown here walking under her own muscle power with computer assistance. Shown with her in the biomedical engineering laboratories of Wright State University is Dr. Jerrold S. Petrofsky, Associate Professor of Engineering and Physiology, Wright State University, Dayton, Ohio.

Computer graphics are used by a U.S. mining corporation to evaluate the potential of exploration opportunities.

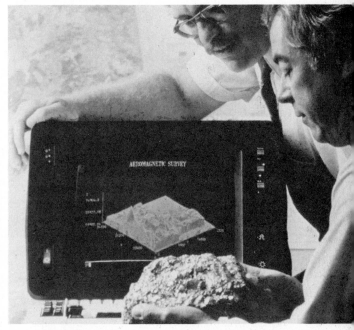

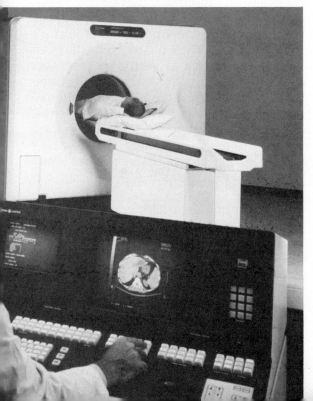

This computed tomography system combines an advanced X-ray scanning system with a powerful computer to permit doctors to study virtually any portion of the human anatomy.

ELECTRONIC DEVICES

The age of electronics effectively began with the invention of the three-element vacuum tube, or audion, by Lee De Forest in 1906. In 1973, the centennial year of De Forest's birth, the United States commemorated his pioneering work with this stamp.

In this chapter you will gain an understanding of:

✓ the role of the vacuum tube in the growth of electronics
✓ the thermionic emission of electrons
✓ the diode as a voltage rectifier
✓ the triode as a voltage amplifier
✓ the performance of a cathode-ray tube
✓ the function of a photomultiplier tube
✓ the semiconductor diode
✓ the doping process of forming P- and N-type semiconductor materials
✓ the P-N junction and its diode properties
✓ the role of P-N junctions in the performance of junction transistors
✓ some basic concepts of transistor circuits

VACUUM TUBES

22.1 Electronics, a Branch of Physics   The influence of *electronics* has become so commonplace that its role in our daily activities goes largely unnoticed. Many household appliances are electronically programmed to perform a sequence of timed operations at the push of a button. An increasing number of technical and industrial processes are controlled electronically by computers. Sophisticated electronic devices and circuits have an indispensable role in the advances now being achieved in medical science, communications, computer technology, space exploration, and automation—to name just a few. *Electronics is the branch of physics that is concerned with the emission, behavior, and effects of electrons in devices and with the utilization of these devices.*

The emergence of electronics as an important branch of physics began with the development of the *vacuum tube.* Vacuum tubes made possible such innovations as radio broadcasting, long-distance communications, sound motion pictures, public address systems, television, radar, electronic computers, and industrial automation.

Today, semiconductor devices have replaced vacuum tubes in many diverse applications of electronic circuits. However, some knowledge of vacuum tubes and basic vacuum-tube electronics is an important adjunct to the study of semiconductor electronics.

22.2 Vacuum-Tube Applications It may be useful to think of vacuum tubes as electron valves in electronic circuits. They have been designed to perform many different functions:

1. As *rectifiers*, they convert alternating current to direct current.

2. As *mixers*, they combine separate signals to produce a different signal.

3. As *detectors*, they separate the useful component from a complex signal.

4. As *amplifiers*, they increase the strength of a signal.

5. As *oscillators*, they convert a direct current to an alternating current of a desired frequency.

6. As *wave-shapers*, they change a voltage waveform into a desired shape for a special use.

A vacuum tube usually contains a *cathode* as a source of electrons, an *anode*, commonly called a *plate*, that attracts electrons from the cathode, and one or more *grids* for controlling the flow of electrons between cathode and plate. These electrodes are enclosed in a highly evacuated, gastight envelope. Modified forms of the vacuum tube may contain only a cathode and a plate or two independent sets of electrodes in a single envelope. In some tubes, a small amount of a particular gas is introduced to obtain special operating characteristics.

Common tubes are classified as *diodes, triodes, tetrodes,* and *pentodes,* according to the number of electrodes. High-vacuum tubes are known as *hard* tubes, and those containing a gas under very low pressure are known as *soft* tubes. Special tubes have special names. For example, the cathode-ray tube is a visual-indicating tube used in cathode-ray oscilloscopes. It serves also as the picture tube in television sets, the indicator tube in radar equipment, and the visual-display tube in computers.

22.3 Vacuum Tube Development 1. *The Edison effect.* While experimenting with the first electric lamp in 1883, Thomas Edison (1847–1931) was plagued by the frequent burning out of the carbon filament and the accompanying black deposit inside the bulb. In an effort to correct this difficulty Edison sealed a metal plate inside the lamp near the filament and connected it through a galvanometer to the filament battery. See Figure 22-2. He observed a deflection on the galvanometer when the plate was connected to the positive terminal of the battery. But no deflection occurred when the plate was connected to the negative terminal. Edison recorded these observations and continued his lamp experiments without attempting to explain the phenomenon.

Vacuum tubes are also referred to as "electron tubes." In Great Britain they are called "valves."

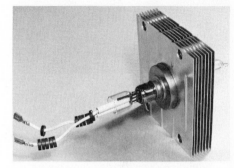

Figure 22-1. Modern vacuum tubes are designed to perform a variety of highly specialized functions. Shown here is a magnetron oscillator, the radiation generator in a microwave oven. More powerful magnetrons serve as transmitter tubes in pulsed radar systems.

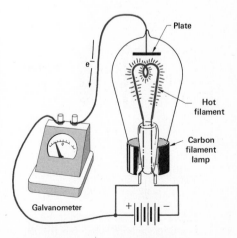

Figure 22-2. The Edison effect.

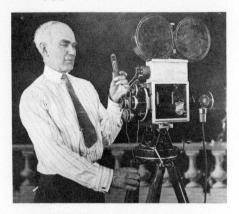

Figure 22-3. Lee De Forest was one of the pioneers of the development of wireless telegraphy in America. Among his many inventions was the triode, known as the audion, which made long-distance telephone and radio communications possible. De Forest also invented a system for recording sound on film which made sound motion pictures possible.

Several years later the English physicist Sir Joseph Thomson (1856–1940) discovered the electron and provided the explanation for the *Edison effect*. When the plate in Figure 22-2 was made positive, electrons escaping from the heated carbon filament were attracted through the evacuated space to the plate. The resulting electron current in the galvanometer circuit produced the meter deflection that Edison had observed. When the plate was made negative, on the other hand, the escaping electrons were repelled from the plate and there was no electron current in the galvanometer circuit.

2. *The Fleming valve.* Near the end of 1901 "wireless telegraphy," the forerunner of modern radio communications, was successfully used to transmit radio signals over long distances. In these early days of radio, crystals of semiconductor materials were used as receivers to *detect,* or recover, the message from the transmitted signals. These crystals were unsatisfactory in many ways. A better method of detection was clearly needed.

In 1904 the English physicist J. A. Fleming (1849–1945) patented the first *diode,* called the *Fleming valve.* He used the Edison effect to develop a crude detector for radio signals. Fleming's valve was so insensitive that it found little immediate application, and yet it was an important link in the evolution of the vacuum tube.

3. *The De Forest triode.* In 1906 the American inventor Lee De Forest (1873–1961) developed a vacuum tube that amplified feeble radio signals that neither the Fleming valve nor the crystal detector was sensitive enough to detect.

De Forest placed a third electrode, consisting of a *grid* of fine wire, between the filament and plate of the diode, as shown in Figure 22-4. He found that a small variation in voltage applied to the grid produced a large variation of current in the plate circuit of the tube. Thus the feeble radio signals from an antenna could be amplified by De Forest's *triode.* The amplified signals could then be successfully detected by the Fleming valve to provide the varying d-c signal necessary to operate headphones. The success of the De Forest triode led to the development of modern vacuum tubes.

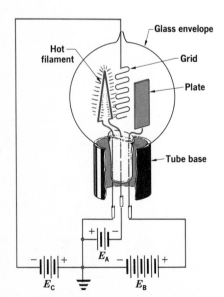

Figure 22-4. De Forest's three-element vacuum tube, the triode.

22.4 Thermionic Emission Electric conduction involves a transfer of some kind of charged particles. Metallic substances conduct electricity because they have many free electrons. In a tube containing a gas under low pressure, ionized gas molecules are responsible for conduction. The number of gas particles in a highly evacuated tube is so small, however, that conduction by gas ionization is virtually insignificant. Conduction in a high vacuum involves the introduction of charged particles into the electric field in the evacuated space between two electrodes. The charged particles released into the electric field are electrons. These electrons are released from a metallic surface by a process known as *thermionic emission.*

The free electrons of a metallic conductor are in continuous motion with speeds that increase with temperature. To escape from the surface of a conductor, an electron must do work to overcome the attractive forces at the surface of a material. At ordinary temperatures, the kinetic energies of electrons are not large enough to exceed the work function of the surface; the electrons cannot escape.

At a *high* temperature the average kinetic energy of the free electrons is large and a relatively large number will have enough energy to escape through the surface of the material. *Thermionic emission is the emission of electrons from a hot surface.* It is analogous to the evaporation of molecules from the surface of water.

The rate at which electrons escape from an emitting surface increases as the temperature of the emitter is raised. In the photoelectric effect, photons of light supply the energy required to eject electrons through the surface barrier of the photosensitive material. In thermionic emission, it is thermal energy that produces this same effect.

The high temperatures that are needed for satisfactory thermionic emission in vacuum tubes limit the number of suitable emitters to such substances as *tungsten, thoriated tungsten,* and certain *oxide-coated metals.* Tungsten and thoriated tungsten operate at very high temperatures and are connected directly to a source of filament current. These *directly heated cathodes* are known as *filament cathodes,* or simply *filaments.* They are used as electron emitters in large transmitting tubes.

Oxide-coated emitters consist of a metal such as nickel coated with a mixture of barium and strontium oxides, which in turn are covered with a surface layer of metallic barium and strontium. The electrons are emitted from this metallic surface layer. Such emitters can be heated to operating temperature either directly as a filament cathode or indirectly by radiation from an incandescent tungsten filament. *Indirectly heated cathodes* are commonly called *heater cathodes,* or simply *cathodes.* See Figure 22-6. Since practically all receiving tubes in use today have oxide-coated emitters, we shall concern ourselves principally with vacuum tubes of this type.

Figure 22-5. Vacuum tubes vary in their appearance because different designs are manufactured for different circuit requirements. This vacuum tube, designed for high-power applications, has radiation fins to dissipate heat.

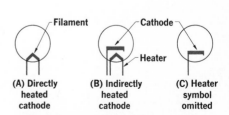

(A) Directly heated cathode

(B) Indirectly heated cathode

(C) Heater symbol omitted

Figure 22-6. Circuit symbols for both directly and indirectly heated cathodes.

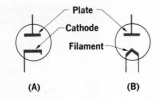

Figure 22-7. Circuit symbols for diodes with an indirectly heated cathode in **(A)** and a directly heated cathode in **(B)**.

When the plate charge is positive, electrons are accelerated toward the plate by the intervening electric field.

22.5 Diode Characteristics The simplest vacuum tubes are diodes that contain two elements—a cathode and a plate. Diode circuit symbols are shown in Figure 22-7. If the plate is made positive with respect to the cathode, the diode conducts because electrons flow from the cathode to the plate inside the tube and from the plate through the plate circuit to the cathode outside the tube. If the plate is negative with respect to the cathode, the diode does not conduct and there is no plate current in the external circuit. Thus a diode acts as a one-way valve for electrons; it allows electrons to flow in one direction through the circuit, but not in the other.

As electrons are emitted from the hot cathode, the space about the cathode becomes negatively charged. This negatively charged space, called the *space charge*, opposes the escape of additional electrons from the cathode. The extent of electron emission depends on the temperature of the cathode.

At low positive plate voltages, only those electrons near the plate are attracted to it, so a small plate current is produced. As the plate voltage is increased, greater numbers of electrons are attracted and the plate current is increased while the space charge is reduced.

In an ordinary circuit resistance the proportionality of current to voltage is described by Ohm's law. The resistor is said to be a linear, or *ohmic*, element in the circuit. In the case of the diode, the plate current is not proportional to the plate voltage. The diode and other such devices are said to be nonlinear, or nonohmic, elements in the circuit.

A diode circuit from which we can determine the plate current, I_P, as a function of plate voltage, E_P, for different filament temperatures is shown in Figure 22-8(A). The family of I_P vs. E_P curves is shown in Figure 22-8(B).

Figure 22-8. A diode circuit and characteristic curves showing the plate current as a function of plate voltage at different filament temperatures.

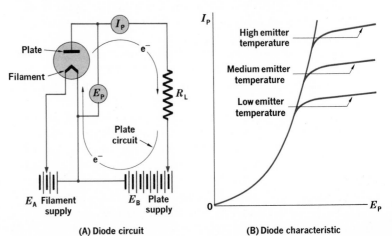

(A) Diode circuit   (B) Diode characteristic

Diode tubes can be classified as *power* diodes or *signal* diodes. The difference is primarily one of size and power-handling capabilities. Power diodes are used as rectifiers in power-supply circuits and must be large enough to dissipate the heat generated during their operation. Signal diodes are small and handle signals of negligible power. They are used principally as detectors and wave-shapers. See Figure 22-9. They are frequently enclosed in the same envelope with triodes and pentodes.

22.6 The Triode Amplifier   De Forest perfected the first *triode* amplifier by inserting a third electrode between the cathode and plate of a diode. This electrode consists of an open spiral of fine wire that presents a negligible physical barrier to electrons flowing from cathode to plate. However, a small potential difference between this electrode and the cathode has an important controlling effect on the flow of electrons from cathode to plate, and thus on the current in the plate circuit. For this reason the third electrode is called a *control grid*. We may think of a triode as an adjustable electron valve. The electrode configuration of a triode and the triode circuit symbol are shown in Figure 22-10.

If the control grid is made sufficiently negative with respect to the cathode, all electrons are repelled toward the cathode. In this condition the plate current is zero and the triode is said to be *cut off.* The grid-to-cathode voltage is called the *grid bias. The smallest negative grid voltage for a given plate voltage that causes the tube to cease to conduct is the* **cut-off bias.** This design characteristic of vacuum tubes containing control grids is important in the circuit applications of these tubes.

The effects of different grid-bias voltages on the plate current of a triode are shown in Figure 22-11. If a negative grid bias less than the cut-off value is used, some electrons pass through the grid and reach the plate to provide some magnitude of plate current. As the negative grid bias is

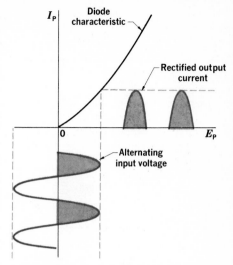

Figure 22-9. A diode characteristic used to show the rectifying action of a power diode.

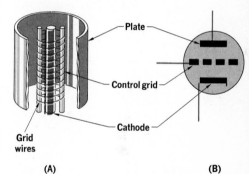

Figure 22-10. Sketch of the electrode configuration of a triode (**A**) and the triode circuit symbol (**B**).

Figure 22-11. The effects of different control-grid bias voltages on the plate current of a triode.

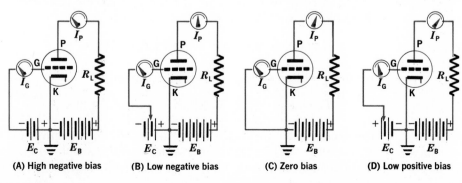

(A) High negative bias   (B) Low negative bias   (C) Zero bias   (D) Low positive bias

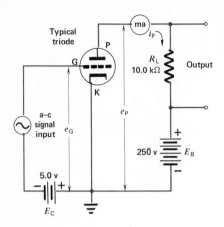

Figure 22-12. A basic triode amplifier circuit.

There is no d-c current in the grid circuit because the grid bias is negative with respect to the cathode. See Figure 22-11(B).

The sum of the potential drops around a series circuit must equal the applied emf.

reduced toward zero, the plate current increases accordingly. At zero bias, the plate current is fairly large and depends primarily on the plate voltage. The performance of the triode is now essentially the same as that of a diode of similar construction.

If, on the other hand, the grid is made positive with respect to the cathode, it no longer repels the electrons of the space charge but accelerates them in the direction of the plate. The plate current is increased by this action; however, *some electrons are attracted to the grid and produce an appreciable current in the grid circuit.* Grid current causes power dissipation in the grid circuit, which represents a waste of power. Thus the grid is normally maintained at some negative potential with respect to that of the cathode.

A basic triode amplifier circuit is shown in Figure 22-12. Small variations in grid potential cause larger variations in plate current than those caused by similar variations in plate potential. Thus voltage amplification across the circuit is possible. A small a-c voltage across the input circuit produces an amplified a-c voltage across the output circuit.

With batteries E_B and E_C in the circuit of Figure 22-12, a d-c current is in the plate circuit of the triode. Observe that the d-c plate circuit consists of the plate-supply battery E_B, the cathode-to-plate resistance in the triode (an inherent value characteristic of the triode used), and the plate-load resistance R_L.

The plate current I_P produces a drop in potential $I_P R_L$ across the load resistance. Consequently, the d-c plate potential E_P is less than E_B by the magnitude of $I_P R_L$.

$$E_P = E_B - I_P R_L$$

In Figure 22-12, E_B is $25\overline{0}$ v, E_C is -5.0 v, and R_L is 10.0 kΩ. With no a-c signal on the grid, typical d-c operating conditions of the amplifier circuit can be established by assuming a d-c plate current I_P of 10.0 ma.

A d-c current of 10.0 ma in R_L produces a drop in potential across R_L of $10\overline{0}$ v.

$$I_P R_L = 10.0 \text{ ma} \times 10.0 \text{ k}\Omega = 10\overline{0} \text{ v}$$

The d-c plate potential E_P is therefore $15\overline{0}$ v.

$$E_P = E_B - I_P R_L = 25\overline{0} \text{ v} - 10\overline{0} \text{ v} = 15\overline{0} \text{ v}$$

These circuit values of I_P, E_P, and $I_P R_L$ represent the steady-state (d-c) operating conditions of the amplifier circuit. These steady-state values are shown in Figure 22-13.

Suppose we now apply an a-c signal of ±1.0 v across the grid circuit of Figure 22-12. This a-c voltage is imposed on the steady-state grid bias of −5.0 v. It gives an a-c component e_G that varies between −4.0 v and −6.0 v *at the frequency of the a-c signal.*

When the input signal is +1.0 v, e_G will be −4.0 v.

$$e_G = -5.0 \text{ v} + 1.0 \text{ v} = -4.0 \text{ v}$$

The plate current will *increase* because the grid bias is *less negative.* Assuming that the plate current increases to 15.0 ma, $i_P R_L$ will *increase* to $15\overline{0}$ v.

$$i_P R_L = 15.0 \text{ ma} \times 10.0 \text{ k}\Omega = 15\overline{0} \text{ v}$$

Therefore, e_P will *decrease* to $10\overline{0}$ v.

$$e_P = 25\overline{0} \text{ v} - 15\overline{0} \text{ v} = 10\overline{0} \text{ v}$$

Compare these results with the first half-period e_G, i_P, and e_P waveforms in Figure 22-13.

When the input signal is −1.0 v, e_G will be −6.0 v

$$e_G = -5.0 \text{ v} + (-1.0 \text{ v}) = -6.0 \text{ v}$$

and the plate current will *decrease* because the grid is *more negative.* If i_P decreases to 5.0 ma, $i_P R_L$ will *decrease* to $5\overline{0}$ v.

$$i_P R_L = 5.0 \text{ ma} \times 10.0 \text{ k}\Omega = 5\overline{0} \text{ v}$$

As a result, e_P will *increase* to $20\overline{0}$ v.

$$e_P = 25\overline{0} \text{ v} - 5\overline{0} \text{ v} = 20\overline{0} \text{ v}$$

Compare these results with the second half-period e_G, i_P, and e_P waveforms in Figure 22-13.

Figure 22-13 shows that a variation of ±1.0 v at the grid produces a variation of ±$5\overline{0}$ v at the plate; voltage amplification or "gain" across the amplifier circuit is $5\overline{0}$. Note that a change in grid voltage in a positive sense results in a change in plate voltage in a negative sense. The output voltage is opposite in *phase* to that of the input voltage. The a-c voltage gain across the amplifier circuit is accompanied by a voltage phase inversion.

22.7 The Cathode-Ray Tube The cathode-ray tube is a special kind of vacuum tube used in test equipment, radar, television receivers, and computers to present a visual display of information. Electrons are emitted from a cathode, accelerated to a high velocity, and brought to focus on a fluorescent screen. This screen is a translucent coating on the inside surface of the glass face of the tube. It *fluoresces,* emitting visible light at the point where the electron beam strikes.

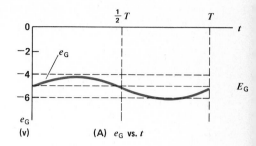

(A) e_G vs. t

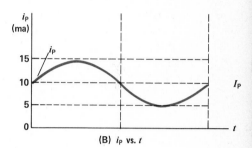

(B) i_P vs. t

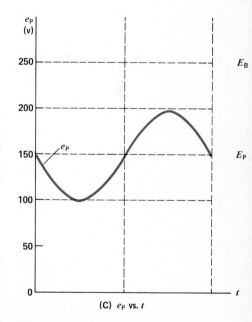

(C) e_P vs. t

Figure 22-13. Grid voltage, plate current, and plate voltage variations of a vacuum triode amplifier showing phase inversion between e_G and e_P.

When bombarded by electrons, fluorescent substances emit light of characteristic colors. All fluorescent substances have some *phosphorescence,* or afterglow, the persistence of which depends on the material and the energy of the bombarding electrons. Cathode-ray tubes that are used in test equipment such as the oscilloscope have short-persistence screens that emit predominantly green light. The picture tubes of television receivers have long-persistence screens.

The electron beam is moved over the fluorescent screen in response to the information signals being presented. This means that the information, in the form of electric signals, must be used to deflect the beam from its quiescent path within the tube.

An electron has negligible inertia since its mass is approximately 10^{-27} gram. Thus a beam of electrons can be deflected back and forth across the screen of a cathode-ray tube at very high speeds in response to weak signals. In a television receiver, for example, the electron beam sweeps across the face of the picture tube 15,750 times each second at a speed of the order of 24,000 km per hour.

There are two general methods of deflecting the electron beam in cathode-ray tubes: *electrostatic deflection* and *electromagnetic deflection.*

1. Electrostatic deflection systems. A cathode-ray tube employing electrostatic deflection is shown in Figure 22-14. The electrode arrangement that produces a focused beam of electrons is called the *electron gun.* It consists of an indirectly heated *cathode* capable of high emission, a *control grid* that regulates the intensity of the electron beam, a *focusing anode,* and an *accelerating anode.* The two anodes are maintained at high positive potentials with respect to the cathode. They produce an electric field that acts as an *electrostatic lens* to converge the electron paths at a point on the screen.

Figure 22-14. The construction of a cathode-ray tube using electrostatic deflection.

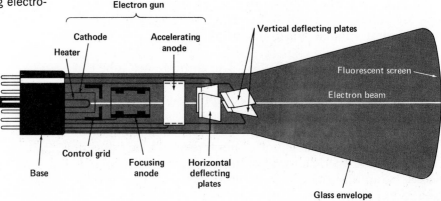

Deflection of the beam is accomplished by two pairs of *deflecting plates* set at right angles to each other. Signal voltages applied to these plates establish electrostatic fields through which the electrons move. The electron beam is subjected to deflecting forces in both the horizontal and vertical planes. Since electrons have negligible inertia, their response to these forces is practically instantaneous in the direction of the resultant force. Electrostatic deflection is generally used in the small cathode-ray tubes of test equipment.

2. Electromagnetic deflection. Large cathode-ray tubes have electromagnetic deflection systems. The beam can be focused either electrostatically, as shown in Figure 22-14, or magnetically by means of an external focusing coil. Beam control is accomplished with coils formed into a deflection yoke placed on the neck of the tube.

The most common color picture tubes have three electron guns, one for each primary-color phosphor. These color phosphors are on a tricolor, phosphor-dot screen. The electron beams scan the screen to produce the primary colors of light on the face of the tube. The structure of a typical color television tube is shown in Figure 22-15. Cathode-ray tubes with more than three electron guns have been developed for use in highly specialized electronic equipment.

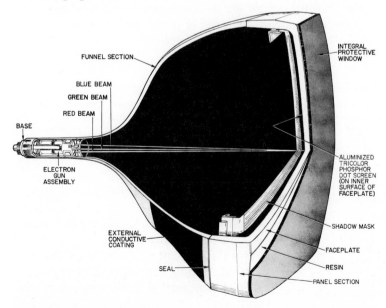

Figure 22-15. A cutaway view of a television color picture tube.

22.8 Photomultiplier Tube Secondary emission, the liberation of electrons from a metal surface by the impact of a high-speed electron, can be used in conjunction with

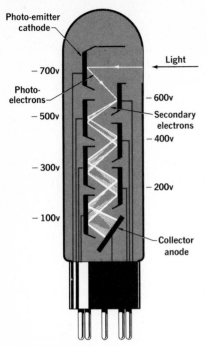

Figure 22-16. A photomultiplier tube with the six stages of secondary emission.

the photoelectric effect to detect light of very low intensity. This special form of photoelectric cell is known as a *photomultiplier tube.*

A diagram of a photomultiplier tube is shown in Figure 22-16. The number of photoelectrons emitted by the light-sensitive cathode in a given time is proportional to the intensity of the incident light. These photoelectrons are attracted to the first secondary-emission electrode, which is 100 volts positive with respect to the emitter cathode. On impact, additional electrons are released by secondary emission.

Successive secondary-emission electrodes, each more positive (less negative) than the one preceding it, contribute additional electrons as the electron stream progresses toward the collector anode. Thus a meager photoemission from a faint incident light can develop into an avalanche of electrons by the time it arrives at the collector plate. If each electron impact produces **n** secondary electrons, a photomultiplier tube with **x** secondary-emission electrodes will supply n^x electrons to the collector plate for each photoelectron emitted.

Suppose a faint light incident on the photoemission cathode of the tube shown in Figure 22-16 will produce *five* photoelectrons per millisecond. Also suppose that each electron impact will produce *four* secondary electrons. Since the tube has *six* secondary-emission stages, the number of electrons arriving at the collector plate will be of the order of 2×10^4 per millisecond.

$$5n^x = 5 \times 4^6 = 2 \times 10^4$$

This equation suggests the extraordinary gain that can be achieved with a photomultiplier tube compared to that of a regular photoelectric cell.

Questions
GROUP A

1. What device is primarily responsible for the development of electronics as a branch of engineering?
2. What discovery by Joseph Thomson provided the explanation for the Edison effect?
3. What significant advantage was realized with De Forest's triode that had not been possible with previously developed diodes?
4. (a) Define thermionic emission.

(b) What type of emitter is commonly used in receiving tubes?
5. Why is the grid of a triode called a control grid?
6. What is meant when a vacuum tube is said to be cut off?
7. Why is the control grid of a vacuum tube normally maintained at some negative potential with respect to the cathode?

GROUP B

8. Suggest two possible reasons why Edison did not offer an explanation

for the phenomenon known as the Edison effect.

9. Explain voltage inversion between the control grid and plate in a vacuum-tube circuit.

10. What is the electron gun of a cathode-ray tube?

11. What is the basic difference between the electrostatic and electromagnetic deflecting systems used in cathode-ray tubes?

12. (a) Explain why the first secondary-emission electrode of a photomultiplier tube is maintained at a positive (less negative) potential with respect to the photoemission electrode.
(b) Explain why each successive secondary-emission electrode is at a positive potential with respect to the preceding electrode?

TRANSISTORS

22.9 Crystal Diodes Metals are good *conductors* of electric currents. Many nonmetallic materials are classed as *insulators* because their resistivities are enormous compared with those of metals. For most practical purposes they are considered to be *nonconductors*. Some elements, including germanium, silicon, and selenium, and certain compounds, such as copper oxide, lead sulfide, and zinc oxide, are in a class of crystals called *semiconductors*. These substances have resistivities that fall between those of conductors and insulators.

Refer to the resistivity spectrum of Figure 17-23.

One of the devices from the early days of radio that is found in modern electronic circuits is the *crystal diode*. The rectifying property of silicon or galena (lead sulfide) crystals was used in early receivers before the development of the diode tube. During World War II, improved crystal diodes were developed for use in radar receivers. They are now used extensively in many types of circuits to perform a variety of specialized functions. They have the advantage of small size, and they require no heater or filament power.

A modern crystal diode can consist of a tiny wafer of silicon or germanium and a platinum *catwhisker*, all contained in a sealed capsule. See Figure 22-17. The crystal acts as the cathode of the diode and the catwhisker acts as the anode. The diode characteristic depends on the unique

Figure 22-17. A crystal diode and its circuit symbol.

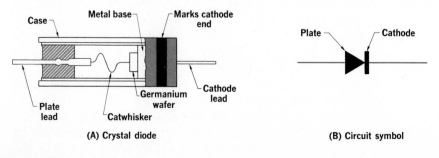

(A) Crystal diode

(B) Circuit symbol

property of a semiconductor that permits a large flow of electrons in the *forward* direction with a small applied voltage, and only a very small flow of electrons in the *reverse* direction, even at a much larger applied voltage.

22.10 Transistor Development

After nearly a decade of basic research in the physics of semiconductors, physicists discovered how to produce a semiconductor device that could *amplify.* This device, known as a *transistor,* was developed by a team of physicists at the Bell Telephone Laboratories in 1948. Transistors are tiny, rugged structures that require very little power. The development of a family of transistors that can perform the functions of vacuum tubes has made possible the miniaturization of electronic circuits. See Figure 22-18.

The first transistors were known as *point-contact* transistors. These devices were replaced by *junction* transistors. Junction transistors can be further classified into several broad categories depending on the forming process and performance characteristics. Another addition to the semiconductor family is the *field-effect* transistor.

As the physicist's knowledge of semiconductor materials expands, new transistor configurations with improved frequency response and power-handling capabilities continue to appear. The key to the transistor effect is the introduction of specific impurities into silicon and germanium crystals to form **P**- and **N**-type semiconductor materials. The process is referred to as *doping.*

22.11 P- and N-Type Semiconductors

Silicon and germanium are quite similar in their structure and chemical behavior. The atoms of both elements have *four* valence electrons bound in the same way in their respective crystals. In its transistor functions, germanium is the more versatile semiconductor. It has a diamond cubic crystal lattice in which each atom is bonded to four neighboring atoms through shared electrons, as illustrated in Figure 22-19. This arrangement binds the four valence electrons of each atom so that pure germanium would appear to be a nonconductor. *Germanium acquires the diode property of rectification and the transistor property of amplification through the presence of trace quantities of certain impurities in the crystal structure.* Two types of impurities are important: one is known as a *donor,* the other as an *acceptor.*

Antimony, arsenic, and phosphorus are typical donor elements. The atoms of each have *five* valence electrons. When minute traces of phosphorus are added to germanium, each phosphorus atom joins the crystal lattice by *donating* one electron to the crystal structure. Four of the

Figure 22-18. Semiconductor components are produced in a variety of circuit designs. The coin (dime) suggests component size.

The controlled addition of donor or acceptor atoms to a semiconductor material during crystal growth is referred to as "doping" the semiconductor.

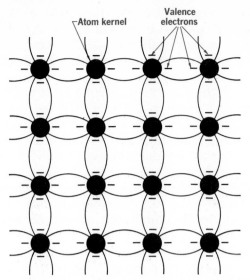

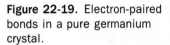

Valence electrons

Figure 22-19. Electron-paired bonds in a pure germanium crystal.

five valence electrons are paired, but the fifth electron is relatively free to wander through the lattice like the free electrons of a metallic conductor. The detached electron leaves behind a phosphorus atom with a unit-positive charge bound into the crystal lattice. Germanium with this kind of crystal structure is known as *N-type,* or *electron-rich,* germanium because most of its conductivity results from *negative* charge carriers. **N-type germanium** *consists of germanium to which are added equal numbers of free electrons and bound positive charges so that the net charge is zero.* The added negative charge carriers (the free electrons) greatly increase conductivity over that of the pure semiconductor. An **N**-type semiconductor crystal is illustrated in Figure 22-20(A).

In N-type semiconductors the charge carriers are negatively charged electrons.

Figure 22-20. The **N**-type germanium crystal illustrated in **(A)** has donor atoms and free electrons in equal numbers. The **P**-type germanium crystal in **(B)** has acceptor atoms and free (positive) holes in equal numbers.

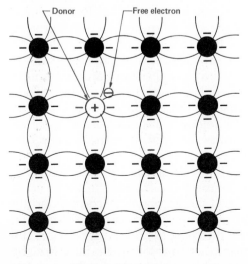

(A)   N-type semiconductor

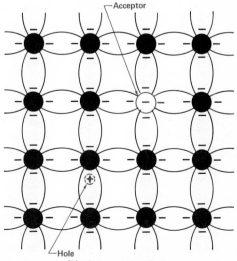

(B)   P-type semiconductor

Atoms with *three* valence electrons, such as those of aluminum, boron, and gallium, will act as acceptors. When minute traces of aluminum are added to germanium, each aluminum atom joins the crystal lattice by *accepting* a single electron from a neighboring germanium atom. This leaves a *hole* in the electron-pair bond from which the electron is acquired.

We can think of this hole as the equivalent of a *positive charge* since it acts as a trap into which an electron can fall. As an electron fills the hole, it leaves another hole behind, into which another electron can fall. *In effect* then, the hole (positive charge) detaches itself and becomes free to move in the opposite sense to that of the electron, leaving behind the aluminum atom with a unit negative charge bound into the crystal lattice. Germanium with this crystal structure is called *P-type*, or *hole-rich*, germanium because most of the conductivity results from *positive* charge carriers. ***P-type germanium*** *consists of germanium to which are added equal numbers of free positive holes and bound negative charges so that the net charge is zero.* The added positive charge carriers (the holes) greatly increase conductivity over that of the pure semiconductor. A **P**-type semiconductor crystal is illustrated in Figure 22-20(B).

In P-type semiconductors the charge carriers are, in effect, positively charged holes.

22.12 The P-N Junction

As a germanium crystal is formed, donor and acceptor atoms can be added in such a way that a definite boundary is created between **P**-type and **N**-type regions of the crystal. At this boundary some free electrons from the **N** region drift into the **P** region and a corresponding number of holes from the **P** region drift into the **N** region. An electric field, $\mathcal{E}$, is established across the boundary layer as it becomes devoid of electron carriers on the **N** side and of hole carriers on the **P** side. In the resulting equilibrium state the boundary layer becomes a kind of electric barrier. It is called the ***P-N*** *junction*. The electric barrier of the **P-N** junction is shown in Figure 22-21.

The crystal as a whole has properties of a diode and is called a ***P-N*** *junction diode*. The hole-charge carriers of the **P** region cannot pass through the electric barrier of the **P-N** junction to reach the **N** region. The electron-charge carriers of the **N** region cannot pass through the electric barrier to reach the **P** region.

A small potential difference impressed across the diode, as in Figure 22-22(A), enables the free electrons to cross the junction and pass into the **P** crystal. A corresponding number of holes cross into the **N** crystal. Recall that the apparent movement of holes is in the opposite sense to the actual movement of electrons that fall in the holes of the

Review electric fields in Sections 16.9 and 16.16.

The P-N junction is the meeting zone of oppositely doped regions of a semiconductor crystal.

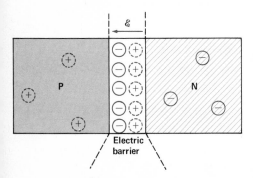

Figure 22-21. The electric barrier of the **P-N** junction.

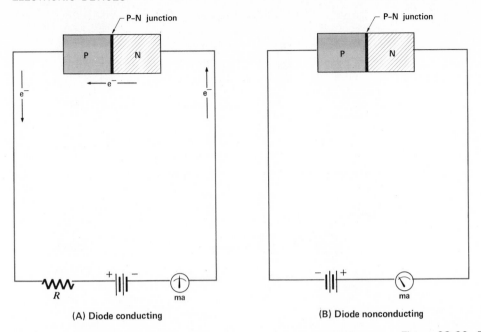

(A) Diode conducting   (B) Diode nonconducting

Figure 22-22. The **P-N** junction is the rectifying element of semiconductor crystals.

P-type crystal. Thus an electron current is established across the **P-N** junction and in the external circuit. The current-limiting resistance R is in the circuit to protect the conducting diode from excessive current.

If the battery connections are reversed, as in Figure 22-22(B), the free electrons in the **N** crystal and the free holes of the **P** crystal are attracted away from the **P-N** junction. The junction region is left without current carriers; consequently, there is no conduction across the junction and no current in the circuit.

The junction has the diode characteristic of permitting unidirectional electron flow with ease when a small voltage is applied in the proper sense. Thus the **P-N** junction is the rectifying element of semiconductor crystals.

22.13 Two Types of Junction Transistors Junction transistors consist of two **P-N** junctions back to back with a thin **P**-type or **N**-type region formed between them. If the center region is composed of **P**-type material, the transistor is designated **N-P-N**. The physical diagram of an **N-P-N** transistor and its circuit symbol are illustrated in Figure 22-23(A).

The **N**-type material shown at the left of the **P**-type center is the *emitter*. It is normally biased negatively with respect to the **P**-region, which is called the *base*. The **N**-type material on the right of the base is the *collector*. It is biased positively with respect to the base. The basic circuit of the **N-P-N** transistor is shown in Figure 22-24(A).

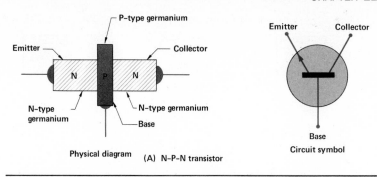

Figure 22-23. A comparison of **N-P-N** and **P-N-P** junction transistors.

Transistors in which the center region is composed of **N**-type semiconductor material are designated **P-N-P**. A diagram of a **P-N-P** transistor and its circuit symbol are shown in Figure 22-23(B).

The basic circuit of **P-N-P** transistors is given in Figure 22-24(B). Observe that it is the same as that for **N-P-N** transistors except that the battery connections are reversed.

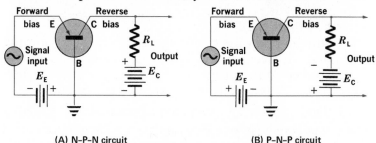

Figure 22-24. Basic **N-P-N** and **P-N-P** junction transistor circuits. Compare these circuits with that of the vacuum triode circuit of Figure 22-12.

In either **N-P-N** or **P-N-P** transistors, the emitter is biased to supply charges—electrons or holes—to be acquired by the collector through the two junctions of the base. An **N**-type emitter is biased negatively with respect to the base to supply electrons to its **N-P** junction and then to the **P**-type base. A **P**-type emitter is biased positively to withdraw electrons from its **P-N** junction and thus from the **N**-type base. This statement is equivalent to saying that the **P**-type emitter supplies holes to its **P-N** junction and then to the **N**-type base. Both emitter types are said to have *forward* bias. The emitter circuits of Figure 22-24(A) and Figure 22-24(B) have forward-bias voltages applied by batteries E_E.

In the transistor circuit symbol, the arrowhead at the emitter is conventionally drawn to indicate the direction of the positive hole current. *Electron current is in the opposite direction.* This arrowhead in the transistor circuit symbol provides a way of showing which type of transistor is represented in a circuit diagram. In the **P-N-P** circuit symbol, note that the arrowhead at the emitter *points toward the base*. It points *away from the base* in the **N-P-N** circuit symbol.

An **N**-type collector is biased positively with respect to the base to attract electrons from its **N-P** junction and from the **P**-type base. A **P**-type collector is biased negatively to transfer electrons to its **P-N** junction and into the **N**-type base. Again, this is equivalent to saying it attracts holes from its **P-N** junction and thus from the **N**-type base. Both collector types are said to have *reverse* bias. The collector circuits of Figure 22-24(A) and of Figure 22-24(B) have reverse-bias voltages applied by batteries E_C.

The forward bias on emitters and the reverse bias on collectors have the effect of producing an electron current in the **N-P-N** transistor from emitter to base in the emitter circuit and from base to collector in the collector circuit. Forward emitter bias and reverse collector bias in the **P-N-P** transistor produce an electron current from collector to base and from base to emitter.

The base section serves to isolate the emitter input circuit from the collector output circuit of a transistor. This isolation results from the barrier voltages established at the two junctions. The barrier polarities at these two junctions are reversed because the base has an **N-P** junction at one side and a **P-N** junction at the other.

Forward bias at the emitter junction overcomes the barrier voltage and enables the emitter to supply charge to the base. The reversed polarity of the barrier voltage at the collector junction, augmented by the reverse collector bias, aids the transfer of charge through the junction to the collector. Transistor performance is achieved when the collector current is controlled by the emitter current in the base.

22.14 Transistor Amplification   The transistor circuits shown in Figure 22-24 are *common-base* amplifiers. Small signals introduced into the emitter circuit are supplemented by energy from the collector battery and larger signals are transferred from the collector circuit. This effect is accomplished in spite of the fact that the collector current is slightly smaller than the emitter current.

Amplification in the common-base transistor amplifier is due to the ratio of output impedance to input impedance, which may be of the order of 1000:1. Low emitter impedance is obtained by the *forward* biasing of the emitter junction, that is, the junction between the emitter section and the base section. As stated in Section 22.12, the effect is to reduce the electric field across the emitter junction. High collector impedance is obtained by the *reverse* biasing of the junction between the collector section and the base section. This effect is to increase the electric field across the collector junction.

Only 1% to 5% of the emitter current passes through the base leg of the circuit. The remaining 99% to 95% passes through the collector circuit. This current can pass through the collector circuit because of the thinness of the base section and the consequent ease with which positive holes from the emitter diffuse across it to be collected.

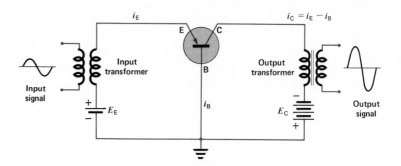

Figure 22-25. A P-N-P common-base amplifier.

A typical **P-N-P** common-base amplifier circuit is shown in Figure 22-25. A small signal current i_E in the emitter circuit results in a low input power $i_E^2 Z_E$ because the input impedance Z_E is low. The slightly smaller collector current $(i_C = i_E - i_B)$ yields a higher output power $i_C^2 Z_C$ because of the high output impedance Z_C. The *power gain* of the circuit is the ratio of the output power to the input power.

$$P_{gain} = \frac{P_{out}}{P_{in}}$$

Suppose the input impedance of the amplifier is $5\overline{0}$ Ω, the a-c emitter current is 2.0 ma, the collector current is 1.9 ma, and the output impedance is $1\overline{0}$ kΩ.

The power input is

$$P_{in} = i_E^2 Z_E = (2.0 \text{ ma})^2 \times 5\overline{0} \text{ }\Omega$$
$$P_{in} = 0.20 \text{ mw}$$

The power output is

$$P_{out} = i_C^2 Z_C = (1.9 \text{ ma})^2 \times 1\overline{0} \text{ k}\Omega$$
$$P_{out} = 36 \text{ mw}$$

The power gain is

$$P_{gain} = \frac{P_{out}}{P_{in}} = \frac{36 \text{ mw}}{0.20 \text{ mw}} = 180$$

A **P-N-P** transistor is used in Figure 22-25. An **N-P-N** transistor would serve just as well with the bias-voltage polarities reversed. In some circuits the two types of junction transistors are equivalent. In other circuits one type of junction transistor may have small advantages over the other.

The common-base transistor circuit that we have been considering is one of three general circuit configurations for junction transistors. Others are the common-emitter and the common-collector configurations. These configurations for both **N-P-N** and **P-N-P** transistors are shown in Figure 22-26. Observe that in all circuits the emitters have forward bias to provide low impedance and the collectors have reverse bias to provide high impedance.

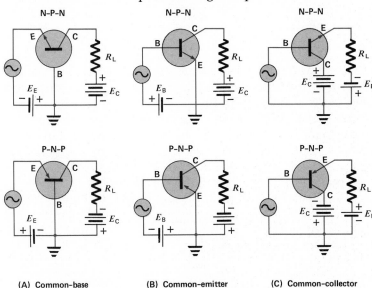

(A) Common-base   (B) Common-emitter   (C) Common-collector

Figure 22-26. Circuit configurations for **N-P-N** and **P-N-P** junction transistors.

The amplifier output signal in Figure 22-25 is *in phase* with the input signal. The common-base amplifier *does not* produce phase inversion. However, the common-emitter amplifier *does* produce a phase shift, as does a vacuum-tube amplifier.

The amplifying property of a transistor is the result of current changes that occur in the different regions of the structure when proper voltages are applied. Consequently, the transistor is often referred to as a current amplifier. More precisely, it is a *current-operated device* in much the same way that a vacuum triode is a voltage-operated device. A transistor amplifier can produce a voltage gain, a current gain, or a power gain, depending on the circuit configuration.

When the operation of a transistor is compared with that of a vacuum triode, the emitter is equivalent to the cathode, the base to the control grid, and the collector to the plate. These relationships are illustrated in Figure 22-27. The emitter and base form the signal input circuit for a transistor amplifier in the same way that the grid and cathode form the input circuit of the vacuum-triode amplifier. The collector is the output section of the transistor just as the plate is the output of the vacuum triode.

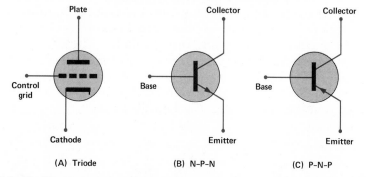

Figure 22-27. Transistor elements compared with those of a vacuum triode. (A) Triode circuit symbol. (B) **N-P-N** transistor circuit symbol. (C) **P-N-P** transistor circuit symbol.

(A) Triode   (B) N-P-N   (C) P-N-P

22.15 Transistor Characteristics A small signal applied to the input circuit of a transistor can be amplified because the magnitude of collector current in its output circuit is controlled by the current in its input circuit. The performance of a transistor is indicated by its family of characteristic curves. These curves describe the behavior of the transistor in a specific circuit configuration.

Typical output characteristics for an **N-P-N** transistor in a common-base configuration are shown in Figure 22-28. Each curve shows how the collector current I_C varies with the output voltage V_{CB}, measured between the collector and base terminals, for a given value of emitter current I_E.

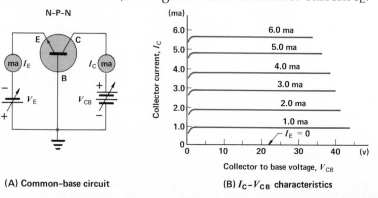

Figure 22-28. Typical common-base output characteristic curves for an **N-P-N** transistor. (A) Common-base test circuit. (B) $I_C - V_{CB}$ characteristics.

(A) Common-base circuit   (B) I_C-V_{CB} characteristics

The family of curves is produced by using different values of emitter current.

These curves show that the collector current is independent of the collector voltage but is dependent on the magnitude of emitter current. Variations in I_E produce variations in I_C. At a constant value of V_{CB}, the ratio of a change in I_C to a change in I_E expresses the *current gain* of the common-base circuit. It is expressed as a positive number without dimension and is denoted by the Greek letter α (alpha).

$$\alpha = \frac{\Delta I_C}{\Delta I_E} \; (V_{CB} \text{ constant})$$

Referring to Figure 22-28(B), suppose the emitter current I_E is varied from 2.0 ma to 4.0 ma (a change of 2.0 ma) while the voltage V_{CB} is held constant at 15 v. Reading from the collector current axis, I_C will be found to vary from 1.9 ma to 3.8 ma (a change of 1.9 ma). The current gain can be determined from these data as follows:

$$\alpha = \frac{\Delta I_C}{\Delta I_E} = \frac{1.9 \text{ ma}}{2.0 \text{ ma}} = 0.95$$

The forward current gain, or alpha characteristic, of a common-base circuit is always slightly less than unity. Typical values of α range from 0.95 to 0.99. The closer α is to unity, the better the quality of the transistor.

Figure 22-29 shows a typical family of characteristic curves of transistor performance in a common-emitter configuration. Collector current I_C is plotted as a function of output voltage V_{CE}, measured between collector and emitter, for each value of base current I_B.

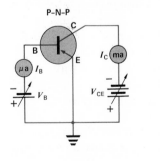

(A) Common-emitter circuit

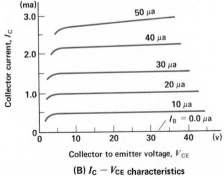

(B) $I_C - V_{CE}$ characteristics

Figure 22-29. Typical common-emitter output characteristic curves for a P-N-P transistor.
(A) Common-emitter test circuit.
(B) $I_C - V_{CE}$ characteristics.

Base currents are very small compared to emitter and collector currents. In the common-emitter circuit, small changes in I_B produce large variations in I_C. At a constant value of V_{CE}, the ratio of a change in I_C to a change in I_B expresses the *current gain* for the common-emitter circuit.

This current-gain factor is denoted by the Greek letter β (beta).

$$\beta = \frac{\Delta I_C}{\Delta I_B} \; (V_{CE} \text{ constant})$$

Using the typical curves of Figure 22-29(B), we will assume that the base current I_B is varied from $2\bar{0}$ μa to $4\bar{0}$ μa (a change of $2\bar{0}$ μa) while V_{CE} is held constant at $2\bar{0}$ v. Projecting to the collector current axis, we find that I_C will vary from 1.0 ma to 2.3 ma (a change of 1.3 ma). The current gain, or β characteristic, of this common-emitter circuit will be

$$\beta = \frac{\Delta I_C}{\Delta I_B} = \frac{1.3 \times 10^{-3} \text{ a}}{2.0 \times 10^{-5} \text{ a}} = 65$$

Some typical values of β range from 20 to 200. Either the common-base or the common-emitter transistor circuit will provide amplification of an input signal. With a low-impedance load, where high *current gain* is desired, the common-emitter circuit is preferred. With a high-impedance load, where large *power gain* is desired, the common-base circuit will serve as well and is even preferable.

The common-collector circuit is essentially the reverse of the common-emitter circuit. This configuration can be recognized in Figure 22-26. The common-collector circuit has a high input impedance and a low output impedance and a voltage gain of less than unity. Thus it has little value as an amplifier but can be used for impedance-matching between a high-impedance source and a low-impedance load.

22.16 The Photovoltaic Effect Energy-conversion processes are described in Section 17.3 as sources of electric currents. For example, heat is the source of electric current in thermoelectric action. Mechanical stress is the source of current in piezoelectric action. Chemical reactions are the source of current in fuel-cell action. These processes are characterized by a capability for *direct* conversion of thermal, mechanical, or chemical energy to electric energy.

Semiconductor materials can provide the direct conversion of radiant energy to electric energy under suitable conditions. The process is known as the *photovoltaic effect*. If a semiconductor **P-N** junction is designed to absorb incident light, a potential difference can be developed across the junction as a consequence of photovoltaic action. This potential difference can generate an electron current in a load resistance connected across the junction. Thus the

P-N junction can act as a source of emf while exposed to light radiations.

The silicon *solar cell* is a photovoltaic converter of considerable interest. The components of a solar-cell photovoltaic circuit are shown schematically in Figure 22-30. The cell consists of a thin base layer of **P**-type silicon crystal over which a very thin (about 0.5 μm) **N**-type layer is formed. The **P-N** junction is a barrier layer with its electric field situated where the **P**- and **N**-type materials are joined. The external circuit of the cell consists of the resistance load R_L connected across the **P-N** junction at terminals in contact with the **P** and **N** surfaces.

The P- and N-type semiconductor sections of a photovoltaic cell can just as well be the reverse of the arrangement shown in Figure 22-30.

Review the formation of the P-N junction in Section 22.12.

Figure 22-30. Photovoltaic action in a silicon solar cell.

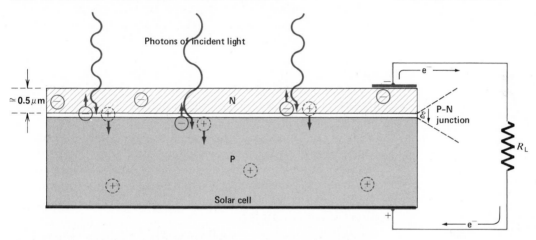

Photons of light radiations with energies corresponding to wavelengths shorter than approximately 10^{-6} m can dislodge electrons from silicon atoms in the crystal lattice of the solar cell. Reference to the radiation spectrum of Figure 12-8 shows that this minimum energy radiation ($\lambda = 10^{-6}$ m) lies in the near infrared region of the electromagnetic spectrum. Figure 12-8 also suggests that roughly half of the sun's radiation, including the visible and ultraviolet regions, has photon energies that can be converted to electric energy by the silicon solar cell.

When light is incident on the exposed surface of the solar cell, most of the photons pass through the very thin **N** layer and are absorbed by atoms of the crystal lattice in the vicinity of the **P-N** junction. The energy *hf* of an absorbed photon is transferred to an electron, which is then ejected from the atom, leaving behind a positive hole. In this way, light photons generate *electron-hole pairs* in the region of the junction layer. The electric field of the **P-N** junction forces electrons that have not recombined into the **N** layer and a corresponding number of holes into the **P** layer. These electrons are excess negative charge carriers

$\lambda = 10^{-6}$ m = 1 μm (1 micron) = 10^4 Å identifies radiation in the near infrared region of the solar spectrum.

in the **N** layer and flow through the external circuit from the **N** terminal to the **P** terminal. They then combine with the excess holes in the **P** layer. Thus the illuminated solar cell acts as a source of emf and of electron current in its external circuit.

Batteries of solar cells provide electric energy for space vehicles. Earth-bound solar batteries serve many purposes; for example, they supply energy for telephone lines and recharge storage batteries that replace the solar batteries during hours of darkness. The silicon solar cell has an important role in the on-going search for practical and efficient ways to harness solar energy.

A photovoltaic cell of the type usually used in photographic exposure meters consists of a junction formed by a film of copper oxide deposited on metallic copper, or of a film of selenium deposited on metallic iron. Incident light photons pass through the selenium film to the iron. Electrons are moved from the iron into the selenium layer and an emf is generated across the junction. The iron becomes the positive electrode and the selenium the negative electrode of the photovoltaic cell. If a sensitive galvanometer is connected across these electrodes, the emf produces an electron current in the meter that is proportional to the illumination on the cell. By calibrating the galvanometer in light values, proper exposure conditions can be indicated directly by the meter.

The photoelectric effect is described in Section 12.6.

Photovoltaic action is a type of photoelectric effect in the sense that electron ejection from a parent atom is initiated by a light photon. The distinction between photoelectric action and photovoltaic action is that the former requires a vacuum environment and an external source of emf in the photoemission circuit. The latter does not require a vacuum environment, and it generates its own emf. Both types of cells are commonly referred to as *photocells*.

A photoelectric circuit is shown in Figure 17-5.

Questions
GROUP A

1. What is the property of a semiconductor crystal known as the diode characteristic?
2. Why is **N**-type germanium referred to as electron-rich germanium?
3. Why is **P**-type germanium referred to as hole-rich germanium?
4. What is the **P-N** junction of a semiconductor?

5. How are the two types of junction transistors designated?
6. Draw the circuit symbol for an **N-P-N** transistor. Label the emitter, base, and collector.
7. Draw the circuit symbol for a **P-N-P** transistor. Label the emitter, base, and collector.
8. What is the significance of the arrowhead at the emitter in a transistor circuit symbol?

9. Which sections of a junction transistor are equivalent to the cathode, control grid, and plate of a vacuum triode?
10. What are the different circuit configurations used for transistor service?

GROUP B

11. Distinguish between a germanium diode and a germanium transistor.
12. Distinguish between a **P-N-P** transistor and an **N-P-N** transistor.
13. The emitter of an **N-P-N** transistor is biased negatively and the emitter of a **P-N-P** transistor is biased positively with respect to the base. In both cases they are described as forward bias. Explain.
14. What is the alpha characteristic of a transistor and to what transistor configuration does it apply?
15. Why must the current gain in a common-base transistor circuit always be less than unity?
16. Which transistor circuit configuration most closely resembles the basic vacuum triode circuit? Explain why.
17. To what does the beta characteristic of a transistor refer?
18. How can you account for amplification in a common-base transistor circuit when the collector current in the output circuit can never be as large as the emitter current in the input circuit?
19. Photovoltaic action is a type of photoelectric effect. How do the two actions differ?
20. Suggest a reason for the exposed layer of a solar cell being so thin that most of the incident photons will penetrate the layer and be absorbed by atoms in or very near the **P-N** junction layer.

SUMMARY

High-vacuum tubes are identified as diodes, triodes, tetrodes, and pentodes based on the number of electrodes present. Low-vacuum tubes contain a gas under reduced pressure. Diodes are used as detectors for radio signals and as rectifiers for converting alternating current to direct current. They are classed as signal or power diodes.

Electrons are emitted from the cathode of a vacuum tube by a process of thermionic emission. Electron emission is a function of emitter temperature. Emitters can be either directly or indirectly heated.

Voltage and power amplification are possible in triode circuits because a small signal on the grid circuit is capable of controlling a large current in the plate circuit. Conduction in a vacuum tube results from the thermionic emission of electrons from the cathode and the attraction of the electrons to the positively charged plate.

The tube bias voltage between the control grid and the cathode establishes the operating condition of the tube. Bias voltages are produced by three methods. The a-c voltage gain across a vacuum-tube amplifier is accompanied by a voltage phase inversion.

P- and N-type semiconductors are formed by doping with trace quantities of acceptor and donor atoms respectively. N-type semiconductors are electron-rich materials. The free electrons are the negative charge carriers. P-type semiconductors are hole-rich materials. The holes act as positive charge carriers. The P-N junction acts as an electric barrier in the meeting zone of oppositely doped regions of a semiconductor crystal. The direction of the electric field across the junction is from the N side to the P side.

Junction transistors consist of two P-N junctions back to back with thin P or N regions in between. They are designated N-P-N and P-N-P. Transistor elements correspond functionally to the elements of a vacuum triode. In transistor circuits the emitter has a forward-bias voltage and the collector has a reverse-bias voltage. Three circuit configurations are used: common base, common emitter, and common collector.

The transistor is a current-operated device. The amplifying property is the result of current changes that occur in the different transistor regions when proper voltages are applied to the transistor elements. The performance of a transistor is indicated by its family of characteristic curves. The forward current gain of a common-base configuration is called its alpha characteristic. It is the ratio of the change in collector current to the change in emitter current for a constant collector-to-base voltage. The current gain of a common-emitter configuration is called its beta characteristic. It is expressed as the ratio of the change in collector current to the change in base current for a constant collector-to-emitter voltage.

The photovoltaic effect can provide the direct conversion of radiant energy to electric energy. Photovoltaic action is a type of photoelectric effect. The solar cell is a semiconductor diode in which the semiconductor crystal is doped in a special way. The construction and P-N junction provide the energy conversion effect when light strikes the cell's surface.

VOCABULARY

acceptor element
amplifier
base
cathode-ray tube
collector
cut-off bias
diode
donor element
doping
Edison effect
electronics

electron-rich
   semiconductor
electrostatic lens
emitter
forward bias
hole-rich semiconductor
N-P-N transistor
N-type semiconductor
photovoltaic effect
P-N junction

P-N-P transistor
P-type semiconductor
rectifier
reverse bias
secondary emission
solar cell
space charge
thermionic emission
transistor
triode

ATOMIC STRUCTURE

Ernest Rutherford grew up in New Zealand and won scholarships to the university in that country and to Cambridge University in England. In 1911, Rutherford confirmed the nuclear model of the atom with the alpha-scattering experiments that are symbolized on this stamp.

THE ELECTRON

23.1 Subatomic Particles Toward the end of the 19th century, evidence indicating that atoms were not the ultimate particles of matter began to appear. A simple way of demonstrating the existence of subatomic particles is shown in Figure 23-1. In each photo, the glass tube contains hydrogen gas under low pressure. The conductor leading to the center of the tube is connected to the negative terminal of a direct-current source of electricity. The conductor at the top of the right bulb in each picture is connected to the positive terminal. When a high voltage is impressed across the tubes, two beams shoot out in opposite directions from the hollow electrode in the center. The two beams have different colors.

Since the tubes contain only hydrogen atoms, the differently colored beams indicate that the hydrogen atom has at least two different parts. The fact that the beams flow in opposite directions shows that the two parts of the atom have opposite electric charges. One beam is negatively charged since it passes into the part of the tube where the positive electrode is located. The other beam is positively charged since it is attracted through the negative electrode at the center and into the left side of the tube.

The magnets in the pictures tell still more about the parts of the hydrogen atom. In the top picture, the magnet is held near the positive beam. The effect on the beam is

I n this chapter you will gain an understanding of:

➤ the determination of the mass, size, and charge of the electron
➤ the wave-particle nature of the electron
➤ the relative positions of the atomic nucleus and electron shells
➤ the discovery and nature of the proton and the neutron
➤ the nature of isotopes
➤ the relationship between nuclear binding force and nuclear mass defect

The different colors in the tube also indicate that the electrons of the hydrogen atoms have been excited to different energy levels. Refer to Section 12.11.

Figure 23-1. Evidence for subatomic particles. The dumbbell-shaped apparatus contains the gas, hydrogen. When d-c voltage is applied, hydrogen atoms separate into streams of oppositely charged particles. This shows that hydrogen atoms are composed of at least two different particles.

not very pronounced. In the bottom picture, the magnet is held near the negative beam. This time the bending effect is more noticeable. This suggests that the particles composing the two beams have differences in addition to that of opposite charge. Further studies show that the main reason for the difference in the way the beams react to magnetism is that the positive particles have much larger masses than do the negative particles.

With this and other experiments, scientists have developed a comprehensive theory of atomic structure. Various parts of this theory will be presented in this and the following chapters. It is important to remember, however, that all these discussions are based on current knowledge of atomic structure. Revisions are frequently necessary as new investigations and new interpretations are made. What sort of picture of the atom will finally emerge from this research no one can fully predict. It is certain that the atom is made up of a number of still smaller particles that are arranged in various complex ways. These subatomic particles interact with each other to make up various atoms.

23.2 Discovery of the Electron The first subatomic particle to be identified and studied by scientists was the electron. Actually, the discovery of this particle was a by-product of other investigations. Toward the end of the 19th century physicists were using tubes similar to the one just described to study the conduction of electricity through various gases. When such a tube was filled with air (at atmospheric pressure), electricity would not flow between the electrodes, even under high voltages. When some of the air was removed, however, electricity began to flow from one electrode to the other through the rarefied air. The air in the tube glowed with a purplish light. When still more air was pumped from the tube, the purplish light faded and the glass walls of the tube glowed with a greenish fluorescence.

Additional work with evacuated tubes showed that the greenish glow on the glass walls was produced by the particles emitted from the cathode, or negative electrode. Figure 23-2(A) shows how this was deduced. With the electrodes placed as shown in the figure, the walls of the tube glowed everywhere except where the cross inside the tube cast a shadow on the face of the tube. The shadow is in a direct line with the cross and the cathode. This means that the green glow is produced by something coming from the cathode in straight lines. These so-called *cathode rays* are invisible to the eye unless they produce a visible

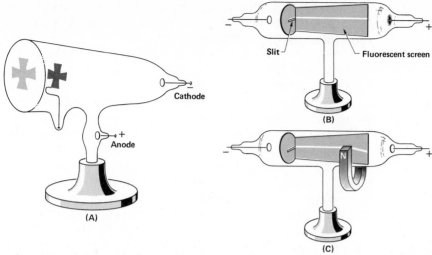

Slit — — Fluorescent screen

(B)

(C)

effect on some other substance, such as a glow on the glass walls of the tube. In addition, these cathode rays can be stopped by substances like the cross inside the tube.

In 1895, the French scientist Jean Perrin (1870–1942) discovered another property of cathode rays. He arranged an evacuated tube as shown in Figure 23-2(B). The slit in the tube forms a broad beam of cathode rays that strikes the fluorescent screen and causes it to glow. The screen is placed at an angle so that the beam is visible along the entire length of the screen. Ordinarily, the beam would trace a straight horizontal path on the screen. But when Perrin placed a magnet in the position shown in Figure 23-2(C), the beam was deflected downward. When he reversed the poles of the magnet, the beam was deflected upward. Since light beams are not deflected by magnets, Perrin concluded that cathode rays were really streams of negative electricity coming from the cathode.

The modern television picture tube makes use of many of the principles discovered many years ago by scientists working with cathode rays. The face of the picture tube glows when cathode rays strike it. In color television, different chemicals are used on the face of the tube so that multicolored glows are produced. The position of the picture on the tube is adjusted by magnets surrounding the neck of the tube inside the set.

Other scientists continued the investigation of cathode rays begun by Perrin. Among these were the English physicists Sir William Crookes (1832–1919) and Sir Joseph Thomson. Thomson was able to show that cathode-ray particles are much lighter than atoms and are present in all forms of matter. He arrived at this latter conclusion by observing that the nature of the particles did not change

Figure 23-2. Properties of cathode rays. The shadow cast by the cross in (A) shows that cathode rays travel in straight lines. The coated screen in (B) shows the ability of cathode rays to produce fluorescence. In (C) the negative charge of cathode rays is identified by means of a magnet.

The word "electron" is derived from a Greek word meaning "beaming sun."

when he changed the composition of the cathode or of the gas in his tubes. Thus Thomson was the first to discover a subatomic particle—the *electron.*

23.3 Measuring the Mass of the Electron

It is obviously impossible to use a laboratory balance to determine the mass of something as small as an electron. Even if you could pile as many as 10^{13} electrons on the balance, the total mass would still amount to only 10^{-17} kg. An indirect method must therefore be used. Thomson developed such a method in a famous series of experiments in 1897. The basic parts of this experiment are shown in Figure 23-3. The cathode rays (electrons) emanating from the electrode at the left are accelerated and focused by the positively charged disk that has a small hole in its center. The potential difference, V, between the cathode and the disk determines the kinetic energy of each electron since $eV = \frac{1}{2}mv^2$, where e is the electric charge on a single electron. A second disk further narrows the electron beam.

The beam then passes between two oppositely charged plates on its way to a fluorescent screen. These charged plates produce an electric field. Surrounding the neck of the tube is a magnetic field that is set at right angles to the electric field. In the arrangement of Figure 23-3, the electron beam is under the influence of a downward force due to the electric field and an upward force due to the magnetic field. Both forces depend on the charge of the electrons, but the force due to the magnetic field also depends on the speed of the electrons. By varying these forces until they exactly balance each other, the speed of the electrons, v, can be determined and used in $eV = \frac{1}{2}mv^2$. Two unknowns, e and m, still remain. Though neither e nor m can be determined individually by this method, their ratio e/m can be calculated from the measured values of the potential difference, the electric field, and the magnetic field. In other words, Thomson was able to calculate the quantity of electric charge per unit mass of electrons. This quantity is expressed in coulombs per kilogram of electrons. The presently accepted value for the ratio is 1.758803×10^{11} c/kg.

This measurement is a step in the right direction, but it still does not tell us the mass of an electron. If the charge on a single electron could be measured, then the charge-to-mass ratio could be used to calculate the mass. Similarly, if the mass could be determined, the ratio would tell us the charge. In 1912, the American physicist Robert A. Millikan succeeded in measuring the charge on the elec-

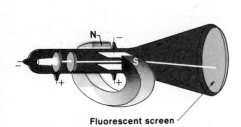

Figure 23-3. The Thomson experiment. By means of an evacuated tube such as this, Thomson found that cathode radiation particles are subatomic in mass and are present in all forms of matter.

tron; the problem was solved. His experiment is described in the next section. First we shall take a look at a completely different way of measuring the electron's mass.

Millikan's work with photoelectric emissions is described in Section 12.7.

As mentioned in Chapter 1, Einstein's equation, $E = mc^2$, states the relationship between mass and energy. Consequently, if an electron could be converted into energy, Einstein's equation could be used to calculate the original mass of the electron from the resulting energy. Conversely, if a form of energy such as a beam of X rays could be converted into an electron, the mass could again be calculated from the known energy of the X rays. Figure 23-4 shows the results of such an experiment. As the diagram to the right of the photograph indicates, X rays had entered from the left and then changed into a pair of electrons, one with a positive charge, called a positron, and one with a negative charge. The bubble chamber in which the photograph was taken was in a magnetic field that caused the two particles to spiral off in opposite directions after their formation; the positron spiraled away from the camera while the electron spiraled toward it.

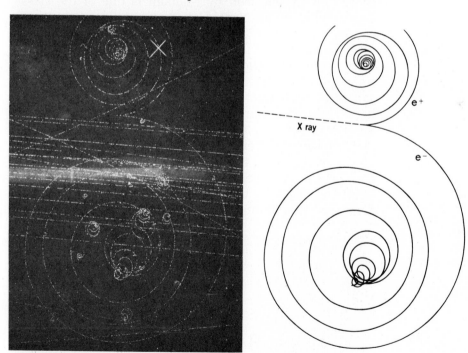

Repeated experiments have shown that in order to produce an electron pair, X rays must have an energy value of 1.64×10^{-13} joule. Since X rays with this energy produce two particles (a single electron is never produced), and since we assume that the two particles have equal masses,

Figure 23-4. Pair production, as photographed in a particle detector. X rays entering from the left were converted into a positron-electron pair.

the mass of each particle is equivalent to half of this energy, or 8.2×10^{-14} joule. Solving Einstein's equation for m and substituting,

$$m = \frac{E}{c^2} = \frac{8.2 \times 10^{-14}\,\text{j}}{(3.0 \times 10^8\,\text{m/s})^2}$$

$$m = 9.1 \times 10^{-31}\,\text{kg}$$

In Section 1.14 it was mentioned that the mass of an object varies with its velocity. In the case of the positron and electron in Figure 23-4, the two particles have velocities so low that they can be considered almost at rest. Thus the mass computed above is called the *rest mass*. In terms of atomic mass units, the rest mass of the electron has been measured as $0.00054858026\,u$ or $\frac{1}{1837}$ of the mass of an atom of hydrogen. Obviously electrons do not account for a very large share of the mass of substances. Other subatomic particles must be responsible for most of the mass. These particles are identified later in this chapter.

23.4 The Electronic Charge The determination of the charge on the electron by Millikan is one of the classic experiments of physics. A diagram of the apparatus is shown in Figure 23-5. Charged capacitor plates were placed inside a metal box in which temperature and pressure conditions could be held constant. The top plate had a small hole in the center. The plates were connected to a source of high electric potential that could be varied at will. The plates were parallel and the distance between them was carefully measured. From the voltage and the plate spacing, the electric field, $\mathcal{E}_q$, could be computed. The space between the plates was illuminated by a spotlight. Through a telemicroscope, a combination of two instruments used for viewing small objects at intermediate distances, this illuminated space could be viewed.

Figure 23-5. Millikan's oil drop experiment. When the oil drop is stationary, the upward force of attraction of the charged plates equals the downward force of gravitation.

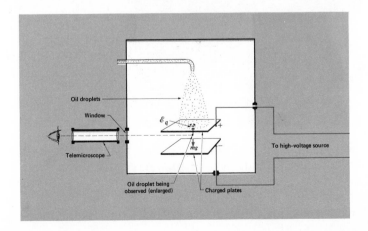

Oil droplets

Window

$\mathcal{E}_q$

mg

Telemicroscope

To high-voltage source

Oil droplet being observed (enlarged)

Charged plates

A mist of light oil was sprayed into the region above the plates. Eventually a single oil drop fell through the opening in the top capacitor plate. The spray was then turned off so that this single drop could be studied as it fell between the plates. When the plates were uncharged, the drop fell through the field of view with a constant terminal velocity that was dependent on the drop's weight, mg. By means of crosshairs in the eyepiece of the telemicroscope, this velocity could be accurately measured. When the voltage was turned on, however, the behavior of the oil drop changed. The drop slowed its descent, stopped in midair, or even began to rise. These variations in behavior depended upon the weight of the drop, the amount of charge it acquired in the oil-spraying process, and the strength of the electric field between the capacitor plates.

Consider the case of a drop that was suspended motionless between the two plates. The force of gravity on the drop was exactly balanced by the force of the electric field between the two plates. In order for this to happen there had to be an electric charge on the oil drop upon which the electric field could exert a force. Hence the field strength between the capacitor plates needed to suspend the drop was a measure of the electric charge on the drop. This observation did not tell how many charged particles were on the drop, however, and so the value of the unit charge was still unknown.

Millikan studied the behavior of thousands of oil drops in the apparatus. By directing X rays at the drops, he was able to change the charge on a drop in midair as a result of the photoelectric effect. Finally he was able to arrive at two important conclusions:

1. Even though the equipment was capable of detecting much smaller values, the charges on the oil drops never fell below a certain minimum value.

2. All charges were integral multiples of the minimum charge.

Millikan correctly concluded that the minimum charge he had measured was the electronic charge—the charge on a single electron. This important fundamental unit of physics has the presently accepted value of $1.6021892 \times 10^{-19}$ c. The mass of the electron could now be computed by dividing the charge by the charge-to-mass ratio found by Thomson.

$$m = \frac{e}{e/m} = \frac{1.6021892 \times 10^{-19} \text{ c}}{1.758803 \times 10^{11} \text{ c/kg}}$$

$$m = 9.109543 \times 10^{-31} \text{ kg}$$

See Section 23.3 for the value of e/m.

This result agrees with the mass of the electron as determined by the bubble chamber method described earlier. It is, however, much more precise.

The sizes of various atoms are given in Section 7.3.

23.5 Size of the Electron   Since the electron is a subatomic particle, it must be smaller than the smallest atom, which is only about 1 Å in diameter. Suppose a narrow beam of electrons is directed through a thin metal foil and then allowed to strike a fluorescent screen. (The interatomic distances in the foil can be quite accurately measured by X rays.) If the electrons are very small, they can be expected to pass straight through the spaces between the foil atoms and to produce a bright narrow spot on the screen. Some of the electrons will come close enough to an atom in the foil to be partly deflected (scattered) by the atom, producing a fuzziness around the spot on the screen. A few may even come so close to hitting atoms head-on as to be strongly scattered, still further broadening the spot on the screen. Careful measurements of such scattering could be used to estimate the size of the electron. The actual result of such experiments, however, is a series of concentric rings in addition to a central spot. Figure 23-6 is a photograph of the image produced in such an experiment. The central spot is understandable, but where do the rings come from? In Chapter 15 we saw that a diffraction grating produces several discrete images of the incident light. The grating is a one-dimensional array of equally spaced lines. In a similar sense, the metal foil is a three-dimensional array of regularly spaced atoms. This should produce a set of discrete images that are symmetrical around a center spot. This is indeed the case when X rays, which are short wavelength photons, pass through a crystal. When X rays pass through a metal foil, the pattern will appear very much like Figure 23-6. In other words, the attempt to measure the size of the electron produced the unexpected evidence that the electron has wave properties!

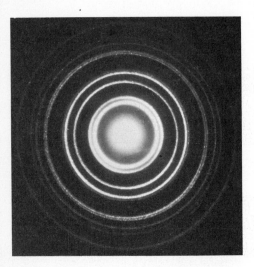

Figure 23-6. An electron diffraction pattern made with an electron microscope of a gold film about 40 Å thick. The pattern shows the rings characteristic of a crystalline substance.

Further experiments with electron patterns have shown that the patterns vary not only with the type of metal foil through which the electrons are fired but also with the momentum of the electron beam. If a description of the electron is given in terms of its wavelength, and if this wavelength is computed from the changing ring patterns, we must conclude that the size of the electron varies with its momentum.

In the measurements of Thomson and Millikan the electron behaved as a charged particle in an electric or mag-

This is a form of the wave-particle duality of matter that was mentioned in Section 1.14.

netic field. Now we see that it behaves as a wave. Does this mean that an electron is sometimes a particle and sometimes a wave? No. It means that *an electron sometimes exhibits particle characteristics and sometimes wave characteristics, depending upon the nature of the experiment and the measuring instruments used.* Thus an electron may at times be described as a wave and at times as a particle. Actually, it is neither a particle nor a wave—it is an electron.

23.6 Motions of the Electron   Figures 23-7 and 23-8 are typical representations of hydrogen and carbon atoms. It is important to note that the electrons are not shown as either particles or waves; they are shown as rings of varying thickness around the center of each atom. Electrons are negatively charged. Hence a neutral atom must have a positive charge in its nucleus. If the electrons were not in motion, the force of attraction between these opposite charges would immediately produce a collapse of the atom. Our knowledge of atomic and nuclear sizes and of chemical and nuclear reactions precludes the possibility of the existence of collapsed atoms in ordinary matter. The force of attraction between the electron and the nucleus provides a centripetal force sufficient to hold the electron in its orbit.

The wave-particle duality of the electron makes it difficult to describe its path around the nucleus. Each electron has a definite amount of energy that determines the *orbital* in which it moves. An orbital is not an orbit or path in the sense that the orbit of a planet is. An orbital is a probable pattern of movement characteristic of the energy of the electron. Groups of orbitals in an atom are usually called *shells.* Shells are frequently designated by letters of the alphabet beginning with K and continuing sequentially to Q in the most complex elements. They also may be given numbers from 1 to 7. The number of orbitals in a shell is the square of the shell number (1st shell, 1 orbital; 2nd shell, 4 orbitals; 3rd shell, 9 orbitals, etc.). The maximum number of electrons that can occupy an orbital is 2; thus the maximum number of electrons that can occupy a shell is two times the number of orbitals in the shell.

The aggregate of electrons is referred to as the *electron cloud.* The characteristics of the electron cloud about an atom give the atom its volume and prevent the interpenetration of one atom by another under ordinary conditions.

Experiments have also shown that, in addition to orbiting the nucleus, the electron spins on its own axis.

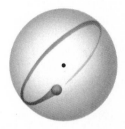

Figure 23-7. A hydrogen atom has a nucleus consisting of one proton. One electron moves about this nucleus in the K shell. Electron paths are not as definite as these diagrams show.

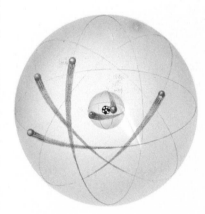

Figure 23-8. A carbon atom has a nucleus consisting of 6 protons and 6 neutrons. Two electrons are in the K shell and 4 electrons are in the L shell.

The evidence and consequence of electron spin were mentioned in Section 19.2.

Questions

GROUP A

1. (a) Why are cathode rays so called? (b) What is the relationship between cathode rays and electrons?
2. (a) Who is credited with the discovery of the electron? (b) Describe his experiments and conclusions.
3. (a) Describe two ways in which the mass of the electron can be determined. (b) Which way is more precise?
4. In what two ways was the charge placed on the oil drops in Millikan's experiment?
5. What difficulties are encountered in measuring the size of the electron?
6. Is an electron a particle or a wave? Explain your answer.
7. What is the relationship between the orbital of an electron and an electron shell?
8. (a) How does the mass of an electron compare with the total mass of a hydrogen atom? (b) What does this signify?

THE NUCLEUS

23.7 Discovery of the Atomic Nucleus In order to be electrically neutral, an atom must contain a positive charge of the same magnitude as the total negative charge of its electrons. But what is the exact location of the positive charge? At the time of the discovery of the electron, there were a number of theories about this. Some scientists suggested that the atom has a uniform structure in which the electrons are distributed evenly throughout a positively charged mass: something like the bits of fruit in a fruitcake. Others thought that the positive charge was concentrated at the center and that the electrons surrounded this positive core.

To test the validity of these theories, in 1911 the English physicist Ernest Rutherford conducted a series of experiments with particles emitted by radioactive materials. As will be explained more fully in Chapter 24, certain elements emit rays and particles that can be studied and identified as being subatomic in nature. One of these particles, called an *alpha particle,* was found to be the same as an atom of helium with its electrons missing. In other words, an alpha particle is a subatomic particle with a positive charge.

In his experiment, Rutherford used alpha particles that were emitted by a piece of radioactive polonium contained in a lead box, as shown in Figure 23-9. These alpha particles have a velocity of 1.60×10^7 m/s. This gave Rutherford subatomic "bullets" of known mass, charge, and velocity with which to explore the structure of other atoms.

Rutherford directed the alpha particles at thin metal foils of various kinds. First he used gold because it can be hammered into extremely thin sheets. The thickness of the foil was about 10^{-7} m, equivalent to only a few hundred

An alpha particle is a helium nucleus.

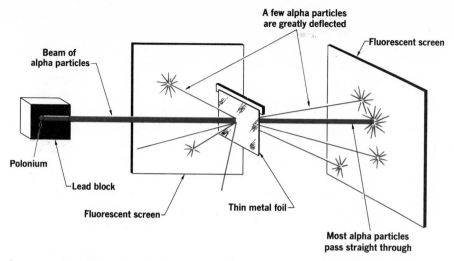

A few alpha particles are greatly deflected

Fluorescent screen

Beam of alpha particles

Polonium

Lead block

Fluorescent screen

Thin metal foil

Most alpha particles pass straight through

layers of gold atoms. Most of the alpha particles passed straight through the foil. This indicated that the metal foil was very porous to the alpha particles even though the atoms of the metal were tightly packed into many layers. Some of the alpha particles were deflected from their paths, however, at angles ranging from a very slight deflection to a direct rebound. The large deflections were a surprise. Could a massive alpha particle be deflected up to 180° by electrons of much smaller masses found inside the atom or by a positive charge spread out uniformly within the atom? In Rutherford's words: "It was about as credible as if you fired a 15-inch shell at a piece of tissue paper and it came back and hit you."

The answer must be that the positive charge within the atom is not uniformly distributed as had been thought by some scientists. Instead, it is concentrated in a dense *atomic nucleus* that is able to deflect alpha particles the way a brick wall would deflect a tennis ball. Furthermore, the number of alpha particles rebounding compared with the number passing through the foil provided a clue to the size of the atomic nucleus compared with the size of the entire atom. Rutherford's students sat for hours in a darkened room to count and measure the deflections on a fluorescent screen. They found that about one out of every 8000 alpha particles was deflected by more than 90°.

From this and subsequent evidence, scientists have been able to determine the radius of the largest atomic nucleus at about 10^{-14} m, or 10^{-4} Å. This is less than 1/10,000th of the diameter of the smallest atom, and less than 1/1,000,000,000,000th of its volume. No wonder that alpha particles, which are themselves atomic nuclei, are able to pass through the atomic structure of metal foils with only an occasional collision! This situation is like

Figure 23-9. Rutherford's alpha scattering experiment. Fluorescent screens show whether the alpha particles from a sample of polonium are deflected when they are directed at a metal foil.

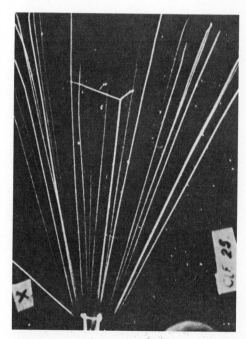

Figure 23-10. Alpha particle tracks in a sample of chlorine gas. Most of the particles are not scattered, but occasionally a particle is deflected through a large angle.

shooting at widely separated peas with an air rifle from the opposite side of a large football stadium.

Yet this tiny nucleus accounts for more than 99.95% of the mass of the entire atom. Thus the density of the atomic nucleus is extremely high: about 10^{13} times the density of lead. One cubic centimeter of closely packed atomic nuclei would have a mass of 10,000,000 metric tons! Astrophysicists believe that some stars are made of very concentrated nuclei, and the electrons of the composite atoms have been forced into the nuclei by intense pressures.

Astronomers call a mass that consists only of atomic nuclei a "black hole" because not even light can escape from its concentrated gravitational field.

23.8 The Proton The existence of positive charges in the atom was known even before the time of Rutherford's experiments. If a perforated disk is used as a cathode in a discharge tube, luminous rays are seen coming through the hole on the side opposite the anode, as in Figure 23-1. In 1895, Perrin tested this radiation with magnets and with electrically charged plates and showed that it consists of particles with a positive charge. In 1907, Thomson named them *positive rays*.

It was found that the properties of the positive rays, unlike those of cathode rays, depend on the nature of the gas in the tube. The charge-to-mass ratio of the particles composing positive rays indicated that they are charged particles formed from atoms of the gas in the tube. No positive particles with a charge-to-mass ratio similar to that of the electron were found. The lightest particle found in positive rays has the mass of a hydrogen atom and carries a charge equal in magnitude but opposite in sign to that carried by the electron. This positive particle was assumed to be a hydrogen atom from which one electron had been removed.

The measurement of the amount of positive charge in the nuclei of various elements was also the work of Rutherford and his colleagues. They developed equations stating that the scattering of alpha particles depends on three factors: the thickness of the metal foil, the velocity of the alpha particles, and the amount of positive charge in the atomic nuclei. Experiments soon showed that the first two of these relationships agreed remarkably with the results as predicted by the equations. So it could be assumed that the third relationship, scattering vs. relative nuclear charge, was also true. Tests were made with many different elements, and it was found that *the nuclear charge of the atoms of a given element is always the same and is characteristic of that element.*

The smallest nuclear charge is that of the hydrogen nucleus. In 1920, the name **proton** was given to the nuclear particle carrying this unit positive charge. The number of protons, and therefore the amount of positive charge (measured in units of electron charge), is called the ***atomic number.*** Atomic numbers range from 1 for hydrogen to over 100 for the most massive elements. The positive charge of a proton has the same magnitude as the negative charge of an electron. Consequently the atomic number of an element also gives the number of electrons surrounding the nucleus of neutral atoms.

The atomic number of an element is the number of protons in its nucleus.

The mass of the proton has been found to be $1.6726485 \times 10^{-27}$ kg, or $1.007276470\ u$ on the atomic mass scale. This is $\frac{1836}{1837}$ of the mass of a hydrogen atom. Thus the mass of a hydrogen atom is contained almost entirely in its nucleus.

23.9 The Neutron   If an atom consisted entirely of electrons and protons and if the mass of the atom were almost entirely contained in its protons, then it should be possible to predict the atomic mass of an element from its atomic number. For example, since the atomic number of oxygen is 8, the atomic mass of oxygen should be eight times the atomic mass of hydrogen, which consists of a single proton and single electron. Similarly, the atomic mass of uranium ought to be 92. But this is not the case. The atomic mass of the most abundant form of oxygen is about 16 and that of the most common form of uranium is about 238. The same discrepancy shows up in all the elements except ordinary hydrogen. What explanation can be given to resolve this discrepancy?

One possibility is that the excess mass is due to protons whose positive charge is neutralized by electrons located within the nucleus. However, such an arrangement presents a number of complex problems that rule it out as a possible solution. For instance, the spin of such nuclear electrons would produce effects that disagree with observed nuclear behavior.

The answer was provided by the English physicist James Chadwick (1891–1974). In 1930 physicists found that when certain elements, such as beryllium or boron, are bombarded with alpha particles from the radioactive element polonium, a radiation with very high penetrating power is obtained. Two years later Chadwick discovered that this highly penetrating radiation is able to knock protons out of a block of paraffin, a compound containing carbon and hydrogen atoms. Figure 23-11 is a schematic diagram of Chadwick's experiment. The radiation that produces this proton emission is not a beam of X rays because it can be shown that X rays would require unusually high energies to expel protons from paraffin. Chadwick reasoned that the radiation is a stream of neutral particles.

In 1935 Chadwick received the Nobel Prize in Physics for his discovery of the neutron.

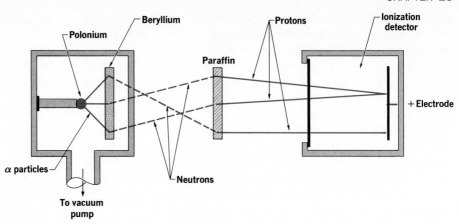

Figure 23-11. Chadwick's neutron experiment. Particles ejected from beryllium were in turn able to knock protons out of a sample of paraffin. The intermediate particles were found to have no charge but to have a mass similar to that of the proton. Chadwick called them neutrons.

Calculations showed that the particles have masses similar to that of the proton.

Additional experiments indicated that the same neutral particles can be released from a great variety of elements. This means that the neutral particles are regular components of atomic nuclei. Chadwick called these particles *neutrons.* They have a mass of $1.6749543 \times 10^{-27}$ kg (very slightly higher than that of the proton), or 1.008665012 u on the atomic mass scale. The two particles are also quite similar in size.

An interesting result of recent research with neutrons is shown in Figure 23-12. The bottom photo, which was made with the help of neutrons, shows more inner detail in the opaque toy locomotive than does the X-ray photo in the center. The neutrograph is made by passing high-intensity neutrons through the locomotive to a metallic converter, where they release the rays that expose the photographic film. Industrial and medical applications of neutron radiography are being developed.

23.10 Isotopes Experiments show that the chemical properties of an element are related to its atomic number, Z. For hydrogen $Z = 1$, for carbon $Z = 6$, and for uranium $Z = 92$. The total number of protons and neutrons in the nucleus of an atom is equal to the mass number, A. Using the symbol N to represent the number of neutrons,

$$A = Z + N$$

Protons and neutrons are referred to collectively as *nucleons.* Thus in the formula above, A equals the number of nucleons.

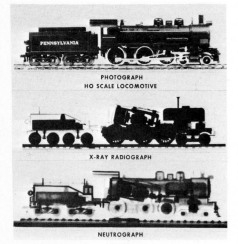

Figure 23-12. Neutrons can be used to take pictures of opaque objects, revealing more inner detail than is obtainable with X rays.

All atoms of an element have the same chemical properties, but their masses may be different. The number of neutrons in the atomic nucleus of a given element may vary. Atoms of a given element that have different masses are called *isotopes.* **Isotopes** *of an element contain the same number of protons but a different number of neutrons. Each dif-*

ferent variety of atom as determined by the composition of its nucleus is called a **nuclide.** Nuclides having the same atomic number are isotopes. There are three isotopes of hydrogen: *protium* (ordinary hydrogen), 1 proton, no neutrons; *deuterium*, 1 proton, 1 neutron; and *tritium*, 1 proton, 2 neutrons. Isotopes of all chemical elements either occur naturally or have been produced artificially.

In general, nuclei with odd atomic numbers exist naturally in only one or two isotopic forms. Those with even atomic numbers exist in several isotopic forms; some elements have as many as eleven variations. Figure 23-13 is a graph of the nuclei of the naturally occurring light nuclides in which Z is plotted as the abscissa and N as the ordinate. Notice that Z and N are equal in many of the lightest nuclides, but that the neutrons outnumber the protons as the mass number increases. This fact is reflected in the properties of the elements, some of which will be discussed in the next chapter.

Naturally occurring elements are mixtures of isotopes in quite definite proportions. Hydrogen consists of 99.985% protium, 0.015% deuterium, and less than 0.001% tritium. *The **gram-atomic weight** of an element is the mass in grams of one mole of naturally occurring atoms of an element.* Hydrogen has a gram-atomic weight of 1.00797 g and carbon has a gram-atomic weight of 12.01115 g. Similarly, the gram-atomic weights of all naturally occurring elements are mixed numbers.

23.11 The Mass Spectrograph The masses of ionized atoms can be measured with great precision in a device called a *mass spectrograph.* There are many different types of mass spectrographs, but in all types the operation depends on the slight difference in mass between isotopes. One type of spectrograph is shown in Figure 23-14.

A beam of ions enters from the ion gun and passes into a region of crossed electric and magnetic fields that allows only the ions with a specific velocity to pass through the slit **S.** The velocity of the ions is inversely proportional to their masses. (Ions with other velocities are deflected and blocked by the chamber wall around the slit.) As the ions enter the upper chamber they are influenced by another magnetic field perpendicular to the plane of the page. This magnetic field causes the ions to move in a circular path. If the ions have the same mass-charge ratio, they will all describe the same arc and will strike the photographic plate at the same place. However, if the beam entering the slit contains ions of various masses, the less massive ions will be deflected more than the more massive ones and will strike the photographic plate at different places. If the

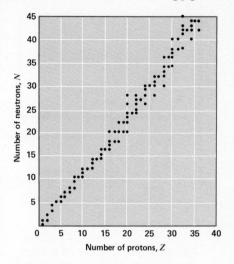

Figure 23-13. Nuclei of the stable isotopes of the light elements. Note that the lightest nuclides have equal numbers of neutrons and protons, but in the heavier nuclides there are more neutrons than protons.

The gram-atomic weight is the mass of a naturally occurring mole of an element.

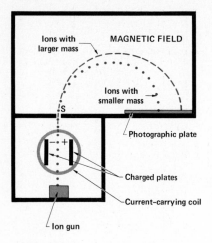

MAGNETIC FIELD

Ions with larger mass

Ions with smaller mass

S

Photographic plate

+

Charged plates

Current-carrying coil

Ion gun

Figure 23-14. Diagram of a mass spectrograph. The magnetic field deflects the less massive ions more than the more massive ones.

The nuclear mass defect of a nucleus is the difference between its mass and the sum of the masses of its uncombined constituent particles.

photographic plate is replaced by collecting vessels, the mass spectrograph can be used to separate isotopes.

23.12 Nuclear Binding   A very strong force holds protons and neutrons together in the nucleus. This force is not electrostatic because the similarly charged protons would then repel each other. The force cannot be gravitational either because calculations show that it would be much too weak. It is a completely unique force that acts only within the extremely small distances between nucleons. Experiments have shown that this *nuclear binding force* is effective to a distance of 2.0×10^{-15} m, or 2.0×10^{-5} Å. This is less than the diameter of a proton, which is 2.6×10^{-5} Å.

The magnitude of the nuclear binding force has been investigated in several ways. One way is to study the scattering pattern that results when nucleons are directed at each other, very much like the scattering experiments of Rutherford and Chadwick. Another way is to compare the mass of a nucleus with the sum of the masses of the uncombined nucleons that make up the nucleus. In each case, the magnitude of the nuclear force is derived from measurements of the energy involved.

Careful measurements have shown that *the mass of a nucleus is always less than the sum of the uncombined masses of its constituent particles. This difference in mass is called the **nuclear mass defect.*** The reason for the nuclear mass defect is found in Einstein's equation $E = mc^2$. When a nucleus is formed, energy is released. This is similar to the energy release that occurs when atoms combine to form a molecule. Einstein's equation states that a decrease in energy must be accompanied by a corresponding decrease in mass since the two quantities are directly proportional. This decrease in mass is the nuclear mass defect.

The nuclear mass defect can be used to compute the total *nuclear binding energy*. Binding energy is equivalent to the energy released during the formation of a nucleus and is the energy that must be applied to the nucleus in order to break it apart. Let us consider the formation of a helium nucleus (alpha particle) as an example. It contains 2 protons and 2 neutrons. Each proton has a mass of 1.007276 *u*, and each neutron has a mass of 1.008665 *u*. The mass of a helium nucleus is 4.001509 *u*. The nuclear mass defect can now be computed as follows:

$$
\begin{aligned}
\text{2 protons} &= 2 \times 1.007276 \ u = 2.014552 \ u \\
\text{2 neutrons} &= 2 \times 1.008665 \ u = \underline{2.017330 \ u} \\
&\qquad\qquad\qquad\qquad\quad\ 4.031882 \ u \\
\text{helium nucleus} &= \underline{4.001509 \ u} \\
\textbf{nuclear mass defect} &= \underline{0.030373 \ u}
\end{aligned}
$$

In nuclear physics, nuclear binding energy is expressed in electron volts (ev). *An **electron volt** is the energy required to move an electron between two points that have a potential difference of one volt.* The mega electron volt (Mev) is also used. A single atomic mass unit is the equivalent of 931 Mev. Hence the binding energy of the helium nucleus is

$$0.030373 \, u \times 931 \text{ Mev}/u = 28.3 \text{ Mev}$$

If the nuclear binding energy is divided by the number of nucleons, the binding energy per nucleon is obtained. For helium, this is 28.3 Mev/4 = 7.1 Mev/nucleon. The graph in Figure 23-15 shows the variation of binding energy per nucleon with the mass number. Note that elements of low or high mass number have lower binding energies per nucleon than those of intermediate mass number. The binding energy per nucleon is low for the lightest nuclei because the nucleons are all on the surface of the nucleus and are not held on all sides by other nucleons. This geometry reduces the binding energy per nucleon. As the number of nucleons increases, more and more of them are completely surrounded and the binding energy per nucleon increases. Beyond the maximum of 8.7 Mev/nucleon, the repelling force between the protons begins to cancel the binding force. Finally a point is reached at which the nucleus begins to fall apart.

One atomic mass unit has a nuclear binding energy equivalent to 931 mega electron volts.

The disintegration of a nucleus is a form of nuclear reaction. This is the topic of the next chapter.

Figure 23-15. Graph showing the relationship between mass number and binding energy per nucleon.

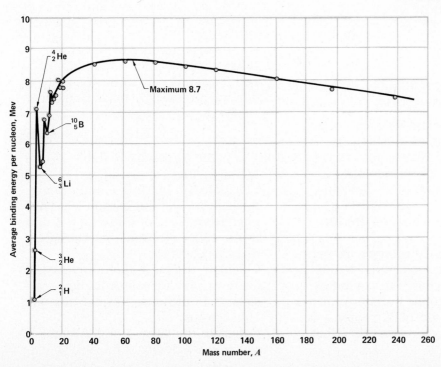

Questions

GROUP A

1. Compare the relative masses, charges, and sizes of the following particles: molecule, atom, electron, proton, neutron, nucleus.
2. Distinguish between atomic mass and gram-atomic weight.
3. (a) What do the isotopes of an element have in common? (b) How do they differ?
4. Define the following: (a) A, (b) Å, (c) Z, (d) N, (e) u, (f) Mev.
5. Copy and complete the following table on a separate sheet of paper.

Nuclide	Z	A	N	Binding energy/nucleon (Mev/A)
deuterium	1			
carbon-14	6			
oxygen-18	8			
sodium-23	11			
sulfur-32	16			
argon-40	18			
uranium-235	92			
uranium-238	92			

GROUP B

6. (a) How did Rutherford obtain alpha particles? (b) How did he detect them after they had passed through the metal foils?
7. (a) What did the experiments of Perrin and Rutherford have in common? (b) Of Rutherford and Chadwick?
8. How did Rutherford know that the alpha particles that rebounded did not rebound from the electrons in the metal foils?
9. What is the general relationship between the atomic number of an atom and the number of its isotopes?
10. (a) What is meant by nuclear mass defect? (b) What is nuclear binding energy? (c) How are they related?
11. What element has the highest Mev/nucleon?

Problems

GROUP A

Note: Consult Appendix B, Table 23, for the atomic masses of nuclides.

1. (a) Calculate the mass in grams of an atom of an isotope of iron that has an atomic mass of 55.9349 u. (b) What is the mass number of this isotope?
2. The mass of an atom of an unstable isotope of neon is 3.818×10^{-23} g. (a) What is its atomic mass? (b) Its mass number?
3. A calcium ion has 20 protons, 20 neutrons, and 18 electrons. What is the magnitude of its charge in coulombs?
4. (a) What is the atomic number of a zinc atom composed of 30 protons, 34 neutrons, and 30 electrons? (b) What is the mass number of the atom?
5. What particles, and how many of each, make up an atom of silver (atomic number 47, mass number 109)?

GROUP B

Note: Consult Appendix B, Table 23, for the atomic masses of nuclides.

6. Calculate the binding energy per nucleon in Mev for carbon-12. This nuclide consists of 6 protons, 6 neutrons, and 6 electrons. Its atomic mass is 12.00000.
7. Calculate the binding energy per nucleon in Mev for sulfur-32 that consists of 16 protons, 16 neutrons, and 16 electrons.
8. Calculate the gram-atomic weight for

chlorine if naturally occurring chlorine consists of 75.4% Cl-35 and 24.6% Cl-37.

9. Naturally occurring magnesium consists of 78.6% Mg-24, 10.1% Mg-25, and 11.3% Mg-26. Calculate the gram-atomic weight.

10. An electron pair is produced from X rays with a total energy of 1.5 Mev. If the positron and electron each receive half of the extra energy, calculate what the resulting velocity of each particle would be if we could disregard relativity.

SUMMARY

The electron was the first subatomic particle to be identified by scientists. It was found to have a mass of 9.1×10^{-31} kg and a negative charge of 1.6×10^{-19} c. The electron exhibits both particle and wave characteristics. It orbits the nucleus of the atom in a path determined by its energy.

Almost all the mass of an atom is contained in its nucleus, which takes up only a very small percentage of the space occupied by an atom. The positive charge of the nucleus is due to protons. The magnitude of the charge of a proton is the same as that of an electron. The number of protons in a neutral atom is equal to the number of orbiting electrons. This number is known as the atomic number.

The atomic nucleus also contains uncharged particles called neutrons. The mass of a neutron is slightly greater than the mass of a proton. Protons and neu-

trons are collectively called nucleons, and their total number is the mass number of the atom. Isotopes are atoms with the same atomic number but different mass numbers. Each variety of an atom as determined by its nuclear composition is called a nuclide. The gram-atomic weight of an element is the average atomic mass of one mole of naturally occurring atoms of an element. The atomic weight scale is based on the carbon-12 isotope. Nuclides are separated by a mass spectrograph.

The mass of a nucleus is always less than the combined mass of its constituent particles by an amount called the nuclear mass defect. The nuclear mass defect is equivalent to the nuclear binding energy. Binding energies are expressed in mega electron volts. Elements with intermediate mass numbers have higher binding energies per nucleon than do elements with low or high mass numbers.

VOCABULARY

alpha particle
atomic mass unit
atomic number
cathode rays
electron
electron shell
electron volt

gram-atomic weight
isotope
neutron
nuclear binding force
nuclear mass defect
nucleon

nucleus
nuclide
oil drop experiment
pair production
proton
rest mass

$\overset{\displaystyle\frown}{24}$ 　NUCLEAR REACTIONS

Marie Sklodowska Curie discovered the elements polonium and radium in a series of experiments performed in Paris in collaboration with her husband, Pierre. The Curies jointly won the Nobel Prize in Physics in 1903, and Madame Curie received the chemistry prize in 1911. She was the first person to receive the Nobel prize twice.

In this chapter you will gain an understanding of:

▷ the characteristic effects of radioactive nuclides on various substances
▷ the nature of alpha, beta, and gamma emissions
▷ the writing and the balancing of nuclear equations
▷ the distinction between fission and fusion reactions
▷ the requirements for a nuclear chain reaction
▷ nuclear reactors and the uses of radioisotopes

TYPES OF NUCLEAR REACTIONS

24.1 Discovery of Radioactivity At the time that Thomson was conducting his experiments with cathode rays, other scientists were gathering evidence that the atom could be subdivided. Some of this new evidence showed that certain atoms disintegrate by themselves. In 1896 Henri Becquerel (1852–1908) discovered this phenomenon while investigating the properties of *fluorescent* minerals. Fluorescent minerals glow after they have been exposed to strong light. Becquerel used photographic plates to record this fluorescence.

One of the minerals Becquerel worked with was a uranium compound. During a day when it was too cloudy to expose his mineral samples to direct sunlight, Becquerel stored some of the compound in the same drawer with the photographic plates. When he later developed these same plates for use in his experiments, he discovered that they were fogged. What could have produced this fogging? The plates were wrapped tightly before being used, so the fogging could not be due to stray light. Also, only the plates that were in the drawer with the uranium compound were fogged. Becquerel reasoned that the uranium compound must give off a type of radiation that could penetrate heavy paper and affect photographic film.

Elements that emit such radiation are called *radioactive* and possess the property of *radioactivity*. **Radioactivity** is *the spontaneous breakdown of an unstable atomic nucleus with the emission of particles and rays*. At the suggestion of Becquerel, Pierre (1859–1906) and Marie (1867–1934) Curie investigated uranium and its various ores. They found that all uranium ores were radioactive. However, one of them, pitchblende, was four times as radioactive as might be expected from the amount of uranium in it. Further studies showed that this extra radiation was actually due to the presence of two previously unknown elements, *polonium* and *radium*, both of which were intensely radioactive. The elements polonium and radium were discovered and named by the Curies.

Other radioactive elements have since been discovered or produced. All of the naturally occurring elements with atomic numbers greater than 83 are radioactive. Also, a few naturally radioactive isotopes of elements with atomic numbers smaller than 83 are known. Many artificial radioactive nuclides have been produced and put to use in different ways, as we shall see later in this chapter.

24.2 Nature of Radioactivity   All radioactive nuclides have certain common characteristics.

1. Their radiations affect the emulsion on a photographic film. Even though photographic film is wrapped in heavy black paper and kept in the dark, some of the radiations from radioactive nuclides penetrate the wrapping and affect the film. When the film is developed, a black spot can be seen where the invisible radiations struck the film. Some of the radiations penetrate wood, flesh, *thin* sheets of metal, and, as shown in Figure 24-3, even thick sheets of glass.

2. Their radiations ionize the surrounding air molecules. The radiations from radioactive nuclides knock out electrons from the atoms of the gas molecules in the air surrounding the radioactive material. This process leaves the gas molecules with a positive charge. An atom or a group of atoms having an electric charge is called an ion. The production of ions is termed ionization. An electroscope can detect ionized gas molecules in the air. As we noted in Section 7.23, an ionized gas is called plasma.

3. Their radiations make certain compounds fluoresce. Radiations from radioactive nuclides produce bright flashes of light when they strike certain compounds. The combined effect of these flashes is a fluorescence, or glow, given off by the affected material. For example, radium compounds added to zinc sulfide cause the zinc sulfide to glow.

4. Their radiations have special physiological effects. Radiations from natural sources can destroy the germinating power of plant seeds, kill bacteria, and even injure and kill large animals. Burns from radioactive materials heal with great difficulty and may sometimes be fatal.

Figure 24-1. Henri Becquerel, the French physicist who discovered radioactivity.

In 1903, Becquerel and the Curies shared the Nobel Prize in Physics for their work in radioactivity.

Figure 24-2. Marie and Pierre Curie discovered the radioactive elements radium and polonium.

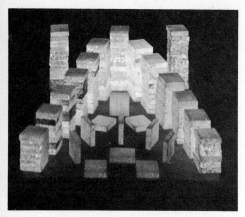

Figure 24-3. Self-portrait of a radioactive substance. The only source of illumination for this time exposure, which was taken through a 3-foot-thick heavy-glass window, was the natural radioactivity of the cesium-137 in the blocks of cesium chloride.

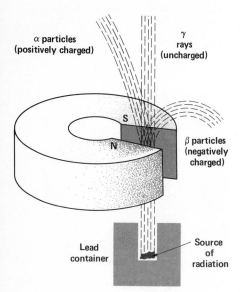

Figure 24-4. Types of radioactivity. Alpha and beta particles are deflected by a magnet, but gamma rays are not.

5. They undergo radioactive decay. The atoms of all radioactive elements continually decay into simpler atoms and simultaneously emit radiations. However, it is not possible to predict when a specific atom will decay. The rate of decay may be described by a rule that states how long it will take for half of the atoms in a given sample to decay. The accuracy of the rule is greater for larger samples. The time associated with radioactive decays of a certain type is called the *half-life* of the nuclide. ***Half-life** is the length of time during which, on the average, half of a given number of atoms of a radioactive nuclide decays.* For example, the half-life of radium-226 is 1620 years. This means that one-half of a given sample of radium-226 atoms can be expected to decay into simpler atoms in 1620 years.

*24.3 **Types of Natural Radioactivity*** In the course of his alpha-scattering experiments, Rutherford discovered that the radiations from radioactive materials could be separated into three distinct types by means of a strong magnet. A diagram of this separation is shown in Figure 24-4.

1. α (alpha) particles. These are composed of two protons and two neutrons; hence they are helium nuclei. This conclusion can be proved by collecting large quantities of alpha particles in a partially evacuated tube and passing an electric discharge through the tube. The alpha particles acquire electrons from the residual air molecules in the tube. The spectrum of the resulting substance is identical with the spectrum of helium. Alpha particles have two positive electric charges, speeds about one-tenth the speed of light, and masses about four times that of the hydrogen atom. Hence they are deflected only slightly by the field of a magnet. Their penetrating power is not very great. They can be stopped by a thin piece of aluminum foil or by a thin sheet of paper.

2. β (beta) particles. These are electrons, just like cathode rays. They have single negative charges and may travel with nearly the speed of light. Their mass is only a small fraction of the mass of alpha particles. Hence, even though they have only half as much charge, they are deflected much more by a magnetic field and in the opposite direction. The high speed of beta particles makes them much more penetrating than alpha particles. It was the beta particles from the uranium compound that fogged Becquerel's photographic plates.

3. γ (gamma) rays. These are high-energy photons. They are the same kind of radiation as visible light, but of much shorter wavelength and thus higher frequency. Gamma rays are produced by energy transitions in the nucleus that do not change the composition of the nuclide. They are the most penetrating radiations given off by radioactive elements. They are not electrically charged and are not deflected by a magnetic field.

24.4 Nuclear Symbols and Equations Changes in matter can be classified into three types: *physical, chemical,* and *nuclear.* When a physical change occurs, the identity of the substance is not lost. When water freezes and forms ice, the composition of its molecules is not changed; if we heat the ice it changes back into water. When sugar dissolves in water it undergoes a physical change; if the water evaporates, the sugar remains behind. *In a **physical change** the composition of the substance is not changed.* The substance consists of the same molecules, atoms, or ions that it did before the change occurred.

When a chemical change occurs, new substances are formed from the atoms or ions of the original substances. These new substances have their own distinct chemical and physical properties. The rusting of iron and the souring of milk are examples of chemical changes. *In a **chemical change** the composition of the substance is changed, and new substances with new properties are produced.* The type and number of atoms or ions remain the same, but the atoms or ions are rearranged to form new substances. A chemical change does not alter the nuclei of the interacting atoms.

A nuclear change is similar to a chemical change in that new substances with new properties are formed. However, *in a **nuclear change** the new materials are formed by changes in the identities of the atoms themselves.* The radioactive decay of radium is a nuclear change, as are all forms of alpha, beta, and gamma emission.

Both chemical and nuclear changes are usually represented by equations. In equations, symbols or formulas are used to represent the particles that enter into a reaction and that are produced by it. You are probably familiar with the symbols of many of the chemical elements. In working with nuclear equations, you must know how to use the symbols for some of the subatomic particles and radiations. The most basic of these are given in Table 24-1.

Except for the gamma-ray symbol, all the symbols in the table have subscripts and superscripts. The subscript designates the electric charge of the particle. The superscript denotes the number of nucleons in the particle. These charge and nucleon designations are also used with the symbols of all other atomic nuclei presented in nuclear equations.

In a chemical equation, the same atoms appear on both

Table 24-1
SYMBOLS FOR SUBATOMIC PARTICLES

Symbol	Particle
$^1_0 n$	neutron
$^1_1 H$	proton (hydrogen nucleus)
$^0_{-1} e$	electron (beta particle)
$^0_{+1} e$	positron (positive electron)
$^4_2 He$	alpha particle (helium nucleus)
γ	gamma ray

Note: In the case of the proton and the alpha particle the atomic symbols for hydrogen and helium are meant to designate only the nuclei of these atoms.

In a nuclear change the atomic number of the nucleus is changed.

The symbols of other subatomic particles are given in Chapter 25.

sides of the equation in equal numbers. This is true because the identity of the atoms is not changed. In a nuclear equation, however, the same atomic symbols do not necessarily appear on both sides. The factors that will remain constant in a nuclear equation are the total number of nucleons and the arithmetic sum of the electric charges on each side.

The following example will illustrate this.

$$^{14}_{7}\text{N} + {}^{4}_{2}\text{He} \rightarrow {}^{17}_{8}\text{O} + {}^{1}_{1}\text{H}$$

In this equation, a nitrogen nucleus (charge, 7; nucleons, 14) reacts with an alpha particle (charge, 2; nucleons, 4) to form an oxygen nucleus (charge, 8; nucleons, 17) and a proton (charge, 1; nucleon, 1). There is a total of 9 electric charges and 18 nucleons on each side of the equation.

24.5 Radioactive Decay It was noted in Section 24.1 that the atoms of all naturally occurring elements with atomic numbers greater than 83 are unstable and decay spontaneously into lighter particles. This is also true of several naturally occurring nuclides and a great many artificially made nuclides with atomic numbers below 83. As a general rule, an atomic nucleus is stable only if the ratio of its neutrons to its protons is about 1 for the light elements and about $1\frac{1}{2}$ for the heavier elements. When there are more neutrons than this, a *nuclear transformation* will probably take place.

Figure 24-5 shows a series of nuclear transformations beginning with uranium-238 and ending with lead-206. Each step of the series can be represented by a nuclear equation in which nucleons and charge are conserved. For instance the equation for the first step in the chart may be written as

$$^{238}_{92}\text{U} \rightarrow {}^{234}_{90}\text{Th} + {}^{4}_{2}\text{He}$$

This reaction is called *alpha decay* because an alpha particle is emitted.

The next two steps in the series are examples of *beta decay*. In the first step, a neutron in the thorium nucleus ejects a beta particle and becomes a proton. This transformation increases the nuclear charge by one, and the nucleus becomes a protactinium nucleus.

$$^{234}_{90}\text{Th} \rightarrow {}^{234}_{91}\text{Pa} + {}^{0}_{-1}\text{e}$$

The chart also shows the half-life of each nuclide in the decay series. You will notice that there is a great variation in these values, from small fractions of a second to millions of years. In each case the exact length of the half-life

The fact that natural radioactive decay stops with lead makes that element a good shielding material for experiments in radioactivity.

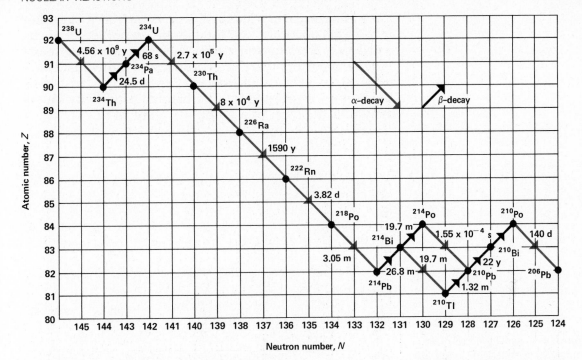

Figure 24-5. The uranium-238 disintegration series. Note that in a beta decay, the atomic number increases by one but the mass number remains constant.

is related to the energy change that accompanies the transformation. As a rule, the shorter the half-life of the nuclide, the greater is the kinetic energy of the alpha or beta particle it emits.

From the definition of half-life it is obvious that the radioactivity of a sample of material depends on the total number of nuclei present. The ratio between the number of nuclei decaying per unit time and the total number of original nuclei is known as the *decay constant*. If the half-life of a nuclide is known, its decay constant can be calculated by the following equation:

$$\lambda = \frac{0.693}{T_{1/2}}$$

In this equation λ (the Greek letter lambda) designates the decay constant in reciprocal seconds, 1/s or s^{-1}, if the half-life $T_{1/2}$ is given in seconds. For example, for $^{226}_{88}Ra$, $T_{1/2} = 1620$ years and

$$\lambda = \frac{0.693}{1620 \text{ y} \times 365 \text{ d/y} \times 86{,}400 \text{ s/d}}$$

$$\lambda = 1.36 \times 10^{-11} \text{ s}^{-1}$$

In 1.000 kg of radium there are 2.665×10^{24} nuclei. Consequently, the number of radium nuclei decaying per second is 2.665×10^{24} nuclei $\times 1.36 \times 10^{-11} \text{ s}^{-1} = 3.62 \times 10^{13}$ nuclei/s. Observe that this equation does not indicate *when* a particular nucleus will decay. Instead it says that in a certain amount of time a definite proportion of nuclei will decay. A relationship of this kind represents the statistical behavior of a very large number of individual situations. It uses the mathematical methods of statistics or the theory of probability. The half-life probability law is shown in the form of a graph in Figure 24-6.

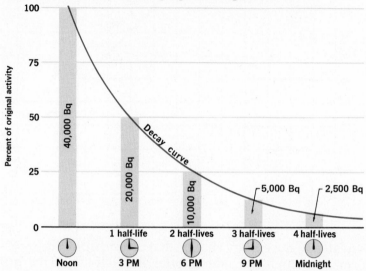

Figure 24-6. Activity curve of a radioactive nuclide with a half-life of three hours and an original strength of 40,000 Bq.

One radioactive disintegration per second is called a *becquerel* (Bq). A sample of radioactive material that emits 3.7×10^{10} Bq is said to have a strength of one *curie* (Ci). Since the curie is a very large unit, the millicurie (mCi) and microcurie (μCi) are frequently used. For example, a source with a strength of 4.00 μCi emits ($4.00 \times 10^{-6})(3.7 \times 10^{10}$), or 1.5×10^5 particles per second.

The curie is not an SI unit, but it is widely used to express the strength of radioactive nuclides.

The *gray* (Gy) is the unit used to measure the amount of absorbed radiation. *One gray is equal to the absorption of one joule of radioactive energy per kilogram of absorbing matter.* Radioactive dosage is described in terms of the *sievert* (Sv). A sievert has the energy equivalent of a gray.

24.6 Nuclear Bombardment The study of radioactive decay led scientists to believe that transformations that would produce a different element could be achieved by adding protons to the nucleus. The first successful transformation of this kind was carried out by Rutherford in 1919. A diagram of his apparatus is shown in Figure 24-7. The container was filled with nitrogen gas. A radioactive source emitted alpha particles. A silver foil thick enough to absorb any alpha particles not absorbed by the nitrogen gas was used. A zinc sulfide screen recorded the scintillations of any particles with enough energy to pass through

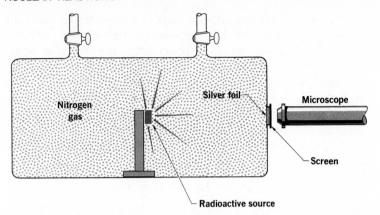

Figure 24-7. Rutherford's transformation experiment. Bombardments from the radioactive sample ejected protons from the nitrogen gas in the chamber. The protons produced scintillations on a fluorescent screen.

the silver foil. These scintillations could be observed through the microscope.

Rutherford observed scintillations on the screen when the chamber was filled with nitrogen and other light gases. However, there were no scintillations when he used oxygen and other heavy gases such as carbon dioxide. He concluded that low density gases were changed by the alpha-particle bombardment with the accompanying emission of a high-energy particle such as a proton. The nuclear equation in Section 24.4 represents the transformation produced by the alpha bombardment of nitrogen. The detection of oxygen gas in the container after the alpha bombardment of pure nitrogen helps to verify the validity of the equation.

The alpha bombardment of nitrogen will result in the emission of protons. Other bombardments produce other subatomic particles. In 1932 J. D. Cockcroft and E. T. S. Walton bombarded lithium with high-speed protons.

$$\ce{^{7}_{3}Li} + \ce{^{1}_{1}H} \rightarrow \ce{^{4}_{2}He} + \ce{^{4}_{2}He} + \text{energy}$$

In this case all the products have lower mass numbers than the original target material. In addition there is the release of a considerable amount of binding energy. Actually, adjustments in binding energy are involved in all nuclear reactions. Cockcroft and Walton found that the measured energy in their experiments agreed closely with the theoretical value as derived from Einstein's equation, $E = mc^2$.

The method of computing binding energy is described in Section 23.12.

Another nuclear bombardment, also carried out in 1932, led to the discovery of the neutron. The nuclear equation for Chadwick's experiment is

Neutrons are described in Section 23.9.

$$\ce{^{9}_{4}Be} + \ce{^{4}_{2}He} \rightarrow \ce{^{12}_{6}C} + \ce{^{1}_{0}n} + \text{energy}$$

Many similar bombardments have since been found to produce neutrons. Since the neutron is not electrically charged, it penetrates atomic nuclei more easily than do charged particles like the proton or the alpha particle. Hence the neutron has become an important nuclear "bullet." In fact one of the goals in neutron bombardment is to get the neutrons to move *slowly* enough to produce a transformation. Fast neutrons may be slowed down by passage through materials called *moderators,* composed of elements of low atomic weight. Deuterium oxide or graphite are useful moderators. The equation for a typical neutron absorption is

$$^{10}_{5}\text{B} + ^{1}_{0}\text{n} \rightarrow ^{7}_{3}\text{Li} + ^{4}_{2}\text{He} + \textbf{energy}$$

Bombardment reactions are usually designated by the symbols for the incident and emitted subatomic particles. For example, the preceding equation is called a (n, α) reaction. In bombardment reactions, the mass number of the target nucleus is increased or decreased by several units. Sometimes, however, a nucleus will split into two large segments when struck by a neutron moving at just the right speed.

24.7 Fission   When a neutron is captured by $^{238}_{92}\text{U}$, the following reaction takes place:

$$^{238}_{92}\text{U} + ^{1}_{0}\text{n} \rightarrow ^{239}_{92}\text{U}$$

This new isotope of uranium is unstable, and so it emits a beta particle.

$$^{239}_{92}\text{U} \rightarrow ^{239}_{93}\text{Np} + ^{0}_{-1}\text{e}$$

$^{239}_{93}\text{Np}$ has a half-life of 2.3 days and decays by emitting another beta particle.

$$^{239}_{93}\text{Np} \rightarrow ^{239}_{94}\text{Pu} + ^{0}_{-1}\text{e}$$

Neptunium and plutonium were the first two artificial *transuranium* elements that were discovered. *A transuranium element is one with an atomic number greater than 92, the atomic number of uranium.* At the time of this writing, sixteen artificial transuranium elements have been confirmed. All of them are radioactive, and some of them exist for only a small fraction of a second.

Uranium-238 accounts for 99.3% of naturally occurring uranium. 0.7% is uranium-235. There are also traces of uranium-234. When $^{235}_{92}\text{U}$ absorbs a neutron, the reaction is quite different from that described above for $^{238}_{92}\text{U}$. Instead of emitting alpha or beta particles, the $^{235}_{92}\text{U}$ may split into segments of intermediate mass. *The splitting of a heavy nucleus into nuclei of intermediate mass is called* **fission.** During the process of fission, neutrons are emitted and a large

The fission of U-235 yields products with intermediate atomic numbers. Fission into two equal fragments is very unlikely, as Figure 24-8 shows.

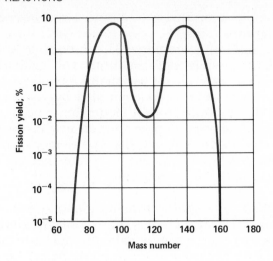

Figure 24-8. Graph of nuclides produced by the fission of the element uranium-235. The fission products vary in mass number from 72 to 161. The most probable products have mass numbers of 95 and 138.

amount of energy is released. One equation for the fission of $^{235}_{92}U$ is

$$^{235}_{92}U + ^{1}_{0}n \rightarrow ^{138}_{56}Ba + ^{95}_{36}Kr + 3^{1}_{0}n + \textbf{energy}$$

The energy involved in this reaction can be computed from the atomic masses of the particles as given in Appendix B, Table 23.

$$
\begin{aligned}
^{235}_{92}U &= 235.0439 \quad u \\
^{1}_{0}n &= \underline{1.008665} \; u \\
& 236.0526 \quad u \\
\\
^{138}_{56}Ba &= 137.9050 \quad u \\
^{95}_{36}Kr &= 94.9 u \\
3 \times ^{1}_{0}n &= \underline{3.025995} \; u \\
& 235.8 u
\end{aligned}
$$

nuclear mass defect = 236.0526 u − 235.8 u = 0.3 u

binding energy released = 0.3 u × 931 Mev/u = 300 Mev

The above calculation has been simplified by using the masses of the neutral atoms given in the table, rather than the masses of the nuclei by themselves. No error is introduced in this case because the masses of the electrons cancel. When a beta particle or a positron is emitted, however, the masses of the nuclei must be used in the calculation.

Note that the atomic mass of $^{95}_{36}Kr$ is not known with the same precision as the masses of the other particles in this reaction. This uncertainty is due to the fact that the atoms of the nuclide are so unstable that it is very difficult to measure their mass. Since the mass of krypton-95 is given in three significant figures, the mass defect and binding energy can be computed with a precision of only one significant figure.

Uranium-238 and uranium-235 can both be split with fast neutrons, while slow neutrons will split only the uranium-235 atoms. Hence fission in a mixture of uranium isotopes can be selectively controlled by regulating the speed of the neutrons with a moderator.

The first successful fission reaction was carried out in 1939. It was later discovered that the element plutonium also undergoes fission and produces more neutrons when bombarded with slow neutrons.

24.8 Fusion The graph of binding energy per nucleon in Section 23.12 suggests that a great deal of binding energy could be released by combining nuclei of the light elements into nuclei of medium mass. In fact, we could expect energies of about 7 Mev/nucleon as compared with about 1 Mev/nucleon for fission reactions. *A reaction in which light nuclei combine to form a nucleus with a larger mass number is called fusion.* Because fusion can take place only under extremely high temperature conditions, the process is also called a *thermonuclear reaction.*

One of the first fusion reactions to be accomplished was the combination of two deuterons (nuclei of the isotope of hydrogen, deuterium).

$$^2_1H + {}^2_1H \rightarrow {}^3_2He + {}^1_0n + 3.3 \text{ Mev}$$

The total energy released in this reaction is less than that of uranium fission, but the energy per nucleon is much greater. In other words, less material is required to produce the energy. In 1938 the American astrophysicist Hans Bethe (b. 1906) suggested that the tremendous and long-lasting rate of energy production of the sun and other stars is due to nuclear fusion. The sufficiently high temperatures and pressures of a star bring about the fusion of hydrogen nuclei to form nuclei of helium, with the accompanying release of energy. The net result of such a reaction is

$$4^1_1H \rightarrow {}^4_2He + 2_{+1}^{0}e + 25.7 \text{ Mev}$$

The intermediate steps that take place in the thermonuclear reactions in the stars are not completely understood. Some scientists believe that the reaction is somewhat different in different parts of the interior of a star. But it is almost certain that several other atomic nuclei, especially those of oxygen and nitrogen, play an important role. Both oxygen and nitrogen have been identified in the sun and other stars.

Figure 24-9. Hans Bethe, the American scientist who proposed the thermonuclear fusion theory of energy production in the stars.

In 1967, Bethe won the Nobel Prize in Physics for his work on energy production in stars.

Estimates of the sun's energy output indicate that about 6×10^{11} kg of hydrogen are converted into helium *every second*, with a mass defect of 4×10^9 kg *per second*. It is comforting to know that the sun has a mass of 2×10^{30} kg, and so there is no immediate danger that its mass will disappear.

24.9 Cosmic Rays   When a charged electroscope is exposed to the air, it will slowly discharge, even though no radioactive material is present. This indicates that there are always some ions in the atmosphere. Experiments have shown that most of this ionization is caused by high-energy particles coming into the atmosphere at great speeds from outer space. These particles are *cosmic rays*.

Cosmic rays have great penetrating power. Their intensity varies with latitude, indicating that they are charged and are affected by the earth's magnetic field. Scientists believe that cosmic rays consist largely of the nuclei of elements of low atomic weight, protons (hydrogen nuclei) being the most abundant type. Other nuclei, ranging up to those as heavy as iron and beyond, have also been detected in the upper atmosphere. A few of these particles from outer space reach the surface of the earth, but most of them collide with the gas particles in the upper atmosphere and produce showers of secondary particles. In effect, cosmic rays are furnishing the nuclear physicist with a constant supply of bombarding nuclei that would otherwise not be available for study.

The term "cosmic ray" was coined by the American physicist Robert Millikan.

Scientists believe that gigantic magnetic fields in space, in galaxies, and near certain stars may be influential in producing cosmic rays. Some cosmic radiation is also thought to originate from highly energetic explosions on certain stars, including the flares that erupt from our sun.

Questions
GROUP A

1. (a) How was radioactivity discovered? (b) By whom was it discovered?
2. (a) Define radioactivity. (b) Which elements are naturally radioactive?
3. List four effects of radioactive radiation.
4. Give another name for (a) alpha particles, (b) beta particles, (c) gamma rays.
5. In a nuclear equation, (a) what do the letter symbols represent, (b) what do the subscripts indicate, (c) what do the superscripts indicate?
6. (a) What is a nuclear transformation? (b) Describe two ways in which it can occur.
7. (a) Define half-life. (b) How is half-life used to find the decay constant of a radioactive nuclide? (c) Define the various units associated with radioactivity.
8. What important conclusion was drawn from the experimental work of Cockcroft and Walton?

9. Why are neutrons more effective than other particles for bombarding nuclei?

10. (a) What is the purpose of a nuclear moderator? (b) List two substances that are used as moderators.

11. (a) How many chemical elements are known today? (b) How many of them occur naturally?

12. (a) Write the equation for the reaction that is believed to be responsible for the energy of the stars. (b) How is the mass of a star affected by this process?

13. What quantities are balanced in a balanced nuclear equation?

14. (a) What are cosmic rays? (b) What is known about their source?

GROUP B

15. How can a magnet be used to identify the radiations from a radioactive substance?

16. List three differences between gamma rays and other radioactive emissions.

17. (a) What is the source of the energy released in a nuclear reaction? (b) Give a specific example.

18. How does the stability of a nucleus vary with (a) mass, (b) the proton-neutron ratio?

19. (a) What is the difference between the average life of a radioactive nuclide and its half-life? (b) What is the relationship between the average life of a nuclide and its half-life?

20. (a) Suggest a way in which $^{200}_{80}Hg$ (mercury) might be changed into $^{196}_{79}Au$ (gold). (b) Write nuclear equations for the reactions that are required.

21. Write the nuclear equations for the alpha decay of (a) $^{234}_{90}Th$, (b) $^{234}_{92}U$, (c) $^{214}_{83}Bi$, (d) $^{210}_{84}Po$.

22. Write nuclear equations showing the preparation of $^{239}_{94}Pu$ from $^{238}_{92}U$.

23. (a) Which nuclides have the smallest binding energy per nucleon? (b) Which have the largest binding energy per nucleon? (c) How does the binding energy per nucleon affect the stability of a nucleus?

24. Describe the difference in the results when $^{235}_{92}U$ and $^{238}_{92}U$ absorb neutrons.

25. Explain, using a specific example, whether the half-life of an element tells when its nucleus will disintegrate.

26. When a radioactive nucleus disintegrates, the products have kinetic energy even though the original nucleus was at rest. What is the source of this energy?

Problems

GROUP A

Note: Consult Appendix B, Table 23 for the atomic masses of nuclides.

1. How much energy in Mev is evolved in the decay of a $^{222}_{86}Rn$ nucleus?

2. Calculate the energy evolved when a $^{214}_{82}Pb$ nucleus decays.

3. What is the decay constant of $^{222}_{86}Rn$ that has a half-life of 3.82 da?

4. Determine the decay constant of $^{238}_{92}U$ if its half-life is 4.49×10^9 years.

GROUP B

5. (a) What is the radioactivity in millicuries of $10\bar{0}$ g of pure $^{232}_{90}Th$? Its half-life is 1.4×10^{10} yr, and it decays to $^{228}_{88}Ra$. (b) What is the rate of energy evolution?

6. (a) Calculate the radioactivity of 5.00×10^{-3} g of $^{222}_{86}Rn$. (b) What is the rate of energy evolution?

7. In which of the following reactions is a greater proportion of the fuel elements converted into energy: $^{239}_{94}Pu + ^{1}_{0}n \rightarrow ^{137}_{52}Te + ^{100}_{42}Mo + 3^{1}_{0}n +$ energy, or $2^{2}_{1}H \rightarrow ^{4}_{2}He +$ energy?

8. If the sun continues producing energy at its present rate, what fraction of the sun's mass would disappear in the next 1000 years?

9. The earth receives 3×10^{22} joules of energy from the sun per day. If this energy is expressed as mass, what is the proportional increase in the earth's mass in one year? (The mass of the earth is presently 6×10^{24} kg.)

10. Compute the mass defect and binding energy per nucleon of $^{13}_{6}C$.

11. (a) Compute the mass defect and binding energy per nucleon of $^{12}_{7}N$. (b) Compare the stability of $^{13}_{6}C$ and $^{12}_{7}N$.

USES OF NUCLEAR ENERGY

24.10 Chain Reactions In a fission reaction the target nucleus may break up in a great many different ways. The reaction involving uranium-235 in Section 24.7, for example, is only one of the ways in which this nucleus may react under neutron bombardment. In most cases, however, two or three neutrons are emitted in the process. This is an example of a *chain reaction. In a **chain reaction** the material or energy that starts a reaction is also one of the products and can cause similar reactions.* Figure 24-10 shows the sequence of events in a typical chain reaction that involves uranium-235. The first controlled chain reaction was carried out at the University of Chicago in 1942 under the direction of the Italian physicist Enrico Fermi (1901–1954).

In 1938, Fermi received the Nobel Prize in Physics for his work in neutron bombardment.

Figure 24-10. A nuclear chain reaction. When a neutron strikes a U-235 nucleus, it is absorbed. This makes the U-235 nucleus unstable, causing it to split into two lighter nuclei called fission products. Heat and two or three other neutrons are also released. These neutrons can strike other U-235 nuclei to produce additional fissions.

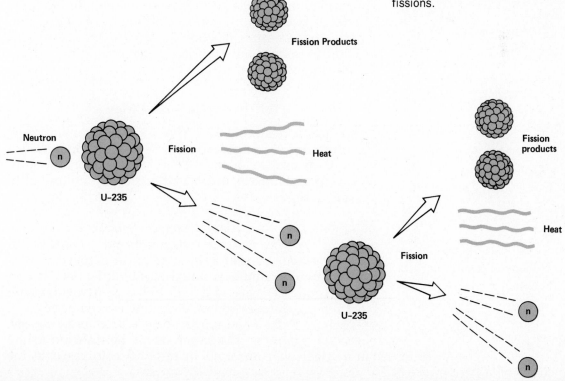

Fission Products

Neutron

Fission

Heat

U-235

n

n

U-235

Fission

Fission products

Heat

n

n

If the material surrounding a fission reaction has the right dimensions and characteristics, the reaction is self-sustaining. *The amount of a particular fissionable material required to make a fission reaction self-sustaining is called the **critical mass.*** If the critical mass is exceeded and if the emitted neutrons are not absorbed by nonfissionable material, the reaction runs out of control and a nuclear explosion results. For example, suppose that two neutrons from each fission are able to produce further fission reactions and that 10^{-8} second elapses between successive reactions. In one millionth of a second, 100 successive reactions can take place. The total energy released is 200 Mev/nucleus $\times 2^{100}$ reactions, or 4×10^{19} j. This is an enormous amount of energy.

Actually, the chain reaction continues until all the fissionable nuclei have split or until the neutrons no longer strike the fissionable material. This is what happens in an uncontrolled chain reaction such as the explosion of a nuclear warhead. Two subcritical masses are placed a short distance from each other inside the warhead. To detonate it, the two masses are brought together quickly by means of explosive charges. The resulting mass exceeds the critical mass and an uncontrolled chain reaction takes place.

A thermonuclear bomb, sometimes called a hydrogen bomb or H-bomb, produces energy by a fusion reaction. The formation of alpha particles and tremendous amounts of energy from a compound of lithium and deuterium, $^6_3\text{Li}^2_1\text{H}$, is one possible reaction in a hydrogen bomb. This reaction is started by subjecting the compound to extremely high temperature and pressure and using a fission reaction as a detonator.

$$^6_3\text{Li}^2_1\text{H} \rightarrow 2^4_2\text{He} + 22.4 \text{ Mev}$$

This reaction results in 22.4 Mev for 8 nucleons, or 2.80 Mev/nucleon, as compared with 186 Mev for 236 nucleons in uranium fission, or only 0.788 Mev/nucleon. In addition, a thermonuclear bomb is theoretically unlimited in possible size.

24.11 Nuclear Reactors A **nuclear reactor** *is a device in which the controlled fission of certain substances is used to produce new substances and energy.* One of the earliest reactors using natural uranium as a fuel was built in 1943 at Oak Ridge, Tennessee. It had a lattice-type construction with blocks of graphite forming the framework and acting as a moderator. Rods of uranium were placed between the blocks of graphite. Control rods of neutron-absorbing boron steel were inserted in the lattice to regulate the number of free neutrons. The reactor contained a critical mass of uranium.

Two types of reactions occur in this kind of reactor. Neutrons cause $^{235}_{92}U$ nuclei to undergo fission. Fast neutrons from this fission are slowed down by passage through the graphite. Some neutrons strike other $^{235}_{92}U$ nuclei and continue the chain reaction. Other neutrons strike $^{238}_{92}U$ nuclei and initiate the changes that produce plutonium. Because great quantities of heat energy are liberated, the reactor is cooled continuously by air blown through tubes in the lattice. The rate of the nuclear reactions is controlled by insertion or removal of control rods.

*A **breeder reactor** is one in which a fissionable material is produced at a greater rate than the fuel is consumed.* The fuel is a mixture of $^{238}_{92}U$ and $^{239}_{94}Pu$. Neutrons from the fission of $^{239}_{94}Pu$ strike the $^{238}_{92}U$ nuclei and convert them to $^{239}_{92}U$. These nuclei decay to $^{239}_{94}Pu$ by double beta decay, as described in Section 24.7. Thus $^{239}_{94}Pu$ is produced at the same time that it is being consumed.

Figure 24-11 illustrates the events in a breeder reactor. Since uranium-238 is much more plentiful than uranium-235, a breeder reactor can extract much larger amounts of

The nuclear equations for these changes are given in Section 24.7.

Figure 24-11. Reactions in a breeder reactor. When a neutron strikes a Pu-239 nucleus, the nucleus splits into lighter fission products and releases heat along with additional neutrons. Some of these neutrons strike other Pu-239 nuclei to continue the chain reaction. Other neutrons strike U-238 nuclei, where they are absorbed to produce Pu-239. This produces additional fuel for the reactor.

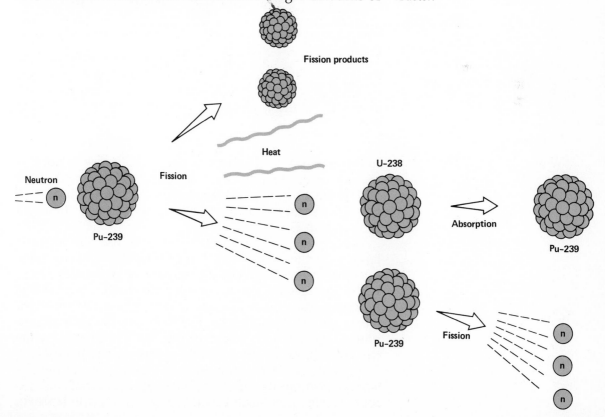

energy from natural uranium than can a reactor that utilizes the lighter isotope.

24.12 Nuclear Power A kilogram of uranium, when completely utilized in fission reactions, has about the same fuel value as 3,000,000 kg of coal or 12,000,000 kg of oil burned in the conventional way. Obviously, nuclear power is an important substitute for the world's dwindling supplies of fossil fuels. The challenge is to produce it safely and to dispose of radioactive wastes without damaging the environment.

As shown in Figure 1-1, less than 5% of the energy used in the United States is presently obtained from nuclear power plants.

As Figure 24-12 shows, the basic difference between a nuclear power plant and a conventional one is in the way in which the steam that turns the turbines is produced. In an ordinary plant, the steam comes from a boiler fired with coal, oil, or gas. In a nuclear plant, the steam is generated by the heat released during the fission process in the reactor.

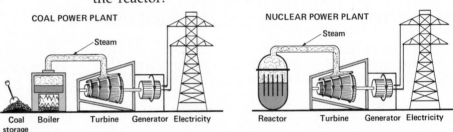

Figure 24-12. A nuclear power plant differs from other power plants only in the way in which steam is generated for the turbines.

The reactor in a nuclear power plant must be able to sustain a chain reaction. Several factors determine whether a chain reaction will continue. The relationship among these factors is described by the following equation:

$$k = \frac{P}{A + L}$$

In this equation k is known as the *multiplication factor*, P is the rate of production of neutrons in the reactor, A is the rate of fission-producing absorption, and L is the rate of leakage. When $k = 1$, the reactor is said to be *critical* and the reaction is self-sustaining.

The power production of a nuclear reactor depends largely on the rate of neutron production. Neutron production, in turn, reduces the amount of fissionable material in the reactor. Hence, to keep the reactor critical it is necessary to reduce the rate of absorption, A, or to increase the production of neutrons, P. The former is accomplished by gradually withdrawing the control rods, while the latter requires the addition of fissionable material.

Natural uranium contains only about 0.7% of the fissionable isotope U-235. Consequently, when natural uranium is used as a reactor fuel, the rate of neutron leakage,

L, is so high that the reactor can never become critical. *L* can be reduced, however, by using a moderator to slow down the neutrons so that they are more likely to strike U-235 nuclei. Another way to make the reactor critical is to use fuel that has been processed to increase the concentration of U-235.

A typical "neutron history" in a nuclear reactor is given in Table 24-2. It shows the relative number of neutrons involved in other events for every 100 neutrons that produce new fissions. Thus of a total of 256 neutrons, only the two that are absorbed by the control rods are available for increasing the power of the reactor. When the control rods are withdrawn, of course, these two neutrons will raise the multiplication factor of the reactor.

The first nonexplosive application of nuclear power was in the American submarine *Nautilus*, which was launched in 1954. Two years later, the first large-scale nuclear power plant was placed in operation in Shippingport, Pennsylvania. It had an output of 60,000 kilowatts of electric power. Table 24-3 lists the nuclear power plants in operation or under construction in various countries in 1981. A nuclear plant with a capacity of nearly 3,500,000 kw is shown in Figure 24-13.

The problem of releasing fusion energy in controlled amounts is being studied in a number of laboratories around the world. At Princeton University's Plasma Physics Laboratory, fusion reactions are produced in so-called *tokamaks* (a name derived from the Russian acronym for "toroidal magnetic chamber"). See Figure 24-14. In a tokamak, plasma is contained in an evacuated, doughnut-shaped stainless steel vessel. Magnetic fields, generated by massive coils located around the outside of the chamber, prevent the high-temperature plasma from striking the vessel walls. The objective is to hold the ions within the vacuum vessel long enough, pack them together densely enough, and heat them to a high enough temperature so that net fusion energy can be produced.

In 1980, a Princeton tokamak reached an ion temperature mark of 82 million °C. This high temperature was achieved with average electron densities of about 1.5×10^{13} particles per cubic centimeter and energy confinement times of about 25 milliseconds. In a reactor that went into operation at Princeton in 1982 it is expected that plasma temperatures of 100 million °C will be attained with the heavy isotopes of hydrogen, deuterium and tritium. At the 100 million °C level, the fusion power produced in the plasma is expected to equal the power required to maintain the plasma temperature.

Table 24-2 NEUTRONS IN A REACTOR	
Number of neutrons	**History**
100	produce new fissions
90	captured by U-238
20	captured by U-235
30	absorbed by moderator
5	absorbed by housing
9	escape from reactor
2	absorbed by control rods
256	

Figure 24-13. Browns Ferry Nuclear Plant in northern Alabama. Research on the effect of heated water from the plant on aquatic life is carried out in the six long channels in the foreground.

Figure 24-14. With this magnetic-confinement fusion reactor, temperatures much higher than at the sun's core have been attained at Princeton University.

Table 24-3
NUCLEAR POWER PLANTS

Country	Plants in operation	Plants planned or under construction	Power output (Mw)
Argentina	1	2	1,627
Austria	—	1	692
Belgium	3	4	5,450
Brazil	—	3	3,116
Bulgaria	3	1	1,760
Canada	10	14	15,332
Czechoslovakia	2	8	4,400
Egypt	—	2	1,800
Finland	3	1	2,160
France	29	30	52,468
Germany	17	19	31,967
Hungary	4	10	3,884
Italy	3	3	3,276
Japan	25	10	25,072
Korea	1	8	7,427
Mexico	—	2	1,308
Netherlands	2	—	495
Pakistan	1	—	125
Philippines	—	1	620
Poland	—	2	880
Rumania	—	3	1,840
South Africa	—	2	1,844
Spain	5	12	15,557
Sweden	9	3	9,410
Switzerland	4	3	4,947
Taiwan	4	2	4,924
United Kingdom	32	10	14,388
United States	78	80	147,767
USSR	31	10	24,795
Yugoslavia	1	—	615

Another approach to controlled nuclear fusion is to direct lasers or beams of heavy ions at targets consisting of hydrogen isotopes. The reaction forces of these collisions implode the target to the densities and temperatures necessary for the fusion of the hydrogen nuclei. A device of this kind is called an inertial-confinement-fusion reactor.

24.13 Radioisotopes *A **radioisotope** is a radioactive isotope of an element.* Some radioisotopes are the products of natural radioactive decay, as in the case of the uranium decay series in Section 24.5. Since the advent of the nuclear reactor, however, a far greater number of radioisotopes has been prepared artificially. The reactor produces

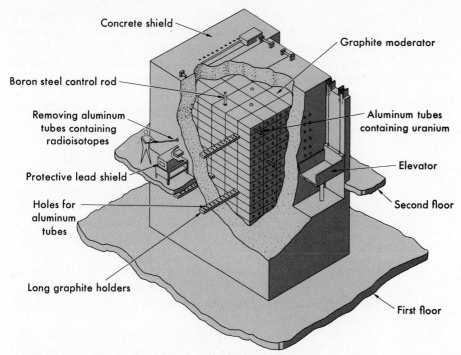

Concrete shield

Graphite moderator

Boron steel control rod

Removing aluminum tubes containing radioisotopes

Aluminum tubes containing uranium

Protective lead shield

Elevator

Holes for aluminum tubes

Second floor

Long graphite holders

First floor

Figure 24-15. Cutaway view of a uranium reactor used for the preparation of radioisotopes. Note the thick concrete shield that is used to protect the technicians from radiation.

radioisotopes during the fission process. In addition, radioisotopes can be prepared by inserting elements into the reactor while it is operating. Radioisotopes of all the elements have been prepared in this way.

One way in which an element becomes radioactive in a reactor is by absorbing a neutron. For example, $^{13}_{6}C$ will turn into $^{14}_{6}C$, which is a radioisotope. In other cases, elements will absorb a neutron and immediately emit a proton. This process lowers the atomic number of the element but does not change its mass number. For example, natural sulfur may change into radiophosphorus.

$$^{32}_{16}S + ^{1}_{0}n \rightarrow ^{32}_{15}P + ^{1}_{1}H$$

A large reactor can release neutrons at the rate of 10^{14} neutrons/cm^2 s. When the reactor is used to produce radioisotopes, it must be run at a rate that is somewhat higher than the rate needed for a chain reaction by itself. The extra neutrons are then available for radioisotope production. Special control and shielding arrangements make it possible to insert and remove substances without shutting down the reactor.

It is often desirable to incorporate radioisotopes into specific compounds before they are used. This can be done by chemically combining the radioisotope with other elements. The *radiocompound* that results is said to be tagged, or *labeled*. Complex compounds that cannot be tagged by direct combination can sometimes be labeled through the biological processes of plants and animals. Radioisotopes are injected into or fed to an animal and then removed from the blood or tissue in the form of radiocompounds. Similar procedures can be used with plants.

The many uses of radioisotopes can be divided into three categories:

1. Effects of radioisotope radiations on materials
2. Effects of materials on radioisotope radiations
3. Tracing materials with radioisotope radiations

In the first of these categories, the radioisotope is used as a source of radiation, just as X rays and radium are used. The target material consists of the substance that is to be changed or destroyed by the radiation. In this way cancer can be treated and food and drugs can be sterilized to retard spoilage. Plastic can be irradiated to change its properties and static electricity can be reduced or eliminated by irradiating the air surrounding the source of the static electricity.

In the second category of radioisotope utilization, the effect of the material on the radiation yields information about the material. Radiation is beamed at a target and the amount that is transmitted or reflected is recorded on photographic film or with some other detection device. Radiations from radioisotopes that are given to a patient can be used to trace or diagnose physiological processes. This method can also be used to measure the thickness of a moving sheet of metal or other material, to analyze the internal structure of a metal casting, to measure the level of a liquid inside a closed container, and even to sort out materials on a moving conveyor.

In the third category of radioisotope utilization, a radioisotope acts as a *tracer* that makes it possible to follow the course of a chemical or biological process. The target and the radioisotope are intimately mixed or combined. The radioisotope then serves as the label that indicates the location of the successive compounds or materials with which the radioisotope becomes associated. Detection instruments or photographic film again serve as measuring devices. The material that is labeled and traced may be food that is consumed by a human being, feed that is converted into milk inside the body of a cow, water that is running out of an undiscovered underground leak, atoms that are rearranged to form a complex molecule, wax that is worn away from the hood of a polished car, or any one of a great many other research materials or industrial products. Figure 24-16 shows how tracers are used to

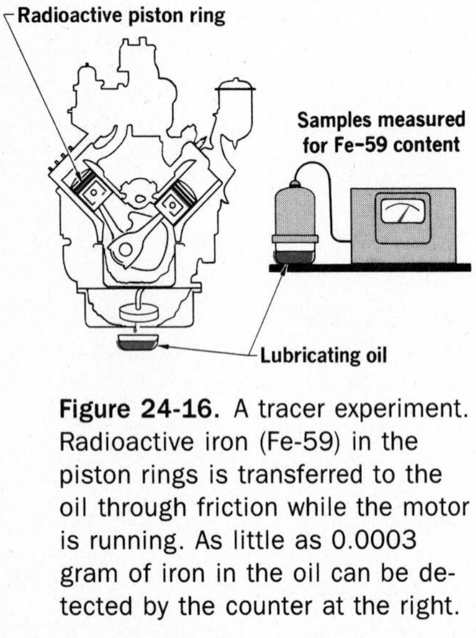

Radiotherapy is an important tool in the fight against cancer.

Figure 24-16. A tracer experiment. Radioactive iron (Fe-59) in the piston rings is transferred to the oil through friction while the motor is running. As little as 0.0003 gram of iron in the oil can be detected by the counter at the right.

study the performance of piston rings in an automobile engine.

An interesting use of a natural radioisotope is the process known as *carbon-14 dating,* which was developed by the American chemist W. F. Libby (1908–1980) in 1952. $^{14}_{7}N$ atoms in the atmosphere are constantly being bombarded by radiations from outer space and converted into the radioisotope $^{14}_{6}C$. In the process of photosynthesis, living plants absorb $^{14}_{6}C$ (as carbon dioxide) along with the more common and nonradioactive $^{12}_{6}C$. When the plant dies, it no longer replaces the carbon atoms in its cells with carbon atoms from the atmosphere and the intake of $^{14}_{6}C$ stops. The amount of radioactivity in the plant gradually decreases due to the fact that the half-life of $^{14}_{6}C$ is 5600 years. Measurements show that a gram of carbon from a living plant has an activity of 16 disintegrations/min. Consequently, a sample showing an activity of 8 disintegrations/g min is assumed to be 5600 years old. A sample with 4 disintegrations/ g min is 11,200 years old, etc. See Figure 24-17. The precision of carbon-14 dating falls off rapidly after several half-life periods, although objects up to about 50,000 years old can be dated in this way.

Other nuclear dating methods are used to learn the age of rocks on the earth and the moon. Moon rocks as old as 4.5 billion years have been dated. It is believed that the solar system itself is not much older than this.

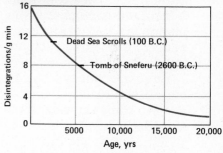

Figure 24-17. Carbon-14 dating. The age of organic compounds can be determined by measuring the remaining amount of radioactive carbon. The Dead Sea Scrolls, a portion of which is pictured above, were dated in this manner.

Questions
GROUP A

1. (a) What conditions are necessary for a chain reaction? (b) What is meant by critical mass?
2. Describe the two types of reactions in a nuclear reactor.
3. (a) How does a nuclear reaction produce electric power? (b) What is meant by a critical reactor? (c) How can natural uranium be used in a critical reactor?
4. How does a breeder reactor differ from conventional reactors?
5. What are the similarities and differences between a nuclear power plant and a fossil-fuel plant?
6. What is the distinction between a magnetic-confinement and an inertial-confinement fusion reactor?
7. (a) What are radioisotopes? (b) How are they made?
8. How could radioisotopes be used to determine the effectiveness of a fertilizer?
9. Estimate the age of a carbon sample showing an activity of 3 disintegrations/g min.

SUMMARY

Radioactivity is the spontaneous decay of atomic nuclei with the emission of particles and rays. It is a property of all elements with atomic numbers greater than 83. Radioactive elements have these characteristics in common: effect on photographic film, ionization of air, production of fluorescence in certain compounds, special physiological effects, and radioactive decay.

The radiations from radioactive elements consist of alpha particles (helium nuclei), beta particles (high-speed electrons), and gamma rays (high-energy photons). In a nuclear change, new materials are formed by changes in the identity of the atoms themselves.

There are four types of energy-producing nuclear reactions: radioactive decay, nuclear bombardment, fission, and fusion. The rate at which a radioactive nuclide decays is indicated by its half-life. Radioactivity is also described in terms of becquerels, curies, and grays. In nuclear bombardment, a nuclide breaks into a large and a small fragment. In fission, two large fragments are produced. In fusion, two small fragments combine to form a larger one. More energy is produced from equal masses of fuel by fusion reactions than by fission reactions.

A chain reaction yields as a product the material or energy that started the reaction. The amount of material required to sustain a fission reaction is called the critical mass. The isotopes U-235 and Pu-239 capture slow neutrons and undergo fission, thereby producing neutrons to continue the chain reaction.

Nuclear reactors produce heat energy for power plants, new fissionable materials, and radioisotopes. The effects of these radioisotopes on certain materials and the effects of certain materials on radioisotopes are used in research. Radioisotopes are also used to trace materials in chemical or biological processes. Carbon-14 dating is an example of a process that employs the use of a natural radioisotope to learn the age of an ancient object.

VOCABULARY

alpha particle
becquerel
beta particle
breeder reactor
chain reaction
chemical change
cosmic ray

critical mass
curie
fission
fusion
gamma ray
gray

nuclear change
nuclear reactor
physical change
radioactivity
radioisotope
sievert

HIGH-ENERGY PHYSICS

The Nobel Prize was established by the Swedish industrialist, Alfred Nobel, who invented dynamite. Nobel's will specified that the awards should go to those who "have conferred the greatest benefit on mankind." In recent years, the majority of Nobel prizes in physics have been awarded for studies of nuclear particles and interactions.

QUANTUM MECHANICS

25.1 The Uncertainty Principle   As seen in Section 12.9, the quantum theory states that light exhibits particle-like properties in addition to its wave-like properties. This means that light energy is transferred in discrete quantities, or quanta. Applying this idea to the atom, Bohr concluded that the electron is not free to assume any orbit whatever; rather, the size and shape of the electron orbit are governed by the quantum theory. An electron cannot move into a lower energy orbit except through the loss of energy exactly equal to the energy difference between the initial and the final orbit. An electron cannot move into a higher energy orbit unless it gains enough energy. An electron must stay in its orbit unless sufficient energy is gained or lost so that an abrupt orbital transition can take place. This process can occur spontaneously.

Bohr's model was remarkably successful in explaining the spectral lines of low-atomic-numbered elements. But the model ran into trouble when it was used to explain the behavior of helium atoms or atoms with large atomic numbers. Then, too, small but important differences were found between the predictions of the Bohr model regarding atomic spectra and the results of experiments based on the predictions. Attempts were made to account for these differences by suggesting that electron orbits are elliptical rather than circular and by using relativistic mass (a method that was successful in solving a similar problem

In this chapter you will gain an understanding of:

▷ the uncertainty and exclusion principles of electron behavior
▷ quantum numbers in the description of electron motion
▷ various types of particle accelerators
▷ the principles of operation of particle detection devices
▷ the classification of subatomic particles as baryons, mesons, leptons, and bosons
▷ the four basic interactions between particles of matter
▷ the conservation laws of particle physics
▷ recent developments in the search for quarks and unified field theories

Figure 25-1. Werner Heisenberg, the German physicist, formulated the uncertainty principle, which provides a basis for quantum mechanics.

In 1932 Heisenberg received the Nobel Prize in Physics for his work in quantum mechanics.

with the orbit of the planet Mercury). These explanations resolved only some of the discrepancies.

There were other problems with the Bohr atom. The effect of the spin of the electron (Sections 19.2 and 23.6) on its orbital motion had not been taken into account by Bohr. Such spinning sets up electromagnetic forces that must be considered in computing the total energy of the orbiting electron.

The most serious blow to the Bohr model of the atom came in 1927 with the announcement of the *uncertainty principle* by the German physicist Werner Heisenberg (1901–1976). The **uncertainty principle** states that *it is impossible to specify simultaneously the exact position of an object and its momentum.* Other pairs of physical quantities related to the motion of an object, such as time and energy, are also subject to the uncertainty principle. For bodies of ordinary mass, the inherent uncertainty in any such simultaneous determination is smaller than errors in measurement, and so the uncertainty effect is not observed. With electrons, however, the effect is pronounced. This principle means that it is possible to determine where an electron is at a given time, but it is not possible to determine its exact velocity at that same time. The reason is that the method used to determine the position of the electron, short-wavelength radiation, changes the electron's momentum. Likewise, to determine an electron's momentum, an instrument that changes the electron's direction of motion is used, and so the determination of the electron's position is prevented.

25.2 *Quantum Numbers* To understand the description of the atom according to the quantum theory, it is necessary to be familiar with the idea of *quantum numbers.* As used in atomic physics, a **quantum number** is *a number that describes the allowable value of certain physical quantities.* For example, quantum numbers are used in special equations for computing the energy of a particle, such as the electron. *The branch of physics that deals with the behavior of particles whose specific properties are given by quantum numbers is called* **quantum mechanics.**

The quantum-theory model of the atom is three-dimensional. The motion of the electron is not restricted to a single plane, as in the Bohr atom. Within the limits prescribed by its quantum numbers, the electron describes an orbit that completely encloses the nucleus.

The first quantum number for the electron describes the radius of its orbit. It is called the *principal quantum number* and is designated by the letter *n*. The principal quantum number may have the integral values 1, 2, 3, 4, etc., which correspond to the similarly numbered energy levels of the Bohr atom. For example, an electron with a principal quantum number 1 is in the first energy level, an electron with the number 2 is in the second level, and so on.

The equation for finding the energy of an electron from its principal quantum number is

$$E = -\frac{me^4}{8\epsilon_0^2 h^2 n^2}$$

where E is the energy of the electron, m is its mass, e is the electronic charge, ϵ_0 (the Greek letter epsilon) is a constant with the value 8.854×10^{-12} $c^2/n\ m^2$, h is Planck's constant, and n is the principal quantum number. The energy is expressed as a negative quantity to show that the electron is held by the nucleus.

Thus far the use of the principal quantum number yields exactly the same results for the orbits of electrons that Bohr obtained for his atomic model. But here the similarity ends. A second quantum number, called the *angular momentum quantum number* or *secondary quantum number*, describes the magnitude of the angular momentum of the orbiting electron. This quantum number is designated by the letter l. Unlike energy, which is a scalar quantity, angular momentum is a vector with both magnitude and direction. The equation for the magnitude of the angular momentum is

$$L = \sqrt{l(l + 1)}\,\frac{h}{2\pi}$$

where L is the angular momentum, l is the angular momentum quantum number, and h is Planck's constant. l may have integral values ranging from zero to $n - 1$ and is therefore limited by the principal quantum number. In effect, a high value of l means that the electron's orbit is circular. The lower the value of l, the more elliptical the orbit is.

The magnitude of the angular momentum in a magnetic field is described by a third quantum number, the *magnetic quantum number*. It is designated as m_l and can have integral values ranging from $-l$ through zero to $+l$. The equation using m_l is

$$L_z = m_l\frac{h}{2\pi}$$

in which L_z is the component of the angular momentum that is parallel to the magnetic field.

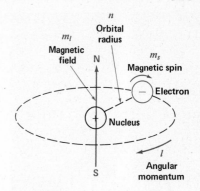

Figure 25-2. Quantum numbers describe the position and motion of an electron.

Figure 25-3. The Zeeman effect. The single line at the top is the 3779-Å line of the element curium. At the bottom is the spectrum of the same line in a strong magnetic field.

In 1945 Pauli received the Nobel Prize in Physics for his discovery of the exclusion principle.

Since m_l can have several values for all values of n except $n = 1$, the energy levels of an atom are split into two or more sublevels when the atom is subjected to a magnetic field. This effect is called the *Zeeman effect,* after the Dutch physicist Pieter Zeeman (1865–1943). The Zeeman effect can be detected in the spectra of excited atoms, as is shown in the photo of Figure 25-3.

In the 1920s it was discovered that many pairs of spectral lines could be described in terms of a model in which electrons spin on their axes. This electron spin model also harmonizes with the relativistic explanation of the behavior of matter. The spin of the electron is designated by the *spin quantum number, s,* which has the value $\frac{1}{2}$. The equation for the magnitude of spin angular momentum is

$$L_s = \sqrt{s(s + 1)}\frac{h}{2\pi}$$

The component of spin angular momentum that is parallel to a magnetic field is given by the equation

$$L_s = m_s\frac{h}{2\pi}$$

where m_s is the *spin magnetic quantum number* and can have the values $+\frac{1}{2}$ and $-\frac{1}{2}$. The $+$ and $-$ signs indicate that the electrons are spinning in opposite directions.

The assignment of four quantum numbers to each electron leads to the logical question of how the quantum numbers of any one electron compare with the numbers of other electrons. Does it ever happen that two electrons have the same set of numbers? The answer to this question is provided by the *exclusion principle* of the Austrian physicist Wolfgang Pauli (1900–1958). The **exclusion principle** states that *no two electrons in an atom can have the same set of quantum numbers.* The number of different energy levels that could be computed if there were no restrictions on the quantum numbers associated with each electron is far greater than those revealed by spectral evidence.

With the help of the four quantum numbers and the exclusion principle, it is possible to determine the electron configuration of any particular atom. Electron configurations for several elements are given in Table 25-1. In each case, the number of electrons in a particular sublevel is found by using the permissible values of each quantum number. For example, aluminum has 13 electrons. When $n = 1$, l equals 0, m_l equals 0, and m_s can have values of $+\frac{1}{2}$ and $-\frac{1}{2}$. In other words, two electrons can be listed in the column headed $n = 1$. When $n = 2$, l can be 0 or 1. When $l = 0$, m_l equals 0 and m_s can be $+\frac{1}{2}$ and $-\frac{1}{2}$, as before.

Table 25-1
ELECTRON CONFIGURATION OF THE LIGHT ELEMENTS

Atomic number	Symbol	Element	Electron configuration of atom						
			$n=1$	$n=2$		$n=3$			$n=4$	
			$l=0$	$l=0$	$l=1$	$l=0$	$l=1$	$l=2$	$l=0$	$l=1$
1	H	hydrogen	1							
2	He	helium	2							
3	Li	lithium	2	1						
4	Be	beryllium	2	2						
5	B	boron	2	2	1					
6	C	carbon	2	2	2					
7	N	nitrogen	2	2	3					
8	O	oxygen	2	2	4					
9	F	fluorine	2	2	5					
10	Ne	neon	2	2	6					
11	Na	sodium	2	2	6	1				
12	Mg	magnesium	2	2	6	2				
13	Al	aluminum	2	2	6	2	1			
14	Si	silicon	2	2	6	2	2			
15	P	phosphorus	2	2	6	2	3			
16	S	sulfur	2	2	6	2	4			
17	Cl	chlorine	2	2	6	2	5			
18	Ar	argon	2	2	6	2	6			
19	K	potassium	2	2	6	2	6		1	
20	Ca	calcium	2	2	6	2	6		2	
21	Sc	scandium	2	2	6	2	6	1	2	
22	Ti	titanium	2	2	6	2	6	2	2	
23	V	vanadium	2	2	6	2	6	3	2	
24	Cr	chromium	2	2	6	2	6	5	1	
25	Mn	manganese	2	2	6	2	6	5	2	
26	Fe	iron	2	2	6	2	6	6	2	
27	Co	cobalt	2	2	6	2	6	7	2	
28	Ni	nickel	2	2	6	2	6	8	2	
29	Cu	copper	2	2	6	2	6	10	1	
30	Zn	zinc	2	2	6	2	6	10	2	
31	Ga	gallium	2	2	6	2	6	10	2	1
32	Ge	germanium	2	2	6	2	6	10	2	2
33	As	arsenic	2	2	6	2	6	10	2	3
34	Se	selenium	2	2	6	2	6	10	2	4
35	Br	bromine	2	2	6	2	6	10	2	5
36	Kr	krypton	2	2	6	2	6	10	2	6

Figure 25-4. Wolfgang Pauli, the Austrian-born physicist, developed the exclusion principle, which states that no two electrons in an atom can have the same set of quantum numbers.

Hence two electrons are listed under $l = 0$ in this column. When $l = 1$, m_l can be $+1$ or 0 or -1. For each of these three values of m_l, there are again two values of m_s. In other words, there are six possible combinations of quantum numbers when $l = 1$. When $n = 3$ and $l = 0$, two values of m_s are again possible. This tally leaves only one more electron to account for in the aluminum atom; this final electron is listed under $l = 1$.

Notice that in the elements beyond argon, electrons may appear in the fourth energy level before the third level is complete. In such cases, the sublevel for which $n = 4$ has a lower energy than the remaining sublevels for which $n = 3$. For atoms with atomic numbers beyond 18, therefore, it is not simply a matter of successively filling in sublevels.

Quantum mechanics allows only a definite number of orbiting electrons for each value of the principal quantum number. For $n = 1$, it is 2 electrons, for $n = 2$, it is 8 electrons, and so on. When an atom has the maximum number of electrons for a particular electron shell (or subshell for heavy atoms), it is said to have a *complete shell*. In most cases, atoms are stable when they have complete shells.

The stability of an atom is described in terms of *ionization energy*. **Ionization energy** *is the energy required to remove an electron from an atom.* Obviously, the ionization energy must be equal to the energy of the electron as described by its quantum numbers, and it must also have the same sign. As n increases, the energy of the electron decreases. Consequently, the outermost electrons are usually the easiest to remove from an atom. Figure 25-5 shows the ionization energies of the outermost electrons of the elements. While there are variations in ionization energies within electron energy levels, you will notice that the ionization energies for completed levels (atomic numbers 2, 10, 18, 36, 54, and 86) decrease as the size of the atom increases.

25.3 Matter Waves   A final word of explanation about the equations of the quantum theory is necessary at this

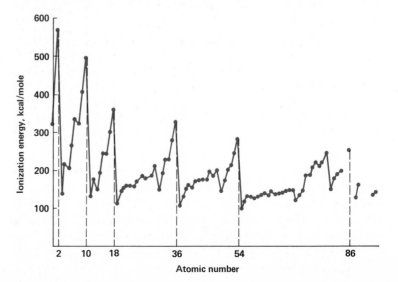

Figure 25-5. Ionization energies of the outermost electrons of the elements.

point. You will notice that Planck's constant, h, is used in each of the equations. In Section 12.11, we saw that h is associated with wave motion. The use of this same constant in describing the behavior of subatomic particles suggests that the particles have wave-like characteristics. This wave-particle duality was mentioned in the very first chapter and in the discussion of light. Now it is important to emphasize it once more. About 1924 the French physicist Louis de Broglie suggested that *all* matter possesses wave characteristics and that these matter waves become longer as the momentum of a particle of matter decreases.

In 1929 de Broglie received the Nobel Prize in Physics for his discovery of the wave nature of electrons.

In Section 12.11, the wavelength of a matter wave was given by the equation

$$\lambda = \frac{h}{mv}$$

where λ is the wavelength, h is Planck's constant, m is the mass of the particle, and v is its velocity. Since h has a very small value (6.63×10^{-34} j s), it can be seen from the equation that m likewise must be extremely small before λ can become large enough to be detectable. This is why the wave properties of, let us say, a moving baseball are completely undetectable, while the wavelength of the electron is long enough to consider in the modern view of the atom.

It must not be assumed, however, that a de Broglie wave is like the physical disturbance of the water in a ripple tank. A de Broglie wave is not a motion, but rather a property of matter. Matter waves can exist in a vacuum in the same way that light waves can. They permeate space in the way a gravitational or magnetic field influences the area surrounding a mass or a magnet. In the final analysis, it is not possible to construct an analogy for matter waves. Furthermore, it should be recognized that de Broglie waves, like so many other descriptions in science, represent a particular model of the universe. As with all models in the history of science, this one can also be modified and improved as our understanding of the universe keeps growing.

Figure 25-6. Louis de Broglie, the French physicist who proposed the wave nature of matter.

Questions

GROUP A

1. (a) Why is it impossible to determine simultaneously the position and velocity of an electron? (b) What is the name of this principle? (c) Who first postulated it?

2. (a) What is quantum mechanics? (b) How is it used to explain atomic structure?

3. (a) Define the four quantum numbers of an electron. (b) What quantum number always has only two values? (c) Why?
4. (a) What is the Zeeman effect? (b) How can spectra be used to tell whether a specific example of the Zeeman effect is the result of the angular momentum of an electron or of electron spin?
5. (a) What is meant by the exclusion principle? (b) Who postulated it? (c) What led him to this conclusion?
6. How does the ionization energy of an electron affect the stability of an atom?
7. Why is the wavelength of a moving baseball undetectable while that of a much smaller and less massive electron is noticeable?
8. (a) How do matter waves differ from water waves? (b) Describe a model for de Broglie waves.

Problems
GROUP A

1. List the quantum numbers for each electron in the following atoms: (a) helium, (b) sodium, (c) argon, (d) magnesium.
2. (a) Compute the energy of an electron with a principal quantum number of 3. Use SI units. The mass and charge of the electron and the value of Planck's constant are given in Appendix B, Table 4. (b) Convert this energy to electron volts, and compare it with the ionization energy of the outer electron of the neutral sodium atom.

GROUP B

3. (a) Find the wavelength of a baseball with a mass of 0.15 kg and a velocity of 26 m/s. (b) Compare this wavelength with that of an electron moving at 5.5×10^6 m/s.
4. Show that the number of electrons that can exist in a particular shell is given by the expression $2n^2$, in which n is the principal quantum number.

PARTICLE ACCELERATORS

25.4 Providing High-Energy Particles As we saw in Chapter 24, nuclear reactions can be caused by particles and radiations from naturally radioactive substances. But the subatomic research that can be conducted in this way is limited by the energy of the radioactive emissions. To provide nuclear "bullets" that have greater penetrating power, it is necessary to accelerate them. In the following sections, we shall describe several particle accelerators, among which are the largest scientific devices ever built.

25.5 Van de Graaff Generators One of the first particle accelerators was developed in 1931 by the American physicist R. J. Van de Graaff (1901–1967). As Figure 25-7 illustrates, a Van de Graaff generator projects electrons onto an insulated moving belt as a result of repulsion from a strongly negative electrode. The electrons ride on the belt to the other end of the generator, where they are picked off by a collector and transferred to a metal sphere. Since an electric charge always resides on the *outside* of a hollow conductor, the inside of the sphere is able to receive more charges. Thus a large negative charge is built up on the sphere.

Conversely, a large positive charge may be built up on the sphere by removing electrons from the belt by use of a strongly positive electrode. The large difference of electric potential thus obtained can be used to accelerate electrons or ions. Since the high charge on a Van de Graaff generator leaks off into the atmosphere if the air is humid, such generators are sometimes built under a large, pear-shaped shell that is filled with dry air under pressure to keep the charge on the sphere. Placing the generator in a vertical position helps to prevent the discharge of particles to the ground.

To provide maximum energy to particles, Van de Graaff generators are sometimes used in series with linear accelerators. The world's most powerful Van de Graaff installation is a pair of end-to-end generators at the Brookhaven National Laboratory in New York. The generators are well insulated to prevent grounding. This installation is capable of accelerating hydrogen ions to an energy of 30 Mev. Although this is not as high as the energies produced in other types of large accelerators, the Van de Graaff machines have certain other advantages. For example, the particle beam can be more precisely focused, energy variations are more easily controlled, and a wide range of particles can be employed.

25.6 Circular Accelerators Shortly after the development of the Van de Graaff generator, it was found that charged particles can be accelerated to very high velocities by driving them in a circular path with the aid of electromagnets. A variety of such accelerators have since been built.

1. Cyclotrons. In 1932 the American physicist E. O. Lawrence (1901–1958) developed a circular accelerator in which various kinds of charged particles could be used. He called it a *cyclotron.* It consists of a large cylindrical box placed between the poles of an electromagnet, as shown in Figure 25-8. The box is exhausted until a very high vacuum exists inside. Charged particles, such as protons or deuterons, are fed into the center of the box. Inside the box are two hollow, D-shaped electrodes called *dees* that are connected to a source of high-voltage electricity.

When the cyclotron is in operation, the electric charge on the dees is reversed very rapidly by an oscillator. The action of the field of the electromagnet causes the charged particles inside the cylindrical box to move in a semicircle in a fixed period of time. The potential applied to the dees

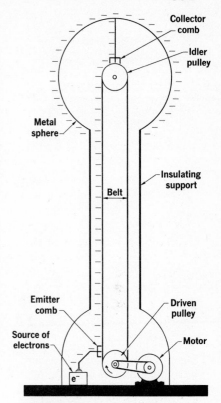

Figure 25-7. Schematic diagram of a Van de Graaff generator (top). Shown at the bottom are two such generators in tandem at Brookhaven National Laboratory. (The second generator is barely visible in the rear.)

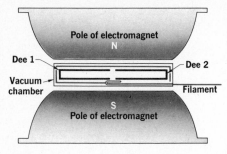

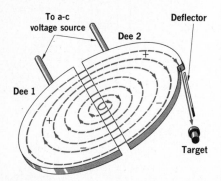

Figure 25-8. Schematic diagram of a cyclotron.

Figure 25-9. The inventor of the betatron, Donald Kerst, is shown with his original accelerator, a 2.5-Mev model built in 1940. In the background is a 340-Mev betatron used at the University of Illinois.

is adjusted to reverse itself in the same time period. Thus at each transit across the gap between the dees, the ions gain energy from the electric field. While the ions are inside the dees, their velocity is constant because there is no electric field on the inside. But at each transit, the ions move faster and approach the outside of the box. When they reach the outer rim of the box, they are deflected toward the target.

The energy of particles accelerated in a conventional cyclotron may reach 15 Mev. To obtain higher values, it is necessary to modify the accelerator to deal with the relativistic increase in mass of the particles as their velocity increases. This increased mass keeps the particles from gaining enough speed to reach the end of the dees by the time the oscillating voltage reverses. To overcome this difficulty, it is necessary to vary the oscillating frequency. Then as the mass of the particles increases and their acceleration decreases, the frequency of the oscillator can be reduced. The changes in voltage of the field are timed to coincide with the slower acceleration of the more massive particles. This type of accelerator is called a *synchrocyclotron*. It can boost the energy of particles to about 800 Mev.

2. Betatrons. As its name implies, the betatron is used to accelerate beta particles (electrons). The betatron uses the principle of the transformer; the primary is a huge electromagnet and the secondary is the stream of electrons that are being accelerated. The electrons are enclosed in a circular, evacuated tube called the *doughnut*. After the electrons have acquired maximum energy, they are directed toward a metal plate. The impact of the electrons on the metal plate produces high-energy X rays that are used for nuclear studies and for industrial and medical applications. In effect, the betatron is a large X-ray machine, producing energies up to about 300 Mev. See Figure 25-9. Beyond this range, electrons rapidly lose energy by electromagnetic radiation.

3. Synchrotrons. A newer type of particle accelerator is the *synchrotron*. Like the cyclotron, it operates on the principle of accelerating particles by making them move with increasing velocity in a circular path. However, the path of the particle is confined to a doughnut (as in the betatron) by varying both the oscillating voltage and the magnetic field.

The energies attainable in a synchrotron are theoretically unlimited. At the present time, the installation at the Fermi National Accelerator Laboratory (Fermilab) in Batavia, Illinois, is the world's most powerful accelerator. It consists of four different kinds of accelerators in series. See Figure 25-10. The circular path of the synchrotron in the final stage is 6.6 km long! By using superconducting magnets in this synchrotron, it is possible to obtain proton energies of 1000 Gev (1 Tev) and velocities greater than 99.999% of the velocity of light. High-energy synchrotrons are also called *bevatrons* or *cosmotrons*.

4. *Intersecting storage rings.* The circular accelerators mentioned thus far produce particles by the collision of beams of high-energy particles with stationary targets. The disadvantage of this arrangement is that most of the beam's energy is dissipated in pushing against the target material; only a small amount of the energy is available for nuclear reactions.

In the intersecting storage accelerator, particles are made to collide with each other head-on as they move in opposite directions. To increase the number of collisions, the two beams are first concentrated in *storage rings*. When the density of particles in the storage rings is sufficiently high, the beams are allowed to collide.

Large intersecting storage accelerators are located in Hamburg, West Germany; Geneva, Switzerland; and at Stanford University in California. ISABELLE, another intersecting storage accelerator, is under construction at the Brookhaven National Laboratory in New York. ISABELLE will contain storage rings that intersect in six places. See Figure 25-11. More than 1000 superconducting magnets will guide the beams inside the rings. Designed to study

Figure 25-10. Sections of the particle accelerator at Fermilab. The voltage multiplier at the far left supplies an initial acceleration of protons to 750 kev. Second from left is a view of the linear accelerator that drives the protons up to 200 Mev. At the top right is part of the magnet ring of the booster accelerator that raises the energy to 8 Gev. The main accelerator tunnel, a section of which is shown at the bottom right, is 6.6 km long. Protons emerge from this final stage of the accelerator with energies of more than 500 Gev.

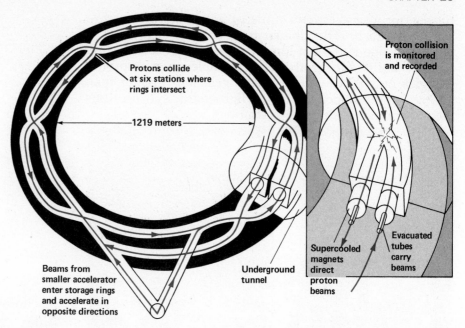

Figure 25-11. Schematic diagram of the intersecting storage accelerator under construction at the Brookhaven National Laboratory. Almost all the energy of the proton beams is converted into particle production in the head-on collisions in this accelerator.

collisions between protons, ISABELLE will set a new record for accelerator energies—over 800 Gev. Because of the efficiency of intersecting accelerators, this is thirteen times greater than the available energy of the most powerful nonintersecting accelerator in operation today. Storage rings are also being added to the accelerator installation at Fermilab.

Since intersecting storage rings are best used with protons, however, and since the total number of collisions per second is not very large compared with that in other accelerators, intersecting storage accelerators cannot be used for all experiments.

25.7 Linear Accelerators A linear accelerator is a device that moves charged particles to high velocities along a straight path, in contrast to the circular arrangements in cyclotrons and synchrotrons. Linear accelerators are used individually and in conjunction with other accelerators.

A simplified diagram of a linear accelerator is shown in Figure 25-12. A series of hollow *drift tubes* is mounted in a long, evacuated chamber. Charged particles are fed into one end of this arrangement and are accelerated by means of a high-frequency alternating voltage applied to the drift tubes. At any given instant, successive tubes have opposite charges. The alternating voltage is timed in such a way that the particle is repelled by the tube it is leaving and attracted by the tube it is approaching. In this way, the particle is accelerated every time it crosses the gap be-

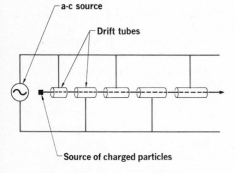

Figure 25-12. Schematic diagram of a linear accelerator. Charged particles accelerate as they are attracted by the alternately charged drift tubes.

tween two tubes. Because of this acceleration, the drift tubes are successively longer so that the particles will always be in a gap when the voltage is reversed.

A 3.3-km-long linear accelerator is located near Palo Alto, California. It is pictured in Figure 25-13. Inside a copper tube in a concrete tunnel buried 7.6 m underground, electrons are accelerated to 40 Gev. As mentioned previously, some circular accelerators are limited in their ability to accelerate electrons because of electromagnetic radiation losses. This difficulty is not encountered in linear accelerators. In addition, the beam of particles can be more easily controlled in a linear accelerator than in a circular one. A linear accelerator is also more accessible than a circular one, and it can provide a higher frequency of impacts on the target material.

Figure 25-13. The world's largest linear accelerator. Electrons are injected at the far end of the 3.3-km-long structure (upper right). Particle production and analysis take place in the two buildings at the lower left.

Questions

GROUP A

1. (a) How does a Van de Graaff generator accelerate particles? (b) What type of particles can it *not* accelerate? (c) Why not?
2. (a) In what units is a particle accelerator rated? (b) Define the unit. (c) Give the ratings of a large Van de Graaff generator, a large cyclotron, and a large synchrotron.
3. Distinguish between a cyclotron and a synchrotron.
4. (a) What is the function of a drift tube in a linear accelerator? (b) Why are drift tubes made successively longer inside the accelerator?

5. One particle accelerator can accelerate electrons to a speed of 99.9% of the speed of light. Another machine can bring protons up to a speed of 10% of the speed of light. In terms of particle mass, which machine requires more energy?
6. (a) How does a cyclotron accelerate a proton? (b) How does it keep the proton inside the machine? (c) Why is it impossible for a cyclotron to accelerate neutrons?
7. Why is an intersecting storage accelerator more efficient than a single-beam accelerator?
8. Why are magnets unnecessary in a linear accelerator?

DETECTION INSTRUMENTS

25.8 Types of Detectors   Subatomic particles are far too small to be observed and measured directly. Most of these particles have extremely short half-lives. This, together with the high speed at which they usually travel, means that the time available to study them is very short. As is true of the construction of high-energy accelerators, the development of instruments capable of detecting such elusive events is one of the most impressive accomplishments of modern technology. Some particle detectors are able to photograph the effects of subatomic particles that spend

Figure 25-14. A gold leaf electroscope. Radioactive materials are placed in the hinged drawer at the bottom.

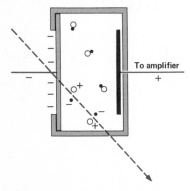

To amplifier
+

Figure 25-15. An ionization chamber. The passage of a subatomic particle (represented by the arrow) ionizes the gas in the chamber along its path. The negative ions are then attracted to the anode on the right side of the chamber.

as little as 10^{-11} second in the instruments!

In the following sections we will describe several low-energy detection instruments that have been available to physicists for many years. Then we shall examine several high-energy detectors, including the computerized electronic bubble chamber.

25.9 Low-Energy Detectors *1. The electroscope.* One of the first instruments to be used to detect and measure radioactivity is the electroscope, shown in Figure 25-14. A negatively charged electroscope becomes discharged when ions in the air take electrons from the electroscope. Similarly, a positively charged electroscope becomes discharged as it takes electrons from ions in the air. Thus the rate at which an electroscope is discharged is a measure of the number of ions in the air near the electroscope.

Madame Curie used the electroscope to study the radioactivity of uranium ore. She found that some ores discharged the electroscope more rapidly than others. This finding suggested the presence of an undiscovered, highly radioactive element in some of the ores and led to the isolation of radium.

2. The spinthariscope. The spinthariscope (derived from the Greek word meaning "spark") is another instrument that was used in early studies of radioactivity. In this device, alpha and beta particles from radioactive materials or nuclear collisions strike a zinc sulfide screen. Where they strike the screen, they produce a momentary flash of light that can be observed through a magnifier. By counting the number of flashes in a given period of time, the intensity of the radiation can be determined.

3. The scintillation counter. As its name implies, the scintillation counter registers the flashes of light produced when subatomic particles hit a fluorescent screen. In this respect the scintillation counter is similar to both the spinthariscope and a television picture tube. However, the scintillation counter also provides information about the energy, and thus the identity of the particles it detects. This is done by having the screen of the counter coupled with a photomultiplier tube. (See Section 22.8.)

4. The ionization chamber. The ionization chamber, shown in Figure 25-15, contains fixed electrodes. When the gas in the chamber becomes ionized because of the passage of a high-speed particle or ray, the electrons and ions drift to the electrodes. The electrons drift to the positive plate while the positively charged ions drift to the negative plate.

The charge produced by a single passing particle is much too small to activate the circuit in an ionization chamber, and so an amplifier is used in conjunction with the instrument. The resulting impulse can then be used to trigger a light or some other recording device.

5. *The Geiger tube.* The principle of the ionization chamber is also used in the Geiger tube, which is named after the German physicist Hans Geiger (1882–1945). A hollow cylinder acts as the negative electrode and a thin wire down the center is the positive electrode. The tube is filled with a gas at low pressure, and a potential difference of about 1000 volts is applied to the electrodes. This potential difference is slightly less than the voltage required to ionize the gas and enables it to conduct an electric current between the electrodes.

When a charged particle or gamma ray enters the Geiger tube, it releases electrons from the atoms in its path and produces ions, as shown in Figure 25-16. These electrons are then attracted to the positively charged wire. As they move toward the wire, they collide with more gas atoms to produce additional free electrons and positive ions. Eventually, the surge of electrons toward the positive electrode is large enough to register as a detectable current flowing through the tube. Thus the Geiger tube acts as its own amplifier for the detection of nuclear events. The Geiger tube is generally used in conjunction with external circuits to activate flashing lights, a loudspeaker, or a counting device.

6. *The cloud chamber.* A device that makes the paths of ionizing particles clearly visible is the cloud chamber. It was invented by the British physicist Charles Wilson (1869–1959) in 1911. Its principle of operation is quite simple, and an operating cloud chamber can be assembled with a minimum of material and effort. Figure 25-17 shows a simple, homemade cloud chamber.

The chamber is a plastic or glass container. It is resting on a block of dry ice. A piece of dark cloth is saturated with alcohol and placed around the inside of the container near the top. A radioactive source is placed in the eye of a needle that is suspended from a cork placed on the lid of the container.

In the chamber, alcohol evaporates from the cloth and condenses again as it reaches the cold region at the bottom of the chamber. Just above the floor of the chamber there is a region where the alcohol vapor does not condense unless there are "seeds" around which drops of alcohol can form. (The situation is similar to a cloud that can be

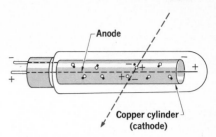

Figure 25-16. A Geiger tube. The ions produced by the subatomic particle (arrow) permit the passage of a flow of electrons between the cathode and the anode.

Figure 25-17. A simple cloud chamber. The radioactive source is placed in the eye of the needle that is suspended from the cork.

made to form rain by "seeding" it with a powdered chemi-cal.) If there is no dust in the chamber, the only seeds available are ions produced by the passage of charged par-ticles or rays. The resulting trail of alcohol droplets is then visible against the black bottom of the chamber and can be photographed.

The device we have just described is called a *diffusion cloud chamber*. Others, including the chamber developed by Wilson, rely on pressure changes instead of low tem-peratures to produce condensation. It is possible to use many other liquid-vapor arrangements to produce the tracks. In all cloud chambers, it is the ionizing effect of high-speed particles that makes a visible trail possible.

Figure 25-18. Impressions of sub-atomic particles on photographic film. An antiproton (p⁻) entered from the left and produced a shower of eight other particles when it collided with a proton. This is the first photo of the reac-tion between a proton and its antiparticle. Other particles are identified by their Greek symbols.

25.10 Photographic Film In addition to its use with detection instruments, photographic film can be used di-rectly to record the passage of subatomic particles. When charged particles or gamma rays strike photographic film, they change the chemicals in the emulsion in much the same way that a ray of light changes them. When the film is developed, the passage of the particle or ray is recorded as a spot. If a stack of photographic plates is exposed to a nuclear event, the event is recorded as a series of streaks through the emulsions, as shown in Figure 25-18.

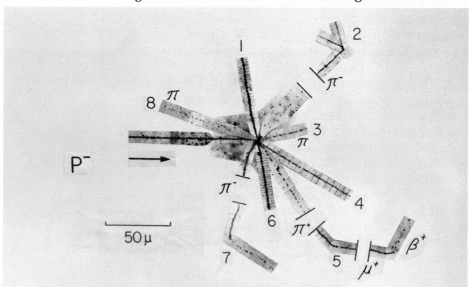

25.11 Liquid Bubble Chambers In a cloud chamber, nuclear events are visible because of the condensation of a substance like alcohol. In a bubble chamber, exactly the opposite principle is used. A container is filled with a liq-uid, and the tracks of particles become visible as a series of bubbles when the liquid boils.

One substance used in bubble chambers is liquid hydrogen. As long as the temperature and pressure conditions are right, the hydrogen stays in liquid form. If the pressure is suddenly lowered, however, the hydrogen boils violently. If high-speed charged particles are allowed to pass through the hydrogen at this same instant, ions are formed. The hydrogen boils a few thousandths of a second sooner around these ions than it does in the rest of the chamber. Carefully timed photos, such as the one in Figure 25-19, are used to record the resulting bubble trails.

The advantage of the bubble chamber over the cloud chamber is that more nuclear collisions take place in the dense liquid of the bubble chamber. This difference makes it possible to study particles and rays having speeds that are much higher than those that produce recordable events in a cloud chamber.

Figure 25-19. This photograph shows tracks of subatomic particles in a liquid hydrogen bubble chamber. Two charged particles arise from the collision shown at the upper left. One of these undergoes an additional collision at lower left center. The straight tracks were made by protons, some of which collided with electrons of the hydrogen atoms in the chamber. These electrons move in spiral paths under the influence of the strong magnetic field that is acting on the bubble chamber.

25.12 Electronic Bubble Chambers In 1962, the American physicist S. J. Lindenbaum (b. 1925) and his colleagues at Brookhaven National Laboratory developed a new technique that resulted in an electronic version of the bubble chamber. This detector, an exterior view of which is shown in Figure 25-20, is almost universally used today in high-energy physics experiments. The Lindenbaum detector consists of a series of picket-fence wire planes immersed in a gas such as argon. When a charged particle passes near one of the wires, it causes a signal that is fed into a computer. The computer then reconstructs and analyzes the events in the chamber in three dimensions. Such electronic chambers are more than 10,000 times faster than other particle detectors, and they have better resolution and identification capabilities.

Figure 25-20.

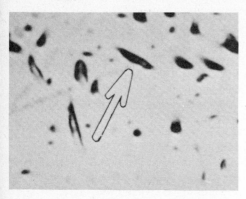

Figure 25-21. By etching fragments of lunar rocks collected by American astronauts, scientists have discovered that the rocks are marked with tracks left by cosmic rays. The arrow points to a typical cosmic-ray track.

25.13 Solid-State Detectors When certain nonconducting solids are exposed to charged nuclear particles, tracks that can be seen with an electron microscope are produced in the solids. The tracks are regions of distortion of the crystal lattices of the solids. Etching the solids with acid makes the tracks more permanent and more readily visible.

Cosmic rays and various radioactive decay products have left tracks in naturally occurring solids since the origin of the solar system. See Figure 25-21. Through the etching technique just described, scientists have examined these fossil tracks in rocks, in ancient artifacts, in meteorites, and in lava from the ocean floor. Important clues about the earth's history have been gathered from these studies. The same technique was used on lunar samples.

One of the first solid-state detectors made was a thin sheet of mica that was used by two British scientists in 1959 in connection with a fission experiment. Glass and certain plastics have also been used as detectors. Plastics are particularly sensitive to slow protons and alpha particles.

Each solid used as a solid-state detector has a threshold below which no tracks are produced. This property makes the detector particularly suited for the study of high-energy particles; the dense background of light-particle tracks that is present in other types of detectors is eliminated by the threshold property of the solid-state detector.

Another type of solid-state detector, the *junction detector*, is based on the transistor principle. See Figure 25-22. An applied bias removes all free carriers from the detector. When charged particles enter the device, ionization is produced in the base and an output signal results. In essence, the junction detector is a solid-state ionization chamber. However, the junction detector is smaller in size, requires less power, has a faster response, is completely insensitive to magnetic fields, and produces an output signal that is directly proportional to the energy of the incident particle.

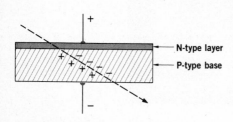

Figure 25-22. A junction detector. The passage of a charged particle (represented by the arrow) ionizes a region of the **P**-type base and produces an output signal.

Questions
GROUP A

1. (a) Which detection instrument was the first to record the effects of nuclear radiation? (b) Describe the event.
2. (a) What is the derivation of the term "spinthariscope"? (b) How does this instrument detect subatomic particles?
3. Distinguish between a scintillation counter and a spinthariscope.

4. List one similarity and one difference between a Geiger tube and an electroscope.
5. Describe the construction of a simple cloud chamber.
6. (a) What substance is commonly used in a bubble chamber? (b) Why?
7. (a) How does an electronic bubble chamber work? (b) List two ways in which it is superior to a liquid bubble chamber.

8. Explain the property of solid-state detectors that makes them particularly suited for the study of high-energy particles.

SUBATOMIC REACTIONS

25.14 Classification of Subatomic Particles   Through the use of the detection devices and particle accelerators just described, physicists have released or produced more than two hundred different subatomic particles. Almost all these particles are unstable, and many of them have extremely short half-lives. Nevertheless, these fleeting fragments of matter are essential pieces in the puzzle of nuclear structure. In the remainder of this chapter, we shall look at the picture of the atom that is emerging from these investigations.

More than one hundred years ago, scientists recognized relationships among the chemical elements that made it possible to arrange them in groups according to their properties. This arrangement is called the *periodic table of the chemical elements.* As more and more subatomic particles were discovered, it became evident that a similar grouping could be made of the subatomic particles. In 1957 such a table was proposed by the American physicist Murray Gell-Mann (b. 1929). Working independently, the Japanese scientist Kazuhiko Nishijima (b. 1926) also proposed such a table.

Table 25-2 shows the main features of the Gell-Mann/Nishijima table of subatomic particles. Although not all known particles in each category are listed, enough of them are shown so that the purpose of the classification is evident.

The names and symbols of subatomic particles are taken from Greek words and letters. The names refer to the masses: *baryon* (*berry*-on) means heavy, *meson* (*mez*-on) means medium, and *lepton* (*lep*-ton) means small. Baryons and mesons are also called *hadrons* (*had*-rons); hadron comes from a Greek word meaning bulky. *Bosons* (*boz*-ons) are named after the Indian physicist Satyendra Bose (1894–1974). The four major divisions of the table were first based largely on mass. But, this distinction could not be strictly applied as new particles were discovered.

The masses of the subatomic particles are given as the energy equivalents (in Mev) of the rest masses. Neutrinos, photons, and gravitons are given a mass of zero because they do not exist as stationary entities.

The main reasons for the construction of ever-larger particle accelerators are to study the interrelationships between nuclear particles and to explore the possibility that

The first useful periodic table of the chemical elements was devised by the Russian chemist Dmitri Mendeleyev (1834–1907).

In 1969 Gell-Mann received the Nobel Prize in Physics for his work with subatomic particles.

Baryon numbers, lepton numbers, and hypercharge will be explained in Section 25.16.

Table 25-2
SUBATOMIC PARTICLES

Category	Name	Symbol	Antiparticle	Mass (Mev)	Half-life (s)
baryons	omega hyperon	Ω^-	Ω^+	1676	1.3×10^{-10}
	xi hyperon	Ξ^- Ξ^0	$\overline{\Xi^-}$ $\overline{\Xi^0}$	~1320 ~1310	1.2×10^{-10} 2.1×10^{-10}
	sigma hyperon	Σ^- Σ^0 Σ^+	$\overline{\Sigma^-}$ $\overline{\Sigma^0}$ $\overline{\Sigma^+}$	~1190 ~1190 ~1190	1.0×10^{-10} $\sim 10^{-20}$ 5.5×10^{-11}
	lambda hyperon	Λ^0	$\overline{\Lambda^0}$	1115	1.8×10^{-10}
	neutron	n	$\overline{n}$	940	700
	proton	p	$\overline{p}$	938	stable*
mesons	eta meson	η^0	$\overline{\eta^0}$	549	$\sim 10^{-20}$
	kaon (kappa meson)	K^0 K^+	$\overline{K^0}$ $\overline{K^+}$	498 494	8.5×10^{-7} 6.4×10^{-9} 3.5×10^{-8}
	pion (pi meson)	π^+ π^0	π^- π^0	140 135	1.0×10^{-8} 0.6×10^{-16}
leptons	muon (mu meson)	μ^-	μ^+	105.6	1.5×10^{-6}
	electron	e^-	e^+	0.5	stable
	mu neutrino	ν_μ	$\overline{\nu_\mu}$	0	stable
	e-neutrino	ν_e	$\overline{\nu_e}$	0	stable
bosons	photon	γ	γ	0	stable
	graviton	g	g	0	stable

*See Section 25.18.

The electron-positron pair that was discussed in Section 23.3 is an example of a particle and its antiparticle.

they are, in turn, composed of still simpler particles. Discoveries made with high-energy accelerators often require the revision of existing theories about the structure of matter. These studies of the building blocks of the universe are often referred to as "modern physics."

For many years, nuclear physicists have speculated on the interesting idea that every particle in the universe has a mirror-image counterpart. The table of subatomic particles suggests this idea by listing an antiparticle for each particle. When a particle and its antiparticle collide, all physical quantities associated with the particle or antiparticle (mass, charge, etc.) disappear and pure energy is left.

In 1981, scientists in Geneva, Switzerland, produced the first proton-antiproton collisions in a particle accelerator.

A substance composed of antiparticles is called **antimatter.** The collision of matter with antimatter results in the annihilation of matter. Such a collision illustrates the interchangeability of matter and energy. Scientists believe that large quantities of antimatter may exist in the universe in the form of stars or even entire galaxies (or antistars and antigalaxies, to be more exact).

25.15 Particle Interactions   As we saw in Section 4.1, a force produces a change in the state of motion of an object. In quantum mechanics, the broader concept of "interaction" replaces the Newtonian concept of force. *An* **interaction** *is any change in the amount or quantum numbers of particles that are near each other.* For example, an interaction occurs when an electron and a positron collide and change into two photons. An interaction is not merely another word for force. In an interaction, the identity of the particles colliding is changed as new particles are created and annihilated. The intermediate particles of an interaction are sometimes called "carriers."

1. *Strong interactions.* The nature of the strong interaction was discussed in Section 23.12. The strong interaction holds the particles of the nucleus together against the electromagnetic repulsion of the nucleus' similarly charged particles. The strong interaction is independent of charge. In other words, the attraction between two neutrons is the same as that between a proton and a neutron or between two protons.

Although the strong interaction is the strongest of all interactions, it has a very limited range. It is not effective beyond a distance of 2.0×10^{-5} Å. The carrier of the strong interaction is believed to be the meson; there is a constant exchange of mesons among the baryons of a stable nucleus. When the baryons are separated by the intervention of a high-energy particle from an accelerator, the carrier mesons are released.

The production and exchange of particles during an interaction is usually represented by a *Feynman diagram,* named after the American physicist Richard Feynman (b. 1918). In Figure 25-23, a Feynman diagram illustrates the reaction that takes place during the close approach of a proton and a neutron, the reaction in which a meson is emitted and absorbed. This carrier meson transfers the energy that is responsible for the change in momentum of the two baryons. The explanation of the strong "force" by meson exchange is the main reason the process is now

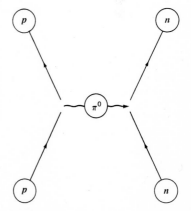

Figure 25-23. A Feynman diagram. In this interaction, a neutral pi meson is exchanged during the close approach of a proton and a neutron. The energy transferred by the carrier meson changes the momenta of the two baryons.

called an interaction. In 1935 the Japanese physicist Hideki Yukawa (1907–1981) proposed the theory of the existence of energy-carrying mesons to explain the strong interaction. But mesons were not actually detected until 1947.

2. Electromagnetic interactions. These interactions differ from other interactions in several ways. (1) Electromagnetic interactions can either attract or repel the objects on which they are acting. The other three interactions are attractive only. (2) Electromagnetic interactions can act in directions other than along a straight line between objects. (3) The electromagnetic interactions depend on the relative velocities of particles. This is not true of other interactions. The carrier of the electromagnetic interaction is the photon.

Electromagnetism keeps electrons in orbit around the nucleus. The same interaction is also responsible for the bonding between atoms to form molecules and between molecules to form larger substances. Nevertheless, the electromagnetic interaction is less than 10^{-2} as strong as the strong interaction.

The elastic collisions that were discussed in Section 6.14 are examples of electromagnetic interactions between the orbital electrons of the atoms that make up the colliding objects.

3. Weak interactions. This is the most mysterious of the known interactions between particles. As the study of the nucleus progressed and the number of subatomic particles grew, a "strange" principle was discovered. Some particles were always produced in pairs, such as lambda hyperon and K meson, and these particles had much longer half-lives than was expected.

To explain the behavior of these so-called strange particles, the "weak interaction" was postulated. Its ratio to the strong interaction is about 10^{-13} to 1, and so it affects only leptons. Beta decay is an example of a weak interaction.

The carrier for weak interactions is the W particle. Discovery of the W particle was announced in 1982 by scientists working with the intersecting storage accelerator in Geneva, Switzerland.

4. Gravitational interactions. Gravitation was the first of the interactions to be studied, and yet it is the weakest of them all. Its ratio to the strong interaction is only 10^{-38} to 1. Gravitation has no known distance limitations, and so its effects are easily observed in the behavior of large objects on the earth and in the motions of celestial bodies. Within the atomic nucleus, the effect of gravitational interactions is negligible in comparison with the other interactions that are present.

The carrier for the gravitational interaction has been named the *graviton*. Like the photon, its rest mass is thought to be zero. Because they are very weak, gravity waves are difficult to detect. Scientists at the University of Maryland used 1000-kg aluminum cylinders as receivers for gravity waves from certain massive stars, but the results were inconclusive. In 1978, astrophysicists from the University of Massachusetts announced that they had verified the existence of gravity waves after four years of observations with a large radio telescope in Puerto Rico. See Figure 25-24. Aimed at a pair of massive stars that are orbiting each other, the telescope recorded a cumulative delay of 4×10^{-4}s in the orbital period, which is the amount predicted by the theory of relativity if gravity waves are draining energy from the two stars.

25.16 Subatomic Conservation Laws Repeatedly in your study of physics, the importance of the conservation principles has been emphasized. The conservation of mass-energy, the conservation of linear and angular momentum, the conservation of electric charge—these laws are basic to an understanding of all reactions above the atomic scale. Similar laws hold true in subatomic interactions. We have already seen in Section 24.4, for example, that both mass-energy and electric charge are conserved in nuclear reactions. As other subatomic particles were discovered, additional conservation laws were formulated.

1. *Conservation of baryons.* When a baryon decays or reacts with another particle, the number of baryons is the same on both sides of the equation. For example, if a neutron is left by itself, it will decay as follows:

$$n^0 \rightarrow p + e^- + \bar{\nu}$$
$$\text{Baryon No.} = +1 = +1 \quad +0 \quad +0$$

The neutron and proton are both baryons, while the electron and antineutrino have baryon numbers of zero. Consequently, the baryon number of the equation is +1 on each side.

The conservation of baryons keeps the universe from deteriorating into mesons and leptons. If such a deterioration could happen, atoms as we know them would no longer exist and matter would become a rather homogeneous mixture of a few kinds of low-mass particles.

2. *Conservation of leptons.* The decay of the neutron illustrates another nuclear conservation principle. The lepton number of the neutron and proton is zero. The lepton number of the electron is +1. The lepton number of the

Figure 25-24. This 305-meter receiver near Arecibo, Puerto Rico, is the world's largest single radio telescope. It was used to confirm the existence of gravity waves in an experiment that extended over a 4-year period.

antineutrino is −1, since it is an antiparticle. Consequently, the arithmetic sum of the lepton numbers is the same on each side of the equation. This principle holds true for every interaction in which leptons are involved.

$$n^0 \rightarrow p + e^- + \bar{\nu}$$

Lepton No. $= 0 = 0 \quad +1 \quad -1$

Lepton conservation explains the necessity of pair production, which was considered "strange" when it was first discovered. When a lepton is created in an interaction, a particle with the opposite lepton number must also be created in order to conserve the lepton value of the reaction. Similarly, a lepton can be annihilated only by interacting with a particle of opposite lepton value.

The conservation of baryons and leptons does not prevent the change of a particle to another particle within the same category, however. In neutron decay, one kind of baryon (the neutron) changed into another kind (the proton). The decay of the antimuon is another example.

$$\mu^+ \rightarrow e^+ + \nu - \bar{\nu}$$

Lepton No. $= -1 = -1 \quad +1 \quad -1$

3. *Conservation of hypercharge.* Some baryons and some mesons have a property called *hypercharge.* Hypercharge can be either positive or negative. The hypercharge sign of a particle is the opposite of the hypercharge sign of its antiparticle. Hypercharge is conserved in strong and electromagnetic interactions, but it is not conserved in weak interactions.

Consider the following interaction involving hypercharge:

$$\pi^+ + p \rightarrow K^+ + \Lambda^0$$

Hypercharge $= 0 \quad +1 = +1 \quad +0$

The identity of the original particles has changed completely, but the sum of the hypercharges is +1 on each side of the reaction. (The sum of the baryon numbers is also +1 on each side.)

4. *Conservation of parity.* The distribution of matter in space is often described in terms of *parity. The **law of parity** states that for every process in nature there is a mirror-image process that is indistinguishable from the original process.* Another way to state this law is to say that there is no absolute and independent way of describing right and left, clockwise and counterclockwise, in the universe. Any experiment devised to establish one of these directions is also matched by an equally probable event in the opposite direction.

Parity experiments deal largely with the spin of nuclear particles. Spin is also responsible for the direction in which particles are emitted in subatomic interactions. The conservation of parity stipulates that particles will be emitted in opposite directions in equal numbers since nature shows no preference for spin direction. But in 1956 a series of experiments based on predictions by the Chinese physicists Tsung-Dao Lee (b. 1926) and Chen Ning Yang (b. 1922) showed that this was not the case. In the case of weak interactions involving beta decay, parity was not conserved!

Experiments were specifically designed to test the conservation of parity in other interactions. The question was whether parity is still conserved in a larger sense when particles are replaced by their antiparticles in subatomic interactions. In 1966 physicists at the Brookhaven National Laboratory in New York discovered a violation of parity with antiparticles as well. So perhaps there is an absolute sense of direction in the universe after all.

25.17 Quarks In 1963 Murray Gell-Mann and the American physicist George Zweig (b. 1937) proposed the theory that hadrons are made up of still smaller particles called *quarks*, each having a charge of $\pm\frac{1}{3}$ or $\pm\frac{2}{3}$. According to this theory, baryons are composed of three quarks each and mesons contain two quarks each. Three different quarks were postulated with the designations of "up," "down," and "strange." For each of these three quarks there is also an antiquark.

In 1974 two subnuclear particles were discovered that have unusually large masses and long half-lives. These so-called J/psi particles could not be explained with only three quarks, and so three more were eventually added to the list. These were called "charm," "truth," and "beauty." Also, in order to avoid possible violations of the exclusion principle in combining two and three quarks within a hadron, each quark is assigned one of three "colors": "red," "blue," and "green." (The "color" of a quark is the subnuclear counterpart of the quantum number of an electron.) In keeping with the terminology of the colors of visible light, the three "colors" of antiquarks are "cyan," "yellow," and "magenta."

The carrier between quarks is the *gluon*. It is the exchange of "color" between quarks by gluons that provides the binding force between quarks within a hadron. Strangely enough, the gluon-grip between quarks is thought to increase, rather than decrease, with distance. That is probably why present particle accelerators have

Figure 25-25. The Chinese physicists C. N. Yang and T. D. Lee shared the Nobel Prize in Physics in 1957 for their hypothesis that parity is not conserved in all nuclear reactions.

The word "quark" does not have a specific meaning. It appears in the novel Finnegans Wake *by James Joyce.*

Figure 25-26. The American physicists Richard Feynman (left) and Murray Gell-Mann. They won the Nobel Prize in Physics in 1965 and 1969, respectively, for their work with subatomic particles.

not been able to isolate individual quarks. According to one theory, there may even be as many as eight different gluons, each with a characteristic "color."

25.18 Unified Field Theories   In 1974 it was found that the subatomic conservation laws are not completely inviolate at high energies. At 300 Gev, collisions of electrons and positrons produced hadrons and vice versa. This suggests that the weak interaction is not a constant and that under certain conditions it may become stronger than the electromagnetic interaction.

One explanation of this phenomenon is that colliding leptons exchange a so-called Z particle. This interaction is known as a *neutral weak current* (since there is no change in the charges of the colliding particles).

The discovery of neutral weak currents is an important step toward a *unified field theory*, which combines two or more of the interactions in the universe into a single concept, much as Maxwell's electromagnetic theory combined electricity and magnetism in the nineteenth century. A unified field theory proposed in 1967 combined the weak and electromagnetic interaction, and the neutral weak current helps to substantiate that concept.

Figure 25-27. The four known interactions and their carriers. Unified field theories are attempts to combine two or more of these phenomena into a single concept.

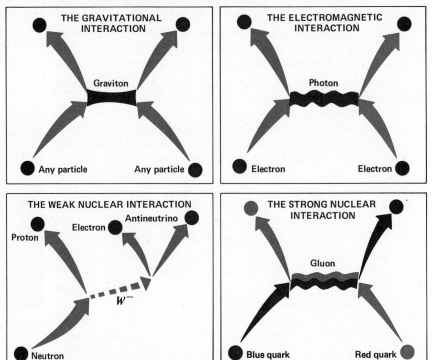

Efforts are presently being made to include the strong interaction in a unified field theory (or *grand* unified field theory, as it has been called). Virtually every version of such a theory predicts the startling idea that the proton is unstable! The half-life of the proton, in this scheme, is more than 10^{30} years. Experiments to detect the decay of a proton are being carried out in several countries; and in 1982, physicists in Europe announced that such an event was recorded in a 134-ton iron detector inside a tunnel under the Alps. Even with this massive shielding, however, it is possible that the recorded nuclear event was produced by a cosmic ray and not by the spontaneous decay of a proton.

*The acronym **GUTS** has been used in discussing grand unified theories.*

A successful grand unified field theory would combine the three nuclear interactions, but it would still not include gravitation. Until such a supertheory is found and successfully tested, no one really knows what the ultimate model of matter and its interactions will be, or even if such a model can ever be formulated. But one thing is certain: the search for new knowledge in high-energy physics requires keen and open minds and a close correlation between theory and technology.

According to theory, an average of one proton will decay in every human being during a lifetime.

Figure 25-28. The American physicists Sheldon Glashow and Steven Weinberg at a news conference after they won the 1979 Nobel Prize in Physics for their unified field theory of nuclear interactions. The third recipient was Abdus Salam of Pakistan.

Questions
GROUP A

1. (a) Who proposed the table of subatomic particles? (b) What other scientific table does it resemble?
2. Define: (a) baryon, (b) meson, (c) lepton, (d) hadron, (e) boson.
3. (a) What is antimatter? (b) What happens when a particle collides with its antiparticle?
4. (a) List the four types of interactions in the universe in order of their increasing magnitude. (b) Name the carrier of each interaction. (c) Which of these interactions are effective only within the size range of the atomic nucleus?
5. Distinguish between a Newtonian force and a subatomic interaction.
6. (a) What are Feynman diagrams? (b) How are they used to describe subatomic interactions?
7. (a) List four kinds of conservation in subatomic interactions. (b) List two properties of matter that are not always conserved on the subatomic scale.

8. How does the conservation of baryons prevent the collapse of matter?
9. What is the relationship between symmetry and parity?
10. What is meant by a unified field theory?
11. Discuss the quark theory. Why, when, and by whom was it proposed?
12. List the various types of quarks.

13. How does the quark theory avoid violations of the exclusion principle?
14. Distinguish between a unified field theory and a grand unified field theory.
15. What is the role of the gluon in the strong interaction?
16. Why are proton-decay experiments carried out underground?

SUMMARY

Modern atomic models take into account the uncertainty principle, which states that the exact position and momentum of an object cannot be simultaneously specified. The allowable values of physical quantities in modern atomic theory are given in terms of quantum numbers. The first quantum number for the electron describes its orbital radius. The second quantum number gives its angular momentum. The third quantum number specifies its direction in terms of a magnetic field. The fourth quantum number designates the spin of the electron. The exclusion principle states that no two electrons in an atom may have the same set of quantum numbers.

The stability of an electron is described in terms of ionization energy, which is the energy required to remove an electron from its orbit. All the equations for quantum numbers include Planck's constant. This indicates that all matter has wave characteristics.

Particle accelerators fall into three main categories: Van de Graaff accelerators, circular accelerators, and linear accelerators. In Van de Graaff generators, energies of up to 20 Mev may be produced. Betatrons, cyclotrons, synchrotrons, and intersecting storage rings are types of circular accelerators. The world's largest accelerator is a synchrotron, in which proton energies of 1000 Gev are achieved after several preliminary stages of acceleration in other devices. Intersecting storage accelerators are more efficient than other types of circular accelerators. Linear accelerators avoid the radiation losses encountered in circular accelerators.

Subatomic particles can be detected and studied because they are able to ionize certain substances or to produce visible paths. Low-energy detectors include the electroscope, the spinthariscope, the scintillation counter, the ionization chamber, the Geiger tube, and the cloud chamber.

The bubble chamber detects high-energy particles by producing momentary trails of hydrogen or helium gas, which are then photographed. Electronic detectors use computers to analyze nuclear events in three dimensions. Solid-state detectors are more selective than bubble chambers. Some solid-state detectors work on the transistor principle.

The number of identified subatomic particles now exceeds two hundred. They can be classified into baryons, mesons, and leptons. Baryons and mesons are also called hadrons. Each particle has an antiparticle with opposite charge.

The four known subatomic interactions are strong, electromagnetic, weak, and gravitational. In strong interactions, the number of baryons and leptons remains constant. Hypercharge and parity are not conserved in weak interactions.

Hadrons are thought to consist of quarks. Six kinds of quarks, each having one of three "colors," have been postulated. Efforts are also being made to combine the four known interactions in the universe into a grand unified field theory.

VOCABULARY

antimatter
antiparticle
baryon
betatron
bevatron
boson
bubble chamber
cloud chamber
cosmotron
cyclotron
dees
drift tubes
electromagnetic interaction
electroscope
energy level
exclusion principle
Feynman diagram
Geiger tube
gravitational interaction

graviton
hypercharge
interaction
ionization chamber
ionization energy
junction detector
law of conservation of
   baryons
law of conservation of
   hypercharge
law of conservation of
   leptons
lepton
linear accelerator
matter wave
meson
neutral weak current
orbital
photon

principle of parity
quantum mechanics
quantum number
quantum theory
quark
solid-state detector
spinthariscope
storage ring
strangeness
strong interaction
synchrocyclotron
synchrotron
uncertainty principle
unified field theory
Van de Graaff generator
W particle
weak interaction
Zeeman effect

Mathematics Refresher

1. Introduction.

The topics selected for this Mathematics Refresher are those that sometimes trouble physics students. You may want to study this section before working on certain types of physics problems. The presentation here is not as detailed as that given in mathematics textbooks, but there is sufficient review to help you perform certain important mathematical operations.

The following references are given for your convenience in reviewing these topics:

Section 2.8 defines *significant figures* and gives rules for their notation.

Section 2.9 explains the *scientific notation* system.

Section 2.12 describes the *orderly procedure* you should use *in problem solving*, and the proper method of handling units in computations.

2. Conversion of a fraction to a decimal.

Divide the numerator by the denominator, expressing the answer in the required number of significant figures.

Example

Convert $\frac{45}{85}$ to a decimal.

Solution

Dividing 45 by 85, we find the equivalent decimal to be 0.53, rounded to two significant figures.

$$
\begin{array}{r}
0.529 \\
85\overline{)45.000} \\
42\,5 \\
\overline{2\,50} \\
1\,70 \\
\overline{800} \\
765
\end{array}
$$

3. Calculation of percentage.

Uncertainty always exists in measurements of physical phenomena. Uncertainty exists because measuring instruments have limited precision. The instruments you use in laboratory experiments generally have less precision than those used by the physicists who obtained the values found in tables. *The precision of any measurement is limited to the measurement detail that the instrument is capable of providing.* In physics, an error is defined as the difference between experimentally obtained data and the accepted values for these data. In experimental work, the *relative error* (percentage error) is usually more meaningful than the *absolute error* (the actual difference between an observed value and its accepted value). Hence, as a physics student you must be able to calculate percentages and percentage errors, and convert fractions to percentages and percentages to decimals. Examples of these calculations are given in the following sections.

4. Conversion of fractions to percentages.

To convert a fraction to a percentage, divide the numerator by the denominator and multiply the quotient by 100%.

Example
Express $\frac{11}{13}$ as a percentage.

Solution
$\frac{11}{13} \times 100\% = 85\%$, rounded to two significant figures.

5. Conversion of percentages to decimals.

To convert a percentage to a decimal, move the decimal point two places to the left, and remove the percent sign.

Example
Convert 62.5% to a decimal.

Solution
If we express 62.5% as a fraction, it becomes 62.5/100. If we actually perform the indicated division, we obtain 0.625 as the decimal equivalent. Observe that the only changes have been to move the decimal point two places to the left, and remove the percent sign.

6. Calculation of relative error.

$$\text{Relative Error} = \frac{\text{Absolute Error}}{\text{Accepted Value}} \times 100\%$$

where the *absolute error* is the difference between the *observed value* and the *accepted value*.

Example
In a laboratory experiment carried out at 20.0°C, a student found the speed of sound in air to be 329.8 m/s. The accepted value at this temperature is 343.5 m/s. What was the relative error?

Solution
Absolute Error = 343.5 m/s − 329.8 m/s
= 13.7 m/s

$$\text{Relative Error} = \frac{13.7 \text{ m/s}}{343.5 \text{ m/s}} \times 100\% = 3.99\%$$

7. Proportions.

Many physics problems involving temperature, pressure, and volume relationships of gases may be solved by using proportions.

Example
Solve the proportion $\frac{V}{225 \text{ mL}} = \frac{273°\text{C}}{298°\text{C}}$ for V.

Solution
Multiply both sides of the equation by 225 mL, the denominator of V, which yields
$$V = \frac{225 \text{ mL} \times 273°\text{C}}{298°\text{C}}. \text{ Solving, } V = 206 \text{ mL}.$$

8. Fractional Equations.

The equations for certain problems involving lenses, mirrors, and electric resistances produce fractional equations where the unknown is in the denominator. To solve such equations, clear them of fractions by multiplying each term by the lowest common denominator. Then isolate the unknown and complete the solution.

Example

Solve $\dfrac{1}{d_o} + \dfrac{1}{d_i} = \dfrac{1}{f}$ for f.

Solution

The lowest common denominator of d_o, d_i, and f is $d_o d_i f$. Multiplying the fractional equation by this product, we obtain

$$\frac{d_o d_i f}{d_o} + \frac{d_o d_i f}{d_i} = \frac{d_o d_i f}{f} \quad \text{or} \quad d_i f + d_o f = d_o d_i$$

Thus, $f(d_o + d_i) = d_o d_i$ and $f = \dfrac{d_o d_i}{d_o + d_i}$

9. Equations.

When an equation is used in solving a problem and the unknown quantity is not the one isolated, the equation should be solved algebraically to isolate the unknown quantity required. Then the values (with units) of the known quantitites can be substituted and the indicated operations performed.

Example

From the equation for potential energy, $E_p = mgh$, we are to calculate h.

Solution

Before substituting known values for E_p, m, and g, the unknown term h is isolated and expressed in terms of E_p, m, and g. This is accomplished by dividing both sides of the basic equation by mg.

$$\frac{E_p}{mg} = \frac{mgh}{mg} \quad \text{or} \quad h = \frac{E_p}{mg}$$

If E_p is given in joules, m in kilograms, and g in meters/second^2, the unit of h is

$$h = \frac{j}{kg\ m/s^2} = \frac{kg\ m^2/s^2}{kg\ m/s^2} = m \qquad \text{Thus } h \text{ is expressed in meters.}$$

10. Laws of exponents.

Operations involving exponents may be expressed in general fashion as $a^m \times a^n = a^{m+n}$; $a^m \div a^n = a^{m-n}$; $(a^m)^n = a^{mn}$; $\sqrt[n]{a^m} = a^{m/n}$. For example:

$$x^2 \times x^3 = x^5 \quad 10^5 \div 10^{-3} = 10^8 \quad (t^2)^3 = t^6 \quad \sqrt[4]{4^2} = 4^{2/4} = 4^{1/2} = 2$$

11. Quadratic equation.

The two roots of the quadratic equation $ax^2 + bx + c = 0$ in which a does not equal zero are given by the quadratic equation

$$x = \frac{-b \pm \sqrt{b^2 - 4ac}}{2a}$$

12. *Triangles.*

In physics, triangles are used to solve problems about forces and velocities.

1. *30°—60°—90° right triangle.* It is useful to remember that the hypotenuse of such a triangle is twice as long as the side opposite the 30° angle. The length of the side ópposite the 60° angle is $l\sqrt{3}$, where l is the length of the side opposite the 30° angle.

2. *45°—45°—90° right triangle.* The sides opposite the 45° angles are equal. The length of the hypotenuse is $l\sqrt{2}$, where l is the length of a side.

3. *Trigonometric functions.* Trigonometric functions are ratios of the lengths of sides of a right triangle and depend on the magnitude of one of its acute angles. Using right triangle **ABC**, below, we define the following trigonometric functions of $\angle$**A**:

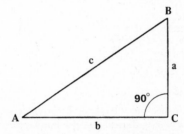

$$\text{sine } \angle \mathbf{A} = \frac{\mathbf{a}}{\mathbf{c}}$$

$$\text{cosine } \angle \mathbf{A} = \frac{\mathbf{b}}{\mathbf{c}}$$

$$\text{tangent } \angle \mathbf{A} = \frac{\mathbf{a}}{\mathbf{b}}$$

These functions are usually abbreviated as sin **A**, cos **A**, and tan **A**. Appendix B, Table 6, gives values of trigonometric functions.

Example

In a right triangle the length of one side is 25.3 cm and the length of the hypotenuse is 37.6 cm. Find the angle between these two sides.

Solution

Using the designations in the triangle shown above, **b** = 25.3 cm and **c** = 37.6 cm. You are to find $\angle$**A**. Hence you will use the trigonometric function cos **A** = **b/c**, or cos **A** = 25.3 cm/37.6 cm = 0.673. From Appendix B, Table 6, the cosine of 47.5° is 0.676 and the cosine of 48.0° is 0.669. The required angle is 3/7 × 0.5°, or 0.2°, greater than 47.5°. Thus $\angle$**A** = 47.7°.

Example

In a right triangle, one angle is 23.8° and the length of the adjacent side (not the hypotenuse) is 43.2 cm. Find the length of the other side.

Solution

Again using the designations in the above triangle, $\angle$**A** = 23.8° and **b** = 43.2 cm. The function involving these values and the unknown side, **a**, is tan **A** = **a/b**. Solving for **a**: **a** = **b** tan **A**. From Appendix B, Table 6, tan 23.8° = 0.441. Thus **a** = 43.2 cm × 0.441 = 19.1 cm.

4. *Sine law and cosine law.* For any triangle **ABC**, below, the sine law and the cosine law enable us to calculate the magnitudes of the remaining sides and angles if the magnitudes of one side and of any other two parts are given.

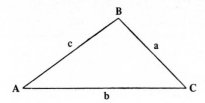

$$\text{Sine law: } \frac{a}{\sin A} = \frac{b}{\sin B} = \frac{c}{\sin C}$$

$$\text{Cosine law: } a = \sqrt{b^2 + c^2 - 2bc \cos A}$$
$$b = \sqrt{c^2 + a^2 - 2ca \cos B}$$
$$c = \sqrt{a^2 + b^2 - 2ab \cos C}$$

13. Circles.

The circumference c of a circle of radius r or diameter d is

$$c = 2\pi r = \pi d$$

The area A of a circle of radius r or diameter d is

$$A = \pi r^2 = \pi d^2/4 = \tfrac{1}{4}\pi d$$

14. Cylinders.

The volume V of a right circular cylinder of height h and radius r or diameter d is

$$V = \pi r^2 h = \pi d^2 h/4 = \tfrac{1}{4}\pi d^2 h$$

15. Spheres.

The surface area A of a sphere of radius r or diameter d is

$$A = 4\pi r^2 = \pi d^2$$

The volume V of a sphere of radius r or diameter d is

$$V = \tfrac{4}{3}\pi r^3 = \tfrac{1}{6}\pi d^3$$

16. Logarithms.

The common logarithm of a number is the exponent or the power to which 10 must be raised in order to obtain the given number. A logarithm is composed of two parts: the *characteristic*, or integral part; and the *mantissa*, or decimal part. The characteristic of the logarithm *of any whole or mixed number* is one less than the number of digits to the left of its decimal point. The characteristic of the logarithm *of a decimal fraction* is always negative and is numerically one greater than the number of zeros immediately to the right of the decimal point. Mantissas are always positive and are read from tables such as Appendix B, Table 24. Proportional parts are used when numbers having four significant figures are involved. In determining the mantissa, the decimal point in the original number is ignored since its position is indicated by the characteristic.

Logarithms are exponents and follow the laws of exponents:

Logarithm of a product = sum of the logarithms of the factors

Logarithm of a quotient = logarithm of the dividend minus logarithm of the divisor

To find the number whose logarithm is given, determine the digits in the number from the table of mantissas. The characteristic indicates the position of the decimal point.

Example

Find the logarithm of 35.76.

Solution

There are two digits to the left of the decimal point. Therefore the characteristic of the logarithm of 35.76 is one less than two, or 1. To find the mantissa, ignore the decimal point and look up 3576 in Appendix B, Table 24. Read down the *n* column to 35. Follow this row across to the 7 column, where you will find 5527. This is the mantissa of 3570. Similarly, the mantissa for 3580 is 5539. To calculate the mantissa for 3576, take 6/10 of the difference between 5527 and 5539, or 7, and add it to 5527. Thus, the required mantissa is 5534, and the complete logarithm of 35.76 is 1.5534.

Example

Find the logarithm of 0.4692.

Solution

There are no zeros immediately to the right of the decimal point, hence the characteristic of the logarithm of 0.4692 is −1. However, only the characteristic is negative; the mantissa is positive and is found from Appendix B, Table 24, to be 6714. The complete logarithm of 0.4692 may be written as $\bar{1}.6714$, or 0.6714 − 1, or 9.6714 − 10.

Example

Find the logarithm of (1) 357,600,000   (2) 0.0000003576

Solution

(1) $357,600,000 = 3.576 \times 10^8$ log = 8.5534
(2) $0.0000003576 = 3.576 \times 10^{-7}$ log = $\bar{7}$.5534, or 0.5534 − 7, or 3.5534 − 10

Note: Trigonometric functions and logarithms can also be obtained with "scientific" calculators.

Appendix A—Equations

(The number indicates the section in which the equation is introduced.)

1.9 Mass Density

$$\text{mass density} = \frac{\text{mass}}{\text{volume}}$$

1.14 Mass-Energy Relationship

$$E = mc^2$$

E is energy; m is mass; c is speed of light.

2.6 Relative Error

$$E_r = \frac{E_a}{A} \times 100\%$$

E_r is relative error; E_a is absolute error; A is accepted value.

2.11 Vector addition—The resultant of two vectors acting at an angle of between $0°$ and $180°$ upon a given point is equal to the diagonal of a parallelogram of which the two vectors are sides.

3.2 Speed

$$\text{Average Speed} = \frac{\text{displacement}}{\text{elapsed time}}$$

3.3 Velocity

$$v_{av} = \frac{d_f - d_i}{t_f - t_i} = \frac{\Delta d}{\Delta t}$$

v_{av} is average velocity; d_i is initial position at time t_i; d_f is final position at time t_f; Δd is displacement; Δt is elapsed time.

3.5 Acceleration

$$a_{av} = \frac{v_f - v_i}{t_f = t_i} = \frac{\Delta v}{\Delta t}$$

a_{av} is average accelaration; v_i is initial velocity at time t_i; v_f is final velocity at time t_f; Δv is change in velocity; Δt is elapsed time.

3.6 Uniformly Accelerated Motion

$$v_f = v_i + a\Delta t \quad \text{or} \quad v_f = v_i + g\Delta t$$

v_f is final velocity; v_i is initial velocity; a is acceleration or g is free-fall acceleration; Δt is elapsed time.

3.6 Accelerated Motion

$$\Delta d = v_i\Delta t + \tfrac{1}{2}a\Delta t^2 \quad \text{or} \quad \Delta d = v_i\Delta t + \tfrac{1}{2}g\Delta t^2$$

Δd is displacement; v_i is initial velocity; Δt is elapsed time; a is acceleration or g is free-fall acceleration.

3.6 Accelerated Motion

$$v_f = \sqrt{v_i^2 + 2a\Delta d} \quad \text{or} \quad v_f = \sqrt{v_i^2 + 2g\Delta d}$$

v_f is final velocity; v_i is initial velocity; a is acceleration or g is free-fall acceleration; Δd is displacement.

3.9 Newton's Second Law of Motion

$$F = ma$$

F is force; m is mass; a is acceleration.

3.9 Force and Acceleration on Bodies of Known Weight

$$F = \frac{F_w a}{g}$$

F is force; F_w is weight; a is acceleration; g is free-fall acceleration.

3.9 Weight and Mass

$$F_w = mg$$

F_w is weight; m is mass; g is free-fall acceleration.

3.11 Law of Universal Gravitation

$$F = \frac{G\,m_1 m_2}{d^2}$$

F is force of attraction; G is gravitational constant; m_1 and m_2 are masses of bodies; d is distance between their centers of mass.

3.12 Acceleration of Gravity

$$g = \frac{Gm_e}{d^2}$$

g is free-fall acceleration; G is gravitational constant; m_e is mass of earth; d is distance from center of earth.

4.7 Coefficient of Friction

$$\mu = \frac{F_f}{F_N}$$

μ is coefficient of friction; F_f is force of friction; F_N is force normal to surface.

5.2 Centripetal Force

$$F_c \quad \frac{mv^2}{r}$$

F_c is centripetal force; m is mass; v is velocity; r is radius of path.

5.3 Motion in Vertical Circle

$$v_{min} = \sqrt{rg}$$

v_{min} is critical velocity; r is radius of vertical circle; g is free fall acceleration.

5.6 Angular Velocity

$$\omega = \frac{\Delta\theta}{\Delta t}$$

ω is angular velocity; $\Delta\theta$ is angular displacement; Δt is elapsed time.

5.7 Angular Acceleration

$$\alpha = \frac{\Delta\omega}{\Delta t}$$

α is angular acceleration; $\Delta\omega$ is angular velocity; Δt is elapsed time.

5.7 Accelerated Rotary Motion

$$\omega_f = \omega_i + \alpha\Delta t$$
$$\Delta\theta = \omega_i \Delta t + \tfrac{1}{2}\alpha\Delta t^2$$
$$\omega_f = \sqrt{\omega_i^2 + 2\alpha\theta}$$

ω_f is final angular velocity; ω_i is initial angular velocity; α is angular acceleration; Δt is elapsed time; $\Delta\theta$ is angular displacement.

5.8 Rotational Inertia

$$T = I\alpha$$

T is torque; I is rotational inertia; α is angular acceleration.

5.8 Torque

$$T = Fr$$

T is torque; F is tangential force; r is distance from pivot point to point of application of force.

5.12 Pendulum

$$T = 2\pi\sqrt{\frac{l}{g}}$$

T is period; l is length; g is free-fall acceleration.

6.1 Work

$$W = F\Delta d$$

W is work; F is force; Δd is displacement.

6.3 Work (Rotary Motion)

$$W = T\Delta\theta$$

W is work; T is torque; $\Delta\theta$ is angular displacement.

6.4 Efficiency

$$\text{Efficiency} = \frac{W_{out}}{W_{in}}$$

W_{out} is work output; W_{in} is work input.

6.5 Power

$$P = \frac{F\Delta d}{\Delta t}$$

P is power; F is force; Δd is displacement; Δt is elapsed time.

6.6 Power (Rotary Motion)

$$P = T\omega$$

P is power; T is torque; ω is angular velocity.

6.7 Gravitational Potential Energy

$$E_p = mg\Delta h$$

E_p is gravitational potential energy; m is mass; g is free-fall acceleration; Δh is vertical distance.

6.8 Kinetic Energy

$$E_K = \tfrac{1}{2}mv^2$$

E_K is kinetic energy; m is mass; v is velocity.

6.9 Kinetic Energy (Rotary Motion)

$$E_K = \tfrac{1}{2}I\omega^2$$

E_K is kinetic energy; I is rotational inertia; ω is angular velocity.

6.10 Potential Energy (Elastic)

$$E_p = \tfrac{1}{2}k(\Delta d)^2$$

E_p is potential energy; k is elastic constant; Δd is displacement.

6.12 Impulse and Change of Momentum

$$F\Delta t = m\Delta v$$

F is force; Δt is elapsed time; the product $F\Delta t$ is impulse; m is mass; Δv is change of velocity; the product $m\Delta v$ is the change of momentum.

6.14 Elastic Collisions in One Dimension

$$mv_i + m'v_i' = mv_f + m'v_f'$$

m is mass of first body, v_i its initial velocity, v_f its final velocity; m' is mass of second body, v_i' its initial velocity, v_f' its final velocity.

6.15 Angular Impulse and Change of Angular Momentum

$$T\Delta t = I\omega_f - I\omega_i$$

T is torque; Δt is elapsed time; the product $T\Delta t$ is angular impulse; I is moment of inertia; ω_f is final angular velocity; ω_i is initial angular velocity; the product $I\omega$ is angular momentum.

7.11 Hooke's Law

$$Y = \frac{Fl}{\Delta lA}$$

Y is Young's modulus; F is distorting force; l is original length; Δl is change in length; A is cross-sectional area.

8.2 Kelvin Temperature

$$°K = °C + 273°$$

$°K$ is Kelvin temperature; $°C$ is Celsius temperature.

8.4 Linear Expansion

$$\Delta l = \alpha l\,\Delta T$$

Δl is change in length; α is coefficient of linear expansion; l is original length; ΔT is change in temperature.

8.5 Volume Expansion

$$\Delta V = \beta V\,\Delta T$$

ΔV is change in volume; β is coefficient of volume expansion; V is original volume; ΔT is change in temperature.

8.7 Charles' Law

$$\frac{V}{T_K} = \frac{V'}{T_K'}$$

V is original volume; T_K is original Kelvin temperature; V' is new volume; T_K' is new Kelvin temperature; provided pressure is constant.

8.8 Boyle's Law

$$pV = p'V'$$

p is original pressure; V is original volume; p' is new pressure; V' is new volume; provided temperature is constant.

8.8 Variation of Gas Density with Pressure

$$\frac{D}{D'} = \frac{p}{p'}$$

D is gas density at pressure p; D' is gas density at pressure p'.

8.9 Boyle's and Charles' Laws Combined

$$\frac{pV}{T_K} = \frac{p'V'}{T_K'}$$

p, V, and T_K are original pressure, volume, and Kelvin temperature, respectively; p', V', and T_K' are new pressure, volume, and Kelvin temperature, respectively.

8.10 Ideal Gas Equation

$$pV = nRT_K$$

p is pressure; V is volume; n is number of moles; R is universal gas constant; T_K is Kelvin temperature.

8.11 Heat Capacity

$$\text{heat capacity} = Q / \Delta T$$

Q is quantity of heat; ΔT is change in temperature.

8.12 Specific Heat

$$c = \frac{Q}{m \Delta T}$$

c is specific heat; Q is quantity of heat; m is mass of material; ΔT is change in temperature.

8.13 Heat Exchange

$$Q_1 = Q_g$$

Q_1 is heat lost; Q_g is heat gained.

9.6 Ideal Heat Engine Efficiency

$$\text{Efficiency} = \frac{T_1 - T_2}{T_1} \times 100\%$$

T_1 is input Kelvin temperature; T_2 is exhaust Kelvin temperature.

9.8 Change in Entropy

$$\Delta S = \Delta Q / T$$

ΔS is change in entropy; ΔQ is heat added or removed; T is Kelvin temperature.

10.6 Wave Equation

$$v = f \lambda$$

v is wave speed; f is frequency; λ is wavelength.

11.7 Intensity Level

$$\beta = 10 \log \frac{I}{I_0}$$

β is intensity level in db; I is intensity of sound; I_0 is intensity of threshold of hearing.

11.9 Doppler effect

1. Source moving toward stationary listener.

$$f_{LF} = f_s \frac{v}{v - v_s}$$

2. Source moving away from stationary listener.

$$f_{LB} = f_s \frac{v}{v + v_s}$$

3. Listener moving toward stationary source.

$$f_{LC} = f_s \frac{v + v_{LC}}{v}$$

4. Listener moving away from stationary source.

$$f_{LO} = f_s \frac{v - v_{LO}}{v}$$

f_{LF}, f_{LB}, f_{LC}, f_{LO} represent frequency of sound reaching the listener in the different listener-source conditions stated; f_s is the frequency of the source; v is the velocity of sound in the medium; v_s is the velocity of the source; v_{LC} is the closing velocity of the listener; v_{LO} is the opening velocity of the listener.

11.13 Laws of Strings

1. Law of Lengths

$$\frac{f}{f'} = \frac{l'}{l}$$

f and f' are frequencies corresponding to lengths l and l'.

2. Law of Diameters

$$\frac{f}{f'} = \frac{d'}{d}$$

f and f' are frequencies corresponding to diameters d and d'.

3. Law of Tensions

$$\frac{f}{f'} = \frac{\sqrt{F}}{\sqrt{F'}}$$

f and f' are frequencies corresponding to tensions F and F'.

4. Law of Densities

$$\frac{f}{f'} = \frac{\sqrt{D'}}{\sqrt{D}}$$

f and f' are frequencies corresponding to densities D and D'.

11.15 Resonance in Tubes

1. Closed Tube

$$\lambda = 4(l + 0.4d)$$

λ is wavelength; l is length of tube; d is diameter of tube.

2. Open Tube

$$\lambda = 2(l + 0.8d)$$

λ is wavelength; l is length of tube; d is diameter of tube.

11.16 Beats

$$f_{av} = \tfrac{1}{2}(f_l + f_h)$$

f_{av} is difference frequency; f_l is lower frequency; f_h is higher frequency.

12.9 Photon Energy

$$E = hf$$

E is energy; h is Planck's constant (6.63×10^{-34} joule sec); f is frequency.

12.10 Photoelectric Equation

$$\tfrac{1}{2}mv^2{}_{max} = hf - w$$

$\tfrac{1}{2}mv^2{}_{max}$ is maximum kinetic energy of photoelectrons; h is Planck's constant; f is frequency of impinging radiation; w is work function of emitting material.

12.11 Photon Wavelength

$$\lambda = \frac{h}{mc}$$

λ is photon wavelength; h is Planck's constant; c is velocity of light; mc is photon momentum.

12.11 Wavelength of Particle Having a Velocity v

$$\lambda = \frac{h}{mv}$$

λ is particle wavelength; h is Planck's constant; m is mass of particle; v is velocity; (mv is particle momentum).

12.17 Illumination

1. Uniformly Illuminated Surface

$$E = \frac{\Phi}{A}$$

E is illumination; Φ is luminous flux; A is area uniformly illuminated.

2. Surface Perpendicular to Luminous Flux

$$E = \frac{I}{r^2}$$

E is illumination; I is intensity of source; r is the distance from source to surface perpendicular to the beam.

3. Surface at Any Angle to Luminous Flux

$$E = \frac{I \cos \theta}{r^2}$$

E is illumination; I is intensity of source; θ is the angle which the light beam makes with the normal to the illuminated surface; r is the distance from source to surface.

13.12 Focal Length of Mirrors and Lenses

$$\frac{1}{f} = \frac{1}{d_o} + \frac{1}{d_i}$$

f is focal length; d_o is object distance; d_i is image distance.

13.12 Images in Mirrors and Lenses

$$\frac{h_i}{h_o} = \frac{d_i}{d_o}$$

h_o is object size; h_i is image size; d_o is object distance; d_i is image distance.

14.3 Snell's Law

$$n = \frac{\sin i}{\sin r}$$

n is index of refraction; i is the angle of incidence; r is the angle of refraction.

14.12 Simple Magnifier

$$M = \frac{25 \text{ cm}}{f} \text{ (approx.)}$$

f is focal length of lens in cm.

14.13 Compound Microscope

$$M = \frac{25 \text{ cm} \times l}{f_e \times f_o} \text{ (approx.)}$$

l is length of tube; f_e is focal length of eyepiece lens; f_o is focal length of objective lens in cm.

15.4 Diffraction Grating Equation

$$\lambda = \frac{d \sin \theta_n}{n}$$

λ is wavelength; d is grating constant; θ is diffraction angle; n is the order of image.

16.8 Coulomb's Law of Electrostatics

$$F = k \frac{Q_1 Q_2}{d^2}$$

F is force between two point charges; k is a proportionality constant; Q_1 and Q_2 are the two charges; d is distance separating the charges.

16.9 Electric field intensity

$$\mathscr{E} = \frac{F}{q}$$

$\mathscr{E}$ is the electric field intensity; F is force acting on a charge; q is quantity of charge.

16.10 Potential Difference

$$V = \frac{W}{q}$$

V is potential difference; W is work done in moving a charge; q is quantity of charge moved.

16.14 Capacitance of a Capacitor

$$C = \frac{Q}{V}$$

C is capacitance of a capacitor; Q is charge on either plate; V is potential difference between the plates.

16.17 Capacitors in Parallel

$$C_T = C_1 + C_2 + C_3 + \text{etc.}$$

C_T is total capacitance; C_1, C_2, C_3, etc. are separate capacitances connected in parallel.

16.17 Capacitors in Series

$$\frac{1}{C_T} = \frac{1}{C_1} + \frac{1}{C_2} + \frac{1}{C_3} + \text{etc.}$$

C_T is total capacitance; C_1, C_2, C_3, etc. are separate capacitances connected in series.

17.1 Electric Current

$$I = \frac{Q}{t}$$

I is current; Q is quantity of charge; t is time.

17.6 Ohm's Law of Resistance

1. Entire Circuit Including a Source of emf

$$E = IR \quad \text{or} \quad E = IZ$$

E is emf of source; I is current in the circuit; R is resistance of the circuit; Z is impedance of a-c circuit.

2. Any Part of a Circuit That Does Not Include a Source of emf

$$V = IR \quad \text{or} \quad V = IZ$$

V is potential difference across a part of the circuit; I is the current in that part of the circuit; R is the resistance of that part of the circuit; Z is the impedance of that part of an a-c circuit.

17.8 Resistances in Series

$$R_T = R_1 + R_2 + R_3 + \text{etc.}$$

R_T is the total resistance; R_1, R_2, R_3, etc. are separate resistances connected in series.

17.9 Resistances in Parallel

$$\frac{1}{R_{eq}} = \frac{1}{R_1} + \frac{1}{R_2} + \frac{1}{R_3} + \text{etc.}$$

R_{eq} is equivalent resistance; R_1, R_2, R_3, etc. are separate resistances connected in parallel.

17.11 Law of Resistance

$$R = \rho \frac{l}{A}$$

R is resistance; ρ is resistivity, a proportionality constant; l is length; A is cross-sectional area.

18.3 Joule's Law

$$Q = \frac{I^2 R t}{J}$$

Q is quantity of heat energy; I is current; R is resistance; t is time; J is the mechanical equivalent of heat.

18.4 Electric Power

1. Total Power Consumed in a Circuit

$$P_T = I_T^2 R_T \quad \text{or} \quad P_T = EI \cos \phi$$

P_T is total power; I_T is total current; R_T is total resistance; E is emf; ϕ is phase angle between current and voltage in an a-c circuit.

2. Power Expended in the Circuit Load

$$P_L = I^2 R_L \quad \text{or} \quad P_L = I^2 Z_L$$

P_L is load power; I is current; R_L is load resistance; Z_L is load impedance of an a-c circuit.

3. Power Dissipated in a Source of emf

$$P_r = I^2 r$$

P_r is power dissipated in the internal resistance of source; I is current; r is internal resistance of source.

18.9 Faraday's Laws of Electrolysis

$$m = zIt$$

m is mass; z is the electrochemical equivalent; I is current; t is time.

19.4 Magnetic Flux Density

$$B = \Phi / A$$

B is magnetic flux density; Φ is magnetic flux; A is area.

20.3 Induced emf *(Coil in a Magnetic Field)*

$$E = -N \frac{\Delta \Phi}{\Delta t}$$

E is induced emf; N is number of turns; $\Delta\Phi/\Delta t$ is the change in flux linkage in a given interval of time.

20.4 Induced emf *(Conductor in a Magnetic Field)*

$$E = Blv$$

E is induced emf; B is flux density of the magnetic field; l is length of conductor; v is velocity of conductor across magnetic field.

20.9 Instantaneous Voltage

$$e = E_{\max} \sin \theta$$

e is instantaneous voltage; $E_{\max}$ is maximum voltage; θ is displacement angle, the angle between the plane of the conducting loop and the perpendicular to the magnetic flux.

20.9 Instantaneous Current

$$i = I_{\max} \sin \theta$$

i is instantaneous current; $I_{\max}$ is maximum current; θ is displacement angle, the angle between the plane of the conducting loop and the perpendicular to the magnetic flux.

20.14 The Motor Effect

$$T = Fw \cos \alpha$$

T is torque; F is magnetic force acting on either conductor of loop; w is width of conducting loop; α is the angle between the plane of the loop and the magnetic flux.

20.18 Mutual Inductance

$$M = \frac{-E_S}{\Delta I_P / \Delta t}$$

M is mutual inductance; E_S is average induced emf across the secondary; $\Delta I_P / \Delta t$ is time rate of change of current in the primary.

20.19 Self-inductance

$$L = \frac{-E}{\Delta I / \Delta t}$$

L is self-inductance; E is the average emf induced across the coil; $\Delta I / \Delta t$ is time rate of change of current in the coil.

20.20 Inductors in Series

1. No Mutual Inductance Between Inductors

$$L_T = L_1 + L_2 + L_3 + \text{etc.}$$

L_T is total inductance; L_1, L_2, L_3, etc. are separate inductances connected in series.

2. *Mutual Inductance Between Two Inductors*

$$L_T = L_1 + L_2 \pm 2M$$

L_T is total inductance; L_1 and L_2 are inductances connected in series; M is mutual inductance between them.

20.20 Inductors in Parallel

$$\frac{1}{L_T} = \frac{1}{L_1} + \frac{1}{L_2} + \frac{1}{L_3} + \text{etc.}$$

L_T is total inductance; L_1, L_2, L_3, etc. are separate inductances connected in parallel so that each is unaffected by the magnetic field of the other.

20.21 Transformer

1. *Voltage Ratio*

$$\frac{V_S}{V_P} = \frac{N_S}{N_P}$$

V_S and V_P are secondary and primary terminal voltages; N_S/N_P is the turns ratio, secondary to primary.

2. *Current Ratio*

$$\frac{I_P}{I_S} = \frac{N_S}{N_P}$$

I_P and I_S are primary and secondary currents; N_S/N_P is the turns ratio, secondary to primary.

21.4 Power Factor

$$\text{pf} = \cos \phi$$

pf is power factor; ϕ is phase angle between voltage and current.

21.5 Inductive Reactance

$$X_L = 2\pi f L$$

X_L is inductive reactance; f is frequency; L is inductance.

21.8 Capacitive Reactance

$$X_C = \frac{1}{2\pi f C}$$

X_C is capacitive reactance; f is frequency; C is capacitance.

21.9 Impedance

$$Z = \sqrt{R^2 + X^2} \;\underline{|\arctan X/R}$$

Z is impedance; R is resistance; X is reactance $(X_L - X_C)$.

21.12 Resonance

$$f_r = \frac{1}{2\pi \sqrt{LC}}$$

f_r is resonant frequency; L is inductance; C is capacitance.

22.15 Transistor Characteristics

1. *Current gain, common-base circuit*

$$\alpha = \frac{\Delta I_C}{\Delta I_E} \; (V_{CB} \text{ constant})$$

α is current gain; ΔI_C is change in collector current; ΔI_E is change in emitter current.

2. *Current gain, common-emitter circuit*

$$\beta = \frac{\Delta I_C}{\Delta I_B} \; (V_{CE} \text{ constant})$$

β is current gain; ΔI_C is change in collector current; ΔI_B is change in base current.

23.10 Mass Number

$$A = Z + N$$

A is mass number of an atom; Z is atomic number; N is number of neutrons.

24.5 Decay Constant

$$\lambda = \frac{0.693}{T_{1/2}}$$

λ is decay constant; $T_{1/2}$ is half-life.

25.2 Energy of Electron

$$E = -\frac{me^4}{8\epsilon_0^2 h^2 n^2}$$

E is energy; m is mass; e is electronic charge; ϵ_0 is a constant; h is Planck's constant; n is principal quantum number.

Appendix B—Tables

Table 1 PREFIXES OF THE METRIC SYSTEM

Factor	Prefix	Symbol
10^{18}	exa	E
10^{15}	peta	P
10^{12}	tera	T
10^9	giga	G
10^6	mega	M
10^3	kilo	k
10^2	hecto	h
10	deka	da
10^{-1}	deci	d
10^{-2}	centi	c
10^{-3}	milli	m
10^{-6}	micro	μ
10^{-9}	nano	n
10^{-12}	pico	p
10^{-15}	femto	f
10^{-18}	atto	a

Table 2 GREEK ALPHABET

Greek letter	Greek name	English equivalent	Greek letter	Greek name	English equivalent
A α	alpha	ä	N ν	nu	n
B β	beta	b	Ξ ξ	xi	ks
Γ γ	gamma	g	O o	omicron	o
Δ δ	delta	d	Π π	pi	p
E ϵ	epsilon	e	P ρ	rho	r
Z ζ	zeta	z	Σ σ	sigma	s
H η	eta	ā	T τ	tau	t
Θ θ	theta	th	Y υ	upsilon	ü, ōō
I ι	iota	ē	Φ ϕ	phi	f
K κ	kappa	k	X χ	chi	h
Λ λ	lambda	l	Ψ ψ	psi	ps
M μ	mu	m	Ω ω	omega	ō

Table 3 PHYSICAL CONSTANTS

Quantity	Symbol	Value
atmospheric pressure, normal	atm	1.01325×10^5 n/m^2
atomic mass unit	u	$1.6605655 \times 10^{-27}$ kg
Avogadro number	N_A	6.022045×10^{23}/mole
charge to mass ratio for electron	e/m_e	1.7588047×10^{11} c/kg
electron rest mass	m_e	9.109534×10^{-31} kg
		5.4858026×10^{-4} u
electron volt	ev	1.60210×10^{-19} j
electrostatic constant	k	8.987×10^9 n m^2/c^2
elementary charge	e	$1.6021892 \times 10^{-19}$ c
faraday	f	9.648456×10^4 c/mole
gas constant, universal	R	6.236×10^4 mm cm^3/mole °K
		8.20568×10^{-2} L atm/mole °K
		8.3143×10^0 j/mole °K
gravitational acceleration, standard	g	9.80665×10^0 m/s^2
mechanical equivalent of heat	J	4.1868×10^0 j/cal
molar volume of ideal gas at STP	V_m	2.24136×10^1 L/mole
neutron rest mass	m_n	$1.6749543 \times 10^{-27}$ kg
		1.008665012×10^0 u
Planck's constant	h	6.626176×10^{-34} j s
proton rest mass	m_p	$1.6726485 \times 10^{-27}$ kg
		1.007276470×10^0 u
speed of light in a vacuum	c	2.99792458×10^8 m/s
speed of sound in air at STP	v	3.3145×10^2 m/s
temperature of the triple point of water	—	2.7316×10^2 °K (0.01 °C)
universal gravitational constant	G	6.6720×10^{-11} n m^2/kg^2

Table 4 SELECTED PHYSICAL QUANTITIES AND MEASUREMENT UNITS

Physical quantity	Quantity symbol	Measurement unit	Unit symbol	Unit dimensions
		Fundamental (base) units		
length	l	meter	m	m
mass	m	kilogram	kg	kg
time	t	second	s	s
electric charge*	Q	coulomb	c	c
temperature	T	degree Kelvin	°K	°K
amount of substance	n	mole	mol	mol
luminous intensity	I	candle	cd	cd
		Derived units		
acceleration	a	meter per second per second	m/s^2	m/s^2
area	A	square meter	m^2	m^2
capacitance	C	farad	f	$c^2\,s^2/kg\,m^2$
density	D	kilogram per cubic meter	kg/m^3	kg/m^3
electric current	I	ampere	a	c/s
electric field intensity	$\mathscr{E}$	newton per coulomb	n/c	$kg\,m/c\,s^2$
electric resistance	R	ohm	Ω	$kg\,m^2/c^2\,s$
emf	E	volt	v	$kg\,m^2/c\,s^2$
energy	E	joule	j	$kg\,m^2/s^2$
force	F	newton	n	$kg\,m/s^2$
frequency	f	hertz	hz	s^{-1}
heat	Q	joule	j	$kg\,m^2/s^2$
illumination	E	lumen per square meter	lm/m^2	cd/m^2
inductance	L	henry	h	$kg\,m^2/c^2$
luminous flux	Φ	lumen	lm	cd
magnetic flux	Φ	weber	wb	$kg\,m^2/c\,s$
magnetic flux density	B	weber per square meter	wb/m^2	$kg/c\,s$
potential difference	V	volt	v	$kg\,m^2/c\,s^2$
power	P	watt	w	$kg\,m^2/s^3$
pressure	p	newton per square meter	n/m^2	$kg/m\,s^2$
velocity	v	meter per second	m/s	m/s
volume	V	cubic meter	m^3	m^3
work	W	joule	j	$kg\,m^2/s^2$

*The SI fundamental (base) electric unit is the ampere. For pedagogical reasons the coulomb of electric charge is introduced in this book as the fundamental unit and the ampere as a derived unit having the dimensions, coulomb per second.

Table 5 CONVERSION FACTORS

Length
$1 \text{ m} = 10^{-3} \text{ km} = 10^2 \text{ cm} = 10^3 \text{ mm} = 10^6 \text{ } \mu\text{m} = 10^9 \text{ nm} = 10^{10} \text{ Å}$
$1 \text{ } \mu = 1 \text{ } \mu\text{m} = 10^{-6} \text{ m} = 10^{-4} \text{ cm} = 10^{-3} \text{ mm} = 10^3 \text{ m}\mu = 10^3 \text{ nm} = 10^4 \text{ Å}$
$1 \text{ m}\mu = 1 \text{ nm} = 10^{-9} \text{ m} = 10^{-7} \text{ cm} = 10^{-6} \text{ mm} = 10 \text{ Å}$
$1 \text{ Å} = 10^{-10} \text{ m} = 10^{-8} \text{ cm} = 10^{-4} \text{ } \mu = 10^{-1} \text{ m}\mu = 10^{-1} \text{ nm}$

Area
$1 \text{ m}^2 = 10^{-6} \text{ km}^2 = 10^4 \text{ cm}^2 = 10^6 \text{ mm}^2$

Volume
$1 \text{ m}^3 = 10^{-9} \text{ km}^3 = 10^3 \text{ L} = 10^6 \text{ cm}^3$
$1 \text{ L} = 10^3 \text{ mL} = 10^3 \text{ cm}^3 = 10^{-3} \text{ m}^3$

Angular
$1° = 1.74 \times 10^{-2} \text{ radian} = 2.78 \times 10^{-3} \text{ revolution}$
$1 \text{ radian} = 57.3° = 1.59 \times 10^{-1} \text{ revolution}$
$1 \text{ revolution} = 360° = 6.28 \text{ radians}$

Mass
$1 \text{ kg} = 10^3 \text{ g} = 10^6 \text{ mg} = 6.02 \times 10^{26} \text{ } u$
$1 \text{ g} = 10^{-3} \text{ kg} = 10^3 \text{ mg} = 6.02 \times 10^{23} \text{ } u$
$1 \text{ } u = 1.66 \times 10^{-24} \text{ g} = 1.66 \times 10^{-21} \text{ mg} = 1.66 \times 10^{-27} \text{ kg}$

Time
$1 \text{ hr} = 60 \text{ min} = 3.6 \times 10^3 \text{ s}$
$1 \text{ min} = 60 \text{ s} = 1.67 \times 10^{-2} \text{ hr}$
$1 \text{ s} = 1.67 \times 10^{-2} \text{ min} = 2.78 \times 10^{-4} \text{ hr}$

Velocity
$1 \text{ km/hr} = 10^3 \text{ m/hr} = 16.7 \text{ m/min} = 2.78 \times 10^{-1} \text{ m/s}$
$1 \text{ m/min} = 10^2 \text{ cm/min} = 1.67 \times 10^{-2} \text{ m/s} = 1.67 \text{ cm/s}$
$1 \text{ m/s} = 10^{-3} \text{ km/s} = 3.6 \text{ km/hr} = 10^2 \text{ cm/s}$

Acceleration
$1 \text{ cm/s}^2 = 10^{-2} \text{ m/s}^2 = 10^{-5} \text{ km/s}^2$
$1 \text{ m/s}^2 = 10^2 \text{ cm/s}^2 = 10^{-3} \text{ km/s}^2$
$1 \text{ km/hr/s} = 10^3 \text{ m/hr/s} = 2.78 \times 10^{-1} \text{ m/s}^2 = 2.78 \times 10^1 \text{ cm/s}^2 = 2.78 \times 10^2 \text{ mm/s}^2$

Pressure
$1 \text{ atm} = 760.00 \text{ mm Hg} = 1.013 \times 10^5 \text{ n/m}^2 = 1.013 \times 10^5 \text{ Pa} = 1.013 \times 10^2 \text{ kPa}$
$1 \text{ n/m}^2 = 10 \text{ dynes/cm}^2 = 9.87 \times 10^{-6} \text{ atm} = 1 \text{ Pa} = 10^{-3} \text{ kPa} = 7.50 \times 10^{-3} \text{ mm Hg}$

Energy
$1 \text{ j} = 2.39 \times 10^{-1} \text{ cal} = 2.39 \times 10^{-4} \text{ kcal} = 2.78 \times 10^{-7} \text{ kw hr} = 6.25 \times 10^{18} \text{ ev}$
$1 \text{ cal} = 10^{-3} \text{ kcal} = 4.19 \text{ j} = 1.16 \times 10^{-6} \text{ kw hr}$
$1 \text{ kcal} = 10^3 \text{ cal} = 4.19 \times 10^3 \text{ j} = 1.16 \times 10^{-3} \text{ kw hr}$
$1 \text{ ev} = 10^{-6} \text{ Mev} = 1.60 \times 10^{-19} \text{ j}$
$1 \text{ kw hr} = 10^3 \text{ w hr} = 3.6 \times 10^3 \text{ kw s} = 3.6 \times 10^6 \text{ w s} = 8.6 \times 10^5 \text{ cal}$
$1 \text{ w s} = 2.78 \times 10^{-4} \text{ w hr} = 2.78 \times 10^{-7} \text{ kw hr}$

Mass-Energy
$1 \text{ j} = 1.11 \times 10^{-17} \text{ kg} = 1.11 \times 10^{-14} \text{ g} = 6.69 \times 10^9 \text{ } u$
$1 \text{ ev} = 1.07 \times 10^{-9} \text{ } u = 1.78 \times 10^{-33} \text{ g}$
$1 u = 1.49 \times 10^{-10} \text{ j} = 931 \text{ Mev} = 9.31 \times 10^8 \text{ ev}$
$1 \text{ kg} = 9.00 \times 10^{16} \text{ j}$

Table 6 NATURAL TRIGONOMETRIC FUNCTIONS

Angle (°)	Sine	Cosine	Tangent	Angle (°)	Sine	Cosine	Tangent
0.0	0.000	1.000	0.000				
0.5	0.009	1.000	0.009	23.0	0.391	0.921	0.424
1.0	0.017	1.000	0.017	23.5	0.399	0.917	0.435
1.5	0.026	1.000	0.026	24.0	0.407	0.914	0.445
2.0	0.035	0.999	0.035	24.5	0.415	0.910	0.456
2.5	0.044	0.999	0.044	25.0	0.423	0.906	0.466
3.0	0.052	0.999	0.052	25.5	0.431	0.903	0.477
3.5	0.061	0.998	0.061	26.0	0.438	0.899	0.488
4.0	0.070	0.998	0.070	26.5	0.446	0.895	0.499
4.5	0.078	0.997	0.079	27.0	0.454	0.891	0.510
5.0	0.087	0.996	0.087	27.5	0.462	0.887	0.521
5.5	0.096	0.995	0.096	28.0	0.470	0.883	0.532
6.0	0.104	0.995	0.105	28.5	0.477	0.879	0.543
6.5	0.113	0.994	0.114	29.0	0.485	0.875	0.554
7.0	0.122	0.992	0.123	29.5	0.492	0.870	0.566
7.5	0.131	0.991	0.132	30.0	0.500	0.866	0.577
8.0	0.139	0.990	0.141	30.5	0.508	0.862	0.589
8.5	0.148	0.989	0.149	31.0	0.515	0.857	0.601
9.0	0.156	0.988	0.158	31.5	0.522	0.853	0.613
9.5	0.165	0.986	0.167	32.0	0.530	0.848	0.625
10.0	0.174	0.985	0.176	32.5	0.537	0.843	0.637
10.5	0.182	0.983	0.185	33.0	0.545	0.839	0.649
11.0	0.191	0.982	0.194	33.5	0.552	0.834	0.662
11.5	0.199	0.980	0.204	34.0	0.559	0.829	0.674
12.0	0.208	0.978	0.213	34.5	0.566	0.824	0.687
12.5	0.216	0.976	0.222	35.0	0.574	0.819	0.700
13.0	0.225	0.974	0.231	35.5	0.581	0.814	0.713
13.5	0.233	0.972	0.240	36.0	0.588	0.809	0.726
14.0	0.242	0.970	0.249	36.5	0.595	0.804	0.740
14.5	0.250	0.968	0.259	37.0	0.602	0.799	0.754
15.0	0.259	0.966	0.268	37.5	0.609	0.793	0.767
15.5	0.267	0.964	0.277	38.0	0.616	0.788	0.781
16.0	0.276	0.961	0.287	38.5	0.622	0.783	0.795
16.5	0.284	0.959	0.296	39.0	0.629	0.777	0.810
17.0	0.292	0.956	0.306	39.5	0.636	0.772	0.824
17.5	0.301	0.954	0.315	40.0	0.643	0.766	0.839
18.0	0.309	0.951	0.325	40.5	0.649	0.760	0.854
18.5	0.317	0.948	0.335	41.0	0.656	0.755	0.869
19.0	0.326	0.946	0.344	41.5	0.663	0.749	0.885
19.5	0.334	0.943	0.354	42.0	0.669	0.743	0.900
20.0	0.342	0.940	0.364	42.5	0.676	0.737	0.916
20.5	0.350	0.937	0.374	43.0	0.682	0.731	0.932
21.0	0.358	0.934	0.384	43.5	0.688	0.725	0.949
21.5	0.366	0.930	0.394	44.0	0.695	0.719	0.966
22.0	0.375	0.927	0.404	44.5	0.701	0.713	0.983
22.5	0.383	0.924	0.414	45.0	0.707	0.707	1.000

Table 6   NATURAL TRIGONOMETRIC FUNCTIONS (cont'd)

Angle (°)	Sine	Cosine	Tangent	Angle (°)	Sine	Cosine	Tangent
45.5	0.713	0.701	1.018	68.0	0.927	0.375	2.475
46.0	0.719	0.695	1.036	68.5	0.930	0.366	2.539
46.5	0.725	0.688	1.054	69.0	0.934	0.358	2.605
47.0	0.731	0.682	1.072	69.5	0.937	0.350	2.675
47.5	0.737	0.676	1.091	70.0	0.940	0.342	2.747
48.0	0.743	0.669	1.111	70.5	0.943	0.334	2.824
48.5	0.749	0.663	1.130	71.0	0.946	0.326	2.904
49.0	0.755	0.656	1.150	71.5	0.948	0.317	2.983
49.5	0.760	0.649	1.171	72.0	0.951	0.309	3.078
50.0	0.766	0.643	1.192	72.5	0.954	0.301	3.172
50.5	0.772	0.636	1.213	73.0	0.956	0.292	3.271
51.0	0.777	0.629	1.235	73.5	0.959	0.284	3.376
51.5	0.783	0.622	1.257	74.0	0.961	0.276	3.487
52.0	0.788	0.616	1.280	74.5	0.964	0.267	3.606
52.5	0.793	0.609	1.303	75.0	0.966	0.259	3.732
53.0	0.799	0.602	1.327	75.5	0.968	0.250	3.867
53.5	0.804	0.595	1.351	76.0	0.970	0.242	4.011
54.0	0.809	0.588	1.376	76.5	0.972	0.233	4.165
54.5	0.814	0.581	1.402	77.0	0.974	0.225	4.331
55.0	0.819	0.574	1.428	77.5	0.976	0.216	4.511
55.5	0.824	0.566	1.455	78.0	0.978	0.208	4.705
56.0	0.829	0.559	1.483	78.5	0.980	0.199	4.915
56.5	0.834	0.552	1.511	79.0	0.982	0.191	5.145
57.0	0.839	0.545	1.540	79.5	0.983	0.182	5.396
57.5	0.843	0.537	1.570	80.0	0.985	0.174	5.671
58.0	0.848	0.530	1.600	80.5	0.986	0.165	5.976
58.5	0.853	0.522	1.632	81.0	0.988	0.156	6.314
59.0	0.857	0.515	1.664	81.5	0.989	0.148	6.691
59.5	0.862	0.508	1.698	82.0	0.990	0.139	7.115
60.0	0.866	0.500	1.732	82.5	0.991	0.131	7.596
60.5	0.870	0.492	1.767	83.0	0.992	0.122	8.144
61.0	0.875	0.485	1.804	83.5	0.994	0.113	8.777
61.5	0.879	0.477	1.842	84.0	0.994	0.104	9.514
62.0	0.883	0.470	1.881	84.5	0.995	0.096	10.38
62.5	0.887	0.462	1.921	85.0	0.996	0.087	11.43
63.0	0.891	0.454	1.963	85.5	0.997	0.078	12.71
63.5	0.895	0.446	2.006	86.0	0.998	0.070	14.30
64.0	0.899	0.438	2.050	86.5	0.998	0.061	16.35
64.5	0.903	0.431	2.097	87.0	0.999	0.052	19.08
65.0	0.906	0.423	2.145	87.5	0.999	0.044	22.90
65.5	0.910	0.415	2.194	88.0	0.999	0.035	28.64
66.0	0.914	0.407	2.246	88.5	1.000	0.026	38.19
66.5	0.917	0.399	2.300	89.0	1.000	0.017	57.29
67.0	0.921	0.391	2.356	89.5	1.000	0.009	114.6
67.5	0.924	0.383	2.414	90.0	1.000	0.000	. . .

Table 7 TENSILE STRENGTH OF METALS

Metal	Tensile strength (n/m^2)
aluminum wire	2.4×10^8
copper wire, hard drawn	4.8×10^8
iron wire, annealed	3.8×10^8
iron wire, hard drawn	6.9×10^8
lead, cast or drawn	2.1×10^8
platinum wire	3.5×10^8
silver wire	2.9×10^8
steel (minimum)	2.8×10^8
steel wire (maximum)	$32 \ \ \times 10^8$

Table 8 YOUNG'S MODULUS

Metal	Elastic modulus (n/m^2)
aluminum, 99.3%, rolled	6.96×10^{10}
brass	9.02×10^{10}
copper, wire, hard drawn	11.6×10^{10}
gold, pure, hard drawn	7.85×10^{10}
iron, cast	$9.1 \ \times 10^{10}$
iron, wrought	$19.3 \ \times 10^{10}$
lead, rolled	1.57×10^{10}
platinum, pure, drawn	$16.7 \ \times 10^{10}$
silver, hard drawn	7.75×10^{10}
steel, 0.38% C, annealed	$20.0 \ \times 10^{10}$
tungsten, drawn	$35.5 \ \times 10^{10}$

Table 9 RELATIVE DENSITY AND VOLUME OF WATER

The mass of one cm^3 of water at 4 °C is taken as unity.
The values given are numerically equal to the absolute density in g/mL.

T (°C)	D (g/cm^3)	V (cm^3/g)	T (°C)	D (g/cm^3)	V (cm^3/g)
−10	0.99815	1.00186	45	0.99025	1.00985
−5	0.99930	1.00070	50	0.98807	1.01207
0	0.99987	1.00013	55	0.98573	1.01448
+1	0.99993	1.00007	60	0.98324	1.01705
2	0.99997	1.00003	65	0.98059	1.01979
3	0.99999	1.00001	70	0.97781	1.02270
4	1.00000	1.00000	75	0.97489	1.02576
5	0.99999	1.00001	80	0.97183	1.02899
+10	0.99973	1.00027	85	0.96865	1.03237
15	0.99913	1.00087	90	0.96534	1.03590
20	0.99823	1.00177	95	0.96192	1.03959
25	0.99707	1.00294	100	0.95838	1.04343
30	0.99567	1.00435	110	0.9510	1.0515
35	0.99406	1.00598	120	0.9434	1.0601
40	0.99224	1.00782	150	0.9173	1.0902

Table 10 SURFACE TENSION OF VARIOUS LIQUIDS

Liquid	In contact with	Temp. (°C)	Surface tension (n/m)
carbon disulfide	vapor	20	3.233×10^{-2}
carbon tetrachloride	vapor	20	2.695×10^{-2}
ethyl alcohol	air	0	2.405×10^{-2}
ethyl alcohol	vapor	20	2.275×10^{-2}
water	air	20	7.275×10^{-2}

Table 11 MASS DENSITY OF GASES

Gas	(T = 0 °C; p = 760 mm Hg) Formula	Density (g/L)
acetylene	C_2H_2	1.17910
air, dry, CO_2 free		1.29284
ammonia	NH_3	0.77126
argon	Ar	1.78364
chlorine	Cl_2	3.214
carbon dioxide	CO_2	1.9769
carbon monoxide	CO	1.25004
ethane	C_2H_6	1.3562
helium	He	0.17846
hydrogen	H_2	0.08988
hydrogen chloride	HCl	1.6392
methane	CH_4	0.7168
neon	Ne	0.89990
nitrogen	N_2	1.25036
oxygen	O_2	1.42896
sulfur dioxide	SO_2	2.9262

Table 12 EQUILIBRIUM VAPOR PRESSURE OF WATER

Temp. (°C)	Pressure (mm Hg)	Temp. (°C)	Pressure (mm Hg)	Temp. (°C)	Pressure (mm Hg)
0	4.6	25	23.8	90	525.8
5	6.5	26	25.2	95	633.9
10	9.2	27	26.7	96	657.6
15	12.8	28	28.3	97	682.1
16	13.6	29	30.0	98	707.3
17	14.5	30	31.8	99	733.2
18	15.5	35	42.2	100	760.0
19	16.5	40	55.3	101	787.5
20	17.5	50	92.5	103	845.1
21	18.7	60	149.4	105	906.1
22	19.8	70	233.7	110	1074.6
23	21.1	80	355.1	120	1489.1
24	22.4	85	433.6	150	3570.5

Table 13   HEAT CONSTANTS

Material	Specific heat (cal/g C°)	Melting point (°C)	Boiling point (°C)	Heat of fusion (cal/g)	Heat of vaporization (cal/g)
alcohol, ethyl	0.581 (25°)	−115	78.5	24.9	204
aluminum	0.214 (20°)	660.2	2467	94	2520
	0.217 (0–100°)				
	0.220 (20–100°)				
	0.225 (100°)				
ammonia, liquid	1.047 (−60°)	−77.7	−33.35	108.1	327.1
liquid	1.125 (20°)				
gas	0.523 (20°)				
brass (40% Zn)	0.0917	900			
copper	0.0924	1083	2595	49.0	1150
glass, crown	0.161				
iron	0.1075	1535	3000	7.89	1600
lead	0.0305	327.5	1744	5.47	207
mercury	0.0333	−38.87	356.58	2.82	70.613
platinum	0.0317	1769	3827 ± 100	27.2	
silver	0.0562	960.8	2212	26.0	565
tin	0.0543 (25°)	232	2270	14.4	520
tungsten	0.0322	3410 ± 10	5927	43	
water	1.00		100.00		538.7
ice	0.530	0.00		79.71	
steam	0.481				
zinc	0.0922	419.4	907	23.0	420

Table 14   COEFFICIENT OF LINEAR EXPANSION

(Increase in length per unit length per Celsius degree)

Material	Coefficient ($\Delta l/l$ C°)	Temperature (°C)	Material	Coefficient ($\Delta l/l$ C°)	Temperature (°C)
aluminum	23.8 × 10⁻⁶	20–100	invar (nickel steel)	0.9 × 10⁻⁶	20
brass	19.30 × 10⁻⁶	0–100	lead	29.40 × 10⁻⁶	18–100
copper	16.8 × 10⁻⁶	25–100	magnesium	26.08 × 10⁻⁶	18–100
glass, tube	8.33 × 10⁻⁶	0–100	platinum	8.99 × 10⁻⁶	40
crown	8.97 × 10⁻⁶	0–100	rubber, hard	8̄0 × 10⁻⁶	20–60
Pyrex	3.3 × 10⁻⁶	20–300	quartz, fused	0.546 × 10⁻⁶	0–800
gold	14.3 × 10⁻⁶	16–100	silver	18.8 × 10⁻⁶	20
ice	50.7 × 10⁻⁶	−10–0	tin	26.92 × 10⁻⁶	18–100
iron, soft	12.10 × 10⁻⁶	40	zinc	26.28 × 10⁻⁶	10–100
steel	10.5 × 10⁻⁶	0–100			

The coefficient of cubical expansion may be taken as three times the linear coefficient.

Table 15 COEFFICIENT OF VOLUME EXPANSION

(Increase in volume per unit volume per C° at 20 °C)

Liquid	Coefficient ($\Delta V/V\ C°$)
acetone	14.87×10^{-4}
alcohol, ethyl	$11.2\ \times 10^{-4}$
benzene	12.37×10^{-4}
carbon disulfide	12.18×10^{-4}
carbon tetrachloride	12.36×10^{-4}
chloroform	12.73×10^{-4}
ether	16.56×10^{-4}
gasoline	$10.8\ \times 10^{-4}$
glycerol	5.05×10^{-4}
mercury	1.82×10^{-4}
petroleum	9.55×10^{-4}
turpentine	9.73×10^{-4}
water	2.07×10^{-4}

Table 16 HEAT OF VAPORIZATION OF SATURATED STEAM

Temp. (°C)	Pressure (mm Hg)	Heat of vaporization (cal/g)
95	634.0	541.9
96	657.7	541.2
97	682.1	540.6
98	707.3	539.9
99	733.3	539.3
100	760.0	538.7
101	787.5	538.1
102	815.9	537.4
103	845.1	536.8
104	875.1	536.2
105	906.1	535.6
106	937.9	534.9

Table 17 SPEED OF SOUND

Substance	Density (g/L)	Velocity (m/s)	$\Delta v/\Delta T$ (m/s C°)
GASES (STP)			
air, dry	1.293	331.45	0.59
carbon dioxide	1.977	259	0.4
helium	0.178	965	0.8
hydrogen	0.0899	1284	2.2
nitrogen	1.251	334	0.6
oxygen	1.429	316	0.56
LIQUIDS (25 °C)	(g/cm³)		
acetone	0.79	1174	
alcohol, ethyl	0.79	1207	
carbon tetrachloride	1.595	926	
glycerol	1.26	1904	
kerosene	0.81	1324	
water, distilled	0.998	1497	
water, sea	1.025	1531	
SOLIDS (thin rods)			
aluminum	2.7	5000	
brass	8.6	3480	
brick	1.8	3650	
copper	8.93	3810	
cork	0.25	500	
glass, crown	2.24	4540	
iron	7.85	5200	
lucite	1.18	1840	
maple (along grain)	0.69	4110	
pine (along grain)	0.43	3320	
steel	7.85	5200	

Table 18 INDEX OF REFRACTION

(λ = 5893 Å; Temperature = 20 °C except as noted)

Material	Refractive index
air, dry (STP)	1.00029
alcohol, ethyl	1.360
benzene	1.501
calcite	1.6583
	1.4864
canada balsam	1.530
carbon dioxide (STP)	1.00045
carbon disulfide	1.625
carbon tetrachloride	1.459
diamond	2.4195
glass, crown	1.5172
flint	1.6270
glycerol	1.475
ice	1.310
lucite	1.50
quartz	1.544
	1.553
quartz, fused	1.45845
sapphire (Al_2O_3)	1.7686
	1.7604
water, distilled	1.333
water vapor (STP)	1.00025

Table 19 ELECTROCHEMICAL EQUIVALENTS

Element	g-at wt (g)	Ionic charge (oxidation number)	z (g/c)
aluminum	27.0	+3	0.0000933
cadmium	112	+2	0.000580
calcium	40.1	+2	0.000208
chlorine	35.5	−1	0.000368
chromium	52.0	+6	0.0000898
chromium	52.0	+3	0.000180
copper	63.5	+2	0.000329
copper	63.5	+1	0.000658
gold	197	+3	0.000680
gold	197	+1	0.00204
hydrogen	1.01	+1	0.0000105
lead	207	+4	0.000536
lead	207	+2	0.00107
magnesium	24.3	+2	0.000126
nickel	58.7	+2	0.000304
oxygen	16.0	−2	0.0000829
potassium	39.1	+1	0.000405
silver	108	+1	0.00112
sodium	23.0	+1	0.000238
tin	119	+4	0.000308
tin	119	+2	0.000617
zinc	65.4	+2	0.000339

Table 20 RESISTIVITY

(Temperature = 20 °C)

Material	Resistivity (Ω cm)	Melting point (°C)
advance	48 × 10⁻⁶	1190
aluminum	2.824 × 10⁻⁶	660
brass	7.00 × 10⁻⁶	900
climax	87 × 10⁻⁶	1250
constantan (Cu 60, Ni 40)	49 × 10⁻⁶	1190
copper	1.724 × 10⁻⁶	1083
german silver (Cu 55, Zn 25, Ni 20)	33 × 10⁻⁶	1100
gold	2.44 × 10⁻⁶	1063
iron	1̄0 × 10⁻⁶	1535
magnesium	4.6 × 10⁻⁶	651
manganin (Cu 84, Mn 12, Ni 4)	44 × 10⁻⁶	910
mercury	95.783 × 10⁻⁶	−39
monel metal	42 × 10⁻⁶	1300
nichrome	115 × 10⁻⁶	1500
nickel	7.8 × 10⁻⁶	1452
nickel silver (Cu 57, Ni 43)	49 × 10⁻⁶	1190
platinum	1̄0 × 10⁻⁶	1769
silver	1.59 × 10⁻⁶	961
tungsten	5.6 × 10⁻⁶	3410

Table 21 PROPERTIES OF COPPER WIRE

	(Temperature, 20 °C) American Wire Gauge (B&S) — for any metal		for copper only	
Gauge number	Diameter	Cross section	Resistance	
	(mm)	(mm^2)	(Ω/km)	(m/Ω)
0000	11.68	107.2	0.1608	6219
000	10.40	85.03	0.2028	4932
00	9.266	67.43	0.2557	3911
0	8.252	53.48	0.3224	3102
1	7.348	42.41	0.4066	2460
2	6.544	33.63	0.5027	1951
3	5.827	26.67	0.6465	1547
4	5.189	21.15	0.8152	1227
5	4.621	16.77	1.028	972.9
6	4.115	13.30	1.296	771.5
7	3.665	10.55	1.634	611.8
8	3.264	8.366	2.061	485.2
9	2.906	6.634	2.599	384.8
10	2.588	5.261	3.277	305.1
11	2.305	4.172	4.132	242.0
12	2.053	3.309	5.211	191.9
13	1.828	2.624	6.571	152.2
14	1.628	2.081	8.258	120.7
15	1.450	1.650	10.45	95.71
16	1.291	1.309	13.17	75.90
17	1.150	1.038	16.61	60.20
18	1.024	0.8231	20.95	47.74
19	0.9116	0.6527	26.42	37.86
20	0.8118	0.5176	33.31	30.02
21	0.7230	0.4105	42.00	23.81
22	0.6438	0.3255	52.96	18.88
23	0.5733	0.2582	66.79	14.97
24	0.5106	0.2047	84.21	11.87
25	0.4547	0.1624	106.2	9.415
26	0.4049	0.1288	133.9	7.486
27	0.3606	0.1021	168.9	5.922
28	0.3211	0.08098	212.9	4.697
29	0.2859	0.06422	268.5	3.725
30	0.2546	0.05093	338.6	2.954
31	0.2268	0.04039	426.9	2.342
32	0.2019	0.03203	538.3	1.858
33	0.1798	0.02540	678.8	1.473
34	0.1601	0.02014	856.0	1.168
35	0.1426	0.01597	1079	0.9265
36	0.1270	0.01267	1361	0.7347
37	0.1131	0.01005	1716	0.5827
38	0.1007	0.007967	2164	0.4621
39	0.08969	0.006318	2729	0.3664
40	0.07987	0.005010	3441	0.2906

Table 22 THE CHEMICAL ELEMENTS

Name of element	Symbol	Atomic number	Atomic weight	Name of element	Symbol	Atomic number	Atomic weight
actinium	Ac	89	227.0278	lead	Pb	82	207.2
aluminum	Al	13	26.98154	lithium	Li	3	6.941
americium	Am	95	[243]	lutetium	Lu	71	174.967
antimony	Sb	51	121.75	magnesium	Mg	12	24.305
argon	Ar	18	39.948	manganese	Mn	25	54.9380
arsenic	As	33	74.9216	mendelevium	Md	101	[258]
astatine	At	85	[210]	mercury	Hg	80	200.59
barium	Ba	56	137.33	molybdenum	Mo	42	95.94
berkelium	Bk	97	[247]	neodymium	Nd	60	144.24
beryllium	Be	4	9.01218	neon	Ne	10	20.179
bismuth	Bi	83	208.9804	neptunium	Np	93	237.0482
boron	B	5	10.81	nickel	Ni	28	58.70
bromine	Br	35	79.904	niobium	Nb	41	92.9064
cadmium	Cd	48	112.41	nitrogen	N	7	14.0067
calcium	Ca	20	40.08	nobelium	No	102	[259]
californium	Cf	98	[251]	osmium	Os	76	190.2
carbon	C	6	12.011	oxygen	O	8	15.9994
cerium	Ce	58	140.12	palladium	Pd	46	106.4
cesium	Cs	55	132.9054	phosphorus	P	15	30.97376
chlorine	Cl	17	35.453	platinum	Pt	78	195.09
chromium	Cr	24	51.996	plutonium	Pu	94	[244]
cobalt	Co	27	58.9332	polonium	Po	84	[209]
copper	Cu	29	63.546	potassium	K	19	39.0983
curium	Cm	96	[247]	praseodymium	Pr	59	140.9077
dysprosium	Dy	66	162.50	promethium	Pm	61	[145]
einsteinium	Es	99	[254]	protactinium	Pa	91	231.0359
erbium	Er	68	167.26	radium	Ra	88	226.0254
europium	Eu	63	151.96	radon	Rn	86	[222]
fermium	Fm	100	[257]	rhenium	Re	75	186.207
fluorine	F	9	18.998403	rhodium	Rh	45	102.9055
francium	Fr	87	[223]	rubidium	Rb	37	85.4678
gadolinium	Gd	64	157.25	ruthenium	Ru	44	101.07
gallium	Ga	31	69.72	rutherfordium	Rf	104	[261]
germanium	Ge	32	72.59	samarium	Sm	62	150.4
gold	Au	79	196.9665	scandium	Sc	21	44.9559
hafnium	Hf	72	178.49	selenium	Se	34	78.96
hahnium	Ha	105	[263]	silicon	Si	14	28.0855
helium	He	2	4.00260	silver	Ag	47	107.868
holmium	Ho	67	164.9304	sodium	Na	11	22.98977
hydrogen	H	1	1.0079	strontium	Sr	38	87.62
indium	In	49	114.82	sulfur	S	16	32.06
iodine	I	53	126.9045	tantalum	Ta	73	180.9479
iridium	Ir	77	192.22	technetium	Tc	43	98.9062
iron	Fe	26	55.847	tellurium	Te	52	127.60
krypton	Kr	36	83.80	terbium	Tb	65	158.9254
lanthanum	La	57	138.9005	thallium	Tl	81	204.37
lawrencium	Lr	103	[260]	thorium	Th	90	232.0381

Table 22 THE CHEMICAL ELEMENTS (cont'd)

Name of element	Symbol	Atomic number	Atomic weight	Name of element	Symbol	Atomic number	Atomic weight
thulium	Tm	69	168.9342	vanadium	V	23	50.9414
tin	Sn	50	118.69	xenon	Xe	54	131.30
titanium	Ti	22	47.90	ytterbium	Yb	70	173.04
tungsten	W	74	183.85	yttrium	Y	39	88.9059
uranium	U	92	238.029	zinc	Zn	30	65.37
				zirconium	Zr	40	91.22

A value given in brackets denotes the mass number of the isotope of longest known half-life. The atomic weights of most of these elements are believed to have no error greater than ±1 of the last digit given.

Table 23 MASSES OF SOME NUCLIDES

Element	Symbol	Atomic mass* (u)**	Element	Symbol	Atomic mass* (u)**
hydrogen	$^1_1$H	1.007825	sodium	$^{23}_{11}$Na	22.9898
deuterium	$^2_1$H	2.0140	magnesium	$^{24}_{12}$Mg	23.98504
helium	$^3_2$He	3.01603		$^{25}_{12}$Mg	24.98584
	$^4_2$He	4.00260		$^{26}_{12}$Mg	25.98259
lithium	$^6_3$Li	6.01512	sulfur	$^{32}_{16}$S	31.97207
	$^7_3$Li	7.01600	chlorine	$^{35}_{17}$Cl	34.96885
beryllium	$^6_4$Be	6.0197		$^{37}_{17}$Cl	36.96590
	$^8_4$Be	8.0053	potassium	$^{39}_{19}$K	38.96371
	$^9_4$Be	9.01218		$^{41}_{19}$K	40.96184
boron	$^{10}_5$B	10.0129	krypton	$^{95}_{36}$Kr	94.9
	$^{11}_5$B	11.00931	molybdenum	$^{100}_{42}$Mo	99.9076
carbon	$^{12}_6$C	12.00000	silver	$^{107}_{47}$Ag	106.90509
	$^{13}_6$C	13.00335		$^{109}_{47}$Ag	108.9047
nitrogen	$^{12}_7$N	12.0188	tellurium	$^{137}_{52}$Te	136.91
	$^{14}_7$N	14.00307	barium	$^{138}_{56}$Ba	137.9050
	$^{15}_7$N	15.00011	lead	$^{214}_{82}$Pb	213.9982
oxygen	$^{16}_8$O	15.99491	bismuth	$^{214}_{83}$Bi	213.9972
	$^{17}_8$O	16.99914	polonium	$^{218}_{84}$Po	218.0089
	$^{18}_8$O	17.99916	radon	$^{222}_{86}$Rn	222.0175
fluorine	$^{19}_9$F	18.99840	radium	$^{228}_{88}$Ra	228.0303
neon	$^{20}_{10}$Ne	19.99244	thorium	$^{232}_{90}$Th	232.0382
	$^{22}_{10}$Ne	21.99138	uranium	$^{235}_{92}$U	235.0439
				$^{238}_{92}$U	238.0508
			plutonium	$^{239}_{94}$Pu	239.0522

*Atomic mass of neutral atom is given.

**1 atomic mass unit (u) = $1.6605655 \times 10^{-27}$ kg.

Table 24 FOUR-PLACE LOGARITHMS

n	0	1	2	3	4	5	6	7	8	9
10	0000	0043	0086	0128	0170	0212	0253	0294	0334	0374
11	0414	0453	0492	0531	0569	0607	0645	0682	0719	0755
12	0792	0828	0864	0899	0934	0969	1004	1038	1072	1106
13	1139	1173	1206	1239	1271	1303	1335	1367	1399	1430
14	1461	1492	1523	1553	1584	1614	1644	1673	1703	1732
15	1761	1790	1818	1847	1875	1903	1931	1959	1987	2014
16	2041	2068	2095	2122	2148	2175	2201	2227	2253	2279
17	2304	2330	2355	2380	2405	2430	2455.	2480	2504	2529
18	2553	2577	2601	2625	2648	2672	2695	2718	2742	2765
19	2788	2810	2833	2856	2878	2900	2923	2945	2967	2989
20	3010	3032	3054	3075	3096	3118	3139	3160	3181	3201
21	3222	3243	3263	3284	3304	3324	3345	3365	3385	3404
22	3424	3444	3464	3483	3502	3522	3541	3560	3579	3598
23	3617	3636	3655	3674	3692	3711	3729	3747	3766	3784
24	3802	3820	3838	3856	3874	3892	3909	3927	3945	3962
25	3979	3997	4014	4031	4048	4065	4082	4099	4116	4133
26	4150	4166	4183	4200	4216	4232	4249	4265	4281	4298
27	4314	4330	4346	4362	4378	4393	4409	4425	4440	4456
28	4472	4487	4502	4518	4533	4548	4564	4579	4594	4609
29	4624	4639	4654	4669	4683	4698	4713	4728	4742	4757
30	4771	4786	4800	4814	4829	4843	4857	4871	4886	4900
31	4914	4928	4942	4955	4969	4983	4997	5011	5024	5038
32	5051	5065	5079	5092	5105	5119	5132	5145	5159	5172
33	5185	5198	5211	5224	5237	5250	5263	5276	5289	5302
34	5315	5328	5340	5353	5366	5378	5391	5403	5416	5428
35	5441	5453	5465	5478	5490	5502	5514	5527	5539	5551
36	5563	5575	5587	5599	5611	5623	5635	5647	5658	5670
37	5682	5694	5705	5717	5729	5740	5752	5763	5775	5786
38	5798	5809	5821	5832	5843	5855	5866	5877	5888	5899
39	5911	5922	5933	5944	5955	5966	5977	5988	5999	6010
40	6021	6031	6042	6053	6064	6075	6085	6096	6107	6117
41	6128	6138	6149	6160	6170	6180	6191	6201	6212	6222
42	6232	6243	6253	6263	6274	6284	6294	6304	6314	6325
43	6335	6345	6355	6365	6375	6385	6395	6405	6415	6425
44	6435	6444	6454	6464	6474	6484	6493	6503	6513	6522
45	6532	6542	6551	6561	6571	6580	6590	6599	6609	6618
46	6628	6637	6646	6656	6665	6675	6684	6693	6702	6712
47	6721	6730	6739	6749	6758	6767	6776	6785	6794	6803
48	6812	6821	6830	6839	6848	6857	6866	6875	6884	6893
49	6902	6911	6920	6928	6937	6946	6955	6964	6972	6981
50	6990	6998	7007	7016	7024	7033	7042	7050	7059	7067
51	7076	7084	7093	7101	7110	7118	7126	7135	7143	7152
52	7160	7168	7177	7185	7193	7202	7210	7218	7226	7235
53	7243	7251	7259	7267	7275	7284	7292	7300	7308	7316
54	7324	7332	7340	7348	7356	7364	7372	7380	7388	7396

Table 24 FOUR-PLACE LOGARITHMS (cont'd)

n	0	1	2	3	4	5	6	7	8	9
55	7404	7412	7419	7427	7435	7443	7451	7459	7466	7474
56	7482	7490	7497	7505	7513	7520	7528	7536	7543	7551
57	7559	7566	7574	7582	7589	7597	7604	7612	7619	7627
58	7634	7642	7649	7657	7664	7672	7679	7686	7694	7701
59	7709	7716	7723	7731	7738	7745	7752	7760	7767	7774
60	7782	7789	7796	7803	7810	7818	7825	7832	7839	7846
61	7853	7860	7868	7875	7882	7889	7896	7903	7910	7917
62	7924	7931	7938	7945	7952	7959	7966	7973	7980	7987
63	7993	8000	8007	8014	8021	8028	8035	8041	8048	8055
64	8062	8069	8075	8082	8089	8096	8102	8109	8116	8122
65	8129	8136	8142	8149	8156	8162	8169	8176	8182	8189
66	8195	8202	8209	8215	8222	8228	8235	8241	8248	8254
67	8261	8267	8274	8280	8287	8293	8299	8306	8312	8319
68	8325	8331	8338	8344	8351	8357	8363	8370	8376	8382
69	8388	8395	8401	8407	8414	8420	8426	8432	8439	8445
70	8451	8457	8463	8470	8476	8482	8488	8494	8500	8506
71	8513	8519	8525	8531	8537	8543	8549	8555	8561	8567
72	8573	8579	8585	8591	8597	8603	8609	8615	8621	8627
73	8633	8639	8645	8651	8657	8663	8669	8675	8681	8686
74	8692	8698	8704	8710	8716	8722	8727	8733	8739	8745
75	8751	8756	8762	8768	8774	8779	8785	8791	8797	8802
76	8808	8814	8820	8825	8831	8837	8842	8848	8854	8859
77	8865	8871	8876	8882	8887	8893	8899	8904	8910	8915
78	8921	8927	8932	8938	8943	8949	8954	8960	8965	8971
79	8976	8982	8987	8993	8998	9004	9009	9015	9020	9025
80	9031	9036	9042	9047	9053	9058	9063	9069	9074	9079
81	9085	9090	9096	9101	9106	9112	9117	9122	9128	9133
82	9138	9143	9149	9154	9159	9165	9170	9175	9180	9186
83	9191	9196	9201	9206	9212	9217	9222	9227	9232	9238
84	9243	9248	9253	9258	9263	9269	9274	9279	9284	9289
85	9294	9299	9304	9309	9315	9320	9325	9330	9335	9340
86	9345	9350	9355	9360	9365	9370	9375	9380	9385	9390
87	9395	9400	9405	9410	9415	9420	9425	9430	9435	9440
88	9445	9450	9455	9460	9465	9469	9474	9479	9484	9489
89	9494	9499	9504	9509	9513	9518	9523	9528	9533	9538
90	9542	9547	9552	9557	9562	9566	9571	9576	9581	9586
91	9590	9595	9600	9605	9609	9614	9619	9624	9628	9633
92	9638	9643	9647	9652	9657	9661	9666	9671	9675	9680
93	9685	9689	9694	9699	9703	9708	9713	9717	9722	9727
94	9731	9736	9741	9745	9750	9754	9759	9763	9768	9773
95	9777	9782	9786	9791	9795	9800	9805	9809	9814	9818
96	9823	9827	9832	9836	9841	9845	9850	9854	9859	9863
97	9868	9872	9877	9881	9886	9890	9894	9899	9903	9908
98	9912	9917	9921	9926	9930	9934	9939	9943	9948	9952
99	9956	9961	9965	9969	9974	9978	9983	9987	9991	9996

GLOSSARY

abscissa. The value corresponding to the horizontal distance of a point on a graph from the Y axis. The X coordinate.

absolute deviation. The difference between a single measured value and the average of several measurements made in the same way.

absolute error. The actual difference between a measured value and its accepted value.

absolute zero. The temperature of a body at which the kinetic energy of its molecules is at a minimum; $0\,^\circ$K or $-273.16\,^\circ$C.

absorption spectrum. A continuous spectrum interrupted by dark lines or bands that are characteristic of the medium through which the radiation has passed.

acceleration. Time rate of change of velocity.

acceptor. An element with three valence electrons per atom which when added to a semiconductor crystal provides electron "holes" in the lattice structure of the crystal.

accuracy. Closeness of a measurement to the accepted value for a specific physical quantity; expressed in terms of error.

adhesion. The force of attraction between unlike molecules.

adiabatic process. A thermal process in which no heat is added to or removed from a system.

alpha (α) particle. A helium-4 nucleus, especially when emitted from the nucleus of a radioactive atom.

alternating current. An electric current that has one direction during one part of a generating cycle and the opposite direction during the remainder of the cycle.

ammeter. An electric meter designed to measure current.

ampere. The unit of electric current; one coulomb per second.

amplifier. A device consisting of one or more vacuum tubes (or transistors) and associated circuits, used to increase the strength of a signal.

amplitude. The maximum displacement of a vibrating particle from its equilibrium position.

angle of incidence. The angle between the incident ray and the normal drawn to the point of incidence.

angle of reflection. The angle between the reflected ray and the normal drawn to the point of incidence.

angle of refraction. The angle between the refracted ray and the normal drawn to the point of refraction.

Ångstrom. A unit of linear measure equal to 10^{-10} m.

angular acceleration. The time rate of change of angular velocity.

angular impulse. The product of a torque and the time interval during which it acts.

angular momentum. The product of the rotational inertia of a body and its angular velocity.

angular velocity. The time rate of change of angular displacement.

anode. (1) The positive electrode of an electric cell. (2) The positive electrode or plate of an electronic tube. (3) The electron-poor electrode.

antimatter. A substance composed of antiparticles.

antiparticle. A counterpart of a subatomic particle having opposite properties (except for equal mass).

aperture. Any opening through which radiation may pass. The diameter of an opening that admits light to a lens or mirror.

apparent power. The product of the effective values of alternating voltage and current.

arc tangent. The inverse function to the tangent. Symbol: arctan or $\tan^{-1}$. Interpretation: "An angle whose tangent is _____."

armature. A coil of wire formed around an iron or steel core that rotates in the magnetic field of a generator or motor.

atom. The smallest particle of an element that can exist either alone or in combination with other atoms of the same or other elements.

atomic mass unit. One-twelfth of the mass of carbon-12, or $1.6605655 \times 10^{-27}$ kg.

atomic number. The number of protons in the nucleus of an atom.

atomic weight. The weighted average of the atomic masses of an element's isotopes based on their relative abundance.

audio signal. The alternating voltage proportional to the sound pressure produced in an electric circuit.

average velocity. Total displacement divided by elapsed time.

back emf. An induced emf in the armature of a motor that opposes the applied voltage.

band spectrum. An emission spectrum consisting of fluted bands of color. The spectrum of a substance in the molecular state.

barometer. A device used to measure the pressure of the atmosphere.

baryon. A subatomic particle with a large rest mass, e.g., the proton.

basic equation. An equation that relates the unknown quantity with known quantities in a problem.

basic law of electrostatics. Similarly charged objects repel each other. Oppositely charged objects attract each other.

beam. Several parallel rays of light considered collectively.

beat. The interference effect resulting from the superposition of two waves of slightly different frequencies propagating in the same direction. The amplitude of the resultant wave varies with time.

becquerel. The rate of radioactivity equal to one disintegration per second.

beta (β) particle. An electron emitted from the nucleus of a radioactive atom.

betatron. A device that accelerates electrons by means of the transformer principle.

bevatron. A high-energy synchrotron.

binding energy. Energy that must be applied to a nucleus to break it up.

boiling point. The temperature at which the vapor pressure of a liquid equals the pressure of the atmosphere.

boson. A subatomic particle with zero charge and rest mass, e.g., the photon.

Boyle's law. The volume of a dry gas varies inversely with the pressure exerted upon it, provided the temperature is constant.

breeder reactor. A nuclear reactor in which a fissionable material is produced at a greater rate than the fuel is consumed.

Brownian movement. The irregular and random movement of small particles suspended in a fluid, known to be a consequence of the thermal motion of fluid molecules.

bubble chamber. Instrument used for making the paths of ionizing particles visible as a trail of tiny bubbles in a liquid.

calorie. The quantity of heat equal to 4.19 joules.

calorimeter. A heat-measuring device consisting of nested metal cups separated by an air space.

candle. The unit of luminous intensity of a light source.

capacitance. The ratio of the charge on either plate of a capacitor to the potential difference between the plates.

capacitive reactance. Reactance in an a-c circuit containing capacitance which causes a lagging voltage.

capacitor. A combination of conducting plates separated by layers of a dielectric that is used to store an electric charge.

capillarity. The elevation or depression of liquids in small-diameter tubes.

cathode. (1) The negative electrode of an electric cell. (2) The electron-emitting electrode of an electronic tube. (3) The electron-rich electrode.

cathode rays. Particles emanating from a cathode; electrons.

Celsius scale. The temperature scale using the ice point as 0° and the steam point as 100°, with 100 equal divisions, or degrees, between; formerly the centigrade scale.

center of curvature. The center of the sphere of which the mirror or lens surface forms a part.

center of gravity. The point at which all of the weight of a body can be considered to be concentrated.

centrifugal force. Force that tends to move the particles of a rotating object away from the center of rotation.

centripetal acceleration. Acceleration directed toward the center of a circular path.

centripetal force. The force that produces centripetal acceleration.

chain reaction. A reaction in which the material or energy that starts the reaction is also one of the products and can cause similar reactions.

Charles' law. The volume of a dry gas is directly proportional to its Kelvin temperature, providing the pressure is constant.

chemical change. A change in which new substances with new properties are formed.

chemical equivalent. The quantity of an element, expressed in grams, equal to the ratio of its atomic weight to its valence.

chromatic aberration. The nonfocusing of light of different colors.

circular motion. Motion of a body along a circular path.

cloud chamber. A chamber in which charged subatomic particles appear as trails of liquid droplets.

coefficient of area expansion. The change in area per unit area of a solid per degree change in temperature.

coefficient of cubic expansion. The change in volume per unit volume of a solid or liquid per degree change in temperature.

coefficient of linear expansion. The change in length per unit length of a solid per degree change in temperature.

coefficient of sliding friction. The ratio of the force needed to overcome sliding friction to the normal force pressing the surfaces together.

coherence. The property of two wave trains with identical wavelengths and a constant phase relationship.

cohesion. The force of attraction between like molecules.

color. The visual perception of light associated with its frequency or wavelength.

commutator. A split ring in a d-c generator, each segment of which is connected to an end of a corresponding armature loop.

complementary colors. Two colors that combine to form white light.

complete vibration. Back-and-forth motion of an object describing simple harmonic motion.

component. One of the several vectors that can be combined geometrically to yield a resultant vector.

composition of forces. The combining of two or more component forces into a single resultant force.

compression. The region of a longitudinal wave in which the distance separating the vibrating particles is less than their equilibrium distance.

concave. Surface with center of curvature on the same side as the observer.

concave lens. A lens that diverges parallel light rays (assuming the outside refractive index to be smaller).

concave mirror. A mirror that converges parallel light rays incident on its surface.

concurrent forces. Forces with lines of action that pass through the same point.

condensation. The change of phase from a gas or vapor to a liquid.

conductance. The reciprocal of the ohmic resistance.

conductor. A material through which an electric charge is readily transferred.

conservative forces. Forces for which the law of conservation of mechanical energy holds true; gravitational forces and elastic forces.

continuous spectrum. A spectrum without dark lines or bands or in which there is an uninterrupted change from one color to another.

converging lens. A lens that is thicker in the middle than it is at the edge and bends incident parallel rays toward a common point.

convex. Surface with center of curvature on the opposite side from the observer.

convex lens. A lens that converges parallel light rays (assuming the outside refractive index to be smaller).

convex mirror. A mirror that diverges parallel light rays incident on its surface.

cosmic rays. High-energy nuclear particles apparently originating from outer space.

cosmotron. A high-energy synchrotron.

coulomb. The quantity of electricity equal to the charge on 6.25×10^{18} electrons.

Coulomb's law of electrostatics. The force between two point charges is directly proportional to the product of their magnitudes and inversely proportional to the square of the distance between them.

Coulomb's law of magnetism. The force between two magnetic poles is directly proportional to the strengths of the poles and inversely proportional to the square of their distance apart.

couple. Two forces of equal magnitude acting in opposite directions in the same plane, but not along the same line.

crest. A region of upward displacement in a transverse wave.

critical angle. That limiting angle of incidence in the optically denser medium that results in an angle of refraction of $90°$.

critical mass. The amount of a particular fissionable material required to make a fission reaction self-sustaining.

critical point. The upper limit of the temperature-pressure curve of a substance.

critical pressure. The pressure needed to liquefy a gas at its critical temperature.

critical temperature. The temperature to which a gas must be cooled before it can be liquefied by pressure.

critical velocity. Velocity below which an object moving in a vertical circle will not describe a circular path.

curie. The quantity of any radioactive nuclide that has a disintegration rate of 3.7×10^{10} becquerels.

current sensitivity. Current per unit scale division of an electric meter.

cut-off bias. The smallest negative grid voltage, for a given plate voltage, that causes a vacuum tube to cease to conduct.

cut-off frequency. A characteristic threshold frequency of incident light below which, for a given material, the photoelectric emission of electrons ceases.

cut-off potential. A negative potential on the collector of a photoelectric cell that reduces the photoelectric current to zero.

cycle. A series of changes produced in sequence that recur periodically.

cyclotron. A device for accelerating charged atomic particles by means of D-shaped electrodes.

damping. The reduction in amplitude of a wave due to the dissipation of wave energy.

decay constant. The ratio between the number of nuclei decaying per second and the total number of nuclei.

decibel. A unit of sound intensity level. The smallest change of sound intensity that the normal human ear can detect.

declination. The angle between magnetic north and the true north from any surface location; also called *variation*.

dees. The electrodes of a cyclotron.

density. See *mass density*.

derived unit. A unit of measure that consists of combinations of fundamental units.

dew point. The temperature at which a given amount of water vapor will exert equilibrium vapor pressure.

diamagnetism. The property of a substance whereby it is feebly repelled by a strong magnet.

dichroism. A property of certain crystalline substances in which one polarized component of incident light is absorbed and the other is transmitted.

dielectric. An electric insulator. A nonconducting medium.

dielectric constant. The ratio of the capacitance with a particular material separating the plates of a capacitor to the capacitance with a vacuum between the plates.

diffraction. The spreading of a wave disturbance into a region behind an obstruction.

diffraction angle. The angle that a diffracted wavefront forms with the grating plane.

diffraction grating. An optical surface, either transmitting or reflecting, with several thousand equally spaced and parallel grooves ruled in it.

diffusion. (1) The penetration of one type of particle into a mass of a second type of particle. (2) The scattering of light by irregular reflection.

dimensional analysis. The performance of indicated mathematical operations in a problem with the measurement units alone.

diode. A two-terminal device that will conduct electric current more easily in one direction than in the other.

direct current. An essentially constant-value current in which the movement of charge is in only one direction.

direct proportion. The relation between two quantities whose graph is a straight line.

dispersion. The process of separating polychromatic light into its component wavelengths.

displacement. (1) A change of position in a particular direction. (2) Distance of a vibrating particle from the midpoint of its vibration.

dissipative forces. Forces for which the law of conservation of mechanical energy does not hold true; frictional forces.

distillation. The evaporation of volatile materials from a liquid or solid mixture and their condensation in a separate vessel.

diverging lens. A lens that is thicker at the edge than it is in the middle and bends incident parallel rays so that they appear to come from a common point.

domain. A microscopic magnetic region composed of a group of atoms whose magnetic fields are aligned in a common direction.

donor. A substance with five valence electrons per atom which when added to a semiconductor crystal provides free electrons in the lattice structure of the crystal.

Doppler effect. The change observed in the frequency with which a wave from a given source reaches an observer when the source and the observer are in relative motion.

double refraction. The separation of a beam of unpolarized light into two refracted plane-polarized beams by certain crystals such as quartz and calcite.

drift tubes. Charged cylinders used to accelerate charged subatomic particles in a linear accelerator.

ductility. The property of a metal that enables it to be drawn through a die to form a wire.

eddy currents. Closed loops of induced current set up in a piece of metal when there is relative motion between the metal and a magnetic field. The eddy currents are in such direction that the resulting magnetic forces oppose the relative motion.

Edison effect. The emission of electrons from a heated metal in a vacuum.

effective value of current. The magnitude of an alternating current that in a given resistance produces heat at the same average rate as that magnitude of steady direct current.

efficiency. The ratio of the useful work output of a machine to total work input.

elastic collision. A collision in which ob-

jects rebound from each other without a loss of kinetic energy.

elastic limit. The condition in which a substance is on the verge of becoming permanently deformed.

elastic potential energy. The potential energy in a stretched or compressed elastic object.

elasticity. The ability of an object to return to its original size or shape when the external forces producing distortion are removed.

electric current. The rate of flow of charge past a given point in an electric circuit.

electric field. The region in which a force acts on an electric charge brought into the region.

electric field intensity. The force per unit positive charge at a given point in an electric field.

electric ground. (1) A conductor connected with the earth to establish zero (ground) potential. (2) A common return to an arbitrary zero potential.

electrification. The process of charging a body by adding or removing electrons.

electrochemical cell. A cell in which chemical energy is converted to electric energy by a spontaneous electron-transfer reaction.

electrochemical equivalent. The mass of an element, in grams, deposited by one coulomb of electric charge.

electrode. A conducting element in an electric cell, electronic tube, or semiconductor device.

electrolysis. The conduction of electricity through a solution of an electrolyte or through a fused ionic compound, together with the resulting chemical changes.

electrolyte. A substance whose solution conducts an electric current.

electrolytic cell. A cell in which electric energy is converted to chemical energy by means of an electron-transfer reaction.

electromagnetic induction. The process

by which an emf is set up in a conducting circuit by a changing magnetic flux linked by the circuit.

electromagnetic interaction. The interaction that keeps electrons in orbit and forms bonds between atoms and molecules.

electromagnetic waves. Transverse waves having an electric component and a magnetic component, each being perpendicular to the other and both perpendicular to the direction of propagation.

electromotive force. See *emf.*

electron. A negatively charged subatomic particle having a rest mass of 9.109534×10^{-31} kg and a charge of $1.6021892 \times 10^{-19}$ c.

electron shell. A region about the nucleus of an atom in which electrons move and which is made up of electron orbitals.

electron volt. The energy required to move an electron between two points that have a difference of potential of one volt.

electronics. The branch of physics concerned with the emission, behavior, and effects of electrons.

electroscope. A device used to observe the presence of an electrostatic charge.

elementary colors. The six regions of color in the solar spectrum observed by the dispersion of sunlight: red, orange, yellow, green, blue, and violet.

elongation strain. The ratio of the increase in length to the unstretched length.

emf. The energy per unit charge supplied by a source of electric current.

emission spectrum. A spectrum formed by the dispersion of light from an incandescent solid, liquid, and gas.

endothermic. Referring to a process that absorbs energy.

energy. The capacity for doing work.

energy level. One of a series of discrete energy values that characterize a phys-ical system governed by quantum rules.

entropy. (1) The internal energy of a system that cannot be converted to mechanical work. (2) The property that describes the disorder of a system.

equilibrant force. The force that produces equilibrium.

equilibrium. The state of a body in which there is no change in its motion.

equilibrium position. Midpoint of the path of an object describing simple harmonic motion.

equilibrium vapor pressure. The pressure exerted by vapor molecules in equilibrium with a liquid.

evaporation. The change of phase from a liquid to a gas or vapor.

exclusion principle. No two electrons in an atom can have the same set of quantum numbers.

exothermic. Referring to a process that liberates energy.

external combustion engine. A heat engine in which the fuel burns outside the cylinder or turbine chamber.

farad. The unit of capacitance; one coulomb per volt.

faraday. The quantity of electricity (96,500 coulombs) required to deposit one chemical equivalent of an element.

Faraday's first law. The mass of an element deposited during electrolysis is proportional to the quantity of charge that passes through the electrolytic cell.

Faraday's second law. The mass of an element deposited during electrolysis is proportional to the chemical equivalent of that element.

ferromagnetism. The property of a substance by which it is strongly attracted by a magnet.

Feynman diagram. A diagram showing the production and exchange of particles during a subatomic interaction.

first law of photoelectric emission. The rate of emission of photoelectrons is directly proportional to the intensity of the incident light.

first law of thermodynamics. When heat is converted to another form of energy, or when another form of energy is converted to heat, there is no loss of energy.

fission. The splitting of a heavy nucleus into nuclei of intermediate mass.

flash tube. The ionization tube that emits the light in a chemical laser.

Fleming valve. The first vacuum-tube diode.

fluorescence. The emission of light during the absorption of radiation from another source.

flux. Flow.

f-number. The ratio of the focal length of a lens to the effective aperture.

focal length. The distance between the principal focus of a lens or mirror and its optical center or vertex.

focal plane. The plane perpendicular to the principal axis of a converging lens or mirror and containing the principal focus.

focus. A point at which light rays meet or from which rays of light appear to diverge.

force. (1) A physical quantity that can affect the motion of an object. (2) A measure of the momentum gained per second by an accelerating body.

force of gravity. See *gravity*.

forced vibration. Vibration that is due to the application of a periodic force, and not to the natural vibrations of the system.

forward bias. Voltage applied to a semiconductor P-N junction that increases the electron current across the junction.

fractional distillation. The process of separating the components of a liquid mixture by means of differences in their boiling points.

frame of reference. Any system for specifying the precise location of objects in space.

freezing point. The temperature at which a liquid changes to a solid.

frequency. Number of vibrations, oscillations, or cycles per unit time.

friction. A force that resists the relative motion of objects that are in contact with each other.

fuel cell. An electrochemical cell in which the chemical energy of continuously supplied fuel is converted into electric energy.

fundamental. The lowest frequency produced by a musical tone source. That harmonic component of a wave which has the lowest frequency.

fundamental unit. Any one of seven basic units of measure.

fusion. (1) The change of phase from a solid to a liquid; melting. (2) A reaction in which light nuclei combine, forming a nucleus with a larger mass number.

galvanometer. An instrument used to measure minute electric currents.

gamma (γ) ray. High-energy photon emitted from the nucleus of a radioactive atom.

Geiger tube. Ion-sensitive instrument used for the detection of subatomic particles.

gravitational field. Region of space in which each point is associated with a value of gravitational acceleration.

gravitational force. The mutual force of attraction between particles of matter.

gravitational interaction. The interaction between particles of matter that has no known distance limitations, but is the weakest interaction of all.

gravitational potential energy. Potential energy acquired by an object when it is moved against gravity.

graviton. The carrier for the gravitational interaction.

gravity. The force of gravitation on an

object on or near the surface of a celestial body.

grid. An element of an electronic tube. An electrode used to control the flow of electrons from the cathode to the plate.

grid bias. The grid-to-cathode voltage.

half-life. The length of time during which, on the average, half of a large number of radioactive nuclides decay.

harmonics. The fundamental and the tones whose frequencies are whole-number multiples of the fundamental.

heat. Thermal energy in the process of being added to, or removed from, a substance.

heat capacity. The quantity of heat needed to raise the temperature of a body one degree.

heat engine. Any device that converts heat energy into mechanical work.

heat of fusion. The heat required per unit mass to change a substance from solid to liquid at its melting point.

heat of vaporization. The heat required per unit mass to change a substance from liquid to vapor at its boiling point.

heat pump. A device that absorbs heat from a cool environment and gives it off to a region of higher temperature.

heat sink. A reservoir that absorbs heat without a significant increase in temperature.

henry. The unit of inductance; one henry of inductance is present in a circuit when a change in the current of 1 ampere per second induces an emf of 1 volt.

Hooke's law. Below the elastic limit, strain is directly proportional to stress.

hyperbola. Graph of an inverse proportion.

hypercharge. A property of some baryons and leptons that is conserved in strong and electromagnetic interactions but not in weak interactions.

hypothesis. A plausible solution to a problem.

ice point. The melting point of ice when in equilibrium with water saturated with air at standard atmospheric pressure.

ideal gas. A theoretical gas consisting of infinitely small molecules that exert no forces on each other; also called perfect gas.

illumination. The luminous flux per unit area of a surface.

image. The optical counterpart of an object formed by lenses or mirrors.

impedance. (1) The ratio of applied wave-producing force to resulting displacement velocity of a wave-transmitting medium. (2) The ratio of sound pressure to volume displacement at a given surface in a sound-transmitting medium. (3) The ratio of the effective voltage to the effective current in an a-c circuit.

impedance matching. A technique used to ensure maximum transfer of energy from the output of one circuit to the input of another.

impulse. The product of a force and the time interval during which it acts.

index of refraction. The ratio of the speed of light in a vacuum to its speed in a given matter medium.

induced magnetism. Magnetism produced in a ferromagnetic substance by the influence of a magnetic field. Magnetization.

inductance. The property of an electric circuit by which a varying current induces a back emf in that circuit or a neighboring circuit.

induction. The process of charging one body by bringing it into the electric field of another charged body.

inductive reactance. Reactance in an a-c circuit containing inductance, which causes a lagging current.

inelastic collision. A collision in which

the colliding objects stick together after impact.

inertia. The property of matter that opposes any change in its state of motion.

inertial frame of reference. A nonaccelerating frame of reference in which Newton's first law holds true.

infrared light. Electromagnetic waves longer than those of visible light and shorter than radio waves.

infrasonic range. Vibrations in matter below 20 cycles/second.

instantaneous current. The magnitude of a varying current at any instant of time.

instantaneous velocity. Short displacement divided by elapsed time. Slope of the line that is tangent to a velocity graph at a given point.

instantaneous voltage. The magnitude of a varying voltage at any instant of time.

insulator. A material through which an electric charge is not readily transferred.

intensity level. The logarithm of the ratio of the intensity of a sound to the intensity of the threshold of hearing.

interaction. Any change in the amount or quantum numbers of particles that are near each other.

interface. A surface that forms the boundary between two phases or systems.

interference. (1) The superposing of one wave on another. (2) The mutual effect of two beams of light, resulting in a reduction of energy in certain areas and an increase of energy in others.

internal combustion engine. A heat engine in which the fuel burns inside the cylinder or turbine chamber.

internal energy. Total potential and kinetic energy of the molecules and atomic particles of a substance.

intersecting storage ring. An accelerator in which particles collide as they move in opposite directions.

inverse photoelectric effect. The emission of photons of radiation from a material when bombarded with high-speed electrons.

inverse proportion. The relation between two quantities whose product is a constant and whose graph is a hyperbola.

ion. An atom or a group of atoms having an electric charge.

ionization chamber. A device used to detect the passage of charged rays or particles by their ionizing effect on a gas.

ionization energy. The energy required to remove an electron from an atom.

irregular reflection. Scattering. Reflection in many different directions from an irregular surface.

isothermal process. A thermal process that takes place at constant temperature.

isotopes. Atoms whose nuclei contain the same number of protons but different numbers of neutrons.

joule. The unit of work; the product of a force of one newton acting through a distance of one meter.

Joule's law. The heat developed in a conductor is directly proportional to the resistance of the conductor, the square of the current, and the time the current is maintained.

junction detector. A solid-state device based on the transistor principle that is used to detect the passage of charged particles.

Kelvin scale. The scale of temperature having a single fixed point, the temperature of the triple point of water, which is assigned the value 273.16 °K.

kilogram. A unit of mass in the metric system; one of the seven fundamental units.

kilowatt hour. A unit of electric energy equal to 3.6×10^6 w s.

kinetic energy. Energy possessed by an object because of its motion.

kinetic theory. The molecules of matter are continuously in motion and the collisions between molecules are perfectly elastic.

Kirchhoff's first law. The algebraic sum of the currents at any circuit junction is equal to zero.

Kirchhoff's second law. The algebraic sum of all changes in potential occurring around any loop in a circuit equals zero.

laser. An acronym for light amplification by stimulated emission of radiation.

law. A statement that describes a natural phenomenon; a principle.

law of conservation of baryons. When a baryon decays or reacts with another particle, the number of baryons is the same on both sides of the equation.

law of conservation of energy. The total quantity of energy in a closed system is constant.

law of conservation of hypercharge. Hypercharge is conserved in strong and electromagnetic interactions, but not in weak interactions.

law of conservation of leptons. In a reaction involving leptons, the arithmetic sum of the lepton numbers is the same on each side of the equation.

law of conservation of mechanical energy. The sum of the potential and kinetic energies of an ideal energy system is constant.

law of conservation of momentum. When no net external forces are acting on an object, the total vector momentum of the object remains constant.

law of entropy. A natural process always takes place in such a direction as to increase the entropy of the universe.

law of heat exchange. In any heat transfer system, the heat lost by hot materials equals the heat gained by cold materials.

Lenz's law. An induced current is in such direction that its magnetic property opposes the change by which the current is induced.

lepton. A subatomic particle with a small rest mass, e.g., the electron.

line of flux. A line so drawn that a tangent to it at any point indicates the direction of the magnetic field.

line of force. A line so drawn that a tangent to it at any point indicates the direction of the electric field.

line spectrum. A spectrum consisting of monochromatic slit images having wavelengths characteristic of the atoms present in the source.

linear accelerator. A device for accelerating particles in a straight line through many stages of small potential difference.

liquefaction. The change to the liquid phase. The condensation of a gas to a liquid.

liter. A special name for the cubic decimeter. Symbol: L.

longitudinal wave. A wave in which the vibrations are parallel to the direction of propagation of the wave.

loop. A midpoint of a vibrating segment of a standing wave.

loudness. The sensation that depends principally on the intensity of sound waves reaching the ear.

lumen. The unit of luminous flux; the luminous flux on a unit surface all points of which are at unit distance from a point source of one candle.

luminous. Visible because of the light emitted by its oscillating particles.

luminous flux. The part of the total energy radiated per unit of time from a luminous source that is capable of producing the sensation of sight.

machine. A device that multiplies force at the expense of distance or that multiplies distance at the expense of force.

magnetic field. A region in which a magnetic force can be detected.

magnetic field intensity. The force exerted by a magnetic field on a unit N pole situated in the field.

magnetic flux. Lines of flux through a region of a magnetic field, considered collectively.

magnetic flux density. The magnetic flux through a unit area normal to the magnetic field; also called magnetic induction.

magnetic force. A force associated with motion of electric charges.

magnetosphere. A region of the upper atmosphere in which the motion of charged particles is governed primarily by the magnetic field of the earth.

magnification. The ratio of the image distance to the object distance; the ratio of the image size to the object size.

malleability. The property of a metal that enables it to be hammered or rolled into sheets.

mass. A measure of the quantity of matter; a fundamental physical quantity.

mass density. Mass per unit volume of a substance.

mass number. (1) The sum of the number of protons and neutrons in the nucleus of an atom. (2) The integer nearest to the atomic mass.

mass spectrograph. Instrument used to determine the mass of ionized particles.

matter. Anything that has the properties of mass and inertia.

matter wave. A property of matter that is directly proportional to Planck's constant and inversely proportional to mass and velocity.

mechanical equivalent of heat. The conversion factor that relates heat units to work units; 4.19 j/cal.

mechanical wave. A wave that originates in the displacement of a portion of an elastic medium from its normal position, causing it to oscillate about an equilibrium position.

medium. Any region through which a wave disturbance propagates. Me-

chanical waves require a matter medium. Electromagnetic waves propagate through a vacuum and various matter media.

melting point. The temperature at which a solid changes to a liquid.

meniscus. The crescent-shaped surface at the edge of a liquid column.

meson. A subatomic particle with a rest mass intermediate between that of a lepton and a baryon; the carrier of the strong interaction.

meter. A unit of length in the metric system equivalent to 1,650,763.73 wavelengths of the orange-red light emitted by krypton-86. One of the seven fundamental units of measure.

metric system. A system of measurement that is based on decimal multiples and subdivisions.

moderator. A material that slows down neutrons.

mole. Amount of substance containing the Avogadro number of particles such as atoms, molecules, ions, electrons, etc.

molecule. The smallest chemical species of a substance that is capable of stable independent existence.

momentum. The product of the mass and velocity of a moving body.

monochromatic light. Light composed of a single color.

mutual inductance. The ratio of the induced emf in one circuit to the rate of change of current in the coil of another circuit.

neutral weak current. A subatomic reaction in which leptons collide without change in the charges of the colliding particles.

neutron. A neutral subatomic particle having a mass of 1.674943×10^{-27} kg.

newton. The unit of force; a derived unit having the dimensions kg m/s^2. The force required to accelerate a one-kilogram mass at a rate of one meter per second each second.

Newton's first law of motion. A body at rest or in uniform motion in a straight line will remain at rest or in the same uniform motion unless acted upon by an external force; also called the law of inertia.

Newton's law of universal gravitation. The force of attraction between any two particles of matter in the universe is directly proportional to the product of their masses and inversely proportional to the square of the distance between their centers of mass.

Newton's second law of motion. The acceleration of a body is directly proportional to the net force exerted on the body, is inversely proportional to the mass of the body, and has the same direction as the net force; also called the law of acceleration.

Newton's third law of motion. If one body exerts a force on a second body, then the second body exerts a force equal in magnitude and opposite in direction on the first body; also called the law of interaction.

node. A point of no disturbance of a standing wave.

noise. Sound produced by irregular vibrations in matter which is unpleasant to the listener.

noninertial frame of reference. An accelerating frame of reference in which Newton's first law of motion does not hold true.

normal. A line drawn perpendicular to a line or surface.

N-type germanium. "Electron-rich" germanium consisting of equal numbers of free electrons and bound positive charges so that the net charge is zero.

nuclear binding force. The force that acts within the small distances between nucleons.

nuclear change. A change in the identity of atomic nuclei.

nuclear mass defect. The arithmetic difference between the mass of a nucleus and the larger sum of its uncombined constituent particles.

nuclear reactor. A device in which the controlled fission of certain substances is used to produce new substances and energy.

nucleon. A proton or neutron in the nucleus of an atom.

nucleus. The positively charged dense central part of an atom.

nuclide. An atom of a particular mass and of a particular element.

octave. The interval between a given musical tone and one with double or half the frequency.

ohm. The unit of electric resistance; one volt per ampere.

Ohm's law. The ratio of the emf applied to a closed circuit to the current in the circuit is a constant.

optical center. The point in a thin lens through which the secondary axes pass.

optical density. A property of a transparent material that is a measure of the speed of light through it.

orbital. The probability pattern of position of an electron about the nucleus of an atom.

order of magnitude. A numerical approximation to the nearest power of ten.

ordinate. The value corresponding to the vertical distance of a point on a graph from the X axis. The Y coordinate.

oscilloscope. A cathode-ray tube with associated electronic circuits that enable external voltages to deflect the electron beam of the cathode-ray tube simultaneously along both horizontal and vertical axes.

parallel circuit. An electric circuit in which two or more components are connected across two common points in the circuit so as to provide separate conducting paths for the current.

parallelogram method. The graphic method of finding the resultant of two vectors that do not act along a straight line.

paramagnetism. The property of a substance by which it is feebly attracted by a strong magnet.

pendulum. A body suspended so that it can swing back and forth about an axis.

penumbra. The partially illuminated part of a shadow.

period. (1) The time for one complete cycle, vibration, revolution, or oscillation. (2) The time required for a single wavelength to pass a given point.

periodic motion. Motion repeated in each of a succession of equal time intervals.

permeability. The property of a material by which it changes the flux density in a magnetic field from its value in air.

phase. (1) A condition of matter. (2) In any periodic phenomenon, a number that describes a specific stage within each oscillation. (3) The angular relationship between current and voltage in an a-c circuit. (4) The number of separate voltage waves in a commercial a-c supply.

phase angle. (1) Of any periodic function, the angle obtained by multiplying the phase by 360 if the angle is to be expressed in degrees, or by 2π if in radians. (2) The angle between the voltage and current vectors.

phasor. A representation of the concepts of magnitude and direction in a reference plane; a rotating vector.

photoelastic. Pertaining to certain materials that become double refracting when strained.

photoelectric effect. The emission of electrons by a substance when illuminated by electromagnetic radiation of sufficiently short wavelength.

photoelectrons. Electrons emitted from a light-sensitive material when it is illuminated with light of sufficiently short wavelength.

photometer. An instrument used for comparing the intensity of a light source with that of a standard source.

photometry. The quantitative measurement of visible radiation from light sources.

photon. A quantum of light energy; the carrier of the electromagnetic interaction.

photovoltaic effect. The generation of a potential difference across a P-N junction as a consequence of the absorption of incident light of appropriate frequency.

physical change. A change in which the composition and identifying properties of a substance remain unchanged.

physical quantity. A measurable aspect of the universe, such as length.

physics. The science that deals with the relationships between matter and energy.

piezoelectric effect. The property of certain natural and synthetic crystals to develop a potential difference between opposite surfaces when subjected to a mechanical stress, and conversely.

pitch. The identification of a certain sound with a definite tone; depends on the frequency which the ear receives.

pivot point. The point from which the lengths of all torque arms are measured.

Planck's constant. A fundamental constant in nature that determines what values are allowed for physical quantities in quantum mechanics; $h = 6.63 \times 10^{-34}$ j s.

plasma. A gas that is capable of conducting an electric current.

plate. The anode of an electronic tube.

P-N junction. The boundary between P- and N-type materials in a semiconductor crystal.

polarized light. Light radiations in which the vibrations of all light waves present are confined to planes parallel to each other.

polarizing angle. A particular angle of incidence at which polarization of reflected light is complete.

polychromatic light. Light composed of several colors.

positive rays. Rays coming through holes in a cathode on the side opposite the anode in a discharge tube. Positively charged ions.

potential difference. The work done per unit charge as a charge is moved between two points in an electric field.

potential energy. Energy that is the result of the position of an object.

potential gradient. The change in potential per unit distance.

power. The time rate of doing work.

power factor. The cosine of the phase angle between current and voltage in an a-c circuit.

precession. The motion that results from the application of a torque that tends to displace the axis of rotation of a rotating object.

precision. The agreement between the numerical values of two or more measurements made in the same way and expressed in terms of deviation; the reproducibility of measured data.

pressure. Force per unit area.

primary. A transformer winding that carries current and normally induces a current in one or more secondary windings.

primary cell. An electrochemical cell in which the reacting materials must be replaced after a given amount of energy has been supplied to the external circuit.

primary colors. Colors in terms of which all other colors may be described or from which all other colors may be evolved by mixtures.

primary pigments. The complements of the primary colors.

principal axis. (1) A line drawn through the center of curvature and the vertex of a curved mirror. (2) A line drawn through the center of curvature and the optical center of a lens.

principal focus. A point at which rays parallel to the principal axis converge or from which they diverge after reflection or refraction.

principle. See *law.*

principle of parity. For every process in nature there is a mirror-image process which is indistinguishable from the original process.

propagate. To travel through a material or space.

propagation. The act of propagating. The action of traveling through a material or space.

property. A measurable aspect of matter, e.g., mass and inertia.

proton. A positively charged subatomic particle having a mass of $1.6726485 \times 10^{-27}$ kg and a charge equal and opposite to that of the electron.

P-type germanium. "Hole-rich" germanium consisting of equal numbers of free positive holes and bound negative charges so that the net charge is zero.

pulse. A single nonrepeated disturbance.

quality. The property of sound waves that depends on the number of harmonics and their prominence.

quantum. An elemental unit of energy; a photon of energy *hf.*

quantum mechanics. The branch of physics that deals with the behavior of particles whose specific properties are given by quantum numbers.

quantum number. One of a set of notations used to characterize a discrete value that a quantized variable is allowed to assume.

quantum theory. A unifying theory based on the concept of the subdivision of radiant energy into discrete quanta (photons) and applied to the studies of structure at the atomic and molecular levels.

quark. A hypothetical subatomic particle of which all other subatomic particles are composed.

quiescent. A steady-state condition. The operating condition of an electronic circuit when no input signal is applied.

radian. A unit of angular measurement. The angle that, when placed with its vertex at the center of a circle, subtends on the circumference an arc equal in length to the radius of the circle. Approximately 57.3°.

radio waves. Also called Hertzian waves. Electromagnetic radiations produced by rapid reverses of current in a conductor.

radioactivity. The spontaneous breakdown of an atomic nucleus with the emission of particles and rays.

radioisotope. An isotope of an element that is radioactive.

rarefaction. The region of a longitudinal wave in which the vibrating particles are farther apart than their equilibrium distance.

ray. A single line of light from a luminous point. A line showing the direction of propagation of light.

reactance. The nonresistive opposition to current in an a-c circuit.

reaction motor. A heat engine whose acceleration is produced by the thrust of exhaust gases.

real image. An image formed by actual rays of light.

rectifier. A device for changing alternating current to direct current.

rectilinear propagation. Traveling in a straight line.

reflectance. The ratio of the light reflected from a surface to the light falling on it, expressed in percentage.

reflection. The return of a wave from the boundary of a medium.

refraction. The bending of a wave disturbance as it passes obliquely from one medium into another in which the disturbance has a different velocity.

regelation. The melting of a substance under pressure and the refreezing after the pressure is released.

regular reflection. Reflection from a polished surface in which scattering effects are negligible.

relative deviation. Percentage average deviation of a set of measurements.

relative error. Percentage absolute error of a set of measurements.

relative humidity. The ratio of the water vapor pressure in the atmosphere to the equilibrium vapor pressure at a given temperature.

relativistic mass. The mass of an object in motion with respect to the observer.

residual magnetism. Magnetism retained in a magnet after the magnetizing field has been removed.

resistance. The ratio of the potential difference across a conductor to the magnitude of current in it.

resistivity. A proportionality constant that relates the length and cross-sectional area of a given electric conductor to its resistance, at a given temperature.

resolution of forces. The resolving of a single force into component forces acting in given directions on the same point.

resonance. (1) The inducing of vibrations of a natural rate by a vibrating source having the same frequency. (2) The condition in an a-c circuit in which the inductive reactance and capacitive reactance are equal.

rest mass. The mass of an object not in motion.

resultant. A vector representing the sum of several vector components.

resultant force. The single force that has the same effect as two or more forces applied simultaneously at the same point.

reverse bias. Voltage applied to a semiconductor P-N junction that reduces the electron current across the junction.

rheostat. A variable resistance.

root-mean-square (rms) current. The effective value of an alternating current; the square root of the mean of the instantaneous values squared.

rotary motion. Motion of a body about an internal axis.

rotational equilibrium. The state of a body in which the sum of all the clockwise torques in a given plane equals the sum of all the counterclockwise torques about a pivot point.

rotational inertia. The property of a rotating object that resists changes in its angular velocity.

scalar quantity. A quantity that is completely specified by a magnitude.

scientific notation. A positive number expressed in the form of $M \times 10^n$ in which M is a number between 1 and 10 and n is an integral power of 10.

scintillation counter. A device that counts the impacts of charged subatomic particles on a fluorescent screen by means of a photomultiplier tube.

second. A unit of time; equivalent to 9,192,631,770 vibrations of cesium-113. One of the seven fundamental units of measure.

second law of photoelectric emission. The kinetic energy of photoelectrons is independent of the intensity of the incident light.

second law of thermodynamics. It is not possible for an engine to transfer heat from one body to another at a higher temperature unless work is done on the engine.

secondary. A transformer output winding in which the current is due to inductive coupling with another winding called the primary.

secondary axis. Any line other than the principal axis drawn through the center of curvature of a mirror or the optical center of a lens.

secondary emission. Emission of electrons as a result of the bombardment of an electrode by high-velocity electrons.

selectivity. The property of a tuned circuit that discriminates between signal voltages of different frequencies.

self-inductance. The ratio of the induced emf across a coil to the rate of change of current in the coil.

series circuit. An electric circuit in which the components are arranged to provide a single conducting path for current.

series resonance. A condition in which the impedance of a series circuit containing resistance, inductance, and capacitance is equal to the resistance of the circuit and the voltage across the circuit is in phase with the current.

shear strain. The ratio of the amount of deformation of the side of a body to the length of the side.

short circuit. An electric circuit through a negligible resistance that usually shunts a normal load and overloads the circuit.

significant figures. Those digits in an observed quantity (measurement) that are known with certainty plus the first digit that is uncertain.

simple harmonic motion. Motion in which the acceleration is proportional to the displacement from an equilibrium position and is directed toward that position.

solar spectrum. The band of colors produced when sunlight is dispersed by a prism.

solenoid. A long helically wound coil of insulated wire.

solid state detector. A device used to detect the passage of charged subatomic particles by their crystal-distorting or ionizing effects on a nonconducting or semiconducting solid.

solidification. The change of phase from a liquid to a solid.

sonometer. A device, consisting of two or more wires or strings stretched over a sounding board, used for testing the frequency of strings and for showing how they vibrate.

sound. The series of disturbances in matter to which the human ear is sensitive. Also similar disturbances in matter above and below the normal range of human hearing.

sound intensity. The rate at which sound energy flows through a unit area.

space charge. The negative charge in the space between the cathode and plate of a vacuum tube.

spark chamber. A device used to detect the passage of charged subatomic particles by the light flashes they trigger.

specific gravity. The ratio of the mass density of a substance to that of water.

specific heat. The heat capacity of a material per unit mass.

spectroscope. Optical instrument used for the study of spectra.

speed. Time rate of motion.

spherical aberration. The failure of parallel rays to meet at a single point on a spherical surface after reflection or refraction.

spinthariscope. A device used to detect subatomic particles by the light flashes they produce on a zinc sulfide screen.

standard pressure. The pressure exerted by $76\bar{0}$ mm of mercury at 0 °C.

standard temperature. 0 °C; 273 °K.

standing wave. The resultant of two wave trains of the same wavelength, frequency, and amplitude traveling in opposite directions through the same medium.

static electricity. Electricity at rest.

steam point. The boiling point of water at standard atmospheric pressure.

steradian. The ratio of the intercepted surface area of a sphere to the square of the radius. A unit of solid angle.

storage cell. An electrochemical cell in which the reacting materials are regenerated by the use of a reverse current from an external source.

strain. The relative amount of distortion produced in a body under stress.

stress. The distorting force per unit area.

strong nuclear interaction. The interaction that holds the particles of the nucleus together and is independent of charge.

sublimation. The change of a solid to a gaseous phase without passing through the liquid phase.

superconductivity. The condition of zero resistivity below the transition temperature of a substance.

supercooling. The process of cooling a substance below its normal phase-change point without a change of phase.

superposition. Combining the displacements of two or more waves vectorially to produce a resultant displacement.

surface tension. The tendency of a liquid surface to contract; the measure of this tendency in newtons per meter.

sympathetic vibration. See *resonance (1)*.

synchrotron. A particle accelerator in which the oscillating frequency varies.

technology. The application of science to human needs and goals.

temperature. The physical quantity that is proportional to the average kinetic energy of translation of particles in matter.

temporary magnet. A magnet produced by induction.

tensile strength. The force required to break a rod or wire of unit cross-sectional area.

theory. A plausible explanation of an observed event, supported experimentally and confirmed by experiments designed to test predictions based upon the explanation.

thermal energy. The total potential and kinetic energy associated with the random motions of the particles of a material.

thermionic emission. The liberation of

electrons from the surface of a hot body.

thermocouple. An electric circuit composed of two dissimilar metals whose junctions are maintained at different temperatures.

thermodynamics. Study of quantitative relationships between heat and other forms of energy.

thermoelectric effect. The production of an electron current in a closed circuit consisting of two dissimilar metals as a result of the emf developed when the two junctions are maintained at different temperatures.

thermonuclear reaction. Nuclear fusion.

third law of photoelectric emission. Within the region of effective frequencies, the maximum kinetic energy of photoelectrons varies directly with the difference between the frequency of the incident light and the cut-off frequency.

thought experiment. An idealized experiment that cannot be performed under actual conditions.

threshold of hearing. The intensity of the faintest sound audible to the average human ear, 10^{-16} w/cm^2 at 10^3 hz.

threshold of pain. For audible frequencies of sound, an intensity level above which pain results in the average human ear.

tolerance. Degree of precision obtainable with a measuring instrument.

torque. Product of a force and the effective length of its torque arm.

torque arm. The perpendicular distance between the line of action of the torque producing force and the axis of rotation.

total reflection. The reflection of light at the boundary of two transparent media when the angle of incidence exceeds the critical angle.

transformer. A device for changing an alternating voltage from one potential to another.

transistor. A semiconductor device used as a substitute for vacuum tubes in electronic applications.

transition temperature. A specific temperature at which the resistivity of some materials drops suddenly to zero.

translational equilibrium. The state of a body in which there are no unbalanced forces acting on it.

transuranium elements. Elements with atomic number greater than 92.

transverse wave. A wave in which the vibrations are at right angles to the direction of propagation of the wave.

triode. Vacuum tube consisting of a plate, grid, and cathode.

triple point. The single condition of temperature and pressure at which the solid, liquid, and vapor phases of a substance can coexist in stable equilibrium.

trough. A region of downward displacement in a transverse wave.

tuning fork. A metal two-prong fork that produces a sound of a definite pitch.

ultrasonic range. Vibrations in matter above 20,000 vibrations/second.

ultraviolet light. Electromagnetic radiations of shorter wavelength than visible light but longer than X rays.

umbra. The part of a shadow from which all light rays are excluded.

uncertainty principle. It is impossible to specify simultaneously both the position of an object and its momentum.

unified field theory. The principle that all forces in the universe are part of a single concept.

unit magnetic pole. One that repels an exactly similar pole placed one centimeter away with a force of one dyne.

universal gas constant. Constant of proportionality, R, in the ideal gas equation.

universal gravitational constant. Constant of proportionality in Newton's law of universal gravitation; $G = 6.67 \times 10^{-11} \text{ nm}^2/\text{kg}^2$.

Van de Graaff generator. A particle accelerator that transfers a charge from an electron source to an insulated sphere by means of a moving belt composed of an insulating material.

vapor. The gaseous phase of a substance that exists as a liquid or solid under normal conditions.

vaporization. The change in phase from a solid or a liquid to a gas.

variation. See *declination*.

vector quantity. A quantity that is completely specified by a magnitude and a direction.

velocity. Speed in a particular direction.

vertex. The center of a curved mirror.

virtual image. An image that only appears to be formed by rays of light.

viscosity. The ratio of shear stress to the rate of change of shear strain in a liquid or gas.

volt. The unit of potential difference. The potential difference between two points in an electric field such that one joule of work moves a charge of one coulomb between these points.

voltage sensitivity. Voltage per unit scale division of an electric instrument.

voltaic cell. A device that changes chemical into electric energy by the action of two dissimilar metals immersed in an electrolyte.

voltmeter. An instrument used to measure the difference of potential between two points in an electric circuit.

volume strain. The ratio of the decrease in volume to the volume before stress is applied.

$W^{\pm}$ **particle.** The theoretical carrier for weak subatomic interactions.

watt. The unit of power; one joule per second.

wavelength. In a periodic wave, the distance between consecutive points of corresponding phase.

weak nuclear interaction. The interaction that produces pairs of particles that have unusually long half-lives.

weber. The unit of magnetic flux.

weight. The measure of the gravitational force acting on a substance.

Wheatstone bridge. Instrument used for the measurement of electric resistance.

work. The product of a displacement and the force in the direction of the displacement.

work function. The minimum energy required to remove an electron from the surface of a material and send it into field-free space.

working equation. The equation derived from a basic equation as an expression of the unknown quantity in a problem directly in terms of the known quantities stated in the problem. The working equation is the mathematical expression of the solution to the problem.

X rays. Invisible electromagnetic radiations of great penetrating power.

Young's modulus. The ratio of stress to strain in a solid.

Zeeman effect. The splitting of atomic energy levels into two or more sublevels by means of a magnetic field as seen in spectral lines under these conditions.

INDEX

PHOTO CREDITS

Cover   NASA

Chapter 1   p. 1 Swiss Philatelic Society; p. 2 IBM; p. 3 (both) General Electric Company; p. 4 Argonne National Laboratory; p. 7 (both) Alex Mulligan; p. 8 (both) Holt, Rinehart & Winston by Russell Dian; p. 12 John G. Zimmerman; p. 13 Deutsches Museum, Munich; p. 14 Courtesy, The Archive, California Institute of Technology

Photo Essay   CAREERS p. 18 (left) I.B.M.; (top right) NASA; (bottom right) HRW photo by Luis Fernandez; p. 19 (top left) I.B.M.; (bottom left) I.B.M.; (right) A.I.P. Niels Bohr Library

Chapter 2   p. 20 Hungarian Philatelic Society (top) p. 20 c Tom McHugh, 1981/Photo Researchers, Inc.; p. 22 National Bureau of Standards; p. 23 Bureau International des Poids et Mesures; p. 24 National Museum of American History, Smithsonian Institute-Division of Mechanisms; p. 25 Alex Mulligan; p. 27 Cenco; p. 29 Hale Observatories

Chapter 3   p. 40 Italian Philatelic Society; p. 41 Project Physics, HRW; p. 54 © Hatton, 1980/ Woodfin Camp & Associates; p. 56 The Bettmann Archive/Painting-Kneller; p. 58 (both) Loomis Dean, Life Magazine © 1954. Time, Inc.; p. 66 Jet Propulsion Laboratory, Pasadena, CA.

Chapter 4   p. 70 Hungarian Philatelic Society; p. 71 FMC Corporation Construction Equipment; p. 81 Austrian Press and Information Service

Chapter 5   p. 93 Philatelic Society of the Netherlands; p. 94 PSSC Physics, D.C. Heath & Co.; p. 99 Department of the Navy

Chapter 6   p. 115 West German Philatelic Society; p. 128 CSC Scientific; p. 131 Dr. Harold Edgerton, Massachusetts Institute of Technology; p. 132 NASA; p. 133 Ealing Corporation; p. 137 Henry Hirschfeld (both); p. 139 PSSC Physics, D.C. Heath & Co.

Chapter 7   p. 141 Philatelic Society of Greece; p. 142 Brown Brothers; p. 143 Jutta Schwab

Mueller; p. 146 United States Steel Corporation; p. 152 Ken Chen; p. 153 Ken Chen; p. 155 Courtesy, Citicorp; p. 156 John King; p. 157 Dick Saunders/Ebony Magazine; p. 162 NASA

Photo Essay   ASTROPHYSICS p. 164 (left) Fred Seward, Center for Astrophysics Harvard College Observatory, Smithsonian Astrophysical Observatory; (right) © Ian Berry/Magnum; p. 165 (a) © California Institute of Technology; (b) NASA; (c) NASA; (top right) Jet Propulsion Lab; (bottom right) NASA

Chapter 8   p. 166 Swedish Philatelic Society; p. 168 National Bureau of Standards; p. 173 Alex Mulligan

Chapter 9   p. 200 Massachusetts Philatelic Society; p. 201 The Bettmann Archive; p. 208 AIP, Niels Bohr Library; p. 211 G. Hunter/National Film Board of Canada; p. 213 Westinghouse; p. 215 Courtesy, Detroit Diesel Allison; p. 218 Courtesy, NASA/HRW photo by Russell Dian (top), (bottom) NASA

Chapter 10   p. 222 West German Philatelic Society; p. 223 (left) © Alexander Lowry/Photo Researchers, Inc. (right) Kay Chernush; p. 231 © Devon Jacklin; p. 233 (left) Fundamental Photographs from the Granger Collection; p. 234 (left) PSSC Physics, D.C. Heath & Company (right) Education Development Center; p. 235 PSSC Physics, D.C. Heath & Company; p. 236 Education Development Center; p. 237 PSSC Physics, D.C. Heath & Company; p. 240 (top) PSSC Physics, D.C. Heath & Company (bottom) PSSC Physics, D.C. Heath & Company; p. 242 (all) Fundamental Photographs from the Granger Collection; p. 246 Education Development Center

Chapter 11   p. 252 Canadian Philatelic Society; p. 253 College Physics, HRW; p. 255 Bell Telephone Laboratories; p. 264 Education Development Center; p. 267 Ken Chen; p. 268 & 269 (all) (C.G. Conn. Ltd.; p. 270 Ken Chen

Chapter 12   p. 280 Mexican Philatelic Society; p. 286 Master and Fellows of Trinity College,

Cambridge; p. 287 The Bettmann Archive; p. 288 U.P.I.; p. 293 AIP, Niels Bohr Library; p. 295 Danish Information Office; p. 298 © Howard Sochurek/Woodfin Camp & Associates; p. 306 National Bureau of Standards; p. 311 Stansi Scientific Company

Chapter 13   p. 315 California Philatelic Society; p. 323 General Electric Company; p. 326 (top) Wide World Photos (bottom) Courtesy, Roland Cormier

Chapter 14   p. 332 Philatelic Society of Pakistan; p. 333 General Electric Company; p. 334 Bausch & Lomb; p. 337 Alex Mulligan; p. 343 (bottom) Bausch & Lomb; p. 344 (bottom) HSS, Inc., Bedford, Mass.; p. 353 Bausch & Lomb; Color Plate VI(A) © Ken Garrett/Woodfin Camp & Associates; C.P. VI(B) Bausch & Lomb; C.P.

Chapter 15   p. 362 Swedish Philatelic Society; p. 363 Foundations of Physics, HRW; p. 366 (both) American Optical Company; p. 367 Bausch & Lomb; p. 376 Photo by George Garard with permission of Henry Semat, as published in Fundamentals of Physics 4th ed., HRW; p. 377 (left) Bruce Roberts/Photo Researchers, Inc. (right) NASA; p. 378 Carl Zeiss, Inc.

Chapter 16   p. 380 United States Postal Service; p. 394 General Electric Corporation; p. 399 © Adam Woolfit/Woodfin Camp & Associates; p. 400 Cardwell Condenser Corporation

Chapter 17   p. 410 Philatelic Society of France; p. 418 NASA; p. 424 Burndy Library, Norwalk, Conn.; p. 441 Cenco

Photo Essay   METROLOGY p. 446 Grumman Aerospace Corporation; p. 447 NASA

Chapter 18   p. 448 United States Postal Service; p. 449 General Electric Company; p. 450 Bethlehem Steel; p. 460 Royal Institution of Great Britain

Chapter 19   p. 465 Philatelic Society of Denmark; p. 466 Walter Dawn; p. 469 (both) Bell Telephone Laboratories; p. 481 Brookhaven National Laboratory; p. 482 Sargent-Welch Scientific Company; p. 483 Leeds & Northrup

Chapter 20   p. 490 West German Philatelic Society; p. 505 Public Service Electric and Gas

Company; p. 513 General Electric Company; p. 516 The New York Public Library; p. 521 General Electric Company

Chapter 21   p. 522 Italian Philatelic Society; p. 526 The Granger Collection; p. 549 Ken Chen

Photo Essay   COMPUTERS p. 552 (top) NOAA National Weather Service; (bottom) Apple Computers; p. 553 (top left & right) Wright State University, Dayton, Ohio; (bottom left) General Electric (bottom right) I.B.M.

Chapter 22   p. 554 United States Postal Service; p. 555 Amana; p. 556 The New York Public Library; p. 557 Westinghouse Electric Corporation; p. 566 Ken Chen

Chapter 23   p. 581 Philatelic Society of New Zealand; p. 582 (both) Fritz Goro/Life Magazine © Time, Inc.; p. 585 (left) Brookhaven National Laboratory; p. 586 Margot Bergmann, Polytechnic Institute of Brooklyn; p. 594 General Electric Company

Chapter 24   p. 600 Philatelic Society of France; p. 601 (top) The Bettmann Archive (bottom) French Embassy Press & Information Division; p. 602 National Laboratory, Oak Ridge, Tennessee; p. 610 Cornell University; p. 617 (top) Tennessee Valley Authority (bottom) Princeton University Plasma Physics Laboratory; p. 621 Zionist Archive and Library

Chapter 25   p. 623 Philatelic Society of Sénégal; p. 624 AIP, Niels Bohr Library; p. 626 University of California, Lawrence Radiation Laboratory; p. 627 The Granger Collection; p. 629 AIP, Niels Bohr Library Bainbridge Collection; p. 631 (bottom) Brookhaven National Laboratory; p. 632 (bottom) University of Illinois; p. 633 (all) Fermi National Accelerator Laboratory; p. 635 Roland Quintero; p. 636 (top) Sargent-Welch Scientific Company; p. 637 Education Development Center; p. 638 Brookhaven National Laboratory; p. 639 (top) University of California, Lawrence Radiation Laboratory (bottom) Brookhaven National Laboratory; p. 640 U.S. Atomic Energy Commission; p. 645 Barrett Gallagher; p. 647 (top) AIP, Niels Bohr Library Courtesy, Alan Richards (bottom) Michael R. Dressler; p. 649 Courtesy of the Harvard University News Office